T0255719

Graduate Texts in Physics

Graduate Texts in Physics

Graduate Texts in Physics publishes core learning/teaching material for graduate- and advanced-level undergraduate courses on topics of current and emerging fields within physics, both pure and applied. These textbooks serve students at the MS- or PhD-level and their instructors as comprehensive sources of principles, definitions, derivations, experiments and applications (as relevant) for their mastery and teaching, respectively. International in scope and relevance, the textbooks correspond to course syllabi sufficiently to serve as required reading. Their didactic style, comprehensiveness and coverage of fundamental material also make them suitable as introductions or references for scientists entering, or requiring timely knowledge of, a research field.

More information about this series at http://www.springer.com/series/8431

Ervin B. Podgoršak

Radiation Physics for Medical Physicists

Third Edition

With 232 Figures, 98 Tables

 Springer

Ervin B. Podgoršak
Faculty of Medicine, Department of
  Oncology and Medical Physics Unit
McGill University
Montreal, QC
Canada

ISSN 1868-4513   ISSN 1868-4521   (electronic)
Graduate Texts in Physics
ISBN 978-3-319-79781-6   ISBN 978-3-319-25382-4   (eBook)
DOI 10.1007/978-3-319-25382-4

This Springer imprint is published by Springer Nature
The registered company is Springer International Publishing AG Switzerland

Mariani za razumevanje in pomoč

Preface

This book is intended as a textbook for a radiation physics course in academic medical physics graduate programs as well as a reference book for candidates preparing for certification examinations in medical physics subspecialties. The book may also be of interest to many professionals, not only physicists, who in their daily occupations deal with various aspects of medical physics or radiation physics and have a need or desire to improve their understanding of radiation physics.

Medical physics is a rapidly growing specialty of physics, concerned with the application of physics to medicine, mainly but not exclusively in the application of ionizing radiation to diagnosis and treatment of human disease. In contrast to other physics specialties, such as nuclear physics, condensed matter physics, and high-energy physics, studies of modern medical physics attract a much broader base of professionals, including graduate students in medical physics; medical residents and technology students in radiation oncology and diagnostic imaging; students in biomedical engineering; and students in radiation safety and radiation dosimetry educational programs. These professionals have diverse background knowledge of physics and mathematics, but they all have a common need to improve their knowledge and understanding of the physical concepts that govern the application of ionizing radiation in diagnosis and treatment of disease.

Numerous textbooks that cover the various subspecialties of medical physics are available, but they generally make a transition from the elementary basic physics directly to the intricacies of the given medical physics subspecialty. The intent of this textbook is to provide the missing link between the elementary physics and the physics of the subspecialties of medical physics.

This book deals mainly with theoretical aspects of the discussed subject material, with some examples added for clarification. However, a recently published text book "Compendium to Radiation Physics for Medical Physicists: 300 Problems and Solutions" covers the first 14 of the 17 chapters of this book with 300 solved problems. The reader of this book who would like to reinforce the material of this book through pertinent examples is encouraged to peruse the "Compendium" book in conjunction with this book.

Most of the subjects covered in this textbook can be found discussed in greater detail in many other specialized physics texts, such as nuclear physics, quantum mechanics, and modern physics. However, these texts are aimed at students in a specific physics specialty, giving more in-depth knowledge of the particular specialty but providing no evident link with medical physics and radiation physics. Some of these important specialized texts are listed in the bibliography at the end of this book for the benefit of readers who wish to attain a better insight into subjects discussed in this book. To recognize the importance of relevant history for understanding of modern physics and medical physics, Appendix C provides short biographies on scientists whose work is discussed in this book.

I am indebted to my colleagues in the Medical Physics Department of the McGill University Health Centre for their encouragement, approval, and tolerance of my concentrating on the book during the past several years. I am greatly indebted to my former students and/or colleagues Dr. Geoffrey Dean, Dr. François DeBlois, Dr. Slobodan Dević, Michael D.C. Evans, Marina Olivares, William Parker, Horacio Patrocinio, Dr. Matthew B. Podgorsak, and Dr. Jan P. Seuntjens who helped me with discussions on specific topics as well as with advice on how to present certain ideas to make the text flow better. I also appreciate constructive comments by Prof. José M. Fernandez-Varea from the University of Barcelona and Prof. Pedro Andreo from the University of Stockholm and the Karolinska University Hospital.

Special thanks are due to my colleague Dr. Wamied Abdel-Rahman, not only for helpful discussions of the subject matter, but also for his skillful drawing of figures presented in the book and for significant contributions to Chaps. 7 and 12. Secretarial help from Ms. Margery Knewstubb and Ms. Tatjana Nišić is very much appreciated.

I received my undergraduate physics education at the University of Ljubljana in Slovenia. I would like to thank the many teachers from the University of Ljubljana who introduced me to the beauty of physics and provided me with the knowledge that allowed me to continue my studies in the USA and Canada.

My sincere appreciation is due to my former teachers and mentors Profs. John R. Cameron and Paul R. Moran from the University of Wisconsin in Madison and Profs. Harold E. Johns and John R. Cunningham from the University of Toronto who introduced me to medical physics; a truly rewarding profession that brings together one's love of physics and compassion for patients.

Finally, I gratefully acknowledge that the completion of this book could not have been accomplished without the support and encouragement of my spouse Mariana. Especially appreciated are her enthusiasm for the project and her tolerance of the seemingly endless hours I spent on the project during the past several years.

Montreal   Ervin B. Podgoršak
October 2015

Medical Physics: A Specialty and Profession

Medical Physics and Its Subspecialties

Medical physics is a branch of physics concerned with the application of physics to medicine. It deals mainly, but not exclusively, with the use of ionizing radiation in diagnosis and treatment of human disease.

Diagnostic procedures involving ionizing radiation use relatively low-energy x-rays in the 100 kV range (*diagnostic radiology*) or γ-rays (*nuclear medicine* also known as *molecular imaging*); therapeutic procedures involving ionizing radiation most commonly use high-energy megavoltage x-rays and γ-rays or megavoltage electrons (*radiotherapy* also known as *radiation therapy*, *radiation oncology*, and *therapeutic radiology*).

Other applications of physics in diagnosis of disease include the use of nuclear magnetic resonance in anatomic, functional, and spectroscopic magnetic resonance imaging (MRI); ultrasound (US) in imaging; bioelectrical investigations of the brain (*electroencephalography*) and heart (*electrocardiography*); biomagnetic investigations of the brain (*magnetoencephalography*); and infrared radiation in *thermography*. Physicists are also involved in the use of heat for cancer therapy (*hyperthermia*), in applications of lasers for surgery, and in medical informatics.

During the past three decades medical physics has undergone a tremendous evolution, progressing from a branch of applied science on the fringes of physics into an important mainstream discipline that can now be placed on equal footing with other more traditional branches of physics such as nuclear physics, particle physics, and condensed matter physics. Since the number of new jobs in medical physics is growing faster than the number of jobs in other specialties of physics, universities are under much pressure to develop new graduate programs in medical physics or to expand their existing medical physics programs.

The medical physics specialty covers several diverse areas of medicine. It is therefore customary for medical physicists to concentrate and work on only one of the four specific subspecialties of medical physics:

1. *Diagnostic radiology physics* dealing with diagnostic imaging with x-rays, ultrasound, and magnetic resonance.
2. *Nuclear medicine physics* also referred to as *molecular imaging physics* dealing with diagnostic imaging using radionuclides.
3. *Radiotherapy physics* or *radiation oncology physics* dealing with treatment of cancer with ionizing radiation.
4. *Health physics* dealing with the study of radiation hazards and radiation protection.

Brief History of Use of Ionizing Radiation in Medicine

The study and use of ionizing radiation started with three momentous discoveries: *x-rays* by Wilhelm Röntgen in 1895, *natural radioactivity* by Henri Becquerel in 1896, and *radium* by Pierre Curie and Marie Curie-Skłodowska in 1898. Since then, ionizing radiation has played an important role in atomic and nuclear physics where it ushered in the era of quantum mechanics, provided the impetus for development of radiology and radiotherapy as medical specialties and medical physics as a specialty of physics. In addition, ionizing radiation also proved useful in many other diverse areas of human endeavor, such as in industry, power generation, waste management, and security services.

The potential benefit of x-ray use in medicine for imaging and treatment of cancer was recognized within a few weeks of Röntgen's discovery of x-rays. Two new medical specialties: radiology and radiotherapy evolved rapidly, both relying heavily on physicists for routine use of radiation as well as for development of new techniques and equipment.

Initially, most technological advances in medical use of ionizing radiation were related to: (1) improvements in efficient x-ray beam delivery; (2) development of analog imaging techniques; (3) optimization of image quality with concurrent minimization of delivered dose; and (4) increase in beam energies for radiotherapy.

During the past two decades, on the other hand, most developments in radiation medicine were related to integration of computers in imaging, development of digital diagnostic imaging techniques, and incorporation of computers into therapeutic dose delivery with high-energy linear accelerators. Radiation dosimetry and treatment planning have also undergone tremendous advances in recent years: from development of new absolute and relative dosimetry techniques to improved theoretical understanding of basic radiation interactions with human tissues, and to the introduction of Monte Carlo techniques in the determination of dose distributions resulting from penetration of ionizing radiation into tissue.

Currently, ionizing radiation is used in medicine for imaging in diagnostic radiology and nuclear medicine, for treatment of cancer in radiotherapy, for blood irradiation to prevent transfusion-associated graft versus host disease, and for sterilization of single-use medical devices. The equipment used for modern imaging and radiotherapy is very complex and requires continuous maintenance, servicing, and calibration to ensure optimal as well as safe performance. Optimal performance in imaging implies acquisition of optimized image quality for lowest possible patient radiation dose, while in radiotherapy it implies a numerically and spatially accurate dose delivery to the prescribed target. Optimal performance can only be achieved with services provided by engineers who maintain and service the equipment, and medical physicists who deal with calibration of equipment, in vitro and in vivo dose measurement, dosimetry and treatment planning, as well as quality assurance of image acquisition and patient dose delivery.

Educational Requirements for Medical Physicists

Pioneers and early workers in medical physics came from traditional branches of physics. By chance or choice they ended up working in nuclear medicine, radiology, or radiotherapy, and through on-the-job training developed the necessary skills and knowledge required for work in medical environment. In addition to clinical work, they also promoted medical physics as science as well as profession, and developed graduate medical physics educational programs, first through special medical physics courses offered as electives in physics departments and more recently through independent, well-structured medical physics academic programs that lead directly to graduate degrees in medical physics.

Since medical physicists occupy a responsible position in the medical environment, they are required to have a broad background of education and experience. The requirement for basic education in physics and mathematics is obvious, but the close working relationship of medical physicists with physicians and medical scientists also requires some familiarity with basic medical sciences, such as anatomy, physiology, genetics, and biochemistry.

Today's sophistication of modern medical physics and the complexity of the technologies applied to diagnosis and treatment of human disease by radiation demand a stringent approach to becoming a member of the medical physics profession. Currently, the most common path to a career in medical physics is academic progression, through a B.Sc. degree in one of the physical sciences but preferably in Physics, to an M.Sc. degree in Medical Physics, and then to a Ph.D. degree (doctorate) in Medical Physics.

The minimum academic requirement for a practicing medical physicist is an M.Sc. degree in Medical Physics, and this level is adequate for physicists who are mainly interested in clinical and service responsibilities. However, medical physicists working in academic environments should possess a Ph.D. degree in Medical Physics.

Academic education alone does not make a medical physicist. In addition to academic education, practical experience with medical problems and medical equipment is essential, and this may be acquired through on-the-job clinical education or, preferably, through a structured two-year traineeship (also referred to as internship or residency) program in a hospital after graduation with M.Sc. or Ph.D. degree in Medical Physics.

Because medical physicists work in health care and their work directly or indirectly affects patient safety and well-being, standards for their didactic and clinical education, work, and professional conduct are set and maintained by various professional bodies through the processes of educational accreditation, professional certification, professional licensure, and requirement for continuing education. Accreditation is usually given to institutions, while certification and licensure are given to individual professionals. As the profession of medical physics matures, these processes are becoming more and more stringent to ensure quality of work and protection of patients undergoing diagnostic or therapeutic procedures using ionizing radiation.

In relation to the medical physics profession, *accreditation*, *certification*, and *licensure* are defined as follows:

- *Educational accreditation* represents an attestation by an appropriate accreditation agency that the accredited educational program (M.Sc. and/or Ph.D.) or clinical residency education program, offered by a given educational institution, typically a university incorporating a medical school, provides quality education in medical physics and meets applicable standards.
- *Professional certification* in medical physics like in other professions is obtained from a national or international professional society and attests that the certified medical physicist is able to competently execute a job or task in the area covered by the certification. The certification is usually attained through a rigorous examination process run by an appropriate national medical physics organization or medical organization. With respect to expiry there are two types of professional certification: (1) lifetime and (2) more commonly, time-limited typically to 5 or 10 years. For continued certification, the time-limited certificates must be renewed before expiry, usually through a rigorous "maintenance of certification" (MOC) process involving continuing education.
- While professional certification attests to the competence of a professional in certain field, it does not confer legal right to practice. Certain jurisdictions require by law that medical physicists demonstrate competence before they are legally allowed to practice. This legal requirement is referred to as *licensure* and the examination process is often similar to that used for professional certification except that professional certification is conferred by a professional organization while licensure is conferred by a government agency. Thus, certification and licensure deal with the same professional issues but differ in legal status they confer.

Accreditation of Medical Physics Educational Programs

Many universities around the world offer academic and clinical educational programs in medical physics. To achieve international recognition for its graduates, a medical physics educational program should hold accreditation by an international accreditation body that attests to the program's meeting rigorous academic and clinical standards in medical physics. Currently, there are two such international bodies, the *Commission on Accreditation of Medical Physics Educational Programs* (CAMPEP) and the *International Medical Physics Certification Board* (IMPCB).

The CAMPEP – www.campep.org/, was founded in the late 1980s and is currently sponsored by five organizations: the American Association of Physicists in Medicine (AAPM), the American College of Radiology (ACR), American Society for Radiation Oncology (ASTRO), Canadian Organization of Medical Physicists (COMP), and the Radiological Society of North America (RSNA). In September 2015, 139 medical physics education programs (49 academic and 90 residency programs) were accredited by the CAMPEP.

The IMPCB – www.impcb.org/ was formed in 2010 by 11 charter member organizations in medical physics with the following main objectives:

(i) To support the practice of medical physics through a certification program in accordance with IOMP guidelines.
(ii) To establish the infrastructure requirement and assessment procedures for the accreditation of medical physics certification programs in accordance with the requirements of IOMP guidelines.

Toward the objectives, the IMPCB, in collaboration with the IOMP and the IAEA has started to build models to develop national certification programs and established requirements for successful completion of the certification process.

Certification of Medical Physicists

Several national professional medical physics organizations certify the competence of medical physicists. The certification is obtained through passing a rigorous written and oral examination that can be taken by candidates who possess M.Sc. or Ph.D. degree in medical physics and have completed an accredited residency in medical physics. Currently, the residency requirement is relaxed and a minimum of two years of work experience in medical physics after graduation with M.Sc. or Ph.D. degree in Medical Physics is also accepted, because of the shortage of available residency positions. However, in the future, graduation from an accredited medical physics academic program as well as graduation from an accredited medical physics residency education program will likely become mandatory for admission to write the medical physics certification examination.

The medical physics certification attests to the candidate's competence in the delivery of patient care in one of the subspecialties of medical physics. The requirement that its medical physics staff be certified provides a medical institution with the necessary mechanism to ensure that high standard medical physics services are given to its patients.

Appointments and Areas of Activities

Medical physicists are involved in four areas of activity: (1) *Clinical service and consultation*; (2) *Research and development*; (3) *Teaching and mentoring*; and (4) *Administration*. They are usually employed in hospitals and other medical care facilities. Frequently, the hospital is associated with a medical school and medical physicists are members of its academic staff. In many non-teaching hospitals, physicists hold professional appointments in one of the clinical departments and are members of the professional staff of the hospital. Larger teaching hospitals usually employ a number of medical physicists who are organized into medical physics departments that provide medical physics services to clinical departments.

Medical Physics Organizations

Medical physicists are organized in national, regional, and international medical physics organizations. The objectives of these organizations generally are to advance the medical physics practice and profession through:

- Promoting medical physics education and training.
- Promoting the advancement in status and stature of the medical physics profession.
- Lobbying for a formal national and international recognition of medical physics as a profession.
- Holding regular meetings and conferences as well as publishing journals, proceedings, reports, and newsletters to disseminate scientific knowledge and discuss professional issues of interest to medical physicists.
- Improving the scientific knowledge and technical skills of physicists working in medicine.
- Sponsoring accreditation commissions for academic and residency programs in medical physics, and organizing certification programs as well as maintenance of certification and continuing education programs for medical physicists.
- Developing professional standards and quality assurance procedures for applications of physics in medicine.
- Fostering collaborations with other medical physics organizations as well as other related scientific and professional organizations.

The *International Organization for Medical Physics* (IOMP) is the largest medical physics organization representing 18,000 medical physicists worldwide and 80 adhering national member organizations (www.iomp.org). The organization was founded in 1963, largely through the efforts of the UK-based *Hospital Physicists' Association* (HPA) which was the first national body of medical physicists in the world. The four national founding members of the IOMP were the UK, USA, Sweden, and Canada.

According to the inaugural statutes of the IOMP, each interested country was to join the IOMP through a National Committee for Medical Physics which was to coordinate medical physics interests within its own country and select delegates to represent it at the general meeting of the IOMP.

In retrospect, the requirements for joining the IOMP through a National Committee proved to have been successful in stimulating medical physicists in many countries to first form their national organization and then through the national organization join the IOMP. During the first 50 years of its existence, the IOMP grew from its original four sponsors to 80 national sponsoring organizations spanning all inhabited continents of the world.

The IOMP speaks on professional and scientific issues of interest to the world community of medical physicists. In particular, it sponsors the *World Congress on Medical Physics and Biomedical Engineering* that is held every 3 years and attracts several thousand medical physicists and biomedical engineers from around the world.

The IOMP also sponsors regular conferences on medical physics with specific objectives aiming at developing local medical physics services, strengthening the links among the regional medical physicists, and promoting the medical physics profession in regions and countries where holding a large world congress would not be feasible.

The following journals are recognized as official publications of the IOMP:

1. Physics in Medicine and Biology
2. Physiological Measurement
3. Medical Physics
4. Journal of Applied Clinical Medical Physics
5. Medical Physics International

In addition to representing 80 adhering national medical physics organizations, the IOMP also represents six regional medical physics federations. The six existing regional federations are as follows:

- Asia-Oceania Federation of Organizations for Medical Physics (AFOMP)
- European Federation of Organisations for Medical Physics (EFOMP)
- Latin American Medical Physics Association (Associação Latino-americana de Fisica Medica – ALFIM)
- Southeast Asian Federation for Medical Physics (SEAFOMP)
- Federation of African Medical Physics Organisations (FAMPO)
- Middle East Federation of Organisations for Medical Physics (MEFOMP).

From the list above we note that all continents except North America are covered by one or more regional medical physics organizations. North America consisting of only three countries, on the other hand, is covered well by the American Association of Physicists in Medicine (AAPM). The AAPM, despite being formed in 1958 as a U.S. national medical physics organization, can also be considered an international organization open to all practicing medical physicists irrespective of their country of work. Most Canadian medical physicists, in addition to supporting their national Canadian medical physics organizations, also belong to the AAPM and so do many Mexican medical physicists. It is thus reasonable to assume that the AAPM, in addition to playing the role of one of the 80 national IOMP sponsoring countries, also plays the role of a regional organization covering North America. For reasons of language, Mexico is one of the 11 countries sponsoring the ALFIM.

The *European Federation of Organizations for Medical Physics* (EFOMP) – www.efomp.org/ was founded in 1980 in London by 18 European Community members to serve as an umbrella organization to all national member medical physics organizations in Europe. It became the largest regional medical physics organization in the world, currently covering 39 national medical physics organizations representing over 5000 medical physicists from the 28 European Union (EU) countries and 12 countries adjacent to the EU.

The EFOMP's mission is:

1. To harmonize and advance medical physics at an utmost level both in its professional, clinical, and scientific expression throughout Europe.
2. To strengthen and make more effective the activities of the national member organizations by bringing about and maintaining systematic exchange of professional and scientific information, by the formulation of common policies, and by promoting education and training programs in medical physics.

The EFOMP accomplishes its mission by:

- Organizing congresses, meetings, and special courses.
- Publishing the journal "Physica Medica: *The European Journal of Medical Physics*" (EJMP), sponsoring four other scientific journals, and publishing an electronic bulletin *The European Medical Physics News*.
- Harmonizing European education and training in medical physics.
- Improving the profession and practice of medical physics in Europe.
- Encouraging the formation of organizations for medical physics where such organizations do not exist.
- Making recommendations on the appropriate general responsibilities, organizational relationships, and roles of medical physicists.

The *Asia-Oceania Federation of Organizations for Medical Physics* (AFOMP) – www.afomp.org/ was founded in 2000 and is currently sponsored by 16 regional countries. The federation publishes a newsletter and works closely with its member organizations as well as with the IOMP on professional and educational issues of interest to medical physicists in the region. It also organizes a regional medical physics congress on a yearly basis, cosponsors the *Journal of the Australasian College of Physical Scientists and Engineers in Medicine*, and is developing an accreditation program for academic medical physics programs as well as a certification program for medical physicists.

The *American Association of Physicists in Medicine* (AAPM), founded in 1958 (www.aapm.org), is the most prominent and by far the largest national medical physics organization in the world. It has over 8000 members, many of them from countries other than the US, making the AAPM the most international of the national medical physics organizations.

According to the AAPM *"The mission of the Association is to advance the practice of physics in medicine and biology by encouraging innovative research and development, disseminating scientific and technical information, fostering the education and professional development of medical physicists, and promoting the highest quality medical services for patients"*.

The AAPM is a very active organization involved in promotion of medical physics through:

- Sponsoring monthly scientific journals *Medical Physics* and *Journal of Applied Clinical Medical Physics*.
- Conducting annual summer schools on relevant medical physics or clinical subjects.
- Publishing task group reports and summer school proceedings.
- Publishing a bimonthly newsletter about AAPM activities and items of interest to AAPM members. It contains timely information and serves as a forum for debate about professional and educational issues of interest to AAPM members.
- Hosting a virtual library that contains many of the continuing education courses presented at the AAPM annual meetings.
- Providing a variety of educational and training programs in cooperation with medical physics organizations throughout the world.
- Sponsoring the Commission on Accreditation of Medical Physics Educational Programs (CAMPEP) with three other organizations.

Medical Physics Around the World

Of the 80 national medical physics organizations that are sponsoring the IOMP, most have a relatively small number of members, as evident from the average number of members for the 80 countries that amounts to about 200. There are only a few countries that have national organizations exceeding 500 members: USA 8000; UK 1700; Germany 1100; India 850; Japan 780; Italy 730; Canada 700; and Spain 510.

Since it is reasonable to assume that the IOMP represents the majority of practicing medical physicists worldwide, one may estimate the current average concentration of medical physicists in the world at 2.5 per million population.

However, there are large variations in this concentration from one region to another and from one country to another. For example, developed countries employ from 10 to 20 medical physicists per million population, while there are several developing countries with no medical physicists and many countries with less than 1 medical physicist per million population.

Considering the rapidly evolving technological base in modern medicine which requires an ever-increasing technical input from physicists, engineers, and technicians, the potential for growth in demand for medical physicists in the future is obvious. This is especially so for developing countries with many of them making serious efforts to modernize their health care services and in dire need of a significant improvement in their technological health care base not only in terms of equipment but also in terms of trained personnel to operate and maintain it.

Career in Medical Physics

Many academic and clinical educational programs are now available to an aspiring medical physicist for entering the medical physics profession. The ideal educational and professional steps are as follows:

1. *Undergraduate B.Sc. degree in Physics* (typical duration: 4 years)
2. *Graduate degree* (M.Sc. and/or Ph.D.) *in Medical Physics* from an accredited medical physics program. Typical duration of an M.Sc. program is 2 years; typical duration of Ph.D. program is 3 years or more after M.Sc. studies.
3. *Residency in medical physics* from an accredited residency program in medical physics. Typical duration of a residency program for a resident holding M.Sc. or Ph.D. degree in Medical Physics is 2 years.
4. Successful completion of a national *certification examination* in one of the four subspecialties of medical physics (as soon as possible upon completion of residency).

In principle, becoming a medical physicist through the four steps listed above is feasible; however, in practice, the steps are still somewhat difficult to follow because of the relatively low number of accredited academic and residency programs in medical physics. The number of these programs is growing, however. We are now in a transition period and within a decade, progression through the four steps listed above is likely to become mandatory for physicists entering the medical physics profession, similarly to the physicians entering the medical profession. The sooner broad-based didactic and clinical education through accredited educational programs in medical physics to become the norm, the better it will be for the medical physics profession and for patients the profession serves.

A career in medical physics is very rewarding and the work of medical physicists is interesting and versatile. A characteristic of modern societies is their ever-increasing preoccupation with health. Research in cancer and heart disease is growing yearly and many new methods for diagnosis and therapy are physical in nature, requiring the special skills of medical physicists not only in research but also in the direct application to patient care.

Undergraduate students with a strong background in science in general and physics in particular who decide upon a career in medical physics will find their studies of medical physics interesting and enjoyable, their employment prospects after completion of studies excellent, and their professional life satisfying and rewarding.

Physics played an important role in the development of imaging and treatment of disease with ionizing radiation and provided the scientific base, initially for the understanding of the production of radiation, its interaction with matter, and its measurement, and, during recent years, for the technological development of equipment used for imaging and delivery of radiation dose. The importance of modern technology and computerization in radiation medicine has been increasing steadily and dramatically, resulting in extremely sophisticated, efficient, and accurate equipment that is also very costly.

Medical physicists, with their scientific education in general and their under-standing of modern imaging and radiation therapy in particular, are well placed to play an important role in safe, efficient, and cost-effective use of high technology in diagnosis and treatment of disease with radiation, be it as part of an engineering team that designs the equipment or part of a medical team that purchases the equipment and uses it for patient care. Medical physicists also get involved in general health technology assessment as part of a biomedical engineering team that offers impartial and objective evaluation of medical devices to ensure that they meet appropriate standards of safety, quality, and performance both technically and clinically.

As part of medical team, medical physicists are involved with writing specifi-cations for high technology equipment before it is purchased, with negotiating conditions for its purchase, with organizing its acceptance and commissioning upon delivery by the vendor, as well as with organizing the equipment maintenance, servicing and calibration upon its acceptance and commissioning. Medical physi-cists also deal with governmental regulatory agencies and ensure that hospitals and clinics meet regulatory requirements to make the use of radiation in diagnosis and treatment of disease as safe as possible for both the patients and staff.

International Organizations

whose mission statements fully or partially address radiation protection, use of ionizing radiation in medicine, and promotion of medical physics:

American Association of Physicists in Medicine (AAPM), Alexandria, VA, USA	www.aapm.org
Asia-Oceania Federation of Organizations for Medical Physics (AFOMP)	www.afomp.org
European Federation of Organisations for Medical Physics (EFOMP), York, UK	www.efomp.org
European Society for Therapeutic Radiology and Oncology (ESTRO), Brussels, Belgium	www.estro.org
International Atomic Energy Agency (IAEA), Vienna, Austria	www.iaea.org
International Commission on Radiological Protection (ICRP), Stockholm. Sweden	www.icrp.org
International Commission on Radiation Units and Measurements (ICRU), Bethesda, Maryland, USA	www.icru.org
International Electrotechnical Commission (IEC), Geneva, Switzerland	www.iec.ch
International Organisation for Standardization (ISO), Geneva, Switzerland	www.iso.org
International Organisation for Medical Physics (IOMP)	www.iomp.org
International Radiation Protection Association (IRPA), Fontenay-aux-Roses, France	www.irpa.net
International Society of Radiology (ISR), Bethesda, Maryland, USA	www.isradiology.org
Pan American Health Organisation (PAHO), Washington, D.C., USA	www.paho.org
Radiological Society of North America (RSNA), Oak Brook, IL, USA	www.rsna.org
World Health Organization (WHO), Geneva, Switzerland	www.who.int

Contents

About the Author

Ervin B. Podgoršak, Ph.D., C.M. is Professor Emeritus of Medical Physics at McGill University in Montreal, Canada. He was born in Vienna, Austria and grew up in Ljubljana, Slovenia where he earned his Dipl. Ing. degree in Technical Physics from the University of Ljubljana. He pursued graduate work in physics at the University of Wisconsin in Madison, Wisconsin, USA receiving his M.Sc. degree in Physics under Prof. John R. Cameron in 1970 and Ph.D. in Physics under Prof. Paul R. Moran in 1973. He then specialized in medical and clinical physics as a post-doctoral fellow under Drs. Harold E. Johns and John R. Cunningham at the Ontario Cancer Institute and the University of Toronto in Canada. In 1975 he joined McGill University in Montreal, Canada and remained there until his retirement in 2010 from positions of Professor of Medical Physics, Director of McGill Academic and Residency programs in Medical Physics, and Director of Medical Physics Department at the McGill University Health Centre (MUHC). For his scientific and educational contributions to Medical Physics he received the Coolidge Award from the American Association of Physicists in Medicine (AAPM) in 2006, Gold Medal from the Canadian Organization of Medical Physicists (COMP) in 2008 and Kirkby Memorial Medal from the Canadian Association of Physicists (CAP) in 2011. In 2014 he was appointed by the Canadian Government to Membership (C.M.) in the Order of Canada.

Roman Letter Symbols

a	Acceleration; radius of atom; vertex–center distance for a hyperbola; specific activity; annum (year)
a	Year (annum)
a_{max}	Maximum specific activity
a_0	Bohr radius (0.5292 Å)
a_{TF}	Thomas–Fermi atomic radius
a_{theor}	Theoretical specific activity
A	Ampere (SI unit of current)
A	Vector function
A	Atomic mass number; Richardson thermionic constant
Å	Ångström (unit of length or distance: 10^{-10} m)
$\mathcal{A}$	Activity
$\mathcal{A}_D$	Daughter activity
$\mathcal{A}_P$	Parent activity
$\mathcal{A}_{sat}$	Saturation activity
$\mathcal{A}_{max}$	Maximum activity
b	Barn (unit of area: 10^{-24} cm^2)
b	Impact parameter
b_{max}	Maximum impact parameter
b_{min}	Minimum impact parameter
B_{col}	Atomic stopping number in collision stopping power
B	Build-up factor in broad beam attenuation
$\mathcal{B}$	Magnetic field
B	Boron atom
B_{rad}	Parameter in radiation stopping power
B_q	Becquerel (SI unit of activity)
c	Speed of light in vacuum (3×10^8 m/s)
c_n	Speed of light in medium
C	Coulomb (unit of electric charge); carbon atom
C_0	Collision stopping power constant (0.3071 MeV·cm^2/mol)
C_i	Curie (old unit of activity: 3.7×10^{10} s^{-1} = 3.7×10^{10} Bq)
C_K	K-shell correction for stopping power

C_M	Nuclear mass correction factor
C_v	Electric field correction factor
d	Day, deuteron
d	Distance; spacing
D	Daughter nucleus
D	Dose; characteristic distance in two-particle collision
$D_{\alpha-N}$	Distance of closest approach (between α-particle and nucleus)
D_{eff}	Effective characteristic scattering distance
$D_{\alpha-N}$	Effective characteristic scattering distance of closest approach between α-particle and nucleus
D_{e-a}	Effective characteristic scattering distance between electron and atom
D_{e-e}	Effective characteristic distance between the electron and orbital electron
D_{e-N}	Effective characteristic distance between electron and nucleus
D_{ex}	Exit dose
D_s	Surface dose
e	Electron charge (1.6×10^{-19} C)
e^-	Electron
e^+	Positron
e	Base of natural logarithm (2.7183...)
erf(x)	Error function
eV	Electron volt (unit of energy: 1.6×10^{-19} J)
$e\phi$	Work function
E	Energy
$\mathcal{E}$	Electric field
E_{ab}	Energy absorbed
$\bar{E}_{ab}$	Mean energy absorbed
E_B	Binding energy of electron in atom or neucleon in nucleus
E_{col}	Energy lost through collisions
E_i	Initial total energy of charged particle
$\mathcal{E}_{in}$	Electric field for incident radiation
E_K	Kinetic energy
$(E_K)_0$	Initial kinetic energy of charged particle
$(E_K)_{crit}$	Critical kinetic energy
$(E_K)_D$	Recoil kinetic energy of daughter
$(E_K)_f$	Final kinetic energy
$(E_K)_i$	Initial kinetic energy
$(E_K)_{IC}$	Kinetic energy of conversion electron
$(E_K)_{max}$	Maximum kinetic energy
$(E_K)_n$	Kinetic energy of incident neutron
$(E_K)_{thr}$	Threshold kinetic energy
E_n	Allowed energy state (eigenvalue)
E_0	Rest energy

$\mathcal{E}_{out}$	Electric field for scattered radiation
E_p	Barrier potential
E_R	Rydberg energy
E_{rad}	Energy radiated by charged particle
E_{thr}	Threshold energy
E_{tr}	Energy transferred
$\bar{E}_{tr}$	Average energy transferred
E_v	Photon energy; energy of neutrino
$\bar{E}_{tr}^{PP}$	Mean energy transferred from photons to charged particles in pair production
$\bar{E}_{tr}^{C}$	Mean energy transferred from photons to electrons in Compton effect
$\bar{E}_{tr}^{PE}$	Mean energy transferred from photons to electrons in photoeffect
E_β	Energy of beta particle
$(E_\beta)_{max}$	Maximum total energy of electron or positron in β decay
E_γ	Energy of gamma photon
$(E_\gamma)_{thr}$	Threshold energy for pair production
$(\mathcal{E}_z)_0$	Amplitude of electric field in uniform wave guide
f	Function; theoretical activity fraction; branching fraction in radioactive decay
$f(x)$	Function of independent variable x
$\bar{f}_{PE}$	Mean fraction of energy transferred from photons to electrons in photoelectric effect
$\bar{f}_C$	Mean fraction of energy transferred from photons to electrons in Compton effect
$(\bar{f}_C)_{max}$	Maximum energy transfer fraction in Compton effect
$\bar{f}_{PP}$	Mean fraction of energy transferred from photons to charged particles in pair production
$\bar{f}_{ab}$	Total mean energy absorption fraction
$\bar{f}_{tr}$	Total mean energy transfer fraction
f_{spin}	Spin correction factor
f_{recoil}	Recoil correction factor
fm	Femtometer (10^{-15} m)
F	Fluorine
F	Force
F_{coul}	Coulomb force
$F(K)$	Form factor
$F(x, Z)$	Atomic form factor
F_{KN}	Klein–Nishina form factor
F_L	Lorentz force
F_n	Neutron kerma factor
F^+	Stopping power function for positrons
F^-	Stopping power function for electrons
g	Gram (unit of mass: 10^{-3} kg)

$\bar{g}$	Mean radiation fraction
$\bar{g}_A$	Mean in-flight radiation fraction
$\bar{g}_B$	Mean bremsstrahlung fraction
$\bar{g}_i$	Mean impulse ionization fraction
G	Granddaughter nucleus
G	Newtonian gravitational constant
G_y	Gray (SI unit of kerma and dose: 1 J/kg)
h	Planck constant $(6.626 \times 10^{-34}\,\text{J·s})$, hour
h	Hour (1 h = 60 min = 3600 s)
H	Hydrogen
H	Equivalent dose; Hamiltonian operator
Hz	Unit of frequency (s^{-1})
$\hbar$	Reduced Planck constant $(h/2\pi)$
I	Electric current; mean ionization/excitation potential; beam intensity; radiation intensity
I_0	Initial photon beam intensity
j	Current density; quantum number in spin–orbit interaction
J	Joule (SI unit of energy)
$J_m(x)$	Bessel function of order m
k	Wave number, free space wave number, Boltzmann constant, effective neutron multiplication factor in fission chain reaction
k_g	Wave guide wave number (propagation coefficient)
k_g	Kilogram (SI unit of mass)
$k(K_\alpha)$	Wave number for K_α transition
kVp	Kilovolt peak (in x-ray tubes)
k^*	Ratio σ_P/σ_D in neutron activation
k_i	Initial wave vector
k_j	Final wave vector
K	$n = 1$ allowed shell (orbit) in an atom; Kelvin temperature; potassium
K	Wave vector
K	Kerma; capture constant in disk-loaded waveguide
K_{col}	Collision kerma
K_{rad}	Radiation kerma
K_α	Characteristic transition from L shell to K shell
l	Length
L	$n = 2$ allowed shell (orbit) in an atom
L	Angular momentum
L	Angular momentum vector
ℓ	Orbital quantum number; distance; path length
m	Meter (SI unit of length or distance)
m	Mass; magnetic quantum number; decay factor in parent–daughter–granddaughter decay; activation factor in nuclear activation; integer in Bragg relationship

m_e	Electron rest mass $(0.5110\ \text{MeV}/c^2)$
m_{e^-}	Electron rest mass
m_{e^+}	Positron rest mass
m_ℓ	Magnetic quantum number
m_n	Neutron rest mass $(939.6\ \text{MeV}/c^2)$
m_0	Rest mass of particle
m_p	Proton rest mass $(938.3\ \text{MeV}/c^2)$
m_α	Rest mass of α-particle
$m(v)$	Relativistic mass m at velocity v
m^*	Modified activation factor
M	$n = 3$ allowed shell (orbit) in an atom
$\mathbf{M}_{if}$	Matrix element
M	Mass of heavy nucleus
M_u	Molar mass constant
MeV	Megaeletron volt (unit of energy: 10^6 eV)
MHz	Megahertz (unit of frequency: 10^6 Hz)
MV	Megavoltage (in linacs)
$M(Z, A)$	Nuclear mass in atomic mass units
$\mathcal{M}(Z, A)$	Atomic mass in atomic mass units
Mu	Muonium
n	Neutron
nm	Nanometer (unit of length or distance: 10^{-9} m)
$\mathbf{n}$	Unit vector
n	Principal quantum number
$n^\square$	Number of atoms per volume
N	$n = 4$ allowed shell (orbit) in an atom; nitrogen, Newton (SI unit of force)
$N_m(x)$	Neumann function (Bessel function of second kind) of order m
N	Number of radioactive nuclei; number of experiments in central limit theorem; number or monoenergetic electrons in medium
N_a	Number of atoms
N_A	Avogadro number $(6.022 \times 10^{23}\ \text{atom/mol})$
N_e	Number of electrons
$N_{v/m}$	Number of specific nuclei per unit mass of tissue
O	Oxygen
OER	Oxygen enhancement ratio
p	Proton
p	Momentum
p_e	Electron momentum
p_ν	Photon momentum
$\mathbf{p}_i$	Initial particle momentum vector
$\mathbf{p}_f$	Final particle momentum vector
P	Parent nucleus
P	Power; probability

$\bar{P}$	Mean power
Pa	Pascal, SI derived unit of pressure ($1\,Pa = 1\,N/m^2$)
P_j	Probability for photoelectric effect, if it occurs, to occur in the j subshell
$P(\varepsilon, Z)$	Pair production function
Ps	Positronium
P_K	Fraction of photoelectric interactions that occur in the K shell
$P(x)$	Probability density function
q	Charge
Q	Charge; nuclear reaction energy; Q value
$\bar{Q}$	Expectation (mean) value of physical quantity Q
$[Q]$	Operator associated with the physical quantity Q
Q_{EC}	Decay energy (Q value) for electron capture
Q_{IC}	Decay energy (Q value) for internal conversion
Q_α	Decay energy (Q value) for alpha decay
Q_β	Decay energy (Q value) for beta decay
$Q(x)$	Standard cumulative distribution function
r	Radius vector; separation between two interacting particles, radius, of curvature
$\mathbf{r}$	Radius vector
rad	Old unit of absorbed dose (100 erg/g); radian
rem	Old unit of equivalent dose
r_e	Classical electron radius (2.818 fm)
r_n	Radius of the nth allowed Bohr orbit
$\bar{r}$	Average electron radius
R	Roentgen (unit of exposure: $2.58 \times 10^{-4}\,C/kg_{air}$)
RBE	Relative biological effectiveness
R	Radial wave function; radius (of nucleus); reaction rate; distance of closest approach
$\bar{R}$	Mean range
R_{CSDA}	Continuous slowing down approximation range
R_H	Rydberg constant for hydrogen ($109678\,cm^{-1}$)
R_{max}	Maximum penetration depth
R_0	Nuclear radius constant
$R_{\alpha-N}$	Distance of closest approach between the α-particle and nucleus in a non-direct hit collision
R_∞	Rydberg constant assuming an infinite nuclear mass ($109737\,cm^{-1}$)
R_{50}	Depth of the 50% depth dose in water for electron beam
s	Second (unit of time)
s	Spin quantum number
S	Mass stopping power
$\mathbf{S}$	Poynting vector
$\bar{S}$	Mean total mass stopping power
S_{col}	Mass collision stopping power

$\bar{S}_{col}$	Mean collision stopping power
S_{in}	Poynting vector of incident radiation
$\bar{S}_{in}$	Mean Poynting vector of incident radiation
S_{out}	Poynting vector of scattered radiation
$\bar{S}_{out}$	Mean Poynting vector of scattered radiation
S_{rad}	Mass radiation stopping power
S_{tot}	Total mass stopping power
Sv	Sievert (SI unit of equivalent dose)
$S(x, Z)$	Incoherent scattering function
t	Triton
t	Time; thickness of absorber in mass scattering power
t_{max}	Characteristic time in nuclear decay series or nuclear activation
$t_{1/2}$	Half-life
T	Temperature; linear scattering power; temporal function
T/ρ	Mass scattering power
TE	Transverse electric mode
TM	Transverse magnetic mode
torr	Non-SI unit of pressure defined as 1/760 of a standard atmosphere (1 torr = 1 mm Hg)
u	Unified atomic mass constant $(931.5 \, \text{MeV}/c^2)$
u	Particle velocity after collision; EM field density
U	Uranium atom
U	Applied potential
v	Velocity
v_{el}	Electron velocity
v_{thr}	Threshold velocity in Čerenkov effect
v_{en}	Velocity of energy flow
v_{gr}	Group velocity
v_n	Velocity of electron in nth allowed orbit
v_{ph}	Phase velocity
v_α	Velocity of α-particle
V	Volt (unit of potential difference); potential operator
V	Volume; potential energy
V_N	Nuclear potential
$V_{TF}(r)$	Thomas–Fermi potential
V_{FNS}	Potential energy for finite nuclear size
V_{Yuk}	Yukawa potential
$\mathcal{V}$	Volume
ν	Variance
w_R	Radiation weighting factor
w_C	Relative weight of Compton effect
w_{PE}	Relative weight of photoelectric effect
w_{PP}	Relative weight of pair production
W	Transmitted particle in weak interaction; tungsten atom

W_{el}	Electric energy stored per unit length
W_{if}	Transition (reaction) rate
W_{mag}	Magnetic energy stored per unit length
W	Watt (unit of power)
x	Momentum transfer variable $(x = \sin(\theta/2)/\lambda$; normalized time $x = t/t_{1/2}$; horizontal axis in 2D and 3D Cartesian coordinate system; coordinate in Cartesian coordinate system; abscissa axis
x_f	Particle final position
x_i	Particle initial position
x_0	Target thickness
x_{01}	First zero of the zeroth order Bessel function $(x_{01} = 2.405)$
$\bar{x}$	Mean free path; mean value of variable x
$(x_D)_{max}$	Maximum normalized characteristic time of the daughter
$x_{1/10}$	Tenth-value layer
$x_{1/2}$	Half-value layer
$^A_Z X$	Nucleus with symbol X, atomic mass number A and atomic number Z
X	Exposure
X_0	Target thickness; radiation length
$\bar{X}_{PE}(j)$	Mean fluorescence emission energy
y	Vertical axis in 2D Cartesian coordinate system; coordinate in Cartesian coordinate system; ordinate axis
Y	Radiation yield; activation yield
y_P	Normalized activity
$(y_D)_{max}$	Maximum normalized daughter activity
Y_D	Radioactivation yield of the daughter
$Y[(E_K)_0, Z]$	Radiation yield
y_P	Normalized parent activity
z	Atomic number of the projectile; depth in phantom; coordinate in Cartesian coordinate system; applicate axis
z_{max}	Depth of dose maximum
Z	Atomic number
Z_{eff}	Effective atomic number
Z^0	Transmitted particle in weak interaction

Greek Letter Symbols

α — Fine structure constant (1/137); ratioσ_P/σ_D; nucleus of helium atom (alpha particle)

α_{IC} — Internal conversion factor

β — Normalized particle velocity(v/c)

β^+ — Beta plus particle (positron)

β^- — Beta minus particle (electron)

γ — Photon originating in a nuclear transition; ratio of total to rest energy of a particle; ratio of total to rest mass of a particle

δ — Polarization (density effect) correction for stopping power; delta particle (electron); duty cycle for linear accelerators

Δ — Energy threshold for restricted stopping power

ε — Eccentricity of hyperbola; normalized photon energy: $\varepsilon = h\nu/(m_e c^2)$; Planck energy

ε^* — Ratio λ_D^*/λ_D in nuclear activation

ε_0 — Electric constant (electric permittivity of vacuum): $8.85 \times 10^{-12}\,\text{A·s/(V·m)}$

θ — Scattering angle for a single scattering event; scattering angle of projectile in projectile/target collision; scattering angle of photon in Compton and Rayleigh scattering

$\overline{\theta^2}$ — Mean square scattering angle for single scattering

θ_{cer} — Čerenkov characteristic angle

θ_{max} — Characteristic angle in bremsstrahlung production; maximum scattering angle

θ_{min} — Minimum scattering angle

θ_R — Characteristic angle for Rayleigh scattering

Θ — Scattering angle for multiple scattering

$\overline{\Theta^2}$ — Mean square scattering angle for multiple scattering

η — Pair production parameter; maximum energy transfer fraction in nuclear collision; energy boundary between hard and soft collision; fluorescence efficiency

κ	Linear attenuation coefficient for pair production
$_a\kappa$	Atomic attenuation coefficient for pair production
κ/ρ	Mass attenuation coefficient for pair production
λ	Wavelength; separation constant; decay constant; de Broglie wavelength of particle
λ_C	Compton wavelength
$(\lambda)_c$	Cutoff wavelength in uniform waveguide
λ_D	Decay constant of daughter
λ_D^*	Modified decay constant
λ_{min}	Duane–Hunt short wavelength cutoff
λ_P	Decay constant of parent
Λ	Separation constant
μ	Linear attenuation coefficient; reduced mass
μ_{ab}	Linear energy absorption coefficient
μ_{eff}	Effective attenuation coefficient
μ_H	Reduced mass of hydrogen atom
μ_m	Mass attenuation coefficient
μ_{tr}	Linear energy transfer coefficient
μ/ρ	Mass attenuation coefficient
μ_0	Magnetic constant (magnetic permeability of vacuum): $4\pi \times 10^{-7}$ (V·s)/(A·m)
(μ_{ab}/ρ)	Mass energy absorption coefficient
(μ_{tr}/ρ)	Mass energy transfer coefficient
$_a\mu$	Atomic attenuation coefficient
$_e\mu$	Electronic attenuation coefficient
μm	Unit of length or distance (10^{-6} m)
ν	Frequency
ν_{eq}	Photon frequency at which the atomic cross sections for Rayleigh scattering and Compton scattering are equal
ν_e	Electronic neutrino
ν_{orb}	Orbital frequency
ν_{trans}	Transition frequency
ν_μ	Muonic neutrino
ξ	Ratio between daughter and parent activities at time t; Thomas–Fermi atomic radius constant; absorption edge parameter in photoelectric effect
ξ_j	Absorption edge parameter for subshell j
π	Pi meson (pion)
π^+	Positive pi meson (pion)
π^-	Negative pi meson (pion)
ρ	Density; energy density
$\rho(E_f)$	Density of final states
σ	Cross section; linear attenuation coefficient; standard deviation
σ_{rad}	Cross section for emission of bremsstrahlung
σ_C	Compton cross section (attenuation coefficient)

σ_C^{KN}	Klein–Nishina cross section for Compton effect
$_a\sigma_C$	Atomic attenuation coefficient (cross section) for Compton effect
$_a\sigma_R$	Atomic attenuation coefficient (cross section) for Rayleigh scattering
$_a\sigma_{Th}$	Atomic attenuation coefficient (cross section) for Thomson scattering
$_e\sigma_C$	Electronic attenuation coefficient for Compton effect
σ_D	Daughter cross section in particle radioactivation
σ_R	Rayleigh cross section (linear attenuation coefficient)
σ_{Ruth}	Cross section for Rutherford scattering
σ_{Th}	Thomson cross section (linear attenuation coefficient)
$_a\sigma$	Atomic cross section (in cm^2/atom)
$_e\sigma$	Electronic cross section (in cm^2/electron)
$[\sigma(z)]^2$	Spatial spread of electron pencil beam
τ	Linear attenuation coefficient for photoelectric effect; normalized electron kinetic energy; mean (average) life
$_a\tau$	Atomic attenuation coefficient for photoelectric effect
τ/ρ	Mass attenuation coefficient for photoelectric effect
ϕ	Angle between radius vector and axis of symmetry on a hyperbola; recoil angle of the target in projectile/target collision; neutron recoil angle in elastic scattering on nucleus; recoil angle of the electron in Compton scattering
φ	Particle fluence
$\dot{\varphi}$	Particle fluence rate
χ	Homogeneity factor
ψ	Wavefunction (eigenfunction) depending on spatial coordinates; energy fluence
Ψ	Wavefunction depending on spatial and temporal coordinates
ω	Fluorescence yield; angular frequency
ω_c	Cutoff angular frequency in acceleration waveguide
ω_{cyc}	Cyclotron frequency fluorescence yield for K-shell transition
ω_K	Fluorescence yield for K-shell transition
Ω	Solid angle

Acronyms

AAPM	American Association of Physicists in Medicine
ACR	American College of Radiology
AFOMP	Asia-Oceania Federation of Organizations for Medical Physics
ALFIM	Associaçåo Latino-americana de Fisica Medica
ART	Adaptive radiotherapy
BNCT	Boron neutron capture therapy
BNL	Brookhaven National Laboratory
CAMPEP	Commission on Accreditation of Medical Physics Educational Programs
CCD	Charge-coupled device
CCPM	Canadian College of Physicists in Medicine
CODATA	Committee on Data for Science and Technology
CPA	Charged particle activation
CPE	Charged particle equilibrium
CSDA	Continuous slowing down approximation
CT	Computerized tomography
CVD	Chemical vapor deposition
CNT	Carbon nanotube
DT	Deuterium-tritium
EC	Electron capture
EFOMP	European Federation of Organisations of Medical Physics
EM	Electromagnetic
EGS	Electron-gamma shower
EE	Exoelectron emission
FAMPO	Federation of African Medical Physics Organizations
FDG	Fluoro-deoxy-glucose
FE	Field emission
FNS	Finite nuclear size
FWHM	Full width at half maximum
HDR	High dose rate
HPA	Hospital Physicists' Association

HVL	Half-value layer
IAEA	International Atomic Energy Agency
IC	Internal conversion, Ionization chamber
ICRP	International Commission on Radiation Protection
ICRU	International Commission on Radiation Units and Measurements
IE	Ionization energy
IGRT	Image-guided radiotherapy
IMPCB	International Medical Physics Certification Board
IMRT	Intensity modulated radiotherapy
IOMP	International Organization for Medical Physics
IP	Ionization potential
IUPAC	International Union of Pure and Applied Chemistry
KN	Klein–Nishina
LDR	Low dose rate
LET	Linear energy transfer
LIC	Liquid ionization chamber
LINAC	Linear accelerator
MC	Monte Carlo
MEFOMP	Middle East Federation of Organizations for Medical Physics
MFP	Mean free path
MLC	Multi-leaf collimator
MOC	Maintenance of certification
MRI	Magnetic resonance imaging
MV	Megavoltage
NA	Nuclear activation
NDS	Nuclear data section
NE	Neutron emission
NIST	National Institute of Standards and Technology
NNDC	National Nuclear Data Center
NPP	Nuclear Pair Production
NRC	National Research Council
NTCP	Normal tissue complication probability
OER	Oxygen enhancement ratio
OSL	Optically stimulated luminescence
PE	Photoelectric, Proton emission
PET	Positron emission tomography
PMT	Photomultiplier tube
PP	Pair production
PPM	Parts per million
PPS	Pulses per second
RBE	Relative biological effectiveness
RF	Radiofrequency
RT	Radiotherapy
SAD	Source-axis distance
SF	Spontaneous fission

SI	Système International
SEAFOMP	Southeast Asian Federation for Medical Physics
SLAC	Stanford Linear Accelerator Center
SRA	Synchrotron radiation angioplasty
SSD	Source-skin distance, source-surface distance
STP	Standard temperature and pressure
TCP	Tumor control probability
TL	Thermoluminescence
TVL	Tenth-value layer
TE	Transverse electric
TM	Transverse magnetic
TP	Triplet Production
UK	United Kingdom
US	Ultrasound

First Solvay Conference on Physics

Photograph on next page shows many of the outstanding physicists that shaped modern physics during the first decades of the 20th century. The photograph was taken during the First Solvay International Conference on Physics (*Conseil de Physique de Solvay*) held in Brussels in the autumn of 1911 under the chairmanship of Hendrik A. Lorentz and under the sponsorship of Ernest Solvay. The subject of the conference was the Theory of Radiation and Quanta (*La théorie du rayonnement et les quanta*) and the discussion centered on problems with having two approaches to physics: classical physics and the nascent quantum theory.

The Solvay International Conferences on Physics are held every three to five years. Organized by the International Solvay Institutes for Physics and Chemistry located in Brussels, they are noted for their role in stimulating advances in atomic and quantum physics. The key to the success of the "Conseils Solvay" is in their organization: an international scientific committee is in charge of defining the general theme of the conference and of selecting the chair person who then sets up the program and invites the selected individual participants. The emphasis during the conference is on discussions of a particular scientific problem rather than on presentations.

The International Solvay Institutes were founded by Ernest Solvay, a Belgian chemist, industrialist, and philanthropist who in the 1860s developed the ammonia process for the manufacture of sodium carbonate, widely used in the production of glass and soap. In 1863 Solvay founded a company Solvay & Cie and during his lifetime transformed it into a global company currently headquartered in Brussels and specializing in pharmaceuticals, chemicals, and plastics.

Solvay was also a man of liberal and progressive social ideals. After becoming wealthy through his company Solvay & Cie, he founded several scientific institutes as well as charitable foundations. To date 26 Solvay International Conferences on Physics were held; however, in terms of notable participants, the first one that was held in Brussels in 1911 seems to have been of greatest historical significance and also carries the distinction of having been the first international physics conference ever organized.

From left to right seated at the table:

Hermann Walther **Nernst**; Marcel Louis **Brillouin**; Ernest **Solvay**; Hendrik Antoon **Lorentz** (chairman); Emil Gabriel **Warburg**; Jean Baptiste **Perrin**; Wilhelm **Wien**; Marie **Sklodowska-Curie**; Henri **Poincaré**

From left to right standing:

Robert **Goldschmidt**; Max **Planck**; Heinrich **Rubens**; Arnold **Sommerfeld**; Frederick **Lindemann**; Maurice **DeBroglie**; Martin **Knudsen**; Friedrich **Hasenöhrl**; Georges **Hostelet**; Edouard **Herzen**; James Hopwood **Jeans**; Ernest **Rutherford**; Heike **Kamerlingh-Onnes**; Albert **Einstein**; Paul **Langevin**.

Photograph taken by Benjamin Couprie © International Institutes for Physics and Chemistry. Courtesy: Emilio Segrè Visual Archives, American Institute of Physics. Reproduced with Permission.

Chapter 1
Introduction to Modern Physics

This chapter provides an introduction to modern physics and covers basic elements of atomic, nuclear, relativistic, and quantum physics as well as electromagnetic theory. These elements form the background knowledge that is required for a study of medical radiation physics. The first few pages of this chapter present lists of basic physical constants, of important derived physical constants, and of milestones in modern physics and medical physics. These lists would normally be relegated to appendices at the end of the book; however, in this textbook they are given a prominent place at the beginning of the book to stress their importance to modern physics as well as to medical physics.

Scan ©American Institute of Physics

© Springer International Publishing Switzerland 2016
E.B. Podgoršak, *Radiation Physics for Medical Physicists*,
Graduate Texts in Physics, DOI 10.1007/978-3-319-25382-4_1

After introducing the basic physical constants and the derived physical constants of importance in modern physics and medical physics, the chapter spells out the rules governing physical quantities and units and introduces the classification of natural forces, fundamental particles, ionizing radiation in general as well as directly and indirectly ionizing radiation. Next the basic definitions for atomic and nuclear structure are given and the concepts of the physics of small dimensions (quantum physics) as well as large velocities (relativistic physics) are briefly reviewed.

A short introduction to particle-wave duality is given, wave mechanics is briefly discussed, and Fermi second golden rule and the Born collision formula are introduced. After a brief discussion of Maxwell equations and the Poynting theorem, the chapter concludes with a discussion of the normal probability distribution described by the continuous probability density function.

Medical physics is intimately related to modern physics and most milestone discoveries in modern physics were rapidly translated into medical physics and medicine, as evident from the list of milestones in medical physics provided in Sect. 1.3. Medical physics is a perfect and long-standing example of translational research where basic experimental and theoretical discoveries are rapidly implemented into benefiting humanity through improved procedures in diagnosis and treatment of disease.

A thorough understanding of the basics presented in this chapter will facilitate the readers' study of subsequent chapters and enhance their appreciation of the nature, importance, and history of medical physics as it relates to the use of ionizing radiation in diagnosis and treatment of human disease.

1.1   Fundamental Physical Constants

Currently the best source of data on fundamental physical constants is the *Committee on Data for Science and Technology* (CODATA), an inter-disciplinary scientific committee of the International Council for Science (ICSU) with headquarters in Paris, France. The ICSU's membership comprises 121 national organizations and 32 international scientific unions.

The CODATA Task Group on Fundamental Constants (www.codata.org) was established in 1969 and its purpose is to periodically provide the scientific and technological communities with an internationally accepted set of values of fundamental physical constants for worldwide use. The mission of CODATA is: "*To strengthen international science for the benefit of society by promoting improved scientific and technical data management and use*" and its membership comprises 23 full national members, 2 associate members, and 17 international scientific unions. The committee publishes "The DATA Science Journal", a peer-reviewed, open access electronic journal featuring papers on management of data and databases in science and technology.

The data below (rounded off to four significant figures) were taken from the recent CODATA set of values issued in 2010 and easily available from the web-site supported by the National Institute of Science and Technology (NIST) in Washington, DC, USA (http://physics.nist.gov/cuu/Constants/)

Avogadro constant ....$N_A = 6.022 \times 10^{23}$ mol^{-1}

Speed of light in vacuum .... $c = 2.998 \times 10^8$ m/s $\approx 3 \times 10^8$ m/s

Atomic mass constant .... u $= 1.661 \times 10^{-27}$ kg $= 931.5$ MeV/c^2

Elementary charge .... $e = 1.602 \times 10^{-19}$ C

Electron rest mass .... $m_e = 9.109 \times 10^{-31}$ kg $= 0.5110$ MeV/c^2

Positron rest mass .... $m_e = 9.109 \times 10^{-31}$ kg $= 0.5110$ MeV/c^2

Proton rest mass .... $m_p = 1.673 \times 10^{-27}$ kg $= 1.007$ u $= 938.3$ MeV/c^2

Neutron rest mass .... $m_n = 1.675 \times 10^{-27}$ kg $= 1.009$ u $= 939.6$ MeV/c^2

Planck constant .... $h = 6.626 \times 10^{-34}$ J·s $= 4.136 \times 10^{-15}$ eV·s

Reduced Planck constant .... $\hbar = \dfrac{1}{2\pi}h = 1.055 \times 10^{-34}$ J·s

Boltzmann constant .... $k = 1.381 \times 10^{-23}$ J·K^{-1} $= 0.8617 \times 10^{-4}$ eV·K^{-1}

Electric constant .... $\varepsilon_0 = 8.854 \times 10^{-12}$ F·m^{-1} $= 8.854 \times 10^{-12}$ C/(V·m)

Magnetic constant .... $\mu_0 = 4\pi \times 10^{-7}$ N·A^{-2}/(A·m) ≈ 12.57 (V·s)/(A·m)

Newtonian gravitation constant .... $G = 6.674 \times 10^{-11}$ m^3·kg^{-1}·s^{-2}

Proton/electron mass .... $m_p/m_e = 1836.1$

Specific charge of electron .... $e/m_e = 1.759 \times 10^{11}$ C·kg^{-1}

Planck constant/electron charge .... $h/e = 4.136 \times 10^{-15}$ V·s

Alpha particle mass ...... $m_\alpha = 6.645 \times 10^{-27}$ kg $= 4.0015$ u $= 3727.4$ MeV/c^2

Elementary charge / Planck constant .... $e/h = 2.418 \times 10^{14}$ A·J^{-1}

1.2   Derived Physical Constants and Relationships

- **Speed of light** in vacuum:

$$c = \frac{1}{\sqrt{\varepsilon_0 \mu_0}} \approx 3 \times 10^8 \text{ m/s} \tag{1.1}$$

- **Reduced Planck constant × speed of light in vacuum**:

$$\hbar c = \frac{h}{2\pi} c = 197.3 \text{ MeV·fm} = 197.3 \text{ eV·nm} \approx 200 \text{ MeV·fm} \tag{1.2}$$

- **Bohr radius constant** [see (3.4)]

$$a_0 = \frac{\hbar c}{\alpha \, m_e c^2} = \frac{4\pi\varepsilon_0}{e^2} \frac{(\hbar c)^2}{m_e c^2} = 0.5292 \text{ Å} = 0.5292 \times 10^{-10} \text{ m} \tag{1.3}$$

- **Fine structure constant** [see (3.6)]

$$\alpha = \frac{e^2}{4\pi\varepsilon_0} \frac{1}{\hbar c} = \frac{\hbar c}{a_0 m_e c^2} = 7.297 \times 10^{-3} \approx \frac{1}{137} \tag{1.4}$$

- **Rydberg energy** [see (3.8)]

$$E_R = \frac{1}{2} m_e c^2 \alpha^2 = \frac{1}{2} \left[\frac{e^2}{4\pi\varepsilon_0} \right]^2 \frac{m_e c^2}{(\hbar c)^2} = 13.61 \text{ eV} \tag{1.5}$$

- **Rydberg constant** [see (3.11)]

$$R_\infty = \frac{E_R}{2\pi\hbar c} = \frac{m_e c^2 \alpha^2}{4\pi\hbar c} = \frac{1}{4\pi} \left[\frac{e^2}{4\pi \, \varepsilon_0} \right]^2 \frac{m_e c^2}{(\hbar c)^3} = 109737 \text{ cm}^{-1} \tag{1.6}$$

- **Classical electron radius** [see (6.43), (6.60), (7.31), (7.41), and (7.89)]

$$r_e = \frac{e^2}{4\pi\varepsilon_0 m_e c^2} = 2.818 \text{ fm} = 2.818 \times 10^{-15} \text{ m} \tag{1.7}$$

- **Compton wavelength** of the electron [see (7.44)]

$$\lambda_c = \frac{h}{m_e c} = \frac{2\pi\hbar c}{m_e c^2} = 0.02426 \text{ Å} = 2.426 \times 10^{-12} \text{ m} \tag{1.8}$$

- **Thomson classical cross section** for free electron [see (7.41)]

$$\sigma_{Th} = \frac{8\pi}{3} r_e^2 = 0.6653 \text{ b} = 0.6653 \times 10^{-24} \text{ cm}^2 \tag{1.9}$$

- **Collision stopping power constant** [see (6.43)]

$$C_0 = 4\pi N_A \left(\frac{e^2}{4\pi\varepsilon_0} \right)^2 \frac{1}{m_e c^2} = 4\pi N_A r_e^2 m_e c^2 = 0.3071 \text{ MeV·cm}^2 \text{·mol}^{-1} \tag{1.10}$$

1.3 Milestones in Modern Physics and Medical Physics

x-rays	Wilhelm Konrad **Röntgen**	1895
Natural radioactivity	Antoine-Henri **Becquerel**	1896
Electron	Joseph John **Thomson**	1897
Radium-226	Pierre **Curie**, Marie **Skłodowska-Curie**	1898
Alpha particle	Ernest **Rutherford**	1899
Energy quantization	Max **Planck**	1900
Special theory of relativity	Albert **Einstein**	1905
Photoelectric effect	Albert **Einstein**	1905
Characteristic x-rays	Charles **Barkla**	1906
Alpha particle scattering	Hans **Geiger**, Ernest **Marsden**	1909
Atomic model	Ernest **Rutherford**	1911
Thermionic emission	Owen W. **Richardson**	1911
Electron charge	Robert **Millikan**	1911
Model of hydrogen atom	Neils **Bohr**	1913
Tungsten filament for x-ray tubes	William D. **Coolidge**	1913
Energy quantization	James **Franck**, Gustav **Hertz**	1914
Proton	Ernest **Rutherford**	1919
x-ray scattering (Compton effect)	Arthur H. **Compton**	1922
Exclusion principle	Wolfgang **Pauli**	1925
Quantum wave mechanics	Erwin **Schrödinger**	1926
Wave nature of the electron	Clinton J. **Davisson**, Lester H. **Germer**	1927
Cyclotron	Ernest O. **Lawrence**	1931
Neutron	James **Chadwick**	1932
Positron	Carl D. **Anderson**	1932
Artificial radioactivity	Irène **Joliot-Curie**, Frédéric **Joliot**	1934
Čerenkov radiation	Pavel A. **Čerenkov**, Sergei I. **Vavilov**	1934
Uranium fission	**Meitner, Frisch, Hahn, Strassmann**	1939
Betatron	Donald W. **Kerst**	1940
Spontaneous fission	Georgij N. **Flerov**, Konstantin A. **Petržak**	1940
Nuclear magnetic resonance	Felix **Bloch**, Edward **Purcell**	1946
Cobalt-60 machine	Harold E. **Johns**	1951
Recoil-less nuclear transition	Rudolf L. **Mössbauer**	1957
Gamma Knife	Lars **Leksell**	1968
Computerized tomography (CT)	Godfrey N. **Hounsfield**, Alan M. **Cormack**	1971
Magnetic resonance imaging (MRI)	Paul C. **Lauterbur**, Peter **Mansfield**	1973
Positron emission tomography (PET)	Michael **Phelps**	1973

1.4   Physical Quantities and Units

1.4.1   Rules Governing Physical Quantities and Units

A physical quantity is defined as quantity that can be used in mathematical equations of science and technology. It is characterized by its numerical value (magnitude) and associated unit. The following rules apply in general:

- Symbols for physical quantities are set in *italic (sloping)* type, while symbols for units are set in roman (upright) type.

    For example: $m = 21$ kg; $E = 15$ MeV; $K = 180$ cGy.

- The numerical value and the unit of a physical quantity must be separated by space.

    For example: 21 kg, *not* 21kg; 15 MeV, *not* 15MeV.

- Superscripts and subscripts used with physical quantities are in italic type if they represent variables, quantities, or running numbers; they are in roman type if they are descriptive.

    For example: N_X (exposure calibration coefficient with X a quantity), *not* N_X; $\sum_{i=0}^{n} X_i$, where i and n represent running numbers, *not* $\sum_{i=0}^{n} X_i$; $_a\mu_{tr}$ where a and tr are descriptive subscripts, *not* $_a\mu_{tr}$; U_{max} *not* U_{max}.

- A space or half-high dot is used to signify multiplication of units.

    For example: 15 m/s or $15 \, \text{m·s}^{-1}$ or $15 \, \text{m s}^{-1}$, *not* $15 \, \text{ms}^{-1}$.

- It must be clear to which unit a numerical value belongs and which mathematical operation applies to the specific quantity.

    For example:
    $10 \, \text{cm} \times 15 \, \text{cm}$, *not* 10×15 cm;
    1 MeV to 10 MeV or (1 to 10) MeV, *not* 1 MeV–10 MeV and *not* 1 to 10 MeV;
    $100 \, \text{cGy} \pm 2 \, \text{cGy}$ or (100 ± 2) cGy, *not* 100 ± 2 cGy;
    $80\% \pm 10\%$ or $(80 \pm 10)\%$, *not* $80 \pm 10\%$;
    $210 \times (1 \pm 10\%)$ cGy, *not* 210 cGy $\pm 10\%$.

1.4.2   The SI System of Units

The currently used metric system of units is known as the *Système International d'Unités* (International System of Units) with the international abbreviation SI. The system is founded on base units for *seven basic physical quantities*. All other physical quantities and units are derived from the seven base quantities and units. The system also defines standard prefixes to the unit names and symbols to form decimal multiples of fundamental and derived units with special names. Currently 20 agreed upon prefixes are in use.

The seven base quantities and their units are:

Length ℓ .... meter (m)
Mass m .... kilogram (kg)
Time t .... second (s)
Electric current I .... ampere (A)
Temperature T .... kelvin (K)
Amount of substance .... mole (mol)
Luminous intensity .... candela (cd)

Examples of basic and derived physical quantities and their units are given in Table 1.1. The Système International obtains its international authority from the Meter Convention that was endorsed in 1875 by 17 countries; the current membership stands at 48 countries.

While six of the seven basic physical quantities and their units seem straightforward, the quantity "amount of substance" and its unit the mole (mol) cause many conceptual difficulties for students. The SI definition of the mole is as follows: *"One mole is the amount of substance of a system which contains as many elementary entities as there are atoms (unbound, at rest, and in ground state) in 0.012 kg of carbon-12"*.

Table 1.1 The main basic physical quantities and several derived physical quantities with their units in Système International (SI) and in radiation physics

Physical quantity	Symbol	Unit in SI	Units commonly used in physics	Conversion
Length	ℓ	m	nm, Å, fm	$1\,\text{m} = 10^9\,\text{nm} = 10^{10}\,\text{Å} = 10^{15}\,\text{fm}$
Mass	m	kg	MeV/c^2	$1\,\text{MeV}/c^2 = 1.778 \times 10^{-30}\,\text{kg}$
Time	t	s	ms, μs, ns, ps	$1\,\text{s} = 10^3\,\text{ms} = 10^6\,\mu\text{s} = 10^9\,\text{ns}$ $= 10^{12}\,\text{ps}$
Current	I	A	mA, μA, nA, pA	$1\,\text{A} = 10^3\,\text{mA} = 10^6\,\mu\text{A} = 10^9\,\text{nA}$
Temperature	T	K		$T\,(\text{in K}) = T\,(\text{in}\,^\circ\text{C}) + 273.16$
Mass density	ρ	kg/m^3	g/cm^3	$1\,\text{kg/m}^3 = 10^{-3}\,\text{g/cm}^3$
Current density	j	A/m^2		
Velocity	v	m/s		
Acceleration	**a**	m/s^2		
Frequency	ν	Hz		$1\,\text{Hz} = 1\,\text{s}^{-1}$
Electric charge	q	C	e	$1\,\text{e} = 1.602 \times 10^{-19}\,\text{C}$
Force	**F**	N		$1\,\text{N} = 1\,\text{kg·m·s}^{-2}$
Pressure	p	Pa	760 torr = 101.3 kPa	$1\,\text{Pa} = 1\,\text{N·m}^{-2} = 7.5 \times 10^{-3}\,\text{torr}$
Momentum	**p**	N·s		$1\,\text{N·s} = 1\,\text{kg·m·s}^{-1}$
Energy	E	J	eV, keV, MeV	$1\,\text{eV} = 1.602 \times 10^{-19}\,\text{J} = 10^{-3}\,\text{keV}$
Power	P	W		$1\,\text{W} = 1\,\text{J/s} = 1\,\text{V·A}$

Table 1.2 Non-SI unit of quantities of importance to general physics as well as medical physics

Physical quantity	Unit	Symbol	Value in SI units
Time	Day	d	1 d = 24 h = 86400 s
	Hour	h	1 h = 60 min = 3600 s
	Minute	min	1 min = 60 s
Angle	Degree	o	$1° = (\pi/180)$ rad
Energy	Electron volt[a]	eV	1 eV = 1.6×10^{-19} J
Mass	Unified atomic mass unit[b]	u	1 u = 931.5 MeV/c^2

[a]The electron volt (eV) is defined as the kinetic energy acquired by an electron with charge $e = 1.602 \times 10^{-19}$ C passing through a potential difference of 1 V in vacuum
[b]The unified atomic mass unit (u) or dalton (Da) is a unit of atomic or molecular mass. It is equal to 1/12 of the mass of an unbound carbon-12 atom, at rest in its ground state. (for more detail see Sect. 1.13)

The following additional features should be noted:

- The term "elementary entity" is defined as atom, molecule, ion, electron, or some other particle and represents the smallest component of a substance which cannot be broken down further without altering the nature of the substance.
- When referring to "mole of a substance" it is important to specify the elementary entity under consideration, such as atom, molecule, ion, other particle, etc.
- The number of elementary entities in a mole is by definition the same for all substances and is equal to a universal constant called the Avogadro constant N_A with unit mol^{-1}, as discussed in Sect. 1.13.2.

1.4.3 Non-SI Units

Certain units are not part of the SI system of units but, despite being outside the SI, are important and widely used with the SI system. Some of the important non-SI units are listed in Table 1.2.

1.5 Classification of Forces in Nature

Four distinct forces are observed in the interaction between various types of particles. These forces, in decreasing order of strength, are the *strong force, electromagnetic (EM) force, weak force*, and *gravitational force* with relative strengths of 1, 1/137, 10^{-6}, and 10^{-39}, respectively. The four fundamental forces, their source, and their transmitted particle are listed in Table 1.3. As far as the range of the four fundamental forces is concerned, the forces are divided into two groups: two forces are infinite range force and two are very short-range force:

Table 1.3 The four fundamental forces in nature, their source, their transmitted particle, and their relative strength normalized to 1 for the strong force

Force	Source	Transmitted particle	Relative strength
Strong	Strong charge	Gluon	1
EM	Electric charge	Photon	1/137
Weak	Weak charge	W^+, W^-, and Z^0	10^{-6}
Gravitational	Energy	Graviton	10^{-39}

1. The range of the EM and gravitational force is infinite ($1/r^2$ dependence where r is the separation between two interacting particles).
2. The range of the strong and weak force is extremely short (of the order of a few femtometers).

Each force results from a particular intrinsic property of the particles, such as strong charge for the strong force, electric charge for the EM force, weak charge for the weak force, and energy for the gravitational force:

- *Strong charge* enables the strong force transmitted by mass-less particles called gluons and resides in particles referred to as quarks.
- *Electric charge* enables the electromagnetic force transmitted by photons and resides in charged particles such as electrons, positrons, protons, etc.
- *Weak charge* enables the weak force transmitted by particles called W and Z^0, and resides in particles called quarks and leptons.
- *Energy* enables the gravitational force transmitted by a hypothetical particle called graviton.

1.6 Classification of Fundamental Particles

Two classes of fundamental particles are known: *hadrons* and *leptons*.

1. *Hadrons* are particles that exhibit strong interactions. They are composed of quarks with a fractional electric charge ($\frac{2}{3}$ or $-\frac{1}{3}$) and characterized by one of three types of strong charge called color (red, blue, green). There are six known quarks: up, down, strange, charm, top, and bottom. Two classes of hadrons are known: mesons and baryons.

   - Mesons are unstable subatomic particles consisting of a quark and antiquark bound together by strong force. They have rest masses that fall between the rest mass of the electron ($m_e = 0.511$ MeV/c^2) and the rest mass of the proton ($m_p = 938.3$ MeV/c^2).
   - Baryons have a rest mass equal to or greater than the proton rest mass. Proton and neutron as well as many more exotic heavy particles belong to the baryon

group. All baryons with the exception of proton are unstable and decay into products that include a proton as the end product.

2. *Leptons* are particles that do not interact strongly. Electron e, muon μ, tau τ and their corresponding neutrinos ν_e, ν_μ, and ν_τ are in this category.

1.7 Classification of Radiation

Radiation is classified into two main categories, as shown in Fig. 1.1: *non-ionizing* and *ionizing*, depending on its ability to ionize matter. The ionization energy (IE), also known as ionization potential (IP), of atoms is defined as the minimum energy required for ionizing an atom and is typically specified in electron volts (eV). In nature IE ranges from a few electron volts ($\sim$4 eV) for alkali elements to 24.6 eV for helium (noble gas) with IE for all other atoms lying between the two extremes.

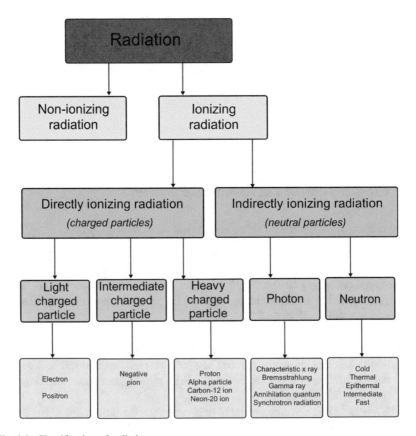

Fig. 1.1 Classification of radiation

- *Non-ionizing radiation* cannot ionize matter because its energy is lower than the ionization energy of atoms or molecules of the absorber. The term non-ionizing radiation thus refers to all types of electromagnetic radiation that do not carry enough energy per quantum to ionize atoms or molecules of the absorber. Near ultraviolet radiation, visible light, infrared photons, microwaves, and radio waves are examples of non-ionizing radiation.
- *Ionizing radiation* can ionize matter either directly or indirectly because its quantum energy exceeds the ionization potential of atoms and molecules of the absorber. Ionizing radiation has many practical uses (see Sect. 1.8.3) in industry, agriculture, and medicine but it also presents a health hazard when used carelessly or improperly. Medical physics is largely, but not exclusively, based on the study and use of ionizing radiation in medicine; health physics deals with health hazards posed by ionizing radiation and with safety issues related to use of ionizing radiation.

1.8 Classification of Ionizing Radiation

Ionizing radiation is classified into two distinct categories according to the mode of ionization and also into two categories according to the density of ionization it produces in the absorber.

1.8.1 Directly and Indirectly Ionizing Radiation

According to the mode of ionization there are two categories of ionizing radiation, directly ionizing and indirectly ionizing:

1. *Directly ionizing radiation*: Comprises charged particles (electrons, protons, α-particles, heavy ions) that deposit energy in the absorber through a direct one-step process involving Coulomb interactions between the directly ionizing charged particle and orbital electrons of the atoms in the absorber.
2. *Indirectly ionizing radiation*: Comprises neutral particles (photons such as x-rays and γ-rays, neutrons) that deposit energy in the absorber through a two-step process as follows:

   - In the first step a charged particle is released in the absorber (photons release either electrons or electron/positron pairs, neutrons release protons or heavier ions).
   - In the second step, the released charged particles deposit energy to the absorber through direct Coulomb interactions with orbital electrons of the atoms in the absorber.

Both directly and indirectly ionizing radiations are used in diagnosis and treatment of disease. The branch of medicine that uses ionizing radiation in treatment of disease is called radiotherapy, therapeutic radiology or radiation oncology. The branch of medicine that uses ionizing radiation in diagnosis of disease is called medical imaging and is usually divided into two categories: diagnostic radiology based on use of x-rays for imaging and nuclear medicine now often referred to as molecular imaging based on use of radionuclides for imaging.

1.8.2 Low LET and High LET Radiation

The ionization density produced by ionizing radiation in tissue depends on the linear energy transfer (LET) of the ionizing radiation beam. The LET is defined as the mean amount of energy that a given ionizing radiation imparts to absorbing medium (such as tissue) per unit path length and is used in radiobiology and radiation protection to specify the quality of an ionizing radiation beam. According to the density of ionization produced in the absorber there are two distinct categories of ionizing radiation:

1. Low LET (also referred to as *sparsely ionizing*) radiation.
2. High LET (also referred to as *densely ionizing*) radiation.

In contrast to stopping power (see Chap. 6) which focuses attention on the energy loss by an energetic charged particle moving through an absorber, the LET focuses on the linear rate of energy absorption by the absorbing medium as the charged particle traverses the absorber. The LET is measured in keV/μm with 10 keV/μm separating the low LET (sparsely ionizing) radiation from the high LET (densely ionizing) radiation. Table 1.4 gives a list of various low LET and high LET ionizing radiation beams and the LETs they produce in tissue.

Table 1.4 LET values for various low LET radiation beams (left hand side of table) and high LET radiation beams (right hand side of table)

Low LET radiation	LET (keV/μm)	High LET	LET (keV/μm)
x-rays: 250 kVp	2	Electrons: 1 keV	12.3
γ-rays: Co-60	0.3	Neutrons: 14 MeV	12
x-rays: 3 MeV	0.3	Protons: 2 MeV	17
Electrons: 10 keV	2.3	Carbon ions: 100 MeV	160
Electrons: 1 MeV	0.25	Heavy ions	100–2000

LET of 10 keV/μm separates low LET radiation from high LET radiation

1.8.3 Use of Ionizing Radiation

The study and use of ionizing radiation started with the discovery of x-rays by Wilhelm Röntgen in 1895 and the discovery of natural radioactivity by Henri Becquerel in 1896. Since then, ionizing radiation played an important role in atomic and nuclear physics where it ushered in the era of modern physics and in many diverse areas of human endeavor, such as medicine, industry, power generation, weapon production, waste management, and security services. Concurrently with the development of new practical uses of ionizing radiation, it became apparent that ionizing radiation can cause somatic and genetic damage to biologic material including human tissue. For safe use of ionizing radiation it is thus imperative that the users have not only a clear understanding of the underlying physics but also of the biological hazards posed by ionizing radiation.
Ionizing radiation is used in the following areas:

1. In *medicine* where it is used for: (i) imaging in diagnostic radiology and nuclear medicine; (ii) treatment of cancer in radiotherapy; (iii) blood irradiation to prevent transfusion-associated graft versus host disease; and (iv) sterilization of single use medical devices.
2. In *nuclear reactors* where it is used for: (i) basic nuclear physics research; (ii) production of radionuclides used in medicine and industry; and (iii) electric power generation.
3. In *industrial radiography* where it is used for nondestructive inspection of welds in airplane manufacturing as well as inspection of welds in gas and oil pipelines.
4. In *well logging* where it is used to obtain information about the geologic media and recoverable hydrocarbon zones through which a borehole has been drilled.
5. In *insect pest control* where insects made sterile by a high radiation dose are released into the wild to control and eradicate insect pests.
6. In *security services* where it is used for screening of cargo and luggage as well as for mail sanitation mainly against the anthrax bacterium.
7. In *food production* where it is used for irradiation of foods such as meat, poultry, fish, spices, fresh fruit, vegetables, and grains to: (i) kill bacteria, viruses, parasites, and mold; (ii) slow the ripening process; (iii) prevent sprouting; and (iv) extend shelf life.
8. In *waste management* where hospital waste and domestic sewage sludge are irradiated with the objective to kill pathogenic microorganisms and disease-causing bacteria before release into the environment.
9. In *chemical industry* where radiation processing produces a chemical modification of industrial materials such as polymers (polyethylene) and crude rubber used in vulcanized tires.
10. In *production of weapons* based on fission and fusion for military purpose.

1.9 Classification of Directly Ionizing Radiation

Most directly ionizing radiations have been found suitable for use in external beam radiotherapy; however, their usage varies significantly from one particle to another, as result of physical and economic considerations. Generally, with regard to radiotherapy, directly ionizing radiations are divided into two categories: (i) electron therapy with megavoltage electron beams and (ii) hadron therapy with hadron beams.

- Electrons interact with absorber atoms mainly through Coulomb interactions with atomic orbital electrons experiencing collision (ionization) loss and with atomic nuclei experiencing radiation loss sometimes also referred to as bremsstrahlung loss.
- Hadrons, with the exception of neutrons that fall into the indirectly ionizing radiation category, interact with absorber atoms through Coulomb interactions with atomic orbital electrons experiencing collision loss as well as through strong interactions with atomic nuclei (referred to as nuclear reactions).

Electrons have been used in routine radiotherapy for treatment of superficial lesions for the past 50 years, while proton beams, the most common hadron beams used in external beam radiotherapy, have only recently been used on a wider, albeit still limited, scale. Heavier hadrons, such as carbon-12, are still considered an experimental modality available in only a few institutions around the world.

Electron beams are produced relatively inexpensively in clinical linear accelerators (linacs). Proton beams, on the other hand, are produced in a cyclotron or synchrotron and these machines are significantly more sophisticated and expensive in comparison with linacs. Much work is currently being done on alternative means for proton beam generation with the goal to design compact machines that will fit into a treatment bunker similarly to the current experience with clinical linacs. Laser-based proton generating methods currently hold most promise for eventual use in compact, inexpensive, and practical proton machines.

1.9.1 Electrons

Electrons play an important role in medical physics and, because of their relatively small mass are considered light charged particles. *Joseph J. Thomson* discovered electrons in 1897 while studying the electric discharge in a partially evacuated Crookes tube (see Sect. 14.4.1). They are used directly as beams for cancer therapy, are responsible for the dose deposition in media by photon and electron beams, and they govern the experimental and theoretical aspects of radiation dosimetry. With regard to their mode of production, electrons fall into the following categories:

- Electrons released in medium by photoelectric effect are referred to as *photoelectrons*.

- Electrons released in medium by Compton effect are referred to as *Compton* or recoil *electrons*.
- Electrons produced in medium by pair production interactions in the field of the nucleus or in the field of an orbital electron are referred to as *pair production electrons*.
- Electrons emitted from nuclei by β^- decay are referred to as *beta particles* or *beta rays*.
- Electrons produced by linear accelerators (linacs), betatrons or microtrons for use in radiotherapy with kinetic energies typically in the range from 4 MeV to 30 MeV are referred to as *megavoltage electrons*.
- Electrons produced through Auger effect are referred to as *Auger electrons, Coster–Kronig electrons,* or *super Coster–Kronig electrons*.
- Electrons produced through internal conversion are referred to as *internal conversion electrons*.
- Electrons produced by charged particle collisions are of interest in radiation dosimetry and are referred to as *delta* (δ) *rays*.
- Electrons released from metallic surface in thermionic emission are referred to as *thermions*.

1.9.2 Positrons

The positron or antielectron is the antiparticle of an electron with same mass 0.511 MeV/c^2 and spin (1/2) as the electron and charge (1.602×10^{-19} C) equal in magnitude but opposite in sign to that of the electron. In 1928 *Paul Dirac* was the first to postulate positron's existence and in 1932 *Carl D. Anderson* discovered it as the first evidence of antimatter in his experimental study of cosmic rays. There are three ways for generating positrons: (1) positron emission beta decay, (2) nuclear pair production, and (3) triplet production:

1. Positrons emitted from nuclei by β^+ radioactive decay are used in positron emission tomography (PET) and referred to as *beta particles* or *beta rays*.
2. Positrons produced by nuclear pair production and triplet production are referred to as *pair production positrons* and play an important role in interactions of high-energy photons with absorbing medium.

1.9.3 Heavy Charged Particles

For use in radiotherapy heavy charged particles are defined as particles such as proton and heavier ions with mass exceeding the electron mass. They are produced through acceleration of nuclei or ions in cyclotrons, synchrotrons or heavy particle linacs.

Heavy charged particles of importance in nuclear physics and also potentially useful in medicine for treatment of disease are:

- *Proton* is the nucleus of a hydrogen-1 ($^1_1$H) atom. The hydrogen-1 atom, a stable isotope of hydrogen with a natural abundance of 99.985%, is called protium or light hydrogen and consists of a nucleus (proton) and one electron.
- *Deuteron,* the nucleus of a hydrogen-2 ($^2_1$H) atom, consists of one proton and one neutron bound together with a total binding energy of 2.225 MeV or 1.1125 MeV/nucleon. The hydrogen-2 atom, a stable isotope of hydrogen with a natural abundance of 0.015%, is called deuterium or heavy hydrogen and consists of a nucleus (deuteron) and one electron.
- *Triton,* the nucleus of a hydrogen-3 ($^3_1$H) atom, consists of one proton and two neutrons bound together with a total binding energy of 8.48 MeV or 2.83 MeV/nucleon. The hydrogen-3 atom, a radioactive isotope of hydrogen with a half-life $t_{1/2}$ of 12.32 years, is called tritium and consists of a nucleus (triton) and one electron.
- *Helion,* the nucleus of a helium-3 ($^3_2$He) atom, consists of two protons and one neutron bound together with a total binding energy of 7.72 MeV or 2.57 MeV/nucleon. The helium-3 atom, a stable isotope of helium with a natural abundance of ~0.00014%, consists of a nucleus (helion) and two electrons.
- *Alpha particle,* the nucleus of a helium-4 ($^4_2$He) atom, consists of two protons and two neutrons bound together with a total binding energy of 28.3 MeV or 7.075 MeV/nucleon. The helium-4 atom, a stable isotope of helium with a natural abundance of ~99.99986%, consists of a nucleus (alpha particle) and two electrons.

The basic atomic and nuclear properties of heavy charged particles and atoms listed above are summarized in Appendix A and in Table 1.5. Appendix A also lists the atomic mass, nuclear rest energy, nuclear total binding energy, and binding energy per nucleon.

Table 1.5 Basic properties of heavy charged particles used in nuclear physics and medicine

ATOM				NUCLEUS			
Designation	Symbol	Natural abundance (%)	Name	Name	Protons	Neutrons	Nuclear stability
Hydrogen-1	$^1_1$H	99.985	Protium	Proton	1	0	Stable
Hydrogen-2	$^2_1$H	0.015	Deuterium	Deuteron	1	1	Stable
Hydrogen-3	$^3_1$H	–	Tritium	Triton	1	2	Radio-active
Helium-3	$^3_2$He	0.00014	Helium-3	Helion	2	1	Stable
Helium-4	$^4_2$He	99.99986	Helium-4	Alpha particle	2	2	Stable

For the nuclear structures, the table lists the special name as well as the number of protons and neutrons; for associated atomic structures the table lists the symbol, natural abundance and the special name

As they penetrate into an absorber, energetic heavy charged particles lose energy through Coulomb interactions with orbital electrons of the absorber. Just before the heavy charged particle has expended all of its kinetic energy, its energy loss per unit distance traveled increases drastically and this results in a high dose deposition at that depth in the absorber. This high dose region appears close to the particle's range in the absorber and is referred to as the *Bragg peak*. The depth of the Bragg peak in tissue depends on the mass and incident energy of the charged particle (see Sect. 1.12.4).

In contrast to heavy charged particles listed above, heavier charged particles are nuclei or ions of heavier atoms such as carbon-12 ($^{12}_{6}C$), nitrogen-14 ($^{14}_{7}N$), or neon-20 ($^{20}_{10}Ne$). They are generated with cyclotrons and synchrotrons for general use in nuclear and high-energy physics but are also used for radiotherapy in a few highly specialized institutions around the world. They offer some advantages over charged particle radiotherapy with proton beams; however, equipment for their production is very expensive to build and operate, and advantages of their use for general radiotherapy are still not clearly established.

1.9.4 Pions

Pions π also called π mesons belong to a group of short-lived subatomic particles called mesons. They are either neutral (π^0) or come with positive (π^+) or negative (π^-) electron charge and their rest mass is about $273m_e$ with $m_e = 0.511$ MeV/c^2 the rest mass of the electron. Pions do not exist in free state in nature; they reside inside the nuclei of atoms and, based on their mass, were identified as the quanta of the strong interaction. They can be ejected from the nucleus in nuclear reactions by bombarding target nuclei with energetic electrons or protons.

Of the three pion types, the negative π mesons (π^-) have been used for radiotherapy, since, by virtue of their negative charge, they produce the so-called "pion stars" in irradiated nuclei. As negative pions penetrate an absorber, they lose energy, similarly to heavy charged particles, through Coulomb interactions with orbital electrons of the absorber. However, close to their range in the absorber, they not only exhibit a Bragg peak, they are also drawn into a nucleus of the absorber. This nuclear penetration makes the absorbing nucleus unstable and causes it to break up violently into smaller energetic fragments. These fragments fly apart and deposit a significant amount of energy within a short distance from the point of the nuclear reaction. The effect is called a "pion star" and it accentuates the normal Bragg peak dose distribution. In the past pions showed great promise for use in radiotherapy; however, during recent years the studies of pions were largely abandoned in favor of heavy charged particles such as protons.

1.10   Classification of Indirectly Ionizing Photon Radiation

Indirectly ionizing photon radiation consists of three categories of photon: ultraviolet (UV), x-ray, and γ-ray. While UV photons are of some limited use in medicine, imaging and treatment of disease are carried out with photons of higher energy such as x-rays and γ-rays. With regard to their origin, these photons fall into five categories, all of them discussed in detail in subsequent chapters of this book. These five photon categories are:

- *Gamma rays*: photons resulting from nuclear shell transitions (see Sect. 11.7).
- *Annihilation quanta*: photons resulting from positron–electron annihilation (see Sect. 7.6.10).
- *Characteristic (fluorescence) x-rays*: photons resulting from electron transitions between atomic shells (see Sect. 4.1).
- *Bremsstrahlung x-rays*: photons resulting from Coulomb interaction between energetic electrons and positrons with atomic nuclei of absorber (see Sect. 4.2).
- *Synchrotron radiation* (also known as *magnetic bremsstrahlung* and *cyclotron radiation*): photons resulting from charged particles (electrons, positrons, protons, etc.) moving through a magnetic field (e.g., in storage rings, see Sect. 4.3).

1.11   Radiation Quantities and Units

Accurate measurement of radiation is very important in any medical use of radiation, be it for diagnosis or treatment of disease. In diagnosis one must optimize the image quality so as to obtain the best possible image quality with the lowest possible radiation dose to the patient to minimize the risk of radiation induced morbidity. In radiotherapy the prescribed dose must be delivered accurately and precisely to maximize the tumor control probability (TCP) and to minimize the normal tissue complication probability (NTCP). In both instances the risk of morbidity includes acute radiation effects (radiation injury) as well as late effects such as induction of cancer and genetic damage.

Several quantities and units were introduced for the purpose of quantifying radiation and the most important of these are listed in Table 1.6. Also listed in Table 1.6 are the definitions for the various quantities and the relationships between the old units and the SI units for these quantities.

- *Exposure X* is related to the ability of photons to ionize air. Its unit roentgen (R) is defined as charge of 2.58×10^{-4} C of either sign produced per kilogram of air.
- *Kerma K* (acronym for kinetic energy released in matter) is defined for indirectly ionizing radiations (photons and neutrons) as energy transferred to charged particles per unit mass of the absorber.
- *Dose D* is defined as energy absorbed per unit mass of absorbing medium. Its SI unit gray (Gy) is defined as 1 J of energy absorbed per kilogram of absorbing medium.

Table 1.6 Radiation quantities, radiation units, and conversion between old and SI units

Quantity	Definition	SI unit	Old unit	Conversion
Exposure X	$X = \dfrac{\Delta Q}{\Delta m_{\text{air}}}$	$2.58 \times \dfrac{10^{-4}\,\text{C}}{\text{kg air}}$	$1\,\text{R} = \dfrac{1\,\text{esu}}{\text{cm}^3\,\text{air}_{\text{STP}}}$	$1\,\text{R} = 2.58 \times \dfrac{10^{-4}\,\text{C}}{\text{kg air}}$
Kerma K	$K = \dfrac{\Delta E_{\text{tr}}}{\Delta m}$	$1\,\text{Gy} = 1\,\dfrac{\text{J}}{\text{kg}}$	–	–
Dose D	$D = \dfrac{\Delta E_{\text{ab}}}{\Delta m}$	$1\,\text{Gy} = 1\,\dfrac{\text{J}}{\text{kg}}$	$1\,\text{rad} = 100\,\dfrac{\text{erg}}{\text{g}}$	$1\,\text{Gy} = 100\,\text{rad}$
Equiv. dose H	$H = D w_{\text{R}}$	$1\,\text{Sv}$	$1\,\text{rem}$	$1\,\text{Sv} = 100\,\text{rem}$
Activity $\mathcal{A}$	$\mathcal{A} = \lambda N$	$1\,\text{Bq} = 1\,\text{s}^{-1}$	$1\,\text{Ci} = 3.7 \times 10^{10}\,\text{s}^{-1}$	$1\,\text{Bq} = \dfrac{1\,\text{Ci}}{3.7 \times 10^{10}}$

where

ΔQ	is the charge of either sign collected.
Δm_{air}	is the mass of air.
ΔE_{tr}	is energy transferred from indirectly ionizing particles to charged particles in absorber.
ΔE_{ab}	is absorbed energy.
Δm	is the mass of medium.
w_{R}	is the radiation weighting factor.
λ	is the decay constant.
N	is the number of radioactive atoms.
R	stands for roentgen.
Gy	stands for gray.
Sv	stands for sievert.
Bq	stands for becquerel.
Ci	stands for curie.
STP	stands for standard temperature (273.2 K) and standard pressure (101.3 kPa)

- *Equivalent dose H* is defined as the dose multiplied by a radiation-weighting factor w_{R}. The SI unit of equivalent dose is sievert (Sv).
- *Activity $\mathcal{A}$* of a radioactive substance is defined as the number of nuclear decays per time. Its SI unit is becquerel (Bq) corresponding to one decay per second.

1.12 Dose Distribution in Water for Various Radiation Beams

Dose deposition in water is one of the most important characteristics of the interaction of radiation beams with matter. This is true in general radiation physics and even more so in medical physics, where the dose deposition properties in tissue govern both the diagnosis of disease with radiation (*imaging physics*) as well as treatment of disease with radiation (*radiotherapy physics*).

Imaging with ionizing radiation is limited to the use of x-ray beams in *diagnostic radiology* and γ-ray beams in *nuclear medicine*, while in *radiotherapy* the use of radiation is broader and covers essentially all ionizing radiation types ranging from x-rays and γ-rays through electrons to neutrons, protons and heavier charged particles.

In diagnostic radiology imaging one is interested in the radiation beam that propagates and is transmitted through the patient, while in nuclear medicine imaging (now referred to as molecular imaging) one is interested in the radiation beam that emanates from the patient. In radiotherapy, on the other hand, one is interested in the energy deposited in the patient by a radiation source that is located either outside of the patient (*external beam radiotherapy*) or inside the tumor (*brachytherapy*).

When considering the dose deposition in tissue by radiation beams, four beam categories are usually defined: two categories (*photons* and *neutrons*) for indirectly ionizing radiation and two categories (*electrons* and *heavy charged particles*) for directly ionizing radiation. Typical dose distributions in water for the four categories are displayed in Fig. 1.2, normalized to 100% at the dose maximum (percentage depth doses) for various radiation types and energies: for indirectly ionizing radiation in (a) for *photons* and in (b) for *neutrons* and for directly ionizing radiation in (c) for

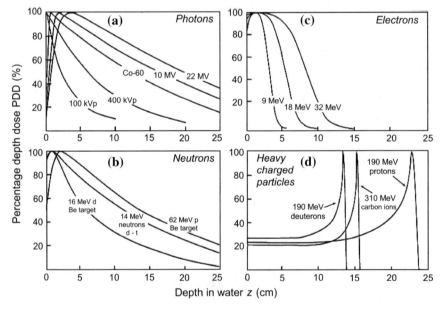

Fig. 1.2 Percentage depth dose against depth in water for radiation beams of various types and energies. Parts **a** and **b** are for *indirectly ionizing radiation*: in (**a**) for photon beams in the range from 100 kVp to 22 MV and in (**b**) for various neutron beams. Parts **c** and **d** are for *directly ionizing radiation*: in (**c**) for electron beams in the range from 9 MeV to 32 MeV and in (**d**) for heavy charged particle beams (190 MeV protons, 190 MeV deuterons, and 310 MeV carbon ions)

electrons and in (d) for *heavy charged particles* (protons, deuterons, and carbon-12 ions). It is evident that the depth dose characteristics of radiation beams depend strongly upon beam type and energy. However, they also depend in a complex fashion on other beam parameters, such as field size, source–patient distance, etc. In general, indirectly ionizing radiation exhibits exponential-like attenuation in absorbing media, while directly ionizing radiation exhibits a well-defined range in absorbing media.

Of the four beam categories of Fig. 1.2, photon beams in the indirectly ionizing radiation category and electron beams in the directly ionizing radiation category are considered conventional beams, well understood and readily available for radiotherapy in all major medical institutions around the world. On the other hand, neutron beams in the indirectly ionizing radiation category and heavy ions including protons in the directly ionizing radiation category remain in the category of special beams, available in only a limited number of institutions around the world, despite having been in use for the past five decades. These beams offer some advantages in treatment of certain malignant diseases; however, in comparison with conventional beams, they are significantly more complicated to use as well as to maintain and their infrastructure and operating costs are also considerably higher, currently precluding a widespread clinical use.

The special beams (neutrons and protons) provide certain advantages when used in treatment of selected tumor types; however, their choice and prescribed dose must account not only for the physical beam characteristics but also for the biological effects associated with radiation beams: the relative biological effectiveness (RBE) and the oxygen enhancement ratio (OER).

Since the biological effect of a dose of radiation depends on its LET, knowing the LET of a given radiation beam is important when prescribing a tumoricidal dose in radiotherapy. As the LET of radiation increases, the ability of the radiation to produce biological damage also increases. The relative biological effectiveness (RBE) is used for comparison of the dose of test radiation to the dose of standard radiation to produce the same biological effect. The RBE varies not only with the type of radiation but also with the type of cell or tissue, biologic effect under investigation, dose, dose rate and fractionation. In general, the RBE increases with LET to reach a maximum of 3–8 at very high LET of $\sim$200 keV/μm and then it decreases with further increase in LET.

The presence or absence of molecular oxygen within a cell influences the biological effect of ionizing radiation: the larger is the cell oxygenation above anoxia, the larger is the biological effect of ionizing radiation. The effect is quite dramatic for low LET (sparsely ionizing) radiations, while for high LET (densely ionizing) radiations it is much less pronounced. The oxygen enhancement ratio (OER) is defined as the ratio of doses without and with oxygen (hypoxic vs. well oxygenated cells) to produce the same biological effect. For low LET radiations, such as x-rays, γ-rays and electrons, OER equals about 3, while for high LET radiations such as neutrons it is about 1.5.

1.12.1   Dose Distribution in Water for Photon Beams

A photon beam propagating through air or vacuum is governed by the inverse-square law; a photon beam propagating through a patient, on the other hand, is not only affected by the inverse-square law but also by attenuation and scattering of the photon beam inside the patient. The three effects make the dose deposition in a patient a complicated process and its determination a complex task.

Typical dose distributions for several photon beams in the energy range from 100 kVp to 22 MV are shown in Fig. 1.2a. Several important points and regions of the absorbed dose curves may be identified. The beam enters the patient on the surface where it delivers a certain surface dose D_s. Beneath the surface, the dose first rises rapidly, reaches a maximum value at a depth z_{max}, and then decreases almost exponentially until it reaches an exit dose value at the patient's exit point. The depth of dose maximum z_{max} is proportional to the beam energy and amounts to 0 for superficial (50 kVp to 100 kVp) and orthovoltage (100 kVp to 300 kVp) beams; 0.5 cm for cobalt-60 γ-rays; 1.5 cm for 6 MV beams; 2.5 cm for 10 MV beams; and 4 cm for 22 MV beams.

The relatively low surface dose for high-energy photon beams (referred to as the *skin sparing effect*) is of great importance in radiotherapy for treatment of deep-seated lesions without involvement of the skin. The tumor dose can be concentrated at large depths in the patient concurrently with delivering a low dose to patient's skin that is highly sensitive to radiation and must be spared as much as possible when it is not involved in the disease.

The dose region between the surface and the depth of dose maximum z_{max} is called the dose *build-up region* and represents the region in the patient in which the dose deposition rises with depth as a result of the range of secondary electrons released in tissue by photon interactions with the atoms of tissue. It is these secondary electrons released by photons that deposit energy in tissue (indirect ionization). The larger is the photon energy, the larger are the energy and the range of secondary electrons and, consequently, the larger is the depth of dose maximum.

It is often assumed that at depths of dose maximum and beyond electronic equilibrium is achieved; however, a better term is transient electronic equilibrium because of the effects of photon beam attenuation as the photon beam penetrates into a patient. Electronic equilibrium or, more generally, charged particle equilibrium (CPE) exist for a volume if each charged particle of a given type and energy leaving the volume is replaced by an identical particle of the same type and energy entering the volume.

1.12.2   Dose Distribution in Water for Neutron Beams

Neutron beams belong to the group of indirectly ionizing radiation, but rather than releasing electrons like photons do, they release protons or heavier ions that subsequently deposit their energy in absorbing medium through Coulomb interactions with electrons and nuclei of the absorber.

As shown in Fig. 1.2b, the dose deposition characteristics in water by neutrons are similar to those of photon beams. Neutron beams exhibit a relatively low surface dose although the skin sparing effect is less pronounced than that for photon beams. They also exhibit a dose maximum beneath the skin surface and an almost exponential decrease in dose beyond the depth of dose maximum. Similarly to photons, the dose build up region depends on neutron beam energy; the larger is the energy, the larger is the depth of dose maximum.

For use in radiotherapy, neutron beams are produced either with cyclotrons or neutron generators (see Sect. 9.6). In a cyclotron, protons or deuterons are accelerated to kinetic energies of 50 MeV to 80 MeV and strike a thick beryllium target to produce fast neutrons that are collimated into a clinical neutron beam. The neutron beams produced in the beryllium target have beam penetration and build up characteristics similar to those of 4 MV to 10 MV megavoltage x-ray beams.

Less common in clinical use are neutron generators (see Sect. 9.6.3) in which deuterons are accelerated to 250 keV and strike a tritium target to produce a 14 MeV neutron beam which exhibits penetration characteristics similar to those obtained for a cobalt-60 teletherapy γ-ray beam.

Producing physical depth dose characteristics that are similar to those produced by megavoltage photon beams, neutrons offer no advantage over photons in this area. However, neutrons are high LET (densely ionizing) particles in comparison with low LET (sparsely ionizing) photon radiation and produce more efficient cell kill per unit dose. The high LET of neutron beams produces RBE > 1 which means that, to achieve the same biological effect, a lower neutron dose is required compared to the photon dose.

Moreover, in comparison with photons, neutrons by virtue of their high LET are more efficient in killing hypoxic cells in comparison to well-oxygenated cells. The oxygen enhancement ratio (OER) of neutrons is 1.5 compared to an OER of 3 for photons. Thus, neutrons offer no physical advantage over photons and it is the biological advantage of neutron beams that makes neutrons attractive for use in radiotherapy despite the increased complexity of producing and using them clinically.

1.12.3 Dose Distribution in Water for Electron Beams

Electrons are directly ionizing radiations that deposit their energy in tissue through Coulomb interactions with orbital electrons and nuclei of the absorber atoms. Megavoltage electron beams represent an important treatment modality in modern radiotherapy, often providing a unique option for treatment of superficial tumors that are less than 5 cm deep. Electrons have been used in radiotherapy since the early 1950s, first produced by betatrons and then by linear accelerators. Modern high-energy linear accelerators used in radiotherapy typically provide, in addition to two megavoltage x-ray beam energies, several electron beams with energies from 4 MeV to 25 MeV.

As shown in Fig. 1.2c, the electron beam percentage depth dose curve plotted against depth in patient exhibits a relatively high surface dose (of the order of 80%) and then builds up to a maximum dose at a certain depth referred to as the electron beam depth dose maximum z_{max}. Beyond z_{max} the dose drops off rapidly, and levels off at a small low-level dose component referred to as the *bremsstrahlung* tail. The bremsstrahlung component of the electron beam is the photon contamination of the beam that results from radiation loss experienced by incident electrons as they penetrate the various machine components, air, and the patient. The higher is the energy of the incident electrons, the higher is the bremsstrahlung contamination of the electron beam.

Several parameters are used to describe clinical electron beams, such as the most probable energy on the patient's skin surface; mean electron energy on the patient's skin surface; and the depth at which the absorbed dose falls to 50% of the maximum dose. The depth of dose maximum generally does not depend on beam energy, contrary to the case for photon beams; rather, it is a function of machine design. On the other hand, the beam penetration into tissue clearly depends on incident electron beam energy; the higher is the energy, the more penetrating is the electron beam, as is evident from Fig. 1.2c.

1.12.4 Dose Distribution in Water for Heavy Charged Particle Beams

Heavy charged particle beams fall into the category of directly ionizing radiation depositing their energy in tissue through Coulomb interactions with orbital electrons of the absorber. As they penetrate into tissue, heavy charged particles lose energy but, in contrast to electrons, do not diverge appreciably from their direction of motion and therefore exhibit a distinct range in tissue. This range depends on the incident particle's kinetic energy and mass.

Just before the heavy charged particle reaches its range in the absorbing medium, its energy loss per unit distance traveled increases dramatically and this results in a high dose deposition referred to as Bragg peak (see Sect. 1.9.3). As indicated in Fig. 1.2d, the Bragg peak appears close to particle's range, is very narrow, defines the maximum dose deposited in tissue, and characterizes all heavy charged particle dose distributions.

Because of their large mass compared to the electron mass, heavy charged particles lose their kinetic energy only interacting with orbital electrons of the absorber; they do not lose any appreciable amount of energy through bremsstrahlung interactions with absorber nuclei.

1.12.5 Choice of Radiation Beam and Prescribed Target Dose

The choice of radiation beam and dose prescription in treatment of disease with radiation depends on many factors such as:

1. Medical patient-related and physician-related factors:

   - Tumor type and histology.
   - Tumor location in the patient.
   - Location of sensitive structures and healthy tissues in the vicinity of the target.
   - Patient's tolerance of treatment.
   - Any previous radiation treatment.
   - Physician's training and experience.

2. Availability of equipment for diagnostic imaging and dose delivery.
3. Physical parameters of the radiation beam to be used in treatment:

   - Depth dose characteristics, governed by machine design, beam energy, field size, and other machine parameters.
   - Density of ionization produced in tissue by the radiation beam to be used in treatment.

4. Biological factors produced in tissue by the radiation beam:

   - Relative biological effectiveness (RBE).
   - Oxygen enhancement ratio (OER).

Based on machine-related physical factors, *superficial tumors* are treated, depending on their size, with single beams of:

1. Superficial x-rays (50 kVp to 80 kVp) produced in superficial x-ray machines.
2. Orthovoltage x-rays (100 kVp to 300 kVp) produced in orthovoltage x-ray machines.
3. Electrons in the energy range from 4 MeV to 25 MeV produced in clinical linear accelerators.

Deep-seated tumors, on the other hand, are treated with multiple megavoltage beams from cobalt-60 teletherapy machine or linear accelerator in the energy range from 4 MV to 25 MV. Over the past decade there have been significant advances in technology and techniques used to plan and deliver precision radiotherapy. The patient's path through a radiotherapy department involves several steps, some of them not involving the patient directly but all of them important for a favorable treatment outcome. In short, the steps are as follows:

1. Definition of target and collection of patient data with diagnostic imaging techniques.
2. Treatment planning and, if required, fabrication of treatment accessories.
3. Prescription of target dose and dose fractionation.

4. Dose delivery, typically with multiple fractions at one fraction per day for a duration of several weeks.
5. Follow up at regular intervals.

The aim of modern dose delivery is to optimize the tumor control probability (TCP) by delivering as high as possible dose to the target and causing as little as possible morbidity for the patient by keeping the normal tissue complication probability (NTCP) as low as possible. Therefore, accurate knowledge of the tumor (target) location in the body, as well as the target's shape and volume are very important parameters of modern radiotherapy. This knowledge is usually acquired with diagnostic imaging which involves appropriate fusion of patient data collected with two or more imaging modalities.

Traditional treatment planning is carried out by matching radiation fields with target dimensions and subsequently calculating the resulting dose distribution, while modern treatment planning is carried out by prescribing a suitable dose distribution and subsequently calculating the intensity modulated fields required to achieve the prescribed dose distribution. The traditional treatment planning is now referred to as forward planning; planning with beam intensity modulation is called inverse planning and dose delivery using intensity modulated fields is called intensity modulated radiotherapy (IMRT). In principle, the IMRT optimizes the dose delivery to the patient by conforming the prescribed dose distribution to the target volume; however, in this process one assumes that the target location is accurately known and that the position and shape of the target do not change during the treatment (intra-treatment motion) and from one fractionated treatment to another (inter-treatment motion).

To have a better control of the target motion problem recent development in dose delivery technology introduced the so-called image guided radiotherapy (IGRT) which merged imaging and dose delivery into one machine, allowing accurate positioning of the target into the radiation beam. The most recent development is adaptive radiotherapy (ART) which enables target shape and position verification not only before and after treatment but also during the dose delivery process.

1.13 Basic Definitions for Atomic Structure

The constituent particles forming an atom are *protons*, *neutrons* and *electrons*. Protons and neutrons are known as *nucleons* and form the nucleus of the atom. The following definitions apply for atomic structure:

- *Atomic number* Z: number of protons and number of electrons in an atom.
- *Atomic mass number* A: number of nucleons in an atom, i.e., number of protons Z plus number of neutrons N in an atom; i.e., $A = Z + N$.
- *Atomic mass* $\mathcal{M}$: expressed in unified atomic mass units u, where 1 u is equal to one twelfth of the mass of the carbon-12 atom (unbound, at rest, and in ground state) or 931.5 MeV/c^2. The atomic mass $\mathcal{M}$ is smaller than the sum of individual masses of constituent particles because of the intrinsic energy associated with

binding the particles (nucleons) within the nucleus (see Sect. 1.15). On the other hand, the atomic mass $\mathcal{M}$ is larger than the nuclear mass M because the atomic mass $\mathcal{M}$ includes the mass contribution by Z orbital electrons while the nuclear mass M does not. The binding energy of orbital electrons to the nucleus is ignored in the definition of the atomic mass $\mathcal{M}$.

- While for carbon-12 the atomic mass $\mathcal{M}$ is exactly 12 u, for all other atoms $\mathcal{M}$ in u does not exactly match the atomic mass number A. However, for all atomic entities A (an integer) and $\mathcal{M}$ are very similar to one another and often the same symbol (A) is used for the designation of both.
- Number of atoms N_a per mass of an element is given as

$$\frac{N_a}{m} = \frac{N_A}{A},$$ (1.11)

where N_A is the Avogadro number discussed in Sect. 1.13.2.

- Number of electrons per volume of an element is

$$Z\frac{N_a}{V} = \rho \, Z\frac{N_a}{m} = \rho \, Z\frac{N_A}{A}.$$ (1.12)

- Number of electrons per mass of an element is

$$Z\frac{N_a}{m} = Z\frac{N_A}{A}.$$ (1.13)

Note that $(Z/A) \approx 0.5$ for all elements with one notable exception of hydrogen for which $(Z/A) = 1$. Actually, Z/A slowly decreases from 0.5 for low Z elements to 0.4 for high Z elements. *For example*: Z/A for helium-4 is 0.5, for cobalt-60 it is 0.45, for uranium-235 it is 0.39.

1.13.1 Mean Atomic Mass (Standard Atomic Weight)

Most of the naturally occurring elements are mixtures of several stable isotopes, each isotope with its own relative natural abundance. For a given chemical element one stable isotope usually predominates; however, natural elements generally consist of atoms of same atomic number Z but of various different atomic mass numbers A as a result of different numbers of neutrons N. The *mean atomic mass* $\mathcal{M}$ of an element is often referred to as the *standard atomic weight* of an element and is given as the mean atomic mass of all stable isotopes of the element, accounting for the isotopes' natural relative abundance and relative atomic mass.

For example:

- Natural carbon ($Z = 6$) consists of two stable isotopes, carbon-12 with a natural abundance of 98.93% and relative atomic mass of 12.0000 u and carbon-13 with a natural abundance of 1.07% and relative atomic mass of 13.003355 u. The mean atomic mass (standard atomic weight) of carbon $\bar{\mathcal{M}}(C)$ is determined as follows

$$\bar{\mathcal{M}}(C) = 0.9893 \times 12.0000 \text{ u} + 0.0107 \times 13.003355 \text{ u} = 12.0107 \text{ u}. \quad (1.14)$$

- Natural iridium ($Z = 77$) consists of two stable isotopes, iridium-191 with a natural abundance of 37.3% and relative atomic mass of 190.960591 u and iridium-193 with a natural abundance of 62.7% and relative atomic mass of 192.962924 u. The mean atomic mass (standard atomic weight) of iridium $\bar{\mathcal{M}}(Ir)$ is determined as follows

$$\bar{\mathcal{M}}(Ir) = 0.373 \times 190.960591 \text{ u} + 0.627 \times 192.962924 \text{ u} = 192.216 \text{ u}.$$
$$(1.15)$$

- Natural iron is a slightly more complicated example containing four stable isotopes with the following relative abundances and relative atomic masses:

  Iron-54: 5.845% and 53.9396148 u
  Iron-56: 91.754% and 55.9349421 u
  Iron-57: 2.119% and 56.9353987 u
  Iron-58: 0.282% and 57.9332805 u

After accounting for the relative abundances and atomic masses for the four iron isotopes we get the following mean atomic mass (standard atomic weight) for natural iron: $\bar{\mathcal{M}}(Fe) = 55.845$ u.

1.13.2 Atomic Mass Constant and the Mole

Atomic mass constant u or Dalton (Da) previously known as *unified atomic mass unit* (amu) is related to the macroscopic SI base unit of mass, the kilogram (kg), through the Avogadro constant (Avogadro number) N_A, defined as the number of atoms ($6.022141293 \times 10^{23}$) at rest and in their ground state contained in exactly 12 g (12×10^{-3} kg) of carbon-12. Since 12 g of carbon-12 is also defined as a mole (mol) of carbon-12, we can state that $N_A = 6.022141293 \times 10^{23}$ mol^{-1}.

Since 1 u by definition equals to the mass of 1/12 of the carbon-12 atom and since 12 g of carbon-12 by definition contains Avogadro number ($6.022141293 \times 10^{23}$) of atoms, we conclude that the mass of one carbon-12 atom equals to $(12 \text{ g})/N_A$ and that the relationship between the *atomic mass constant* u and the SI mass unit *kilogram* is

$$1\,\mathrm{u} = \frac{1}{12} \times \frac{12\,\mathrm{g\cdot mol^{-1}}}{N_A} = \frac{10^{-3}\,\mathrm{kg}}{\mathrm{mol}}\frac{\mathrm{mol}}{6.022141293 \times 10^{23}} = 1.660538922 \times 10^{-27}\,\mathrm{kg} \tag{1.16}$$

In terms of energy we can express the mass 1 u in MeV/c^2 to get

$$
\begin{aligned}
1\,\mathrm{u} &= \frac{1\,\mathrm{u}c^2}{c^2} \\
&= \frac{(1.660538922 \times 10^{-27}\,\mathrm{kg}) \times (2.99792458 \times 10^8\,\mathrm{m\cdot s^{-1}})^2}{c^2 \times 1.602176565 \times 10^{-13}} \times \frac{\mathrm{MeV}}{\mathrm{J}} \\
&= 931.494060\,\mathrm{MeV}/c^2 \approx 931.5\,\mathrm{MeV}/c^2,
\end{aligned}
\tag{1.17}
$$

where we use the following 2010 CODATA constants available from the NIST:

Avogadro constant:	$N_A = 6.022141293 \times 10^{23}\,\mathrm{mol^{-1}}$	(1.18)
Atomic mass constant:	$\mathrm{u} = 1.660538922 \times 10^{-27}\,\mathrm{kg}$	(1.19)
Speed of light in vacuum:	$c = 2.9979458 \times 10^8\,\mathrm{m\cdot s^{-1}}$	(1.20)
Elementary charge (electron):	$e = 1.602176565 \times 10^{-19}\,\mathrm{C}$	(1.21)

The mass in grams of a chemical element equal to the mean atomic mass (standard atomic weight) $\bar{\mathcal{M}}(\mathrm{X})$ of an element X is referred to as a *mole* of the element and contains exactly Avogadro number of atoms. In general, Avogadro constant N_A is given as the number of entities per mole, where the entity can be atoms per mole, molecules per mole, ions per mole, electrons per mole, etc. The mean atomic mass number $\bar{\mathcal{M}}$ of all elements is thus defined such that a mass of $\bar{\mathcal{M}}$ grams of the element contains exactly Avogadro number of atoms.

For example: Cobalt has only one stable isotope, cobalt-59, so that the mean atomic mass $\bar{\mathcal{M}}(\mathrm{Co})$ of natural cobalt is the atomic mass of cobalt-59 at 58.9332 u. Thus:

- 1 mol of natural cobalt is 58.9332 g of natural cobalt.
- A mass of 58.9332 g of natural cobalt contains 6.022×10^{23} cobalt atoms.

As far as the unit of atomic mass is concerned, we have three related options. The atomic mass can be expressed in one of the following three formats:

- Without a unit when it represents the ratio between the mass of a given element and the unified atomic mass unit u ($\frac{1}{12}$ of the mass of the carbon-12 atom). In this case the atomic mass is dimensionless and expresses the magnitude of the atomic mass relative to the standard mass that is $\frac{1}{12}$ of the mass of the carbon-12 atom.
- In units of the unified atomic mass unit u where u represents $\frac{1}{12}$ of the mass of the carbon-12 atom.
- In units of g/mol when the mean atomic mass (standard atomic weight) is multiplied by the molar mass constant $M_u = 1$ g/mol.

For example, for the carbon-12 atom we can state that its relative atomic mass is 12.000, or that its atomic mass is 12.000 u, or that its molar mass is 12.000 g/mol. For elemental carbon we can state that its mean atomic mass (standard atomic weight) is 12.0107, or that its mean atomic mass (standard atomic weight) is 12.0107 u, or that its molar atomic weight is 12.0107 g/mol.

1.13.3   Mean Molecular Mass (Standard Molecular Weight)

If we assume that the mass of a molecule is equal to the sum of the masses of the atoms that make up the molecule, then for any molecular compound there are N_A molecules per mole of the compound where the *mole* in grams is defined as the sum of the mean atomic masses of the atoms making up the molecule. Moreover, a mole of a molecular compound contains Avogadro number N_A of molecules.

The standard molecular weight of a molecular compound is calculated from the molecule's chemical formula and the standard atomic weights for the atoms constituting the molecule. *For example*:

- Water molecule contains two atoms of hydrogen [$\bar{\mathcal{M}}(H) = 1.00794$] and one atom of oxygen [$\bar{\mathcal{M}}(O) = 15.9994$]. The standard molecular weight of water is:

$$\bar{\mathcal{M}}(H_2O) = 2 \times \bar{\mathcal{M}}(H) + \bar{\mathcal{M}}(O) = 2 \times 1.00794 \text{ u} + 15.9994 \text{ u} = 18.0153 \text{ u}$$
(1.22)

  and one mole of water that contains N_A molecules is 18.0153 g of water.
- The molecule of carbon dioxide contains one atom of carbon [$\bar{\mathcal{M}}(C) = 12.0107$] and two atoms of oxygen [$\bar{\mathcal{M}}(O) = 15.9994$]. The standard molecular weight of carbon dioxide CO_2 is:

$$\bar{\mathcal{M}}(CO_2) = \bar{\mathcal{M}}(C) + 2 \times \bar{\mathcal{M}}(O) = 12.0107 \text{ u} + 2 \times 15.9994 \text{ u} = 44.0095 \text{ u}$$
(1.23)

  and one mole of carbon dioxide that contains N_A molecules is 44.0095 g of carbon dioxide.

In the calculation of the standard molecular weight of a water molecule above we used the mean values for the standard atomic weights to account for traces of deuterium, oxygen-17, and oxygen-18 in natural water molecules and found 18.0153 for the molecular weight of water. However, the most common water molecule will contain hydrogen-1 (protium) and oxygen-16 and will thus have a slightly lower molecular weight amounting to 18.0106 as a result of protium atomic mass of 1.00783 and oxygen-16 atomic mass of 15.9949.

Similarly, we get a molecular weight of carbon dioxide as 43.9898 for a typical molecule containing carbon-12 and oxygen-16 in contrast to 44.0095 that we calculated as the mean value after accounting for traces of carbon-13, oxygen-17 and oxygen-18 in natural carbon dioxide.

In the first approximation assuming that atomic mass equals to the atomic mass number A, we get 18 g for the mole of water and 44 g for the mole of carbon dioxide, very similar to the values obtained in (1.22) and (1.23), respectively.

1.14 Basic Definitions for Nuclear Structure

Most of the atomic mass is concentrated in the atomic nucleus consisting of Z protons and $(A - Z)$ neutrons, where Z is the atomic number and A the atomic mass number of a given nucleus. Proton and neutron have nearly identical rest masses; the proton has positive charge, identical in magnitude to the negative electron charge and the neutron has no charge and is thus neutral.

In nuclear physics the convention is to designate a nucleus with symbol X as $_{Z}^{A}X$, where A is the atomic mass number and Z the atomic number. *For example*: The cobalt-60 nucleus is identified as $_{27}^{60}Co$, the radium-226 nucleus as $_{88}^{226}Ra$, and the uranium-235 nucleus as $_{92}^{235}U$.

There is no basic relation between the atomic mass number A and the atomic number Z in a nucleus, but the empirical relationship

$$Z = \frac{A}{1.98 + 0.0155A^{2/3}} \tag{1.24}$$

provides a good approximation for stable nuclei. Protons and neutrons are commonly referred to as *nucleons*, have identical strong attractive interactions, and are bound in the nucleus with the strong force. As discussed in Sect. 1.5, in contrast to electrostatic and gravitational forces that are inversely proportional to the square of the distance between two particles, the strong force between two nucleons is a very short-range force, active only at distances of the order of a few femtometers. At these short distances the strong force is the predominant force exceeding other forces by many orders of magnitude, as shown in Table 1.3. With regard to relative values of atomic number Z and atomic mass number A of nuclei, the following conventions apply:

- An element may be composed of atoms that all have the same number of protons, i.e., have the same atomic number Z, but have a different number of neutrons, i.e., have different atomic mass numbers A. Such atoms of identical Z but differing A are called *isotopes* of a given element.
- The term isotope is often misused to designate nuclear species. For example, cobalt-60, cesium-137, and radium-226 are not isotopes, since they do not belong to the same element. Rather than isotopes, they should be referred to as *nuclides*. On the other hand, it is correct to state that deuterium (with nucleus called deuteron) and tritium (with nucleus called triton) are heavy isotopes of hydrogen or that cobalt-59 and cobalt-60 are isotopes of cobalt. Thus, the term *radionuclide* should

be used to designate radioactive species; however, the term radioisotope is often used for this purpose.

- The term *nuclide* refers to all atomic forms of all elements. The term *isotope* is narrower and only refers to various atomic forms of a single chemical element.
- In addition to being classified into isotopic groups (common atomic number Z), nuclides are also classified into groups with common atomic mass number A (*isobars*) and common number of neutrons (*isotones*). For example, cobalt-60 and nickel-60 are isobars with 60 nucleons each ($A = 60$); hydrogen-3 (tritium) and helium-4 are isotones with two neutrons each ($A - Z = 2$).
- If a nucleus exists in an excited state for some time, it is said to be in an isomeric (metastable) state. *Isomers* are thus nuclear species that have common atomic number Z and atomic mass number A. *For example*, technetium-99m is an isomeric state of technetium-99 and cobalt-60m is an isomeric state of cobalt-60.

1.15 Nuclear Binding Energies

The sum of masses of the individual components of a nucleus that contains Z protons and $(A - Z)$ neutrons is larger than the actual mass of the nucleus. This difference in mass is called the mass defect (deficit) Δm and its energy equivalent $\Delta m c^2$ is called the total binding energy E_B of the nucleus. The *total binding energy* E_B of a nucleus can thus be defined as:

1. The positive work required to disassemble a nucleus into its individual components: Z protons and $(A - Z)$ neutrons.

or

2. The energy liberated when Z protons and $(A - Z)$ neutrons are brought together to form the nucleus.

The *binding energy per nucleon* (E_B/A) in a nucleus (i.e., the total binding energy of a nucleus divided by the number of nucleons) varies with the number of nucleons A and is of the order of ~ 8 MeV/nucleon. It may be calculated from the energy equivalent of the mass deficit Δm as follows:

$$\frac{E_B}{A} = \frac{\Delta m c^2}{A} = \frac{Z m_p c^2 + (A - Z) m_n c^2 - M c^2}{A}, \tag{1.25}$$

where

A	is the atomic mass number.
M	is the nuclear mass in atomic mass units u.
$m_p c^2$	is the proton rest energy (938.3 MeV).
$m_n c^2$	is the neutron rest energy (939.6 MeV).

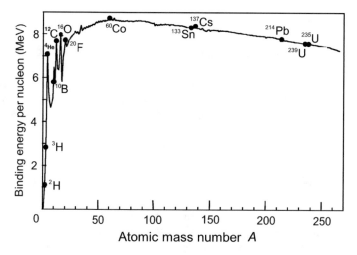

Fig. 1.3 Binding energy per nucleon in MeV/nucleon against atomic mass number A

As shown in Fig. 1.3, the binding energy per nucleon E_B/A against the atomic mass number A exhibits the following characteristics:

1. For $1 \leq A \leq 4$ the binding energy per nucleon rises rapidly from 1.1 MeV per nucleon for deuteron ($^2_1$H) through 2.8 MeV and 2.6 MeV per nucleon for triton (^{3}H) and helium-3 ($^3_2$He), respectively, to 7.1 MeV per nucleon for helium-4 ($^4_2$He). The nucleus of the helium-4 atom is the α-particle.
2. For $4 \leq A \leq 28$, E_B/A fluctuates and exhibits peaks for nuclides in which A is a multiple of four.
3. For $28 < A < 60$, E_B/A rises slowly with increasing A to reach a peak value of 8.8 MeV per nucleon for $A \approx 60$ (iron, cobalt, nickel).
4. For A exceeding 60, E_B/A falls monotonically from 8.8 MeV/nucleon to reach 7.5 MeV per nucleon for uranium-238.

The larger is the binding energy per nucleon (E_B/A) of an atom, the larger is the stability of the atom. Thus the most stable nuclei in nature are the ones with $A \approx 60$. Nuclei of light elements (small A) are generally less stable than the nuclei with $A \approx 60$ and the heaviest nuclei (large A) are also less stable than the nuclei with $A \approx 60$.

The peculiar shape of the E_B/A against A curve suggests two methods for converting mass into energy: (1) *fusion* of nuclei at low A and (2) *fission* of nuclei at large A, as discussed in greater detail in Sect. 12.8.

- Fusion of two nuclei of very small mass, e.g., $^2_1$H $+ ^3_1$H $\rightarrow ^4_2$He $+$ n, will create a more massive nucleus and release a certain amount of energy. Experiments using controlled nuclear fusion for production of energy have so far not been successful; however, steady progress in fusion research is being made in various laboratories around the world. It is reasonable to expect that in the future controlled fusion will become possible and will result in a relatively clean and abundant means for sustainable power generation.

- Fission of elements of large mass, e.g., $^{235}_{92}U$ + n, will create two lower mass and more stable nuclei and lose some mass in the form of kinetic energy. Nuclear fission was observed first in 1934 by *Enrico Fermi* and described correctly by *Otto Hahn, Fritz Strassman, Lise Meitner*, and *Otto Frisch* in 1939. In 1942 at the University of Chicago Enrico Fermi and colleagues carried out the first controlled chain reaction based on nuclear fission (see Sect. 12.7).

1.16 Nuclear Models

Several models of the nucleus have been proposed; all phenomenological and none of them capable of explaining completely the complex nature of the nucleus, such as its binding energy, stability, radioactive decay, etc. The two most successful models are the *liquid-drop model* that accounts for the nuclear binding energy and the *shell model* that explains nuclear stability.

1.16.1 Liquid-Drop Nuclear Model

The *liquid-drop* nuclear model, proposed by *Niels Bohr* in 1936, treats the nucleons as if they were molecules in a spherical drop of liquid. Scattering experiments with various particles such as electrons, nucleons and α-particles reveal that to a first approximation nuclei can be considered spherical with essentially constant density.

The radius R of a nucleus with atomic mass number A is estimated from the following expression

$$R = R_0 \sqrt[3]{A}, \tag{1.26}$$

where R_0 is the nuclear radius constant equal to 1.25 fm.

Using (1.26) we estimate the density of the nucleus with mass M and volume $\mathcal{V}$ as

$$\rho = \frac{M}{\mathcal{V}} \approx \frac{Am_p}{(4/3)\,\pi R^3} = \frac{m_p}{(4/3)\,\pi R_0^3} \approx 1.5 \times 10^{14} \text{ g·cm}^{-3}, \tag{1.27}$$

where m_p is the rest mass of the proton (938.3 MeV/c^2).

Based on the liquid drop model of the nucleus the nuclear binding energy was split into various components, each with its own dependence on the atomic number Z and atomic mass number A. Four of the most important components of the nuclear binding energy are:

1. *Volume correction* Since the binding energy per nucleon E_B/A is essentially constant, as shown in Fig. 1.3, the total nuclear binding energy is linearly proportional to A.
2. *Surface correction* Nucleons on the surface of the liquid-drop have fewer neighbors than those in the interior of the drop. The surface nucleons will reduce the

total binding energy by an amount proportional to R^2, where R is the nuclear radius proportional to $A^{1/3}$, as given in (1.26). Thus the surface effect correction is proportional to $A^{2/3}$.

3. *Coulomb repulsion correction* accounts for the Coulomb repulsion among protons in the nucleus. The repulsive energy reduces the total binding energy and is proportional to $Z(Z-1)$, the number of proton pairs in the nucleus, and inversely proportional to R, i.e., inversely proportional to $A^{1/3}$.

4. *Neutron excess correction* reduces the total binding energy and is proportional to $(A-2Z)^2$ and inversely proportional to A.

The total nuclear binding energy E_B is then written as follows

$$E_B = C_1 A - C_2 A^{2/3} - C_3 \frac{Z(Z-1)}{A^{1/3}} - C_4 \frac{(A-2Z)^2}{A}. \qquad (1.28)$$

Equation (1.28) is referred to as the *Weizsächer semi-empirical binding energy formula* in which the various components are deduced theoretically but their relative magnitudes are determined empirically to match the calculated results with experimental data. The constants in (1.28) were determined empirically and are given as follows:

$$C_1 \approx 16\,\text{MeV}; \quad C_2 \approx 18\,\text{MeV}; \quad C_3 \approx 0.7\,\text{MeV}; \quad \text{and} \quad C_4 \approx 24\,\text{MeV}.$$

1.16.2 Shell Structure Nuclear Model

Experiments have shown that the number of nucleons the nucleus contains affects the stability of nuclei. The general trend in binding energy per nucleon E_B/A, as shown in Fig. 1.3, provides the E_B/A maximum at around $A = 60$ and then drops for smaller and larger A. However, there are also considerable variations in stability of nuclei depending on the parity in the number of protons and neutrons forming a nucleus.

In nature there are 280 nuclides that are considered stable with respect to radioactive decay. Some 60% of these stable nuclei have an even number of protons and an even number of neutrons (even-even nuclei); some 20% have an even–odd configuration and a further 20% have and odd even configuration. Only four stable nuclei are known to have an odd–odd proton/neutron configuration. A conclusion may thus be made that an even number of protons or even number of neutrons promotes stability of nuclear configurations.

When the number of protons is: 2, 8, 20, 28, 50, 82 or the number of neutrons is: 2, 8, 20, 28, 50, 82, 126, the nucleus is observed particularly stable and these numbers are referred to as *magic numbers*. Nuclei in which the number of protons as well as the number of neutrons is equal to a magic number belong to the most stable group of nuclei.

The existence of magic numbers stimulated development of a nuclear model containing a nuclear shell structure in analogy with the atomic shell structure configuration of electrons. In the nuclear shell model, often also called the independent particle model, the nucleons are assumed to move in well-defined orbits within the nucleus in a field produced by all other nucleons. The nucleons exist in quantized energy states of discrete energy that can be described by a set of quantum numbers, similarly to the situation with electronic states in atoms.

The ground state of a nucleus constitutes the lowest of the entire set of energy levels and, in contrast to atomic physics where electronic energy levels are negative, in nuclear physics the nuclear ground state is set at zero and the excitation energies of the respective higher bound states are shown positive with respect to the ground state.

To raise the nucleus to an excited state an appropriate amount of energy must be supplied. On de-excitation of a nucleus from an excited state back to the ground state a discrete amount of energy will be emitted.

1.17   Physics of Small Dimensions and Large Velocities

At the end of the nineteenth century physics was considered a completed discipline within which most of the natural physical phenomena were satisfactorily explained. However, as physicists broadened their interests and refined their experimental techniques, it became apparent that classical physics suffered severe limitations in two areas:

1. Dealing with dimensions comparable to small atomic dimensions.
2. Dealing with velocities comparable to the speed of light.

Modern physics handles these limitations in two distinct, yet related, sub-specialties: *quantum physics* and *relativistic physics*, respectively:

1. *Quantum physics* extends the range of application of physical laws to small atomic dimensions of the order of 10^{-10} m (radius a of atom), includes classical laws as special cases when dimension $\gg a$, and introduces the Planck constant h as a universal constant of fundamental significance. *Erwin Schrödinger, Werner Heisenberg*, and *Max Born* are credited with developing quantum physics in the mid 1920s.
2. *Relativistic physics* extends the range of application of physical laws to large velocities v of the order of the speed of light in vacuum c (3×10^8 m/s), includes classical laws as special cases when $v \ll c$, and introduces c as a universal physical constant of fundamental significance. The protagonist of relativistic physics was *Albert Einstein* who formulated the special theory of relativity in 1905.

1.18 Planck Energy Quantization

Modern physics was born in 1900 when *Max Planck* presented his revolutionary idea
of energy quantization of physical systems that undergo simple harmonic oscillations.
Planck energy ε quantization is expressed as

$$\varepsilon = nh\nu, \tag{1.29}$$

where

n is the quantum number ($n = 0, 1, 2, 3 \ldots$).
h is a universal constant referred to as the Planck constant.
ν is the frequency of oscillation.

The allowed energy states in a system oscillating harmonically are continuous in
classical models, while in the Planck model they consist of discrete allowed quantum
states with values $nh\nu$, where n is a non-negative integer quantum number. Planck
used his model to explain the spectral distribution of thermal radiation emitted by
a blackbody defined as an entity that absorbs all radiant energy incident upon it.
All bodies emit thermal radiation to their surroundings and absorb thermal radiation
from their surroundings; in thermal equilibrium the rates of thermal emission and
thermal absorption are equal.

Planck assumed that sources of thermal radiation are harmonically oscillating
atoms possessing discrete vibrational energy states. When an oscillator jumps from
one discrete quantum energy state E_1 to another energy state E_2 where $E_1 > E_2$, the
energy difference $\Delta E = E_1 - E_2$ is emitted in the form of a photon with energy $h\nu$,
i.e.,

$$\Delta E = E_1 - E_2 = h\nu = \frac{hc}{\lambda}, \tag{1.30}$$

where

h is the Planck constant.
ν is the frequency of the photon.
c is the speed of light in vacuum.
λ is the wavelength of the photon.

Thus, according to Planck law, radiation such as light is emitted, transmitted, and
absorbed in discrete energy quanta characterized by the product of frequency ν and
Planck constant h. Planck's postulate of energy quantization leads to the atomic model
with its angular momentum quantization introduced by *Niels Bohr* in 1913 and to
quantum wave mechanics developed by *Erwin Schrödinger* in 1926. The so-called
Schrödinger equation, used extensively in atomic, nuclear, and solid-state physics,
is a wave equation describing probability waves (wave functions) that govern the
motion of small atomic particles. The equation has the same fundamental importance
to quantum mechanics as Newton laws have for large dimension phenomena in
classical mechanics.

1.19   Quantization of Electromagnetic Radiation

Electromagnetic (EM) radiation incident on metallic surface may eject charged particles from the surface, as first observed by *Heinrich Hertz* in 1887. *Joseph Thomson* proved that the emitted charged particles were electrons and *Albert Einstein* in 1905 explained the effect by proposing that EM radiation was quantized similarly to the quantization of oscillator levels in matter introduced by *Max Planck* in 1900 (see Sect. 1.26).

A quantum of EM radiation, called a *photon*, has the following properties:

- It is characterized by frequency ν and wavelength $\lambda = c/\nu$ where c is the speed of light in vacuum.
- It carries energy $h\nu$ and momentum $p_\nu = h/\lambda$ where h is Planck constant.
- It has zero rest mass (see Sect. 1.21).

In a metal the outer electrons move freely from atom to atom and behave like a gas with a continuous spectrum of energy levels. To release an electron from a metal a minimum energy, characteristic of the given metal and referred to as the work function $e\phi$, must be supplied to the electron. Einstein postulated that the maximum kinetic energy $(E_K)_{max}$ of the electron ejected from the surface of a metal by a photon with energy $h\nu$ is given by the following expression

$$(E_K)_{max} = h\nu - e\phi. \tag{1.31}$$

The maximum kinetic energy of the ejected electrons depends on the incident photon energy $h\nu$ and the work function $e\phi$ of the metal but does not depend on the incident radiation intensity. The effect of electron emission from metallic surfaces was named the photoelectric effect and its explanation by Einstein on the basis of quantization of EM radiation is an important contribution to modern physics. Notable features of the surface photoelectric effect are as follows:

- Electrons can be ejected from a metallic surface by the photoelectric effect only when the incident photon energy $h\nu$ exceeds the work function $e\phi$ of the metal, i.e., $h\nu > e\phi$.
- The photoelectric effect is a quantum phenomenon: a single electron absorbs a single photon; the photon disappears and the electron is ejected with a certain kinetic energy.
- The typical magnitude of the work function $e\phi$ for metals is of the order of a few electron volts (e.g., aluminum: 4.3 eV; cesium: 2.1 eV; cobalt: 5.0 eV; copper: 4.7 eV; iron: 4.5 eV; lead: 4.3 eV; uranium: 3.6 eV), as given in the *Handbook of Chemistry and Physics*. The work function is thus of the order of energy $h\nu$ of visible photons ranging from 1.8 eV (700 nm) to 3 eV (400 nm) and near ultraviolet photons ranging in energy from 3 eV (400 nm) to 10 eV (80 nm).
- The surface photoelectric effect is related to the atomic photoelectric effect in which high-energy photons with energies exceeding the binding energy of orbital electrons eject electrons from atomic shells (see Sect. 7.5) rather than from metallic surfaces.

1.20 Special Theory of Relativity

The special theory of relativity, introduced in 1905 by *Albert Einstein*, extends the range of physical laws to large velocities and deals with transformations of physical quantities from one inertial frame of reference to another.

An *inertial frame of reference* implies motion with uniform velocity. The two postulates of the special relativity theory are:

1. The laws of physics are identical in all inertial frames of reference.
2. The speed of light in vacuum c is a universal constant independent of the motion of the source.

Albert A. Michelson and *Edward W. Morley* in 1887 showed that the speed of light c is a universal constant independent of the state of motion of the source or observer. Einstein, with his special theory of relativity, explained the results of the Michelson–Morley experiment and introduced, in contrast to classical *Galilean transformations*, special transformations referred to as *Lorentz transformations* to relate measurements in one inertial frame of reference to measurements in another inertial frame of reference.

When the velocities involved are very small, the Lorentzian transformations simplify to the classical Galilean transformations, and the relativistic relationships for physical quantities transform into classical Newtonian relationships. Galilean and Lorentzian transformations have the following basic characteristics:

- Galilean and Lorentzian transformations relate the spatial and time coordinates x, y, z, and t in a stationary frame of reference to coordinates x', y', z', and t' in a reference frame moving with a uniform speed v in the x direction as follows:

Galilean transformation		*Lorentzian transformation*	
$x' = x - vt$	(1.32)	$x' = \gamma(x - vt)$	(1.33)
$y' = y$	(1.34)	$y' = y$	(1.35)
$z' = z$	(1.36)	$z' = z$	(1.37)
$t' = t$	(1.38)	$t' = \gamma\left(t - \dfrac{xv}{c^2}\right)$	(1.39)

$$\text{where} \quad \gamma = \frac{1}{\sqrt{1 - \dfrac{v^2}{c^2}}} \tag{1.40}$$

- For $v \ll c$ the Lorentzian transformation reduces to the Galilean transformation since $\gamma \approx 1$. The specific form of the Lorentzian transformation is a direct consequence of $c = \text{const}$ in all frames of reference.

- Einstein also showed that atomic and subatomic particles, as they are accelerated to a significant fraction of the speed of light c, exhibit another relativistic effect, an increase in mass as a result of the mass-energy equivalence stated as $E = mc^2$, where m and E are the mass and energy, respectively, of the particle. A corollary to the second postulate of relativity is that no particle can move faster than the speed of light c in vacuum.

- *Conservation of energy and momentum:*

  – In *classical mechanics* where $v \ll c$, the momentum given as $p = m_0 v$ and the kinetic energy given as $E_K = \frac{1}{2} m_0 v^2$ are conserved in all collisions (m_0 is the mass of the particle at $v = 0$).
  – In *relativistic mechanics* where $v \approx c$, the momentum $p = mv = \gamma m_0 v$ and the total energy $E = m_0 c^2 + E_K$ are conserved in all collisions.

1.21 Important Relativistic Relations

In relativistic mechanics the mass of a particle is not a conserved quantity, since it depends on the velocity of the particle and may be converted into kinetic energy. The reverse transformation is also possible and energy may be converted into matter.

1.21.1 Relativistic Mass

Newton classical equation of motion is preserved in relativistic mechanics, i.e.,

$$\mathbf{F} = \frac{d\mathbf{p}}{dt}, \tag{1.41}$$

where $\mathbf{p}$ is the momentum of a particle acted upon by force $\mathbf{F}$. The momentum $\mathbf{p}$ is proportional to the velocity v of a particle through the relationship

$$\mathbf{p} = m\mathbf{v}, \tag{1.42}$$

where m is the mass of the particle, dependent on the magnitude of the particle velocity v, i.e., $m = m(v)$.

The mass $m(v)$ is referred to as the *relativistic mass* of a particle and is given by Einstein's expression (see Fig. 1.4 and Table 1.7) as follows

$$m(v) = \frac{m_0}{\sqrt{1 - \dfrac{v^2}{c^2}}} = \frac{m_0}{\sqrt{1 - \beta^2}} = \gamma m_0 \tag{1.43}$$

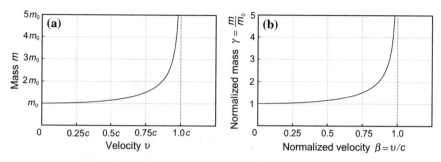

Fig. 1.4 Particle mass m as a function of its velocity v. A plot of m against v of (1.43) in **a** and a plot of γ against β of (1.44) in **b**

Table 1.7 Mass against velocity according to (1.44)

$(v/c) = \beta$	0	0.1	0.25	0.5	0.75	0.9	0.99	0.999	0.9999
$(m/m_0) = \gamma$	1.000	1.005	1.033	1.155	1.512	2.294	7.089	22.37	70.71

or

$$\frac{m(v)}{m_0} = \frac{1}{\sqrt{1 - \dfrac{v^2}{c^2}}} = \frac{1}{\sqrt{1 - \beta^2}} = \gamma, \tag{1.44}$$

where

m_0 is the mass of a particle at $v = 0$, referred to as the particle *rest mass*.
c is the speed of light in vacuum, a universal constant.
β is v/c.
γ is the Lorentz factor expressed as $(1 - \beta^2)^{-1/2}$ or $[1 - (v/c)^2]^{-1/2}$.

1.21.2   Relativistic Force and Relativistic Acceleration

In *classical physics* the Newton second law of mechanics is given as follows

$$\mathbf{F} = \frac{d\mathbf{p}}{dt} = m_0 \frac{dv}{dt} = m_0 \mathbf{a}, \tag{1.45}$$

indicating that the acceleration **a** is parallel to force **F**, and that mass m_0 is constant.

In *relativistic physics* the acceleration **a** is not parallel to the force **F** at large velocities because the speed of a particle cannot exceed c, the speed of light in vacuum. The force **F**, with the mass m a function of particle velocity v as given in (1.43), can be written as

$$\mathbf{F} = \frac{d\mathbf{p}}{dt} = \frac{d(mv)}{dt} = m\frac{dv}{dt} + v\frac{dm}{dt} \tag{1.46}$$

and

$$\mathbf{F} = \frac{d\mathbf{p}}{dt} = \frac{d(\gamma m_0 v)}{dt} = \gamma m_0 \frac{dv}{dt} + m_0 v\frac{d\gamma}{dt} = \gamma m_0 \frac{dv}{dt} + m_0 v\frac{\gamma^3 v}{c^2}\frac{dv}{dt}, \tag{1.47}$$

where

$$\frac{d\gamma}{dt} = \frac{1}{\left[1 - \dfrac{v^2}{c^2}\right]^{3/2}}\frac{v}{c^2}\frac{dv}{dt} = \frac{\gamma^3 v}{c^2}\frac{dv}{dt}. \tag{1.48}$$

The acceleration $\mathbf{a} = dv/dt$ will be determined by obtaining a dot product of the force **F** and velocity v as follows

$$\mathbf{F}{\cdot}v = \gamma m_0 v\frac{dv}{dt} + \frac{m_0\gamma^3 v^3}{c^2}\frac{dv}{dt} = \gamma m_0 v\frac{dv}{dt}\left(1 + \gamma^2\beta^2\right) = \gamma^3 m_0 v\frac{dv}{dt}. \tag{1.49}$$

Inserting (1.49) into (1.47) gives the following result for the relativistic force **F**

$$\mathbf{F} = \gamma m_0 \frac{dv}{dt} + \frac{\mathbf{F}{\cdot}v}{c^2}v = \gamma m_0 \frac{dv}{dt} + (\mathbf{F}{\cdot}\boldsymbol{\beta})\,\boldsymbol{\beta}. \tag{1.50}$$

Solving (1.50) for $\mathbf{a} = dv/dt$ gives the relativistic relationship for the acceleration **a**

$$\mathbf{a} = \frac{dv}{dt} = \frac{\mathbf{F} - (\mathbf{F}{\cdot}\boldsymbol{\beta})\,\boldsymbol{\beta}}{\gamma m_0}. \tag{1.51}$$

For velocities $v \ll c$, where $\beta \to 0$ and $\gamma \to 1$, the relativistic expression (1.51) for acceleration **a** transforms into Newton's classical result $\mathbf{a} = dv/dt = \mathbf{F}/m_0$ with **a** parallel to **F**.

1.21.3   Relativistic Kinetic Energy

The expression for the relativistic kinetic energy $E_K = E - E_0$, where $E = mc^2$ is the total energy of the particle and $E_0 = m_0 c^2$ is its rest energy, is derived below.

   The particle of rest mass m_0 is initially at rest at the initial position x_i and moves under the influence of force F to its final position x_f. The work done by force F is the kinetic energy E_K of the particle calculated using the integration of (1.46) and the following steps:

1.

$$E_K = \int\limits_{x_i}^{x_f} F \, dx = \int\limits_{x_i}^{x_f} \left(m \frac{dv}{dt} + v \frac{dm}{dt} \right) dx \qquad (1.52)$$

2. Multiply (1.43) by c, square the result, and rearrange the terms to obtain

$$m^2 c^2 - m^2 v^2 = m_0^2 c^2. \qquad (1.53)$$

3. Differentiate (1.53) with respect to time t to obtain

$$c^2 \frac{d(m^2)}{dt} - \frac{d}{dt}(m^2 v^2) = 0 \qquad (1.54)$$

4. Equation (1.54), after completing the derivatives, gives

$$2c^2 m \frac{dm}{dt} - 2m^2 v \frac{dv}{dt} - 2v^2 m \frac{dm}{dt} = 0. \qquad (1.55)$$

5. After dividing (1.55) by $2mv$ we obtain the following expression

$$\frac{c^2}{v} \frac{dm}{dt} = m \frac{dv}{dt} + v \frac{dm}{dt}. \qquad (1.56)$$

The expression for E_K in (1.52) using (1.56) can now be written as follows

$$E_K = c^2 \int\limits_{x_i}^{x_f} \frac{1}{v} \frac{dm}{dt} dx = c^2 \int\limits_{m_0}^{m} dm = mc^2 - m_0 c^2 = E - E_0, \qquad (1.57)$$

since dx/dt is the particle velocity v by definition and the masses m_0 and m correspond to particle positions x_i and x_f, respectively.

Inserting (1.43) into (1.57) results in the following expression for the relativistic kinetic energy E_K

$$E_K = mc^2 - m_0 c^2 = \gamma m_0 c^2 - m_0 c^2 = (\gamma - 1) m_0 c^2$$

$$= (\gamma - 1) E_0 = \left(\frac{1}{\sqrt{1 - \frac{v^2}{c^2}}} - 1 \right) E_0, \qquad (1.58)$$

in contrast with the well known classical expression

Fig. 1.5 Classical
expression expression for
kinetic energy (1.59) and
relativistic expression
expression for kinetic energy
(1.58) both normalized to
electron rest mass m_0c^2 and
plotted against normalized
velocity v/c

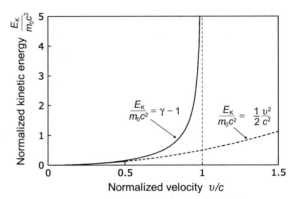

The relativistic expression (1.58) and the classical expression (1.59) are plotted in Fig. 1.5, with the kinetic energy normalized to rest energy $[E_K/(m_0c^2)]$ on the ordinate (y) axis and the velocity normalized to the speed of light (v/c) on the abscissa (x) axis. At low particle velocity where $(v \ll c)$, the two expressions coincide; however, at high velocities v the relativistic expression (1.58), as a result of increase in mass m with velocity v, increases rapidly to attain an infinite value at $v = c$ or at $v/c = 1$, while, for the classical expression (1.59) in which the mass remains constant with increasing velocity v, the ratio $E_K/(m_0c^2)$ attains a value of 0.5 for $v = c$.

$$E_K = \frac{m_0 v^2}{2}.$$ (1.59)

1.21.4   Total Relativistic Energy as a Function of Momentum

The expression for the total relativistic energy E as a function of momentum p is as follows

$$E = \sqrt{E_0^2 + p^2 c^2}.$$ (1.60)

Equation (1.60) is obtained from Einstein expression for the relativistic mass given in (1.43) as follows:

1. Square the relationship for the relativistic mass m of (1.43), multiply the result by c^4, and rearrange the terms to obtain

$$m^2 c^4 - m^2 c^2 v^2 = m_0^2 c^4.$$ (1.61)

2. Equation (1.61) can be written as

$$E^2 - p^2 c^2 = E_0^2$$ (1.62)

or

$$E = \sqrt{E_0^2 + p^2 c^2},\qquad (1.63)$$

using the common relativistic relationships for the total energy E, rest energy E_0, and momentum p, i.e., $E = mc^2$, $E_0 = m_0 c^2$, and $p = mv$.

The following two relationships are also often used in relativistic mechanics:

1. The particle momentum p using (1.57) and (1.63) for the kinetic energy E_K and total energy E, respectively, can be expressed as

$$p = \frac{1}{c}\sqrt{E^2 - E_0^2} = \frac{1}{c}\sqrt{E_K^2 + 2E_K E_0} = \frac{E_K}{c}\sqrt{1 + \frac{2E_0}{E_K}}.\qquad (1.64)$$

2. The particle speed v is, in terms of its total energy E and momentum p, given as

$$\frac{v}{c} = \frac{mvc}{mc^2} = \frac{pc}{E}.\qquad (1.65)$$

It is easy to show that the two expressions for momentum p, given in (1.42) and (1.64) and restated below, are equivalent to one another. From (1.42) and (1.64) we get

$$p = mv = \gamma m_0 v = \frac{m_0 v}{\sqrt{1 - \beta^2}}\qquad (1.66)$$

and

$$p = \frac{1}{c}\sqrt{E^2 - E_0^2} = \frac{1}{c}\sqrt{\gamma^2 m_0^2 c^4 - m_0^2 c^4} = m_0 c\sqrt{\gamma^2 - 1},\qquad (1.67)$$

respectively, and since

$$\gamma^2 - 1 = \frac{1}{1 - \beta^2} - 1 = \frac{\beta^2}{1 - \beta^2},\qquad (1.68)$$

we can express the momentum p of (1.67) as

$$p = m_0 c\sqrt{\gamma^2 - 1} = \frac{m_0 c \beta}{\sqrt{1 - \beta^2}} = \gamma m_0 v = mv,\qquad (1.69)$$

proving the equivalence of the two expressions (1.42) and (1.64) for momentum p.

1.21.5 Taylor Expansion and Classical Approximations for Kinetic Energy and Momentum

The Taylor expansion of a function $f(x)$ about $x = a$ is given as follows

$$f(x) = f(a) + (x - a)\frac{df}{dx}\bigg|_{x=a} + \frac{(x - a)^2}{2!}\frac{d^2f}{dx^2}\bigg|_{x=a} + \cdots + \frac{(x - a)^n}{n!}\frac{d^nf}{dx^n}\bigg|_{x=a}.$$

$$(1.70)$$

The Taylor expansion into a series given by (1.70) is particularly useful when one can neglect all but the first two terms of the series. For example, the first two terms of the Taylor expansion of the function $f(x) = (1 \pm x)^n$ about $x = 0$ for $x \ll 1$ are given as follows

$$f(x) = (1 \pm x)^n \approx 1 \pm nx. \qquad (1.71)$$

- The approximation of (1.71) is used in showing that, for small velocities where $v \ll c$ or $v/c \ll 1$, the relativistic kinetic energy E_K of (1.57) transforms into the well-known classical relationship $E_K = \frac{1}{2}m_0v^2$

$$E_K = E - E_0 = m_0c^2\left(\frac{1}{\sqrt{1 - \dfrac{v^2}{c^2}}} - 1\right)$$

$$\approx m_0c^2\left\{1 + \frac{1}{2}\frac{v^2}{c^2} - \cdots - 1\right\} = \frac{m_0v^2}{2}. \qquad (1.72)$$

- Another example of the use of Taylor expansion of (1.71) is the classical relationship for the momentum $p = m_0v$ that, for $v \ll c$, i.e., $v/c \ll 1$, is obtained from the relativistic relationship of (1.64) as follows

$$p = \frac{1}{c}\sqrt{E^2 - E_0^2} = \frac{1}{c}\sqrt{m_0^2c^4\left(\frac{1}{1 - \dfrac{v^2}{c^2}} - 1\right)} = \frac{m_0c^2}{c}\sqrt{\left(1 - \frac{v^2}{c^2}\right)^{-1} - 1}$$

$$\approx m_0c\sqrt{1 + \frac{v^2}{c^2} + \cdots - 1} = m_0v. \qquad (1.73)$$

1.21.6 Relativistic Doppler Shift

The speed of light emitted from a moving source is equal to c, a universal constant, irrespective of the source velocity. While the speed of the emitted photons equals c, the energy, wavelength, and frequency of the emitted photons all depend on the velocity of the moving source. The energy shift resulting from a moving source in comparison with the stationary source is referred to as the Doppler shift and the following conditions apply:

- When the source is moving toward the observer, the measured photon frequency and energy increase while the wavelength decreases (*blue Doppler shift*).
- When the source is moving away from the observer, the measured photon frequency and energy decrease while the wavelength increases (*red Doppler shift*).

1.22 Particle–Wave Duality

For electromagnetic radiation, the energy E_ν of a photon is given by Planck law as

$$E_\nu = h\nu = h\frac{c}{\lambda},\tag{1.74}$$

where

h is Planck constant.
c is the speed of light in vacuum.
ν is photon frequency.
λ is the wavelength of the photon.

The photon energy E_ν can be also written in terms of the photon momentum p_ν as

$$E_\nu = p_\nu c,\tag{1.75}$$

using (1.63) and recognizing that the rest mass of the photon is zero. Merging (1.74) and (1.75), the photon momentum p_ν is given as

$$p_\nu = \frac{E_\nu}{c} = \frac{h\nu}{c} = \frac{h}{\lambda},\tag{1.76}$$

highlighting the particle–wave duality of electromagnetic radiation, since both (1.74) and (1.76) contain within their structure a wave concept through wavelength λ and frequency ν as well as a particle concept through energy E_ν and momentum p_ν.

1.22.1 De Broglie Equation and de Broglie Wavelength

Following the recognition of particle–wave duality of electromagnetic radiation, *Louis de Broglie* in 1924 postulated a similar property for particles in particular and matter in general, namely a characterization with wavelength λ and momentum p related to one another through the following expression, already applicable to electromagnetic radiation, as shown in (1.74) and (1.75)

$$\lambda = \frac{h}{p}. \tag{1.77}$$

In relation to particles, (1.77) is referred to as de Broglie equation and λ is referred to as de Broglie wavelength of a particle. Using (1.64) and (1.77) we can calculate de Broglie wavelength λ of a particle as a function of its kinetic energy E_K to get

$$\lambda = \frac{h}{p} = \frac{2\pi\hbar c}{E_K\sqrt{1 + \dfrac{2E_0}{E_K}}}. \tag{1.78}$$

De Broglie wavelength λ of a particle can also be expressed as a function of its velocity v or normalized velocity β using (1.42) and (1.43) to get

$$\lambda = \frac{h}{p} = \frac{h}{mv} = \frac{2\pi\hbar c}{\gamma m_0 c^2 \beta} = \frac{2\pi\hbar c}{m_0 c^2} \frac{\sqrt{1 - \beta^2}}{\beta}. \tag{1.79}$$

In Fig. 1.6 we plot (1.78) representing de Broglie wavelength λ for a given particle against the particle's kinetic energy E_K for an electron, proton and α-particle with solid curves and for a photon of energy $h\nu$ with the dashed line. Figure 1.7 displays (1.79), a plot of de Broglie wavelength λ as a function of the normalized velocity β for an electron, proton, and α-particle.

Fig. 1.6 Plot of (1.78) representing de Broglie wavelength λ against kinetic energy E_K for electron, proton, and alpha particle with *solid curves* and for photon with a *dashed curve*

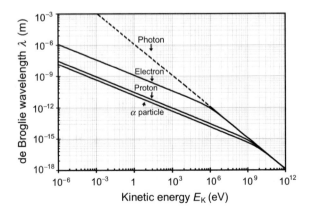

Fig. 1.7 Plot of (1.79) representing de Broglie wavelength λ against the normalized velocity v/c for electron, proton, and alpha particle

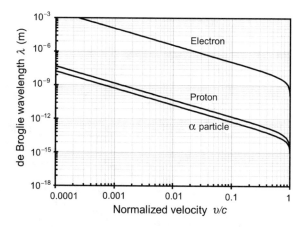

The following observations can be made from Figs. 1.6 and 1.7:

- At large kinetic energies, where $E_K \gg E_0$ and total energy $E \approx E_K$, de Broglie wavelength λ converges to $2\pi \hbar c / E_K$ coinciding with the photon line, so that all particles have the same de Broglie wavelength at a given energy.
- At low kinetic energies, as $E_K \to 0$, de Broglie wavelength λ approaches ∞ as $(2\pi \hbar c) / \sqrt{2E_0 E_K}$; the larger is E_0, the slower is the approach of λ to ∞ with a decreasing kinetic energy.
- For macroscopic objects that are moving with practical speed of the order of 250 m/s (airplane) or less de Broglie wavelength λ is extremely small with an order of magnitude of 10^{-34} m, some 16 orders of magnitude smaller than the highest resolution achievable experimentally.

1.22.2 Davisson–Germer Experiment

In 1927 *Clinton J. Davisson* and *Lester H. Germer* confirmed experimentally the wave nature of electrons by bombarding a nickel target with electrons and measuring the intensity of electrons scattered from the target. The target was in the form of a regular crystalline alloy that was formed through a special annealing process. The beam of electrons was produced by thermionic emission from a heated tungsten filament. The electrons were accelerated through a relatively low variable potential difference U that enabled the selection of the incident electron kinetic energy E_K. The scattered electrons were collected with a Faraday cup and their intensity was measured with a galvanometer.

Davisson and Germer discovered that for certain combinations of electron kinetic energy E_K and angle of incidence ϕ the intensity of scattered electrons exhibited maxima, similar to the scattering of x-rays from a crystal with a crystalline plane separation d that follows the Bragg relationship. Bragg has shown that for x-rays from two successive planes to interfere constructively their path lengths must differ

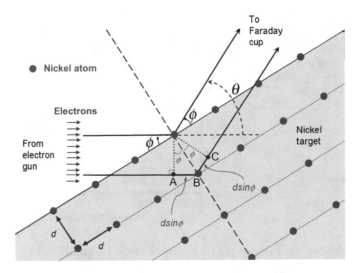

Fig. 1.8 Schematic representation of the Davisson–Germer experiment of elastic electron scattering on a nickel single crystal target. The electrons are produced in an electron gun and scattered by the nickel crystalline structure that has atom spacing d and acts as a reflection grating. The maximum intensity of scattered electrons occurs at Bragg angle of incidence ϕ as a result of constructive interference from electron matter waves following the Bragg relationship of (1.80). The scattering angle θ is the angle between the incident and scattered wave and equals to 2ϕ

by an integral number m of wavelengths λ. The first order diffraction maximum for $m = 1$ is the most intense.

Figure 1.8 shows the similarity between the Bragg experiment in x-ray scattering on a crystal with plane separation d and Davisson–Germer experiment of electron wave scattering on nickel atoms also arranged in crystalline planes with separation d. Constructive interference between waves scattered from two planes occurs when

$$\overline{AB} + \overline{BC} = 2d \sin \phi = m\lambda, \tag{1.80}$$

where

ϕ is the angle of incidence (Bragg angle) of the incident wave equal to the angle of reflection.

λ is the wavelength of the incident wave that may be a monoenergetic x-ray wave in Bragg experiment or matter (electron) wave in Davisson–Germer experiment.

Davisson and Germer determined the wavelength λ of electrons from the known atomic separation d and the measured Bragg angle ϕ at which the electron intensity exhibited a maximum. They found that the wavelength λ calculated with (1.80) agreed well with electron wavelength λ_e calculated from de Broglie relationship

$$\lambda_e = \frac{h}{p} = \frac{h}{m_e v} = \frac{2\pi \hbar c}{\sqrt{2m_e c^2 eU}}, \tag{1.81}$$

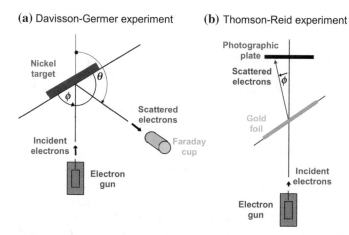

Fig. 1.9 Schematic representation of two seminal experiments to confirm the wave nature of electrons with diffraction patterns: **a** Davisson–Germer experiment using electron scattering on nickel crystalline structure and **b** Thomson–Reid experiment using electron scattering on a gold foil. In (**a**) electrons are collected with a Faraday cup, in (**b**) they are collected on a photographic plate

where v is the velocity of electrons determined from the classical kinetic energy relationship $E_K = \frac{1}{2} m_e v^2 = eU$ with U the applied potential of the order of 10 V to 100 V. The Davisson–Germer experiment, shown schematically in Fig. 1.9a, unequivocally demonstrated the diffraction of electrons and with it the wave nature of the electron.

1.22.3 Thomson–Reid Experiment

At about the same time as Davisson and Germer, *George P. Thomson* and his assistant A. Reid carried out an experiment that also confirmed de Broglie contention that matter can behave as waves. In contrast to Davisson and Germer who measured with a Faraday cup the intensity of electrons scattered from a nickel absorber, Thomson and Reid passed a collimated electron beam with kinetic energy of the order of 50 keV through a thin gold or aluminum foil (thickness of about 1000 Å and measured the transmitted electron intensity with a photographic plate.

Because of the crystalline structure of the metallic foil and the wave nature of the electrons, the intensity of electrons measured with film appeared in the form of concentric circles governed by the Bragg condition (1.80). The diameter of each ring is proportional to de Broglie wavelength of the electron and inversely proportional to the speed of the electron, similar to the results found in Davisson–Germer experiment. The Thomson–Reid experiment is shown schematically in Fig. 1.9b.

1.22.4   General Confirmation of Particle–Wave Duality

Davisson–Germer and Thomson–Reid experiments confirmed that electrons behave as waves under certain conditions, and other experimentalists have subsequently tested and confirmed the universal character of the de Broglie postulate by observing similar diffraction results for hydrogen and helium atoms as well as for neutron beams.

The experimental discovery of electron and neutron diffraction was a very important finding in support of quantum mechanics. The wavelength λ associated with a particle is called its de Broglic wavelength and is defined as the Planck constant divided by the particle momentum p, as shown in (1.77).

The experimentally determined *particle–wave duality* suggests that both models can be used for particles as well as for photon radiation. However, for a given measurement only one of the two models will apply. For example, in the case of photon radiation, the Compton effect is explained with the particle model, while the diffraction of x-rays is explained with the wave model. On the other hand, the charge-to-mass ratio e/m_e of the electron implies a particle phenomenon, while electron diffraction suggests wave-like behavior.

Both Davisson and Thomson shared the 1937 Nobel Prize in Physics for their discovery of the diffraction of electrons by crystals. It is interesting to note that Joseph J. Thomson, the father of George Thomson, discovered the electron as particle in 1897, while George Thomson confirmed the wave nature of the electron in 1927.

1.23   Matter Waves

1.23.1   Introduction to Wave Mechanics

Associated with any particle is a matter wave, as suggested by the de Broglie relationship of (1.77). This matter wave is referred to as the particle's wave function $\Psi(z, t)$ for one-dimensional problems or $\Psi(x, y, z, t)$ for three-dimensional problems and contains all the relevant information about the particle. Quantum mechanics, developed by *Erwin Schrödinger* as wave mechanics and *Werner Heisenberg* as matrix mechanics between 1925 and 1929, is a branch of physics that deals with the properties of wave functions as they pertain to particles, nuclei, atoms, molecules, and solids.

The main characteristics of wave mechanics are as follows:

- The theory has general application to microscopic systems and includes Newton theory of macroscopic particle motion as a special case in the macroscopic limit.
- The theory specifies the laws of wave motion that the particles of any microscopic system follow.

- The theory provides techniques for obtaining the wave functions for a given microscopic system.
- It offers means to extract information about a particle from its wave function.

The main attributes of wave functions $\Psi\,(z, t)$ are:

- Wave functions are generally but not necessarily complex and contain the imaginary number i.
- Wave functions cannot be measured with any physical instrument.
- Wave functions serve in the context of the Schrödinger wave theory but contain physical information about the particle they describe.
- Wave functions must be single-valued and continuous functions of z and t to avoid ambiguities in predictions of the theory.

The information on a particle can be extracted from a complex wave function $\Psi\,(z, t)$ through a postulate proposed by *Max Born* in 1926 relating the probability density $dP\,(z, t)\,/dz$ in one dimension and $dP\,(x, y, z, t)\,/dV$ in three dimensions with the wave functions $\Psi\,(z, t)$ and $\Psi\,(x, y, z, t)$, respectively, as follows

$$\frac{dP\,(z, t)}{dz} = \Psi^*\,(z, t) \cdot \Psi\,(z, t) \tag{1.82}$$

and

$$\frac{dP\,(x, y, z, t)}{dV} = \Psi^*\,(x, y, z, t) \cdot \Psi\,(x, y, z, t), \tag{1.83}$$

where

Ψ^* is the complex conjugate of the wave function Ψ.
V stands for volume.

The probability density is real, non-negative and measurable. In one-dimensional wave mechanics, the total probability of finding the particle somewhere along the z axis in the entire range of the z axis is equal to one, if the particle exists. We can use this fact to define the following *normalization condition*

$$\int\limits_{-\infty}^{+\infty} \frac{dP\,(z, t)}{dz} dz = \int\limits_{-\infty}^{+\infty} \Psi^*\,(z, t)\,\Psi\,(z, t)\,dz = 1. \tag{1.84}$$

Similarly, in three-dimensional wave mechanics, the normalization expression is written as

$$\int\limits_{-\infty}^{+\infty} \frac{dP(x, y, z, t)}{dV} dV = \int\limits_{-\infty}^{+\infty}\int\limits_{-\infty}^{+\infty}\int\limits_{-\infty}^{+\infty} \Psi^*(x, y, z, t)\Psi(x, y, z, t)dV = 1, \tag{1.85}$$

where the volume integral extends over all space and represents a certainty that the particle will be found somewhere (unit probability). Any one-dimensional wave function $\Psi\,(z, t)$ that satisfies (1.84) is by definition normalized. Similarly, any three-dimensional wave function $\Psi\,(x, y, z, t)$ that satisfies (1.85) is also normalized.

While the normalization condition expresses certainty that a particle, if it exists, will be found somewhere, the probability that the particle will be found in any interval $a \leq z \leq b$ is obtained by integrating the probability density $\Psi^* \cdot \Psi$ from a to b as follows

$$P = \int_a^b \Psi^* \cdot \Psi \, d\mathcal{V}. \tag{1.86}$$

1.23.2   Quantum Mechanical Wave Equation

The particulate nature of photons and the wave nature of matter are referred to as the wave–particle duality of nature. The waves associated with matter are represented by the wave function $\Psi\,(x, y, z, t)$ that is a solution to a quantum mechanical wave equation. This wave equation cannot be derived directly from first principles of classical mechanics; however, it must honor the following four conditions:

1. It should respect the *de Broglie postulate* relating the wavelength λ of the wave function with the momentum p of the associated particle

$$p = \frac{h}{\lambda} = \hbar k, \tag{1.87}$$

   where k is the wave number defined as $k = 2\pi/\lambda$.
2. It should respect *Planck law* relating the frequency ν of the wave function with the total energy E of the particle

$$E = h\nu = \hbar\omega, \tag{1.88}$$

   where ω is the angular frequency of the wave function.
3. It should respect the relationship expressing the total energy E of a particle of mass m as a sum of the particle's kinetic energy $E_K = p^2\,/\,(2m)$ and potential energy V, i.e.,

$$E = \frac{p^2}{2m} + V. \tag{1.89}$$

4. It should be linear in $\Psi\,(z, t)$ which means that any arbitrary linear combination of two solutions for a given potential energy V is also a solution to the wave equation.

While the wave equation cannot be derived directly, we can determine it for a free particle in a constant potential and then generalize the result to other systems and other potential energies. The free particle wave function in one dimension $\Psi(z, t)$ can be expressed as follows

$$\Psi(z, t) = C\, e^{i(kz - \omega t)}, \tag{1.90}$$

where $(kz - \omega t)$ is the phase of the wave with $k = 2\pi/\lambda$ the wave number and $\omega = 2\pi\nu$ the angular frequency of the wave.

We now determine the partial derivatives $\partial/\partial z$ and $\partial/\partial t$ of the wave function to obtain

$$\frac{\partial \Psi(z, t)}{\partial z} = ikC\, e^{i(kz - \omega t)} = ik\Psi(z, t) = i\frac{p}{\hbar}\Psi(z, t) \tag{1.91}$$

and

$$\frac{\partial \Psi(z, t)}{\partial t} = -i\omega C\, e^{i(kz - \omega t)} = -i\omega\Psi(z, t) = i\frac{E}{\hbar}\Psi(z, t). \tag{1.92}$$

Equations (1.91) and (1.92) can now be written as follows

$$p\Psi(z, t) = -i\hbar\frac{\partial}{\partial z}\Psi(z, t) \tag{1.93}$$

and

$$E\Psi(z, t) = i\hbar\frac{\partial}{\partial t}\Psi(z, t), \tag{1.94}$$

where $(-i\hbar\partial/\partial z)$ and $(i\hbar\partial/\partial t)$ are differential operators for the momentum p and total energy E, respectively.

Equations (1.93) and (1.94) suggest that multiplying the wave function $\Psi(z, t)$ by a given physical quantity, such as p and E in (1.93) and (1.94), respectively, has the same effect as operating on $\Psi(z, t)$ with an operator that is associated with the given physical quantity. As given in (1.89), the total energy E of the particle with mass m is the sum of the particle's kinetic and potential energies. If we now replace p and E in (1.89) with respective operators, given in (1.93) and (1.94), respectively, we obtain

$$-\frac{\hbar^2}{2m}\frac{\partial^2}{\partial x^2} + V = i\hbar\frac{\partial}{\partial t}. \tag{1.95}$$

Equation (1.95) represents two new differential operators; the left hand side operator is referred to as the hamiltonian operator [H], the right hand side operator is the operator for the total energy E. When the two operators of (1.95) are applied to a free particle wave function $\Psi(z, t)$ we get

$$-\frac{\hbar^2}{2m}\frac{\partial^2\Psi\,(z,t)}{\partial z^2} + V\Psi\,(z,t) = i\hbar\frac{\partial\Psi\,(z,t)}{\partial t}. \tag{1.96}$$

Equation (1.96) was derived for a free particle moving in a constant potential V; however, it turns out that the equation is valid in general for any potential energy $V(z,t)$ and is referred to as the time-dependent Schrödinger equation with $V(z,t)$ the potential energy describing the spatial and temporal dependence of forces acting on the particle of interest. The time-dependent Schrödinger equation is thus in the most general three dimensional form written as follows

$$-\frac{\hbar^2}{2m}\nabla^2\Psi(x,y,z,t) + V(x,y,z,t)\Psi(x,y,z,t) = i\hbar\frac{\partial\Psi\,(x,y,z,t)}{\partial t}. \tag{1.97}$$

1.23.3 Time-Independent Schrödinger Equation

In most physical situations the potential energy $V(z,t)$ only depends on the spatial coordinate z, i.e., $V(z,t) = V(z)$ and then the time-dependent Schrödinger equation can be solved with the method of separation of variables. The wave function $\Psi\,(z,t)$ is written as a product of two functions $\psi\,(z)$ and $T\,(t)$, one depending on the spatial coordinate z only and the other depending on the temporal coordinate t only, i.e.,

$$\Psi(z,t) = \psi(z)\,T(t)\,. \tag{1.98}$$

Inserting (1.98) into the time-dependent wave equation given in (1.96) and dividing by $\psi(z)T(t)$ we get

$$-\frac{\hbar^2}{2m}\frac{1}{\psi(z)}\frac{\partial^2\psi\,(z)}{\partial z^2} + V(z) = i\hbar\frac{1}{T(t)}\frac{\partial T\,(t)}{\partial t}. \tag{1.99}$$

Equation (1.99) can be valid in general only if both sides, the left hand side that depends on z alone and the right hand side that depends on t alone, are equal to a constant, referred to as the separation constant Λ. We now have two ordinary differential equations: one for the spatial coordinate z and the other for the temporal coordinate t given as follows

$$-\frac{\hbar^2}{2m}\frac{d^2\psi\,(z)}{dz^2} + V(z)\psi(z) = \Lambda\psi(z) \tag{1.100}$$

and

$$\frac{dT(t)}{dt} = -\frac{i\Lambda}{\hbar}T(t). \tag{1.101}$$

The solution to the temporal equation (1.101) is

$$T(t) = \exp\left(-i\frac{\Lambda}{\hbar}t\right), \tag{1.102}$$

representing simple oscillatory function of time with angular frequency $\omega = \Lambda/\hbar$. According to de Broglie and Planck, the angular frequency must also be given as $E/\hbar$, where E is the total energy of the particle.

We can now conclude that the separation constant Λ equals the total particle energy E and obtain from (1.101) the following solution to the temporal equation

$$T(t) = \exp\left(-i\frac{E}{\hbar}t\right) = e^{-i\omega t}. \tag{1.103}$$

Recognizing that $\Lambda = E$ we can write (1.100) as

$$-\frac{\hbar^2}{2m}\frac{d^2\psi(z)}{dz^2} + V(z)\psi(z) = E\psi(z) \tag{1.104}$$

and obtain the so-called *time-independent Schrödinger wave equation* for the potential $V(z)$.

The essential problem in quantum mechanics is to find solutions to the time-independent Schrödinger equation for a given potential energy V, generally only depending on spatial coordinates. The solutions are given in the form of:

1. Physical wave functions $\psi(x, y, z)$ referred to as *eigenfunctions*.
2. Allowed energy states E referred to as *eigenvalues*.

The time independent Schrödinger equation does not include the imaginary number i and its solutions, the eigenfunctions, are generally not complex. Since only certain functions (eigenfunctions) provide physical solutions to the time-independent Schrödinger equation, it follows that only certain values of E, referred to as *eigenvalues,* are allowed. This results in discrete energy values for physical systems and it also results in energy quantization.

Many mathematical solutions are available as solutions to wave equations. However, to serve as a physical solution, an eigenfunction $\psi(z)$ and its derivative $d\psi/dz$ must be:

1. *Finite*
2. *Single valued*
3. *Continuous*

Corresponding to each eigenvalue E_n is an eigenfunction $\psi_n(z)$ that is a solution to the time-independent Schrödinger equation for the potential $V_n(z)$. Each eigenvalue is also associated with a corresponding wave function $\Psi(z, t)$ that is a solution to the time-dependent Schrödinger equation and can be expressed as

$$\Psi(z, t) = \psi(z)e^{-i\frac{E}{\hbar}t}. \tag{1.105}$$

1.23.4 Measurable Quantities and Operators

As the term implies, a measurable quantity is any physical quantity of a particle that can be measured. Examples of measurable physical quantities are: position z, momentum p, kinetic energy E_K, potential energy V, total energy E, etc.

In quantum mechanics an operator is associated with each measurable quantity. The operator allows for a calculation of the average (expectation) value of the measurable quantity, provided that the wave function of the particle is known.

The *expectation value* (also referred to as the average or mean value) $\overline{Q}$ of a physical quantity Q, such as position z, momentum p, potential energy V, and total energy E of a particle, is determined as follows provided that the particle's wave function $\Psi(z, t)$ is known

$$\overline{Q} = \int \Psi^*(z, t)[Q]\Psi(z, t)\mathrm{d}z, \tag{1.106}$$

where $[Q]$ is the operator associated with the physical quantity Q. A listing of most common measurable quantities in quantum mechanics and their associated operators is given in Table 1.8. Two entries are given for the momentum, kinetic energy, and Hamiltonian operators: in one dimension and in three dimensions.

The quantum uncertainty ΔQ for any measurable quantity Q is given as

$$\Delta Q = \sqrt{\overline{Q^2} - \bar{Q}^2}, \tag{1.113}$$

where $\bar{Q}^2$ is the square of the expectation value of the quantity Q and $\overline{Q^2}$ is the expectation value of Q^2. The following conditions apply for ΔQ:

- When $\Delta Q = 0$, the measurable quantity Q is said to be sharp and all measurements of Q yield identical results.
- In general, $\Delta Q > 0$ and repeated measurements result in a distribution of measured points.

1.23.5 Transition Rate and the Fermi Second Golden Rule

Many physical interactions that particles undergo can be described and evaluated with the help of the transition rate between the initial state i and the final state f of

Table 1.8 Several measurable quantities and their associated operators used in quantum mechanics

Measurable	Symbol	Associated operator		Symbol	Equation
Position	z	z		$[z]$	(1.107)
Momentum	p	$-i\hbar\dfrac{\partial}{\partial z}$	or $\quad -i\hbar\nabla$	$[p]$	(1.108)
Potential energy	V	V		$[V]$	(1.109)
Kinetic energy	E_K	$-\dfrac{\hbar^2}{2m}\dfrac{\partial^2}{\partial z^2}$	or $\quad -\dfrac{\hbar^2}{2m}\nabla^2$	$[E_K]$	(1.110)
Hamiltonian	H	$-\dfrac{\hbar^2}{2m}\dfrac{\partial^2}{\partial z^2}+V$	or $\quad -\dfrac{\hbar^2}{2m}\nabla^2+V$	$[H]$	(1.111)
Total energy	E	$i\hbar\dfrac{\partial}{\partial t}$		$[E]$	(1.112)

Two entries are given for the momentum, kinetic energy, and Hamiltonian operators, one-dimensional on the left and 3-dimensional on the right

the particular system consisting of the particle and the potential operator acting on it. In general, the transition rate (also called reaction rate) W_{if} is equal to the transition probability per unit time and depends upon:

1. Coupling between the initial and final states.
2. Density of the final states $\rho(E_f)$ defined as the number of levels per energy interval.

The transition (reaction) rate W_{if} is usually expressed with the following relationship referred to as the *Fermi second golden rule*

$$W_{if} = \frac{2\pi}{\hbar}\,|\mathbf{M}_{if}|^2\,\rho(E_f),\tag{1.114}$$

with $\mathbf{M}_{if}$ denoting the amplitude of the transition matrix element (probability amplitude) and describing the dynamics of the particular interaction. $\mathbf{M}_{if}$ is usually expressed in an integral form as

$$\mathbf{M}_{if} = \iiint \psi_f^* V \psi_i \mathrm{d}\mathcal{V},\tag{1.115}$$

where

ψ_i is the wave function of the initial state.
ψ_f^* is the complex conjugate of the wave function of the final state.
V is the potential which operates on the initial state wave function ψ_i, i.e., is the potential operator which couples the initial and final eigenfunctions of the system.

Fermi second golden rule is derived from time dependent perturbation theory and is used widely in atomic, nuclear, and particle physics to deal with a wide variety of interactions, such as atomic transitions, beta decay, gamma decay, particle scattering (Rutherford, Mott) and nuclear reactions. At relatively low energies, all these interactions also play an important role in medical physics and radiation dosimetry.

1.23.6 Particle Scattering and Born Collision Formula

Much of the current knowledge in atomic, nuclear, and particle physics comes from analyses of scattering experiments carried out with various particles as projectile interacting with a target represented by atoms, nuclei, or other particles. An important parameter for describing a given scattering interaction is the differential cross section $d\sigma/d\Omega$ for the scattering interaction. The differential cross section is defined as the probability of scattering into a given solid angle $d\Omega$ and in quantum mechanics it is expressed as a transition rate, highlighting the most important use of the Fermi second golden rule stated in (1.114)

$$\frac{d\sigma}{d\Omega} = \frac{2\pi}{\hbar} |\mathbf{M}_{if}|^2 \rho(E_f),$$

(1.116)

where the matrix element $\mathbf{M}_{if}$ for the scattering interaction is given in (1.115).

The differential cross section for a scattering process can be calculated with many degrees of sophistication, and the result of the calculation is accepted as valid when it agrees with the experimental result. Some of the issues to be considered and accounted for in the calculations are:

1. Classical versus relativistic mechanics.
2. Elastic versus inelastic scattering.
3. Point-like versus finite projectile.
4. Point-like versus finite scatterer.
5. No spin versus spin of projectile.
6. No spin versus spin of scatterer.
7. Shape of the coupling potential.

The matrix element $\mathbf{M}_{if}$ is usually calculated using the Born approximation under the following assumptions:

1. Only a single scattering event occurs.
2. Initial and final states of the particle undergoing the scattering event are described by plane waves.

The wave function ψ_i of the plane wave for the initial particle to undergo scattering is given by

$$\psi_i(\mathbf{r}) = C \exp(i\mathbf{k}_i\mathbf{r}) = C \exp\left[\frac{i\mathbf{p}_i\mathbf{r}}{\hbar}\right],$$

(1.117)

where

$\mathbf{k}_i$ is the initial wave vector of the particle (projectile) before interaction, related to the incident particle momentum $\mathbf{p}_i$ through $\mathbf{k}_i = \mathbf{p}_i/\hbar$.

$\mathbf{p}_i$ is the initial momentum of the particle before scattering.

C is a normalization constant obtained from the normalization equation (see (1.85)) $\int \psi_i^* \psi_i \, d\mathcal{V} = 1$.

Similarly, the wave function ψ_f of the final plane wave is given as

$$\psi_f(\mathbf{r}) = C \exp(i\mathbf{k}_f\mathbf{r}) = C \exp\left[\frac{i\mathbf{p}_f\mathbf{r}}{\hbar}\right], \qquad (1.118)$$

where

$\mathbf{k}_f$ is the final wave vector of the particle (projectile) after the scattering event, related to the final particle momentum $\mathbf{p}_f$ through $\mathbf{k}_f = \mathbf{p}_f/\hbar$.

$\mathbf{p}_f$ is the final momentum of the particle after scattering.

C is a normalization constant obtained from the normalization equation (see (1.85)) $\int \psi_f^* \psi_f \, d\mathcal{V} = 1$.

Figure 1.10 shows the general relationship between the initial momentum of the particle $\mathbf{p}_i = \hbar\mathbf{k}_i$ before scattering and the final momentum of the particle $\mathbf{p}_f = \hbar\mathbf{k}_f$ after scattering. The momentum transfer $\Delta\mathbf{p}$ from the particle to the scatterer is given as

$$\Delta\mathbf{p} = \mathbf{p}_i - \mathbf{p}_f \qquad (1.119)$$

(a) *Inelastic scattering*

$p_i > p_f$

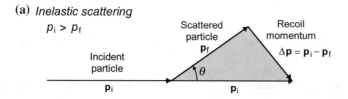

(b) *Elastic scattering*

$p_i = p_f$

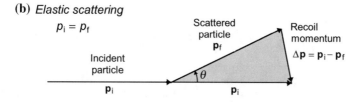

Fig. 1.10 General relationship between the initial momentum of the particle $\mathbf{p}_i$ before scattering and the final momentum of the particle $\mathbf{p}_f$ after scattering. $\Delta\mathbf{p}$ is the momentum transfer from the incident particle to the scatterer and θ is the scattering angle. Part **a** is for general inelastic scattering with Δp given by (1.121); part **b** is for elastic scattering with Δp given by (1.123)

and we define the wave vector $\mathbf{K}$ as

$$\mathbf{K} = \frac{\Delta \mathbf{p}}{\hbar} = \frac{\mathbf{p}_i}{\hbar} - \frac{\mathbf{p}_f}{\hbar} = \mathbf{k}_i - \mathbf{k}_f. \tag{1.120}$$

Using the law of cosines we express the magnitudes of $\Delta \mathbf{p}$ and $\mathbf{K}$ as follows

$$\Delta p = |\Delta \mathbf{p}| = \sqrt{|\mathbf{p}_i|^2 + |\mathbf{p}_f|^2 - 2\,|\mathbf{p}_i|\,|\mathbf{p}_f|\cos\theta} = \sqrt{p_i^2 + p_f^2 - 2p_i p_f \cos\theta} \tag{1.121}$$

and

$$K = |\mathbf{K}| = \frac{1}{\hbar}\sqrt{|\mathbf{p}_i|^2 + |\mathbf{p}_f|^2 - 2\,|\mathbf{p}_i|\,|\mathbf{p}_f|\cos\theta} = \frac{1}{\hbar}\sqrt{p_i^2 + p_f^2 - 2p_i p_f \cos\theta}, \tag{1.122}$$

where θ, as shown in Fig. 1.10, is the scattering angle.

When the scattering process is elastic, the magnitude of the initial momentum is equal to the magnitude of the final momentum, i.e., $|\mathbf{p}_i| = |\mathbf{p}_f| = p$ and (1.121) and (1.122) yield

$$\Delta p = |\Delta \mathbf{p}| = \sqrt{2p^2 - 2p^2 \cos\theta} = \sqrt{2p^2 (1 - \cos\theta)} = 2p \sin\frac{\theta}{2}. \tag{1.123}$$

and

$$K = |\mathbf{K}| = \frac{\Delta p}{\hbar} = \frac{2p}{\hbar}\sin\frac{\theta}{2} = \frac{2}{\lambda}\sin\frac{\theta}{2}. \tag{1.124}$$

The matrix element (scattering amplitude) $\mathbf{M}_{if}$ for spherically symmetrical central scattering potential $V(r)$ is in general written as

$$\mathbf{M}_{if} = C^2 \iiint e^{-i\mathbf{k}_f \cdot \mathbf{r}} V\, e^{+i\mathbf{k}_i \cdot \mathbf{r}}\, d\mathcal{V} = C^2 \iiint V(r)\, e^{+i\mathbf{K} \cdot \mathbf{r}}\, d\mathcal{V} = V(\mathbf{K}) \tag{1.125}$$

using (1.120) for $\mathbf{K}$. The right hand integral in (1.125) is the Fourier transform of the central potential $V(r)$ representing $V(r)$ in the momentum space. For the central potential $V(r)$ the matrix element $\mathbf{M}_{if}$ can now be simplified to get

$$\mathbf{M}_{if} = C^2 \iiint V(r)\, e^{+i\mathbf{K} \cdot \mathbf{r}}\, d\mathcal{V} = C^2 \int_0^{2\pi} \int_0^{\pi} \int_0^{\infty} V(r)\, e^{iKr} r^2 \sin\Theta\, dr\, d\Theta\, d\phi$$

$$= 2\pi C^2 \int_0^{\infty} r^2 V(r) \left\{ \int_{-1}^{+1} e^{iKr \cos\Theta}\, d\cos\Theta \right\} dr. \tag{1.126}$$

First, we deal with the integral inside the curly bracket of (1.126) to get

$$\int_{-1}^{+1} e^{iKr\cos\Theta}\, d(\cos\Theta) = \left[\frac{e^{iKr\cos\Theta}}{iKr}\right]_{\cos\Theta=-1}^{\cos\Theta=1} = \left[\frac{e^{iKr} - e^{-iKr}}{iKr}\right] = \frac{2\sin(Kr)}{Kr},$$

(1.127)

and finally we have the following expression for the matrix element $\mathbf{M}_{if}$

$$\mathbf{M}_{if} = 4\pi C^2 \int_0^\infty r^2 V(r)\,\frac{\sin(Kr)}{Kr}\,dr.$$

(1.128)

The differential scattering cross section $d\sigma/d\Omega$ (1.116) is now given as

$$\frac{d\sigma}{d\Omega} = \frac{2\pi}{\hbar}|\mathbf{M}_{if}|^2\,\rho(E_f) = \left\{\frac{2\pi}{\hbar}\left(4\pi C^2\right)^2\right\}\rho(E_f)\left|\int_0^\infty r^2 V(r)\,\frac{\sin(Kr)}{Kr}\,dr\right|^2$$

$$= \left|\frac{2m}{\hbar^2}\int_0^\infty r^2 V(r)\,\frac{\sin(Kr)}{Kr}\,dr\right|^2.$$

(1.129)

Equation (1.129) is referred to as the *Born collision formula* and is valid for elastic scattering brought about by a spherically symmetrical central scattering potential $V(r)$. The derivation of the density of final states $\rho(E_f)$ is cumbersome but can be found in standard quantum mechanics and nuclear physics texts.

1.24 Uncertainty Principle

In classical mechanics the act of measuring the value of a measurable quantity does not disturb the quantity; therefore, the position and momentum of an object can be determined simultaneously and precisely. However, when the size of the object diminishes and approaches the dimensions of microscopic particles, it becomes impossible to determine with great precision at the same instant both the position and momentum of particles or radiation nor is it possible to determine the energy of a system in an arbitrarily short time interval.

Werner Heisenberg in 1927 proposed the *uncertainty principle* that limits the attainable precision of measurement results. The uncertainty principle covers two distinct components:

One component (*momentum-position uncertainty principle*) deals with the simultaneous measurement of the position z and momentum p_z of a particle and limits the attainable precision of z and p_z measurement to the following product

$$\Delta z \Delta p_z \geq \frac{1}{2}\hbar, \tag{1.130}$$

where

Δz is the uncertainty on z.
Δp_z is the uncertainty on p_z.

Thus, there are no limits on the precision of individual z and p_z measurements. However, in a simultaneous measurement of z and p_z the product of the two uncertainties cannot be smaller than $\frac{1}{2}\hbar$ where $\hbar$ is the reduced Planck constant $\left(\hbar = \frac{1}{2\pi}h\right)$. Thus, if z is known precisely $(\Delta z = 0)$, then we cannot know p_z, since $(\Delta p_z = \infty)$. The reverse is also true: if p_z is known exactly $(\Delta p_z = 0)$, then we cannot know z, since $\Delta z = \infty$.

The other component (*energy-time uncertainty principle*) deals with the measurement of the energy E of a system and the time interval Δt required for the measurement. Similarly to the $(\Delta z, \Delta p_z)$ situation, the Heisenberg uncertainty principle states the following

$$\Delta E \Delta t \geq \frac{1}{2}\hbar, \tag{1.131}$$

where

ΔE is the uncertainty in the energy determination.
Δt is the time interval taken for the measurement.

Classical mechanics sets no limits on the precision of measurement results and allows a deterministic prediction of the behavior of a system in the future. Quantum mechanics, on the other hand, limits the precision of measurement results and thus allows only probabilistic predictions of the system's behavior in the future.

1.25   Complementarity Principle

In 1928 *Niels Bohr* proposed the *principle of complementarity* postulating that any atomic scale phenomenon for its full and complete description requires that both its wave and particle properties be considered and determined. This is in contrast to macroscopic scale phenomena where particle and wave characteristics (e.g., billiard ball as compared to water wave) of the same macroscopic phenomenon are mutually incompatible rather than complementary.

The Bohr principle of complementarity is thus valid only for atomic size processes and asserts that these processes can manifest themselves either as waves or as particles (corpuscles) during a given experiment, but never as both during the same experiment. However, to understand and describe fully an atomic scale physical process the two types of properties must be investigated with different experiments, since both properties complement rather than exclude each other.

The most important example of this *particle–wave duality* is the photon, a massless particle characterized with energy, frequency and wavelength. However, in certain experiments such as in Compton effect the photon behaves like a particle; in other experiments such as double-slit diffraction it behaves like a wave.

Another example of the particle–wave duality are the wave-like properties of electrons as well as heavy charged particles and neutrons that manifest themselves through diffraction experiments (Davisson–Germer and Thomson–Reid experiments; see Sect. 1.22.2 and 1.22.3, respectively).

1.26   Emission of Electrons from Material Surface: Work Function

Emission of electrons from the surface of a solid material into vacuum is an important phenomenon governed by the so-called work function $e\phi$ defined as the minimum energy that must be supplied to an electron to remove it from the surface of a given material. For condensed matter $e\phi$ is of the order of a few electron volts and presents an effective surface barrier preventing electrons from leaving the material under normal circumstances. However, electrons can be liberated from the material surface into vacuum through various effects such as, for example:

1. Energy equal to or exceeding the work function $e\phi$ can be supplied to surface electrons by photons of energy $h\nu$ larger than $e\phi$, typically in the visible or near ultraviolet region. The electrons obtain sufficient energy to overcome the surface potential barrier and can leave the metal surface. The effect is referred to as the surface photoelectric effect or photoemission and Albert Einstein is credited with explaining the effect theoretically in 1905 on the basis of quantization of electromagnetic radiation.

2. Heating a metal to temperature above 1000 °C increases the kinetic energy of electrons and enables these electrons to overcome the potential barrier and leave the metal surface. Emission of electrons under this condition is referred to as thermionic emission (see Sect. 1.27) and forms the basis for production of electrons with hot cathode in Coolidge x-ray tubes, electron guns of linear accelerators, and many other modern sources of electrons.

3. Placing a material into a very strong electric field may deform the material potential barrier and allow unexcited electrons to escape through the surface barrier from the condensed material into vacuum. This leakage or tunneling of electrons through the potential barrier is referred to as field emission (FE) and, as discussed in Sect. 1.28, has found use in electron microscopes, flat panel displays and, more recently has shown promise in electron sources based on cold cathode manufactured with carbon nanotubes.

4. In radiation dosimetry, weak electron emission from pre-irradiated condensed matter dosimeters (phosphors) is referred to as exoelectron emission (EE) and can be stimulated by heating of the phosphor to get thermally stimulated EE or by exposing the phosphor with visible or ultraviolet light to get optically stimulated EE.

1.27   Thermionic Emission

Thermionic emission, a very important phenomenon in medical physics, is defined as the flow of charge carriers from the surface of a solid or over some kind of potential barrier, facilitated by supplying thermal energy to the solid. Charge carriers so released from the solid are called thermions and the science dealing with the phenomenon is called thermionics. The most common practical example of thermionic emission is the emission of electrons from a hot metal cathode into vacuum, as used in filaments of Coolidge x-ray tubes, electron guns of linear accelerators, and so-called thermionic diodes.

Many physicists have contributed to the science of thermionics, most notably *Owen W. Richardson* who in 1928 received the Nobel Prize in Physics for his work and the discovery of the law governing the phenomenon. Saul Dushman demonstrated the modern form of the law governing the thermionic emission and the law is now referred to as the Richardson–Dushman equation. It expresses the relationship between the current density j in ampere per meter of electrons emitted from the metal and the absolute temperature T in kelvin of the metal as

$$j = A_R T^2 \, e^{-\frac{e\phi}{kT}}, \tag{1.132}$$

where

A_R is the Richardson constant with theoretical value of 1.2×10^6 A·m^{-2}·K^{-2}.
$e\phi$ is the work function of the metal.
k is Boltzmann constant (1.381×10^{-23} J·K^{-1} = 0.8617×10^{-4} eV·K^{-1}).

From (1.132) it is evident that thermionic emission is controlled by three characteristics of the emitter: (1) its temperature; (2) its material composition (work function); and (3) its surface area. The current density j rises rapidly with temperature T of the emitter and decreases with an increase in the work function $e\phi$. The work function is defined, similarly to the definition of the atomic ionization energy, as:

• Ionization of a free atom involves removing one of its outer shell electrons by giving it an energy which is equal or exceeds its binding energy and is referred to as the ionization energy of the atom. The range of atomic ionization energy in nature is from a few electron volts for alkali elements to 24.6 eV for the noble gas helium.

- To remove an electron from a solid consisting of an array of atoms an electron in the conduction band must be supplied a minimum energy referred to as the work function typically ranging from 2 eV to 5 eV for most elements. The magnitude of the work function for a metal is usually about a half of the ionization potential of the free atom of the metal.

In the absence of external electric field electrons leaving a heated metal in vacuum form a cloud surrounding the emitter. The cloud, referred to a space charge, represents an electronic equilibrium in which the number of electrons governed by (1.132) that leave the emitter and enter the cloud is equal to the number of electrons that are attracted back to the emitter. As the temperature of the emitter increases, the number of electrons in the space charge also increases because of the increased electron emission, as predicted by (1.132).

In practical use of thermionic emission, the thermionic emitter is not only heated but is also immersed in an external electric field. This field enhances the emission current density j if it has the same sign as the emitted thermion charge and diminishes it if the signs are opposite. In practice, one is interested in enhancing j with external electric field. This corresponds to lowering the work function $e\phi$ by $e\,\Delta\phi$ where $\Delta\phi$ is proportional to $\mathcal{E}^{1/2}$ where $\mathcal{E}$ is the electric field applied externally, such as, for example, the field applied between the cathode and anode of a Coolidge x-ray tube. The Richardson–Dushman equation (1.132) then reads

$$ j = A_R T^2 \, e^{-\frac{e\phi - e\,\Delta\phi}{kT}}. \tag{1.133} $$

The effect of lowering the work function of a metal by applying an external electric field is called the Schottky effect and (1.133) is referred to as the Schottky equation. In typical x-ray tubes the external electric field is such that all electrons emitted from the filament are attracted to the anode and no space charge around the filament is present. At very high external electric fields of the order of 10^8 V/m quantum tunneling begins to contribute to the emission current and the effect is referred to as field emission (see Sect. 1.28.2).

1.28 Tunneling

The particle–wave duality is highlighted in discussions of potential wells and potential barriers, both important phenomena in quantum and wave mechanics; the potential wells attract and trap particles, potential barriers reflect or transmit them. While medical physics and clinical physics rarely deal with quantum and wave mechanics, there are several physical phenomena of importance to radiation physics and, by extension, to medical physics that can only be explained through wave-mechanical reasoning. *Tunneling*, for example, is a purely wave-mechanical phenomenon that is used in explaining α decay and field emission: two important effects in radiation physics.

In addition, there are several other phenomena of importance in electronics that can be explained invoking tunneling such as, for example, in the periodic inversion of the ammonia molecule NH_3, used as standard in atomic clocks, and in a semiconductor device called tunnel diode that is used for fast switching in electronic circuits.

A *classical particle* incident on a square barrier will pass the barrier only if its kinetic energy E_K exceeds the barrier potential E_P. If $E_P > E_K$, the classical particle is reflected at the barrier and no transmission occurs.

A *quantum-mechanical particle* incident on a square barrier has access to regions on both sides of the barrier, irrespective of the relative magnitudes of kinetic energy E_K and the barrier potential E_P. A matter wave is associated with the particle and it has a non-zero magnitude on both sides of the barrier as well as inside the barrier. The wave penetrates and traverses the barrier even when $E_P > E_K$, clearly contravening classical physics but conforming to the rules of wave mechanics. The non-zero probability for finding the particle on the opposite side of the barrier indicates that the particle may tunnel through the barrier or one may say that the particle undergoes the tunneling effect. In tunneling through a barrier, the particle behaves as a pure wave inside the barrier and as a pure particle outside the barrier.

1.28.1   Alpha Decay Tunneling

Alpha decay is considered a tunneling phenomenon in which α-particles with kinetic energies between 4 MeV and 9 MeV tunnel through a potential barrier of the order of 30 MeV. The tunneling theory of the α decay was proposed by *George Gamow* in 1928 (see Sect. 11.2). Inside the parent nucleus (atomic number Z) the α-particle is free yet confined to the nuclear potential well by the strong nuclear force. The dimension of the well is of the order of few femtometers; once the α-particle is beyond this distance from the center of the parent nucleus, it only experiences Coulomb repulsion between its charge $2e$ and the charge of the daughter nucleus $(Z - 2)\,e$.

A classical α-particle with $E_K < 9$ MeV cannot overcome a potential barrier with $E_P > 30$ MeV. On the other hand, a α-particle with wave-like attributes may tunnel through the potential barrier and escape the parent nucleus through this purely quantum-mechanical phenomenon.

1.28.2   Field Emission Tunneling

Emission of electrons from a solid into vacuum under the influence of a strong electric field aimed in a direction to accelerate electrons away from the surface is referred to as field emission (FE). Unlike thermionic emission (see Sect. 1.27), field emission does not depend on temperature of the material. The effect is purely quantum-mechanical and is attributed to wave-mechanical tunneling of electrons through the surface potential barrier which is affected by the strong electric field so as to facilitate the tunneling process. Since electric field rather than heat is used to induce FE, the effect is often referred to as cold cathode emission, in contrast to thermionic emission which is stimulated by heat and consequently referred to as hot cathode emission.

In FE as well as in the Schottky effect the surface potential barrier is effectively lowered by the applied electric field. However, in Schottky effect the electrons surmount the barrier while in TE they tunnel through it.

Field emission has been known from the early days of quantum mechanics as a clear example of electron tunneling through a sufficiently thin potential barrier. The effect is also known as Fowler–Nordheim tunneling in honor of the physicists who in 1928 were the first to study it experimentally and theoretically. The tunneling probability shows exponential dependence on the tunneling distance which is inversely proportional to the electric field. The Fowler–Nordheim expression for the current density j in FE exhibits a functional dependence on electric field that is similar to the functional dependence of current density j on temperature T for thermionic emission given in (1.132).

Field emission occurs at surface points where the local electric field is extremely high, typically of the order of 1 V/nm or 10^9 V/m. These high electric fields are generated by applying relatively low voltages to needlelike metal tips with minute radius of curvature r of the order of 100 nm or less. Since the electric field equals $\sim V/r$, its magnitude for small r is very large.

Field emission has found practical application in solid-state electronic components such as tunnel diodes, high-resolution electron microscopy, high-resolution flat panel display, and many other electronic devices using an electron source. During recent years, cold cathodes based on carbon nanotubes (CNT) have shown promise in x-ray tube technology where they could serve as an alternative to Coolidge's hot cathode design (see Sect. 14.4.3).

1.29 Maxwell Equations

The basic laws of electricity and magnetism can be summarized by four Maxwell equations. The equations may be expressed in integral form or in differential form and the two forms are linked through two theorems of vector calculus: *Gauss divergence theorem* and *Stokes theorem*. In radiation physics and medical physics Maxwell equations play an important role in the understanding of bremsstrahlung production, in waveguide theory of particle acceleration, and in the theory of ionization chamber operation.

For a vector function **A** and volume V bounded by a surface S the two vector calculus theorems are given as follows:

Gauss theorem

$$\iiint_{\mathcal{V}} \nabla \cdot \mathbf{A} \, d\mathcal{V} = \iiint_{\mathcal{V}} \text{div} \, \mathbf{A} \, d\mathcal{V} = \oiint_{\mathcal{S}} \mathbf{A} \cdot d\mathcal{S} \qquad (1.134)$$

Stokes theorem

$$\iint_{\mathcal{S}} (\nabla \times \mathbf{A}) \cdot d\mathcal{S} = \iint_{\mathcal{S}} \text{curl} \, \mathbf{A} \cdot d\mathcal{S} = \oint_{\ell} \mathbf{A} \cdot d\boldsymbol{\ell} \qquad (1.135)$$

The four Maxwell equations (integral form on the right, differential form on the left) are given as follows:

1. *Maxwell–Gauss equation for electricity* (also known as Gauss law for electricity)

$$\nabla \cdot \boldsymbol{\mathcal{E}} = \frac{\rho}{\varepsilon_0} \qquad \oiint_{\mathcal{S}} \boldsymbol{\mathcal{E}} \cdot d\mathcal{S} = \frac{1}{\varepsilon_0} \iiint_{\mathcal{V}} \rho \, d\mathcal{V} = \frac{q}{\varepsilon_0} \qquad (1.136)$$

2. *Maxwell–Gauss equation for magnetism* (also known as Gauss law for magnetism)

$$\nabla \cdot \boldsymbol{\mathcal{B}} = 0 \qquad \oiint_{\mathcal{S}} \boldsymbol{\mathcal{B}} \cdot d\mathcal{S} = 0. \qquad (1.137)$$

3. *Maxwell–Faraday equation* (also known as Faraday law of induction)

$$\nabla \times \boldsymbol{\mathcal{E}} = -\frac{\partial \boldsymbol{\mathcal{B}}}{\partial t} \qquad \oint_{\ell} \boldsymbol{\mathcal{E}} \cdot d\boldsymbol{\ell} = -\frac{\partial}{\partial t} \iint_{\mathcal{S}} \boldsymbol{\mathcal{B}} \cdot d\mathcal{S} = -\frac{\partial \phi_{\text{mag}}}{\partial t} \qquad (1.138)$$

4. *Maxwell–Ampère equation* (also known as Ampère circuital law)

$$\nabla \times \boldsymbol{\mathcal{B}} = \mu_0 \mathbf{j} + \frac{1}{c^2} \frac{\partial \boldsymbol{\mathcal{E}}}{\partial t} \qquad \oint_{\ell} \boldsymbol{\mathcal{B}} \cdot d\boldsymbol{\ell} = \mu_0 I + \frac{1}{c^2} \int_{\mathcal{S}} \boldsymbol{\mathcal{E}} \cdot d\mathcal{S}, \qquad (1.139)$$

where

$\boldsymbol{\mathcal{E}}$ is the electric field in volt per meter (V/m).
$\boldsymbol{\mathcal{B}}$ is the magnetic field in tesla T where $1 \text{ T} = 1 \text{ V·s/m}^2$.
ρ is the total charge density in coulomb per cubic meter (C/m^3).
$\mathbf{j}$ is the total current density in ampère per square meter (A/m^2).
ε_0 is the permeability of vacuum (electric constant).
μ_0 is are the permittivity of vacuum (magnetic constant).
q is the total charge enclosed by the Gaussian surface $\mathcal{S}$ in coulombs (C).
I is the total current passing through the surface $\mathcal{S}$ in ampères (A).

The four Maxwell equations combined with the *Lorentz force* and *Newton second law of motion* provide a complete description of the classical dynamics of interacting charged particles and electromagnetic fields. The Lorentz force $\mathbf{F_L}$ for charge q moving in electric field $\mathcal{E}$ and magnetic field $\mathcal{B}$ with velocity $\boldsymbol{v}$ is given as follows

$$\mathbf{F_L} = q\,(\mathcal{E} + \boldsymbol{v}\times\mathcal{B})\,. \tag{1.140}$$

Maxwell equations boosted the theory of electromagnetic fields in a similar fashion to the boost classical mechanics received from Newton laws. However, classical mechanics has subsequently been shown deficient at small dimensions on atomic and nuclear scale where quantum physics applies and at large velocities of the order of the speed of light in vacuum where relativistic physics applies. Maxwell equations, on the other hand, survived subsequent developments in physics related to quantum and relativistic mechanics and remain as valid today as they were when Maxwell introduced them some 150 years ago. With the theory of electromagnetic field, Maxwell equations succeeded in unifying electricity, magnetism, and photons on a broad spectrum ranging in frequency from radio waves to gamma rays.

1.30 Poynting Theorem and Poynting Vector

In 1884 English physicist *John Henry Poynting* used the Lorentz equation for a moving charge in an electromagnetic (EM) field and Maxwell equations for electromagnetism to derive a theorem that expresses the conservation of energy for EM fields. The theorem relates the rate of change of the energy u stored in the EM field and the energy flow expressed by the Poynting vector $\mathbf{S}$.

An electromagnetic field interacts with a charged particle q traveling with velocity $\boldsymbol{v}$ through the Lorentz force $\mathbf{F_L}$

$$\mathbf{F_L} = \frac{\mathrm{d}\,(m\boldsymbol{v})}{\mathrm{d}t} = q\,(\mathcal{E} + \boldsymbol{v}\times\mathcal{B})\,. \tag{1.141}$$

Multiply (1.141) with velocity $\boldsymbol{v}$ to obtain an energy relationship

$$\boldsymbol{v}{\cdot}\mathbf{F_L} = \boldsymbol{v}\frac{\mathrm{d}\,(m\boldsymbol{v})}{\mathrm{d}t} = \frac{1}{2}\frac{\mathrm{d}m v^2}{\mathrm{d}t} = q\boldsymbol{v}{\cdot}\mathcal{E}, \tag{1.142}$$

where only the electric field contributes to the particle's energy, since for the magnetic filed $\boldsymbol{v}{\cdot}\,(\boldsymbol{v}\times\mathcal{B}) = 0$. We now multiply (1.142) with the particle density n to get

$$n\frac{\mathrm{d}}{\mathrm{d}t}\frac{m v^2}{2} = \frac{\mathrm{d}E_K}{\mathrm{d}t} = nq\boldsymbol{v}{\cdot}\mathcal{E} = \mathbf{j}{\cdot}\mathcal{E}, \tag{1.143}$$

where

E_K is the kinetic energy of the ensemble of charged particles.
$\mathbf{j}$ is the current density $\mathbf{j} = nq\boldsymbol{v}$.

Next, we use the Ampère–Maxwell equation (1.139) to express the current density $\mathbf{j}$ in terms of electric field $\boldsymbol{\mathcal{E}}$ and magnetic field $\boldsymbol{\mathcal{B}}$.

$$\mathbf{j}\cdot\boldsymbol{\mathcal{E}} = \frac{1}{\mu_0}\boldsymbol{\mathcal{E}}\cdot\nabla\times\boldsymbol{\mathcal{B}} - \varepsilon_0\frac{\partial}{\partial t}\frac{\boldsymbol{\mathcal{E}}^2}{2}, \tag{1.144}$$

where ε_0 and μ_0 are the electric permittivity of vacuum and magnetic permeability of vacuum, respectively. Using the vector identity

$$\nabla\cdot(\boldsymbol{\mathcal{E}}\times\boldsymbol{\mathcal{B}}) = \boldsymbol{\mathcal{B}}\cdot\nabla\times\boldsymbol{\mathcal{E}} - \boldsymbol{\mathcal{E}}\cdot\nabla\times\boldsymbol{\mathcal{B}} \tag{1.145}$$

we express (1.144) as

$$\mathbf{j}\cdot\boldsymbol{\mathcal{E}} = -\frac{1}{\mu_0}\nabla\,(\boldsymbol{\mathcal{E}}\times\boldsymbol{\mathcal{B}}) - \frac{1}{\mu_0}\boldsymbol{\mathcal{B}}\cdot\nabla\times\boldsymbol{\mathcal{E}} - \varepsilon_0\frac{\partial}{\partial t}\frac{\boldsymbol{\mathcal{E}}^2}{2}. \tag{1.146}$$

Inserting (1.138), the Faraday–Maxwell equation, into (1.146) we get

$$\mathbf{j}\cdot\boldsymbol{\mathcal{E}} = -\nabla\left(\boldsymbol{\mathcal{E}}\times\frac{\boldsymbol{\mathcal{B}}}{\mu_0}\right) - \frac{\partial}{\partial t}\left\{\frac{\varepsilon_0\boldsymbol{\mathcal{E}}^2}{2} + \frac{\boldsymbol{\mathcal{B}}^2}{2\mu_0}\right\} = -\nabla\cdot\boldsymbol{\mathcal{S}} - \frac{\partial u}{\partial t}, \tag{1.147}$$

where

u is the electromagnetic field energy density

$$u = \varepsilon_0\frac{\boldsymbol{\mathcal{E}}^2}{2} + \frac{\boldsymbol{\mathcal{B}}^2}{2\mu_0}. \tag{1.148}$$

$\mathbf{S}$ is the Poynting vector

$$\mathbf{S} = \boldsymbol{\mathcal{E}}\times\frac{\boldsymbol{\mathcal{B}}}{\mu_0}, \tag{1.149}$$

representing the energy flow with dimensions energy/(area×time) or power/area.

The Poynting theorem can also be expressed in integral form where the integration is carried out over an arbitrary volume $\mathcal{V}$

$$-\int_{\mathcal{V}} \mathbf{j}\cdot\boldsymbol{\mathcal{E}}\,\mathrm{d}\mathcal{V} = \int_{\mathcal{V}} \left[\frac{\mathrm{d}u}{\mathrm{d}t} + \nabla\cdot\boldsymbol{\mathcal{S}}\right]\mathrm{d}\mathcal{V}. \tag{1.150}$$

The conservation of energy is with Poynting theorem thus expressed as follows: the time rate of change of electromagnetic (EM) energy within a given volume $\mathcal{V}$ added to the energy leaving the given volume per unit time equals to the negative value of the total work done by the EM fields on the ensemble of charged particles encompassed by the given volume.

1.31 Normal Probability Distribution

Random variation in natural processes most commonly follows the probability distribution generally known in mathematics as the normal probability distribution but also referred to as Gaussian distribution in physics and "bell curve" in social science. The function describing the normal distribution has a long tradition in mathematics and physics. De Moivre used it in 18th century as an approximation to the binomial distribution, Laplace used it to study measurement errors, and Gauss used it in his analysis of astronomical data.

In general, the normal probability distribution is described by the following continuous *probability density function $P(x)$*

$$P(x) = \frac{1}{\sigma\sqrt{2\pi}}e^{-\frac{(x-\bar{x})^2}{2\sigma^2}}, \tag{1.151}$$

where

σ is the standard deviation related to the width of the "bell curve".
$\bar{x}$ is the mean value of x also called the expectation value of x.

The following general characteristics apply to the probability distribution function $P(x)$ given by (1.151)

1. $P(x)$ has a peak at $\bar{x}$ and is symmetric about $\bar{x}$.
2. $P(x)$ is unimodal.
3. $P(x)$ extends from $-\infty$ to $+\infty$.
4. The area under $P(x)$ equals to 1. i.e., $\int\limits_{-\infty}^{+\infty} P(x)dx = 1$.
5. $P(x)$ is completely specified by the two parameters $\bar{x}$ and σ.
6. $P(x)$ follows the empirical rule which states that (see Fig. 1.11):

   – 68.3% of data will fall within 1 standard deviation σ of the mean $\bar{x}$.
   – 95.5% of the data will fall within 2σ of the mean $\bar{x}$.
   – 99.7% of the data will fall within 3σ of the mean $\bar{x}$.

7. The inflection points on the $P(x)$ curve occur one standard deviation from the mean, i.e., at $\bar{x} - \sigma$ and $\bar{x} + \sigma$.

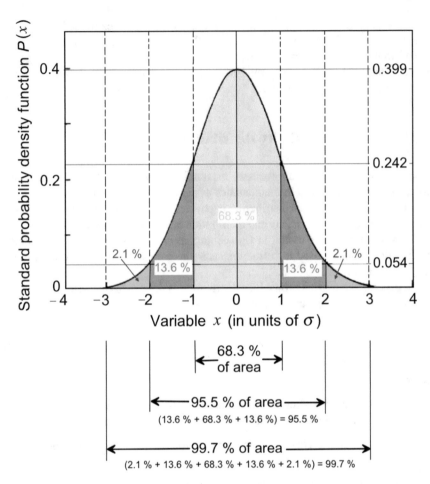

Fig. 1.11 Standard probability density function $P(x)$ of (1.152) with the variable x plotted in units of σ, the standard deviation of the mean. The total area under the curve from $x = -\infty$ to $x = +\infty$ is equal to 1. The *yellow* area under the probability curve represents 68.3% of the total area, i.e., $\frac{1}{\sqrt{2\pi}} \int\limits_{-1}^{+1} e^{-\frac{x^2}{2}} \, dx = 0.683$; the two *brown* areas represent 13.6% of the total area each, i.e.,

$$\frac{1}{\sqrt{2\pi}} \int\limits_{-2}^{+2} e^{-\frac{x^2}{2}} \, dx = 2 \times 0.136 + 0.683 = 0.955$$

1.31.1 Standard Probability Density Function

For the special case of mean value $\bar{x} = 0$ and standard deviation $\sigma = 1$, the probability distribution is called the standard normal distribution and the probability density function of (1.151) simplifies into the *standard probability density function*

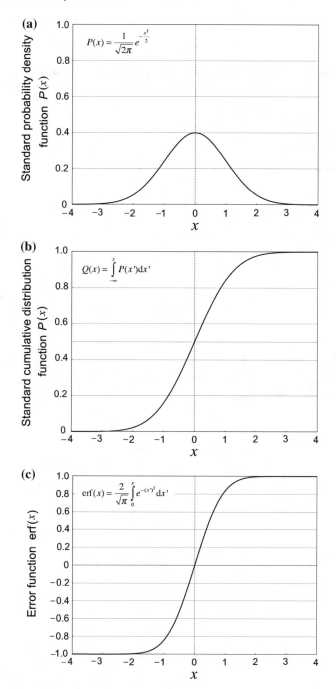

Fig. 1.12 Part **a** Standard probability density function $P(x)$ of (1.152) for the mean or expectation value $\bar{x} = 0$; standard deviation $\sigma = 1$, and variance $v = 1$. Part **b** Standard cumulative distribution function $Q(x)$ of (1.158) for the mean or expectation value $\bar{x} = 0$; standard deviation $\sigma = 1$, and variance $v = 1$. Part **c** Error function erf(x) of (1.159)

$$P(x) = \frac{1}{\sqrt{2\pi}} e^{-\frac{x^2}{2}}. \tag{1.152}$$

The standard probability density function $P(x)$ of (1.152) is plotted in Figs. 1.11 and 1.12a. It exhibits the following notable characteristics:

1. For $x = 0$, $P(0) = 1/\sqrt{2\pi} = 0.399$.
   For $x = 1$, $P(1) = (1/\sqrt{2\pi})e^{-0.5} = 0.242$.
   For $x = 2$, $P(2) = (1/\sqrt{2\pi})e^{-2} = 0.054$.

2. *Total area* under the $P(x)$ curve from $x = -\infty$ to $x = +\infty$

$$\int_{-\infty}^{+\infty} P(x)dx = \frac{1}{\sqrt{2\pi}} \int_{-\infty}^{+\infty} e^{-\frac{x^2}{2}} dx = \frac{2}{\sqrt{2\pi}} \int_{0}^{+\infty} e^{-\frac{x^2}{2}} dx = 1. \tag{1.153}$$

3. *Mean value* or *expectation value* of x

$$\bar{x} = \frac{1}{\sqrt{2\pi}} \int_{-\infty}^{+\infty} x e^{-\frac{x^2}{2}} dx = 0. \tag{1.154}$$

4. *Variance* v of x is a parameter giving a measure of the dispersion of a set of data points around the mean value. For the normal probability distribution the variance is defined as follows

$$v(x) = \overline{x^2} = \frac{1}{\sqrt{2\pi}} \int_{-\infty}^{+\infty} x^2 e^{-\frac{x^2}{2}} dx = \frac{4}{\sqrt{\pi}} \int_{0}^{+\infty} u^2 e^{-u^2} du = 1. \tag{1.155}$$

5. *Standard deviation* σ of the mean $\bar{x}$ is the square root of the variance v

$$\sigma = \sqrt{v} = 1. \tag{1.156}$$

1.31.2 Cumulative Distribution Function

A probability distribution can also be characterized with the general cumulative distribution function $Q(x)$ expressed as follows

$$Q(x) = \int_{-\infty}^{x} P(x')dx' = \frac{1}{\sigma\sqrt{2\pi}} \int_{-\infty}^{x} e^{-\frac{(x'-\bar{x})^2}{2\sigma^2}} dx' = \frac{1}{2}\left[1 + \mathrm{erf}\frac{x-\bar{x}}{\sigma\sqrt{2}}\right], \tag{1.157}$$

while the standard cumulative distribution function is the general cumulative distribution function of (1.157) evaluated with $\bar{x} = 0$ and $\sigma = 1$

$$Q(x) = \int_{-\infty}^{x} P(x')dx' = \frac{1}{\sqrt{2\pi}} \int_{-\infty}^{x} e^{-\frac{(x')^2}{2}} dx' = \sqrt{\frac{2}{\pi}} \int_{-\infty}^{\frac{x}{\sqrt{2}}} e^{-\left(\frac{x'}{\sqrt{2}}\right)^2} d\frac{x'}{\sqrt{2}}$$

$$= \frac{1}{\sqrt{\pi}} \int_{-\infty}^{0} e^{-\left(\frac{x'}{\sqrt{2}}\right)^2} d\frac{x'}{\sqrt{2}} + \frac{1}{2} \frac{2}{\sqrt{\pi}} \int_{0}^{\frac{x}{\sqrt{2}}} e^{-\left(\frac{x'}{\sqrt{2}}\right)^2} d\frac{x'}{\sqrt{2}} = \frac{1}{2} \left[1 + \mathrm{erf} \frac{x}{\sqrt{2}} \right].$$

$$\tag{1.158}$$

The standard cumulative distribution function is plotted in Fig. 1.12b and exhibits the following notable properties:
1. $Q(x) = 0$ for $x = -\infty$.
2. $Q(x) = 0.5$ for $x = 0$.
3. $Q(x) = 1$ for $x = \infty$.

1.31.3 Error Function

In (1.157) and (1.158) "erf" denotes a special function of sigmoid shape called the error function defined as

$$\mathrm{erf}(x) = \frac{2}{\sqrt{\pi}} \int_{0}^{x} e^{-(x')^2} dx', \tag{1.159}$$

with the following notable features: $-1 \leq \mathrm{erf}(x) \leq +1$; $\mathrm{erf}(0) = 0$; $\mathrm{erf}(\infty) = 1$; $\mathrm{erf}(-x) = -\mathrm{erf}(x)$; and $\mathrm{erf}(-\infty) = -1$.

The error function $\mathrm{erf}(x)$ is plotted in Fig. 1.12c and its tabulated values are readily available in standard mathematical tables.

Monte Carlo Treatment Planning in Radiotherapy

In modern societies cancer became the most significant health care problem surpassing heart disease as the leading cause of death and potential years of life lost. The cancer rate is slowly but steadily growing mainly because of population aging, and currently stands at about 4500 new cancer patients per million population per year.

Radiotherapy is an important modality of cancer therapy and over 50% of all cancer patients receive radiation treatment either as the primary component of their treatment or as adjuvant therapy. Many crucial steps affect the final outcome of treatment and physics plays a pivotal role in many of these steps, such as in: (1) Imaging for cancer diagnosis, target localization, virtual simulation, and treatment planning and (2) Delivery of the prescribed dose either with external beam radiotherapy or with brachytherapy involving sealed internal sources.

Currently, linear accelerators (linacs) are the most important high technology machines for dose delivery in modern radiotherapy. Algorithms for calculation of dose distribution in the irradiated patient (treatment planning) are an important component of radiotherapy. They evolved during the past few decades from calculations that relied heavily on analytic, semi-analytic and empirical algorithms based on dose measurements in water to the current stage, where we are on the verge of using patient-specific Monte Carlo algorithm-based dose distribution calculation as the most accurate method for predicting the dose distribution in the patient. Accurate and reliable Monte Carlo algorithms have been available for some time, the main problem, however, was adequate computing power to allow fast, reliable, and efficient dose calculations. With the current rapid increase in computing power one can expect that routine treatment planning for radiotherapy will be carried out with Monte Carlo-based dose calculation engines in the near future.

Dose distributions can be calculated accurately only if the calculation adequately models: (1) the radiation source, (2) beam geometry, and (3) patient anatomy. Future Monte Carlo engines are expected to allow a single simulation involving radiation transport through the linac as well as the patient; to save on calculation time current techniques all share the common approach of separating calculations into two parts: (1) source and beam geometry and (2) patient.

Figure on next page shows results of a Monte Carlo simulation of a typical linac head (Varian, Clinac 2100 C) operating in the 6 MeV clinical electron mode. About 100 histories are displayed in addition to the main beam-forming components of the linac. The 6 MeV electron pencil beam produced in the accelerator waveguide of the linac is transmitted through the beryllium exit window and strikes the scattering foil to increase the beam cross sectional area. The scattered beam then passes through the monitor chamber and the field mirror and is shaped into the clinical beam by the photon collimator and the electron applicator (cone).

Figure on next page: Courtesy of Dr. David W.O. Rogers, Carleton University, Ottawa, Canada. Reproduced with permission.

Chapter 2
Coulomb Scattering

This chapter deals with various types of elastic scattering interactions that heavy and light charged particles can have with atoms of an absorber. The interactions fall into the general category of Coulomb interactions and the chapter starts with a discussion of the intriguing Geiger–Marsden experiment of alpha particle scattering on thin gold foils. The experiment is of great historical importance and its results have lead to Rutherford's ingenious conclusion that most of the atom is empty space and that most of the atomic mass is concentrated in the atomic nucleus. The kinematics of the α-particle scattering is discussed in detail and the differential and total cross section concept for scattering is introduced for Rutherford scattering and expanded to other types of Coulomb scattering.

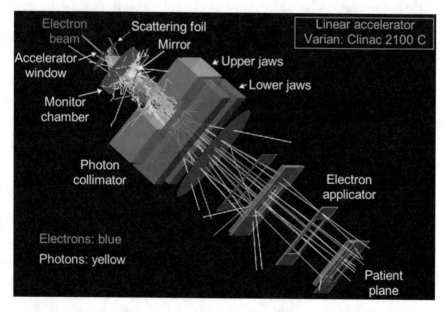

© Springer International Publishing Switzerland 2016
E.B. Podgoršak, *Radiation Physics for Medical Physicists*,
Graduate Texts in Physics, DOI 10.1007/978-3-319-25382-4_2

The chapter continues with a discussion of the Mott electron-nucleus scattering and introduces correction factors for electron spin, nuclear recoil, and the finite size of the nucleus to achieve better agreement with measured data. A brief discussion of the form factor representing the Fourier transform of the nuclear charge density follows and the chapter continues with a general discussion of elastic scattering of charged particles. The chapter concludes with a discussion of the characteristic scattering distance, scattering cross section and mean square scattering angle for various scattering events occurring on single scattering centers (single scattering) as well as the mean square scattering angle and mass scattering power for multiple scattering.

2.1 General Aspects of Coulomb Scattering

Coulomb scattering is a general term used to describe elastic Coulomb interactions between two charged particles: an energetic projectile and a target. Much of the knowledge in atomic, nuclear, and particle physics has been derived from various Coulomb scattering experiments, starting with the famous Geiger and Marsden experiment of 1909 in which α-particles were scattered on gold nuclei. Based on the angular distribution of the scattered α-particles, measured by Geiger and Marsden, Rutherford concluded that most of the atomic mass and the positive atomic charge are concentrated in the atomic nucleus which is at least four orders of magnitude smaller than the size of the atom.

The Rutherford model of the atom revolutionized physics in particular and science in general. Since then other Coulomb-type scattering experiments were carried out, typically using energetic protons or electrons as projectiles bombarding atomic nuclei or orbital electrons with the objective to learn more about the atomic and nuclear structure.

It is now well understood that in order for a particle to be useful as a nuclear probe, its de Broglie wavelength (Sect. 1.22.1) must be of the order of the nuclear size which is currently estimated with the relationship $R = R_0 \sqrt[3]{A}$ given in (1.26) with R the nuclear radius, A the atomic mass number and R_0 the nuclear radius constant (1.25 fm). As shown in Sect. 1.22.1, the de Broglie wavelength of a particle can be expressed as a function of the particle's kinetic energy E_K as

$$\lambda = \frac{2\pi\hbar c}{E_K \sqrt{1 + \dfrac{2E_0}{E_K}}} \begin{cases} \approx \dfrac{2\pi\hbar c}{\sqrt{2E_0 E_K}} & \text{for } E_K \ll E_0 = m_0 c^2, \\[2ex] \approx \dfrac{2\pi\hbar c}{E_K} \approx \dfrac{2\pi\hbar c}{E} & \text{for } E_K \gg E_0 = m_0 c^2. \end{cases} \qquad (2.1)$$

In Fig. 1.6 we show the de Broglie wavelength λ against kinetic energy E_K for electrons, protons, and α-particles. Typical nuclear size is of the order of 10 fm and the de Broglie wavelength λ of 10 fm is attained at kinetic energies E_K of

2 MeV for α-particles, 10 MeV for protons, and 130 MeV for electrons. Electrons with kinetic energies above 200 MeV can serve as an excellent probe for nuclear studies, not only because these electrons possess a suitable de Broglie wavelength but also because they are point-like and experience only Coulomb interactions with the nuclear constituents even when they penetrate the nucleus. This is in contrast to heavy charged particles which upon penetration of the nucleus will undergo strong interactions in addition to Coulomb interactions, making the analysis of experimental results difficult and cumbersome.

In the first approximation electron scattering on a nucleus can be treated like Rutherford scattering; however, when doing so, several other interactions are ignored, such as: spin effects in magnetic interactions; energy transfer to the nucleus of the scatterer (target recoil); relativistic and quantum effects; and effects of the finite size of the nucleus. Modern scattering theories now account for these additional interactions; however, they are still based on principles enunciated 100 years ago in Manchester by Geiger, Marsden, and Rutherford.

Scattering of α-particles on atomic nuclei is referred to as Rutherford scattering in honor of Rutherford's contribution to the understanding of the scattering process as well as the structure of the atom. In addition to the Rutherford scattering of α-particles on atomic nuclei (see Sect. 2.2), the most notable other Coulomb elastic scattering phenomena are:

- Scattering of high energy electrons on atomic nuclei referred to as Mott scattering (Sect. 2.5).
- Scattering of electrons on atomic orbital electrons referred to as Møller scattering.
- Scattering of positrons on atomic orbital electrons referred to as Bhabha scattering.
- Multiple scattering involving any one of the above listed scattering types and referred to as Molière multiple scattering (Sect. 2.7).

2.2 Geiger–Marsden Experiment

In 1909 *Hans Geiger* and *Ernest Marsden* in collaboration with *Ernest Rutherford* carried out an experiment studying the scattering of 5.5 MeV α-particles on a thin gold foil with a thickness of the order of 10^{-6} m. They obtained the α-particles from radon-222, a natural α-particle emitter, collimated them into a small pencil beam, and directed the beam in vacuum onto a thin gold foil in which scattering occurred. The scattered α-particles were detected by counting with a low-power microscope the scintillations produced in a zinc sulphide (ZnS) receptor (area: 1 cm^2) that could be rotated around the foil at a given distance from the source. The alpha particle counter was invented several years before by *William Crookes* and called the spinthariscope. The Geiger–Marsden experiment, shown schematically in Fig. 2.1,

Fig. 2.1 Schematic diagram
of the Geiger–Marsden
experiment in the study of
α-particle scattering on gold
nuclei in a thin gold foil. Θ is
the total scattering angle for
α-particles upon traversing
the 1 μm thick gold foil and
undergoing a large number
of scattering interactions

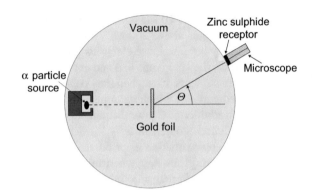

seems rather mundane; however, its peculiar and unexpected results had a profound
effect on modern physics in particular and on science in general.

2.2.1 Thomson Model of the Atom

In 1898 *Joseph J. Thomson*, who is also credited with the discovery of the electron
in 1897, proposed an atomic model in which the mass of the atom is distributed
uniformly over the volume of the atom with a radius of the order of 1 Å and negatively
charged electrons are dispersed uniformly within a continuous spherical distribution
of positive charge. The electrons form rings and each ring can accommodate a certain
upper limit in the number of electrons and then other rings begin to form. With this
ring structure Thomson could in principle account for the periodicity of chemical
properties of elements. A schematic representation of the Thomson's atomic model,
often referred to as the *"plum-pudding model"*, is given in Fig. 2.2a, suggesting the
following features:

- In the ground state of the atom the electrons are fixed at their equilibrium positions
  and emit no radiation.
- In an excited state of the atom the electrons oscillate about their equilibrium
  positions and emit radiation through dipole oscillations by virtue of possessing
  charge and being continuously accelerated or decelerated (Larmor relationship).

According to the Thomson atomic model the angular distribution of a pencil beam of
α-particles scattered in the gold foil in the Geiger–Marsden experiment is Gaussian
and given by the following expression (for derivation see Sect. 2.7.5)

$$N(\Theta)\mathrm{d}\Theta = \frac{2\Theta N_0}{\overline{\Theta^2}}e^{-\frac{\Theta^2}{\overline{\Theta^2}}}\,\mathrm{d}\Theta, \tag{2.2}$$

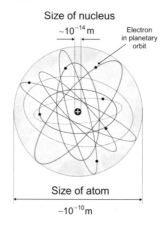

Fig. 2.2 Schematic diagram of two atomic models: **a** Thomson "plum-pudding" model of 1898 in which the electrons are uniformly distributed in a sea of positive atomic charge and **b** Rutherford nuclear model in which the electrons revolve in empty space around the nucleus that is positively charged and contains most of the atomic mass. The size of the nucleus with diameter of the order of 10^{-14} m is at least 4 orders of magnitude smaller than the size of the Rutherford atom with diameter of the order of 10^{-10} m. The size of the Thomson atom is of the order of 10^{-10} m, similar to the size of Rutherford atom

where

Θ is the scattering angle of the α-particle after it passes through the gold foil (note: the α-particle undergoes $\sim 10^4$ interactions as a result of a foil thickness of 10^{-6} m and an approximate atomic diameter of 10^{-10} m.

$N(\Theta)\,d\Theta$ is the number of α-particles scattered within the angular range of Θ to $\Theta + d\Theta$.

N_0 is the number of α-particles striking the gold foil.

$\overline{\Theta^2}$ is the mean square net deflection experimentally determined to be of the order of 3×10^{-4} rad^2, i.e., $\sqrt{\overline{\Theta^2}} \approx 1°$.

Geiger and Marsden found that more than 99% of the α-particles incident on the gold foil were scattered at angles less than 3° and that their distribution followed a Gaussian shape given in (2.2); however, they also found that one in $\sim 10^4$ α-particles was scattered with a scattering angle Θ exceeding 90°. This implied a measured probability of 10^{-4} for scattering with scattering angle $\Theta > 90°$, in drastic disagreement with the probability of 10^{-3500} predicted by the theory based on the Thomson atomic model, as shown in (2.3) below.

According to the Thomson atomic model the probability for α-particle scattering with $\Theta > 90°$ (i.e., with a scattering angle Θ between $\frac{1}{2}\pi$ and π) is calculated by integrating (2.2) from $\frac{1}{2}\pi$ to π as follows

$$\frac{N\left(\Theta > \frac{\pi}{2}\right)}{N_0} = \frac{\displaystyle\int_{\pi/2}^{\pi} N\left(\Theta\right) d\Theta}{N_0} = -\int_{\frac{\pi}{2}}^{\pi} e^{-\frac{\Theta^2}{\overline{\Theta^2}}}\, d\left(-\frac{\Theta^2}{\overline{\Theta^2}}\right) \qquad (2.3)$$

$$= -e^{-\frac{\Theta^2}{\overline{\Theta^2}}}\Bigg|_{\frac{\pi}{2}}^{\pi} = -e^{-\left\{\frac{180°}{1°}\right\}^2} + e^{-\left\{\frac{90°}{1°}\right\}^2} = e^{-90^2} \approx 10^{-3500},$$

where we use the experimentally determined value of $1°$ for the root mean square angle $\sqrt{\overline{\Theta^2}}$.

2.2.2 Rutherford Model of the Atom

At the time of the Geiger–Marsden experiment, the Thomson atomic model was the prevailing atomic model based on the assumption that the positive charges and the negative (electron) charges of an atom were distributed uniformly over the atomic volume ("plum-pudding" model) to make the atom neutral on the outside. The theoretical result of 10^{-3500} for the probability of α-particle scattering with a scattering angle greater than $90°$ on a gold foil consisting of Thomson atoms is an extremely small number in comparison with the result of 10^{-4} obtained experimentally by Geiger and Marsden. This discrepancy between experiment and theory highlighted a serious problem with the Thomson atomic model and stimulated *Ernest Rutherford* to propose a completely new atomic model that agreed better with experimental results obtained by Geiger and Marsden. The two main features of the Rutherford model are as follows:

1. Mass and positive charge of the atom are concentrated in the nucleus the size of which is of the order of 10^{-15} m $= 1$ fm.
2. Negatively charged electrons revolve about the nucleus in a cloud, the radius of which is of the order of 10^{-10} m $= 1$ Å.

The two competing atomic models are depicted schematically in Fig. 2.2. Contrary to the Thomson "plum-pudding" atomic model, essentially all mass of the Rutherford atom is concentrated in the atomic nucleus that is also the seat of the positive charge of the atom and has a radius of the order of 10^{-15} m, as shown schematically in Fig. 2.2b. As shown in (1.27), the density of the nucleus with mass M is enormous with an order of magnitude of 1.5×10^{14} g·cm^{-3}.

As for the atomic electrons, Rutherford proposed that they are distributed in a spherical cloud on the periphery of the atom with a radius of the order of 10^{-10} m; however, he did not speculate on the rules governing the motion of electrons in an atom. It was *Niels Bohr* who soon thereafter expanded the Rutherford model by proposing four postulates, one of them dealing with quantization of electron angular momentum, which allowed him to derive from first principles the electron planetary motion in one-electron structures (See Sect. 3.1). Rutherford, a superb experimental physicist, and Bohr, an extremely gifted theoretical physicist, are credited with developing the currently accepted atomic model which in their honor is referred to as the Rutherford–Bohr atomic model.

2.3 Rutherford Scattering

2.3.1 Kinematics of Rutherford Scattering

Based on his model and five additional assumptions, Rutherford derived the kinematics for the scattering of α-particles on gold nuclei using basic principles of classical mechanics. The five additional assumptions are as follows:

1. Scattering of α-particles on gold nuclei is elastic.
2. The mass of the gold nucleus M is much larger than the mass of the α-particle m_α, i.e., $M \gg m_\alpha$.
3. Scattering of α-particles on atomic electrons is negligible because $m_\alpha \gg m_e$, where m_e is the electron mass.
4. The α-particle does not penetrate the nucleus (no nuclear reactions).
5. The classical relationship for the kinetic energy E_K of the α-particle, i.e., $E_K = \frac{1}{2}m_\alpha v^2$, is valid, where v_α is the velocity of the α-particle.

Rutherford used concepts of classical mechanics in his derivation of the kinematics of α-particle scattering. To show that this was an acceptable approach we determine the speed of 5.5 MeV α-particles used in the Geiger–Marsden experiment. The speed v_α of the α-particles relative to the speed of light c in vacuum for 5.5 MeV α-particles can be calculated using either the classical relationship (1.59) or the relativistic relationship of (1.58). Note that $m_\alpha c^2 = 3727$ MeV:

1. The *classical calculation* is done using the classical expression for the kinetic energy E_K of the α-particle

$$E_K = \frac{1}{2}m_\alpha v_\alpha^2 = \frac{1}{2}m_\alpha c^2 \left\{ \frac{v_\alpha^2}{c^2} \right\}. \tag{2.4}$$

Solve (2.4) for v_α/c to obtain

$$\frac{v_\alpha}{c} = \sqrt{\frac{2E_K}{m_\alpha c^2}} = \sqrt{\frac{2 \times 5.5 \text{ MeV}}{3727 \text{ MeV}}} = 0.0543. \tag{2.5}$$

2. The *relativistic calculation* is carried out using the relativistic expression (1.58) for the kinetic energy E_K of the α-particle

$$E_K = \frac{m_\alpha c^2}{\sqrt{1 - \left(\dfrac{v_\alpha}{c}\right)^2}} - m_\alpha c^2. \tag{2.6}$$

Solve (2.6) for v_α/c to obtain

$$\frac{v_\alpha}{c} = \sqrt{1 - \frac{1}{\left(1 + \dfrac{E_K}{m_\alpha c^2}\right)^2}} = \sqrt{1 - \frac{1}{\left(1 + \dfrac{5.5}{3727}\right)^2}} = 0.0543 \tag{2.7}$$

The relativistic calculation of (2.6) and classical calculation of (2.4) give identical results since the velocity of the α-particle is much smaller than c, the speed of light in vacuum, or $(v_\alpha/c) \ll 1$, for α-particles with kinetic energy E_K of the order of a few million electron volts. Rutherford's use of the simple classical relationship rather than the correct relativistic expression for the kinetic energy of the naturally occurring α-particles was thus justified. Note that all naturally occurring α-particles have kinetic energy of the order of a few million electron volts, so the use of classical mechanics is appropriate for all naturally occurring α-particles.

The interaction between the α-particle (charge ze) and the nucleus (charge Ze) is a repulsive Coulomb interaction between two positive point charges, and, as result, the α-particle follows a hyperbolic trajectory, as shown schematically in Fig. 2.3. Note that θ represents the scattering angle in a single α-particle interaction with one nucleus, whereas Θ of (2.2) represents the scattering angle resulting from the α-particle traversing the thin gold foil and undergoing some 10^4 interactions while traversing the foil.

For a single α-particle interaction depicted in Fig. 2.3 the nucleus is in the outer focus of the hyperbola because of the repulsive interaction between the α-particle and the nucleus. For an interaction between two charges of opposite sign (for example, energetic electron interacting with atomic nucleus) the Coulomb interaction is attractive and the trajectory of the projectile is also a hyperbola but the target resides in the inner focus of the hyperbola.

Two important parameters of Coulomb scattering are the impact parameter b and the scattering angle θ. As shown in Fig. 2.3:

- Impact parameter b is defined as the perpendicular distance between the initial velocity vector v_i of the projectile and the center of the target it is approaching.
- Scattering angle θ is defined as the angle between the initial momentum vector $\mathbf{p}_i$ and the final momentum vector $\mathbf{p}_f$.

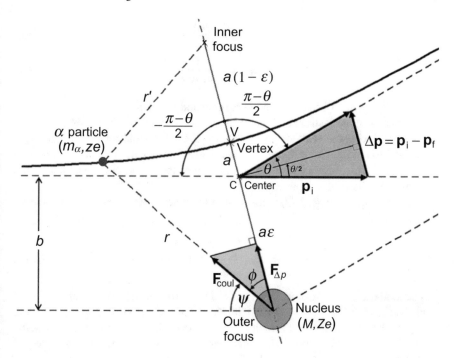

Fig. 2.3 Schematic diagram for scattering of an α-particle on a nucleus: θ is the scattering angle; b the impact parameter; $\Delta\mathbf{p}$ the change in α-particle momentum; v_i the initial velocity of the α-particle; and $\mathbf{p}_i$ the initial momentum of the α-particle. The trajectory of the α-particle is a hyperbola as result of the repulsive Coulomb interaction between the α-particle and the nucleus. The nucleus is in the outer focus of the hyperbolic trajectory of the α-particle. The two asymptotes of the hyperbola intersect at the center C of the hyperbola. The axis of the hyperbola connecting inner focus, center C, and outer focus intersects the hyperbola at vertex V on the hyperbolic trajectory of the particle.

2.3.2 Distance of Closest Approach in Head-on Collision Between α-Particle and Nucleus

A special case of Rutherford scattering occurs when $b = 0$ corresponding to the α-particle being on a direct-hit trajectory. Considering the classical conservation of energy in a direct hit α-particle elastic scattering event, the following points can be made:

1. The total energy $E(r)$ of the α-particle–nucleus system consists of two components: kinetic energy $E_K(r)$ of the α-particle and the repulsive Coulomb potential energy $E_P(r)$ where

$$E_K(r) = \frac{m_\alpha v^2}{2}, \tag{2.8}$$

$$E_P(r) = \frac{zZe^2}{4\pi\varepsilon_0}\frac{1}{r}. \tag{2.9}$$

and r is the distance between the α-particle and the nucleus.

2. The scattering is elastic. The kinetic energy E_K of the α-particle does not remain constant during the scattering process; however, the initial kinetic energy $(E_K)_i$ is equal to the final kinetic energy $(E_K)_f$ since the nucleus is assumed to remain stationary. This means that the final velocity of the α-particle v_f is equal to the initial velocity of the α-particle v_i.

3. In general, total energy $E(r)$ is the sum of the kinetic energy $E_K(r)$ and potential energy $E_P(r)$

$$E(r) = (E_K)_i = E_K(r) + E_P(r) = E_K(r) + \frac{zZe^2}{4\pi\varepsilon_0}\frac{1}{r}. \tag{2.10}$$

4. The total energy $E(r)$ at any distance $r > D_{\alpha-N}$ from the nucleus equals the initial kinetic energy $(E_K)_i$ of the α-particle, since $E_P(r = \infty) \to 0$. As the α-particle approaches the nucleus, its velocity v_α and kinetic energy $E_K(r)$ diminish and the repulsive potential energy $E_P(r)$ increases with the sum of the two always equal to the initial kinetic energy $(E_K)_i$ of the α-particle.

5. In its approach toward the nucleus the α-particle eventually stops at a distance from the nucleus $D_{\alpha-N}$, defined as the distance of closest approach. At $r = D_{\alpha-N}$ the α-particle kinetic energy $E_K(r = D_{\alpha-N})$ is zero, and the total energy $E(r)$ equals the potential energy $E_P(r = D_{\alpha-N})$ that is expressed as

$$E\left(r = D_{\alpha-N}\right) = (E_K)_i = E_K\left(r = D_{\alpha-N}\right) + E_P\left(r = D_{\alpha-N}\right)$$

$$= 0 + \frac{zZe^2}{4\pi\varepsilon_0}\frac{1}{D_{\alpha-N}}. \tag{2.11}$$

The distance of closest approach $D_{\alpha-N}$ between the α-particle with $(E_K)_i = 5.5$ MeV and a gold nucleus $(Z = 79)$ in a direct hit scattering event is determined from (2.11) as follows

$$D_{\alpha-N} = \frac{zZe^2}{4\pi\varepsilon_0}\frac{1}{E_K} = \frac{2\times79\times e\times1.6\times10^{-19}\ \text{C·V·m}}{4\pi\times8.85\times10^{-12}\ \text{C}\times5.5\times10^6\ \text{eV}} = 41.3\ \text{fm} \tag{2.12}$$

or

$$D_{\alpha-N} = \frac{zZ\hbar c\alpha}{E_K} = \frac{2\times79\times197.3\ \text{MeV·fm}}{137\times5.5\ \text{MeV}} \approx 41.3\ \text{fm}$$

For naturally occurring α-particles interacting with nuclei of atoms the distance of closest approach $D_{\alpha-N}$ exceeds the radius R of the nucleus. Thus, the α-particle does not penetrate the nucleus and no nuclear reaction occurs. For example, as shown in (2.12), $D_{\alpha-N}$ for the Geiger–Marsden experiment with 5.5 MeV α-particle scattering on gold nuclei is 41.3 fm compared to the gold nucleus radius determined from (1.26) as

$$R = R_0 \sqrt[3]{A} = 1.25 \text{ fm } \sqrt[3]{197} \approx 7.3 \text{ fm}, \tag{2.13}$$

where R_0 is the nuclear radius constant equal to 1.25 fm, as discussed in Sect. 1.16.1.

2.3.3 General Relationship Between Impact Parameter and Scattering Angle

The general relationship between the impact parameter b and the scattering angle θ may be derived most elegantly by determining two independent expressions for the change in momentum Δp of the scattered α-particle. The momentum transfer is along the symmetry line that bisects the angle $\pi - \theta$, as indicated in Fig. 2.3. The magnitude of the repulsive Coulomb force F_{Coul} acting on the α-particle is given by

$$F_{Coul} = \frac{zZe^2}{4\pi\varepsilon_0} \frac{1}{r^2}, \tag{2.14}$$

where

r is the distance between the α-particle and the nucleus M,
z is the atomic number of the α-particle (for helium $z = 2$ and $A = 4$),
Z is the atomic number of the absorber (for gold $Z = 79$ and $A = 197$).

Since the component of the force F_{coul} in the direction of the momentum transfer is $F_{\Delta p} = F_{Coul} \cos \phi$, the momentum transfer (impulse of force) Δp may be written as the time integral of the force component $F_{\Delta p}$ as follows

$$\Delta p = \int_{-\infty}^{\infty} F_{\Delta p} \, dt = \int_{-\infty}^{\infty} F_{Coul} \cos \phi \, dt = \frac{zZe^2}{4\pi\varepsilon_0} \int_{-\frac{\pi-\theta}{2}}^{\frac{\pi-\theta}{2}} \frac{\cos \phi}{r^2} \frac{dt}{d\phi} d\phi$$

$$= \frac{zZe^2}{4\pi\varepsilon_0} \int_{-\frac{\pi-\theta}{2}}^{\frac{\pi-\theta}{2}} \frac{\cos \phi}{wr^2} d\phi, \tag{2.15}$$

where

ϕ is the angle between the radius vector r and the bisector, as also shown in
       Fig. 2.3,

$\dfrac{\mathrm{d}t}{\mathrm{d}\phi}$ is the inverse of the angular frequency ω.

The angular frequency $\omega = \mathrm{d}\phi/\mathrm{d}t = v/r$ can be expressed as a function of the impact parameter b, initial α-particle velocity v_i, and radius vector r by invoking conservation of angular momentum $\mathbf{L}$, where $\mathbf{L}$ is in general defined as

$$\mathbf{L} = \mathbf{r} \times \mathbf{p} = \mathbf{r} \times m_\alpha \mathbf{v}. \tag{2.16}$$

With the help of Fig. 2.4 we now express $|\mathbf{L}| = L$, the magnitude of the angular momentum $\mathbf{L}$, for two different points (V and B) on the α-particle hyperbolic trajectory. Point V is at the vertex of the hyperbola and point B is at a very large distance from the nucleus where the α-particle position defines the impact parameter b. The angular momentum L at point B is given as

$$|\mathbf{L}| = L = r m_\alpha v_i \sin \psi = m_\alpha v_i b, \tag{2.17}$$

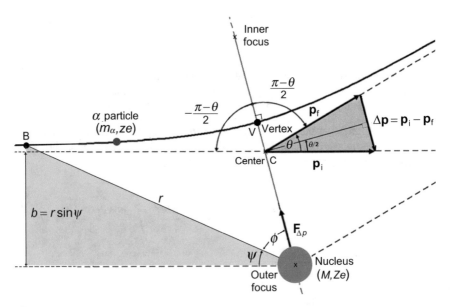

Fig. 2.4 Geometry for determination of the angular momentum $\mathbf{L}$ for two different points on the α-particle trajectory. Point V is for the α-particle located in the vertex of the hyperbola and point B is for the α-particle located at a very large distance from the nucleus

while for the vertex point V, where v and $\mathbf{r}$ are perpendicular to each other and $v = \omega r$, it is

$$|\mathbf{L}| = L = |\mathbf{r} \times m_\alpha \mathbf{v}| = m_\alpha r v \sin 90° = m_\alpha \omega r^2. \tag{2.18}$$

Using the conservation of angular momentum L, we merge (2.17) and (2.18) to get

$$L = m_\alpha v_i b = m_\alpha \omega r^2, \tag{2.19}$$

and the following expression for the angular frequency ω

$$\omega = \frac{v_i b}{r^2}, \tag{2.20}$$

with v_i the initial velocity of the α-particle at $r = \infty$. Since the scattering is elastic, the kinetic energy will be conserved in the scattering interaction, so it follows that the final α-particle velocity v_f will be equal to the initial α-particle velocity ($v_f = v_i$).

After inserting (2.20) into (2.15) we get a simple integral for Δp with the following solution

$$\Delta p = \frac{zZe^2}{4\pi\varepsilon_0} \frac{1}{v_i b} \int_{-\frac{\pi-\theta}{2}}^{\frac{\pi-\theta}{2}} \cos\phi \, d\phi = \frac{zZe^2}{4\pi\varepsilon_0} \frac{1}{v_\infty b} \{\sin\phi\}_{-\frac{\pi-\theta}{2}}^{+\frac{\pi-\theta}{2}} = 2\frac{zZe^2}{4\pi\varepsilon_0} \frac{1}{v_i b} \cos\frac{\theta}{2}. \tag{2.21}$$

With the help of the momentum vector diagram, given in Figs. 2.3 and 2.4, the momentum transfer Δp may also be written as

$$\Delta p = 2p_i \sin\frac{\theta}{2} = 2m_\alpha v_i \sin\frac{\theta}{2}. \tag{2.22}$$

Combining (2.21) and (2.22) we obtain the following expressions for the impact parameter b

$$b = \frac{zZe^2}{4\pi\varepsilon_0 m_\alpha v_i^2} \cot\frac{\theta}{2} = \frac{1}{2}\frac{zZe^2}{4\pi\varepsilon_0}\frac{1}{E_K}\cot\frac{\theta}{2}$$

$$= \frac{1}{2}D_{\alpha-N}\cot\frac{\theta}{2} = \frac{1}{2}D_{\alpha-N}\sqrt{\frac{1+\cos\theta}{1-\cos\theta}}, \tag{2.23}$$

with the use of:

1. Classical relationship for the kinetic energy of the α-particle ($E_K = \frac{1}{2}m_\alpha v_i^2$), since $v_i \ll c$.
2. Definition of $D_{\alpha-N}$ as the distance of closest approach between the α-particle and the nucleus in a "direct-hit" head-on collision for which the impact parameter $b = 0$, the scattering angle $\theta = \pi$, and $D_{\alpha-N} = zZe^2/(4\pi\varepsilon_0 E_K)$ from (2.12).

2.3.4 Hyperbolic Trajectory and Distance of Closest Approach

Equations for the hyperbolic trajectory of an alpha particle interacting with a nucleus can be derived from the diagram given in Fig. 2.3 and the simple rule governing the hyperbola with the target in the outer focus because of the repulsive interaction between the projectile (α-particle) and the target (nucleus)

$$r - r' = 2a, \tag{2.24}$$

where

a is the distance between the vertex V and the center C of the hyperbola.
r is the distance between the point of interest on the hyperbola and the outer focus.
r' is the distance between the point of interest on the hyperbola and the inner focus.

The parameters of the hyperbola, such as a, r, and r', are defined in Fig. 2.3 and the locations of the inner focus, vertex V, center C, and outer focus on the hyperbola axis are also indicated in Fig. 2.3. Solving (2.24) for r' and squaring the result, we get the following expression for $(r')^2$

$$(r')^2 = r^2 - 4ar + 4a^2. \tag{2.25}$$

Using the law of cosines in conjunction with Fig. 2.3, we express $(r')^2$ as

$$(r')^2 = r^2 - 4a\varepsilon r \cos\phi + 4a^2\varepsilon^2, \tag{2.26}$$

where ε is the eccentricity of the hyperbola.

Subtracting (2.26) from (2.25) and solving for $r(\phi)$, we now obtain the following general equation for the hyperbolic trajectory of the α-particle

$$r(\phi) = \frac{a(\varepsilon^2 - 1)}{\varepsilon \cos\phi - 1}. \tag{2.27}$$

Three separate special conditions are of interest with regard to (2.27):

1. $r = \infty$ for determining the eccentricity ε.
2. $\phi = 0$ for determining the general distance of closest approach $R_{\alpha-N}$.
3. $\theta = \pi$ for determining the distance of closest approach in a direct hit that results in the shortest distance of closest approach defined as $D_{\alpha-N}$ in (2.12).

Eccentricity ε is determined as follows:
For $r = \infty$ the angle ϕ equals to $\frac{1}{2}(\pi - \theta)$ and, to get $r = \infty$, the denominator in (2.27) $\left[\varepsilon \cos[\frac{1}{2}(\pi - \theta)] - 1\right]$ must equal to zero, resulting in the following relationship for the eccentricity ε

$$\varepsilon \cos \frac{\pi - \theta}{2} - 1 = \varepsilon \sin \frac{\theta}{2} - 1 = 0 \quad \text{or} \quad \varepsilon = \frac{1}{\sin \frac{\theta}{2}}. \tag{2.28}$$

Distance of closest approach $R_{\alpha-N}$ between the α-particle and the nucleus in a non-direct hit collision ($\theta < \pi$ and $\phi = 0$) is from (2.27) given as

$$R_{\alpha-N} = r(\phi = 0) = \frac{a(\varepsilon^2 - 1)}{\varepsilon - 1} = a(1 + \varepsilon) = a\left\{1 + \frac{1}{\sin \frac{\theta}{2}}\right\}. \tag{2.29}$$

The result $R_{\alpha-N} = a(1 + \varepsilon)$ can also be obtained directly from Fig. 2.3 by recognizing that the distance between the outer focus and vertex V of the hyperbola equals to $(a\varepsilon + a)$.

Distance of closest approach in a direct-hit collision, $D_{\alpha-N}$ ($b = 0$; $\theta = \pi$) can now from (2.29) with $\theta = \pi$ be written as

$$D_{\alpha-N} = R_{\alpha-N}(\theta = \pi) = 2a, \tag{2.30}$$

from where it follows that a, the distance between vertex V and center C of the hyperbola, (see Figs. 2.3 and 2.4) is equal to $\frac{1}{2}D_{\alpha-N}$. This allows us to express $R_{\alpha-N}$ of (2.29) as a function of the direct-hit distance of closest approach $D_{\alpha-N}$ or as a function of the impact parameter b using the relationship (2.23) between $D_{\alpha-N}$ and b

$$R_{\alpha-N} = a\left\{1 + \frac{1}{\sin \frac{\theta}{2}}\right\} = \frac{D_{\alpha-N}}{2}\left\{1 + \frac{1}{\sin \frac{\theta}{2}}\right\} = b\frac{1 + \sin \frac{\theta}{2}}{\cos \frac{\theta}{2}} = b\frac{\cos \frac{\theta}{2}}{1 - \sin \frac{\theta}{2}}. \tag{2.31}$$

2.3.5 Hyperbola in Polar Coordinates

In *polar coordinates* (r, φ) the hyperbolic α-particle trajectory may be expressed as

$$\frac{1}{r} = \frac{1}{b} \sin \psi + \frac{a}{b^2} (\cos \psi - 1), \tag{2.32}$$

with parameters a, b, and ψ defined in Fig. 2.3. Note that ψ and ϕ are different angles and that the following relationship for angles ψ, ϕ, and θ applies

$$\psi + \phi = \left| \frac{\pi - \theta}{2} \right|. \tag{2.33}$$

It can be shown that the general expressions (2.32) and (2.27) defining the hyperbola are equivalent.

2.4 Cross Sections for Rutherford Scattering

2.4.1 Differential Cross-Section for Rutherford Scattering: Classical Derivation

The differential cross section $d\sigma_{Ruth}/d\Omega$ for Rutherford scattering into a solid angle $d\Omega = 2\pi \sin \theta \, d\theta$ that corresponds to an angular aperture between θ and $\theta + d\theta$ (equivalent to impact parameters between b and $b - db$), assuming the azimuthal distribution to be isotropic, is the area of a ring with mean radius b and width db

$$d\sigma_{Ruth} = 2\pi b \, db = 2\pi \frac{b}{\sin \theta} \sin \theta \left| \frac{db}{d\theta} \right| d\theta. \tag{2.34}$$

Recognizing that

$$d\Omega = 2\pi \sin \theta \, d\theta, \tag{2.35}$$

expressing $\sin \theta$ as

$$\sin \theta = 2 \sin \frac{\theta}{2} \cos \frac{\theta}{2}, \tag{2.36}$$

and, using (2.23) for the impact parameter b to determine $|db/d\theta|$ as

$$\left| \frac{db}{d\theta} \right| = \frac{D_{\alpha-N}}{4} \frac{1}{\sin^2 \frac{\theta}{2}}, \tag{2.37}$$

we obtain from (2.34) combined with (2.35)–(2.37) the following expression for $d\sigma_{Ruth}/d\Omega$, the differential Rutherford cross section

$$\frac{d\sigma_{Ruth}}{d\Omega} = \left(\frac{D_{\alpha-N}}{4}\right)^2 \frac{1}{\sin^4\frac{\theta}{2}} = \left(\frac{D_{\alpha-N}}{4}\right)^2 \frac{1}{(1-\cos\theta)^2}. \qquad (2.38)$$

Inserting the expression for $D_{\alpha-N}$ of (2.12) into (2.38) and using the definition of the fine structure constant $\alpha = e^2 (4\pi\varepsilon_0\hbar c)^{-1}$, we can express the Rutherford differential cross section as

$$\frac{d\sigma_{Ruth}}{d\Omega} = \left(\frac{zZ\hbar c}{4}\right)^2 \left(\frac{\alpha}{E_K}\right)^2 \frac{1}{\sin^4\frac{\theta}{2}}, \qquad (2.39)$$

allowing us to conclude that the Rutherford differential scattering cross section $|d\sigma_{Ruth}/d\Omega|$ is:

1. Proportional to the atomic number z of the projectile and the atomic number Z of the target.
2. Proportional to the electromagnetic coupling (fine structure) constant α^2. Thus, the electromagnetic force is governed by photon exchange between the α-particle and the nucleus.
3. Inversely proportional to the square of the initial kinetic energy E_K of the α-particle.
4. Inversely proportional to the fourth power of $\sin(\theta/2)$ arising from the $1/r^2$ variation of the Coulomb repulsive force in effect between the α-particle and the nucleus.

At small scattering angles θ, where $\sin\left(\frac{1}{2}\theta\right) \approx \frac{1}{2}\theta$, the differential Rutherford cross section (2.38) can be simplified to read

$$\frac{d\sigma_{Ruth}}{d\Omega} = \frac{D_{\alpha-N}^2}{\theta^4}. \qquad (2.40)$$

Since most of the Rutherford scattering occurs for $\theta \ll 1$ rad and even at $\theta = \frac{1}{2}\pi$ the small angle result is within 30% of the general Rutherford expression, it is reasonable to use the small angle approximation of (2.40) at all angles for which the unscreened point-Coulomb field expression is valid. Departures from the point Coulomb field approximation appear for large and small angles θ, corresponding to small and large impact parameters b, respectively, and resulting from α-particle penetration of the nucleus and nuclear field screening by orbital electrons, respectively.

2.4.2 Differential Cross Section for Rutherford Scattering (Quantum-Mechanical Derivation)

The Rutherford differential cross section $d\sigma_{Ruth}/d\Omega$ of (2.38) was derived classically; however, it can also be derived quantum-mechanically in a short and simple manner by using the Fermi second golden rule, discussed in Sect. 1.23.5, and the Born collision formula, discussed in Sect. 1.23.6. The Born collision formula was introduced in a general form in (1.129) and can be written for Rutherford scattering in terms of the spherically symmetric Coulomb nuclear potential $V_N(r)$ playing the role of the potential operator $V(r)$

$$V(r) = V_N(r) = \left(\frac{zZe^2}{4\pi\varepsilon_0} \right) \frac{1}{r}. \tag{2.41}$$

For Rutherford scattering, (1.129) is expressed as follows

$$\frac{d\sigma_{Ruth}}{d\Omega} = \left| \frac{2m_\alpha}{\hbar^2} \int_0^\infty r^2 \frac{zZe^2}{4\pi\varepsilon_0} \frac{\sin(Kr)}{K^2r^2} d(Kr) \right|^2$$

$$= \left(\frac{D_{\alpha-N}}{4} \right)^2 \frac{1}{\sin^4\frac{\theta}{2}} \left| \int_0^\infty \sin(Kr) \, d(Kr) \right|^2, \tag{2.42}$$

after inserting the expression for nuclear Coulomb potential $V_N(r)$ given in (2.41), expression for K given in (1.124), and the expression for $D_{\alpha-N}$ given in (2.12).

The value of the integral in (2.42) poses a problem at its upper limit, since it oscillates about zero there. This problem can be obviated by accounting for screening effects or simply by substituting into (2.42) the nuclear Coulomb potential $V_N(r)$ of (2.41) by a Yukawa type potential $V_{Yuk}(r)$ where

$$V_{Yuk}(r) = V_N(r) e^{-\eta r} = \frac{zZe^2}{4\pi\varepsilon_0} \frac{1}{r} e^{-\eta r}, \tag{2.43}$$

to get

$$\frac{d\sigma_{Ruth}}{d\Omega} = \left(\frac{D_{\alpha-N}}{4} \right)^2 \frac{1}{\sin^4\frac{\theta}{2}} \left| K \int_0^\infty e^{-\eta r} \sin(Kr) \, dr \right|^2, \tag{2.44}$$

with η a positive constant which is set to zero upon solving (2.44). The integral in (2.44) in the limit $\eta \to 0$ gives

$$\lim_{\eta\to 0}\int_0^\infty e^{-\eta r}\sin(Kr)\,dr = \lim_{\eta\to 0}\left[e^{-\eta r}\frac{-\eta\sin Kr - K\cos Kr}{\eta^2 + K^2}\right]_0^\infty$$

$$= \lim_{\eta\to 0}\frac{K}{\eta^2 + K^2} = \frac{1}{K}, \tag{2.45}$$

and (2.44) then gives the standard well known result for the Rutherford differential cross section derived classically in Sect. 2.4.1 and presented in (2.38) as

$$\frac{d\sigma_{\text{Ruth}}}{d\Omega} = \left(\frac{D_{\alpha-N}}{4}\right)^2\frac{1}{\sin^4\dfrac{\theta}{2}}. \tag{2.46}$$

2.4.3 Screening of Nuclear Potential by Orbital Electrons

At large impact parameters b (i.e., at small scattering angles θ) the screening effects of the atomic orbital electrons cause the potential felt by the α-particle to fall off more rapidly than the $1/r$ Coulomb point-source potential. It is convenient to account for electron screening of the nuclear potential with the Thomas–Fermi statistical model of the atom in which the *Thomas–Fermi atomic potential* is given as

$$V_{\text{TF}}(r) \approx \frac{zZe^2}{4\pi\varepsilon_0}\frac{1}{r}e^{-\frac{r}{a_{\text{TF}}}}. \tag{2.47}$$

In (2.47) a_{TF} is the Thomas–Fermi atomic radius expressed as

$$a_{\text{TF}} = \frac{\xi a_0}{\sqrt[3]{Z}}, \tag{2.48}$$

where

ξ is the Thomas–Fermi atomic radius constant,
a_0 is the Bohr atomic radius $\left(a_0 = 0.53\text{ Å}\right)$, discussed in Sect. 3.1.1,
Z is the atomic number of the atom.

The Thomas–Fermi radius a_{TF} represents a fixed fraction of all atomic electrons or, more loosely, the radius of the atomic electron cloud that effectively screens the nucleus. The nuclear screening implies that, with a decreasing scattering angle θ, the scattering cross-section will flatten off at small angles θ to a finite value at $\theta = 0$ rather than increasing as θ^{-4} and exhibiting a singularity at $\theta = 0$. The constant ξ in (2.48) calculated from the Thomas–Fermi atomic model has a value of 0.885, while Jackson recommends a value of 1.4 as a better description of a general range of atomic and nuclear phenomena.

For our purposes $\xi \approx 1$ is a good and simple approximation to yield the following expression for the Thomas–Fermi radius a_{TF}

$$a_{TF} \approx \frac{a_0}{\sqrt[3]{Z}}, \tag{2.49}$$

suggesting that the effective radius of the atomic electron charge cloud decreases with an increasing atomic number Z as $1/\sqrt[3]{Z}$, decreasing from $\sim a_0$ for low Z to $\sim 0.2a_0$ for high Z elements. At first glance this result seems surprising considering that the radius of atoms increases with Z, as shown in (3.39). However, the radii of lower level atomic shells are inversely proportional to Z and this in turn results in a decreasing effective charge radius a_{TF} with increasing Z.

The Fermi second golden rule (Sect. 1.23.5) can be used in conjunction with the Born approximation (Sect. 1.23.6) to calculate $d\sigma_{Ruth}/d\Omega$ for very small scattering angles θ approaching 0 where (2.38) and (2.40) exhibit a singularity and predict $d\sigma_{Ruth}/d\Omega = \infty$, an obviously unacceptable result. Using the Thomas–Fermi potential of (2.47) for the potential $V(r)$, the differential cross section for Rutherford scattering is expressed as

$$\frac{d\sigma_{Ruth}}{d\Omega} = \left| \frac{2m_\alpha}{\hbar^2} \int_0^\infty r^2 V_{TF}(r) \frac{\sin(Kr)}{Kr} dr \right|^2 = \left| \frac{2m_\alpha}{\hbar^2} \frac{zZe^2}{4\pi\varepsilon_0} \int_0^\infty e^{-\frac{r}{a_{TF}}} \frac{\sin(Kr)}{Kr} dr \right|^2. \tag{2.50}$$

In standard tables of integrals we find the following solution for the integral in (2.50)

$$\int_0^\infty e^{-ax} \sin(bx)\, dx = -\left[\frac{e^{-ax}}{a^2+b^2}[a\sin(bx) + b\cos(bx)] \right]_{x=0}^{x=\infty} \tag{2.51}$$

and with its help we evaluate the integral in (2.50) as

$$\int_0^\infty e^{-\frac{r}{a_{TF}}} \sin(Kr)\, dr = \frac{K}{\frac{1}{a_{TF}^2} + K^2} = \frac{1}{K\left[1 + \frac{1}{(Ka_{TF})^2}\right]}. \tag{2.52}$$

The differential cross section for the Rutherford scattering $d\sigma_{Ruth}/d\Omega$ is now expressed as

$$\frac{d\sigma_{Ruth}}{d\Omega} = \left| \frac{2m_\alpha}{\hbar^2} \frac{zZe^2}{4\pi\varepsilon_0} \frac{1}{K^2} \frac{1}{1 + \frac{1}{(Ka_{TF})^2}} \right|^2. \tag{2.53}$$

The term $[1 + (Ka_{TF})^{-2}]^{-2}$ may be regarded as a correction factor to the standard differential scattering cross section (2.46) for $\theta \to 0$ where the Thomas–Fermi screening of the simple point-source Coulomb nuclear potential becomes important. The product Ka_{TF} using the expression for K of (1.124) is now given as

$$Ka_{TF} = \frac{2pa_{TF}}{\hbar} \sin \frac{\theta}{2} \tag{2.54}$$

and for a typical Rutherford scattering experiment using naturally emitted α-particles on a gold foil amounts to $\sim 10^5 \sin(\theta/2)$. Thus, unless the scattering angle is very small, the correction factor $[1 + (Ka_{TF})^{-2}]^{-2}$ is equal to 1 and (2.53) transforms into the simple Rutherford relationship given in (2.38). Equation (2.53) is then simplified to read

$$\frac{d\sigma_{Ruth}}{d\Omega} = \left| \frac{2m_\alpha}{\hbar^2} \frac{zZe^2}{4\pi\varepsilon_0} \frac{1}{K^2} \right|^2 = \left| \frac{2m_\alpha}{\hbar^2} \frac{zZe^2}{4\pi\varepsilon_0} \frac{\hbar^2}{4p^2 \sin^2(\theta/2)} \right|^2$$

$$= \left(\frac{D_{\alpha-N}}{4} \right)^2 \frac{1}{\sin^4 \dfrac{\theta}{2}} \tag{2.55}$$

and shows that the Rutherford scattering formula for a point-charge Coulomb field approximation can also be derived through quantum mechanical reasoning using the Fermi second golden rule and the Born approximation but neglecting any magnetic interaction involving spin effects.

2.4.4 Minimum Scattering Angle

We now return to (2.53) to show that for very small scattering angles θ it provides a finite result for $d\sigma_{Ruth}/d\Omega$ in contrast to the singularity exhibited by (2.55). The general differential cross section $d\sigma_{Ruth}/d\Omega$ including the small-θ correction factor $[1 + (Ka_{TF})^{-2}]^{-2}$ is

$$\frac{d\sigma_{Ruth}}{d\Omega} = \left(\frac{D_{\alpha-N}}{4} \right)^2 \frac{1}{\sin^4 \dfrac{\theta}{2}} \frac{1}{\left[1 + \dfrac{1}{(Ka_{TF})^2} \right]^2} \approx \frac{D_{\alpha-N}^2}{\theta^4} \frac{1}{\left[1 + \left(\dfrac{\hbar}{pa_{TF}\theta} \right)^2 \right]^2}, \tag{2.56}$$

after introducing the expression for K given in (2.54) and using the approximation $\sin \theta \approx \theta$ for small scattering angles θ.

Next we introduce the concept of θ_{min}, the minimum cutoff scattering angle for a given scattering experiment. Using the expressions for p and a_{TF} given by (1.64) and (2.49), respectively, we define θ_{min} as

$$\theta_{min} = \frac{\hbar}{pa_{TF}} = \frac{\hbar \sqrt[3]{Z}}{pa_0} = \frac{\hbar c \sqrt[3]{Z}}{a_0 \sqrt{E_K (E_K + 2E_0)}}, \tag{2.57}$$

where

E_K is the kinetic energy of the α-particle,
E_0 is the rest energy of the α-particle (3727.4 MeV).

Quantum-mechanically, based on Heisenberg uncertainty principle of (1.130), we define the minimum cutoff angle θ_{min} (also referred to as the Born screening angle) as follows: When the classical trajectory of the incident particle is localized within $\Delta z \approx a_{TF}$, the corresponding uncertainty on the transverse momentum of the particle is $\Delta p \geq \hbar/a_{TF}$, resulting in

$$\theta_{min} = \frac{\Delta p}{p} \approx \frac{\hbar}{p a_{TF}} = \frac{\lambda}{a_{TF}}. \tag{2.58}$$

For small scattering angles θ including $\theta = 0$ the differential scattering cross section $d\sigma_{Ruth}/d\Omega$ given in (2.56) simplifies to

$$\frac{d\sigma_{Ruth}}{d\Omega} = \frac{D^2_{\alpha-N}}{\left[\theta^2 + \theta^2_{min}\right]^2}, \tag{2.59}$$

and converges to the following finite value for $\theta = 0$

$$\frac{d\sigma_{Ruth}}{d\Omega} = \frac{D^2_{\alpha-N}}{\theta^4_{min}}. \tag{2.60}$$

2.4.5   Effect of the Finite Size of the Nucleus

At relatively large scattering angles θ the differential cross section $d\sigma_{Ruth}/d\Omega$ is smaller than that predicted by (2.38) because of the finite size of the nucleus. Approximating the charge distribution of the atomic nucleus by a volume distribution inside a sphere of radius R results in the following electrostatic potentials $V(r)$ for regions inside and outside the nucleus

$$V(r) = \frac{zZe^2}{4\pi\varepsilon_0 R}\left(\frac{3}{2} - \frac{1}{2}\frac{r^2}{R^2}\right) \quad \text{for} \quad r < R \text{ (inside the nucleus)}, \tag{2.61}$$

$$V(r) = \frac{3}{8}\frac{zZe^2}{\pi\varepsilon_0 R} \quad \text{for} \quad r = 0 \text{ (at the center of the nucleus)}, \tag{2.62}$$

$$V(r) = \frac{zZe^2}{4\pi\varepsilon_0 R} \quad \text{for} \quad r = R \text{ (at the edge of the nucleus)}, \tag{2.63}$$

$$V(r) = \frac{zZe^2}{4\pi\varepsilon_0 r} \quad \text{for} \quad r > R \text{ (outside the nucleus)}. \tag{2.64}$$

For use in the Fermi golden rule in conjunction with the Born approximation the four functions above can be approximated with the following single function $V_{FNS}(r)$ approximating the effects of the finite nuclear size (FNS) and covering the whole region of r from 0 to ∞

$$V_{FNS}(r) = \frac{zZe^2}{4\pi\varepsilon_0}\frac{1}{r}\left(1 - e^{-\frac{2r}{R}}\right), \tag{2.65}$$

where R is the nominal radius of the nucleus calculated from $R = R_0\sqrt[3]{A}$, given in (1.26).

In Fig. 2.5 the potential energy V_{FNS} is plotted against r, the distance from the center of the nucleus, for the gold nucleus. It converges to $(2zZe^2)/(4\pi\varepsilon_0 R)$ at $r = 0$ and provides a reasonable and continuous approximation both inside the finite nucleus where $r \leq R$ and outside the nucleus for $r > R$ where the point source Coulomb approximation holds.

For comparison, also shown in Fig. 2.5 are the Coulomb point source potential (dashed curve) and the finite source potential assuming a uniform charge distribution inside the nuclear sphere with radius R (light solid curve). At $r = 0$ the Coulomb source potential exhibits a singularity and the finite source with uniform charge density converges to $(3zZe^2)/(8\pi\varepsilon_0 R)$, as shown in (2.62). For the gold nucleus V_{FNS} at $r = 0$ converges to 64.9 MeV and the field assuming a uniform charge distribution inside the nucleus converges to 48.7 MeV, as shown in Fig. 2.5.

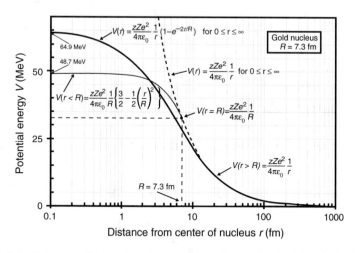

Fig. 2.5 Potential energy $V(r)$ against distance r from the center of gold nucleus with radius $R = 7.3$ fm. Three different potential energies are plotted: (1) Point source nuclear potential $V_N(r)$ of (2.41) shown with dashed curve exhibiting singularity at $r = 0$; (2) Potential for uniform charge distribution inside nuclear sphere of (2.61) shown with light solid curve and converging to $(3zZe^2)/(8\pi\varepsilon_0 R) = 48.7$ MeV at $r = 0$; and (3) Exponential function potential $V_{FNS}(r)$ of (2.65) approximating potential inside and outside the finite size nucleus, shown with heavy solid curve, and converging to $(zZe^2)/(2\pi\varepsilon_0 R) = 64.9$ MeV at $r = 0$

Inserting (2.65) into the Born approximation of (1.129) results in the following integral for $d\sigma_{\text{Ruth}}/d\Omega$

$$
\begin{aligned}
\frac{d\sigma_{\text{Ruth}}}{d\Omega} &= \left| \frac{2m_\alpha}{\hbar^2} \int_0^\infty r^2 V_{\text{FNS}}(r) \frac{\sin(Kr)}{Kr} dr \right|^2 \\
&= \left| \frac{2m_\alpha}{\hbar^2} \frac{zZe^2}{4\pi\varepsilon_0 K} \int_0^\infty \left(1 - e^{-\frac{2r}{R}} \right) \sin(Kr)\, dr \right|^2 \\
&= \left| \frac{2m_\alpha}{\hbar^2} \frac{zZe^2}{4\pi\varepsilon_0 K} \left\{ \int_0^\infty \sin(Kr)\, dr - \int_0^\infty e^{-\frac{2r}{R}} \sin(Kr)\, dr \right\} \right|^2 . \quad (2.66)
\end{aligned}
$$

The integrals in the curly bracket of (2.66) are calculated using (2.51) to get

$$
\begin{aligned}
\left\{ \int_0^\infty \sin(Kr)\, dr - \int_0^\infty e^{-\frac{2r}{R}} \sin(Kr)\, dr \right\} \\
= \left[-\frac{\cos(Kr)}{K} + \frac{e^{-\frac{2r}{R}} [2R^{-1} \sin(Kr) + K\cos(Kr)]}{\frac{4}{R^2} + K^2} \right]_{r=0}^{r=\infty} \\
= \frac{1}{K} - \frac{K}{\frac{4}{R^2} + K^2} = \frac{1}{K \left(1 + \frac{K^2 R^2}{4} \right)} . \quad (2.67)
\end{aligned}
$$

2.4.6 Maximum Scattering Angle

The differential cross section $d\sigma_{\text{Ruth}}/d\Omega$ after inserting (2.67) into (2.66) is given as

$$
\begin{aligned}
\frac{d\sigma_{\text{Ruth}}}{d\Omega} &= \left| \frac{2m_\alpha}{\hbar^2} \frac{zZe^2}{4\pi\varepsilon_0} \frac{1}{K^2} \frac{1}{\left(1 + \frac{K^2 R^2}{4} \right)} \right|^2 \\
&= \left(\frac{D_{\alpha-N}}{4} \right)^2 \frac{1}{\sin^4 \frac{\theta}{2}} \frac{1}{\left[1 + \left(\frac{pR \sin\left(\frac{1}{2}\theta\right)}{\hbar} \right)^2 \right]^2}
\end{aligned}
$$

$$= \left(\frac{D_{\alpha-N}}{4}\right)^2 \frac{1}{\sin^4 \frac{\theta}{2}} \frac{1}{\left[1 + \left(\frac{\sin\left(\frac{1}{2}\theta\right)}{\theta_{max}}\right)^2\right]^2}, \qquad (2.68)$$

after first inserting the expression for $K = 2(p/\hbar)\sin(\theta/2)$ of (1.124) and then defining the maximum (cutoff) scattering angle θ_{max}, beyond which the scattering cross section falls significantly below the $\sin^{-4}\left(\frac{1}{2}\theta\right)$ expression, as

$$\theta_{max} = \frac{\hbar}{pR} = \frac{\hbar c}{R_0 \sqrt[3]{A}\sqrt{E_K(E_K + 2E_0)}}, \qquad (2.69)$$

where again

E_K is the kinetic energy of the α-particle,
E_0 is the rest energy of the α-particle (3727.4 MeV).

The maximum cutoff scattering angle θ_{max} can be defined quantum-mechanically based on the Heisenberg uncertainty principle of (1.130) as follows: When the classical trajectory of the incident particle is localized within $\Delta z \approx R$, the corresponding uncertainty on the transverse momentum of the particle is $\Delta p \geq \hbar/R$, leading to

$$\theta_{max} = \frac{\Delta p}{p} \approx \frac{\hbar}{pR} = \frac{\lambda}{R} = \frac{\lambda}{2\pi R}, \qquad (2.70)$$

where λ is the de Broglie wavelength of the incident α-particle and we assume that $\theta_{max} \ll 1$.

2.4.7 General Relationships for Differential Cross Section in Rutherford Scattering

In each Rutherford collision the angular deflections obey the Rutherford expression with cutoff at θ_{min} and θ_{max} given by (2.57) and (2.69), respectively. The typical value for $\hbar/p$ in the two expressions can be estimated for α-particles with a typical kinetic energy of 5.5 MeV as follows

$$\frac{\hbar}{p} = \frac{\hbar c}{\sqrt{E_K(E_K + 2E_0)}} \approx \frac{197.3 \text{ MeV·fm}}{\sqrt{5.5(5.5 + 2\times3727.4)} \text{ MeV}} \approx 1 \text{ fm}, \qquad (2.71)$$

where we use the expression for p given in (1.64). Inserting the value for $\hbar/p \approx 1$ fm into (2.58) and (2.69), respectively, for a typical α-particle kinetic energy of 5.5 MeV, combined with appropriate values for $a_{TF} = 0.123\times10^5$ fm and $R = 7.3$ fm, results in the following angles θ_{min} and θ_{max} for the gold atom

$$\theta_{min} = \frac{\hbar}{pa_{TF}} = \frac{\hbar c}{pc}\frac{\sqrt[3]{Z}}{a_0} \approx (1 \text{ fm})\frac{\sqrt[3]{79}}{0.5292 \times 10^5 \text{ fm}} \approx 8.1 \times 10^{-5} \text{ rad} \quad (2.72)$$

and

$$\theta_{max} = \frac{\hbar}{pR} = \frac{\hbar c}{pc}\frac{1}{R_0\sqrt[3]{A}} \approx (1 \text{ fm})\frac{1}{(1.25 \text{ fm})\sqrt[3]{197}} \approx 0.14 \text{ rad.} \quad (2.73)$$

We now see that the Rutherford scattering of α-particles on nuclei is governed by the following stipulation: $\theta_{min} \ll \theta_{max} \ll 1$, justifying our assumptions in (2.58) and (2.70) that both cutoff angles are much smaller than 1. We also note that the ratio $\theta_{max}/\theta_{min}$ is independent of α-particle kinetic energy E_K but depends on the atomic number Z and mass number A of the scatterer and is given as

$$\frac{\theta_{max}}{\theta_{min}} = \frac{a_{TF}}{R} \approx \frac{a_0}{R_0\sqrt[3]{ZA}} = \frac{0.5292 \times 10^5 \text{ fm}}{1.25 \text{ fm}}\frac{1}{\sqrt[3]{ZA}} \approx \frac{0.423 \times 10^5}{\sqrt[3]{ZA}}. \quad (2.74)$$

From (2.74) we estimate that the ratio $\theta_{max}/\theta_{min}$ ranges from $\sim 5 \times 10^4$ for low atomic number Z scatterers to $\sim 1.5 \times 10^3$ for high atomic number Z scatterers, since $\sqrt[3]{ZA}$ ranges from 1 at low Z to about 30 at high Z. We may thus conclude that $\theta_{max}/\theta_{min} \gg 1$ for all elements. For gold, the material used in Geiger–Marsden experiment, $\sqrt[3]{ZA}$ amounts to $\sim 1.76 \times 10^3$.

The differential cross section $d\sigma_{Ruth}/d\Omega$ for Rutherford scattering of 5.5 MeV α-particles on gold nuclei is plotted in Fig. 2.6 in the form $\left(D_{\alpha-N}^{-2}\right)d\sigma_{Ruth}/d\Omega$ against

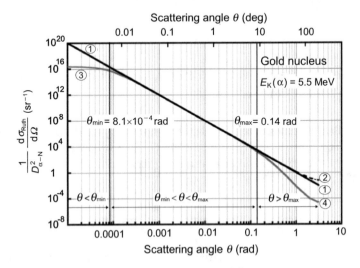

Fig. 2.6 Differential Rutherford scattering cross section $[(1/D_{\alpha-N}^2) \times (d\sigma_{Ruth}/d\Omega)]$ plotted against the scattering angle θ for 5.5 MeV α-particles interacting with gold. The minimum and maximum scattering angles $\theta_{min} = 8.1 \times 10^{-5}$ rad and $\theta_{max} = 0.14$ rad, respectively, are identified. For $\theta \to 0$ the value of the ordinate approaches $(1/\theta_{min}^4) \approx 2.32 \times 10^{16} \text{ (rad)}^{-4}$

the scattering angle θ in the range from 10^{-5} rad to π. As calculated in (2.72) and (2.73), $\theta_{min} = 8.1 \times 10^{-5}$ rad and $\theta_{max} = 0.14$ rad, respectively.

Three distinct regions can be identified on the graph: small θ; intermediate θ; and large θ.

1. In the intermediate region $\theta_{min} \ll \theta \ll \theta_{max}$ where $\theta_{min} \ll \theta_{max} \ll 1$, the simple Rutherford differential scattering expressions given by (2.38) and (2.40) apply, resulting in a straight line [curves (1) and (2)] on the log–log plot

$$\frac{1}{D_{\alpha-N}^2} \frac{d\sigma_{Ruth}}{d\Omega} = \frac{1}{16} \frac{1}{\sin^4(\theta/2)} \approx \frac{1}{\theta^4}. \tag{2.75}$$

2. In the small angle θ region ($\theta < \theta_{min}$), as a result of nuclear screening and after applying the Thomas–Fermi atomic model, the differential cross section is given by

$$\frac{1}{D_{\alpha-N}^2} \frac{d\sigma_{Ruth}}{d\Omega} = \frac{1}{\left(\theta^2 + \theta_{min}^2\right)^2}, \tag{2.76}$$

resulting in curve (3) in Fig. 2.6 and converging to a finite value of $\theta_{min}^{-4} = 2.32 \times 10^{16}$ rad^{-4} for $\theta = 0$.

3. In the large angle θ region where $\theta > \theta_{max}$, (2.40) represented by curve (1) is still linear, while (2.38) results in curve (2). A correction for finite nuclear size and nuclear penetration of the scattered particle lowers the value of the differential cross section from the value predicted by the simple Rutherford equation and results from (2.68) in

$$\frac{1}{D_{\alpha-N}^2} \frac{d\sigma_{Ruth}}{d\Omega} = \frac{1}{16} \frac{1}{\sin^4(\theta/2)} \frac{1}{\left(1 + \dfrac{\sin^2(\theta/2)}{\theta_{max}^2}\right)^2}, \tag{2.77}$$

shown as curve (4) in Fig. 2.6.

2.4.8 Total Rutherford Scattering Cross Section

The total cross section for Rutherford scattering can be approximated by using the small angle approximation and integrating (2.59) over the complete solid angle to obtain

$$\sigma_{Ruth} = \int \frac{d\sigma_{Ruth}}{d\Omega} d\Omega = 2\pi \int_0^{\theta_{max}} \frac{d\sigma_{Ruth}}{d\Omega} \sin\theta \, d\theta \approx 2\pi D_{\alpha-N}^2 \int_0^{\theta_{max}} \frac{\theta \, d\theta}{\left(\theta^2 + \theta_{min}^2\right)^2}$$

$$= \pi D_{\alpha-N}^2 \int_0^{\theta_{max}} \frac{d\left(\theta^2 + \theta_{min}^2\right)}{\left(\theta^2 + \theta_{min}^2\right)^2} = \pi D_{\alpha-N}^2 \left\{ \frac{1}{\theta_{min}^2} - \frac{1}{\theta_{max}^2 + \theta_{min}^2} \right\}$$

$$= \pi D_{\alpha-N}^2 \frac{1}{\theta_{min}^2} \left\{ 1 - \frac{1}{1 + \left(\frac{\theta_{max}}{\theta_{min}}\right)^2} \right\}. \tag{2.78}$$

In each Rutherford collision the angular deflections obey the Rutherford expression with cutoffs at θ_{min} and θ_{max} given by (2.72) and (2.73), respectively. The typical value for $\hbar/p$ in the expressions for θ_{min} and θ_{max} was estimated in (2.71) for α-particles with a typical kinetic energy of 5.5 MeV as 1 fm while θ_{min} and θ_{max} for gold atom were estimated in (2.72) and (2.73) as 8.1×10^{-5} rad and 0.14 rad, respectively. The cutoff angles θ_{min} and θ_{max} thus satisfy the Rutherford condition stipulating that $\theta_{min} \ll \theta_{max} \ll 1$ and, since also $\theta_{max}/\theta_{min} \gg 1$, the total cross section for Rutherford scattering given in (2.78) can be simplified, after inserting (2.12) and (2.57), to read

$$\sigma_{Ruth} \approx \frac{\pi D_{\alpha-N}^2}{\theta_{min}^2} = \pi a_{TF}^2 \left(\frac{D_{\alpha-N}}{(\hbar/p)}\right)^2 = \pi a_{TF}^2 \left\{ \frac{2zZe^2}{4\pi\varepsilon_0 \hbar v_i} \right\}^2. \tag{2.79}$$

The parameters of (2.79) are as follows:

a_{TF} is the Thomas–Fermi atomic radius

Z is the atomic number of the absorber foil,

z is the atomic number of the α-particle,

v_i is the initial velocity of the α-particle, equal to the final velocity of the α-particle,

$D_{\alpha-N}$ is the distance of closest approach between the α-particle and nucleus in a direct-hit head-on collision ($b = 0$).

For the Geiger–Marsden experiment with 5.5 MeV α-particles scattered on a 1 μm thick gold foil we calculate the following total scattering cross section

$$\sigma_{Ruth} = \frac{\pi D_{\alpha-N}^2}{\theta_{min}^2} = \frac{\pi(41 \times 10^{-13} \text{ cm})^2}{(8.1 \times 10^{-5})^2} = 8.05 \times 10^9 \text{ b}. \tag{2.80}$$

2.4.9 Mean Square Scattering Angle for Single Rutherford Scattering

Rutherford scattering is confined to very small angles and for energetic α-particles $\theta_{max} \ll 1$ rad. An α-particle traversing a gold foil will undergo a large number of small angle θ scatterings and emerge from the foil with a small cumulative angle Θ that represents a statistical superposition of a large number of small angle deflections.

Large angle scattering events, on the other hand, are rare and a given α-particle will undergo at most only one such rare scattering event while traversing the gold foil. As discussed in Sect. 2.2, Geiger and Marsden found that only about 1 in 10^4 α-particles traverses the 1 μm thick gold foil with a scattering angle Θ exceeding $90°$. The range of Rutherford angular scattering is thus divided into two distinct regions:

1. Single scattering events with large angle θ.
2. Multiple scattering events resulting in a small cumulative angle Θ.

In the multiple-scatter region, the mean square angle for single scattering $\overline{\theta^2}$ is

$$\overline{\theta^2} = \frac{\int \theta^2 \dfrac{d\sigma_{Ruth}}{d\Omega} d\Omega}{\int \dfrac{d\sigma_{Ruth}}{d\Omega} d\Omega} = \frac{\int \theta^2 \dfrac{d\sigma_{Ruth}}{d\Omega} d\Omega}{\sigma_{Ruth}}. \tag{2.81}$$

The denominator in (2.81) is the total Rutherford scattering cross section σ_{Ruth}, given in (2.79). It is proportional to the square of the distance of closest approach $(D_{\alpha-N})^2$ and inversely proportional to θ^2_{min}. The integral in the numerator of (2.81) is in the small angle approximation ($\sin\theta \approx \theta$) calculated as follows

$$\int \theta^2 \frac{d\sigma_{Ruth}}{d\Omega} d\Omega = 2\pi D^2_{\alpha-N} \int_0^{\theta_{max}} \frac{\theta^2 \sin\theta \, d\theta}{\left(\theta^2 + \theta^2_{min}\right)^2} \approx 2\pi D^2_{\alpha-N} \int_0^{\theta_{max}} \frac{\theta^3 d\theta}{\left(\theta^2 + \theta^2_{min}\right)^2}$$

$$= \pi D^2_{\alpha-N} \int_0^{\theta_{max}} \frac{\left(\theta^2 + \theta^2_{min}\right) d\left(\theta^2 + \theta^2_{min}\right)}{\left(\theta^2 + \theta^2_{min}\right)^2} - \pi D^2_{\alpha-N} \int_0^{\theta_{max}} \frac{\theta^2_{min} d\left(\theta^2 + \theta^2_{min}\right)}{\left(\theta^2 + \theta^2_{min}\right)^2}$$

$$= \pi D^2_{\alpha-N} \left\{ \ln\left(\theta^2 + \theta^2_{min}\right) + \frac{\theta^2_{min}}{\theta^2 + \theta^2_{min}} \right\}_0^{\theta_{max}}$$

$$= \pi D^2_{\alpha-N} \left\{ \ln\left(1 + \frac{\theta^2_{max}}{\theta^2_{min}}\right) + \frac{\theta^2_{min}}{\theta^2_{max} + \theta^2_{min}} - 1 \right\}. \tag{2.82}$$

The mean square angle $\overline{\theta^2}$ of (2.81) for a single scattering event, after incorporating the Rutherford total scattering cross section given in (2.79), is then given by the following relationship

$$\overline{\theta^2} = \theta^2_{min} \ln\left(1 + \frac{\theta^2_{max}}{\theta^2_{min}}\right) - \frac{\theta^2_{min}\theta^2_{max}}{\theta^2_{min} + \theta^2_{max}} = \theta^2_{min} \ln\left(1 + \frac{\theta^2_{max}}{\theta^2_{min}}\right) - \left(\frac{1}{\theta^2_{min}} + \frac{1}{\theta^2_{max}}\right)^{-1}. \tag{2.83}$$

The expression in (2.83) can be simplified using Rutherford scattering condition stipulating that $\theta_{min} \ll \theta_{max} \ll 1$ to obtain

$$\overline{\theta^2} \approx 2\,\theta_{min}^2\,\ln\frac{\theta_{max}}{\theta_{min}}. \tag{2.84}$$

For the Geiger–Marsden experiment with 5.5 MeV α-particles scattered on a gold foil we calculate the following mean square angle for single Rutherford scattering

$$\overline{\theta^2} \approx 2\theta_{min}^2\,\ln\frac{\theta_{max}}{\theta_{min}} = 2\times\left(8.1\times10^{-5}\,\text{rad}\right)^2\ln\frac{0.14}{8.1\times10^{-5}} = 9.8\times10^{-8}\,(\text{rad})^2, \tag{2.85}$$

resulting in the following root mean square scattering angle

$$\sqrt{\overline{\theta^2}} = 3.13\times10^{-4}\,\text{rad}. \tag{2.86}$$

2.4.10 Mean Square Scattering Angle for Multiple Rutherford Scattering

Since the successive scattering collisions are independent events, the *central-limit theorem* of statistics (see Sect. 2.7.1) shows that for a large number $n > 20$ of such collisions, the distribution in angle will be Gaussian around the forward direction [see (2.2)] with a cumulative mean square scattering angle $\overline{\Theta^2}$ related to the mean square scattering angle $\overline{\theta^2}$ for a single scattering event given in (2.83). The cumulative mean square angle $\overline{\Theta^2}$ and the mean square angle $\overline{\theta^2}$ for a single scattering event are related as follows

$$\overline{\Theta^2} = n\overline{\theta^2} \tag{2.87}$$

where n, the number of scattering events, is

$$n = \frac{N_a}{V}\sigma_{\text{Ruth}}t = \rho\frac{N_A}{A}\sigma_{\text{Ruth}}t = \pi\rho\frac{N_A}{A}\frac{D_{\alpha-N}^2}{\theta_{min}^2}t. \tag{2.88}$$

In (2.88) the parameters are as follows:

σ_{Ruth}	is the total Rutherford cross section given by (2.78),
N_a/V	is the number of atoms per volume equal to $\rho N_A/A$,
ρ	is the density of the foil material,
t	is the thickness of the foil,
A	is the atomic mass number,
N_A	is the Avogadro number ($N_A = 6.023\times10^{23}$ mol^{-1}),

$D_{\alpha-N}$ is the distance of closest approach between the α-particle and the nucleus in a direct hit interaction where $b = 0$ [see (2.12)],

θ_{min} is the cutoff angle defined in (2.57).

The mean square angle $\overline{\Theta^2}$ of the Gaussian distribution after combining (2.85), (2.87) and (2.88) is then given by

$$\overline{\Theta^2} = 2\pi\rho\frac{N_A}{A}t\,D^2_{\alpha-N}\,\ln\frac{\theta_{max}}{\theta_{min}}, \tag{2.89}$$

indicating that the mean square angle $\overline{\Theta^2}$ for multiple Rutherford scattering increases linearly with the foil thickness t. Inserting the expressions for θ_{min} and θ_{max} of (2.57) and (2.69), respectively, into (2.89), we now get the following expression for the mean square angle $\overline{\Theta^2}$ in Rutherford scattering

$$\overline{\Theta^2} = 2\pi\rho\frac{N_A}{A}t\,D^2_{\alpha-N}\,\ln\frac{1.4a_0}{R_0\sqrt[3]{AZ}} = 2\pi\rho\frac{N_A}{A}t\left\{\frac{zZe^2}{4\pi\varepsilon_0 E_K}\right\}^2\ln\frac{1.4a_0}{R_0\sqrt[3]{AZ}}, \tag{2.90}$$

where $a_0 = 0.5292$ Å and $R_0 = 1.25$ fm are the Bohr radius constant of (3.4) and the nuclear radius constant of (1.26), respectively.

For the Geiger–Marsden experiment with 5.5 MeV α-particles scattered on a gold foil we calculate the following mean square angle for multiple Rutherford scattering

$$\begin{aligned}
\overline{\Theta^2} &= 2\pi\rho\frac{N_A}{A}t\,D^2_{\alpha-N}\,\ln\frac{\theta_{max}}{\theta_{min}}\\
&= 2\pi\times 19.3\ (\text{g/cm}^3)\frac{6.022\times 10^{23}\ (\text{mol})^{-1}}{197\ (\text{g/mol})}(10^{-4}\ \text{cm})\times(41\times 10^{-13}\ \text{cm})^2\\
&\quad\times\ln\frac{0.14}{8.1\times 10^{-5}} = 46.4\times 10^{-4}\ (\text{rad})^2,
\end{aligned} \tag{2.91}$$

resulting in the following root mean square scattering angle for multiple scattering

$$\sqrt{\overline{\Theta^2}} = 0.068\ \text{rad} = 3.9°. \tag{2.92}$$

2.4.11 Importance of the Rutherford Scattering Experiment

Tables 2.1, 2.2 and 2.3 summarize the parameters of the Geiger–Marsden α-particle scattering experiment, listing the important parameters of the α-particles; the gold atom; and Rutherford scattering, respectively, based on expressions derived in this section. All data are calculated for Rutherford scattering of 5.5 MeV α-particles on gold nuclei.

Table 2.1 Properties of α-particles used in the Geiger–Marsden experiment

	Properties of α-particles	
Atomic number	$z = 2$	
Rest energy	$E_0 = m_\alpha c^2 = 3727.4 \text{ MeV}$	
Kinetic energy	$E_K = E - E_0 = 5.5 \text{ MeV}$	
Normalized velocity	$\dfrac{v_\alpha}{c} = \sqrt{1 - \dfrac{1}{\left(1 + \dfrac{E_K}{m_\alpha c^2}\right)^2}} = 0.054$	(2.7)
Momentum	$p = \dfrac{1}{c}\sqrt{E^2 - E_0^2} = \dfrac{1}{c}\sqrt{E_K^2 + 2E_K E_0} = 202.6 \text{ MeV}/c$	(1.64)
Reduced Planck constant divided by momentum	$\dfrac{\hbar}{p} = \dfrac{\hbar c}{pc} = \dfrac{197.3 \text{ MeV·fm}}{202.6 \text{ MeV}} = 0.974 \text{ fm} \approx 1 \text{ fm}$	(2.71)

Table 2.2 Properties of gold atom of importance in Rutherford scattering

	Properties of gold atom $^{197}_{79}\text{Au}$	
Atomic number	$Z = 79$	
Atomic mass number	$A = 197$	
Density	$\rho = 19.3 \text{ g/cm}^3$	
Thomas–Fermi radius	$a_{TF} = \dfrac{a_0}{\sqrt[3]{Z}} = \dfrac{0.5292 \text{ Å}}{\sqrt[3]{79}} = 0.123 \text{ Å}$	(2.49)
Nuclear radius	$R = R_0 \sqrt[3]{A} = (1.25 \text{ fm})\sqrt[3]{197} = 7.3 \text{ fm}$	(2.13)
Thickness of gold foil	$t = 10^{-4} \text{ cm} = 1 \text{ μm}$	

The α-particle scattering experiment on a thin gold foil conducted by Hans Geiger and Ernest Marsden under the guidance of Ernest Rutherford seems rather mundane, yet it is one of the most important experiments in the history of physics. Nature provided Geiger and Marsden with ideal conditions to probe the nucleus with radon-222 α-particles with kinetic energy of 5.5 MeV.

The radon-222 α-particles allowed penetration of the atom but their energy was neither too large to cause nuclear penetration and associated nuclear reactions nor large enough to require relativistic treatment of the α-particle velocity. Since artificial nuclear reactions and the relativistic mechanics were not understood in 1909 when the Geiger–Marsden experiment was carried out, Rutherford would not be able to solve with such elegance the atomic model question, if the kinetic energy of the α-particles used in the experiment was much larger than 5.5 MeV thereby causing penetration of the gold nucleus or much smaller than 5.5 MeV thereby preventing penetration of the atom.

Table 2.3 Parameters of Geiger–Marsden experiment for α-particles with kinetic energy of 5.5 MeV undergoing Rutherford scattering on a 1 μm thick gold foil

	Parameters of Rutherford scattering	
Distance of closest approach	$D_{\alpha-N} = \dfrac{zZe^2}{4\pi\varepsilon_0}\dfrac{1}{E_K} = 41$ fm	(2.12)
Minimum scattering angle	$\theta_{min} = \dfrac{\hbar}{p}\dfrac{1}{a_{TF}} = \dfrac{1 \text{ fm}}{0.123\times 10^5 \text{ fm}}$ $= 8.1\times 10^{-5}$ rad	(2.72)
Maximum scattering angle	$\theta_{max} = \dfrac{\hbar}{p}\dfrac{1}{R} = \dfrac{1 \text{ fm}}{7.3 \text{ fm}} \approx 0.14$ rad	(2.73)
Ratio $\dfrac{\theta_{max}}{\theta_{min}}$	$\dfrac{\theta_{max}}{\theta_{min}} = \dfrac{0.14}{8.1\times 10^{-5}} = 1.766\times 10^3$	(2.74)
Differential Rutherford cross section at $\theta = 0$	$\left.\dfrac{d\sigma_{Ruth}}{d\Omega}\right|_{\theta=0} = \dfrac{D_{\alpha-N}^2}{\theta_{min}^4} = 3.9\times 10^{17}$ b/sr	(2.76)
Rutherford cross section	$\sigma_{Ruth} = \pi\dfrac{D_{\alpha-N}^2}{\theta_{min}^2} = 8.05\times 10^9$ b	(2.79)
Mean square scattering angle for single scattering	$\overline{\theta^2} \approx 2\theta_{min}^2 \ln\dfrac{\theta_{max}}{\theta_{min}} = 9.8\times 10^{-8}$ (rad)^2	(2.80)
Root mean square angle for single scattering	$\sqrt{\overline{\theta^2}} = 3.13\times 10^{-4}$ rad	(2.86)
Mean square scattering angle for multiple scattering	$\overline{\Theta^2} = 2\pi\rho\dfrac{N_A}{A}tD_{\alpha-N}^2 \ln\dfrac{\theta_{max}}{\theta_{min}} = n\overline{\theta^2}$ $= 46.4\times 10^{-4}$ (rad)^2	(2.89)
Root mean square angle for multiple scattering	$\sqrt{\overline{\Theta^2}} = 0.068$ rad $= 3.9°$	(2.92)
Number of scattering events	$n = \pi\rho\dfrac{N_A}{A}\dfrac{D_{\alpha-N}^2}{\theta_{min}^2}t \approx 47500$	(2.88)

Geiger–Marsden experiment provided the stimulus for development of nuclear physics and will remain forever on the short list of milestones in physics. It also served as the first known method for estimation of the upper limit of nuclear size through the calculation of the distance of closest approach $D_{\alpha-N}$ but was soon eclipsed by new and more sophisticated scattering experiments that are now used for this purpose. However, the basic principles of the original technique are still used in the so-called Rutherford backscattering spectroscopy (RBS) which is an analytical tool used in materials science for determining structure and composition of materials by measuring backscattering of a beam of high energy ions (protons or helium ions) accelerated in a linear accelerator.

2.5 Mott Scattering

In comparison with heavy charged particles, energetic electrons are much better suited for studies of nuclear size and charge distribution. However, to obtain agreement with experimental results, the theoretical treatment of the scattering process must go beyond the rudimentary Rutherford–Coulomb point-source scattering approach and account for various other parameters such as:

1. Electron spin
2. Relativistic effects
3. Quantum effects
4. Recoil of the nucleus
5. Nuclear spin
6. Finite size of the nucleus

Accounting for these additional parameters refines the scattering theory beyond the level achieved by Rutherford but also makes it significantly more complex. For example, the finite size of the nucleus implies that the target is not a point charge but consists of its own structure containing protons and neutrons which, in turn, have their own constituents referred to as quarks.

At low electron energies where the electron does not penetrate the nucleus the electron scattering by the nucleus can be described with the standard Rutherford-type scattering formula [see (2.39)]

$$
\frac{d\sigma_{\text{Ruth}}}{d\Omega} = \left(\frac{D_{e-N}}{4}\right)^2 \frac{1}{\sin^4 \dfrac{\theta}{2}} = \left(\frac{D_{e-N}}{2}\right)^2 \frac{1}{(1 - \cos \theta)^2} = \left(\frac{Z\alpha\hbar c}{4E_K}\right)^2 \frac{1}{\sin^4 \dfrac{\theta}{2}},
$$

$$(2.93)$$

where

$$
D_{e-N} = \frac{Ze^2}{4\pi\varepsilon_0 \dfrac{1}{2}mv^2}
$$

$$(2.94)$$

is here referred to as the effective characteristic distance for the electron–nucleus scattering (see Sect. 2.6.2) in contrast to the distance of closest approach $D_{\alpha-N}$ used in Rutherford scattering of α-particles, as discussed in Sect. 2.3.2. In the expression for D_{e-N} of (2.93), m is the total mass of the incident electron in contrast to m_e which is the rest mass of the electron ($m = m_e/\sqrt{1 - (v/c)^2}$), and v is the velocity of the incident electron.

At very high electron energies (above 100 MeV) electrons are highly relativistic and two corrections to the simple Rutherford-type formula (2.93) are required: correction for electron spin and correction for nuclear recoil.

2.5.1 Correction for Electron Spin

The effect of the electron magnetic moment introduces to the Rutherford relationship for electron scattering given in (2.93) a spin correction factor expressed as

$$f_{\text{spin}} = 1 - \beta^2 \sin^2 \frac{\theta}{2}, \qquad (2.95)$$

which, for relativistic electrons where $\beta = v/c \to 1$, simplifies to

$$f_{\text{spin}} \approx 1 - \sin^2 \frac{\theta}{2} = \cos^2 \frac{\theta}{2} = \frac{1 + \cos \theta}{2}. \qquad (2.96)$$

For relativistic electrons ($v \approx c$) the spin correction factor f_{spin} does not depend on the kinetic energy E_K of the incident electron but depends on the scattering angle θ and, as shown in Fig. 2.7, ranges from $f_{\text{spin}} = 1$ for $\theta = 0$ through $f_{\text{spin}} = 0.85$ for $\theta = 45°$, $f_{\text{spin}} = 0.5$ for $\theta = 90°$ and $f_{\text{spin}} = 0.146$ for $\theta = 135°$ to $f_{\text{spin}} = 0$ for $\theta = 180°$. Thus, at small scattering angles θ the electron spin effects are negligible, while at large scattering angles they significantly decrease the differential scattering cross section from that given by the Rutherford expression of (2.93), essentially disallowing electron backscattering at $\theta = 180°$.

Figure 2.8 plots, for a point-like Coulomb scattering source, the differential cross section for electron–nucleus scattering without spin correction in curve (1) and with spin correction in curve (2). The following expressions are plotted:

$$\frac{16}{D_{e-N}^2} \frac{d\sigma_{\text{Mott}}}{d\Omega} = \frac{16}{D_{e-N}^2} \frac{d\sigma_{\text{Ruth}}}{d\Omega} f_{\text{spin}} = \frac{1}{\sin^4 \dfrac{\theta}{2}} \times \frac{1 + \cos \theta}{2}, \qquad (2.97)$$

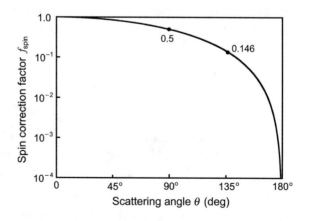

Fig. 2.7 Spin correction factor f_{spin} of (2.96) against scattering angle θ for electron–nucleus (Mott) scattering

Fig. 2.8 Normalized Mott differential scattering cross section $d\sigma_{Mott}/d\Omega$ against scattering angle θ. Curve (1) is the Rutherford component without electron spin correction (i.e., $f_{spin} = 1$); curve (2) is for the Rutherford component corrected for the electron spin effect given as $f_{spin} = \cos^2\left(\frac{1}{2}\theta\right) = \frac{1}{2}(1 + \cos\theta)$

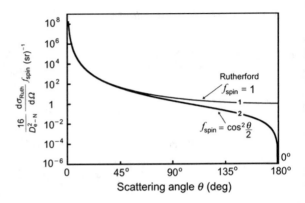

where

curve (1)   is without spin correction or $f_{spin} = 1$ independent of θ

curve (2)   is with spin correction given in Fig. 2.7 and (2.96) as $f_{spin} = \cos^2(\theta/2)$.

2.5.2 Correction for Recoil of the Nucleus

The nuclear recoil correction factor f_{recoil} is given as the ratio between the kinetic energy of the scattered (recoil) electron E'_K and the kinetic energy of the incident electron E_K. The kinetic energy of the scattered electron E'_K is determined from considerations of conservation of energy and momentum during the scattering process. The considerations resemble the derivation of scattered photon energy and recoil electron kinetic energy in Compton effect (see Sect. 7.3.3).

The schematic diagram of the scattering process is shown in Fig. 2.9 where an incident electron with momentum **p** and kinetic energy E_K is scattered, essentially elastically, through a scattering angle θ to end with momentum **p**$'$ and kinetic energy E'_K. To be useful as a nuclear probe and to have a relatively small de Broglie wavelength (of the order of 10 fm) the electron must be of sufficiently high kinetic energy

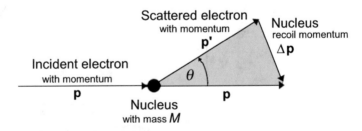

Fig. 2.9 Schematic representation of electron–nucleus (Mott) scattering

and is thus relativistic. The conservation of energy during the scattering process is written as follows

$$Mc^2 + E_K + m_e c^2 = \Delta E_K + Mc^2 + E'_K + m_e c^2 \tag{2.98}$$

or

$$E_K = \Delta E_K + E'_K, \tag{2.99}$$

where

ΔE_K is the recoil kinetic energy transferred from the incident electron to the nucleus,

Mc^2 is the rest energy of the nucleus,

$m_e c^2$ is the rest energy of the incident electron.

Using the law of cosines on the vector diagram of Fig. 2.9, the conservation of momentum $\mathbf{p} = \mathbf{p'} + \Delta\mathbf{p}$, with an assumption that in elastic scattering $|\mathbf{p}| \approx |\mathbf{p'}|$, can be stated as follows

$$|\Delta\mathbf{p}|^2 = |\mathbf{p}|^2 + |\mathbf{p'}|^2 - 2\,|\mathbf{p}|\,|\mathbf{p'}|\cos\theta \approx 2\,|\mathbf{p}|^2\,(1 - \cos\theta), \tag{2.100}$$

where $\Delta\mathbf{p}$ is the recoil momentum of the nucleus. The recoil kinetic energy ΔE_K is given as

$$\Delta E_K \approx \frac{|\Delta\mathbf{p}|^2}{2M} = \frac{|\mathbf{p}|^2\,(1 - \cos\theta)}{M} = \frac{E_K^2}{Mc^2}\left(1 + \frac{2m_e c^2}{E_K}\right)(1 - \cos\theta), \tag{2.101}$$

using the expression for the incident electron momentum magnitude $|\mathbf{p}| = p$ of (1.64) given as

$$p = \frac{1}{c}\sqrt{E_K^2 + 2E_K m_e c^2} = \frac{E_K}{c}\sqrt{1 + \frac{2m_e c^2}{E_K}}. \tag{2.102}$$

Recognizing that in the electron scattering experiment $m_e c^2 \ll E_K \ll Mc^2$ and $E'_K = E_K - \Delta E_K$, we now get the following expression for the recoil correction f_{recoil}

$$f_{\text{recoil}} = \frac{E'_K}{E_K} = \frac{1}{1 + \dfrac{E_K}{Mc^2}(1 - \cos\theta)} = \frac{1}{1 + \dfrac{2E_K}{Mc^2}\sin^2\dfrac{\theta}{2}}. \tag{2.103}$$

The recoil correction factor f_{recoil} depends on the kinetic energy E_K of the incident electron, the rest mass of the scattering nucleus Mc^2, and the scattering angle θ. For small scattering angles $f_{\text{recoil}} \approx 1$ irrespective of E_K and then, for a given E_K/Mc^2, it decreases with θ increasing from 0 to 180°. Since generally $Mc^2 \gg E_K$, unless we

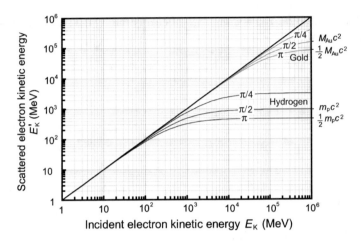

Fig. 2.10 Scattered electron kinetic energy $E'_K = f_{recoil} E_K$ against the incident electron kinetic energy E_K for Mott scattering on hydrogen and gold nuclei for four different scattering angles $(0, \frac{1}{4}\pi, \frac{1}{2}\pi,$ and $\pi)$. The recoil correction f_{recoil} is given in (2.103)

are dealing with very low atomic number scatterer and very high incident electron kinetic energy, it is reasonable to assume that $f_{recoil} \approx 1$.

Figure 2.10 plots the relationship between the kinetic energy of the scattered electron E'_K and the kinetic energy of the incident electron E_K for four scattering angles $(0, \frac{1}{4}\pi, \frac{1}{2}\pi,$ and $\pi)$ of Mott scattering on hydrogen and gold nuclei in the kinetic energy range from 1 MeV to 10^6 MeV. In the electron kinetic energy of interest in medical physics (up to 30 MeV), the kinetic energy of the scattered electron E'_K is equal to the kinetic energy of the incident electron E_K for all scattering materials and all scattering angles θ. This implies that $f_{recoil} = 1$ for all situations of interest in medical physics.

From Fig. 2.10 we arrive at several other conclusions, of little interest in medical physics but relevant to high energy physics:

1. $f_{recoil} = 1$ for $\theta = 0$ at all kinetic energies of the incident electron from 0 to ∞.
2. For backscattered electron $(\theta = \pi)$, its kinetic energy saturates at $\frac{1}{2}Mc^2$ where M is the rest mass of the recoil nucleus. This results in $f_{recoil} \to 0$ but happens only at very large incident electron kinetic energies, way outside of the energy region of interest in medical physics.
3. Similarly, for side-scattered electron $(\theta = \frac{1}{2}\pi)$, its kinetic energy saturates at Mc^2 at very high incident electron kinetic energy.
4. The findings in points (2) and (3) are similar to relationships observed in Compton scattering (see Sect. 7.3.3) except that in Compton scattering the recoil particle is an electron which has a significantly smaller rest energy than a nucleus. This makes the recoil of the Compton electron of great importance to medical physics, since a significant fraction of the incident photon energy is transferred to the recoil electron in the photon energy range of interest in medical physics.

2.5.3 Differential Cross Section for Mott Scattering of Electrons on Point-Like Atomic Nuclei

Accounting for the spin correction of (2.96) and the nuclear recoil correction of (2.103) we now write the Mott expression for the differential cross section in electron–nucleus scattering as

$$\frac{d\sigma_{\text{Mott}}}{d\Omega} = \frac{d\sigma_{\text{Ruth}}}{d\Omega} f_{\text{spin}} f_{\text{recoil}} = \frac{d\sigma_{\text{Ruth}}}{d\Omega} \left\{ \cos^2 \frac{\theta}{2} \right\} \times \frac{1}{1 + \frac{E_K}{Mc^2}(1 - \cos\theta)}, \quad (2.104)$$

where $d\sigma_{\text{Ruth}}/d\Omega$ is the Rutherford electron–nucleus scattering formula given in (2.93) and valid at very low electron kinetic energies. The most important component of (2.104) is the Rutherford component; the product of the two corrections to the Rutherford component (the electron spin quantum effect f_{spin} and the nuclear recoil f_{recoil}) is of the order of unity except when the scattering angle θ is close to $180°$ or when the kinetic energy of the incident electron is very large.

2.5.4 Hofstadter Correction for Finite Nuclear Size and the Form Factor

Figure 2.11 shows, for scattering of 125 MeV electrons on gold nuclei, several differential cross sections plotted against the scattering angle θ:

1. Curve (R) represents the calculated simple Rutherford differential cross section (2.93) assuming a point-like Coulomb field and ignoring the electron spin effects as well as nuclear recoil.
2. Curve (M) represents the Mott differential scattering cross section (2.104) assuming a point-like Coulomb source and incorporating corrections for electron spin (f_{spin}) and nuclear recoil ($f_{\text{recoil}} \approx 1$).
3. Data points represent data that *Robert Hofstadter* measured in the early 1960s at Stanford University. While at small scattering angles Hofstadter's measurements agree with the Mott theory, for scattering angles θ exceeding $45°$ the measured points show significantly lower values than the theory, and the discrepancy increases with increasing θ.

Hofstadter carried out extensive experimental and theoretical studies of electron–nucleus scattering and for this work received a Nobel Prize in Physics in 1961. He explained the discrepancy between his measured data and Mott theory of Fig. 2.11 by expanding the Mott expression of (2.104) to account for the finite size of the nucleus using a form factor $F(K)$ correction. The experimental differential cross section for elastic electron–nucleus scattering then becomes expressed as

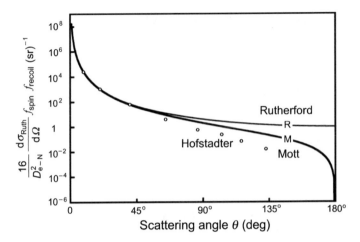

Fig. 2.11 Elastic scattering of 125 MeV electrons on gold nuclei. Curve (R) is for data calculated with Rutherford equation (2.93) without spin or nuclear recoil correction. Curve (M) is for Rutherford equation incorporating spin correction of (2.96) and nuclear recoil correction of (2.103). Data points are Hofstadter's measured data

$$\frac{d\sigma_{exp}}{d\Omega} = \frac{d\sigma_{Mott}}{d\Omega} |F(K)|^2 = \frac{d\sigma_{Ruth}}{d\Omega} \left\{ \cos^2 \frac{\theta}{2} \right\} \times \frac{1}{1 + \frac{E_K}{Mc^2}(1 - \cos\theta)} |F(K)|^2,$$

$$(2.105)$$

where K is proportional to the momentum transferred from incident electron to the nucleus, or

$$K = |\mathbf{K}| = \frac{1}{\hbar}\sqrt{|\mathbf{p}_i|^2 + |\mathbf{p}_f|^2 - 2|\mathbf{p}_i||\mathbf{p}_f|\cos\theta} = \frac{1}{\hbar}2p\sin\frac{\theta}{2} = \frac{2}{\lambda}\sin\frac{\theta}{2}. \quad (2.106)$$

The form factor $F(K)$ represents a Fourier transform of the nuclear charge density distribution $\rho(r)$ assumed to be spherically symmetric. In the Born approximation (see Sect. 1.23.6), $F(K)$ is expressed as

$$F(K) = \int\int\int \rho(r) e^{i\mathbf{Kr}} d\mathcal{V} = \int_0^\infty \int_0^\pi \int_0^{2\pi} \rho(r) e^{iKr\cos\theta} r^2 \, dr \sin\theta \, d\theta \, d\phi$$

$$= 2\pi \int_0^\infty r^2 \rho(r) \left\{ \int_{-1}^1 e^{iKr\cos\theta} d(\cos\theta) \right\} dr$$

$$= 2\pi \int_0^\infty r^2 \rho(r) \frac{e^{iKr} - e^{-iKr}}{iKr} dr = 4\pi \int_0^\infty r^2 \rho(r) \frac{\sin Kr}{Kr} dr, \quad (2.107)$$

with the normalization

$$\int \rho(r)\mathrm{d}\mathcal{V} = \int\limits_{0}^{\infty}\int\limits_{-1}^{+1}\int\limits_{0}^{2\pi} r^2\rho(r)\mathrm{d}\phi\,\mathrm{d}(\cos\theta)\mathrm{d}r = 4\pi\int\limits_{0}^{\infty}\rho(r)r^2\mathrm{d}r = 1. \qquad (2.108)$$

The magnitude of the form factor $F(K)$ is determined experimentally by comparing the measured cross section to the Mott cross section for point-like nucleus. The measurements are carried out for fixed electron beam energy at various scattering angles θ, i.e., at various values of $|\mathbf{K}| = K$. In practice, however, $F(K)$ can be measured only over a limited range of momentum transfer $|\mathbf{K}|/\hbar$ so that a full functional dependence of $F(K)$ cannot be determined for use in inverse Fourier transform which would yield the nuclear charge distribution $\rho(r)$

$$\rho(r) = \frac{1}{(2\pi)^3}\int F(K)e^{-i\mathbf{K}\mathbf{r}}\mathrm{d}\mathcal{V}. \qquad (2.109)$$

Much effort has been spent on experimental determination of nuclear size and charge distribution. The current consensus is that nuclei are not charged spheres with a sharply defined surface. Rather, the nuclear charge density $\rho(r)$ can be described by a Fermi function with two parameters (α and β) both of the order of 1 fm

$$\rho(r) = \frac{\rho(0)}{1 + e^{(r-\alpha)/\beta}}. \qquad (2.110)$$

As a guide to nuclear size, the nucleus is commonly approximated as a homogeneously charged sphere with radius R given as

$$R = R_0\sqrt[3]{A}, \qquad (2.111)$$

where A is the atomic mass number and R_0 is the nuclear radius constant amounting to 1.25 fm, as discussed in Sect. 1.16.1.

2.6 General Aspects of Elastic Scattering of Charged Particles

Most interactions of energetic charged particles as they traverse an absorber can be characterized as elastic Coulomb scattering between an energetic charged particle and the atoms of the absorber. The charged particles of interest in medical physics are either light charged particles such as electrons and positrons or heavy charged particles such as protons, α-particles, and heavier ions. Negative pions π^- were included in the group of charged particles as intermediate mass particles; however, interest in their use in radiotherapy has waned during the past 20 years with the advent of proton radiotherapy machines.

Charged particles can have elastic scattering interactions with orbital electrons as well as nuclei of the absorber atoms. The Coulomb force between the charged particle and the orbital electron or the nucleus of the absorber governs the elastic collisions and is either attractive or repulsive depending on the polarity of the interacting charged particles. In either case the trajectory of the projectile is a hyperbola: for an attractive Coulomb force the target is in the inner focus of the hyperbola; for a repulsive Coulomb force the target is in the outer focus of the hyperbola. An elastic collision between an α-particle and a nucleus of an absorber is shown schematically in Fig. 2.3 (Rutherford scattering) in Sect. 2.3.1; an elastic collision between a heavy charged particle and an orbital electron is shown schematically in Fig. 6.3 in Sect. 6.4.1.

Various investigators worked on theoretical aspects of elastic scattering of charged particles, most notably: *Rutherford* with *Geiger* and *Marsden* on α-particle scattering; *Mott* on electron–nucleus scattering as well as on non-relativistic electron–orbital electron scattering; *Møller* on relativistic electron–orbital electron scattering; *Bhabha* on positron–orbital electron scattering; and *Molière* on multiple scattering.

As shown in previous sections of this chapter, Rutherford scattering theory forms the basis for all charged particle single scattering theories. However, various corrections must be applied to Rutherford's formalism when moving from a discussion of classical α-particle scattering on an infinite-mass gold nucleus to a discussion of relativistic electrons scattered on finite size absorber nuclei. To highlight the various different projectiles, scattering centers, and corrections, in addition to Rutherford scattering, we speak of Mott scattering, Møller scattering, Bhabha scattering, Hofstadter scattering, etc. in single scattering events and of Molière scattering when we consider the composite effect of scattering on a large number of scattering centers.

The particle interactions in absorbers are characterized by various parameters that describe single and multiple scattering events:

1. For *single scattering* we define the differential and total scattering cross section, effective characteristic distance, and mean square scattering angle.
2. For *multiple scattering* we define the radiation length, mean square scattering angle, and mass scattering power.

2.6.1 Differential Scattering Cross Section for a Single Scattering Event

The differential scattering cross section $d\sigma/d\Omega$ for a single scattering event between two charged particles was discussed in relation to Rutherford scattering in Sect. 2.3. In the small scattering angle θ approximation where $\sin(\frac{1}{2}\theta) \approx \frac{1}{2}\theta$, the differential scattering cross section based on Rutherford's seminal work is in general expressed as

$$\frac{d\sigma}{d\Omega} = \frac{D^2}{(\theta^2 + \theta_{min}^2)^2},$$ (2.112)

where θ_{min} is a cutoff angle is a cutoff angle defined as the minimum angle below minimum angle below which the unscreened point Coulomb field expression is no longer valid; D is a scattering parameter generally referred to as the *characteristic scattering distance*, such as, for example, $D_{\alpha-N}$ defined as the distance of closest approach between the α-particle and the nucleus in Rutherford scattering.

2.6.2 Characteristic Scattering Distance

Each elastic scattering event between two particles (energetic projectile and stationary target) can be characterized by a scattering parameter referred to as the characteristic scattering distance D. This distance depends on the nature of the specific scattering event as well as on the physical properties of the scattered particle and the atomic number Z of the scattering material. The differential scattering cross section of (2.38) was derived for Rutherford scattering of α-particles on gold nuclei in Sect. 2.4.1 and is a good approximation for scattering of both heavy and light charged particles, as long as the characteristic scattering distance D, appropriate for the particular scattering event, is used in the calculations.

Characteristic Scattering Distance for Rutherford Scattering

In *Rutherford scattering* of a α-particle (projectile) on a nucleus (target) the characteristic scattering distance D, as shown in (2.12) and (2.30), is the distance of closest approach $D_{\alpha-N}$ between the α-particle and the nucleus in a direct-hit (head on) collision ($b = 0$, $\theta = \pi$)

$$D_{\alpha-N} = \frac{zZe^2}{4\pi\varepsilon_0} \frac{1}{(E_K)_i} = \frac{zZe^2}{4\pi\varepsilon_0} \frac{1}{\frac{m_\alpha v_\alpha^2}{2}} = \frac{2zZe^2}{4\pi\varepsilon_0} \frac{1}{p_\alpha v_\alpha},$$ (2.113)

where

z	is the atomic number of the α-particle,
Z	is the atomic number of the absorber atom,
$(E_K)_i$	is the initial kinetic energy of the α-particle,
m_α	is the mass of the α-particle,
v_α	is the initial and final velocity of the α-particle,
p_α	is the momentum of the α-particle.

Characteristic Scattering Distance for Electron–Nucleus Scattering

In *electron* (projectile)–*nucleus* (target) *elastic scattering* the characteristic scattering distance D_{e-N}, similarly to (2.113), is given as follows (note that $z = 1$ for the electron)

$$D_{\text{e-N}} = \frac{Ze^2}{4\pi\varepsilon_0} \frac{1}{\frac{mv^2}{2}} = \frac{2Ze^2}{4\pi\varepsilon_0} \frac{1}{pv} = \frac{2Ze^2\sqrt{1-\beta^2}}{4\pi\varepsilon_0(m_{\text{e}}c^2\beta^2)} = \frac{2Zr_{\text{e}}\sqrt{1-\beta^2}}{\beta^2}, \quad (2.114)$$

where

m is the total mass of the electron, i.e., $m = m_{\text{e}}/\sqrt{1-\beta^2} = \gamma m_{\text{e}}c^2$.
m_{e} is the rest mass of the electron.
β is the velocity of the electron normalized to c, i.e., $\beta = v/c$.
v is the velocity of the electron.
p is the momentum of the electron.
Z is the atomic number of the absorber.
r_{e} is the classical radius of the electron (2.82 fm).

Characteristic Scattering Distance for Electron–Orbital Electron Scattering

In *electron* (projectile)–*orbital electron* (target) *scattering* the characteristic scattering distance $D_{\text{e-e}}$, similarly to (2.114), is given by (note that $Z = 1$ for orbital electron)

$$D_{\text{e-e}} = \frac{e^2}{4\pi\varepsilon_0} \frac{1}{\frac{mv^2}{2}} = \frac{2e^2}{4\pi\varepsilon_0} \frac{1}{pv} = \frac{2e^2\sqrt{1-\beta^2}}{4\pi\varepsilon_0(m_{\text{e}}c^2\beta^2)} = \frac{2r_{\text{e}}\sqrt{1-\beta^2}}{\beta^2}, \quad (2.115)$$

where

m is the *total mass* of the electron, i.e., $m = m_{\text{e}}/\sqrt{1-\beta^2} = \gamma m_{\text{e}}c^2$.
m_{e} is the rest mass of the electron.
β is the velocity of the electron normalized to c, i.e., $\beta = v/c$.
v is the velocity of the electron.
p is the momentum of the electron.
r_{e} is the classical radius of the electron (2.82 fm).

Characteristic Scattering Distance for Electron–Atom Scattering

The *characteristic scattering distance* $D_{\text{e-a}}$ for electron (projectile) scattering on absorber atoms (target) has two components: the *electron–nucleus* (e–N) component of (2.114) and the *electron–orbital electron* (e–e) component of (2.115). The differential cross section for elastic electron scattering on atoms of an absorber consists of the sum of the differential electron–nucleus cross section and Z times the differential electron–orbital electron cross section, i.e.,

$$\left.\frac{d\sigma}{d\Omega}\right|_{\text{e-a}} = \left.\frac{d\sigma}{d\Omega}\right|_{\text{e-N}} + Z \left.\frac{d\sigma}{d\Omega}\right|_{\text{e-e}} = \frac{D_{\text{e-a}}^2}{(\theta^2 + \theta_{\min}^2)^2}, \quad (2.116)$$

where $D_{\text{e-a}}$ is the characteristic scattering distance for *electron-atom* elastic scattering given as

$$D_{e-a}^2 = D_{e-N}^2 + ZD_{e-e}^2. \qquad (2.117)$$

The characteristic scattering distance D_{e-a} is determined from (2.117) after inserting (2.114) and (2.115) to get

$$D_{e-a} = \sqrt{D_{e-N}^2 + ZD_{e-e}^2} = \frac{e^2}{4\pi\varepsilon_0} \frac{\sqrt{Z(Z+1)}}{\frac{mv^2}{2}} = \frac{2e^2}{4\pi\varepsilon_0} \frac{\sqrt{Z(Z+1)}}{pv}$$

$$= \frac{2r_e\sqrt{Z(Z+1)}\sqrt{1-\beta^2}}{\beta^2} = \frac{2r_e\sqrt{Z(Z+1)}}{\gamma\beta^2}, \qquad (2.118)$$

where

m is the *total mass* of the electron, i.e., $m = m_e / \sqrt{1 - \beta^2} = \gamma m_e$.
m_e is the rest mass of the electron.
β is the velocity of the electron normalized to c, i.e., $\beta = v_e/c$.
v_e is the velocity of the electron.
p is the momentum of the electron.
Z is the atomic number of the absorber.
r_e is the classical radius of the electron (2.82 fm).

A summary of characteristic scattering distances D for four elastic Coulomb scattering events including scattering of: (1) α-particle on nucleus (Rutherford scattering); (2) electron on nucleus (Mott scattering); (3) electron on atomic orbital electron; and (4) electron on atom is given in Table 2.4.

Table 2.4 Characteristic scattering distances D for four elastic Coulomb scattering events

Elastic Coulomb scattering	Characteristic scattering distance D	
α-particle–nucleus (Rutherford)	$D_{\alpha-N} = \dfrac{zZe^2}{4\pi\varepsilon_0 \frac{m_\alpha v_\alpha^2}{2}} = \dfrac{zZe^2}{4\pi\varepsilon_0 (E_K)_i}$	See (2.12) and (2.113)
Electron–nucleus (Mott)	$D_{e-N} = \dfrac{Ze^2}{4\pi\varepsilon_0 \frac{mv_e^2}{2}} = \dfrac{2Zr_e\sqrt{1-\beta^2}}{\beta^2}$	See (2.94) and (2.114)
Electron–orbital electron	$D_{e-e} = \dfrac{e^2}{4\pi\varepsilon_0 \frac{mv_e^2}{2}} = \dfrac{2r_e\sqrt{1-\beta^2}}{\beta^2}$	See (2.115)
Electron–atom	$D_{e-a} = \dfrac{e^2\sqrt{Z(Z+1)}}{4\pi\varepsilon_0 \frac{mv_e^2}{2}} = \dfrac{2r_e\sqrt{Z(Z+1)}\sqrt{1-\beta^2}}{\beta^2}$	See (2.116)

Note that m in D_{e-N}, D_{e-e}, and D_{e-a} stands for the **total mass** of the electron and not the rest mass of the electron

2.6.3 Minimum and Maximum Scattering Angles

The minimum and maximum scattering angles θ_{min} and θ_{max}, respectively, are angles where the deviation from point Coulomb nuclear field becomes significant. These departures from the point Coulomb field approximation appear at very small and very large scattering angles θ, corresponding to very large and very small impact parameters b, respectively.

At very small angles θ the screening of the nuclear charge by atomic orbital electrons decreases the differential cross section and at large angles θ the finite nuclear size or nuclear penetration by the charged particle decreases the differential cross section, as discussed for Rutherford scattering in Sects. 2.4.4 and 2.4.6, respectively.

As evident from Figs. 2.3 and 6.3, the relationship governing the change of momentum Δp in elastic scattering is given as follows

$$\sin \frac{\theta}{2} = \frac{\Delta p}{2p_i}, \tag{2.119}$$

where

θ is the scattering angle,

p_i is the particle initial momentum at a large distance from the scattering interaction.

In the small angle θ approximation, we get the following simple relationship from (2.22) and (2.119) recognizing that $\sin \theta \approx \theta$

$$\theta \approx \frac{\Delta p}{p_i}. \tag{2.120}$$

As shown in (2.57) and (2.69), θ_{min} and θ_{max}, respectively, are given by the following quantum-mechanical expressions

$$\theta_{min} \approx \frac{\Delta p}{p_i} \approx \frac{\hbar}{a_{TF}p_i} = \frac{\hbar}{p_i} \frac{\sqrt[3]{Z}}{a_0} = \frac{\hbar c}{a_0} \frac{\sqrt[3]{Z}}{\sqrt{E_K(E_K + 2E_0)}}$$
$$\approx \frac{3.723 \times 10^{-3} \text{ MeV} \sqrt[3]{Z}}{\sqrt{E_K(E_K + 2E_0)}} \tag{2.121}$$

and

$$\theta_{max} \approx \frac{\Delta p}{p_i} \approx \frac{\hbar}{Rp_i} = \frac{\hbar}{p_i R_0 \sqrt[3]{A}} = \frac{\hbar c}{R_0 \sqrt[3]{A}} \frac{1}{\sqrt{E_K(E_K + 2E_0)}}$$
$$\approx \frac{1.578 \times 10^2 \text{ MeV}}{\sqrt[3]{A}\sqrt{E_K(E_K + 2E_0)}}, \tag{2.122}$$

where

p_i is the initial momentum of the charged particle.

a_{TF} is the Thomas–Fermi atomic radius equal to $\sim a_0 Z^{-1/3}$ with a_0 the Bohr radius constant and Z the atomic number of the absorber, as given in (2.49).

a_0 is the Bohr radius constant defined in (3.4).

R is the radius of the nucleus equal to $R_0 A^{1/3}$ with R_0 the nuclear radius constant ($R_0 = 1.25$ fm), as discussed in Sect. 1.16.1.

E_K is the initial kinetic energy of the charged particle related to the initial momentum of the charged particle through (1.64).

E_0 is the rest energy of the charged particle.

A is the atomic mass number of the absorber.

Figure 2.12a shows the maximum scattering angle θ_{max} against kinetic energy E_K given in (2.122) in the range from 1 keV to 1000 MeV for electron and α-particle elastic scattering in carbon, aluminum, copper, silver, and lead. Figure 2.12b shows the minimum scattering angle θ_{min} given in (2.121) under same conditions as those

Fig. 2.12 Maximum scattering angle θ_{max} in (**a**) and minimum scattering angle θ_{min} in (**b**) against kinetic energy E_K for electrons and α-particles scattered on carbon, aluminum, copper, silver, and gold

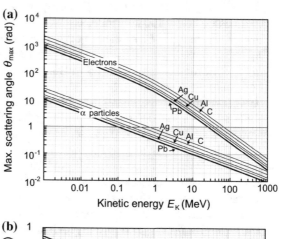

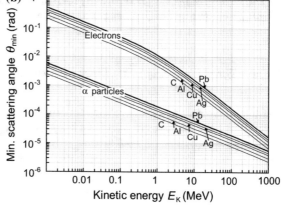

in Fig. 2.12a. Based on (2.121) and (2.122) as well as Fig. 2.12 we now make the following observations about the minimum and maximum scattering angles θ_{min} and θ_{max}, respectively:

1. In general, θ_{min} and θ_{max} depend on the kinetic energy E_K and rest energy E_0 of the elastically scattered projectile as well as the atomic number Z and atomic mass number A of the target. However, the ratio $\theta_{max}/\theta_{min}$ is independent of the incident particle physical properties and depends solely on the atomic number Z and the atomic mass A of the absorber target as follows

$$\frac{\theta_{max}}{\theta_{min}} = \frac{a_0}{R_0 \sqrt[3]{A}\sqrt[3]{Z}} \approx \frac{0.5292 \text{ Å}}{1.25 \times 10^{-5} \text{ Å}\sqrt[3]{AZ}} \approx \frac{0.423 \times 10^5}{(AZ)^{1/3}} = \frac{\text{Const}}{\sqrt[3]{AZ}}. \quad (2.123)$$

2. For a given E_K the maximum scattering angle θ_{max} is inversely proportional to $Z^{1/3}$ since $\theta_{max} \propto A^{-1/3}$ and $A \approx 2Z$ and the minimum scattering angle θ_{min} is proportional to $Z^{1/3}$.
3. For kinetic energies E_K of the projectile much smaller than its rest energy E_0, or $E_K \ll E_0$, both θ_{min} and θ_{max} for a given target are proportional to $1/\sqrt{E_K}$, as shown in Fig. 2.12 in the whole E_K energy range for α-particles and at kinetic energies E_K below 100 keV for electrons.
4. For kinetic energies E_K of the projectile much larger than its rest energy E_0, or $E_K \gg E_0$, both θ_{min} and θ_{max} for a given target are proportional to $1/E_K$, as shown in Fig. 2.12 for electrons in the energy range above 10 MeV.
5. For Rutherford scattering of 5.5 MeV α-particles on gold nucleus Au-197 (Geiger–Marsden experiment) we obtain from (2.121) a minimum scattering angle θ_{min} of 8.1×10^{-5} rad, as given in (2.72), and from (2.122) a maximum scattering angle of θ_{max} of 0.14 rad, as given in (2.73), in agreement with the general condition that $\theta_{min} \ll \theta_{max} \ll 1$.
6. For 10 MeV electrons scattered on gold-197, on the other hand, we find significantly larger θ_{min} from (2.121) and θ_{max} from (2.122) at 1.5×10^{-3} rad and 2.6 rad, respectively. However, we may still assume that $\theta_{min} \ll \theta_{max}$. Note: For θ_{max} calculated from (2.122) larger than unity, θ_{max} is usually set equal to 1.
7. The factor $(AZ)^{1/3}$ ranges from unity for hydrogen to ~ 28 for high atomic number absorbers such as uranium with $Z = 92$ and $A = 235$.

2.6.4   Total Cross Section for a Single Scattering Event

The total cross section σ for a single scattering event, similarly to the discussion of Rutherford cross section given in Sect. 2.4.8, is approximated as follows using the small angle approximation $\sin \theta \approx \theta$:

$$\sigma = \int \frac{d\sigma}{d\Omega} d\Omega \approx 2\pi D^2 \int_0^{\theta_{max}} \frac{\theta \, d\theta}{(\theta^2 + \theta_{min}^2)^2} = \pi D^2 \int_0^{\theta_{max}} \frac{d(\theta^2 + \theta_{min}^2)}{(\theta^2 + \theta_{min}^2)^2}$$

$$= -\pi D^2 \left[\frac{1}{\theta^2 + \theta_{min}^2} \right]_0^{\theta_{max}} = \pi D^2 \left\{ \frac{1}{\theta_{min}^2} - \frac{1}{\theta_{max}^2 + \theta_{min}^2} \right\}$$

$$= \pi D^2 \frac{1}{\theta_{min}^2} \left\{ 1 - \frac{1}{1 + (\theta_{max}/\theta_{min})^2} \right\}. \tag{2.124}$$

Since $\theta_{max}/\theta_{min} \gg 1$ even for very high atomic number materials, we can simplify the expression for total cross section σ to read

$$\sigma \approx \frac{\pi D^2}{\theta_{min}^2}, \tag{2.125}$$

where

D is the effective characteristic distance discussed in Sect. 2.6.2,
θ_{min} is the minimum scattering angle discussed in Sect. 2.6.3.

2.6.5 Mean Square Scattering Angle for Single Scattering

The mean square scattering angle for a single scattering event $\overline{\theta^2}$ is defined by the following general relationship

$$\overline{\theta^2} = \frac{\int_0^{\theta_{max}} \theta^2 \frac{d\sigma}{d\Omega} d\Omega}{\int_0^{\theta_{max}} \frac{d\sigma}{d\Omega} d\Omega} = \frac{2\pi}{\sigma} \int_0^{\theta_{max}} \theta^2 \frac{d\sigma}{d\Omega} \sin\theta \, d\theta, \tag{2.126}$$

where

$d\sigma/d\Omega$ is the differential cross section for the single scattering event, given in (2.112),
σ is the total cross section for the single scattering event [see (2.124) and (2.125)],
θ is the scattering angle for the single scattering event,
θ_{max} is the maximum scattering angle calculated from (2.122). It is taken as the actual calculated value when the calculated θ_{max} is smaller than 1 and is taken as unity when the calculated θ_{max} exceeds 1.

The mean square angle $\overline{\theta^2}$ for a single scattering event may be approximated in the small angle approximation as follows

$$
\overline{\theta^2} = \frac{2\pi D^2}{\sigma} \int\limits_0^{\theta_{max}} \frac{\theta^3 \, d\theta}{(\theta^2 + \theta_{min}^2)^2} = \frac{\pi D^2}{\sigma} \int\limits_0^{\theta_{max}} \frac{(\theta^2 + \theta_{min}^2) \, d(\theta^2 + \theta_{min}^2)}{(\theta^2 + \theta_{min}^2)^2}
$$

$$
- \frac{\pi D^2}{\sigma} \int\limits_0^{\theta_{max}} \frac{\theta_{min}^2 \, d(\theta^2 + \theta_{min}^2)}{(\theta^2 + \theta_{min}^2)^2}
$$

$$
= \frac{\pi D^2}{\sigma} \left\{ \ln\left(1 + \frac{\theta_{max}^2}{\theta_{min}^2}\right) - \frac{1}{1 + (\theta_{min}/\theta_{max})^2} \right\} \tag{2.127}
$$

or, after inserting the expression for σ given in (2.125)

$$
\overline{\theta^2} = \theta_{min}^2 \left\{ \ln\left(1 + \frac{\theta_{max}^2}{\theta_{min}^2}\right) - \frac{1}{1 + (\theta_{min}/\theta_{max})^2} \right\}
$$

$$
= \theta_{min}^2 \ln\left(1 + \frac{\theta_{max}^2}{\theta_{min}^2}\right) - \frac{\theta_{min}^2 \theta_{max}^2}{\theta_{min}^2 + \theta_{max}^2}, \tag{2.128}
$$

with θ_{min} minimum scattering angle defined in (2.121) and θ_{max} largest angle to be still considered a small angle in single scattering and defined in (2.122). At low energies θ_{max} calculated from (2.122) may exceed 1 rad and the maximum scattering angle is then taken as $\theta_{max} \approx 1$ rad.

The ratio $\theta_{max}/\theta_{min}$ is independent of particle kinetic and total energy and depends only on the atomic number Z and the atomic mass number A of the absorber, as shown in (2.123). Since, in addition $\theta_{max} \gg \theta_{min}$, we can simplify (2.128) to read

$$
\overline{\theta^2} \approx 2\,\theta_{min}^2 \, \ln \frac{\theta_{max}}{\theta_{min}}. \tag{2.129}
$$

After inserting (2.123) into (2.129) and assuming that $A \approx 2Z$ we get the following approximation for the mean square scattering angle $\overline{\theta^2}$ for single scattering

$$
\overline{\theta^2} \approx 2\theta_{min}^2 \, \ln\left(\frac{\sqrt{0.423 \times 10^5}}{\sqrt[3]{\sqrt{2}}\sqrt[3]{Z}}\right)^2 = 4\,\theta_{min}^2 \, \ln\left[183 Z^{-1/3}\right], \tag{2.130}
$$

with the minimum scattering angle θ_{min} given in (2.121).

2.7 Molière Multiple Elastic Scattering

Multiple or compound Coulomb scattering results from a large number of single scattering events that a charged particle will experience as it moves through an absorber. These single scattering events are independent and statistically random processes governed by a Rutherford-type Coulomb interaction and confined to a very small scattering angle θ with respect to the direction of incidence. In honor of the German theoretical physicist Gert Molière who carried out much of the initial theoretical work on multiple scattering, this type of scattering is often referred to as Molière multiple scattering.

As discussed for standard Rutherford scattering in Sect. 2.3, a particle traversing a thin metallic foil will experience a large number of Coulomb interactions with nuclei of the absorber and these interactions will generally produce only small angle deflections. The cumulative effect of these independent interactions will be a superposition of a large number of random deflections resulting in the particle emerging through the foil: (1) at a small cumulative scattering angle Θ, and (2) at a mean scattering angle $\overline{\Theta}$ with respect to the incident direction of zero for a beam of particles striking the foil.

The angular distribution of particles transmitted through a foil is Gaussian in shape and centered round the direction of the incident particles, reflecting the cumulative action of a large number of independent small-angle scattering interactions. This was shown by (2.2) for α-particle scattering on gold nuclei.

The measured angular distributions of charged particles emerging through a foil show excellent agreement with a Gaussian distribution at small cumulative scattering angles Θ but also exhibit a higher tail than the Gaussian distribution at large scattering angles. This discrepancy at large scattering angles is attributed to the effect of rare large-angle single scattering events which were first explained by Rutherford as follows: In its travel through the foil a charged particle may experience a close encounter with a scattering center and this hard collision will result in a large angle deflection, possibly amounting to 180°. These large-angle Coulomb scattering interactions are extremely rare yet not negligible and occur with a typical frequency of about one such interaction per several thousand particles transmitted through a thin foil. The following conclusions can now be made:

1. A charged particle traversing a foil will have numerous soft interactions with scattering centers in the absorber that are random and independent from one another. These interactions result in small individual deflections from the incident direction as well as in a small cumulative scattering angle Θ.
2. One in several 1000 particles of the particle beam traversing a foil will have a hard interaction (close encounter) with a scattering center resulting in a large-angle deflection. Because of the very small probability for a hard collision, only one such large angle deflection can occur to a given charged particle. All large-angle deflections are therefore attributed to one single-scattering event for a given charged particle.

3. The angular distribution of charged particles traversing a foil thus has three regions:

   a. Small angle Θ region governed by a Gaussian distribution resulting from Molière multiple scattering.
   b. Large angle single-scatter region produced by a small fraction of particles striking the foil and resulting from single hard collisions between a charged particle and a scattering center.
   c. Intermediate region between the multiple scatter small-angle region and the single-scatter large-scattering angle region referred to as the region of *plural scattering*. The plural scattering distribution enables the transition from the multiple scattering region into the single scattering region.

The mean square angle $\overline{\theta^2}$ for single scattering derived in Sect. 2.6.5 also plays a role in determining the mean square angle $\overline{\Theta^2}$ which governs the Gaussian distribution in Molière multiple scattering distribution, as shown in Sect. 2.7.1.

2.7.1 Mean Square Scattering Angle for Multiple Scattering

The thicker is the absorber and the larger is its atomic number Z, the greater is the likelihood that the incident particle will undergo several single scattering events. For a sufficiently thick absorber the mean number of successive encounters rises to a value that permits a statistical treatment of the process. Generally, 20 collisions are deemed sufficient and we then speak of multiple Coulomb scattering that is characterized by a large succession of small angle deflections symmetrically distributed about the incident particle direction.

The mean square angle for multiple Coulomb scattering $\overline{\Theta^2}$ is calculated from the mean square angle for single scattering $\overline{\theta^2}$ (2.128) with the help of the *central limit theorem* that states the following:

> For a large number N of experiments that measure some stochastic variable X, the probability distribution of the average of all measurements is Gaussian and is centered at $\overline{X}$ with a standard deviation $1/\sqrt{N}$ times the standard deviation of the probability density of X.

Since the successive single scattering collisions in the absorber are independent events, the central limit theorem shows that for a large number $n > 20$ of such collisions the distribution in angle will be Gaussian around the forward direction with a mean square scattering angle $\overline{\Theta^2}$ given as

$$\overline{\Theta^2} = n\overline{\theta^2}, \tag{2.131}$$

where

$\overline{\theta^2}$ is the mean square scattering angle for single scattering given in (2.128),

n is the number of scattering events calculated as follows

$$n = \frac{N_a}{V}\sigma t = \rho \frac{N_A}{A}\sigma t \approx \pi\rho \frac{N_A}{A}\frac{D^2}{\theta^2_{min}}t \qquad (2.132)$$

where we inserted the expression of (2.125) for the cross section and

N_a/V is number of atoms per volume equal to $\rho N_A/A$,

σ is the total cross section for a single scattering event given in (2.124) and (2.125),

t is thickness of the absorber,

ρ is density of the absorber,

N_A is the Avogadro number,

A is the atomic mass number of the absorber.

Incorporating the expression for the mean square angle for single scattering $\overline{\theta^2}$ from (2.128) into (2.131) and using (2.132) for the number of scattering events, the mean square angle for multiple scattering $\overline{\Theta^2}$ can be written as

$$\overline{\Theta^2} = \rho \frac{N_A}{A}\sigma t\theta^2_{min}\left\{\ln\left(1 + \frac{\theta^2_{max}}{\theta^2_{min}}\right) - \frac{1}{1 + \theta^2_{min}/\theta^2_{max}}\right\}, \qquad (2.133)$$

where θ_{min} and θ_{max} are the minimum and maximum scattering angles, respectively, defined in Sect. 2.6.3, and D is the characteristic scattering distance for a particular scattering event, defined in Sect. 2.6.2.

Since $\theta_{max} \gg \theta_{min}$ holds in general, we can simplify (2.133) for heavy charged particle scattering on nuclei of an absorber by inserting (2.125) for the total cross section σ with (2.113) for the characteristic scattering distance D and (2.130) for $\overline{\theta^2}$ to get

$$\overline{\Theta^2} = n\overline{\theta^2} = 4\left(\rho \frac{N_A}{A}\sigma t\right)\theta^2_{min}\ln[183Z^{-1/3}] = 4\pi\rho \frac{N_A}{A}D^2 t\,\ln[183Z^{-1/3}]$$

$$= 4\pi\rho \frac{N_A}{A}\left(\frac{2zZe^2}{4\pi\varepsilon_0 p\upsilon}\right)^2\{\ln[183Z^{-1/3}]\}t. \qquad (2.134)$$

Similarly, for electrons scattered on absorber atoms (nuclei and orbital electrons) we simplify (2.133) by inserting (2.125) for the total cross section σ with (2.118) for the characteristic scattering distance D and (2.130) for $\overline{\theta^2}$ to get

$$\overline{\Theta^2} = n\overline{\theta^2} = \left(\rho \frac{N_A}{A}\sigma t\right)4\theta^2_{min}\,\ln[183Z^{-1/3}]$$

$$= 16\pi\rho \frac{N_A r_e^2 Z(Z+1)}{A\gamma^2\beta^4}\{\ln[183Z^{-1/3}]\}t = 4\pi\frac{\rho}{\alpha X_0\gamma^2\beta^4}t, \qquad (2.135)$$

where X_0 is defined as the radiation length and discussed in Sect. 2.7.2.

From (2.135) we can express the change in the mean square scattering angle $\overline{\Theta^2}$ with propagation distance t in the foil as

$$\frac{d\overline{\Theta^2}}{dt} = 16\pi\rho\frac{N_A r_e^2 Z(Z+1)}{A\gamma^2\beta^4}\ln[183Z^{-1/3}] = 4\pi\frac{\rho}{\alpha X_0\gamma^2\beta^4}. \qquad (2.136)$$

As shown in (2.134) and (2.135), the mean square scattering angle $\overline{\Theta^2}$ for multiple scattering increases linearly with the foil thickness t but, as long as the foil thickness is not excessive, the angular distribution of transmitted particles will remain Gaussian and forward peaked.

2.7.2 Radiation Length

The expressions for the mean square scattering angle $\overline{\Theta^2}$ of (2.135) and (2.136) can be expressed in terms of a distance parameter called the radiation length X_0. This parameter serves as a unit of length, depends on the mass of the charged particle as well as on the atomic number of the absorbing material, and is defined as the mean distance a relativistic charged particle travels in an absorbing medium while its energy, due to radiation loss, decreases to $1/e$ ($\sim$36.8%) of its initial value. X_0 is also defined as 7/9 of the mean free path for pair production by a high energy photon in the absorber.

The radiation length X_0, which usually refers to electrons, is expressed in square centimeters per gram as follows

$$\frac{1}{X_0} = 4\alpha\frac{N_A}{A}Z(Z+1)r_e^2\ln(183Z^{-1/3})$$

$$= 1.4\times10^{-3}\,(\text{cm}^2/\text{mol})\frac{Z(Z+1)}{A}\ln(183Z^{-1/3}), \qquad (2.137)$$

where

α is the fine structure constant (1/137),
N_A is the Avogadro number (6.022×10^{23}/mol),
Z is the atomic number of the absorber,
r_e is the classical electron radius (2.818 fm).

For electrons, values of radiation length X_0 calculated from (2.137) are 24 g/cm^2 (9 cm) in aluminum; 10.2 g/cm^2 (1.1 cm) in copper; and 5.8 g/cm^2 (0.51 cm) in lead.

2.7.3 Mass Scattering Power

As shown in (2.133), the mean square scattering angle for multiple scattering $\overline{\Theta^2}$ increases linearly with the absorber thickness t. A mass scattering power T/ρ can thus be defined for electrons:

1. Either as the mean square angle for multiple scattering $\overline{\Theta^2}$ per mass thickness ρt.
2. Or the increase in the mean square angle $\overline{\Theta^2}$ per unit mass thickness ρt, in analogy with the mass stopping power.

The mass scattering power (T/ρ) is thus expressed as follows

$$\frac{T}{\rho} = \frac{\overline{\Theta^2}}{\rho t} = \frac{d\overline{\Theta^2}}{d(\rho t)} = \frac{N_A}{A}\sigma\theta_{min}^2 \left\{ \ln\left(1 + \frac{\theta_{max}^2}{\theta_{min}^2}\right) - \frac{1}{1 + \theta_{min}^2/\theta_{max}^2} \right\} \quad (2.138)$$

and this result, after inserting (2.125) for the total cross section σ, is usually given as follows (ICRU #35)

$$\frac{T}{\rho} = \pi\frac{N_A}{A}D^2 \left\{ \ln\left(1 + \frac{\theta_{max}^2}{\theta_{min}^2}\right) - 1 + \left[1 + \frac{\theta_{max}^2}{\theta_{min}^2}\right]^{-1} \right\}, \quad (2.139)$$

with D, the effective characteristic scattering distance, discussed in Sect. 2.6.2 for various scattering interactions.

2.7.4 Mass Scattering Power for Electrons

The mass scattering power T/ρ for electrons is determined from the general relationship of (2.139) by inserting (2.118) for the characteristic distance D in electron scattering with nuclei and orbital electrons of the absorber foil to get

$$\frac{T}{\rho} = 4\pi r_e^2\frac{N_A Z(Z+1)}{A\gamma^2\beta^4} \left\{ \ln\left(1 + \frac{\theta_{max}^2}{\theta_{min}^2}\right) - 1 + \left[1 + \frac{\theta_{max}^2}{\theta_{min}^2}\right]^{-1} \right\}. \quad (2.140)$$

The term $\left(\sqrt{1-\beta^2}\right)/\beta^2$ in (2.118) for D can be expressed in terms of the electron kinetic energy E_K and electron rest energy $E_0 = m_e c^2$. We first define the ratio $E_K/(m_e c^2)$ as τ and then use the standard relativistic relationship for the total energy of the electron, i.e.,

$$m_e c^2 + E_K = \frac{m_e c^2}{\sqrt{1-\beta^2}} \quad (2.141)$$

to obtain

$$\sqrt{1-\beta^2} = \frac{1}{\gamma} = \frac{1}{1+\tau} \qquad (2.142)$$

and

$$\beta^2 = \frac{\tau(2+\tau)}{(1+\tau)^2}, \qquad (2.143)$$

resulting in the following expression for the term $\frac{\sqrt{1-\beta^2}}{\beta^2} = \frac{1}{\gamma\beta^2}$

$$\frac{\sqrt{1-\beta^2}}{\beta^2} = \frac{1}{\gamma\beta^2} = \frac{1+\tau}{\tau(2+\tau)}. \qquad (2.144)$$

The mass scattering power T/ρ of (2.139) for electron may then be expressed as follows

$$\frac{T}{\rho} = 4\pi \frac{N_A}{A} r_e^2 Z(Z+1) \left[\frac{1+\tau}{\tau(2+\tau)}\right]^2 \left\{\ln\left(1+\frac{\theta_{max}^2}{\theta_{min}^2}\right) - 1 + \left[1+\frac{\theta_{max}^2}{\theta_{min}^2}\right]^{-1}\right\}. \qquad (2.145)$$

In (2.145), θ_{max} is the cutoff angle resulting from the finite size of the nucleus. In (2.122), the cutoff angle θ_{max} was given by the ratio of the reduced de Broglie wavelength of the electron $\hbar/p_e$ to the nuclear radius R given in (1.26) as $R = R_0\sqrt[3]{A}$ with $R_0 = 1.25$ fm the nuclear radius constant and A the nucleon number or atomic mass number. The electron momentum p_e using (1.60) and (2.142) can be expressed as

$$p_e = \frac{1}{c}\sqrt{E^2 - E_0^2} = \frac{1}{c}\sqrt{E_K(E_K + 2E_0)} = \frac{E_0}{c}\sqrt{\tau(\tau+2)} \qquad (2.146)$$

Recognizing from (1.4) that $\hbar c/E_0 = \alpha a_0$ where $\alpha = 1/137$ is the fine structure constant (3.6) and $a_0 = 0.5292$ Å is the Bohr radius constant (3.4), we now get the following expression for θ_{max}

$$\theta_{max} = \frac{\hbar}{Rp_e} = \frac{\hbar c}{R_0\sqrt[3]{A}E_0\sqrt{\tau(\tau+2)}} = \frac{\alpha a_0}{R_0\sqrt[3]{A}\sqrt{\tau(\tau+2)}}$$
$$= \frac{0.5292\times 10^5 \text{ fm}}{137\times(1.25 \text{ fm}\times\sqrt[3]{A}\sqrt{\tau(\tau+2)}} = \frac{309}{\sqrt[3]{A}\sqrt{\tau(\tau+2)}}, \qquad (2.147)$$

with

β electron velocity normalized to c, the speed of light in vacuum,
A atomic mass number of the absorber,
τ electron kinetic energy normalized to electron rest energy E_0,
E_K kinetic energy of the electron.

The screening angle θ_{min} results from the screening of the nucleus by the atomic orbital electrons and is expressed in (2.121) by the ratio of the reduced de Broglie wavelength of the electron $\hbar/p_e$ [given in (2.71)] to the Thomas–Fermi atomic radius a_{TF} given in (2.49) as $a_{TF} \approx a_0 Z^{-1/3}$, with a_0 the Bohr radius constant of (3.4) and Z the atomic number of the absorber. Recognizing that $\hbar c/(E_0 a_0) = \alpha$, the minimum scattering angle, also known as the *screening angle*, θ_{min} can be expressed as

$$\theta_{min} = \frac{\hbar}{a_{TF} p_e} = \frac{\hbar c \sqrt[3]{Z}}{a_0 E_0 \sqrt{\tau(\tau+2)}} = \frac{\alpha \sqrt[3]{Z}}{\sqrt{\tau(\tau+2)}} = \frac{\sqrt[3]{Z}}{137\sqrt{\tau(\tau+2)}} \qquad (2.148)$$

with

Z atomic number of the absorber,

τ electron kinetic energy normalized to electron rest energy,

α fine structure constant defined in (3.6) as $\hbar c/(a_0 m_e c^2) = 1/137$.

Similarly to the expression in (2.74) and (2.123), the ratio $\theta_{max}/\theta_{min}$ is now given by a simple expression independent of electron rest energy E_0 and kinetic energy E_K

$$\frac{\theta_{max}}{\theta_{min}} = \frac{309 \times 137 A^{-1/3}}{Z^{1/3}} \approx \frac{0.423 \times 10^5}{\sqrt[3]{AZ}}, \qquad (2.149)$$

and ranges from $\theta_{max}/\theta_{min} \approx 0.42 \times 10^5$ for hydrogen through $\sim$3450 for copper, $\sim$2450 for silver down to $\theta_{max}/\theta_{min} \approx 1500$ for uranium-235.

Two features of the mass scattering power T/ρ can be identified:

1. (T/ρ) is roughly proportional to Z. This follows from the $Z(Z+1)/A$ dependence recognizing that $A \approx 2Z$ to obtain $(T/\rho) \propto Z$.
2. (T/ρ) for large electron kinetic energies E_K where $\tau \gg 1$ is proportional to $1/E_K^2$. This follows from $(1+\tau)^2/\{\tau(2+\tau)\}^2 \approx 1/\tau^2$ for $\tau \gg 1$.

A plot of the mass scattering power (T/ρ) for electrons in various materials of interest in medical physics in the electron kinetic energy range from 1 keV to 1000 MeV is given in Fig. 2.13. The mass scattering power (T/ρ) consists of two components: the electron–nucleus (e–N) scattering and the electron–orbital electron (e–e) scattering.

2.7.5 *Fermi-Eyges Pencil Beam Model for Electrons*

Fermi in his study of cosmic radiation derived an analytical solution to the transport equation for energetic charged particles traversing thin foils. He used Molière's small angle multiple scattering approximation and assumed that the energetic cosmic particles lost no energy in the thin foils he used in his experiments. Eyges extended Fermi's work to electron pencil beams traversing absorbing media and accounted for electron energy loss as well as for electron transport through heterogeneous absorbers.

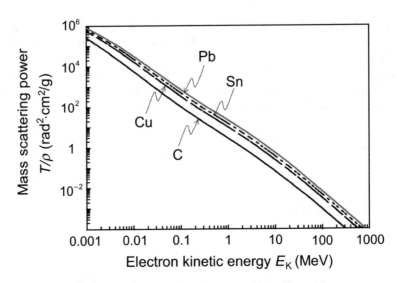

Fig. 2.13 Mass scattering power (T/ρ) against electron kinetic energy E_K for various materials of interest in medical physics

As shown in Fig. 2.14, electrons moving in a pencil beam along the z axis (applicate axis) of a Cartesian coordinate system strike the absorber at the origin $(0,0,0)$ of the coordinate system and undergo multiple scattering interactions as they penetrate into the absorber. After traversing a given thickness of the absorber, each electron emerges in a direction defined by angles Θ and Φ. The projections of the polar angle Θ onto the (x, z) and (y, z) planes are Θ_x and Θ_y, respectively.

The Fermi-Eyges solution to the transport equation gives the probability $P(x, z)\mathrm{d}x$ of finding an electron at depth z in the absorber with a displacement from the original z direction between x and $x + \mathrm{d}x$ on the abscissa and the probability $P(y, z)\mathrm{d}y$ of finding the electron between y and $y + \mathrm{d}y$ on the ordinate. The two probability density functions $P(x, z)\mathrm{d}x$ and $P(y, z)\mathrm{d}y$ are given as follows (see Sect. 1.30)

$$P(x, z)\mathrm{d}x = \frac{1}{\sigma(z)\sqrt{2\pi}}e^{-\frac{x^2}{2[\sigma(z)]^2}}\,\mathrm{d}x \qquad (2.150)$$

and

$$P(y, z)\mathrm{d}y = \frac{1}{\sigma(z)\sqrt{2\pi}}e^{-\frac{y^2}{2[\sigma(z)]^2}}\,\mathrm{d}y, \qquad (2.151)$$

with $\sigma(z)$ representing the standard deviation of the mean as a measure of the width of the distribution at depth z. According to the Fermi-Eyges theory the variance $v(z)$ which by definition is the square of the standard deviation $\sigma(z)$ is expressed as

$$v(z) = [\sigma(z)]^2 = \frac{1}{2}\int_0^z T(z')[z - z']^2\mathrm{d}z', \qquad (2.152)$$

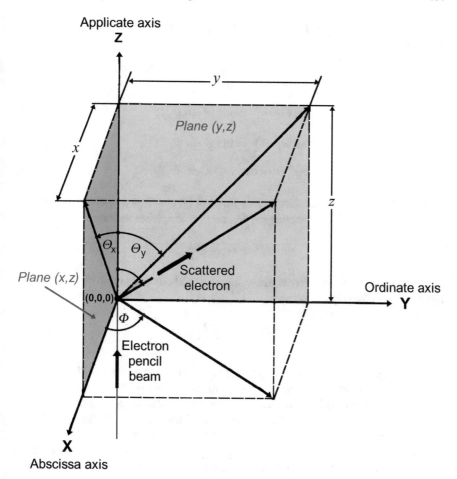

Fig. 2.14 Electrons in a pencil beam moving along applicate (z) axis of a Cartesian coordinate system strike an absorber at the origin of the coordinate system and undergo multiple scattering interactions as they penetrate into the absorber. After traversing a given thickness of the absorber, each electron emerges in a direction defined by angles Θ and Φ

where $T(z')$ is the linear scattering power of the absorber at depth z', evaluated for the mean electron energy at depth z'. The scattering power T was discussed in Sect. 2.7.3 and shown to be proportional to the mean square scattering angle $\overline{\Theta^2}$ in (2.138).

In general, the combined probability $P(x, y, z)\mathrm{d}x\mathrm{d}y$ is a product of the two probability density functions, $P(x, z)\mathrm{d}x$ of (2.150) and $P(y, z)\mathrm{d}y$ of (2.151), expressed as follows

$$P(x, y, z)\mathrm{d}x\mathrm{d}y = [P(x, z)\mathrm{d}x] \times [P(y, z)\mathrm{d}y] = \frac{1}{2\pi[\sigma(z)]^2} e^{-\frac{x^2+y^2}{2[\sigma(z)]^2}} \mathrm{d}x\mathrm{d}y. \qquad (2.153)$$

From Fig. 2.14 we get the following relationships among angles Θ, Θ_x, Θ_y, and Φ and Cartesian coordinates x, y, and z

$$\tan \Theta = \frac{x/\cos \Phi}{z} = \frac{x/\sin \Phi}{z},$$

(2.154)

$$\tan \Theta_x = \frac{x}{z} = \tan \Theta \cos \Phi,$$

(2.155)

$$\tan \Theta_y = \frac{y}{z} = \tan \Theta \sin \Phi,$$

(2.156)

$$\tan^2 \Theta = \tan^2 \Theta_x + \tan^2 \Theta_y.$$

(2.157)

In the small angle approximation where $\sin \Theta \approx \Theta$, $\cos \Theta \approx 1$, and $\tan \Theta \approx \Theta$, the relationship of (2.157) connecting Θ, Θ_x, and Θ_y simplifies to read

$$\Theta^2 = \Theta_x^2 + \Theta_y^2$$

(2.158)

and, since the scattering events are symmetrical about the initial direction z, the following relationships also apply

$$\overline{\Theta_x^2} = \overline{\Theta_y^2} = \frac{1}{2}\overline{\Theta^2}.$$

(2.159)

The final polar angle Θ following multiple scattering events cannot be determined by a simple addition of the polar angles for the individual scattering events because of the Φ component which is present in each single scattering event. The projections Θ_x and Θ_y, however, are additive and this then allows us to apply the central limit theorem stated in Sect. 2.7.1. For the (x, z) plane we define $P(x, \Theta_x, z)d\Theta_x$ as the probability that an electron, after traversing an absorber thickness dz, will be deflected through an angle, the projection of which onto the (x, z) plane will be between Θ_x and $\Theta_x + d\Theta_x$. Similarly, for the (y, z) plane we define $P(y, \Theta_y, z)d\Theta_y$ as the probability that an electron, after traversing an absorber thickness dz, will be deflected through an angle, the projection of which onto the (y, z) plane will be between Θ_y and $\Theta_y + d\Theta_y$.

The two probability functions $P(x, \Theta_x, z)$ and $P(y, \Theta_y, z)$ are Gaussian functions expressed as

$$P(x, \Theta_x, z) = \frac{1}{\sqrt{2\pi\overline{\Theta_x^2}}}e^{-\frac{\Theta_x^2}{2\overline{\Theta_x^2}}} = \frac{1}{\sqrt{\pi\overline{\Theta^2}}}e^{-\frac{\Theta_x^2}{\overline{\Theta^2}}}$$

(2.160)

and

$$P(y, \Theta_y, z) = \frac{1}{\sqrt{2\pi\overline{\Theta_y^2}}}e^{-\frac{\Theta_y^2}{2\overline{\Theta_y^2}}} = \frac{1}{\sqrt{\pi\overline{\Theta^2}}}e^{-\frac{\Theta_y^2}{\overline{\Theta^2}}},$$

(2.161)

where we used (2.159) to modify the two original Gaussian distributions.

Similarly to (2.153), the combined probability $P(x, \Theta_x, y, \Theta_y, z)$ is given as the product of the two probability functions $P(x, \Theta_x, z)$ and $P(y, \Theta_y, z)$ to give

$$
\begin{aligned}
P(x, \Theta_x, y, \Theta_y, z) &= P(x, \Theta_x, z) \times P(y, \Theta_y, z) \\
&= \frac{1}{\sqrt{\pi \overline{\Theta^2}}} e^{-\frac{\Theta_x^2}{\overline{\Theta^2}}} \frac{1}{\sqrt{\pi \overline{\Theta^2}}} e^{-\frac{\Theta_y^2}{\overline{\Theta^2}}} = \frac{1}{\pi \overline{\Theta^2}} e^{-\frac{\Theta^2}{\overline{\Theta^2}}},
\end{aligned} \tag{2.162}
$$

after we use (2.158) for the sum of Θ_x^2 and Θ_y^2.

The discussion of the Thomson model of the atom in Sect. 2.2.1 made use of (2.162) when in (2.2) we estimated $N(\Theta)\mathrm{d}\Theta$, the number of α-particles that are scattered on gold nuclei within the angular range Θ to $\Theta + \mathrm{d}\Theta$, with N_0 representing the number of α-particles striking, and passing through, the gold foil. The fractional number of α-particles scattered into the angular range Θ to $\Theta + \mathrm{d}\Theta$ is expressed as follows

$$
\begin{aligned}
\frac{N(\Theta)\mathrm{d}\Theta}{N_0} &= P(x, \Theta_x, y, \Theta_y, z)\mathrm{d}\Omega = 2\pi P(x, \Theta_x, y, \Theta_y, z)\sin\Theta \mathrm{d}\Theta \\
&\approx 2\pi\Theta \frac{1}{\pi\overline{\Theta^2}} e^{-\frac{\Theta^2}{\overline{\Theta^2}}} \mathrm{d}\Theta = \frac{2\Theta}{\overline{\Theta^2}} e^{-\frac{\Theta^2}{\overline{\Theta^2}}} \mathrm{d}\Theta = e^{-\frac{\Theta^2}{\overline{\Theta^2}}} \mathrm{d}\frac{\Theta^2}{\overline{\Theta^2}},
\end{aligned} \tag{2.163}
$$

where we used the small angle approximation $\sin\Theta \approx \Theta$.

An integration of (2.163) over Θ from 0 to π results in 1, since $\overline{\Theta^2}$, the mean square scattering angle for multiple scattering, is very small. The Θ angular distribution is strongly peaked in the forward (z) direction of the incident pencil electron beam, with $\overline{\Theta}$, the mean scattering angle Θ, equal to 0 and $\sqrt{\overline{\Theta^2}}$, the root mean square angle for multiple scattering of the order of $1°$.

2.7.6 Dose Distribution for Pencil Electron Beam

The dose distribution for a pencil electron beam in absorbing medium is related to the distribution function given by the Fermi-Eyges solution to the Fermi electron transport equation that in three dimensions is expressed as follows

$$
\frac{\partial P}{\partial z} = -\Theta_x \frac{\partial P}{\partial \Theta_x} - \Theta_y \frac{\partial P}{\partial \Theta_y} + \frac{T(x, y, z)}{4} \left(\frac{\partial^2 P}{\partial \Theta_x^2} + \frac{\partial^2 P}{\partial \Theta_y^2} \right), \tag{2.164}
$$

with $T(x, y, z)$ the linear scattering power of the absorber and the probability function P given as a product of two Gaussian probability functions

$$
P = P(x, \Theta_x, y, \Theta_y, z) = P(x, \Theta_x, z) \times P(y, \Theta_y, z). \tag{2.165}
$$

The Fermi-Eyges theory predicts that the dose distribution in the absorber, in a plane perpendicular to the incident direction of the initial pencil electron beam, is represented by a Gaussian distribution. The theory also predicts that the spatial spread of the electron beam in the absorber is an increasing function of depth in the absorber irrespective of the depth. However, experiments show that the spatial spread indeed increases with depth from the absorber surface to about a depth close to 2/3 of the practical electron range, but at larger depths the spatial spread saturates, then decreases, and vanishes at depths greater than the range of electrons in the absorber.

The Fermi-Eyges theory considers only the small angle multiple Coulomb scattering and assumes that the energy of the electron, as it moves through the phantom, is dependent only on depth and that no electrons are absorbed in the scattering medium. This is certainly an improvement over the Fermi assumption of no energy loss of charged particles in the absorber; however, neglecting the electron absorption in the absorber causes significant discrepancy between measurement and Fermi-Eyges theory at depths close to the electron range.

2.7.7 Determination of Electron Beam Kinetic Energy from Measured Mass Scattering Power

The plot of (T/ρ) against electron kinetic energy E_K for kinetic energies in the megavoltage energy range (Fig. 2.13) is essentially linear on a log-log plot resulting in the $(T/\rho) \propto 1/E_K^2$ dependence. The steady $1/E_K^2$ drop of (T/ρ) as a function of increasing E_K suggests a relatively simple means for electron kinetic energy determination from a measurement of the mass scattering power (T/ρ) in a given medium.

The propagation of an electron pencil beam in an absorber is described by a distribution function that is given by the Fermi-Eyges solution to the Fermi differential transport equation. The Fermi-Eyges theory predicts that the dose distribution in a medium on a plane perpendicular to the incident direction of the pencil electron beam is given by a Gaussian distribution with a spatial spread proportional to the variance of the Gaussian distribution.

Equation (2.152) shows that the variance $[\sigma(z)]^2$ of the Gaussian distribution is related to the scattering power $T(z)$ at depth z. In situations where the scattering power $T(z)$ of the absorber is constant in the absorber thickness z (for example, in measurements of spatial spread in air layers z much thinner than the range of electrons in air), (2.152) can be simplified to read

$$[\sigma(z)]^2 = \frac{1}{2} \int_0^z T(z')[z-z']^2 dz' = \frac{1}{2}T(z) \int_0^z [z-z']^2 dz'$$

$$= \frac{1}{2}T(z) \int_0^z [z^2 - 2zz' + (z')^2]dz' = \frac{1}{6}z^3 T(z). \qquad (2.166)$$

In deriving (2.166) the following assumptions are made:

1. Only small angle scattering events are considered.
2. The air layer z is much smaller than the electron range in air.
3. Secondary electrons, set in motion by the electron incident pencil beam, are ignored.
4. The bremsstrahlung contamination of the electron pencil beam is ignored.

Function $[\sigma(z)]^2$ given in (2.166) is a linear function of z^3 with the slope proportional to the mass scattering power (T/ρ), which in turn is a function of electron beam kinetic energy E_K through function τ, as given in (2.145). Thus, from a measurement of $[\sigma(z)]^2$, the spatial spread of an electron pencil beam in air, at several distances z from the pencil beam origin, one first determines (T/ρ) through (2.166) and then determines the electron beam kinetic energy E_K with data tabulated for air or data calculated for air from (2.145).

Ernest Rutherford and Niels Bohr: Giants of Modern Physics

Photographs on next page show stamps issued in honor of physicists *Ernest Rutherford* (1871–1937) and *Niels Bohr* (1885–1962), two scientists credited with developing our current atomic model. According to the Rutherford–Bohr atomic model, most of the atomic mass is concentrated in the positively charged nucleus and the negative electrons revolve in orbits about the nucleus. New Zealand issued the stamp for Rutherford and Denmark for Bohr, both countries honoring their respective native son. Both physicists received Nobel Prizes for their work: Rutherford in 1908 *"for investigations into the disintegration of elements and the chemistry of radioactive substances,"* Bohr in 1922 for *"his services in the investigation of the structure of atoms and the radiation emanating from them."*

Rutherford studied in New Zealand and England, but spent all his professional life first in Canada at McGill University in Montreal (1898–1907) and then in England at the University of Manchester (1908–1919) and at the Cavendish Laboratory in Cambridge (1919–1937). During his nine years at McGill, Rutherford published some 70 papers and worked with Frederick Soddy on the disintegration theory of radioactivity. In his 12 years at the University of Manchester, he collaborated with Geiger and Marsden and, based on their experiments, proposed the currently accepted nuclear model of the atom. In Cambridge, Rutherford continued using α-particles from radium and polonium sources to probe the atom. He collaborated with James Chadwick who in 1932 discovered the neutron and with Charles Wilson, who is best known for the development of the "cloud chamber."

Rutherford is considered one of the most illustrious scientists of all time. His work on the atomic structure parallels Newton's work in mechanics, Darwin's work on evolution, Faraday's work in electricity, Maxwell's work in electromagnetism, and Einstein's work on relativity.

Bohr studied at the Copenhagen University in Denmark and spent most of his professional life there, except for short intervals in 1912 and in 1914–1916 when he worked with Rutherford in Manchester. He built a renowned school of theoretical and experimental physics at the University of Copenhagen and became its first director from its inauguration in 1921 to his death in 1962. The school is now known as the Niels Bohr Institute. Bohr is best known for his introduction of the electron angular momentum quantization into the atomic model that is now referred to as the Rutherford–Bohr atom. He also made numerous other contributions to theoretical physics, most notably with his complementarity principle and the liquid drop nuclear model.

Bohr was also interested in national and international politics and advised Presidents Roosevelt and Truman as well as Prime Minister Churchill on scientific issues in general and nuclear matters in particular. He enjoyed tremendous esteem by physics colleagues, world leaders, and the general public. Among the scientists of the twentieth century only Marie Curie, Albert Einstein, and Ernest Rutherford have reached similar stature and recognition.

Chapter 3
Rutherford–Bohr Model of the Atom

This chapter is devoted to a discussion of the Rutherford–Bohr model of the atom. The two giants of modern physics, Rutherford and Bohr, have not collaborated on the model; however, they both made a major contribution to it; Rutherford by introducing the concept of the atomic nucleus with electrons revolving about the nucleus in a cloud and Bohr by introducing the idea of electron angular momentum quantization and by deriving from first principles the kinematics of the hydrogen atom and one-electron atoms in general.

© Springer International Publishing Switzerland 2016
E.B. Podgoršak, *Radiation Physics for Medical Physicists*,
Graduate Texts in Physics, DOI 10.1007/978-3-319-25382-4_3

The chapter deals first with the hydrogen atom in detail following the steps that Bohr enunciated almost 100 years ago for determining: (i) radii of electron allowed orbits, (ii) velocity of electron in orbit, (iii) binding energy of electron while in allowed orbit, as well as (iv) hydrogen atom spectra.

Next, the successes and limitations of the Rutherford–Bohr model are discussed, the Hartree approximation for multi-electron atoms is introduced, and the experimental confirmation of the validity of the atomic model is presented. The chapter concludes with a discussion of the Schrödinger equation for the ground state of the hydrogen atom providing several sample calculations for the ground state of the hydrogen atom based on the Schrödinger equation.

3.1 Bohr Model of the Hydrogen Atom

In 1913 *Niels Bohr* combined Rutherford's concept of the nuclear atom with Planck's idea of the quantized nature of the radiative process and developed, from first principles, an atomic model that successfully deals with one-electron structures like the hydrogen atom and one-electron ions such as singly ionized helium, doubly ionized lithium, etc. forming a hydrogen-like or hydrogenic structure. The model, known as the Bohr model of the atom, is based on four postulates that combine principles of classical mechanics with the concept of angular momentum quantization.

The *four Bohr postulates* are stated as follows:

1. **Postulate 1**: Electrons revolve about the Rutherford nucleus in well-defined, allowed orbits (referred to as shells). The Coulomb force of attraction $F_{coul} = Ze^2 / \left(4\pi\varepsilon_0 r^2\right)$ between the electrons and the positively charged nucleus is balanced by the centrifugal force $F_{cent} = m_e v^2 / r$, where Z is the number of protons in the nucleus (atomic number); r the radius of the orbit or shell; m_e the electron mass; and v the velocity of the electron in the orbit.

2. **Postulate 2**: While in orbit, the electron does not lose any energy despite being constantly accelerated (this postulate is in contravention of the basic law of nature which states that an accelerated charged particle will lose part of its energy in the form of radiation).

3. **Postulate 3**: The angular momentum $L = m_e v r$ of the electron in an allowed orbit is quantized and given as $L = n\hbar$, where n is an integer referred to as the principal quantum number and $\hbar = h/(2\pi)$ is the reduced Planck constant with h the Planck constant. The simple quantization of angular momentum stipulates that the angular momentum can have only integral multiples of a basic unit which is equal to $\hbar$.

4. **Postulate 4**: An atom or ion emits radiation when an orbital electron makes a transition from an initial allowed orbit with quantum number n_i to a final allowed orbit with quantum number n_f for $n_i > n_f$.

The angular momentum quantization rule simply means that $\hbar$ is the lowest angular momentum available to the electron ($n = 1$, ground state) and that higher n orbits ($n > 1$, excited states) can only have integer values of $\hbar$ for the magnitude of the orbital angular momentum, where n is the *principal quantum number* or the shell number. One-electron atomic structures are now referred to as the Bohr atom, while the atomic model consisting of a nucleus and electrons in planetary orbits about the nucleus is called the Rutherford–Bohr atomic model in honor of Rutherford who introduced the nuclear atomic model and Bohr who explained its kinematics from first principles.

3.1.1 Radius of the Bohr Atom

Assuming that the mass of the nucleus M is much larger than the mass of the electron m_e, i.e., $M \gg m_e$ and that $M \to \infty$, equating the centrifugal force and the Coulomb force on the electron (see Fig. 3.1a)

$$\frac{m_e v^2}{r_n} = \frac{1}{4\pi\varepsilon_0}\frac{Ze^2}{r_n^2} \tag{3.1}$$

and inserting the quantization relationship for the angular momentum L of the electron (third Bohr postulate)

$$L = m_e v_n r_n = m_e \omega_n r_n^2 = n\hbar, \tag{3.2}$$

we get the following relationship for r_n, the radius of the nth allowed Bohr orbit,

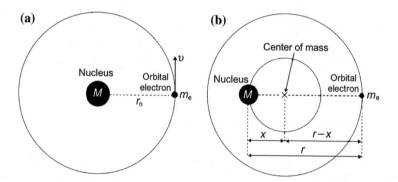

Fig. 3.1 Schematic diagram of the Rutherford–Bohr atomic model. In part **a** electron revolves about the center of the nucleus M under the assumption that $M \to \infty$; in part **b** nuclear mass is finite and both the electron as well as the nucleus revolve about their common center-of-mass

$$r_n = \frac{4\pi\varepsilon_0}{e^2} \frac{(\hbar c)^2}{m_e c^2} \left(\frac{n^2}{Z}\right) = a_0 \left(\frac{n^2}{Z}\right) = (0.5292 \text{ Å}) \times \left(\frac{n^2}{Z}\right), \qquad (3.3)$$

where a_0 is a constant called the Bohr radius of the Bohr one-electron atom and given as

$$a_0 = \frac{4\pi\varepsilon_0}{e^2} \frac{(\hbar c)^2}{m_e c^2} = 0.5292 \text{ Å}. \qquad (3.4)$$

3.1.2 Velocity of the Bohr Electron

Inserting the expression for r_n of (3.3) into (3.2) we obtain the following expression for v_n/c, where v_n is the velocity of the electron in the nth allowed Bohr orbit

$$\frac{v_n}{c} = \frac{n\,\hbar c}{m_e c^2 r_n} = \frac{e^2}{4\pi\varepsilon_0}\frac{1}{\hbar c}\left(\frac{Z}{n}\right) = \alpha\left(\frac{Z}{n}\right) \approx \frac{1}{137}\left(\frac{Z}{n}\right), \qquad (3.5)$$

where α is the so-called fine structure constant expressed as

$$\alpha = \frac{e^2}{4\pi\varepsilon_0}\frac{1}{\hbar c} = \frac{1}{137} \approx 7.3 \times 10^{-3}. \qquad (3.6)$$

Since, as evident from (3.5), the electron velocity in the ground state ($n = 1$) orbit of hydrogen is less than 1% of the speed of light c in vacuum, the use of classical mechanics in one-electron Bohr atom is justifiable. Both Rutherford and Bohr used classical mechanics in their momentous discoveries of the atomic structure and the kinematics of electronic motion, respectively. On the one hand, nature provided Rutherford with an atomic probe in the form of naturally occurring α-particles having just the appropriate energy (few MeV) to probe the atom without having to deal with relativistic effects and nuclear penetration. On the other hand, nature provided Bohr with the hydrogen one-electron atom in which the electron orbital velocity is less than 1% of the speed of light in vacuum so that the electron can be treated with simple classical relationships.

3.1.3 Total Energy of the Bohr Electron

The total energy E_n of the electron when in one of the allowed orbits (shells) with radius r_n is the sum of the electron's kinetic energy E_K and potential energy E_P and can be, as shown by Bohr, expressed as follows

$$E_n = E_K + E_P = \frac{m_e v_n^2}{2} + \frac{Ze^2}{4\pi\varepsilon_0} \int_{\infty}^{r_n} \frac{dr}{r^2} = \frac{1}{2} \frac{Ze^2}{4\pi\varepsilon_0} \frac{1}{r_n} - \frac{Ze^2}{4\pi\varepsilon_0} \frac{1}{r_n}$$

$$= -\frac{1}{2} \frac{Ze^2}{4\pi\varepsilon_0} \frac{1}{r_n} = -\frac{1}{2} \left(\frac{e^2}{4\pi\varepsilon_0} \right)^2 \frac{m_e c^2}{(\hbar c)^2} \left(\frac{Z}{n} \right)^2 = -E_R \left(\frac{Z}{n} \right)^2$$

$$= -(13.61 \text{ eV}) \times \left(\frac{Z}{n} \right)^2. \tag{3.7}$$

Equation (3.7) represents the energy quantization of allowed bound electronic states in a one-electron atom. This energy quantization is a direct consequence of the simple angular momentum quantization $L = n\hbar$ introduced by Bohr. E_R is a constant called Rydberg energy and is expressed as

$$E_R = \frac{1}{2} \left(\frac{e^2}{4\pi\varepsilon_0} \right)^2 \frac{m_e c^2}{(\hbar c)^2} = \frac{1}{2} \alpha^2 m_e c^2 = 13.61 \text{ eV}. \tag{3.8}$$

The energy level diagram for a hydrogen atom is shown in Fig. 3.2. It provides an excellent example of energy level diagram for one-electron structures such as hydrogen, singly ionized helium atom, or doubly ionized lithium atom. The energy levels for hydrogen were calculated from (3.7) using $Z = 1$. The six lowest bound energy levels ($n = 1$ through $n = 6$) of the hydrogen atom according to (3.7) are: -13.61 eV, -3.40 eV, -1.51 eV, -0.85 eV, -0.54 eV, and -0.38 eV.

By convention the following general conditions apply:

1. The negative energy levels of the electron represent discrete allowed electron states bound to the nucleus with a given binding energy E_B.
2. An electron bound to the nucleus in a one-electron configuration can only attain discrete allowed negative energy levels, as predicted by (3.7). Energies of bound states in an atom are negative; however, the binding energy of the electron in a bound state is positive. Thus we say that the energy level of the hydrogen ground state is $E_1 = -13.61$ eV, but the electron is bound in the hydrogen atom with a binding energy $E_B = 13.61$ eV.
3. A stationary free electron, infinitely far from the nucleus has zero kinetic energy.
4. An electron with positive energy is free and moving in a continuum of kinetic energies.
5. The zero energy level separates the continuum of positive free electron kinetic energies from the negative discreet energy levels of bound atomic electrons.
6. Electron in $n = 1$ state is said to be in the ground state; an electron in a bound state with $n > 1$ is said to be in an excited state.
7. Energy must be supplied to an electron in the ground state of a hydrogen atom to move it to an excited state. An electron cannot remain in an excited state; it will move to a lower level shell and the transition energy will be emitted in the form of a photon.

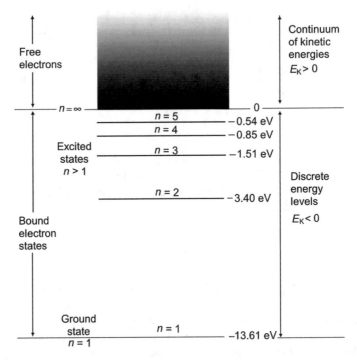

Fig. 3.2 Energy level diagram calculated with (3.7) for the hydrogen atom ($Z = 1$) as example of energy level diagram for one-electron structure. In the ground state ($n = 1$) of hydrogen the electron is bound to the nucleus with a binding energy of 13.61 eV and the first excited state is at -3.40 eV

3.1.4 Transition Frequency and Wave Number

The energy $h\nu$ of a photon emitted as a result of an electronic transition from an initial allowed orbit with $n = n_i$ to a final allowed orbit with $n = n_f$, where $n_i > n_f$, is given by

$$h\nu = E_i - E_f = -E_R Z^2 \left(\frac{1}{n_i^2} - \frac{1}{n_f^2} \right). \tag{3.9}$$

The wave number k of the emitted photon is then given by

$$k = \frac{1}{\lambda} = \frac{\nu}{c} = \frac{E_R}{2\pi\hbar c} Z^2 \left(\frac{1}{n_f^2} - \frac{1}{n_i^2} \right) = R_\infty Z^2 \left(\frac{1}{n_f^2} - \frac{1}{n_i^2} \right)$$

$$= \left(109\,737 \text{ cm}^{-1} \right) \times Z^2 \left(\frac{1}{n_f^2} - \frac{1}{n_i^2} \right), \tag{3.10}$$

where R_∞ is the so-called Rydberg constant expressed as

$$R_\infty = \frac{E_R}{2\pi\hbar c} = \frac{1}{4\pi}\left(\frac{e^2}{4\pi\varepsilon_0}\right)^2 \frac{m_e c^2}{(\hbar c)^3} = \frac{\alpha^2 m_e c^2}{4\pi\hbar c} = 109\,737 \text{ cm}^{-1}. \qquad (3.11)$$

3.1.5 Atomic Spectra of Hydrogen

Photons emitted by excited atoms are concentrated at a number of discrete wavelengths (lines). The hydrogen spectrum is relatively simple and results from transitions of a single electron in the hydrogen atom. Table 3.1 gives a listing for the first six known series of the hydrogen emission spectrum. It also provides the upper limit in electron volts corresponding to the lower limit in ångströms (Å) for each of the six series. The Lyman series is in the ultraviolet region of the photon spectrum, the Balmer series is in the visible region, all the other series are in the infrared region.

3.1.6 Correction for Finite Mass of the Nucleus

A careful experimental study of the hydrogen spectrum has shown that the Rydberg constant for hydrogen is $R_H = 109\,677$ cm^{-1} rather than the $R_\infty = 109\,737$ cm^{-1} value that Bohr derived from first principles. This small discrepancy is of the order of one part in 2000 and arises from Bohr's assumption that the nuclear mass (proton in the case of hydrogen atom) M is infinite and that the electron revolves about a point at the center of the nucleus, as shown schematically in Fig. 3.1a.

When the finite mass of the nucleus M is taken into consideration, both the electron and the nucleus revolve about their common center-of-mass, as shown schematically in Fig. 3.1b. The total angular momentum L of the system is given by the following expression:

$$L = m_e (r - x)^2 \omega + M x^2 \omega, \qquad (3.12)$$

Table 3.1 Characteristics of the first six emission series of the hydrogen atom

Name of series	Spectral range	Final orbit n_f	Initial orbit n_i	Limit of series (eV)	Limit of series (Å)
Lyman	Ultraviolet	1	$2, 3, 4, \ldots, \infty$	13.61	911
Balmer	Visible	2	$3, 4, 5, \ldots, \infty$	3.40	3646
Paschen	Infrared	3	$4, 5, 6, \ldots, \infty$	1.51	8210
Brackett	Infrared	4	$5, 6, 7, \ldots, \infty$	0.85	14584
Pfund	Infrared	5	$6, 7, 8, \ldots, \infty$	0.54	22957
Humphreys	Infrared	6	$7, 8, 9, \ldots, \infty$	0.38	32623

where

r	is the distance between the electron and the nucleus.
x	is the distance between the center-of-mass and the nucleus.
$r - x$	is the distance between the center-of-mass and the electron.
m_e	is the mass of the orbital electron.
M	is the mass of the nucleus.
ω	is the angular frequency of the electron in an allowed orbit.

After introducing the relationship which states that the radii of the orbits (x for the nucleus and $r - x$ for the electron) are in inverse proportion to the masses (M for the nucleus and m_e for the electron)

$$m_e\,(r - x) = Mx \tag{3.13}$$

into (3.12), the angular momentum L for the atomic nucleus–electron system may be written as

$$L = Mx^2\omega + m_e\,(r - x)^2\,\omega = \frac{m_e\,M}{m_e + M}r^2\omega = \mu_M r^2\omega, \tag{3.14}$$

where μ_M is the so-called *reduced mass* of the atomic nucleus–electron system given as

$$\mu_M = \frac{m_e\,M}{m_e + M} = \frac{1}{\dfrac{1}{m_e} + \dfrac{1}{M}} = \frac{m_e}{1 + \dfrac{m_e}{M}}. \tag{3.15}$$

All Bohr relationships, given above for one-electron structures in (3.3) through (3.7) under the assumption that the nuclear mass M is infinite ($M \to \infty$) compared to the electron mass m_e, are also valid for finite nuclear masses M as long as the electron rest mass m_e in these relationships is replaced with the appropriate reduced mass μ_M. Thus, the corrected expression for the radius r_n of the Bohr one-electron atom is

$$r_n = \frac{4\pi\varepsilon_0}{e^2}\frac{\hbar}{\mu_M}\frac{n^2}{Z} = a_0\frac{m_e}{\mu_M}\frac{n^2}{Z} = a_0\left(1 + \frac{m_e}{M}\right)\frac{n^2}{Z} \tag{3.16}$$

and the corrected energy levels E_n are

$$E_n = -\frac{1}{2}\left(\frac{e^2}{4\pi\varepsilon_0}\right)^2\frac{\mu_M c^2}{(\hbar c)^2}\left(\frac{Z}{n}\right)^2 = -E_R\frac{\mu_M}{m_e}\left(\frac{Z}{n}\right)^2 = -E_R\frac{1}{1 + \dfrac{m_e}{M}}\left(\frac{Z}{n}\right)^2. \tag{3.17}$$

The correction accounting for a finite rather than infinite nuclear mass thus amounts to $(1 + m_e/M)$ for r_n and to $(1 + m_e/M)^{-1}$ for E_n. Both are minute and often ignored corrections of the order of 0.05% for hydrogen (proton) and even smaller for larger mass nuclei. From (3.16) and (3.17) it is also evident that the Bohr atom radius

Table 3.2 Nuclear mass correction factor C_M, Rydberg constant R_M, and orbital electron binding energy E_B for the lowest atomic number Z one-electron atoms or ions [$R_\infty = 109\,737$ cm^{-1}; $C_M = 1 + (m_e c^2/Mc^2)$; and $R_M = R_\infty/C_M$]

One-electron atom or ion	Nuclear rest energy (MeV)	$\dfrac{m_e c^2}{Mc^2}$	C_M	R_M	E_B(eV)
Hydrogen	938.3	5.446×10^{-4}	1.0005446	109677	$13.61/C_H$
Deuterium	1876	2.724×10^{-4}	1.0002724	109707	$13.61/C_d$
Tritium	2809	1.819×10^{-4}	1.0001819	109717	$13.61/C_t$
Helium-3	2808	1.821×10^{-4}	1.0001821	109717	$54.44/C_{He-3}$
Helium-4	3727	1.371×10^{-4}	1.0001371	109722	$54.44/C_{He-4}$
Lithium-7	6534	0.782×10^{-4}	1.0000782	109728	$122.5/C_{Li-7}$
∞	∞	0	1.00	R_∞	$13.61 \times Z^2$

r_n scales inversely with the reduced mass μ_M, while the Rydberg energy E_n scales linearly with the reduced mass μ_M.

For the hydrogen atom the reduced mass μ_H of the electron–proton system is very close to the electron mass because the proton is much heavier than the electron $\left(m_p : m_e = 1836 : 1\right)$

$$\mu_H = \frac{m_e m_p}{m_e + m_p} = \frac{m_e}{1 + \dfrac{m_e}{m_p}} = 0.9995 m_e \tag{3.18}$$

and the Rydberg constant R_H is given as

$$R_H = \frac{\mu_H}{m_e} R_\infty = \frac{1}{1 + \dfrac{m_e}{m_p}} R_\infty = \frac{109\,737 \text{ cm}^{-1}}{1 + \dfrac{1}{1836}} = 109\,677 \text{ cm}^{-1}, \tag{3.19}$$

representing a 1 part in 2000 correction, in excellent agreement with the experimental result which was measured for the hydrogen emission spectrum.

Table 3.2 displays, for the lowest atomic number Z one-electron structures, the nuclear mass correction factor C_M, the Rydberg constant R_M, and the binding energy E_B of the orbital electron to the nucleus. It is evident that the finite nuclear mass correction to the Bohr theory is indeed very small amounting to 5 parts per 10^4 for hydrogen and rapidly falling to even lower values with increasing Z.

3.1.7 Positronium, Muonium, and Muonic Atom

In addition to one-electron atoms and ions, several more exotic, short-lived, and unusual "atomic" structures are known whose kinematics can be described using the same concepts as those applied to the Bohr atom. However, to achieve meaningful theoretical results, the use of the appropriate reduced mass μ_M rather than the electron mass m_e is mandatory, because μ_M for these structures can be significantly different from m_e in contrast to the reduced mass of the Bohr atom which is within 0.05% or better of the electron mass.

Examples of these special atom-like structures are: positronium Ps, muonium Mu, and muonic atom. The basic properties of these structures pertinent to the Bohr model are summarized in Table 3.3. These structures are not of much interest in medical physics; however, their Bohr atom-related behavior serves as an excellent example of the relevance of the reduced mass μ_M to the Bohr atomic model.

The *positronium* (Ps) is a semi-stable, hydrogen-like atomic configuration consisting of a positron and electron revolving about their common center-of-mass before the process of annihilation occurs (Sect. 7.6.10). Because it consists of two particles of equal mass, positronium is sometimes considered to be the lightest atom and carries the chemical symbol Ps. Its lifetime is of the order of 10^{-7} s and its reduced mass equals to $0.5 m_e$. Croatian physicist Stjepan Mohorovičić predicted the existence of positronium in 1934 and the Austrian-American physicist Martin Deutsch discovered it in 1951.

Muonium (Mu) is a light, hydrogen-like neutral atom consisting of a positive muon μ^+ and an orbital electron e^-. Physical chemists consider muonium to be a light unstable isotope of hydrogen. It is formed when an energetic positive muon slows down in an absorber and attracts an electron which then revolves about the muon similarly to an orbital electron revolving about a proton in the hydrogen atom. The reduced mass of muonium is within 0.5% of the electron mass.

Muonic atom is an atom in which an electron is replaced by a negative muon μ^- orbiting close to or within the nucleus. Muonic hydrogen is the simplest muonic atom consisting of a proton and negative muon with a reduced mass of $186 m_e$.

Table 3.3 Basic properties of Bohr atom, positronium Ps, muonium Mu, and muonic hydrogen related to Bohr theory for one-electron structures

Atomic system	Bohr atom	Positronium Ps	Muonium Mu	Muonic hydrogen
Constituents	p and e^-	e^+ and e^-	μ^+ and e^-	p and μ^-
	$m_p = 1836\, m_{e^-}$	$m_{e^+} = m_{e^-}$	$m_{\mu^+} = 207\, m_{e^-}$	$m_{\mu^-} = 207\, m_{e^-}$
Reduced mass μ_M	$0.9995\, m_e$	$0.5\, m_e$	$0.995\, m_e$	$186\, m_e$
Radius of orbits r_n	$1.0005\, a_0 n^2$	$2\, a_0 n^2$	$1.005\, a_0 n^2$	$a_0 n^2 / 186$
Energy levels E_n	$-0.9995\, E_R \dfrac{1}{n^2}$	$-0.5\, E_R \dfrac{1}{n^2}$	$-0.995\, E_R \dfrac{1}{n^2}$	$-186\, E_R \dfrac{1}{n^2}$
Ground state E_1 (eV)	-13.61	-6.805	-13.54	-2531.5
Rydberg constant R_M	$0.9995\, R_\infty$	$0.5\, R_\infty$	$0.995\, R_\infty$	$186\, R_\infty$

Muons are unstable elementary particles with a mass of $207m_e$ belonging to the group of leptons which also incorporates the electron, tau, and neutrinos. They are produced in high-energy particle accelerators, have a mean lifetime of 2.2 μs, and decay into two neutrinos and an electron or positron depending on their charge.

3.1.8 Quantum Numbers

The Bohr atomic theory predicts quantized energy levels for the one-electron hydrogen atom that depend only on n, the principal quantum number, since $E_n = -E_R/n^2$, as shown in (3.7). In contrast, the solution of the Schrödinger equation in spherical coordinates for the hydrogen atom (Sect. 3.4) gives three quantum numbers for the hydrogen atom: n, ℓ, and m_ℓ, where

n is the *principal quantum number* with allowed values $n = 1, 2, 3, \ldots$, giving the electron binding energy in shell n as $E_n = -E_R/n^2$.

ℓ is the *orbital angular momentum quantum number* with the following allowed values $\ell = 0, 1, 2, 3, \ldots, n - 1$, giving the electron orbital angular momentum $L = \hbar\sqrt{\ell(\ell+1)}$.

m_ℓ is referred to as the *magnetic quantum number* giving the z component of the orbital angular momentum $L_z = m_\ell\hbar$ and has the following allowed values: $m_\ell = -\ell, -\ell+1, -\ell+2, \ldots, \ell-2, \ell-1, \ell$.

Fig. 3.3 Schematic representation of the Stern–Gerlach experiments. Part **a** shows that silver atoms originate in an oven and are formed into a beam by a collimator before they pass through an inhomogeneous magnet and strike a glass collector plate. **b** Cross section through the magnet in a plane perpendicular to the motion of the silver atoms. **c, d, e** Sketches of the expected beam trace on the receptor plate using classical physics in **c** and **d** and quantum physics in **e**

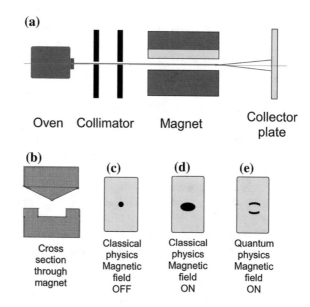

3.1.9 Stern–Gerlach Experiment and Electron Spin

In 1921 an experiment conducted by *Otto Stern* and *Walter Gerlach* has shown that each electron, in addition to its orbital angular momentum **L**, possesses an intrinsic angular momentum referred to as the *spin* **S**. The Stern–Gerlach experiment has shown that electron spin is quantized into two states and provided a major impetus for development of the quantum theory of the atom.

As shown schematically in Fig. 3.3a, the experiment consisted of passing a beam of neutral silver atoms through a strongly inhomogeneous magnetic field with a gradient transverse to their direction of motion. The silver atoms originated in an oven and were formed into a narrow beam by a collimator before traversing the inhomogeneous magnet. After passing through the magnet, the silver atoms were collected on a glass plate. Figure 3.3b shows a cross section through the magnet in a plane perpendicular to the direction of motion of the gold atoms.

Stern and Gerlach designed their experiment with the aim of proving spatial quantization produced by orbiting electrons in an atom. Rather than finding an expected $(2\ell + 1)$ quantization, they actually observed a splitting of the silver beam into only two components. Thus, while in classical mechanics there would be no splitting of the beam, as suggested in Fig. 3.3c and d, the result found by Stern and Gerlach unequivocally proved spatial quantization but was clearly unexpected showing only two silver beam components, as illustrated in Fig. 3.3e.

The Stern–Gerlach experiment eventually lead to a conclusion that electrons, in addition to possessing an orbital angular momentum, also possess an intrinsic angular momentum known as the spin, specified by two quantum numbers: $s = 1/2$ and m_s that can take two values ($1/2$ or $-1/2$). The electron spin is given as $S = \hbar\sqrt{s(s+1)} = \hbar\sqrt{3}/2$ and its z component $S_z = m_s\hbar$. Thus, for each set of the spatial quantum numbers n, ℓ, and m_ℓ that follow from the Schrödinger equation there are two options for the spin quantum number (spin up with $m_s = +1/2$ and spin down with $m_s = -1/2$).

A neutral silver atom has one electron in its outer shell and the intrinsic spin of this unpaired electron causes the silver atom to behave like a small magnet with two poles (magnetic dipole). In moving through a uniform magnetic field this atomic dipole precesses in the field. However, if this atomic dipole moves through an inhomogeneous magnetic field, the magnetic forces acting on each of the two poles are different from one another and this causes a deflection of the atom from its straight trajectory as it traverses the magnet.

If silver atoms possess one of two quantized intrinsic angular momentum states provided by the unpaired outer shell electron, in passing through the inhomogeneous magnetic field, they will first take up their quantized orientations with respect to the field direction and then deviate from their original path following one of two trajectories: one caused by deflection in the direction of the field gradient and the other caused by deflection in direction opposite to the field gradient.

Stern and Gerlach found a clear separation of silver atoms between two traces on the receptor plate and this served as proof of spatial quantization from which the quantization of the intrinsic angular momentum of the electron was inferred. The Stern–Gerlach experiment is one of about a dozen seminal experiments that shaped modern physics during the past century.

3.1.10 Spin–Orbit Coupling

The orbital and spin angular momenta of an electron actually interact with one another. This interaction is referred to as the *spin–orbit coupling* and results in a total electronic angular momentum $\mathbf{J}$ that is the vector sum of the orbital and intrinsic spin components, i.e., $\mathbf{J} = \mathbf{L} + \mathbf{S}$. The following features are notable:

- The total angular momentum $\mathbf{J}$ has the value $J = \hbar\sqrt{j\,(j+1)}$ where the possible values of the quantum number j are: $|\ell - s|$, $|\ell - s + 1|$, ..., $|\ell + s|$, with $s = 1/2$ for all electrons.
- The z component of the total angular momentum has the value $J_z = m_j\hbar$, where the possible values of m_j are: $-j$, $-j + 1$, $-j + 2$, ..., $j - 2$, $j - 1$, j.
- The state of atomic electrons is thus specified with a set of four quantum numbers: n, ℓ, m_ℓ, m_s when there is *no spin–orbit interaction* and n, ℓ, j, m_j when there is *spin–orbit interaction*.

3.1.11 Successes and Limitations of the Bohr Atomic Model

With his four postulates and the innovative idea of angular momentum quantization Bohr provided an excellent extension of the Rutherford atomic model and succeeded in explaining quantitatively the photon spectrum of the hydrogen atom and other one-electron structures such as singly ionized helium, doubly ionized lithium, etc.

According to the Bohr atomic model, each of the six known series of the hydrogen spectrum arises from a family of electronic transitions that all end at the same final state n_f. The Lyman ($n_f = 1$), Brackett ($n_f = 4$), and Pfund ($n_f = 5$) series were not known at the time when Bohr proposed his model; however, the three series were discovered soon after Bohr predicted them with his model. The Humphreys ($n_f = 6$) series has been discovered only recently.

In addition to its tremendous successes, the Bohr atomic model suffers several severe limitations:

1. The orbital electron in revolving about the nucleus is constantly accelerated and by virtue of its charge should lose part of its energy in the form of photons (Larmor law) and spiral into the nucleus. With its assumption that the electron, while in an allowed orbit, emits no photons the Bohr model is in contravention of Larmor law.
2. The model does not predict the relative intensities of the photon emission in characteristic orbital transitions.

3. The model fails to explain the observed fine structure of hydrogen spectral lines where each spectral line is further composed of closely spaced spectral lines.
4. The model does not work quantitatively for multi-electron atoms.
5. The model does not explain the splitting of a spectral line into several lines in a magnetic field (Zeeman effect).
6. The model does not explain the splitting of a spectral line into several lines in an electric field (Stark effect).

The idea of atomic electrons revolving about the nucleus should not be taken too literally; however, the Bohr model for the one-electron structure combined with angular momentum quantization serves as a reasonable intermediate step on the way to more elaborate and accurate theories provided with quantum mechanics and quantum electrodynamics. The Schrödinger quantum theory dispensed with the picture of electrons moving in well defined orbits but the Bohr theory is still often used to provide the first approximation to a particular problem, because it is known to provide reasonable results with mathematical procedures that are, in comparison with those employed in quantum mechanics, significantly simpler and faster.

3.1.12 Correspondence Principle

Niels Bohr postulated that the smallest change in angular momentum L of a particle is equal to $\hbar$ where $\hbar$ is the reduced Planck constant ($2\pi\hbar = h$). This is seemingly in drastic disagreement with classical mechanics where the angular momentum as well as the energy of a particle behave as continuous functions. In macroscopic systems the angular momentum quantization is not noticed because $\hbar$ represents such a small fraction of the angular momentum; on the atomic scale, however, $\hbar$ may be of the order of the angular momentum making the $\hbar$ quantization very noticeable.

The *correspondence principle* proposed by *Niels Bohr* in 1923 states that for large values of the principal quantum number n (i.e., for $n \to \infty$) the quantum and classical theories must merge and agree. In general, the correspondence principle stipulates that the predictions of the quantum theory for any physical system must match the predictions of the corresponding classical theory in the limit where the quantum numbers specifying the state of the system are very large. This principle can be used to confirm the Bohr angular momentum quantization ($L = n\hbar$) postulate as follows.

Consider an electron that makes a transition from an initial orbit $n_i = n$ to a final orbit $n_f = n - \Delta n$, where n is large and $\Delta n \ll n$. The transition energy ΔE and the transition frequency ν_{trans} of the emitted photon are related as follows:

$$\Delta E = E_{initial} - E_{final} = 2\pi\hbar\, \nu_{trans} \tag{3.20}$$

and

$$\nu_{trans} = \frac{\Delta E}{2\pi\hbar}. \tag{3.21}$$

Since n is large, we can calculate ΔE from the derivative with respect to n of the total orbital energy E_n given in (3.7) to obtain

$$\frac{dE_n}{dn} = \frac{d}{dn}\left(-E_R\frac{Z^2}{n^2}\right) = 2E_R\frac{Z^2}{n^3}. \tag{3.22}$$

To get ΔE we express (3.7) as follows

$$\Delta E = 2E_R Z^2 \frac{\Delta n}{n^3} \tag{3.23}$$

resulting in the following expression for the transition frequency ν_{trans}

$$\nu_{trans} = \frac{\Delta E}{2\pi\hbar} = \frac{2E_R Z^2}{2\pi\hbar}\frac{\Delta n}{n^3} = \left\{\frac{Ze^2}{4\pi\varepsilon_0}\right\}^2 \frac{m_e}{2\pi\hbar^3}\frac{\Delta n}{n^3} \tag{3.24}$$

after we insert E_R from (3.8).

Recognizing that the velocity v and angular frequency ω are related through $v = \omega r$, we get from (3.1) the following expression

$$\frac{Ze^2}{4\pi\varepsilon_0} = m_e v^2 r = m_e\omega^2 r^3, \tag{3.25}$$

which, when squared, yields

$$\left\{\frac{Ze^2}{4\pi\varepsilon_0}\right\}^2 = m_e^2\omega^4 r^6. \tag{3.26}$$

The angular momentum was given in (3.2) as

$$L = n\hbar = m_e v r = m_e\omega r^2, \tag{3.27}$$

which, when cubed, gives

$$n^3\hbar^3 = m_e^3\omega^3 r^6. \tag{3.28}$$

Combining (3.26) and (3.28) with (3.24), we get the following expression for the transition frequency ν_{trans}

$$\nu_{trans} = \left\{\frac{Ze^2}{4\pi\varepsilon_0}\right\}^2 \frac{m_e}{2\pi\hbar^3}\frac{\Delta n}{n^3} = \frac{m_e^2\omega^4 r^6 m_e \Delta n}{2\pi m_e^3\omega^3 r^6} = \frac{\omega}{2\pi}\Delta n. \tag{3.29}$$

After incorporating expressions for r_n and v_n given in (3.3) and (3.5), respectively, the classical orbital frequency ν_{orb} for the orbit n is given as

$$\nu_{\text{orb}} = \frac{\omega_n}{2\pi} = \frac{\upsilon_n}{2\pi r_n} = \frac{\alpha c}{2\pi a_0 n^3}. \tag{3.30}$$

It is of note that ν_{trans} of (3.24) equals to ν_{orb} of (3.30) for large values of n and $\Delta n = 1$, confirming the correspondence between quantum physics and classical physics for $n \to \infty$.

We now compare the transition frequency ν_{trans} and orbital frequency ν_{orb} for a small n transition from $n_i = 2$ to $n_f = 1$ in a hydrogen atom ($Z = 1$) and obtain

$$\nu_{\text{orb}}(n = 2) = \frac{\upsilon_2}{2\pi r_2} = \frac{\alpha c}{16\pi a_0} = 8.24 \times 10^{14} \text{ s}^{-1} \tag{3.31}$$

and

$$\nu_{\text{trans}} = \frac{E_2 - E_1}{2\pi \hbar} = \frac{E_R}{2\pi \hbar}\left\{1 - \frac{1}{4}\right\} = \frac{3 E_R}{8\pi \hbar} = \frac{3\alpha c}{16\pi a_0} = 24.72 \times 10^{14} \text{ s}^{-1} = 3\nu_{\text{orb}} \tag{3.32}$$

From (3.31) and (3.32) we note that for low values of n the orbital frequency ν_{orb} differs from the transition frequency ν_{trans}, in contrast to the situation at large n where $\nu_{\text{trans}} = \nu_{\text{orb}}$, as shown by (3.29) and (3.30). Thus, at large n there is agreement between quantum and classical physics, as predicted by the correspondence principle enunciated by *Niels Bohr*, while for low n quantum and classical physics give different results, with $\nu_{\text{trans}} > \nu_{\text{orb}}$.

3.2 Multi-electron Atom

A multi-electron atom of atomic number Z contains a nucleus of charge $+Ze$ surrounded by Z electrons, each of charge $-e$ and revolving in an orbit about the nucleus. The kinematics of individual electron motion and its energy level in a multi-electron atom are governed by:

1. Kinetic energy of the orbital electron.
2. Attractive Coulomb force between the electron and the nucleus.
3. Repulsive Coulomb force exerted on the electron by the other $Z - 1$ atomic electrons.
4. Weak interactions involving orbital and spin angular momenta of orbital electrons.
5. Minor interactions between the electron and nuclear angular momenta.
6. Relativistic effects and the effect of the finite nuclear size.

3.2.1 Exclusion Principle

Wolfgang Pauli in 1925 eloquently answered the question on the values of quantum numbers assigned to individual electrons in a multi-electron atom. Pauli exclusion principle that states: "*In a multielectron atom there can never be more than one electron in the same quantum state*" is important for the understanding of the properties of multi-electron atoms and the periodic table of elements. The following conditions apply:

- According to the Pauli exclusion principle in a multi-electron atom no two electrons can have all four quantum numbers identical. The four quantum numbers are: n, ℓ, j, and m_j.
- The energy and position of each electron in a multi-electron atom are most affected by the principal quantum number n. The electrons that have the same value of n in an atom form a *shell*.
- Within a shell, the energy and position of each electron are affected by the value of the orbital angular momentum quantum number ℓ. Electrons that have the same value of ℓ in a shell form a *sub-shell*.
- The specification of quantum numbers n and ℓ for each electron in a multi-electron atom is referred to as the *electronic configuration* of the atom.
- Pauli exclusion principle confirms the shell structure of the atom as well as the sub-shell structure of individual atomic shells:

  - Number of electrons in sub-shells that are labeled with quantum numbers n, ℓ, m_ℓ: $2(2\ell + 1)$
  - Number of electrons in sub-shells that are labeled with quantum numbers n, ℓ, j: $(2j + 1)$
  - Number of electrons in a shell:

$$2\sum_{\ell=0}^{n=1} (2\ell + 1) = 2n^2 \tag{3.33}$$

The main characteristics of atomic shells and sub-shells are given in Tables 3.4 and 3.5, respectively. The spectroscopic notation for electrons in the K, L, and M shells and associated sub-shells is given in Table 3.6.

Table 3.4 Main characteristics of atomic shells

Principal quantum number n	1	2	3	4	5
Spectroscopic notation	K	L	M	N	O
Maximum number of electrons	2	8	18	32	–

Table 3.5 Main characteristics of atomic subshells

Orbital quantum number ℓ	0	1	2	3
Spectroscopic notation	s	p	d	f
Maximum number of electrons	2	6	10	14

Table 3.6 Spectroscopic notation for electrons in the K, L, M shells of a multi-electron atom

Principal quantum number n	Orbital angular momentum ℓ and total angular momentum j of electron					
	$s_{1/2}$	$p_{1/2}$	$p_{3/2}$	$d_{3/2}$	$d_{5/2}$	f
1	K					
2	L_I	L_{II}	L_{III}			
3	M_I	M_{II}	M_{III}	M_{IV}	M_V	

3.2.2 Hartree Approximation for Multi-electron Atoms

The Bohr theory works well for one-electron structures (hydrogen atom, singly ionized helium, doubly ionized lithium, etc.) but does not apply directly to multi-electron atoms because of the repulsive Coulomb interactions among electrons constituting the atom.

The repulsive Coulomb interactions among electrons constituting the atom disrupt the attractive Coulomb interaction between an orbital electron and the nucleus and make it impossible to predict accurately the potential that influences the kinematics of the orbital electron. *Douglas Hartree* proposed an approximation that predicts the energy levels and radii of multi-electron atoms reasonably well despite its inherent simplicity.

Hartree assumed that the potential seen by a given atomic electron is

$$V(r) = -\frac{Z_{\text{eff}}\, e^2}{4\pi\varepsilon_0}\frac{1}{r}, \qquad (3.34)$$

where

Z_{eff} is the effective atomic number.

$Z_{\text{eff}}\, e$ is an effective charge that accounts for the nuclear charge Ze as well as for the effects of all other atomic electrons.

Hartree's calculations show that in multi-electron atoms the effective atomic number Z_{eff} for K-shell electrons ($n = 1$) has a value of about $Z - 2$. Charge distributions of all other atomic electrons produce a charge of about $-2e$ inside a sphere with the radius of the K shell, partially shielding the K-shell electron from the nuclear charge $+Ze$ and producing an effective charge $Z_{\text{eff}} e = (Z - 2)\, e$.

For outer shell electrons Hartree's calculations show that the effective atomic number Z_{eff} approximately equals n, where n specifies the principal quantum number of the outermost filled shell of the atom in the ground state.

Based on the Bohr one-electron atom model, the Hartree relationships for the radii r_n of atomic orbits (shells) and the energy levels E_n of atomic orbits are similar to the Bohr expressions given for the one-electron structures in (3.3) to (3.7) except that the atomic number Z of Bohr expressions is replaced by effective atomic number Z_{eff} in Hartree expressions. The atomic radius r_n and energy level E_n are thus in the Hartree multi-electron approximation expressed as

$$r_n = \frac{a_0 \, n^2}{Z_{eff}} \tag{3.35}$$

and

$$E_n = -E_R \left(\frac{Z_{eff}}{n}\right)^2 . \tag{3.36}$$

The Hartree approximation for the K-shell ($n = 1$) electrons in multi-electron atoms then results in the following expressions for the K-shell radius and K-shell binding energy

$$r_1 = r_K = \frac{a_0}{Z - 2} \tag{3.37}$$

and

$$E_1 = -E_B(K) \approx -E_R(Z - 2)^2, \tag{3.38}$$

showing that the K-shell radii are inversely proportional to $Z - 2$ and the K-shell binding energies increase as $(Z - 2)^2$.

- K-shell radii range from a low of 0.5×10^{-2} Å for very high atomic number elements to 0.53 Å for hydrogen.
- K-shell binding energies (equal to ionization potentials of K shell) range from 13.61 eV for hydrogen to 9 keV for copper, 33 keV for iodine, 69.5 keV for tungsten, 88 keV for lead, and 115 keV for uranium. For $Z > 30$, (3.38) gives values in good agreement with measured data. For example, the calculated and measured K-shell binding energies for copper are 9.9 keV and 9 keV, respectively, for tungsten 70.5 keV and 69.5 keV, respectively, and for lead 87 keV and 88 keV, respectively.

In Fig. 3.4 we plot, for all natural elements with Z from 1 to 100, the K-shell binding energy E_B (K) which may also be referred to as the ionization potential of the K shell. Measured data are shown with the solid curve and data calculated with the Hartree approximation (3.38) are shown with the dashed curve. The agreement between the two curves is reasonable, indicating that the simple Hartree theory provides a good approximation for estimation of K-shell binding energies of all elements.

Fig. 3.4 K-shell binding energy $E_B(K)$ against atomic number Z for elements from $Z = 1$ to $Z = 100$. *Solid curve* represents measured data, *dashed curve* is for data calculated with the Hartree approximation of (3.38)

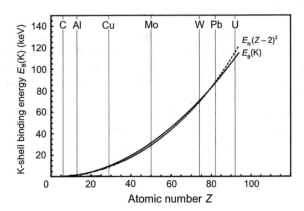

The Hartree approximation for outer shell electrons in multi-electron atoms with ($Z_{eff} \approx n$) predicts the following outer shell radius (radius of atom) and binding energy of outer shell electrons (ionization potential of atom)

$$r_{\text{outer shell}} = \frac{a_0 n^2}{Z_{eff}} \approx na_0 = n \frac{4\pi\epsilon_0}{e^2} \frac{(\hbar c)^2}{m_e c^2} = n \times (0.53 \text{ Å}) \qquad (3.39)$$

and

$$E_{\text{outer shell}} = -E_R \left(\frac{Z_{eff}}{n}\right)^2 \approx -E_R = \frac{1}{2}\alpha^2 m_e c^2 = -13.6 \text{ eV}, \qquad (3.40)$$

with a_0 and E_R the Bohr radius and Rydberg energy, respectively, defined in Sect. 3.1.

The radius of the K-shell constricts with an increasing Z; the radius of the outer-most shell (atomic radius), on the other hand, increases slowly with Z, resulting in a very slow variation of the atomic size with the atomic number Z.

A comparison between the Bohr relationships for one-electron atoms and Hartree relationships for multi-electron atoms is given in Table 3.7. The table compares expressions for the radii of shells, velocities of electrons in shells, energy levels of shells, and wave number for electronic transitions between shells. Two sets of equations are provided for the Bohr theory: a simpler set obtained under the assumption that nuclear mass is infinite ($M \to \infty$) and a set obtained after account is made of the finite nuclear mass.

A simplified energy level diagram for tungsten, a typical multi-electron atom of importance in medical physics for its use as target material in x-ray tubes, is shown in Fig. 3.5. The K, L, M, and N shells are completely filled with their normal allotment of electrons ($2n^2$), the O shell has 12 electrons and the P shell has two electrons. The $n > 1$ shells are actually split into subshells, as discussed in Sect. 3.1.1. In Fig. 3.5 the fine structure of shells is represented by only one energy level that is equal to the average energy of all subshells for a given n.

Table 3.7 Expressions for the radius, velocity, energy, and wave number of atomic structure according to: (1) Bohr one-electron atom model assuming infinite nuclear mass, (2) Bohr one-electron atom model corrected for finite nuclear mass, and (3) Hartree multi-electron model approximation

	One-electron atom Bohr theory		Multi-electron atom Hartree approximation Z_{eff} (for $n=1$) $\approx Z-2$ Z_{eff} (for outer shell) $\approx n$
	Assumption: $M \to \infty$	Corrected for finite nucleus size	
Radius r_n	$r_n = a_0 \dfrac{n^2}{Z}$ $r_1 = \dfrac{a_0}{Z}$ $a_0 = 0.5292\ \text{Å}$	$r_n = a_0 \dfrac{m_e}{\mu_M} \dfrac{n^2}{Z}$ $= a_0 \left(1 + \dfrac{m_e}{M}\right) \dfrac{n^2}{Z}$	$r_n = a_0 \dfrac{n^2}{Z_{\text{eff}}}$ $r_1 = r_K \approx \dfrac{a_0}{Z-2}$ $r_{\text{outer shell}} \approx n a_0$
Velocity v_n	$v_n = \alpha c \dfrac{Z}{n}$ $v_1 = \alpha c Z$ $\alpha = \dfrac{1}{137} \approx 7 \times 10^{-3}$	$v_n = \alpha c \dfrac{Z}{n}$	$v_n = \alpha c \dfrac{Z_{\text{eff}}}{n}$ $v_1 = v_K \approx \alpha c\,(Z-2)$ $v_{\text{outer shell}} \approx \alpha c$
Energy E_n	$E_n = -E_R \left(\dfrac{Z}{n}\right)^2$ $E_1 = -E_R Z^2$ $E_R = 13.61\ \text{eV}$	$E_n = -E_R \dfrac{\mu_M}{m_e} \left(\dfrac{Z}{n}\right)^2$ $= -E_R \dfrac{1}{1 + \dfrac{m_e}{M}} \left(\dfrac{Z}{n}\right)^2$	$E_n = -E_R \left(\dfrac{Z_{\text{eff}}}{n}\right)^2$ $E_1 = E_{K\ \text{shell}}$ $\approx -E_R (Z-2)^2$ $E_{\text{outer shell}} \approx -E_R$
Wave number k	$k = R_\infty Z^2 \left(\dfrac{1}{n_f^2} - \dfrac{1}{n_i^2}\right)$ $R_\infty = 109\,737\ \text{cm}^{-1}$	$k = R_M Z^2 \left(\dfrac{1}{n_f^2} - \dfrac{1}{n_i^2}\right)$ $R_M = \dfrac{\mu_M}{m_e} R_\infty$ $= \dfrac{1}{1 + \dfrac{m_e}{M}} R_\infty$	$k = R_\infty Z_{\text{eff}}^2 \left(\dfrac{1}{n_f^2} - \dfrac{1}{n_i^2}\right)$ $Z_{\text{eff}}(K_\alpha) \approx Z-1$ $k(K_\alpha) \approx \dfrac{3}{4} R_\infty (Z-2)^2$

- A vacancy in a shell with a low quantum number n (inner shell) will result in high-energy transitions in the keV range referred to as *x-ray transitions*.
- A vacancy in a shell with a high quantum number n (outer shell) will result in relatively low-energy transitions (in the eV range) referred to as *optical transitions*.

3.2.3 Periodic Table of Elements

The chemical properties of atoms are periodic functions of the atomic number Z and are governed mainly by electrons with the lowest binding energy, i.e., by outer shell electrons commonly referred to as valence electrons. The periodicity of chemical

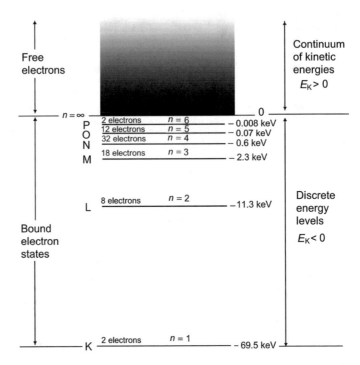

Fig. 3.5 A simplified energy level diagram for the tungsten atom, a typical example of a multi-electron atom

and physical properties of elements (periodic law) was first noticed by *Dmitri Mendeleyev*, who in 1869 produced a periodic table of the then-known elements.

Since Mendeleyev's time the periodic table of elements has undergone several modifications as the knowledge of the underlying physics and chemistry expanded and new elements were discovered and added to the pool. However, the basic principles elucidated by Mendeleyev are still valid today.

In a modern periodic table of elements each element is represented by its chemical symbol and its atomic number. The periodicity of properties of elements is caused by the periodicity in electronic structure that follows the rules of the Pauli exclusion principle (see Sect. 3.2.1).

The periodic table of elements is now most commonly arranged in the form of seven horizontal rows or *periods* and eight vertical columns or *groups*. Elements with similar chemical and physical properties are listed in the same column.

The *periods* in the periodic table are of increasing length as follows:

- Period 1 has two elements: hydrogen and helium.
- Periods 2 and 3 have eight elements each.
- Periods 4 and 5 have 18 elements each.
- Period 6 has 32 elements but lists only 18 entries with entry under lanthanum at $Z = 57$ actually representing the lanthanon series of 15 elements with atomic

numbers from 57 (lanthanum) through 71 listed separately. Synonyms for lanthanon are lanthanide, lanthanoid, rare earth, and rare-earth element.

- Period 7 is still incomplete with 23 elements (ranging in atomic number Z from 87 to 109) but lists only nine entries with entry under actinium at $Z = 89$ actually representing the actinon series of 15 elements with atomic numbers from 89 (actinium) through 103 listed separately. Synonyms for actinon are actinide and actinoid.

The *groups* in the periodic table are arranged into eight distinct groups, each group split into subgroups A and B. Each subgroup has a complement of electrons in the outermost atomic shell (in the range from 1 to 8) that determines its valence, i.e., chemical property, hence the term "valence electron" to designate outer shell electrons.

Table 3.8 gives a simplified modern periodic table of elements with atomic numbers Z ranging from 1 (hydrogen) to 109 (meitnerium). Several groups of elements have distinct names, such as alkali elements (group I.A), alkali earth elements (group II.A), halogens (group VII.A) and noble gases (group VIII.A). Elements of other groups are grouped into transition metals, non-transition metals, non-metals (including halogens of group VII.A), lanthanons, and actinons.

3.2.4 Ionization Energy of Atoms

Ionization energy (IE), also known as ionization potential (IP), of an atom is defined as the energy required for removal of the least bound electron (i.e., the outer shell or valence electron) from the atom. The E_R value predicted by Hartree is only an approximation and it turns out, as shown in Fig. 3.6, that the ionization energies of atoms vary periodically with Z from hydrogen at 13.6 eV to a high value of 24.6 eV for helium down to 4 eV for alkali elements that have only one outer shell (valence) electron.

The highest atomic ionization energy in nature is the ionization energy of the helium atom at 24.6 eV. In contrast, the ionization energy of a singly ionized helium atom $\left(He^+\right)$ can be calculated easily from the Bohr theory using (3.7) with $Z = 2$ and $n = 1$ to obtain an IE of 54.4 eV. This value is substantially higher than the IE for a helium atom because of the two-electron repulsive interaction that lowers the IE in the neutral helium atom.

A plot of the ionization energy IE against atomic number Z, shown in Fig. 3.6, exhibits peaks and valleys in the range from 4 eV to 24.6 eV, with the peaks occurring for noble gases (outer shell filled with a shell-completing complement of orbital electrons such as two electrons for the K shell of helium or eight electrons for the L shell of neon, etc.) and valleys for alkali elements with one solitary electron in the outer shell. Note: the ionization potential of the lead atom is a few electron volts in contrast to the ionization potential of the K shell in lead that is 88 keV.

Table 3.8 Simplified periodic table of elements covering 109 known elements and consisting of 7 periods and 8 groups, each group divided into subgroups A and B

	I. A	II. A	III. B	IV. B	V. B	VI. B	VII. B	VIII. B			I. B	II. B	III. A	IV. A	V. A	VI. A	VII. A	VIII A
1	1 H																	2 He
2	3 Li	4 Be											5 B	6 C	7 N	8 O	9 F	10 Ne
3	11 Na	12 Mg											13 Al	14 Si	15 P	16 S	17 Cl	18 Ar
4	19 K	20 Ca	21 Sc	22 Ti	23 V	24 Cr	25 Mn	26 Fe	27 Co	28 Ni	29 Cu	30 Zn	31 Ga	32 Ge	33 As	34 Se	35 Br	36 Kr
5	37 Rb	38 Sr	39 Y	40 Zr	41 Nb	42 Mo	43 Tc	44 Ru	45 Rh	46 Pd	47 Ag	48 Cd	49 In	50 Sn	51 Sb	52 Te	53 I	54 Xe
6	55 Cs	56 Ba	La	72 Hf	73 Ta	74 W	75 Re	76 Os	77 Ir	78 Pt	79 Au	80 Hg	81 Tl	82 Pb	83 Bi	84 Po	85 At	86 Rn
7	87 Fr	88 Ra	Ac	104 Rf	105 Db	106 Sg	107 Bh	108 Hs	109 Mt									

Alkali metals (group I.A) with one electron in the outer atomic shell:
Lithium Li (3); Sodium Na (11); Potassium K (19); Rubidium Rb (37); Cesium Cs (55); Francium Fr (87).

Alkali earth metals (group II.A) with 2 electrons in the outer atomic shell:
Beryllium Be (4); Magnesium Mg (12); Calcium Ca (20); Strontium Sr (38); Barium Ba (56); Radium Ra (88).

Transition metals with 1 or 2 electrons in the outer atomic shell.

Non-transition metals with 3, 4, 5, or 6 electrons in the outer atomic shell.

Non-metals with 3, 4, 5, 6, or 7 electrons in the outer atomic shell.

Halogens (group VII.A) with 7 electrons in the outer atomic shell
Fluorine F (9); Chlorine Cl (17); Bromine Br (35); Iodine I (53); Astatine At (85).

Noble (Inert) gases (group VIII.A) with 8 electrons in the outer atomic shell
Helium He (4); Neon Ne (10); Argon Ar (18); Krypton Kr(36); Xenon Xe (54); Radon Rn (86).

Lanthanons (lanthanide series from $Z = 57$ to $Z = 71$): **15 elements**

Lanthanum La (57)	Cerium Ce (58)	Praseodymium Pr (59)	Neodymium Nd (60);
Promethium Pm (61)	Samarium Sm (62)	Europium Eu (63)	Gadolinium Gd (64);
Terbium Tb (65)	Dysprosium Dy (66)	Holmium Ho (67)	Erbium Er (68);
Thulium Tm (69)	Ytterbium Yb (70)	Lutetium Lu (71).	

Actinons (actinide series from $Z = 89$ to $Z = 103$): **15 elements**

Actinium Ac (89)	Thorium Th (90)	Protactinium Pr (91)	Uranium U(92)
Neptunium Np (93)	Plutonium Pu (94)	Americium Am (95)	Curium Cm (96);
Berkelium Bk (97)	Californium Cf (98)	Einsteinium Es (99)	Fermium Fm (100)
Mendelevium Md (101)	Nobelium No(102)	Lawrencium Lr(103).	

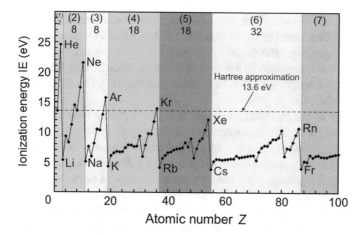

Fig. 3.6 Ionization energy (IE) of atoms against atomic number Z. The noble gases that contain the most stable electronic configurations and the highest ionization potentials are identified, as are the alkali elements that contain the least stable electronic configurations and the lowest ionization potentials with only one valence electron in the outer shell. The numbers in brackets identify the 7 periods, the numbers below the period number indicate the number of atoms in a given period

3.3 Experimental Confirmation of the Bohr Atomic Model

The Rutherford–Bohr atomic model postulates that the total energy of atomic electrons bound to the nucleus is quantized. The binding energy quantization follows from the simple quantization of the electron angular momentum $L = n\hbar$. Direct confirmation of the electron binding energy quantization was obtained from the following three experiments:

1. Measurement of *absorption and emission spectrum* of monoatomic gases
2. *Moseley experiment*
3. *Franck–Hertz experiment*

3.3.1 Emission and Absorption Spectra of Monoatomic Gases

In contrast to the continuous spectra emitted from the surface of solids at high temperatures, the spectra emitted by free excited atoms of gases consist of a number of discrete wavelengths. An electric discharge produces excitations in the gas, and the radiation is emitted when the gas atoms return to their ground state. Correct prediction of line spectra emitted or absorbed by mono-atomic gases, especially hydrogen, serves as an important confirmation of the Rutherford–Bohr atomic model. The following features of emission and absorption spectra of mono-atomic gases are notable:

- The *emission spectrum* is measured by first collimating the emitted radiation by a slit, and then passing the collimated slit-beam through an optical prism or a diffraction grating. The prism or grating breaks the beam into its wavelength spectrum that is recorded on a photographic plate. Each kind of free atom produces its own characteristic emission line, making spectroscopy a useful complement to chemical analysis.
- In addition to the emission spectrum, it is also possible to study the *absorption spectrum* of gases. The experimental technique is similar to that used in measurement of the emission spectrum except that in the measurement of the absorption spectrum a continuous spectrum is made to pass through the gas under investigation. The photographic plate shows a set of unexposed lines that result from the absorption by the gas of distinct wavelengths of the continuous spectrum.
- For every line in the absorption spectrum of a given gas there is a corresponding line in the emission spectrum; however, the reverse is not true. The lines in the absorption spectrum represent transitions to excited states that all originate in the ground state. The lines in the emission spectrum, on the other hand, represent not only transitions to the ground state but also transitions between various excited states. The number of lines in an emission spectrum will thus exceed the number of lines in the corresponding absorption spectrum.

3.3.2 Moseley Experiment

Henry Moseley in 1913 carried out a systematic study of K_α x-rays produced by all then-known elements from aluminum to gold using the Bragg technique of x-ray scattering from a crystalline lattice of a potassium ferrocyanide crystal. The characteristic K_α x-rays (electronic transition from $n_i = 2$ to $n_f = 1$) were produced by bombardment of targets with energetic electrons. The results of Moseley's experiments serve as an excellent confirmation of the Bohr atomic theory.

From the relationship between the measured scattering angle ϕ and the known crystalline lattice spacing d (the Bragg law: $2d \sin \phi = m\lambda$, where m is an integer) Moseley determined the wavelengths λ of K_α x-rays for various elements and observed that the $\sqrt{\nu}$ where ν is the frequency ($\nu = c/\lambda$) of the K_α x-rays was linearly proportional to the atomic number Z. He then showed that all x-ray data could be fitted by the following relationship

$$\sqrt{\nu} = \sqrt{a}(Z - b), \qquad (3.41)$$

where a and b are constants.

The same $\sqrt{\nu}$ versus Z behavior also follows from the Hartree-type approximation that in general predicts the following relationship for the wave number k

$$k = \frac{1}{\lambda} = \frac{\nu}{c} = R_\infty Z_{\text{eff}}^2 \left(\frac{1}{n_f^2} - \frac{1}{n_i^2} \right) \tag{3.42}$$

or

$$\sqrt{\nu} = \sqrt{cR_\infty \left(\frac{1}{n_f^2} - \frac{1}{n_i^2} \right)} Z_{\text{eff}}. \tag{3.43}$$

For K_α characteristic x-rays, where $n_i = 2$ and $n_f = 1$, the Hartree expression gives

$$k\,(K_\alpha) = \frac{3}{4} R_\infty Z_{\text{eff}}^2 = \frac{3}{4} R_\infty (Z - 1)^2. \tag{3.44}$$

Note that in the K_α emission $Z_{\text{eff}} = Z - 1$ rather than $Z_{\text{eff}} = Z - 2$ which is the Z_{eff} predicted by Hartree for neutral multi-electron atoms. In the K_α emission there is a vacancy in the K shell and the L-shell electron making the K_α transition actually sees an effective charge $(Z - 1)\,e$ rather than an effective charge $(Z - 2)\,e$, as is the case for K-shell electrons in neutral atoms.

Moseley also measured several lines belonging to the L characteristic radiation series for various elements and found similar regularities. This result, now referred to as the Moseley law, has been of great importance in substantiating the Rutherford–Bohr atomic theory and highlighted the significance of the atomic number Z of elements as indicator of nuclear charge.

3.3.3 Franck–Hertz Experiment

Direct confirmation that the internal energy states of an atom are quantized came from an experiment carried out by *James Franck* and *Gustav Hertz* in 1914. The experimental set up is shown schematically in Fig. 3.7a.

An evacuated vessel containing three electrodes (cathode, anode, and plate) is filled with low pressure mercury vapor. Electrons are emitted thermio- nically from the heated cathode and accelerated toward the perforated anode by a potential U applied between cathode and anode. Some of the electrons pass through the perforated anode and travel to the plate, provided their kinetic energy upon passing through the perforated anode is sufficiently high to overcome a small retarding potential U_{ret} that is applied between the anode and the plate.

The experiment involves measuring the electron current reaching the plate as a function of the accelerating voltage U. With an increasing potential U the current at the plate increases with U until, at a potential of 4.9 V, it abruptly drops, indicating that some interaction between the electrons and mercury atoms suddenly appears when the electrons attain a kinetic energy of 4.9 eV. The interaction was interpreted as an excitation of mercury atoms with a discrete energy of 4.9 eV; the electron raising an outer shell mercury electron from its ground state to its first excited state and in doing so losing a 4.9 eV portion of its kinetic energy and its ability to overcome the retarding potential U_{ret} between the anode and the plate.

The sharpness of the current drop at 4.9 V indicates that electrons with energy below 4.9 eV cannot transfer their energy to a mercury atom, substantiating the existence of discrete energy levels for the mercury atom. With voltage increase beyond 4.9 V the current reaches a minimum and then rises again until it reaches another maximum at 9.8 V, indicating that some electrons underwent two interactions with mercury atoms. Other maxima at higher multiples of 4.9 V were also observed with careful experiments. Typical experimental results are shown in Fig. 3.7b. In contrast to the minimum excitation potential of the outer shell electron in mercury of 4.9 eV, the ionization potential of mercury is 10.4 eV.

A further investigation showed a concurrent emission of 2530 Å uv-rays by the mercury vapor gas that, according to Bohr model, will be emitted when the mercury atom reverts from its first excited state to the ground state through a 4.9 eV optical transition. The photon energy $E_\nu = 4.9$ eV is given by the standard relationship

$$E_\nu = h\nu = 2\pi\hbar\frac{c}{\lambda} = \frac{12397 \text{ eV·Å}}{\lambda}. \tag{3.45}$$

from which the wavelength of the emitted photon can be calculated as

$$\lambda = \frac{2\pi\hbar c}{E_\nu} = \frac{2\pi \times (1973 \text{ eV·Å})}{4.9 \text{ eV}} = 2530 \text{ Å}. \tag{3.46}$$

Ultraviolet photons with a wavelength of 2536 Å were actually observed accompanying the Franck–Hertz experiment, adding to the measured peaks in the current

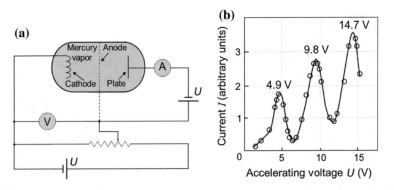

Fig. 3.7 **a** Schematic diagram of the Franck–Hertz experiment; **b** Typical result of the Franck–Hertz experiment using mercury vapor

I versus voltage *U* diagram of Fig. 3.7b another means for the confirmation of the quantization of atomic energy levels.

3.4 Schrödinger Equation for Hydrogen Atom

In solving the Schrödinger equation for a hydrogen or hydrogen-like one-electron atom, a 3-dimensional approach must be used to account for the electron motion under the influence of a central force. The Coulomb potential binds the electron to the nucleus and the coordinate system is chosen such that its origin coincides with the center of the nucleus. To account for the motion of the nucleus we use the reduced mass μ_M of (3.18) rather than the pure electron rest mass m_e in the calculation.

The time-independent Schrödinger wave equation was given in (1.104) as

$$-\frac{\hbar^2}{2\mu_M}\nabla^2\psi + V(r)\psi = E\psi, \tag{3.47}$$

where

$V(r)$ is the potential seen by the electron.
E is the total energy of the electron.
∇^2 is the Laplacian operator in Cartesian, cylindrical or spherical coordinates.
μ_M is the reduced mass of the electron-proton system given in (3.18).

For the hydrogen atom, the potential $V(r)$ is represented by the spherically symmetric Coulomb potential as follows

$$V(r) = -\frac{1}{4\pi\varepsilon_0}\frac{e^2}{r}. \tag{3.48}$$

The Schrödinger wave equation is separable in spherical coordinates (r, θ, ϕ) and for the hydrogen atom it is written by expressing the Laplacian operator in spherical coordinates as follows

$$-\frac{\hbar^2}{2\mu_M}\left\{\frac{1}{r^2}\frac{\partial}{\partial r}\left(r^2\frac{\partial}{\partial r}\right) + \frac{1}{r^2\sin\theta}\frac{\partial}{\partial\theta}\left(\sin\theta\frac{\partial}{\partial\theta}\right) + \frac{1}{r^2\sin^2\theta}\frac{\partial^2}{\partial\phi^2}\right\}\psi(r,\theta,\phi)$$
$$-\frac{e^2}{4\pi\varepsilon_0}\frac{1}{r}\psi(r,\theta,\phi) = E\psi(r,\theta,\phi), \tag{3.49}$$

with (r, θ, ϕ) the spherical coordinates of the electron.

The boundary conditions stipulate that $|\psi|^2$ must be an integrable function. This implies that the wave function $\psi(r, \theta, \phi)$ vanishes as $r \to \infty$, i.e., the condition that $\lim_{r\to\infty} \psi(r, \theta, \phi) = 0$ must hold.

Equation (3.49) can be solved with the method of separation of variables by expressing the function $\psi(r, \theta, \phi)$ as a product of three functions: $R(r)$, $\Theta(\theta)$, and $\Phi(\phi)$; each of the three functions depends on only one of the three spherical variables, i.e.,

$$\psi(r, \theta, \phi) = R(r)\Theta(\theta)\Phi(\phi). \tag{3.50}$$

Inserting (3.50) into (3.49) and dividing by $R(r)\Theta(\theta)\Phi(\phi)$ we get the following expression

$$-\frac{\hbar^2}{2\mu_{\rm M}}\left\{\frac{1}{r^2}\frac{1}{R}\frac{\partial}{\partial r}\left(r^2\frac{\partial R}{\partial r}\right) + \frac{1}{r^2\sin\theta}\frac{1}{\Theta}\frac{\partial}{\partial\theta}\left(\sin\theta\frac{\partial\Theta}{\partial\theta}\right) + \frac{1}{r^2\sin^2\theta}\frac{1}{\Phi}\frac{\partial^2\Phi}{\partial\phi^2}\right\}$$
$$-\frac{e^2}{4\pi\varepsilon_0}\frac{1}{r} = E. \tag{3.51}$$

Separation of variables then results in the following three ordinary differential equations

$$\frac{\mathrm{d}^2\Phi}{\mathrm{d}\phi^2} = -m_\ell\Phi, \tag{3.52}$$

$$-\frac{1}{\sin\theta}\frac{\mathrm{d}}{\mathrm{d}\theta}\left(\sin\theta\frac{\mathrm{d}\Theta}{\mathrm{d}\theta}\right) + \frac{m_\ell^2\Theta}{\sin^2\theta} = \ell(\ell+1)\Theta, \tag{3.53}$$

and

$$\frac{1}{r^2}\frac{\mathrm{d}}{\mathrm{d}r}\left(r^2\frac{\mathrm{d}R}{\mathrm{d}r}\right) + \frac{2\mu_{\rm M}}{\hbar^2}\left(E + \frac{e^2}{4\pi\varepsilon_0}\right)R = \ell(\ell+1)\frac{R}{r^2}, \tag{3.54}$$

with separation constants m_ℓ and $\ell(\ell+1)$, where m_ℓ and ℓ are referred to as the magnetic and orbital quantum numbers, respectively.

Equation (3.54) for $R(r)$ gives physical solutions only for certain values of the total energy E. This indicates that the energy of the hydrogen atom is quantized, as suggested by the Bohr theory, and predicts energy states that are identical to those calculated for the Bohr model of the hydrogen atom. The energy levels E_n calculated from the Schrödinger wave equation, similarly to those calculated for the Bohr atom, depend only on the principal quantum number n; however, the wave function solutions depend on three quantum numbers: n (principal), ℓ (orbital) and m_ℓ (magnetic). All quantum numbers are integers governed by the following rules:

$$n = 1, 2, 3\ldots, \quad \ell = 0, 1, 2, \ldots, n-1, \quad m_\ell = -\ell, -\ell+1, \ldots, (\ell-1), \ell. \tag{3.55}$$

3.4.1 Schrödinger Equation for Ground State of Hydrogen

Equation (3.49) is generally quite complex yielding wave functions for the ground state $n = 1$ of the hydrogen atom as well as for any of the excited states with associated values of quantum numbers ℓ and m_ℓ. However, the ground state of the hydrogen atom can be calculated in a simple fashion as follows.

Since $V(r)$ is spherically symmetric, we assume that solutions to the Schrödinger equation for the ground state of hydrogen will be spherically symmetric which means that the wave function $\psi(r, \theta, \phi)$ does not depend on θ and ϕ, it depends on r alone, and we can write for the spherically symmetric solutions that $\psi(r, \theta, \phi) = R(r)$. The general Schrödinger equation of (3.51) then becomes significantly simpler and, after some rearranging of terms, it is given as follows

$$\frac{d^2 R(r)}{dr^2} + \frac{2}{r}\left\{ \frac{dR(r)}{dr} + \frac{\mu_M}{\hbar^2}\frac{e^2}{4\pi\varepsilon_0}R(r) \right\} + \frac{2\mu_M E}{\hbar^2}R(r) = 0. \qquad (3.56)$$

We can now simplify the Schrödinger equation further by recognizing that for large r the $(1/r)$ term will be negligible and we obtain

$$\frac{d^2 R(r)}{dr^2} - \left[-\frac{2\mu_M E}{\hbar^2} \right] R(r) \approx 0. \qquad (3.57)$$

Next we define the constant $-2\mu_M E/\hbar^2$ as λ^2 and recognize that the total energy E_1 for the ground state of hydrogen will be negative

$$\lambda^2 = -\frac{2\mu_M E_1}{\hbar^2}. \qquad (3.58)$$

The simplified Schrödinger equation is now given as follows

$$\frac{d^2 R(r)}{dr^2} - \lambda^2 R(r) = 0. \qquad (3.59)$$

Equation (3.59) is recognized as a form of the Helmholtz differential equation in one dimension that leads to exponential functions for $\lambda^2 > 0$, to a linear function for $\lambda = 0$, and to trigonometric functions for $\lambda^2 < 0$. Since the total energy E is negative for bound states in hydrogen, λ^2 is positive and the solutions to (3.59) are exponential functions. The simplest exponential solution is

$$R(r) = Ce^{-\lambda r}, \qquad (3.60)$$

with the first derivative expressed as

$$\frac{dR(r)}{dr} = -\lambda Ce^{-\lambda r} = -\lambda R(r). \qquad (3.61)$$

The second derivative of the function $R(r)$ of (3.60) is given as follows

$$\frac{d^2 R(r)}{dr^2} = \lambda^2 C e^{-\lambda r} = \lambda^2 R(r). \tag{3.62}$$

Inserting (3.60) and (3.62) into (3.57) shows that (3.60) is a valid solution to (3.59). We now insert (3.60), (3.61), and (3.62) into (3.56) and get the following expression for the ground state of the hydrogen atom

$$\lambda^2 R(r) + \frac{2}{r}\left\{-\lambda + \frac{\mu_M}{\hbar^2}\frac{e^2}{4\pi\varepsilon_0}\right\}R(r) + \frac{2\mu_M E_1}{\hbar^2}R(r) = 0. \tag{3.63}$$

The first and fourth term of (3.63) cancel out because λ^2 is defined as $\left(-2\mu_M E_1/\hbar^2\right)$ in (3.58). Since (3.63) must be valid for any $\psi\,(r)$, the term in curly brackets equals to zero and provides another definition for the constant λ as follows

$$\lambda = \frac{\mu_M}{\hbar^2}\frac{e^2}{4\pi\varepsilon_0} = \frac{\mu_M c^2}{(\hbar c)^2}\frac{e^2}{4\pi\varepsilon_0}. \tag{3.64}$$

We recognize (3.64) for λ as the inverse of the Bohr radius constant a_0 that was given in (3.4). Therefore, we express $1/\lambda$ as follows

$$\frac{1}{\lambda} = a_0 = \frac{4\pi\varepsilon_0}{e^2}\frac{(\hbar c)^2}{\mu_M c^2} = 0.5292\ \text{Å}. \tag{3.65}$$

Combining (3.58) and (3.64) for the constant λ, we now express the ground state energy E_1 as

$$E_1 = -\frac{1}{2}\frac{\hbar^2}{\mu_M}\frac{1}{a_0^2} = -\frac{1}{2}\left(\frac{e^2}{4\pi\varepsilon_0}\right)^2\frac{\mu_M c^2}{(\hbar c)^2} \simeq -13.61\ \text{eV}. \tag{3.66}$$

The wave function $R(r)$ for the ground state of hydrogen is given in (3.60) in general terms with constants C and λ. The constant λ was established in (3.65) as the inverse of the Bohr radius constant a_0; the constant C we determine from the normalization condition of (1.85) that is given by the following expression

$$\iiint |\psi\,(r)|^2\,d\mathcal{V} = 1, \tag{3.67}$$

with the volume integral extending over all space. The constant C is determined after inserting $\psi(r, \theta, \phi) = R(r)$ given by (3.60) into (3.67) to obtain

$$\iiint |\psi(r)|^2 \, d\mathcal{V} = C^2 \int_0^{2\pi} \int_0^{\pi} \int_0^{\infty} e^{-\frac{2r}{a_0}} r^2 \, d\phi \sin\theta \, d\theta \, dr$$

$$= 4\pi C^2 \int_0^{\infty} r^2 e^{-\frac{2r}{a_0}} \, dr = 4\pi C^2 \frac{1}{4a_0^3} = 1, \qquad (3.68)$$

where the last integral over r is determined from the following recursive formula

$$\int x^n e^{ax} \, dx = \frac{1}{a} x^n e^{ax} - \frac{n}{a} \int x^{n-1} e^{ax} \, dx. \qquad (3.69)$$

The integral over r in (3.68) is equal to $1/(4a_0^3)$ and the constant C is now given as follows

$$C = \pi^{-1/2} a_0^{-3/2}, \qquad (3.70)$$

resulting in the following expression for the wave function $R(r)$ for the ground state of the hydrogen atom

$$\psi_{n,\ell,m_\ell}(r, \theta, \phi) = \psi_{100} = R_1(r) = \frac{1}{\pi^{1/2} a_0^{3/2}} e^{-\frac{r}{a_0}}. \qquad (3.71)$$

The probability density of (1.83) can now be modified to calculate the radial probability density dP/dr as follows

$$\frac{dP}{d\mathcal{V}} = \psi^*(r, \theta, \phi)\psi(r, \theta, \phi) = |\psi(r, \theta, \phi)|^2 \qquad (3.72)$$

and

$$\frac{dP}{dr} = 4\pi r^2 |\psi(r, \theta, \phi)|^2, \qquad (3.73)$$

since $d\mathcal{V} = 4\pi r^2 \, dr$ for the spherical symmetry governing the ground state ($n = 1$) of the hydrogen atom.

The radial probability density dP/dr for the ground state is given as follows, after inserting (3.71) into (3.73)

$$\frac{dP}{dr} = \frac{4r^2}{a_0^3} e^{-\frac{2r}{a_0}} \qquad \text{or} \qquad a_0 \frac{dP}{dr} = \left[\frac{2r}{a_0}\right]^2 e^{-\frac{2r}{a_0}}. \qquad (3.74)$$

A plot of unit-less $a_0 \, (dP/dr)$, given as $4 \, (r/a_0)^2 \exp(-2r/a_0)$, against (r/a_0) for the ground state of the hydrogen atom, is shown in Fig. 3.8. The following observations can be made based on data shown in Fig. 3.8:

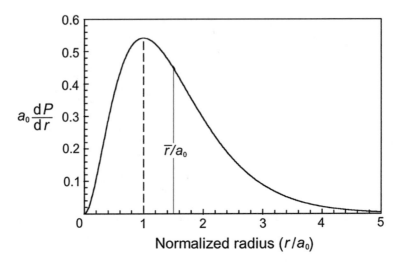

Fig. 3.8 The radial probability density of (3.74) multiplied with the Bohr radius constant a_0 against normalized radius r/a_0 for the ground state electron in the hydrogen atom

1. The radial probability density $dP/dr = 0$ for $r = 0$ and $r = \infty$.
2. dP/dr reaches its maximum at $r = a_0$, highlighting the Schrödinger theory prediction that the ground state electron in hydrogen atom is most likely to be found at $r = a_0$, where a_0 is the Bohr radius constant given in (3.4). One can also obtain this result by calculating d^2P/dr^2 and setting the result equal to zero to obtain the maximum of $a_0\,(dP/dr) = 4e^{-2} = 0.541$ at the normalized radius of $r/a_0 = 1$. Thus, the most probable radius r_p for the electron in the ground state of hydrogen is equal to a_0.
3. Contrary to Bohr theory that predicts the electron in a fixed orbit with $r = a_0$, the Schrödinger theory predicts that there is a finite probability for the electron to be anywhere between $r = 0$ and $r = \infty$. However, the most probable radius for the electron is $r = a_0$.

3.4.2 Sample Calculations for Ground State of Hydrogen

To illustrate the Schrödinger theory further we now carry out a few sample calculations for the ground state of the hydrogen atom based on the Schrödinger equation:

1. The probability that the electron will be found inside the first Bohr radius a_0 is calculated by integrating (3.74) from $r = 0$ to $r = a_0$ to get

$$
P = \frac{4}{a_0^3} \int_0^{a_0} r^2 e^{-\frac{2r}{a_0}}\, dr = \left\{ \frac{4e^{-\frac{2r}{a_0}}}{a_0^3}\left[-\frac{a_0 r^2}{2} - \frac{a_0^2 r}{2} - \frac{a_0^3}{4} \right] \right\}_{r=0}^{r=a_0}
$$

$$
= -\left\{ e^{-\frac{2r}{a_0}}\left[2\left(\frac{r}{a_0}\right)^2 + 2\frac{r}{a_0} + 1 \right] \right\}_{r=0}^{r=a_0} = 1 - 5e^{-2} = 0.323. \quad (3.75)
$$

2. The probability that the electron will be found with radius exceeding a_0 is similarly calculated by integrating (3.74) from $r = a_0$ to $r = \infty$

$$
\begin{aligned}
P &= \frac{4}{a_0^3} \int_{a_0}^{\infty} r^2 e^{-\frac{2r}{a_0}} dr = \left\{ \frac{4e^{-\frac{2r}{a_0}}}{a_0^3} \left[-\frac{a_0 r^2}{2} - \frac{a_0^2 r}{2} - \frac{a_0^3}{4} \right] \right\}_{r=a_0}^{r=\infty} \\
&= -\left\{ e^{-\frac{2r}{a_0}} \left[2\left(\frac{r}{a_0}\right)^2 + 2\frac{r}{a_0} + 1 \right] \right\}_{r=a_0}^{r=\infty} = 5e^{-2} = 0.677. \quad (3.76)
\end{aligned}
$$

3. The probability that the electron will be found with radius between $r = 0$ and $r = \infty$ obviously must be equal to 1 and is given as the sum of (3.75) and (3.76) or calculated directly

$$
\begin{aligned}
P &= \frac{4}{a_0^3} \int_{0}^{\infty} r^2 e^{-\frac{2r}{a_0}} dr = \left\{ \frac{4e^{-\frac{2r}{a_0}}}{a_0^3} \left[-\frac{a_0 r^2}{2} - \frac{a_0^2 r}{2} - \frac{a_0^3}{4} \right] \right\}_{r=0}^{r=\infty} \\
&= -\left\{ e^{-\frac{2r}{a_0}} \left[2\left(\frac{r}{a_0}\right)^2 + 2\frac{r}{a_0} + 1 \right] \right\}_{r=0}^{r=\infty} = 1. \quad (3.77)
\end{aligned}
$$

4. The probability that the orbital electron will be found inside the nucleus (proton) is calculated by integrating (3.74) from $r = 0$ to $r = R$ where R is the proton radius estimated from (1.26) as $R \approx 1.25$ fm. Using $R/a_0 = (1.25 \text{ fm}) / (0.53 \times 10^5 \text{ fm}) = 2.36 \times 10^{-5}$ we get the following probability

$$
\begin{aligned}
P &= \frac{4}{a_0^3} \int_{0}^{R} r^2 e^{-\frac{2r}{a_0}} dr = \left\{ \frac{4e^{-\frac{2r}{a_0}}}{a_0^3} \left[-\frac{a_0 r^2}{2} - \frac{a_0^2 r}{2} - \frac{a_0^3}{4} \right] \right\}_{r=0}^{r=R} \\
&= -\left\{ e^{-\frac{2r}{a_0}} \left[2\left(\frac{r}{a_0}\right)^2 + 2\frac{r}{a_0} + 1 \right] \right\}_{r=0}^{r=R} \\
&= 1 - e^{-\frac{2R}{a_0}} \left\{ 2\left(\frac{R}{a_0}\right)^2 + 2\frac{R}{a_0} + 1 \right\} \\
&\approx 1 - \left[1 - \frac{2R}{a_0} \right] \left[2\left(\frac{R}{a_0}\right)^2 + 2\frac{R}{a_0} + 1 \right] \\
&\approx 2\left(\frac{R}{a_0}\right)^2 = 1.11 \times 10^{-9}. \quad (3.78)
\end{aligned}
$$

5. The average electron radius $\bar{r}$ is calculated using (1.106) and (3.71) to get

$$
\bar{r} = \iiint r \left[R(r)\right]^2 dV = \frac{4}{a_0^3} \int_{0}^{\infty} r^3 e^{-\frac{2r}{a_0}} dr = \frac{4a_0^4 3!}{16a_0^3} = \frac{3}{2} a_0. \quad (3.79)
$$

6. The most probable radius $r_p = a_0$ and the average radius $\bar{r} = 1.5\,a_0$ are not identical because the radial probability density distribution is not symmetrical about its maximum at a_0, as shown in Fig. 3.8. As calculated in (3.75) and (3.76), the area under the dP/dr curve between $r = 0$ and $r = a_0$ amounts to only about one half the area under the curve between $r = a_0$ and $r = \infty$.

7. The expectation value of electron's kinetic energy $\bar{E}_K$ (also referred to as the average or mean kinetic energy of the electron) in the ground state of the hydrogen atom is calculated from (1.106) using $\left[\left(-\hbar^2 / (2m_e)\right)\right] \nabla^2$ for the associated kinetic energy operator, as given in (1.110) for one dimension. The wave function for the ground state of hydrogen ψ is spherically symmetric and given in (3.71). Following (1.106) $\bar{E}_K$ is in general written as

$$\bar{E}_K = -\frac{\hbar^2}{2m_e} \iiint \psi^* \nabla^2 \psi \; d\mathcal{V}. \tag{3.80}$$

In (3.80) the Laplace operator ∇^2 in spherical coordinates operates on spherically symmetrical wave function ψ to give

$$\nabla^2 \psi = \frac{1}{a_0^{3/2} \pi^{1/2}} \frac{1}{r^2} \frac{\partial}{\partial r} \left[r^2 \frac{\partial}{\partial r} e^{-\frac{r}{a_0}} \right] = \frac{1}{a_0^{3/2} \pi^{1/2}} \left[\frac{1}{a_0^2} - \frac{2}{ra_0} \right] e^{-\frac{r}{a_0}}. \tag{3.81}$$

Since $d\mathcal{V}$ for spherical symmetry is given as $4\pi r^2 \, dr$, we write $\bar{E}_K$ as follows

$$
\begin{aligned}
\bar{E}_K &= -\frac{2\hbar^2}{m_e a_0^4} \left\{ \frac{1}{a_0} \int_0^\infty r^2 e^{-\frac{2r}{a_0}} \, dr - 2 \int_0^\infty r e^{-\frac{2r}{a_0}} \, dr \right\} \\
&= -\frac{\hbar^2}{2m_e a_0^2} + \frac{\hbar^2}{m_e a_0^2} = \frac{\hbar^2}{2m_e a_0^2} = \frac{1}{2m_e c^2} \left[\frac{\hbar c}{a_0} \right]^2 \\
&= \frac{1}{2 \times 0.511 \times 10^6 \text{ eV}} \left[\frac{197.3 \times 10^6 \text{ eV·fm}}{0.5292 \times 10^5 \text{ fm}} \right]^2 = 13.6 \text{ eV},
\end{aligned}
\tag{3.82}
$$

and get $+13.6$ eV for the expectation value of electron's kinetic energy.

8. The expectation (mean) value of the potential energy $\bar{E}_P$ can be determined in similar manner to the derivation of the kinetic energy expectation value in (3.82). The associated potential energy operator is given as $[V] = \left[-e^2 / (4\pi\varepsilon_0 r) \right]$ and the hydrogen ground state wave function is spherically symmetrical and given in (3.71). $\bar{E}_P$ is in general expressed as follows

$$\bar{E}_P = \iiint \psi^* [V] \psi \; d\mathcal{V}, \tag{3.83}$$

where $[V]$ is the potential energy operator. For ground state of hydrogen we get

$$\bar{E}_P = -\frac{e^2}{4\pi\varepsilon_0}\frac{4\pi}{\pi a_0^3}\int_0^\infty \frac{e^{-\frac{2r}{a_0}}}{r}r^2 dr = -\frac{e^2}{4\pi\varepsilon_0 a_0}\left[e^{-\frac{2r}{a_0}}\left(\frac{2r}{a_0}-1\right)\right]_0^\infty \quad (3.84)$$

$$= -\frac{e^2}{4\pi\varepsilon_0 a_0} = \frac{1.602\times10^{-19}\ \text{C·eV·m}}{4\pi\times8.85\times10^{-12}\ \text{C}\times0.5292\times10^{-10}\ \text{m}} = -27.2\ \text{eV}$$

for the expectation (mean) value of the potential energy $\bar{E}_P$.

9. The expectation (mean) values of the kinetic energy $\bar{E}_K$ and the potential energy $\bar{E}_P$ of the ground state of hydrogen atom were determined quantum-mechanically in (3.82) and (3.84), respectively, using appropriate ground state wave function (3.71) and appropriate energy operators. The mean kinetic energy $\bar{E}_K = 13.6\ \text{eV}$ is equal to one half of the mean potential energy $\bar{E}_P = -27.2\ \text{eV}$ but with opposite sign. The sum of the two expectation values $\bar{E}_K + \bar{E}_P$ is equal to the total energy E_1 of the electron in the ground state of the hydrogen atom

$$E_1 = \bar{E}_K + \bar{E}_P = 13.6\ \text{eV} - 27.2\ \text{eV} = -13.6\ \text{eV}. \quad (3.85)$$

The quantum-mechanical result (3.85) is in agreement with the value of $-13.6\ \text{eV}$ for E_1 obtained by Bohr, as discussed in Sect. 3.1.3

$$E_1 = E_K + E_P = \frac{m_e\upsilon_1^2}{2} - \frac{e^2}{4\pi\varepsilon_0 r_1} = \frac{1}{2}m_e c^2\alpha^2 - \frac{e^2}{4\pi\varepsilon_0 a_0}$$

$$= \frac{0.511\times10^6\ \text{eV}}{2\times137^2} - \frac{e\times1.6\times10^{-19}\ \text{C·V·m}}{4\pi\times8.85\times10^{-12}\ \text{C}\times0.5292\times10^{-15}\ \text{m}} \quad (3.86)$$

$$= 13.6\ \text{eV} - 27.2\ \text{eV} = -13.6\ \text{eV},$$

where we used (3.3) and (3.4) for $r_1 = a_0 = 0.5292$ Å and (3.5) and (3.6) for $\upsilon_1/c = \alpha = 1/137$.

From Crookes Tube to Coolidge X-Ray Tube

Röntgen's discovery of x-rays in November 1895 ranks among the most important scientific discoveries of all times and no scientific discovery has ever become as rapidly and widely known as did the discovery of x-rays. While the exact mechanism of x-ray production and the nature of x-rays remained a mystery for a number of years after 1895, Röntgen's discovery not only ushered in the era of modern physics, it also opened the door to a revolution in medicine by spawning diagnostic imaging and radiotherapy as two new medical specialties, as well as medical physics as a new specialty of physics. No other physicist except Albert Einstein is as widely known and revered as Röntgen is for contribution to science and medicine in particular and humanity in general. Röntgen received the first Nobel Prize in Physics in 1901 and was honored on postage stamps by many countries, both developed and developing, around the world, as **shown by a few examples on next page**.

The "new kind of rays" that Röntgen discovered and called x-rays behaved in similar fashion to visible light and were capable of exposing photographic film, and, in addition, had the astounding ability to penetrate opaque objects including hands, feet and other parts of the human body. Röntgen made his momentous discovery during a routine investigation of "cathode rays" produced in a discharge tube invented by William Crookes in the 1870s.

The Crookes tube consists of a sealed glass tube filled with gas at relatively low pressure (0.005 Pa to 0.1 Pa) and incorporates two electrodes (cathode and anode) connected to a power supply. A current carried inside the tube by "cathode rays" is produced in the circuit and results in many interesting physical phenomena. Toward the end of nineteenth century the Crookes tube became a very important device and many experimental physicists in laboratories around the world were using it for studies of "cathode rays" in particular and atomic physics in general.

The most notable discoveries that resulted from experiments with Crookes tube are: (1) Röntgen's serendipitous discovery that the Crookes tube, in addition to producing "cathode rays," gives off a new kind of ray which he called x-ray; (2) Thomson's finding in 1897 that "cathode rays" in the Crookes tube are actually negatively charged particles that he called electrons; and (3) Millikan's determination of the charge of the electron with the oil drop experiment in 1913.

The early x-ray tubes used in medicine were essentially of the Crookes tube design (cold cathode) and were plagued by many problems, such as low x-ray output, unreliable and temperamental operation, and severe output fluctuations depending on the pressure of the gas in the tube. Electrons producing x-rays in the anode (target) were generated by positive ions striking the "cold cathode" and produced by an electric discharge in the low-pressure gas inside the tube.

The reliability of x-ray tubes improved significantly when William Coolidge in 1913 introduced the "hot cathode" design that featured high vacuum inside the tube ($\sim 10^{-4}$ Pa) and electron generation through thermionic emission by a heated filament or "hot cathode." A century later, Coolidge x-ray tube design is still in use for all types of modern x-ray tubes in medicine and industry. However, in recent years research with carbon nanotubes has shown promise that field emission from a cold cathode could be used for generation of electrons in an x-ray tube.

Chapter 4
Production of X-Rays

This chapter deals mainly with production of x-rays but it also provides a brief introduction to Čerenkov radiation and synchrotron radiation (magnetic bremsstrahlung). Two types of x-ray are known: *characteristic* (fluorescence) *radiation* and *bremsstrahlung*, and both are important in medical physics, because they are used extensively in diagnostic imaging and in external beam radiotherapy.

© Springer International Publishing Switzerland 2016
E.B. Podgoršak, *Radiation Physics for Medical Physicists*,
Graduate Texts in Physics, DOI 10.1007/978-3-319-25382-4_4

Characteristic x-rays are produced by electronic transitions in atoms triggered by vacancies in inner electronic shells of the absorber atom. Bremsstrahlung x-rays, on the other hand, are produced by inelastic Coulomb interactions between a light energetic charged particle and the nucleus of the absorber atom.

Theoretical and practical aspects of characteristic radiation are discussed by introducing first the Siegbahn and IUPAC notation for designation of electronic levels in atom, followed by a discussion of the fluorescence yield and the Auger effect. A theoretical discussion of bremsstrahlung follows by introducing the Larmor relationship and basic principles of emission of radiation from accelerated charged particles. The chapter concludes with a brief discussion of the Čerenkov radiation and the synchrotron radiation, both of some interest in medical physics and nuclear physics.

4.1 X-Ray Line Spectra

A vacancy in an atomic shell plays an important role in physics and chemistry. Defined as an electron missing from the normal complement of electrons in a given atomic shell, a vacancy can be produced by eight different effects or interactions ranging from various photon–atom interactions through charge particle–atom interactions to nuclear effects. Depending on the nature and energy of the interaction, the vacancy may occur in the outer shell or in one of the inner shells of the atom. A list of the eight effects for production of shell vacancy in an atom is as follows:

1. Photoelectric effect (see Sect. 7.5)
2. Compton scattering (see Sect. 7.3)
3. Triplet production (see Sect. 7.6.1)
4. Charged particle Coulomb interaction with atom (see Sects. 6.3 and 6.11)
5. Internal conversion (see Sect. 11.8)
6. Electron capture (see Sect. 11.6)
7. Positron annihilation (see Sect. 7.6.10)
8. Auger effect (see Sect. 4.1.2)

An atom with a vacancy in its inner shell is in a highly excited state and returns to its ground state through one or several electronic transitions. In each of these transitions an electron from a higher atomic shell fills the shell vacancy and the energy difference in binding energy between the initial and final shell or sub-shell is emitted from the atom in one of two ways:

1. In the form of characteristic (fluorescence) radiation.
2. Radiation-less in the form of Auger electrons, Coster–Kronig electrons or super Coster–Kronig electrons.

4.1.1 Characteristic Radiation

Radiation transitions result in emission of photons that are called *characteristic radiation*, since the wavelength λ and energy $h\nu$ of the emitted photon are characteristic of the atom in which the photon originated. An older term, fluorescence radiation, is still occasionally used to describe the characteristic photons. The set of radiation transition photons emitted from a given atom is referred to as the line spectrum of the atom. *Charles G. Barkla* is credited with the discovery of characteristic x-rays and in 1917 he was awarded the Nobel Prize in Physics for his discovery.

Energy level diagrams for high atomic number x-ray targets are usually drawn using the n, ℓ, j and m_j quantum numbers, as shown with a generic energy level diagram in Fig. 4.1. In addition to dependence on n (main structure), the energy level diagram also exhibits dependence on ℓ and j (fine structure). However, only certain transitions, fulfilling specific selection rules, result in x-rays. The selection rules for allowed characteristic transitions leading to most intense characteristic lines are called the electric dipole selection rules and are stipulated as follows:

$$\Delta\ell = \pm 1 \quad \text{and} \quad \Delta j = 0 \quad \text{or} \quad \pm 1. \tag{4.1}$$

In addition to electric dipole selection rules, magnetic dipole and electric quadrupole selection rules are also known, but they produce significantly weaker characteristic lines. Lines allowed by electric dipole selection rules are referred to as normal x-ray lines; lines not allowed by electric dipole selection rules are referred to as forbidden lines. In Fig. 4.1 only transitions from the M and L to the K shell are shown; allowed transitions with solid lines and a forbidden transition with a *dotted line*.

The energies released through an electronic transition are affected by the atomic number Z of the absorbing atom and by the quantum numbers of the atomic shells involved in the electronic transition. Transitions between outer shells of an atom generally result in optical photons and are referred to as optical transitions (photon energy $h\nu$ is of the order of a few electronvolts); transitions between inner shells of high atomic number elements may result in x-rays and are referred to as x-ray transitions (photon energy $h\nu$ is of the order of 10 keV to 100 keV).

Traditionally, the following conventions introduced by early workers in x-ray spectroscopy, most notably by *Karl M.G. Siegbahn*, have been used in atomic physics:

1. Transitions of electrons to the K shell are referred to as the K lines, to the L shell as L lines, to the M shell as M lines, etc.
2. Transitions from the nearest neighbor shell are designated as α transition; transitions from the second nearest neighbor shell or a higher-level shell are generally designated as β transition; however, other designations may also be used for some of these.
3. Transitions from one shell to another do not all have the same energy because of the fine structure (sub-shells) in the shell levels. The highest energy transition between two shells is usually designated with number 1, the second highest energy with number 2, etc.

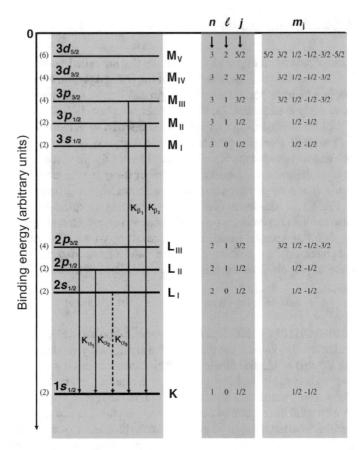

Fig. 4.1 Typical energy level diagram for a high atomic number element showing the K ($n = 1$), L ($n = 2$), and M ($n = 3$) shells with associated sub-shells. The *numbers in brackets* indicate the maximum possible number of electrons in a given sub-shell equal to $(2j + 1)$. K_α and K_β transitions are also shown. The allowed K_α and K_β transitions are shown with *solid lines*, the forbidden K_{α_3} transition is shown with a *dashed line*

4. In Fig. 4.1 the transition K_{α_3} represents a forbidden transition ($\Delta\ell = 0$) from the L to the K shell ($2s_{1/2} \rightarrow 1s_{1/2}$ with $\Delta\ell = 0$ and $\Delta j = 0$).
5. The transition $K_{\beta 1}$ represents an allowed transition from the M shell to the K shell ($3p_{3/2} \rightarrow 1s_{1/2}$ with $\Delta\ell = 1$ and $\Delta j = 1$).
6. Radiation transitions that do not follow the strict electric dipole selection rules of (4.1) may also occur but their intensities are much lower than those of the allowed (normal) transitions.

The original x-ray spectroscopy nomenclature for identification of characteristic (fluorescence) x-rays has been in use since 1920s and is referred to as the *Siegbahn notation* in honor of Karl M.G. Siegbahn who introduced the notation and received the 1924 Nobel Prize in Physics for his research in x-ray spectroscopy. The notation became outdated and cumbersome to use in recent years and the International Union of Pure and Applied Chemistry (IUPAC) in 1991 proposed a new and more systematic nomenclature referred to as the *IUPAC notation*.

The main differences between Siegbahn's notation and the IUPAC notation are as follows:

1. Both notations follow Barkla's designation of electronic shells as the K shell for $n = 1$, L shell for $n = 2$, M shell for $n = 3$, etc.; however, the sub-shells are designated with Roman numerals in the Siegbahn notation (e.g., L_I, L_{II}, L_{III}, etc.) and with Arabic numerals in the IUPAC notation (e.g., L_1, L_2, L_3, etc.).
2. In the Siegbahn notation the subshell origin of a characteristic photon is not identified (e.g., K_{α_1}, K_{α_2}, K_{β_3}, etc.), while in the IUPAC notation both the initial and the final subshell producing the characteristic photon are identified and separated by a hyphen (e.g., $K - L_3$, $K - L_2$, $K - M_2$, etc.). Note that in the IUPAC notation the initial and final states of the electron vacancy rather than those of the electron making the transition are identified.
3. Another advantage of the IUPAC notation is that it coincides with the current nomenclature used in Auger electron emission and photoelectron spectroscopy.

An example of a generic x-ray energy diagram is given in Fig. 4.2 showing a series of K-shell characteristic photons; transitions allowed by electric dipole selection rules of (4.1) are shown with solid lines, transitions forbidden by electric dipole selection rules are shown with dashed lines. For comparison, all electronic transitions to the K shell are identified with both notations: Siegbahn and IUPAC. It is obvious that, of the two notations, the IUPAC notation is more systematic and easier to follow, since it clearly identifies the two sub-shells contributing to the emission of the characteristic photon.

4.1.2 Fluorescence Yield and Auger Effect

Each spontaneous electronic transition from an initial higher level atomic shell (higher quantum number n; lower binding energy E_B) to a final lower level shell (lower n; higher E_B) is characterized by a transition energy ($E_{final} - E_{initial}$) that is:

1. Either emitted in the form of a characteristic (fluorescence) photon.
2. Or transferred to a higher shell electron which is ejected from the atom as an Auger electron.

The phenomenon of emission of Auger electron from an excited atom is called the general Auger effect and actually encompasses three different mechanisms: Auger effect, Coster–Kronig effect, and super Coster–Kronig effect. In Auger effect the primary transition occurs between two shells, in Coster–Kronig and super Coster–Kronig effects the primary transition occurs within two sub-shells of a shell. The three effects are illustrated schematically in Fig. 4.3:

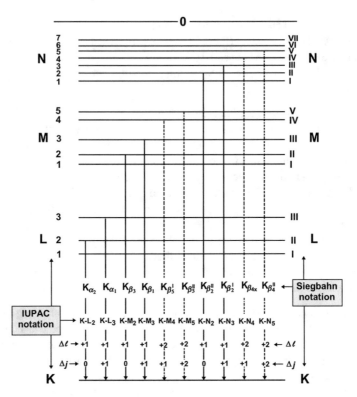

Fig. 4.2 Atomic energy level diagram for a high atomic number element showing the K shell as well as the L, M, and N sub-shells. All possible transitions to the K shell are displayed using the traditional Siegbahn notation (*right hand side*) and the new IUPAC (International Union of Pure and Applied Chemistry) notation (*left hand side*)

1. In Auger effect the primary transition occurs between two shells and the transition energy is transferred to an orbital electron from the initial shell or an even higher-level shell.
2. In Coster–Kronig effect the transition energy originates from two sub-shells of a given shell and is transferred to an electron in another shell. The emitted electron is called a Coster–Kronig electron.
3. In super Coster–Kronig effect the transition energy which, like in (2), originates from two sub-shells of a given shell is transferred to a sub-shell electron within the shell in which the primary transition occurred. The emitted electron is called a super Coster–Kronig electron.

The radiation-less electronic transition is called Auger effect after French physicist *Pierre Auger* who is credited with its discovery in 1925. It is now accepted that Austrian physicist *Lise Meitner* actually discovered the effect in 1923, but the effect continues to be known as the Auger effect.

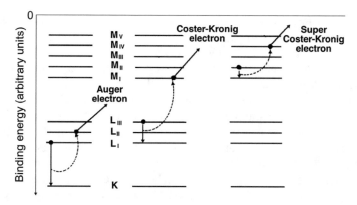

Fig. 4.3 Schematic representation of Auger effect, Coster–Kronig effect, and super Coster–Kronig effect. In Auger effect the electron makes an intershell transition and the transition energy is transferred to the Auger electron. In Coster–Kronig effect the electron makes an intrashell transition and the transition energy is transferred to an electron in a higher shell. In super Coster–Kronig effect the electron makes an intrashell transition and the energy is transferred to an electron in the same shell

By way of example, in Fig. 4.1 a vacancy in the K shell may be filled with an electron from the L_2 sub-shell making the transition energy E_B (K) $- E_B$ (L_2) available either for emission of a characteristic K_2 photon or for emission of an Auger electron, say from an M_3 sub-shell. The energy of the emitted Auger electron is equal to

$$E_K \left(e_{KL_2M_3} \right) = [E_B \text{ (K)} - E_B \text{ (}L_2\text{)}] - E_B \text{ (}M_3\text{)} \tag{4.2}$$

and the Auger electron emitted from the M shell is designated as a $e_{KL_2M_3}$ electron following standard nomenclature for Auger electrons. The subscript stipulates that:

1. The initial vacancy occurred in the K-shell.
2. The K-shell vacancy was filled by an electron from the L_2 sub-shell.
3. The transition energy was transferred to an M_3 electron which was emitted from the atom as Auger electron with kinetic energy given in (4.2).

A vacancy created in an electronic shell or sub-shell will thus be followed by emission of a characteristic (fluorescence) photon or an Auger electron. The branching between the two possible routes is governed by the fluorescence yield ω for the given atom of the absorber and for the given atomic shell of the absorber atom. The fluorescence yield of a given shell is defined as the number of fluorescence (characteristic) photons emitted per vacancy in the shell. It can also be regarded as the probability, after creation of a shell vacancy, of fluorescence photon emission as opposed to Auger electron emission.

Fluorescence yields ω_K, ω_L, and ω_M for K, L, and M shell electron vacancies, respectively, are plotted in Fig. 4.4 against atomic number Z for all elements. The following features of fluorescence yield are noteworthy:

1. A plot of the fluorescence yield ω_K against absorber atomic number Z results in a sigmoid shaped curve with ω_K ranging from $\omega_K = 0$ for low Z elements through $\omega_K = 0.5$ at $Z = 30$ to $\omega_K = 0.96$ at very high Z.
2. For the L-shell vacancy the fluorescence yield ω_L is zero at $Z < 30$ and then rises with Z to reach a value $\omega_L = 0.5$ at $Z = 100$.
3. The fluorescence yield ω_M is zero for all elements with $Z < 60$, and for $Z > 60$ it rises slowly with increasing Z to attain a value $\omega_M \approx 0.05$ for very high Z absorbers, indicating that fluorescence emission from the M shell and higher level electronic shells is essentially negligible for all absorbers, even those with very high atomic number Z.
4. For a given absorber, the higher is the shell level (i.e., the lower is the shell binding energy), the lower is the fluorescence yield ω and, consequently, the higher is the probability for the Auger effect $(1 - \omega)$.

Figure 4.4 also plots the probability for Auger effect for vacancies in the K, L, and M shells. In general, the probability for Auger effect following a given shell vacancy equals to $1 - \omega$, where ω is the fluorescence yield for the given electronic shell.

The exact mechanism of energy transfer in Auger effect is difficult to calculate numerically. In the past, the effect was often considered an internal atomic photoelectric effect and the explanation makes sense energetically. However, two experimental facts contradict this assumption:

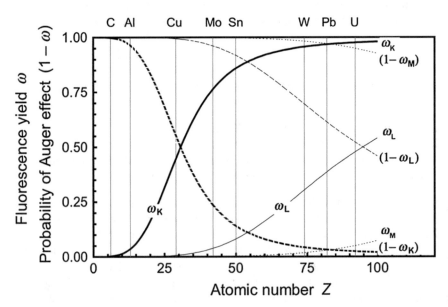

Fig. 4.4 Fluorescence yields ω_K for $h\nu > (E_B)_K$, ω_L for $(E_B)_L < h\nu < (E_B)_K$, and ω_M for $(E_B)_M < h\nu < (E_B)_L$ against atomic number Z. Also shown are probabilities for Auger effect $1 - \omega_K$, $1 - \omega_L$, and $1 - \omega_M$. Data were obtained from Hubbell

1. Auger effect often results from forbidden radiation transitions, i.e., transitions that violate the selection rules for the radiative fluorescence process.
2. Fluorescence yield ω for high atomic number Z materials is significantly larger than that for low atomic number materials; contrary to the well-known photoelectric effect Z dependence that follows a Z^3 behavior (see Sect 7.5).

4.2 Emission of Radiation by Accelerated Charged Particle (Bremsstrahlung Production)

Bremsstrahlung is a type of x-ray produced by light charged particles when they undergo inelastic collisions with nuclei of absorber atoms. Charged particles are characterized by their rest mass, charge, velocity and kinetic energy. With regard to their rest mass, charged particles of interest to medical physics and dosimetry are classified into two categories:

1. *Light charged particles*: electrons e^- and positrons e^+, both capable of producing bremsstrahlung photons because of their relatively small mass.
2. *Heavy charged particles*: protons p, deuterons d, alpha particles α, heavier ions such as Li^+, Be^+, C^+, Ne^+, etc. producing negligible amount of bremsstrahlung photons.

With regard to their velocity, charged particles are classified into three categories:

1. Stationary with $v = 0$
2. Moving with a uniform velocity $v = \text{constant}$
3. Accelerated with an acceleration $\mathbf{a} = d\mathbf{v}/dt$.

Of the three categories, only accelerated light charged particles are capable of producing bremsstrahlung photons even though particles in all three categories, by virtue of being charged, are surrounded by an electric field generated by the charge of the particle. With regard to velocity, only accelerated charged particles lose some of their energy in the form of photon radiation.

4.2.1 Stationary Charged Particle: No Emission of Radiation

A stationary charged particle has an associated constant electric field $\mathcal{E}$ whose energy density ρ is given by

$$\rho = \frac{1}{2}\varepsilon_0\mathcal{E}^2, \tag{4.3}$$

where ε_0 is the electric constant also called permittivity of vacuum [$\varepsilon_0 = 8.85 \times 10^{-12}$ A·s/ (V·m)]. The energy is stored in the field and is not radiated away by the charged particle of charge q. There is no magnetic field associated with a stationary charged particle. The electric field $\mathcal{E}(r)$ produced by a stationary charged particle is governed by the Coulomb law which states that the electric field is proportional to q and inversely proportional to the square of the distance r between the charge q and the point-of-interest P. The electric field thus follows an inverse square law with distance r from q and is isotropic (spherically symmetric)

$$\mathcal{E}(r) = \frac{1}{4\pi\varepsilon_0}\frac{q}{r^3}\mathbf{r} \quad \text{or} \quad \mathcal{E}(r) = \frac{1}{4\pi\varepsilon_0}\frac{q}{r^2}, \tag{4.4}$$

where $\mathbf{r}$ is a vector pointing from particle with charge q to the point-of-interest P. The science that deals with phenomena arising from stationary electric charges is referred to as electrostatics.

4.2.2   Charged Particle Moving with Uniform Velocity: No Emission of Radiation

A charged particle moving with a uniform velocity v has an associated magnetic field $\mathcal{B}$ as well as an electric field $\mathcal{E}$. The energy density ρ is then given by

$$\rho = \frac{1}{2}\varepsilon_0\mathcal{E}^2 + \frac{1}{2\mu_0}\mathcal{B}^2, \tag{4.5}$$

where ε_0 is the electric constant (permittivity of vacuum) and μ_0 is the magnetic constant, also called permeability of vacuum [$\mu_0 = 4\pi \times 10^{-7}$ V . s/ (A·m)]. The energy is stored in the field, moves along with the charged particle, and is not radiated away by the charged particle.

Since a moving charged particle q produces both an electric field and a magnetic field, the two fields are considered together as the electromagnetic (EM) field and the science dealing with these fields is referred to as electromagnetism. The classical approach to electromagnetism was developed toward the end of nineteenth century and is based on the law of Biot–Savart, the Ampère law, and Maxwell equations. In the early 1960s *Edward M. Purcell* introduced the idea of relativistic electromagnetism by explaining electromagnetism using Einstein theory of special relativity and the Lorentz contraction factor $\sqrt{1 - (v/c)^2}$.

To determine $\mathcal{E}$ and $\mathcal{B}$ produced by a moving charged particle, Purcell introduced two inertial reference frames, Σ and Σ'. The charged particle is at rest at the origin of Σ', and Σ' is moving along the x axis of the reference frame Σ with uniform velocity v. At time $t = 0$ the origins of the two reference frames coincide. The charged particle is at rest in Σ', thus in the reference frame Σ' it produces an electric field $\mathcal{E}$ governed by Coulomb law given in (4.4) but it produces no magnetic field $\mathcal{B}$. In the reference frame Σ, on the other hand, we have a moving charge, implying a current with an associated magnetic field $\mathcal{B}$ in addition to an electric field $\mathcal{E}$.

Texts on electromagnetism and relativistic electrodynamics give the following results for $\mathcal{E}$ and $\mathcal{B}$ produced by charge q moving with uniform velocity v

$$\mathcal{E} = \frac{q}{4\pi\varepsilon_0} \frac{\mathbf{r}}{r^3} \frac{1 - \beta^2}{(1 - \beta^2 \sin^2 \theta)^{3/2}} \quad \text{or} \quad \mathcal{E} = \frac{q}{4\pi\varepsilon_0 r^2} \frac{1 - \beta^2}{\left(1 - \beta^2 \sin^2 \theta\right)^{3/2}} \quad (4.6)$$

and

$$\mathcal{B} = \frac{\boldsymbol{\upsilon} \times \mathcal{E}}{c^2} = \frac{q}{4\pi\varepsilon_0 c^2} \frac{\boldsymbol{\upsilon} \times \mathbf{r}}{r^3} \frac{1 - \beta^2}{(1 - \beta^2 \sin^2 \theta)^{3/2}} \quad \text{or}$$

$$\mathcal{B} = \frac{q\beta \sin \theta}{4\pi\varepsilon_0 c r^2} \frac{1 - \beta^2}{\left(1 - \beta^2 \sin^2 \theta\right)^{3/2}}, \quad (4.7)$$

where

q is the charge of the charged particle, identical in both reference frames.
$\mathbf{r}$ is the radius vector connecting the origin of Σ with the point-of-interest P.
θ is the angle between $\boldsymbol{\upsilon}$ and $\mathbf{r}$ in the reference frame Σ.
β is υ/c with c the speed of light in vacuum according to the standard notation used in relativistic physics.

The electric field $\mathcal{E}$ of (4.6) is equal to the Coulomb field of (4.4) produced by a stationary charge q, corrected for charge motion with a field correction factor C_υ which depends on the magnitude of velocity υ and on the angle θ

$$C_\upsilon = \frac{1 - \beta^2}{\left(1 - \beta^2 \sin^2 \theta\right)^{3/2}}. \quad (4.8)$$

Equation (4.6) shows that the electric field is radial, it diverges radially from the charge, and its intensity is proportional to $1/r^2$. However, because of the effect of the field correction factor C_υ, the field is not the same in all directions unlike the case with the Coulomb law for stationary charge given in (4.4).

Figure 4.5 shows a plot of the field correction factor C_υ against angle θ in the range from 0 to π for four values of β (0, 0.5, 0.75, and 0.9). The following special features are noted:

1. For $\upsilon = 0$ or $\beta = 0$ the expressions for $\mathcal{E}$ and $\mathcal{B}$ of (4.6) and (4.7), respectively, reduce to expressions for stationary charge q where $\mathcal{E}$ is given by (4.4) and $\mathcal{B} = 0$. Moreover, we note that for $\upsilon = 0$ or ($\beta = 0$) the field correction factor $C_\upsilon = 1$ for all angles θ confirming the Coulomb law and isotropic electric field $\mathcal{E}$ for a stationary charge q.
2. For $\upsilon \ll c$ or $\beta \ll 1$ expressions (4.6) and (4.7), respectively, reduce to classical expressions

$$\mathcal{E} = \frac{q\mathbf{r}}{4\pi\varepsilon_0 r^3} \quad (4.9)$$

Fig. 4.5 Field correction factor C_v of (4.8) against angle θ for normalized velocity β of 0, 0.5, 0.75, and 0.9. For a given β the minimum C_v occurs at $\theta = 0$ and $\theta = \pi$ and equals to $1 - \beta^2$. Maximum C_v occurs at $\theta = \frac{1}{2}\pi$ and equals to $1/\sqrt{1 - \beta^2}$

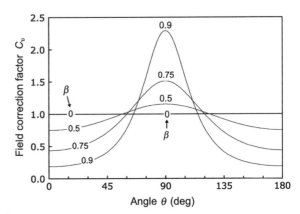

and

$$\mathcal{B} = \frac{q\boldsymbol{v} \times \mathbf{r}}{4\pi\varepsilon_0 c^2 r^3}, \tag{4.10}$$

referred to as Coulomb law (4.4) and Biot–Savart law, respectively.

3. For $\theta = 0$ in the direction of charge motion and $\theta = \pi$ in the direction opposite to charge motion, the field correction factor C_v has a minimum value for a given β, is smaller than 1 for all $\beta < 1$, and simplifies to

$$C_v(\theta = 0) = C_v\,(\theta = \pi) = (C_v)_{\min} = 1 - \beta^2. \tag{4.11}$$

This suggests a constriction of the electric field along the direction of motion, since $1 - \beta^2 < 1$ for all β except for the case of stationary charged particle for which $\beta = 0$ and the electric field is given by the Coulomb law of (4.4).

4. For $\theta = \pi/2$ in directions perpendicular to the direction of motion, the field correction factor C_v has a maximum value for a given β, exceeds 1 for all $\beta < 1$, and simplifies to

$$C_v\,(\theta = \pi/2) = (C_v)_{\max} = \frac{1}{\sqrt{1 - \beta^2}} = \gamma, \tag{4.12}$$

where γ is defined in Sect. 1.2.1. This suggests an increase in the electric field $\mathcal{E}$ in comparison with electric field produced by a stationary charged particle of charge q, since $1/\sqrt{1 - \beta^2} = \gamma > 1$ for all β except for $\beta = 0$.

5. For $\beta = 0$ (stationary particle) the field correction factor C_v equals to 1 for all angles θ. For $\beta < 1$ the field correction factor C_v goes through 1 around angles θ of 60° and 120° for all β.

In Fig. 4.6 we show the field correction factor C_v of (4.8) in a polar diagram with radius representing C_v and the angle representing angle θ for various velocities β

Fig. 4.6 Electric field $\mathcal{E}$ produced by a charged particle q moving with uniform velocity v in vacuum for four velocities β of 0 (stationary particle), 0.5, 0.75, and 0.9. At $\beta = 0$ the field is isotropic and governed by Coulomb law; for relativistic particles the field constricts in the direction of motion and in the direction opposite to direction of motion and the field expands in directions perpendicular to direction of motion. The plot is essentially a polar diagram of the field correction factor C_v of (4.8)

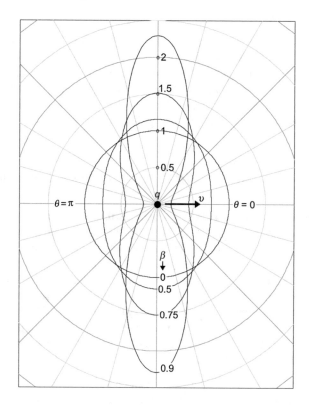

ranging from $\beta = 0$ (stationary particle) through $\beta = 0.5$ and $\beta = 0.75$ to $\beta = 0.9$. For $\theta = 0$ and $\theta = \pi$, C_v is given by (4.11); for $\theta = \frac{1}{2}\pi$ it is given by (4.12); for any arbitrary angle θ it is given by (4.8). We can also regard the graph of Fig. 4.6 as a display of the electric field $\mathcal{E}$ associated with the charged particle and normalized to 1 for the field of the particle when it is stationary ($v = 0$). For the stationary particle the electric field is governed by the Coulomb relationship and is spherically symmetric. For charged particle moving with velocity v the field is cylindrically symmetric in planes perpendicular to the direction of motion; however, in planes containing the direction of motion the electric field distribution is no longer isotropic and depends on angle θ and velocity v.

For a moving charged particle the fields in Fig. 4.6 are shown corrected and normalized to value of 1 of the stationary field. As discussed above, at low (classical) velocities v the electric field $\mathcal{E}$ produced by the charged particle is isotropic and follows the inverse square law. However, as the velocity of the charged particle increases and approaches c, the speed of light in vacuum, the electric field is constricted in the forward and backward direction and increases in a direction at right angles to the particle motion. The electric field is contracted by a factor $(1 - \beta^2)$ in the direction of flight of the particle, whereas it is enhanced by a factor $\gamma = 1/\sqrt{1 - \beta^2}$ in the transverse direction.

The electric field distortion for moving charged particles is of importance in collision stopping power calculations (Chap. 6). As a charged particle passes through an absorber, it sweeps out a cylinder throughout which its electric field is capable of transferring energy to orbital electrons of the absorber. The radius of this cylinder increases with increasing parameter γ as the charged particle velocity v increases, allowing more orbital electrons to be affected by the charged particle, thereby increasing the energy loss of the charged particle in a given absorber.

4.2.3 Accelerated Charged Particle: Emission of Radiation

For an accelerated charged particle the non-static electric and magnetic fields cannot adjust themselves in such a way that no energy is radiated away from the charged particle. As a result, an accelerated or decelerated charged particle emits some of its kinetic energy in the form of photons referred to as bremsstrahlung radiation. The electric and magnetic fields associated with accelerated charged particle are calculated from the so-called Lienard–Wiechert potentials. To determine the fields at time t the potentials must be evaluated for an earlier time (called the retarded time), with the charged particle at the appropriate retarded position on its trajectory.

The electric field $\mathcal{E}$ and the magnetic field $\mathcal{B}$ of an accelerated charged particle have two components:

1. *Local* (or near) *velocity field* component which falls off as $1/r^2$.
2. *Far* (or radiation) *acceleration field* component which falls off as $1/r$.

At large distances r of interest in medical physics and dosimetry the $1/r$ radiation component dominates and the $1/r^2$ near field component may be ignored, since it approaches zero much faster than the $1/r$ component. The energy loss by radiation is thus determined by the far field components of the electric field $\mathcal{E}$ and the magnetic field $\mathcal{B}$. The far field components of $\mathcal{E}$ and $\mathcal{B}$ are given as follows

$$\mathcal{E} = \frac{q}{4\pi\varepsilon_0}\frac{\mathbf{r}\times(\mathbf{r}\times\dot{\boldsymbol{v}})}{c^2 r^3} \quad \text{or} \quad \mathcal{E} = \frac{1}{4\pi\varepsilon_0}\frac{q}{c^2}\frac{\dot{v}\sin\theta}{r} \qquad (4.13)$$

and

$$\mathcal{B} = \frac{q}{4\pi\varepsilon_0}\frac{\dot{\boldsymbol{v}}\times\mathbf{r}}{c^3 r^2} = \frac{\mu_0}{4\pi}\frac{q}{c}\frac{\dot{\boldsymbol{v}}\times\mathbf{r}}{r^2} \quad \text{or} \quad \mathcal{B} = \frac{q}{4\pi\varepsilon_0}\frac{\dot{v}\sin\theta}{c^3 r} = \frac{\mu_0}{4\pi}\frac{q}{c}\frac{\dot{v}\sin\theta}{r} = \frac{\mathcal{E}}{c}, \qquad (4.14)$$

where

$\mathbf{r}$ is the radius vector connecting the charged particle with the point of observation.
$\dot{\boldsymbol{v}}$ is the acceleration vector of the charged particle.
$\dot{v}$ is the magnitude of the acceleration vector.
q is the charge of the charged particle,

θ is the angle between $\mathbf{r}$ and $\mathbf{v}$.

c is the speed of light in vacuum.

The $\mathcal{E}$ and $\mathcal{B}$ fields propagate outward with velocity c and form the electromagnetic (EM) radiation (bremsstrahlung) emitted by the accelerated charged particle. The energy density ρ of the emitted radiation is given by

$$\rho = \frac{1}{2}\varepsilon_0 \mathcal{E}^2 + \frac{1}{2\mu_0}\mathcal{B}^2 = \varepsilon_0 \mathcal{E}^2 = \frac{\mathcal{B}^2}{\mu_0} = \frac{\mathcal{E}\mathcal{B}}{\mu_0 c}, \tag{4.15}$$

noting that $\mathcal{B} = \mathcal{E}/c$ and $c^2 = 1/(\varepsilon_0 \mu_0)$ in vacuum.

4.2.4 Intensity of Radiation Emitted by Accelerated Charged Particle

The intensity of the emitted radiation is defined as the energy flow per unit area A per unit time t and is given by the vector product $\mathcal{E} \times \mathcal{B}/\mu_0$, known as the Poynting vector $\mathbf{S}$ (see Sect. 1.30), where

$$\mathbf{S} = \frac{\mathcal{E} \times \mathcal{B}}{\mu_0}. \tag{4.16}$$

After using (4.13) and (4.14) and recognizing that $\mathcal{E}$ and $\mathcal{B}$ are perpendicular to one another, we obtain the following relationship for the magnitude of the Poynting vector

$$S = |\mathbf{S}| = \frac{\mathcal{E}\mathcal{B}}{\mu_0} = \varepsilon_0 c \mathcal{E}^2 = \frac{1}{16\pi^2 \varepsilon_0} \frac{q^2 a^2}{c^3} \frac{\sin^2 \theta}{r^2}. \tag{4.17}$$

The following characteristics of the emitted radiation intensity are notable:

1. Emitted radiation intensity $S(r, \theta)$ is linearly proportional to: q^2, square of particle's charge; a^2, square of particle's acceleration; and $\sin^2 \theta$.
2. Emitted radiation intensity $S(r, \theta)$ is inversely proportional to r^2, reflecting an inverse square law behavior.
3. Emitted radiation intensity $S(r, \theta)$ exhibits a maximum at right angles to the direction of motion where $\theta = \frac{1}{2}\pi$. No radiation is emitted in the forward direction ($\theta = 0$) or in the backward direction ($\theta = \pi$).

4.2.5 Power Emitted by Accelerated Charged Particle Through Electromagnetic Radiation (Classical Larmor Expression)

The power P (energy per unit time) emitted by the accelerated charged particle in the form of bremsstrahlung radiation is obtained by integrating the intensity $S(r, \theta)$ over the area A. Recognizing that $dA = r^2 d\Omega = 2\pi r^2 \sin \theta \, d\theta$ we obtain

$$P = \frac{dE}{dt} = \int S(r, \theta) \, dA = \int S(r, \theta) r^2 \, d\Omega = 2\pi \int_0^\pi S(r, \theta) r^2 \sin \theta \, d\theta$$

$$= -\frac{2\pi}{16\pi^2 \varepsilon_0} \frac{q^2 a^2}{c^3} \int_0^\pi \sin^2 \theta \, d(\cos\theta)$$

$$= -\frac{q^2 a^2}{8\pi\varepsilon_0 c^3} \int_0^\pi (1 - \cos^2 \theta) \, d(\cos \theta) = \frac{1}{6\pi\varepsilon_0} \frac{q^2 a^2}{c^3}. \tag{4.18}$$

Equation (4.18) is the classical Larmor relationship predicting that the power P emitted in the form of bremsstrahlung radiation by an accelerated charged particle is proportional to:

1. q^2, square of particle's charge.
2. a^2, square of particle's acceleration.

The Larmor expression represents one of the basic laws of nature and is of great importance to radiation physics. It can be expressed as follows: *"Any time a charged particle is accelerated or decelerated it emits part of its kinetic energy in the form of bremsstrahlung photons."*

As shown by (4.18), the power emitted in the form of radiation depends on $(qa)^2$ where q is the particle charge and a is its acceleration. The question arises on the efficiency for x-ray production for various charged particles of mass m and charge ze. As charged particles interact with an absorber, they experience Coulomb interactions with orbital electrons (charge e) and nuclei (charge Ze) of the absorber.

Bremsstrahlung is only produced through inelastic Coulomb interactions between a charged particle and the nucleus of the absorber. The acceleration a produced in this type of Coulomb interaction can be evaluated through equating the Newton force with the Coulomb force

$$ma = \frac{zeZe}{4\pi\varepsilon_0 r^2}, \tag{4.19}$$

from where it follows that

$$a \propto \frac{zZe^2}{m}. \tag{4.20}$$

Thus, acceleration a experienced by a charged particle interacting with absorber nuclei is linearly proportional with:

1. Charge of the charged particle ze
2. Charge of the absorber nucleus Ze

and inversely proportional to:

1. Mass m of the charged particle
2. Square of the distance between the two interacting particles r^2.

Since it is proportional to a^2, as shown in (4.18), the power of bremsstrahlung production is inversely proportional to m^2, the square of the particle's mass. Thus, a proton, by virtue of its relatively large mass m_p in comparison with the electron mass m_e, $(m_p/m_e = 1836)$ will produce much less bremsstrahlung radiation than does an electron, specifically about $(m_p/m_e)^2 \approx 4 \times 10^6$ times less. The radiation stopping power for electrons in comparison to that for protons is over six orders of magnitude larger at the same velocity and in the same absorbing material (see Sect. 6.2).

1. As a result of the inverse m^2 dependence, a heavy charged particle traversing a medium loses energy only through ionization (collision) losses and its radiation losses are negligible. The collision losses occur in interactions of the heavy charged particle with orbital electrons of the medium. The total stopping power for heavy charged particle is then given by the collision stopping power and the radiation stopping power is ignored, i.e., $S_{tot} = S_{col}$ and $S_{rad} \approx 0$.
2. Light charged particles, on the other hand, undergo collision as well as radiation loss, since they interact with both the orbital electrons and the nuclei of the absorber. The total stopping power for light charged particles is then a sum of the collision stopping power and the radiation stopping power, i.e., $S_{tot} = S_{col} + S_{rad}$.
3. As established in 1915 by *William Duane* and *Franklin L. Hunt*, the incident light particle can radiate an amount of energy which ranges from zero to the incident particle kinetic energy E_K

$$E_K = h\nu_{max} = 2\pi \frac{\hbar c}{\lambda_{min}}, \tag{4.21}$$

producing a sharp cutoff at the short-wavelength end of the continuous bremsstrahlung spectrum (Duane–Hunt law).

4.2.6 Relativistic Larmor Relationship

Recognizing that $\mathbf{a} = \dot{\upsilon} = \dot{\mathbf{p}}/m$ we can extend the classical Larmor result for bremsstrahlung power P of (4.18) to relativistic velocities and obtain

$$P = \frac{dE}{dt} = \frac{1}{6\pi\varepsilon_0} \frac{q^2}{m^2 c^3} \left| \frac{d\mathbf{p}}{dt} \right| \cdot \left| \frac{d\mathbf{p}}{dt} \right|. \tag{4.22}$$

For the special case of linear motion (e.g., in a linear accelerator waveguide) the emitted power P is given as follows

$$P = \frac{dE}{dt} = \frac{1}{6\pi\varepsilon_0}\frac{q^2}{m^2c^3}\left(\frac{dp}{dt}\right)^2 = \frac{1}{6\pi\varepsilon_0}\frac{q^2}{m^2c^3}\left(\frac{dE}{dx}\right)^2, \tag{4.23}$$

noting that the rate of change of momentum (dp/dt) is equal to the change in energy of the particle per unit distance (dE/dx).

4.2.7   Relativistic Electric Field Produced by Accelerated Charged Particle

The velocity v of the charged particle affects the electric field $\mathcal{E}$ and, as $\beta = v/c$ increases, the electric field $\mathcal{E}$ becomes tipped forward and increases in magnitude as predicted by an expression differing from the classical result of (4.13) by a factor $1/(1 - \beta\cos\theta)^{5/2}$

$$\mathcal{E}(r,\theta) = \frac{1}{4\pi\varepsilon_0}\frac{q}{c^2}\frac{\dot{v}}{r}\frac{\sin\theta}{\sqrt{(1-\beta\cos\theta)^5}}. \tag{4.24}$$

As a result, the emitted radiation intensity which is equal to the magnitude $S(r,\theta)$ of the Poynting vector $\mathbf{S}$ also becomes tipped forward

$$S(r,\theta) = \varepsilon_0 c\mathcal{E}^2 = \frac{1}{16\pi^2\varepsilon_0}\frac{q^2a^2}{c^3r^2}\frac{\sin^2\theta}{(1-\beta\cos\theta)^5}. \tag{4.25}$$

Note that at classical velocities where $\beta \to 0$, (4.24) and (4.25) revert to the classical relationships, given in (4.13) and (4.17), respectively. The emitted radiation intensity of (4.25) has the following notable properties:

1. Intensity $S(r,\theta)$ is in general proportional to $\sin^2\theta/(1-\beta\cos\theta)^5$. In classical mechanics where $\beta \to 0$, the radiation intensity is proportional to $\sin^2\theta$ and the maximum intensity occurs at $\theta = \frac{1}{2}\pi$.
2. As β increases, the radiation intensity becomes more and more forward-peaked; however, the intensities for the forward direction $(\theta = 0)$ and the backward direction $(\theta = \pi)$ are still equal to zero, similarly to the classical situation.
3. The function $\sin^2\theta/(1-\beta\cos\theta)^5$ that governs the radiation intensity distribution $S(r,\theta)$ of (4.25) is plotted in Fig. 4.7 for $\beta = 0.006$ (classical result for $v \to 0$) and for $\beta = 0.941$. For electrons $(m_ec^2 = 0.511\text{ MeV})$ these two β values correspond to kinetic energies of 10 eV and 1 MeV, respectively.
4. Note that in Fig. 4.7 the maximum values of both β distributions are normalized to 1. In reality, as shown in Table 4.1, if the maximum value for the $\beta = 0.006$ distribution is 1, then, for the $\beta = 0.941$ distribution, it is more than four orders of magnitude larger at 1.44×10^4.

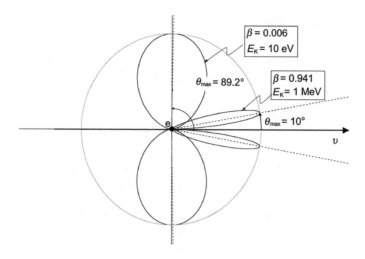

Fig. 4.7 Radiation intensity distributions for two accelerated electrons; one with $\beta = 0.006$ corresponding to an electron kinetic energy of 10 eV and θ_{max} of 89.2° and the other with $\beta = 0.941$ corresponding to an electron kinetic energy of 1 MeV and θ_{max} of 10°. Both distributions are normalized to 1 at θ_{max}. The actual ratio of radiation intensities at $\theta_{max} = 89.2°$ and $\theta_{max} = 10°$ is 1 versus 1.44×10^4, as shown in Table 4.1

Table 4.1 Various parameters for bremsstrahlung production by electrons with kinetic energy E_K

E_K (MeV)	β^a	γ^b	θ_{max}^c	$CS(r, \theta_{max})^d$
10^{-5}	0.006	1.00002	89.2°	1.0000
10^{-4}	0.020	1.0002	87.2°	1.0025
10^{-3}	0.063	1.002	81.2°	1.024
10^{-2}	0.195	1.02	64.4°	1.263
10^{-1}	0.548	1.20	35.0°	6.47
1	0.941	2.96	10.0°	1.44×10^4
10	0.999	20.4	1.4°	1.62×10^{11}
10^2	0.9999	70.71	0.4°	1.64×10^{15}

[a] $\beta = \dfrac{v}{c} = \sqrt{1 - \dfrac{1}{\left(1 + \dfrac{E_K}{m_e c^2}\right)^2}}$, where $m_e c^2 = 0.511$ MeV

[b] $\gamma = \dfrac{1}{\sqrt{1 - \beta^2}}$, where γ and β are defined in (1.44)

[c] $\theta_{max} = \arccos\left\{\dfrac{1}{3\beta}\left(\sqrt{1 + 15\beta^2} - 1\right)\right\}$, where β is given in [a]

[d] $CS(r, \theta_{max}) = \dfrac{\sin^2\theta_{max}}{(1 - \beta\cos\theta_{max})^5}$, where $C = \left(\dfrac{e^2 a^2}{16\pi^2\varepsilon_0 c^3 r^2}\right)^{-1}$

4.2.8   Characteristic Angle

From (4.25) it is evident that, as β increases, the emitted radiation intensity $S(r, \theta)$ becomes more forward-peaked, and its peak intensity that occurs at a characteristic angle θ_{max} also increases.

The characteristic angle θ_{max} is determined as follows:

Set $dS(r, \theta)/d\theta \,|_{\theta=\theta_{max}} = 0$, where $S(r, \theta)$ is given in (4.25), to obtain

$$\frac{2 \sin \theta_{max} \cos \theta_{max}}{(1 - \beta \cos \theta_{max})^5} - \frac{5\beta \sin^3 \theta_{max}}{(1 - \beta \cos \theta_{max})^6} = 0. \tag{4.26}$$

Equation (4.26) yields a quadratic equation for $\cos \theta_{max}$, given as follows

$$3\beta \cos^2 \theta_{max} + 2 \cos \theta_{max} - 5\beta = 0. \tag{4.27}$$

The physically relevant solution to the quadratic equation (4.27) is

$$\cos \theta_{max} = \frac{1}{3\beta} \left(\sqrt{1 + 15\beta^2} - 1 \right), \tag{4.28}$$

resulting in the following expression for θ_{max}

$$\theta_{max} = \arccos \left\{ \frac{1}{3\beta} \left(\sqrt{1 + 15\beta^2} - 1 \right) \right\}. \tag{4.29}$$

The limiting values for θ_{max} are as follows:

1. In the classical region $\beta \to 0$, resulting in $\theta_{max} \to \frac{1}{2}\pi$, as shown below:

$$\begin{aligned} \lim_{\beta \to 0} \theta_{max} &= \lim_{\beta \to 0} \arccos \left\{ \frac{1}{3\beta} \left(\sqrt{1 + 15\beta^2} - 1 \right) \right\} \\ &= \lim_{\beta \to 0} \arccos \left\{ \frac{1}{3\beta} \left(1 + \frac{15}{2}\beta^2 + \cdots - 1 \right) \right\} \\ &= \lim_{\beta \to 0} \arccos \left\{ \frac{5}{2}\beta \right\} = \arccos 0 = \frac{\pi}{2}. \end{aligned} \tag{4.30}$$

2. In the relativistic region $\beta \to 1$, resulting in $\theta_{max} \to 0$

$$\lim_{\beta \to 1} \theta_{max} = \lim_{\beta \to 1} \arccos \left\{ \frac{1}{3\beta} \left(\sqrt{1 + 15\beta^2} - 1 \right) \right\} = \arccos 1 = 0. \tag{4.31}$$

3. In the extreme relativistic region, where $\beta \to 1$ and the Lorentz factor $\gamma = 1/\sqrt{1 - \beta^2} \to \infty$, we show below that θ_{max} of (4.29) approaches zero as $1/(2\gamma)$. The Lorentz factor γ can be expressed as

$$\beta^2 = 1 - \frac{1}{\gamma^2} \quad \text{or} \quad \beta = \sqrt{1 - \frac{1}{\gamma^2}} \tag{4.32}$$

allowing us to write $1/\beta$ as

$$\frac{1}{\beta} = \lim_{\gamma \to \infty} \left(1 - \frac{1}{\gamma^2}\right)^{-1/2} \approx 1 + \frac{1}{2\gamma^2}. \tag{4.33}$$

and $\cos \theta_{max}$ of (4.28) as

$$\cos \theta_{max} = \lim_{\beta \to 1} \left(1 - \frac{1}{2}\theta_{max}^2 + \frac{1}{24}\theta_{max}^4 - \cdots\right) = \lim_{\beta \to 1} \sum_{n=0}^{\infty} (-1)^n \frac{x^{2n}}{(2n)!} \approx 1 - \frac{1}{2}\theta_{max}^2 \tag{4.34}$$

Inserting (4.32), (4.33), and (4.34) simplifies (4.28) to read

$$1 - \frac{1}{2}\theta_{max}^2 \approx \frac{1}{3}\left(1 + \frac{2}{2\gamma^2}\right)\left(\sqrt{1 + 15\left(1 - \frac{1}{\gamma^2}\right)} - 1\right)$$

$$\approx \frac{1}{3}\left(1 + \frac{2}{2\gamma^2}\right)3\left(1 - \frac{5}{8\gamma^2}\right) \approx 1 - \frac{1}{8\gamma^2} = 1 - \frac{1}{2}\frac{1}{(2\gamma)^2}, \tag{4.35}$$

showing that in the extreme relativistic region where $\beta \to 1$ and $\gamma \to \infty$, the characteristic angle θ_{max} approaches zero as $1/(2\gamma)$, i.e.,

$$\theta_{max} = \arccos\left\{\frac{1}{3\beta}\left(\sqrt{1 + 15\beta^2} - 1\right)\right\} \approx \frac{1}{2\gamma} = \frac{1}{2}\sqrt{1 - \beta^2}. \tag{4.36}$$

The two functions for θ_{max} given in (4.29) and (4.36) are plotted against electron kinetic energy E_K in Fig. 4.8. In the extreme relativistic region where $E_K > 1$ MeV the two functions coincide well and $1/(2\gamma)$ provides an excellent and simple expression for θ_{max}. However, for kinetic energies below 1 MeV the two functions diverge with decreasing E_K and approach $\frac{1}{2}\pi$ rad (90°) and 0.5 rad (28.7°), respectively, for $E_K \to 0$.

Table 4.1 lists parameters β, γ, θ_{max} and $S(r, \theta_{max})$ for bremsstrahlung production by electrons and positrons with kinetic energies between 10^{-5} MeV and 10^2 MeV.

The entry $CS(r, \theta_{max})$ in Table 4.1 highlights the significant increase in the bremsstrahlung photon distribution at $\theta = \theta_{max}$ and confirms the rapid increase in x-ray production efficiency with an increase in electron (or positron) kinetic energy.

Parameters β and θ_{max}, given in Table 4.1, are also plotted against the electron kinetic energy E_K in Figs. 4.9 and 4.10, respectively. For very low kinetic energies E_K (classical region) $\beta \approx 0$ and $\theta_{max} \approx 90°$. As E_K increases, β rises and

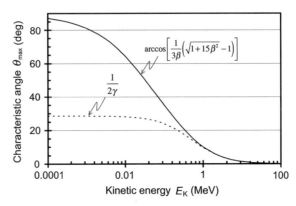

Fig. 4.8 Functions $\theta_{max} = \arccos\left\{(3\beta)^{-1}\left(\sqrt{1+15\beta^2}-1\right)\right\}$ given in (4.29) and $\theta_{max} \approx 1/(2\gamma)$ given in (4.36) plotted against electron kinetic energy E_K

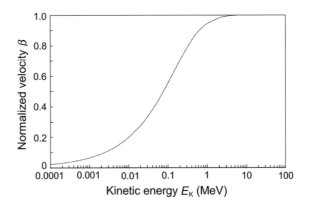

Fig. 4.9 Normalized electron velocity β against the kinetic energy E_K of the electron

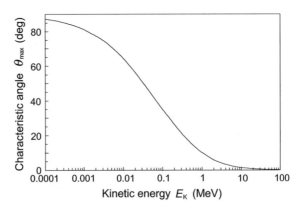

Fig. 4.10 Characteristic angle θ_{max} against kinetic energy E_K of the electron

asymptotically approaches 1 for very high E_K, while θ_{max} decreases with increasing E_K and asymptotically approaches $0°$ for very high E_K. In the orthovoltage x-ray range $\theta_{max} \approx 40°$; in the megavoltage x-ray range $\theta_{max} \approx 5°$.

4.2.9 Electromagnetic Fields Produced by Charged Particles

Generally, charged particles are surrounded by electric and magnetic fields determined by several parameters such as the charge q and velocity v of the charged particle, the distance r of the point of interest from the charged particle, as well as the angle θ between the radius vector $\mathbf{r}$ and velocity v.

The production of bremsstrahlung is governed by the acceleration $\mathbf{a} = dv/dt$ and the rest mass m_0 of the charged particle. A stationary charged particle ($v = 0$) or a charged particle moving with constant velocity v emits no energy in the form of bremsstrahlung photons which means that it experiences no radiation loss. For bremsstrahlung emission to occur, a charged particle must be of relatively small rest mass, such as electron and positron, and must be subjected to acceleration or deceleration. A summary of electric and magnetic fields associated with charged particles under various velocity conditions is provided in Table 4.2.

Table 4.2 Electric field $\mathcal{E}$ and magnetic field $\mathcal{B}$ associated with a charged particle of charge q moving with: (1) velocity $v = 0$; (2) constant velocity v; or (3) with acceleration $\mathbf{a} = dv/dt = \dot{v}$

	Electric field $\mathcal{E}$	Magnetic field $\mathcal{B}$	Bremsstrahlung possible?
Velocity $v = 0$	$\mathcal{E} = \dfrac{q}{4\pi\varepsilon_0}\dfrac{\mathbf{r}}{r^3}$	$\mathcal{B} = 0$	NO
Acceleration $\mathbf{a} = 0$. See (4.14)	$\mathcal{E} = \dfrac{q}{4\pi\varepsilon_0}\dfrac{1}{r^2}$	$\mathcal{B} = 0$	
Velocity $v = $ constant	$\mathcal{E} = \dfrac{q}{4\pi\varepsilon_0}\dfrac{\mathbf{r}}{r^3}C_v$	$\mathcal{B} = \dfrac{q}{4\pi\varepsilon_0 c^2}\dfrac{v\times\mathbf{r}}{r^3}C_v$	NO
Acceleration $\mathbf{a} = 0$. See (4.6) and (4.7)	$\mathcal{E} = \dfrac{q}{4\pi\varepsilon_0}\dfrac{1}{r^2}C_v$	$\mathcal{B} = \dfrac{q}{4\pi\varepsilon_0 c}\dfrac{\beta\sin\theta}{r^2}C_v$	
Velocity $v \neq $ constant	$\mathcal{E} = \dfrac{q}{4\pi\varepsilon_0}\dfrac{\mathbf{r}\times(\mathbf{r}\times\dot{v})}{c^2 r^3}$	$\mathcal{B} = \dfrac{q}{4\pi\varepsilon_0}\dfrac{\dot{v}\times r}{c^3 r^2}$	YES
Acceleration $\mathbf{a} = dv/dt = \dot{v} \neq 0$. See (4.13) and (4.14)	$\mathcal{E} = \dfrac{q}{4\pi\varepsilon_0}\dfrac{\dot{v}\sin\theta}{c^2 r}$	$\mathcal{B} = \dfrac{q}{4\pi\varepsilon_0}\dfrac{\dot{v}\sin\theta}{c^3 r} = \dfrac{\mathcal{E}}{c}$	

The field correction factor C_v is given in (4.8) as $C_v = (1 - \beta^2)(1 - \beta^2 \sin^2\theta)^{-3/2}$ where β is the velocity v of the charged particle normalized to c, the speed of light in vacuum

4.3 Synchrotron Radiation

Synchrotron radiation refers to electromagnetic radiation emitted by charged particles following a curved trajectory in free space under the influence of a magnetic
field. The phenomenon was first observed in 1947 in synchrotrons (hence the term
synchrotron radiation) which, as discussed in Sect. 14.5.4, are accelerators that accelerate charged particles in circular orbits to very high relativistic energies. Since the
effect occurs under the influence of a magnetic field that keeps the particles in a
circular trajectory, it is sometimes called *magnetic bremsstrahlung*.

Electrons as well as heavier charged particles may produce the synchrotron radiation. The radiation can be considered:

1. Either an unnecessary nuisance causing energy losses when the objective is to
   attain high kinetic energies of charged particles in circular accelerators.
2. Or an extraordinary dedicated source of intense, short duration, x-ray or ultraviolet
   pulses that can be exploited as a tool to study structure of matter on an atomic,
   molecular, and cellular scale or to devise ultra fast imaging studies in cardiology.

Originally, research on synchrotron radiation was conducted as a sideline to particle acceleration, recently, however, special sources of synchrotron radiation called
storage rings were built with the specific purpose to produce and exploit synchrotron
radiation (Sect. 14.5.5).

The magnetic field exerts a Lorentz force on the charged particle perpendicularly
to the particle's direction of motion, causing particle's acceleration and, according
to the Larmor relationship of (4.18), emission of photons. Larmor relationship of
(4.18) for power P radiated by particle of charge q accelerated with acceleration a is
given as follows

$$P = \frac{1}{6\pi\varepsilon_0} \frac{q^2 a^2}{c^3}. \tag{4.37}$$

For a classical particle in circular motion with radius R, the acceleration is simply the
centrifugal acceleration v^2/R, where v is the velocity of the particle. For a relativistic
particle with velocity $v \to c$ and mass $m = \gamma m_0$, where m_0 is the particle's rest mass,
in circular motion in a circular accelerator with radius R, the acceleration is similarly
obtained from

$$F = m_0 a = \frac{\mathrm{d}p}{\mathrm{d}t'}, \tag{4.38}$$

where

p is the relativistic momentum of the particle: $p = mv = \gamma m_0 v$.
t' is the proper time in the particle's reference frame given as: $t' = t/\gamma = t\sqrt{1-\beta^2}$.

Neglecting the rate of change of γ with time t, the acceleration a can now be written
as

$$a = \frac{1}{m_0}\frac{dp}{dt'} = \frac{\gamma}{m_0}\frac{d\,(\gamma m_0 v)}{dt} = \gamma^2\frac{dv}{dt} = \gamma^2\frac{v^2}{R}. \tag{4.39}$$

The power radiated from a relativistic particle according to Larmor relationship of (4.18) and acceleration a of (4.39) is as follows

$$P = \frac{1}{6\pi\varepsilon_0}\frac{q^2 a^2}{c^3} = \frac{q^2\gamma^4}{6\pi\varepsilon_0 c^3}\frac{v^4}{R^2} = \frac{cq^2\beta^4\gamma^4}{6\pi\varepsilon_0 R^2}. \tag{4.40}$$

Since we know that the particle total energy E is given as $E = \gamma m_0 c^2 = \gamma E_0$, where E_0 is the particle rest energy, we write (4.40) as follows

$$P = \frac{cq^2\beta^4}{6\pi\varepsilon_0 R^2}\left(\frac{E}{E_0}\right)^4. \tag{4.41}$$

For highly relativistic particles, $v \to c$ and energy loss rate is governed by $\gamma^4 = (E/E_0)^4$ when R is fixed for a given accelerator. Equation (4.41) suggests that the larger is the accelerator radius R, the smaller is the rate of energy loss.

The radiation loss ΔE during one complete revolution of a highly relativistic particle ($\beta \approx 1$) is calculated by first determining the duration τ of one revolution as

$$\tau = \frac{2\pi R}{v} \approx \frac{2\pi R}{c}. \tag{4.42}$$

The radiation loss in one revolution is then

$$\Delta E = P\tau = \frac{cq^2}{6\pi\varepsilon_0 R^2}\left(\frac{E}{E_0}\right)^4\frac{2\pi R}{c} = \frac{q^2}{3\varepsilon_0 R}\left(\frac{E}{E_0}\right)^4. \tag{4.43}$$

The radiation energy loss ΔE per turn is inversely proportional to the radius R of the orbit and linearly proportional to $(E/E_0)^4$. For electrons ($q = e$ and $m_0 = m_e = 0.511$ MeV) we get the following expression for ΔE

$$\Delta E = \frac{e^2}{3\varepsilon_0\left(m_e c^2\right)^4}\frac{E^4}{R} = \left\{8.85\times10^{-8}\,\frac{\text{eV·m}}{(\text{MeV})^4}\right\}\frac{E^4}{R}. \tag{4.44}$$

The energy is radiated in a cone centered along the instantaneous velocity of the particle. The cone has a half angle θ_{syn} approximated as (E_0/E). For highly relativistic particles the cone is very narrow and the radiation is emitted in the forward direction similarly to the situation with the bremsstrahlung loss by relativistic particles, discussed in Sect. 4.2.8.

The wavelength distribution of synchrotron radiation follows a continuous spectrum in the x-ray, ultraviolet, and visible region, with the peak emitted wavelength linearly proportional to R and $(E_0/E)^3$.

4.4 Čerenkov Radiation

As discussed in Sect. 4.2.3, a charged particle radiates energy in free space only if accelerated or decelerated; a charged particle in rectilinear uniform velocity motion in free space does not lose any of its kinetic energy in the form of photon radiation. However, if a charged particle moves with uniform rectilinear motion through a transparent dielectric material, part of its kinetic energy is radiated in the form of electromagnetic radiation if the particle velocity v exceeds the phase velocity of light c_n in the particular medium, i.e.,

$$v > c_n = \frac{c}{n}, \tag{4.45}$$

where n is the index of refraction of light in the particular medium.

The phenomenon of visible light emission under these conditions is referred to as Čerenkov radiation and was discovered by *Pavel A. Čerenkov* and *Sergei I. Vavilov* in 1934. The effect is commonly referred to as Čerenkov effect; however, in honor of both discoverers it is often also called the Čerenkov–Vavilov effect.

The emitted Čerenkov radiation does not come directly from the charged particle. Rather, the emission of Čerenkov radiation involves a large number of atoms of the dielectric medium that become polarized by the fast charged particle moving with uniform velocity through the medium. The orbital electrons of the polarized atoms are accelerated by the fields of the charged particle and emit radiation coherently when $v > c_n = c/n$.

Čerenkov radiation is emitted along the surface of a forward directed cone centered on the charged particle direction of motion. The cone is specified with the following relationship

$$\cos \theta_{cer} = \frac{c_n}{v} = \frac{1}{\beta n}, \tag{4.46}$$

where θ_{cer} is the Čerenkov angle defined as the angle between the charged particle direction of motion and the envelope of the cone.

Equation (4.46) suggests that there is a threshold velocity v_{thr} below which no Čerenkov radiation will occur for a given charged particle and absorbing dielectric

$$v_{thr} = \frac{c}{n} = c_n \tag{4.47}$$

or

$$\beta_{thr} = \frac{1}{n} \tag{4.48}$$

The threshold velocity v_{thr} has the following properties:

1. For $v > v_{thr}$ the Čerenkov radiation is emitted with the Čerenkov angle θ_{cer}.
2. For $v < v_{thr}$ no Čerenkov photons are produced.

3. The velocity threshold for Čerenkov radiation in water (index of refraction n also known as refractive index n is 1.33) is $v_{thr} = (1/1.33) \, c = 0.75c$.
4. The velocity threshold of $0.75c$ for water corresponds to kinetic energy threshold

$$(E_K)_{thr} = \frac{m_e c^2}{\sqrt{1 - \beta_{thr}^2}} - m_e c^2 = m_e c^2 \left[\frac{n}{\sqrt{n^2 - 1}} - 1 \right] = 0.264 \, \text{MeV}.$$

(4.49)

Thus, Čerenkov radiation occurs in water for electrons with kinetic energy exceeding 0.264 MeV, i.e., total energy exceeding 0.775 MeV.

For emission of Čerenkov radiation, the number of quanta per wavelength interval $\Delta\lambda$ is proportional to $1/\lambda^2$, favoring the blue end of the visible spectrum. This explains the characteristic bluish glow surrounding the fission core of a swimming-pool nuclear reactor (see introductory figure to Chap. 9) or surrounding the high activity cobalt-60 sources stored in water-filled storage tanks prior to their installation in teletherapy machines. The Čerenkov radiation results from Compton electrons that propagate through water with velocities v exceeding $c/n = 0.75c$.

Other notable characteristics of Čerenkov radiation are as follows:

1. Čerenkov radiation is independent of charged particle mass but depends on particle charge and particle velocity v.
2. Equation (4.46) also shows that there is a maximum angle of emission $(\theta_{cer})_{max}$ in the extreme relativistic limit where $\beta \to 1$

$$(\theta_{cer})_{max} = \arccos (1/n).$$

(4.50)

Thus, for relativistic electrons ($\beta \to 1$) in water ($n = 1.33$), $(\theta_{cer})_{max} = 41.2°$.
3. Čerenkov radiation frequencies appear in the high frequency visible and near visible regions of the electromagnetic spectrum, but do not extend into the x-ray region because for x-rays $n < 1$.
4. Since the refraction index n depends on the wavelength λ of the emitted radiation, the emission angle θ_{cer} for Čerenkov radiation also depends on the frequency of the Čerenkov radiation in addition to depending on the charged particle velocity v.

As a charged particle moves through a dielectric, the total amount of energy appearing as Čerenkov radiation is very small compared to the total energy loss by a charged particle through collision (ionization) and radiation ($\sim$bremsstrahlung) losses. For example, electrons in water lose about 2 MeV·cm^{-1} through collision and bremsstrahlung radiation losses and only about 400 eV·cm^{-1} through Čerenkov radiation losses, i.e., about a factor of 5000 times less. It is obvious that Čerenkov radiation is negligible as far as radiation dosimetry is concerned. However, the Čerenkov–Vavilov effect is used in Čerenkov detectors not only to detect fast moving charged particles but also to determine their energy through a measurement of the Čerenkov angle.

Electron Pencil Beam Penetrating into Water

Figure on next page represents a 1 MeV electron pencil beam consisting of 50 electrons penetrating into a water phantom. The distribution is calculated with the EGS-nrc Monte Carlo code, a package for the Monte Carlo simulation of coupled electron–photon transport that traces the trajectories of the individual incident electrons through their various Coulomb interactions with orbital electrons and nuclei of the water molecules. The code was developed at the National Research Council (NRC) in Ottawa, Canada as an extended and improved version of the EGS4 package that was originally developed at the Stanford Linear Accelerator Center (SLAC).

Interactions of incident electrons with orbital electrons result in collision (ionization) losses of the incident electrons; interactions of incident electrons with nuclei result in elastic scattering (change of direction of motion) and may also result in radiation (bremsstrahlung) losses. The jagged paths in the figure represent incident electron tracks in water; the two straight traces represent bremsstrahlung photons, both escaping the phantom. A careful observer will also be able to discern the tracks of secondary (δ) electrons that are liberated in water by the primary electrons and given sufficient kinetic energy to be able to ionize matter in their own right.

Monte Carlo calculations are a statistical process and their accuracy depends on the number of events included in the calculation. The larger is this number, the better is the accuracy of the calculation and, of course, the longer is the calculation time. With the ever-increasing power and speed of computers, Monte Carlo techniques are becoming important in radiation dosimetry as well as in calculations of dose distributions in patients treated with x-rays, γ-rays, and light or heavy particle beams.

While the current treatment planning techniques are based on a set of measurements carried out in water phantoms, practical Monte Carlo-based treatment planning algorithms that are currently under development in many research centers base the calculations directly on data for a particular patient, thereby, in principle, significantly improving the accuracy and reliability of dose distribution calculations.

Recently, patient specific Monte Carlo-based treatment planning systems have become commercially available; however, their routine implementation in radiotherapy clinics still hinges on many factors, such as adequate modeling of radiation sources; solving several experimental problems involving tissue inhomogeneities; answering many important clinical questions; updating the dose calculation algorithms; and improving the computing hardware. It is expected that in the near future incorporation of predictive biological models for tumor control and normal tissue complication into Monte Carlo-based dose calculation engines will form the standard approach to radiotherapy treatment planning.

Figure on this page: Courtesy of Jan Seuntjens, Ph.D., McGill University, Montreal. Reproduced with Permission.

Chapter 5
Two-Particle Collisions

This chapter deals with collisions between two particles characterized by an energetic projectile striking a stationary target. Three categories of projectiles of interest in medical physics are considered: *light charged particles* such as electrons and positrons, *heavy charged particles* such as protons and α-particles, and *neutral particles* such as neutrons. The targets are either atoms as a whole, atomic nuclei, or atomic orbital electrons. The collisions are classified into three categories: (1) Nuclear reactions, (2) Elastic collisions, and (3) Inelastic collisions.

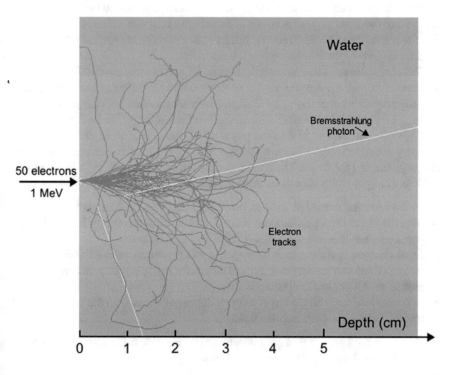

© Springer International Publishing Switzerland 2016
E.B. Podgoršak, *Radiation Physics for Medical Physicists*,
Graduate Texts in Physics, DOI 10.1007/978-3-319-25382-4_5

The many types of interacting particles as well as the various possible categories of interactions result in a wide range of two-particle collisions of interest in nuclear physics and in medical physics. Several parameters, such as the Q value and energy threshold in nuclear reactions, as well as energy transfer in elastic scattering, used in characterization of two-particle collisions are defined in this chapter and determined using considerations of momentum and energy conservation classically as well as relativistically. Many of these parameters play an important role in radiation dosimetry through their effects on stopping powers, as discussed in Chap. 6. They also play an important role in the production of radioactive nuclides, as discussed in Chap. 12.

5.1 Collisions of Two Particles: General Aspects

A common problem in nuclear physics and radiation dosimetry is the collision of two particles in which a projectile with mass m_1, velocity v_1 and kinetic energy $(E_K)_1$ strikes a stationary target with mass m_2 and velocity $v_2 = 0$. The probability or cross section for a particular collision as well as the collision outcome depend on the physical properties of the projectile (mass, charge, velocity, kinetic energy) and the stationary target (mass, charge).

As shown schematically in Fig. 5.1, the collision between the projectile and the target in the most general case results in an intermediate compound that subsequently decays into two reaction products: one of mass m_3 ejected with velocity v_3 at an angle θ to the incident projectile direction, and the other of mass m_4 ejected with velocity v_4 at an angle ϕ to the incident projectile direction.

Targets are either atoms as a whole, atomic nuclei, or atomic orbital electrons. In an interaction with a projectile, targets are assumed to be stationary and they interact with the projectile either through a Coulomb interaction when both the projectile and the target are charged or through a direct collision when the projectile is not charged.

Projectiles of interest in medical physics fall into one of three categories, each category characterized by its own specific mechanism for the interaction between the projectile and the target. The three categories of projectile are: (1) heavy charged particle, (2) light charged particle, and (3) neutron:

1. *Heavy charged particles*, such as protons, α-particles, and heavy ions, interact with the target through Coulomb interactions. Typical targets for heavy charged particles are either atomic nuclei or atomic orbital electrons.
2. *Light charged particles*, such as electrons and positrons, interact with the target through Coulomb interactions. Typical targets for light charged particles are either atomic nuclei or atomic orbital electrons.
3. *Neutrons* interact with the target through direct collisions with the target. Typical targets for neutrons are atomic nuclei.

Two-particle collisions are classified into three categories: (1) *nuclear reactions*, (2) *elastic collisions*, and (3) *inelastic collisions*.

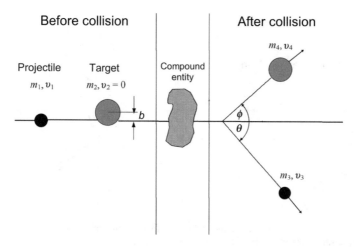

Fig. 5.1 Schematic representation of a two-particle collision of a projectile (*incident particle*) with mass m_1, velocity v_1, momentum p_i and kinetic energy $(E_K)_i$ striking a stationary target with mass m_2 and velocity $v_2 = 0$. An intermediate compound entity is produced temporarily and it decays into two reaction products, one of mass m_3 ejected with velocity v_3 at an angle θ to the incident projectile direction, and the other of mass m_4 ejected with velocity v_4 at an angle ϕ to the incident projectile direction

1. *Nuclear reactions*, shown schematically in Fig. 5.1 and discussed in Sect. 5.2, represent the most general case of a two-particle collision of a projectile m_1 with a target m_2 resulting in two reaction products, m_3 and m_4, that differ from the initial products m_1 and m_2.

   - In any nuclear reaction a number of physical quantities must be conserved, most notably: *charge, linear momentum* and *mass-energy*.
   - In addition, the sum of atomic numbers Z and the sum of atomic mass numbers A for before and after the collision must also be conserved, i.e.,

$$\sum Z \text{ (before collision)} = \sum Z \text{ (after collision)}$$

   and

$$\sum A \text{ (before collision)} = \sum A \text{ (after collision)}.$$

2. *Elastic scattering* is a special case of a two-particle collision in which:

   - The products after the collision are identical to the products before the collision, i.e., $m_3 = m_1$ and $m_4 = m_2$.
   - The total kinetic energy and momentum before the collision are equal to the total kinetic energy and momentum, respectively, after the collision.
   - A minute and generally negligible fraction of the initial kinetic energy of the projectile is transferred to the target.

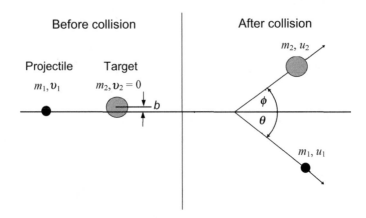

Fig. 5.2 Schematic diagram of an elastic collision between a projectile with mass m_1 and velocity v_1 striking a stationary target m_2. The projectile is scattered with a *scattering angle* θ; the target recoils with a *recoil angle* ϕ. The impact parameter is b. After the collision the velocity of the projectile m_1 is u_1; the velocity of the target m_2 is u_2

Two-particle elastic scattering is shown schematically in Fig. 5.2. The energy transfer in elastic collisions is discussed in Sect. 5.3; the cross sections for single and multiple elastic scattering of two charged particles are discussed in Sect. 2.6

3. In *inelastic scattering* of the projectile m_1 on the target m_2, similarly to elastic scattering, the reaction products after collision are identical to the initial products, i.e., $m_3 = m_1$ and $m_4 = m_2$; however, the incident projectile transfers a portion of its kinetic energy to the target in the form of not only kinetic energy but also in the form of an intrinsic excitation energy E^*. The excitation energy E^* may represent:

- Nuclear excitation of the target.
- Atomic excitation or ionization of the target.
- Emission of bremsstrahlung by the projectile.

As a result of the various types of projectiles and targets as well as several categories of two-particle collisions, many different two-particle interactions are possible. The interactions of interest in medical physics and radiation dosimetry are summarized in Table 5.1.

5.2 Nuclear Reactions

Two-particle collisions between the projectile m_1 and target m_2 resulting in products m_3 and m_4 are referred to as *nuclear reactions* and are governed by conservation of total energy and momentum laws. As shown in Table 5.1, the projectile can be a heavy charged particle, a light charged particle, or a neutron.

Table 5.1 Collisions between various projectiles and targets of interest in medical physics and radiation dosimetry; examples of interactions resulting from specific combinations of projectiles and targets are listed immediately below the table

Projectile	Heavy charged particle		Light charged particle		Neutron	
Target	Nucleus	Electron	Nucleus	Electron	Nucleus	Electron
Nuclear reaction $m_2(m_1, m_3)m_4$	Yes[a]	No	Yes[e]	No	Yes[j]	No
Elastic scattering $m_2(m_1, m_1)m_2$	Yes[b]	No	Yes[f]	Yes[h]	Yes[k]	No
Inelastic scattering $m_2(m_1, m_1)m_2^*$	Yes[c]	Yes[d]	Yes[g]	Yes[i]	Yes[l]	No

Heavy charged particle interactions with nuclei of the target:

a. *Nuclear reaction* precipitated by a heavy charged particle projectile m_1 striking a nucleus m_2 resulting in products m_3 and m_4
   Example: Deuteron bombarding nitrogen-14, resulting in nitrogen-15 and a proton:

$$^{14}_{7}\text{N(d, p)}^{15}_{7}\text{N}.$$

b. *Elastic Coulomb collision* of heavy charged particle with atomic nucleus
   Example: Rutherford scattering of α-particle on gold nucleus:

$$^{197}_{79}\text{Au}(\alpha, \alpha)^{197}_{79}\text{Au}.$$

c. *Inelastic collision* of heavy charged particle with nucleus
   Example: Nuclear excitation, resulting in excited nucleus which reverts to its ground state through emission of a γ-ray photon:

$$^{A}_{Z}\text{X}(\alpha, \alpha)^{A}_{Z}\text{X}^* \Rightarrow {}^{A}_{Z}\text{X}^* \rightarrow {}^{A}_{Z}\text{X} + \gamma.$$

Heavy charged particle interactions with orbital electrons of the target:

d. *Inelastic collision* of heavy charged particle with atomic orbital electron
   Example: Excitation or ionization of an atom.

Light charged particle interactions with nuclei of the target:

e. *Nuclear reaction* precipitated by an energetic light charged particle striking the nucleus
   Examples: (e,n) and (e,p) nuclear reactions.
f. *Elastic collision* between a light charged particle and atomic nucleus
   Example: Radiation-less scattering of electrons on the nuclei of the target.
g. *Inelastic collision* between a light charged particle and atomic nucleus
   Example: Bremsstrahlung production by electrons or positrons undergoing a Coulomb interaction with an atomic nucleus (radiation loss).

Light charged particle interactions with orbital electrons of the target:

h. *Elastic collision* between a light charged particle and an orbital electron
   Example: Ramsauer effect in which an electron of very low kinetic energy (below 100 eV) undergoes an elastic collision with an atomic orbital electron.
i. *Inelastic collision* between a light charged particle and atomic orbital electron
   Example 1: Electron-orbital electron interaction resulting in atomic excitation or ionization (hard and soft collisions).
   Example 2: Positron annihilation leaving atom in an ionized state coinciding with emission of two γ annihilation quanta (see Sect. 7.6.10).

Neutron interactions with nuclei of the target:

j. *Nuclear reaction* caused by neutron colliding with atomic nucleus
   Example 1: Neutron capture by a stable target or neutron activation of a stable target (see Sect. 8.4), resulting in radioactive isotope of same species as the stable target

$$^{59}_{27}\text{Co(n, }\gamma)^{60}_{27}\text{Co}.$$

   Example 2: Spallation and nuclear fission for high atomic number targets.
k. *Elastic collision* between neutron and atomic nucleus (see Sect. 9.2.1)
l. *Inelastic collision* between neutron and atomic nucleus (see Sect. 9.2.2)
   Example: Nuclear excitation

$$^{A}_{Z}\text{X}(n, n)^{A}_{Z}\text{X}^* \Rightarrow {}^{A}_{Z}\text{X}^* \rightarrow {}^{A}_{Z}\text{X} + \gamma.$$

The collision leading to a nuclear reaction is shown schematically in Fig. 5.1 with the projectile m_1 moving with velocity v_1 and kinetic energy $(E_K)_1$ striking a stationary target m_2. An intermediate compound entity is produced temporarily that decays into two reaction products, m_3 and m_4, ejected with velocities v_3 at angle θ and v_4 at angle ϕ, respectively.

5.2.1   Conservation of Momentum in Nuclear Reaction

The conservation of momentum in a two-particle nuclear collision is expressed through the vector relationship

$$m_1 v_1 = m_3 v_3 + m_4 v_4, \tag{5.1}$$

that can be resolved into a component along the incident direction and a component perpendicular to the incident direction to obtain

$$m_1 v_1 = m_3 v_3 \cos \theta + m_4 v_4 \cos \phi \tag{5.2}$$

and

$$0 = m_3 v_3 \sin \theta - m_4 v_4 \sin \phi, \tag{5.3}$$

where the angles θ and ϕ are defined in Fig. 5.1 and v_1, v_3, and v_4 are magnitudes of velocity vectors v_1, v_3, and v_4, respectively.

5.2.2   Conservation of Energy in Nuclear Reaction

The total energy of the projectile m_1 and target m_2 before the interaction (collision) must equal to the total energy of reaction products m_3 and m_4 after the collision

$$\left\{ m_{10} c^2 + (E_K)_i \right\} + \left(m_{20} c^2 + 0 \right) = \left\{ m_{30} c^2 + (E_K)_3 \right\} + \left\{ m_{40} c^2 + (E_K)_4 \right\}, \tag{5.4}$$

where

$m_{10} c^2$	is the rest energy of the projectile.
$m_{20} c^2$	is the rest energy of the target.
$m_{30} c^2$	is the rest energy of the reaction product m_3.
$m_{40} c^2$	is the rest energy of the reaction product m_4.
$(E_K)_i$	is the kinetic energy of the projectile (incident particle).
$(E_K)_3$	is the kinetic energy of the reaction product m_3.
$(E_K)_4$	is the kinetic energy of the reaction product m_4.

Inserting into (5.4) the so-called Q value for the collision in the form

$$Q = \left(m_{10}c^2 + m_{20}c^2\right) - \left(m_{30}c^2 + m_{40}c^2\right),\tag{5.5}$$

we get the following relationship for the conservation of energy

$$(E_K)_i + Q = (E_K)_3 + (E_K)_4.\tag{5.6}$$

Each two-particle collision possesses a characteristic Q value that can be either positive, zero, or negative.

- For $Q > 0$, the collision is *exothermic* (also called exoergic) and results in release of energy.
- For $Q = 0$, the collision is termed *elastic*.
- For $Q < 0$, the collision is termed *endothermic* (also called endoergic) and, to take place, it requires an energy transfer from the projectile to the target.

In (5.5) the Q value for a nuclear reaction is determined by subtracting the sum of nuclear rest energies of reaction products after the reaction $\sum_{i,\text{after}} M_i c^2$ from the sum of nuclear rest energies of reactants (projectile and target) before the reaction $\sum_{i,\text{before}} M_i c^2$, or

$$Q = \sum_{i,\text{before}} M_i c^2 - \sum_{i,\text{after}} M_i c^2.\tag{5.7}$$

If atomic masses rather than nuclear masses are used in calculations of Q values for nuclear reactions, in many instances the electron masses cancel out, so that there is no difference in the end result. However, in situations where electron masses do not cancel out, special care must be taken when using atomic masses to account for all electrons involved in the interaction.

The Q value for a nuclear reaction can also be determined with the help of nuclear binding energy E_B by subtracting the sum of nuclear binding energies of reactants before the interaction $\sum_{i,\text{before}} E_B(i)$ from the sum of nuclear binding energies of reaction products after the interaction $\sum_{i,\text{after}} E_B(i)$, or

$$Q = \sum_{i,\text{after}} E_B(i) - \sum_{i,\text{before}} E_B(i).\tag{5.8}$$

Similar approach can be taken in calculating Q values for spontaneous nuclear decay, as shown in Chap. 11, by using nuclear rest masses, atomic rest masses, or binding energies of the parent, daughter, and other products specific to the particular decay.

5.2.3 Threshold Energy for Nuclear Reactions

An exothermic reaction can occur spontaneously; an endothermic reaction cannot take place unless the projectile has a kinetic energy exceeding *threshold energy*.

- The threshold energy is defined as the smallest total energy E_{thr} or the smallest kinetic energy $(E_K)_{thr}$ of the projectile at which an endothermic collision can still occur.
- The threshold energy for an endothermic collision is determined through the use of the so-called invariant

$$E^2 - p^2c^2 = \text{invariant}, \tag{5.9}$$

where

E is the total energy before the collision and total energy after the collision.
p is the total momentum before and the total momentum after the collision.
c is the speed of light in vacuum.

The invariant is valid for both the *laboratory coordinate system* and for the *center-of-mass coordinate system* and, for convenience, the conditions before the collision are written for the laboratory system while the conditions after the collision are written for the center-of-mass system.

The conditions for before and after the collision are written as follows:

1. **Before collision**:

   Total energy before:

$$E_{thr} + m_{20}c^2 = \sqrt{m_{10}^2c^4 + p_1^2c^2} + m_{20}c^2, \tag{5.10}$$

   where E_{thr} is the total threshold energy of the projectile.

   Total momentum before: p_1

2. **After collision**:

   Total energy after in the center-of-mass system: $m_{30}c^2 + m_{40}c^2$

   Total momentum after in the center-of-mass system: 0

   The invariant of (5.9) for before and after the collision then gives

$$E^2 - p^2c^2 = \left(\sqrt{m_{10}^2c^4 + p_1^2c^2} + m_{20}c^2\right)^2 - p_1^2c^2 = \left(m_{30}c^2 + m_{40}c^2\right)^2 - 0. \tag{5.11}$$

Solving for $E_{\text{thr}} = \sqrt{m_{10}^2 c^4 + p_1^2 c^2}$ results in the following expression for the total threshold energy

$$E_{\text{thr}} = \frac{\left(m_{30}c^2 + m_{40}c^2\right)^2 - \left(m_{10}^2 c^4 + m_{20}^2 c^4\right)}{2m_{20}c^2}. \tag{5.12}$$

Noting that $E_{\text{thr}} = (E_{\text{K}})_{\text{thr}} + m_{10}c^2$, where $(E_{\text{K}})_{\text{thr}}$ is the threshold kinetic energy of the projectile, we get the following expression for $(E_{\text{K}})_{\text{thr}}$

$$(E_{\text{K}})_{\text{thr}} = \frac{\left(m_{30}c^2 + m_{40}c^2\right)^2 - \left(m_{10}c^2 + m_{20}c^2\right)^2}{2m_{20}c^2}. \tag{5.13}$$

The threshold kinetic energy $(E_{\text{K}})_{\text{thr}}$ of the projectile given in (5.13) may now be written in terms of the nuclear reaction Q value as follows:

1. First, we note that from (5.5) for the Q value we can write the following expression

$$\left(m_{30}c^2 + m_{40}c^2\right)^2 = \left(m_{10}c^2 + m_{20}c^2\right)^2 + Q^2 - 2Q\left(m_{10}c^2 + m_{20}c^2\right). \tag{5.14}$$

2. Inserting the relationship of (5.14) into (5.13) we obtain

$$(E_{\text{K}})_{\text{thr}} = -Q\left[\frac{m_{10}c^2 + m_{20}c^2}{m_{20}c^2} - \frac{Q}{2m_{20}c^2}\right] \approx -Q\left(1 + \frac{m_{10}}{m_{20}}\right), \tag{5.15}$$

where, since $Q \ll m_{20}c^2$, we can ignore the $Q/2m_{20}c^2$ term in (5.15).

In (5.15) the threshold kinetic energy $(E_{\text{K}})_{\text{thr}}$ of the projectile exceeds the $|Q|$ value by a relatively small amount to account for conservation of both energy and momentum in the collision.

As a special case the invariant of (5.9) may also be used to calculate the threshold photon energy $\left(E_\gamma\right)_{\text{thr}}$ for pair production (see Sect. 7.6.2):

1. In the field of nucleus of rest mass m_{A} (nuclear pair production) as the threshold energy $\left(E_\gamma^{\text{NPP}}\right)_{\text{thr}}$

$$\left(E_\gamma^{\text{NPP}}\right)_{\text{thr}} = 2m_{\text{e}}c^2\left(1 + \frac{m_{\text{e}}}{m_{\text{A}}}\right). \tag{5.16}$$

2. In the field of orbital electron of rest mass m_{e} (electronic pair production also called triplet production) as the threshold energy $\left(E_\gamma^{\text{TP}}\right)_{\text{thr}}$

$$\left(E_\gamma^{\text{TP}}\right)_{\text{thr}} = 4m_{\text{e}}c^2. \tag{5.17}$$

5.3 Two-Particle Elastic Scattering: Energy Transfer

Elastic scattering in a two-particle collision is a special case of a nuclear collision between a projectile m_1 and target m_2:

1. The initial and final products are identical (i.e., $m_3 = m_1$ and $m_4 = m_2$); however, the projectile changes its direction of motion (i.e., is scattered) and the target recoils.
2. Q value for the collision, as given in (5.7), equals zero, i.e., $Q = 0$.
3. A certain amount of kinetic energy (ΔE_K) is transferred from the projectile m_1 to the target m_2. The amount of energy transfer is governed by conservation of the kinetic energy and momentum, and depends on the scattering angle θ of the projectile and the recoil angle ϕ of the target.

Two-particle elastic scattering between projectile m_1 moving with velocity v_1 and a stationary target m_2 is shown schematically in Fig. 5.2, with θ the scattering angle of the projectile, ϕ the recoil angle of the target, and b the impact parameter. After the collision particle m_1 continues with velocity u_1 and the target recoils with velocity u_2.

5.3.1 General Energy Transfer from Projectile to Target in Elastic Scattering

The kinetic energy transfer ΔE_K from projectile m_1 to the target m_2 is determined classically using the conservation of kinetic energy and momentum laws as follows:

Conservation of kinetic energy:

$$(E_K)_i = \frac{1}{2}m_1 v_1^2 = \frac{1}{2}m_1 u_1^2 + \frac{1}{2}m_2 u_2^2, \tag{5.18}$$

where $(E_K)_i$ is the kinetic energy of the projectile (incident particle) m_1.
Conservation of momentum:

$$m_1 v_1 = m_1 u_1 \cos\theta + m_2 u_2 \cos\phi \tag{5.19}$$

and

$$0 = m_1 u_1 \sin\theta - m_2 u_2 \sin\phi, \tag{5.20}$$

where

v_1 is the initial velocity of the projectile m_1.
u_1 is the final velocity of the projectile m_1.
u_2 is the final velocity of the target m_2.

θ is the scattering angle of the projectile m_1.
ϕ is the recoil angle of the target m_2.

Equations (5.19) and (5.20) can, respectively, be written as follows

$$(m_1 v_1 - m_2 u_2 \cos \phi)^2 = m_1^2 u_1^2 \cos^2 \theta \tag{5.21}$$

and

$$m_1^2 u_1^2 \sin^2 \theta = m_1^2 u_1^2 - m_1^2 u_1^2 \cos^2 \theta = m_2^2 u_2^2 \sin^2 \phi, \tag{5.22}$$

Inserting (5.21) into (5.22) gives

$$m_2^2 u_2^2 = m_1^2 u_1^2 - m_1^2 v_1^2 + 2m_1 v_1 m_2 u_2 \cos \phi, \tag{5.23}$$

which, after inserting (5.18) multiplied by $2m_1$, reads

$$m_2^2 u_2^2 = 2m_1 v_1 m_2 u_2 \cos \phi - m_1 m_2 u_2^2$$

or

$$2m_1 v_1 \cos \phi = (m_1 + m_2) u_2. \tag{5.24}$$

Since $\Delta E_K = \frac{1}{2} m_2 u_2^2$, we get the following general expression for the kinetic energy transfer ΔE_K from the projectile (incident particle) m_1 with kinetic energy $(E_K)_i$ to the stationary target m_2

$$\Delta E_K = (E_K)_i \frac{4m_1 m_2}{(m_1 + m_2)^2} \cos^2 \phi, \tag{5.25}$$

where ϕ is the recoil angle of the target m_2, defined graphically in Fig. 5.2.

5.3.2 Energy Transfer in a Two-Particle Elastic Head-on Collision

A head-on (direct hit) elastic collision between two particles is a special elastic collision in which the impact parameter b equals to zero. This results in a maximum possible momentum transfer Δp_{max} and maximum possible energy transfer $(\Delta E_K)_{max}$ from the projectile m_1 to the target m_2.

The head-on two-particle elastic collision is characterized as follows:

1. The impact parameter $b = 0$
2. The target recoil angle $\phi = 0$

3. The projectile scattering angle θ is either 0 or π depending on the relative magnitudes of masses m_1 and m_2

- For $m_1 > m_2$, the scattering angle $\theta = 0$ (forward scattering).
- For $m_1 < m_2$, the scattering angle $\theta = \pi$ (back-scattering).
- For $m_1 = m_2$, the projectile stops and target recoils with $\phi = 0$.

5.3.3 Classical Relationships for a Head-on Collision

Before collision	*After collision*
o → o	o → o →
m_1, v_1 $m_2, v_2 = 0$	m_1, u_1 m_2, u_2

Conservation of momentum:

$$m_1 v_1 + 0 = m_1 u_1 + m_2 u_2. \tag{5.26}$$

Conservation of kinetic energy:

$$\frac{m_1 v_1^2}{2} + 0 = \frac{m_1 u_1^2}{2} + \frac{m_2 u_2^2}{2}. \tag{5.27}$$

The maximum momentum transfer Δp_{max} and the maximum kinetic energy transfer ΔE_{max} from the projectile (incident particle) to stationary target in a head-on collision are given as follows:

- The *maximum momentum transfer* Δp_{max} from the projectile m_1 to the target m_2 is given by:

$$\Delta p_{max} = m_1 v_1 - m_1 u_1 = \frac{2m_1 m_2}{m_1 + m_2} v_1 = \frac{2m_2}{m_1 + m_2} p_i, \tag{5.28}$$

where p_i is the momentum of the projectile m_1.
- The *maximum energy transfer* ΔE_{max} from the projectile m_1 to the stationary target m_2 is given by

$$\Delta E_{max} = \frac{m_1 v_1^2}{2} - \frac{m_1 u_1^2}{2} = \frac{m_2 u_2^2}{2} = \frac{4m_1 m_2}{(m_1 + m_2)^2} (E_K)_i, \tag{5.29}$$

where $(E_K)_i$ is the initial kinetic energy of the projectile (incident particle) m_1. The same result can be obtained from the general relationship given in (5.25) after inserting $\phi = 0$ for the target recoil angle.

5.3.4 Special Cases for Classical Energy Transfer in a Head-on Collision

1. **Projectile mass m_1 much larger than target mass m_2;**

$$m_1 \gg m_2 \rightarrow \Delta E_{max} = \frac{4m_1 m_2}{(m_1 + m_2)^2} (E_K)_i \approx 4\frac{m_2}{m_1} (E_K)_i = 2m_2 v_1^2. \quad (5.30)$$

Example: proton colliding with orbital electron: $m_p \gg m_e$:

$$\Delta E_{max} = \frac{4m_e m_p}{\left(m_e + m_p\right)^2} (E_K)_i \approx 4\frac{m_e}{m_p} (E_K)_i = 2m_e v_1^2. \quad (5.31)$$

Since $4m_e/m_p = 4/1836 \approx 0.002$, we see that in a direct hit between a proton and an electron only about 0.2% of the proton kinetic energy is transferred to the target electron in a single collision.

2. **Projectile mass m_1 much smaller than target mass m_2;**

$$m_1 \ll m_2 \rightarrow \Delta E_{max} = \frac{4m_1 m_2}{(m_1 + m_2)^2} (E_K)_i \approx 4\frac{m_1}{m_2} (E_K)_i. \quad (5.32)$$

Example 1: α-particle colliding with gold nucleus (Au-207): $m_\alpha \ll m_{Au}$ (Rutherford scattering, see Sect. 2.2):

$$\Delta E_{max} = 4\frac{m_\alpha m_{Au}}{(m_\alpha + m_{Au})^2} (E_K)_i \approx 4\frac{m_\alpha}{m_{Au}} (E_K)_i. \quad (5.33)$$

Since $4m_\alpha/m_{Au} \approx 0.08$, we see that in a single direct hit head-on collision only about 8% of the incident α-particle kinetic energy is transferred to the gold target.

Example 2: Neutron colliding with lead nucleus (Pb-207): $m_n \ll m_{pb}$:

$$\Delta E_{max} = 4\frac{m_n m_{Pb}}{(m_n + m_{Pb})^2} (E_K)_i \approx 4\frac{m_n}{m_{Pb}} (E_K)_i. \quad (5.34)$$

Since $4m_n/m_{Pb} \approx 1/50 = 0.02$, we see that in a direct hit only about 2% of the incident neutron kinetic energy is transferred to the lead target. This shows that lead is a very inefficient material for slowing down the neutrons; low atomic number materials are much more suitable for this purpose. Of practical importance here is the use of polyethylene as shielding material for doors in high-energy linac bunkers to shield against neutrons produced in the linac.

3. **Projectile mass m_1 equal to target mass m_2;**

$$m_1 = m_2 \rightarrow \Delta E_{max} = \frac{4m_1 m_2}{(m_1 + m_2)^2} (E_K)_i = (E_K)_i \qquad (5.35)$$

Example: (interaction between two *distinguishable particles*) such as *positron colliding with orbital electron* or *neutron colliding with hydrogen atom*.
In a direct hit between two distinguishable particles of equal mass the whole kinetic energy of the incident particle is transferred to the target in a single hit.

4. **Projectile mass m_1 equal to target mass m_2;**

$$m_1 = m_2 \rightarrow \Delta E_{max} = \frac{1}{2} (E_K)_i \qquad (5.36)$$

Example: (interaction between two *indistinguishable particles*) such as *electron colliding with orbital electron*: after the interaction, the particle with the larger kinetic energy is assumed to be the incident particle; therefore the maximum possible energy transfer is $\frac{1}{2}(E_K)_i$.

5.3.5 Relativistic Relationships for a Head-on Collision

The relationship for ΔE_{max} in (5.29) was calculated classically. The relativistic result given below is similar to the classical result, with m_{10} and m_{20} standing for the rest masses of the projectile (incident particle) m_1 and stationary target m_2, respectively

Before collision   *After collision*

   o $\rightarrow$ o   o $\rightarrow$ o $\rightarrow$

m_{10}, v_1 $m_{20}, v_2 = 0$ m_{10}, u_1 m_{20}, u_2

Conservation of momentum:

$$\gamma \beta m_{10} c + 0 = \gamma_1 \beta_1 m_{10} c + \gamma_2 \beta_2 m_{20} c. \qquad (5.37)$$

Conservation of total energy:

$$\gamma m_{10} c^2 + m_{20} c^2 = \gamma_1 m_{10} c^2 + \gamma_2 m_{20} c^2, \qquad (5.38)$$

with $\beta = \dfrac{v_1}{c}$; $\beta_1 = \dfrac{u_1}{c}$; $\beta_2 = \dfrac{u_2}{c}$

and $\gamma = \dfrac{1}{\sqrt{1 - \beta^2}}$; $\gamma_1 = \dfrac{1}{\sqrt{1 - \beta_1^2}}$; $\gamma_2 = \dfrac{1}{\sqrt{1 - \beta_2^2}}$.

The maximum momentum transfer Δp_{max} and the maximum total energy transfer ΔE_{max} from the projectile to the target in a head-on collision are given relativistically by the following expressions:

- The *maximum momentum transfer* Δp_{max} from the projectile m_{10} to the target m_{20} is given by

$$\Delta p_{max} = \frac{2\,(\gamma m_{10} + m_{20})\,m_{20}}{m_{10}^2 + m_{20}^2 + 2\gamma m_{10}m_{20}}\,p_i, \tag{5.39}$$

where p_i is the momentum of the projectile (incident particle) of rest mass m_{10}.

- The *maximum energy transfer* ΔE_{max} from the projectile m_{10} to the target m_{20} is given by

$$\Delta E_{max} = \frac{2\,(\gamma + 1)\,m_{10}m_{20}}{m_{10}^2 + m_{20}^2 + 2\gamma m_{10}m_{20}}\,(E_K)_i, \tag{5.40}$$

where $(E_K)_i$ is the kinetic energy of the projectile (incident particle) m_{10}.
- The relativistic equations for the maximum momentum transfer of (5.39) and maximum energy transfer of (5.40) transform into the classical Equations (5.28) and (5.29), respectively, for small velocities of the projectile where $\beta \to 0$, corresponding to $\gamma = \left(1 - \beta^2\right)^{-1/2} \to 1$.

5.3.6 Special Cases for Relativistic Energy Transfer in Head-on Collision

1. **Projectile rest mass m_{10} much larger than target rest mass m_{20}; $(m_{10} \gg m_{20})$**

$$\Delta E_{max} = \frac{2\,(\gamma + 1)\,\dfrac{m_{20}}{m_{10}}}{1 + \left[\dfrac{m_{20}}{m_{10}}\right]^2 + 2\gamma \left[\dfrac{m_{20}}{m_{10}}\right]}\,(E_K)_i \approx 2\,(\gamma^2 - 1)\,\frac{m_{20}}{m_{10}}m_{10}c^2$$

$$= 2m_{20}c^2\,\frac{\beta^2}{1 - \beta^2}, \tag{5.41}$$

with the kinetic energy of the projectile m_{10} given as follows

$$(E_K)_i = m_{10}c^2\left[\frac{1}{\sqrt{1 - \beta^2}} - 1\right] = m_{10}c^2\,(\gamma - 1). \tag{5.42}$$

Example: Heavy charged particle with mass $m_{10} = m_p$ (e.g., proton) colliding with an orbital electron with mass $m_{20} = m_e$.

Note that, for the classical case of low velocity v_1 of the incident heavy projectile m_{10}, (5.41) transforms into $\Delta E_{max} \approx 2m_{20}v_1^2$ given in (5.30) with m_{20} the rest mass of the target and v_1 the velocity of the projectile m_{10}.

2. **Projectile rest mass m_{10} much smaller than target rest mass m_{20}; $(m_{10} \ll m_{20})$**

$$\Delta E_{max} = \frac{2(\gamma+1)\dfrac{m_{20}}{m_{10}}}{1+\left[\dfrac{m_{20}}{m_{10}}\right]^2 + 2\gamma\left[\dfrac{m_{20}}{m_{10}}\right]}(E_K)_1 \approx \frac{2(\gamma+1)\dfrac{m_{20}}{m_{10}}}{\dfrac{m_{20}}{m_{10}}\left[\dfrac{m_{20}}{m_{10}}+2\gamma\right]}m_{10}c^2(\gamma-1)$$

$$= \frac{2(\gamma^2-1)}{\dfrac{m_{20}}{m_{10}}+2\gamma}m_{10}c^2 = \frac{2m_{10}c^2}{\dfrac{m_{20}}{m_{10}}+2\gamma}\frac{\beta^2}{1-\beta^2}. \qquad (5.43)$$

Note that, for small velocity $v_1 = c\beta$ of the projectile with rest mass m_{10}, (5.43) transforms into (5.32) as follows

$$\Delta E_{max} = \frac{2m_{10}v_1^2}{\dfrac{m_{20}}{m_{10}}+2\gamma}\frac{1}{1-\beta^2} \approx 4\frac{m_{10}}{m_{20}}\frac{m_{10}v_1^2}{2} = 4\frac{m_{10}}{m_{20}}(E_K)_i. \qquad (5.44)$$

3. **Projectile rest mass m_{10} equal to target rest mass m_{20}; $(m_{10} = m_{20})$** (collision between two *distinguishable particles*)

$$\Delta E_{max} = \frac{2(\gamma+1)m_{10}m_{20}}{m_{10}^2 + m_{20}^2 + 2\gamma m_{10}m_{20}}(E_K)_i = \frac{2(\gamma+1)m_{10}^2}{2(\gamma+1)m_{10}^2}(E_K)_i = (E_K)_i. \quad (5.45)$$

Example: positron (projectile) colliding with an orbital electron (target) [see (5.35)]. The positron stops and the electron moves away with the kinetic energy of the incident positron.

4. **Projectile rest mass m_{10} equal to target rest mass m_{20}; $(m_{10} = m_{20})$** (collision between two *indistinguishable particles*)

$$\Delta E_{max} = \frac{1}{2}(E_K)_i. \qquad (5.46)$$

Example: electron colliding with an orbital electron representing a collision between two indistinguishable particles – after the interaction, the particle with larger kinetic energy is assumed to be the incident particle [see (5.36)].

5.3.7 Maximum Energy Transfer Fraction in Head-on Collision

The general expression for energy transfer ΔE_{max} from the projectile with rest mass energy $m_{10}c^2$ and kinetic energy E_K to the target with rest mass energy $m_{20}c^2$ in a head-on (direct hit) two-particle collision is given in (5.40). The classical limit ($\gamma \to 1$) of (5.40) is given in (5.29). To express the maximum energy transfer fraction $\Delta E_{max}/E_K$ we rewrite (5.40) as follows

$$\eta = \left(\frac{\Delta E_{max}}{E_K}\right)_\eta = \frac{2\,(\gamma+1)\,m_{10}c^2 m_{20}c^2}{\left(m_{10}c^2\right)^2 + \left(m_{20}c^2\right)^2 + 2\gamma m_{10}c^2 m_{20}c^2}$$

$$= \frac{4 m_{10}c^2 m_{20}c^2 + 2 E_K m_{20}c^2}{\left(m_{10}c^2 + m_{20}c^2\right)^2 + 2 E_K m_{20}c^2}, \tag{5.47}$$

where we used $E_K = (\gamma - 1)\,m_{10}c^2$ with γ the standard relativistic velocity factor $\gamma = [1 - (v/c)^2]^{-1/2}$ and v the velocity of the projectile.

The classical limit of (5.47) is determined by taking (5.47) to the limit of $\gamma \to 1$ or assuming that $E_K \ll m_{20}c^2$ to get

$$\eta = \lim_{\gamma \to 1} \frac{\Delta E_K}{E_K} = \frac{4 m_{10}c^2 m_{20}c^2}{\left(m_{10}c^2 + m_{20}c^2\right)^2}. \tag{5.48}$$

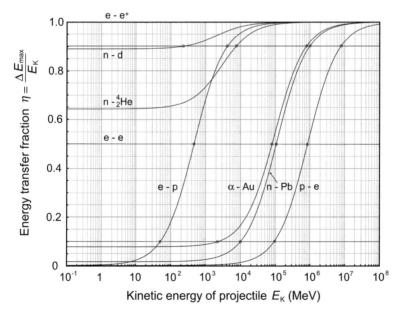

Fig. 5.3 Energy transfer fraction η of (5.47) against kinetic energy E_K of the projectile for various projectile-target combinations in direct hit head-on collision

In Fig. 5.3 we plot the energy transfer fraction η of (5.47) against the projectile kinetic energy E_K for head-on elastic collision of various projectile–target combinations. For all collisions, η ranges from the classical limit of (5.48) to $\eta = 1$ at very large relativistic kinetic energies E_K of the projectile.

A few special features of the η against E_K plot become apparent:

1. Curves for η against E_K are of sigmoid shape ranging from a minimum η given by the classical limit of (5.48) to a maximum of $\eta = 1$ at high E_K.
2. For all head-on elastic collisions, η eventually attains a value of unity at some large value of E_K. This means that at large E_K there is full energy transfer from the projectile to the target.
3. For identical masses of the projectile and the target $\left(m_{10}c^2 = m_{20}c^2\right)$ the maximum energy transfer fraction η equals to 1 for all E_K from the classical range to ∞ when the two particles are distinguishable, such as electron and positron or neutron and proton, for example.
4. For scattering of indistinguishable particles such as electron/electron scattering, for example, the convention is to assume that the scattered particle is the one with

Table 5.2 Kinetic energy $(E_K)_\eta$ of the projectile at which the maximum energy transfer fraction in head-on elastic collision attains values of $\eta = 10\%$, $\eta = 50\%$, and $\eta = 90\%$ for various projectile-target combinations, calculated using (5.49) with appropriate projectile mass m_{10} and target mass m_{20}. The classical limit of η was determined with (5.48)

Projectile	Target	Classical limit of η	Energy $(E_K)_\eta$ in MeV to reach		
			10%	50%	90%
Electron $m_{10} = 0.511$ MeV/c^2	Proton $m_{20} = 938.3$ MeV/c^2	0.002	50	4.7×10^2	4.2×10^3
Proton $m_{10} = 938.3$ MeV/c^2	Electron $m_{20} = 0.511$ MeV/c^2	0.002	9.4×10^4	8.6×10^5	7.7×10^6
Neutron $m_{10} = 939.6$ MeV/c^2	Lead nucleus $m_{20} = 192.8 \times 10^3$ MeV/c^2	0.019	8.7×10^3	9.4×10^4	8.6×10^5
α-particle $m_{10} = 3727.4$ MeV/c^2	Gold nucleus $m_{20} = 183.5 \times 10^3$ MeV/c^2	0.078	2.7×10^3	8.1×10^4	7.9×10^5
Neutron $m_{10} = 939.6$ MeV/c^2	Helium-4 $m_{20} = 3727.4$ MeV/c^2	0.643	–	–	7.5×10^3
Neutron $m_{10} = 939.6$ MeV/c^2	Deuteron $m_{20} = 1875.6$ MeV/c^2	0.889	–	–	220
Positron $m_{10} = 0.511$ MeV/c^2	Electron $m_{20} = 0.511$ MeV/c^2	1.000	–	–	–
Neutron $m_{10} = 939.6$ MeV/c^2	Proton $m_{20} = 938.3$ MeV/c^2	~1.000	–	–	–
Electron $m_{10} = 0.511$ MeV/c^2	Electron $m_{20} = 0.511$ MeV/c^2	0.500	–	–	–
Proton $m_{10} = 938.3$ MeV/c^2	Proton $m_{20} = 938.3$ MeV/c^2	0.500	–	–	–

the higher kinetic energy so that the maximum energy transfer can only amount to $\frac{1}{2}\Delta E_{max}$.

Solving (5.47) for E_K allows us to calculate the projectile kinetic energy $(E_K)_\eta$ at which η attains a predetermined value. In general we get

$$(E_K)_\eta = \frac{\eta(m_{10}c^2 + m_{20}c^2)^2 - 4m_{10}c^2m_{20}c^2}{2m_{20}c^2(1 - \eta)} \tag{5.49}$$

Table 5.2 lists the projectile kinetic energies $(E_K)_\eta$ at which the maximum energy transfer fraction η in a head-on collision attains values of 0.1; 0.5; and 0.9 or 10%; 50%; and 90% for various projectile–target combinations. Of course, for some interactions not all or even none of the above η values are relevant depending on the relative masses of the projectile and the target. The closer are the two masses to one another, the higher is already the classical limit of η and η can only increase with kinetic energy E_K or remain constant at $\eta = 1$ when the two particles have the same mass but are distinguishable from one another. Table 5.2 also gives the classical limits of (5.48) for the various elastic scattering interactions listed.

Lichtenberg Figures

Images on next page are so-called *Lichtenberg figures* in (**a**) calculated using fractal geometry techniques and in (**b**) produced by 10 MeV electrons deposited in a Lucite (acrylic) block.

The first Lichtenberg figures were actually two-dimensional patterns formed in dust on a charged plate in the laboratory of their discoverer, *Georg Christoph Lichtenberg,* an eighteenth century German physicist. The basic principles involved in the formation of these early figures are fundamental to the operation of modern copy machines and laser printers.

Fractal geometry is a modern invention in comparison to the over 2000 year-old Euclidean geometry. Man-made objects usually follow Euclidean geometry shapes and are defined by simple algebraic formulas. In contrast, objects in nature often follow the rules of fractal geometry defined by iterative or recursive algorithms. *Benoît B. Mandelbrot,* a Polish-born French-American mathematician is credited for introducing the term and techniques of fractal geometry in the 1970s. The most striking feature of fractal geometry is the so-called self-similarity implying that the fractal contains smaller components that replicate the whole fractal when magnified. In theory a fractal is composed of an infinite number of ever diminishing components, all of the same shape. In nature, many objects approach the fractal behavior; however, the self-similarity breaks down at some small enough scale and the objects are then called fractal-like. Natural examples of fractal-like shapes have been observed on a wide variety of natural objects such as clouds, mountain ranges, lightning bolts, crystals, trees, coast lines, and snow flakes.

High-voltage *electrical discharges* on the surface or inside insulating materials often result in Lichtenberg figures or patterns. Lucite is usually used as the medium for capturing the Lichtenberg figures, because it has an excellent combination of optical (it is transparent), dielectric (it is an insulator), and mechanical (it is strong, yet easy to machine) properties suitable for highlighting the Lichtenberg effect. Electrons accelerated in a linear accelerator to a speed close to the speed of light are made to strike a Lucite block. They penetrate into the block, come to rest inside the block, and form a plane of excess negative charge. This electron space charge trapped in the block is released either spontaneously or through mechanical stress, and the discharge paths within the Lucite leave permanent records of their passage as they melt and fracture the plastic along the way. The charge exit point appears as a small hole at the surface of the Lucite block. Similar breakdown, albeit on a much larger scale, occurs during a lightning flash as the electrical discharge drains the highly charged regions within storm clouds; however, the discharge in air leaves behind no permanent record of the passage through air.

The fractal tree shown in (**a**) on next page is a typical example of fractal geometry use in calculating the shape of a natural object. An example of a frozen Lichtenberg discharge in Lucite, often referred to as an electrical tree, is shown in (**b**). The similarity between the calculated and the "measured" tree is striking.

(a) *Courtesy of Prof. Volkhard Nordmeier, Technische Universität, Berlin*

(b) *Courtesy of Bert Hickman, Stoneridge Engineering* (www.Teslamania.com)

Chapter 6
Interactions of Charged Particles with Matter

In this chapter we discuss interactions of charged particle radiation with matter. A charged particle is surrounded by its Coulomb electric field that interacts with orbital electrons and the nucleus of all atoms it encounters, as it penetrates into matter. Charged particle interactions with orbital electrons of the absorber result in collision loss, interactions with nuclei of the absorber result in radiation loss. The energy transfer from the charged particle to matter in each individual atomic interaction is generally small, so that the particle undergoes a large number of interactions before its kinetic energy is spent.

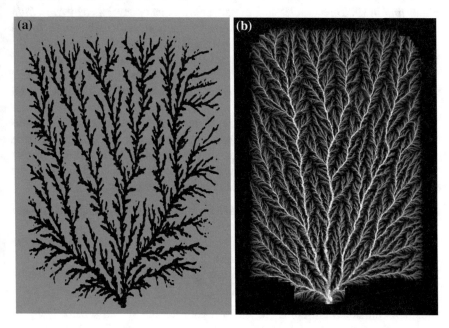

© Springer International Publishing Switzerland 2016
E.B. Podgoršak, *Radiation Physics for Medical Physicists*,
Graduate Texts in Physics, DOI 10.1007/978-3-319-25382-4_6

Stopping power is the parameter used to describe the gradual loss of energy of the charged particle, as it penetrates into an absorbing medium. Two classes of stopping power are known: *collision stopping power* that results from charged particle interaction with orbital electrons of the absorber and *radiation stopping power* that results from charged particle interaction with nuclei of the absorber.

Stopping powers play an important role in radiation dosimetry. They depend on the properties of the charged particle such as its mass, charge, velocity and energy as well as on the properties of the absorbing medium such as its density and atomic number. In addition to stopping powers, other parameters of charged particle interaction with matter, such as the range, energy transfer, mean ionization potential, and radiation yield, are also discussed in this chapter.

6.1 General Aspects of Energy Transfer from Charged Particle to Medium

The discovery of energetic charged particle emission from radioactive materials in 1896 stimulated interest not only in the origin of the emitted particles but also in how they were slowed down as they traversed matter. The theory of stopping power played an important role in the development of atomic and nuclear models starting with the α-particle scattering studies of *Hans Geiger, Ernest Marsden* and *Ernest Rutherford* in 1908 and the classical stopping power theory developed by *Niels Bohr* in 1913, and culminating with the quantum mechanical and relativistic theory of stopping power proposed by *Hans Bethe* in the 1930s and refined by *Ugo Fano* in the 1960s. More recent developments introduced several additional secondary correction factors to increase the accuracy of theoretical stopping power expressions; however, the main theoretical foundations that early workers enunciated decades ago are still valid today.

As a charged particle travels through an absorber, it experiences Coulomb interactions with the nuclei and orbital electrons of absorber atoms. These interactions can be divided into three categories depending on the size of the classical impact parameter b of the charged particle trajectory compared to the classical atomic radius a of the absorber atom with which the charged particle interacts:

1. Coulomb force interaction of the charged particle with the external nuclear field of the absorber atom for $b \ll a$ (bremsstrahlung production).
2. Coulomb force interaction of the charged particle with orbital electron of the absorber atom for $b \approx a$ (hard collision).
3. Coulomb force interaction of the charged particle with orbital electron of the absorber atom for $b \gg a$ (soft collision).

Radiation collision, hard collision, and soft collision are shown schematically in Fig. 6.1, with b the impact parameter of the particle trajectory and a the atomic radius of the absorber atom.

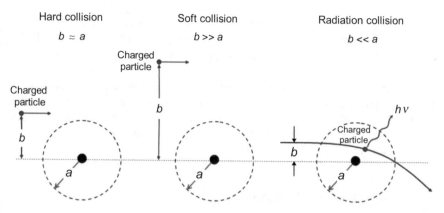

Fig. 6.1 Three different types of collision of a charged particle with an atom, depending on the relative size of the impact parameter b and atomic radius a. Hard (close) collision for $b \approx a$; soft (distant) collision for $b \gg a$; and radiation collision for $b \ll a$

6.1.1   Charged Particle Interaction with Coulomb Field of the Nucleus (Radiation Collision)

When the impact parameter b of a charged particle is much smaller than the radius a of the absorber atom (i.e., $b \ll a$), the charged particle interacts mainly with the nucleus and undergoes either elastic or inelastic scattering possibly accompanied with a change in direction of motion.

The vast majority of these interactions are elastic so that the particle is scattered by the nucleus but loses only an insignificant amount of its kinetic energy to satisfy the conservation of momentum requirement. However, a small percentage of the scattering interactions are inelastic and may result in significant energy loss for the charged particle accompanied by emission of x-ray photons. This type of interaction is called *bremsstrahlung collision*. At a given particle acceleration, the probability for this type of interaction is inversely proportional to the square of the mass of the charged particle, making the bremsstrahlung production for charged particles other then electrons and positrons essentially negligible.

6.1.2   Hard (Close) Collision

When the impact parameter b of a charged particle trajectory is of the order of the radius a of the absorber atom (i.e., $b \approx a$), the charged particle may have a direct Coulomb impact interaction with a single atomic orbital electron and transfer to it a significant amount of energy. The interaction is referred to as *hard* or *close collision*.

The orbital electron leaves the atom as a δ-ray, and is energetic enough to undergo its own Coulomb interactions with absorber atoms. The maximum possible energy transfer from a charged particle to an orbital electron (δ-ray) is discussed in detail in

Sect. 5.3. The number of hard collisions experienced by a charged particle moving in an absorber is generally small; however, the energy transfers associated with hard collisions are relatively large, so that the particle loses roughly 50% of its kinetic energy through hard collisions.

The theories that govern hard collisions depend strongly on the characteristics of charged particles and generally assume that the orbital electron (δ-ray) released through a hard collision is free before and after the interaction, since the kinetic energy transferred to it from the charged particle is much larger than its atomic binding energy.

6.1.3   Soft (Distant) Collision

When the impact parameter b of the charged particle trajectory is much larger than the radius a of the absorber atom (i.e., $b \gg a$), the charged particle interacts with the whole atom and the whole atomic complement of bound electrons. The interaction is called a *soft* or *distant collision*. The energy transfer from the charged particle to a given bound electron is very small; however, the number of these interactions is large, so that approximately 50% of energy loss by a charged particle occurs through these small-energy-transfer interactions that may cause atomic polarization, excitation or ionization through removal of a valence electron. In the energy region of soft collisions the expressions derived with a given theory are valid for all types of charged particles including electrons and positrons.

6.2   General Aspects of Stopping Power

During its motion through an absorbing medium a charged particle experiences a large number of interactions before its kinetic energy is expended. In each interaction the charged particle's path may be altered (*elastic* or *inelastic scattering*) and it may lose some of its kinetic energy that will be transferred to the medium (*collision loss*) or to photons (*radiation loss*). Each of these possible interactions between the charged particle and orbital electrons or the nucleus of the absorber atoms is characterized by a specific cross section (probability) σ for the particular interaction. The energy loss of the charged particle propagating through an absorber depends on the characteristics of the particle as well as the absorber.

The rate of energy loss (typically expressed in MeV) per unit of path length (typically expressed in cm) by a charged particle in an absorbing medium is called the linear stopping power ($-dE/dx$). Dividing the linear stopping power by the density ρ of the absorber results in the mass stopping power S given in units of MeV·cm^2·g^{-1}. The stopping power is a property of the material in which a charged particle propagates.

In general, the average energy loss per unit path length $-dE/d\ell$ experienced by the heavy particle is calculated by multiplying the cross section for a given energy loss σ_{ni} by the energy loss ΔE_{ni} and a summation over all possible individual collisions i

$$-\frac{dE}{d\ell} = \sum_i N_i \sum_n \Delta E_{ni} \sigma_{ni}, \tag{6.1}$$

where N_i is the density of atoms i that can be expressed:

1. Either in number of atoms per unit volume resulting in $-dE/d\ell$ referred to as the linear stopping power $-dE/dx$ and representing energy loss per unit distance traversed in the absorber. The typical units of linear stopping power are MeV/cm or less common keV/μm.
2. Or in number of atoms per unit mass resulting in $-dE/d\ell$ referred to as mass stopping power $S = -(1/\rho)\,dE/dx$ and representing energy loss per g/cm^2 of material traversed in the absorber. The typical unit of mass stopping power is MeV·cm^2·g^{-1}.

With regard to charged particle interaction, two types of stopping power are known:

1. *Radiation stopping power* (also called nuclear stopping power) resulting from charged particle Coulomb interaction with the nuclei of the absorber. Only light charged particles (electrons and positrons) experience appreciable energy loss through these interactions that are usually referred to as bremsstrahlung interactions. For heavy charged particles (protons, α-particles, etc.) the radiation (bremsstrahlung) loss is negligible in comparison with the collision loss.
2. *Collision stopping power* (also called ionization or electronic stopping power) resulting from charged particle Coulomb interactions with orbital electrons of the absorber. Both heavy and light charged particles experience these interactions that result in energy transfer from the charged particle to orbital electrons through impact excitation and ionization of absorber atoms.

The total stopping power S_{tot} for a charged particle of kinetic energy E_K traveling through an absorber of atomic number Z is in general the sum of the radiation (nuclear) stopping power S_{rad} and collision (electronic) stopping power S_{col}, i.e.,

$$S_{tot} = S_{rad} + S_{col}. \tag{6.2}$$

The collision stopping power S_{col} is further subdivided into two components: the soft (distant) collision stopping power S_{col}^{soft} and the hard (close) collision stopping power S_{col}^{hard}

$$S_{col} = S_{col}^{soft} + S_{col}^{hard} \tag{6.3}$$

The total stopping power is thus in general terms expressed as the following sum

$$S_{tot} = S_{rad} + S_{col} = S_{rad} + S_{col}^{soft} + S_{col}^{hard} \tag{6.4}$$

6.3 Radiation (Nuclear) Stopping Power

As shown by the Larmor relationship (4.18), any time a charged particle is accelerated or decelerated part of its kinetic energy is emitted in the form of bremsstrahlung photons. The rate of bremsstrahlung energy dissipation is proportional to a^2 (the square of the charged particle acceleration a) which in turn is proportional to $(zZ/m)^2$ with z and m the atomic number and mass, respectively, of the radiating charged particle, and Z the atomic number of the absorber target.

The bremsstrahlung intensity is thus linearly proportional to $(zZ)^2$ and inversely proportional to m^2. As a consequence of the relatively large mass of heavy charged particles, the bremsstrahlung yield produced by heavy charged particles such as protons and α-particles in comparison with electrons and positrons is insignificant and generally ignored.

Hans Bethe and Walter Heitler have shown in 1930s that the cross section for emission of bremsstrahlung σ_{rad} has the same form in classical and quantum theory and is proportional to

$$\sigma_{rad} \propto \alpha r_e^2 Z^2 \; (cm^2/nucleus) , \tag{6.5}$$

where

α is the fine structure constant $\left[e^2/\left(4\pi\varepsilon_0\hbar c\right) = 1/137\right]$.
r_e is the classical electron radius $\left[e^2/\left(4\pi\varepsilon_0 m_e c^2\right) = 2.818 \text{ fm}\right]$.
Z is the atomic number of the absorber target.

Table 6.1 provides expressions for σ_{rad} for various regions of incident electron kinetic energy $(E_K)_0$ from the classical region where $(E_K)_0 \ll m_e c^2$ all the way to the extreme relativistic region where $(E_K)_0 \gg m_e c^2/\alpha$.

The rate of bremsstrahlung production by light charged particles (electrons and positrons) traveling through an absorber is generally expressed by the mass radiation stopping power S_{rad} (in $MeV \cdot cm^2 \cdot g^{-1}$) given as follows

$$S_{rad} = N_a \, \sigma_{rad} \, E_i, \tag{6.6}$$

where

N_a is the atomic density, i.e., number of atoms per unit mass: $N_a = N/m = N_A/A$.
σ_{rad} is the total cross section for bremsstrahlung production given for various energy regions in Table 6.1.
E_i is the initial total energy of the light charged particle, i.e., $E_i = (E_K)_0 + m_e c^2$.
$(E_K)_0$ is the initial kinetic energy of the light charged particle.

Table 6.1 Total cross section for bremsstrahlung production and parameter B_{rad} for various ranges of electron kinetic energies

Energy range	$\sigma_{rad}\ (cm^2/nucleon)$	$B_{rad} = \sigma_{rad}/(\alpha r_e^2 Z^2)$	
Non-relativistic $(E_K)_0 \ll m_e c^2$	$\dfrac{16}{3}\alpha r_e^2 Z^2$	$\dfrac{16}{3}$	(6.8)
Relativistic $(E_K)_0 \approx m_e c^2$	Complicated power series	–	(6.9)
High-relativistic $m_e c^2 \ll (E_K)_0 \ll \dfrac{m_e c^2}{\alpha Z^{1/3}}$	$8r_e^2 Z^2 \left[\ln\left(\dfrac{E_i}{m_e c^2}\right) - \dfrac{1}{6}\right]$	$8\left[\ln\left(\dfrac{E_i}{m_e c^2}\right) - \dfrac{1}{6}\right]$	(6.10)
Extreme relativistic $(E_K)_0 \gg \dfrac{m_e c^2}{\alpha Z^{1/3}}$	$4\alpha r_e^2 Z^2\left[\ln\dfrac{183}{Z^{1/3}} + \dfrac{1}{18}\right]$	$4\left[\ln\dfrac{183}{Z^{1/3}} + \dfrac{1}{18}\right]$	(6.11)

Inserting σ_{rad} from Table 6.1 into (6.6) we obtain the following expression for S_{rad}

$$S_{rad} = \alpha\, r_e^2 Z^2 \frac{N_A}{A} E_i B_{rad}, \tag{6.7}$$

where B_{rad} is a slowly varying function of Z and E_i, also given in Table 6.1 and determined from $\sigma_{rad}/(\alpha r_e^2 Z^2)$. As shown in Table 6.2, the parameter B_{rad} has a value of $\frac{16}{3}$ for light charged particles in the non-relativistic energy range $(E_K)_0 \ll m_e c^2$; about 6 at $(E_K)_0 = 1$ MeV; 12 at $(E_K)_0 = 10$ MeV; and 15 at $(E_K)_0 = 100$ MeV. *Hans Bethe* and *Walter Heitler* derived (6.7) theoretically. *Martin Berger* and *Stephen Seltzer* have provided extensive tables of S_{rad} for a wide range of absorbing materials. As indicated in (6.7), the mass radiation stopping power S_{rad} is proportional to:

1. $(N_A Z^2/A)$ indicating a proportionality with the atomic number of the absorber Z by virtue of $Z/A \approx 0.5$ for all elements with the exception of hydrogen. The higher is the atomic number Z of the absorber, the larger is the radiation stopping power S_{rad} and the larger is the radiation yield.
2. Total energy E_i [(or kinetic energy $(E_K)_0$ for $(E_K)_0 \gg m_e c^2$] of the light charged particle.
3. Parameter B_{rad} which is a slowly varying function of light charged particle total energy E_i and absorber atomic number Z, as shown in Table 6.2.

Figure 6.2 shows the mass radiation stopping power S_{rad} for electrons in water, aluminum, and lead based on tabulated data obtained from the National Institute of Standards and Technology (NIST). The S_{rad} data are shown with heavy solid curves,

Table 6.2 Parameter B_{rad} for various initial kinetic energies of light charged particles

Kinetic energy	Classical	1 MeV	10 MeV	100 MeV
B_{rad}	~5.3	~6	~10	~15

Fig. 6.2 Mass radiation stopping power S_{rad} for electrons in water, aluminum and lead shown with heavy solid curves against the electron kinetic energy E_K. Mass collision stopping powers S_{col}, discussed in Sect. 6.5, for the same materials are shown with light solid curves for comparison. Data were obtained from the NIST

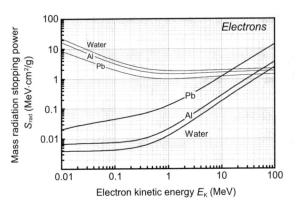

mass collision stopping powers S_{col} (discussed in Sect. 6.5) are shown with light curves for comparison. The radiation stopping power S_{rad} clearly shows an approximate proportionality: (1) to the atomic number Z of the absorber at a given initial kinetic energy of the light charged particle and (2) to initial kinetic energy $(E_K)_0$ of the light charged particle for a given absorber material.

6.4 Collision (Electronic) Stopping Power for Heavy Charged Particles

Energy transfer from energetic heavy charged particles to a medium (absorber) they traverse occurs mainly through Coulomb interactions of the charged particles with orbital electrons of the absorber atoms (collision or electronic loss); inelastic Coulomb interactions between heavy charged particles and nuclei of absorber atoms (radiation loss) are negligible and thus ignored.

Two different approaches were developed to describe a heavy charged particle energy loss to orbital electrons of absorber atoms:

1. Bohr's approach (1913) is in the realm of classical physics and is based on the concept of impact parameter between the particle's trajectory and the absorber nucleus.
2. Bethe's approach (1931) is in the realm of quantum mechanics and relativistic physics, and assumes that the momentum transfer related to the particle's energy loss is quantized.

The basic theories dealing with collision loss of energetic heavy charged particles in absorbing media make the following assumptions:

1. The energetic charged particle is moving through the absorber much faster than the orbital electrons of absorber atoms.
2. The energetic charged particle is much heavier than the energy-absorbing orbital electrons.

3. The energetic charged particle interacts with absorber atoms only through electromagnetic forces; nuclear reactions between the particle and absorber nuclei are not considered.
4. The energetic charged particle loses energy through interactions with orbital electrons of the absorber; elastic and inelastic interactions between the heavy charged particle and nuclei of the absorber are negligible.

6.4.1 Momentum and Energy Transfer from Heavy Charged Particle to Orbital Electron

In 1913 *Niels Bohr* was first to propose a theory describing a heavy charged particle energy loss in an absorbing medium. His classical derivation of the mass collision stopping power S_{col} for a heavy charged particle, such as a proton, is based on the calculation of the momentum change (impulse) Δp of the heavy charged particle having a Coulomb interaction with an orbital electron.

The Coulomb interaction between the heavy charged particle (charge ze and mass M) and the orbital electron (charge e and mass m_e) is shown schematically in Fig. 6.3. The situation here seems similar to that depicted in Fig. 2.3 for Rutherford scattering between an α-particle with mass m_α and gold nucleus with mass M. We must note, however, that in Rutherford scattering $m_\alpha \ll M$, where m_α is the mass of the projectile (α-particle) and M is the mass of the target (nucleus) while the case here is reversed as we have a heavy charged particle with mass M in translational motion and interacting with a stationary orbital electron with mass m_e where $M \gg m_e$. Assuming that the heavy particle is positively charged, the orbital electron is located in the inner focus of the hyperbolic trajectory that, in principle, the heavy charged particle follows. In Rutherford scattering, on the other hand, the α-particle and the positively charged nucleus repel one another and the nucleus is located in the outer focus of the hyperbolic trajectory that the α-particle follows.

The momentum transfer (impulse) Δp is along a line that bisects the angle $\pi - \theta$, as indicated in Fig. 6.3, and the magnitude of Δp is calculated in a similar manner to that used for Rutherford scattering in Sect. 2.3.3

$$\Delta p = \int F_{\Delta p}\, dt = \int_{-\infty}^{\infty} F_{Coul} \cos \phi\, dt. \tag{6.12}$$

The magnitude of the Coulomb force F_{Coul} between the heavy particle of charge ze and the orbital electron of charge $-e$ is

$$F_{Coul} = \frac{ze^2}{4\pi\varepsilon_0} \frac{1}{r^2}, \tag{6.13}$$

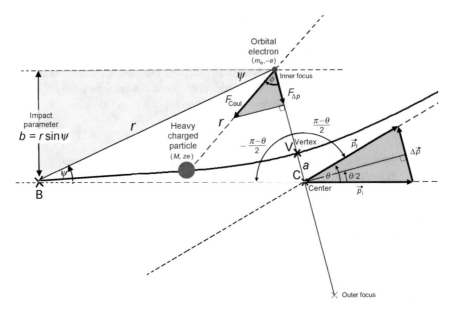

Fig. 6.3 Schematic diagram of a collision between a positively charged heavy particle with mass M and an orbital electron with mass m_e. Since $M \gg m_e$, the scattering angle $\theta \approx 0°$. The scattering angle is shown larger than $0°$ to highlight the principles of the Coulomb collision and aid in the derivation of Δp. The electron is in the inner focus of the hyperbola because of the attractive Coulomb force between the positive heavy charged particle and the negative orbital electron. Point B is for the charged particle at a very large distance from the orbital electron; point V is for the charged particle at the vertex V of the hyperbolic trajectory

Incorporating the expression for the Coulomb force (6.13) into (6.12), the momentum transfer Δp can now be written as

$$\Delta p = \frac{ze^2}{4\pi\varepsilon_0} \int_{-\frac{\pi-\theta}{2}}^{\frac{\pi-\theta}{2}} \frac{\cos\phi}{r^2} \frac{dt}{d\phi} d\phi = \frac{ze^2}{4\pi\varepsilon_0} \int_{-\frac{\pi-\theta}{2}}^{\frac{\pi-\theta}{2}} \frac{\cos\phi}{r^2\omega} d\phi \qquad (6.14)$$

where ϕ is the angle between the radius vector r and the bisector (axis) of the hyperbola, as shown in Fig. 6.3, and ω is the angular frequency defined as $\omega = \frac{d\phi}{dt} = \frac{v}{r}$ with v the initial velocity of the incident charged particle.

The angular momentum L for the elastic collision process between a heavy charged particle M and an electron m_e is in general expressed as

$$L = |\mathbf{L}|\,|\mathbf{r}\times M\mathbf{v}| = rMv\sin\psi = Mr^2\omega\sin\psi \qquad (6.15)$$

where v and $\boldsymbol{v}$ are the incident velocity and velocity vector, respectively, of the charged particle and ψ, as shown in Fig. 6.3, is the angle between the radius vector $\mathbf{r}$ and vector velocity $\boldsymbol{v}$ (see also Fig. 2.4 and Sect. 2.33).

We now use conservation of angular momentum L to express ωr^2 of (6.14) in terms of impact parameter b by determining L for two special points (B and V) on the hyperbola: point B is at a large distance (∞) from the center of the hyperbola and point V is at the vertex of the hyperbola. Note: at point B we have $b = r \sin \psi$ and at point V the angle ψ between $\mathbf{r}$ and $\boldsymbol{v}$ is 90° and $\sin \psi = 1$. We can thus express L_B and L_V as follows

$$L_B = rMv \sin \psi = Mvb \quad \text{and} \quad L_V = rMv = M\omega r^2 \quad (6.16)$$

from where it follows that ωr^2 of (6.14) is, as a result of angular momentum conservation ($L = L_B = L_V$), equal to vb.

The momentum transfer Δp of (6.14) can now be simplified as follows

$$\Delta p = \frac{ze^2}{4\pi\varepsilon_0} \frac{1}{vb} \int_{-\frac{\pi-\theta}{2}}^{\frac{\pi-\theta}{2}} \cos \phi \, d\phi = \frac{ze^2}{4\pi\varepsilon_0} \frac{1}{vb} \{\sin \phi\}_{-(\pi-\theta)/2}^{(\pi-\theta)/2} = 2\frac{ze^2}{4\pi\varepsilon_0} \frac{1}{vb} \cos \frac{\theta}{2}.$$

$$(6.17)$$

Equation (6.17) is identical to the Rutherford expression for Δp in (2.21). However, in the case of a heavy charged particle (ze) interacting with a stationary orbital electron (e), the scattering angle $\theta \approx 0$ and $\cos\left(\frac{1}{2}\theta\right) \approx 1$, resulting in the following simplified expression for Δp

$$\Delta p = 2\frac{ze^2}{4\pi\varepsilon_0} \frac{1}{vb}. \quad (6.18)$$

The energy transferred to the orbital electron from the heavy charged particle for a single interaction with an impact parameter b is

$$\Delta E(b) = \frac{(\Delta p)^2}{2m_e} = 2\left(\frac{e^2}{4\pi\varepsilon_0}\right)^2 \frac{z^2}{m_e v^2 b^2}, \quad (6.19)$$

using the classical expression between kinetic energy $E_K = \frac{1}{2}m_e v^2$ and momentum $p = m_e v$ expressed as $E_K = \frac{1}{2}p^2/m_e$. Note that in (6.19) m_e is the rest mass of the electron (target) and v is the velocity of the heavy charged particle (projectile), so that $\frac{1}{2}m_e v^2$ should not be misconstrued for the kinetic energy of the projectile. From (6.18) and (6.19) we express the impact parameter b as

$$b = 2\left(\frac{e^2}{4\pi\varepsilon_0}\right) \frac{z}{v\Delta p} = \sqrt{\frac{2}{m_e}}\left(\frac{e^2}{4\pi\varepsilon_0}\right) \frac{z}{v[\Delta E(b)]^{1/2}}. \quad (6.20)$$

Following (6.1) we set up a summation over all possible collisions to determine the mass collision stopping power for heavy charged particles

$$
\begin{aligned}
S_{\text{col}} &= -\frac{1}{\rho}\frac{\mathrm{d}E}{\mathrm{d}x} = N_e \int \Delta E(b)\mathrm{d}\sigma_{\text{col}} = N_e \int \Delta E(b)\frac{\mathrm{d}\sigma_{\text{col}}}{\mathrm{d}b}\,\mathrm{d}b \\
&= \int \Delta E(b)\frac{\mathrm{d}\sigma_{\text{col}}}{\mathrm{d}(\Delta E)}\mathrm{d}(\Delta E),
\end{aligned}
\tag{6.21}
$$

where N_e is the electron density (number of electrons per unit mass) of the absorber ($N_e = ZN_A/A$), and $\mathrm{d}\sigma_{\text{col}}/\mathrm{d}b$ as well as $\mathrm{d}\sigma_{\text{col}}/\mathrm{d}(\Delta E)$ are differential cross sections for the collision interaction.

The mass collision stopping power S_{col} is calculated by integrating $\Delta E(b)$ of (6.21) over all possible impact parameters b ranging from $b_{\min}$ to $b_{\max}$ or over all possible energy transfers $\Delta E(b)$ ranging from $\Delta E_{\min}$ to $\Delta E_{\max}$. Intuitively we might have considered integration in (6.21) over all possible impact parameters b from 0 to ∞ or over all possible energy transfers ΔE from 0 to the kinetic energy of the incident charged particle $(E_K)_i$; however, we must account for two physical limitations affecting the energy transfer from a heavy charged particle to an orbital electron which is bound to the atomic nucleus:

1. *Minimum possible energy transfer* $\Delta E_{\min}$ is governed by the ionization and excitation potential of orbital electrons of absorber atoms resulting in a minimum possible energy transfer $\Delta E_{\min}$ below which energy transfer becomes impossible. Since energy transfer ΔE and impact parameter b are inversely proportional, as shown in (6.19), $\Delta E_{\min}$ corresponds to the maximum impact parameter $b_{\max}$ beyond which energy transfer becomes impossible.
2. *Maximum possible energy transfer* $\Delta E_{\max}$ in a head-on collision between the heavy charged particle and an orbital electron is governed by the masses of the two interacting particles and is significantly lower than the incident particle kinetic energy $(E_K)_i$, as discussed in Sect. 5.3. Energy transfer larger than $\Delta E_{\max}$ is physically impossible.

6.4.2 Minimum Energy Transfer and Mean Ionization/Excitation Energy

For large impact parameters b the energy transfer $\Delta E(b)$, calculated from (6.19), may be smaller than the binding energy of the orbital electron or smaller than the minimum excitation energy of the given orbital electron. Thus, no energy transfer is possible for $b > b_{\max}$ where $b_{\max}$ corresponds to a minimum energy transfer $\Delta E_{\min}$, referred to as the mean ionization/excitation energy I of the absorber atom. For a given atom, its mean ionization/excitation energy is always larger than the ionization energy of the atom, since I accounts for all possible atomic ionizations as well as atomic excitations, while the atomic ionization energy (IE), as discussed in Sect. 3.2.4,

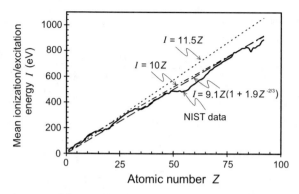

Fig. 6.4 Mean ionization/excitation energy I for elements against atomic number Z. Also plotted are three empirical relationships given in (6.23)–(6.25) that can be used for a rough estimation of I for a given absorber Z. Data were obtained from the ICRU Report 37 and from the NIST

pertains to the energy required to remove the least bound atomic electron (i.e., valence electron in the outer shell).

The mean ionization/excitation energy I corresponds to the minimum amount of energy ΔE_{min} that can be transferred, on average, to an absorber atom in a Coulomb interaction between a charged particle and an orbital electron. Using (6.19), ΔE_{min} is written as

$$\Delta E_{min} = 2 \left(\frac{e^2}{4\pi\varepsilon_o} \right)^2 \frac{z^2}{m_e \upsilon^2 b_{max}^2} = I, \tag{6.22}$$

showing that $\Delta E_{min} \propto 1/b_{max}^2$ and $b_{max} \propto 1/\sqrt{\Delta E_{min}} = 1/\sqrt{I}$.

In general, the mean ionization/excitation energy I cannot be calculated from the atomic theory; however, it can be determined empirically from measured stopping power data compared to data calculated using an appropriate stopping power formula. Current values for I recommended for use by the ICRU and the NIST are plotted in Fig. 6.4 and show a general rise of I with increasing atomic number Z. Figure 6.4 also highlights three empirical relationships that can be used for a rough estimation of I for a given absorber Z. The three approximations are

$$I \text{ (in eV)} = 11.5Z \quad \text{for } Z < 15, \tag{6.23}$$

$$I \text{ (in eV)} = 10Z \quad \text{for } Z > 15, \tag{6.24}$$

$$I \text{ (in eV)} = 9.1Z \left(1 + 1.9Z^{-2/3}\right). \tag{6.25}$$

Typical values of the mean ionization/excitation energy I from the ICRU Report 37 are given in Table 6.3 for several elements and in Table 6.4 for several compounds of interest in medical physics and dosimetry. It is important to note that the mean ionization/excitation energy I only depends on the absorbing medium but does not depend on the type of charged particle interacting with the absorbing medium.

Table 6.3 Mean ionization/excitation energy I for various absorbing materials (from the ICRU Report 37)

Element	H	C	Al	Cu	Ag	W	Pb	Ra	U	Cf
Z	1	6	13	29	47	74	82	88	92	98
I (eV)	19.2	78	167	322	470	727	823	826	890	966

Table 6.4 Mean ionization/excitation energy I for various compounds of interest in medical physics (from the ICRU Report 37)

Compound	I (eV)	Compound	I (eV)
Air (dry)	85.7	Lithium fluoride	94
Water (liquid)	75	Photographic emulsion	331
Water (vapor)	71.6	Sodium iodide	452
Muscle (skeletal)	75.3	Polystyrene	68.7
Bone (compact)	91.9	A-150 plastic	65.1

6.4.3   Maximum Energy Transfer

For small impact parameters b the energy transfer is governed by the maximum energy ΔE_{max} that can be transferred in a single head-on collision, as discussed in Sect. 5.3.2. Classically ΔE_{max} for a head-on collision between a heavy charged particle M with kinetic energy $E_K = \frac{1}{2}Mv^2$ and an orbital electron with mass m_e where $m_e \ll M$, as discussed in Sect. 5.3.4, is given by

$$\Delta E_{max} = \frac{4m_e M}{(m_e + M)^2} E_K \approx 4\frac{m_e}{M} E_K = 4\frac{m_e}{M} \frac{Mv^2}{2} = 2m_e v^2. \qquad (6.26)$$

Equation (6.26) shows that only a very small fraction $(4m_e/M)$ of the heavy charged particle kinetic energy can be transferred to an orbital electron in a single collision (note that $M \gg m_e$). The classical relationship between ΔE_{max} and the minimum impact parameter b_{min} that allows the maximum energy transfer from a heavy charged particle to an orbital electron is

$$\Delta E_{max} = 2\left(\frac{e^2}{4\pi\varepsilon_0}\right)^2 \frac{z^2}{m_e v^2 b_{min}^2} = 2m_e v^2, \qquad (6.27)$$

resulting in $\Delta E_{max} \propto 1/b_{min}^2$ or $b_{min} \propto 1/\sqrt{\Delta E_{max}}$.

In relativistic mechanics the maximum energy transfer ΔE_{max} from a heavy charged particle of rest mass M_0 to orbital electron of rest mass m_e where $m_e \ll M_0$ was in (5.41) given as

$$\Delta E_{max} = 2m_e c^2 \left(\gamma^2 - 1\right) \frac{1}{1 + 2\gamma \dfrac{m_e}{M_0} + \left(\dfrac{m_e}{M_0}\right)^2} \approx 2m_e c^2 \frac{\beta^2}{1 - \beta^2}, \tag{6.28}$$

where β and γ are given by the standard relativistic relationships

$$\beta = \frac{\upsilon}{c}; \quad \gamma = \frac{1}{\sqrt{1 - \beta^2}}; \quad \text{and} \quad \gamma^2 - 1 = \frac{\beta^2}{1 - \beta^2}. \tag{6.29}$$

6.4.4 Classical Derivation of Mass Collision Stopping Power

Following the general definition of the mass collision stopping power given in (6.21), integration over the impact parameter b or energy transfer $\Delta E(b)$ gives, respectively

$$S_{col} = -\frac{1}{\rho} \frac{dE}{dx} = N_e \int \Delta E(b) d\sigma_{col} = 2\pi N_e \int_{b_{min}}^{b_{max}} \Delta E(b) b db \tag{6.30}$$

and

$$S_{col} = -\frac{1}{\rho} \frac{dE}{dx} = N_e \int \Delta E(b) \, d\sigma_{col}$$

$$= 2\pi N_e \int_{\Delta E_{min}}^{\Delta E_{max}} \Delta E(b) \frac{b \, db}{d[\Delta E(b)]} d[\Delta E(b)], \tag{6.31}$$

where we integrate (6.30) from the minimum impact parameter b_{min} to maximum impact parameter b_{max} and (6.31) from minimum energy transfer ΔE_{min} to maximum energy transfer ΔE_{max}, and

N_e is the electron density, i.e., number of electrons per mass of the absorber ($N_e = Z N_A / A$), with Z and A the atomic number and atomic mass number of the absorber, respectively.

$d\sigma_{col}$ is the differential cross section for collision loss expressed as

$$d\sigma_{col} = 2\pi b \, db = 2\pi b \frac{db}{d[\Delta E(b)]} d[\Delta E(b)], \tag{6.32}$$

with b given in (6.20) and the derivative $db/d[\Delta E(b)]$ equal to

$$\frac{db}{d[\Delta E(b)]} = \frac{1}{2} \sqrt{\frac{2}{m_e}} \left(\frac{e^2}{4\pi\varepsilon_0}\right) \frac{z}{\upsilon} [\Delta E(b)]^{-3/2}, \tag{6.33}$$

to give the following expression for $d\sigma_{col}$

$$d\sigma_{col} = 2\pi b\, db = 2\pi b \frac{db}{d[\Delta E(b)]} d[\Delta E(b)]$$

$$= 2\pi \left(\frac{e^2}{4\pi\varepsilon_0}\right)^2 \frac{z^2}{m_e v^2} \frac{d[\Delta E(b)]}{[\Delta E(b)]^2}. \tag{6.34}$$

After incorporating (6.19) into (6.30), the mass collision stopping power S_{col} can be written as

$$S_{col} = 4\pi \frac{ZN_A}{A}\left(\frac{e^2}{4\pi\varepsilon_0}\right)^2 \frac{z^2}{m_e v^2} \int_{b_{min}}^{b_{max}} \frac{db}{b} = 4\pi \frac{ZN_A}{A}\left(\frac{e^2}{4\pi\varepsilon_0}\right)^2 \frac{z^2}{m_e v^2} \ln\frac{b_{max}}{b_{min}}, \tag{6.35}$$

while, after incorporating (6.34) into (6.31), we get the following expression for the mass collision stopping power S_{col}

$$S_{col} = 2\pi \frac{ZN_A}{A}\left(\frac{e^2}{4\pi\varepsilon_0}\right)^2 \frac{z^2}{m_e v^2} \int_{\Delta E_{min}}^{\Delta E_{max}} \frac{d[\Delta E(b)]}{[\Delta E(b)]}$$

$$= 2\pi \frac{ZN_A}{A}\left(\frac{e^2}{4\pi\varepsilon_0}\right)^2 \frac{z^2}{m_e v^2} \ln\frac{\Delta E_{max}}{\Delta E_{min}}. \tag{6.36}$$

Since from (6.22) and (6.27) we write

$$\frac{b_{max}}{b_{min}} = \sqrt{\frac{\Delta E_{max}}{\Delta E_{min}}}, \tag{6.37}$$

and obtain the following expression for S_{col} of (6.35)

$$S_{col} = 4\pi \frac{ZN_A}{A}\left(\frac{e^2}{4\pi\varepsilon_0}\right)^2 \frac{z^2}{m_e v^2} \ln\frac{b_{max}}{b_{min}}$$

$$= 4\pi \frac{ZN_A}{A}\left(\frac{e^2}{4\pi\varepsilon_0}\right)^2 \frac{z^2}{m_e v^2} \ln\sqrt{\frac{\Delta E_{max}}{\Delta E_{min}}}, \tag{6.38}$$

identical to S_{col} in (6.36).

The energy transfer $\Delta E(b)$ from the heavy charged particle to an orbital electron ranges from ΔE_{min} $(b_{max}) = I$ to ΔE_{max} (b_{min}) of (6.27). Insertion of expressions for ΔE_{min} and ΔE_{max} given in (6.22) and (6.27), respectively, into (6.36) or (6.38) results in the following classical expression for mass collision stopping power derived by *Niels Bohr* in 1913

Fig. 6.5 Mass collision stopping powers of *part* **a** aluminum and *part* **b** lead against proton kinetic energy E_K. Curves (*2*) represent experimental data compiled by the NIST, curves (*1*) are calculated with Bohr's stopping power equation (6.39)

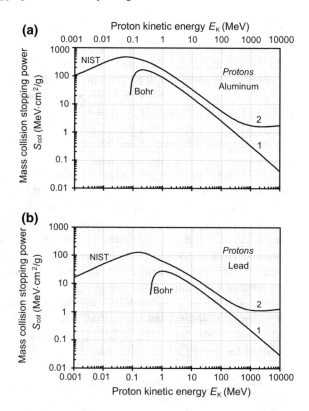

$$S_{col} = 2\pi \frac{Z N_A}{A} \left(\frac{e^2}{4\pi\varepsilon_0} \right)^2 \frac{z^2}{m_e v^2} \ln \frac{2m_e v^2}{I}. \qquad (6.39)$$

Figure 6.5 shows a plot of the mass collision stopping power S_{col} of aluminum in part (a) and lead in part (b) for protons ($z = 1$) in the kinetic energy range from 10^{-3} MeV to 10^4 MeV. Curves (1) are for measured data tabulated by the NIST, curves (1) are for the Bohr classical expression given in (6.39).

In the intermediate energy range from $\sim$300 keV to $\sim$100 MeV both the theory and measurement exhibit the same trends; however, the measurement consistently exceeds the calculated data by a factor of 2. In the low energy range ($E_K < 300$ keV) and the high energy range ($E_K > 100$ MeV) the discrepancy between measured and calculated data is significantly larger, clearly indicating that Bohr classical theory does not provide a realistic description of particle stopping in absorbing media. The reasons for this seem obvious: the Bohr theory ignores quantum mechanical and relativistic effects and treats electron binding effects in a very rudimentary fashion through the mean ionization/excitation energy I.

6.4.5 Bethe Collision Stopping Power

With the advent of quantum mechanics and relativistic physics many scientists attempted to improve the Bohr collision stopping power theory by incorporating the new concepts into the newly developed theories. Most notable were efforts in 1930s by *Hans Bethe* and *Felix Bloch* who developed a new collision stopping power theory based on quantum mechanical and relativistic concepts and, in contrast to Bohr classical theory, achieved excellent agreement between theoretical and experimental data.

In 1931 *Hans Bethe* proposed a theory in which the energy loss of a heavy charged particle traversing an absorber is calculated using the Born approximation applied to the collision of the heavy charged particle with atomic orbital electrons. As suggested in Sect. 1.23.6, the differential cross section for momentum transfer from the heavy charged particle to atomic electrons was proportional to the square of the matrix element defined by the Coulomb interaction between the relevant initial and final states, and the wave functions for the incident and scattered heavy charged particle were described by plane waves.

Bethe divided the possible energy transfers ΔE to an orbital electron into two categories: soft (distant) and hard (close). The energy boundary η between soft and hard collisions was not clearly defined but was generally chosen such that for soft collisions the energy transfer ΔE was smaller than η and for hard collisions ΔE was larger than η. Thus, η was large compared to the binding energies of atomic electrons, yet it was also small enough to allow for all hard collisions, characterized with $\Delta E \geq \eta$, a representation of the charged particle as point source.

Bethe considered average energy losses for the soft and hard collisions separately and obtained the following expressions for the soft and hard mass collision stopping powers

$$S_{\text{col}}^{\text{soft}} = \left(-\frac{1}{\rho}\frac{dE}{dx} \right)_{\Delta E < \eta}^{\text{soft}} = 2\pi \frac{ZN_A}{A} \left(\frac{e^2}{4\pi\varepsilon_0} \right)^2 \frac{z^2}{m_e\beta^2c^2} \left\{ \ln\frac{2m_ec^2\beta^2\eta}{(1-\beta^2)\,I^2} - \beta^2 \right\}$$

$$(6.40)$$

and

$$S_{\text{col}}^{\text{hard}} = -\frac{1}{\rho}\left(\frac{dE}{dx} \right)_{\Delta E > \eta}^{\text{hard}} = 2\pi \frac{ZN_A}{A} \left(\frac{e^2}{4\pi\varepsilon_0} \right)^2 \frac{z^2}{m_e\beta^2c^2} \left\{ \ln\frac{\Delta E_{\max}}{\eta} - \beta^2 \right\}$$

$$= 2\pi \frac{ZN_A}{A} \left(\frac{e^2}{4\pi\varepsilon_0} \right)^2 \frac{z^2}{m_e\beta^2c^2} \left\{ \ln\frac{2m_ec^2\beta^2}{(1-\beta^2)\,\eta} - \beta^2 \right\},$$

$$(6.41)$$

where

I is the mean ionization/excitation energy, discussed in Sect. 6.4.2.

$\Delta E_{\max}$ is the maximum energy transfer from the heavy charged particle to the orbital electron, as given relativistically in (5.4.1) and (6.28).

The expression for S_{col}^{soft} given in (6.40) is valid for all types of charged particles (heavy and light), while the expression for S_{col}^{hard} given in (6.41) is valid only for heavy charged particles and cannot be used for electrons and positrons.

As suggested in (6.3), the mass collision stopping power is given as the sum of the soft (6.40) and hard (6.41) mass collision stopping powers. Adding (6.40) and (6.41) we get

$$
\begin{aligned}
S_{col} = S_{col}^{soft} + S_{col}^{hard} &= 2\pi \frac{Z N_A}{A} \left(\frac{e^2}{4\pi\varepsilon_0} \right)^2 \frac{z^2}{m_e c^2 \beta^2} \\
&\times \left\{ \ln \frac{2m_e c^2 \beta^2 \eta}{(1 - \beta^2) I^2} + \ln \frac{2m_e c^2 \beta^2}{(1 - \beta^2) \eta} - 2\beta^2 \right\} \\
&= 4\pi \frac{Z N_A}{A} \left(\frac{c^2}{4\pi\varepsilon_0} \right)^2 \frac{z^2}{m_e e^2 \beta^2} \left\{ \ln \frac{2m_e c^2}{I} + \ln \frac{\beta^2}{1 - \beta^2} - \beta^2 \right\} \\
&= C_0 \frac{z^2}{A\beta^2} Z \left\{ \ln \frac{2m_e c^2}{I} + \ln \frac{\beta^2}{1 - \beta^2} - \beta^2 \right\}, \qquad (6.42)
\end{aligned}
$$

where the demarcation energy η separating soft from hard collisions falls out of the final equation and C_0 is the collision stopping power constant equal to

$$
C_0 = 4\pi N_A \left(\frac{e^2}{4\pi\varepsilon_0} \right)^2 \frac{1}{m_e c^2} = 4\pi N_A r_e^2 m_e c^2 = 0.3071 \text{ MeV·cm}^2\text{·mol}^{-1}, \qquad (6.43)
$$

with $r_e = e^2 / (4\pi\varepsilon_0 m_e c^2) = 2.818$ fm defined as the classical electron radius.

The expression (6.42) is known as the Bethe mass collision stopping power equation. It is valid for heavy charged particles, such as protons and α-particles, and accounts for quantum mechanical as well as relativistic effects. In the classical limit for $v \ll c$ (6.42) yields

$$
S_{col} = 4\pi \frac{Z N_A}{A} \left(\frac{e^2}{4\pi\varepsilon_0} \right)^2 \frac{z^2}{m_e v^2} \ln \frac{2m v^2}{I} = C_0 \frac{z^2 c^2}{A v^2} Z \ln \frac{2m_e v^2}{I}, \qquad (6.44)
$$

that is non-relativistic yet it accounts for quantum-mechanical effects and is exactly double the result obtained from the Bohr classical stopping power theory in (6.39). Since, as shown in Fig. 6.5, the discrepancy between Bohr theory and measurement in the intermediate energy range was also by a factor of 2, one may surmise that a plot of mass collision stopping power calculated from the Bethe equation (6.42) will yield good agreement with measured data. That this is so is shown in Fig. 6.6 in which we plot the measured (available from the NIST) and calculated mass collision stopping powers S_{col} of aluminum for protons in the kinetic energy range from 1 keV to 10^4 MeV. The calculated data are obtained from two different theories: Bohr classical theory data are calculated from (6.39) and shown by the dashed curve and Bethe theory data are calculated from (6.42) and shown by the solid curve.

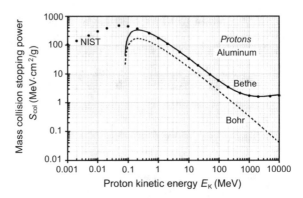

Fig. 6.6 Measured and calculated mass collision stopping powers of aluminium against proton kinetic energy E_K. Experimental data available from the NIST are shown with data points; calculated data are shown with *solid* and *dashed curves*. The *dashed curve* is calculated with the Bohr classical mass collision stopping power equation (6.39); the *solid curve* is calculated with the Bethe mass collision stopping power equation (6.42) accounting for quantum mechanical as well as relativistic effects

As shown in Fig. 6.6, in the intermediate and high energy relativistic range the agreement between experiment and Bethe theory is excellent; however, at low energies the agreement fails. Both the theory and measurement predict a peak in stopping power with decreasing energy, but the measured peak is higher and appears at a lower kinetic energy than the calculated peak. Clearly there must be some other unaccounted for effect that causes this discrepancy. As discussed in Sect. 6.4.6, two corrections (shell correction and density effect correction) are applied to Bethe theory in order to explain this discrepancy at low kinetic energies.

The discrepancy at low kinetic energies notwithstanding, the Bethe collision stopping power formula agrees well with measured data in the intermediate and high energy relativistic region, and clearly predicts the shape of the stopping power curve as well as provides the functional dependence of the many important parameters of the charged particles and absorbing media that influence the stopping power curve.

Shape of the Bethe Mass Collision Stopping Power Curve

As indicated schematically in Fig. 6.7, a plot of the collision stopping power S_{col} against kinetic energy E_K for a heavy charged particle goes through three distinct regions with increasing E_K:

Region 1:   At very low kinetic energies, S_{col} rises with energy and reaches a peak.
Region 2:   In the intermediate energy region beyond the peak, S_{col} decreases as $1/v^2$ or $1/E_K$ of the charged particle until it reaches a broad minimum.
Region 3:   In the relativistic energy region beyond the broad minimum, S_{col} rises slowly with increasing kinetic energy E_K as a result of the relativistic terms $\{\ln \beta^2 - \ln(1 - \beta^2) - \beta^2\}$.

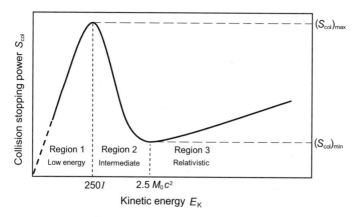

Fig. 6.7 Schematic representation of the shape of the collision stopping power S_{col} as a function of the charged particle kinetic energy E_K. Three regions are shown as the kinetic energy increases from zero: (**1**) low energy region; (**2**) intermediate energy region; and (**3**) relativistic energy region. In region (**1**), S_{col} rises almost linearly and reaches a maximum at about 250 I, where I is the mean ionization/excitation energy of the absorber. In region (**2**), S_{col} decreases as $1/v^2$ or $1/E_K$ where v is the velocity of the charged particle to reach a broad minimum at $\sim 2.5 M_0 c^2$ where $M_0 c^2$ is the rest energy of the charged particle. In region (**3**), S_{col} rises slowly with increasing E_K because of relativistic effects

Maxima and Minima on the Stopping Power Curves

The maximum $S_{col}^{max}(E_K)$ and minimum $S_{col}^{min}(E_K)$ on the stopping power curve (6.42) can be determined by setting $dS_{col}/dE_K = 0$. To simplify the calculation we note that $dS_{col}/dE_K = (dS_{col}/d\beta)\,(d\beta/dE_K)$ and $d\beta/dE_K \neq 0$. We then write the Bethe equation (6.42) as

$$S_{col} = \frac{const}{\beta^2}\left\{ \ln\frac{2m_e c^2}{I} + \ln\beta^2 - \ln(1-\beta^2) - \beta^2 \right\} \tag{6.45}$$

and determine $dS_{col}/d\beta$ as

$$\frac{dS_{col}}{d\beta} = -\frac{2\,const}{\beta^3}\left\{ \ln\frac{2m_e c^2}{I} + \ln\beta^2 - \ln(1-\beta^2) - \beta^2 \right\}$$
$$+ \frac{const}{\beta^2}\left\{ \frac{2}{\beta} + \frac{2\beta}{1-\beta^2} - 2\beta \right\} = 0 \tag{6.46}$$

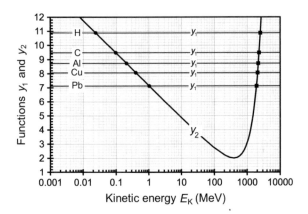

Fig. 6.8 Functions y_1 and y_2 against kinetic energy E_K of the heavy charged particle. Function y_1 is a constant for a given element and is shown for various elements between hydrogen ($Z = 1$) and lead ($Z = 82$). The left intercepts between the two functions indicate the maximum in stopping power according to Bethe collision stopping power equation (6.42), the right intercepts indicate the minimum

or

$$\ln \frac{2m_e c^2}{I} = 1 + \frac{\beta^2}{1 - \beta^2} - \ln \frac{\beta^2}{1 - \beta^2} = \frac{1}{1 - \beta^2} - \ln \frac{\beta^2}{1 - \beta^2}$$

$$= 1 + \frac{E_K}{E_0} \left(2 + \frac{E_K}{E_0} \right) - \ln \left\{ \frac{E_K}{E_0} \left(2 + \frac{E_K}{E_0} \right) \right\}, \qquad (6.47)$$

where we used $E_K / E_0 = 1/\sqrt{1 - \beta^2}$ with E_K and E_0 kinetic energy and rest energy, respectively, of the incident heavy charged particle of rest mass M_0.

We now have two functions, y_1 and y_2, where

$$y_1 = \ln \frac{2m_e c^2}{I} \qquad (6.48)$$

depends on the mean ionization/excitation energy I and

$$y_2 = \frac{1}{1 - \beta^2} - \ln \frac{\beta^2}{1 - \beta^2} = 1 + \frac{E_K}{E_0} \left(2 + \frac{E_K}{E_0} \right) - \ln \left\{ \frac{E_K}{E_0} \left(2 + \frac{E_K}{E_0} \right) \right\} \quad (6.49)$$

depends on the charged particle kinetic energy E_K and rest energy E_0.

A plot of the two functions, shown in Fig. 6.8 for a full range of absorbers from $Z = 1$ to $Z = 92$, for a given absorber results in two intercepts between the two functions; intercept on the left defines E_K for $S_{col}^{max}(E_K)$ and the one on the right defines E_K for $S_{col}^{min}(E_K)$. While the region of E_K for $S_{col}^{max}(E_K)$ seems relatively broad, ranging from 0.02 MeV for hydrogen to 1 MeV for lead, the region of E_K for

$S_{col}^{min}(E_K)$ is very narrow ranging from $\sim 2 \times 10^3$ MeV for uranium to $\sim 2.4 \times 10^3$ MeV for hydrogen. The minimum in the collision stopping power curve is thus essentially independent of absorber material and occurs at $E_K \approx 2.5E_0 = 2.5\,M_0c^2$.

$S_{col}^{max}(E_K)$ is in the low kinetic energy region where $E_K \approx M_0c^2$ and can be evaluated by assuming that $E_K = \frac{1}{2}M_0c^2\beta^2$ or $\beta^2 = 2E_K/(M_0c^2)$. This results in significantly simpler form of function y_2 that can be approximated as $y_2 \approx 1 - \ln \beta^2 \approx 1 - \ln\{2E_K/(M_0c^2)\}$. The (y_1, y_2) intercept that signifies S_{col}^{max} is then determined as

$$\ln \frac{2m_ec^2}{I} = y_1 \equiv y_2 \approx 1 - \ln \frac{2E_K}{M_0c^2} \tag{6.50}$$

or

$$E_K = \frac{e\,M_0c^2}{4m_ec^2} I \approx 1250\,AI, \tag{6.51}$$

where we use $M_0 \approx Am_p$. For the mass of the heavy charged particle, with m_p the proton mass and A the atomic mass number of the heavy charged particle; $m_p/m_e = 1836$; and $e = 2.718$ the base of natural logarithm.

Figure 6.5 shows the proton beam maxima for aluminum and lead at 207.5 keV and 1.03 MeV, respectively, significantly higher than experimental data which show the maxima at 50 keV and 200 keV, respectively. An inspection of stopping power tables for various absorber elements shows that the experimentally measured E_K for $S_{col}^{max}(E_K)$ increases with atomic number Z of the absorber and occurs at approximately $250I$ for all absorber elements.

Dependence of Collision Stopping Power on Particle Charge

S_{col} of (6.42) depends on z^2, the atomic number of the heavy charged particle (projectile). This implies, for example, that mass collision stopping powers of an absorbing medium for deuteron and proton of same velocity β will be the same, i.e., $S_d(\beta) = S_p(\beta)$ while for α-particle and proton of same velocity β they will differ by a factor of 4, i.e., $S_\alpha(\beta) = 4S_p(\beta)$.

Dependence of Collision Stopping Power on Particle Velocity

At low kinetic energies the Bethe equation (6.42) breaks down, as shown in Fig. 6.6, and to achieve better agreement between theory and measurement one must use several correction factors, discussed in Sect. 6.4.6. At intermediate energies S_{col} is governed by the $1/v^2$ (i.e., $1/E_K$) term and decreases rapidly with increasing E_K. In the relativistic energy region S_{col} rises slowly with kinetic energy as a result of the $1/\beta^2$ term which saturates at ~ 1 and the $\{\ln[\beta^2/(1-\beta^2)] - \beta^2\}$ term that slowly increases with increasing kinetic energy.

Dependence of Collision Stopping Power on Particle Mass

As shown in (6.42), S_{col} does **not** depend on the mass of the heavy charged particle. A given absorbing material will have the same collision stopping power for all heavy charged particles of a given velocity and charge ze.

Dependence of Collision Stopping Power on Absorbing Medium

S_{col} of (6.42) depends on the atomic number Z, atomic mass A, and mean ionization/excitation energy I of the absorbing medium through two terms, both of which decrease the mass collision stopping power with an increase in atomic number Z. The first term is the Z/A term of the electron density $N_e = ZN_A/A$ and the second term is the $(-\ln I)$ term.

Since Z/A varies from substance to substance within quite a narrow range (it falls from 0.5 for low Z elements to $\sim$0.4 for high Z elements, with one notable exception of hydrogen for which $Z/A \approx 1$), we note that the mass collision stopping power does not vary much from substance to substance. This means that the energy losses of a given charged particle passing through layers of equal thickness in g/cm^2 are about the same for all substances. One should note, however, that, for a given charged particle kinetic energy, S_{col} of lower atomic number absorber will exceed S_{col} of a higher atomic number absorber.

6.4.6 Corrections to Bethe Collision Stopping Power Equation

As discussed in relation to Fig. 6.6, the Bethe collision equation is not in complete agreement with experimental data and various approximations have been developed to correct the problem, most notably:

(i) the so-called shell correction applicable at relatively low charged particle kinetic energy where the particle velocity is comparable to the velocity of orbital electrons of the absorber atoms.

(ii) Barkas-Bloch correction accounting for departures from the first order Born approximation.

(iii) density effect correction that accounts for the polarization of the absorber medium and is important at high kinetic energies of the charged particle traversing the absorber.

Shell Correction

Bethe's approach, based on the Born approximation, assumes that the velocity of the heavy charged particle is much larger than the velocity of the bound orbital electrons of the absorber. While at high kinectic energies this assumption is correct, at low kinetic energies of the incident charged particle it does not hold. Orbital electrons stop participating in energy transfer from the charged particle when their velocity becomes comparable to the charged particle velocity. This effect causes an overestimate in the mean ionization/excitation energy I at low energies and results in an underestimate in S_{col} calculated from the Bethe equation.

The so-called shell correction term C/Z is introduced to account for the overestimate in I and various theories have subsequently been developed for its determination. Since the K shell electrons are the fastest of all orbital electrons, they are the first to be affected by low particle velocity with decreasing particle velocity. The shell correction is often addressed as the K shell correction, labeled C_K/Z and all possible higher shell corrections are usually ignored.

The correction term C/Z is a function of the absorbing medium and charged particle velocity; however, for the same medium and particle velocity, it is the same for all particles including electrons and positrons.

Polarization or Density Effect Correction

In addition to the shell correction factor, a second correction term δ is applied to the Bethe collision stopping power equation to account for the polarization or density effect in condensed media. The effect influences the soft (distant) collision interactions by polarizing the condensed absorbing medium thereby decreasing the collision stopping power of the condensed medium in comparison with same absorbing medium in gaseous state. For heavy charged particles the density correction is important at relativistic energies and negligible at intermediate and low energies; however, for electrons and positrons it plays a role in stopping power formulas at all energies.

Bethe Collision Stopping Power Equation Incorporating Shell Correction and Density Effect Correction

The Bethe collision stopping power equation incorporating the shell correction and density correction terms is for heavy charged particles written as

$$S_{col} = 4\pi \frac{N_A}{A} \left(\frac{e^2}{4\pi\varepsilon_0} \right)^2 \frac{z^2}{m_e c^2 \beta^2} Z \left\{ \ln \frac{2m_e c^2}{I} + \ln \frac{\beta^2}{1-\beta^2} - \beta^2 - \frac{C}{Z} - \delta \right\},$$

(6.52)

where C/Z and δ are the shell correction and density effect correction, respectively, and all other parameters of the absorber (Z, A, and I) and of the heavy charged particle (β and z) were defined before in (6.42).

6.4.7 Collision Stopping Power Equations for Heavy Charged Particles

The collision stopping power equations stated above for heavy charged particles can be summarized as follows

$$S_{col} = C_0 \frac{z^2}{\beta^2} \frac{Z}{A} B_{col},$$

(6.53)

where

C_0 is the collision stopping power constant defined in (6.43) with the following expression $C_0 = 4\pi N_A r_e^2 m_e c^2 = 0.3071$ MeV·cm^2·mol^{-1}.

z is the atomic number of the heavy charged particle.

β is the heavy charged particle velocity normalized to the speed of light c.

Table 6.5 Expressions for the atomic stopping number B_{col} for various energy ranges of heavy charged particle energy

Derivation of S_{col}	Atomic stopping number B_{col}	
Classical Bohr – 1913	$\left\{ \ln \sqrt{\dfrac{2m_e v^2}{I}} \right\}$	(6.54)
Non-relativistic, quantum-mechanical Bethe – 1931	$\left\{ \ln \dfrac{2m_e v^2}{I} \right\}$	(6.55)
Relativistic, quantum-mechanical Bethe – 1931	$\left\{ \ln \dfrac{2m_e c^2}{I} + \ln \dfrac{\beta^2}{1 - \beta^2} - \beta^2 \right\}$	(6.56)
Relativistic, quantum-mechanical with shell correction (C/Z) and density effect correction (δ)	$\left\{ \ln \dfrac{2m_e c^2}{I} + \ln \dfrac{\beta^2}{1 - \beta^2} - \beta^2 - \dfrac{C}{Z} - \delta \right\}$	(6.57)

Z is the atomic number of the absorbing medium.

A is the atomic mass of the absorber in g/mol.

B_{col} is the so-called *atomic stopping number*.

The atomic stopping number B_{col} is a function of the velocity β of the charged particle and of the atomic number Z of the absorber through the mean ionization/excitation energy I. The form of the expression for B_{col} also depends on the specific approach taken in its derivation, as indicated in Table 6.5.

The units of S_{col} in (6.53) are $MeV \cdot cm^2 \cdot g^{-1}$; the constant C_0 has units of $MeV \cdot cm^2 \cdot mol^{-1}$, and B_{col} has no units. Since the units of A are g/mol, incorporating an appropriate value for A into (6.53) results in proper units for S_{col} in $MeV \cdot cm^2 \cdot g^{-1}$. From the general expression for the mass collision stopping power S_{col} given in (6.53) it is evident that S_{col} for a heavy charged particle traversing an absorber *does not depend on charged particle mass* but depends upon charged particle velocity β and atomic number z as well as absorber atomic number Z, atomic mass A, and mean ionization/excitation energy I.

Figure 6.9 plots the mass collision stopping powers of carbon, aluminum, copper, and lead for protons in the kinetic energy E_K range from 1 keV to 10^4 MeV. The data were obtained from the NIST which at high proton energies uses Bethe collision stopping power equation (6.42) including shell and density correction terms as well as the so-called Barkas and Bloch corrections for deviations from the Born approximation. At low proton energies the NIST data are determined with fitting formulas which are largely based on experimental data. The NIST collision stopping power database uses mean ionization/excitation energies recommended by the ICRU Report 37.

The four curves in Fig. 6.9 exhibit the standard collision stopping power behavior of rising with E_K at low kinetic energies, reaching a peak at between 50 keV and

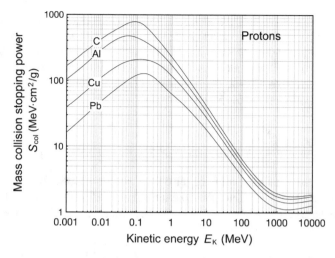

Fig. 6.9 Mass collision stopping powers of carbon, aluminum, copper, and lead against proton beam kinetic energy. Data were obtained from the NIST

200 keV, then falling as $1/v^2$ or $1/E_K$ at intermediate energies, reaching a minimum around $E_K/E_0 \approx 2.5$, and then rising slowly with E_K in the relativistic energy range. In the intermediate and relativistic energy range the mass collision stopping powers at a given kinetic energy E_K are fairly similar; however, it is evident that lower atomic number absorbers have higher mass collision stopping powers because of the effects of the (Z/A) and $(-\ln I)$ terms in the Bethe equation (6.42), as discussed in Sect. 6.4.6.

6.5 Collision Stopping Power for Light Charged Particles

Electron and positron interactions (collisions) with orbital electrons of an absorber differ from those of heavy charged particles in three important aspects. For light charged particles:

1. Relativistic effects become important at relatively low kinetic energies.
2. Collisions with orbital electrons may result in large energy transfers (up to 50% of the incident energy for electrons, up to 100% of the incident energy for positrons). Collisions may also result in elastic and inelastic scattering with large angular deviations.
3. Collisions with nuclei of the absorber may result in bremsstrahlung production (radiation loss, see Sect. 6.3) and, depending on the light charged particle incident energy, radiation loss may actually exceed the collision loss.

In a manner similar to the approach taken with heavy charged particles, S_{col} for electrons and positrons is also calculated separately for soft and hard collisions, and the results are then added to obtain the collision stopping power S_{col}, as indicated in (6.3). The soft collision term ($\Delta E < \eta$) is identical to that obtained for heavy charged particles in (6.40) and the integration also runs from I to η except that for electrons and positrons we use $z = 1$ to obtain

$$S_{col}^{soft} = 2\pi \frac{ZN_A}{A} \left(\frac{e^2}{4\pi\varepsilon_0}\right)^2 \frac{1}{m_e c^2 \beta^2} \left\{ \ln \frac{2m_e c^2 \beta^2 \eta}{(1-\beta^2)I^2} - \beta^2 \right\}. \tag{6.58}$$

On the other hand, the hard collision term ($\Delta E > \eta$) is significantly more complicated for light charged particles than the term given for heavy charged particles in (6.41). The hard collision term for light charged particles is calculated by integrating

$$S_{col}^{hard} = N_e \int_{\eta}^{\Delta E_{max}} \Delta E \, d\sigma_{col} = N_e \int_{\eta}^{\Delta E_{max}} \Delta E \frac{d\sigma_{col}}{d(\Delta E)} d(\Delta E), \tag{6.59}$$

with the differential cross section for electrons determined by *Christian Møller* and for positrons by *Homi J. Bhabha*. The integration runs from η to ΔE_{max} where $\Delta E_{max} = \frac{1}{2} E_K$ for electrons and $\Delta E_{max} = E_K$ for positrons, as discussed in Sect. 5.3.4.

The complete mass collision stopping powers for electrons and positrons, according to the ICRU Report 37, are expressed as follows

$$S_{col} = 2\pi r_e^2 \frac{Z}{A} N_A \frac{m_e c^2}{\beta^2} \left\{ \ln \frac{E_K^2}{I^2} + \ln \left(1 + \frac{\tau}{2}\right) + F^{\pm}(\tau) - \delta \right\}, \tag{6.60}$$

where r_e is the classical electron radius defined in (6.43). In (6.60) the function $F^-(\tau)$ applies to electrons and is given as

$$F^-(\tau) = \left(1 - \beta^2\right)\left[1 + \tau^2/8 - (2\tau + 1)\ln 2\right], \tag{6.61}$$

while the function $F^+(\tau)$ applies to positrons and is given as

$$F^+(\tau) = 2\ln 2 - (\beta^2/12)\left[23 + 14/(\tau+2) + 10/(\tau+2)^2 + 4/(\tau+2)^3\right], \tag{6.62}$$

where

τ is the electron or positron kinetic energy E_K normalized to $m_e c^2$, i.e., $\tau = E_K/(m_e c^2)$, as discussed in Sect. 2.7.4.

β is the electron or positron velocity normalized to c, i.e., $\beta = v/c$.

Fig. 6.10 Mass collision stopping power S_{col} for electrons in water, aluminum and lead against electron kinetic energy. The collision stopping power data are shown with heavy solid curves; the radiation stopping power data of Fig. 6.2 are shown with light solid curves for comparison. Data were obtained from the NIST

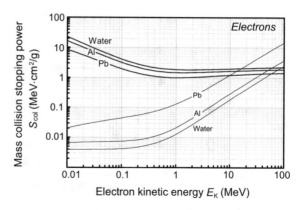

Figure 6.10 shows mass collision stopping powers S_{col} for electrons in water, aluminum and lead with heavy solid lines and, for comparison, mass radiation stopping powers of Fig. 6.2 are shown with light solid lines. Similarly to stopping power behavior for heavy charged particles, the data of Fig. 6.10 show that higher atomic number absorbers have lower S_{col} than lower atomic number absorbers at same electron energies. The dependence of S_{col} on stopping medium results from two factors in the stopping power expression given by (6.60), both lowering S_{col} with an increasing Z of the stopping medium:

1. The factor Z/A makes S_{col} dependent on the number of electrons per unit mass of the absorber. Z/A is 1 for hydrogen; 0.5 for low Z absorbers; then gradually drops to ~0.4 for high Z absorbers.
2. The $(-\ln I)$ term decreases S_{col} with increasing Z, since I increases almost linearly with increasing Z, as shown in (6.23)–(6.25).

6.6 Total Mass Stopping Power

Generally, the total mass stopping power S_{tot} of an absorber is given by the sum of two components: the mass radiation stopping power S_{rad} and the mass collision stopping power S_{col}

$$S_{tot} = S_{rad} + S_{col}. \qquad (6.63)$$

For heavy charged particles the radiation stopping power is negligible ($S_{rad} \approx 0$), thus $S_{tot} = S_{col}$.

For light charged particles both components contribute to the total stopping power; within a broad range of kinetic energies below 10 MeV collision (ionization) losses are dominant ($S_{col} > S_{rad}$); however, the situation is reversed at high kinetic energies where $S_{rad} > S_{col}$. The following features are noted:

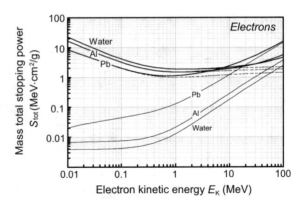

Fig. 6.11 Total mass stopping power S_{tot} for electrons in water, aluminum and lead against the electron kinetic energy shown with heavy solid curves. The mass collision stopping powers and mass radiation stopping powers are shown with dashed curves and light solid curves, respectively, for comparison. The total stopping power of a given material is the sum of the radiation and collision stopping powers. Data were obtained from the NIST. The critical energies $(E_K)_{crit}$ are 100 MeV, 61 MeV, and 10 MeV for water, aluminum, and lead, respectively

1. The crossover between the two modes occurs at a critical kinetic energy $(E_K)_{crit}$ where the two stopping powers are equal, i.e., $S_{rad}[(E_K)_{crit}] = S_{col}[(E_K)_{crit}]$ for a given absorber with atomic number Z
2. The critical kinetic energy $(E_K)_{crit}$ can be estimated from the following relationship

$$(E_K)_{crit} \approx \frac{800 \text{ MeV}}{Z}, \qquad (6.64)$$

and for water, aluminum and lead it amounts to $\sim$100 MeV, $\sim$61 MeV and $\sim$10 MeV, respectively.
3. For high Z absorbers the dominance of radiation loss over collision loss starts at lower kinetic energies than in low Z absorbers. However, even in high Z media, such as lead and uranium, $(E_K)_{crit}$ is at $\sim$10 MeV, well in the relativistic region.
4. The ratio of collision to radiation stopping power (S_{col}/S_{rad}) at a given electron kinetic energy may be estimated from the following

$$\frac{S_{col}}{S_{rad}} = \frac{800 \text{ MeV}}{Z E_K} = \frac{(E_K)_{crit}}{E_K}. \qquad (6.65)$$

Figure 6.11 shows the total mass stopping power S_{tot} for electrons (heavy solid curves) in water, aluminum and lead against the electron kinetic energy E_K. For comparison the radiation and collision components of the total stopping power of Figs. 6.2 and 6.10 are also shown.

6.7 Radiation Yield

The radiation yield (sometimes also referred to as the bremsstrahlung yield) $Y[(E_K)_0]$ of a light charged particle with initial kinetic energy $(E_K)_0$ striking an absorber is defined as that fraction of the initial kinetic energy that is emitted as radiation with energy E_{rad} through the slowing down process of the particle in the absorber. The following features should be noted:

1. The energy E_{rad} is generally emitted in the form of bremsstrahlung, but can also be from positron in-flight annihilation, and in the form of characteristic radiation following impact ionization and impact excitation atomic events.
2. For heavy charged particles $Y\left[(E_K)_0\right] \approx 0$.
3. For light charged particles (electrons and positrons) the radiation yield $Y[(E_K)_0]$ is determined from stopping power data as follows:

$$Y[(E_K)_0] = \frac{\int\limits_0^{(E_K)_0} \dfrac{S_{rad}(E)}{S_{tot}(E)} dE}{\int\limits_0^{(E_K)_0} dE} = \frac{1}{(E_K)_0} \int\limits_0^{(E_K)_0} \frac{S_{rad}(E)}{S_{tot}(E)} dE. \qquad (6.66)$$

4. For positron interactions, in-flight annihilation may also produce photons before the positron energy has been completely expended, so that in general the radiation yield is the sum of two contributions: the bremsstrahlung yield $Y_B[(E_K)_0]$ and the in-flight annihilation yield $Y_A[(E_K)_0]$. However, since the in-flight annihilation yield is much smaller than the bremsstrahlung yield, it is generally ignored in calculation of the radiation yield $Y[(E_K)_0]$ and the term "bremsstrahlung yield" is often incorrectly used to describe the total radiation yield.

Energy E_{rad} radiated per charged particle is from (6.66) given as

$$E_{rad} = (E_K)_0 \cdot Y[(E_K)_0] = \int\limits_0^{(E_K)_0} \frac{S_{rad}(E)}{S_{tot}(E)} dE, \qquad (6.67)$$

while energy E_{col} lost through ionization per charged particle is

$$E_{col} = (E_K)_0 - E_{rad} = (E_K)_0 \cdot \{1 - Y[(E_K)_0]\} = \int\limits_0^{(E_K)_0} \frac{S_{col}(E)}{S_{tot}(E)} dE. \qquad (6.68)$$

The radiation yield $Y[(E_K)_0]$ for various absorbing media from carbon at low Z to uranium at high Z is plotted against incident electron kinetic energy $(E_K)_0$ in Fig. 6.12. For a given incident electron kinetic energy $(E_K)_0$, the radiation yield $Y[(E_K)_0]$ increases with absorber atomic number Z and, for a given absorber atomic number Z, the radiation yield $Y[(E_K)_0]$ increases with increasing incident (initial) electron kinetic energy $(E_K)_0$.

Fig. 6.12 Radiation yield
$Y[(E_K)_0]$ in percent of
incident charged particle
kinetic energy $(E_K)_0$ for
selected absorbing media
from carbon to uranium
against incident electron
kinetic energy $(E_K)_0$. Data
were obtained from the NIST

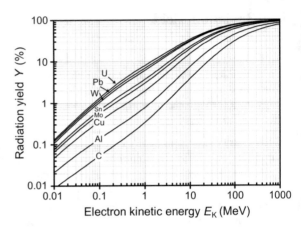

6.8 Range of Charged Particles

The range R of a charged particle in a particular absorbing medium is an experimental concept providing the thickness of an absorber that the particle can just penetrate. It depends on the particle's kinetic energy, mass as well as charge, and on the composition of the absorbing medium. In traversing matter charged particles lose their energy in ionizing and radiation collisions that may also result in significant deflections from their incident trajectory. In addition, charged particles suffer a large number of deflections as a result of elastic scattering.

These scattering effects are much more pronounced for light charged particles (electrons and positrons) in comparison to heavy charged particles because:

1. Heavy charged particles do not experience radiation losses, transfer only small amounts of energy in individual ionizing collisions, and mainly suffer small angle deflections in elastic collisions. Their path through an absorbing medium is thus essentially rectilinear, as shown schematically in Fig. 6.13.
2. Electrons with kinetic energy E_K, on the other hand, can lose energy up to $\frac{1}{2}E_K$ in individual ionizing collisions and energy up to E_K in individual radiation collisions. Since they can also be scattered with very large scattering angles, their path through the absorbing medium is very tortuous, as shown schematically in Fig. 6.13.

Various definitions of range are in common use. For example, the *path length* of a charged particle is the total distance along the particle's actual trajectory until it comes to rest, regardless of the direction of motion. The *projected range*, on the other hand, is the sum of individual path lengths projected onto the incident particle direction. Also used and defined below are the CSDA range R_{CSDA}, maximum range R_{max}, the 50% range R_{50}, practical range R_p, and the therapeutic ranges R_{80} and R_{90}.

Fig. 6.13 Schematic diagram of charged particle penetration into absorbing medium. *Top* Heavy charged particle (proton, α-particle, heavy ion, etc.); *bottom* light charged particle (electron or positron)

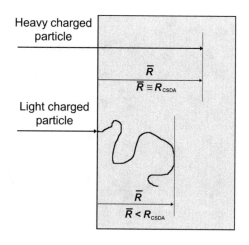

6.8.1 CSDA Range

Most of the collision and radiation interactions individually transfer only minute fractions of the incident particle's kinetic energy, and it is convenient to think of the particle that is moving through an absorber as losing its kinetic energy gradually and continuously in a process often referred to as the "continuous slowing down approximation" (CSDA). *Martin Berger* and *Stephen Seltzer* introduced the CSDA range concept in 1983 and defined the CSDA range R_{CSDA} as

$$R_{CSDA} = \int_{0}^{(E_K)_0} \frac{dE}{S_{tot}(E)}, \tag{6.69}$$

where

R_{CSDA} is the CSDA range (mean path length) of the charged particle in the absorber (typically in $cm^2 \cdot g^{-1}$).

$(E_K)_0$ is the initial kinetic energy of the charged particle.

$S_{tot}(E)$ is the total mass stopping power of the charged particle as a function of the kinetic energy E_K.

The CSDA range is a calculated quantity that represents the mean path length along the particle's trajectory and not necessarily the depth of penetration in a defined direction in the absorbing medium. For heavy charged particles, R_{CSDA} is a very good approximation to the average range $\overline{R}$ of the charged particle in the absorbing medium, because of the essentially rectilinear path of the charged particle (see Fig. 6.13) in the absorbing medium; for light charged particles the CSDA range can be up to twice the average range $\overline{R}$.

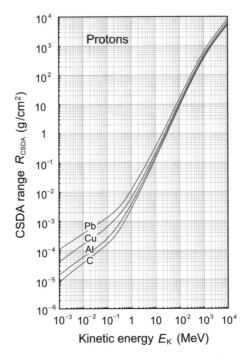

Fig. 6.14 CSDA range of protons against incident kinetic energy $(E_K)_0$ in carbon, aluminum, copper, and lead. Data are from the NIST

6.8.2 Maximum Penetration Depth

The maximum penetration depth R_{max} is defined as the depth in the absorbing medium beyond which no particles are observed to penetrate. For heavy charged particles $R_{max} \approx R_{CSDA}$ for all absorbing media, while for light charged particles $R_{max}/R_{CSDA} \approx 1$ for low Z absorbers and decreases with increasing Z to reach $R_{max}/R_{CSDA} \approx 0.5$ for high Z absorbers.

6.8.3 Range of Heavy Charged Particles in Absorbing Medium

Figure 6.14 gives the CSDA range R_{CSDA} of protons in aluminum, copper, and lead. Because of the decrease in the collision stopping power S_{col} with increasing atomic number Z of the absorber, for a given proton kinetic energy E_K, the range in g/cm^2 increases with the absorber atomic number Z. The CSDA range of heavy charged particles in an absorbing medium can be determined from the following expression

$$R_{CSDA} = \int\limits_{0}^{(E_K)_0} \frac{dE_K}{S_{col}(E_K)}, \tag{6.70}$$

after inserting into (6.69) the relationship $S_{tot} = S_{col}$, since for heavy charged particles S_{rad} is negligible. $(E_K)_0$ in (6.70) is the initial kinetic energy of the incident heavy charged particle. The collision stopping power S_{col} for non-relativistic heavy charged particles is given quantum mechanically by the Bethe equation (6.55) as

$$S_{col} = C_0 \frac{z^2}{\beta^2} \frac{Z}{A} B_{col} = C_1 \frac{z^2}{v^2} \ln \frac{2m_e v^2}{I}, \tag{6.71}$$

where

C_1 is a constant for a given absorber containing C_0 defined in (6.43) as well as the atomic number Z and atomic mass A of the absorber: $C_1 = C_0 c^2 Z/A$.

B_{col} is the atomic stopping number defined in (6.53).

v is the velocity of the heavy charged particle.

m_e is the rest mass of the electron.

For a non-relativistic heavy charged particle of mass M_0, its kinetic energy E_K and velocity v are related through the classical expression $E_K = \frac{1}{2} M_0 v^2$. The CSDA range is then given as

$$R_{CSDA}^{(M_0)} = \frac{1}{C_1 z^2} \int\limits_{0}^{(E_K)_0} \frac{v^2 dE}{\ln \frac{2m_e v^2}{I}} = \frac{2M_0}{C_1 z^2} \int\limits_{0}^{(E_K)_0} \frac{\frac{E_K}{M_0} d\left(\frac{E_K}{M_0}\right)}{\ln\left[C_2 \frac{E_K}{M_0}\right]}, \tag{6.72}$$

where C_2 is a constant for a given absorbing medium: $C_2 = 4m_e/I$. For non-relativistic protons with rest mass m_p and incident kinetic energy $(E_K)_0$, based on (6.72), the CSDA range in a given absorber is given as

$$R_{CSDA}^{(p)}\left[(E_K)_0\right] = \frac{2m_p}{C_1} \int\limits_{0}^{(E_K)_0} \frac{\frac{E_K}{m_p} d\left(\frac{E_K}{m_p}\right)}{\ln\left[C_2 \frac{E_K}{m_p}\right]}. \tag{6.73}$$

The CSDA range for protons in various absorbers has been measured extensively in the past and the data are readily available in tabular as well as graphic form. It turns out that the proton CSDA range $\left(R_{CSDA}^{(p)}\right)$ data can be used to estimate the CSDA range R_{CSDA} of any heavy charged particle following a simple relationship that links the two CSDA ranges.

The CSDA range R_{CSDA} (6.72) of an arbitrary heavy charged particle in a given absorber can be written in terms of the proton CSDA range as follows

$$R_{\text{CSDA}}^{(M_0)}\left[(E_K)_0\right] = \left(\frac{M_0}{m_p}\right)\frac{2m_p}{C_1 z^2}\int_0^{(E_K)_0}\frac{\left(\dfrac{E_K}{m_p}\dfrac{m_p}{M_0}\right)d\left(\dfrac{E_K}{m_p}\dfrac{m_p}{M_0}\right)}{\ln\left[C_2\dfrac{E_K}{m_p}\dfrac{m_p}{M_0}\right]}$$

$$= \frac{1}{z^2}\left(\frac{M_0}{m_p}\right)\left\{\frac{2m_p}{C_1}\int_0^{(E_K')_0}\frac{\dfrac{E_K'}{m_p}\,d\left(\dfrac{E_K'}{m_p}\right)}{\ln\left[C_2\dfrac{E_K'}{m_p}\right]}\right\}$$

$$= \frac{1}{z^2}\left(\frac{M_0}{m_p}\right)R_{\text{CSDA}}^{(p)}[(E_K')_0], \tag{6.74}$$

where we use the substitution

$$E_K' = E_K\frac{m_p}{M_0} \tag{6.75}$$

and

$(E_K)_0$ is the incident kinetic energy of the arbitrary heavy charged particle.
$(E_K')_0$ is the corresponding incident kinetic energy of the proton given as

$$(E_K')_0 = (E_K)_0\frac{m_p}{M_0}. \tag{6.76}$$

Therefore, the general procedure for finding R_{CSDA} of a heavy charged particle of rest mass M_0 and incident kinetic energy $(E_K)_0$ is to find in range tables for protons the $R_{\text{CSDA}}^{(p)}$ at corresponding incident proton kinetic energy of $(E_K')_0 = (E_K)_0\, m_p/M_0$ and then multiplying this value with $M_0/(m_p z^2)$, as shown in (6.74).

For example, in a given absorber the CSDA range of a 20 MeV α-particle is the same as the range of a 5 MeV proton, since, according to (6.74), for the α-particle

$$\frac{M_\alpha}{m_p}\frac{1}{z^2} \approx 1 \quad\text{and}\quad (E_K')_0 = \frac{m_p}{M_\alpha}(E_K)_0 \approx \frac{1}{4}(E_K)_0. \tag{6.77}$$

Thus

$$R_{\text{CSDA}}^{(\alpha)}(20\text{ MeV}) \approx R_{\text{CSDA}}^{(p)}(5\text{ MeV}). \tag{6.78}$$

Another example: a deuteron with kinetic energy of 20 MeV has the same velocity and stopping power as a 10 MeV proton. However, its CSDA range is twice that of a 10 MeV proton, since

$$\frac{M_d}{m_p}\frac{1}{z^2} \approx 2 \quad \text{and} \quad \left(E_K'\right)_0 = (E_K)_0 \frac{m_p}{M_d} \approx \frac{1}{2}(E_K)_0 . \tag{6.79}$$

Thus

$$R_{\text{CSDA}}^{(d)}(20 \text{ MeV}) \approx 2R_{\text{CSDA}}^{(p)}(10 \text{ MeV}) . \tag{6.80}$$

6.8.4 Range of Light Charged Particles (Electrons and Positrons) in Absorbers

For light charged particles, the CSDA range R_{CSDA} exceeds the average range $\overline{R}$ in an absorbing medium, because of the very tortuous path that the light charged particles experience in the absorbing medium (see Fig. 6.13). For low atomic number absorbers the difference is only about 10% to 15%, however, for high Z absorbers, the CSDA range can be up to twice the average range of charged particles in the absorber.

Electrons are used in external beam radiotherapy for treatment of superficial lesions, therefore, accurate knowledge of electron range in water and tissue is important. Since the CSDA range can serve only as a rough guide on the penetration of electron beams into tissue, other more appropriate ranges have been defined for use in radiotherapy, all of them based on measurement of electron depth dose distribution in water.

A typical electron beam depth dose curve (dose versus depth in water, see Sect. 1.12.3) is plotted in Fig. 6.15. The dose distribution is normalized to 100% at the depth of dose maximum z_{max}, and exhibits a relatively high surface dose, a rapid dose fall off beyond z_{max}, and a leveling off of the dose at a low level component referred to as the bremsstrahlung tail. Various electron ranges of interest in radiotherapy and radiation dosimetry, such as R_{80}, R_{90}, R_p, and R_{max}, are identified on the depth dose curve of Fig. 6.15.

The maximum range R_{max} is defined as the depth at which extrapolation of the tail of the depth dose curve meets the bremsstrahlung background. It is the largest penetration depth of electrons in the absorbing medium. The drawback of R_{max} is that it does not provide a well defined measurement point.

The practical range R_p is defined as the depth at which the tangent plotted through the steepest section of the electron depth dose curve intersects with the extrapolation line of the bremsstrahlung background. It is used for the determination of $\overline{E}_K(z)$, the mean electron beam kinetic energy at a depth z in a water phantom, with a relationship proposed by Harder

$$\overline{E}_K(z) = \overline{E}_K(0)\left[1 - \frac{z}{R_p}\right], \tag{6.81}$$

Fig. 6.15 Typical electron beam depth dose curve (dose against depth in water) normalized to 100 at the depth of dose maximum z_{max}. Several ranges of interest in radiotherapy and dosimetry, such as R_{80}, R_{50}, R_p, and R_{max}, are identified on the curve

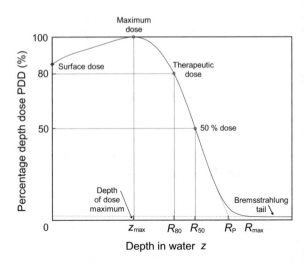

where $\overline{E}_K(0)$ is the mean electron kinetic energy at the water phantom surface at $z = 0$.

The depths R_{90}, R_{80}, and R_{50} on the depth dose curve are defined as depths at which the percentage depth doses beyond the depth of dose maximum z_{max} attain values of 90%, 80%, and 50%, respectively. R_{90} or R_{80} are used for prescription of tumor dose in radiotherapy, while R_{50} is used in radiation dosimetry for determination of the mean electron kinetic energy on the water surface $\overline{E}_K(0)$ which is then in turn used in the Harder relationship of (6.81). The relationship between R_{50} and $\overline{E}_K(0)$ is as follows

$$\overline{E}_K(0) = C R_{50}, \tag{6.82}$$

where C is a constant, for water equal to 2.33 MeV/cm.

6.9   Mean Collision Stopping Power

In radiation dosimetry the main interest is in the energy absorbed per unit mass of the absorbing medium governed by collision losses of charged particles. It is often convenient to characterize a given radiation beam with electrons of only one energy rather than with an electron spectrum $d\phi/dE$ that is present in practice. For example, monoenergetic electrons set in motion with an initial kinetic energy $(E_K)_0$ in an absorbing medium will through their own slowing down process produce a spectrum of electrons in the medium ranging in energy from $(E_K)_0$ down to zero.

The electron spectrum $d\phi(E)/dE$, when ignoring any possible hard collisions, is given as

$$\frac{d\phi(E)}{dE} = \frac{N}{S_{tot}(E)}, \tag{6.83}$$

where

N is the number of monoenergetic electrons of kinetic energy $(E_K)_0$ produced per unit mass in the absorbing medium.

$S_{tot}(E)$ is the total stopping power.

For this electron spectrum, produced by monoenergetic electrons, as shown by Harold E. Johns and John R. Cunningham, we can define a *mean collision stopping power* $\overline{S}_{col}$ as follows

$$\overline{S}_{col}\big[(E_K)_0\big] = \frac{\int_0^{(E_K)_0} \frac{d\phi}{dE} S_{col}(E) dE}{\int_0^{(E_K)_0} \frac{d\phi}{dE} dE}. \tag{6.84}$$

Using (6.69) and (6.83) the integral in the denominator of (6.84) is determined as follows

$$\int_0^{(E_K)_0} \frac{d\phi}{dE} dE = N \int_0^{(E_K)_0} \frac{dE}{S_{tot}(E)} = N \cdot R_{CSDA}. \tag{6.85}$$

Using (6.63), (6.66) and (6.83), the numerator of (6.86) is determined as follows

$$\int_0^{(E_K)_0} \frac{d\phi}{dE} S_{col}(E) dE = N \int_0^{(E_K)_0} \frac{S_{col}(E)}{S_{tot}(E)} dE \tag{6.86}$$

$$= N \int_0^{(E_K)_0} \frac{S_{tot}(E) - S_{rad}(E)}{S_{tot}(E)} dE = N \cdot (E_K)_0 - N \cdot (E_K)_0 \cdot Y[(E_K)_0].$$

The mean collision stopping power $\overline{S}_{col}\big[(E_K)_0\big]$ of (6.84) can now be written as

$$\overline{S}_{col}[(E_K)_0] = (E_K)_0 \frac{1 - Y[(E_K)_0]}{R_{CSDA}}. \tag{6.87}$$

The relationship (6.87) for $\overline{S}_{col}[(E_K)_0]$ could also be stated intuitively by noting that an electron with an initial kinetic energy $(E_K)_0$ will, through traveling the pathlength ℓ equal to R_{CSDA} in the absorbing medium: (1) lose an energy $(E_K)_0 Y[(E_K)_0]$ to bremsstrahlung and (2) deposit an energy $(E_K)_0 \{1 - Y[(E_K)_0]\}$ in the medium. The same would hold for a positron except that it could also lose part of its energy to photon production in in-flight annihilation events.

6.10 Restricted Collision Stopping Power

In radiation dosimetry one is interested in determining the energy transferred to a localized region of interest; however, the use of the mass collision stopping power S_{col} for this purpose may overestimate the dose because S_{col} incorporates both hard and soft collisions. The δ-rays resulting from hard collisions may be energetic enough to carry their kinetic energy a significant distance from the track of the primary particle thereby escaping from the region of interest in which the dose is determined.

The concept of restricted mass collision stopping power L_Δ has been introduced to address this issue by excluding the δ-rays with energies exceeding a suitable threshold value Δ.

The choice of the energy threshold Δ depends on the problem at hand. For dosimetric measurements involving air-filled ionization chambers with a typical electrode separation of 2 mm a frequently used threshold value is 10 keV (Note: the range of a 10 keV electron in air is of the order of 2 mm). For microdosimetric studies, on the other hand, one usually takes 100 eV as a reasonable threshold Δ value.

Of course, to be physically relevant Δ must not exceed ΔE_{max}, the maximum possible energy transfer to orbital electron from the incident particle with kinetic energy E_K in a direct-hit collision. As shown in Sect. 5.3.6, ΔE_{max} equals to $\frac{1}{2}E_K$ for electrons, E_K for positrons, and $2m_e c^2 \beta^2/(1 - \beta^2)$ for heavy charged particles [see (5.46), (5.45), and (5.41), respectively].

For a given kinetic energy E_K of the primary particle the restricted collision stopping power L_Δ is in general smaller than the unrestricted collision stopping power S_{col}; the smaller is the threshold Δ, the larger is the discrepancy. As Δ increases from a very small value, the discrepancy diminishes until at $\Delta = \Delta E_{max}$ the restricted and unrestricted collision stopping powers become equal, i.e., $L_{\Delta=\Delta E_{max}} = S_{col}$, irrespective of E_K.

Figure 6.16 displays the unrestricted collision mass stopping power as well as the restricted collision mass stopping powers with $\Delta = 10$ keV and $\Delta = 100$ keV against kinetic energy E_K for electrons in carbon based on data in the ICRU Report 37. The following observations can now be made:

1. Since energy transfers to secondary electrons are limited to $\frac{1}{2}E_K$, the unrestricted mass collision stopping power S_{col} and restricted mass collision stopping power L_Δ are identical for a given kinetic energy E_K of the electron for kinetic energies lower than or equal to 2Δ. This is indicated in Fig. 6.16 with vertical lines at 20 keV and 200 keV for the threshold values $\Delta = 10$ keV and $\Delta = 100$ keV, respectively.

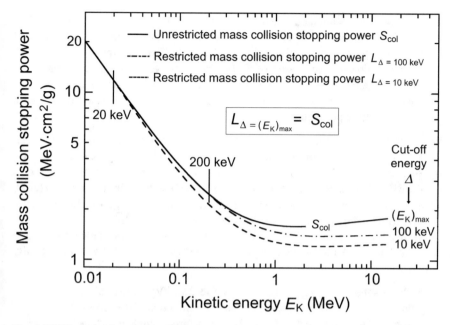

Fig. 6.16 Unrestricted mass collision stopping power S_{col} and restricted mass collision stopping power L_Δ with thresholds $\Delta = 10$ keV and $\Delta = 100$ keV for electrons in carbon against kinetic energy E_K. Data are based on the ICRU Report 37

2. For a given $E_K > 2\Delta$, the smaller is Δ, the larger is the discrepancy between the unrestricted and restricted stopping powers.
3. For given Δ and $E_K > 2\Delta$, the larger is E_K, the larger is the discrepancy between the unrestricted and restricted stopping powers.

6.11 Bremsstrahlung Targets

Bremsstrahlung production is of great importance in medical physics, since the majority of radiation beams used in diagnostic radiology and in external beam radiotherapy are produced through bremsstrahlung interactions of monoenergetic electrons with solid targets. These targets are components of x-ray machines and linear accelerators; the most commonly used radiation-emitting machines for diagnosis and treatment of disease.

An electron that strikes the target with a given kinetic energy will undergo many different interactions with target atoms before it comes to rest and dissipates all of its kinetic energy in the target. As discussed in Sect. 6.1, there are two classes of electron interactions with a target atom:

1. Incident electron interaction with orbital electron of a target atom results mainly in collision impact loss and ionization of the target atom that may be accompanied by an energetic electron referred to as delta ray. The collision loss in an x-ray target is followed by emission of characteristic x-rays and Auger electrons.
2. Incident electron interaction with the nucleus of a target atom results mainly in elastic scattering events but may also result in radiation loss accompanied with bremsstrahlung production.

While bremsstrahlung is a major contributor to the x-ray spectrum at superficial (50 kVp to 100 kVp) and orthovoltage (100 kVp to 350 kVp) energies, it is essentially the sole contributor to the x-ray spectrum at megavoltage energies. With regard to their thickness compared to the average range $\overline{R}$ of electrons in the target material, x-ray targets are either *thin* or *thick*; the thickness of a thin target is much smaller than $\overline{R}$, while the thickness of a thick target is of the order of $\overline{R}$.

As discussed in Sect. 4.2.8, the peak x-ray intensity occurs at a characteristic angle θ_{max} that depends on the kinetic energy of incident electrons. As shown schematically in Fig. 6.17:

1. In the diagnostic energy range (50 kVp to 350 kVp) x-ray tubes are used for production of x-rays and most photons are emitted at 90° from the direction of electron deceleration in the target. The characteristic angle θ_{max} thus equals ∼90° (see Fig. 6.17a).
2. In the megavoltage radiotherapy range (4 MV and above) acceleration waveguides are used for electron acceleration and x-ray production and most photons are emitted in the direction of electron deceleration in the target. The characteristic angle θ_{max} is then ∼0° and the target is referred to as a transmission target (see Fig. 6.17b).

X-ray beams produced in x-ray targets are heterogeneous and contain photons of many energies ranging from 0 to a maximum energy $h\nu_{max}$ which is equal to the

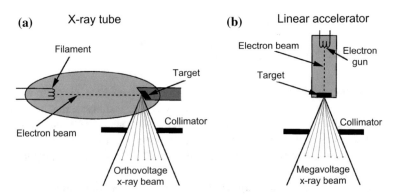

Fig. 6.17 Schematic comparison of x-ray production in the diagnostic radiology orthovoltage range with an x-ray machine (x-rays emitted mostly orthogonally to the electron beam) and in the radiotherapy megavoltage range with a linear accelerator (x-rays emitted mostly in the direction of the electron beam)

kinetic energy E_K of the electrons striking the target. The relationship

$$h\nu_{max} = E_K = eU \tag{6.88}$$

is referred to as the Duane–Hunt law in honor of two American physicists, *William Duane* and *Franklin Hunt*, who discovered the law in 1915. In its original form the Duane–Hunt law stated that there is a sharp upper limit to the x-ray frequencies emitted from an x-ray target stimulated by an impact of electrons and that this frequency ν_{max} is given by the quantum connection, independently of the target material, as

$$\nu_{max} = \frac{c}{\lambda_{min}} = \frac{eU}{h}, \tag{6.89}$$

where

c is the speed of light in vacuum.
λ_{min} is the minimum photon wavelength.
e is the electron charge.
U is the potential difference used in electron acceleration.

The Duane–Hunt law, sometimes called the inverse photoelectric effect, defines the maximum photon energy $h\nu_{max}$ in an x-ray beam produced by thin or thick x-ray targets. However, to describe a heterogeneous x-ray beam we must provide information on a full x-ray spectrum, since the spectrum is affected not only by the maximum photon energy $h\nu_{max}$ but also by the target type (thin or thick) and target atomic number Z.

The spectrum of an x-ray beam is most commonly shown by plotting either:

1. Number of photons per energy interval $[\Delta N/\Delta(h\nu)]$ against photon energy $h\nu$ of the given energy interval bin

or

2. Photon intensity I_ν against photon energy $h\nu$ of a given energy interval bin, where the photon intensity I_ν is proportional to the product $(\Delta N/\Delta E)\cdot h\nu$.

X-ray spectra can also be presented by plotting $(\Delta N/\Delta E)$ or beam intensity I against photon wavelength λ. In this case the Duane–Hunt law is given by specifying the minimum wavelength λ_{min} rather than the maximum frequency ν_{max}, where $\lambda_{min} = c/\nu_{max}$. Since $I_\nu d\nu = I_\lambda d\lambda$ the following relationship also applies

$$I_\lambda = I_\nu \left| \frac{d\nu}{d\lambda} \right| = \left| I_\nu \frac{d}{d\lambda}\left(\frac{c}{\lambda}\right) \right| = \frac{c}{\lambda^2} I_\nu. \tag{6.90}$$

Fig. 6.18 Typical x-ray
intensity spectrum: **a** for thin
target radiation and **b** for
thick target radiation

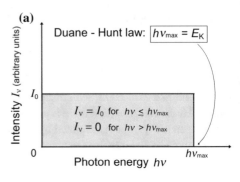

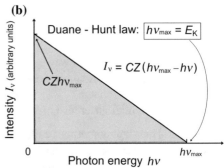

6.11.1   Thin X-Ray Targets

Thin x-ray targets are mainly of theoretical interest and their thickness is very small compared to the range of electrons of given kinetic energy in the target material. By definition, a thin target is so thin that electron striking it:

1. Loses essentially no energy by atomic ionizations.
2. Suffers no significant elastic collisions.
3. Traverses the target without interacting with target atoms or experiences only one bremsstrahlung interaction while traversing the target.

The bremsstrahlung radiation produced in a thin target by electrons of kinetic energy E_K has a constant intensity $I_\nu = I_0$ for $h\nu \leq h\nu_{max}$ and zero intensity $I_\nu = 0$ for $h\nu > h\nu_{max}$, as shown schematically in Fig. 6.18a. The maximum photon energy in the spectrum follows the Duane–Hunt law (6.89). Since the x-ray intensity is proportional to the product $(\Delta N/\Delta h\nu) \cdot h\nu$, it follows that, in comparison with the number of photons of energy $h\nu_{max}$, the x-ray spectrum contains twice as many photons of energy $0.50\,h\nu_{max}$; 4 times as many photons of energy $0.25\,h\nu_{max}$; 10 times as many photons of energy $0.10\,h\nu_{max}$, etc.

6.11.2 Thick X-Ray Targets

Thick x-ray targets have thicknesses of the order of the average range of electrons $\overline{R}$ in the target material. In practice, typical thicknesses are equal to about $1.1\overline{R}$ to satisfy two opposing conditions:

1. To ensure that no electrons that strike the target can traverse the target.
2. To minimize the attenuation of the bremsstrahlung beam produced by electrons that are scattered (decelerated) many times in the target.

Thick target radiation is much more difficult to handle theoretically than thin target radiation; however, in practice most targets used in bremsstrahlung production are of the thick target variety. A typical spectrum of a clinical x-ray beam consists of spectral lines characteristic of the target as well as the filtration material and superimposed onto the continuous bremsstrahlung spectrum. The bremsstrahlung spectrum originates in the x-ray target, while the characteristic spectrum originates not only in the target but also in any filtration placed into the x-ray beam.

Figure 6.19a displays three typical spectra from a tungsten target:

1. Unfiltered 100 kVp bremsstrahlung spectrum with superimposed tungsten K_α and K_β lines.
2. Filtered 100 kVp bremsstrahlung spectrum with superimposed tungsten K_α and K_β lines.
3. Filtered 60 kVp bremsstrahlung spectrum. The K lines are not present in the spectrum because the ionization energy of the K shell in tungsten is 70 keV and the kinetic energy of electrons striking the tungsten target is only 60 keV.

The spectral distribution of thick-target bremsstrahlung can be represented as a superposition of contributions from a large number of thin targets, each thin target traversed by a lower energy monoenergetic electron beam having a lower $h\nu_{max}$ than the previous thin target. This is shown schematically in Fig. 6.19b which depicts a typical thin target bremsstrahlung spectrum (curve 1) and several thick target bremsstrahlung spectra (curves 2, 3, and 4) for an x-ray tube in which 100 keV electrons strike the target. Curve 1 is for a thin target producing constant beam intensity for photon energies from 0 to the kinetic energy of the electrons striking the target (100 keV). Curve 2 represents an unfiltered spectrum (inside the x-ray tube) for a thick target and a superposition of numerous thin target spectra; curve 3 is spectrum for a beam filtered by an x-ray tube window which filters out the low energy photons; and curve 4 is spectrum for a beam filtered by the x-ray tube window and additional filtration.

As indicated in Figs. 6.18b and 6.19b, the beam intensity I_ν for a thick target as a function of photon energy $h\nu$ may be described with an approximate linear relationship

$$I_\nu \approx CZ\left(h\nu_{max} - h\nu\right), \tag{6.91}$$

Fig. 6.19 Typical spectra produced by electrons with kinetic energy of 100 keV striking a tungsten target. *Part a* displays characteristic line spectra superimposed onto the bremsstrahlung spectrum and *part b* displays only the bremsstrahlung spectrum under various conditions: curve *1* is for a thin target, producing a constant intensity beam; curves *2*, *3*, and *4* are for a thick target with curve *2* representing an unfiltered spectrum (inside the x-ray tube), curve *3* representing an x-ray beam filtered by an x-ray tube window, and curve *4* representing an x-ray beam filtered by the x-ray tube window and additional filtration

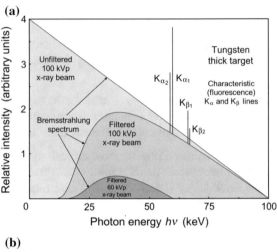

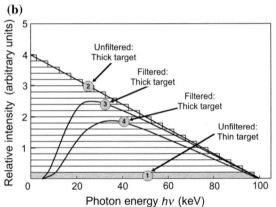

where

C	is a constant.
Z	is the atomic number of the thick target.
$h\nu_{max}$	is the Duane–Hunt photon energy limit.
I_ν	is the beam intensity at photon energy $h\nu$ with a maximum value $CZh\nu_{max}$ at $h\nu = 0$ and a value of zero for $h\nu \geq h\nu_{max}$.

The total intensity I is determined by integrating I_ν over the whole energy range from 0 to $h\nu_{max}$

$$I = CZ \int_0^{h\nu_{max}} (h\nu_{max} - h\nu)\, \mathrm{d}h\nu = \frac{1}{2}CZ\,(h\nu_{max})^2 = \frac{1}{2}CZ\,(eU)^2, \qquad (6.92)$$

where U is the accelerating voltage and eU is the electron kinetic energy.

The total photon intensity I emitted from the x-ray target is thus proportional to:

1. Target atomic number Z.
2. Square of the accelerating potential U.

The average energy E_{rad} radiated by an electron of initial energy $(E_K)_0$ in being stopped in a thick target was given in (6.67) as

$$E_{rad} = (E_K)_0 \, Y\left[(E_K)_0\right] = \int_0^{(E_K)_0} \frac{S_{rad}(E)}{S_{tot}(E)} \, dE. \tag{6.93}$$

In the diagnostic energy range, where $(E_K)_0 \ll m_e c^2$, the mass radiation stopping power S_{rad} is independent of the electron kinetic energy and, from (6.6) combined with Table 6.1, given as

$$S_{rad} = \frac{N_A}{A} \sigma_{rad} E_i = \frac{16}{3} \alpha \, r_e^2 Z^2 E_i = \frac{16}{3} \alpha \, r_e^2 \frac{N_A}{A} Z^2 \left[(E_K)_0 + m_e c^2\right]. \tag{6.94}$$

For $(E_K)_0 \ll m_e c^2$ the mass radiation stopping power S_{rad} of (6.93) is independent of the kinetic energy of the electron and (6.93) may then be simplified to read

$$E_{rad} = S_{rad} \int_0^{(E_K)_0} \frac{dE}{S_{tot}(E)} = S_{rad} R_{CSDA} = \text{const} \frac{N_A}{A} Z^2 R_{CSDA}. \tag{6.95}$$

Since in the low energy range $S_{tot} \approx S_{col}$ and $S_{col} \propto N_A Z/A$, we note that $R_{CSDA} \propto (N_A Z/A)^{-1}$ and the average energy radiated by the electron stopped in a thick target is linearly proportional to the atomic number Z of the target, i.e.,

$$E_{rad} \propto Z f\left[(E_K)_0, Z\right], \tag{6.96}$$

where $f\left[(E_K)_0, Z\right]$ is a slowly varying function of the target atomic number Z.

A few other notable features of bremsstrahlung targets used in diagnostic and therapeutic medical equipment are listed below:

- The integrated intensity of thick-target bremsstrahlung depends linearly on the atomic number Z of the target material. This implies that high Z targets will be more efficient for x-ray production that low Z targets.
- In megavoltage radiotherapy only photons in the narrow cone in the forward direction are used for the clinical beams and the radiation yield in the forward direction is essentially independent of the atomic number Z of the target.
- The thick target bremsstrahlung is linearly proportional to the atomic number Z of the target in the diagnostic energy range where $(E_K)_0 \ll m_e c^2$. This rule will fail when $(E_K)_0$ becomes large enough for the radiation losses to no longer be negligible in comparison with collision losses.

Computerized Tomography Images and Leonardo Da Vinci

Center image on next page is a famous sketch of Vitruvian man by *Leonardo da Vinci* (1452–1519). The drawing is based on geometrical proportions of man proposed by the first century Roman architect Vitruvian, hence the name Vitruvian man. While other great men and women that humanity produced in arts and science generally excelled in only one specific area of art or science, Leonardo da Vinci was a man of enormous talents covering most areas of human endeavor, be it in arts or science; a truly versatile renaissance man. He was active as sculptor, painter, musician, architect, engineer, inventor, and researcher of the human body. Many of da Vinci's drawings of the human body helped doctors of his time to understand better the layout of muscle and bone structures within the body.

Left and right images on next page present *computerized tomography* (CT) images of the human body, representing the most important development that resulted from *Wilhelm Roentgen's* discovery of x-rays in 1895. A CT scanner is a machine that uses an x-ray beam rotating about a specific area of a patient to collect x-ray attenuation data for patient's tissues. It then manipulates these data with special mathematical algorithms to display a series of transverse slices through the patient. The transverse CT data can be reconstructed so as to obtain sagittal sections (shown in the right image on next page) and coronal sections (shown on the left image on next page) through the patient's organs or to obtain digitally reconstructed radiographs. The excellent resolution obtained with a modern CT scanner provides an extremely versatile "non-invasive" diagnostic tool. CT scanners have been in clinical and industrial use since the early 1970s and evolved through five generations, each generation increasingly more sophisticated and faster than the previous one.

Initially, CT scanners have been used for imaging in radiology departments and for target delineation in radiotherapy. Since the early 1990s CT scanners are also an invaluable tool in radiotherapy departments where they form part of a CT-simulator machine which allows not only visualization of internal organs and target delineation but also the so-called virtual simulation and actual treatment planning for radiotherapy.

Three types of detector are used in CT scanners: (1) scintillation detectors (sodium iodide or calcium fluoride) in conjunction with a photomultiplier tube; (2) gas (xenon or krypton) ionization chamber; and (3) semiconductor detectors (cesium iodide) in conjunction with a p–n junction photodiode.

Three-dimensional images can be obtained through three techniques: (1) multiple 2D acquisitions based on a series of sequential scans; (2) spiral (helical) CT with the x-ray source rotating continuously around the patient while simultaneously the patient is translated through the gantry; and (3) cone-beam CT using a 2D detector array in order to measure the entire volume-of-interest during one single orbit of the x-ray source.

Allan Cormack (1924–1998), a South African–American physicist, developed the theoretical foundations that made computerized tomography possible and published his work during 1963–64. His work generated little interest until *Godfrey Hounsfield* (1919–2004), a British electrical engineer, developed a practical model of a CT scanner in the early 1970s. Hounsfield and Cormack received the 1979 Nobel Prize in Physiology and Medicine for their independent invention of the CT scanner.

Chapter 7
Interactions of Photons with Matter

In this chapter we discuss the various types of interaction that photons of energy exceeding the ionization energy of absorber atoms can have with absorbing media. These photons belong to the category of indirectly ionizing radiation and they deposit energy in the absorbing medium through a two-step process: (1) in the first step energy is transferred to an energetic light charged particle and (2) in the second step energy is deposited in the absorbing medium by the charged particle. The energy transferred to charged particles from the interacting photon generally exceeds the energy subsequently deposited in the absorbing medium by the charged particles, because some of the transferred energy may be radiated from the charged particles in the form of photons. Some of the photon interactions are only of theoretical interest and help in the understanding of the general photon interaction phenomena, others are of great importance in medical physics, since they play a fundamental role in imaging, radiotherapy as well as radiation dosimetry.

Coronal CT section   Vitruvian man   Sagittal CT section

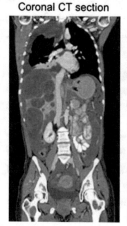

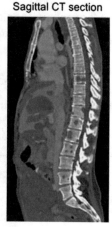

© Springer International Publishing Switzerland 2016
E.B. Podgoršak, *Radiation Physics for Medical Physicists*,
Graduate Texts in Physics, DOI 10.1007/978-3-319-25382-4_7

Depending on their energy and the atomic number of the absorber, photons may interact with an absorber atom as a whole, with the nucleus of an absorber atom, or with an orbital electron of the absorber atom. As far as the photon fate after the interaction with an absorber atom is concerned there are two possible outcomes: (1) photon disappears and a portion of its energy is transferred to light charged particles and (2) photon is scattered. The probability of a particular interaction to occur depends on the photon energy as well as on the density and atomic number of the absorber, and is generally expressed in the form of an interaction cross section.

In this chapter we first discuss the general aspects of photon interaction with matter and define the various attenuation coefficients and beam geometries that are used in describing the penetration of ionizing photon beams in absorbers. In the second part of the chapter we discuss in detail and on a microscopic scale the various interactions that a photon can experience with absorber atoms or their constituents, the nuclei and orbital electrons. For each photon interaction we also discuss the various events that follow the photon interactions. The chapter concludes with a brief discussion of photonuclear reaction, a direct photon–nucleus interaction that is of little importance in medical physics and is added mainly for academic interest.

7.1 General Aspects of Photon Interactions with Absorbers

In penetrating an absorbing medium, photons may experience various interactions with the atoms of the medium. These interactions involve either the nuclei of the absorbing medium or the orbital electrons of the absorbing medium:

1. The interactions with nuclei may be direct photon–nucleus interactions (*photo-disintegration*) or interactions between the photon and the electrostatic field of the nucleus (*pair production*).
2. The photon–orbital electron interactions are characterized as interactions between the photon and either (1) a loosely bound electron (*Thomson scattering, Compton effect, triplet production*) or (2) a tightly bound electron (*photoelectric effect, Rayleigh scattering*).

A *loosely bound electron* is an electron whose binding energy E_B is small in comparison with photon energy $h\nu$, i.e., $E_B \ll h\nu$. An interaction between a photon and a loosely bound electron is considered to be an interaction between a photon and a "free" (i.e., unbound) electron.

A *tightly bound electron* is an electron whose binding energy E_B is comparable to, larger than, or slightly smaller than the photon energy $h\nu$. For a photon interaction to occur with a tightly bound electron, the binding energy E_B of the electron must be of the order of, but slightly smaller than, the photon energy $h\nu$, i.e., $E_B \lesssim h\nu$. An interaction between a photon and a tightly bound electron is considered an interaction between a photon and the atom as a whole.

As far as the photon fate after the interaction with an atom is concerned there are two possible outcomes:

1. *Photon disappears* (i.e., is absorbed completely) and a portion of its energy is transferred to light charged particles (electrons and positrons).
2. *Photon is scattered* and two outcomes are possible:

   a. The resulting photon has the same energy as the incident photon and no light charged particles are released in the interaction.
   b. The resulting scattered photon has a lower energy than the incident photon and the energy excess is transferred to a light charged particle (electron).

The light charged particles (electrons and positrons) released or produced in the absorbing medium through photon interactions will:

1. Either deposit their energy to the medium through Coulomb interactions with orbital electrons of the absorbing medium (collision loss also referred to as ionization loss), as discussed in detail in Sect. 6.3.
2. Or radiate their kinetic energy away in the form of photons through Coulomb interactions with the nuclei of the absorbing medium (radiation loss), as discussed in detail in Sect. 6.2.

The most important parameter used for characterization of x-ray or gamma ray penetration into absorbing media is the linear attenuation coefficient μ. This coefficient depends on energy $h\nu$ of the photon and atomic number Z of the absorber, and may be described as the probability per unit path length that a photon will have an interaction with the absorber.

The functional relationship between the thickness of an absorber and intensity of a photon beam attenuated by the absorber is usually derived using differential calculus, as shown in Sect. 7.1.1. However, the relationship can also be derived without calculus with the help of a thought experiment (Gedanken experiment) shown schematically in Fig. 7.1. A collimated monoenergetic photon beam of energy $h\nu$ strikes a detector and produces a measured intensity I_0. When an absorber of thickness ℓ is placed into the photon beam's path, the measured beam intensity decreases to I_1 which can be expressed as $I_1 = R I_0$ with R the ratio between I_1 and I_0. Since $I_1 < I_0$, it follows that $R < 1$.

When another layer ℓ of the same absorber material is placed into the photon beam, the measured intensity decreases to $I_2 = R I_1 = R^2 I_0$, with three layers to $I_3 = R I_2 = R^2 I_1 = R^3 I_0$, etc., until for a large number n of absorber layers we get $I_n = R I_{n-1} = \cdots = R^{n-1} I_1 = R^n I_0$. Designating the total absorber thickness as x where $x = n\ell$ we can now write

$$I_n = I(x) = R^n I_0 = R^{x/\ell} I_0 \tag{7.1}$$

or

$$\ln \frac{I(x)}{I_0} = x \frac{\ln R}{\ell} = -\mu x, \tag{7.2}$$

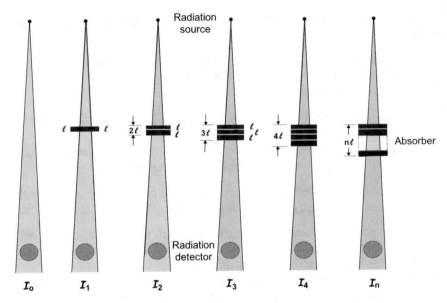

Fig. 7.1 Experimental setup for a simple determination of the attenuation coefficient μ

where, as a consequence of $R < 1$, i.e., $\ln R < 0$, we define the ratio $-(\ln R)/\ell$ as the attenuation coefficient μ. Equation (7.2) for intensity $I(x)$ represents the standard law of exponential attenuation and is usually written in an exponential form as follows

$$I(x) = I_0\, e^{-\mu x}, \tag{7.3}$$

with

I_0 the beam intensity without attenuator.
μ the linear attenuation coefficient.

7.1.1 Narrow Beam Geometry

The attenuation coefficient μ is determined experimentally using the so-called *narrow beam geometry* technique that implies a narrowly collimated source of monoenergtic photons and a narrowly collimated detector. As shown in Fig. 7.2a, a slab of absorber material of thickness x is placed between the source and detector. The absorber decreases the detector signal (intensity which is proportional to the number of photons striking the detector) from $I(0)$ measured without the absorber in place to $I(x)$ measured with absorber thickness x in the beam. A layer of thickness dx' of the absorber reduces the beam intensity by dI and the fractional reduction in intensity, $-dI/I$, is proportional to two parameters:

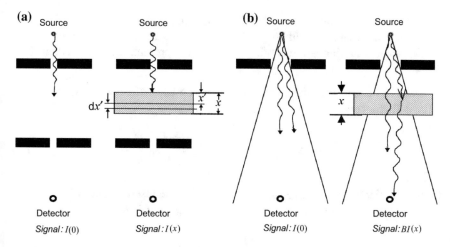

Fig. 7.2 Measurement of photon attenuation in absorbing material: **a** narrow beam geometry; **b** broad beam geometry

1. Linear attenuation coefficient μ (often referred to as the attenuation coefficient μ), measured in units of (length)$^{-1}$, such as m^{-1} or cm^{-1}.
2. Layer thickness dx'.

We can thus write $-dI/I$ as follows

$$-\frac{dI}{I} = \mu dx', \tag{7.4}$$

where the negative sign is used to indicate a decrease in signal I with an increase in absorber thickness x and μ represents the probability that a photon interacts in a unit thickness of absorber traversed. The product $\mu dx'$ represents the probability that a photon interacts in the absorber layer dx'.

After integration of (7.4) over absorber thickness from 0 to x and over intensity from the initial intensity $I(0)$ to intensity $I(x)$ at absorber thickness x, we get

$$\int_{I(0)}^{I(x)} \frac{dI}{I} = -\int_{0}^{x} \mu dx' \quad \text{or} \quad I(x) = I(0)e^{-\int_{0}^{x} \mu dx'}. \tag{7.5}$$

For a homogeneous medium the attenuation coefficient μ is uniform ($\mu = \text{const}$) and (7.5) reduces to the standard exponential relationship valid for monoenergetic photon beams

$$I(x) = I(0)e^{-\mu \int_{0}^{x} dx'} = I(0)e^{-\mu x}. \tag{7.6}$$

Equation (7.6) represents the standard expression of the law of exponential attenuation which applies for narrow beam geometry and:

1. Either a constant attenuation coefficient μ and simple absorption of radiation without any scattering and without production of secondary radiation in the absorber
2. Or, when scattering and secondary radiation are present, they do not contribute to $I(x)$.

Since the exponential function $y = e^z$ can be approximated by the following infinite Taylor series

$$y = e^z = \sum_{n=0}^{\infty} \frac{z^n}{n!} \approx 1 + \frac{z}{1!} + \frac{z^2}{2!} + \frac{z^3}{3!} + \frac{z^4}{4!} + \cdots , \qquad (7.7)$$

for μx sufficiently small, we can approximate (7.6), after inserting into (7.7) $z = -\mu x$ and $y = I(x)/I(0)$, by the first two terms of the series (7.7) to get

$$\frac{I(x)}{I(0)} \approx 1 - \mu x. \qquad (7.8)$$

The smaller is μx in comparison to 1, the more accurate is the approximation (7.8). For example, at $\mu x = 0.1$, approximation (7.8) is within 0.5% of the value calculated with (7.6) (0.900 vs. 0.905); at $\mu x = 0.05$, approximation (7.8) is within 0.1% of the value calculated with (7.6) (0.950 vs. 0.951).

7.1.2 Characteristic Absorber Thicknesses

Several thicknesses of special interest are defined as parameters for monoenergetic photon beam characterization in narrow beam geometry:

1. *First half-value layer* $\left(\text{HVL}_1 \text{ or } x_{1/2}\right)$ is the thickness of a homogeneous absorber that attenuates the narrow beam intensity $I(0)$ to one-half (50%) of the original intensity, i.e., $I(x_{1/2}) = 0.5I(0)$. Half-value layers are often used for characterization of superficial and orthovoltage x-ray beams. The absorbing materials used for this purpose are usually aluminum (for the superficial energy range) and copper (for the orthovoltage energy range). The relationship between the half-value layer $x_{1/2}$ and the attenuation coefficient μ is determined from the basic definition of the half-value layer as follows

$$I(x_{1/2}) = \frac{1}{2}I(0) = I(0)e^{-\mu x_{1/2}}, \qquad (7.9)$$

resulting in

$$\frac{1}{2} = e^{-\mu x_{1/2}} \qquad \text{or} \qquad \mu x_{1/2} = \ln 2 \qquad \text{or} \qquad \text{HVL} = x_{1/2} = \frac{\ln 2}{\mu} \qquad (7.10)$$

2. *Mean free path* (MFP or $\bar{x}$) or *relaxation length* is the thickness of a homogeneous absorber that attenuates the beam intensity $I(0)$ to $1/e = 0.368$ (36.8%) of its original intensity, i.e., $I(\bar{x}) = 0.368I(0)$. The photon mean free path is the average distance a photon of energy $h\nu$ travels through a given absorber before undergoing an interaction. The relationship between the mean free path $\bar{x}$ and the attenuation coefficient μ is determined from the basic definition of the mean free path as follows

$$\bar{x} = \frac{\int\limits_0^\infty xe^{-\mu x}dx}{\int\limits_0^\infty e^{-\mu x}dx} = \frac{\frac{1}{\mu^2}}{\frac{1}{\mu}} = \frac{1}{\mu} \tag{7.11}$$

or

$$I(\bar{x}) = \frac{1}{e}I(0) = I(0)e^{-\mu\bar{x}}, \tag{7.12}$$

resulting in

$$\frac{1}{e} = e^{-\mu\bar{x}} \quad \text{or} \quad \mu\bar{x} = 1 \quad \text{and} \quad \text{MFP} = \bar{x} = \frac{1}{\mu}. \tag{7.13}$$

3. *Tenth-value layer* (TVL or $x_{1/10}$) is the thickness of a homogeneous absorber that attenuates the beam intensity $I(0)$ to one-tenth (10%) of its original intensity, i.e., $I(x_{1/10}) = 0.1I(0)$. Tenth-value layers are used in radiation protection in treatment room shielding calculations. The relationship between the tenth-value layer $x_{1/10}$ and the attenuation coefficient μ is determined from the basic definition of the tenth-value layer as follows

$$I(x_{1/10}) = \frac{1}{10}I(0) = I(0)\,e^{-\mu x_{1/10}}, \tag{7.14}$$

resulting in

$$\frac{1}{10} = e^{-\mu x_{1/10}} \quad \text{or} \quad \mu x_{1/10} = \ln 10 \quad \text{and} \quad \text{TVL} = x_{1/10} = \frac{\ln 10}{\mu}. \tag{7.15}$$

4. *Second half-value layer* (HVL$_2$), measured with the same homogeneous absorber material as the first half value layer (HVL$_1$), is defined as the thickness of the absorber that attenuates the narrow beam intensity from $0.5I(0)$ to $0.25I(0)$. The ratio between HVL$_1$ and HVL$_2$ is called the homogeneity factor χ of the photon beam.

- When $\chi = 1$, the photon beam is monoenergetic such as a cobalt-60 beam with energy of 1.25 MeV or cesium-137 beam with energy of 0.662 MeV.
- When $\chi \neq 1$, the photon beam possesses a spectral distribution.
- For $\chi < 1$ the absorber is hardening the photon beam, i.e., preferentially removing low-energy photons from the spectrum (photoelectric effect region).
- For $\chi > 1$ the absorber is softening the photon beam, i.e., preferentially removing high-energy photons from the spectrum (pair production region).

From (7.10), (7.13), and (7.15), the linear attenuation coefficient μ may be expressed in terms of $x_{1/2}$, $\bar{x}$, and $x_{1/10}$, respectively, as follows

$$\mu = \frac{\ln 2}{x_{1/2}} = \frac{1}{\bar{x}} = \frac{\ln 10}{x_{1/10}}, \tag{7.16}$$

resulting in the following relationships among the characteristic thicknesses

$$x_{1/2} = (\ln 2)\,\bar{x} = \frac{\ln 2}{\ln 10}x_{1/10} \equiv 0.301 x_{1/10}. \tag{7.17}$$

Figure 7.3 is a typical plot of intensity $I(x)$ against absorber thickness x for a narrow, monoenergetic photon beam. The functional relationship $I(x)$ against x is a perfect exponential function expressed in (7.3). The figure also highlights the half-value layer $x_{1/2}$, the mean free path $\bar{x}$, and the tenth-value layer $x_{1/10}$. The various characteristic thicknesses and their effects on photon beam intensity are summarized in Table 7.1.

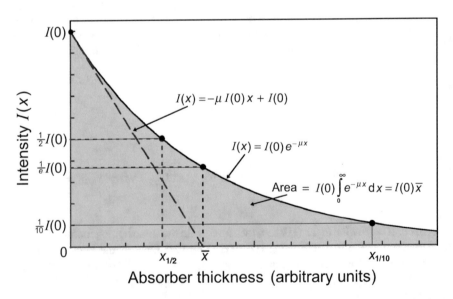

Fig. 7.3 Intensity $I(x)$ against absorber thickness x for monoenergetic photon beam. Half-value layer $x_{1/2}$, mean-free path $\bar{x}$, and tenth-value layer $x_{1/10}$ are identified

Table 7.1 Characteristic absorber thickness and its effects upon beam intensity attenuation

Absorber thickness	$\dfrac{I(x)}{I(0)}$	$100\dfrac{I(x)}{I(0)}$ (%)	Name	Symbol
$(\ln 2)/\mu$	0.500	50.0	Half-value layer	$\text{HVL} = x_{1/2}$
$1/\mu$	0.368	36.8	Mean free path	$\text{MFP} = \bar{x}$
$(\ln 10)/\mu$	0.100	10.0	Tenth-value layer	$\text{TVL} = x_{1/10}$
$3/\mu$	0.050	5.0	Three mean free paths	$3\bar{x}$
$5/\mu$	0.0067	~ 0.7	Five mean free paths	$5\bar{x}$
$7/\mu$	0.0009	~ 0.1	Seven mean free paths	$7\bar{x}$
$9/\mu$	0.00012	~ 0.012	Nine mean free paths	$9\bar{x}$

7.1.3 Other Attenuation Coefficients and Cross Sections

In addition to the *linear attenuation coefficient* μ, three other related attenuation coefficients (often referred to as cross sections) are in use for describing photon beam attenuation characteristics. They are: the mass attenuation coefficient μ_m, atomic attenuation coefficient $_a\mu$, and electronic attenuation coefficient $_e\mu$.

1. **Mass attenuation coefficient** μ_m is defined as the linear attenuation coefficient μ divided by the mass per unit volume of the absorber (absorber mass density) ρ. The mass attenuation coefficient $\mu_m = \mu/\rho$ is independent of absorber density and its SI unit is m^2/kg. The older unit cm^2/g is still often used $\left(1 m^2/kg = 10\,cm^2/g\right)$. When the mass attenuation coefficient is used in (7.3), the thickness of the absorber is expressed in kg/m^2 or g/cm^2 where $1\,kg/m^2 = 10^{-1}\,g/cm^2$.

2. **Atomic attenuation coefficient** $_a\mu$ is defined as linear attenuation coefficient μ divided by the number of atoms N_a per volume $\mathcal{V}$ of the absorber. It can also be defined as the mass attenuation coefficient μ/ρ divided by the number of atoms N_a per mass of the absorber. The SI unit of the atomic attenuation coefficient is $m^2/atom$; however, a smaller unit $cm^2/atom$ is still in common use $\left(1\,m^2/atom = 10^4\,cm^2/atom\right)$. When the atomic attenuation coefficient is used in (7.3), the thickness of the absorber is given in $atom/m^2$ or $atom/cm^2$ where $1\,atom/m^2 = 10^{-4}\,atom/cm^2$.

3. **Electronic attenuation coefficient** $_e\mu$ is defined as the linear attenuation coefficient μ divided by the number of electrons N_e per volume $\mathcal{V}$ of the absorber. It can also be defined as the mass attenuation coefficient μ/ρ divided by the number of electrons N_e per mass of the absorber. The SI unit of the electronic attenuation coefficient is $m^2/electron$; however, a smaller unit $cm^2/electron$ is still in common use. When the electronic attenuation coefficient is used in (7.3), the thickness of the absorber is given in $electron/m^2$ or $electron/cm^2$ where $1\,electron/m^2 = 10^{-4}\,electron/cm^2$.

Since the mass, atomic and electronic attenuation coefficients are measured in units of area per mass, area per atom, and area per electron, respectively, they are often referred to as cross sections following the nomenclature in common use

Table 7.2 Attenuation coefficients and cross sections used in photon attenuation studies

	Symbol	Relationship to μ	SI units	Common units
Linear attenuation coefficient	μ	μ	m^{-1}	cm^{-1}
Mass attenuation coefficient	μ_m	μ/ρ	m^2/kg	cm^2/g
Atomic cross section (attenuation coefficient)	$_a\mu$	$\mu/n^\square$	$m^2/atom$	$cm^2/atom$
Electronic cross section (attenuation coefficient)	$_e\mu$	$\mu/(Zn^\square)$	$m^2/electron$	$cm^2/electron$

in nuclear physics. Based on their definition, the relationship among the various attenuation coefficients or interaction cross sections can be expressed as follows

$$\mu = \rho\mu_m = n^\square {_a\mu} = Zn^\square {_e\mu} \tag{7.18}$$

where

ρ is the mass density of the absorber.

$n^\square$ is the number of atoms N_a per volume $\mathcal{V}$ of the absorber, i.e., $n^\square = N_a/\mathcal{V}$, and $N_a/\mathcal{V} = \rho N_a/m = \rho N_A/A$ with m the mass of the absorber, N_A the Avogadro number of atoms per mole, and A the atomic mass of the absorber in g/mol.

Z is the atomic number of the absorber.

$Zn^\square$ is the number of electrons per volume $\mathcal{V}$ of absorber. i.e., $Zn^\square = \rho Z N_A/A$.

Table 7.2 lists the various attenuation coefficients and cross sections, their relationship to the linear attenuation coefficient μ, their SI units, and their commonly used units.

7.1.4 Energy Transfer Coefficient and Energy Absorption Coefficient

In radiation dosimetry two energy-related coefficients are in use to account for:

1. **Mean energy transferred** from photons to charged particles (electrons and positrons) in a photon–atom interaction (linear energy transfer coefficient μ_{tr} and mass energy transfer coefficient μ_{tr}/ρ).

2. **Mean energy absorbed** in the medium (linear energy absorption coefficient μ_{ab} and mass energy absorption coefficient μ_{ab}/ρ). Note that the linear energy absorption coefficient is commonly labeled as μ_{en} in the literature; however, we use the subscript "ab" for energy absorbed, similarly to the common usage of the subscript "tr" for energy transferred.

Mean energy transferred $\overline{E}_{tr}$ and energy transfer coefficient μ_{tr} as well as mean energy absorbed $\overline{E}_{ab}$ and energy absorption coefficient μ_{ab} are discussed in more detail in Chap. 8.

Linear energy transfer coefficient μ_{tr} with SI unit m^{-1} and more commonly used unit cm^{-1}, is defined as

$$\mu_{tr} = \mu \frac{\overline{E}_{tr}}{h\nu} = \mu \overline{f}_{tr}, \tag{7.19}$$

where

μ is the linear attenuation coefficient dependent on the incident photon energy $h\nu$ and the absorber atomic number Z.

$\overline{E}_{tr}$ is the mean energy transferred from incident photon to kinetic energy of charged particles released or produced in the absorber.

$\overline{f}_{tr}$ is the total mean energy transfer fraction $\left[\overline{f}_{tr} = \overline{E}_{tr}/(h\nu)\right]$, i.e., the total mean fraction of the photon energy $h\nu$ transferred to kinetic energy of charged particles through various interactions between the incident photon and atoms of the absorber.

The mass energy transfer coefficient is defined as the ratio μ_{tr}/ρ where ρ is the mass density of the absorber. The common units of μ_{tr}/ρ are m^2/kg and cm^2/g, related as follows: 1 m^2/kg = 10 cm^2/g.

Linear energy absorption coefficient μ_{ab} with units m^{-1} and cm^{-1} is in a similar manner to (7.19) defined as

$$\mu_{ab} = \mu \frac{\overline{E}_{ab}}{h\nu} = \mu \overline{f}_{ab}, \tag{7.20}$$

where

μ is the linear attenuation coefficient dependent on incident photon energy $h\nu$ and the absorber atomic number Z.

$\overline{E}_{ab}$ is the mean energy transferred from the secondary charged particles to the absorber, i.e., the mean energy absorbed in the absorber.

$\overline{f}_{ab}$ is the total mean energy absorption fraction, i.e., the total mean fraction of the incident photon energy $h\nu$ absorbed in the absorber.

The mass energy absorption coefficient is defined as the ratio μ_{ab}/ρ with ρ the absorber mass density. The common units of μ_{ab}/ρ are m^2/kg and cm^2/g.

Mean energy transferred from photon to secondary charged particles in the absorber, $\overline{E}_{tr}$, can be expressed as a sum of two components: $\overline{E}_{ab}$ defined above and $\overline{E}_{rad}$, the mean energy lost by secondary charged particles and radiated from the secondary charged particles in the form of photons through bremsstrahlung and in-flight annihilation or emitted as fluorescence photons during atomic relaxation after impulse ionization or impulse excitation of absorber atoms.

$\overline{E}_{tr}$, the mean energy transferred from photon to secondary charged particles, is thus expressed as the following sum

$$\overline{E}_{tr} = \overline{E}_{ab} + \overline{E}_{rad}. \tag{7.21}$$

Combining (7.20) and (7.21) we now get the following expression for the mass energy absorption coefficient μ_{ab}/ρ

$$\frac{\mu_{ab}}{\rho} = \frac{\overline{E}_{tr} - \overline{E}_{rad}}{h\nu}\frac{\mu}{\rho} = \frac{\mu_{tr}}{\rho}\left(1 - \frac{\overline{E}_{rad}}{\overline{E}_{tr}}\right) = \frac{\mu_{tr}}{\rho}(1 - \overline{g})$$

$$= \frac{\mu}{\rho}\overline{f}_{tr}(1 - \overline{g}) = \frac{\mu}{\rho}\overline{f}_{ab}, \tag{7.22}$$

where $\overline{g}$ is the mean radiation fraction defined as that fraction of the mean kinetic energy $\overline{E}_{tr}$ which was transferred from the incident photon to charged particles and subsequently radiated as $\overline{E}_{rad}$ from the charged particles in the form of photons.

Mean radiation fraction $\overline{g}$ can thus be expressed as follows

$$\overline{g} = \frac{\overline{E}_{rad}}{\overline{E}_{tr}} = 1 - \frac{\overline{E}_{ab}}{\overline{E}_{tr}} = 1 - \frac{\overline{f}_{ab}}{\overline{f}_{tr}} = 1 - \frac{\dfrac{\mu_{ab}}{\rho}}{\dfrac{\mu_{tr}}{\rho}}. \tag{7.23}$$

As discussed in Sect. 7.10, the mean radiation fraction $\overline{g}$ is the mean value of the radiation yields $Y[(E_K)_0]$ for the spectrum of all electrons and positrons of various starting energies $(E_K)_0$ produced or released in the medium by primary photons.

7.1.5 Broad Beam Geometry

In contrast to the narrow beam geometry that is used in determination of the various attenuation coefficients and cross sections for photon beam attenuation, one can also deal with broad beam geometry in which the detector reading is not only diminished by attenuation of the primary photon beam in the absorber, but is also increased by the radiation scattered from the absorber into the detector. The geometry for a broad beam experiment on photon attenuation in an absorber is shown in Fig. 7.2b.

The signal $I_B(x)$ measured by the detector for an absorber thickness x in broad beam geometry is expressed as follows

$$\frac{I_B(x)}{I_N(x)} = B, \tag{7.24}$$

where

B is the so-called *build-up factor* that accounts for the secondary photons that are scattered from the absorber into the detector.

$I_N(x)$ is narrow beam geometry signal for absorber thickness x given in (7.6) as

$$I_N(x) = I_N(0)e^{-\mu x}, \tag{7.25}$$

with $I_N(0)$ the narrow beam geometry signal measured in the absence of the absorber. The buildup factor B is affected by photon beam energy as well as the geometry, absorber atomic number and thickness, and the quantity measured which can be photon fluence, photon energy fluence, beam intensity, beam exposure, kerma, or dose. For narrow beam geometry $B = 1$, for broad beam geometry B is positive and under certain conditions can amount to a factor of 10 or more. Since in broad beam attenuation photons interacting with the absorber may be scattered into the detector thereby contributing to the measured signal, the apparent attenuation is lower than that obtained under narrow beam conditions and results in an overestimation of the HVL of the beam.

An alternative concept to the buildup factor is the concept of the *mean effective attenuation coefficient* $\bar{\mu}_{\text{eff}}$ expressed, using (7.24) and (7.25), as follows

$$I_B(x) = I_N(x)B = I_N(0)Be^{-\mu x} = I_N(0)e^{-\bar{\mu}_{\text{eff}} x}. \tag{7.26}$$

From (7.26) we get the following expression for $\bar{\mu}_{\text{eff}}$

$$\bar{\mu}_{\text{eff}} = \mu - \frac{\ln B}{x}. \tag{7.27}$$

Broad beam geometry is used in radiation protection for design of treatment room shielding and in beam transport studies. When measuring attenuation coefficients, however, care must be taken to ensure that the build up factor B is unity, i.e., one must use narrow beam geometry, defined in Sect. 7.1.1.

7.1.6 Classification of Photon Interactions with Absorber Atoms

Photons with energy in the ionizing radiation category have several options for interacting with matter. The seven interactions of importance in medical physics and radiation dosimetry are summarized in Table 7.3. The specific interactions are classified as effects and many of them carry the name of their discoverer. The effects listed in Table 7.3 can be classified according to:

1. Type of target (orbital electron or nucleus), as shown in Table 7.4.
2. Type of interaction (photon disappearance or photon scattering), as shown in Table 7.5.
3. Type of particle released (electron or positron), as shown in Table 7.6.

Table 7.3 Most important photon interactions with atoms of the absorber

Interaction	Electronic attenuation coefficient (cross section)	Atomic attenuation coefficient (cross section)	Linear attenuation coefficient
Thomson scattering	$_e\sigma_{Th}$	$_a\sigma_{Th}$	σ_{Th}
Rayleigh scattering	–	$_a\sigma_R$	σ_R
Compton scattering	$_e\sigma_c$	$_a\sigma_c$	σ_c
Photoelectric effect	–	$_a\tau$	τ
Nuclear pair production	–	$_a\kappa_{pp}$	κ_p
Triplet production	$_e\kappa_{tp}$	$_a\kappa_{tp}$	κ_t
Photodisintegration	–	$_a\sigma_{PN}$	σ_{PN}

Table 7.4 Types of targets in photon interactions with atoms (orbital electrons or nuclei)

Photon–orbital electron interactions	Photon–nucleus interactions
With bound electrons	With nucleus directly
Photoelectric effect	Photodisintegration
Rayleigh scattering	
With "free electrons"	With Coulomb field of nucleus
Thomson scattering	Nuclear pair production
Rayleigh scattering	
With Coulomb field of electron	
Triplet productions	

Table 7.5 Types of photon–atom interactions (complete photon absorption or photon scattering)

Complete absorption of photon	Photon scattering
Photoelectric effect	Thomson scattering
Nuclear pair production	Rayleigh scattering
Triplet production	Compton scattering
Photodisintegration	

Table 7.6 Release and production of charged particles in photon interactions with absorber atoms

Result of interaction	Interaction event
1. No charged particles released or produced	Thomson scattering
	Rayleigh scattering
2. Only electrons released	Photoelectric effect
	Compton scattering
3. Electrons and positrons produced and released	Nuclear pair production
	Triplet production

As far as importance to medical physics and radiation dosimetry is concerned, photon interactions with atom of absorber are classified into four categories:

1. *Interactions of major importance*:

    Photoelectric effect.
    Compton scattering by "free" electron.
    Pair production in the field of nucleus (including triplet production).

2. *Interactions of moderate importance*:

    Rayleigh scattering.

3. *Interactions of minor importance*:

    Photonuclear reaction (also known as photonuclear effect).
    Thompson scattering by "free" electron.

4. *Negligible interactions*:

    Thomson scattering by the nucleus.
    Compton scattering by the nucleus.
    Meson production.
    Delbrück scattering (elastic scattering of photon by nuclear Coulomb field-virtual nuclear pair production).

7.2   Thomson Scattering

The scattering of low energy photons $\left(h\nu \ll m_e c^2\right)$ by loosely bound, i.e., essentially "free" electrons of an absorber, is described adequately by non-relativistic classical theory of *Joseph J. Thomson*. Thomson assumed that the incident photon beam set each quasi-free electron of the absorber atom into a forced resonant oscillation and then used classical theory to calculate the cross section for re-emission of the electromagnetic (EM) radiation as a result of the induced dipole oscillation of the electron. This type of photon scattering is now called Thomson scattering (elastic scattering) and implies that energy of scattered photon is equal to energy of incident photon.

The electric fields $\mathcal{E}_{in}$ for the harmonic incident radiation and $\mathcal{E}_{out}$ for the emitted scattered electromagnetic waves [far field, see (4.13)] are given, respectively, by

$$\mathcal{E}_{in} = \mathcal{E}_0 \sin \omega t \tag{7.28}$$

and

$$\mathcal{E}_{out} = \frac{e}{4\pi\varepsilon_0} \frac{\ddot{x} \, \sin \Theta}{c^2 r}, \tag{7.29}$$

where

$\mathcal{E}_0$ is the amplitude of the incident harmonic oscillation.
Θ is the angle between the direction of emission $\mathbf{r}$ and the polarization vector of the incident wave $\mathcal{E}_{in}$.
$\ddot{x}$ is the acceleration of the electron.

The equation of motion for the accelerated electron vibrating about its equilibrium position is

$$m_e\ddot{x} = e\mathcal{E} = e\mathcal{E}_0 \sin\omega t. \tag{7.30}$$

Inserting $\ddot{x}$ from (7.30), the equation of motion for the accelerated electron, into (7.29), we get the following expression for $\mathcal{E}_{out}$

$$\mathcal{E}_{out} = \frac{e^2}{4\pi\varepsilon_0}\frac{\mathcal{E}_0}{m_ec^2}\frac{\sin\omega t\,\sin\Theta}{r} = r_e\mathcal{E}_0\frac{\sin\omega t\,\sin\Theta}{r}, \tag{7.31}$$

where r_e is the classical radius of the electron defined as $r_e = e^2/(4\pi\varepsilon_0m_ec^2) = 2.82\,fm$.

7.2.1 Thomson Differential Electronic Cross Section per Unit Solid Angle

The differential electronic cross section $d_e\sigma_{Th}$ for re-emission of radiation into a solid angle $d\Omega$ in Thomson scattering is by definition given as follows

$$d_e\sigma_{Th} = \frac{\overline{S}_{out}}{\overline{S}_{in}}dA = \frac{\overline{S}_{out}}{\overline{S}_{in}}r^2d\Omega \qquad \text{or} \qquad \frac{d_e\sigma_{Th}}{d\Omega} = r^2\frac{\overline{S}_{out}}{\overline{S}_{in}}. \tag{7.32}$$

The incident and emitted wave intensities are expressed by the time averages of the corresponding Poynting vectors $\overline{S}_{out}$ and $\overline{S}_{in}$, respectively [see (4.17)]

$$\overline{S}_{in} = \varepsilon_0c\overline{\mathcal{E}_{in}^2} = \varepsilon_0c\mathcal{E}_0^2\overline{\sin^2\omega t} = \frac{1}{2}\varepsilon_0c\mathcal{E}_0^2 \tag{7.33}$$

and

$$\overline{S}_{out} = \varepsilon_0c\overline{\mathcal{E}_{out}^2} = \varepsilon_0c\frac{r_e^2\mathcal{E}_0^2\overline{\sin^2\omega t}\,\sin^2\Theta}{r^2} = \frac{\varepsilon_0cr_e^2\mathcal{E}_0^2}{2}\frac{\sin^2\Theta}{r^2}, \tag{7.34}$$

recognizing that $\overline{\sin^2\omega t} = \frac{1}{2}$.

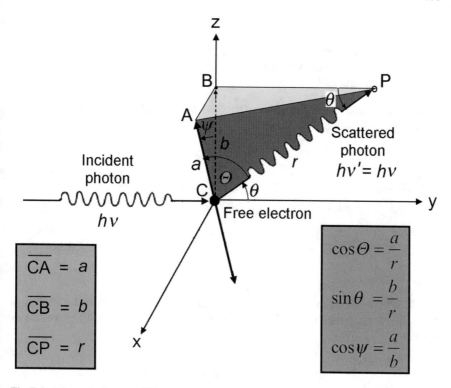

Fig. 7.4 Schematic diagram of Thomson scattering where the incident for unpolarized beam photon with energy $h\nu$ is scattered and emitted with a scattering angle θ. Note that for unpolarized photon beam angles θ and Θ are not coplanar (i.e., they are not in the same plane)

Inserting $\overline{S}_{\text{in}}$ and $\overline{S}_{\text{out}}$ into (7.32) we get the following expression for the differential electronic cross section per unit solid angle $d_e\sigma_{\text{Th}}/d\Omega$ for unpolarized radiation averaged over all possible polarization angles

$$\frac{d_e\sigma_{\text{Th}}}{d\Omega} = r_e^2\,\overline{\sin^2\Theta}. \qquad (7.35)$$

The mean value of $\sin^2\Theta$, i.e., $\overline{\sin^2\Theta}$, for unpolarized radiation may be evaluated using the following relationships

$$\cos\Theta = \frac{a}{r}; \qquad \sin\theta = \frac{b}{r}; \qquad \text{and} \qquad \cos\psi = \frac{a}{b}, \qquad (7.36)$$

where the angles θ, Θ, and ψ as well as the parameters a and b are defined in Fig. 7.4.
Combining expressions given in (7.36) we obtain

$$\cos\Theta = \sin\theta\,\cos\psi, \qquad (7.37)$$

where

θ is the scattering angle defined as the angle between the incident photon
              direction and the scattered photon direction, as shown in Fig. 7.4.

ψ is the polarization angle.

The mean value $\overline{\sin^2 \Theta}$ is now determined by integration over the polarization
angle ψ from 0 to 2π as follows

$$\overline{\sin^2 \Theta} = \frac{\int\limits_0^{2\pi} \sin^2 \Theta \, d\psi}{\int\limits_0^{2\pi} d\psi} = \frac{1}{2\pi} \int_0^{2\pi} \left(1 - \cos^2 \Theta\right) d\psi = 1 - \frac{\sin^2 \theta}{2\pi} \int_0^{2\pi} \cos^2 \psi \, d\psi$$

$$= 1 - \frac{\sin^2 \theta}{2\pi} \left\{ \frac{1}{2} \sin \psi \cos \psi + \frac{1}{2} \psi \right\}_0^{2\pi} = 1 - \frac{1}{2} \sin^2 \theta = \frac{1}{2} \left(1 + \cos^2 \theta\right).$$

$$(7.38)$$

The differential electronic cross section per unit solid angle for Thomson scattering
$d_e\sigma_{Th}/d\Omega$ for unpolarized incident radiation is from (7.35) and (7.38) expressed as
follows

$$\frac{d_e\sigma_{Th}}{d\Omega} = \frac{r_e^2}{2} \left(1 + \cos^2 \theta\right) \qquad \text{in } (cm^2/electron)/steradian, \qquad (7.39)$$

and drawn in Fig. 7.5 against the scattering angle θ in the range from 0 to π. The graph
in part (a) is plotted in the Cartesian coordinate system; that in part (b) shows the
same data in the polar coordinate system. Both graphs show that $d_e\sigma_{Th}/d\Omega$ ranges
from 39.7 mb/(electron·steradian) at $\theta = \pi/2$ to 79.4 mb/(electron·steradian) for
$\theta = 0°$ and $\theta = \pi$.

The differential electronic cross section per unit scattering angle for Thomson
scattering $d_e\sigma_{Th}/d\theta$ gives the fraction of the incident energy that is scattered into a
cone contained between θ and $\theta + d\theta$. The function, plotted in Fig. 7.6 against the
scattering angle θ, is expressed as follows, noting that $d\Omega = 2\pi \sin \theta \, d\theta$

$$\frac{d_e\sigma_{Th}}{d\theta} = \frac{d_e\sigma_{Th}}{d\Omega} \frac{d\Omega}{d\theta} = 2\pi \sin \theta \frac{d_e\sigma_{Th}}{d\Omega} = \pi r_e^2 \sin \theta (1 + \cos^2 \theta). \qquad (7.40)$$

As shown in Fig. 7.6, $d_e\sigma_{Th}/d\theta$ is zero at $\theta = 0$ and $\theta = 180°$, reaches maxima at
$\theta = 55°$ and $\theta = 125°$, and attains a non-zero minimum at $\theta = 90°$. The two maxima
and the non-zero minimum are determined after setting $d^2\sigma_{Th}/d\theta^2 = 0$ and solving
the result for θ.

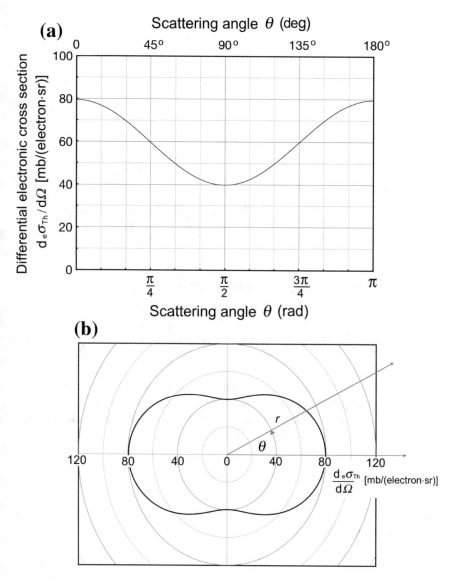

Fig. 7.5 Differential electronic cross section $d_e\sigma_{Th}/d\Omega$ per unit solid angle against the scattering angle θ for Thomson scattering, as given by (7.39). **a** Plotted in Cartesian coordinate system and **b** in polar coordinate system. The units shown are mb/(electron · steradian)

7.2.2 Thomson Total Electronic Cross Section

The total electronic cross section $_e\sigma_{Th}$ for Thomson scattering is obtained by determining the area under the $d_e\sigma_{Th}/d\theta$ curve of Fig. 7.6 or by integrating (7.40) over all scattering angles θ from 0 to π to obtain

Fig. 7.6 Differential electronic cross section $d_e\sigma_{Th}/d\theta$ per unit angle θ for Thomson scattering plotted against the scattering angle θ

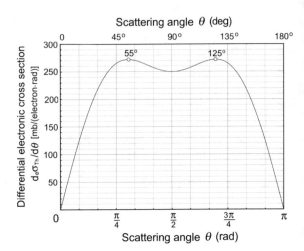

$$_e\sigma_{Th} = \int \frac{d_e\sigma_{Th}}{d\Omega} d\Omega = \frac{r_e^2}{2} \int_0^\pi \left(1 + \cos^2\theta\right) 2\pi \sin\theta \, d\theta$$

$$= -\pi r_e^2 \left\{ \int_0^\pi d\left(\cos\theta\right) + \int_0^\pi \cos^2\theta \, d\left(\cos\theta\right) \right\} = \frac{8\pi}{3} r_e^2 = 0.665 \, \text{b}. \quad (7.41)$$

This is a noteworthy result in that it contains no energy-dependent terms and predicts no change in energy upon re-emission of the electromagnetic radiation. The cross section $_e\sigma_{Th}$ is called the Thomson classical cross section for a free electron and has the same value (0.665 b) for all incident photon energies.

7.2.3 Thomson Total Atomic Cross Section

The atomic cross section for Thomson scattering $_a\sigma_{Th}$ is in terms of the electronic cross section $_e\sigma_{Th}$ given as

$$_a\sigma_{Th} = Z(_e\sigma_{Th}), \quad (7.42)$$

showing a linear dependence upon atomic number Z, as elucidated experimentally for low atomic number elements by *Charles Glover Barkla*, an English physicist who received the Nobel Prize in Physics for his discovery of characteristic x-rays.

For photon energies $h\nu$ exceeding the electron binding energy but small in comparison with the electron rest mass energy m_ec^2, i.e., $E_B \ll h\nu \ll m_ec^2$, the atomic cross section $_a\sigma_{Th}$ measured at small θ approaches the Thomson's value of (7.41). At larger θ and larger photon energies ($h\nu \to m_ec^2$); however, the Thomson classical theory breaks down and the intensity of coherently scattered radiation on free electrons diminishes in favor of incoherently Compton scattered radiation.

7.3 Incoherent Scattering (Compton Effect)

An interaction of a photon of energy $h\nu$ with a loosely bound orbital electron of an absorber is called Compton effect (Compton scattering) in honor of *Arthur Compton* who made the first measurements of photon-"free electron" scattering in 1922. The effect is also known as incoherent scattering. In theoretical studies of the Compton effect an assumption is made that the incident photon interacts with a free and stationary electron. A photon, referred to as a scattered photon with energy $h\nu'$ that is smaller than the incident photon energy $h\nu$, is produced in Compton effect and an electron, referred to as a Compton (recoil) electron, is ejected from the atom with kinetic energy E_K.

A typical Compton effect interaction is shown schematically in Fig. 7.7 for a 1 MeV photon scattered on a "free" (loosely bound) electron with a scattering angle $\theta = 60°$. The scattering angle θ is defined as the angle between the incident photon

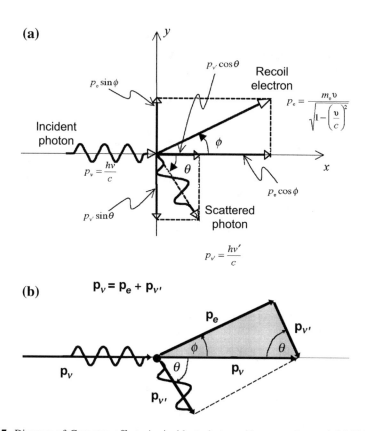

Fig. 7.7 Diagram of Compton effect. An incident photon with energy $h\nu = 1$ MeV interacts with a stationary and free electron. A photon with energy $h\nu'$ is produced and scattered with a scattering angle $\theta = 60°$. The difference between the incident photon energy $h\nu$ and the scattered photon energy $h\nu'$ is given as kinetic energy to the recoil electron. **a** Schematic diagram; **b** vector representation of the effect

direction and the scattered photon direction. It ranges from $\theta = 0°$ (forward scattering) through $90°$ (side scattering) to $\theta = 180°$ (back scattering). The recoil electron angle ϕ is the angle between the incident photon direction and the direction of the recoil Compton electron and it ranges from $\phi = 0$ to $\phi = 90°$.

7.3.1 Compton Wavelength-Shift Equation

The corpuscular nature of the photon is assumed and relativistic conservation of total energy and momentum laws are used in the derivation of the well-known Compton wavelength shift relationship

$$\Delta\lambda = \lambda' - \lambda = \lambda_C (1 - \cos\theta), \tag{7.43}$$

where

λ is the wavelength of the incident photon: $\lambda = 2\pi\hbar c/(h\nu)$.
λ' is the wavelength of the scattered photon: $\lambda' = 2\pi\hbar c/(h\nu')$.
$\Delta\lambda$ is the difference between the scattered and incident photon wavelength, i.e.,
         $\Delta\lambda = \lambda' - \lambda$.
λ_C is the so-called Compton wavelength of the electron defined as

$$\lambda_C = \frac{h}{m_e c} = \frac{2\pi\hbar c}{m_e c^2} = 0.0243 \,\text{Å}. \tag{7.44}$$

Equation (7.43) for the wavelength shift $\Delta\lambda$ can be derived using relativistic relationships for conservation of total energy and momentum in a Compton interaction between the incident photon and a "free" electron.

The *conservation of total energy* is expressed as

$$h\nu + m_e c^2 = h\nu' + E_e, \tag{7.45}$$

where

$m_e c^2$ is the rest energy of the recoil electron.
E_e is the total energy of the recoil electron.

Using (1.58) the total energy of the recoil (Compton) electron is

$$E_e = m_e c^2 + E_K \tag{7.46}$$

and (7.45) can be expressed as

$$h\nu + m_e c^2 = h\nu' + m_e c^2 + E_K \tag{7.47}$$

or

$$hv = hv' + E_K, \tag{7.48}$$

where E_K is the kinetic energy of the recoil electron. Using (1.70) the total recoil electron energy is

$$E_e = \sqrt{(m_e c^2)^2 + p_e^2 c^2} \tag{7.49}$$

and (7.45) can be expressed as

$$hv + m_e c^2 = hv' + \sqrt{(m_e c^2)^2 + p_e^2 c^2}, \tag{7.50}$$

where p_e is the momentum of the recoil electron. Equation (7.50) can be written as

$$(hv - hv') + m_e c^2 = \sqrt{(m_e c^2)^2 + p_e^2 c^2}, \tag{7.51}$$

and, after squaring (7.51), we get the following expression for p_e^2

$$p_e^2 = \frac{1}{c^2} \left\{ (hv)^2 + (hv')^2 - 2(hv)(hv') + 2m_e c^2 (hv - hv') \right\}. \tag{7.52}$$

As shown in Fig. 7.7, the *conservation of momentum* in the direction of the incident photon hv is expressed as

$$p_v = p_{v'} \cos\theta + p_e \cos\phi \quad \text{or} \quad p_e \cos\phi = p_v - p_{v'} \cos\theta \tag{7.53}$$

and in the direction normal to that of the incident photon hv it is expressed as

$$0 = -p_{v'} \sin\theta + p_e \sin\phi \quad \text{or} \quad p_e \sin\phi = p_{v'} \sin\theta, \tag{7.54}$$

where

p_v	is the momentum of the incident photon $p_v = hv/c$.
$p_{v'}$	is the momentum of the scattered photon $p_{v'} = hv'/c$.
p_e	is the momentum of the recoil (Compton) electron, as given in (1.60): $p_e = c^{-1}\sqrt{E_K(E_K + 2m_e c^2)}$.

A summary of parameters for the conservation of total energy and momentum in Compton effect is presented in Table 7.7 in terms of total energy and momentum before and after the Compton interaction.

Table 7.7 Summary of parameters for conservation of total energy and momentum in Compton effect. The incident photon direction coincides with the abscissa (x) axis of the Cartesian coordinate system, as shown in Fig. 7.7

Before Compton interaction	After Compton interaction
Total energy before interaction $h\nu + m_e c^2$	Total energy after interaction $h\nu' + E_K + m_e c^2$
Momentum before interaction (x axis) $\dfrac{h\nu}{c}$	Momentum after interaction (x axis) $\dfrac{h\nu'}{c}\cos\theta + p_e \cos\phi$
Momentum before interaction (y axis) 0	Momentum after interaction (y axis) $\dfrac{h\nu'}{c}\sin\theta - p_e \sin\phi$

We now continue with the derivation of the wavelength shift equation and square (7.53) and (7.54) to get, respectively

$$p_e^2 \cos^2\phi = \left(\frac{h\nu}{c}\right)^2 - 2\frac{h\nu}{c}\frac{h\nu'}{c}\cos\theta + \left(\frac{h\nu'}{c}\right)^2 \cos^2\theta \qquad (7.55)$$

and

$$p_e^2 \sin^2\phi = \left(\frac{h\nu'}{c}\right)^2 \sin^2\theta. \qquad (7.56)$$

Summation of (7.55) and (7.56) eliminates the recoil angle ϕ dependence and we get the following expression for p_e^2

$$p_e^2 = \frac{1}{c^2}\left\{(h\nu)^2 - 2(h\nu)(h\nu')\cos\theta + (h\nu')^2\right\}. \qquad (7.57)$$

Merging expressions (7.52) and (7.57) for p_e^2 results in

$$2(h\nu)(h\nu')\left(1 - \cos\theta\right) = 2m_e c^2(h\nu - h\nu'), \qquad (7.58)$$

that, after division with $2(h\nu)(h\nu')$ and incorporation of $\nu = c/\lambda$ and $\nu' = c/\lambda'$, results in the well known Compton wavelength shift equation for photon interaction with a "free" electron

$$\lambda' - \lambda = \Delta\lambda = \frac{h}{m_e c}(1 - \cos\theta) = \lambda_C(1 - \cos\theta), \qquad (7.59)$$

with λ_C the Compton wavelength of the electron defined in (7.44). Compton derived the wavelength shift equation which carries his name in 1923 and in 1927 he was awarded the Nobel Prize in Physics for his discovery.

The Compton wavelength shift equation (7.59) can also be derived by using a general relationship for p_e^2. In Fig. 7.7b we redraw the schematic diagram of Fig. 7.7a for the Compton effect following the general vector relationship for *conservation of momentum* expressed as

$$\mathbf{p}_\nu = \mathbf{p}_e + \mathbf{p}_{\nu'}. \tag{7.60}$$

Using the law of cosines in conjunction with Fig. 7.7b we now get the following relationship for p_e^2

$$p_e^2 = p_\nu^2 + p_{\nu'}^2 - 2p_\nu p_{\nu'} \cos\theta = \left(\frac{h\nu}{c}\right)^2 + \left(\frac{h\nu'}{c}\right)^2 - 2\frac{h\nu}{c}\frac{h\nu'}{c}\cos\theta. \tag{7.61}$$

Equating expressions for p_e^2 given in (7.52) and (7.61) and using $c = \lambda\nu$ we get the following expression relating the wavelength of the incident photon λ, the wavelength of the scattered photon λ', and the scattering angle θ

$$\frac{1}{\lambda}\frac{1}{\lambda'}\cos\theta = \frac{1}{\lambda}\frac{1}{\lambda'} - \frac{m_e c^2}{h}\left[\frac{1}{\lambda} - \frac{1}{\lambda'}\right], \tag{7.62}$$

that, after multiplication with $\lambda\lambda'$, gives the well known Compton equation for the wavelength shift $\Delta\lambda = \lambda' - \lambda$ in a Compton interaction given in (7.59). The change in wavelength λ governed by the Compton wavelength shift expression given in (7.59) depends only on the scattering angle θ and does not depend on incident photon energy $h\nu$.

7.3.2 Relationship Between Scattering Angle and Recoil Angle

The scattering angle θ and the Compton electron recoil angle ϕ, as defined in Fig. 7.7, are related to one another and their explicit relationship can be derived from the momentum conservation expressions given in (7.53) and (7.54). We first express (7.54) as

$$p_e = p_{\nu'}\frac{\sin\theta}{\sin\phi} \tag{7.63}$$

and insert the resulting (7.63) into (7.53) to get

$$\frac{p_\nu}{p_{\nu'}} = \cos\theta + \sin\theta\cot\phi. \tag{7.64}$$

The ratio $p_\nu / p_{\nu'}$ of (7.64) is from (7.58) equal to

$$\frac{p_\nu}{p_{\nu'}} = \frac{h\nu/c}{h\nu'/c} = 1 + \varepsilon \left(1 - \cos\theta\right), \qquad (7.65)$$

where ε is defined as the incident photon energy $h\nu$ normalized to electron rest mass energy $m_e c^2 = 0.511\,\text{keV}$

$$\varepsilon = \frac{h\nu}{m_e c^2}. \qquad (7.66)$$

Merging (7.64) and (7.65) we get the following relationships between θ and ϕ

$$\cot\phi = \frac{(1+\varepsilon)(1-\cos\theta)}{\sin\theta} = (1+\varepsilon)\,\frac{1 - \cos^2\dfrac{\theta}{2} + \sin^2\dfrac{\theta}{2}}{2\sin\dfrac{\theta}{2}\cos\dfrac{\theta}{2}} = (1+\varepsilon)\tan\frac{\theta}{2}. \qquad (7.67)$$

Recognizing that $\cot\phi = 1/\tan\phi$, we can write (7.67) as

$$\tan\phi = \frac{1}{1+\varepsilon}\cot\frac{\theta}{2} \qquad \text{or} \qquad \cot\frac{\theta}{2} = (1+\varepsilon)\tan\phi. \qquad (7.68)$$

The ϕ versus θ relationships (7.67) and (7.68), plotted in Fig. 7.8 for various values of $\varepsilon = h\nu/(m_e c^2)$, show that for a given θ, the higher is the incident photon energy $h\nu$ or the higher is ε, the smaller is the recoil electron angle ϕ. Figure 7.8 also shows that the range of the scattering angle θ is from 0 to π, while the corresponding range of the recoil electron angle ϕ is limited from $\phi = \pi/2$ for $\theta = 0$ to $\phi = 0$ for

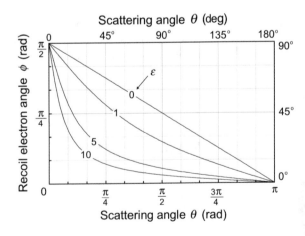

Fig. 7.8 Relationships (7.67) and (7.68) between the electron recoil angle ϕ and photon scattering angle θ

$\theta = \pi$, respectively. The Compton electron recoil angle ϕ is thus confined to the forward hemisphere with respect to the direction of the incident photon, while the photon scattering angle θ ranges between $0°$ for forward scattering through $\frac{1}{2}\pi$ for side scattering to π for backscattering.

7.3.3 Scattered Photon Energy as Function of Incident Photon Energy and Photon Scattering Angle

The Compton wavelength shift equation for $\Delta\lambda$ given in (7.59) leads to a relationship for the energy of the scattered photon $h\nu'$ as a function of the incident photon energy $h\nu$ and scattering angle θ

$$\Delta\lambda = \lambda' - \lambda = \frac{c}{\nu'} - \frac{c}{\nu} = \frac{h}{m_e c}(1 - \cos\theta) = \frac{2h}{m_e c}\sin^2\frac{\theta}{2} \tag{7.69}$$

or

$$\frac{1}{h\nu'} - \frac{1}{h\nu} = \frac{1}{m_e c^2}(1 - \cos\theta) = \frac{2}{m_e c^2}\sin^2\frac{\theta}{2}, \tag{7.70}$$

where we used the trigonometric relationship $\sin^2(\theta/2) = (1 - \cos\theta)/2$. From (7.70) we obtain the following expression for $h\nu'$

$$h\nu' = h\nu\frac{1}{1 + \varepsilon(1 - \cos\theta)} = h\nu\frac{1}{1 + 2\varepsilon\sin^2\frac{\theta}{2}} = h\nu\frac{\dfrac{1}{\sin^2\dfrac{\theta}{2}}}{\dfrac{1}{\sin^2\dfrac{\theta}{2}} + 2\varepsilon}, \tag{7.71}$$

where $\varepsilon = h\nu/(m_e c^2)$, as defined in (7.66).

One can also express the scattered photon energy $h\nu'$ as a function of the electron recoil angle ϕ. From the θ vs. ϕ relationship (7.67) we express $1/\sin^2(\theta/2)$ as a function of ϕ

$$\frac{1}{\sin^2\dfrac{\theta}{2}} = 1 + (1 + \varepsilon)^2\frac{1 - \cos^2\phi}{\cos^2\phi} \tag{7.72}$$

and insert it into (7.71) to get

$$h\nu' = h\nu\frac{(1 + \varepsilon)^2 - \varepsilon(\varepsilon + 2)\cos^2\phi}{(1 + \varepsilon)^2 - \varepsilon^2\cos^2\phi}. \tag{7.73}$$

Fig. 7.9 Scattered photon
energy $h\nu'$ against the
incident photon energy $h\nu$
for various scattering angles
θ in the range from $0°$ to
$180°$

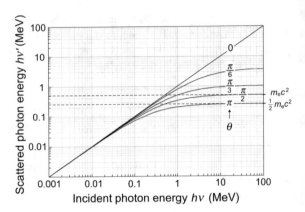

The relationship between $h\nu'$ and $h\nu$ of (7.71) is plotted in Fig. 7.9 for various
scattering angles θ between $0°$ and π (backscattering). The following conclusions
can be made:

1. For $\theta = 0$ corresponding to $\phi = \frac{1}{2}\pi$, the energy of the scattered photon $h\nu'$
   equals the energy of the incident photon $h\nu$, irrespective of $h\nu$. Since in this
   case no energy is transferred to the recoil electron, we are dealing with classical
   Thomson scattering (see Sect. 7.2).
2. For $\theta > 0$, the energy of the scattered photon saturates at high values of $h\nu$;
   the larger is the scattering angle θ, the lower is the saturation value of $h\nu'$ for
   $h\nu \to \infty$.
3. For $\theta = \frac{1}{2}\pi$ corresponding to $\phi = \cot^{-1}(1+\varepsilon)$, the scattered photon energy $h\nu'$,
   after inserting $\theta = \pi/2$ into (7.71) or $\phi = \cot^{-1}(1+\varepsilon)$ into (7.73), is given as

$$h\nu' = \frac{h\nu}{1+\varepsilon}, \tag{7.74}$$

with the following saturation energy $h\nu'_{\text{sat}}$ for $h\nu \to \infty$

$$h\nu'_{\text{sat}}\left[\theta = \frac{\pi}{2}, \phi = \cot^{-1}(1+\varepsilon)\right] = \lim_{h\nu \to \infty} \frac{h\nu}{1+\varepsilon} = \lim_{h\nu \to \infty} \frac{h\nu}{1 + \dfrac{h\nu}{m_e c^2}}$$

$$= \lim_{h\nu \to \infty} \frac{1}{\dfrac{1}{h\nu} + \dfrac{1}{m_e c^2}} = m_e c^2 = 0.511 \, \text{MeV}. \tag{7.75}$$

4. For $\theta = \pi$ corresponding to $\phi = 0$, the scattered photon energy $h\nu'$, after inserting
   $\theta = \pi$ into (7.71) or $\phi = 0$ into (7.73), is given as

$$h\nu' = \frac{h\nu}{1+2\varepsilon}, \tag{7.76}$$

with the following saturation energy $h\nu'_{sat}$ for $h\nu \to \infty$

$$h\nu'_{sat}(\theta = \pi) = \lim_{h\nu \to \infty} \frac{h\nu}{1 + 2\varepsilon} = \lim_{h\nu \to \infty} \frac{h\nu}{1 + \dfrac{2h\nu}{m_ec^2}}$$

$$= \lim_{h\nu \to \infty} \frac{1}{\dfrac{1}{h\nu} + \dfrac{2}{m_ec^2}} = \frac{m_ec^2}{2} = 0.255 \text{ MeV}. \qquad (7.77)$$

5. Results in points (3) and (4) above show that photon scattered with angles θ larger than $\frac{1}{2}\pi$ cannot exceed 511 keV in kinetic energy no matter how high is the incident photon energy $h\nu$. This finding is of practical importance in design of shielding barriers for linear accelerator installations. We also note from (7.77) that the maximum energy of the backscattered photon ($\theta = \pi$) cannot exceed $0.255 \text{ MeV} = \frac{1}{2}m_ec^2$ no matter how high is the incident photon energy $h\nu$.

6. For a given $h\nu$ the scattered photon energy $h\nu'$ will be in the range between $h\nu/(1 + 2\varepsilon)$ for $\theta = \pi$ corresponding to $\phi = 0$ and $h\nu$ for $\theta = 0$ corresponding to $\phi = \frac{1}{2}\pi$, i.e.,

$$\left.\frac{h\nu}{1 + 2\varepsilon}\right|_{\theta=\pi} \le h\nu' \le h\nu|_{\theta=0}. \qquad (7.78)$$

We now define the Compton scatter fraction $f'_C(h\nu, \theta)$ as the ratio between the scattered photon energy $h\nu'$ to the incident photon energy $h\nu$. From (7.71) we note that $f'_C(h\nu, \theta) = h\nu'/(h\nu) = [1 + \varepsilon(1 - \cos\theta)]^{-1}$ and plot the expression with dotted curves in Fig. 7.10 against the scattering angle θ for various incident photon energies $h\nu$ in the range from 10 keV to 100 MeV. The following features are notable:

1. As evident from Fig. 7.9, $h\nu = h\nu'$ for all $h\nu$ at $\theta = 0$ (forward scattering) and this then gives $f'_C(h\nu, \theta)|_{\theta=0} = 1$ for all $h\nu$ from 0 to ∞.
2. For a given $h\nu$, as θ increases, $f'_C(h\nu, \theta)$ gradually decreases from $f'_C(h\nu, \theta)|_{\theta=0} = 1$ and levels off at $f'_C(h\nu, \theta)|_{\theta=\pi} = 1/(1 + 2\varepsilon)$ with $\varepsilon = h\nu/(m_ec^2)$.
3. For a given scattering angle θ, the larger is the incident photon energy $h\nu$, the smaller is the Compton scatter fraction $f'_C(h\nu, \theta)$.

7.3.4 Energy Transfer to Compton Recoil Electron

Kinetic energy of the Compton (recoil) electron $E_K^C(h\nu, \theta)$ depends on photon energy $h\nu$ and photon scattering angle θ. The relationship is determined using conservation of energy expressed in (7.48) to get

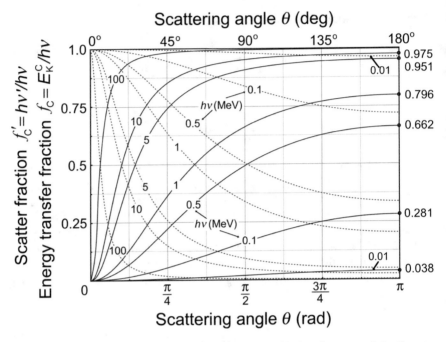

Fig. 7.10 The Compton scatter fraction $f'_C(h\nu, \theta)$ shown with *dotted curves* and the Compton energy transfer fraction $f_C(h\nu, \theta)$ shown with *solid curves* against the scattering angle θ for various incident photon energies $h\nu$ in the range from 10 keV to 100 MeV. Numbers on the vertical scale at $\theta = \pi$ indicate the maximum value of $f_c(h\nu, \theta)$ at $\theta = \pi$

$$
E_K^C(h\nu, \theta) = h\nu - h\nu' = h\nu - h\nu \frac{1}{1 + \varepsilon(1 - \cos\theta)}
$$

$$
= h\nu \frac{\varepsilon(1 - \cos\theta)}{1 + \varepsilon(1 - \cos\theta)} = h\nu \frac{2\varepsilon \sin^2 \dfrac{\theta}{2}}{1 + 2\varepsilon \sin^2 \dfrac{\theta}{2}}, \tag{7.79}
$$

where the scattered photon energy $h\nu'$ as a function of the incident photon energy $h\nu$ is given in (7.71).

The recoil electron kinetic energy of (7.79) can also be expressed as a function of electron recoil angle ϕ similarly to the derivation of the ϕ dependence of the scattered photon energy $h\nu'$ of (7.73). We insert the expression (7.72) for $1/\sin^2(\theta/2)$ into (7.79) and get

$$
E_K^C(h\nu, \phi) = h\nu \frac{2\varepsilon \cos^2 \phi}{(1 + \varepsilon)^2 - \varepsilon^2 \cos^2 \phi}. \tag{7.80}
$$

For a given photon energy $h\nu$ the recoil electron kinetic energy ranges from a minimum value of $(E_K^C)_{min} = 0$ for scattering angle $\theta = 0$ (forward scattering) corresponding to electron recoil angle $\phi = \frac{1}{2}\pi$ (see Fig. 7.8) to a maximum value of

$$(E_K^C)_{max} = E_K^C(h\nu, \theta = \pi) = h\nu\frac{2\varepsilon}{1 + 2\varepsilon} \qquad (7.81)$$

for scattering angle $\theta = \pi$ (backscattering) corresponding to electron recoil angle $\phi = 0$.

The ratio of the kinetic energy of the Compton (recoil) electron $E_K^C(h\nu, \theta)$ to the energy of the incident photon $h\nu$ represents the fraction of the incident photon energy that is transferred to the Compton electron in a Compton effect and is called the Compton energy transfer fraction $f_C(h\nu, \theta)$. Using (7.79) the Compton energy transfer fraction $f_C(h\nu, \theta)$ is expressed as follows

$$f_C(h\nu, \theta) = \frac{E_K^C(h\nu, \theta)}{h\nu} = \frac{\varepsilon(1 - \cos\theta)}{1 + \varepsilon(1 - \cos\theta)}, \qquad (7.82)$$

with the maximum value of the energy transfer fraction $(f_C)_{max}$ given for $\theta = \pi$ as

$$(f_C)_{max} = f_C(h\nu, \theta)|_{\theta=\pi} = \frac{2\varepsilon}{1 + 2\varepsilon}. \qquad (7.83)$$

Figure 7.10 shows, with solid curves, a plot of $f_C(h\nu, \theta)$ against scattering angle θ for various incident photon energies in the range from 10 keV to 100 MeV. The following features are notable:

1. For all $h\nu$, as a result of $h\nu'|_{\theta=0} = h\nu$, the Compton energy transfer fraction $f_C(h\nu, \theta)|_{\theta=0} = 0$.
2. For a given $h\nu$, as the angle θ increases from 0, the energy transfer fraction $f_C(h\nu, \theta)$ increases from 0 and saturates at $2\varepsilon/(1 + 2\varepsilon)$ of (7.83) for $\theta = \pi$ (back-scattering). The values of $(f_C)_{max}$ for a given incident photon energy $h\nu$ are indicated on the graph of Fig. 7.10 at $\theta = \pi$.
3. As a function of $h\nu$, $(f_C)_{max}$ is proportional to $h\nu$ amounting to 0.038 for $h\nu = 0.01$ MeV, 0.281 for $h\nu = 0.1$ MeV, 0.796 for $h\nu = 1$ MeV, 0.975 for $h\nu = 10$ MeV, and 0.997 for $h\nu = 100$ MeV.
4. For a given θ, the larger is the incident photon energy $h\nu$, the larger is the Compton energy transfer fraction $f_C(h\nu, \theta)$.
5. For a given $h\nu$ and a given θ, the sum of the Compton scatter fraction $f_C'(h\nu, \theta)$ and the Compton energy transfer fraction $f_C(h\nu, \theta)$ equals to 1, i.e., $f_C'(h\nu, \theta) + f_C(h\nu, \theta) = 1$.
6. As shown in (7.59), the Compton shift in wavelength $\Delta\lambda$ is independent of the energy of the incident photon $h\nu$. The Compton shift in energy, on the other hand, depends strongly on the incident photon energy $h\nu$. Low-energy photons are scattered with minimal change in energy, while high-energy photons suffer a very large change in energy. The shift in photon energy $h\nu - h\nu'$ is equal to the kinetic energy E_K^C transferred to the Compton recoil electron.

The expression of (7.81) which gives the relationship between the incident photon energy $h\nu$ and the maximum recoil energy of the electron $(E_K^C)_{max}$ can be solved for $h\nu$ after inserting $\varepsilon = h\nu/(m_e c^2)$ to obtain the following quadratic equation for $h\nu$

$$(h\nu)^2 - (E_K^C)_{\text{max}} h\nu - \frac{1}{2}(E_K^C)_{\text{max}} m_e c^2 = 0, \tag{7.84}$$

that has the following physically relevant solution ($h\nu \geq 0$)

$$h\nu = \frac{1}{2}(E_K^C)_{\text{max}} \left\{ 1 + \sqrt{1 + \frac{2m_e c^2}{(E_K^C)_{\text{max}}}} \right\}. \tag{7.85}$$

In Fig. (7.11) we plot, against incident photon energy $h\nu$ the minimum, mean, and maximum Compton recoil electron energies as the minimum, mean, and maximum energy transfer fractions, respectively, obtained by dividing the energies E_K^C with the appropriate photon energy $h\nu$. From the dosimetric point of view, the most important relationship plotted in Fig. 7.11 is the plot of $\overline{f}_C = \overline{E}_K^C/(h\nu)$, the mean fraction of the incident photon energy $h\nu$ transferred to recoil electrons, also referred to as the mean energy transfer fraction for the Compton effect $\overline{f}_C$. It represents the energy transferred from photon to Compton recoil electron averaged over all scattering angles θ.

Data for $\overline{f}_C = \overline{E}_K^C/(h\nu)$ are summarized in Table 7.8 in bold face, showing that the fractional energy transfer to recoil electrons is quite low at low photon

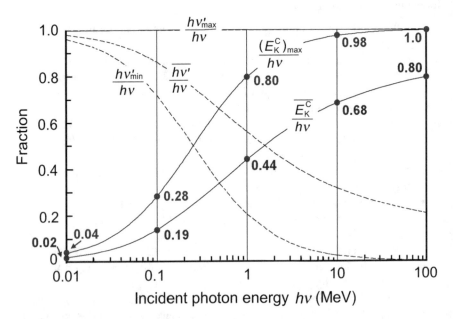

Fig. 7.11 Fraction of incident photon energy $h\nu$ transferred in Compton effect to: Maximum energy of recoil electron: $(E_K^C)_{\text{max}}/(h\nu)$; $\theta = \pi$ [see (7.81)]. Mean energy of recoil electron: $\overline{E}_K^C/(h\nu)$ [see (7.113) below]. Maximum energy of scattered photon: $h\nu'_{\text{max}}/(h\nu)$; $\theta = 0°$ [see (7.86)]. Mean energy of the scattered photon: $h\overline{\nu}'/(h\nu)$ [see (7.87)]. Minimum energy of the scattered photon: $h\nu'_{\text{min}}/(h\nu)$; $\theta = \pi$ [see (7.86)]

Table 7.8 Fractions of the incident photon energy transferred through Compton effect to the maximum electron kinetic energy $\left(E_K^C\right)_{max} / (h\nu)$; mean electron kinetic energy $(\overline{E}_K^C)/ (h\nu)$; maximum scattered photon energy $h\nu'_{max}/ (h\nu)$; mean scattered photon energy $h\overline{\nu}'/ (h\nu)$; and minimum scattered photon energy $h\nu'_{min}/ (h\nu)$

$h\nu$ (MeV)	0.01	0.1	1.0	10.0	100.0
$\varepsilon = h\nu/(m_e^2 c^2)$	0.0196	0.1956	1.956	19.56	195.6
$\left(E_K^C\right)_{max} / (h\nu)$	0.038	0.282	0.796	0.975	0.998
$\overline{f}_C = (\overline{E}_K^C)/ (h\nu)$	**0.019**	**0.139**	**0.440**	**0.684**	**0.796**
$h\nu'_{max}/ (h\nu)$	1.0	1.0	1.0	1.0	1.0
$h\overline{\nu}'/ (h\nu)$	0.981	0.861	0.560	0.316	0.204
$h\nu'_{min}/ (h\nu)$	0.962	0.718	0.204	0.025	0.002

energies ($\overline{f}_C = 0.019$ at $h\nu = 0.01$ MeV) and then slowly rises through $\overline{f}_C = 0.138$ at $h\nu = 0.1$ MeV and $\overline{f}_C = 0.440$ at $h\nu = 1$ MeV to become $\overline{f}_C = 0.684$ at $h\nu = 10$ MeV and $\overline{f}_C = 0.796$ at $h\nu = 100$ MeV. The mean energy transfer fraction $\overline{f}_C = \overline{E}_K^C/(h\nu)$ is discussed further in Sect. 8.2.1.

Figure 7.11 and Table 7.8 also show the *maximum*, *mean*, and *minimum fractions* ($h\nu'_{max}/h\nu$, $h\overline{\nu}'/h\nu$, and $h\nu'_{min}/h\nu$, respectively) of the incident photon energy $h\nu$ given to the scattered photon. The fractions are calculated as follows

$$(f'_C)_{max} = \frac{h\nu'_{max}}{h\nu} = \frac{h\nu'|_{\theta=0}}{h\nu} = 1, \tag{7.86}$$

$$(\overline{f}'_C) = \frac{h\nu'}{h\nu} = 1 - \frac{\overline{E}_K^C}{h\nu} = 1 - \overline{f}_C, \tag{7.87}$$

$$(f'_C)_{min} = \frac{h\nu'_{min}}{h\nu} = \frac{h\nu'|_{\theta=\pi}}{h\nu} = \frac{1}{1+2\varepsilon} = 1 - \frac{(E_K^C)_{max}}{h\nu} = 1 - (f_C)_{max}, \tag{7.88}$$

where $\varepsilon = h\nu/(m_e c^2)$ and $(f_C)_{max}$ was given in (7.83).

7.3.5 Differential Electronic Cross Section for Compton Scattering

The probability or cross section for a Compton interaction between a photon and a "free electron" per unit solid angle is given by an expression derived by *Oskar Klein* and *Yoshio Nishina* in 1928. The differential Klein–Nishina electronic cross section per unit solid angle for Compton effect $d_e\sigma_C^{KN}/d\Omega$ is given in $\left[(cm^2/electron) /steradian\right]$ or in $\left[(m^2/electron) /steradian\right]$ as follows

$$\frac{d_e \sigma_C^{KN}}{d\Omega} = \frac{r_e^2}{2} \left(\frac{\nu'}{\nu}\right)^2 \left\{\frac{\nu'}{\nu} + \frac{\nu}{\nu'} - \sin^2 \theta\right\} = \frac{r_e^2}{2}(1 + \cos^2 \theta) F_{KN} = \frac{d_e \sigma_{Th}}{d\Omega} F_{KN},$$

(7.89)

where

ν	is the frequency of the incident photon.
ν'	is the frequency of the scattered photon.
θ	is the scattering angle of the photon.
r_e	is the classical radius of the electron (2.82 fm).
$F_{KN}(h\nu, \theta)$	is the Klein–Nishina form factor, dependent on incident photon energy $h\nu$ and photon scattering angle θ.
$d_e \sigma_{Th}/d\Omega$	is the differential electronic cross section per unit solid angle for Thomson scattering given in (7.39).

The Klein–Nishina form factor $F_{KN}(h\nu, \theta)$ for a free electron is given as follows

$$F_{KN}(h\nu, \theta) = \frac{1}{[1 + \varepsilon(1 - \cos\theta)]^2}\left\{1 + \frac{\varepsilon^2(1 - \cos\theta)^2}{[1 + \varepsilon(1 - \cos\theta)](1 + \cos^2\theta)}\right\},$$

(7.90)

where $\varepsilon = h\nu/(m_e c^2)$.

The Klein–Nishina form factor F_{KN} is plotted in Fig. 7.12 against the scattering angle θ for various values of the energy parameter ε. For $\varepsilon = 0$ the form factor equals 1 irrespective of the scattering angle θ. As shown in (7.90) and in Fig. 7.12, the form factor F_{KN} is a complicated function of the scattering angle θ and parameter ε. However, it is easy to see that:

1. $F_{KN} \leq 1$ for all θ and ε.   (7.91)

2. $F_{KN} = 1$ for $\theta = 0$ at any ε.   (7.92)

3. $F_{KN} = 1$ for $\varepsilon = 0$ at any θ (Thomson scattering).   (7.93)

From (7.89) we see that the differential electronic cross section for the Compton effect $d_e \sigma_C^{KN}/d\Omega$ when $F_{KN} = 1$ is equal to the Thomson differential cross section $d_e \sigma_{Th}/d\Omega$ of (7.39)

$$\frac{d_e \sigma_C^{KN}}{d\Omega}\bigg|_{F_{KN}=1} = \frac{d_e \sigma_{Th}}{d\Omega} = \frac{r_e^2}{2}(1 + \cos^2 \theta).$$

(7.94)

This result also follows directly from the first part of (7.89) after inserting the relationship $h\nu' = h\nu$ for elastic (Thomson) scattering.

The differential Compton electronic cross section $d_e \sigma_C^{KN}/d\Omega$ of (7.89) is plotted in Fig. 7.13 against the scattering angle θ for various values of ε ranging from $\varepsilon \approx 0$ which results in $F_{KN} = 1$ for all θ (i.e., Thomson scattering) to $\varepsilon = 10$ for which the

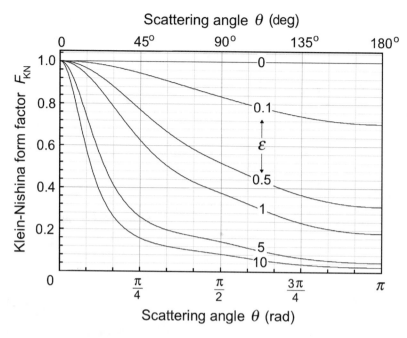

Fig. 7.12 Klein–Nishina atomic form factor for Compton effect F_{KN} against scattering angle θ

F_{KN} causes a significant deviation from the Thomson cross section for all angles θ except for $\theta = 0$.

The data are plotted in Cartesian coordinate system in Fig. 7.13a and in polar coordinate system in Fig. 7.13b to give a better illustration of the Compton scattering phenomenon. The following features are notable:

1. At low ε the probabilities for forward scattering ($\theta = 0$) and backscattering ($\theta = \pi$) are equal and amount to 79.4 mb (Thomson scattering), twice as large as the probability for side scattering ($\theta = \frac{1}{2}\pi$).
2. As energy $h\nu$ increases, the scattering becomes increasingly more forward peaked and backscattering rapidly diminishes. Thus, the probability for backscattering decreases but the probability for forward scattering remains constant at 79.4 mb (Thomson limit).
3. The polar diagram of Fig. 7.13 is sometimes colloquially referred to as the "peanut diagram" to help students remember its shape.

7.3.6 Differential Electronic Cross Section per Unit Scattering Angle

The directional distribution of the Compton scattered photons and recoil electrons is of significant theoretical and practical interest. The differential scattering cross

Fig. 7.13 Differential electronic cross section for Compton effect $d_e\sigma_C^{KN}/d\Omega$ against scattering angle θ for various values of $\varepsilon = h\nu/(m_e c^2)$, as given by (7.89). The differential electronic cross section for Compton effect $d_e\sigma_C^{KN}/d\Omega$ for $\varepsilon = 0$ is equal to the differential electronic cross section for Thomson scattering $d_e\sigma_{Th}/d\Omega$ (see Fig. 7.5). **a** Displays the data in Cartesian coordinate system; **b** in polar coordinate system

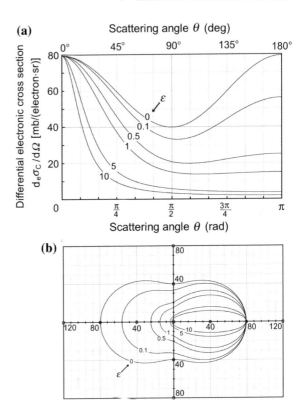

section per unit solid angle $d_e\sigma_C^{KN}/d\Omega$ is given in (7.89) and plotted in Fig. 7.13 for various incident photon energies ε; however, it is also important to consider the directional distribution of scattered photons and recoil electrons in the form of the cross section per unit scattering angle θ and per unit recoil angle ϕ.

The differential electronic cross section per unit scattering angle $d_e\sigma_C^{KN}/d\theta$ is obtained from the differential electronic cross section per unit solid angle $d_e\sigma_C^{KN}/d\Omega$ of (7.89) by recognizing that $d\Omega = 2\pi \sin\theta\, d\theta$

$$\frac{d_e\sigma_C^{KN}}{d\theta} = \frac{d_e\sigma_C^{KN}}{d\Omega}\frac{d\Omega}{d\theta} = \pi r_e^2 F_{KN}(1 + \cos^2\theta)\sin\theta. \qquad (7.95)$$

The differential cross section $d_e\sigma_C^{KN}/d\theta$ is plotted against scattering angle θ in Fig. 7.14 for four values of ε ranging from $\varepsilon = 0$ to $\varepsilon = 10$. The curve for $\varepsilon = 0$ approximates the Thomson cross section discussed in Sect. 7.2, encompasses the largest area, and, similarly to the Thomson cross section, exhibits two maxima (one at $\theta = 55°$ and the other at $\theta = 125°$) and a non-zero minimum at $\theta = 90°$. With increasing ε the area under the $d_e\sigma_C^{KN}/d\theta$ curve diminishes, becomes increasingly asymmetrical, and exhibits a single maximum at increasingly smaller angles θ. Note that the recoil electron data appear only in the forward hemisphere because the electron recoil angle ϕ ranges from $0°$ to $90°$.

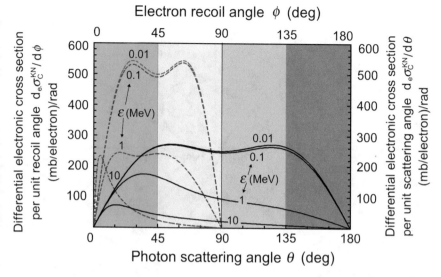

Fig. 7.14 Differential electronic cross section per unit scattering angle θ for Compton effect $d_e\sigma_C^{KN}/d\theta$ against scattering angle θ (solid curves) and differential electronic cross sections per unit recoil angle ϕ for Compton effect $d_e\sigma_C^{KN}/d\phi$ against recoil angle ϕ (dashed curves) for four values of ε (0, 0.1, 1, and 10), the incident photon energy $h\nu$ normalized to the rest mass energy of the electron m_ec^2. The cross sections are drawn on a Cartesian plot

For better visualization, the data of Fig. 7.14 are re-plotted with solid curves in a polar diagram in Fig. 7.15. The increased forward scattering of photons with increase in ε is clearly visible.

7.3.7 Differential Electronic Cross Section per Unit Recoil Angle

For each photon which is scattered with an angle θ, there is an electron which recoils with an angle ϕ. The two angles are related through (7.68) and (7.72), and shown schematically in Fig. 7.7. The differential electronic cross section per unit recoil angle $d_e\sigma_C/d\phi$ is determined from the differential electronic cross section per unit scattering angle $d_e\sigma_C/d\theta$ as follows

$$\frac{d_e\sigma_C^{KN}}{d\phi} = \frac{d_e\sigma_C^{KN}}{d\theta}\frac{d\theta}{d\phi}. \tag{7.96}$$

The derivative $d\theta/d\phi$ is determined from the relationship θ versus ϕ, given in (7.67) and (7.68) and derived in Sect. 7.3.2. Differentiating both sides of (7.67) and (7.68) results, respectively, in

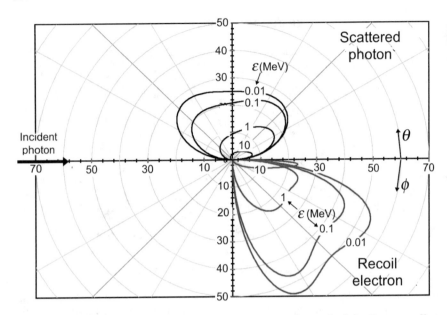

Fig. 7.15 Differential electronic cross section per unit scattering angle θ for Compton effect $d_e\sigma_C^{KN}/d\theta$ against scattering angle θ (*black*) and differential electronic cross sections per unit recoil angle ϕ for Compton effect $d_e\sigma_C^{KN}/d\phi$ against recoil angle ϕ (*red*) for four values of ε (0, 0.1, 1, and 10), the incident photon energy $h\nu$ normalized to the rest mass energy of the electron m_ec^2. The cross sections are drawn on a polar plot; the upper half of the plot shows the distributions for scattered photons, the lower half for recoil electrons

$$-\frac{d\phi}{\sin^2\phi} = (1+\varepsilon)\frac{d\theta}{2\cos^2\dfrac{\theta}{2}} \quad \text{and} \quad \frac{d\phi}{\cos^2\phi} = -\frac{1}{2(1+\varepsilon)}\frac{d\theta}{\sin^2\dfrac{\theta}{2}} \qquad (7.97)$$

or

$$\frac{d\theta}{d\phi} = -\frac{2\cos^2\dfrac{\theta}{2}}{(1+\varepsilon)\sin^2\phi} \quad \text{and} \quad \frac{d\theta}{d\phi} = -\frac{2(1+\varepsilon)\sin^2\dfrac{\theta}{2}}{\cos^2\phi}. \qquad (7.98)$$

The differential electronic cross section per unit recoil angle $d_e\sigma_C/d\phi$, after incorporating (7.95) and (7.98) into (7.96), is expressed with the following two equations

$$\frac{d_e\sigma_C^{KN}}{d\phi} = 2\pi r_e^2 F_{KN}\frac{(1+\cos^2\theta)\sin\theta\cos^2\dfrac{\theta}{2}}{(1+\varepsilon)\sin^2\phi} \qquad (7.99)$$

or

$$\frac{d_e\sigma_C^{KN}}{d\phi} = 2\pi r_e^2 F_{KN}\frac{(1+\varepsilon)(1+\cos^2\theta)\sin\theta\sin^2\dfrac{\theta}{2}}{\cos^2\phi}. \qquad (7.100)$$

and plotted in Fig. 7.14 with *dashed curves*. The angular distribution of electrons exhibits a similar general shape to that for scattered photons, except that the distribution for scattered photons spans a range in θ from $0°$ to $180°$, while the distribution for corresponding recoil electrons spans a range in ϕ from $90°$ to $0°$. The following features in $d_e\sigma_C^{KN}/d\phi$ are noted:

1. The area under the $d_e\sigma_C^{KN}/d\phi$ decreases with increasing ε.
2. At very low incident photon energy $h\nu$ the $d_e\sigma_C^{KN}/d\phi$ curve is symmetrical and exhibits two maxima, similar to the $d_e\sigma_C^{KN}/d\theta$ curve which also exhibits two maxima (at $\theta = 55°$ and $\theta = 125°$, as seen in Thomson scattering) at low $h\nu$. The $55°$ maximum on the photon curve corresponds to a $\phi = 62.5°$ maximum on the recoil electron curve; the $125°$ maximum on the photon curve corresponds to a $\phi = 27.5°$ maximum on the recoil electron curve.
3. With increasing ε, the electron curve becomes more and more asymmetrical and for large ε exhibits only one peak which moves to increasingly smaller angles ϕ.

Electron recoil data of Fig. 7.14 are re-plotted with light solid curves in the polar diagram of Fig. 7.15. The angular distribution of recoil electrons is present only in the forward hemisphere; however, it is zero in forward direction $\phi = 0$ and exhibits maxima at values of ϕ which depend on photon energy $h\nu$; the larger is $h\nu$ or ε, the smaller is the angle at which maximum occurs.

7.3.8 Differential Klein–Nishina Energy Transfer Cross Section

The differential electronic energy transfer coefficient $\left(d_e\sigma_C^{KN}\right)_{tr}/d\Omega$ for the Compton effect is calculated by multiplying the differential electronic cross section $d_e\sigma_C/d\Omega$ of (7.89) with the fractional energy of the Compton recoil electron given in (7.79) to get

$$
\begin{aligned}
\frac{(d_e\sigma_C^{KN})_{tr}}{d\Omega} &= \frac{d_e\sigma_C^{KN}}{d\Omega}\frac{E_K^C}{h\nu} = \frac{d_e\sigma_{Th}}{d\Omega}F_{KN}\frac{\varepsilon(1-\cos\theta)}{1+\varepsilon(1-\cos\theta)} \\
&= \frac{r_e^2}{2}\left(\frac{\nu'}{\nu}\right)^2\left[\frac{\nu'}{\nu}+\frac{\nu}{\nu'}-\sin^2\theta\right]\frac{h\nu-h\nu'}{h\nu} \\
&= \frac{r_e^2}{2}\left(1+\cos^2\theta\right)\frac{\varepsilon(1-\cos\theta)}{[1+\varepsilon(1-\cos\theta)]^3} \\
&\quad \times\left\{1+\frac{\varepsilon^2(1-\cos\theta)^2}{[1+\varepsilon(1-\cos\theta)](1+\cos^2\theta)}\right\},
\end{aligned}\qquad (7.101)
$$

where $\varepsilon = h\nu/(m_ec^2)$, is the incident photon energy normalized to the electron rest energy.

7.3.9 Energy Distribution of Recoil Electrons

The differential electronic Klein–Nishina cross section $d_e\sigma_C^{KN}/dE_K$ expressing the initial energy spectrum of Compton recoil electrons averaged over all scattering angles θ is calculated from the general Klein–Nishina relationship for $d_e\sigma_C^{KN}/d\Omega$ as follows

$$\frac{d_e\sigma_C^{KN}(E_K)}{dE_K} = \frac{d_e\sigma_C^{KN}}{d\Omega}\frac{d\Omega}{d\theta}\frac{d\theta}{dE_K}$$
$$= \frac{\pi r_e^2}{\varepsilon h\nu}\left\{2 - \frac{2E_K}{\varepsilon(h\nu - E_K)} + \frac{E_K^2}{\varepsilon^2(h\nu - E_K)^2} + \frac{E_K^2}{h\nu(h\nu - E_K)}\right\},$$
(7.102)

where

$d_e\sigma_C^{KN}/d\Omega$	is given in (7.89).
$d\Omega/d\theta$	is $2\pi\sin\theta$.
$d\theta/dE_K$	is $(dE_K/d\theta)^{-1}$ with $E_K(\theta)$ given in (7.79).

The differential electronic cross section $d_e\sigma_C^{KN}/dE_K$ of (7.102) is plotted in Fig. 7.16 against the kinetic energy E_K^C of the recoil electron for various values of the incident photon energy $h\nu$ in the range from 0.5 MeV to 10 MeV.

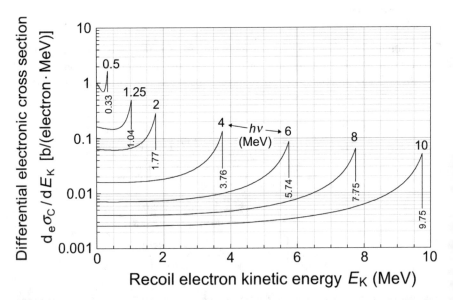

Fig. 7.16 Differential electronic Klein–Nishina cross section per unit kinetic energy $d_e\sigma_c^{KN}/dE_K$ calculated from (7.102) and plotted against the kinetic energy of the Compton recoil electron E_K^C for various incident photon energies $h\nu$ in the range from 0.5 to 10 MeV. For a given photon energy the maximum kinetic energy of the recoil electron in MeV, calculated from (7.81), is indicated

The following features are easily recognized from Fig. 7.16:

- The distribution of kinetic energies given to the Compton recoil electrons is essentially flat from zero almost to the maximum electron kinetic energy $(E_K^C)_{max}$ where a sharp increase in concentration occurs.
- $(E_K^C)_{max}$ is determined using (7.81)

$$(E_K^C)_{max} = \frac{2h\nu\varepsilon}{(1+2\varepsilon)} = h\nu - h\nu'_{min}. \tag{7.103}$$

Since, as shown in (7.77), $h\nu'_{min}$ approaches $\frac{1}{2}m_ec^2$ for high $h\nu$, we note that $(E_K^C)_{max}$ approaches $h\nu - \frac{1}{2}m_ec^2$ as $h\nu \to \infty$.

7.3.10 Total Electronic Klein–Nishina Cross Section for Compton Scattering

The total electronic Klein–Nishina cross section for the Compton scattering on a free electron $_e\sigma_C^{KN}$ [in cm^2/electron] is calculated by integrating the differential electronic cross section per unit solid angle $d_e\sigma_C^{KN}/d\Omega$ of (7.89) over the whole solid angle to get

$$_e\sigma_C^{KN} = \int \frac{d_e\sigma_c^{KN}}{d\Omega} d\Omega = 2\pi r_e^2 \left\{ \frac{1+\varepsilon}{\varepsilon^2} \left[\frac{2(1+\varepsilon)}{1+2\varepsilon} - \frac{\ln(1+2\varepsilon)}{\varepsilon} \right] \right.$$
$$\left. + \frac{\ln(1+2\varepsilon)}{2\varepsilon} - \frac{1+3\varepsilon}{(1+2\varepsilon)^2} \right\}, \tag{7.104}$$

where $\varepsilon = h\nu/(m_ec^2)$. The numerical value of $_e\sigma_C^{KN}$ can also be obtained through a determination of the area under the $d_e\sigma_C^{KN}/d\theta$ curve for a given ε (see Fig. 7.14 – solid curves).

Two extreme cases are of interest, since they simplify (7.104) for $_e\sigma_C^{KN}$:

1. For small incident photon energies $h\nu$ we get the following relationship from (7.104)

$$_e\sigma_C^{KN} = \frac{8\pi}{3} r_e^2 \left(1 - 2\varepsilon + \frac{26}{5}\varepsilon^2 - \frac{133}{10}\varepsilon^3 + \frac{1144}{35}\varepsilon^4 - \cdots \right), \tag{7.105}$$

that for $\varepsilon \to 0$ approaches the classical Thomson result of (7.41), i.e.,

$$_e\sigma_C^{KN}\big|_{\varepsilon\to 0} \approx {_e\sigma_{Th}} = \frac{8\pi}{3} r_e^2 = 0.665\,\text{b}. \tag{7.106}$$

2. For very large incident photon energies $h\nu$, i.e., $\varepsilon \gg 1$, Heitler showed that (7.104) for $\varepsilon \to \infty$ gives

$$_e\sigma_C^{KN} \approx \pi r_e^2 \frac{2\ln(2\varepsilon)+1}{2\varepsilon}. \qquad (7.107)$$

Figure 7.17 shows the Compton electronic cross section $_e\sigma_C^{KN}$ as determined by the Klein–Nishina relationship of (7.104) against the incident photon energy $h\nu$ in the energy range from 0.001 MeV to 1000 MeV. The following features are easily identified:

1. At low photon energies $_e\sigma_C^{KN}$ is approximately equal to the classical Thomson cross section $_e\sigma_{Th}$ which, with its value of 0.665 b, is independent of photon energy [see (7.41)].
2. For intermediate photon energies $_e\sigma_C^{KN}$ decreases gradually with photon energy to read 0.46 b at $h\nu = 0.1$ MeV; 0.21 b at $h\nu = 1$ MeV; 0.05 b at $h\nu = 10$ MeV; and 0.008 b at $h\nu = 100$ MeV.
3. At very high photon energies $h\nu$, the Compton electronic cross section $_e\sigma_c$ attains $1/(h\nu)$ dependence, as shown in (7.107).
4. The Compton electronic cross section $_e\sigma_C^{KN}$ is independent of atomic number Z of the absorber, since in the Compton theory the electron is assumed to be free and stationary, i.e., the electron's binding energy to the atom is assumed to be negligible in comparison with the photon energy $h\nu$.

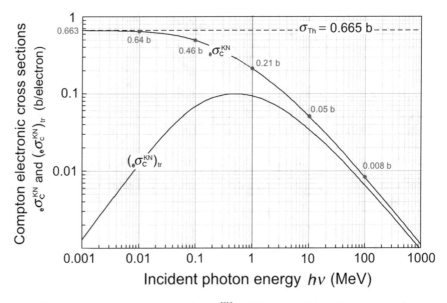

Fig. 7.17 Compton electronic cross section $_e\sigma_C^{KN}$ and Compton electronic energy transfer cross section $(_e\sigma_C^{KN})_{tr}$ for a free electron against incident photon energy $h\nu$ in the energy range from 0.001 MeV to 1000 MeV, determined from Klein–Nishina equations given in (7.104) and (7.108), respectively. For very low photon energies, $_e\sigma_C^{KN} \approx {_e\sigma_{Th}} = 0.665$ b. For very high photon energies $(h\nu \to \infty)$, $(_e\sigma_C^{KN})_{tr} \approx {_e\sigma_C^{KN}}$

7.3.11 Electronic Energy Transfer Cross Section for Compton Effect

The electronic energy transfer cross section $(_e\sigma_C^{KN})_{tr}$ in (cm^2/electron) is obtained by integrating the differential electronic energy transfer cross section $d(_e\sigma_C^{KN})_{tr}/d\Omega$ of (7.101) over all photon scattering angles θ from $0°$ to $180°$ and over recoil angle ϕ from $0°$ to $90°$ to get

$$\frac{(_e\sigma_C^{KN})_{tr}}{2\pi r_e^2} = \frac{2(1+\varepsilon)^2}{\varepsilon^2(1+2\varepsilon)} - \frac{1+3\varepsilon}{(1+2\varepsilon)^2} - \frac{(1+\varepsilon)(2\varepsilon^2-2\varepsilon-1)}{\varepsilon^2(1+2\varepsilon)^2}$$
$$- \frac{4\varepsilon^2}{3(1+2\varepsilon)^3} - \left[\frac{1+\varepsilon}{\varepsilon^3} - \frac{1}{2\varepsilon} + \frac{1}{2\varepsilon^3}\right]\ln(1+2\varepsilon). \qquad (7.108)$$

In addition to the Compton electronic cross section $_e\sigma_C^{KN}$ of (7.104), Fig. 7.17 also shows the energy transfer cross section for the Compton effect $(_e\sigma_C^{KN})_{tr}$ calculated with (7.108) and plotted against the incident photon energy $h\nu$ in the energy range from 0.001 MeV to 1000 MeV. At 0.001 MeV, $(_e\sigma_C^{KN})_{tr}$ is small and of the order of 10^{-3} b/electron; it increases with photon energy $h\nu$ to reach a peak of 0.1 b/electron at $h\nu \approx 700$ keV and then decreases with energy $h\nu$ to $\sim 10^{-3}$ b/electron at $h\nu = 1000$ keV. At high incident photon energies $h\nu$ where $\varepsilon \gg 1$, (7.108) simplifies to

$$(_e\sigma_C^{KN})_{tr} \approx 2\pi r_e^2 \left\{\frac{1}{\varepsilon} - \frac{3}{4\varepsilon} - \frac{1}{2\varepsilon} - \frac{1}{6\varepsilon} + \frac{\ln(2\varepsilon)}{2\varepsilon}\right\} = \pi r_e^2 \frac{2\ln(2\varepsilon)-1.64}{2\varepsilon}. \qquad (7.109)$$

For all incident photon energies $h\nu$, the Compton electronic cross section $_e\sigma_C^{KN}$ exceeds the Compton energy transfer cross section $(_e\sigma_C^{KN})_{tr}$; however, the difference diminishes with increasing $h\nu$. At large photon energies $h\nu$ the difference between $_e\sigma_C$ and $(_e\sigma_C^{KN})_{tr}$ is small and determined by subtracting (7.109) from (7.107) to get

$$_e\sigma_C^{KN} - (_e\sigma_C^{KN})_{tr} \approx \frac{1.32\pi r_e^2}{\varepsilon}. \qquad (7.110)$$

Equation (7.110) shows that as $\varepsilon \to \infty$ the difference between $_e\sigma_C$ and $(_e\sigma_C^{KN})_{tr}$ goes to zero indicating that $(_e\sigma_C^{KN})_{tr} \approx {_e\sigma_C^{KN}}$ for $h\nu \to \infty$.

7.3.12 Mean Energy Transfer Fraction for Compton Effect

Since $(_e\sigma_C^{KN})_{tr}$ and $_e\sigma_C^{KN}$ are related through the following relationship

$$(_e\sigma_C^{KN})_{tr} = {_e\sigma_C^{KN}} \frac{\overline{E}_{tr}^C}{h\nu} = {_e\sigma_C^{KN}} \overline{f}_C, \qquad (7.111)$$

where $\overline{f}_C = \overline{E}_{tr}^C/(h\nu)$ is the mean fraction of the incident photon energy transferred to the kinetic energy of the Compton recoil electron, we can calculate $\overline{E}_K^C/(h\nu)$ from (7.111) as

$$\overline{f}_C = \frac{\overline{E}_K^C}{h\nu} = \frac{({}_e\sigma_C^{KN})_{tr}}{{}_e\sigma_C^{KN}}, \tag{7.112}$$

with $({}_e\sigma_C^{KN})_{tr}$ and ${}_e\sigma_C^{KN}$ given in (7.108) and (7.104), respectively.

Inserting (7.104) and (7.108) into (7.112) gives the following result for the mean fraction of the incident photon energy transferred to the kinetic energy of the recoil electron in Compton effect $\overline{f}_C = \overline{E}_K^C/(h\nu)$

$$\overline{f}_C = \cfrac{\left[\dfrac{2(1+\varepsilon)^2}{\varepsilon^2(1+2\varepsilon)} - \dfrac{1+3\varepsilon}{(1+2\varepsilon)^2} - \dfrac{(1+\varepsilon)(2\varepsilon^2-2\varepsilon-1)}{\varepsilon^2(1+2\varepsilon)^2}\right.}{\left[\dfrac{1+\varepsilon}{\varepsilon^2}\left[\dfrac{2(1+\varepsilon)}{1+2\varepsilon} - \dfrac{\ln(1+2\varepsilon)}{\varepsilon}\right] + \dfrac{\ln(1+2\varepsilon)}{2\varepsilon} - \dfrac{1+3\varepsilon}{(1+2\varepsilon)^2}\right]}} \\ \left.-\dfrac{4\varepsilon^2}{3(1+2\varepsilon)^3} - \left[\dfrac{1+\varepsilon}{\varepsilon^3} - \dfrac{1}{2\varepsilon} + \dfrac{1}{2\varepsilon^3}\right]\ln(1+2\varepsilon)\right]}{}.$$

$$\tag{7.113}$$

At first glance (7.113) looks cumbersome; however, it is simple to use once the appropriate value for ε at a given photon energy $h\nu$ has been established. For example, incident photon of energy $h\nu = 1.022$ MeV results in $\varepsilon = 2$ that, when inserted into (7.113), gives $\overline{f}_C = \overline{E}_K^C/(h\nu) = 0.440$ or mean recoil electron energy of $\overline{E}_K^C = 0.440$ MeV. The mean energy of the corresponding scattered photon is $h\overline{\nu}' = h\nu - \overline{E}_K^C = 0.560$ MeV.

The Compton mean energy transfer fraction $\overline{f}_C = \overline{E}_K^C/(h\nu)$ was plotted as one of several curves in Fig. 7.11. Because of its significance in radiation dosimetry, we plot the Compton mean energy transfer fraction again in Fig. 7.18 in the incident photon energy $h\nu$ range between 0.01 MeV and 100 MeV. On the graph, referred to as "The Compton Graph", we indicate the important anchor points, such as $\overline{f}_C = 0.02; 0.14; 0.44; 0.68;$ and 0.79 at $h\nu = 10$ keV, 100 keV, 1 MeV, 10 MeV, and 100 MeV, respectively. The Compton graph is of much importance in clinical radiation dosimetry because the most important interaction between human tissue and x-rays and γ-rays used in medicine is the Compton interaction.

The plot of $\overline{f}_C = \overline{E}_K^C/(h\nu)$ against incident photon energy $h\nu$ in Fig. 7.18 shows that when low energy photons interact in a Compton process, very little energy is transferred to recoil electrons on the average and most energy goes to the scattered photon resulting in $({}_e\sigma_C^{KN})_{tr} \ll {}_e\sigma_C^{KN}$. On the other hand, when high-energy photons ($h\nu > 10$ MeV) interact in a Compton process, most of the incident photon energy

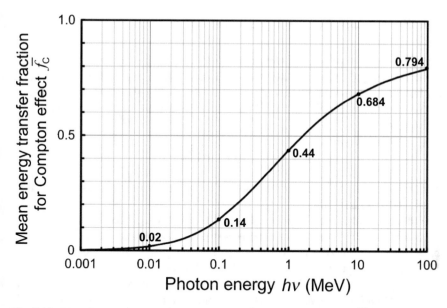

Fig. 7.18 "The Compton Graph" showing the mean energy transfer fraction $\overline{f}_C$ against photon energy $h\nu$ in the energy range from 1 keV to 100 MeV. Anchor points of $\overline{f}_C = 0.02, 014, 0.44, 0.68$, and 0.79 for photon energies $h\nu = 10$ keV, 100 keV, 1 MeV, 10 MeV, and 100 MeV, respectively, calculated from (7.113) are also shown

Table 7.9 Mean energy transfer fractions for Compton effect $\overline{f}_C$ determined from (7.113) and (7.114)

$h\nu$ (MeV)	0.01	0.1	1.0	10	100	1000
ε	0.0196	0.196	1.957	19.6	195.7	1957
$\overline{f}_C$ from (7.113)	0.019	0.139	0.440	0.684	0.796	0.85
$\overline{f}_C$ from (7.114)	–	–	0.293	0.683	0.796	0.85

is given to the recoil electron and relatively little energy is given to the scattered photon resulting in $({}_e\sigma_C^{KN})_{tr} \approx {}_e\sigma_C^{KN}$.

For very high incident photon energies where $\varepsilon \gg 1$, we simplify (7.113) using the ratio of (7.109) and (7.107) to get

$$\overline{f}_C\big|_{\varepsilon \gg 1} \approx \frac{2\ln(2\varepsilon) - 1.64}{2\ln(2\varepsilon) + 1} = \frac{\ln(2\varepsilon) - 0.82}{\ln(2\varepsilon) + 0.5} \tag{7.114}$$

Table 7.9 shows that (7.114) is an excellent approximation for the Compton mean energy transfer fraction $\overline{f}_C$ for incident photon energies $h\nu$ exceeding 10 MeV.

7.3.13 Binding Energy Effects and Corrections

The Compton electronic cross section $_e\sigma_C^{KN}$ of (7.104) and electronic energy transfer cross section $(_e\sigma_C^{KN})_{tr}$ of (7.108) were calculated with Klein–Nishina relationships for free electrons and are plotted in Fig. 7.17. From (7.18) we note that the Compton atomic cross section $_a\sigma_C^{KN}$ is linearly proportional to the Compton electronic cross section $_e\sigma_C^{KN}$

$$_a\sigma_C^{KN} = Z(_e\sigma_C^{KN}), \tag{7.115}$$

where Z, the atomic number of the absorber, is the proportionality constant and $_e\sigma_C^{KN}$ is calculated with (7.104).

When we compare the measured atomic cross sections $_a\sigma_C$ and calculated atomic cross sections $_a\sigma_C^{KN}$ for various absorbers, we find agreement with (7.115) at high photon energies and a discrepancy at low photon energies where the measured atomic cross section is significantly smaller than the calculated $_a\sigma_C^{KN}$. Evidently, at very low incident photon energies the assumption of "free" electron breaks down and the electronic binding energy E_B affects the Compton atomic cross section; the closer is the photon energy $h\nu$ to E_B, the larger is the deviation of the measured atomic cross section $_a\sigma_C$ from that calculated using free-electron Klein–Nishina electronic cross section $_e\sigma_C^{KN}$.

This discrepancy is shown in Fig. 7.19 which displays, for various absorbers ranging from hydrogen to lead, the measured atomic cross sections $_a\sigma_C$ (solid curves) and the calculated Klein–Nishina atomic cross sections $_a\sigma_C^{KN}$ (dashed curves) of (7.115). Two trends are noticed:

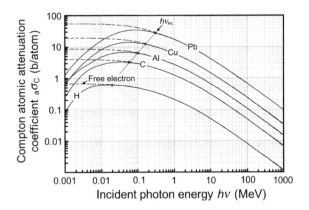

Fig. 7.19 Compton atomic cross section $_a\sigma_C$ plotted against incident photon energy $h\nu$ for various absorbers ranging from hydrogen to lead. The *dashed curves* represent $_a\sigma_C^{KN}$ data calculated with Klein–Nishina free-electron relationship (7.115); the *solid curves* represent the $_a\sigma_C$ data that incorporate the binding effects of orbital electrons. The Klein–Nishina free electron coefficients $_e\sigma_C^{KN}$ for Compton effect are also shown. Equivalent photon energy $h\nu_{eq}$ beyond which the measured cross section $_a\sigma_C$ and calculated cross section $_a\sigma_C^{KN}$ agree are shown with × for a given absorber

1. For a given Z of the absorber, the lower is the incident photon energy $h\nu$, the larger is the discrepancy between the measured $_a\sigma_C$ and the calculated $_a\sigma_C^{KN}$.
2. For a given incident photon energy $h\nu$, the higher is the atomic number Z of the absorber, the more pronounced is the discrepancy.
3. The higher is the atomic number Z of the absorber, the higher is the equivalent photon energy $h\nu_{eq}$ at which the measured $_a\sigma_C$ and the calculated $_a\sigma_C^{KN}$ begin to coincide. *For example*, as shown in Fig. 7.19, $h\nu_{eq}$ for hydrogen is $\sim$20 keV, for aluminum $\sim$100 keV, and for lead $\sim$300 keV. Figure (7.19) also shows that $h\nu_{eq} \propto \sqrt{Z}$.

Incoherent Scattering Function

Various theories have been developed to account for electronic binding effects on Compton atomic cross sections. Most notable is the method developed by *John Hubbell* from the National Institute for Science and Technology (NIST) in Washington, USA.

The binding energy corrections to the Klein–Nishina relationships have usually been treated in the impulse approximation taking into account all orbital electrons of the absorber atom. This involves applying a multiplicative correction function $S(x, Z)$, referred to as the *incoherent scattering function*, to the Klein–Nishina differential atomic cross sections as follows

$$\frac{d_a\sigma_C}{d\Omega} = \frac{d_e\sigma_C^{KN}}{d\Omega} S(x, Z), \tag{7.116}$$

where x, the *momentum transfer variable*, stands for $\sin(\theta/2)/\lambda$, derived in (7.120). The total Compton atomic cross section $_a\sigma_C$ is obtained from the following integral

$$_a\sigma_C = \int S(x, Z) \frac{d_e\sigma_C^{KN}}{d\Omega} d\Omega = \int_{\theta=0}^{\theta=\pi} S(x, Z) d_e\sigma_C^{KN}(\theta), \tag{7.117}$$

where the incoherent scattering function $S(x, Z)$ relates to the properties of the absorber atom and is important for collisions in which the electron momentum p_e is small enough so that the electron has a finite probability for not escaping from the atom.

From Fig. 7.7, in conjunction with the application of the law of cosines on the triangle $(\mathbf{p}_\nu, \mathbf{p}_{\nu'}, \mathbf{p}_e)$, we obtain the following relationship for p_e^2:

$$p_e^2 = p_\nu^2 + p_{\nu'}^2 - 2p_\nu p_{\nu'} \cos\theta \tag{7.118}$$

or

$$p_e = \sqrt{\left(\frac{h\nu}{c}\right)^2 + \left(\frac{h\nu'}{c}\right)^2 - 2\frac{h\nu}{c}\frac{h\nu'}{c}\cos\theta}. \tag{7.119}$$

For small $h\nu$ we know that $h\nu' \approx h\nu$ (see Fig. 7.9) and p_e of (7.119) is approximated as follows:

$$p_e \approx \frac{h\nu}{c}\sqrt{2(1-\cos\theta)} = \frac{h\nu}{c}\sqrt{4\sin^2\frac{\theta}{2}} = 2h\frac{\sin\frac{\theta}{2}}{\lambda} = 2hx, \qquad (7.120)$$

where $x = (\sin\theta/2)/\lambda$ is defined as the *momentum transfer variable* with λ the wavelength of the incident photon.

John Hubbell compiled extensive tables of the incoherent scattering function $S(x, Z)$ and Fig. 7.20 presents Hubbell's data for $S(x, Z)$ plotted against $x = \sin(\theta/2)/\lambda$ for several absorbers in the range from hydrogen to lead. The figure shows that $S(x, Z)$ saturates at Z for relatively large values of x; the higher is Z, the larger is x at which the saturation sets in. With decreasing x, the function $S(x, Z)$ decreases and attains at $x = 0.01$ a value that is less than 1% of its saturation Z value. The following features can be recognized from Fig. 7.20:

1. The electron binding correction is effective only when $S(x, Z) < Z$.
2. For $S(x, Z) = Z$ there is no correction and the Klein–Nishina coefficients $_e\sigma_c^{KN}$ provide correct values for the atomic cross sections $_a\sigma_c$ through the simple relationship $_a\sigma_C = Z(_e\sigma_C^{KN})$.

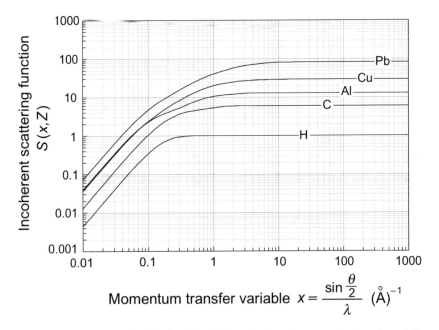

Fig. 7.20 Incoherent scattering function $S(x, Z)$ plotted against the momentum transfer variable x where $x = \sin(\theta/2)/\lambda$ for various absorbers in the range from hydrogen to lead. For large values of x the incoherent scattering function $S(x, Z)$ saturates at Z, the atomic number of the absorber. The saturation values thus are: 1 for hydrogen; 6 for carbon; 29 for copper, and 82 for lead

3. The binding energy correction is only important at photon energies of the order of E_B, and this occurs in the photon energy region where photoelectric effect and Rayleigh scattering are much more probable than is the Compton effect. Thus, ignoring the binding correction on Compton cross sections will only insignificantly affect the determination of the total cross section for photon interactions at relatively low photon energies, since, at these low photon energies, effects other than the Compton effect make a much larger contribution to the total attenuation coefficient than does the Compton effect.

The effects of binding energy corrections on Klein–Nishina differential atomic cross sections per unit scattering angle $d_a \sigma_C^{KN}/d\theta$ are shown in Fig. 7.21 for various incident photons with energies in the range from 1 keV to 10 MeV interacting with hydrogen in part (a), carbon in part (b), and lead in part (c). The solid curves are for simple Klein–Nishina expression given as $d_a \sigma_C^{KN}/d\theta = Z\, d_e \sigma_C^{KN}/d\theta$; the curves accentuated with data points represent the Klein–Nishina electronic cross sections corrected with the incoherent scattering function $S(x, Z)$, i.e., $d_a \sigma_C/d\theta = S(x, Z)\, d_e \sigma_C^{KN}/d\theta$. The following conclusions may be made from Fig. 7.21:

1. For a given absorber Z, the binding energy correction expressed with the incoherent scattering function $S(x, Z)$ is most significant at low photon energies and diminishes with increasing photon energy. For example, in lead the uncorrected and corrected 1 keV curves differ considerably, the 10 keV curves differ less, the 0.1 MeV curves even less, while for 1 MeV and 10 MeV the uncorrected and corrected curves are identical.
2. For a given photon energy $h\nu$, the binding energy correction is more significant for absorbers with higher atomic number Z. For example, the uncorrected and corrected 0.1 MeV curves in hydrogen are identical, for carbon, they are almost identical, and for lead they are significantly different.

Binding Effects and Compton Energy Transfer Fraction

The theory of Compton interaction assumes that the photon interacts with a "free and stationary" orbital electron of the absorber atom. Hence, the Compton energy transfer fraction depends on the incident photon energy $h\nu$ but does not depend on absorber atomic number Z, except for exhibiting a small and generally ignored Z dependence at low photon energies where electron binding to the nucleus affects the "free electron" Compton coefficients, as described by the incoherent scattering function $S(x, Z)$.

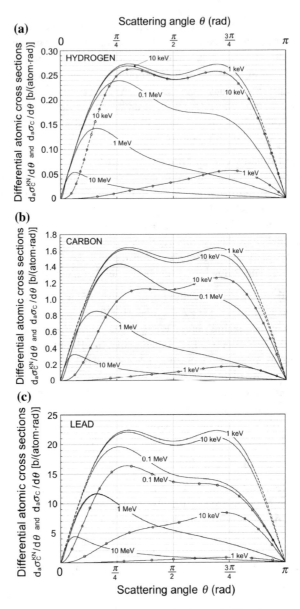

Fig. 7.21 Differential atomic cross section per unit scattering angle θ for Compton effect, $d_a\sigma_C/d\theta$, against scattering angle θ for various photon energies in the range from 1 keV to 10 MeV interacting with hydrogen in (**a**), carbon in (**b**), and lead in (**c**). The *solid curves* represent Klein–Nishina electronic data multiplied by absorber atomic number Z; the curves accentuated with data points represent the Klein–Nishina electronic data corrected with the incoherent scattering function $S(x, Z)$

Table 7.10 "Free electron" mean energy transfer fraction $\overline{f}_C$ (from Figs. 7.18 and Table 7.8) and $\overline{f}_C$ corrected for electron binding effects (data from Steve Seltzer)

$h\nu$ (MeV)	0.01	0.1	1	10	100
$\overline{f}_C$ ("free electron")	0.019	0.139	0.440	0.684	0.796
$\overline{f}_C$ (corrected for electron binding)	0.023	0.148	0.443	0.684	0.796

By way of example, Table 7.10 compares, for lead absorber and various incident photon energies $h\nu$, the Compton mean energy transfer fractions $\overline{f}_C$ for "free electron" in row (2) and corrected for binding energy effects in row (3). The corrected $\overline{f}_C$ accounts for electron binding effects and was calculated by *Steve Seltzer*.

In the MeV photon energy region the agreement between the "free electron" $\overline{f}_C$ and the corrected $\overline{f}_C$ is excellent even for high Z absorbers where the binding effects are the most pronounced, indicating independence of $\overline{f}_C$ from absorber atomic number Z. At low photon energies, on the other hand, there is some Z dependence, yet it is generally ignored because it occurs in the energy range where the photoelectric contribution to the total attenuation coefficient predominates and the Compton effect makes only an insignificant contribution to the total attenuation coefficient.

7.3.14 Compton Atomic Cross Section and Mass Attenuation Coefficient

In the energy region not affected by electron binding effects the Compton atomic cross section $_a\sigma_C^{KN}$ is determined from the electronic cross section of (7.104) using the standard relationship

$$_a\sigma_C^{KN} = Z(_e\sigma_C^{KN}), \qquad (7.121)$$

where Z is the atomic number of the absorber. The Klein–Nishina Compton electronic cross section $_e\sigma_C^{KN}$ is given for free electrons and is thus independent of Z. This makes the atomic attenuation coefficient (cross section) $_a\sigma_c$ linearly dependent on Z, as shown in (7.121).

The Compton mass attenuation coefficient σ_C/ρ is calculated from the Compton atomic cross section $_a\sigma_C^{KN}$ with the standard relationship

$$\frac{\sigma_C^{KN}}{\rho} = \frac{N_A}{A} {}_a\sigma_C^{KN}, \qquad (7.122)$$

which can be expanded using (7.113) to read

$$\frac{\sigma_C^{KN}}{\rho} = \frac{N_A}{A} {}_a\sigma_C^{KN} = \frac{Z N_A}{A} {}_e\sigma_C^{KN} \approx \frac{1}{2} N_A (_e\sigma_C^{KN}). \qquad (7.124)$$

Table 7.11 Compton atomic cross sections $_a\sigma_c$ and mass attenuation coefficients σ_c/ρ at photon energy $h\nu$ of 10 keV for various absorbers

(1)	(2)	(3)	(4)	(5)	(6)	(7)	(8)
Element	Symbol	Atomic number Z	Atomic mass A	$_a\sigma_c$ (b/atom)[a]	$Z \times _e\sigma_C^{KN}$ (b/atom)[b]	$\dfrac{_a\sigma_C}{_a\sigma_C^{KN}}$	σ_c/ρ $(\text{cm}^2/\text{g})^c$
Hydrogen	H	1	1.008	0.60	0.64	0.938	0.0358
Carbon	C	6	12.01	2.70	3.84	0.703	0.0135
Aluminum	Al	13	26.98	4.74	8.33	0.569	0.0106
Copper	Cu	29	63.54	8.15	18.57	0.439	0.0176
Tin	Sn	50	118.69	12.00	32.03	0.375	0.0607
Lead	Pb	82	207.2	15.60	52.52	0.297	0.0153

a. Data are from the NIST
b. $_e\sigma_C^{KN}(h\nu = 10\,\text{keV}) = 0.6405 \times 10^{-24}\,\text{cm}^2/\text{electron} = 0.6405\,\text{b/electron}$

$$\text{c.} \quad \frac{\sigma_C}{\rho} = \frac{N_A}{A}\,_a\sigma_C = \frac{ZN_A}{A}\,_e\sigma_C^{KN} \approx \frac{N_A}{2}\,_e\sigma_C^{KN} = 0.193\,\text{cm}^2/\text{g} \tag{7.123}$$

Table 7.12 Compton atomic cross sections $_a\sigma_c$ and mass attenuation coefficients σ_c/ρ at photon energy $h\nu$ of 1 MeV for various absorbers

(1)	(2)	(3)	(4)	(5)	(6)	(7)	(8)
Element	Symbol	Atomic number Z	Atomic mass A	$_a\sigma_c$ (b/atom)[a]	$Z \times _e\sigma_C^{KN}$ (b/atom)[b]	$\dfrac{_a\sigma_C}{_a\sigma_C^{KN}}$	σ_c/ρ $(\text{cm}^2/\text{g})^c$
Hydrogen	H	1	1.008	0.211	0.211	1.00	0.1261
Carbon	C	6	12.01	1.27	1.27	1.00	0.0636
Aluminum	Al	13	26.98	2.75	2.75	1.00	0.0613
Copper	Cu	29	63.54	6.12	6.12	1.00	0.0580
Tin	Sn	50	118.69	10.5	10.56	0.994	0.0534
Lead	Pb	82	207.2	17.19	17.32	0.992	0.0500

a. Data are from the NIST
b. $_e\sigma_C^{KN}(h\nu = 1\,\text{MeV}) = 0.2112 \times 10^{-24}\,\text{cm}^2/\text{electron} = 0.2112\,\text{b/electron}$

$$\text{c.} \quad \frac{\sigma_C}{\rho} = \frac{N_A}{A}\,_a\sigma_C = \frac{ZN_A}{A}\,_e\sigma_C^{KN} \approx \frac{N_A}{2}\,_e\sigma_C^{KN} = 0.0636\,\text{cm}^2/\text{g} \tag{7.125}$$

Since $Z/A \approx 0.5$ for all elements with the exception of hydrogen for which $Z/A = 1$, σ_C/ρ is essentially independent of Z, as shown in (7.124). In reality, $Z/A = 0.5$ for low atomic number absorbers but with increasing Z the ratio Z/A gradually falls to $Z/A \approx 0.4$ for very high atomic number absorbers, implying a small yet non-negligible Z dependence of σ_C^{KN}/ρ.

Tables 7.11 and 7.12 list the Compton atomic cross section $_a\sigma_C$ and mass attenuation coefficient σ_C/ρ, respectively, for 10 keV and 1 MeV photons interacting with various absorbers in the range from hydrogen to lead. Columns (5) display the atomic cross sections $_a\sigma_C$ incorporating binding energy corrections, while columns (6) display the Klein–Nishina atomic cross sections $_a\sigma_C^{KN} = Z(_e\sigma_C^{KN})$ calculated from (7.121) by a simple multiplication of the electronic Klein–Nishina coefficients $_e\sigma_C^{KN}$ with the absorber atomic number Z.

The two coefficients ($_a\sigma_C$ and $_a\sigma_C^{KN}$) agree well for incident photon energy $h\nu$ of 1 MeV. However, the discrepancy between the two is significant for photon energy of 10 keV, as also evident in Fig. 7.19.

We also note that at $h\nu = 1$ MeV the σ_C/ρ values follow straight from the Klein–Nishina electronic cross sections and are affected only by the specific value for Z/A. This is not the case for σ_c/ρ at $h\nu = 10$ keV since σ_c/ρ at 10 keV is affected not only by Z/A but also by the electronic binding effects that are significant in this energy range for all Z; the larger is Z, the larger is the binding effect, as shown in columns (5) and (6) of Table 7.11. Column (7) of Tables 7.11 and 7.12 gives the ratio between atomic cross sections $_a\sigma_C$ corrected for electron binding effect and the Klein–Nishina atomic cross section $_a\sigma_C^{KN}$. At photon energy $h\nu = 10$ keV the ratio is smaller than 1.00, especially so for high atomic number Z absorbers indicating significant binding effects. On the other hand, at $h\nu = 1$ MeV the ratio is equal to 1.00, except for very high Z where it is 1% lower, indicating absence of the electron binding effect even for very high Z absorbers.

7.3.15 Compton Mass Energy Transfer Coefficient

The Compton mass energy transfer coefficient $(\sigma_C)_{tr}/\rho$ is calculated from the mass attenuation coefficient σ_C/ρ using the standard relationship

$$\frac{(\sigma_C)_{tr}}{\rho} = \frac{\sigma_C}{\rho}\frac{\overline{E}_K^C}{h\nu} = \overline{f}_C\frac{\sigma_C}{\rho}, \qquad (7.126)$$

where $\overline{f}_C$ is the mean energy transfer fraction for the Compton effect given by (7.113) and plotted as "The Compton Graph" in Fig. 7.18 which shows that the Compton mean energy transfer fraction increases with increasing energy from a low value of 0.019 at 10 keV, through 0.440 at 1 MeV, to reach a value of 0.796 at 100 MeV. As a result of the Compton graph behavior we can state that:

1. For low incident photon energies $(\sigma_C)_{tr}/\rho \ll \sigma_C/\rho$.
2. For high incident photon energies $(\sigma_C)_{tr}/\rho \approx \sigma_C/\rho$.

Figure 7.22 shows the $_a\sigma_C$ and $_a\sigma_C^{KN}$ data for lead from Fig. 7.19 and, in addition, it also shows the binding energy effect on the Compton atomic energy transfer coefficients of lead by displaying $(_a\sigma_C)_{tr}$ and $(_a\sigma_C^{KN})_{tr}$ both obtained by multiplying the $_a\sigma_C$ and $_a\sigma_C^{KN}$ data, respectively, with the appropriate Compton mean energy transfer fraction given by (7.113) and plotted in Fig. 7.18.

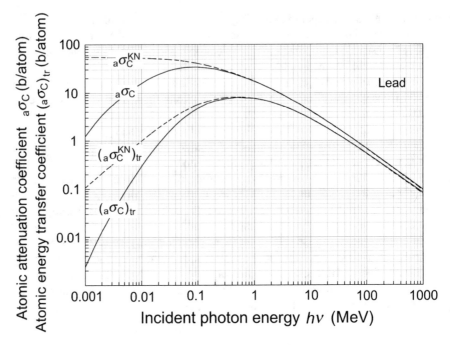

Fig. 7.22 The Compton atomic attenuation coefficient (cross section) for lead of Fig. 7.19 and the Compton atomic energy transfer coefficients for lead; *dashed curves* are Klein–Nishina data for free unbound electrons; *solid curves* are data incorporating electronic binding effects. Data are from the NIST

7.4 Rayleigh Scattering

Rayleigh scattering is an interaction between a photon and absorber atom characterized by photon scattering on bound atomic electrons. The atom is neither excited nor ionized as a result of the interaction and after the interaction the bound electrons revert to their original state. The atom as a whole absorbs the transferred momentum but its recoil energy is very small and the incident photon scattered with scattering angle θ has essentially the same energy as the original photon. The scattering angles are relatively small because the recoil imparted to the atom produces no atomic excitation or ionization.

The Rayleigh scattering is named after physicist *John W. Rayleigh* who in 1900 developed a classical theory for scattering of electromagnetic radiation by atoms. The effect occurs mostly at low photon energies $h\nu$ and for high atomic number Z of the absorber, in the energy region where electron binding effects severely diminish the Compton Klein–Nishina cross sections. As a result of a coherent contribution of all atomic electrons to the Rayleigh (i.e., coherent) atomic cross section, the Rayleigh cross section exceeds the Compton cross section in this energy region.

7.4.1 Differential Atomic Cross Section for Rayleigh Scattering

The differential Rayleigh atomic cross section $d_a\sigma_R/d\Omega$ per unit solid angle is given as follows

$$\frac{d_a\sigma_R}{d\Omega} = \frac{d_e\sigma_{Th}}{d\Omega}\{F(x, Z)\}^2 = \frac{r_e^2}{2}(1 + \cos^2\theta)\{F(x, Z)\}^2, \tag{7.127}$$

where

$d_e\sigma_{Th}/d\Omega$ is the differential Thomson electronic cross section [see (7.39)].
$F(x, Z)$ is the so-called atomic form factor for Rayleigh scattering with the momentum transfer variable $x = \sin(\theta/2)/\lambda$ defined in (7.120).
λ is the wavelength of the incident photon.
Z is the atomic number of the absorber.

Rayleigh differential atomic cross section $d_a\sigma_R/d\theta$ per unit scattering angle θ is

$$\frac{d_a\sigma_R}{d\theta} = \frac{d_a\sigma_R}{d\Omega}\frac{d\Omega}{d\theta} = \frac{r_e^2}{2}(1 + \cos^2\theta)\{F(x, Z)\}^2 2\pi\sin\theta$$

$$= \pi r_e^2\sin\theta(1 + \cos^2\theta)\{F(x, Z)\}^2. \tag{7.128}$$

7.4.2 Form Factor for Rayleigh Scattering

Calculations of the atomic form factor $F(x, Z)$ are difficult and, since they are based on atomic wavefunctions, they can be carried out analytically only for the hydrogen atom. For all other atoms the calculations rely on various approximations and atomic models, such as the Thomas–Fermi, Hartree, or Hartree–Fock model.

The atomic form factor $F(x, Z)$ is equal to Z for small scattering angles θ and approaches zero for large scattering angles θ. Its values are plotted in Fig. 7.23 against the momentum transfer variable $x = \sin(\theta/2)/\lambda$ for various absorbers ranging in atomic number Z from 1 to 82.

Figure 7.24 is a plot of the Rayleigh differential atomic cross section per unit scattering angle $d_a\sigma_R/d\theta$ against the scattering angle θ for hydrogen and carbon, respectively, consisting of a product of the Thomson differential electronic cross section per unit scattering angle $d_e\sigma_{Th}/d\theta$ given in (7.40) and the square of the atomic form factor $F(x, Z)$, as given in (7.128). For comparison the Thomson differential atomic cross section $d_a\sigma_{Th}/d\theta$ is also shown in Fig. 7.24. For hydrogen $d_a\sigma_{Th}/d\theta = d_e\sigma_{Th}/d\theta$, while for carbon $d_a\sigma_{Th}/d\theta = 6 d_e\sigma_{Th}/d\theta$, with both curves symmetrical about $\theta = \frac{1}{2}\pi$.

Fig. 7.23 Atomic form factor $F(x, Z)$ for Rayleigh scattering plotted against the momentum transfer variable $x = \sin(\theta/2)/\lambda$

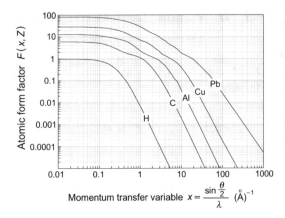

The $d_a\sigma_R/d\theta$ curves for various energies shown in Fig. 7.24 are not symmetrical about $\theta = \frac{1}{2}\pi$ because of the peculiar shape of the atomic form factor $F(x, Z)$ that causes a predominance in forward Rayleigh scattering; the larger the photon energy, the more asymmetrical is the $d_a\sigma_R/d\theta$ curve and the more forward peaked is the Rayleigh scattering. The area under each $d_a\sigma_R/d\theta$ curve gives the total Rayleigh atomic cross section $_a\sigma_R$ for a given photon energy $h\nu$ and absorber atomic number Z.

Fig. 7.24 Differential atomic cross section per unit scattering angle for Rayleigh scattering $d_a\sigma_R/d\theta$, given by (7.128), for incident photon energies of 1, 3, and 10 keV for hydrogen in (**a**) and carbon in (**b**). The Thomson differential atomic cross section $d_a\sigma_{Th}/d\theta$ (similar to the Thomson electronic cross section of Fig. 7.6) for the two absorbing materials is shown by the *dotted curves* for comparison

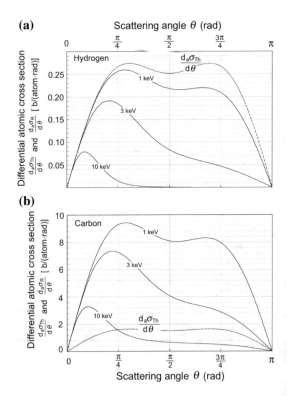

7.4.3 Scattering Angles in Rayleigh Scattering

The angular spread of Rayleigh scattering depends on the photon energy $h\nu$ and the atomic number Z of the absorber. It can be estimated from the following relationship

$$\theta_R \approx 2 \arcsin\left(\frac{0.026Z^{1/3}}{\varepsilon}\right), \tag{7.129}$$

where

θ_R is the characteristic angle for Rayleigh scattering, representing the opening half-angle of a cone that contains 75% of the Rayleigh-scattered photons.

Z is the atomic number of the absorber.

ε is the reduced photon energy, i.e., $\varepsilon = h\nu/(m_ec^2)$.

As suggested by (7.129), the Rayleigh characteristic angle θ_R increases with increasing Z of the absorber for the same $h\nu$ and decreases with increasing photon energy $h\nu$ for the same Z. Table 7.13 lists the angle θ_R for photon energies in the range from 100 keV to 10 MeV and various absorbers (carbon, copper and lead), calculated from (7.129).

The main characteristics of Rayleigh scattering can be summarized as follows:

1. At high photon energies ($h\nu > 1$ MeV) Rayleigh scattering is confined to small angles for all absorbers.
2. At low photon energies $h\nu$, particularly for high Z absorbers, the angular distribution of Rayleigh-scattered photons is much broader. In this energy range the Rayleigh atomic cross section $_a\sigma_R$ exceeds the Compton atomic cross section $_a\sigma_C$ but is nonetheless very small in comparison with the photoelectric atomic cross section $_a\tau$. The Rayleigh atomic cross section $_a\sigma_R$ is therefore often ignored in gamma ray transport as well as in shielding barrier calculations.
3. Rayleigh scattering plays no role in radiation dosimetry, since no energy is transferred to charged particles through Rayleigh scattering. However, the scattering is of interest in diagnostic imaging because it has an adverse effect on image quality.

7.4.4 Atomic Cross Section for Rayleigh Scattering

The Rayleigh atomic cross section $_a\sigma_R$ can be obtained by determining the area under the appropriate $d_a\sigma_R/d\theta$ curve plotted against scattering angle θ, as shown in Fig. 7.24, or it can be calculated by integrating the Rayleigh differential cross section per unit scattering angle $d_a\sigma_R/d\theta$ of (7.128) over all possible scattering angles θ from 0 to π, i.e.,

Table 7.13 The Rayleigh characteristic angle θ_R for various absorbers with atomic number Z and incident photon energy $h\nu$ in the range from 100 keV to 10 MeV, calculated from (7.129)

Absorber	Symbol	Atomic number Z	Incident photon energy $h\nu$ (MeV)				
			0.1	0.5	1	5	10
Carbon	C	6	28°	6°	3°	0.6°	0.3°
Copper	Cu	29	48°	9°	5°	0.9°	0.5°
Lead	Pb	82	70°	13°	7°	1.3°	0.7°

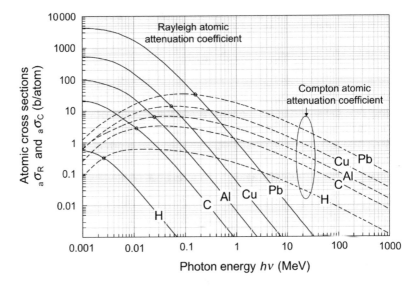

Fig. 7.25 Atomic cross sections for Rayleigh scattering $_a\sigma_R$ shown with *solid curves* and atomic cross sections for Compton scattering $_a\sigma_C$ shown with *dashed curves* against incident photon energy $h\nu$ in the range from 1 keV to 1000 MeV for various absorbers ranging from hydrogen to lead. For very low photon energies $_a\sigma_R$ curves exhibit a plateau with a value of $_e\sigma_{Th}Z^2$ where $_e\sigma_{Th}$ is the energy independent Thomson cross section and Z is the atomic number of the absorber (note that $F(x, Z) \to Z$ for low $h\nu$, i.e., large λ. For a given absorber, the photon energy $h\nu_{eq}$ at which $_a\sigma_R = _a\sigma_C$ is indicated with o. Data are from the NIST

$$_a\sigma_R = \pi r_e^2 \int_0^{\pi} \sin\theta (1 + \cos^2\theta)[F(x, Z)]^2 d\theta. \tag{7.130}$$

Rayleigh atomic cross section $_a\sigma_R$ is shown with solid curves against incident photon energy $h\nu$ in the range from 1 keV to 1000 MeV in Fig. 7.25. For comparison, the figure also shows, with dashed curves, the Compton atomic cross section $_a\sigma_C$ of Fig. 7.19 in the same energy range. The following conclusions may be reached from Fig. 7.25:

1. At low photon energies, $_a\sigma_R$ exceeds $_a\sigma_C$; the higher is the atomic number of the absorber, the larger is the difference. However, at low photon energies both

$_a\sigma_R$ and $_a\sigma_C$ are negligible in comparison with the atomic cross section for the photoelectric effect $_a\tau$, so both are usually ignored in calculations of the total atomic cross section $_a\mu$ for a given absorber at very low photon energies.

2. The photon energy $h\nu_{eq}$ at which the atomic cross sections for Rayleigh and Compton scattering are equal, i.e., $_a\sigma_R = _a\sigma_C$, is proportional to the atomic number Z of the absorber. From Fig. 7.25 we also note that for photon energies exceeding $h\nu_{eq}$ the Rayleigh atomic cross section $_a\sigma_R$ is inversely proportional to $(h\nu)^2$; i.e.,

$$_a\sigma_R \propto \frac{1}{(h\nu)^2}. \tag{7.131}$$

3. In general, as evident from Fig. 7.25, we may also state that, for a given photon energy $h\nu$, the Rayleigh atomic coefficient $_a\sigma_R$ is proportional to Z^2, where Z is the atomic number of the absorber.

7.4.5 Mass Attenuation Coefficient for Rayleigh Scattering

The Rayleigh mass attenuation coefficient σ_R/ρ is determined through the standard relationship

$$\frac{\sigma_R}{\rho} = \frac{N_A}{A} _a\sigma_R. \tag{7.132}$$

Two important conclusions can be made:

1. Since $_a\sigma_R \propto Z^2/(h\nu)^2$ and $A \approx 2Z$, we conclude that $\sigma_R/\rho \propto Z/(h\nu)^2$, where Z and A are the atomic number and mass, respectively, of the absorber.
2. Since no energy is transferred to charged particles in Rayleigh scattering, the energy transfer coefficient for Rayleigh scattering is zero: $(\sigma_R)_{tr} = 0$.
3. At very low photon energies $h\nu$ the Rayleigh atomic cross section $_a\sigma_R$ for a given absorber of atomic number Z exhibits a plateau at $Z^2 _e\sigma_{Th}$ where $_e\sigma_{Th}$ is the energy independent Thomson electronic cross section of 0.665 b/electron given in (7.41).

7.5 Photoelectric Effect

An interaction between a photon and a tightly bound orbital electron of an absorber atom is called photoelectric effect (colloquially often referred to as photoeffect). In the interaction the photon is absorbed completely and the orbital electron is ejected with kinetic energy E_K. The ejected orbital electron is called a photoelectron. The photoelectric interaction between a photon of energy $h\nu$ and a K-shell atomic electron is shown schematically in Fig. 7.26.

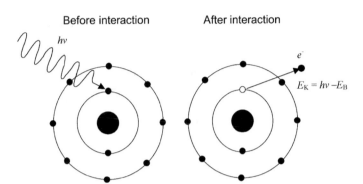

Fig. 7.26 Schematic diagram of the photoelectric effect. A photon with energy $h\nu$ interacts with a K-shell electron. The photon is absorbed completely and the K-shell electron is ejected from the atom as photoelectron with kinetic energy $E_K = h\nu - E_B(K)$, where $E_B(K)$ is the binding energy of the K-shell electron. The vacancy in the K shell is subsequently filled with a higher orbit electron and the energy of the electronic transition will be emitted either in the form of a characteristic (fluorescence) photon or in the form of an Auger electron

In contrast to Compton effect which occurs between photon and a "free (loosely bound) electron", the photoelectric effect occurs between a photon and a "tightly bound" electron. As discussed in Sect. 7.1, the distinction between "loose" and "tight" binding arises from the relative magnitude of photon energy $h\nu$ and electron shell binding energy E_B rather than from an absolute value of $h\nu$ or E_B. Thus, when $E_B \ll h\nu$, the electron is said to be loosely bound or "free" and when $E_B \lesssim h\nu$ the electron is assumed tightly bound.

7.5.1 Conservation of Energy and Momentum in Photoelectric Effect

The requirement for electron tight binding to the atom arises from consideration of total energy and momentum conservation which for photon–free electron photoelectric interaction would be expressed as follows:

1. Total energy before interaction: $h\nu + m_e c^2$ (7.133)

    Total energy after interaction: $E_K + m_e c^2$ (7.134)

    Conservation of energy : $h\nu = E_K$ (7.135)

2. Momentum before interaction: $p_\nu = \dfrac{h\nu}{c}$ (7.136)

    Momentum after interaction [see (1.64)]:

$$p_e = \frac{\sqrt{E^2 - \left(m_e c^2\right)^2}}{c} = \frac{E_K}{c}\sqrt{1 + \frac{2m_e c^2}{E_K}} \qquad (7.137)$$

$$\textit{Conservation of momentum:} \quad \frac{h\nu}{c} = \frac{E_K}{c}\sqrt{1 + \frac{2m_ec^2}{E_K}} \tag{7.138}$$

or, after multiplying (7.138) by c:

$$h\nu = E_K\sqrt{1 + \frac{2m_ec^2}{E_K}}, \tag{7.139}$$

where E, E_K and m_ec^2 are the total energy, kinetic energy, and rest energy of the photoelectron, respectively.

Equations (7.135) and (7.139) clearly contradict one another, since (7.135) states that $h\nu = E_K$ and (7.139) states that $h\nu > E_K$. One concludes that in a photoelectric interaction the photon and free electron alone could not simultaneously conserve the total energy and momentum and therefore photoelectric effect cannot occur between a photon and a free electron. The extra energy and momentum carried by the photon are transferred to a third particle, the parent atom of the photoelectron, but this can happen only when the photoelectron is tightly bound to the parent atom which means that the electron binding energy E_B and the incident photon energy $h\nu$ are of the same order of magnitude, with $h\nu$ slightly exceeding E_B.

The main characteristics of the photoelectric effect are thus as follows:

1. The extra energy and momentum carried by the photon are transferred to the absorbing atom; however, because of the relatively large nuclear mass, the atomic recoil energy is exceedingly small and may be neglected. The kinetic energy E_K of the ejected photoelectron is assumed to be equal to the incident photon energy $h\nu$ less the binding energy E_B of the orbital electron, i.e.,

$$E_K = h\nu - E_B. \tag{7.140}$$

2. When the photon energy $h\nu$ exceeds the K-shell binding energy $E_B(K)$ of the absorber, i.e., $h\nu > E_B(K)$, about 80% of all photoelectric absorptions occur with the K-shell electrons of the absorber and the remaining 20% occur with less tightly bound higher shell electrons.
3. The energy uptake by the photoelectron may be insufficient to bring about its ejection from the atom in a process referred to as atomic ionization but may be sufficient to raise the photoelectron to a higher orbit in a process referred to as atomic excitation.
4. The vacancy that results from the emission of the photoelectron from a given shell will be filled by a higher shell electron and the transition energy will be emitted either as a characteristic (fluorescence) photon or as an Auger electron, the probability for each governed by the fluorescence yield ω, as discussed in Sects. 4.1 and 7.5.7.

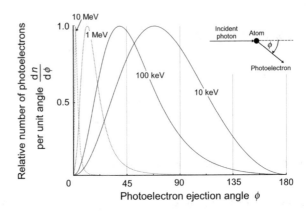

Fig. 7.27 Angular distribution of photoelectrons ejected between two cones with half angles of ϕ and $\phi + d\phi$ for a given incident photon energy $h\nu$ in the incident photon energy $h\nu$ range from 10 keV to 10 MeV. Angle ϕ is the photoelectron emission angle defined as the angle between the incident photon direction and the direction of the emitted photoelectron, as shown in the inset. Graph is based on data from Robley Evans. All peaks in angular distribution are normalized to 1

7.5.2 Angular Distribution of Photoelectrons

The angular distribution of photoelectrons depends on the incident photon energy $h\nu$. The photoelectron emission angle ϕ is defined as the angle between the incident photon direction and the direction of the emitted photoelectron, similarly to the definition of the recoil electron angle ϕ in Compton scattering (see Fig. 7.7). At low $h\nu$ of the order of 10 keV photoelectrons tend to be emitted at angles close to 90° to the incident photon direction, hence in the direction of the electric vector of the incident photon. As $h\nu$ increases, however, the photoelectron emission peak moves progressively to more forward photoelectron emission angles, somewhat akin to the emission of bremsstrahlung photons in electron bremsstrahlung interaction, discussed in Sect. 4.2.8.

Figure 7.27 displays on a Cartesian plot the directional distribution of photoelectron emission for various incident photon energies $h\nu$ in the range from $h\nu = 10$ keV with maximum emission angle $\phi_{\mathrm{max}} \approx 70°$ to $h\nu = 10$ MeV with $\phi_{\mathrm{max}} \approx 2°$. The ordinate ($y$ axis) plots $dn/d\phi$, the relative number of photoelectrons ejected between two cones with half-angles of ϕ and $\phi + d\phi$ for a given incident photon energy $h\nu$. While, for a given $h\nu$, the relative number of photoelectrons per angular interval varies with emission angle ϕ, all photoelectrons irrespective of emission angle ϕ are emitted with the same kinetic energy given in (7.140).

7.5.3 Atomic Cross Section for Photoelectric Effect

The atomic cross section (attenuation coefficient) for the photoelectric effect $_a\tau$ as a function of the incident photon energy $h\nu$ exhibits a characteristic saw-tooth structure in which the sharp discontinuities, referred to as absorption edges, arise whenever the photon energy coincides with the binding energy of a particular electron shell. Since all shells except the K shell exhibit a fine structure, the $_a\tau$ curve plotted against the incident photon energy $h\nu$ also exhibits a fine structure in the L, M, ... etc. absorption edges. Three distinct energy regions characterize the atomic cross section $_a\tau$:

1. Region in the immediate vicinity of absorption edges.
2. Region at some distance from the absorption edge.
3. Region in the relativistic region far from the K absorption edge.

Theoretical predictions for $_a\tau$ in region (1) are difficult and uncertain. For region (2) the atomic attenuation coefficient for K-shell electrons $_a\tau_K$ is given as follows

$$_a\tau_K = \alpha^4 \, (_e\sigma_{Th}) \, Z^n \sqrt{\frac{32}{\varepsilon^7}}, \tag{7.141}$$

where

ε	is the usual normalized photon energy, i.e., $\varepsilon = h\nu/(m_e c^2)$.
α	is the fine structure constant (1/137).
Z	is the atomic number of the absorber.
$_e\sigma_{Th}$	is the Thomson electronic cross section given in (7.41).
n	is the power for the Z dependence of $_a\tau_K$ ranging from $n = 4$ at relatively low photon energies to $n = 4.6$ at high photon energies.

In the relativistic region ($\varepsilon \gg 1$), $_a\tau_K$ is given as follows

$$_a\tau_K = \frac{1.5}{\varepsilon} \alpha^4 Z^5 (_e\sigma_{Th}). \tag{7.142}$$

The following conclusions may be reached with regard to energy and atomic number dependence of $_a\tau_K$:

1. The energy dependence of $_a\tau_K$ is assumed to go as $(1/(h\nu))^3$ at low photon energies $h\nu$ and gradually transforms into $1/(h\nu)$ at high $h\nu$.
2. The energy dependence for regions (2) and (3) can be identified from Fig. 7.28 that displays the atomic cross section for the photoelectric effect $_a\tau$ against incident photon energy for various absorbers ranging from water to lead.
3. Absorption edges are clearly shown in Fig. 7.28, the K absorption edges are identified for aluminum (1.56 keV), copper (8.98 keV) and lead (88 keV). The fine structures of the L and M absorption edges are also displayed.
4. The atomic number Z dependence ($_a\tau \propto Z^n$) of $_a\tau$, where n ranges from 4 to $\sim$5, is also evident from Fig. 7.28.

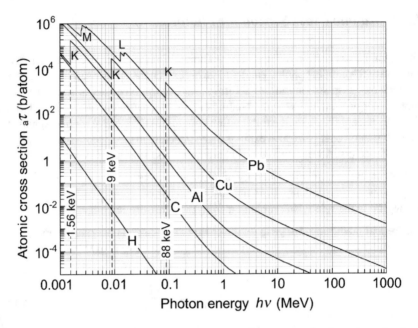

Fig. 7.28 Photoelectric atomic cross section $_a\tau$ against photon energy $h\nu$ for various absorbers. Energies of K-absorption edges are also indicated. Data are from the NIST

7.5.4 Mass Attenuation Coefficient for Photoelectric Effect

The mass attenuation coefficient for the photoelectric effect τ/ρ is calculated from the atomic cross section $_a\tau$ with the standard relationship

$$\frac{\tau}{\rho} = \frac{N_A}{A}\,_a\tau,\tag{7.143}$$

where A and ρ are the atomic number and density, respectively, of the absorber.

7.5.5 Energy Transfer to Charged Particles in Photoelectric Effect

In the photoelectric interaction between a photon of energy $h\nu$ and an absorber atom of atomic number Z the photoelectron is ejected from the atom with a kinetic energy $E_K = h\nu - E_B(j)$, leaving behind a vacancy in the shell or subshell from which it was ejected. $E_B(j)$ stands for the binding energy of the j-subshell electron. Since the photoelectric effect is an interaction between a photon and the whole atom of the absorber, of all orbital electrons available for a photoelectric interaction, the most tightly bound electron has by far the highest probability for interaction with the incident photon. Therefore, photons with energy exceeding the K-shell binding energy of the absorber atom are most likely to interact with K-shell electrons; photons with energy between the L shell and the K shell binding energies are most likely to interact with L shell electrons, etc.

The vacancy that is left behind by the photoelectron is subsequently filled with an electron from a higher-level atomic shell, the resulting vacancy in the higher level shell is filled by another even higher shell electron, and so on until the vacancy migrates ("cascades") to the outer shell of the atom and is filled by a free electron from the environment to neutralize the ion. As discussed in Sect. 4.1, the transition energies, which in total equal the binding energy of the electron that was ejected as the photoelectron, are emitted from the atom:

1. Either in the form of characteristic (fluorescence) photons.
2. Or in the form of Auger electrons including Coster–Kronig and super Coster–Kronig electrons.
3. Or, most generally, in a combination of fluorescence photons and Auger electrons.

The probability for photoelectric effect to occur is governed by the photoelectric attenuation coefficient τ which depends on the absorber atomic number Z and photon energy $h\nu$. As discussed in Sect. 4.1.2, the branching between emission of a fluorescence photon and emission of an Auger electron is governed by the fluorescence yield ω_j for the given atom and for the given atomic shell or subshell j.

Because of the presence of Auger electrons, in addition to the photoelectron which is ejected from the atom with kinetic energy $h\nu - E_B(j)$, the mean energy transferred to charged particles (photoelectron and Auger electrons) in a photoelectric process $\overline{E}_{tr}^{PE}$ for $h\nu > E_B(j)$ is generally somewhere between two possible extremes: $h\nu - E_B(j)$ and $h\nu$, defined as follows:

1. $\overline{E}_{tr}^{PE} = h\nu - E_B(j)$ for $\omega_j = 1$. No Auger electrons are produced in the photoelectric process. The photoelectron is the only charged particle released, and its kinetic energy is $h\nu - E_B(j)$.
2. $\overline{E}_{tr}^{PE} = h\nu$ for $\omega_j = 0$. No characteristic (fluorescence) photons are produced in the photoelectric process so that, in addition to the photoelectron, Auger electrons are also released. The photoelectron leaves the atom with kinetic energy $h\nu - E_B(j)$ and the Auger electrons are released from the atom with a combined kinetic energy equal to $E_B(j)$.
3. In general, $0 < \omega_j < 1$, the mean energy transferred to charged particles in a photoelectric process $\overline{E}_{tr}^{PE}$ is between the two extremes discussed above, i.e., $h\nu - E_B(j) < \overline{E}_{tr}^{PE} < h\nu$, and a combination of characteristic (fluorescence) photons and Auger electrons is released in addition to the photoelectron.

Mean energy transferred to charged particles in a photon interaction with an absorber atom is an important dosimetric quantity, since all or a sizeable portion of this energy will be absorbed in the medium and will contribute to the radiation dose in the medium. In principle, determining the mean energy transferred to charged particles in a photoelectric event $\overline{E}_{\mathrm{tr}}^{\mathrm{PE}}$ is simple: one determines kinetic energy of all Auger, Coster–Kronig and super Coster–Kronig electrons and adds the sum of these energies to the kinetic energy of the photoelectron. In practice, however, because of cascade effects and the large number of Auger electrons released, it is simpler to determine the mean fluorescence emission energy $\overline{X}_{\mathrm{PE}}$ for a given shell vacancy and then subtract $\overline{X}_{\mathrm{PE}}$ from the incident photon energy $h\nu$ to obtain the total mean energy transferred to charged particles (photoelectron and Auger electrons) as

$$\overline{E}_{\mathrm{tr}}^{\mathrm{PE}} = h\nu - \overline{X}_{\mathrm{PE}}. \tag{7.144}$$

For use in radiation dosimetry, the mean photoelectric energy transfer fraction $\overline{f}_{\mathrm{PE}}$ is defined as the mean fraction of the incident photon energy $h\nu$ that is transferred to kinetic energy of secondary charged particles (photoelectron and Auger electrons) released in a photoelectric event, i.e.,

$$\overline{f}_{\mathrm{PE}} = \frac{\overline{E}_{\mathrm{tr}}^{\mathrm{PE}}}{h\nu} = 1 - \frac{\overline{X}_{\mathrm{PE}}}{h\nu}. \tag{7.145}$$

In general, the mean photoelectric fluorescence emission energy $\overline{X}_{\mathrm{PE}}$ is given as

$$\overline{X}_{\mathrm{PE}}(j) = \sum_j P_j \omega_j h\overline{\nu}_j - \sum_j P_j \omega_j \eta_j E_{\mathrm{B}}(j), \tag{7.146}$$

where j stands for:

1. K shell and electronic subshells L_1, L_2, L_3; M_1, M_2, ... for $h\nu \geq E_{\mathrm{B}}$ (K).
2. Electronic subshells L_1, L_2, L_3; M_1, M_2, ... for $E_{\mathrm{B}}(L_1) \leq h\nu < E_{\mathrm{B}}(\mathrm{K})$.
3. Electronic subshells L_2, L_3; M_1, M_2, M_3, ... for $E_{\mathrm{B}}(L_2) \leq h\nu < E_{\mathrm{B}}(L_1)$, etc.

The parameters of (7.146), $P_j, \omega_j, h\overline{\nu}_j, \eta_j$, and $E_{\mathrm{B}}(j)$, are defined below and plotted for the K and L electronic shells in Fig. 7.29 against absorber atomic number Z for all elements from $Z = 1$ to $Z = 100$:

$E_{\mathrm{B}}(j)$ is the binding energy of subshell j electron, by definition equal to the threshold energy for photoelectric interaction between a photon and a subshell j electron.

P_j is the probability for the photoelectric effect, if it occurs, to occur in the j subshell of an absorber atom. Of course, the photon energy $h\nu$ must exceed the threshold energy $E_{\mathrm{B}}(j)$ for the photoelectric event to occur in subshell j. Thus, for $h\nu \geq E_{\mathrm{B}}(j) \rightarrow P_j \neq 0$; for $h\nu < E_{\mathrm{B}}(j) \rightarrow P_j = 0$.

ω_j is the fluorescence yield for subshell j.

$h\overline{\nu}_j$ is the mean fluorescence photon energy representing the energy of the emitted j-subshell fluorescence photon and accounting for all possible fluorescence photons emitted, for their relative intensities, as well as for the resulting cascade fluorescence effects.

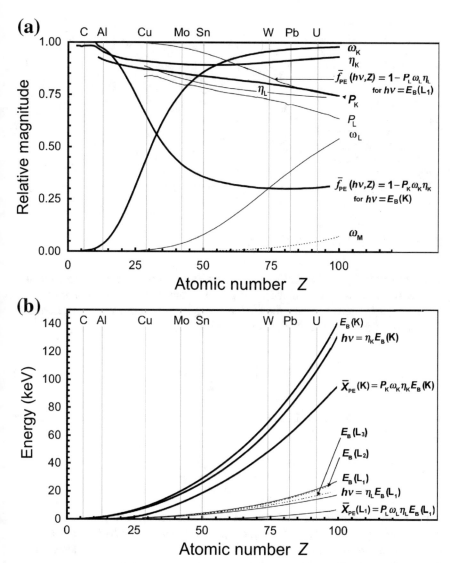

Fig. 7.29 Various atomic parameters relevant to photoelectric effect plotted against atomic number Z of the absorber: **a** fluorescence yields ω_K, ω_L, and ω_m; photoelectric probabilities P_K and P_L; fluorescence efficiencies η_K and η_L, and photoelectric mean energy transfer fractions $\overline{f}_{PE}$ for photon energies $h\nu = E_B\,(K)$ and $h\nu = E_B\,(L_1)$. **b** K-shell and L_1, L_2, L_3-subshell binding energies; mean fluorescence photon energies $h\overline{\nu}_K$ and $h\overline{\nu}_L$; and mean fluorescence emission energies $\overline{X}_{PE}$ (K) and $\overline{X}_{PE}$ (L_1)

η_j is the fluorescence efficiency defined for emission of j-shell fluorescence photon as the mean fraction of the j-subshell binding energy $E_B(j)$ carried by the fluorescence photon.

Fig. 7.30 Photoelectric
mass attenuation coefficient
τ/ρ against incident photon
energy $h\nu$ for lead in the
energy range from 1 to
100 keV covering the K, L,
and M absorption edges.
High and low τ/ρ values at
the K absorption edge as well
as the L and M absorption
subedges are shown to help
in the derivation of the
absorption edge parameter ξ_j

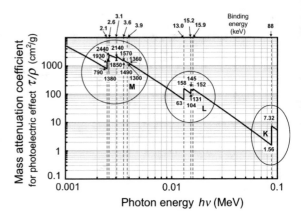

7.5.6 Photoelectric Probability

The probability P_j for the photoelectric effect, if it occurs, to occur in the j subshell
of an absorber atom is determined with the help of the photoelectric mass attenuation
coefficient τ/ρ plotted against the photon energy $h\nu$ to encompass the K, L, and M
absorption edges. In general, P_j is expressed as follows

$$P_j = \left(1 - \sum_{n=0}^{j-1} P_n\right)\xi_j, \qquad (7.147)$$

where $P_0 = 0$; $\sum P_j = 1$; and ξ_j is an absorption edge parameter defined for
subshell j as

$$\xi_j = \frac{(\tau/\rho)_j^H - (\tau/\rho)_j^L}{(\tau/\rho)_j^H}, \qquad (7.148)$$

with H and L designating the high and low values, respectively, of τ/ρ at the K
absorption edge or subedges (L$_1$, L$_2$, etc.). The numerator of (7.148) represents the
photon interactions in subshell j, while the denominator represents interactions in
all subshells including subshell j that are available for interaction.

To illustrate the determination of ξ_j and P_j we plot in Fig. 7.30 the photoelectric
mass attenuation coefficient τ/ρ for lead in the photon energy range from 1 keV to
100 keV. The graph covers the K, L, and M absorption edges with associated subedges
and displays the high and low τ/ρ values for the K absorption edge (7.32 cm^2/g and
1.56 cm^2/g, respectively) as well as for the three L subshells and five M subshells.
Table 7.14 presents the absorption edge parameters ξ_j for the K, L, and M shells of
lead for various photon beam energy ranges. The table also provides a summary of
data required for determination of absorption edge parameters ξ_j for lead. Row (4)
gives the K, L, and M subshell binding energies; row (5) the K, L, and M fluorescence
yields; rows (6) and (7) the high and low τ/ρ values, respectively, at the K shell and

the L and M subshell absorption edges. The table also provides the photoelectric probabilities P_j calculated with (7.148) for various photon energy regions covering photon energies from E_B (M_1) $= 3.9$ keV to $h\nu = \infty$.

As discussed above, for subshells j with binding energy $E_B(j)$ exceeding the photon energy $h\nu$, the photoelectric probability P_j is zero, i.e., for $h\nu < E_B(j) \rightarrow P_j = 0$. On the other hand, for $h\nu \geq E_B(j)$ the photoelectric probability P_j is finite, smaller than 1 and larger than 0. In this energy range the subshell whose binding energy $E_B(j)$ is the closest to $h\nu$ has by far the highest P_j value, so much so that it becomes practical and acceptable to account only for the highest P_j in the calculation of the mean photoelectric fluorescence emission energy $\overline{X}_{PE}$.

By way of example, we calculate the photoelectric probabilities P_K and P_L for photons of energy $h\nu = 100$ keV and $h\nu = 15.5$ keV interacting with lead. Row (4) of Table 7.14 shows that the 100 keV photon energy $h\nu$ exceeds the binding energy of the K-shell electron in lead (88 keV) suggesting that P_K will be the dominant probability for a photoelectric event. Similarly, the photon energy of 15.5 keV is between binding energies of the L_1 (15.9 keV) and L_2 (15.2 keV) subshells indicating that the 15.5 keV photon can interact neither with a K-shell electron ($P_K = 0$) nor with a L_1 subshell electron $\left(P_{L_1} = 0\right)$ but can interact with L_2 and L_3 electrons as well as with all higher shell electrons. Since in this case $P_K = 0$, one expects P_L given as the sum $P_{L_1} + P_{L_2} + P_{L_2}$ to be the dominant probability for the photoelectric interaction.

Equation (7.147) gives the following results for the photoelectric probabilities P_K and P_L for 100 keV and 15.5 keV photons interacting with lead (see Table 7.14):

1. For the $h\nu = 100$ keV photon interacting with lead atom:

$$h\nu > E_B(K) = 88 \text{ keV}$$

$$P_0 = 0 \tag{7.149}$$

$$P_K = (1 - P_0)\,\xi_K = 0.79 \tag{7.150}$$

$$P_{L_1} = [1 - (P_0 + P_K)]\,\xi_{L_1} = (1 - 0.79)\times 0.138 = 0.03 \tag{7.151}$$

$$P_{L_2} = \left[1 - \left(P_0 + P_K + P_{L_1}\right)\right]\xi_{L_2} = (1 - 0.82)\times 0.283 = 0.05 \tag{7.152}$$

$$P_{L_3} = \left[1 - \left(P_0 + P_K + P_{L_1} + P_{L_2}\right)\right]\xi_{L_3} = (1 - 0.87)\times 0.601 = 0.08 \tag{7.153}$$

$$P_L = P_{L_1} + P_{L_2} + P_{L_3} = 0.16 \tag{7.154}$$

2. For $h\nu = 15.5$ keV photon interacting with lead:
$$E_B(L_1) = 15.9 \text{ keV} < h\nu < E_B(L_2) = 15.2 \text{ keV}$$

$$P_K = 0 \tag{7.155}$$

$$P_{L_1} = 0 \tag{7.156}$$

$$P_{L_2} = \left[1 - \left(P_0 + P_K + P_{L_1}\right)\right]\xi_{L_2} = 0.283 \tag{7.157}$$

$$P_{L_3} = \left[1 - \left(P_0 + P_K + P_{L_1} + P_{L_2}\right)\right]\xi_{L_3} = (1 - 0.283)\times 0.601 = 0.431 \tag{7.158}$$

$$P_L = P_{L_1} + P_{L_2} + P_{L_3} = 0.714 \tag{7.159}$$

Table 7.14 Binding energy $E_B(j)$, fluorescence yield ω_j, mass attenuation coefficients $(\tau/\rho)_j^H$ and $(\tau/\rho)_j^L$, absorption edge parameters ξ_j, and photoelectric probability P_j for K, L, and M shell electrons in lead

1	Photon energy range	Shell	K	L			M				
		Subshell	–	L_1	L_2	L_3	M_1	M_2	M_3	M_4	M_5
2		j	1	2	3	4	5	6	7	8	9
3											
4		E_B (keV)	88	15.9	15.2	13.0	3.9	3.6	3.1	2.6	2.5
5		ω_{shell}	0.97		0.39				0.05		
6		$(\tau/\rho)_j^H$	7.32	152	142	158	1360	1570	2140	2440	1380
7		$(\tau/\rho)_j^L$	1.56	131	104	63	1300	1490	1850	1930	790
8		ζ_j	0.79	0.138	0.283	0.601	0.044	0.051	0.136	0.209	0.428
9	$h\nu \geq E_B\,(\text{K})$	P_{subshell}	0.79	0.03	0.05	0.08	0	0	0.01	0.01	0.015
10		P_{shell}	0.79		0.16				0.035		
11	$E_B\,(\text{L}_1) \leq h\nu < E_B\,(\text{K})$	P_{subshell}	0	0.138	0.244	0.370	0.011	0.012	0.030	0.040	0.065
12		P_{shell}	0		0.753				0.159		
13	$E_B\,(\text{L}_2) \leq h\nu < E_B\,(\text{L}_1)$	P_{subshell}	0	0	0.283	0.431	0.013	0.014	0.035	0.047	0.076
14		P_{shell}	0		0.714				0.185		
15	$E_B\,(\text{L}_3) \leq h\nu < E_B\,(\text{L}_2)$	P_{subshell}	0	0	0	0.601	0.018	0.019	0.049	0.065	0.106
16		P_{shell}	0		0.601				0.258		
17	$E_B\,(\text{M}_1) \leq h\nu < E_B\,(\text{L}_3)$	P_{subshell}	0	0	0	0	0.044	0.049	0.123	0.164	0.265
18		P_{shell}	0		0				0.645		

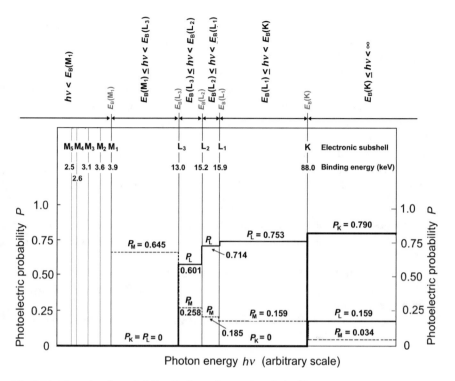

Fig. 7.31 Photoelectric probability P_j for various regions in incident photon energy in the range from $h\nu = E_B\,(M_1) = 3.9$ keV to $h\nu = \infty$

The photoelectric probabilities P_K and P_L calculated for the two photon energies in lead above are plotted in Fig. 7.31 and listed in Table 7.14 together with probabilities for several other regions in photon energy $h\nu$ spanning the absorption edges, as defined for individual subshells in lead. The most important photon energy region is the one for which the photon energy $h\nu$ matches or exceeds the K-shell binding energy $E_B(K)$. In this situation K-shell parameters dominate the photoelectric probabilities and higher shell parameters, for simplicity, are generally ignored. The next photon energy region is $E_B(L_1) \leq h\nu < E_B(K)$ for which the L-shell parameters dominate, etc.

7.5.7 Fluorescence Yield

The fluorescence yield ω_j was discussed in Sect. 4.1.2 and defined as the number of fluorescence (characteristic) photons emitted per vacancy in the given shell or subshell j. It can also be regarded as the probability, after creation of a shell or subshell vacancy, of fluorescence photon emission as opposed to Auger electron emission. Consequently, the probability for emission of an Auger electron following creation of a j-subshell vacancy is given as $(1 - \omega_j)$. The fluorescence yield ω_j is defined for an individual electronic subshell j but is commonly given for a shell as an average value over all subshells forming a shell.

Fluorescence yields ω_K, ω_L, and ω_M following creation of K, L, and M shell electron vacancies, respectively, are plotted in Figs. 4.4 and 7.29a against atomic number Z for all elements from $Z = 1$ to $Z = 100$. Figure 4.4 also plots the probability for emission of Auger electrons $(1 - \omega_j)$ against absorber atomic number Z for vacancies in the K, L, and M shells.

For a given shell, the fluorescence yield ranges from 0 for low atomic number Z absorbers to a maximum value at high atomic number Z; for a given absorber atomic number Z, the fluorescence yield decreases with increasing principal quantum number of a shell. For example:

1. ω_K ranges from $\omega_K = 0$ for absorbers with atomic number below $Z = 10$ through $\omega_K = 0.5$ for $Z = 30$ and saturates at $\omega_K = 0.97$ for absorbers with very high atomic number Z.
2. ω_L ranges from $\omega_L = 0$ for absorbers with atomic number below $Z = 30$ through $\omega_L = 0.25$ for $Z = 70$ and attains a value of $\omega_L \approx 0.5$ for absorbers with very high atomic number Z.
3. ω_M ranges from $\omega_M = 0$ for $Z < 60$ to a low value of $\omega_M = 0.05$ for very high atomic number Z absorbers.

7.5.8 Mean Fluorescence Photon Energy

The determination of the mean energy $h\bar{\nu}_j$ of the j-subshell fluorescence photon is by no means a simple task. These photons result from x-ray and optical transitions governed by well-known selection rules and are emitted with various intensities and energies depending on the subshell in which the allowed x-ray transition originated.

In the 1980s Hubbell determined $h\bar{\nu}_K$ for all elemental absorbers by calculating, for each given absorber, the intensity-weighted mean energy of all possible K-shell fluorescence photons. In 1990s Seltzer refined the calculation by also accounting for all cascade radiation transitions that accompany the migration of the vacancy from the K shell to the outer shell of the absorber atom. The main contribution to $h\bar{\nu}_K$ comes from the intensity-weighted mean energy for the possible K-shell fluorescence photons; however, Seltzer's cascade approach results in additional fluorescence photons from higher-level electronic shells and increases Hubbell's $h\bar{\nu}_K$ values by a small percentage.

By way of an example, we determine the mean K-shell fluorescence energy $h\bar{\nu}_K$ in lead. Figure 7.32 shows for lead absorber an atomic energy level diagram and K-shell fluorescence lines with their relative intensities normalized to 100% for the highest intensity $K - L_3$ transition (in IUPAC notation) or (K_{α_1}) transition (in standard Siegbahn notation). The energy level diagram is displayed using both the original Siegbahn spectroscopic notation as well as the new IUPAC notation, both notations

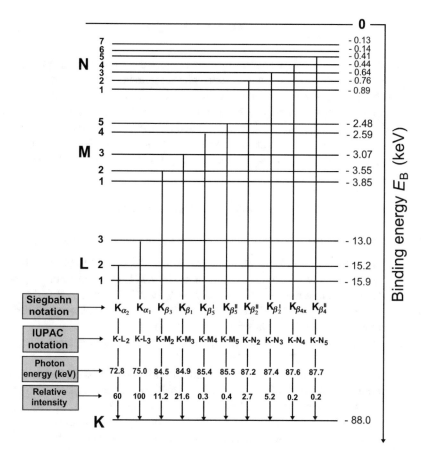

Fig. 7.32 Atomic energy level diagram for the K, L, M, and N shells of lead, illustrating the emission of K-shell fluorescence (characteristic) photons, their energies and relative intensities, normalized to 100% for the $K_{\alpha 1}$ (K − L$_3$) transition. The K-shell characteristic (fluorescence) x-ray photons are identified in the standard Siegbahn notation as well as in the new IUPAC notation

discussed in Sect. 4.1.1. A simple calculation for the lead absorber results in intensity-weighted mean K-shell fluorescence photon energy of 76.5 keV. In addition, the main contribution to cascade fluorescence will come from possible vacancy in L_2 and L_3 subshells, created by the two most probable K_α fluorescence photons.

The L-shell vacancies will be filled by electronic transitions from the M, N or even higher shells creating fluorescence photons or Auger electrons with energies of the order of 12 keV. Since the weight of the two K_α lines, as shown in Fig. 7.32, is ∼80% (relative intensity for the two lines is 160 out of a total of 200) and the fluorescence yield ω_L for lead is 0.4 (see Fig. 7.29a), we estimate the cascade fluorescence contribution as ∼4 keV. Adding this result to the mean K-shell fluorescence photon energy of 76.5 keV, gives a total $h\bar{\nu}_K$ of 80.5 keV resulting in a fluorescence efficiency of $\eta_K = 0.915$ for lead absorber. Note that for lead absorber $E_B(K) = 88$ keV and $h\bar{\nu}_k = \eta_K E_B(K) = 0.915 \times (88\,\text{keV}) = 80.5\,\text{keV}$.

7.5.9   Mean Fluorescence Emission Energy

As indicated in (7.146), the mean photoelectric fluorescence emission energy $\overline{X}_{PE}(j)$ actually emitted in the form of a fluorescence photon following a photoelectric event in the j subshell of an absorber atom is obtained by correcting $h\overline{\nu}_j$ for probabilities P_j and ω_j. For a photon with energy $h\nu$ exceeding the K-shell binding energy $E_B(K)$ of the absorber and ignoring all possible higher shell photoelectric interactions because of their small probability in comparison with the probability for interaction in the K shell, the mean fluorescence emission energy $\overline{X}_{PE}(K)$ emitted after a photoelectric event is approximated as

$$\overline{X}_{PE}(K) = P_K \omega_K h\overline{\nu}_K = P_K \omega_K \eta_K E_B(K), \tag{7.160}$$

where we use only the first term of the summation given in (7.146) and

P_K is the probability for the photoelectric effect, if it occurs, to occur in the K shell of an absorber atom. As shown in Fig. 7.29a, P_K ranges from $P_K \approx 1$ for low atomic number Z absorbers to $P_K \approx 0.75$ for high Z absorbers.

ω_K is the fluorescence yield for the K shell strongly dependent on absorber atomic number Z, as shown in Figs. 4.4 and 7.29a.

$h\overline{\nu}_K$ is the mean K-shell fluorescence photon energy plotted against absorber atomic number Z in Fig. 7.29a.

η_K is the fluorescence efficiency defined for emission of K-shell fluorescence photon as the mean fraction of the K-shell binding energy carried by the fluorescence photon. As shown in Fig. 7.29a, η_K decreases slowly from $\eta_K \approx 0.97$ for low Z absorbers, reaches a broad minimum of $\eta_K \approx 0.9$ at $Z \approx 50$ and then rises slowly to reach $\eta_K = 0.95$ for high Z absorbers.

7.5.10   Mean Photoelectric Energy Transfer Fraction

As suggested by (7.144) and (7.160), the mean energy transferred to charged particles (photoelectron and Auger electrons) in a K shell photoelectric event $\overline{E}_{tr}^{PE}$ is in general given as

$$\overline{E}_{tr}^{PE} = h\nu - \overline{X}_{PE}(K) = h\nu - P_K \omega_K \eta_K E_B(K) \quad \text{for} \quad h\nu > E_B(K), \tag{7.161}$$

allowing us to write the mean photoelectric energy transfer fraction $\overline{f}_{PE}$, valid for $h\nu \geq E_B(K)$, as

$$\overline{f}_{PE} = \frac{\overline{E}_{tr}^{PE}}{h\nu} = 1 - \frac{\overline{X}_{PE}}{h\nu} = 1 - \frac{P_K \omega_K \eta_K E_B(K)}{h\nu}. \tag{7.162}$$

The photoelectric energy transfer fraction $\overline{f}_{PE}$ ranges from $\overline{f}_{PE}[h\nu = E_B(K)] = 1 - P_K\omega_K\eta_K$ to $\overline{f}_{PE}(h\nu \to \infty) = 1$; i.e., $\{\overline{f}_{PE}[E_B(K)] = 1 - P_K\omega_K\eta_K\} \leq \overline{f}_{PE}(h\nu) \leq \{1 = \overline{f}_{PE}(\infty)\}$. Since each absorber Z has its own characteristic $E_B(K)$, it follows that $\overline{f}_{PE}$ depends not only on photon energy $h\nu$ but also on the atomic number Z of the absorber atom.

Figure 7.33a gives a plot of $\overline{f}_{PE}(h\nu, Z)$ against photon energy $h\nu$ for $h\nu > E_B(K)$ and eight selected elements Z ranging from carbon representing low Z absorbers to uranium representing high Z absorbers. The following general conclusions can be made:

1. For all absorbers the fraction $\overline{f}_{PE}(h\nu, Z)$ starts at its lowest value at the K absorption edge where $h\nu = E_B(K)$ and then gradually approaches $\overline{f}_{PE}(h\nu, Z) = 1$ with increasing photon energy $h\nu$.
2. For $\overline{f}_{PE}(Z, h\nu) = 1$, the incident photon energy $h\nu$ for $h\nu > E_B(K)$ is transferred to electrons in full. The photoelectron receives kinetic energy $[h\nu - E_B(K)]$ and the remaining portion of energy amounting to $E_B(K)$ goes either to Auger electrons for low Z absorbers or is essentially negligible in comparison to $h\nu$ for all absorbers at very high photon energies $h\nu$.
3. The dotted curve in Fig. 7.33a represents an envelope of points for the value of $\overline{f}_{PE}(Z, h\nu)$ at the absorption edge, i.e., $\overline{f}_{PE}[h\nu = E_B(K), Z] = 1 - P_K\omega_K\eta_K$ against incident photon energy $h\nu = E_B(K)$. For low Z absorbers $\overline{f}_{PE}[E_B(K)]$ is approximately equal to 1 and it then decreases with Z until it levels off at a broad minimum amounting to $\overline{f}_{PE}[E_B(K)] = 0.3$ for very high Z. The same data for $\overline{f}_{PE}[h\nu = E_B(K), Z]$ are also plotted in Fig. 7.29a against atomic number Z of the absorber.

Equations (7.161) and (7.162) are valid for photon energies $h\nu$ exceeding the absorber K-shell binding energy (K absorption edge energy) $E_B(K)$. We can apply similar methodology to lower energy photons but must recognize that in this energy range K-shell electrons will no longer contribute to photon interaction. Since the L shell contains three subshells (L_1, L_2, and L_3), we limit the energy range to L_1, the L subshell with the highest binding energy. As shown in Fig. 7.31, for energies between the K and L absorption edge, i.e., $E_B(L_1) \leq h\nu < E_B(K)$, the majority of photoelectric interactions will happen with L-shell electrons ($P_K = 0$; $P_L = 0.753$; $P_M = 0.159$) rather than with higher shell electrons and we write the mean photoelectric energy transfer fraction $\overline{f}_{PE}(h\nu, Z)$ as

$$\overline{f}_{PE}(h\nu, Z) = \frac{\overline{E}_{tr}^{PE}}{h\nu} = 1 - \frac{P_L\omega_L h\overline{\nu}_L}{h\nu} = 1 - \frac{P_L\omega_L\eta_L E_B(L_1)}{h\nu}, \qquad (7.163)$$

with parameters P_L, ω_L, and η_L plotted in Fig. 7.29a and $E_B(L_1)$ plotted in Fig. 7.29b.

Figure 7.33b extends the $\overline{f}_{PE}$ range of photon energies into the region between the L and K absorption edges and plots $\overline{f}_{PE}(h\nu, Z)$ for the eight selected elements of Fig. 7.33a for this energy range. The effect of the K absorption edge on $\overline{f}_{PE}(h\nu, Z)$ is now clearly visible, especially for high Z absorbers. For a given absorber, the minimum value in the L shell $\overline{f}_{PE}(h\nu, Z)$ which occurs at $h\nu = E_B(L_1)$ exceeds by

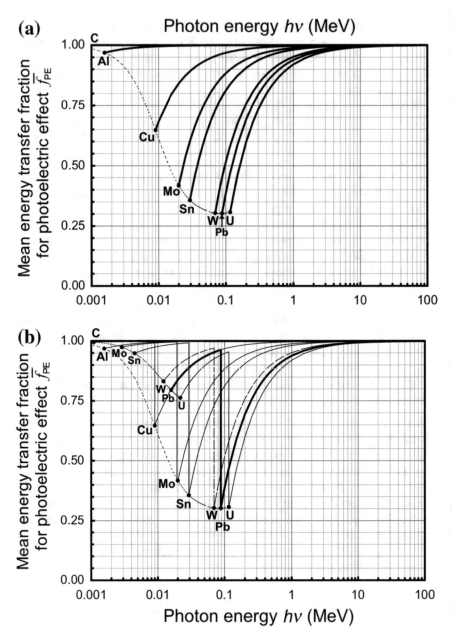

Fig. 7.33 Mean photoelectric energy transfer fraction $\overline{f}_{PE}$ against photon energy $h\nu$ for eight selected absorber atoms ranging from carbon at low Z and uranium at high Z. **a** For photon energy exceeding the K-shell binding energy while **b** extends the energy range downwards to $E_B(L_1)$. The *dashed curves* connect $\overline{f}_{PE}$ points for which $h\nu = E_B(K)$ and $h\nu = E_B(L_1)$

a significant margin the minimum $\overline{f}_{PE}$ value for the K shell, as shown by the two dashed curves connecting the minimum $\overline{f}_{PE}$ values for the K and L shells.

Since, as shown in Fig. 7.29a, the fluorescence yield ω_M is zero for all elements with $Z < 60$ and for $Z > 60$ it rises slowly to attain a value $\omega_M \approx 0.05$ for the very high Z absorbers, one can surmise that:

1. Fluorescence emission from high-level electronic shells is essentially negligible for all absorbers, even those with very high atomic number Z.
2. Extension of $\overline{f}_{PE}(h\nu, Z)$ data to energies below $E_B(L_1)$ will result in $\overline{f}_{PE}(h\nu, Z) \approx 1$ for all absorbers and all photon energies $h\nu < E_B(L_1)$.
3. The mean photoelectric energy transfer fraction $\overline{f}_{PE}(h\nu, Z)$, plotted in Fig. 7.33b and covering photon energies exceeding the binding energy of the L_1 subshell, can be used for all photon energies $h\nu$ and absorber atomic numbers Z. For photon energies below $E_B(L_1)$ for a given absorber, it is reasonable to use the approximation $\overline{f}_{PE}(h\nu, Z) \approx 1$.

A summary of relevant atomic photoelectric parameters is given in Table 7.15 for eight selected absorbers ranging from carbon to uranium. Data in rows (3) through (9) are relevant for photon energies exceeding the K-shell binding energy $E_B(K)$; data in rows (10) through (16) are relevant for photon energies between $E_B(L_1)$ and $E_B(K)$. Row (10) of the table lists $\overline{f}_{PE}[E_B(K), Z]$ for photon energy $h\nu$ equal to the K-shell electron binding energy $E_B(K)$ or K-absorption edge energy, representing the lowest value attained by $\overline{f}_{PE}$ for a given Z. Row (16) of the table lists $\overline{f}_{PE}[E_B(L_1), Z]$ for photon energy equal to the L_1 subshell binding energy $E_B(L_1)$ or L_1 absorption edge energy.

7.5.11 Mass Energy Transfer Coefficient for Photoelectric Effect

The mass energy transfer coefficient for the photoelectric effect $(\tau_K)_{tr}/\rho$ for $h\nu \geq E_B(K)$ is given by the following relationship

$$\frac{(\tau_K)_{tr}}{\rho} = \frac{\tau}{\rho} \frac{(\overline{E}_K^{PE})_{tr}}{h\nu} = \frac{\tau}{\rho}\overline{f}_{PE} = \frac{\tau}{\rho} \frac{h\nu - P_K\omega_K\eta_K E_B(K)}{h\nu}$$
$$= \frac{\tau}{\rho}\left(1 - \frac{P_K\omega_K\eta_K E_B(K)}{h\nu}\right), \tag{7.164}$$

where $\overline{f}_{PE} = 1 - P_K\omega_K\eta_K E_B(K)/(h\nu)$ is the mean photoelectric energy transfer fraction, providing the fraction of the incident photon energy transferred from the photon to charged particles (photoelectron and Auger electrons) in a photoelectric process. All other parameters of (7.164) were defined in Sect. 7.5.5.

Table 7.15 Various photoelectric parameters for the K and L shells of eight selected elements ranging from low atomic number carbon to high atomic number uranium

	Element	C	Al	Cu	Mo	Sn	W	Pb	U
1.	Atomic number Z	6	13	29	42	50	74	82	92
2.									
3.	P_K	1.0	0.911	0.871	0.852	0.839	0.803	0.788	0.767
4.	ω_K	0.002	0.036	0.446	0.764	0.859	0.957	0.968	0.976
5.	η_K	0.982	0.953	0.909	0.895	0.892	0.906	0.915	0.925
6.	$E_B(K)$ (keV)	0.284	1.56	8.98	20.0	29.2	69.5	88.0	115.6
7.	$h\bar{v}_K = \eta_K E_B(K)$ (keV)	0.279	1.49	8.16	17.9	26.1	63.0	80.5	106.9
8.	$\bar{X}_{PE}(K) = P_K \omega_K h\bar{v}_K$ (keV)	0	0.05	3.17	11.7	18.8	48.4	61.4	80.1
9.	$\bar{f}_{PE}[E_B(K)] = 1 - P_K \omega_K \eta_K$	1.0	0.969	0.647	0.417	0.357	0.304	0.302	0.307
10.	P_L	–	–	0.883	0.827	0.803	0.761	0.752	0.742
11.	ω_L	0	0	0.006	0.039	0.081	0.304	0.386	0.478
12.	η_L	–	–	0.839	0.799	0.783	0.728	0.706	0.669
13.	$E_B(L_1)$ (keV)	0.013	0.118	1.1	2.87	4.47	12.1	15.9	21.8
14.	$h\bar{v}_L = \eta_L E_B(L_1)$ (keV)	0	0	0.92	2.29	3.50	8.80	11.2	14.6

(continued)

Table 7.15 (continued)

		C	Al	Cu	Mo	Sn	W	Pb	U
1.	Element	C	Al	Cu	Mo	Sn	W	Pb	U
2.	Atomic number Z	6	13	29	42	50	74	82	92
15.	$\overline{X}_{PE}(L) = P_L \omega_L h\overline{\nu}_L$ (keV)	0	0	0.005	0.074	0.227	2.03	3.25	5.17
16.	$\overline{f}_{PE}[E_B(L_1)] = 1 - P_L\omega_L\eta_L$	1.0	1.0	0.996	0.974	0.949	0.832	0.795	0.763

P_K and $P_L]$ Photoelectric probability for the K shell and L shell, respectively; also defined as fraction of all photoelectric interactions that occur in K shell and L shell, respectively.

ω_K and ω_L Fluorescence yield for the K shell and the L shell, respectively.

η_K and η_L Fluorescence efficiency for the K shell and the L shell, respectively.

$E_B(K)$ Binding energy of orbital electron in the K shell.

$E_B(L_1)$ Binding energy of orbital electron in the L1 subshell $h\overline{\nu}_K$ Mean fluorescence photon energy for all allowed fluorescence x-ray transitions including cascade effects following formation of a K-shell vacancy.

$h\overline{\nu}_L$ Mean fluorescence photon energy for all allowed fluorescence x-ray transitions including cascade effects following formation of a L-shell vacancy.

$\overline{X}_{PE}(K)$ Mean fluorescence emission energy for K-shell photoelectric process.

$\overline{X}_{PE}(L)$ Mean fluorescence emission energy for L-shell photoelectric process.

$\overline{f}_{PE}[E_B(K)]$ Photoelectric mean energy transfer fraction for incident photon energy $h\nu = E_B(K)$.

$\overline{f}_{PE}[E_B(L_1)]$ Photoelectric mean energy transfer fraction for incident photon energy $h\nu = E_B(L_1)$.

The following quantities that were already plotted in Fig. 7.29 are tabulated: photoelectric probabilities P_K and P_L; fluorescence efficiencies η_K and η_L; binding energies $E_B(K)$ and $E_B(L)$; mean fluorescence photon energies $h\overline{\nu}_K$ and $h\overline{\nu}_L$; mean fluorescence emission energies $\overline{X}_{PE}(K)$ and $\overline{X}_{PE}(L)$; and mean photoelectric energy transfer fractions $\overline{f}_{PE}$ for photon energies $h\nu = E_B(K)$ and $h\nu = E_B(L_1)$

7.6 Pair Production

When the incident photon energy $h\nu$ exceeds $2m_ec^2 = 1.02$ MeV, with m_ec^2 the rest energy of electron and positron, the production of an electron–positron pair in conjunction with a complete absorption of the photon by absorber atom becomes energetically possible. For the effect to occur, three quantities must be conserved: *energy*, *charge*, and *momentum*.

For $h\nu > 2m_ec^2$, total energy and charge can be conserved even if pair production occurred in free space. However, to conserve the linear momentum simultaneously with total energy and charge, the effect cannot occur in free space; it can only occur in the Coulomb field of a collision partner (either atomic nucleus or orbital electron) that can take up a suitable fraction of the momentum carried by the photon. The energy distribution and angular distribution of electrons and positrons in pair production are complex functions of the incident photon energy $h\nu$ and absorber atomic number Z.

7.6.1 Conservation of Energy, Momentum and Charge in Pair Production

Before the pair production interaction the incident photon has energy $E_\nu = h\nu > 2m_ec^2$ and momentum $p_\nu = h\nu/c$. In the interaction an electron–positron pair is produced with a total energy $E_{\text{pair}} = 2\gamma m_ec^2$ and total momentum $p_{\text{pair}} = 2\gamma m_e\upsilon$. A summary of parameters for before and after pair production interaction in free space is given in Table 7.16.

Based on data in Table 7.16 the total energy conservation in "free space" pair production would be expressed as

$$E_\nu = h\nu \equiv E_{\text{Pair}} = 2\gamma m_ec^2, \tag{7.165}$$

while the conservation of momentum would be written as follows

$$p_\nu = \frac{h\nu}{c} \equiv p_{\text{pair}} = 2\gamma m_e\upsilon = 2\gamma m_ec^2\frac{\upsilon}{c^2} = E_{\text{pair}}\frac{\upsilon}{c^2} = E_\nu\frac{\upsilon}{c^2} = p_\nu\frac{\upsilon}{c}. \tag{7.170}$$

Table 7.16 Parameters of pair production interaction in "free space"

Before the pair production interaction in "free space"		After the pair production interaction in "free space"	
Total energy before interaction		Total energy after interaction	
$E_\nu = h\nu$	(7.166)	$E_{\text{Pair}} = 2\gamma m_ec^2$	(7.167)
Momentum before the interaction		Momentum after the interaction	
$p_\nu = \frac{h\nu}{c}$	(7.168)	$p_{\text{Pair}} = 2\gamma m_e\upsilon = E_{\text{Pair}}\frac{\upsilon}{c^2}$	(7.169)
Total charge before the interaction		Total charge after the interaction	
0		0	

Equation (7.170) is contradictory and shows that both total energy and momentum cannot be conserved simultaneously. Since the particle velocity v is always smaller than c, it follows that p_ν, the momentum before the pair production interaction, is always larger than p_{pair}, the total momentum after the pair production interaction. The photon possesses momentum excess that cannot be absorbed by the electron–positron pair, therefore, it must be absorbed by a collision partner, be it the atomic nucleus or an orbital electron of the absorber. Thus, in an absorber atom two collision partners are available for absorbing the extra photon momentum: the atomic nucleus and orbital electrons:

1. When the extra momentum is absorbed by the atomic nucleus of the absorber, the recoil energy, as a result of the relatively large nuclear mass, is exceedingly small and the effect is described as the standard pair production (usually referred to as *nuclear pair production*). Two particles (electron and positron) leave the interaction site.
2. When an orbital electron of the absorber picks up the extra photon momentum, the recoil energy of the orbital electron may be significant and the effect is described as the pair production in the field of electron, i.e., electronic pair production better known as *triplet production*. Three particles (two electrons and a positron) leave the interaction site.

The total charge before the pair production interaction is zero and the total charge after interaction is also zero. The charge conservation rule is thus satisfied in pair production. The nuclear pair production and triplet production interactions are shown schematically in Fig. 7.34, nuclear pair production in part (a), triplet production in part (b).

7.6.2 Threshold Energy for Nuclear Pair Production and Triplet Production

In contrast to other common photon interactions, such as photoelectric effect, Rayleigh scattering and Compton scattering, pair production exhibits a clear threshold energy below which the effect cannot happen. The threshold energy is derived following the procedure described in detail in Sect. 5.2.3 that is based on the invariant: $E^2 - p^2c^2 = $ invariant where E and p are the total energy and total momentum, respectively, before and after the interaction.

For pair production in the field of the nucleus (**nuclear pair production**: NPP) the conditions for before the interaction (in the laboratory system) and for after the interaction (in the center-of-mass system) are written, as shown in Table 7.17, with $(h\nu)_{\text{thr}}^{\text{NPP}}$ the threshold photon energy for nuclear pair production and $m_A c^2$ the rest energy of the nucleus, the interaction partner.

The invariant $E^2 - p^2c^2 = $ invariant for before and after the nuclear pair production event is now written as

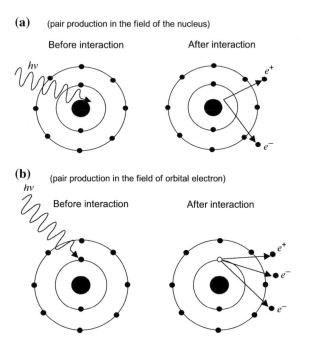

(a)   (pair production in the field of the nucleus)

Before interaction   After interaction

(b)   (pair production in the field of orbital electron)

hv

Before interaction   After interaction

Fig. 7.34 Schematic representation of pair production: **a** nuclear pair production in the Coulomb field of the absorber nucleus and **b** electron pair production (triplet production) in the Coulomb field of an orbital electron

Table 7.17 Parameters for photon energy threshold calculation in nuclear pair production using the invariant $E^2 - p^2c^2 =$ invariant

Before the pair production interaction (in laboratory coordinate system)		After the pair production interaction (in center-of-mass coordinate system)	
Total energy before interaction $(h\nu)_{\text{thr}}^{\text{NPP}} + m_A c^2$	(7.171)	Total energy after interaction $(m_A c^2 + 2m_e c^2)$	(7.172)
Momentum before interaction $\dfrac{(h\nu)_{\text{thr}}^{\text{NPP}}}{c}$	(7.173)	Momentum after interaction 0	

$$\left\{(h\nu)_{\text{thr}}^{\text{NPP}} + m_A c^2\right\}^2 - \left(\frac{(h\nu)_{\text{thr}}^{\text{NPP}}}{c}\right)^2 c^2 = \left(m_A c^2 + 2m_e c^2\right)^2 - 0, \qquad (7.177)$$

resulting in the following expression for pair production threshold energy $E_{\text{thr}}^{\text{NPP}} = (h\nu)_{\text{thr}}^{\text{NPP}}$

$$E_{\text{thr}}^{\text{NPP}} = (h\nu)_{\gamma\text{thr}}^{\text{NPP}} = 2m_e c^2 \left(1 + \frac{m_e c^2}{m_A c^2}\right) = (1.022\,\text{MeV}) \times \left(1 + \frac{m_e c^2}{m_A c^2}\right). \tag{7.178}$$

Table 7.18 Parameters for photon energy threshold calculation in triplet production using the invariant $E^2 - p^2c^2 =$ invariant

Before the triplet production interaction (in laboratory coordinate system)		After the triplet production interaction (in center-of-mass coordinate system)	
Total energy before interaction		Total energy after interaction	
$(h\nu)_{\text{thr}}^{\text{TP}} + m_ec^2$	(7.174)	$3m_ec^2$	(7.175)
Momentum before interaction		Momentum after interaction	
$\dfrac{(h\nu)_{\text{thr}}^{\text{TP}}}{c}$	(7.176)	0	

Threshold energy for pair production in the field of the nucleus is $2m_ec^2[1 + (m_ec^2)/(m_Ac^2)]$ with $2m_ec^2$ accounting for materialization of the electron–positron pair and the minute energy of $2(m_ec^2)^2/(m_Ac^2)$ expended for nuclear recoil in order to conserve the momentum of the incident photon $(h\nu/c)$. In the first approximation we can use $(h\nu)_{\text{thr}}^{\text{NPP}} \approx 2m_ec^2$, since the ratio m_ec^2/m_Ac^2 is very small, indicating that the recoil energy of the absorber nucleus, which enabled the pair production event, is exceedingly small.

For **triplet production** (TP) the conditions for before the interaction (in the laboratory system) and for after the interaction (in the center-of-mass system) are written, as shown in Table 7.18, with $(h\nu)_{\text{thr}}^{\text{TP}}$ the threshold photon energy for triplet production and m_ec^2 the rest mass of the orbital electron, the interaction partner.

Note: The total energy after the triplet interaction is $3m_ec^2$ to account for rest energies of the orbital electron which enables the interaction as well as for the electron–positron pair that is produced in the interaction (materialization).

The invariant $E^2 - p^2c^2 =$ invariant for before and after the triplet production event is now written as

$$\left\{(h\nu)_{\text{thr}}^{\text{TP}} + m_ec^2\right\}^2 - \left[(h\nu)_{\text{thr}}^{\text{TP}}\right]^2 = \left(3m_ec^2\right)^2 - 0, \qquad (7.179)$$

resulting in the following expression for the triplet production threshold energy $E_{\text{thr}}^{\text{TP}}$

$$E_{\text{thr}}^{\text{TP}} = (h\nu)_{\text{thr}}^{\text{TP}} = 4m_ec^2 = 2.044 \,\text{MeV}. \qquad (7.180)$$

The threshold energy for triplet production is $4m_ec^2$ consisting of $2m_ec^2$ for materialization of the electron–positron pair and $2m_ec^2$ for recoil energy of the three participating particles with each obtaining kinetic energy of $\frac{2}{3}m_ec^2$.

7.6.3 Energy Distribution of Electrons and Positrons in Nuclear Pair Production and Triplet Production

In both the nuclear pair production and in triplet production the energy converted into mass (materialization) is $2m_ec^2$, so energy conservation is expressed as:

1. For pair production, ignoring the minute kinetic energy transferred to recoil energy of the nucleus

$$h\nu = 2m_ec^2 + (E_K)_{e^-} + (E_K)_{e^+} \qquad (7.181)$$

2. For triplet production

$$h\nu = 2m_ec^2 + (E_K)_{e^-} + (E_K)_{e^+} + (E_K)_{\text{orb.el.}}, \qquad (7.182)$$

where $(E_K)_{e^-}$ and $(E_K)_{e^+}$ are kinetic energies of the electron and positron, respectively, of the electron–positron pair produced in the nuclear pair production and triplet production. $(E_K)_{\text{orb.el.}}$ is the kinetic energy of the orbital electron which, with its Coulomb field, enabled the triplet production event and was subsequently ejected from the absorber atom.

The total kinetic energy transferred to charged particles (electron and positron) in nuclear pair production is

$$(E_K^{\text{NPP}})_{\text{tr}} = h\nu - 2m_ec^2, \qquad (7.183)$$

ignoring the minute recoil energy of the nucleus. Generally, the electron and the positron do not receive equal kinetic energies but their average is given as

$$\overline{E}_K^{\text{NPP}} = \frac{h\nu - 2m_ec^2}{2}. \qquad (7.184)$$

The mean energy transferred to each charged particle released in the absorber is thus $\frac{1}{2}(h\nu - 2m_ec^2)$ in nuclear pair production and $\frac{1}{3}(h\nu - 2m_ec^2)$ in triplet production. The actual energy distribution among the charged particles released or produced in the absorber follows very broad general guidelines:

1. For nuclear pair production, all distributions of the available energy $(h\nu - 2m_ec^2)$ are almost equally probable except for the extreme case in which one particle would obtain all the available energy and the other particle none of it.
2. For triplet production, F. Perrin, who also was the first to derive the threshold energy of $4m_ec^2$ for triplet production, showed that the kinetic energy of each of the three particles released lies within limits defined by the following expression

$$E_K = \frac{\varepsilon^2 - 2\varepsilon - 2 \pm \varepsilon\sqrt{\varepsilon(\varepsilon - 4)}}{2\varepsilon + 1} m_ec^2, \qquad (7.185)$$

where ε, like in the Compton effect, is defined as the photon energy $h\nu$ normalized to electron rest mass energy m_ec^2

$$\varepsilon = \frac{h\nu}{m_ec^2}. \qquad (7.186)$$

3. For triplet production threshold energy of $h\nu = 4m_ec^2$ Perrin's equation reduces to $E_K = \frac{2}{3}m_ec^2$, meaning that each of the three particles carries $1/3$ of the total energy $(2m_ec^2)$ available for particle recoil. *Note:* Of the $4m_ec^2$ threshold energy, $2m_ec^2$ goes for materialization of the electron–positron pair and the remaining $2m_ec^2$ goes into combined recoil energy of the three particles.

4. *For example,* for a 20 MeV photon Perrin's equation (7.185) predicts that the kinetic energies of each one of the three particles released will lie between limits of 0.68 keV and 18.7 MeV.

7.6.4 Angular Distribution of Charged Particles in Pair Production

Like the energy distribution, the angular distribution of electrons and positrons produced in pair production is a complex function of incident photon energy $h\nu$ and absorber atomic number Z. *Walter Heitler* and *Hans Bethe* are credited with carrying out the seminal theoretical work on this problem. With increasing incident photon energy $h\nu$ the distribution of charged particles is peaked increasingly in the forward direction. For very high photon energies with $\varepsilon = h\nu/(m_ec^2) \gg 1$, the mean angle $\bar{\theta}$ of positron and electron emission is of the order of $\bar{\theta} \approx 1/\varepsilon$ with respect to the direction of the incident photon, resulting in a forward peaked angular distribution of electrons and positrons.

7.6.5 Nuclear Screening

For very high photon energies $(h\nu > 20\,\text{MeV})$ significant contribution to the pair production cross section may come from interaction points that lie outside the orbit of K shell electrons. The Coulomb field in which the pair production occurs is thus reduced because of the screening of the nucleus by the two K-shell electrons, thereby requiring a screening correction in theoretical calculations. *Hans Bethe* and *Walter Heitler* used the Thomas–Fermi atomic model (see Sect. 2.4.3) in their theory of pair production to describe the electrostatic potential resulting from the nuclear charge in conjunction with the K shell electrons producing the screening effect.

7.6.6 Atomic Cross Section for Pair Production

The theoretical derivations of atomic cross sections for pair production $_a\kappa$ are very complicated, some based on Born approximation, others not, some accounting for nuclear screening and others not. In general, the atomic cross sections for pair production in the field of a nucleus or orbital electron appear as follows

$$_a\kappa = \alpha r_e^2 Z^2 P(\varepsilon, Z), \tag{7.187}$$

where

α is the fine structure constant ($\alpha = 1/137$).
r_e is the classical electron radius [$r_e = e^2/(4\pi\varepsilon_0 m_e c^2) = 2.818\,\mathrm{fm}$].
Z is the atomic number of the absorber.
$P(\varepsilon, Z)$ is a complicated function of the photon energy $h\nu$ and atomic number Z of the absorber, as given in Table 7.19.

It is evident from (7.187) through (7.191) and from Table 7.19 that, aside from the effects of nuclear screening by K-shell electrons, the atomic cross section for nuclear pair production $_a\kappa_{NPP}$ is proportional to Z^2, while the atomic cross section for triplet production $_a\kappa_{TP}$ is linearly proportional to Z. In general, the relationship between $_a\kappa_{NPP}$ and $_a\kappa_{TP}$ is given as follows

$$\frac{_a\kappa_{NPP}}{_a\kappa_{TP}} = \eta Z, \tag{7.192}$$

where η is a parameter, depending only on $h\nu$, and, according to *Robley Evans*, equal to 2.6 at $h\nu = 6.5\,\mathrm{MeV}$, 1.2 at $h\nu = 100\,\mathrm{MeV}$, and approaching unity as $h\nu \to \infty$. This indicates that the atomic cross section for triplet production $_a\kappa_{TP}$ is at best about 30% of the pair production cross section $_a\kappa_{NPP}$ for $Z = 1$ and less than 1% for high Z absorbers.

Since $_a\kappa_{NPP}$, the atomic cross section for pair production in the field of the atomic nucleus, exceeds significantly $_a\kappa_{TP}$, the atomic cross section for triplet production, as shown in Fig. 7.35 for two absorbing materials: carbon with $Z = 6$ and lead with $Z = 82$, both the nuclear pair production and the triplet production contributions are

Table 7.19 Characteristics of atomic cross section for pair production in the field of the nucleus (nuclear pair production) and in the field of an orbital electron (electronic pair production also called triplet production), according to Marmier and Sheldon

Field	Energy range	$P(\varepsilon, Z)$	Comment	
Nucleus	$1 \ll \varepsilon \ll 1/(\alpha Z^{1/3})$	$\dfrac{28}{9}\ln 2\varepsilon - \dfrac{218}{27}$	No screening	(7.188)
Nucleus	$\varepsilon \gg 1/(\alpha Z^{1/3})$	$\dfrac{28}{9}\ln\dfrac{183}{Z^{1/3}} - \dfrac{2}{27}$	Complete screening	(7.189)
Nucleus	Outside the limits above but $\varepsilon > 4$	$\dfrac{28}{9}\ln 2\varepsilon - \dfrac{218}{27} - 1.027$	No screening	(7.190)
Electron	$\varepsilon > 4$	$\dfrac{1}{Z}\left(\dfrac{28}{9}\ln 2\varepsilon - 11.3\right)$	No screening	(7.191)

Fig. 7.35 Atomic cross sections for nuclear pair production $_a\kappa_{NPP}$ (*solid curves*) and for triplet production (electronic pair production) $_a\kappa_{TP}$ (*dotted curves*) against incident photon energy $h\nu$ for carbon and lead. Data are from the NIST

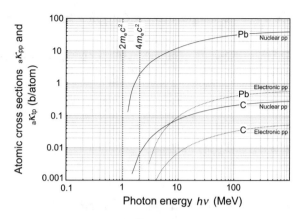

Fig. 7.36 Atomic cross section for general pair production (including nuclear pair production and triplet production) $_a\kappa$ against incident photon energy $h\nu$ for various absorbers in the range from hydrogen to lead. Data are from the NIST

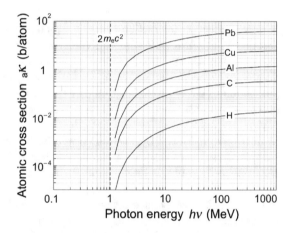

usually combined and their sum given under the header of general pair production as follows

$$_a\kappa = {}_a\kappa_{NPP} + {}_a\kappa_{TP} = {}_a\kappa_{NPP}\left\{1 + \frac{1}{\eta Z}\right\}, \qquad (7.193)$$

where the electronic effects (triplet production) are accounted for with a correction term $1/(\eta Z)$. This term is equal to zero for $h\nu < 4m_ec^2$, where $4m_ec^2$ is the threshold energy for triplet production.

Atomic cross sections for general pair production $_a\kappa$ including both the nuclear pair production and the triplet production are plotted in Fig. 7.36 for various absorbers ranging from hydrogen to lead. The increase of $_a\kappa$ with incident photon energy $h\nu$ and with atomic number Z of the absorber is evident.

7.6.7 Mass Attenuation Coefficient for Pair Production

The mass attenuation coefficient for pair production κ/ρ is calculated from the atomic cross section $_a\kappa$ with the standard relationship

$$\frac{\kappa}{\rho} = \frac{N_A}{A}\,_a\kappa,\qquad(7.194)$$

where A and ρ are the atomic mass and density, respectively, of the absorber.

7.6.8 Energy Transfer to Charged Particles in Nuclear Pair Production and Triplet Production

In pair production interaction between a photon of energy $h\nu$ and a nucleus of absorber atom with atomic number Z (*nuclear pair production*), an electron–positron pair is produced in the field of the nucleus, as shown in Fig. 7.34a. The incident photon disappears in the interaction: $2m_ec^2$ ($=1.022$ MeV) of its energy $h\nu$ is expended for production of the electron–positron pair (materialization) and $(h\nu - 2m_ec^2)$ of its energy is shared as kinetic energy between the two particles of the electron–positron pair.

In *triplet production interaction* between a photon of energy $h\nu$ and an orbital electron of absorber atom with atomic number Z (electronic pair production), an electron–positron pair is produced in the field of an orbital electron, as shown in Fig. 7.34b. The incident photon disappears in the interaction: $2m_ec^2$ ($= 1.022$ MeV) of its energy $h\nu$ is expended for production of the electron–positron pair (materialization) and $(h\nu - 2m_ec^2)$ of its energy is shared between the triplet formed by the electron–positron pair and the orbital electron.

For both the nuclear pair production and the triplet production the mean energy transfer $\overline{E}_{tr}^{PP}$ to charged particles is

$$\overline{E}_{tr}^{PP} = h\nu - 2m_ec^2.\qquad(7.195)$$

The mean energy transfer fraction for general pair production including nuclear pair production and triplet production $\overline{f}_{PP}$ is given as

$$\overline{f}_{PP} = \frac{\overline{E}_{tr}^{PP}}{h\nu} = 1 - \frac{2m_ec^2}{h\nu}\qquad(7.196)$$

and plotted in Fig. 7.37. For incident photon energy $h\nu < 2m_ec^2 = 1.022$ MeV, $\overline{f}_{PP} = 0$; with photon energy $h\nu$ increasing above 1.022 MeV, $\overline{f}_{PP}$ first rises rapidly through $\overline{f}_{PP} = 0.5$ at $h\nu \approx 2$ MeV, $\overline{f}_{PP} = 0.66$ at $h\nu \approx 3$ MeV, and $\overline{f}_{PP} = 0.9$ at $h\nu \approx 10$ MeV, asymptotically approaches $\overline{f}_{PP} = 1$ as $[1 - \text{const}/(h\nu)]$ and

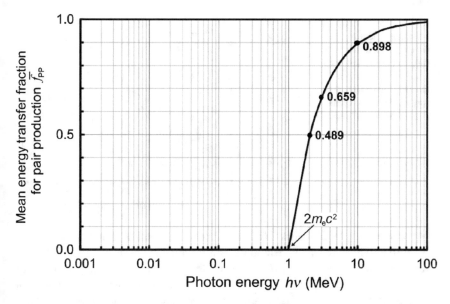

Fig. 7.37 The mean pair production energy transfer fraction $\overline{f}_{PP}$ against photon energy $h\nu$

saturates at $\overline{f}_{PP} \approx 1$ for $h\nu > 100\,\text{MeV}$. Like the mean Compton energy transfer fraction $\overline{f}_C\,(h\nu)$, the mean pair production energy transfer fraction $\overline{f}_{PP}$ depends on photon energy and does not depend on the atomic number Z of the absorber. The anchor points $[\overline{f}_{PP} = 0.5;\ 0.66;\ 0.9$ at $h\nu$ (MeV) $= 2;\ 3;\ 10$, respectively] for the $\overline{f}_{PP}$ diagram are also shown in Fig. 7.37.

7.6.9 Mass Energy Transfer Coefficient for Pair Production

The mass energy transfer coefficient for pair production κ_{tr}/ρ is proportional to the mass attenuation coefficient for pair production κ/ρ through the mean pair production energy transfer fraction $\overline{f}_{PP}$ plotted in Fig. 7.37 and given in (7.196)

$$\frac{\kappa_{tr}}{\rho} = \frac{\kappa}{\rho}\overline{f}_{PP} = \frac{\kappa}{\rho}\left(1 - \frac{2m_e c^2}{h\nu}\right) = \frac{\kappa}{\rho}\left(1 - \frac{2}{\varepsilon}\right), \qquad (7.197)$$

where ε, as before, is the photon energy $h\nu$ normalized to the electron rest energy $m_e c^2$.

Figure 7.38 shows a comparison between the mass attenuation coefficient κ/ρ (dashed curve) and the mass energy transfer coefficient κ_{tr}/ρ (solid curve) against photon energy $h\nu$ for carbon (low Z absorber) and lead (high Z absorber). The following features are apparent:

Fig. 7.38 Mass energy
transfer coefficient κ_{tr}/ρ
(*solid curves*) and mass
attenuation coefficient κ/ρ
(*dashed curves*) for pair
production against photon
energy $h\nu$ for carbon and
lead. Data are from the NIST

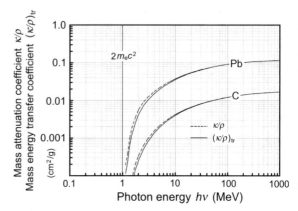

1. At photon energies $h\nu$ below 1.022 MeV both coefficients κ/ρ and κ_{tr}/ρ are equal
   to zero as a result of the threshold for pair production of $2m_ec^2 = 1.022$ MeV.
2. For photon energies $h\nu$ increasing from the threshold value, κ/ρ at first significantly exceeds κ_{tr}/ρ because of the low value of the mean pair production
   energy transfer fraction $\overline{f}_{PP}$ at relatively low photon energies of the order of a
   few megaelectron volt.
3. As shown in Fig. 7.37, however, $\overline{f}_{PP}$ rises rapidly with $h\nu$, reaching a value of
   $\overline{f}_{PP} = 0.9$ at $h\nu = 10$ MeV and asymptotically approaching $\overline{f}_{PP} = 1$ for photon
   energies exceeding 100 MeV. This implies that with increasing photon energy $h\nu$
   the two coefficients converge to the same value.
4. At photon energies $h\nu$ exceeding 10 MeV, $\overline{f}_{PP} \approx 1$ and the mass energy transfer
   coefficient for pair production κ_{tr}/ρ is equal to the mass attenuation coefficient
   for pair production κ/ρ.

7.6.10 *Positron Annihilation*

The positron e^+ is an antiparticle to electron e^-. The two particles have identical
rest masses and rest mass energies, $(m_{e^+}c^2 = m_{e^-}c^2 = m_ec^2 = 0.511$ MeV), and
charges that are opposite in sign but equal in magnitude $(1.6\times10^{-19}$ C); electrons are
negative, positrons positive. Electrons were discovered in 1897 by *Joseph J. Thomson*
while he was carrying out experiments with a Crookes tube; positrons in 1932 by
Carl Anderson during his study of cosmic ray tracks in a Wilson cloud chamber. Of
interest in medical physics are positrons produced by:

1. Energetic photons undergoing pair production or triplet production (important in
   radiation dosimetry and health physics).
2. Beta plus $\left(\beta^+\right)$ decay used in positron emission tomography (PET) imaging.

Positronium

Energetic positrons move through an absorbing medium and lose their kinetic energy
through collision and radiation losses in Coulomb interactions with orbital electrons

and nuclei, respectively, of the absorber. Eventually, each positron collides with an electron of the absorber and the two annihilate directly or they annihilate through an intermediate step forming a metastable hydrogen-like structure (see Sect. 3.1.7) called positronium (Ps). The positron and electron of the positronium revolve about their common center-of-mass in discrete orbits that are subjected to Bohr quantization rules with the reduced mass equal to one half of the electron rest mass and the lowest energy state with a binding energy of $\frac{1}{2}E_R = 6.8\,\text{eV}$. The process of positron–electron annihilation is an inverse to pair production with the total mass before annihilation transformed into one, two, or three photons.

Standard Positron–Electron Annihilation

The most common electron–positron annihilation occurs after the positron lost all of its kinetic energy and undergoes annihilation with an orbital electron of the absorber. The electron is considered stationary and free. The annihilation results in two photons (annihilation quanta) of energy $m_e c^2 = 0.511$ MeV each and moving in opposite directions (at nearly $180°$ to one another), ensuring conservation of total charge (zero), total energy ($2m_e c^2 = 1.02$ MeV), and total momentum (zero).

In-Flight Annihilation

A less common event (of the order of 2% of all positron–electron annihilation interactions) is the in-flight annihilation between a positron with non-zero kinetic energy E_K and either a tightly bound electron or a "free" electron.

When the electron is tightly bound to the nucleus, the nucleus can pick up the recoil momentum, and annihilation-in-flight produces only one photon with energy equal to the sum of the total positron energy E_{e^+} (which in turn is the sum of its rest energy $m_{e^+}c^2$ and its kinetic energy $E_K^{e^+}$) and rest energy of the electron $m_{e^-}c^2$

$$h\nu = E_{e^+} + m_{e^-}c^2 = \left(E_K^{e^+} + m_{e^+}c^2\right) + m_{e^-}c^2. \tag{7.198}$$

When the annihilation electron is essentially free, the in-flight annihilation results in two photons, $h\nu_1$ and $h\nu_2$, moving from the annihilation point each with its own emission angle (θ and ϕ, respectively) with respect to the incident positron direction. Energies of the two photons are governed by emission angles θ and ϕ as well as by the principles of energy and momentum conservation, as indicated in Fig. 7.39 and Table 7.20.

The relationships among the energies of emitted photons and emission angles are determined from the following three expressions for:

1. *Conservation of energy*

$$E_K^{e^+} + 2m_e c^2 = h\nu_1 + h\nu_2 \tag{7.202}$$

2. *Conservation of momentum (x-axis component)*

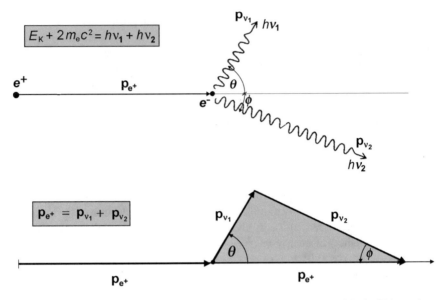

Fig. 7.39 Two-photon in-flight positron annihilation. The kinetic energy of the incident positron is $E_K^{e^+}$, its momentum is $p_e = \sqrt{E_K^{e^+}(E_K^{e^+} + 2m_ec^2)}$. Photon 1 is emitted with energy $h\nu_1$ and momentum $\mathbf{p}_{\nu_1}$; photon 2 is emitted with energy $h\nu_2$ and momentum $\mathbf{p}_{\nu_2}$. The photon emission angles are θ and ϕ measured with respect to the direction of the incoming positron. The x components of $\mathbf{p}_{\nu_1}$ and $\mathbf{p}_{\nu_2}$ are $h\nu_1 c^{-1} \cos\theta$ and $h\nu_2 c^{-1} \cos\phi$, respectively. The y components of $\mathbf{p}_{\nu_1}$ and $\mathbf{p}_{\nu_2}$ are $h\nu_1 c^{-1} \sin\theta$ and $h\nu_2 c^{-1} \sin\theta$, respectively

Table 7.20 Total energy and momentum conservation for in-flight annihilation

Total energy before annihilation	Total energy after annihilation	
$E_K^{e^+} + m_{e^+}c^2 + m_{e^-}c^2$	$h\nu_1 + h\nu_2$	(7.199)
Momentum before annihilation (x axis)	Momentum after annihilation (x axis)	
$p_{e^+} = \dfrac{1}{c}\sqrt{E^2 - (m_{e^+}c^2)^2}$	$\dfrac{h\nu_1}{c}\cos\theta + \dfrac{h\nu_2}{c}\cos\phi$	(7.200)
$\quad = \dfrac{E_K^{e^+}}{c}\sqrt{1 + \dfrac{2m_{e^+}c^2}{E_K^{e^+}}}$		
Momentum before annihilation (y axis)	Momentum after annihilation (y axis)	
0	$\dfrac{h\nu_1}{c}\sin\theta - \dfrac{h\nu_2}{c}\sin\phi$	(7.201)

$$\frac{E_K^{e^+}}{c}\sqrt{1 + \frac{2m_{e^+}c^2}{E_K^{e^+}}} = \frac{h\nu_1}{c}\cos\theta + \frac{h\nu_2}{c}\cos\phi \tag{7.203}$$

3. *Conservation of momentum* (y-axis component)

$$0 = \frac{h\nu_1}{c}\sin\theta - \frac{h\nu_2}{c}\sin\phi \tag{7.204}$$

To simplify the calculation process we rewrite (7.202)–(7.204) as follows

$$h\nu_1 = E - h\nu_2 \tag{7.205}$$
$$h\nu_2 \cos\phi = A - h\nu_1 \cos\theta \tag{7.206}$$
$$h\nu_2 \sin\phi = h\nu_1 \sin\theta, \tag{7.207}$$

where

$$A = E_K \sqrt{1 + \frac{2m_e c^2}{E_K}} \tag{7.208}$$

and

$$E = E_K + 2m_e c^2. \tag{7.209}$$

Next we square equations (7.206) and (7.207), add the two resulting equations, and insert (7.206) to get

$$h\nu_2 = \frac{A^2 + E^2 - 2AE\cos\theta}{2(E - A\cos\theta)}. \tag{7.210}$$

Inserting (7.210) into (7.205) results in

$$h\nu_1 = \frac{E^2 - A^2}{2(E - A\cos\theta)}. \tag{7.211}$$

In a similar manner, eliminating θ from (7.205)–(7.207) results in the following expressions for $h\nu_1$ and $h\nu_2$ as a function of ϕ

$$h\nu_1 = \frac{A^2 + E^2 - 2AE\cos\phi}{2(E - A\cos\phi)} \tag{7.212}$$

and

$$h\nu_2 = \frac{E^2 - A^2}{2(E - A\cos\phi)}. \tag{7.213}$$

The general relationship between θ and ϕ can be stated from (7.207) as

$$\frac{h\nu_1}{h\nu_2} = \frac{\sin\phi}{\sin\theta} \tag{7.214}$$

or from (7.210) and (7.211) as

$$\frac{h\nu_1}{h\nu_2} = \frac{E^2 + A^2 - 2EA\cos\phi}{E^2 - A^2} \tag{7.215}$$

and from (7.212) and (7.213) as

$$\frac{h\nu_1}{h\nu_2} = \frac{E^2 - A^2}{E^2 + A^2 - 2EA\cos\theta}. \tag{7.216}$$

Merging (7.215) and (7.216) results in the following relationship linking θ and ϕ

$$\frac{\cos\theta + \cos\phi}{1 + \cos\theta\cos\phi} = \frac{2EA}{E^2 + A^2}. \tag{7.217}$$

A special case of two-photon in-flight annihilation occurs when one photon of energy $h\nu_1$ is moving in the direction of the incoming positron and the other photon of energy $h\nu_2$ is moving in a direction opposite to the incoming positron. The scattering angles thus are $\theta = 0$ and $\phi = 180°$ and inserting $\theta = 0$ into (7.210) and (7.211) or inserting $\phi = 0$ into (7.212) and (7.213) results in the following simple expressions for $h\nu_1$ and $h\nu_2$

$$h\nu_1 = \frac{1}{2}(E + A) = \frac{1}{2}\left[E_K + 2m_ec^2 + F_K\sqrt{1 + \frac{2m_ec^2}{E_K}}\right] \tag{7.218}$$

and

$$h\nu_2 = \frac{1}{2}(E - A) = \frac{1}{2}\left[E_K + 2m_ec^2 - E_K\sqrt{1 + \frac{2m_ec^2}{E_K}}\right]. \tag{7.219}$$

In the high relativistic region where $E_K \gg m_ec^2$ we can use the approximation

$$\sqrt{1 + \frac{2m_ec^2}{E_K}} \approx 1 + \frac{m_ec^2}{E_K} \tag{7.220}$$

to get a further simplification for $h\nu_1$ and $h\nu_2$

$$h\nu_1 \approx E_K + \frac{3}{2}m_ec^2 \tag{7.221}$$

and

$$h\nu_2 \approx \frac{1}{2}m_ec^2. \tag{7.222}$$

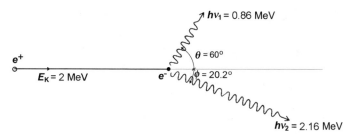

Fig. 7.40 In-flight annihilation of a positron of kinetic energy 2 MeV with a "free electron". One photon leaves the interaction site with angle $\theta = 60°$

By way of example we calculate the parameters for in-flight annihilation of a 2 MeV kinetic energy positron with a stationary and "free electron". We assume that one of the two photons produced moves away with angle $\theta = 60°$, as defined in Fig. 7.40, and calculate the angle ϕ for the other photon as well as the energies $h\nu_1$ and $h\nu_2$ of the two annihilation photons.

We first calculate the two parameters A and E from (7.208) and (7.209) as follows

$$A = E_K \sqrt{1 + \frac{2m_ec^2}{E_K}} = 2.459 \, \text{MeV}, \tag{7.223}$$

$$E = E_K + 2m_ec^2 = 3.022 \, \text{MeV}. \tag{7.224}$$

Next we use (7.217) to determine the angle ϕ for the second photon

$$\frac{\cos\theta + \cos\phi}{1 + \cos\theta\cos\phi} = \frac{0.5 + \cos\phi}{1 + 0.5\cos\phi} = \frac{2EA}{A^2 + E^2} = 0.979 \tag{7.225}$$

or

$$\cos\phi = 0.938 \quad \text{and} \quad \phi = \arccos 0.938 = 20.2° \tag{7.226}$$

Energies $h\nu_1$ and $h\nu_2$ of the two photons produced in the annihilation process are calculated from (7.210) through (7.213)

$$h\nu_1 = \frac{E^2 - A^2}{2(E - A\cos\theta)} = \frac{A^2 + E^2 - 2AE\cos\phi}{2(E - A\cos\phi)} = 0.86 \, \text{MeV} \tag{7.227}$$

and

$$h\nu_2 = \frac{A^2 + E^2 - 2AE\cos\theta}{2(E - A\cos\theta)} = \frac{E^2 - A^2}{2(E - A\cos\phi)} = 2.16 \, \text{MeV}. \tag{7.228}$$

We can verify the results for ϕ, $h\nu_1$, and $h\nu_2$ by using (7.205)–(7.207), and (7.214)

$$h\nu_1 + h\nu_2 = E = E_K + 2m_e c^2 = 3.02\,\text{MeV}, \tag{7.229}$$

$$A = E_K \sqrt{1 + \frac{2m_e c^2}{E_K}} = h\nu_1 \cos\theta + h\nu_2 \cos\phi = 2.46\,\text{MeV}, \tag{7.230}$$

$$h\nu_1 \sin\theta = h\nu_2 \sin\phi = 0.745, \tag{7.231}$$

$$\frac{h\nu_1}{h\nu_2} = 0.398 = \frac{\sin\phi}{\sin\theta}. \tag{7.232}$$

The in-flight annihilation process calculated in this example is plotted schematically in Fig. 7.40.

7.7 Photonuclear Reactions (Photodisintegration)

Photonuclear reaction occurs in a direct interaction between an energetic photon and an absorber nucleus causing nuclear disintegration. Two other names are often used for the effect: "photodisintegration" and "nuclear photoelectric effect". In photonuclear reaction the nucleus absorbs a photon and the most likely result of such an interaction is the emission of a single neutron through a (γ, n) reaction, even though emissions of charged particles such as protons or alpha particles, gamma rays, more than one neutron, or fission fragments (photofission) are also possible but much less likely to occur. Neutrons produced in photonuclear reactions are referred to as *photoneutrons*.

While the photonuclear reactions do not play a role in general photon attenuation studies, they are of considerable importance in shielding calculations whenever photon energies exceed the photonuclear reaction threshold. Neutrons produced through the (γ, n) photonuclear reactions are usually far more penetrating than the photons that produced them. In addition, the daughter nuclei resulting from the (γ, n) reaction may be radioactive and the neutrons, through subsequent neutron capture, may produce radioactivity in the irradiation facility, adding to radiation hazard in the facility. This raises concern over the induced radioactivity in clinical high-energy linear accelerator installations (above 10 MV) and stimulates a selection of appropriate machine components to decrease the magnitude and half-life of the radioactivation. It also sets forth requirements for adequate treatment room ventilation to expel the nitrogen-13 and oxygen-15 produced in the room (typical air exchanges in treatment rooms are of the order of six to eight per hour).

Photonuclear reactions are endothermic (endoergic), thus for the reaction to occur the incident photon must possess minimum or threshold energy to be able to trigger the reaction. The threshold energy represents the separation energy of a neutron from the nucleus that is of the order of 8 MeV or more for most nuclei, except for the deuteron ($^2_1$H) and berillium-9 ($^9_4$Be) where it is at 2.225 MeV and 1.665 MeV, respectively.

In photodisintegration of a deuteron a photon with energy exceeding the deuteron binding energy strikes a deuteron d and breaks it into its constituent nucleons: proton p and neutron n. The reaction is written as follows

$$d + \gamma = n + p + Q \qquad \text{or} \qquad d(\gamma, n)p, \qquad (7.233)$$

where Q is the so-called Q value of the photonuclear reaction (see Sect. 5.2.2).

Chadwick discovered the deuteron photodisintegration in 1935 and the reaction was used as early confirmation of the conversion of energy to mass as predicted by the theory of relativity. The reaction is still of great importance to nuclear physics because the deuteron is the simplest bound nucleus known, just like the hydrogen atom is of great importance to atomic physics as the simplest atomic structure known.

7.7.1 Cross Section for Photonuclear Reaction

The cross sections for photonuclear reactions vary as a function of photon energy as well as the absorber nucleus. The most notable features of the cross section for nuclear absorption of energetic photons are the reaction threshold and the so-called "giant resonance", both depending on the absorbing material.

Above the threshold photon energy the cross section gradually increases, reaches a broad peak referred to as giant resonance, and then decreases with a further increase in photon energy. The giant resonance peak is centered at about 23 MeV for low atomic number Z absorbers and at about 12 MeV for high Z absorbers. The only exceptions again are the two reactions $^2H(\gamma, n)^1H$ and $^9Be(\gamma, n)2\alpha$ that have giant resonance peaks at much lower energy.

The full-width-at-half-maximum (FWHM) in the giant resonance cross sections typically ranges from about 3 MeV to 9 MeV and it depends on the detailed properties of absorber nuclei. Table 7.21 provides various parameters of the "giant (γ, n) resonance" cross section for selected absorbers:

1. With increasing atomic number Z of the absorber the resonance peak energy steadily decreases from 23 MeV for carbon-12 (^{12}C) to 12.2 MeV for uranium-235 (^{235}U).
2. The magnitude of the atomic cross section for photodisintegration $_a\sigma_{PN}$, even at the resonance peak energy $h\nu_{max}$, is relatively small in comparison with the sum of competing "electronic" cross sections and amounts to only a few percent of the total "electronic" cross section. As a result, $_a\sigma_{PN}$ is usually neglected in photon attenuation studies in medical physics.
3. High-energy electrons are also capable of triggering nuclear reactions that release neutron and protons; however, the probabilities for electron-nucleus reactions are smaller than those for photonuclear reactions.
4. Cross sections for photonuclear (γ, n) reactions, generally in the millibarn range, are much smaller than cross sections for neutron activation (n, γ) reactions.

Table 7.21 Photonuclear (γ, n) giant resonance cross section parameters for selected absorbers

Absorber	Threshold energy (MeV)	Resonance peak energy $h\nu_{max}$ (MeV)	Resonance FWHM (MeV)	Percent of total electronic cross section at $h\nu_{max}$
^{2}H	2.225	5	10.0	–
^{12}C	18.7	23.0	3.6	5.9
^{27}Al	13.1	21.5	9.0	3.9
^{63}Cu	10.8	17.0	8.0	2.0
^{208}Pb	7.4	13.6	3.8	2.7
^{235}U	6.1	12.2	7.0	2.4

(Data are from Hubbell)

7.7.2 Threshold Energy for Photonuclear Reaction

We can determine the Q value and threshold kinetic energy $(E_\gamma^{PN})_{thr}$ for a photonuclear reaction, as discussed in general in Sect. 5.2, using the rest energy method or the binding energy method:

- In the rest energy method we subtract the sum of nuclear rest energies of reaction products after the reaction $\sum\limits_{i,\text{after}} M_i c^2$ from the sum of nuclear rest energies of reactants (target and projectile) before the reaction $\sum\limits_{i,\text{before}} M_i c^2$.
- In the binding energy method we subtract the sum of nuclear binding energies of reactants before the interaction $\sum\limits_{i,\text{before}} E_B(i)$ from the sum of nuclear binding energies of reaction products after the interaction $\sum\limits_{i,\text{after}} E_B(i)$.

By way of example, we now determine the threshold energy for the photonuclear (γ, n) reactions on deuteron, beryllium-9 and lead-208. We first determine the reaction Q value and then use (5.15) to determine the threshold energy E_γ^{PN}. The appropriate data for rest energies and binding energies are provided in Appendix A.

1. *Rest energy method:* $Q = \sum\limits_{i,\text{before}} M_i c^2 - \sum\limits_{i,\text{after}} M_i c^2$, as shown in (5.7)

$$Q(\text{d}, \gamma) = M(\text{d})c^2 + 0 - [m_n c^2 + m_p c^2]$$
$$= 1875.6128 \text{ MeV} - [939.5654 \text{ MeV} + 938.2720 \text{ MeV}]$$
$$= -2.225 \text{ MeV} \tag{7.234}$$
$$Q(^9_4\text{Be}, \gamma) = M(^9_4\text{Be})c^2 + 0 - [m_n c^2 + M(^8_4\text{Be})c^2]$$
$$= 8392.7499 \text{ MeV} - [939.5654 \text{ MeV} + 7454.8500 \text{ MeV}]$$
$$= -1.666 \text{ MeV} \tag{7.235}$$

$$Q(^{208}_{82}\text{Pb}, \gamma) = M(^{208}_{82}\text{Pb})c^2 + 0 - [m_n c^2 + M(^{207}_{82}\text{Pb})c^2]$$
$$= 193687.0956 \text{ MeV} - [939.5654 \text{ MeV} + 192754.8983 \text{ MeV}]$$
$$= -7.37 \text{ MeV} \tag{7.236}$$

2. *Binding energy method:* $Q = \sum\limits_{i,\text{after}} E_B(i) - \sum\limits_{i,\text{before}} E_B(i)$, as shown in (5.8)

$$Q(d, \gamma) = 0 - E_B(d) = -2.225 \text{ MeV} \tag{7.237}$$
$$Q(^9_4\text{Be}, \gamma) = E_B(^8_4\text{Be}) - E_B(^9_4\text{Be})$$
$$= 56.4996 \text{ MeV} - 58.1650 \text{ MeV} = -1.666 \text{ MeV} \tag{7.238}$$
$$Q(^{208}_{82}\text{Pb}, \gamma) = E_B(^{207}_{82}\text{Pb}) - E_B(^{208}_{82}\text{Pb})$$
$$= 1629.0779 \text{ MeV} - 1636.4457 \text{ MeV}$$
$$= -7.368 \text{ MeV} \tag{7.239}$$

As expected, for a given photonuclear reaction, both the rest energy method and the binding energy method give the same result for the Q value. The threshold energy E_γ^{PN} can be determined from the Q value using (5.15) which provides a general relationship between the threshold kinetic energy $(E_K)_{\text{thr}}$ of the projectile (positive) and the Q value for an endothermic nuclear reaction (negative). In general, $(E_K)_{\text{thr}}$ exceeds the absolute value of Q by a relatively small amount to account for conservation of energy and momentum in the collision. However, in the case of photonuclear reaction, the projectile is a photon with zero rest mass, resulting in the absolute Q value and the threshold energy being equal to one another, i.e., $(E_\gamma^{PN})_{\text{thr}} \approx |Q|$, as shown in (5.15) by setting $m_{10} = 0$.

Photonuclear threshold energies for the examples given above are thus as follows: $E_\gamma^{PN}(d) = 2.225 \text{ MeV}$, $E_\gamma^{PN}(^9_4\text{Be}) = 1.666 \text{ MeV}$, and $E_\gamma^{PN}(^{208}_{82}\text{Pb}) = 7.368 \text{ MeV}$, as given in Table 7.21.

Interaction of Photon with Absorber

Photons with energy exceeding the ionization energy of atoms are called *indirectly ionizing radiation* because they transfer energy to an absorber by an indirect process. This indirect process may be through one of several effects available to the photon for interaction with an absorber atom, such as the photoelectric effect, Compton effect, nuclear pair production, and triplet production. In each of these effects a portion of the photon energy is transferred to energetic light charged particles (electron or positron) which travel through the absorber and lose their energy through direct Coulomb interactions with nuclei and orbital electrons of absorber atoms. The energy is thus transferred from photon to absorber through an indirect process of first transferring energy to light charged particles which in turn transfer part of their energy to the absorber.

In inelastic Coulomb interactions with nuclei, bremsstrahlung radiation is produced which may escape the volume of interest in the absorber; in Coulomb interactions with orbital electrons of the absorber, the absorber atoms are left ionized or excited and the ionization and excitation energy may be absorbed locally in the absorber. A clear understanding of energy transfer and energy absorption in photon interactions with an absorber is essential for the understanding of radiation dosimetry of external photon beams and for development of new dosimetry techniques.

The **diagram on next page** illustrates the energy transfer process from photon to absorber by showing a simulated history of a 10 MeV photon interacting with lead absorber. The photon penetrates the lead absorber and at point A undergoes a nuclear pair production interaction producing an electron–positron pair. The pair production electron and positron share energy of $10\,\mathrm{MeV} - 2m_e c^2$ and produce their own track as they move through the absorber and dissipate their energy through Coulomb interactions with orbital electrons and nuclei of absorber atoms. Most of these interactions are either elastic collisions of the charged particle with absorber nuclei causing changes in direction of motion or are Coulomb interactions of the charged particle with orbital electrons causing ionization and excitation of absorber atoms.

Three other possible energy-loss mechanisms are also shown in the diagram: (1) bremsstrahlung interactions at points B in the electron track and positron track resulting from an inelastic collision between the energetic charged particle and absorber nucleus; (2) production of a delta ray electron at point C; and (3) in-flight annihilation interaction of the energetic positron with an orbital electron of the absorber at point D.

At points B the bremsstrahlung photons are produced with relatively low energy and are absorbed in the absorber. The delta ray electron produced at point C is an electron energetic enough to produce ionization and excitation of absorber atoms in its own right and is eventually absorbed in the absorber.

Of the two annihilation quanta produced at point D, one subsequently disappears in a photoelectric interaction at point E and the other has a Compton interaction at point F. Both the photoelectron released in the photoelectric effect at point E and the Compton electron released in Compton effect at point F are absorbed locally; the scattered photon released in the Compton effect at point F escapes the lead absorber.

Chapter 8
Energy Transfer and Energy Absorption in Photon Interactions with Matter

This chapter presents a discussion of the important mechanisms involved in energy transfer and energy absorption in an absorber irradiated with an external photon beam. First, the energy transfer mechanism is introduced, the components of the energy transfer fraction are described, the mass energy transfer coefficient is defined, and the mean energy transferred from photon to light charged particles is calculated. The next section of the chapter deals with energy absorption which is the fraction of the energy transferred from photons to energetic charged particles that is subsequently absorbed in the irradiated medium.

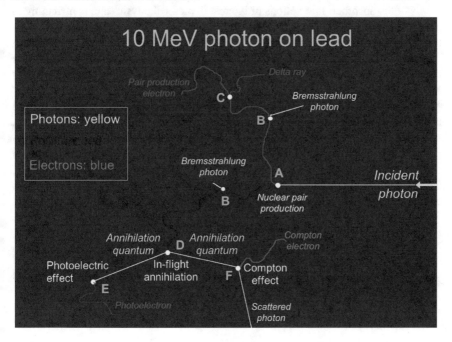

© Springer International Publishing Switzerland 2016
E.B. Podgoršak, *Radiation Physics for Medical Physicists*,
Graduate Texts in Physics, DOI 10.1007/978-3-319-25382-4_8

The difference between the energy transferred and energy absorbed is attributed to energy radiated from the charged particles in the form of photons. The mean radiation fraction and its three components (bremsstrahlung, in-flight annihilation, and impulse ionization) are defined and used to determine the total mean energy absorption fraction, as well as the mass energy absorption coefficient and the mean energy absorbed in absorbing medium.

The chapter concludes with a discussion of attenuation coefficients of compounds and mixtures and gives two specific examples of photon beam interaction with absorber: (1) Interaction of 2 MeV photons with lead absorber and (2) Interaction of 8 MeV photons with copper absorber.

8.1   Macroscopic Attenuation Coefficient

As discussed in Sect. 7.13, four types of attenuation coefficient are in use: linear, mass, atomic, and electronic. The four coefficients are related to one another through physical properties of the absorber (density ρ, atomic number Z, and atomic mass A), as given in (7.18). The attenuation coefficient of monoenergetic photons in a specific absorber depends on photon energy $h\nu$ and absorber atomic number Z.

In general, the macroscopic attenuation coefficient represents a sum of attenuation coefficients for all individual interactions that a photon of energy $h\nu$ may have with atoms of the absorber. Interactions of interest in medical physics and contributing to the attenuation coefficient are the photoelectric effect, Rayleigh scattering, Compton scattering, and pair production including triplet production. Other less common effects, such as Delbrück scattering, nuclear photoelectric effect, and photonuclear reactions, are of interest in nuclear physics but usually ignored in medical physics and radiation dosimetry.

For an absorber of density ρ, atomic number Z, and atomic mass A, we write the relationships for the linear attenuation coefficient μ, mass attenuation coefficient μ_m, and atomic attenuation coefficient (cross section) $_a\mu$, as a sum of contributions μ_i from the four individual effects. Note: the electronic attenuation coefficient $_e\mu_i$ is only defined for photon interactions with "free" electrons (such as Thomson scattering, Compton effect, and triplet production) and is most commonly used in Compton effect.

$$\mu = \sum_i \mu_i = \tau + \sigma_R + \sigma_c + \kappa, \tag{8.1}$$

$$\mu_m = \frac{\mu}{\rho} = \sum_i \left(\frac{\mu}{\rho}\right)_i = \frac{1}{\rho}(\tau + \sigma_R + \sigma_C + \kappa) = \frac{\tau}{\rho} + \frac{\sigma_R}{\rho} + \frac{\sigma_C}{\rho} + \frac{\kappa}{\rho}, \tag{8.2}$$

$$_a\mu = \frac{1}{\rho}\frac{A}{N_A}\mu = {_a}\tau + {_a}\sigma_R + {_a}\sigma_C + {_a}\kappa = \frac{1}{\rho}\frac{A}{N_A}(\tau + \sigma_R + \sigma_C + \kappa),$$

$$= {_a}\tau + {_a}\sigma_R + {_a}\sigma_C + {_a}\kappa \tag{8.3}$$

$$_e\sigma_C = Z^{-1}{_a}\sigma_C \quad \text{and} \quad \sigma_C = \rho\frac{N_A}{A}{_a}\sigma_C = \rho\frac{N_A}{A}Z_e\sigma_C \tag{8.4}$$

where

τ, τ/ρ, and $_a\tau$	are the linear, mass, and atomic attenuation coefficient, respectively, for photoelectric effect,
σ_R, σ_R/ρ and $_a\sigma_R$	are the linear, mass, and atomic attenuation coefficient, respectively, for Rayleigh scattering,
σ_C, σ_C/ρ, and $_a\sigma_C$	are the linear, mass, and atomic attenuation coefficient, respectively, for Compton effect,
κ, κ/ρ, and $_a\kappa$	are the linear, mass, and atomic attenuation coefficient, respectively, for pair production (including triplet).

The fundamental interactions of photons with matter are generally expressed with cross section on atomic scale or cross section per atom for individual effects. The mass attenuation coefficient which is proportional to the atomic cross section is considered more fundamental than the linear attenuation coefficient because, in contrast with the linear attenuation coefficient, the mass attenuation coefficient is independent of the actual mass density as well as physical state of the absorber.

In Fig. 8.1 we show the total mass attenuation coefficient μ/ρ for carbon as example of low atomic number absorber in part (a) and lead as example of high atomic number absorber in part (b), plotted against photon energy $h\nu$. In addition to μ/ρ which represents the sum of the individual coefficients for the photoelectric effect, Rayleigh scattering, Compton effect, and pair production, the coefficients for the individual components are also shown. Also shown are the absorption edges for the lead absorber; the absorption edges for the carbon absorber are not visible, because they occur off-scale at energies below 1 keV. The following general conclusions can be made from Fig. 8.1:

1. For all absorber materials the photoelectric effect is the predominant mode of photon interaction with the absorber at low photon energies.
2. At intermediate photon energies and low atomic numbers Z the Compton effect mass coefficient σ_C/ρ predominates and makes the largest contribution to the total mass attenuation coefficient μ/ρ.
3. The width of the region of Compton scattering predominance depends on the atomic number Z of the absorber; the lower is Z, the broader is the Compton scattering predominance region. For water and tissue this region ranges from $\sim$20 keV up to $\sim$20 MeV, indicating that for most of radiotherapy the most important interaction of photon beams with tissues is the Compton scattering.
4. The pair production dominates at photon energies $h\nu$ above 10 MeV and at high atomic numbers Z of the absorber.
5. In all energy regions the Rayleigh scattering mass coefficient σ_R/ρ plays only a secondary role in comparison with the other three coefficients.

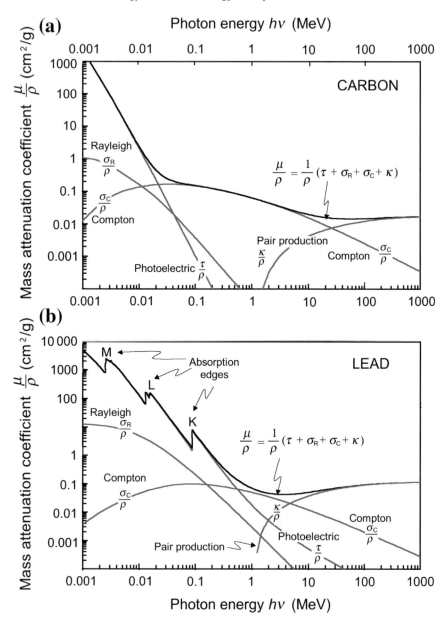

Fig. 8.1 Mass attenuation coefficient μ/ρ against photon energy $h\nu$ in the range from 1 keV to 1000 MeV for carbon in **a** and lead in **b**. In addition to the total coefficient μ/ρ, the individual coefficients for photoelectric effect, Rayleigh scattering, Compton scattering, and pair production (including triplet production) are also shown. The mass attenuation coefficient μ/ρ is the sum of the coefficients for individual effects, i.e., $\mu/\rho = (\tau + \sigma_R + \sigma_C + \kappa)/\rho$

8.2 Energy Transfer from Photons to Charged Particles in Absorber

The total mean energy transfer fraction $\bar{f}_{\mathrm{tr}}(h\nu, Z)$ is defined as the fraction of photon energy $h\nu$ that is transferred to kinetic energy of charged particles produced or released in the absorber during various possible photon interactions with absorber atoms. It is generally expressed as a sum of four components, each component representing a specific effect that contributes to photon attenuation in the absorber (photoelectric effect, Rayleigh scattering, Compton effect, and pair production). The total mean energy transfer fraction $\bar{f}_{\mathrm{tr}}(h\nu, Z)$ is given as follows

$$
\begin{aligned}
\bar{f}_{\mathrm{tr}} = \sum_i \frac{\mu_i}{\mu} \bar{f}_i &= \left\{ \frac{\tau}{\mu} \bar{f}_{\mathrm{PE}} + \frac{\sigma_{\mathrm{R}}}{\mu} \bar{f}_{\mathrm{R}} + \frac{\sigma_{\mathrm{C}}}{\mu} \bar{f}_{\mathrm{C}} + \frac{\kappa}{\mu} \bar{f}_{\mathrm{PP}} \right\} \\
&= \sum_i w_i \bar{f}_i = \left\{ w_{\mathrm{PE}} \bar{f}_{\mathrm{PE}} + w_{\mathrm{R}} \bar{f}_{\mathrm{R}} + w_{\mathrm{C}} \bar{f}_{\mathrm{C}} + w_{\mathrm{PP}} \bar{f}_{\mathrm{PP}} \right\} \qquad (8.5) \\
&= \frac{\bar{E}_{\mathrm{tr}}}{h\nu} = \sum_i \frac{\mu_i}{\mu} \frac{\bar{E}_{\mathrm{tr}}^i}{h\nu} = \frac{1}{h\nu} \left\{ w_{\mathrm{PE}} \bar{E}_{\mathrm{tr}}^{\mathrm{PE}} + w_{\mathrm{R}} \bar{E}_{\mathrm{tr}}^{\mathrm{R}} + w_{\mathrm{C}} \bar{E}_{\mathrm{tr}}^{\mathrm{C}} + w_{\mathrm{PP}} \bar{E}_{\mathrm{tr}}^{\mathrm{PP}} \right\},
\end{aligned}
$$

with

w_i the relative weight of given effect i for photon energy $h\nu$ and absorber atomic number Z defined as the ratio between the linear attenuation coefficient μ_i for the given effect i and the total linear attenuation coefficient μ that, as given in (8.1), is the sum of individual linear attenuation coefficients μ_i (see Sect. 8.2.2).

$\bar{f}_i$ the mean energy transfer fraction for effect i representing the mean energy transfer fractions $\bar{f}_{\mathrm{PE}}, \bar{f}_R = 0, \bar{f}_C$ and $\bar{f}_{\mathrm{pp}}$ for photoelectric effect, Rayleigh scattering, Compton effect, and pair production, respectively, discussed in detail in Sects. 7.5.10, 7.4.3, 7.3.12, and 7.6.8, respectively. Note that $\bar{f}_R = 0$ for all photon energies $h\nu$ and all absorber atomic numbers Z (see Sect. 8.2.1).

$\bar{E}_{\mathrm{tr}}^i$ the mean energy transferred from photon to charged particles for interaction i, representing one of the three effects: photoelectric effect, Compton scattering, and general pair production that includes nuclear and electronic pair production, discussed in detail in Sects. 7.5.10, 7.4.3, 7.3.12, and 7.6.8, respectively.

8.2.1 General Characteristics of Mean Energy Transfer Fractions

Of the five effects listed above (photoelectric, Rayleigh, Compton, nuclear pair production, and triplet production), Rayleigh scattering does not transfer any energy to charged particles, so it can be ignored in the context of energy transfer. Three of the remaining four effects (photoelectric, Compton, and triplet production) produce vacancies in atoms of the absorber, and these vacancies engender either characteristic photons or Auger electrons or both. In general, therefore, kinetic energy of the Auger electrons must be added to the kinetic energy of primary particles produced in photon interactions when the mean energy transfer fractions $\bar{f}_{PE}, \bar{f}_C$, and $\bar{f}_{TP}$ for the photoelectric effect, Compton effect, and triplet production, respectively, are determined. These energy transfer fractions are then expressed as follows

$$\bar{f}_{PE} = \frac{\overline{E}_{tr}^{PE}}{h\nu} = \frac{h\nu - \overline{X}_{PE}}{h\nu} = \frac{h\nu - \sum_j P_j \omega_j h\bar{\nu}_j}{h\nu} = 1 - \frac{\sum_j P_j \omega_j h\bar{\nu}_j}{h\nu} \tag{8.6}$$

$$\bar{f}_C = \frac{\overline{E}_{tr}^{C}}{h\nu} = \frac{h\nu - h\bar{\nu}' - \overline{X}_C}{h\nu} = 1 - \frac{h\bar{\nu}' + \overline{X}_C}{h\nu} \tag{8.7}$$

$$\bar{f}_{TP} = \frac{\overline{E}_{tr}^{TP}}{h\nu} = \frac{h\nu - 2m_e c^2 - \overline{X}_{TP}}{h\nu} = 1 - \frac{2m_e c^2 + \overline{X}_{TP}}{h\nu}, \tag{8.8}$$

where $\overline{X}_{PE}, \overline{X}_C$, and $\overline{X}_{TP}$ are the mean fluorescence emission energies for the photoelectric effect, Compton effect, and triplet production, respectively and $h\bar{\nu}'$ is the mean energy of the scattered photon.

For the photoelectric effect the mean fluorescence emission energy $\overline{X}_{PE}$ and its components have been discussed in detail in Sect. 7.5.9 and the treatment of $\overline{X}_C$ as well as $\overline{X}_{TP}$ would in principle be similar; however, a closer look at $\overline{X}_C$ and $\overline{X}_{TP}$ reveals that their influence upon the mean energy transfer is significantly different from that discussed in Sect. 7.5.9 for the photoelectric effect.

For the Compton effect, the mean fluorescence emission energy $\overline{X}_C$ in (8.7) is different from the photoelectric $\overline{X}_{PE}$ in (8.6) and (7.146) because of the difference in the distribution of electronic vacancies produced in the absorber atom by the two effects. In contrast to the photoelectric interaction that is by far the most likely to happen with the most tightly bound available orbital electron of the absorber, a Compton interaction is equally probable with any one of the whole electronic complement of the absorber atom. Moreover, since the majority of atomic electrons reside in outer atomic shells, a Compton interaction is more probable with an outer shell electron, resulting in relatively low energy fluorescence photons $\overline{X}_C$ that can be ignored in comparison with the incident photon energy $h\nu$.

For triplet production, in view of the relatively high threshold for triplet production ($4m_e c^2 = 2.044$ MeV) in comparison with the mean energy of fluorescence photons $\overline{X}_{TP}$, these photons, similarly to the situation with the Compton effect, can be ignored and the mean energy transferred in triplet production, like that in nuclear pair production, is expressed as

$$\overline{E}_{tr}^{PP} = h\nu - 2m_e c^2, \tag{8.9}$$

where $\overline{E}_{tr}^{PP}$ accounts for both the nuclear pair production and electronic pair production (triplet production).

The mean fluorescence emission energies $\overline{X}_C$ and $\overline{X}_{TP}$ in (8.7) and (8.8), respectively, can thus be ignored thereby significantly simplifying the expressions for $\overline{f}_C$ and $\overline{f}_{TP}$ that can now be written as

$$\overline{f}_C = \frac{\overline{E}_{tr}^C}{h\nu} = \frac{h\nu - h\overline{\nu}'}{h\nu} = 1 - \frac{h\overline{\nu}'}{h\nu}, \tag{8.10}$$

as already seen in (7.133) and "The Compton graph" in Fig. 7.18, and

$$\overline{f}_{TP} = \frac{\overline{E}_{tr}^{TP}}{h\nu} = \frac{h\nu - 2m_ec^2}{h\nu} = 1 - \frac{2m_ec^2}{h\nu}, \tag{8.11}$$

as shown for general pair production in Fig. 7.37.

For comparison, Fig. 8.2 displays the three non-zero energy transfer fractions $\overline{f}_{PE}$, $\overline{f}_C$ and $\overline{f}_{PP}$ together on one graph (note that $\overline{f}_R = 0$) and the following general characteristics are notable:

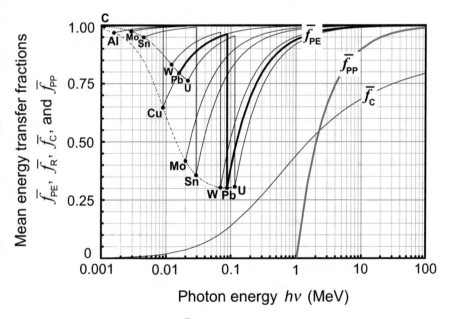

Fig. 8.2 Mean energy transfer fractions $\overline{f}_i$ plotted against photon energy $h\nu$ for the four main photon interactions with absorber atoms: $\overline{f}_{PE}$ for photoelectric effect, $\overline{f}_R$ for Rayleigh scattering, $\overline{f}_C$ for Compton effect, and $\overline{f}_{PP}$ for pair production. The photoelectric energy transfer fraction $\overline{f}_{PE}$ depends on photon energy $h\nu$ and on absorber atomic number Z, and is plotted for eight selected absorber atoms ranging from carbon to uranium. The energy transfer fractions for the other three effects: $\overline{f}_R, \overline{f}_C$, and $\overline{f}_{PP}$ do not depend on absorber atomic number Z, therefore one curve for each effect covers all absorber atomic numbers. The *dashed curves* connect $\overline{f}_{PE}$ points for which $h\nu = E_B(K)$ and $h\nu = E_B(L_1)$. Note that $\overline{f}_R = 0$ for all photon energies $h\nu$ and all absorber atomic numbers Z

Dependence of Mean Energy Transfer Fractions on Photon Energy $h\nu$

1. All three energy transfer fractions depend upon photon energy $h\nu$ and increase with increasing $h\nu$ with the exception of photoelectric discontinuities at the K and L absorption edges where the photoelectric fraction $\bar{f}_{PE}$ drops significantly. All three fractions converge to 1 at large $h\nu$ with $\bar{f}_{PE}$ displaying the fastest convergence to 1, followed by $\bar{f}_{PP}$.
2. Both $\bar{f}_{PE}$ and $\bar{f}_{PP}$ converge to 1 as $[1 - C/(h\nu)]$ where C is a constant $(2m_ec^2)$ for pair production and a parameter depending on the atomic number Z of the absorber for the photoelectric effect. The mean Compton energy transfer fraction $\bar{f}_C$ also increases with $h\nu$ but displays a slow convergence to 1, so that at the very high photon energy $h\nu$ of 1000 MeV it still amounts to only $\bar{f}_C = 0.85$.

Dependence of Mean Energy Transfer Fractions on Absorber Atomic Z

1. The pair production mean energy transfer fraction $\bar{f}_{PP}$ does not depend on the atomic number Z of the absorber.
2. The Compton mean energy transfer fraction $\bar{f}_C$ exhibits a slight, generally negligible, atomic number Z dependence as a result of electron binding effects.
3. In contrast to $\bar{f}_C$ and $\bar{f}_{PP}$, the photoelectric mean energy transfer fraction $\bar{f}_{PE}$, in addition to depending on $h\nu$, also depends on the atomic number Z of the absorber; the higher is Z, the lower is $\bar{f}_{PE}$ at the K absorption edge and the slower is the convergence to 1 with increasing $h\nu$.
4. For a given Z, the lowest value in $\bar{f}_{PE}$ is attained at the K absorption edge.
5. For photon energies $h\nu$ matching the K absorption edge energy of the absorber, $\bar{f}_{PE}[E_B(K)] = P_K\omega_K\eta_K$ ranges from $\bar{f}_{PE} = 1$ for low Z absorbers down to $\bar{f}_{PE} \approx$ 0.3 for high Z absorbers.

Range of Mean Energy Transfer Fractions

1. For a given absorber Z, $\bar{f}_{PE}$ ranges from its lowest value at the K absorption edge $[h\nu = E_B(K)]$ to $\bar{f}_{PE} = 1$ both at low photon energies where $h\nu < E_B(M)$ and at high photon energies where $h\nu > 10$ MeV.
2. For all absorbers, the range of $\bar{f}_C$ is from $\bar{f}_C < 0.02$ for $h\nu < 10$ keV to $\bar{f}_C > 0.85$ for $h\nu > 1000$ MeV.
3. For all absorbers, the range of $\bar{f}_{pp}$ is from $\bar{f}_{pp} = 0$ for $h\nu \le 1.02$ MeV through $\bar{f}_{PP} = 0.5$ at $h\nu = 2$ MeV to $\bar{f}_{pp} \approx 1$ for $h\nu > 10$ MeV.

8.2.2   Relative Weights for Individual Effects

The relative weights w_i of photoelectric effect w_{PE}, Rayleigh scattering w_R, Compton scattering w_C, and pair production w_{PP}, are plotted against photon energy $h\nu$ in Fig. 8.3 for eight selected absorbers ranging from carbon to uranium.

For all elements, on semilog photon energy diagram, w_{PE} and w_{PP} exhibit a sigmoid shape, while w_R and w_C exhibit a bell shape. The relative weight w_{PE} shows discontinuities at absorption edges where the photon energy equals the binding energy of a given atomic shell. As a result of the discontinuities in w_{PE}, the relative weights w_R and w_C also exhibit discontinuities at same photon energies. The relative weights w_i for the eight absorbers of Fig. 8.3 are re-plotted in Fig. 8.4 for $h\nu \geq E_B(K)$: in part (a) for the photoelectric effect; in part (b) for Rayleigh scattering; in part (c) for Compton effect; and in part (d) for pair production.

For *photoelectric effect*, the relative weight w_{PE} is approximately equal to 1 at relatively low photon energies $h\nu$ of the order of the K-shell binding energy $E_B(K)$. With increasing photon energy, w_{PE} decreases and asymptotically approaches zero. It attains a 50% point at $\sim$20 keV for low Z absorbers, at $\sim$100 keV for intermediate Z absorbers, and at $\sim$0.8 MeV for high Z absorbers. For low Z absorbers, $w_{PE} \approx 0$ for photon energies exceeding 0.1 MeV, for intermediate Z absorbers for photon energies above 1 MeV, and for high Z absorbers for photon energies above 10 MeV.

For *Rayleigh scattering*, w_R follows a bell shaped distribution and reaches a peak of only about 10% of the total attenuation coefficient; at $\sim$20 keV for low Z absorbers, at $\sim$100 keV for intermediate Z absorbers, and at $\sim$500 keV for high Z absorbers. Thus, the relative weight for Rayleigh scattering w_R amounts at best to only about 0.1 for all photon energies and absorbers.

For *Compton scattering*, w_C also follows a bell shaped distribution and it peaks at $\sim$2 MeV for all absorbers. The distribution is broad and peaks at $w_C = 1$ for low Z absorbers, while it is narrow and peaks at $w_C \approx 0.7$ for high Z absorbers. The higher is the atomic number Z of the absorber, the narrower is the w_C distribution and the lower is its peak value.

For *pair production*, w_{PP} exhibits a sigmoid curve shape starting at the pair production threshold photon energy of 1.022 MeV. Beyond 1.022 MeV the distribution rises rapidly with photon energy $h\nu$ to reach a saturation value of $w_{PP} = 1$ at high photon energies; the higher is the absorber atomic number Z, the lower is the energy at which w_{PP} attains saturation. The point of 50% saturation occurs at photon energy of $\sim$25 MeV for low Z absorbers and at only $\sim$5 MeV for high Z absorbers.

8.2.3 Regions of Predominance for Individual Effects

In the photon energy range 1 keV $\leq h\nu \leq$ 1000 MeV, of interest in medical physics and dosimetry, the predominance of a given effect i is indicated for photon energy regions where w_i is close to unity ($w_i \approx 1$). For example, $w_{PE} \approx 1$ at low photon energies; $w_C = 1$ (for low Z absorbers) at intermediate photon energies; and $w_{PP} = 1$ at high photon energies. From Figs. 8.3 and 8.4 we can also speculate on atomic number Z dependence; the broader is the energy region where $w_i \approx 1$, the broader is the region of predominance of effect i. For example, for high Z absorbers, w_{PE} and w_{PP} regions are broad and the w_C region is narrow, while for low Z absorbers the w_C region is broad and the w_{PE} as well as w_{PP} regions are narrow.

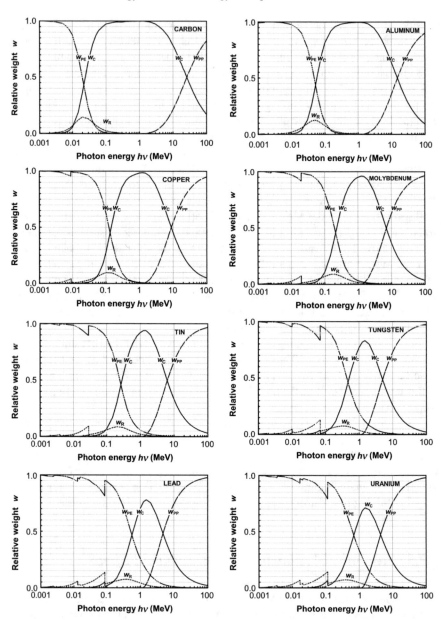

Fig. 8.3 Relative weights w_i plotted for eight selected absorber atoms ranging from carbon to uranium against photon energy $h\nu$ for the four main photon interactions with absorber atoms: photoelectric effect w_{PE}, Rayleigh scattering w_R, Compton effect w_C, and pair production w_{PP}. Data were calculated using the NIST XCOM database

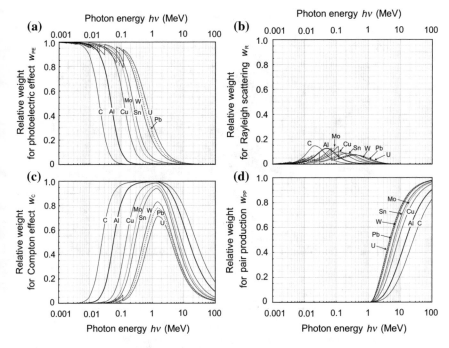

Fig. 8.4 Relative weights w_i for the four main photon interactions with atoms of absorber against photon energy $h\nu$ for eight selected absorber atoms from carbon to uranium. **a** Relative weight of photoelectric effect w_{PE}; **b** relative weight of Rayleigh scattering w_R; **c** relative weight of Compton effect w_C; and **d** relative weight of pair production w_{PP}. Data were calculated using the NIST XCOM database

The areas of predominance of individual effects are customarily shown on a $h\nu$ versus Z diagram displaying points where the attenuation coefficients for photoelectric effect and Compton effect are equal, i.e., $\tau = \sigma_C$, and points where the attenuation coefficients for Compton effect and pair production are equal, i.e., $\sigma_C = \kappa$, as shown in Fig. 8.5a.

We expand this approach in the $(h\nu, Z)$ diagram shown in Fig. 8.5b and connect specific points which have relative percentage weights w_i of 10%, 50%, and 90% for photoelectric effect, Compton scattering, and pair production. The pairs of curves for ($w_C = 50\%$ and $w_{PP} = 50\%$) coincide and so do pairs for ($w_C = 90\%$ and $w_{PP} = 10\%$), and ($w_C = 10\%$ and $w_{PP} = 90\%$) because in their energy range the relative contributions from the photoelectric effect and Rayleigh scattering to the total attenuation coefficients are zero. On the other hand, at low photon energies the pairs of curves for ($w_{PE} = 50\%$ and $w_C = 50\%$); ($w_{PE} = 90\%$ and $w_C = 10\%$); and ($w_{PE} = 10\%$ and $w_C = 90\%$) do not quite coincide because of a small contribution from Rayleigh scattering to the total attenuation coefficient.

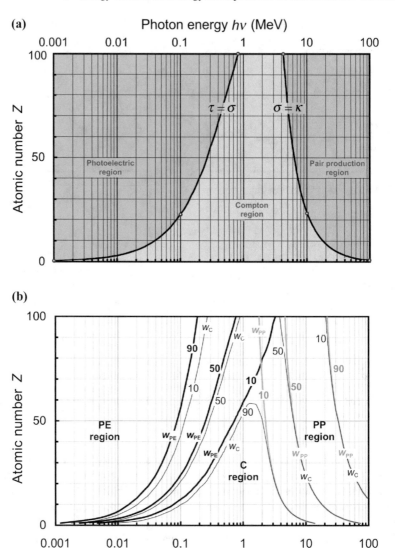

Fig. 8.5 Representation of the relative predominance of the three main processes of photon interaction with absorber atom: photoelectric effect τ, Compton effect σ_C, and pair production κ in a $(h\nu, Z)$ diagram where $h\nu$ is photon energy and Z is the absorber atomic number. The two curves in **a** connect points where photoelectric and Compton cross sections are equal ($\tau = \sigma_C$) shown by the curve on the *left* and Compton and pair production cross sections are equal ($\sigma_C = \kappa$) shown by the curve on the *right*. Anchor points for the two curves are: $h\nu = 0.001$ MeV, $Z \approx 0$; $h\nu = 0.1$ MeV, $Z = 22$; 0.8 MeV, 100; 4 MeV, 100; 10 MeV, 22; and 100 MeV, 0. In **b** we connect specific points which have relative percentage weights of 10%, 50% and 90% for the photoelectric effect (PE), Compton effect (C), and pair production (PP). Data were calculated using the NIST XCOM database

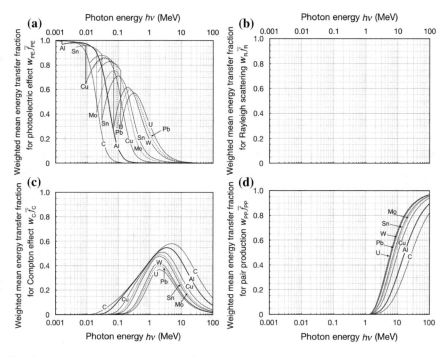

Fig. 8.6 Weighted mean energy transfer fractions $w_i f_i$ for the four main photon interactions with absorber atoms plotted for eight selected absorber atoms against photon energy $h\nu$. Part **a** is for photoelectric effect $(w_{PE}\bar{f}_{PE})$; **b** for Rayleigh scattering $(w_R \bar{f}_R)$; **c** for Compton effect $(w_C \bar{f}_C)$; and **d** for pair production $(w_{PP}\bar{f}_{PP})$

8.2.4 Mean Weighted Energy Transfer Fractions

In Fig. 8.6 we plot for eight selected absorbers the product $w_i \bar{f}_i$ which represents the mean weighted energy transfer fractions: for the photoelectric effect $w_{PE}\bar{f}_{PE}$; for Rayleigh scattering $w_R \bar{f}_R$; for the Compton effect $w_C \bar{f}_C$; and for pair production $w_{PP}\bar{f}_{PP}$. The curves are similar to those plotted for the relative weights w_i in Fig. 8.4, except that their shapes are clearly affected by the individual mean energy transfer fractions $\bar{f}_i$ given in Fig. 8.2. Most notably $w_R \bar{f}_R$ is zero at all photon energies because $\bar{f}_R = 0$ at all energies.

Similarly to average energy transfer fractions $\bar{f}_i$, the product $w_i \bar{f}_i$ is at low photon energies large for the photoelectric effect and zero for pair production, and the roles for $w_{PE}\bar{f}_{PE}$ and $w_{PP}\bar{f}_{PP}$ are reversed at high photon energies. The product $w_C \bar{f}_C$ is bell shaped and equals zero at small photon energies as well as at very large photon energies. It reaches a maximum at intermediate photon energies of the order of a few MeV (the point of maximum actually ranges from $\sim$2 MeV for high Z absorbers to $\sim$5 MeV for low Z absorbers). The maximum in $w_C \bar{f}_C$ ranges from a high value of $\sim$0.6 for low Z absorbers to a low value of $\sim$0.35 for high Z absorbers.

8.2.5   Total Mean Energy Transfer Fraction

As indicated in (8.5), the total mean energy transfer fraction $\bar{f}_{tr}$ is a sum of four mean weighted energy transfer fractions: photoelectric $w_{PE}\bar{f}_{PE}$, Rayleigh $w_R\bar{f}_R$, Compton $w_C\bar{f}_C$, and pair production $w_{PP}\bar{f}_{PP}$. Figure 8.7 shows $\bar{f}_{tr}$ and its three non-zero individual components for two absorbers: carbon as example of a low atomic number Z absorber in part (a) and lead as example of high Z absorber in part (b). The resulting total mean energy transfer fractions $\bar{f}_{tr}$ for the two elements are plotted with heavy solid curves, while the photoelectric, Compton, and pair production components $w_{PE}\bar{f}_{PE}$, $w_C\bar{f}_C$, and $w_{PP}\bar{f}_{PP}$, respectively, are plotted in the background with light solid curves.

For low Z absorbers, such as carbon in Fig. 8.7a, the total mean energy transfer fraction $\bar{f}_{tr}$ is clearly governed by the photoelectric component $w_{PE}\bar{f}_{PE}$ at low photon energies ($h\nu < 100$ keV), by the Compton component $w_C\bar{f}_C$ at intermediate photon energies of the order of 1 MeV, and by the pair production component $w_{PP}\bar{f}_{PP}$ at very high photon energies ($h\nu > 100$ MeV). For high Z absorbers, such as lead in Fig. 8.7b, there is no energy range where $w_C\bar{f}_C$ would be the sole contributor to $\bar{f}_{tr}$, yet, this is the case for the photoelectric component $w_{PE}\bar{f}_{PE}$ which governs $\bar{f}_{tr}$ at low photon energies and the pair production component $w_{PP}\bar{f}_{PP}$ which governs $\bar{f}_{tr}$ at very high photon energies.

In Fig. 8.8 we show the total mean energy transfer fraction $\bar{f}_{tr}$ for eight selected absorber elements ranging from carbon to uranium. The curves for carbon and lead have already been shown in Fig. 8.7; however, in Fig. 8.8, in order to avoid clutter, the individual four components forming $\bar{f}_{tr}$, namely: $w_{PE}\bar{f}_{PE}$, $w_R\bar{f}_R$, $w_C\bar{f}_C$, and $w_{PP}\bar{f}_{PP}$, as given in (8.5), are not shown separately. Figure 8.8a is for photon energy $h\nu$ exceeding the K shell binding energy $E_B(K)$ for each given absorber; Fig. 8.8b is for photon energy $h\nu$ exceeding the L_1 subshell binding energy $E_B(L_1)$.

The following are general features of the total mean energy transfer fraction $\bar{f}_{tr} = \bar{E}_{tr}/(h\nu) = w_{PE}\bar{f}_{PE} + w_R\bar{f}_R + w_C\bar{f}_C + w_{PP}\bar{f}_{PP}$ of (8.5):

1. For low atomic number Z absorbers such as carbon, $\bar{f}_{tr}$ is close to unity ($\bar{f}_{tr} \approx 1$) at low and very high photon energies and reaches a minimum at intermediate photon energies of the order of 100 keV. The minimum value and its photon energy position increase with atomic number Z of the absorber.
2. For high atomic number Z absorbers such as lead, $\bar{f}_{tr}$ has a minimum value when $h\nu$ equals $E_B(K)$, and exhibits with increasing photon energy $h\nu$ a peak and valley, and then asymptotically approaches $\bar{f}_{tr} = 1$ at very high photon energies. The positions of the peak and valley increase with atomic number Z of the absorber.

8.2.6   Mass Energy Transfer Coefficient

Once the total mean energy transfer fraction $\bar{f}_{tr}(h\nu, Z)$ is known, we can use (7.19) to determine the mass energy transfer coefficient μ_{tr}/ρ from the mass attenuation coefficient μ/ρ. The total mean energy transfer fraction $\bar{f}_{tr}$ was given in Fig. 8.7 for

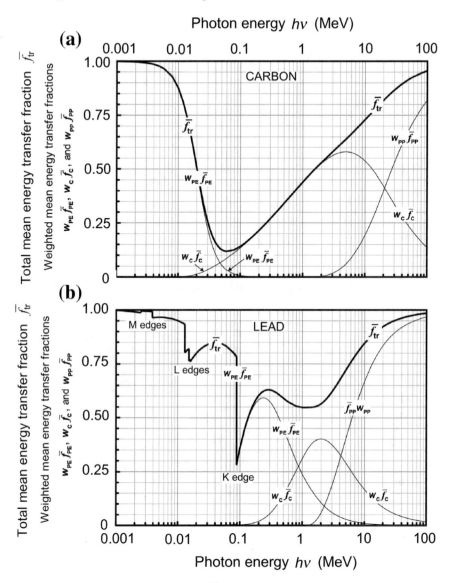

Fig. 8.7 Total mean energy transfer fraction $\overline{f}_{tr}$ against photon energy $h\nu$ for carbon in **a** and lead in **b**. The three components of $\overline{f}_{tr}$; namely the weighted mean energy transfer fractions for the photoelectric effect $w_{PE}\overline{f}_{PE}$, Compton effect $w_C\overline{f}_C$, and pair production $w_{PP}\overline{f}_{PP}$, are plotted with *fine line curves* in the background

carbon and lead and in Fig. 8.8 for eight selected absorbers ranging in atomic number Z from carbon to uranium. In Fig. 8.9 we plot the three components of (7.19): $\overline{f}_{tr}$, μ/ρ, and μ_{tr}/ρ, in the photon energy range from 1 keV to 100 MeV for two absorbers: carbon and lead as representatives of low Z and high Z materials, respectively.

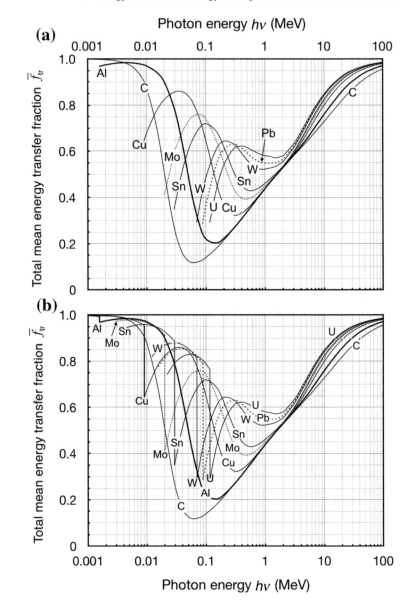

Fig. 8.8 Total mean energy transfer fraction $\overline{f}_{tr}$ against photon energy $h\nu$ for eight selected absorber atoms from carbon to uranium. Part **a** is for $h\nu$ exceeding the K-shell binding energy of the absorber atom, i.e., $h\nu \geq E_B(K)$; **b** is for $h\nu$ exceeding the binding energy of the L_1 subshell, i.e., $h\nu \geq E_B(L_1)$

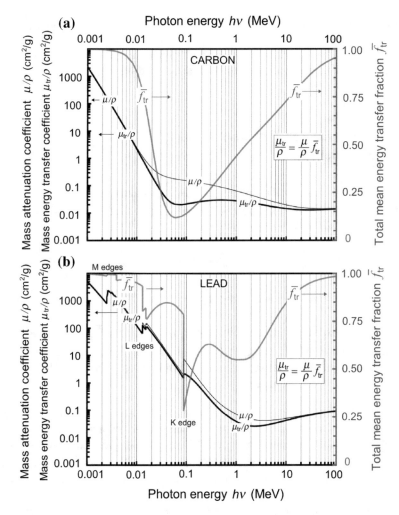

Fig. 8.9 Mass attenuation coefficient μ/ρ from the NIST XCOM database and the calculated mass energy transfer coefficient μ_{tr}/ρ against photon energy $h\nu$ for carbon in **a** and lead in **b**. Values of the two coefficients are given in cm²/g in the *left hand scale*. The total mean energy transfer fraction $\overline{f}_{tr}$ of (8.5) which links μ_{tr}/ρ with μ/ρ is shown in the background with values shown on the *right hand scale*. The attenuation coefficient μ/ρ is plotted with the *fine dark line*, the energy transfer coefficient μ_{tr}/ρ with the *heavy dark line*, and the total energy transfer fraction $\overline{f}_{tr}$ with the *light grey line*

For both absorbers of Fig. 8.9 the mean energy transfer fraction $\overline{f}_{tr}$ equals 1 at low and high photon energies, implying that in these energy ranges the mass energy transfer coefficient μ_{tr}/ρ equals the mass attenuation coefficient μ/ρ. At intermediate photon energies, on the other hand, the total mean energy transfer fraction $\overline{f}_{tr}$ is small and for high Z absorbers it is also structured in a complicated fashion because of the influence of the K and L absorption edges. In the intermediate photon energy range, μ_{tr}/ρ can be significantly (up to a factor of 10) smaller than μ/ρ.

An accurate determination of the total mean energy transfer fraction $\overline{f}_{tr}$ is thus of great importance in radiation dosimetry, since the kerma and ultimately the dose are related to μ_{tr}/ρ which is obtained by multiplying the total mean energy transfer fraction $\overline{f}_{tr}$ with mass attenuation coefficient μ/ρ, as given in (7.19).

8.2.7 Mean Energy Transferred from Photon to Charged Particles

Equation (8.5) gives the total mean energy transfer fraction $\overline{f}_{tr}(h\nu, Z)$ and also allows us to determine the mean energy transferred from photon to secondary charged particles $\overline{E}_{tr}$ using the following relationship

$$\overline{E}_{tr} = h\nu \sum_i w_i \overline{f}_i = \left(w_{PE}\overline{f}_{PE} + w_R\overline{f}_R + w_C\overline{f}_C + w_{PP}\overline{f}_{PP} \right) h\nu$$

$$= w_{PE}\overline{E}_{tr}^{PE} + w_R\overline{E}_{tr}^{R} + w_C\overline{E}_{tr}^{C} + w_{PP}\overline{E}_{tr}^{PP} = \overline{f}_{tr}h\nu. \qquad (8.12)$$

$\overline{E}_{tr}$ consists of three non-zero components: photoelectric $w_{PE}\overline{E}_{tr}^{PE}$, Compton $w_C\overline{E}_{tr}^{C}$, and pair production $w_C\overline{E}_{tr}^{PP}$. In Fig. 8.10 we plot the three components against photon energy $h\nu$ for eight selected absorbers ranging from carbon to uranium. The following general conclusions can be made:

1. The photoelectric component $w_{PE}\overline{E}_{tr}^{PE} = w_{PE}\overline{f}_{PE}h\nu$ of the total mean energy transfer $\overline{E}_{tr}$ is equal to $h\nu$ at very low photon energies $h\nu$; exhibits discontinuities at absorption edges; reaches a peak and then drops with increasing photon energy $h\nu$. The higher is the absorber Z, the higher is the peak in $w_{PE}\overline{E}_{tr}^{PE}$ and the higher is the photon energy $h\nu$ at which the peak occurs, ranging from $\sim$15 keV for carbon to $\sim$1 MeV for uranium.
2. The Compton component $w_C\overline{E}_{tr}^{C} = w_C\overline{f}_Ch\nu$ of the total mean energy transfer $\overline{E}_{tr}$ rises with photon energy $h\nu$ at low $h\nu$ and saturates at high $h\nu$; the higher is the absorber atomic number Z, the lower is the saturation value of $w_C\overline{E}_{tr}^{C}$ amounting to $\sim$1.5 MeV for uranium absorber and to $\sim$20 MeV for carbon absorber.
3. The pair production component $w_{PP}\overline{E}_{tr}^{PP} = w_{PP}\overline{f}_{PP}h\nu$ of the total mean energy transfer $\overline{E}_{tr}$ rises steeply with photon energy $h\nu$ starting at 1.022 MeV (threshold for pair production) and converges to $w_{PP}\overline{E}_{tr}^{PP} = h\nu$ for very high photon energies $h\nu$. The larger is the absorber atomic number Z, the faster is the $w_{PP}\overline{E}_{tr}^{PP}$ convergence to the $\overline{E}_{tr} = h\nu$ curve.
4. The photoelectric component $w_{PE}\overline{E}_{tr}^{PE} = w_{PE}\overline{f}_{PE}h\nu$ predominates at low $h\nu$, the Compton component $w_C\overline{E}_{tr}^{PE} = w_C\overline{f}_Ch\nu$ at intermediate $h\nu$, and the pair production component $w_{PP}\overline{E}_{tr}^{PP} = w_{PP}\overline{f}_{PP}h\nu$ at high $h\nu$.

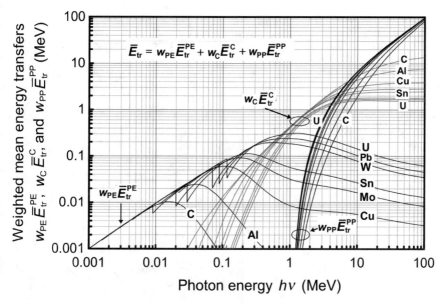

Fig. 8.10 Weighted mean energy transfers $w_i E_{tr}^i$ for photoelectric effect, Compton effect, and pair production for eight selected absorber atoms ranging from carbon (*low atomic number*) to uranium (*high atomic number*) plotted against photon energy $h\nu$

5. The Compton component $w_C \overline{E}_{tr}^C = w_C \overline{f}_C h\nu$ predominates for photon energies $h\nu$ between ~50 keV and ~20 MeV for low Z absorbers and between ~1 MeV and ~5 MeV for high Z absorbers. For photon energies $h\nu$ below these lower limits, the photoelectric component $w_{PE} \overline{E}_{tr}^{PE} = w_{PE} \overline{f}_{PE} h\nu$ predominates; for photon energies $h\nu$ exceeding the upper limit, the pair production component $w_{PP} \overline{E}_{tr}^{PP} = w_{PP} \overline{f}_{PP} h\nu$ predominates. In Fig. 8.11 we plot with the heavy solid line the sum of the three components: $w_{PE} \overline{E}_{tr}^{PE}$, $w_C \overline{E}_{tr}^C$, and $w_{PP} \overline{E}_{tr}^{PP}$ resulting in $\overline{E}_{tr}$, the mean energy transferred from photons to secondary charged particles against photon energy $h\nu$ for carbon in part (a) and lead in part (b). The three components of $\overline{E}_{tr}$ from Fig. 8.10 are shown with light curves in the background. At very low $h\nu$ and very high $h\nu$, $\overline{E}_{tr}$ equals the photon energy $h\nu$ for all absorbers, since in these two energy regions the total mean energy transfer fraction $\overline{f}_{tr}$ equals to 1. In the intermediate photon energy range where $\overline{f}_{tr} < 1$ the mean energy transferred to charged particles $\overline{E}_{tr}$ is smaller than $h\nu$ by up to an order of magnitude for low Z absorbers and by up to a factor of 3 for high Z absorbers. The maximum deviation of $\overline{E}_{tr}$ from $h\nu$ occurs at photon energy $h\nu$ of the order of 50 keV for low Z absorbers and 100 keV for high Z absorbers.

Fig. 8.11 $\overline{E}_{tr}$, the mean energy transferred from photon to charged particles in carbon absorber, in (a) and lead absorber in (b) plotted against photon energy $h\nu$. The three components of $\overline{E}_{tr}$ (photoelectric $w_{PE}\overline{E}_{tr}^{PE}$; Compton $w_C\overline{E}_{tr}^{C}$; and pair production $w_{PP}\overline{E}_{tr}^{PP}$) are shown with *fine line curves* in the background

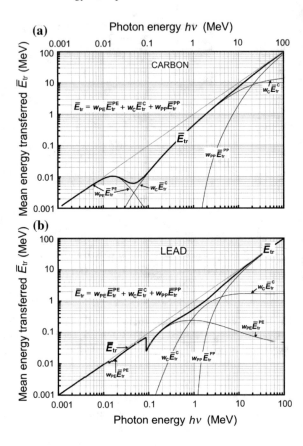

8.3   Energy Absorption

As shown in Sect. 7.1.4, the mass energy absorption coefficient μ_{ab}/ρ can be determined as follows:

1. From the mass attenuation coefficient μ/ρ by multiplying μ/ρ with total mean energy absorption fraction $\overline{f}_{ab}$, discussed in Sect. 8.3.2

$$\frac{\mu_{ab}}{\rho} = \frac{\mu}{\rho}\frac{\overline{E}_{ab}}{h\nu} = \frac{\mu}{\rho}\overline{f}_{ab} \tag{8.13}$$

2. From the mass energy transfer coefficient μ_{tr}/ρ by multiplying μ_{tr}/ρ with $(1-\overline{g})$

$$\frac{\mu_{ab}}{\rho} = \frac{\mu}{\rho}\frac{\overline{E}_{ab}}{h\nu} = \frac{\mu}{\rho}\frac{\overline{E}_{tr}-\overline{E}_{rad}}{h\nu} = \frac{\mu}{\rho}\left\{\frac{\overline{E}_{tr}}{h\nu}\left[1-\frac{\overline{E}_{rad}}{\overline{E}_{tr}}\right]\right\}$$
$$= \frac{\mu_{tr}}{\rho}(1-\overline{g}) = \overline{f}_{tr}\frac{\mu}{\rho}(1-\overline{g}) = \overline{f}_{ab}\frac{\mu}{\rho}, \tag{8.14}$$

where $\bar{g}$ is the mean radiation fraction discussed in Sect. 8.3.1 and given by the ratio $\bar{E}_{rad}/\bar{E}_{tr}$ with $\bar{E}_{rad}$ and $\bar{E}_{tr}$ defined in Sect. 7.14.

Thus, before we can address the mass energy absorption coefficient μ_{ab}/ρ we must define and determine the mean radiation fraction $\bar{g}$ and the total mean energy absorption fraction $\bar{f}_{ab}$.

8.3.1 Mean Radiation Fraction

The mean radiation fraction $\bar{g}$ is defined as the mean fraction of the energy that is transferred through photon interactions from photon to energetic electrons and positrons in the absorbing medium (see Sect. 8.2) and subsequently lost by these secondary charged particles through various radiation processes as the charged particles move through the absorber and come to rest in the absorber. Mathematically $\bar{g}$ is expressed by the following expression

$$\bar{g} = \frac{\bar{E}_{rad}}{\bar{E}_{tr}} = \frac{\bar{E}_{tr} - \bar{E}_{ab}}{\bar{E}_{tr}} = 1 - \frac{\bar{E}_{ab}}{\bar{E}_{tr}} = 1 - \frac{\bar{f}_{ab}}{\bar{f}_{tr}} = 1 - \frac{\mu_{ab}/\rho}{\mu_{tr}/\rho}, \tag{8.15}$$

where

$\bar{E}_{tr}$ is the mean energy transferred from the interacting photon to secondary light charged particles (electrons and positrons) that are released or produced through photon interactions in the absorber ($\bar{E}_{tr} = \bar{E}_{ab} + \bar{E}_{rad}$),

$\bar{E}_{ab}$ is the mean energy deposited in the absorbing medium by the light secondary charged particles (electrons and positrons) as they travel through the absorber,

$\bar{E}_{rad}$ is the mean energy radiated from the light secondary charged particles (electrons and positrons) as they travel through the absorber and come to rest in the absorber.

As discussed in Sect. 6.7 in conjunction with the radiation yield $Y[(E_K)_0]$, the radiation processes by which the energetic secondary charged particles may lose energy as they travel through the absorber are:

1. *Bremsstrahlung interaction* by electrons and positrons while they travel through the absorbing medium (see Sect. 4.2.3). This is the predominant radiation-emitting interaction experienced by the secondary charged particles.
2. *In-flight annihilation* process experienced by positrons. This process is less important than bremsstrahlung, but is generally not negligible.
3. *Production of fluorescence radiation* resulting from electron and positron impact ionization and impact excitation of atoms of the absorbing medium. This process is usually neglected in comparison to bremsstrahlung.

Thus, the mean radiation fraction $\bar{g}$ depends on incident photon energy $h\nu$ as well as on the absorber atomic number Z and generally consists of three components:

$$\bar{g} = \bar{g}_B + \bar{g}_A + \bar{g}_I, \tag{8.16}$$

where $\bar{g}_B$, $\bar{g}_A$, and $\bar{g}_I$ stand for the mean bremsstrahlung, mean in-flight annihilation, and mean impulse ionization fractions, respectively.

Figure 8.12 shows graphs of the mean radiation fraction $\bar{g}$ and its two main components ($\bar{g}_B$ and $\bar{g}_A$) against photon energy $h\nu$ for eight selected absorbers ranging from carbon to uranium. The mean radiation fraction $\bar{g}$ for incident photon energies ranging from 0.1 MeV to 1000 MeV was calculated with the standard g/EGSnrcMP Monte Carlo user code. In the code the charged particles released by photon interactions in the absorber are tracked and the energy fraction lost to radiation through bremsstrahlung interactions, positron in-flight annihilation, and emission of fluorescence x-rays is calculated as the charged particles slow down from their initial kinetic energy to rest in the absorber or to the point of positron in-flight annihilation in the absorber.

Figure 8.12a shows $\bar{g}$ and $\bar{g}_B$ (note linear ordinate scale) and Fig. 8.12b shows $\bar{g}$ and $\bar{g}_A$ (note logarithmic ordinate scale). In Fig. 8.12a the $\bar{g}$ and $\bar{g}_B$ curves for a given absorber are very close to one another, with $\bar{g}$ only slightly exceeding $\bar{g}_B$ confirming that $\bar{g}_B$ is the predominant contributor to $\bar{g}$ and validating the frequently used approximation in which $\bar{g}_B$ is assumed to represent the total mean radiation fraction $\bar{g}$. The mean radiation fraction $\bar{g}$ is often loosely referred to as the mean bremsstrahlung fraction even though strictly speaking it should always be referred to as the mean radiation fraction.

Several features of the mean radiation fraction $\bar{g}$ and its two components $\bar{g}_B$ and $\bar{g}_A$ become apparent from Fig. 8.12:

1. For a given photon energy $h\nu$ the mean radiation fraction $\bar{g}$ increases with absorber atomic number Z.
2. For a given absorber atomic number Z the mean radiation fraction $\bar{g}$ increases with photon energy $h\nu$ and saturates at $\bar{g} = 1$ at very high photon energies $h\nu$.
3. The mean bremsstrahlung fraction $\bar{g}_B$ is the sole contributor to $\bar{g}$ for photon energies $h\nu$ below 1.022 MeV, which, as shown in Sect. 7.6, is the threshold energy for pair production.
4. For $h\nu > 1.022$ MeV pair production becomes possible and both the mean bremsstrahlung fraction $\bar{g}_B$ and the mean in-flight annihilation fraction $\bar{g}_A$ contribute to $\bar{g}$; however, the mean bremsstrahlung fraction $\bar{g}_B$ predominates at all photon energies and exceeds the mean in-flight annihilation fraction $\bar{g}_A$ by at least an order of magnitude.
5. At photon energies $h\nu > 10$ MeV the mean bremsstrahlung fraction $\bar{g}_B$ increases almost linearly with $h\nu$ until it saturates at $\bar{g}_B \approx 1$ for extremely high photon energies.

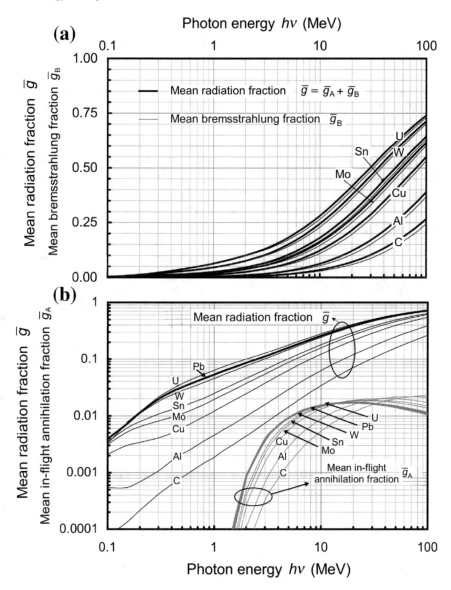

Fig. 8.12 Mean radiation fraction $\bar{g}$ against photon energy $h\nu$ for eight selected absorbers ranging from carbon to uranium. Part **a** displays with *heavy solid curves* the total mean radiation fraction $\bar{g}$ and with *light solid curves* its bremsstrahlung component $\bar{g}_B$ for the various absorbers (note linear scale for $\bar{g}$ and $\bar{g}_B$). Part **b** displays the total mean radiation fraction $\bar{g}$ and its in-flight annihilation component $\bar{g}_A$ (note the logarithmic scale for $\bar{g}$ and $\bar{g}_A$). Data were calculated using the g/EGSnrcMP code obtained from the NRC, Ottawa, Canada

6. The mean in-flight annihilation fraction $\overline{g}_A$ first rises rapidly with photon energy between 1 MeV and 10 MeV and saturates around $\overline{g} \approx 0.02$ at photon energy $h\nu \approx 20\,\mathrm{MeV}$ for high Z absorbers and at photon energy around $h\nu \approx 100\,\mathrm{MeV}$ for low Z absorbers.

7. For low Z absorbers and photon energies below 1 MeV the mean radiation fraction $\overline{g}$ is negligible ($\overline{g} < 0.001$); for electrons produced by cobalt-60 γ-rays in water $\overline{g} = 0.003$.

The mean radiation fraction $\overline{g}$ and the radiation yield $Y[(E_K)_0]$ for light charged particles, discussed in Sect. 6.7, are different, yet related, quantities. The radiation fraction $\overline{g}$ is the mean value of radiation yields $Y[(E_K)_0]$ for all electrons and positrons of various initial energies $(E_K)_0$ present in the spectrum of light charged particles released or produced in absorbing medium by either monoenergetic photons or by a spectrum of photons. On the one hand, the radiation yield $Y[(E_K)_0]$ is defined, as given in (6.66), for monoenergetic electrons and positrons with initial energy $(E_K)_0$, while, on the other hand, $\overline{g}$ is the mean radiation fraction calculated for a spectrum of electrons and positrons released in the medium by photons of energy $h\nu$.

The graph of $\overline{g}$ versus $h\nu$ of Fig. 8.12 can be linked to the graph of the radiation yield $Y[(E_K)_0]$ versus incident electron kinetic energy $(E_K)_0$ of Fig. 6.12 by determining, for a given photon energy $h\nu$, three quantities: $\overline{E}_{tr}$, $(\overline{E}_K)_0$, and $Y[(E_K)_0]$:

1. $\overline{E}_{tr}$, mean energy transferred to light charge carriers (electrons and positrons) through photon interactions with absorber atoms.

2. $(\overline{E}_K)_0$, mean of the initial kinetic energies acquired by the charged particles (note: in pair production two particles, an electron and a positron are produced in each interaction and in the first approximation we assume that they share the energy in equal proportions. Thus, $(\overline{E}_K)_0 < \overline{E}_{tr}$).

3. $Y[(\overline{E}_K)_0]$ for mean initial kinetic energy $(\overline{E}_K)_0$ from the $Y[(E_K)_0]$ versus $(E_K)_0$ diagram given in Fig. 6.12.

4. An assumption can then be made that $Y[(\overline{E}_K)_0, Z] \approx \overline{g}(h\nu, Z)$.

8.3.2 Total Mean Energy Absorption Fraction

The total mean energy absorption fraction $\overline{f}_{ab}$ is determined from the total mean energy transfer fraction $\overline{f}_{tr}$ of (8.5) as follows

$$\overline{f}_{ab} = \overline{f}_{tr}(1 - \overline{g}), \qquad (8.17)$$

with $\overline{g}$ the mean radiation fraction plotted in Fig. 8.12 and discussed in Sect. 8.3.1.

In Fig. 8.13 we plot $\overline{f}_{ab}$ with heavy solid curves against photon beam energy $h\nu$ for two absorbers: carbon in part (a) and lead in part (b). In the background we plot, with light solid curves, the mean energy transfer fraction $\overline{f}_{tr}$ and its three components $w_{PE}\overline{f}_{PE}$, $w_C\overline{f}_C$, and $w_{PP}\overline{f}_{PP}$ as well as the mean radiation fraction $\overline{g}$ and the function $(1 - \overline{g})$ for the two absorbers.

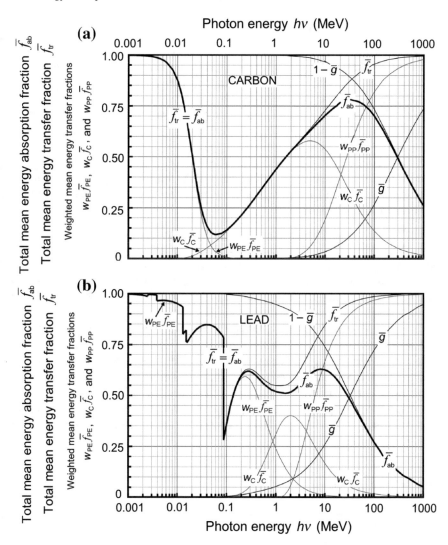

Fig. 8.13 Total mean energy absorption fraction $\bar{f}_{ab}$ against photon energy $h\nu$ for carbon in **a** and lead in **b**. Components of $\bar{f}_{ab}$, namely the weighted mean energy transfer fractions for the photoelectric effect $w_{PE}\bar{f}_{PE}$, Compton effect $w_C\bar{f}_C$, and pair production $w_{PP}\bar{f}_{PP}$ forming the total mean energy transfer fraction $\bar{f}_{tr}$ as well as the radiation fraction $\bar{g}$ and the function $(1 - \bar{g})$ are plotted with *fine line curves* in the background, while the mean energy absorption fraction $\bar{f}_{ab}$ is plotted with the *heavy black line*

The following observations can be made:

1. For relatively low photon energies $h\nu$ (below 1 MeV for low Z absorber and below 200 keV for high Z absorber) where $\overline{g} \approx 0$ and $(1 - \overline{g}) \approx 1$, the mean energy transfer fraction $\overline{f}_{tr}$ and the mean energy absorption fraction $\overline{f}_{ab}$ are equal.
2. At higher photon energies $h\nu$ (above 1 MeV for low Z absorber and above 200 keV for high Z absorber) $\overline{f}_{tr}$ and $\overline{f}_{ab}$ diverge as $h\nu$ increases, with $\overline{f}_{tr}$ asymptotically approaching $\overline{f}_{tr} = 1$, as shown in Fig. 8.13, and $\overline{f}_{ab}$ asymptotically approaching $\overline{f}_{ab} = 0$ at high photon energies $h\nu$.
3. Since the mean radiation fraction $\overline{g}$ is proportional to absorber atomic number Z resulting in inverse proportionality with Z of the function $1 - \overline{g}$, the mean energy absorption fraction $\overline{f}_{ab}$ is at high photon energy $h\nu$ inversely proportional to the absorber atomic number Z.

Figure 8.14 shows the total mean energy absorption fraction $\overline{f}_{ab}$ for eight selected absorbers from carbon to uranium and for photon energies exceeding the L_1 absorption edge energy for a given absorber. At relatively low photon energies $\overline{f}_{ab}$ is equal to the mean energy transfer fraction $\overline{f}_{tr}$ and both exhibit the fine structure caused by absorption edges. At photon energies $h\nu$ above 1 MeV for high Z absorbers and above 10 MeV for low Z absorbers, $\overline{f}_{ab}$ and $\overline{f}_{tr}$ start to deviate from one another. While $\overline{f}_{tr}$ continues to increase with increasing $h\nu$ and attains a value of 1 at very high photon energies $h\nu$, as shown in Fig. 8.8, $\overline{f}_{ab}$ attains a local maximum and then falls to 0 with increasing $h\nu$ as a result of $\overline{g} \to 1$. The energy $h\nu$ at which the local peak in $\overline{f}_{ab}$ occurs is inversely proportional to the absorber atomic number Z appearing at $h\nu = 10$ MeV for high Z absorbers and at $h\nu = 30$ MeV for low Z absorbers. The magnitude of $\overline{f}_{ab}$ at the peak energy $h\nu$ is also inversely proportional with Z, amounting to $\overline{f}_{ab} \approx 0.6$ for high Z absorbers and increasing with Z to reach $\overline{f}_{ab} \approx 0.8$ for low Z absorbers.

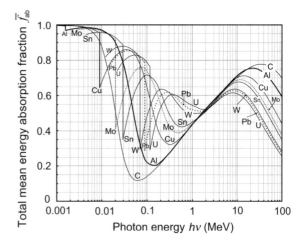

Fig. 8.14 Total mean energy absorption fraction $\overline{f}_{ab}$ against photon energy $h\nu$ for eight selected absorber atoms from carbon to uranium. The data are plotted for photon energy $h\nu$ exceeding the binding energy of the L_1 subshell of the absorber atom

8.3.3 Mass Energy Absorption Coefficient

The mass energy absorption coefficient μ_{ab}/ρ is shown for two absorbers: carbon and lead in Fig. 8.15 with the heavy solid line curve. Shown with fine line curves are the mass energy transfer coefficient μ_{tr}/ρ and the mass attenuation coefficient μ/ρ from Fig. 8.9. In addition, in Fig. 8.15 we also show in the background with light curves the total mean energy transfer fraction $\bar{f}_{tr}$, the function $1 - \bar{g}$, and the total mean energy absorption fraction $\bar{f}_{ab}$ for carbon in part (a) and for lead in part (b).

Based on Fig. 8.15a for carbon and Fig. 8.15b for lead we conclude that:

1. For low Z absorbers and photon energies $h\nu$ below 10 MeV, $\bar{f}_{ab} = \bar{f}_{tr}$ and $1 - \bar{g} \approx 1$ resulting in $\mu_{ab}/\rho \approx \mu_{tr}/\rho$.
2. For high Z absorbers and photon energies $h\nu$ below 100 keV, $\bar{f}_{ab} = \bar{f}_{tr}$ and $1 - \bar{g} \approx 1$ resulting in $\mu_{ab}/\rho = \mu_{tr}/\rho$.
3. With increasing photon energies above 20 MeV for low Z absorbers and above 2 MeV for high Z absorbers, μ_{ab}/ρ starts to deviate from μ_{tr}/ρ because of an increasing mean radiation fraction $\bar{g}$ with photon energy $h\nu$. As the mean radiation fraction $\bar{g}$ approaches 1, μ_{ab}/ρ approaches 0 while μ_{tr}/ρ approaches μ/ρ.

8.3.4 Mean Energy Absorbed in Absorbing Medium

From (8.5) and (8.17) we get the following expression for $\bar{E}_{ab}$, the mean energy absorbed in medium

$$\bar{E}_{ab} = \left(w_{PE}\bar{f}_{PE} + w_C\bar{f}_C + w_{PP}\bar{f}_{PP}\right) \times (1 - \bar{g})h\nu = \bar{f}_{ab}h\nu = \bar{E}_{tr}(1 - \bar{g}). \quad (8.18)$$

$\bar{E}_{ab}$ is plotted against photon energy $h\nu$ in Fig. 8.16 with a heavy line curve for two absorbers: carbon in part (a) and lead in part (b). The fine line curves in the background represent the mass energy transfer coefficient μ_{tr}/ρ, while the heavy grey line curve represents the function $(1 - \bar{g})$ with $\bar{g}$ the mean radiation fraction. The following general conclusions can be made:

1. For all absorbers, $\bar{E}_{tr}$ and $\bar{E}_{ab}$ are equal at relatively low photon energies $h\nu$ where the radiation fraction $\bar{g}$ is negligible.
2. For all absorbers $\bar{E}_{ab} < \bar{E}_{tr}$ at high photon energies. The two quantities start to diverge with increasing $h\nu$ at $h\nu = 10$ MeV for low Z absorbers and $h\nu = 1$ MeV for high Z absorbers.
3. As $\bar{g} \to 1$ at very high photon energies $h\nu$, the energy $\bar{E}_{ab}$ absorbed in the absorber approaches 0.

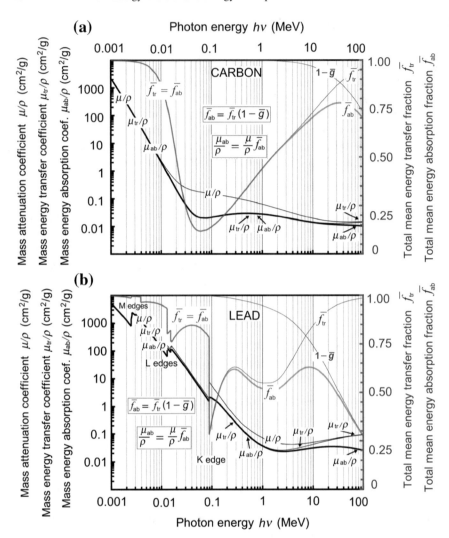

Fig. 8.15 Mass attenuation coefficient μ/ρ from the NIST XCOM database, calculated mass energy transfer coefficient μ_{tr}/ρ, and calculated mass energy absorption coefficient μ_{ab}/ρ, all against photon energy $h\nu$ for carbon in **a** and lead in **b**. Values of the three coefficients are given in cm²/g in the *left hand scale*. The total mean energy transfer fraction $\overline{f}_{tr}$ and the total mean energy absorption fraction $\overline{f}_{ab}$ as well as the function $(1 - \overline{g})$ which links $\overline{f}_{ab}$ with $\overline{f}_{tr}$ are plotted in the background with their values shown on the *right hand scale*. The mass attenuation coefficient μ/ρ and the mass energy transfer coefficient μ_{tr}/ρ are plotted with the *fine black line*; the mass energy absorption coefficient μ_{ab}/ρ with the *heavy black line*; and the total energy transfer fraction $\overline{f}_{tr}$, the total mean energy absorption fraction $\overline{f}_{ab}$, and the function $(1 - \overline{g})$ with the *light grey line*

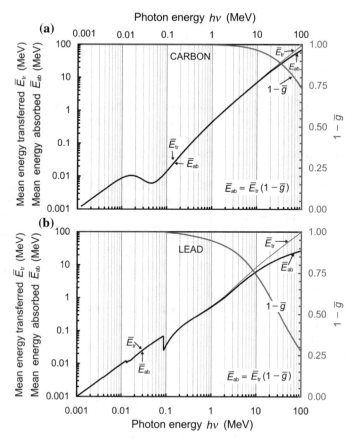

Fig. 8.16 $\overline{E}_{tr}$, mean energy transferred from photon to charged particles, and $\overline{E}_{ab}$, mean energy absorbed in the absorber, for carbon in **a** and lead in **b** plotted against photon energy $h\nu$. The function $(1 - \overline{g})$ which relates $\overline{E}_{tr}$ and $\overline{E}_{ab}$ is plotted in the background

8.4 Coefficients of Compounds and Mixtures

At a given photon energy $h\nu$, the attenuation coefficients μ, energy transfer coefficients μ_{tr}, and energy absorption coefficients μ_{ab} for a compound or mixture of elements are approximated by a summation of a weighted mean of its constituents, as follows

$$\mu = \sum_j w_j \mu_j, \tag{8.19}$$

$$\mu_{tr} = \sum_j w_j (\mu_{tr})_j, \tag{8.20}$$

$$\mu_{ab} = \sum_j w_j (\mu_{ab})_j, \tag{8.21}$$

where

w_j	is the proportion by weight of the jth constituent element.
μ_j	is the attenuation coefficient of the jth constituent element.
$(\mu_{tr})_j$	is the energy transfer coefficient of the jth constituent element.
$(\mu_{ab})_j$	is the energy absorption coefficient of the jth constituent element.

By way of a few **examples**, we determine the proportion by weight w_j of the constituent elements for water, polystyrene, and Lucite and calculate, for 10 MeV photons, the mass attenuation coefficient of water, the mass energy transfer coefficient of polystyrene, and the mass energy absorption coefficient of Lucite. A discussion of mean atomic mass and mean molecular mass is given in Sect. 1.14. Attenuation coefficient data for constituent atoms as well as a summary of results are given in Table 8.1.

1. **Water** H_2O with constituent elements hydrogen (mean atomic mass $\overline{\mathcal{M}}_H = 1.00794$ u) and oxygen (mean atomic mass $\overline{\mathcal{M}}_O = 15.9994$ u) has mean molecular mass $\overline{\mathcal{M}}_{H_2O} = 18.0153$ u. Mass attenuation coefficient of water $(\mu/\rho)^{h\nu}_{H_2O}$ is approximated as follows

$$\left(\frac{\mu}{\rho}\right)^{h\nu}_{H_2O} = \frac{2 \times 1.00794}{18.0153} \left(\frac{\mu}{\rho}\right)^{h\nu}_H + \frac{15.9994}{18.0153} \left(\frac{\mu}{\rho}\right)^{h\nu}_O$$

$$= 0.1119 \left(\frac{\mu}{\rho}\right)^{h\nu}_H + 0.8881 \left(\frac{\mu}{\rho}\right)^{h\nu}_O \tag{8.22}$$

$$\left(\frac{\mu}{\rho}\right)^{10\,MeV}_{H_2O} = 0.1119 \times 0.0325 \ (cm^2/g) + 0.8881 \times 0.0209 \ (cm^2/g)$$

$$= 0.0222 \ cm^2/g. \tag{8.23}$$

2. **Polystyrene** $(C_8H_8)_x$ with constituent elements carbon (mean atomic weight $\overline{\mathcal{M}}_C = 12.0107$ u) and hydrogen (mean atomic mass $\overline{\mathcal{M}}_H = 1.00794$ u) has

Table 8.1 Mass attenuation coefficient μ/ρ, mass energy transfer coefficient μ_{tr}/ρ, and mass energy absorption coefficient μ_{ab}/ρ for 10 MeV photons in various low atomic number absorbers

Coefficient	Hydrogen	Carbon	Oxygen	Water	Polystyrene	Lucite
μ/ρ (cm^2/g)	0.0325	0.0196	0.0209	**0.0222**	0.0206	0.0210
μ_{tr}/ρ (cm^2/g)	0.0227	0.0143	0.0154	0.0162	**0.0150**	0.0153
μ_{ab}/ρ (cm^2/g)	0.0225	0.0138	0.0148	0.0157	0.0145	**0.0148**

Results from the three examples [(8.23), (8.25) and (8.27)] are shown in bold

a molecular mass $\overline{\mathcal{M}}_{C_8H_8} = 104.149$ u. Mass energy transfer coefficient of polystyrene $(\mu_{tr}/\rho)_{C_8H_8}^{h\nu}$ is approximated as

$$\left(\frac{\mu_{tr}}{\rho}\right)_{C_8H_8}^{h\nu} = \frac{8 \times 12.0107}{104.149} \left(\frac{\mu_{tr}}{\rho}\right)_C^{h\nu} + \frac{8 \times 1.00794}{104.149} \left(\frac{\mu_{tr}}{\rho}\right)_H^{h\nu}$$

$$= 0.9226 \left(\frac{\mu_{tr}}{\rho}\right)_C^{h\nu} + 0.0774 \left(\frac{\mu_{tr}}{\rho}\right)_H^{h\nu} \tag{8.24}$$

$$\left(\frac{\mu_{tr}}{\rho}\right)_{C_8H_8}^{10\ MeV} = 0.9226 \times 0.0143\ (cm^2/g) + 0.0774 \times 0.0227\ (cm^2/g)$$

$$= 0.0150\ cm^2/g. \tag{8.25}$$

3. **Lucite** $(C_5H_8O_2)_x$ with constituent elements carbon (mean atomic weight $\overline{\mathcal{M}}_C = 12.0107$ u), hydrogen (mean atomic weight $\overline{\mathcal{M}}_H = 1.00794$ u), and oxygen (atomic mass $\overline{\mathcal{M}}_O = 15.9994$ u) has mean molecular weight $\overline{\mathcal{M}}_{C_5H_8O_2} = 100.1158$ u. Mass energy absorption coefficient of Lucite $(\mu_{ab}/\rho)_{C_5H_8O_2}^{h\nu}$ is approximated as follows

$$\left(\frac{\mu_{ab}}{\rho}\right)_{C_5H_8O_2}^{h\nu} = \frac{5 \times 12.0107}{100.1158} \left(\frac{\mu_{ab}}{\rho}\right)_C^{h\nu} + \frac{8 \times 1.00794}{100.1158} \left(\frac{\mu_{ab}}{\rho}\right)_H^{h\nu}$$

$$+ \frac{2 \times 15.9994}{100.1158} \left(\frac{\mu_{ab}}{\rho}\right)_O^{h\nu} = 0.5998 \left(\frac{\mu_{ab}}{\rho}\right)_C^{h\nu}$$

$$+ 0.0805 \left(\frac{\mu_{ab}}{\rho}\right)_H^{h\nu} + 0.3196 \left(\frac{\mu_{ab}}{\rho}\right)_O^{h\nu} \tag{8.26}$$

$$\left(\frac{\mu_{ab}}{\rho}\right)_{C_5H_8O_2}^{10\ MeV} = 0.5998 \times 0.0138\ (cm^2/g)$$

$$+ 0.0805 \times 0.0225\ (cm^2/g)$$

$$+ 0.3196 \times 0.0148\ (cm^2/g) = 0.0148\ cm^2/g. \tag{8.27}$$

4. **Air** In radiation dosimetry, air is commonly used as the radiation sensitive medium in ionization chambers for determination of exposure and dose in various tissue equivalent media. It thus represents an important mixture of gases and, as given in Table 8.2, consists of molecules of nitrogen (N_2), oxygen (O_2), and carbon dioxide (CO_2), as well as argon atoms (Ar) in the following respective proportions by volume and number: 78.08%, 20.95%, 0.93%, and 0.03%. This translates to respective proportions by weight of 75.8%, 22.6%, 0.93%, and 0.03%. The mass attenuation coefficient of air $(\mu/\rho)_{air}^{h\nu}$ at photon energy $h\nu$, and similarly the mass energy transfer coefficient and mass energy absorption coefficient, are approximated as follows

Table 8.2 Properties of dry air of importance in radiation dosimetry

Gas	Molecular mass (u)	Percent by volume	Partial pressure (kPa)	Percent by weight
Nitrogen (N_2)	28.013	78.08	79.1	75.8
Oxygen (O_2)	31.999	20.95	21.2	22.6
Argon (Ar)	39.948	0.93	0.94	0.93
Carbon dioxide	44.000	0.03	0.03	0.03
		100	101.3	100

$$\left(\frac{\mu}{\rho}\right)^{h\nu}_{air} = 0.758\left(\frac{\mu}{\rho}\right)^{h\nu}_{N_2} + 0.226\left(\frac{\mu}{\rho}\right)^{h\nu}_{O_2} + 0.0093\left(\frac{\mu}{\rho}\right)^{h\nu}_{Ar} + 0.0003\left(\frac{\mu}{\rho}\right)^{h\nu}_{CO_2}.$$

$$(8.28)$$

8.5 Effects Following Photon Interactions with Absorber

In photoelectric effect, Compton effect, and triplet production vacancies are produced in atomic shells through ejection of orbital electrons:

1. For orthovoltage and megavoltage photons used in diagnosis and treatment of disease with radiation, the shell vacancies occur mainly in inner atomic shells of the absorber.
2. Nuclear pair production, Rayleigh scattering and photodisintegration do not produce shell vacancies.
3. As discussed in detail in Sect. 4.1, vacancies in inner atomic shells are not stable; they are followed by emission of characteristic (fluorescence) x-rays or Auger electrons depending on the fluorescence yield of the absorbing material and cascade to the outer shell of the ion. The ion eventually attracts an electron from its surroundings and becomes a neutral atom.
4. Pair production and triplet production are followed by annihilation of the positron with an orbital electron of the absorber, most commonly producing two annihilation quanta of 0.511 MeV each and moving at 180° to each other. Annihilation of a positron before it expended all of its kinetic energy is referred to as in-flight annihilation and may produce photons exceeding 0.511 MeV in energy (Sect. 7.6.10) as well as angular deflections significantly different from 180°.

A list of the major photon interactions with absorber atoms identifying the interactions that result in orbital shell vacancy is provided in Table 8.3. The table also identifies the type of charged particle produced in each given photon interaction as well as the name used to designate the charged particle.

Table 8.3 Production of electron shell vacancy and charged particles produced in various photon interactions with absorber atoms

Photon interaction with absorber	Electron shell vacancy produced	Charged particles released or produced in the photon interaction with absorber
Thomson scattering	No	None
Compton effect	Yes	Compton (recoil) electron
Rayleigh scattering	No	None
Photoelectric effect	Yes	Photoelectron
Pair production	No	Electron–positron pair
Triplet production	Yes	Electron–positron pair + orbital electron
Photodisintegration	No	Photoneutron, proton, etc.

A vacancy produced in inner atomic shell migrates to outer shell and the excess energy is emitted in the form of characteristic (fluorescence) photons or Auger electrons

8.6 Summary of Photon Interactions

As is evident from discussions in Chap. 7 and this chapter, photons have numerous options for interaction with absorber atoms. The probabilities for interaction in general depend on the incident photon energy $h\nu$ and the atomic number Z of the absorber.

While over a dozen different photon interactions are known in nuclear physics, six of these are of importance to medical physics because they govern:

1. The physics of attenuation and scattering of photon beams by tissues of importance in imaging physics and radiation dosimetry.
2. The physics of energy transfer from photons to light charged particles in an absorber and the ultimate energy absorption in irradiated tissues. This is of importance in radiation dosimetry, treatment planning, clinical dose prescription, and dose delivery.
3. Less importantly, physics of neutron production which poses a potential health hazard to patients and staff involved with the use of high energy linacs in treatment of cancer with radiation.

The six modes of photon interactions with absorber atoms of relevance to medical physics are:

1. Photoelectric effect.
2. Rayleigh scattering.
3. Compton effect.
4. Nuclear pair production.
5. Electronic pair production (also known as triplet production).
6. Photonuclear reactions (also known as photodisintegration).

The six modes of photon interactions are discussed in detail in Chap. 7 and this chapter, and their most important characteristics are:

1. Reiterated in the next five sections (Sect. 8.6.1 through Sect. 8.6.5). Note that the discussions of the nuclear pair production and the electronic pair production are combined under the header "pair production".
2. Shown schematically in Figs. 8.17 and 8.18.
3. Summarized in Table 8.4.

Figure 8.17 shows the basic features of the six photon interaction modes, indicating clearly that:

1. Four of the six modes (photoelectric effect, Rayleigh scattering, Compton effect, and electronic pair production) are interactions between a photon and orbital electron of an absorber atom.
2. Three of the six modes (photoelectric effect, Compton effect, and electronic pair production) produce a vacancy in the absorber atom by causing ejection of the orbital electron with which the photon interacts.
3. Two of the six modes (nuclear pair production and photonuclear reaction) are interactions between a photon and nucleus of an absorber atom.

Figure 8.17 also indicates the photon energy $h\nu$ range for the six modes of interaction as well as the products that appear after the particular interaction. Kinetic energies E_K of the charged particles released during the interaction and of importance in radiation dosimetry are also indicated.

Figure 8.18 provides information similar to that of Fig. 8.17 but to a greater detail, as it also presents the effects that follow the individual photon interactions, such as:

1. Emission of characteristic x-ray photons following a vacancy produced in atomic shell of an absorber atom.
2. Emission of Auger electrons following a vacancy produced in atomic shell of an absorber atom.
3. Production of ionization and excitation of absorber atoms by the energetic charged particles produced in the initial photon interactions.
4. Production of delta rays by the energetic charged particles produced in the initial photon interactions.
5. Bremsstrahlung production by the energetic charged particles produced in the initial photon interactions.
6. Production of annihilation quanta by positrons generated in the nuclear pair production and the electronic pair production.
7. Redirection of scattered photons, characteristic photons, annihilation photons, and bremsstrahlung photons to start a new photon interaction cycle in the absorber.

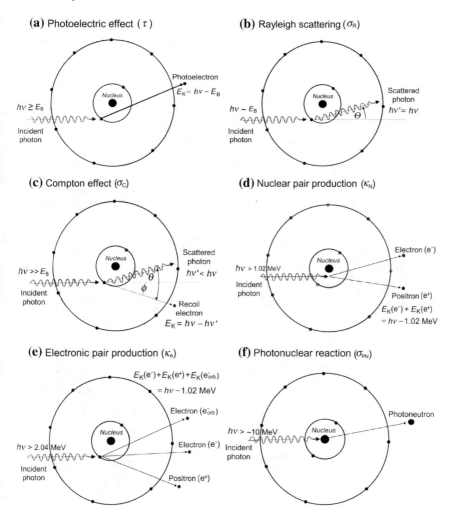

Fig. 8.17 Schematic diagrams of the six most important modes of photon interaction with atoms of absorber: **a** Photoelectric effect; **b** Rayleigh scattering; **c** Compton effect; **d** Electronic pair production (triplet production); **e** Nuclear pair production; **f** Photodisintegration. The first four modes represent photon interactions with a K-shell orbital electron; the last two modes **e** and **f** represent photon interactions with the nucleus of the absorber atom. The electronic and nuclear pair production modes are usually handled together under the header "pair production", and photonuclear reactions are usually ignored, so that often in medical physics an assumptions is made that there are only four important modes of photon interaction with absorber atoms: photoelectric effect, Rayleigh scattering, Compton effect, and pair production

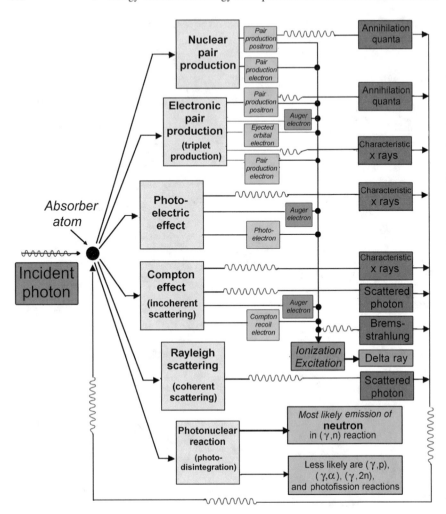

Fig. 8.18 Schematic diagram of the six modes available to incident photons for interaction with absorber atoms of importance to medical physics: photonuclear reactions, Rayleigh scattering, Compton effect, photoelectric effect, electronic pair production, and nuclear pair production (*Note* the modes are not listed in any particular order of importance). Also shown are the particles produced or released in the absorber during a given photon interaction as well as the effects that follow a given photon interaction, such as emission of characteristic radiation, Auger electrons, annihilation quanta, and bremsstrahlung photons. Figure also indicates that the secondary photons (Rayleigh- and Compton-scattered photons, characteristic photons, annihilation quanta, and bremsstrahlung photons) can start their own photon interaction cycle in the absorber

Table 8.4 Main characteristics of photoelectric effect, Rayleigh scattering, Compton effect, and pair production

	Photoelectric effect	Rayleigh scattering	Compton effect	Pair production
Photon interaction	With whole atom (bound electron)	With bound electrons	With free electron	With nuclear Coulomb field
Mode of photon interaction	Photon disappears	Photon scattered	Photon scattered	Photon disappears
Energy dependence	$\dfrac{1}{(h\nu)^3}$	$\dfrac{1}{(h\nu)^2}$	Decreases with energy	Increases with energy
Threshold energy	Shell binding energy	No	Shell binding energy	$\sim 2m_e c^2$
Linear attenuation coefficient	τ	σ_R	σ_C	κ
Atomic coef. dependence on Z	$_a\tau \propto Z^4$	$_a\sigma_R \propto Z^2$	$_a\sigma_C \propto Z$	$_a\kappa \propto Z^2$
Mass coefficient dependence on Z	$\dfrac{\tau}{\rho} \propto Z^3$	$\dfrac{\sigma_R}{\rho} \propto Z$	Independent of Z	$\dfrac{\kappa}{\rho} \propto Z$
Particles released in absorber	Photoelectron	None	Compton (recoil) electron	Electron–positron pair
Mean energy transferred to charged part's	$h\nu - \sum_j P_j \omega_j h\bar{\nu}_j$	0	$\overline{E}_{tr}^{C}$ (see Fig. 7.18)	$h\nu - 2m_e c^2$
Fraction of energy $h\nu$ transferred	$1 - \dfrac{\sum_j P_j \omega_j h\bar{\nu}_j}{h\nu}$	0	$\dfrac{\overline{E}_{tr}^{C}}{h\nu}$	$1 - \dfrac{2m_e c^2}{h\nu}$
Subsequent effect	Characteristic x-ray, Auger effect	None	Characteristic x-ray, Auger effect	Positron annihilation radiation
Predominant energy region for water	<20 keV	–	20 keV–20 MeV	>20 MeV
Predominant energy region for lead	<500 keV	–	500 keV–5 MeV	>5 MeV

8.6.1 Photoelectric Effect

1. The photoelectric effect (sometimes also referred to as photoeffect) is an interaction between a photon with energy $h\nu$ and a tightly bound orbital electron of an absorber atom. The interaction is thus between a photon and an absorber atom as a whole. The electron is ejected from the atom and referred to as a photoelectron.

2. A tightly bound orbital electron is defined as an orbital electron with binding energy E_B either larger than $h\nu$ or of the order of $h\nu$. For $E_B > h\nu$ the photoeffect cannot occur; for $h\nu > E_B$ the photoelectric effect is possible. The closer is $h\nu$ to E_B, the larger is the probability for photoelectric effect to happen, provided, of course, that $h\nu$ exceeds E_B. At $h\nu = E_B$ the probability abruptly drops and exhibits the so-called absorption edge.

3. With increasing incident photon energy $h\nu$, the atomic, linear, and mass photo-electric attenuation coefficients decrease from their absorption edge value approximately as $1/(h\nu)^3$.

4. The atomic photoelectric attenuation coefficient $_a\tau$ varies approximately as Z^5 for low Z absorbers and as Z^4 for high Z absorbers.

5. The mass photoelectric attenuation coefficient $\tau_m = \tau/\rho$ varies approximately as Z^4 for low Z absorbers and as Z^3 for high Z absorbers.

6. In water and tissue $\overline{E}_{tr}^{PE}$, the mean energy transferred to electrons (photoelectrons and Auger electrons) is equal to $\overline{E}_{ab}^{PE}$, the mean energy absorbed in the medium because the radiation fraction $\overline{g}$ is negligible; i.e., $\overline{g} \approx 0$.

7. Furthermore, in water and tissue $\overline{E}_{tr}^{PE}$ is approximately equal to the photon energy $h\nu$ because the fluorescence yield ω_K is approximately equal to zero. Thus in water and tissue the following relationship holds for the photoelectric effect: $\overline{E}_{tr}^{PE} = \overline{E}_{ab}^{PE} \approx h\nu$.

8.6.2 Rayleigh Scattering

1. Rayleigh scattering is an interaction between a photon with energy $h\nu$ and the whole atom. All orbital electrons contribute to the scattering event and the phenomenon is referred to as coherent scattering because the photon is scattered by the constructive action of the tightly bound electrons of the whole atom.

2. The photon leaves the point of interaction with the incident energy $h\nu$ intact but is redirected through a small scattering angle. Since no energy is transferred to charged particles, Rayleigh scattering plays no role in radiation dosimetry; however, it is of some importance in imaging physics because the scattering event has an adverse effect on image quality.

3. The atomic Rayleigh attenuation coefficient $_a\sigma_R$ decreases approximately as $1/(h\nu)^2$ and is approximately proportional to Z^2 of the absorber.

4. Even at very small incident photon energy $h\nu$, the Rayleigh component of the total attenuation coefficient is small and amounts to only a few percent of the total attenuation coefficient.

8.6.3   Compton Effect

1. Compton effect (often referred to as Compton scattering) is an interaction between a photon with energy $h\nu$ and a free orbital electron.
2. A free electron is defined as an orbital electron whose binding energy E_B is much smaller than the photon energy $h\nu$; i.e., $h\nu \gg E_B$.
3. In each Compton interaction a scattered photon and a free electron (referred to as Compton or recoil electron) are produced. The sum of the scattered photon energy $h\nu'$ and the Compton recoil electron kinetic energy E_K is equal to the incident photon energy $h\nu$. The relative distribution of the two energies depends on the incident photon energy $h\nu$ and on the angle of emission (scattering angle θ) of the scattered photon.
4. The electronic and mass Compton attenuation coefficients $_e\sigma_C$ and σ_C/ρ, respectively, are essentially independent of the atomic number Z of the absorber.
5. The atomic and mass Compton attenuation coefficients $_a\sigma_C$ and σ_C/ρ, respectively, decrease with increasing incident photon energy $h\nu$.
6. The atomic Compton attenuation coefficient $_a\sigma$ is linearly proportional to the atomic number Z of the absorber.
7. The average fraction of the incident photon energy $h\nu$ transferred to recoil electron increases with $h\nu$ (see *The Compton Graph* in Figs. 7.11 and 7.18). At low photon energies the Compton energy transfer coefficient $(\sigma_c)_{tr}$ is much smaller than the Compton attenuation coefficient σ_C; i.e., $(\sigma_c)_{tr} \ll \sigma_c$. At high photon energies, on the other hand, $(\sigma_C)_{tr} \approx \sigma_C$.
8. In water and tissue the Compton process is the predominant mode of photon interaction in the wide photon energy range from $\sim$20 keV to $\sim$20 MeV (see Fig. 8.5).

8.6.4   Pair Production

1. Pair production is an interaction between a relatively high energy photon and the Coulomb field of either a nucleus or orbital electron. The photon disappears and an electron–positron pair is produced. The process is an example of mass-energy equivalence and is sometimes referred to as materialization.
2. Pair production in the field of a nucleus is referred to as nuclear pair production and has threshold energy of $\sim 2m_e c^2 = 1.022$ MeV.
3. Pair production in the Coulomb field of an orbital electron of the absorber is referred to as electronic pair production or triplet production. The process is much less probable than nuclear pair production and has threshold energy of $4m_e c^2 = 2.044$ MeV. The photon disappears and three light charged particles are released: the original orbital electron and the electron–positron pair produced in the interaction.

4. Contributions from the nuclear pair production and the electronic pair production are usually combined into one contribution referred to as pair production.
5. The probability for pair production interaction first increases rapidly with the incident photon energy $h\nu$ for photon energies above the threshold energy and then eventually saturates at very high photon energies resulting in a logarithmic type curve.
6. The atomic pair production attenuation coefficient $_a\kappa$ varies approximately as Z^2 of the absorber.
7. The mass pair production coefficient $\kappa_m = \kappa/\rho$ varies approximately linearly with the atomic number Z of the absorber.
8. The average energy transferred from the incident photon $h\nu$ to charged particles in both the nuclear pair production and the electronic pair production, $\overline{E}_{tr}^{PP}$, is $h\nu - 2m_e c^2 = 1.022$ MeV.

8.6.5 Photonuclear Reactions

1. Photonuclear reactions (also called photodisintegration or nuclear photoelectric effect) are direct interactions between an energetic photon with energy $h\nu$ and a nucleus of the absorber atom. The nucleus absorbs the photon and the most likely result of such an interaction is the emission of a single neutron from the nucleus through a (γ, n) reaction.
2. The threshold energy for photonuclear reactions is of the order of ~ 8 MeV for all nuclides with two notable exceptions of the deuteron at 2.22 MeV and beryllium-9 at 1.67 MeV.
3. Cross sections for photonuclear reactions exhibit a broad peak "giant resonance" centered at about 23 MeV for low atomic number Z absorbers and at about 12 MeV for high Z absorbers. The full-width-at-half-maximum (FWHM) typically ranges from ~ 3 MeV to ~ 9 MeV.
4. Atomic cross sections for photonuclear reactions $_a\sigma_{PN}$ even at the peak of the giant resonance amounts to only about a few per cent of the sum of the "electronic cross sections", i.e., $_a\sigma_{PN} \ll {_a\tau} + {_a\sigma_R} + {_a\sigma_C} + {_a\kappa}$. For this reason, the photonuclear reactions are generally ignored in medical physics.

8.7 Sample Calculations

To provide a practical summary of the material presented in Chap. 7 and this chapter we investigate with two examples the various interactions that photons can have with absorbers. In Sect. 8.7.1 we present the simple relationships that govern photoelectric, Rayleigh, Compton, and pair production interactions of monoenergetic 2 MeV photons with a lead absorber. We also determine various attenuation coefficients as well as the mass energy transfer coefficient and the mass energy absorption coefficient.

In Sect. 8.7.2 we deal with interactions of monoenergetic 8 MeV photons with copper absorber. In this example, we use the basic relationships presented in this chapter to estimate the appropriate attenuation coefficients, energy transferred to charged particles, energy absorbed by the absorber, as well as the mass energy transfer coefficient and the mass energy absorption coefficient.

8.7.1 Example 1

Interaction of 2 MeV Photon with Lead Absorber

For monoenergetic 2 MeV *photons interacting with lead* ($Z = 82$; $A = 207.2$ g/mol; $\rho = 11.36$ g/cm^3) linear attenuation coefficients for the photoelectric effect, coherent scattering, Compton effect, and pair production are: $\tau = 0.055$ cm^{-1}, $\sigma_R = 0.008$ cm^{-1}, $\sigma_C = 0.395$ cm^{-1}, and $\kappa = 0.056$ cm^{-1}, respectively. The mean energy transferred to charged particles is $\overline{E}_{tr} = 1.13$ MeV and the mean energy absorbed in lead is $\overline{E}_{ab} = 1.04$ MeV. Compare results with data tabulated in the literature.

Determine

(1) Linear attenuation coefficient μ.
(2) Mass attenuation coefficient μ_m.
(3) Atomic attenuation coefficient $_a\mu$.
(4) Mass energy transfer coefficient μ_{tr}/ρ.
(5) Mean radiation fraction $\overline{g}$.
(6) Mass energy absorption coefficient μ_{ab}/ρ.

1. **Linear attenuation coefficient** μ [see (8.1)] is the sum of the four individual components: photoelectric, Rayleigh, Compton, and pair production including nuclear and electronic pair production

$$\mu = \tau + \sigma_R + \sigma_C + \kappa = (0.055 + 0.008 + 0.395 + 0.056) \text{ cm}^{-1} = 0.514 \text{ cm}^{-1}$$
$$(8.29)$$

2. **Mass attenuation coefficient** μ_m [see (8.2)]

$$\mu_m = \frac{\mu}{\rho} = \frac{0.514 \text{ cm}^{-1}}{11.36 \text{ g/cm}^3} = 0.0453 \text{ cm}^2/\text{g}. \qquad (8.30)$$

3. **Atomic attenuation coefficient** (cross section) $_a\mu$ [see (8.3)]

$$_a\mu = \left\{ \frac{\rho N_A}{A} \right\}^{-1} \mu$$

$$= \frac{207.2 \text{ (g/mol) } 0.514 \text{ cm}^{-1}}{11.36 \text{ g/cm}^3 \ 6.022 \times 10^{23} \text{ (atom/mol)}} = 1.56 \times 10^{-23} \text{ cm}^2/\text{atom} \quad (8.31)$$

4. **Mass energy transfer coefficient** μ_{tr}/ρ

$$\frac{\mu_{tr}}{\rho} = \frac{\overline{E}_{tr}}{h\nu} \frac{\mu}{\rho} = \frac{1.13 \text{ MeV} \times 0.0453 \text{ cm}^2/\text{g}}{2 \text{ MeV}} = 0.0256 \ \frac{\text{cm}^2}{\text{g}}. \quad (8.32)$$

The mass energy transfer coefficient μ_{tr}/ρ can also be determined using (8.5) with the appropriate mean energy transfer fractions $\overline{f}_i$

$$\overline{f}_{tr} = \sum_i \frac{\mu_i}{\mu} \overline{f}_i = \left\{ \frac{\tau}{\mu} \overline{f}_{PE} + \frac{\sigma_R}{\mu} \overline{f}_R + \frac{\sigma_C}{\mu} \overline{f}_C + \frac{\kappa}{\mu} \overline{f}_{PP} \right\} = \frac{\overline{E}_{tr}}{h\nu}. \quad (8.33)$$

The appropriate mean energy transfer fractions can be found in Fig. 8.2

$$\overline{f}_{PE} = \frac{\overline{E}_{tr}^{PE}}{h\nu} = \frac{h\nu - P_K \omega_K h\overline{\nu}_K}{h\nu} = 1 - \frac{P_K \omega_K h\overline{\nu}_K}{h\nu} = 0.965 \quad (8.34)$$

$$\overline{f}_R = \frac{\overline{E}_{tr}^R}{h\nu} = 0 \quad (8.35)$$

$$\overline{f}_C = \frac{\overline{E}_{tr}^C}{h\nu} = 0.53 \quad (8.36)$$

$$\overline{f}_{PP} = \frac{\overline{E}_{tr}^{PP}}{h\nu} = \frac{h\nu - 2m_e c^2}{h\nu} = 1 - \frac{2m_e c^2}{h\nu} = 0.49, \quad (8.37)$$

resulting in the following mass energy transfer coefficient μ_{tr}/ρ

$$\frac{\mu_{tr}}{\rho} = \frac{\overline{E}_{tr}}{h\nu} \frac{\mu}{\rho} = \left\{ \frac{\tau}{\mu} \overline{f}_{PE} + \frac{\sigma_R}{\mu} \overline{f}_R + \frac{\sigma_C}{\mu} \overline{f}_C + \frac{\kappa}{\mu} \overline{f}_{PP} \right\} \frac{\mu}{\rho}$$

$$= \frac{1}{11.36 \text{ g·cm}^{-3}} (0.055 \times 0.965 + 0 + 0.395 \times 0.53 + 0.056 \times 0.50) \text{ cm}^{-1}$$

$$= 0.0256 \ \frac{\text{cm}^2}{\text{g}}, \quad (8.38)$$

in excellent agreement with the result obtained in (8.32). We can now also verify the mean energy transferred from 2 MeV photons to charged particles (electrons and positrons) in the lead absorber stated as 1.13 MeV in the assignment above. Using the mean energy transfer fractions of (8.34) through (8.37) we get the following result for $\overline{E}_{tr}$

$$\overline{E}_{tr} = \overline{f}_{tr} h\nu = \frac{\tau}{\mu}\overline{E}_{tr}^{PE} + \frac{\sigma_R}{\mu}\overline{E}_{tr}^{R} + \frac{\sigma_C}{\mu}\overline{E}_{tr}^{C} + \frac{\kappa}{\mu}\overline{E}_{tr}^{PP}$$

$$= 0.107 \times 1.93 \text{ MeV} + 0.016 \times 0 + 0.769 \times 1.06 \text{ MeV}$$

$$+ 0.109 \times 0.98 \text{ MeV} = 1.13 \text{ MeV}. \tag{8.39}$$

5. **Mean radiation fraction** $\overline{g}$

$$\overline{g} = \frac{\overline{E}_{rad}}{\overline{E}_{tr}} = \frac{\overline{E}_{tr} - \overline{E}_{ab}}{\overline{E}_{tr}} = 1 - \frac{\overline{E}_{ab}}{\overline{E}_{tr}} = 1 - \frac{1.04 \text{ MeV}}{1.13 \text{ MeV}} = 0.08, \tag{8.40}$$

in excellent agreement with the radiation fraction of 0.08 we get for 2 MeV photons in lead from Fig. 8.12.

6. **Mass energy absorption coefficient** μ_{ab}/ρ

$$\frac{\mu_{ab}}{\rho} = \frac{\mu_{en}}{\rho} = \frac{\overline{E}_{ab}}{h\nu}\frac{\mu}{\rho} = \frac{1.04 \text{ MeV} \times 0.0453 \text{ cm}^2/\text{g}}{2 \text{ MeV}} = 0.0236 \frac{\text{cm}^2}{\text{g}} \tag{8.41}$$

or from (7.22) and (8.15)

$$\frac{\mu_{ab}}{\rho} = \frac{\overline{E}_{tr} - \overline{E}_{rad}}{h\nu}\frac{\mu}{\rho} = \frac{\mu_{tr}}{\rho}\left(1 - \frac{\overline{E}_{rad}}{\overline{E}_{tr}}\right) = \frac{\mu_{tr}}{\rho}(1 - \overline{g}) = \frac{\mu}{\rho}\overline{f}_{tr}(1 - \overline{g}) = \frac{\mu}{\rho}\overline{f}_{ab}$$

$$= 0.0256 \frac{\text{cm}^2}{\text{g}}(1 - 0.08) = 0.0235 \frac{\text{cm}^2}{\text{g}}, \tag{8.42}$$

in good agreement with (8.41). The inverse relationship also holds, as expected

$$\overline{g} = 1 - \frac{\mu_{ab}/\rho}{\mu_{tr}/\rho} = 1 - \frac{0.0236 \text{ cm}^2/\text{g}}{0.0256 \text{ cm}^2/\text{g}} = 0.08. \tag{8.43}$$

Summary of results is shown schematically in Fig. 8.19. A 2 MeV photon in lead will, on the average, transfer 1.13 MeV to charged particles, while the scattered photon will have an energy of 0.87 MeV.

- $\overline{E}_{tr} = 1.13$ MeV: Mean energy transferred to charged particles (electrons and positrons).
- $h\nu' = 0.87$ MeV: Mean energy scattered through Rayleigh and Compton scattering.

Of the 1.13 MeV of energy transferred to charged particles in the lead absorber:

- $\overline{E}_{ab} = 1.04$ MeV: Mean energy absorbed in lead.
- $h\nu'' = 0.09$ MeV: Mean energy re-emitted through bremsstrahlung radiation loss.

The radiation fraction $\overline{g}$ for 2 MeV photons in lead is 0.08.

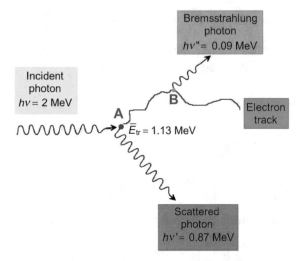

Fig. 8.19 Schematic diagram for general photon interactions with an atom. In this example a 2 MeV photon $h\nu$ interacts with a lead atom. An individual 2 MeV photon, as it encounters a lead atom at point A, may interact with the atom through photoelectric effect, Rayleigh scattering, Compton effect or pair production, or it may not interact at all. However, for a large number of 2 MeV photons striking lead, we may state that on the average: • 1.13 MeV will be transferred at point A to charged particles (mainly to energetic electrons, but possibly also to positrons if the interaction is pair production); • 0.87 MeV will be scattered through Rayleigh and Compton scattering ($h\nu'$). Of the 1.13 MeV transferred to charged particles: • 1.04 MeV will be absorbed in lead over the fast charged particle tracks, and • 0.09 MeV will be emitted in the form of bremsstrahlung photons ($h\nu''$)

8.7.2   Example 2

Interaction of 8 MeV Photon with Copper Absorber

Monoenergetic photons with energy $h\nu = 8$ MeV [$\varepsilon = h\nu/(m_ec^2) = 15.66$] interact with a copper absorber ($Z = 29$; $A = 63.54$ g/mol; $\rho = 8.96$ g/cm^3). Using only the relationships and graphs given in Chap. 7 and this chapter, **determine** the following quantities:

1. Atomic cross section $_a\mu$.
2. Mass attenuation coefficient μ_m.
3. Linear attenuation coefficient μ.
4. Mean energy transferred to charged particles $\overline{E}_{tr}$.
5. Mass energy transfer coefficient μ_{tr}/ρ.
6. Mean radiation fraction $\overline{g}$.
7. Mean energy radiated by charged particles as bremsstrahlung.
8. Mean energy absorbed in the copper absorber $\overline{E}_{ab}$.
9. Mass energy absorption coefficient μ_{ab}/ρ.

1. To determine the total **atomic cross section** $_a\mu$ we first calculate the individual atomic cross sections for photoelectric effect $_a\tau$, Compton effect $_a\sigma_C$, and pair production $_a\kappa$. The total atomic cross section $_a\mu$ will be the sum of the three individual atomic cross sections. We ignore the atomic cross sections for Rayleigh scattering $_a\sigma_R$ and for photonuclear reactions $_a\sigma_{PN}$ because they are very small in comparison with the photoelectric, Compton and pair production cross sections.

Photoelectric Effect

Since $\varepsilon \gg 1$, we use (7.142) to estimate $_a\tau$ for K-shell electrons in copper and get

$$
_a\tau_K = \frac{1.5}{\varepsilon}\alpha^4 Z^5 {_e}\sigma_{Th} = \frac{1.5}{15.66}\frac{29^5}{137^4}0.665\,\frac{b}{atom} = 3.71\times10^{-3}\,\frac{b}{atom}
$$
$$
\approx 0.004\ b/atom. \tag{8.44}
$$

Compton Effect

We use the Klein–Nishina relationship for the electronic cross section $_e\sigma_C^{KN}$, given in (7.104), and then calculate $_a\sigma_C^{KN}$ from $_a\sigma_C^{KN} = Z({_e}\sigma_C^{KN})$

$$
_e\sigma_C^{KN} = 2\pi r_e^2\left\{\frac{1+\varepsilon}{\varepsilon^2}\left[\frac{2(1+\varepsilon)}{1+2\varepsilon} - \frac{\ln(1+2\varepsilon)}{\varepsilon}\right] + \frac{\ln(1+2\varepsilon)}{2\varepsilon} - \frac{1+3\varepsilon}{(1+2\varepsilon)^2}\right\}
$$
$$
= 2\pi(2.818\times10^{-15}\ m)^2(0.068\times0.809 + 0.111 - 0.046)
$$
$$
= 0.0599\ b/electron. \tag{8.45}
$$

The Compton atomic cross section $_a\sigma_C^{KN}$ is calculated from the electronic cross section $_e\sigma_C^{KN}$ as follows [see (7.121)]

$$
_a\sigma_C^{KN} = Z({_e}\sigma_C^{KN}) = 29\ (electron/atom)\times0.0599\ b/electron
$$
$$
= 1.737\ b/atom. \tag{8.46}
$$

Pair Production

Since the photon energy of 8 MeV is significantly above the nuclear pair production threshold of 1.02 MeV and also above the triplet production threshold of 2.04 MeV, both effects (nuclear pair production and triplet production) may occur and then contribute to the total atomic cross section $_a\mu$.

To determine the atomic pair production cross section we use (7.191) to get

$$
_a\kappa_{PP} = \alpha r_e^2 Z^2 P_{PP}(\varepsilon, Z). \tag{8.47}
$$

We assume that $1 \ll \varepsilon \ll 1/(\alpha Z^{1/3})$, where for our example $\varepsilon = 15.66$ and $1/(\alpha Z^{1/3}) = 44.6$, and use (7.187) to determine $P_{PP}(\varepsilon, Z)$ as follows

$$P_{PP}(\varepsilon, Z) = \frac{28}{9} \ln(2\varepsilon) - \frac{218}{27} = 10.73 - 8.07 = 2.65. \qquad (8.48)$$

The atomic cross-section for nuclear pair production $_a\kappa_{NPP}$ is now calculated from (7.191) as follows

$$_a\kappa_{NPP} = \alpha r_e^2 Z^2 P_{PP}(\varepsilon, Z) = \frac{7.94 \times 10^{-2} \times 29^2 \times 2.65}{137} \frac{b}{atom}$$

$$= 1.292 \text{ b/atom}. \qquad (8.49)$$

To account for the *triplet production* contribution we use (7.193) with $\eta = 2.5$ to get the following result for the total pair production atomic cross section $_a\kappa$

$$_a\kappa = {_a\kappa_{NPP}} \left\{ 1 + \frac{1}{\eta Z} \right\} = 1.292 \frac{b}{atom} \left\{ 1 + \frac{1}{2.5 \times 29} \right\} = 1.310 \text{ b/atom}. \qquad (8.50)$$

Two observations can now be made:

- For 8 MeV photons interacting with copper, triplet production contributes only of the order of 1.5 % to the total atomic pair production cross section.
- The atomic cross sections $_a\sigma_C$ and $_a\kappa$ for Compton scattering and pair production, respectively, are similar to one another. This can actually be surmised from Fig. 8.5 that shows the loci of points $(Z, h\nu)$ for which $_a\tau = {_a\sigma_C}$ and $_a\sigma_C = {_a\kappa}$. The point $(Z = 29, \ h\nu = 8 \text{ MeV})$ is very close to the $_a\sigma_C = {_a\kappa}$ curve and thus will possess similar atomic cross-sections $_a\sigma_C$ and $_a\kappa$.

1. **Total atomic cross section** $_a\mu$ is the sum of the cross sections for individual non-negligible effects, as given in (8.1)

$$_a\mu = {_a\tau} + {_a\sigma_R} + \sigma_C + {_a\kappa} = (0.004 + 0 + 1.737 + 1.310) \text{ b/atom}$$

$$= 3.051 \text{ b/atom} \qquad (8.51)$$

2. **Mass attenuation coefficient** μ_m is calculated, as suggested in (8.2), from

$$\mu_m = \frac{\mu}{\rho} = {_a\mu} \frac{N_A}{A} = 3.051 \frac{b}{atom} \frac{6.022 \times 10^{23} \text{ atom/mol}}{63.54 \text{ g/mol}} = 0.0289 \frac{cm^2}{g}.$$

$$(8.52)$$

3. **Linear attenuation coefficient** μ is determined by multiplying μ_m with the absorber density ρ to get [see (8.2)]

$$\mu = \rho \mu_m = 8.96 \frac{g}{cm^3} 0.0289 \frac{cm^2}{g} = 0.259 \text{ cm}^{-1}. \qquad (8.53)$$

4. **Mean energy** $\overline{E}_{tr}$ *transferred* from photons to charged particles is determined using (8.12)

$$\overline{E}_{tr} = \sum_i w_i \overline{E}_{tr}^i = w_{PE}\overline{E}_{tr}^{PE} + w_C\overline{E}_{tr}^C + w_{PP}\overline{E}_{tr}^{PP} = \overline{f}_{tr}h\nu, \qquad (8.54)$$

where

$\overline{f}_{tr}$ is the total mean energy transfer fraction defined in (8.5).
w_i is the relative weight of individual effects i [see (8.5)].
$\overline{E}_{tr}^i$ is the mean energy transferred from photon to charged particles for effect i.

The parameters w_i and $\overline{E}_{tr}^i$ are given as follows

$$w_{PE} = \frac{_a\tau}{_a\mu} = \frac{0.004}{3.051} = 1.3 \times 10^{-3}, \qquad (8.55)$$

$$w_C = \frac{_a\sigma_c}{_a\mu} = \frac{1.737}{3.051} = 0.57, \qquad (8.56)$$

$$w_{PP} = \frac{_a\kappa}{_a\mu} = \frac{1.310}{3.051} = 0.43, \qquad (8.57)$$

and

$$\overline{E}_{tr}^{PE} = h\nu - P_K\omega_K\overline{h\nu}_K = 8 \text{ MeV} - 0.5 \times 0.85 \times 7.7 \times 10^{-3} \text{ MeV} \approx 8 \text{ MeV} \qquad (8.58)$$

(see Fig. 7.29 for values of P_K, ω_K, and $\overline{h\nu}_K$).

$$\overline{E}_{tr}^C = 0.67 \times 8 \text{ MeV} \approx 5.36 \text{ MeV}, \qquad (8.59)$$

(see "The Compton Graph" in Fig. 7.18 or 8.2)

$$\overline{E}_{tr}^{PP} = h\nu - 2m_ec^2 = 8 \text{ MeV} - 1.02 \text{ MeV} = 6.98 \text{ MeV}. \qquad (8.60)$$

(for values of mean energy transfer fraction see Fig. 7.37 or 8.2).

Inserting into (8.54) the weights w_i and mean energy transfers $\overline{E}_{tr}^i$ for the three individual effects, we now calculate the mean energy transferred from 8 MeV photons to charged particles in copper

$$\overline{E}_{tr} = 1.3 \times 10^{-3} \times 8 \text{ MeV} + 0.57 \times 5.36 \text{ MeV} + 0.43 \times 7 \text{ MeV}$$
$$= \sim 0 + 3.06 \text{ MeV} + 3.01 \text{ MeV} = 6.07 \text{ MeV}. \qquad (8.61)$$

5. **Mass energy transfer coefficient** μ_{tr}/ρ is determined from the following expression [see (7.19)]

$$\frac{\mu_{tr}}{\rho} = \frac{\overline{E}_{tr}}{h\nu}\frac{\mu}{\rho} = \frac{6.07}{8}0.0289\,\frac{cm^2}{g} = 0.0219\,cm^2/g. \tag{8.62}$$

6. **Mean radiation fraction** $\overline{g}$ can be read directly from Fig. 8.12 which plots $\overline{g}$ versus photon energy $h\nu$ and for 8 MeV monoenergetic photons yields $\overline{g} = 0.1$. As discussed in Sect. 8.3.1, the mean radiation fraction $\overline{g}$ can also be estimated through first determining $(\overline{E}_K)_0$, the mean initial kinetic energy of charged particles produced in the absorber by photon interactions with absorber atoms, and then finding the radiation yield $Y(\overline{E}_K)_0$ for light charged particles of initial energy $(\overline{E}_K)_0$ (see Fig. 6.12). For our example of 8 MeV photons interacting with copper absorber we proceed as follows:

- In general, the mean radiation fraction $\overline{g}$ represents the mean radiation yield $Y(E_K)_0$ for the spectrum of charged particles released by 8 MeV photons in the copper absorber.
- This charged particle spectrum is composed of recoil Compton electrons with mean energy of 5.36 MeV [as determined in (8.59)] as well as electrons and positrons produced in pair production with mean energy of 0.5×7 MeV = 3.5 MeV [see (8.60)]. Photoelectrons are ignored, because of the low probability for the photoelectric effect at photon energy of 8 MeV.
- The actual spectrum of charged particles released by 8 MeV photons in the copper absorber can only be determined reliably by Monte Carlo calculations. In the first approximation, however, we assume that all charged particles are produced with monoenergetic initial kinetic energies $(\overline{E}_K)_0$.
- The mean energy transferred to charged particles $\overline{E}_{tr}$ exceeds $(\overline{E}_K)_0$, the mean of the initial energies acquired by charged particles that are set in motion in the absorber, because in pair production two charged particles with a combined energy of 6.98 MeV are set in motion and the initial average energy for each of the two charged particles is only 3.5 MeV rather than $\sim$7 MeV.
- **Mean initial energy** $(\overline{E}_K)_0$ of all charged particles released in copper by 8 MeV photons is thus given as

$$(\overline{E}_K)_0 = \overline{E}_{tr}\frac{_a\sigma + _a\kappa}{_a\sigma + 2_a\kappa} = 6.07\,MeV\frac{1.737 + 1.310}{1.737 + 2 \times 1.310} = 4.25\,MeV. \tag{8.63}$$

- The radiation yield $Y(\overline{E}_K)_0$, given in Fig. 6.12, can be equated with the radiation fraction $\overline{g}$ to get $\overline{g} \approx 0.08$ for 4.25 MeV light charged particles (electrons and positrons), in reasonable agreement with $\overline{g} = 0.1$ obtained directly from Fig. 8.12.

7. **Mean energy** $\overline{E}_{rad}$ radiated by charged particles as bremsstrahlung and to some degree as in-flight annihilation photons is given by (7.23) as

$$\overline{E}_{rad} = \overline{g} \times \overline{E}_{tr} = 0.1 \times 6.07 \text{ MeV} = 0.61 \text{ MeV}. \tag{8.64}$$

8. **Mean energy** $\overline{E}_{ab}$ **absorbed** in the copper absorber is given as the difference between the mean energy transferred from photon to electrons as well as positrons and the mean energy radiated in the form of photons by electrons and positrons

$$\overline{E}_{ab} = \overline{E}_{tr} - \overline{E}_{rad} = 6.07 \text{ MeV} - 0.61 \text{ MeV} = 5.46 \text{ MeV}. \tag{8.65}$$

9. **Mass energy absorption coefficient** μ_{ab}/ρ is calculated from (7.20)

$$\frac{\mu_{ab}}{\rho} = \frac{\mu}{\rho} \frac{\overline{E}_{ab}}{h\nu} = 0.0289 \frac{\text{cm}^2}{\text{g}} \frac{5.46}{8} = 0.0197 \text{ cm}^2/\text{g}. \tag{8.66}$$

Mass energy absorption coefficient μ_{ab}/ρ may also be calculated from the mass energy transfer coefficient μ_{tr}/ρ and the radiation fraction $\overline{g}$ as follows [see (7.22) and (8.14)]

$$\frac{\mu_{ab}}{\rho} = \frac{\mu_{tr}}{\rho}(1 - \overline{g}) = 0.0219 \frac{\text{cm}^2}{\text{g}}(1 - 0.1) = 0.0197 \text{ cm}^2/\text{g}. \tag{8.67}$$

In summary, we determined for 8 MeV photons interacting with a copper absorber that on the average:

- $\overline{E}_{tr} = 6.07$ MeV: Mean energy transferred to charged particles (electrons and positrons).
- $h\nu' = 1.93$ MeV: Mean energy scattered through Rayleigh and Compton scattering.
- $\overline{E}_{ab} = 5.46$ MeV: Mean energy absorbed in copper;
- $h\nu'' = 0.61$ MeV: Mean energy radiated in the form of bremsstrahlung.
- Atomic cross section $_a\mu$, mass attenuation coefficient μ_m, and linear attenuation coefficient μ for 8 MeV photons in copper are estimated as 3.051 b/atom; 0.0289 cm^2/g; and 0.259 cm^{-1}, respectively.
- Mass energy transfer coefficient μ_{tr}/ρ and mass energy absorption coefficient μ_{ab}/ρ are estimated as 0.0219 cm^2/g and 0.0197 cm^2/g, respectively.
- Radiation fraction $\overline{g}$ for 8 MeV photons in copper is $\sim$0.1.

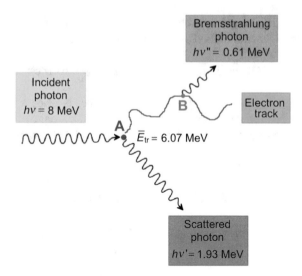

Fig. 8.20 Schematic diagram for general photon interactions with an atom. In this example an 8 MeV photon $h\nu$ interacts with a copper atom. An individual 8 MeV photon, as it encounters a copper atom at point A, may interact with the atom through photoelectric effect, Rayleigh scattering, Compton effect, or pair production, or, of course, it may not interact at all. However, for a large number of 8 MeV photons striking copper, we may state that on the average: • 6.07 MeV will be transferred at point A to charged particles (mainly to fast energetic electrons, but possibly also to positrons if the interaction is pair production). • 1.93 MeV will be scattered through Rayleigh and Compton scattering $(h\nu')$. Of the 6.07 MeV transferred to charged particles: • 5.46 MeV will be absorbed in lead over the fast charged particle tracks. • 0.61 MeV will be emitted in the form of bremsstrahlung photons $(h\nu'')$. • Mean energies transferred to charged particles in a photoelectric process, Rayleigh scattering, Compton scattering, and pair production are: $\sim$8 MeV; 0; 5.36 MeV; and 6.98 MeV, respectively

Table 8.5 Comparison of results of Example 2 (Sect. 8.7.2) with data tabulated by Johns and Cunningham (J&C), Attix, and the NIST

	J&C	Attix	NIST	Our estimate
$_a\tau$ (b/atom)	0.002	0.003	0.002	0.004
$_a\sigma$ (b/atom)	1.74	1.74	1.76	1.74
$_a\kappa$ (b/atom)	1.45	1.49	1.50	1.31
μ_m (cm^2/g)	0.030	0.031	0.031	0.029
μ_{ab}/ρ (cm^2/g)	0.022	–	0.021	0.019
$\bar{E}_{tr}$ (MeV)	6.08	–	–	6.07
$\bar{E}_{ab}$ (MeV)	5.51	–	–	5.46
$\bar{E}_{rad}$ (MeV)	0.57	–	–	0.61

A summary of results for 8 MeV photons interacting with copper absorber is shown schematically in Fig. 8.20. A comparison of our estimates with tabulated values in books by Johns and Cunningham as well as Attix and data provided by the NIST is shown in Table 8.5. Our estimates based on expressions and figures given in Chap. 7 and this chapter are in reasonable, although not perfect, agreement with tabulated data. This shows that measurements still provide the gold standard, as theories cannot give a perfect picture; however, the theories give general trends and help with understanding of the underlying physics.

Čerenkov Radiation in a Nuclear Reactor

Photograph on next page shows *Čerenkov blue radiation* from the reactor core during operation of a TRIGA Mark II nuclear reactor at Kansas State University (KSU) in Manhattan, Kansas. The KSU reactor, in operation since 1961, is a swimming pool reactor, so called because it sits near the bottom of a large concrete pool of water. Currently licensed to operate at thermal power of 250 kW, it serves as an excellent tool for nuclear research and training. It also provides special services, such as production of radionuclides for industry and medicine, neutron activation analysis, neutron radiography, and material irradiation.

Nuclear reactors are based on fission chain reactions that are self-sustained by using some of the fission-produced neutrons to induce new fissions. Each fission event liberates energy of the order of 200 MeV that is distributed among the fission fragments, fission neutrons, β-particles from the β decay of radioactive fission fragments, γ-rays, and neutrinos.

The important components of a nuclear reactor core are: reactor fuel, most commonly uranium-235; moderator that slows down to thermal energies the fission-produced fast neutrons; and control rods that very efficiently absorb neutrons. The position of the control rods in the reactor core affects the number of neutrons available to induce fission thereby controlling the fission rate, reactor power, and reactor shut down.

While no particle can exceed the speed of light in vacuum, in a given medium it is quite possible for charged particles to propagate with velocities that are larger than the speed of light in that medium. When a nuclear reactor is in operation, many fission fragments emit high-energy β-particles and these particles may travel at velocities larger than the speed of light in water. Water molecules line up along the path of particles, and, as they return to their normal random orientations, energy is released in the form of visible (blue) and ultraviolet photons. This type of radiation, produced only when a particle moves faster than the speed of light in a given medium, is called Čerenkov radiation and is named after the Russian scientist who in 1934 was the first to study the phenomenon in depth.

Unlike fluorescence and atomic emission spectra that have characteristic spectral peaks, Čerenkov radiation is continuous and its intensity is proportional to the frequency (inversely proportional to wavelength), resulting in the predominantly blue emission visible to the naked eye and even more emission in the ultraviolet region of the photon spectrum, invisible to the human eye. The blue glow in the reactor continues for some time after the reactor has been shut down because of the β decay of fission products that continues to produce energetic β-particles. The Čerenkov effect, discussed in Sect. 4.4, is analogous to the sonic boom in acoustics when an object exceeds the speed of sound in air. Nuclear reactors and nuclear chain reaction are discussed in Sect. 12.8.

Photo: *Courtesy of Kansas State University, TRIGA Mark II Nuclear Reactor. Reproduced with Permission.*

Chapter 9
Interactions of Neutrons with Matter

Neutrons, by virtue of their neutrality, are indirectly ionizing radiation exhibiting a quasi-exponential penetration into an absorber and depositing energy in the absorber through a two-step process: (1) energy transfer to heavy charged particles and (2) energy deposition in the absorber through Coulomb interactions of these charged particles with atoms of the absorber. As they penetrate into matter, neutrons may undergo elastic and inelastic scattering as well as trigger nuclear reactions, such as neutron capture, spallation, and fission.

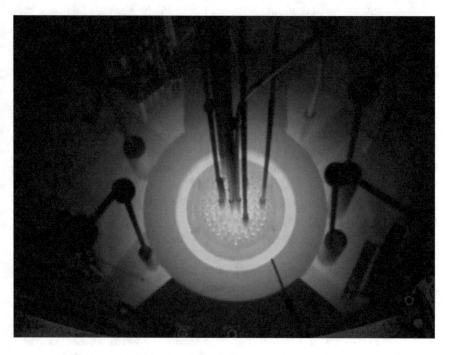

© Springer International Publishing Switzerland 2016   429
E.B. Podgoršak, *Radiation Physics for Medical Physicists*,
Graduate Texts in Physics, DOI 10.1007/978-3-319-25382-4_9

Two distinct categories of neutrons are of direct importance in medical physics: *thermal neutrons* used in boron-neutron capture therapy (BNCT) and *fast neutrons* used in external beam radiotherapy. Indirectly, thermal neutrons play an important role in production of radionuclide sources that are used in external beam radiotherapy, in brachytherapy as well as in nuclear medicine imaging. A nuclear reactor and two types of thermal neutron interaction are used for this purpose: (1) neutron activation of suitable target material and (2) fission reaction induced by thermal neutrons in fissile target materials.

Several parameters used for describing neutron fields and neutron dose deposition in absorbers are defined and discussed in this chapter. Also discussed are several radiotherapy techniques based on neutron beams, machines for production of neutron beams in radiotherapy, and an efficient source of neutrons for use in brachytherapy, the californium-252.

9.1   General Aspects of Neutron Interactions with Absorbers

Neutrons, similarly to photons, may penetrate an absorber without interacting or they may undergo various interactions with the absorber. In contrast to photons, however, neutrons interact mostly with the nuclei of the absorber and have only minor interactions with orbital electrons of the absorber.

Neutron beams, similarly to photon beams, belong to the category of indirectly ionizing radiation beams, both types transferring energy to absorbing medium through an intermediate step in which energy is transferred to a charged particle (protons and heavier nuclei in the case of neutrons; electrons and positrons in the case of photons).

The secondary heavy charged particles released in a medium traversed by neutrons have a very short range in the medium ensuring charged particle equilibrium. Since no bremsstrahlung x-rays are generated by charged particles put in motion by neutrons, the absorbed dose for neutron beams is equal to kerma at any point in the neutron field.

In terms of their kinetic energy E_K, neutrons are classified into several categories:

1. *Ultracold neutrons* with $E_K < 2 \times 10^{-7}$ eV
2. *Very cold neutrons* with 2×10^{-7} eV $\leq E_K \leq 5 \times 10^{-5}$ eV
3. *Cold neutrons* with 5×10^{-5} eV $\leq E_K \leq 0.025$ eV
4. **Thermal neutrons** with $E_K \approx 0.025$ eV
5. **Epithermal neutrons** with 1 eV $< E_K < 1$ keV
6. *Intermediate neutrons* with 1 keV $< E_K < 0.1$ MeV
7. **Fast neutrons** with $E_K > 0.1$ MeV.

Of the seven categories listed above, only thermal, epithermal, and fast neutrons are used in medicine and are thus of interest in medical physics. Note that the velocity of an ultracold neutron with a kinetic energy of 2×10^{-7} eV is

$\sim$6 m/s $\left(v/c \approx 2 \times 10^{-8}\right)$; of a thermal neutron with a kinetic energy of 0.025 eV it is $\sim$2200 m/s $\left(v/c \approx 7 \times 10^{-6}\right)$; and of a fast neutron it is 1.4×10^7 m/s $(v/c \approx 0.05)$.

9.2   Neutron Interactions with Nuclei of the Absorber

Neutrons by virtue of being neutral particles can approach a target nucleus without any interference from a Coulomb repulsive or attractive force, since they, unlike protons and electrons, are not affected by nuclear charge. Once in close proximity to the target nucleus, neutrons can interact with it through the short range attractive nuclear potential and trigger various nuclear reactions.

There are five principal processes by which neutrons interact with the nuclei of the absorber:

1. Elastic scattering
2. Inelastic scattering
3. Neutron capture
4. Nuclear spallation
5. Nuclear fission

The probability (cross section) for these different types of interaction varies with the kinetic energy of the neutron and with the physical properties of the nuclei of the absorber.

9.2.1   Elastic Scattering

In elastic scattering a neutron collides with a nucleus of mass M that recoils with an angle ϕ with respect to the neutron initial direction of motion, as shown schematically in Fig. 5.2 and discussed in Sect. 5.3 for general two-particle elastic scattering. Kinetic energy and momentum are conserved in the interaction.

For a neutron with mass m_n and initial kinetic energy $(E_K)_i$, the kinetic energy ΔE_K transferred to the nucleus is in general given as shown in (5.25)

$$\Delta E_K = (E_K)_i \, \frac{4 m_n M}{(m_n + M)^2} \cos^2 \phi. \tag{9.1}$$

The maximum possible energy transfer $(\Delta E_K)_{\text{max}}$ is attained in a head-on collision for which $\phi = 0°$ (see Sect. 5.3.3)

$$(\Delta E_K)_{\text{max}} = \Delta E_K|_{\phi=0} = (E_K)_i \, \frac{4 m_n M}{(m_n + M)^2}. \tag{9.2}$$

The average kinetic energy $\overline{\Delta E}_K$ transferred to the recoil nucleus is

$$\overline{\Delta E}_K = \frac{1}{2}(\Delta E_K)_{max} = \frac{1}{2}(E_K)_i\frac{4m_nM}{(m_n+M)^2} = 2(E_K)_i\frac{m_nM}{(m_n+M)^2}. \qquad (9.3)$$

The kinetic energy of the scattered neutron, $(E_K)_f$, in a head-on collision is equal to

$$(E_K)_f = (E_K)_i - (\Delta E_K)_{max} = (E_K)_i\left(\frac{m_n-M}{m_n+M}\right)^2, \qquad (9.4)$$

while $(\bar{E}_K)_f$, the average energy attained by the scattered neutron, is

$$(\overline{E}_K)_f = (E_K)_i - \overline{\Delta E}_K = (E_K)_i\frac{m_n^2+M^2}{(m_n+M)^2}. \qquad (9.5)$$

Thus, for example, if the target nucleus is hydrogen (nucleus is a proton with mass m_p), then $M = m_p \approx m_n$ and the neutron will transfer on the average one half of its initial kinetic energy to the proton (see (9.5)), while the maximum energy transferred to the proton equals to the initial neutron energy $(E_K)_i$ (see (9.2)). The recoil proton will then travel a short distance through the absorbing medium and rapidly transfer its kinetic energy to the medium through Coulomb interactions with the nuclei and orbital electrons of the medium.

The transfer of neutron's energy to the absorbing medium is much less efficient when $m_n \ll M$; the larger is M, the less efficient is the energy transfer, as evident from (9.2). For example, as shown in Sect. 5.3.4, (9.2) predicts an only 2% fractional energy transfer from a neutron colliding head-on with a lead nucleus, compared to a 100% energy transfer in a head-on neutron-proton collision. This, of course, has implications for shielding against neutron radiation in high-energy linear accelerator installations, where low atomic number materials are used in neutron barriers for shielding against neutrons produced by high-energy photons.

9.2.2   Inelastic Scattering

In inelastic scattering the neutron n is first captured by the nucleus and then re-emitted as neutron n' with a lower energy and in a direction that is different from the incident neutron direction. The nucleus is left in an excited state and will de-excite by emitting high-energy γ-rays. This process is illustrated by the following relationship

$$n + {}^A_Z X \rightarrow {}^{A+1}_Z X^* \rightarrow {}^A_Z X^* + n' \Rightarrow {}^A_Z X^* \rightarrow {}^A_Z X + \gamma, \qquad (9.6)$$

where

${}^A_Z X$ is the stable target nucleus.

$^{A+1}_{\ \ Z}X^*$ is an unstable compound nucleus.

$^A_ZX^*$ is an excited target nucleus.

9.2.3 Neutron Capture

Neutron capture is a term used to describe a nuclear reaction in which a thermal neutron bombards a nucleus leading to the emission of a proton or γ-ray. Two of these interactions are of particular importance in tissue: $^{14}N(n, p)^{14}C$ and $^1H(n, \gamma)^2H$ and one interaction, $^{113}Cd(n,\gamma)^{114}Cd$, is of importance in shielding against thermal neutrons.

A cadmium filter with a thickness of 1 mm absorbs essentially all incident thermal neutrons with energies below 0.5 eV, but readily transmits neutrons with energies exceeding 0.5 eV. The cross section for neutron capture plotted against neutron kinetic energy exhibits a broad resonance with a peak at 0.178 eV. At the resonance peak energy the cross section for neutron capture by natural cadmium (12% abundance of cadmium-113) is 7800 b, while pure cadmium-113 has a cross section of $\sim 64 \times 10^3$ b.

Often neutron bombardment of a stable target is carried out in a nuclear reactor with the intent of producing a radioactive nuclide (radionuclide) for industrial or medical purposes. When the main interest in the reaction is the end product, the reaction is termed neutron activation. Of interest in medical physics is the neutron activation process in general and in particular when it is used for production of cobalt-60 sources for radiotherapy, iridium-192 sources for brachytherapy, and molybdenum-99 radionuclide for nuclear medicine diagnostic imaging procedures. The neutron activation process is discussed in greater detail in Sect. 12.6.

9.2.4 Spallation

Spallation is defined as fragmentation of a target into many smaller components as a result of impact or stress. Consequently, nuclear spallation is defined as disintegration of a target nucleus into many small residual components such as α-particles and nucleons (protons and neutrons) upon bombardment with a suitable projectile such as light or heavy ion beams or neutrons. Nuclear spallation can also occur naturally in earth's atmosphere as a result of exposure of nuclides to energetic cosmic rays such as protons.

An example of spallation is as follows

$$^{16}_{8}O + n \rightarrow 3\alpha + 2p + 3n. \tag{9.7}$$

Most of the energy released from the spallation process is carried away by the heavier fragments that deposit their energy in the absorber locally. On the other hand, neutrons and de-excitation γ-rays produced in spallation carry their energy to a remote location. Spallation can be used for production of radionuclides and for generation of neutron beams in spallation neutron generators.

9.2.5   Nuclear Fission Induced by Neutron Bombardment

Fission is a particular type of neutron interaction produced by bombardment of certain very high atomic number nuclei ($Z \geq 92$) by thermal or fast neutrons. The target nucleus fragments into two daughter nuclei of lighter mass and the fission process is accompanied with production of several fast neutrons. Nuclei that are capable of undergoing fission are called fissionable nuclei in general; nuclei that undergo fission with thermal neutrons are called fissile nuclei. Fission fragments combined with the nuclei that are subsequently formed through radioactive decay of fission fragments are called fission products.

Since neutrons are produced as a by-product of nuclear fission, the initial fission reaction may be followed by other fission reactions, resulting in a self-sustained nuclear chain reaction and a substantial release of energy. Controlled chain reactions are used in nuclear reactors for research and educational purposes as well as for power generation. Three fissile nuclides have been used in nuclear reactors: one is naturally occurring uranium-235 and the other two are artificially produced uranium-233 and plutonium-239. Nuclear fission and the nuclear chain reaction are discussed in more detail in Sects. 12.7 and 12.8.

Two typical examples of fission reaction triggered by thermal neutron bombarding a uranium-235 nucleus are as follows

$$^{235}_{92}\text{U} + \text{n} = {}^{89}_{36}\text{Kr} + {}^{144}_{56}\text{Ba} + 3\text{n} + \text{energy} \ (\sim 180 \ \text{MeV}) \tag{9.8}$$

and

$$^{235}_{92}\text{U} + \text{n} = {}^{100}_{38}\text{Sr} + {}^{134}_{54}\text{Xe} + 2\text{n} + \text{energy}. \tag{9.9}$$

9.3   Neutron Kerma

Like in other applications of ionizing radiation in medicine, dosimetry of neutron beams is very important for achieving the desired treatment outcome. Since neutrons are indirectly ionizing particles, they are detected by measuring the ionizing particles that are released in the absorbing medium through interactions of neutrons with absorber nuclei. The most common interactions are: (n, α), (n, p), and (n, γ); the most common neutron detectors are: gas-filled ionization chambers, scintillation detectors, thermoluminescent dosimeters, track detectors, and radiographic film. Neutron fields are usually described in terms of fluence $\varphi(E_K)$ rather than energy fluence ψ, as is usually the case with photon fields. For a monoenergetic neutron beam of fluence φ in cm^{-2} undergoing a specific type of interaction i with a particular atom at a point in medium, the kerma K_i in a small mass m is expressed as

$$K_i = \varphi \sigma_i \frac{N}{m} \left(\overline{\Delta E_K}\right)_i , \qquad (9.11)$$

where

σ_i is the cross section for the particular interaction i.

N is the number of target atoms in mass m with $N/m = N_A/A$.

$\left(\overline{\Delta E_K}\right)_i$ is the mean energy transferred from neutrons to charged particles through the particular interaction i.

The product $\sigma_i N/m$ summed over all possible neutron interactions is the mass attenuation coefficient μ/ρ for neutrons in the absorbing medium. Following the convention used for photon beams, the mass energy transfer coefficient μ_{tr}/ρ for neutrons is defined as follows

$$\frac{\mu_{tr}}{\rho} = \frac{\mu}{\rho} \frac{\overline{\Delta E_K}}{E_K}, \qquad (9.12)$$

where $\overline{\Delta E_K}/E_K$ is the fraction of the neutron incident energy transferred to charged particles.

The total kerma K accounting for all possible interactions is

$$K = \sum_i \varphi \, \sigma_i \frac{N}{m} \overline{\Delta E_K} = \varphi \frac{\mu}{\rho} \overline{\Delta E_K} = \varphi \frac{\mu_{tr}}{\rho} E_K, \qquad (9.13)$$

where E_K is the kinetic energy of the monoenergetic neutron beam and the mass energy transfer coefficient (μ_{tr}/ρ) is given in units of cm^2·g.

9.4 Neutron Kerma Factor

The product $(\mu_{tr}/\rho)E_K$ in (9.13), defined as the *neutron kerma factor* F_n with units of cGy·cm^2 or cGy per n/cm^2, is tabulated for neutrons instead of the mass energy transfer coefficient (μ_{tr}/ρ). Figure 9.1 provides the neutron kerma factor F_n against neutron kinetic energy for various materials of interest in medical physics (hydrogen, water, tissue, carbon, oxygen, and nitrogen).

From (9.13) for monoenergetic neutrons we get the following expression for the neutron kerma K

$$K = \varphi(F_n)_{E_K,Z}, \qquad (9.14)$$

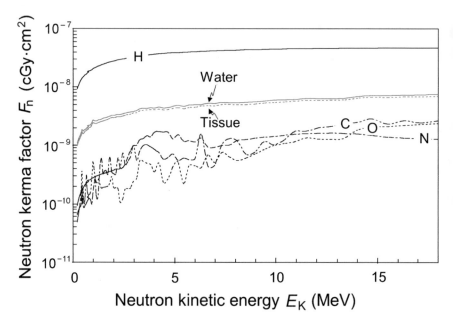

Fig. 9.1 Neutron kerma factor F_n against neutron kinetic energy E_K for various materials of interest in medical physics. Data were obtained from the NIST

where

φ is the fluence of monoenergetic neutrons of kinetic energy E_K.

$(F_n)_{E_K, Z}$ is the neutron kerma factor F_n in $J \cdot cm^2 \cdot g^{-1}$ for neutrons of kinetic energy E_K in the absorber with atomic number Z.

For neutron beams characterized with an *energy spectrum* $\varphi'(E_K)$ of particle fluence, kerma K is expressed as follows

$$K = \int_{0}^{(E_K)_{\max}} \varphi'(E_K)\,(F_n)_{E_K, Z}\,dE_K, \qquad (9.15)$$

where $(E_K)_{\max}$ is the maximum neutron kinetic energy in the continuous neutron spectrum with the differential fluence distribution $\varphi'(E_K)$.

An average value for the neutron kerma factor F_n for the spectrum of neutrons $\varphi'(E_K)$ is given as

$$\overline{(F_n)}_{\varphi'(E_K), Z} = \frac{K}{\varphi} = \frac{\displaystyle\int_{0}^{(E_K)_{\max}} \varphi'(E_K)(F_n)_{E_K, Z}\,dE_K}{\displaystyle\int_{0}^{E_{K_{\max}}} \varphi'(E_K)\,dE_K}. \qquad (9.16)$$

9.5 Neutron Dose Deposition in Tissue

Neutrons, by virtue of their neutrality, similarly to photons, deposit dose in tissue through a two-step process:

1. Energy transfer to heavy charged particles, such as protons and heavier nuclei in tissue.
2. Energy deposition in tissue by heavy charged particles through Coulomb interactions of the charged particles with atoms of tissue.

Similarly to photons, the nature of neutron interactions with tissue depends on the kinetic energy of neutrons; however, the options available for neutron interactions are not as varied as those for photons (see Chap. 7). For neutrons there are only two energy ranges to consider:

1. *Thermal neutron energy* of the order of 0.025 eV.
2. *Epithermal, intermediate* and *fast neutrons*.

9.5.1 Thermal Neutron Interactions in Tissue

Thermal neutrons undergo two possible interactions with nuclei of tissue:

1. *Neutron capture* by nitrogen-14 $\left(^{14}_{7}\text{N}\right)$ nucleus that produces carbon-14 $\left(^{14}_{6}\text{C}\right)$ and a proton. The cross section for the $^{14}_{7}\text{N}$ (n, p) $^{14}_{6}\text{C}$ reaction is $\sigma_{\text{N-14}} = 1.84$ b/atom.
2. *Neutron capture* by hydrogen-1 $\left(^{1}_{1}\text{H}\right)$ nucleus (proton) that produces a deuterium nucleus (deuteron) and a γ photon. The cross section for reaction $^{1}_{1}\text{H}$ (n, γ)$^{2}_{1}\text{H}$ is $\sigma_{\text{H-1}} = 0.33$ b/atom.

According to the ICRU and the ICRP, the human tissue composition in percent by mass is: $\sim$10% for hydrogen-1 and $\sim$3% for nitrogen-14. The data for oxygen-16 and carbon-12, the other two abundant constituents of tissue, are $\sim$75% and $\sim$12%, respectively.

The kerma deposited in muscle tissue per unit neutron fluence φ is from (9.9) given as follows

$$\frac{K}{\varphi} = \sigma \left(\frac{N_t}{m}\right) \overline{\Delta E_K},\tag{9.17}$$

where

σ	is the thermal neutron cross section for the specific nuclear reaction.
$\overline{\Delta E_K}$	is the average energy transfer in the nuclear reaction.
(N_t/m)	is the number of specific nuclei, such as nitrogen-14 or hydrogen-1, per unit mass of tissue.

Thermal Neutron Capture in Nitrogen-14 in Tissue

The kinetic energy released by thermal neutron capture in nitrogen-14 is determined by calculating the change in total nuclear binding energy between the nitrogen-14 nucleus ($E_B = 104.66$ MeV) and the carbon-14 nucleus ($E_B = 105.29$ MeV). Since the total binding energy of carbon-14 exceeds that of nitrogen-14 by 0.63 MeV, we note that the energy released to charged particles in thermal neutron capture by the nitrogen-14 nucleus is 0.63 MeV. This energy is shared as kinetic energy between the proton and the carbon-14 nucleus in the inverse proportion of their masses, since both nuclei carry away the same momenta, but in opposite directions. Thus, the proton receives a kinetic energy of 0.58 MeV; the carbon-14 atom kinetic energy of 0.05 MeV.

The number of nitrogen-14 atoms per gram of tissue, $(N_t/m)_{\text{N-14}}$, is determined as follows:

1. 1 mol of N-14 contains N_A atoms of N-14.
2. 1 g of N-14 contains (N_A/A) atoms of N-14 where $A = 14.01$ g·mol^{-1}.
3. 1 g of tissue contains 0.03 g of N-14 atoms, i.e., $0.03 \times (N_A/A)$ atoms of N-14, therefore $(N_t/m)_{\text{N-14}} = 1.3 \times 10^{21}$ atom/g.

The kerma K per unit thermal neutron fluence φ for the $^{14}_{7}\text{N}$ (n, p) $^{14}_{6}\text{C}$ reaction is thus equal to

$$
\frac{K}{\varphi} = \sigma_{\text{N}-14} \left(\frac{N_t}{m} \right)_{\text{N}-14} \overline{\Delta E_K}
$$

$$
= \left(1.84 \times 10^{-28} \frac{\text{m}^2}{\text{atom}} \right) \times \left(0.13 \times 10^{23} \frac{\text{atom}}{\text{g}} \right) \times (0.63 \text{ MeV}) \times \left(1.602 \times 10^{-13} \frac{\text{J}}{\text{MeV}} \right)
$$

$$
= 2.41 \times 10^{-20} \frac{\text{J}}{\text{g}} \cdot \frac{\text{m}^2}{\text{neutron}} = 2.41 \times 10^{-17} \text{ Gy} \cdot \left(\frac{\text{neutron}}{\text{m}^2} \right)^{-1}. \tag{9.18}
$$

Thermal Neutron Capture in Hydrogen-1 in Tissue

Despite a lower cross section for capture in hydrogen compared to nitrogen, thermal neutrons have a much larger probability for being captured by hydrogen than by nitrogen in tissue because in number of atoms per gram of tissue (concentration) hydrogen surpasses nitrogen with a ratio of ~45 to 1.

In the $^{1}_{1}\text{H}$ (n, γ) $^{2}_{1}\text{H}$ reaction a γ photon is produced and the binding energy difference between a proton $E_B = 0$ and deuteron ($E_B = 2.225$ MeV) is 2.225 MeV. Neglecting the recoil energy of the deuteron, we assume that the γ photon receives the complete available energy of 2.225 MeV, i.e., $E_\gamma = 2.225$ MeV.

The number of hydrogen-1 atoms per gram of tissue, $(N_t/m)_{\text{H-1}}$ is determined as follows:

1. 1 mol of H-1 contains N_A atoms of H-1.
2. 1 g of H-1 contains (N_A/A) atoms of H-1.
3. 1 g of tissue contains 0.1 g of H-1 atoms, i.e., $0.1 \times (N_A/A)$ atoms of H-1, therefore $(N_t/m)_{\text{H-1}} = 6 \times 10^{22}$ atom/g $\approx 45 \times (N_t/m)_{\text{N-14}}$.

The energy transfer to γ photons per thermal neutron fluence φ and per mass of tissue m for the ${}_1^1\mathrm{H}\,(n, \gamma){}_1^2\mathrm{H}$ nuclear reaction is given as follows, again using (9.11)

$$\frac{E_\gamma}{m\varphi} = \frac{E_\gamma/m}{\varphi} = \sigma_{\mathrm{H}-1}\left(\frac{N_t}{m}\right)_{\mathrm{H}-1} \Delta E_\gamma$$

$$= \left(0.33\times 10^{-28}\,\frac{\mathrm{m}^2}{\mathrm{atom}}\right) \times \left(6\times 10^{22}\,\frac{\mathrm{atom}}{\mathrm{g}}\right) \times (2.225\,\mathrm{MeV}) \times \left(1.602\times 10^{-13}\,\frac{\mathrm{J}}{\mathrm{Mev}}\right)$$

$$= 7.05\times 10^{-19}\,\frac{\mathrm{J}}{\mathrm{g}}\cdot\frac{\mathrm{m}^2}{\mathrm{neutron}} = 7.05\times 10^{-16}\,\mathrm{Gy}\cdot\left(\frac{\mathrm{neutron}}{\mathrm{m}^2}\right). \qquad (9.19)$$

The result of (9.19) represents the energy per unit neutron fluence and per unit mass of tissue that is transferred to γ photons. The amount of this energy that actually contributes to the kerma in tissue depends on the fraction of this energy that is transferred from the γ photons to electrons in tissue. This fraction depends on the size of the tissue mass: for a small size mass most of the γ photons may escape; for a large mass all photons might be absorbed.

The human body is intermediate in size, so most of the γ photons produced through the ${}_1^1\mathrm{H}\,(n, \gamma){}_1^2\mathrm{H}$ reaction are absorbed in the body, making the ${}_1^1\mathrm{H}\,(n, \gamma){}_1^2\mathrm{H}$ reaction the main contributor to kerma and dose delivered to humans from thermal neutrons. The ${}_1^1\mathrm{H}\,(n, \gamma){}_1^2\mathrm{H}$ reaction also dominates the kerma production in tissue for epithermal neutrons, since the body acts as moderator for thermalizing the neutrons.

9.5.2 Interactions of Intermediate and Fast Neutrons with Tissue

For neutrons with kinetic energies above $100\,\mathrm{eV}$ (upper end epithermal, intermediate, and fast neutrons) by far the most important interaction is the elastic scattering with nuclei of tissue, most importantly with hydrogen-1.

As given in Sect 5.3, the following expressions govern the elastic collisions by two particles:

1. The kinetic energy transfer ΔE_K from the neutron with mass m_n to tissue nucleus with mass M, as derived in (5.25), is

$$\Delta E_\mathrm{K} = \frac{4m_\mathrm{n}M}{(m_\mathrm{n} + M)^2}\,(E_\mathrm{K})_\mathrm{n}\cos^2\phi, \qquad (9.20)$$

where

$(E_\mathrm{K})_\mathrm{n}$ is the kinetic energy of the incident neutron.
ϕ is the recoil angle of the target M nucleus.

2. The maximum kinetic energy transfer $(\Delta E_\mathrm{K})_\mathrm{max}$, occurs for $\phi = 0$ and is given as follows

Table 9.1 Parameters of tissue constituents relevant to neutron absorption and scattering

	Abundance (% by mass)	Abundance (#atoms/g of tissue)	Abundance (relative to hydrogen)	$\overline{\Delta E}_K$ in % of $(E_K)_n$
Hydrogen-1	10	6.0×10^{22}	1	50
Carbon-12	75	3.8×10^{22}	0.63	14
Nitrogen-14	3	1.3×10^{21}	0.022	12
Oxygen-16	12	4.5×10^{21}	0.075	11

Table 9.2 Two predominant interactions of neutrons depositing dose in tissue and their regions of predominance

	Reaction	$\overline{\Delta E}_K$ (MeV)	K/φ (Gy per neutron/m^2)
$(E_K)_n < 100$ eV	$^{14}_{7}\text{N}(n, p)^{14}_{6}\text{C}$	0.63	2.4×10^{-17}
$(E_K)_n > 100$ eV	$^{1}_{1}\text{H}(n, n')^{1}_{1}\text{H}'$	$0.5(E_K)_n$	$0.5\sigma_{el}(E_K)_n$

$$(\Delta E_K)_{max} = \frac{4m_n M}{(m_n + M)^2}(E_K)_n. \tag{9.21}$$

3. The average energy transfer by elastic scattering from a neutron to tissue nucleus M is given as follows

$$\overline{\Delta E}_K = \frac{4m_n M}{(m_n + M)^2}(E_K)_n \overline{\cos^2 \phi} = \frac{2m_n M}{(m_n + M)^2}(E_K)_n$$

$$= \frac{1}{2}(\Delta E_K)_{max}. \tag{9.22}$$

The average energy $\overline{\Delta E}_K$ transferred to recoil nucleus M in tissue in elastic scattering depends on the nuclear mass M and ranges from $0.5\,(E_K)_n$ for hydrogen-1; through $0.14\,(E_K)_n$ for carbon-12; $0.12\,(E_K)_n$ for nitrogen-14, to $0.11\,(E_K)_n$ for oxygen.

4. Of the possible contributors to energy transfer to nuclei in tissue, hydrogen-1 is the most efficient, since it not only provides the largest number of atoms per tissue mass, it also transfers, on the average, the largest amount of energy (50%) from the neutron to the scattering nucleus per each elastic scattering event, as shown in Table 9.1.

5. The dependence of the kerma factor (K/φ, kerma per unit fluence) on neutron energy is essentially split into two regions, one for neutron energy below 100 eV and the other for neutron energy above 100 eV, as summarized in Table 9.2.

9.6 Neutron Beams in Medicine

Of the four main energy categories of neutrons (thermal, epithermal, intermediate and fast, as listed in Sect. 9.1) three categories: thermal, epithermal and fast neutrons are used in radiotherapy; thermal and epithermal neutrons for *boron neutron capture therapy* (BNCT) and fast neutrons for *external beam radiotherapy* and *brachytherapy*. Fast neutrons are also used in medicine for in-vivo neutron activation analysis and in neutron radiography but these applications are still research oriented and not used routinely in patient care.

9.6.1 Boron Neutron Capture Therapy (BNCT)

The boron neutron capture therapy (BNCT) irradiation technique relies on the exceptionally high thermal neutron cross section ($\sigma = 3840$ barn) of the boron-10 nuclide. Exposed to thermal neutrons, boron-10 undergoes the following nuclear reaction:

$$^{10}_5\text{B} + \text{n} \rightarrow {}^7_3\text{Li} + \alpha + Q \,(2.79\,\text{MeV}), \tag{9.23}$$

where n represents a thermal neutron, α an alpha particle, and ${}^7_3\text{Li}$ the lithium-7 nucleus.

As discussed in Sect. 5.2.2, the Q value for the reaction of (9.23) is calculated using either the nuclear rest energies for the nuclei or the total binding energies for the nuclei as follows

$$\begin{aligned} Q &= M({}^{10}_5\text{B})c^2 + m_\text{n}c^2 - \left\{ M({}^7_3\text{Li})c^2 + m_\alpha c^2 \right\} \\ &= (9324.4362 + 939.5654)\,\text{MeV} - (6533.8329 + 3727.3791)\,\text{MeV} \\ &= 2.79\,\text{MeV} \end{aligned} \tag{9.24}$$

or

$$\begin{aligned} Q &= E_\text{B}({}^7_3\text{Li}) + E_\text{B}(\alpha) - E_\text{B}({}^{10}_5\text{B}) \\ &= (39.24459 + 28.29569 - 64.75071)\,\text{MeV} = 2.79\,\text{MeV}, \end{aligned} \tag{9.25}$$

where $m_\text{n}c^2$ is the neutron rest energy equal to 939.5654 MeV. The nuclear masses M and binding energies E_B for the nuclides of (9.23) are given in Appendix A.

Both methods give a reaction Q value of 2.79 MeV that is shared between a γ photon (0.48 MeV) produced by an excited lithium-7 nucleus and reaction products lithium-7 and the α-particle. The 2.31 MeV kinetic energy (2.79 MeV to 0.48 MeV) is shared between the two reaction products in the inverse proportion of their masses, i.e., lithium-7 carries away an energy of 0.84 MeV; the α-particle 1.47 MeV. The range of these reaction products in tissue is of the order of 6 μm to 10 μm which is of the order of a typical cell diameter.

By virtue of their relatively large masses, both reaction products are densely ionizing particles that can produce significant radiation damage on the cellular level during their short travel through tissue. In addition, the cellular damage produced by these densely ionizing particles depends much less on the presence of oxygen than is the case with standard sparsely ionizing beams, such as x-rays, γ-rays, and electrons.

Boron-10 and thermal neutrons have no direct effect on tissue components, since: (1) boron is not toxic to humans and (2) thermal neutrons cannot produce ionization in tissue components because their kinetic energy ($\sim$0.025 eV) is much lower than the ionization energy of all atoms (range 5 eV to 24.5 eV see, Sect. 3.2.4). However, boron-10 and thermal neutrons interact to produce nuclear fission fragments with a shared energy of 2.31 MeV, as shown in (9.24) and (9.25).

The basic premise behind the BNCT is that: (1) bio-molecules labeled with boron-10 can be administered safely to the patient and selectively concentrated in a rapidly growing tumor and (2) as the tumor is exposed to thermal neutrons, a higher dose will be delivered to the tumor than to the adjacent normal tissue. This is so because the tumor contains the boron-10 nuclide, while the surrounding tissues do not, at least not to the same extent. The therapeutic effect achieved depends on the achieved concentration of boron-10 in the target as well as on the fluence of thermal neutrons. Since, as discussed in Sect. 9.5.1, the thermal neutron cross sections for hydrogen (0.33 b) and nitrogen (1.8 b) in tissue are at least three orders of magnitude smaller than that of boron-10 (3840 b), we may assume that most of the target dose is delivered by the relatively heavy fission products resulting from the boron-10 nuclides interacting with thermal neutrons.

In theory the idea behind the BNCT is logical and simple; however, in practice the technique is still considered experimental despite more than half a century that has already been spent on its development by various research groups. Most attempts with the use of the BNCT are concentrated on treatment of malignant brain tumors, and essentially all practical aspects of dose delivery with the BNCT are wrought with difficulties. The most serious of these difficulties are:

1. Boron-10 is difficult to concentrate in the tumor.
2. Thermal neutrons of sufficient fluence rate (of the order of 10^{12} cm^{-2}·s^{-1}) can only be obtained from a nuclear reactor and reactors are not readily available for this kind of purpose nor are they located close to hospitals.
3. Thermal neutrons have very poor penetration into tissue, exhibiting negligible skin sparing and a rapid dose fall-off with depth in tissue (50% dose at $\sim$2 cm depth in tissue).
4. The thermal neutron beam produced in a nuclear reactor is contaminated with γ photons and the dosimetry of the mixed neutron/γ-ray fields is problematic.

Despite difficulties, there are several research groups around the world (Japan, USA, and Europe) working with great enthusiasm on making the BNCT more clinically useful, yet so far success was limited.

9.6.2 Radiotherapy with Fast Neutron Beams

In contrast to the BNCT, radiotherapy with fast neutrons is quite advanced, practiced in several centers around the world, and accepted as a viable, albeit uncommon, alternative to standard radiotherapy with photon and electron beams. In comparison with photon and electron beams, the main attraction of fast neutron beams is their much lower oxygen enhancement ratio (OER); the main drawback is their significantly more difficult and more expensive means of production.

The OER is defined as the ratio of doses without and with oxygen (hypoxic vs. well-oxygenated cells) to produce the same biological effect. The OER for electrons and photons (sparsely ionizing radiations) is about 2–3 while for neutrons (densely ionizing radiation) it is only about 1.5. This means that treatment of anoxic tumors with neutrons is much less affected by the absence of oxygen than is the standard treatment with photons or electrons.

The depth dose distributions produced in tissue by fast neutron beams exhibit similar characteristics to those of photon beams (see Sect. 1.12 and Fig. 1.2). The dose maximum occurs at a depth beneath the surface and depends on beam energy; the larger is the energy, the larger is the depth of dose maximum and the more penetrating is the neutron beam. The skin sparing effect is present, yet less pronounced than in photon beams of similar penetration. As a rough comparison one can state that in terms of tissue penetration, a 14 MeV neutron beam is equivalent to a cobalt-60 γ-ray beam.

In contrast with the production of clinical photon beams, there are three major technical difficulties associated with generating clinical fast neutron beams:

1. Beam intensity to achieve sufficiently high dose rate.
2. Beam energy to attain sufficient penetration into tissue.
3. Beam collimation to minimize dose to healthy tissues and to minimize total body dose to the patient.

9.6.3 Machines for Production of Clinical Fast Neutron Beams

Two types of machine are used for production of clinical fast neutron beams:

1. Neutron generator operated at about 250 kV.
2. Cyclotron operated at 15 MeV to 75 MeV.

Deuterium–Tritium (DT) Neutron Generator

In a DT neutron generator a beam of deuterons $\left(d = {}^2_1H\right)$ is accelerated to a few hundred keV and directed onto a tritium $\left(t = {}^3_1H\right)$ target thereby producing the following exothermic nuclear reaction:

$$d + t \rightarrow \alpha + n + Q \, (17.59 \text{ MeV}) \quad \text{or} \quad {}^2_1\text{H} + {}^3_1\text{H} = {}^4_2\text{He} + n + Q \, (17.59 \text{ MeV}) . \tag{9.26}$$

As described in Sect. 5.2.2, the Q value for the d–t reaction, Q_{d-t}, is calculated by using the nuclear binding energy method or the nuclear rest energy method. Data for nuclides of (9.26) are given in Appendix A. The binding energy calculation is as follows

$$\begin{aligned} Q_{d-t} &= E_B(\alpha) - \{E_B(d) + E_B(t)\} \\ &= 28.29569 \text{ MeV} - \{2.22458 + 8.48182\} \text{ MeV} = 17.59 \text{ MeV}. \quad (9.27) \end{aligned}$$

Same Q value result can be obtained by accounting for nuclear masses for the nuclei in (9.26)

$$\begin{aligned} Q_{d-t} &= \left[m_d c^2 + m_t c^2 \right] - \left[m_\alpha c^2 + m_n c^2 \right] \\ &= \left\{ \left[M({}^2_1\text{H}) + M({}^3_1\text{H}) \right] - \left[M({}^4_2\text{He}) + m_n \right] \right\} c^2 \quad (9.28) \\ &= \left\{ [1875.6128 + 2808.9209] - [3727.3791 + 939.5654] \right\} \text{ MeV} \\ &= 17.59 \text{ MeV}. \end{aligned}$$

The reaction energy Q_{d-t} of 17.6 MeV is shared between the neutron n and the α-particle in inverse proportions to their masses, resulting in neutron kinetic energy of 14.05 MeV and α-particle kinetic energy of 3.54 MeV.

DT neutron generators are relatively inexpensive; however, they have difficulties producing stable beams of sufficient intensity because of problems with the tritium target. Since at their best, the DT neutron generators produce beams that are only equivalent in penetration to cobalt-60 γ-ray beams and have significantly lower outputs than a standard cobalt unit, they are not serious contenders for delivery of routine radiotherapy treatments. However, as discussed in Sect. 1.12.2, neutron beams with their lower oxygen enhancement ratio (OER) have a certain biological advantage over photon beams in treatment of poorly oxygenated tumors.

Fast Neutron Beams from Cyclotrons

Cyclotrons provide practical means for production of clinical neutron beams, in addition to their use in production of clinical heavy charged particle beams and in production of radionuclides for use in industry and medicine. Ernest Lawrence, the inventor of the cyclotron in 1930, was in 1938, in collaboration with Robert Stone, the first to use external neutron beams in treatment of cancer. The beams were produced with a cyclotron, a cyclic accelerator discussed in Sect. 14.5.2.

While the initial treatment results were encouraging, many patients developed unacceptable late complications and external beam neutron therapy was discontinued for many years. However, during the hiatus, it became obvious that the complications experienced by the first neutron patients resulted from an overdose caused by a poor understanding of the radiobiology of neutron beams rather than from any inherent deleterious property of neutron beams.

Neutron therapy with a more careful prescription of target dose and fractionation was restarted in the second half of 1960s and during the past 40 years external beam neutron therapy has become an accepted, albeit still somewhat esoteric, cancer treatment modality.

The most common and efficient approach for production of clinical neutron beams with cyclotrons is to accelerate protons (p) or deuterons (d) in the energy range from 50 MeV to 70 MeV onto a beryllium-9 target. This results in neutron spectra that are characteristic of the particular nuclear reaction used, with the maximum neutron energy in the spectrum given as the sum of the incident charged particle kinetic energy and reaction Q value for the particular reaction that produces the neutrons.

Beryllium-9, the only stable isotope of beryllium, is chosen as target material for its neutron production efficiency as well as for its suitable mechanical and thermal properties. At the same incident particle energy, deuterons in comparison to protons will produce more neutrons in the beryllium target because some of the neutrons are produced through deuteron stripping reaction in addition to the standard compound nucleus formation. Consequently, currents required to achieve reasonable neutron dose rates with deuteron acceleration are typically 5 times lower than those required with proton acceleration.

The two nuclear reactions involving energetic protons and deuterons striking beryllium thick targets produce neutron beams with beam penetration and build-up characteristics that are similar to those produced by 4 MV to 8 MV x-ray beams. They are as follows

$$p + {}^9_4\text{Be} \rightarrow {}^9_5\text{B} + n + Q \, (-1.85 \text{ MeV}), \tag{9.29}$$

$$d + {}^9_4\text{Be} \rightarrow {}^{10}_5\text{B} + n + Q \, (4.36 \text{ MeV}). \tag{9.30}$$

The reaction Q values for the two reactions can be determined either with the nuclear binding energy E_B method or with the nuclear rest energy Mc^2 method. The appropriate data are given in Appendix A.

For the reaction (9.29) the nuclear binding energy method gives

$$Q = E_B({}^9_5\text{B}) - E_B({}^9_4\text{Be}) = (56.31445 - 58.16497) \text{ MeV} = -1.85 \text{ MeV}, \tag{9.31}$$

while the nuclear rest energy method gives the same result

$$\begin{aligned} Q &= \left[m_p c^2 + M({}^9_4\text{Be})c^2 \right] - \left[M({}^9_5\text{B})c^2 + m_n c^2 \right] \\ &= [938.2703 + 8392.7499] \text{ MeV} - [8393.3069 + 939.5654] \text{ MeV} \\ &= -1.85 \text{ MeV}. \end{aligned} \tag{9.32}$$

In a similar manner we get the following result for the (9.30) reaction

$$\begin{aligned} Q &= E_B({}^{10}_5\text{B}) - [E_B(d) + E_B({}^9_4\text{B})] \\ &= 64.7507 \text{ MeV} - [2.2246 \text{ MeV} + 58.1650 \text{ MeV}] = 4.36 \text{ MeV} \end{aligned} \tag{9.33}$$

and

$$Q = \left[m_d c^2 + M(^9_4\text{Be})c^2\right] - \left[m_n c^2 + M(^{10}_5\text{B})c^2\right]$$
$$= [1875.6128 + 8392.7499] \text{ MeV} - [939.5654 + 9324.4362] \text{ MeV} \quad (9.34)$$
$$= 4.36 \text{ MeV}.$$

Note that in (9.31) and (9.33) the proton and neutron are elementary particles with no binding energy, the deuteron in (9.33), on the other hand, consists of a proton and neutron bound together with a total binding energy $E_B(\text{d}) = 2.22458$ MeV or a binding energy per nucleon of 1.1123 MeV.

9.6.4   Californium-252 Neutron Source

Californium (Cf) is a synthetic radioactive transuranic element in the actinide series with an atomic number Z of 98 and 20 known radioisotopes. Of these only Cf-252, as an intense neutron emitter, is of commercial interest and was found useful in a wide range of specialized areas of science, industry, and medicine, such as the study of fission, neutron activation analysis, neutron radiography, well logging, nuclear reactor start up, and brachytherapy of cancer.

Californium-252 is produced by bombarding actinide oxide target rods in a nuclear reactor with a very high neutron fluence rate of the order of 10^{15} cm$^{-2}\cdot$s^{-1} for 12 months or more. Next, the heavy element components are separated and purified, and, finally, the californium fraction is separated, purified, and encapsulated into portable sealed sources for use in science, industry, and medicine. The main characteristics of Cf-252 are as follows:

1. Cf-252 decays with a half-life $(t_{1/2})_\alpha$ of 2.73 years through α decay (Sect. 11.2) into curium-248. About 96.9% of all Cf-252 decays occur through the α decay.
2. About 3.1% of all Cf-252 decays occur through spontaneous fission (Sect. 11.9). The half-life $(t_{1/2})_{SF}$ for spontaneous fission (SF) of Cf-252 is 85.5 years.
3. The effective half-life $(t_{1/2})_{eff}$ of Cf-252, accounting for both possible decays (α and SF), is 2.645 years, as determined following the relationship of (10.78) for the decay constant resulting from branching decay

$$\frac{1}{(t_{1/2})_{eff}} = \frac{1}{(t_{1/2})_\alpha} + \frac{1}{(t_{1/2})_{SF}}. \quad (9.35)$$

The monthly decay of Cf-252 is 2.2% of the initial mass.
4. In each spontaneous fission event on average 3.8 neutrons are produced per fission, amounting to a neutron production rate of 2.34×10^6 µg$^{-1}\cdot$s^{-1}, as determined by multiplying the neutron factor $f_n = 3.8$ with the specific activity for the spontaneous fission a_{SF}

$$f_n a_{SF} = \frac{\ln 2}{(t_{1/2})_{SF}} \frac{N_A}{A}$$

$$= 3.8 \frac{0.693}{85.5 \times 365 \times 24 \times 3.6 \times 10^3 \text{ s}} \frac{6.022 \times 10^{23}}{252 \text{ g}}$$

$$= 2.34 \times 10^6 \text{ s}^{-1} \cdot \mu\text{g}^{-1}. \tag{9.36}$$

Production rate of 2.34×10^6 neutrons per microgram per second or 140 million neutrons per microgram per minute classify Cf-252 as a very intense neutron source for use in industry and medicine. Industrial sources with mass exceeding 100 mg of Cf-252 approach neutron intensities produced by nuclear reactors and are used for neutron radiography.

5. Specific activity of Cf-252 is 2×10^7 Bq/μg (540 Ci/g) determined using the effective half-life $(t_{1/2})_{\text{eff}} = 2.645$ y in conjunction with (10.2) and (10.13).
6. Neutron spectrum emitted by Cf-252 is similar to that of a fission reactor, with a Maxwellian energy distribution (average energy of 2.1 MeV and most probable energy of $\sim$0.7 MeV).

Industrial sources contain up to 50 mg of Cf-252 emitting of the order of 10^{11} neutrons per second. High dose rate (HDR) brachytherapy requires about 500μg of Cf-252 per source and emits $\sim$$10^9$ neutrons per second. Current technology results in source diameters of the order of 3 mm; adequate for intracavitary brachytherapy but not suitable for interstitial brachytherapy. Smaller dimension (miniature) sources are likely to be produced in the near future, making the Cf-252 brachytherapy more practical and more widely available. Standard HDR brachytherapy is carried out with iridium-192 sources that emit a spectrum of γ-rays with an effective energy of $\sim$400 keV. The advantage of neutron irradiation is that neutron therapy is significantly more effective than conventional photon therapy in treatment of hypoxic (oxygen deficient) malignant disease.

9.6.5 In-vivo Neutron Activation Analysis

In-vivo analysis of human body elements by activation with fast neutrons provides means for quantitative evaluation of elemental and chemical human body composition for studies in human physiology as well as for clinical diagnosis and treatment of a variety of diseases and disorders. Elements that are of main interest in in-vivo neutron activation are calcium, nitrogen, carbon, potassium and sodium. Measurements of calcium body content are related to diagnosis of osteopenia and osteoporosis; measurements of nitrogen content to nutritional status of the body.

Neutron capture by an atom results in an unstable excited compound nucleus. This nucleus either decays into ground state through emission of a "prompt" gamma ray or it decays by emission of a nucleon followed by "delayed" gamma emission. Both the prompt as well as delayed gamma emissions are used in in-vivo neutron activation analysis. Prompt gamma rays are measured during the irradiation procedure; delayed gamma rays are measured after irradiation with neutrons.

The clinical neutron activation technique is non-destructive and based on two steps:

1. Bilateral irradiation of the patient with fast neutrons to obtain uniform activation throughout the body.
2. Subsequent measurement of the induced activity in a whole body counter with large thallium-activated sodium iodide or germanium lithium-drifted detectors.

The whole body dose received by the patient undergoing neutron activation should be as low as possible and with modern equipment amounts to only about 0.001 Gy, typically consisting of a neutron and gamma ray component. Since the introduction of the in-vivo neutron activation technique in the early 1960s, all known sources of fast neutrons have been used for clinical neutron activation of the body: nuclear reactors, DT generators, cyclotrons, and (α, n) sources. Radionuclide neutron sources offer the least expensive and most convenient approach.

9.7 Neutron Radiography

X-ray and γ-ray radiography became indispensable imaging tools in medicine, science and industry; however, radiography with more exotic particles such as protons and neutrons is also being developed.

Neutron radiography (NR) is a non-invasive imaging technique similar to industrial radiography with x-rays and γ-rays except that, instead of x-ray and γ-ray transmission through an object (absorber), it uses attenuation of a thermal neutron beam in a test object (absorber). In contrast to transmission of photons through an absorber that is characterized by photon interactions with orbital electrons of the absorber and is governed by atomic number Z of the absorber, transmission of neutrons through an absorber is characterized by neutron interactions with the nuclei of the absorber and is governed by the neutron cross sections of the absorber nuclei.

While for x-ray and γ-ray beams the attenuation coefficient of absorbers increases monotonically (if we ignore absorption edges) with absorber atomic number Z, for neutron beams the attenuation coefficient (cross section Σ) of absorbers cannot be correlated with Z of absorber and seems to fall in a random fashion on a Σ versus Z plot. In contrast to x-rays, neutrons are attenuated strongly by some low Z materials such as hydrogen, lithium, boron, and cadmium but penetrate many high Z materials such as iron, titanium, and lead with relative ease.

Elements with similar atomic numbers exhibit very similar x-ray attenuation characteristics and yet may have markedly different neutron attenuation characteristics. Moreover, even two isotopes of the same element can have significantly different neutron cross sections. Organic materials and water are clearly visible in neutron radiographs because of their hydrogen content, while many structural materials such as aluminum and iron or x-ray shielding materials such as lead are nearly transparent. This suggests that x-ray imaging and NR have potential to complement each other, since NR can provide information not available with standard x-ray and γ-ray imaging.

NR is currently mainly used in industry for non-destructive testing of specimens, well logging, inspection of cargo, and quality control of industrial products. Use of NR in medicine, while feasible in principle, currently requires a relatively high equivalent dose precluding routine use on patients.

Neutron sources for NR are divided into three main categories: radionuclide neutron sources, particle accelerators, and nuclear reactor:

(1) The best-known example of *radionuclide neutron source* is californium-52 (Cf-252), introduced in Sect. 9.6.4 for use in brachytherapy. Its main attributes are: (i) relatively high neutron output, (ii) low mean emitted neutron energy, and (iii) relatively small size.

(2) As discussed in Sect. 9.6.3, two machines are used for production of fast neutron beams: *neutron generator* and *cyclotron*. Both accelerate heavy charged particles into a suitable target in which fast neutrons are produced.

(3) *Nuclear reactor* (see Sect. 12.8.2) produces a spectrum of neutrons through fission reaction in uranium-235 and can be used for electrical power generation, nuclear and medical research, production of radionuclides, and as source of thermal as well as fast neutrons.

Neutron sources listed above all produce fast neutrons that must be thermalized for use in NR. To attain thermal neutron energy fast neutrons from a neutron source are transmitted through a moderator consisting of a hydrogenous material such as water or polystyrene. The thermal neutrons so produced are then transmitted through the test object to form the latent neutron image and are then made to strike a neutron converter screen that transforms the neutron latent image into a visible object image through a two-step process. In the first step neutrons undergo a nuclear reaction (typically neutron capture or spallation) in the screen coated with a neutron absorbing material such as gadolinium or boron. In the second step the nuclear reaction products generated in the first step (such as γ-rays, conversion electrons, or α-particles) are detected and converted into a visible image of the object. This latter step is achieved using film, imaging plate or digital camera.

Cobalt-60 Teletherapy

The two figures on next page depict a cobalt-60 teletherapy machine: actual modern machine manufactured by Best Theratronics in Ottawa, Canada is shown on the left and a schematic diagram of the machine on the right. The schematic diagram of a cobalt machine was presented on a Canadian postage stamp issued in 1988 by the Canada Post Corporation in honor of *Harold E. Johns* (1915–1997), a Canadian medical physicist and the inventor of the cobalt-60 teletherapy machine.

The cobalt machine was developed in Canada in the 1950s for use in cancer therapy. It was the first truly practical and widely available megavoltage cancer therapy machine in the world and incorporates a radioactive cobalt-60 source that is characterized with features suitable for external beam radiotherapy, such as high gamma ray energy, relatively long half-life, and high specific activity.

The cobalt-60 source is produced in a nuclear reactor by irradiating stable cobalt-59 nuclides with thermal neutrons. The cobalt-60 source decays with a half-life of 5.26 years to nickel-60 with emission of beta particles (electrons) and two gamma rays (1.17 MeV and 1.33 MeV) per each disintegration for use in cancer therapy, as shown schematically on the stamp.

Most modern cobalt machines are arranged on a rotatable gantry so that the source may rotate about a horizontal axis referred to as the machine isocenter axis. The source-axis distance (SAD) typically is either 80 cm or 100 cm depending on the machine design. The isocentric source mounting allows the use of the isocentric treatment technique in which the radiation beam is directed toward the patient from various directions thereby concentrating the radiation dose in the target volume and spreading the dose to healthy tissues over a larger volume to minimize damage to healthy tissues.

During the past two decades the linear accelerator (linac) eclipsed the cobalt machine and became the most widely used radiation-producing machine in modern radiotherapy. Compared to cobalt units, linacs offer higher beam energies that result in better skin sparing effect and more effective penetration into tissue; higher output dose rates that result in shorter treatment times; electron beams in addition to photon beams for treatment of superficial lesions; and a possibility for beam intensity modulation that provides optimal dose distributions in the target volume.

Despite the technological and practical advantages of linacs over cobalt-60 machines, the latter still occupy an important place in the radiotherapy armamen-tarium, mainly because of significantly lower capital, installation and maintenance costs in comparison with linacs. Moreover, the design of modern cobalt machines offers many of the features that until lately were in the sole domain of linacs, such as large source-axis distance, high output, dynamic wedge, independent jaws, and a multileaf collimator (MLC). In the developing world, the cobalt-60 machines, owing to their relatively low costs, simplicity of design, and ease of operation, are likely to play an important role in cancer therapy for the foreseeable future.

Chapter 10
Kinetics of Radioactive Decay

Radioactivity also known as radioactive decay, nuclear transformation, and nuclear disintegration is a spontaneous process by which an unstable parent nucleus emits a particle or electromagnetic radiation and transforms into a more stable daughter nucleus that may or may not be stable. An unstable daughter nucleus will decay further in a decay series until a stable nuclear configuration is reached. The radioactive decay is governed by the formalism based on the definition of activity and the radioactive decay constant. Henri Becquerel discovered the process of natural radioactivity in 1896 and soon thereafter in 1898 Pierre Curie and Marie Skłodowska-Curie discovered radium and polonium and coined the term "radioactivity" to describe the emission of "emanations" from unstable natural elements. Frédéric Joliot and Irène Joliot-Curie discovered artificial radioactivity in 1934.

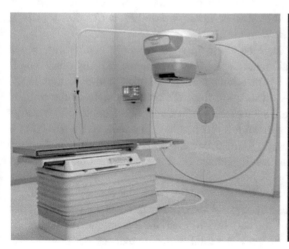

© Springer International Publishing Switzerland 2016
E.B. Podgoršak, *Radiation Physics for Medical Physicists*,
Graduate Texts in Physics, DOI 10.1007/978-3-319-25382-4_10

In this chapter we discuss the general aspects of radioactive decay and show that the decay follows first order kinetics. We first address the simple decay of a radioactive parent into a stable daughter and then discuss the more complicated radioactive series decay of parent producing an unstable daughter which decays into a granddaughter nuclide which may or may not be stable. The various equilibria between the daughter and parent are discussed and a generalized approach to series decay kinetics is presented. The chapter concludes with discussion of special situations such as many-component radioactive series decay described with Bateman equations; branching decay; and decay of mixtures of several radionuclides in a sample.

10.1   General Aspects of Radioactivity

Radioactivity or radioactive decay, discovered in 1986 by *Henri Becquerel*, is a process by which an unstable parent nucleus transforms spontaneously into one or several daughter nuclei that are more stable than the parent nucleus by having larger binding energies per nucleon than does the parent nucleus. The daughter nucleus may also be unstable and will decay further through a chain of radioactive decays until a stable nuclear configuration is reached. Radioactive decay is usually accompanied by emission of energetic particles or gamma ray photons or both that may be used in science, industry, agriculture, and medicine.

Nuclear decay, also called *radioactive decay, nuclear disintegration, nuclear transformation*, and *nuclear transmutation* is a statistical phenomenon that is commonly described by the following characteristics:

- The exponential laws that govern nuclear decay and growth of radioactive substances were first formulated by *Ernest Rutherford* and *Frederick Soddy* in 1902 and then refined by *Harry Bateman* in 1910.
- A radioactive substance containing atoms of same structure is often referred to as radioactive nuclide or radionuclide. Radioactive atoms, like any other atomic structure, are characterized by the atomic number Z and atomic mass number A.
- Radioactive decay involves a transition from the quantum state of the original nuclide (parent) to a quantum state of the product nuclide (daughter). The energy difference between the two quantum levels involved in a radioactive transition is referred to as the decay energy Q. The decay energy is emitted either in the form of electromagnetic radiation (usually gamma rays) or in the form of kinetic energy of the reaction products.
- The mode of radioactive decay depends upon the particular nuclide involved.
- All radioactive decay processes are governed by the same general formalism that is based on the definition of the activity $\mathcal{A}(t)$ and on a characteristic parameter for each radioactive decay process: the total radioactive *decay constant* λ with dimensions of reciprocal time usually in s^{-1}.
- The decay constant λ is independent of the age of the radioactive atom and is essentially independent of physical conditions such as temperature, pressure, and chemical state of the atom's environment. Careful measurements have shown that λ

can actually depend slightly on the physical environment. For example, at extreme pressure or at extremely low temperature the technetium-99m radionuclide shows a fractional change in λ of the order of 10^{-4} in comparison to the value at room temperature (293 K) and standard pressure (101.3 kPa).

- The total *radioactive decay constant* λ multiplied by a time interval that is much smaller than $1/\lambda$ represents the probability that any particular atom of a radioactive substance containing a large number $N(t)$ of identical radioactive atoms will decay (disintegrate) in that time interval. An assumption is made that λ is independent of the physical environment of a given atom.
- *Activity* $A(t)$ of a radioactive substance containing a large number $N(t)$ of identical radioactive atoms represents the total number of decays (disintegrations) per unit time and is defined as a product between $N(t)$ and λ, i.e.,

$$A(t) = \lambda N(t). \tag{10.1}$$

- The *SI unit of activity* is the becquerel (Bq) given as $1\text{ Bq} = 1\text{ s}^{-1}$. The becquerel and hertz both correspond to s^{-1}, but hertz refers to frequency of periodic motion, while becquerel refers to activity.
- The old unit of activity, the curie (Ci), was initially defined as the activity of 1 g of radium-226 and given as $1\text{ Ci} = 3.7 \times 10^{10}\text{ s}^{-1}$. The activity of 1 g of radium-226 was subsequently measured to be $3.665 \times 10^{10}\text{ s}^{-1}$; however, the definition of the curie was kept at $3.7 \times 10^{10}\text{ s}^{-1}$. The current value of the activity of 1 g of radium-226 is thus 0.988 Ci or 3.665×10^{10} Bq.
- Bq and Ci are related as follows:
  $1\text{ Bq} = 2.703 \times 10^{-11}\text{ Ci or } 1\text{ Ci} = 3.7 \times 10^{10}\text{ Bq}.$
- *Specific activity* a is defined as activity A per unit mass M, i.e.,

$$a = \frac{A}{M} = \frac{\lambda N}{M} = \frac{\lambda N_A}{A}, \tag{10.2}$$

where N_A is the Avogadro number ($6.022 \times 10^{23}\text{ mol}^{-1}$).
- The specific activity a of a radioactive atom depends on the decay constant λ and on the atomic mass number A of the radioactive atom. The units of specific activity are Bq/kg (SI unit) and Ci/g (old unit). The relationship between the two units is as follows

$$1\frac{\text{Ci}}{\text{g}} = \frac{3.7 \times 10^{10}}{10^{-3}}\frac{\text{Bq}}{\text{kg}} = 3.7 \times 10^{13}\frac{\text{Bq}}{\text{kg}} = 37\frac{\text{TBq}}{\text{kg}} \tag{10.3}$$

or

$$1\frac{\text{Bq}}{\text{kg}} = \frac{1}{3.7 \times 10^{13}}\frac{\text{Ci}}{\text{g}} = 2.703 \times 10^{-14}\frac{\text{Ci}}{\text{g}}. \tag{10.4}$$

10.2   Decay of Radioactive Parent into a Stable Daughter

The simplest form of radioactive decay is characterized by a radioactive parent nucleus P decaying with decay constant λ_P into a stable daughter nucleus D:

$$P \xrightarrow{\lambda_P} D. \tag{10.5}$$

The rate of depletion of the number of radioactive parent nuclei $N_P(t)$ is equal to the activity $\mathcal{A}_P(t)$ at time t, i.e.,

$$\frac{dN_P(t)}{dt} = -\mathcal{A}_P(t) = -\lambda_P N_P(t). \tag{10.6}$$

The fundamental differential equation of (10.6) for $N_P(t)$ can be rewritten in general integral form to get

$$\int_{N_P(0)}^{N_P(t)} \frac{dN_P(t)}{N_P} = -\int_0^t \lambda_P dt, \tag{10.7}$$

where $N_P(0)$ is the number of radioactive nuclei at time $t = 0$.

Assuming that λ_P is constant, we can write (10.7) as follows

$$\ln \frac{N_P(t)}{N_P(0)} = -\lambda_P t \tag{10.8}$$

or

$$N_P(t) = N_P(0)\, e^{-\lambda_P t}. \tag{10.9}$$

The activity of parent nuclei P at time t may now be expressed as follows

$$\mathcal{A}_P(t) = \lambda_P N_P(t) = \lambda_P N_P(0)\, e^{-\lambda_P t} = \mathcal{A}_P(0)\, e^{-\lambda_P t}, \tag{10.10}$$

where $\mathcal{A}_P(0) = \lambda_P N_P(0)$ is the initial activity of the radioactive substance.

The decay law of (10.10) applies to all radioactive nuclides irrespective of their mode of decay; however, the decay constant λ is different for each radioactive nuclide P and is the most important defining characteristic of a radioactive nuclide. When more than one mode of decay is available to a radioactive nucleus (branching), the total decay constant λ is the sum of the partial decay constants λ_i applicable to each mode

$$\lambda = \sum_i \lambda_i. \tag{10.11}$$

Half-life $(t_{1/2})_P$ of a radioactive substance P is that time during which the number of radioactive nuclei of the substance decays to half of the initial value $N_P(0)$ present at time $t = 0$. We can also state that in the time of one half-life the activity $\mathcal{A}_P(t)$ of a radioactive substance diminishes to one half of its initial value, i.e.,

$$N_P[t = (t_{1/2})_P] = \frac{1}{2}N_P(0) = N_P(0)e^{-\lambda_P(t_{1/2})_P}$$

or

$$\mathcal{A}_P[t = (t_{1/2})_P] = \frac{1}{2}\mathcal{A}_P(0) = \mathcal{A}_P(0)e^{-\lambda_P(t_{1/2})_P} \tag{10.12}$$

From (10.12) we obtain the following relationship between the decay constant λ_P and the half-life $(t_{1/2})_P$

$$\lambda_P = \frac{\ln 2}{(t_{1/2})_P} = \frac{0.693}{(t_{1/2})_P}. \tag{10.13}$$

The actual lifetime of any radioactive nucleus can vary from 0 to ∞; however, for a large number N_P of parent nuclei we can define the *average (mean)* life τ_P of a radioactive parent substance P that equals the sum of lifetimes of all individual atoms divided by the initial number of radioactive nuclei. The average (mean) life thus represents the average life expectancy of all nuclei in the radioactive substance P at time $t = 0$; i.e.,

$$\mathcal{A}_P(0)\tau_P = \mathcal{A}_P(0) \int_0^\infty e^{-\lambda_P t} \, dt = \frac{\mathcal{A}_P(0)}{\lambda_P} = N_P(0), \tag{10.14}$$

Decay constant λ_P and mean life τ_P are related through the following expression

$$\tau_P = \frac{1}{\lambda_P}. \tag{10.15}$$

Mean life τ_P can also be defined as the time required for the number of radioactive atoms or their activity to fall to $1/e = 0.368$ of its initial value $N_P(0)$ or initial activity $\mathcal{A}_P(0)$, respectively. Mean life τ_P and half-life are related as follows

$$\tau_P = \frac{1}{\lambda_P} = \frac{(t_{1/2})_P}{\ln 2} = 1.44 \, (t_{1/2})_P. \tag{10.16}$$

Average (mean) life τ_P can also be determined using the standard method for finding the average of a continuous variable

$$\tau_P = C \int_0^\infty t e^{-\lambda_P t} \, dt = \frac{\int_0^\infty t e^{-\lambda_P t} \, dt}{\int_0^\infty e^{-\lambda_P t} \, dt} = \frac{\frac{1}{\lambda_P^2}}{\frac{1}{\lambda_P}} = \frac{1}{\lambda_P}, \tag{10.17}$$

where C is a normalization constant expressed as

$$C = \left[\int_0^\infty e^{-\lambda_P t}\, dt \right]^{-1} . \tag{10.18}$$

A typical example of a radioactive decay for initial condition $\mathcal{A}_P(t = 0) = \mathcal{A}_P(0)$ is shown in Fig. 10.1 with a plot of parent activity $\mathcal{A}_P(t)$ against time t, i.e.,

$$\mathcal{A}_P(t) = \mathcal{A}_P(0)\, e^{-\lambda_P t} . \tag{10.19}$$

The following properties of the radioactive decay curve are notable:

1. Area under the activity $\mathcal{A}_P(t)$ versus time t curve for $0 \leq t \leq \infty$ is given as

$$\int_0^\infty \mathcal{A}_P(t)\, dt = \mathcal{A}_P(0) \int_0^\infty e^{-\lambda_P t}\, dt = \frac{\mathcal{A}_P(0)}{\lambda_P} = \mathcal{A}_P(0)\tau_P = N_P(0) \tag{10.20}$$

and the result equals the initial number of radioactive nuclei at time $t = 0$.

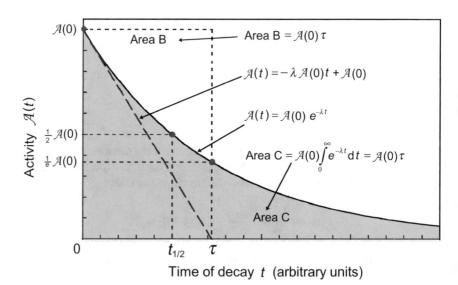

Fig. 10.1 Activity $\mathcal{A}_P(t)$ plotted against time t for a simple decay of a radioactive parent P into a stable daughter D. The activity follows the relationship given in (10.10). The concepts of half-life $(t_{1/2})_P$ and mean life τ_P are also illustrated. The area under the exponential decay curve from 0 to ∞ is equal to $\mathcal{A}_P(0)\tau_P$ where $\mathcal{A}_P(0)$ is the initial activity of the parent nuclei. The slope of the tangent to the decay curve at $t = 0$ is equal to $-\lambda_P\mathcal{A}_P(0)$ and this tangent crosses the abscissa axis at $t = \tau_P$

2. Total number of radioactive nuclei present at any time $t > 0$ is simply the activity $\mathcal{A}_P(t)$ multiplied by the mean life τ_P.
3. The concept of half-life $(t_{1/2})_P$ is shown in Fig. 10.1 as the time in which the activity $\mathcal{A}_P(t)$ drops from $\mathcal{A}_P(0)$ to $0.5\mathcal{A}_P(0)$.
4. The concept of mean life τ_P is shown in Fig. 10.1 as the time in which the activity $\mathcal{A}_P(t)$ drops from $\mathcal{A}_P(0)$ to $0.368\mathcal{A}_P(0) = e^{-1}\mathcal{A}_P(0)$.
5. Area $\mathcal{A}_P(0)\tau_P$ is shown in Fig. 10.1 by a rectangle with sides $\mathcal{A}_P(0)$ and τ_P. If the initial activity $\mathcal{A}_P(0)$ could remain constant for mean life τ_P all atoms would be transformed by the time $t = \tau_P$ at which the activity would abruptly drop to zero.
6. In general, the slope of the tangent to the decay curve at time t is given as

$$\frac{\mathrm{d}\mathcal{A}_P(t)}{\mathrm{d}t} = -\lambda_P\mathcal{A}_P(0)e^{-\lambda_P t}, \tag{10.21}$$

while the initial slope at $t = 0$ is equal to $-\lambda_P\mathcal{A}_P(0)$.
7. The linear function, with the slope equal to $-\lambda_P\mathcal{A}_P(0)$ and the ordinate intercept at time $t = 0$ equal to $\mathcal{A}_P(0)$, is

$$\mathcal{A}_P(t) = -\lambda_P\mathcal{A}_P(0)t + \mathcal{A}_P(0) \tag{10.22}$$

and represents the tangent to the decay curve at $t = 0$. It serves as a good approximation for the activity $\mathcal{A}_P(t)$ versus t relationship when $t \ll \tau_P$, i.e.,

$$\mathcal{A}_P(t) \approx \mathcal{A}_P(0)\{1 - \lambda_P t\} = \mathcal{A}_P(0)\left\{1 - \frac{t}{\tau_P}\right\} \tag{10.23}$$

and results in $\mathcal{A}_P(t) = 0$ at $t = \tau_P$, in contrast to (10.19) that predicts $\mathcal{A}_P(t) = 0$ only at $t \to \infty$.

10.3   Radioactive Series Decay

Section 10.2 dealt with the radioactivity of a sample containing only one parent radionuclide subjected to only one radioactive decay mode leading to a stable daughter. This simple radioactive decay was described with (10.9) and (10.19). The radioactivity of a sample is significantly more complex if the sample consists of two or more components, such as in the case of:

1. Radioactive decay series where a radioactive parent P decays with decay constant λ_P into a daughter D that in turn is radioactive and decays with a decay constant λ_D into a stable granddaughter G

$$P \xrightarrow{\lambda_P} D \xrightarrow{\lambda_D} G. \tag{10.24}$$

2. Complicated decay chain in which the granddaughter and several of its progeny are also radioactive.
3. Radionuclide in the sample is subjected to more than one mode of decay. This type of decay is referred to as branching decay.
4. Sample consists of more than one radioactive species in a mixture of independent activities.

Radioactive decay series is much more common than the simple radioactive decay. It forms a decay chain starting with the parent radionuclide and moves through several generations to eventually end with a stable nuclide. We first analyse the simple chain P $\rightarrow$ D $\rightarrow$ G where both the parent and daughter are radioactive and the granddaughter is not radioactive. We then generalize the discussion into larger chains in Sect. 10.6 and address the question of branching decay in Sect. 10.7 as well as heterogeneous samples in Sect. 10.8.

10.3.1   Parent → Daughter → Granddaughter Relationships

The rate of change dN_D/dt in the number of daughter nuclei D is equal to the supply of new daughter nuclei D through the decay of P given as $\lambda_P N_P(t)$ and the loss of daughter nuclei D from the decay of D to G given as $[-\lambda_D N_D(t)]$, i.e.,

$$\frac{dN_D}{dt} = \lambda_P N_P(t) - \lambda_D N_D(t) = \lambda_P N_P(0)e^{-\lambda_P t} - \lambda_D N_D(t), \qquad (10.25)$$

where $N_P(0)$ is the initial number of parent nuclei at time $t = 0$.

The parent P follows a straightforward radioactive decay process with the initial condition $N_P(t = 0) = N_P(0)$, as described by (10.9)

$$N_P(t) = N_P(0)e^{-\lambda_P t}. \qquad (10.26)$$

We are now interested in obtaining the functional relationship for the number of daughter nuclei $N_D(t)$ assuming an initial condition that at $t = 0$ there are no daughter nuclei D present. The initial condition for the number of daughter nuclei N_D is thus as follows

$$N_D(t = 0) = N_D(0) = 0. \qquad (10.27)$$

The general solution of the differential equation given by (10.25) will be of the form

$$N_D(t) = N_P(0)\{pe^{-\lambda_P t} + de^{-\lambda_D t}\}, \qquad (10.28)$$

where p and d are constants to be determined using the following four steps:

1. Differentiate (10.28) with respect to time t to obtain

$$\frac{dN_D}{dt} = N_P(0)\left\{-p\lambda_P e^{-\lambda_P t} - d\lambda_D e^{-\lambda_D t}\right\}. \tag{10.29}$$

2. Insert (10.28) and (10.29) into (10.25) and rearrange the terms to get

$$e^{-\lambda_P t}\left\{-p\lambda_P - \lambda_P + p\lambda_D\right\} = 0. \tag{10.30}$$

3. The factor in curly brackets of (10.30) must be equal to zero to satisfy the equation for all values of t, yielding the following expression for the constant p

$$p = \frac{\lambda_P}{\lambda_D - \lambda_P}. \tag{10.31}$$

4. The coefficient d depends on the initial condition for N_D, i.e., $N_D(t = 0) = 0$ and may now be determined from (10.28) as

$$p + d = 0 \tag{10.32}$$

or after inserting (10.31)

$$d = -p = -\frac{\lambda_P}{\lambda_D - \lambda_P}. \tag{10.33}$$

The number of daughter nuclei $N_D(t)$ of (10.28) may now be written as follows

$$N_D(t) = N_P(0)\frac{\lambda_P}{\lambda_D - \lambda_P}\left\{e^{-\lambda_P t} - e^{-\lambda_D t}\right\}. \tag{10.34}$$

Recognizing that the activity of the daughter $\mathcal{A}_D(t)$ is $\lambda_D N_D(t)$ we now write the daughter activity $\mathcal{A}_D(t)$ as

$$\mathcal{A}_D(t) = \frac{N_P(0)\lambda_P\lambda_D}{\lambda_D - \lambda_P}\left\{e^{-\lambda_P t} - e^{-\lambda_D t}\right\} = \mathcal{A}_P(0)\frac{\lambda_D}{\lambda_D - \lambda_P}\left\{e^{-\lambda_P t} - e^{-\lambda_D t}\right\}$$

$$= \mathcal{A}_P(0)\frac{1}{1 - \frac{\lambda_P}{\lambda_D}}\left\{e^{-\lambda_P t} - e^{-\lambda_D t}\right\} = \mathcal{A}_P(t)\frac{\lambda_D}{\lambda_D - \lambda_P}\{1 - e^{-(\lambda_D - \lambda_P)t}\},$$

$$\tag{10.35}$$

where

$\mathcal{A}_D(t)$ is the activity at time t of the daughter nuclei equal to $\lambda_D N_D(t)$.
$\mathcal{A}_P(0)$ is the initial activity of the parent nuclei present at time $t = 0$.
$\mathcal{A}_P(t)$ is the activity at time t of the parent nuclei equal to $\lambda_P N_P(t)$.

10.3.2 Characteristic Time

Equation (10.35) represents several general expressions for the activity $\mathcal{A}_D(t)$ of the daughter nuclei D and predicts a value of zero for $\mathcal{A}_D(t)$ at $t = 0$ (initial condition) and at $t = \infty$ (when all nuclei of the parent P and daughter D have decayed). This suggests that $\mathcal{A}_D(t)$ will pass through a maximum at a specified characteristic time $(t_{max})_D$ for $\lambda_P \neq \lambda_D$. The characteristic time $(t_{max})_D$ is determined by setting $d\mathcal{A}_D/dt = 0$ at $t = (t_{max})_D$ and solving for $(t_{max})_D$ to get

$$\lambda_P e^{-\lambda_P (t_{max})_D} = \lambda_D e^{-\lambda_D (t_{max})_D} \tag{10.36}$$

and

$$(t_{max})_D = \frac{\ln \dfrac{\lambda_P}{\lambda_D}}{\lambda_P - \lambda_D}. \tag{10.37}$$

Equation (10.37), governed by the initial conditions at $t = 0$

$$\mathcal{A}_P(t = 0) = \mathcal{A}_P(0) \qquad \text{and} \qquad \mathcal{A}_D(t = 0) = 0, \tag{10.38}$$

may also be expressed in terms of half-lives $(t_{1/2})_P$ and $(t_{1/2})_D$ as well as in terms of mean-lives τ_P and τ_D for the parent P nuclei and daughter D nuclei, respectively, as

$$(t_{max})_D = \frac{\ln \dfrac{(t_{1/2})_D}{(t_{1/2})_P}}{(\ln 2)\left\{\dfrac{1}{(t_{1/2})_P} - \dfrac{1}{(t_{1/2})_D}\right\}} = \frac{(t_{1/2})_P (t_{1/2})_D}{(t_{1/2})_D - (t_{1/2})_P} \frac{\ln \dfrac{(t_{1/2})_D}{(t_{1/2})_P}}{\ln 2} \tag{10.39}$$

and

$$(t_{max})_D = \frac{\ln \dfrac{\tau_D}{\tau_P}}{\dfrac{1}{\tau_D} - \dfrac{1}{\tau_P}} = \frac{\tau_P \tau_D}{\tau_P - \tau_D} \ln \frac{\tau_P}{\tau_D}. \tag{10.40}$$

10.4 General Form of Daughter Activity

Equations (10.37), (10.39) and (10.40) show that $(t_{max})_D$ is positive and real, irrespective of the relative values of λ_P and λ_D, except for the case of $\lambda_P = \lambda_D$ for which $\mathcal{A}_D(t)$ in (10.35) is not defined.

At $t = (t_{max})_D$ we get from (10.35) that $\mathcal{A}_P[(t_{max})_D] = \mathcal{A}_D[(t_{max})_D]$, i.e., the activities of the parent and daughter nuclei are equal and the condition referred to as the *ideal equilibrium* is met. The term "ideal equilibrium" was coined by *Robley Evans* to distinguish this instantaneous condition from other types of equilibrium (transient and secular) that are defined for the relationship between the parent and daughter activity under certain special conditions.

- For $0 < t < (t_{max})_D$, the activity of parent nuclei $\mathcal{A}_P(t)$ always exceeds the activity of the daughter nuclei $\mathcal{A}_D(t)$, i.e., $\mathcal{A}_D(t) < \mathcal{A}_P(t)$.
- For $(t_{max})_D < t < \infty$, the activity of the daughter nuclei $\mathcal{A}_D(t)$ always exceeds the activity of the parent nuclei $\mathcal{A}_P(t)$, i.e., $\mathcal{A}_D(t) > \mathcal{A}_P(t)$.

Equation (10.35), describing the daughter activity $\mathcal{A}_D(t)$, can be written in a general form covering all possible physical situations. This is achieved by introducing variables x, y_P, and y_D as well as a decay factor m defined as follows:

1. x: time t normalized to half-life of parent nuclei $(t_{1/2})_P$

$$x = \frac{t}{(t_{1/2})_P},$$ (10.41)

2. y_P: parent activity $\mathcal{A}_P(t)$ normalized to $\mathcal{A}_P(0)$, the parent activity at $t = 0$

$$y_P = \frac{\mathcal{A}_P(t)}{\mathcal{A}_P(0)} = e^{-\lambda_P t} \quad \text{[see (10.10) and (10.26)]}$$ (10.42)

3. y_D: daughter activity $\mathcal{A}_D(t)$ normalized to $\mathcal{A}_P(0)$, the parent activity at $t = 0$

$$y_D = \frac{\mathcal{A}_D(t)}{\mathcal{A}_P(0)},$$ (10.43)

4. m: *decay factor* defined as the ratio of the two decay constants, i.e., λ_P/λ_D

$$m = \frac{\lambda_P}{\lambda_D} = \frac{(t_{1/2})_D}{(t_{1/2})_P}.$$ (10.44)

Insertion of x, y_D, and m into (10.35) results in the following expression for y_D, the daughter activity $\mathcal{A}_D(t)$ normalized to the initial parent activity $\mathcal{A}_P(0)$, as defined in (10.43)

$$y_D = \frac{1}{1-m} \left\{ e^{-x \ln 2} - e^{-\frac{x}{m} \ln 2} \right\} = \frac{1}{1-m} \left\{ \frac{1}{2^x} - \frac{1}{2^{x/m}} \right\}.$$ (10.45)

Equation (10.45) for y_D as a function of x has physical meaning for all positive values of m except for $m = 1$ for which y_D is not defined. However, since (10.45) gives $y_D = 0/0$ for $m = 1$, we can apply the L'Hôpital rule and determine the appropriate function for y_D as follows

$$y_D(m = 1) = \lim_{m \to 1} \frac{\dfrac{d}{dm}\left\{\dfrac{1}{2^x} - \dfrac{1}{2^{x/m}}\right\}}{\dfrac{d}{dm}(1 - m)} = \lim_{m \to 1} \frac{-2^{-\frac{x}{m}}\ln 2\left\{\dfrac{x}{m^2}\right\}}{-1} = (\ln 2)\frac{x}{2^x}.$$

(10.46)

Similarly, (10.19) for the parent activity $\mathcal{A}_P(t)$ can be written in terms of variables x and y_P as follows

$$y_P = e^{-\lambda_P t} = e^{-x\ln 2} = \frac{1}{2^x},$$

(10.47)

where x was given in (10.41) as $x = t/(t_{1/2})_P$ and $y_P = \mathcal{A}_P(t)/\mathcal{A}_P(0)$ is the parent activity normalized to the parent activity at time $t = 0$.

The characteristic time $(t_{max})_D$ can now be generalized to $(x_D)_{max}$ by using (10.41) to get the following expression

$$(x_D)_{max} = \frac{(t_{max})_D}{(t_{1/2})_P}.$$

(10.48)

Three approaches can now be used to determine $(x_D)_{max}$ for y_D in (10.45)

1. Set $\dfrac{dy_D}{dx} = 0$ at $x = (x_D)_{max}$ and solve for $(x_D)_{max}$ to get

$$\frac{dy_D}{dx}\bigg|_{x=(x_D)_{max}} = \frac{\ln 2}{1-m}\left\{-2^{-x} + \frac{1}{m}2^{-\frac{x}{m}}\right\}\bigg|_{x=(x_D)_{max}} = 0.$$

(10.49)

Solving (10.49) for $(x_D)_{max}$ we finally get

$$(x_D)_{max} = \frac{m}{m-1}\frac{\log m}{\log 2} = \frac{m}{m-1}\frac{\ln m}{\ln 2}.$$

(10.50)

For $m = 1$ (10.50) is not defined; however, since it gives $(x_D)_{max} = 0/0$, we can apply the L'Hôpital rule to get $(x_D)_{max}|_{m \to 1}$ as follows

$$(x_D)_{max}|_{m \to 1} = \lim_{m \to 1} \frac{\dfrac{d(m\ln m)}{dm}}{\ln 2\dfrac{d(m-1)}{dm}} = \lim_{m \to 1} \frac{1+\ln m}{\ln 2} = \frac{1}{\ln 2} = 1.44.$$

(10.51)

Thus, $(x_D)_{max}$ is calculated from (10.50) for any positive m except for $m = 1$. For $m = 1$, (10.51) gives $(x_D)_{max} = 1.44$.

2. Insert (10.41) and (10.44) into (10.37) for $(t_{max})_D$ and solve for $(x_D)_{max}$ to get the result given in (10.50).
3. Recognize that when $x = (x_D)_{max}$ the condition of ideal equilibrium applies for (10.43), i.e., $y_P[(x_D)_{max}] = y_D[(x_D)_{max}]$. Insert $x = (x_D)_{max}$ into (10.45) and

(10.47), set $y_P[(x_D)_{max}] = y_D[(x_D)_{max}]$, and solve for $(x_D)_{max}$ to get the result of (10.50).

In Fig. 10.2 we plot (10.45) for y_D against x using various values of the decay factor m in the range from 0.1 to 10. The function plotted with the dashed curve for $m = 1$ is the function given in (10.46). For comparison we also plot y_P of (10.47) against x.

All y_D curves of Fig. 10.2 start at the coordinate system origin at $(0,0)$, rise with x, reach a peak at $(x_D)_{max}$, as given in (10.50), and then decay with an increasing x. The smaller is m, the steeper is the initial rise of y_D, i.e., the larger is the initial slope of y_D. The initial slope and its dependence on m can be determined from the derivative dy_P/dx of (10.49) by setting $x = 0$ to get

$$\left.\frac{dy_D}{dx}\right|_{x=0} = \frac{\ln 2}{1-m}\left\{-2^{-x} + \frac{1}{m}2^{-\frac{x}{m}}\right\}\Bigg|_{x=0} = \frac{\ln 2}{m}. \qquad (10.52)$$

Noting that $x = t/(t_{1/2})_P$, $m = \lambda_P/\lambda_D$, and $y_D = \mathcal{A}_D(t)/\mathcal{A}_P(0)$, we can link the data of Fig. 10.2 with physical situations that occur in nature in the range $0.1 < m < 10$. Of course, the m region can be expanded easily to smaller and larger values outside the range shown in Fig. 10.2.

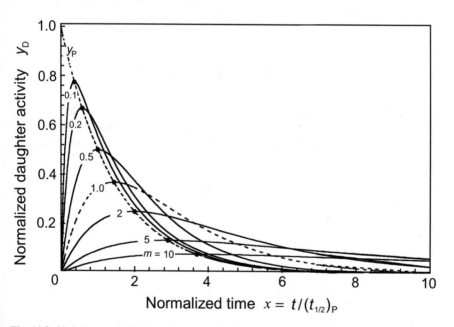

Fig. 10.2 Variable y_D of (10.45) against variable x for various values of the decay parameter m in the range from 0.1 to 10. The *dashed curve* is for y_P of (10.47) against x. The parameter $(x_D)_{max}$ shown by *dots* on the y_P curve is calculated from (10.50). Values for $(y_D)_{max}$ are obtained with (10.53). The *dashed* y_D curve is for decay factor $m = 1$ and is calculated from (10.46)

As indicated with dots on the y_p curve in Fig. 10.2, $(y_D)_{max}$, the maxima in y_D for a given m, occur at points $(x_D)_{max}$ where the y_D curves cross over the y_P curve. The $(x_D)_{max}$ values for a given m can be calculated from (10.50) and $(y_D)_{max}$ for a given m can be calculated simply by determining $y_P(x)$ at $x = (x_D)_{max}$ with $y_P(x)$ given in (10.47). We thus obtain the following expression for $(y_D)_{max}$

$$(y_D)_{max} = y_P(x_D)_{max} = 2^{\left(\frac{m}{1-m}\right)\frac{\ln m}{\ln 2}} = \frac{1}{2^{(x_D)_{max}}} \equiv e^{\frac{m}{1-m}\ln m} = e^{-(\ln 2)(x_D)_{max}}, \quad (10.53)$$

where $(x_D)_{max}$ was given by (10.50). Equation (10.53) is valid for all positive m with the exception of $m = 1$. We determine $(y_D)_{max}$ for $m = 1$ by applying the L'Hôpital rule to (10.53) to get

$$(y_D)_{max}|_{m=1} = \lim_{m \to 1} 2^{\frac{\frac{d}{dm}(m\ln m)}{\frac{d}{dm}(1-m)\ln 2}} = \lim_{m \to 1} 2^{\frac{\ln m + 1}{-\ln 2}} = 2^{-\frac{1}{\ln 2}} = e^{-1} = 0.368. \quad (10.54)$$

As shown in (10.53), $(y_D)_{max}$ and $(x_D)_{max}$ are related through a simple exponential expression plotted in Fig. 10.3 and also given by (10.47) with $x = (x_D)_{max}$ and $y_P = (y_D)_{max}$. Figures 10.4 and 10.5 show plots of $(y_D)_{max}$ and $(x_D)_{max}$, respectively, against m as given by (10.53) and (10.50), respectively, for positive m except for $m = 1$. The $m = 1$ values of $(x_D)_{max}$ and $(y_D)_{max}$, equal to $1/\ln 2$ and $1/e$, respectively, were calculated from (10.51) to (10.54), respectively.

10.5 Equilibria in Parent–Daughter Activities

In many Parent P $\to$ Daughter D $\to$ Granddaughter G relationships after a certain time t the parent and daughter activities reach a constant ratio that is independent of a further increase in time t. This condition is referred to as *radioactive equilibrium* and

Fig. 10.3 Parameter $(y_D)_{max}$ against parameter $(x_D)_{max}$, as given in (10.53)

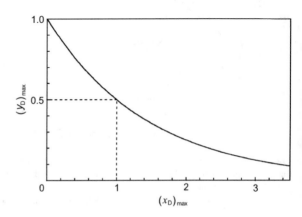

Fig. 10.4 Parameter $(y_D)_{max}$ against decay factor m calculated from (10.53) for all $m > 0$ except for $m = 1$. The value of $(y_D)_{max}$ for $m = 1$ is calculated from (10.54)

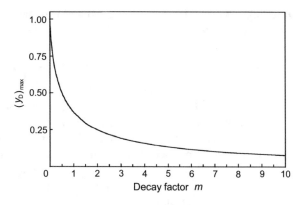

Fig. 10.5 Parameter $(x_D)_{max}$ against decay factor m calculated from (10.50) for all $m > 0$ except for $m = 1$. The value of $(x_D)_{max}$ for $m = 1$ is calculated from (10.51)

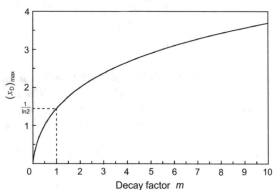

can be analyzed further by examining the behavior of the activity ratio $\mathcal{A}_D(t)/\mathcal{A}_P(t)$ obtained from (10.35) as

$$\frac{\mathcal{A}_D(t)}{\mathcal{A}_P(t)} = \frac{\lambda_D}{\lambda_D - \lambda_P}\{1 - e^{-(\lambda_D - \lambda_P)t}\} = \frac{1}{1 - \dfrac{\lambda_P}{\lambda_D}}\{1 - e^{-(\lambda_D - \lambda_P)t}\}$$

$$= \frac{1}{1 - \dfrac{(t_{1/2})_D}{(t_{1/2})_P}}\left\{1 - e^{-(\ln 2)\left[\dfrac{1}{(t_{1/2})_D} - \dfrac{1}{(t_{1/2})_P}\right]}\right\} \tag{10.55}$$

for the two initial conditions:

1. $\mathcal{A}_P(t = 0) = \mathcal{A}_P(0) = \lambda_P N_P(0),$ (10.56)
2. $\mathcal{A}_D(t = 0) = \mathcal{A}_D(0) = 0.$ (10.57)

Inserting the decay factor m of (10.44) and variable x of (10.41) into (10.55) and defining parameter ξ as $\xi = \mathcal{A}_D(t)/\mathcal{A}_P(t) = y_D/y_P$, we write (10.55) as follows

$$\xi(x) = \frac{\mathcal{A}_D(t)}{\mathcal{A}_P(t)} = \frac{1}{1-m}\{1 - e^{-(\ln 2)\frac{1-m}{m}x}\} \equiv \frac{1}{1-m}\{1 - 2^{\frac{m-1}{m}x}\}. \qquad (10.58)$$

The $\xi(x)$ expression of (10.58) is valid for all positive m except for $m = 1$ for which it is not defined. However, we can determine the $\xi(x)$ functional relationship for $m = 1$ by applying the L'Hôpital rule to convert the indeterminate form of (10.58) into a determinate form which allows the evaluation of the $m \to 1$ limit

$$\xi(m = 1) = \lim_{m\to 1}\frac{\dfrac{d}{dm}\left\{1 - 2^{\frac{m-1}{m}x}\right\}}{\dfrac{d(1-m)}{dm}} = \lim_{m\to 1}\frac{\left\{-2^{\frac{m-1}{m}x}\ln 2\left[\dfrac{x}{m} - \dfrac{m-1}{m^2}x\right]\right\}}{-1}$$

$$= x \ln 2. \qquad (10.59)$$

Equation (10.59) shows that $\xi(x)$ for $m = 1$ is a linear function of x, as shown in Fig. 10.6 in which we plot $\xi(x)$ for various values of m in the range from 0.1 to 10. The $m = 1$ linear equation actually separates two distinct regions for the variable ξ: (1) region where $m > 1$ and (2) region where $0 < m < 1$.

1. For the $m > 1$ region, we write (10.58) as follows

$$\xi = \frac{1}{m-1}\left\{e^{\frac{m-1}{m}x\ln 2} - 1\right\}. \qquad (10.60)$$

Note that ξ rises exponentially with x, implying that the ratio $\mathcal{A}_D(t)/\mathcal{A}_P(t)$ also increases with time t and thus no equilibrium between $\mathcal{A}_P(t)$ and $\mathcal{A}_D(t)$ will ensue with an increasing time t. The exponential behavior of $\xi(x)$ is clearly shown in Fig. 10.6 with the dashed curves for $m > 1$ in the range $1 < m < 10$.

Fig. 10.6 Variable $\xi = \mathcal{A}_D(t)/\mathcal{A}_P(t) = y_D/y_P$ against normalized time x for several decay factors m in the range from 0.1 to 10 calculated from (10.58) except for $\xi(m = 1)$ which gives a linear function calculated in (10.59)

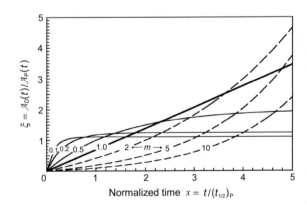

2. For the $0 < m < 1$ region, (10.58) suggests that the exponential term diminishes with increasing x and exponentially approaches zero. This means that at large x the parameter ξ approaches a constant value that is independent of x and is equal to $1/(1-m)$. Under these conditions the parent activity $\mathcal{A}_P(t)$ and daughter activity $\mathcal{A}_D(t)$ are said to be in *transient equilibrium*, and are governed by the following relationship

$$\xi = \frac{\mathcal{A}_D(t)}{\mathcal{A}_P(t)} = \frac{y_D}{y_P} = \frac{1}{1-m} = \frac{1}{1 - \dfrac{\lambda_P}{\lambda_D}} = \frac{\lambda_D}{\lambda_D - \lambda_P}. \qquad (10.61)$$

3. After initially increasing, the daughter activity $\mathcal{A}_D(t)$ goes through a maximum and then decreases at the same rate as the parent activity $\mathcal{A}_P(t)$ and the two activities are related through (10.61). As m decreases the daughter and parent activities at relatively large times t become increasingly more similar, since, as $m \to 0, \xi \to 1$. This represents a special case of transient equilibrium ($\lambda_D \gg \lambda_P$, i.e., $m \to 0$) and in this case the parent and daughter are said to be in *secular equilibrium*. Since in secular equilibrium $\xi = 1$, the parent and daughter activities are approximately equal, i.e., $\mathcal{A}_P(t) \approx \mathcal{A}_D(t)$ and the daughter decays with the same rate as the parent.

Equations (10.55) and (10.59) are valid in general, irrespective of the relative magnitudes of λ_P and λ_D; however, as indicated above, the ratio $\mathcal{A}_D(t)/\mathcal{A}_P(t)$ falls into four distinct categories that are clearly defined by the relative magnitudes of λ_P and λ_D. The four categories are discussed below.

10.5.1 Daughter Longer-Lived Than Parent

Half-life of the daughter exceeds that of the parent: $(t_{1/2})_D > (t_{1/2})_P$ or $\lambda_D < \lambda_P$. We write the ratio $\mathcal{A}_D(t)/\mathcal{A}_P(t)$ of (10.55) as follows

$$\xi(x) = \frac{\mathcal{A}_D(t)}{\mathcal{A}_P(t)} = \frac{\lambda_D}{\lambda_P - \lambda_D}\{e^{(\lambda_P-\lambda_D)t} - 1\}. \qquad (10.62)$$

Since the decay factor m exceeds 1, we note that the ratio $\mathcal{A}_D/\mathcal{A}_P$ increases exponentially with time t. Therefore, no equilibrium between the parent activity $\mathcal{A}_P(t)$ and the daughter activity $\mathcal{A}_D(t)$ will be reached for any time t.

10.5.2 Equal Half-Lives of Parent and Daughter

Half-lives of parent and daughter are equal: $(t_{1/2})_D = (t_{1/2})_P$ or $\lambda_D = \lambda_P$.

The condition is mainly of theoretical interest as no such example has been observed in nature yet. The ratio $A_D(t)/A_P(t)$ is given as a linear function, given in (10.59). As shown in Fig. 10.6, the condition $m = 1$ separates the region of no equilibrium where $m > 1$ from the region of transient and secular equilibrium where $m < 1$.

10.5.3   Daughter Shorter-Lived Than Parent: Transient Equilibrium

Half-life of the daughter is shorter from that of parent: $(t_{1/2})_D < (t_{1/2})_P$ or $\lambda_D > \lambda_P$.

The activity ratio $A_D(t)/A_P(t)$ at large t becomes a constant equal to $\lambda_D/(\lambda_D - \lambda_P)$ and is then independent of t and larger than unity, i.e.,

$$\frac{A_D(t)}{A_P(t)} \approx \frac{\lambda_D}{\lambda_D - \lambda_P} = \text{const} > 1. \tag{10.63}$$

The constancy of the ratio $A_D(t)/A_P(t)$ at large t implies a *transient equilibrium* between $A_P(t)$ and $A_D(t)$. The ratio $A_D(t)/A_P(t)$ of (10.55) can be written in terms of the characteristic time t_{max} by inserting into (10.55) a new variable $t = nt_{max}$ with t_{max} given in (10.37) to get the following expression for $A_D(t)/A_P(t)$

$$\frac{A_D(t)}{A_P(t)} = \frac{\lambda_D}{\lambda_D - \lambda_P}\{1 - e^{-(\lambda_D - \lambda_P)nt_{max}}\} = \frac{\lambda_D}{\lambda_D - \lambda_P}\{1 - e^{-n \ln \frac{\lambda_D}{\lambda_P}}\}$$

$$= \frac{\lambda_D}{\lambda_D - \lambda_P}\left\{1 - \left(\frac{\lambda_P}{\lambda_D}\right)^n\right\}. \tag{10.64}$$

Equation (10.64) allows us to estimate the required value of n to bring the ratio $A_D(t)/A_P(t)$ to within a certain percentage p of the saturation value of $\lambda_D/(\lambda_D - \lambda_P)$ in transient equilibrium. This simply implies that the following relationship must hold

$$\left[\frac{\lambda_P}{\lambda_D}\right]^n = \frac{p}{100}, \tag{10.65}$$

or, after solving for n,

$$n = \frac{\ln \dfrac{100}{p}}{\ln \dfrac{\lambda_D}{\lambda_P}}. \tag{10.66}$$

For example, the activity ratio $\xi = A_D(t)/A_P(t)$ will reach 90%, 98%, 99%, and 99.9% of its saturation value; i.e., p is 10%, 2%, 1%, and 0.1%, respectively, for values of n equal to $2.3/\ln(\lambda_D/\lambda_P)$; $3.9/\ln(\lambda_D/\lambda_P)$; $4.6/\ln(\lambda_D/\lambda_P)$; and $6.9/\ln(\lambda_D/\lambda_P)$.

10.5.4 Daughter Much Shorter-Lived Than Parent: Secular Equilibrium

Half-life of the daughter is much shorter than that of the parent: $(t_{1/2})_D \ll (t_{1/2})_P$ or $\lambda_D \gg \lambda_P$. The ratio of daughter activity $\mathcal{A}_D(t)$ and parent activity $\mathcal{A}_P(t)$, i.e., $\mathcal{A}_D(t)/\mathcal{A}_P(t)$ of (10.55) simplifies to

$$\frac{\mathcal{A}_D(t)}{\mathcal{A}_P(t)} \approx 1 - e^{-\lambda_D t}. \tag{10.67}$$

For relatively large time $t \gg t_{\max}$, (10.67) becomes equal to unity or

$$\frac{\mathcal{A}_D(t)}{\mathcal{A}_P(t)} \approx 1. \tag{10.68}$$

The activity of the daughter $\mathcal{A}_D(t)$ very closely approximates that of its parent $\mathcal{A}_P(t)$, i.e., $\mathcal{A}_D(t) \approx \mathcal{A}_P(t)$, and they decay together at the rate of the parent. This special case of transient equilibrium in which the daughter and parent activities are essentially identical, i.e., $\mathcal{A}_D(t) \approx \mathcal{A}_P(t)$, is called *secular equilibrium*.

10.5.5 Conditions for Parent–Daughter Equilibrium

Figure 10.6 is a plot of $\xi(x)$ the ratio of the daughter activity $\mathcal{A}_D$ to parent activity $\mathcal{A}_P$ against x, the time normalized to the parent half-life $(t_{1/2})_P$. The linear function $x \ln 2$ of (10.59) for $m = 1$, where m is the decay factor of (10.44), clearly separates the equilibrium (transient and secular) region where $m < 1$ from the no-equilibrium region where $m > 1$.

The four regions of the decay factor m are summarized in Table 10.1. The regions with $m \to 0$ and $0 < m < 1$ provide secular and transient equilibrium, respectively, between the parent and daughter activities; the regions for $m = 1$ and $m > 1$ provide no equilibrium between the parent and daughter activities.

An example of transient equilibrium of importance in medical physics is the beta minus decay of molybdenum-99 to technetium-99m (Tc-99m) with a half-life of 66 h in comparison to the half-life of the Tc-99m isomeric state of 6 h. The decay to Tc-99m occurs in 86% of disintegrations, the remaining 14% of disintegrations go to other excited states of Tc-99 which promptly decay through gamma ray emission to ground state. The production and use of Tc-99m for nuclear medicine imaging is discussed in Sect. 12.9.

An excellent example of secular equilibrium is the alpha decay of radium-226 (Ra-226) into radon-222 (Rn-222) discussed in Sect. 11.2.2. The half-life of Ra-226 is 1602 years compared to Rn-222 half-life of 3.82 days.

Table 10.1 Four special regions for the decay factor m between $m = 0$ and $m = \infty$

Decay factor m	Relative value	Equilibrium	Relationship for $\xi = \dfrac{A_D(t)}{A_P(t)}$	
$m \approx 0$	$\lambda_D \gg \lambda_P$	Secular	$\xi = 1$	
$0 < m < 1$	$\lambda_D > \lambda_P$	Transient	$\xi = \dfrac{1}{1-m} = \dfrac{\lambda_D}{\lambda_D - \lambda_P}$	(10.70)
$m = 1$	$\lambda_D = \lambda_P$	No	$\xi = \dfrac{t \ln 2}{(t_{1/2})_P}$	(10.71)
$m > 1$	$\lambda_D < \lambda_P$	No	$\xi = \dfrac{1}{m-1}\left\{ e^{\frac{m-1}{m}\frac{t \ln 2}{(t_{1/2})_P}} - 1 \right\}$	(10.72)

The region where $m \to 0$ results in secular equilibrium between the parent and daughter activities and the region where $0 < m < 1$ results in transient equilibrium between the parent and daughter activities. Regions where $m = 1$ and $m > 1$ do not result in equilibrium between the parent and daughter activities

10.6 Bateman Equations for Radioactive Decay Chain

The laws of spontaneous radioactive decay are independent of the radiation emitted in the radioactive decay process. Since the number of radioactive nuclides in a radioactive sample is in general very large, it can be treated as a continuous variable and its behavior can be evaluated with standard methods of calculus. The exponential laws of radioactive decay were first formulated by Ernest Rutherford and Frederick Soddy in 1902 to explain the results of their experiments on the thorium series of radionuclides.

The P $\to$ D $\to$ G radioactive decay chain discussed in Sect. 10.3 can be extended to a general chain of decaying nuclei with an arbitrary number of radioactive chain links by using equations proposed by U.K. mathematician *Harry Bateman* in 1910. The general radioactive chain is as follows

$$N_1 \to N_2 \to N_3 \to \cdots \to N_{i-1} \to N_i \qquad (10.69)$$

and Bateman's initial condition stipulates that at $t = 0$ only $N_1(0)$ parent nuclei are present, while all other descendent nuclei are not present yet, i.e.,

$$N_2(0) = N_3(0) = \cdots = N_{i-1}(0) = N_i(0) = 0. \qquad (10.73)$$

Bateman equations are given as a set of equations that give the number of atoms $N_i(t)$ of each nuclide of a radioactive decay chain produced after a given time t recognizing that at $t = 0$ (initial condition) only a given number of parent nuclei were present. The number of nuclei $N_i(t)$ is given as follows

$$N_i(t) = C_1 e^{-\lambda_1 t} + C_2 e^{-\lambda_2 t} + C_3 e^{-\lambda_3 t} + \cdots + C_i e^{-\lambda_i t}, \qquad (10.74)$$

where C_i are constants given as follows

$$C_1 = N_1(0) \frac{\lambda_1 \lambda_2 \cdots \lambda_{i-1}}{(\lambda_2 - \lambda_1)(\lambda_3 - \lambda_1) \cdots (\lambda_i - \lambda_1)}, \tag{10.75}$$

$$C_2 = N_1(0) \frac{\lambda_1 \lambda_2 \cdots \lambda_{i-1}}{(\lambda_1 - \lambda_2)(\lambda_3 - \lambda_2) \cdots (\lambda_i - \lambda_2)}, \tag{10.76}$$

$$C_i = N_1(0) \frac{\lambda_1 \lambda_2 \cdots \lambda_{i-1}}{(\lambda_1 - \lambda_i)(\lambda_3 - \lambda_i) \cdots (\lambda_{i-1} - \lambda_i)}. \tag{10.77}$$

10.7 Mixture of Two or More Independently Decaying Radionuclides in a Sample

An unknown mixture of two or more independently decaying radionuclides, each with its own half-life and decay constant, will produce a composite decay curve that does not result in a straight line when plotted on a semi-logarithmic plot, unless, of course, all radionuclides have identical or very similar half-lives. In principle, the decay curves of the individual radionuclides can be resolved graphically, if their half-lives differ sufficiently and if at most three radioactive components are present.

Figure 10.7 illustrates this for a mixture of two radionuclides: nuclide A with short half-life and nuclide B with long half-life. The solid curve represents the measured decay curve (activity) for the mixture with the two components A and B. For large time t, the short-lived component A is essentially gone and the composite activity curve follows the decay of the long-lived radionuclide B.

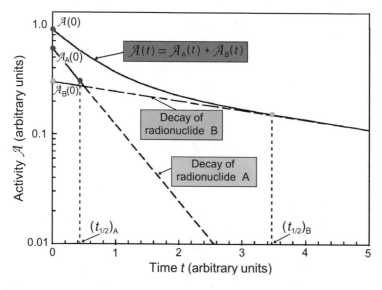

Fig. 10.7 Decay curves shown on a semi-logarithmic plot for a mixture of two radionuclides: short-lived A and long-lived B (*solid curve*). The *dashed lines* are individual decay curves for radionuclides A and B

The initial activities and half-lives of the nuclides A and B can be determined graphically as follows:

1. The first step is to carry out a linear extrapolation to time $t = 0$ of the long-time portion of the composite curve (region where the curve becomes linear on a semi-logarithmic plot). This gives the decay curve for nuclide B and the initial activity of nuclide B at $t = 0$.
2. The second step is to obtain the decay curve for the short-lived component A by subtracting the straight-line curve B from the composite curve. This results in another straight line on the semi-logarithmic plot, this time for nuclide B, and gives the initial activity of nuclide A at $t = 0$.
3. Half-lives for components A and B may be determined from the individual linear decay data for radionuclides A and B. The two radionuclides may then be identified through the use of tabulated half-lives for the known natural and artificial radionuclides.
4. Experimental uncertainties in measured data generally preclude handling systems of more than three components, and even only two-component curves may be difficult to resolve if their decay constants differ by less than a factor of two.

10.8   Branching Decay and Branching Fraction

In many instances decay of a radionuclide can proceed by more than one mode of decay; for example beta plus and beta minus decay or alpha and beta decay, etc., and the radionuclide is said to undergo branching decay to two or more different daughter nuclides. Another avenue for branching decay is a parent decaying to different energy states of the same daughter.

Most of branching decays offer a choice of only two branches; however, examples of more than two possible branches are known. In general, the overall decay constant λ_P for the parent decay is the sum of the partial decay constants $(\lambda_P)_i$ for each possible mode of decay i

$$\lambda_P = \sum_{i=1}^{N} (\lambda_P)_i, \qquad (10.78)$$

with N the number of available decay modes.

The ratio between the decay rate for an individual decay mode i to the total decay rate is referred to as f_i, the branching fraction for mode i. The branching fraction can also be defined as the ratio λ_i / λ where λ_i is the partial decay constant for mode i and λ is the overall decay constant. In the common situation of only two possible branches, one often defines the branching ratio as the ratio between the two partial decay constants or the ratio between the two branching fractions for the two decay modes.

An *example of branching decay into two different daughter nuclides*, important for medical physics, is discussed in Sect. 11.6.4 for iridium-192 (Ir^{192}) that may decay through β^- decay into platinum-192 (Pt-192) with a branching fraction of 95.2% or through electron capture decay into osmium-192 (Os-192) with a branching fraction of 4.8%. The overall decay constant of Ir-192 ($t_{1/2} = 73.83$ d) is 1.0866×10^{-7} s^{-1} resulting in the following partial decay constants:

1. For the beta minus radioactive decay of Ir-102 into Pt-192: $\lambda_{Ir \to Pt} = 0.952\lambda_{Ir\text{-}192}$ $= 1.0344 \times 10^{-7}$ s^{-1}
2. For the beta plus radioactive decay of Ir-192 into Os-192: $\lambda_{Ir \to Os} = 0.048\lambda_{Ir\text{-}192}$ $= 0.0522 \times 10^{-7}$ s^{-1}

An *example of branching decay into different energy states of daughter nucleus* is the production of technetium-99m radionuclide from molybdenum-99 for nuclear medicine imaging. The parent radionuclide Mo-99 undergoes branching beta minus decay with half-life of $t_{1/2} = 65.94$ h and decay constant $\lambda = 1.2166 \times 10^{-7}$ s^{-1} into:

1. Metastable Tc-99m with a branching fraction of 86% resulting in a partial decay constant $\lambda_{Mo \to Tc\text{-}99m} = 0.86\lambda_{Mo} = 1.0463 \times 10^{-7}$ s^{-1}
2. Standard excited states of Tc-99 which promptly decay by γ-ray emission into the Tc-99 ground state.

Nuclear Landscape and the Chart of Nuclides

Nuclear data for stable and radioactive nuclides are usually compiled in a graphic form referred to as the "*Chart of Nuclides*" or the *Segrè Chart* in honor of Emilio Segrè who proposed the graphic layout in the 1940s. The chart of nuclides is drawn in such a way that each nuclide is assigned a unique pixel position on a Cartesian diagram in which the proton number Z is plotted on the ordinate (y) axis and the neutron number N on the abscissa (x) axis. A typical example of a Segrè chart is the so-called "Karlsruher Nuklidkarte" (Karlsruhe Chart of the Nuclides) issued by the Joint Research Center of the European Commission in Karlsruhe, Germany.

The figure on next page shows a condensed version of the Karlsruhe Chart of Nuclides including all known nuclides from $Z = 1$ to $Z = 118$. In addition, the figure also shows regions of possible, but to date not yet discovered radionuclides. Pixels representing known as well as unknown yet theoretically feasible nuclides form an elongated island oriented from southwest to northeast, with the stable nuclides (shown with black pixels) forming the backbone (mountain range) in the central part of the island.

The south shore of the island is occupied by known neutron-rich radionuclides and a vast "terra incognita" region formed by not yet discovered neutron-rich radionuclides (shown in dark grey color). The south shore is delineated by the so-called neutron drip line (white line on graph) beyond which neutrons are no longer bound to the nucleus.

The north shore of the island is populated by known proton-rich radionuclides as well as by a small number of yet undiscovered proton-rich radionuclides and delineated by the so-called proton drip line beyond which protons are no longer bound to the nucleus. The north-eastern tip of the island is populated by very heavy nuclides delineated by the spontaneous fission drip line (shown with black curve).

The actual position of the shore (drip lines), especially in the high Z region, is somewhat arbitrary and depends on various theories and a few sketchy experimental data. The position of the neutron and proton drip lines is governed by interplay in the nucleus between the nuclear strong attractive force in effect among all nucleons and the repulsive Coulomb force in effect among protons. The proton drip line is reached when the binding energy of the outermost proton in a nucleus becomes zero, and, similarly, the neutron drip line is reached when the binding energy of the outermost neutron becomes zero. Since neutrons are not affected by the Coulomb force, the neutron drip line is much farther away from the "mountain range" of stable nuclides and the "terra incognita" region for neutron-rich radionuclides is much larger than that for proton-rich radionuclides.

Source of the consolidated Karlsruhe Nuclear Chart: © *European Communities, 2009. Courtesy of Dr. Joseph Magill: Joint Research Centre of the European Commission, Karlsruhe, Germany. The "Karlsruher Nuklidkarte" is available on-line at*: www.nucleonica.net.

Chapter 11
Modes of Radioactive Decay

Henri Becquerel's discovery of natural radioactivity in 1896 opened a whole new world of physics and introduced new and exciting opportunities for physics research that eventually developed into important branches of modern physics such as nuclear physics and particle physics. While the early investigators explained the macroscopic kinetics of radioactive decay soon after 1896 starting with the work of Marie Skłodowska-Curie and Pierre Curie, Ernest Rutherford and Frederick Soddy, it took several decades until the various radioactive decay modes were fully understood on a microscopic scale.

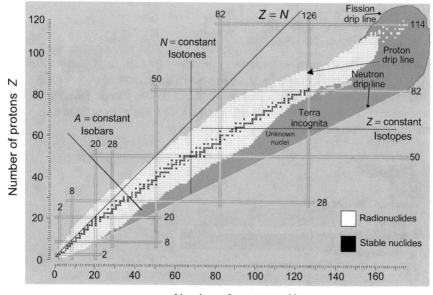

© Springer International Publishing Switzerland 2016
E.B. Podgoršak, *Radiation Physics for Medical Physicists*,
Graduate Texts in Physics, DOI 10.1007/978-3-319-25382-4_11

In this chapter the various radioactive decay modes are presented with a special emphasis on specific aspects of radioactive decay that are of importance to medical physics. In addition to standard modes of radioactive decay, such as alpha, beta and gamma decay, the chapter also includes proton and neutron decay as well as spontaneous fission as interesting examples of spontaneous decay despite their limited importance to medical physics. The chapter concludes with a discussion of the Segrè chart of the nuclides which presents an orderly catalog of all known stable as well as radioactive nuclear species, provides useful basic data for all known nuclides, and indicates the possible decay paths for radionuclides.

11.1   Introduction to Radioactive Decay Processes

Radioactive nuclides, either naturally occurring or artificially produced by nuclear reactions, are unstable and strive to reach more stable nuclear configurations through various processes of spontaneous radioactive decay that involve transformation to a more stable nuclide and emission of energetic particles. General aspects of spontaneous radioactive decay may be discussed using the formalism based on the definitions of activity $\mathcal{A}$ and decay constant λ without regard for the actual microscopic processes that underlie the radioactive disintegrations. A closer look at radioactive decay processes shows that they are divided into six main categories:

1. Alpha (α) decay
2. Beta (β) decay
3. Gamma (γ) decay
4. Spontaneous fission (SF)
5. Proton emission (PE) decay
6. Neutron emission (NE) decay

β decay actually encompasses three decay processes (β^-, β^+, and electron capture) and γ decay encompasses two (γ decay and internal conversion).

There are many spontaneous radioactive decay modes that an unstable nucleus may undergo in its quest for reaching a more stable nuclear configuration. On a microscopic scale the nine most important modes are:

1. α decay
2. β^- decay
3. β^+ decay
4. Electron capture (EC)
5. γ decay
6. Internal conversion (IC)
7. Spontaneous fission (SF)
8. Proton emission (PE) decay
9. Neutron emission (NE) decay

Nuclear transformations are usually accompanied by emission of energetic particles (charged particles, neutral particles, photons, etc.). The particles released in the various decay modes are as follows:

- Alpha (α) particles in α decay,
- Electrons in β^- decay,
- Positrons in β^+ decay,
- Neutrinos in β^+ decay,
- Antineutrinos in β^- decay,
- γ-rays in γ decay,
- Atomic orbital electrons in internal conversion,
- Neutrons in spontaneous fission and in neutron emission decay,
- Heavier nuclei in spontaneous fission,
- Protons in proton emission decay.

In each nuclear transformation a number of physical quantities must be conserved. The most important of these quantities are:

1. *Total energy*
2. *Momentum*
3. *Charge*
4. *Atomic number*
5. *Atomic mass number (number of nucleons).*

The total energy of particles released by the transformation process is equal to the net decrease in the rest energy of the neutral atom, from parent P to daughter D. The disintegration (decay) energy, often referred to as the Q value of the radioactive decay, is defined as follows

$$Q = \{M(\mathrm{P}) - [M(\mathrm{D}) + m]\}\, c^2, \tag{11.1}$$

where $M(\mathrm{P})$, $M(\mathrm{D})$, and m are the nuclear rest masses (in unified atomic mass units u) of the parent, daughter, and emitted particles, respectively. The energy equivalent of u is 931.5 MeV.

Often atomic masses rather than nuclear masses are used in calculations of Q values of radioactive decay. In many decay modes the electron masses cancel out, so that it makes no difference if atomic or nuclear masses are used in (11.1). On the other hand, there are situations where electron masses do not cancel out (e.g., in β^+ decay) and there special care must be taken to account for all electrons involved when atomic rest masses are used in (11.1).

For radioactive decay to be energetically possible the Q value must be greater than zero. This means that spontaneous radioactive decay processes release energy and are called exoergic or exothermic. For $Q > 0$, the energy equivalent of the Q value is shared as kinetic energy between the particles emitted in the decay process and the daughter product. Since the daughter has a much larger mass than the other emitted particles, the kinetic energy acquired by the daughter is usually negligibly small.

In light (low atomic number Z) elements nuclear stability is achieved when the number of neutrons N and the number of protons Z is approximately equal ($N \approx Z$). As the atomic number Z increases, the N/Z ratio for stable nuclei increases from 1 at low Z elements to about 1.5 for heavy stable elements.

- If a nucleus has a N/Z ratio too high for nuclear stability, it has an excess number of neutrons and is called neutron-rich. It decays through conversion of a neutron into a proton and emits an electron and anti-neutrino. This process is referred to as β^- decay. If the N/Z ratio is extremely high, a direct emission of a neutron is possible.
- If a nucleus has a N/Z ratio that is too low for nuclear stability, it has an excess number of protons and is called proton-rich. It decays through conversion of a proton into a neutron and emits a positron and a neutrino (β^+ decay). Alternatively, the nucleus may capture an orbital electron, transform a proton into a neutron and emit a neutrino (electron capture). A direct emission of a proton is also possible, but less likely, unless the nuclear imbalance is very high.

11.2 Alpha Decay

Alpha (α) decay was the first mode of radioactive decay detected and investigated during the 1890s. It played a very important role in early modern physics experiments that led to the currently accepted Rutherford–Bohr atomic model (see Chaps. 2 and 3) and is characterized by a nuclear transformation in which an unstable parent nucleus P attains a more stable nuclear configuration (daughter D) through ejection of an α-particle. This α-particle is a helium-4 nucleus that has, with a binding energy of 7 MeV/nucleon, a very stable configuration.

While α decay was well known since the discovery of natural radioactivity by *Henri Becquerel* in 1896 and α-particles were already used as nuclear probes by *Hans Geiger* and *Ernest Marsden* in 1909, its exact nature was finally unraveled much later in 1928 by *George Gamow*.

In α decay the number of protons and neutrons is conserved by producing a $^4_2$He nucleus (α-particle) and lowering the parent's A and Z by 4 and 2, respectively, i.e.,

$$^A_Z P \rightarrow {}^{A-4}_{Z-2}D + {}^4_2He = {}^{A-4}_{Z-2}D + \alpha. \tag{11.2}$$

- When an α-particle is emitted by the radioactive parent (Z, A) nucleus, the atomic number Z of the parent decreases by 2 and it sheds two orbital electrons from its outermost shell to become a neutral daughter atom ($Z - 2, A - 4$).
- The energetic α-particle slows down in moving through the absorber medium and captures two electrons from its surroundings to become a neutral $^4_2$He atom.
- Typical kinetic energies of α-particles released by naturally occurring radionuclides are between 4 MeV and 9 MeV, corresponding to a range in air of about 1 cm to 10 cm, respectively, and in tissue of about 10^{-3} cm and 10^{-2} cm, respectively.

The Coulomb barrier that an α-particle experiences on the surface of the parent nucleus is of the order of 30 MeV; thus, classically, an α-particle with a kinetic energy of few MeV cannot overcome the barrier. However, the quantum mechanical effect of tunneling (see Sect. 1.28.1) gives the α-particle a certain finite probability for tunneling through the potential barrier and escaping the parent nucleus P that transforms into the daughter nucleus D. Thus, positive decay energy Q_α and the quantum mechanical effect of tunneling make the α decay possible.

11.2.1 Decay Energy in Alpha Decay

The decay energy Q_α released in α decay appears as kinetic energy shared between the α-particle and the daughter nucleus and is given as follows

$$Q_\alpha = \left\{ \mathcal{M}(P) - \left[\mathcal{M}(D) + \mathcal{M}\left({}_2^4He\right) \right] \right\} c^2$$
$$= \{ M(P) - [M(D) + m_\alpha] \} c^2, \tag{11.3}$$

where $\mathcal{M}(P)$, $\mathcal{M}(D)$, and $\mathcal{M}({}_2^4He)$ are the *atomic rest masses* and $M(P)$, $M(D)$ and m_α are the *nuclear rest masses* of the parent, daughter, and α-particle, respectively.

Since neither the total number of protons nor the total number of neutrons changes in the α decay, Q_α can also be expressed in terms of binding energies E_B of the parent, daughter, and helium nucleus, as follows

$$Q_\alpha = E_B(D) + E_B(\alpha) - E_B(P), \tag{11.4}$$

where

$E_B(D)$ is the total binding energy of the daughter D nucleus.
$E_B(\alpha)$ is the total binding energy of the α-particle (28.3 MeV).
$E_B(P)$ is the total binding energy of the parent P nucleus.

The definition of nuclear binding energy E_B is given in (1.25). For α decay to be possible, Q_α must be positive. This implies that the combined total binding energies of the daughter nucleus and the α-particle nucleus must exceed the total binding energy of the parent nucleus. Or, similarly, this implies that the rest mass of the parent nucleus $M(P)$ must exceed the combined rest masses of the daughter nucleus $M(D)$ and the α-particle m_α.

Two entities are produced in α decay: the α-particle and the daughter product. For decay of the parent nucleus at rest this implies that the α-particle and the daughter will acquire momenta p equal in magnitude but opposite in direction and kinetic energies equal to $(E_K)_\alpha = p^2 / (2m_\alpha)$ for the α-particle and $(E_K)_D = p^2 / [2M(D)]$ for the daughter.

The most prominent general features of α decay are as follows:

- α decay occurs commonly in nuclei with $Z > 82$ because, in this range of atomic number Z, decay energies Q_α given by (11.3) or (11.4) are positive and of the order of $\sim$4 MeV to $\sim$9 MeV.
- The $Q_\alpha > 0$ results mainly from the high total binding energy of the $_2^4$He nucleus (28.3 MeV) that is significantly higher than for nuclei of $_2^3$He (helion), $_1^3$H (triton), $_1^2$H (deuteron) for which spontaneous ejection from parent nuclei energetically is not feasible.
- Ejection of a heavy nucleus from the parent nucleus is energetically possible (large Q value); however, the effect of tunneling through the potential barrier is then also much more difficult for the heavy nucleus in comparison with tunneling for the α-particle.
- Emission of heavy particles from parent nuclei with $Z > 92$ is possible and represents a mode of radioactive decay referred to as *spontaneous fission*. This process competes with the α decay and is discussed in Sect. 11.9.

The total decay energy Q_α must be positive for α decay to occur and is written as follows

$$Q_\alpha = (E_K)_\alpha + (E_K)_D = \frac{p^2}{2m_\alpha} + \frac{p^2}{2M(D)} = \frac{p^2}{2m_\alpha}\left\{1 + \frac{m_\alpha}{M(D)}\right\}$$

$$= (E_K)_\alpha\left\{1 + \frac{m_\alpha}{M(D)}\right\}. \tag{11.5}$$

Since $m_\alpha \ll M(D)$, the α-particle recoils with a much higher kinetic energy than the daughter. Thus, the α-particle acquires a much larger fraction of the total disintegration energy Q_α than does the daughter.

From (11.5) we determine $(E_K)_\alpha$, the kinetic energy of the α-particle, as

$$(E_K)_\alpha = \frac{Q_\alpha}{1 + \dfrac{m_\alpha}{M(D)}}. \tag{11.6}$$

After inserting Q_α from (11.3) we get

$$(E_K)_\alpha = \frac{M(P)c^2 - M(D)c^2 - m_\alpha c^2}{1 + \dfrac{m_\alpha}{M(D)}}$$

$$\approx \left[M(P)c^2 - M(D)c^2 - m_\alpha c^2\right]\left\{\frac{A_P - 4}{A_P}\right\}$$

$$= Q_\alpha\left\{\frac{A_P - 4}{A_P}\right\} \approx Q_\alpha\left\{1 - \frac{4}{A_P}\right\}, \tag{11.7}$$

where A_P is the atomic mass number of the parent nucleus; $(A_P - 4)$ is the atomic mass number of the daughter nucleus; and $m_\alpha/M(D) \approx 4/(A_P - 4)$.

The kinetic energy $(E_K)_D$ of the recoil daughter nucleus, on the other hand, is given as follows

$$(E_K)_D = Q_\alpha - (E_K)_\alpha = Q_\alpha - Q_\alpha \left\{ 1 - \frac{4}{A_P} \right\} = 4\frac{Q_\alpha}{A_P}. \qquad (11.8)$$

11.2.2 Alpha Decay of Radium-226 into Radon-222

For historical reasons, the most important example of radioactive decay in general and α decay in particular is the decay of radium-226 with a half-life of 1602 years into radon-222 which in itself is radioactive and decays by α decay into polonium-218 with a half-life of 3.824 days:

$$^{226}_{88}\text{Ra} \rightarrow {}^{222}_{86}\text{Rn} + \alpha. \qquad (11.9)$$

Radium-226 is the sixth member of the naturally occurring uranium series starting with uranium-238 and ending with stable lead-206. It was discovered in 1898 by *Marie Skłodowska-Curie* and *Pierre Curie* and was used for therapeutic purposes almost immediately after its discovery, either as an external (sealed) source of radiation or as an internal (open) source.

The external use of radium-226 and radon-222 focused largely on treatment of malignant disease. In contrast, internal use of these two radionuclides was spread over the whole spectrum of human disease between 1905 through the 1930s and was based on ingestion of soluble radium salts, inhalation of radon gas, or drinking water charged with radon.

When radium-226 is used as a sealed source, the radon-222 gas cannot escape and a build up occurs of the seven daughter products that form the radium-226 series. Some of these radionuclides undergo α decay, others undergo β decay with or without emission of γ-rays. The γ-ray spectrum consists of discrete lines ranging in energy from 0.18 MeV to 2.2 MeV producing a photon beam with an effective energy close to that of cobalt-60 ($\sim$1.25 MeV). The encapsulation of the source is thick enough to absorb all α- and β-particles emitted by radium-226 and its progeny; however, the encapsulation cannot stop the γ-rays and this makes radium-226 sealed sources useful in treatment of cancer with radiation.

Before the advent of cobalt-60 and cesium-137 teletherapy machines in 1950s all radionuclide based external beam radiotherapy machines made use of radium-226. They were called *teleradium* machines, contained up to 10 g of radium-226 and were very expensive because of the tedious radium-226 manufacturing process. They were also very inefficient because of the low inherent specific activity of radium-226 ($0.988 \text{ Ci}\cdot\text{g}^{-1} = 3.665 \times 10^{10} \text{ Bq}\cdot\text{s}^{-1}$) and self-absorption of γ radiation in the source.

Widespread availability of external beam radiotherapy only started in the 1950s with the invention of the cobalt-60 teletherapy machine in Canada. On the other hand, radium-226 proved very practical for use in brachytherapy where sources are placed into body cavities or directly implanted into malignant lesions for a specific time. While radium-based brachytherapy was very popular in the past century, modern brachytherapy is now carried out with other radionuclides (e.g., iridium-192, cesium-137, iodine-125, etc.) that do not pose safety hazards associated with the radon-222 gas that may leak through damaged radium-226 source encapsulation.

The decay energy Q_α for the α decay of radium-226 is calculated either (1) using atomic rest masses $\mathcal{M}$, as shown in (11.3), or (2) using nuclear rest masses M, as also shown in (11.3), or (3) nuclear binding energies E_B, as given in (11.4). All required nuclear and atomic data are provided in Appendix A.

1. Decay energy Q_α with appropriate atomic rest masses $\mathcal{M}$ and (11.3) is given as

$$
\begin{aligned}
Q_\alpha &= \left\{ \mathcal{M}\left(^{226}_{88}\text{Ra}\right) - \mathcal{M}\left(^{222}_{86}\text{Rn}\right) - \mathcal{M}\left(^{4}_{2}\text{He}\right) \right\} c^2 \\
&= (226.025403\ \text{u} - 222.017571\ \text{u} - 4.002603\ \text{u}) \times 931.494\ \text{MeV/u} \\
&= 0.005229\ \text{u} \times 931.494\ \text{MeV/u} = 4.87\ \text{MeV} > 0. \qquad (11.10)
\end{aligned}
$$

2. Decay energy Q_α with appropriate nuclear masses M and (11.3) is calculated as

$$
\begin{aligned}
Q_\alpha &= \left\{ M\left(^{226}_{88}\text{Ra}\right) - M\left(^{222}_{86}\text{Rn}\right) - m_\alpha \right\} c^2 \qquad (11.11) \\
&= (210496.3482 - 206764.0985 - 3727.3791)\ \text{MeV} = 4.87\ \text{MeV} > 0,
\end{aligned}
$$

3. Decay energy Q_α using appropriate binding energies E_B and (11.4) gives the same result as (11.10) and (11.11)

$$
\begin{aligned}
Q_\alpha &= \left\{ E_B\left(^{222}_{86}\text{Rn}\right) + E_B\left(^{4}_{2}\text{He}\right) - E_B\left(^{226}_{88}\text{Ra}\right) \right\} \\
&= (1708.185 + 28.296 - 1731.610)\ \text{MeV} = 4.87\ \text{MeV}. \qquad (11.12)
\end{aligned}
$$

The kinetic energy $(E_K)_\alpha$ of the α-particle is given from (11.7) as

$$
(E_K)_\alpha = Q_\alpha \left(\frac{A_P - 4}{A_P} \right) = 4.87\ \text{MeV}\ \frac{222}{226} = 4.78\ \text{MeV}, \qquad (11.13)
$$

while 0.09 MeV goes into the recoil kinetic energy $(E_K)_D$ of the $^{222}_{86}\text{Rn}$ atom, as calculated from (11.8)

$$
(E_K)_D = Q_\alpha - (E_K)_\alpha = 0.09\ \text{MeV} = \frac{4 Q_\alpha}{A_P} = \frac{4 \times 4.87\ \text{MeV}}{226} = 0.09\ \text{MeV}. \qquad (11.14)
$$

Fig. 11.1 Energy level diagram for the α decay of radium-226 into radon-222. The relative mass-energy levels for the ground states of the two nuclides are calculated from the respective atomic masses of the two radionuclides given in Appendix A

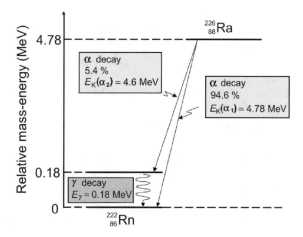

Figure 11.1 shows an energy level diagram for radium-226 decaying through α decay into radon-222. A closer look at the decay scheme of $^{226}_{88}$Ra, shown in Fig. 11.1, paints a slightly more complicated picture than suggested by (11.2) with two α lines emitted; one with $(E_K)_{\alpha_1} = 4.78$ MeV emitted in 94.6% of decays and the other with $(E_K)_{\alpha_2} = 4.60$ MeV emitted in 5.4% of the decays. The 4.78 MeV transition ends at the ground state of $^{222}_{86}$Rn; the 4.60 MeV transition ends at the first excited state of $^{222}_{86}$Rn that instantaneously decays to the ground state through emission of a 0.18 MeV γ-ray (γ decay; see Sect. 11.7).

The decay energy Q_α of 4.87 MeV is thus shared between the α-particle (4.78 MeV) and the recoil daughter (0.09 MeV). The α-particle, because of its relatively small mass in comparison with the daughter mass picks up most of the decay energy; the magnitudes of the momenta for the two decay products are of course equal, and the momenta are opposite in direction to one another.

11.3 Beta Decay

11.3.1 General Aspects of Beta Decay

The term β decay encompasses modes of radioactive decay in which the atomic number Z of the parent nuclide changes by one unit (± 1), while the atomic mass number A remains constant. Thus, the number of nucleons and the total charge are both conserved in the β decay processes and the daughter D can be referred to as an isobar of the parent P.

Three processes fall into the category of β decay:

1. *Beta minus* (β^-) *decay* with the following characteristics: $Z \rightarrow Z+1$; $A = \text{const.}$

$$n \rightarrow p + e^- + \bar{\nu}_e, \qquad\qquad {}^A_Z P \rightarrow {}^A_{Z+1} D + e^- + \bar{\nu}_e. \qquad (11.15)$$

A neutron-rich radioactive nucleus transforms a neutron into proton and ejects an electron e^- and an electron antineutrino $\bar{\nu}_e$. Free (extranuclear) neutrons actually decay into protons through the β^- decay process with a life-time τ of 11.24 min. This decay is possible since the neutron rest mass exceeds that of the proton $(m_n > m_p)$.

2. *Beta plus* $\left(\beta^+\right)$ *decay* with the following characteristics: $Z \rightarrow Z - 1$; $A =$ const.

$$p \rightarrow n + e^+ + \nu_e, \qquad\qquad {}^A_Z P \rightarrow {}^A_{Z-1}D + e^+ + \nu_e. \qquad (11.16)$$

A proton-rich radioactive nucleus transforms a proton into neutron and ejects a positron e^+ and an electron neutrino. Free (extranuclear) protons cannot decay into neutrons through a β^+ decay process because the rest mass of the proton is smaller than that of the neutron $(m_p < m_n)$.

3. *Electron capture* with the following characteristics: $Z \rightarrow Z - 1$; $A =$ const.

$$p + e^- = n + \nu_e, \qquad\qquad {}^A_Z P + e^- = {}^A_{Z-1}D + \nu_e. \qquad (11.17)$$

A proton-rich radioactive nucleus captures an inner shell orbital electron (usually K shell), transforms a proton into a neutron, and ejects a neutrino.

In many cases, β decay of a parent nucleus does not lead directly to the ground state of the daughter nucleus; rather it leads to an *unstable* or even *metastable excited state* of the daughter. The excited state de-excites through emission of γ-rays or through emission of internal conversion electrons (see Sect. 11.8). Of course, the orbital shell vacancies produced by the electron capture or internal conversion process will be followed by emission of discrete characteristic photons or Auger electrons, as is the case with all atomic shell vacancies no matter how they are produced. A detailed discussion of these atomic phenomena is given in Chap. 4. Of course, β decay can only take place when the binding energy of the daughter nucleus E_B (D) exceeds the binding energy of the parent nucleus E_B(P).

11.3.2 Beta Particle Spectrum

For a given β decay, similarly to the situation in α decay, the β-decay energy is uniquely defined by the difference in mass-energy between the parent and daughter nuclei. However, in contrast to the α decay where the energy of the emitted α-particles is also uniquely defined, the β-particles emitted in β decay are not monoenergetic, rather they exhibit a continuous spectral kinetic energy distribution with only the maximum kinetic energy $(E_e)_{max}$ corresponding to the β decay energy.

This apparent contravention of the energy conservation law was puzzling physicists for many years until in 1930 *Wolfgang Pauli* postulated the existence of the neutrino to explain the continuous spectrum of electrons emitted in β decay. In 1934 *Enrico Fermi* expanded on Pauli's neutrino idea and developed a theory of β^- and

β^+ decay. The theory includes the neutrino or the antineutrino as the third particle sharing the available decay energy and momentum with the β-particle and the recoil nucleus. With the emission of a third particle, the neutrino or antineutrino, the momentum and energy can be conserved in β decay.

The neutrino and antineutrino are assumed to have essentially zero rest mass and are moving with the speed of light c. They are also assumed to have only weak interactions with atoms of the absorber and are thus extremely difficult to detect. Their charge is equal to zero. It is obvious that detection of an essentially massless, momentless, uncharged relativistic particle that only experiences weak interactions with matter is extremely difficult. Nonetheless, several techniques were devised to detect the elusive neutrino particle experimentally and thus prove correct Fermi's contention about its existence in β decay. The existence of the neutrino was finally proven experimentally in 1956.

Typical shapes of β^- and β^+ spectra are shown in Fig. 11.2. In general, the spectra exhibit low values at small kinetic energies, reach a maximum at a certain kinetic energy, and then decrease with kinetic energy until they reach zero at a maximum energy $(E_{e^\pm})_{max}$ that corresponds to the β decay energy Q_β, if we neglect the small recoil energy acquired by the daughter nucleus.

The shapes of β^- and β^+ spectra differ at low kinetic energies owing to the charge of the β-particles: electrons in β^- decay are attracted to the nucleus; positrons in β^+ decay are repelled by the nucleus. The charge effects cause an energy shift to lower energies for electrons and to higher energies for positrons, as is clearly shown in Fig. 11.2.

For use in internal dosimetry calculations of β sources the effective energy $(E_e)_{eff}$ of β decay spectra are usually estimated as

$$(E_\beta)_{eff} \approx \frac{1}{3}(E_\beta)_{max}. \tag{11.18}$$

Fig. 11.2 Typical β-particle energy spectra for β^- and β^+ decay normalized to the maximum energy of the β-particle

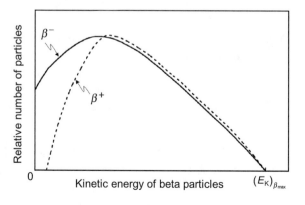

11.3.3 Daughter Recoil in Beta Minus and Beta Plus Decay

In a β^- and β^+ beta decay event the daughter nucleus recoils with a kinetic energy $(E_K)_D$ ranging from 0 to a maximum value.

1. The recoil kinetic energy of the daughter decay product is zero, i.e., $(E_K)_D = 0$, when the electron and antineutrino in β^- decay or positron and neutrino in β^+ decay are emitted with the same momentum but in opposite directions.
2. The maximum recoil kinetic energy $[(E_K)_D]_{max}$ of the daughter nucleus occurs when either one of the two decay particles (electron or antineutrino in β^- decay; positron or neutrino in β^+ decay) is emitted with the maximum available kinetic energy $(E_\beta)_{max}$. The β-decay energy Q_β is expressed as

$$Q_\beta = [(E_K)_D]_{max} + (E_\beta)_{max}. \tag{11.19}$$

The maximum recoil kinetic energy of the daughter $[(E_K)_D]_{max}$ is determined using the laws of energy and momentum conservation and accounting for the relativistic mass changes of the β-particle (electron or positron). A common name for electron and positron in β decay is β-particle.

1. The β-particle momentum $p_{e^\pm} = \gamma m_{e^\pm} \upsilon_{e^\pm}$ is equal to the daughter momentum $p_D = M(D)\upsilon_D$, where $m_{e^\pm}$ and $M(D)$ are the rest masses of the β-particle and daughter nucleus, respectively; $\upsilon_{e^\pm}$ and υ_D are the velocities of the β-particle and the daughter nucleus, respectively; and $\gamma = (1 - \beta^2)^{-1/2}$ with $\beta = \upsilon/c$.
2. Maximum recoil kinetic energy $[(E_K)_D]_{max}$ of the daughter nucleus is calculated classically as $[(E_K)_D]_{max} = \frac{1}{2}M(D)\upsilon_D^2$; the maximum kinetic energy of the β-particle is given relativistically as $(E_\beta)_{max} = (\gamma - 1) m_{e^\pm} c^2$.

Since $p_{e^\pm} = p_D$, we get

$$\upsilon_D = \gamma \frac{m_{e^\pm}}{M(D)} \upsilon_{e^\pm}. \tag{11.20}$$

Inserting (11.20) into the equation for the daughter kinetic energy $[(E_K)_D]_{max} = \frac{1}{2}M(D)\upsilon_D^2$, we obtain

$$[(E_K)_D]_{max} = \frac{M(D)\upsilon_D^2}{2} = \gamma^2 \frac{m_{e^\pm}^2 \upsilon_{e^\pm}^2}{2M(D)} = \gamma^2 \beta^2 \frac{(m_{e^\pm}c^2)^2}{2M(D)c^2} = \frac{\beta^2}{1 - \beta^2} \frac{(m_{e^\pm}c^2)^2}{2M(D)c^2}. \tag{11.21}$$

From the relationship $(E_\beta)_{max} = (\gamma - 1) m_{e^\pm} c^2$ we calculate the expression for $\beta^2/(1 - \beta^2)$ to obtain

$$\frac{\beta^2}{1 - \beta^2} = \frac{2(E_\beta)_{max}}{m_{e^\pm}c^2} + \left\{ \frac{(E_\beta)_{max}}{m_{e^\pm}c^2} \right\}^2. \tag{11.22}$$

Inserting (11.22) into (11.21) we obtain the following expression for maximunm recoil kinetic energy $[(E_K)_D]_{max}$ of the daughter nucleus

$$[(E_K)_D]_{max} = \frac{\beta^2}{1 - \beta^2} \frac{(m_{e^\pm}c^2)^2}{2M(D)c^2} = \frac{m_{e^\pm}c^2}{M(D)c^2}(E_\beta)_{max} + \frac{(E_\beta)^2_{max}}{2M(D)c^2}. \quad (11.23)$$

The daughter recoil kinetic energy $[(E_K)_D]_{max}$ is usually of the order of 10 eV to 100 eV; negligible in comparison to the kinetic energy of the β-particle, yet sufficiently high to be able to cause atomic rearrangements in neighboring molecules in biological materials.

Decay energy Q_β of (11.19) is now given as follows

$$Q_\beta = [(E_K)_D]_{max} + (E_\beta)_{max} = (E_\beta)_{max} \left\{ 1 + \frac{m_{e^\pm}c^2 + \frac{1}{2}(E_\beta)_{max}}{M(D)c^2} \right\}, \quad (11.24)$$

showing that in β^- and β^+ decay by far the larger energy component is the component shared between the β-particle and neutrino, since these two particles in general share the energy $(E_\beta)_{max}$; the recoil kinetic energy given to the daughter is extremely small and may be neglected, unless, of course, we are interested in calculating it, so that we may determine the local damage produced by the daughter atom in biological materials.

11.4 Beta Minus Decay

11.4.1 General Aspects of Beta Minus Decay

Several radionuclides decaying by beta minus (β^-) decay are used in medicine for external beam radiotherapy and brachytherapy. The parent nuclide decays by β^- decay into an excited daughter nuclide that instantaneously or through a metastable decay process decays into its ground state and emits the excitation energy in the form of γ-ray photons. These photons are then used for radiotherapy.

The most important characteristics of radionuclides used in external beam radiotherapy are:

1. *High gamma ray energy*
2. *High specific activity*
3. *Relatively long half-life*
4. *Large specific air-kerma rate constant*

Of the over 3000 natural or artificial radionuclides known, only a few are suitable for use in radiotherapy and of these practically only cobalt-60, with its high photon energy (1.17 MeV and 1.33 MeV), high practical specific activity $(a_{Co} \approx 300$ Ci·g$^{-1})$, and a relatively long half-life (5.26 years), meets the source requirements for external beam radiotherapy.

11.4.2   Beta Minus Decay Energy

The β^- decay can occur to a neutron-rich unstable parent nucleus when the mass $M(Z, A)$ of the parent nucleus exceeds the mass $M(Z+1, A)$ of the daughter nucleus by more than one electron rest mass m_e. The decay energy Q_{β^-} for the β^- decay process is, in terms of *nuclear mass* M, given as

$$Q_{\beta^-} = \{M(Z, A) - [M(Z + 1, A) + m_e]\}\, c^2. \tag{11.25}$$

Adding and subtracting $Zm_e c^2$ to the right-hand side of (11.25) and neglecting the electron binding energies to the nucleus we obtain

$$\begin{aligned} Q_{\beta^-} &= \{M(Z, A) + Zm_e - [M(Z + 1, A) + m_e + Zm_e]\}\, c^2 \\ &= \{\mathcal{M}(Z, A) - \mathcal{M}(Z + 1, A)\}\, c^2, \end{aligned} \tag{11.26}$$

where $\mathcal{M}(Z, A)$ and $\mathcal{M}(Z + 1, A)$ represent the *atomic masses* of the parent and daughter, respectively, noting that

$$\mathcal{M}(Z, A) = M(Z, A) + Zm_e \tag{11.27}$$

and

$$\mathcal{M}(Z + 1, A) = M(Z + 1, A) + (Z + 1)m_e. \tag{11.28}$$

For the β^- decay to occur the atomic mass of the parent $\mathcal{M}(Z, A)$ must exceed the atomic mass of the daughter $\mathcal{M}(Z + 1, A)$; i.e., $\mathcal{M}(Z, A) > \mathcal{M}(Z + 1, A)$.

The atomic rest energy difference between the parent and daughter provides the energy released in a β^- decay event, most generally consisting of:

1. Energy of the emitted electron.
2. Energy of the antineutrino.
3. Energy of the emitted γ-ray photons or conversion electrons with characteristic x-rays and Auger electrons.
4. Recoil kinetic energy of daughter nucleus (small and negligible).

11.4.3   Beta Minus Decay of Free Neutron into Proton

Neutrons, as subatomic particles, along with protons that are also subatomic particles, make up the atomic nucleus held together by the strong force. Neutrons are stable while they reside in a stable nucleus with a balanced number of protons and neutrons. An unstable nucleus with an excessive number of neutrons may undergo several types of nuclear decay such as α decay, β^- decay, neutron emission decay, and spontaneous fission.

A free (extra-nuclear) neutron is not stable; it transforms into a proton p, electron e^-, and electron antineutrino $\bar{\nu}_e$. This spontaneous disintegration of a free neutron n with a half-life of 618 s (10.3 min) is the simplest example of β^-decay

$$n \to p + e^- + \bar{\nu}_e + Q_{\beta^-} (0.7824 \text{ MeV}). \tag{11.29}$$

The β^- process is energetically possible, since it fulfills the requirement which, as spelled out in Sect 11.4.2, states that the parent nucleus rest energy $M(Z, A)c^2$ should exceed the daughter nucleus rest energy $M(Z + 1, A)c^2$ by more than one electron rest energy $m_e c^2$, where for neutron decay $M(Z, A)c^2 = m_n c^2 = 939.5654 \text{ MeV}$, $M(Z + 1, A)c^2 = m_p c^2 = 938.2720 \text{ MeV}$, and $m_e c^2 = 0.5110 \text{ MeV}$.

The data for neutron, proton, and electron rest energies are given in Appendix A and show that the neutron rest energy $m_n c^2$ exceeds the proton rest energy $m_p c^2$ by 1.2934 MeV, an amount larger than the electron rest energy of 0.5110 MeV.

According to (11.25) the decay energy $Q_{\beta^-}^n$ (Q value) for the β^- decay of the neutron is given as follows

$$Q_{\beta^-}^n = m_n c^2 - m_p c^2 - m_e c^2$$
$$= 939.5654 \text{ MeV} - 938.2720 \text{ MeV} - 0.5110 \text{ MeV} = 0.7824 \text{ MeV}. \tag{11.30}$$

Decay energy $Q_{\beta^-}^n$ of (11.30) will be shared between the three particles produced in the β^- decay. As given in (11.19), the maximum recoil kinetic energy of the proton $[(E_K)_p]_{max}$ will occur when either the electron e^- or the electron antineutrino $\bar{\nu}_e$ is emitted with the maximum kinetic energy $(E_\beta)_{max}$. According to (11.19) and (11.24) we can write the neutron decay energy $Q_{\beta^-}^n$ in the form of a quadratic equation for $(E_\beta)_{max}$ as

$$Q_{\beta^-}^n = [(E_K)_p]_{max} + (E_\beta)_{max} = (E_\beta)_{max}\left[1 + \frac{m_e c^2}{m_p c^2}\right] + \frac{[(E_\beta)_{max}]^2}{2 m_p c^2}, \tag{11.31}$$

that has the following physically relevant solution

$$(E_\beta)_{max} = m_p c^2 \left\{-\left[1 + \frac{m_e c^2}{m_p c^2}\right] + \sqrt{\left[1 + \frac{m_e c^2}{m_p c^2}\right]^2 + \frac{2 Q_{\beta^-}^n}{m_p c^2}}\right\}$$

$$= (938.2720 \text{ MeV})\left\{-\left[1 + \frac{0.5110}{938.2720}\right]\right.$$

$$\left. + \sqrt{\left[1 + \frac{0.5110}{938.2720}\right]^2 + \frac{2 \times 0.7824}{938.2720}}\right\}$$

$$= 0.7817 \text{ MeV}. \tag{11.32}$$

Even in the event where the electron receives the maximum possible kinetic energy $(E_\beta)_{max} = 0.7817$ MeV, the maximum recoil energy of the proton $(E_K)_p^{max}$ in β^- decay of a free neutron is very small because the proton is much heavier than the electron $(m_p c^2/(m_e c^2) = 1836)$. Thus, the decay energy of 0.7824 MeV is shared between the electron $(E_\beta)_{max} = 0.7817$ MeV and the proton $(E_K)_p^{max} = Q_{\beta^-}^n - (E_\beta)_{max} \approx 7 \times 10^{-4}$ MeV.

Equation (11.32) can be simplified by recognizing that the term $2Q_{\beta^-}^n / (m_p c^2)$ is very small. Applying Taylor expansion to the square root term of (11.32)

$$
\begin{aligned}
(E_\beta)_{max} &= m_p c^2 \left\{ -\left[1 + \frac{m_e c^2}{m_p c^2}\right] + \sqrt{\left[1 + \frac{m_e c^2}{m_p c^2}\right]^2 + \frac{2Q_{\beta^-}^n}{m_p c^2}} \right\} \\
&= m_p c^2 \left\{ \left[1 + \frac{m_e c^2}{m_p c^2}\right] \left[-1 + \sqrt{1 + \frac{2Q_{\beta^-}^n}{m_p c^2}\left[1 + \frac{m_e c^2}{m_p c^2}\right]^{-2}}}\right] \right\} \\
&\approx m_p c^2 \left\{ \left[1 + \frac{m_e c^2}{m_p c^2}\right] \left[-1 + 1 + \frac{Q_{\beta^-}^n}{m_p c^2}\left[1 + \frac{m_e c^2}{m_p c^2}\right]^{-2}\right] \right\} \\
&= Q_{\beta^-}^n \left[1 + \frac{m_e c^2}{m_p c^2}\right]^{-1} = 0.7824 \left[1 + \frac{0.511}{738.272}\right] \text{ MeV} = 0.7819 \text{ MeV}
\end{aligned}
$$

(11.33)

results in a simpler equation than (11.32) and gives a similar maximum electron energy $(E_\beta)_{max}$ to that of (11.32).

11.4.4 Beta Minus Decay of Cobalt-60 Into Nickel-60

For medical physics an important β^- decay example is the decay of unstable cobalt-60 radionuclide with a half-life of 5.26 years into an excited nickel-60 nuclide that decays instantaneously into its ground state with emission of two γ-ray photons of energies 1.173 MeV and 1.332 MeV, as shown schematically in Fig. 11.3.

$$
{}_{27}^{60}\text{Co} \rightarrow {}_{28}^{60}\text{Ni} + e^- + \bar{\nu}_e + Q_{\beta^-} \text{ (2.82 MeV)} .
$$

(11.34)

Cobalt-60 is used as a radiation source in teletherapy machines applied for external beam radiotherapy. Typical cobalt-60 source activities are of the order of 200 TBq to 400 TBq. There are several thousand cobalt units in operation around the world and Canada is a major producer of these units and cobalt-60 sources. In the past cobalt-60 was also used in brachytherapy sources, however, its use for this purpose has largely been abandoned with the introduction of iridium-192 sources.

The decay energy Q_{β^-} for the Co-60 β^- decay into Ni-60 is calculated with (11.26) and appropriate atomic mass data given in Appendix A as follows

Fig. 11.3 Decay scheme for the β^- decay of cobalt-60 into nickel-60. The relative mass-energy levels for the ground states of the two nuclides are calculated from atomic masses given in Appendix A

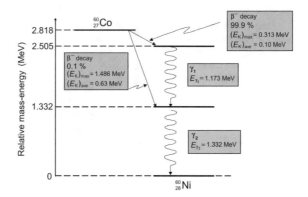

$$Q_{\beta^-} = \left\{ \mathcal{M}\left(^{60}_{27}\text{Co}\right) - \mathcal{M}\left(^{60}_{28}\text{Ni}\right) \right\} c^2$$
$$= \{59.933822\,\text{u} - 59.930791\,\text{u}\}\,931.494\,\text{MeV/u} = 2.82\,\text{MeV}. \quad (11.35)$$

Of course, we can also calculate the decay energy Q_{β^-} for the cobalt-60 β^- decay using (11.25) in conjuction with appropriate data available in Appendix A. The calculated Q_{β^-} of 2.82 MeV is shown in Fig. 11.3 as the energy difference between the ground states of cobalt-60 and nickel-60. There are two β^- decay channels:

1. 99.9% of decays proceed from Co-60 to the second excited state of Ni-60 with maximum and effective electron energy of 0.313 MeV and 0.1 MeV, respectively.
2. Only 0.1% of decays proceed from Co-60 to the first excited state of Ni-60 with maximum and effective electron energy of 1.486 MeV and 0.63 MeV, respectively.

The excited nickel-60 nucleus attains its ground state through emission of γ-ray photons, as discussed further in Sect. 11.7 on γ decay.

11.4.5 Beta Minus Decay of Cesium-137 Into Barium-137

Another example of β^- decay of interest in medical physics is the decay of cesium-137 into barium-137 with a half-life of 30.07 years (see Fig. 11.4):

$$^{137}_{55}\text{Cs} \rightarrow ^{137}_{56}\text{Ba} + \text{e}^- + \overline{\nu}_{\text{e}} + Q_{\beta^-}\,(1.176\,\text{MeV}). \quad (11.36)$$

Decay energy Q_{β^-} for the decay of Cs-137 into Ba-137 by β^- decay is calculated with (11.26) and appropriate atomic mass data from Appendix A as follows

$$Q_{\beta^-} = \left\{ \mathcal{M}\left(^{137}_{55}\text{Cs}\right) - \mathcal{M}\left(^{137}_{56}\text{Ba}\right) \right\} c^2$$
$$= \{136.907084\,\text{u} - 136.905821\,\text{u}\}\,931.494\,\text{MeV/u} = 1.176\,\text{MeV}. \quad (11.37)$$

Fig. 11.4 Decay scheme for β^- decay of cesium-137 into barium-137. The relative mass-energy levels for the ground states of the two nuclides are calculated from atomic masses listed in Appendix A

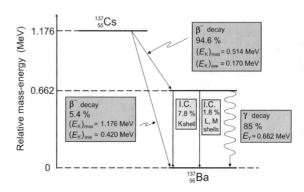

The result of (11.37) can be obtained using (11.25) and appropriate nuclear mass data from Appendix A. The calculated Q_{β^-} of 1.176 MeV is shown in Fig. 11.4 as the energy difference between the ground states of cesium-137 and barium-137. There are two β^- decay channels:

1. 94.6% of β^- decays proceed from Cs-137 to the excited state of Ba-137m with maximum electron energy of 0.514 MeV. The Ba-137m is a metastable state that decays with a 2.552 min half-life to the ground state of Ba-137 with emission of a 0.662 MeV γ-ray photon. The maximum electron energy of 0.514 MeV added to the γ-ray energy of 0.662 MeV results in decay energy of 1.176 MeV, as calculated in (11.37).

2. 5.4% of β^- decays proceed directly from Cs-137 to the ground state of Ba-137 with maximum electron energy of 1.176 MeV.

Cesium-137 has been used in the past as external beam source as well as brachytherapy source. Its use in external beam radiotherapy has been abandoned in favor of more practical cobalt-60 teletherapy and its use in brachytherapy has been abandoned in favor of iridium-192; however, it is still used in blood irradiation machines.

11.5   Beta Plus Decay

11.5.1   General Aspects of the Beta Plus Decay

The beta plus $\left(\beta^+\right)$ decay is characterized by the production of positrons that appear in a spectral distribution with maximum positron energy specific to the particular β^+ decay. As in the β^- decay, the daughter recoil kinetic energy in β^+ decay is essentially negligible. Radionuclides undergoing β^+ decay are often called positron emitters and are used in medicine for functional imaging with a special imaging technique called positron emission tomography (PET). The main characteristics of PET imaging are as follows:

- PET provides information on metabolic function of organs or tissues by detecting how cells process certain compounds such as, for example, glucose. Cancer cells metabolize glucose at a much higher rate than normal tissues. By detecting increased radiolabelled glucose metabolism with a high degree of sensitivity, PET identifies cancerous cells, even at an early stage when other imaging modalities may miss them.
- In a PET study one administers a positron-emitting radionuclide by injection or inhalation. The radionuclide circulates through the bloodstream to reach a particular organ. The positrons emitted by the radionuclide have a very short range in tissue and undergo annihilation with an available atomic orbital electron (see Sect. 7.6.10). This process generally results in emission of two γ photons called annihilation quanta, each with energy of 0.511 MeV, moving away from the point of production in nearly opposite directions.
- The radionuclides used in PET studies are produced by bombardment of an appropriate stable nuclide with protons from a cyclotron (see Sect. 12.10) thereby producing positron-emitting radionuclides that are subsequently attached to clinically useful biological markers. The most commonly used positron emitting radionuclides are: carbon-11, nitrogen-13, oxygen-15, fluorine-18 and rubidium-82.
- Fluorine-18 radionuclide attached to the biological marker deoxyglucose forms the radiopharmaceutical fluorodeoxyglucose (FDG) that is the most commonly used tracer in studies involving glucose metabolism in cancer diagnosis.

11.5.2 Decay Energy in Beta Plus Decay

The β^+ decay can occur to a proton-rich unstable parent nucleus where the mass $M(Z, A)$ of the parent nucleus exceeds the mass $M(Z - 1, A)$ of the daughter nucleus by more than one positron mass m_e. Decay energy Q_{β^+} for the β^+ decay process is given as

$$Q_{\beta^+} = \{M(Z, A) - [M(Z - 1, A) + m_e]\}\, c^2 \qquad (11.38)$$

in terms of *nuclear masses M*.

Adding and subtracting $Zm_e c^2$ to the right-hand side of (11.38) and neglecting the electron binding energies to the nucleus we obtain

$$
\begin{aligned}
Q_{\beta^+} &= \{M(Z, A) + Zm_e - [M(Z - 1, A) + m_e + Zm_e]\}\, c^2 \\
&= \{\mathcal{M}(Z, A) - [\mathcal{M}(Z - 1, A) + 2m_e]\}\, c^2, \qquad (11.39)
\end{aligned}
$$

where $\mathcal{M}(Z, A)$ and $\mathcal{M}(Z - 1, A)$ represent the *atomic masses* of the parent and the daughter, respectively.

We note that the relationship between *atomic* and *nuclear masses* of parent and daughter, ignoring the binding energies of orbital electrons, are

$$\mathcal{M}(Z, A) = M(Z, A) + Zm_e \tag{11.40}$$

and

$$\mathcal{M}(Z - 1, A) = M(Z - 1, A) + (Z - 1)m_e. \tag{11.41}$$

For β^+ decay to occur the atomic mass of the parent $\mathcal{M}(Z, A)$ must exceed the atomic mass of the daughter $\mathcal{M}(Z - 1, A)$ by more than two electron rest masses, or in rest energies

$$\mathcal{M}(Z, A)c^2 > \mathcal{M}(Z - 1, A)c^2 + 2m_e c^2, \tag{11.42}$$

where $m_e c^2$ is the electron rest energy of 0.5110 MeV.

11.5.3 Beta Plus Decay of Nitrogen-13 into Carbon-13

An example for a simple β^+ decay is the decay of nitrogen-13 into carbon-13 with a half-life of 10 min. Nitrogen-13 is a proton-rich radionuclide produced in a cyclotron. The decay scheme is shown in Fig. 11.5 and the basic equation for the decay is as follows

$$^{13}_{7}\text{N} \rightarrow {}^{13}_{6}\text{C} + e^+ + \nu_e + Q_{\beta^+}(1.2 \text{ MeV}). \tag{11.43}$$

The decay energy Q_{β^+} for the β^+ decay of nitrogen-13 into carbon-13 is calculated using (11.39) as follows, with the atomic masses for the two nuclides listed in Appendix A

$$
\begin{aligned}
Q_{\beta^+} &= \left\{ \mathcal{M}(^{13}_{7}\text{N}) - \left[\mathcal{M}(^{13}_{6}\text{C}) + 2m_e \right] \right\} c^2 \\
&= (13.005739 \text{ u} - 13.003355 \text{ u})c^2 - 2m_e c^2 \\
&= 0.002384 \text{ u} \times 931.494 \text{ MeV/u} - 2 \times 0.5110 \text{ MeV} \\
&= 2.221 \text{ MeV} - 1.022 \text{ MeV} = 1.2 \text{ MeV}.
\end{aligned}
\tag{11.44}
$$

The energy difference between the ground state of nitrogen-13 and carbon-13 is 2.22 MeV; however, only 2.22 MeV $- 2m_e c^2 = 1.2$ MeV is available for the maximum energy of the positron. The same result of $Q_{\beta^+} = 1.2$ MeV is obtained if (11.38) is used with appropriate nuclear masses from Appendix A.

Ammonia is the substance that can be labeled with the nitrogen-13 radionuclide for use in functional imaging with positron emission tomography (PET) scanning. The nitrogen-13 labeled ammonia is injected intravenously and is mainly used in cardiac imaging for diagnosis of coronary artery disease and myocardial infarction. It is also occasionally used for liver and brain imaging.

Fig. 11.5 Decay scheme for β^+ decay of nitrogen-13 into carbon-13. The relative mass-energy levels of the ground states of the two nuclides are calculated from atomic masses listed in Appendix A

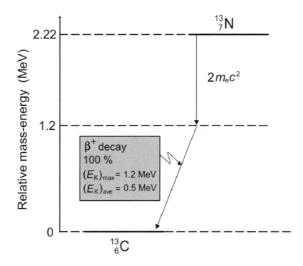

11.5.4 Beta Plus Decay of Fluorine-18 into Oxygen-18

The β^+ decay of fluorine-18 into oxygen-18 with a half-life of 110 min is an important practical example of the β^+ decay. Fluorodeoxyglucose (FDG) labeled with radionuclide fluorine-18 is a sugar compound that can be injected intravenously into a patient for use in positron emission tomography (PET) functional imaging. Based on demonstrated areas of increased glucose metabolism the FDG PET scan:

1. Can detect malignant disease.
2. Can distinguish benign from malignant disease.
3. Can be used for staging of malignant disease.
4. Can be used for monitoring response to therapy of malignant disease.

The decay energy Q_{β^+} for the β^+ decay of fluorine-18 into oxygen-18 is calculated with (11.39) as follows

$$
\begin{aligned}
Q_{\beta^+} &= \left\{ \mathcal{M}(^{18}_9\text{F}) - \left[\mathcal{M}(^{18}_8\text{O}) + 2m_e \right] \right\} c^2 \\
&= (18.000938 \text{ u} - 17.999160 \text{ u})\, c^2 - 2m_e c^2 \qquad (11.45) \\
&= 0.001778 \text{ u} \times 931.494 \text{ MeV/u} - 2 \times 0.511 \text{ MeV} \\
&= 1.656 \text{ MeV} - 1.022 \text{ MeV} = 0.634 \text{ MeV}.
\end{aligned}
$$

The energy difference between the ground states of fluorine-18 and oxygen-18 is 1.66 MeV; however, only 1.66 MeV $- 2m_e c^2 = 0.638$ MeV is available for the maximum energy of the positron, as shown schematically in Fig. 11.6 and in (11.46)

$$
^{18}_9\text{F} \rightarrow \,^{18}_8\text{O} + e^+ + \nu_e + Q_{\beta^+}(0.638 \text{ MeV}). \qquad (11.46)
$$

Fig. 11.6 Decay scheme for β^+ decay of fluorine-18 into oxygen-19. The relative mass- energy levels for the ground states of the two nuclides are calculated from atomic masses listed in Appendix A

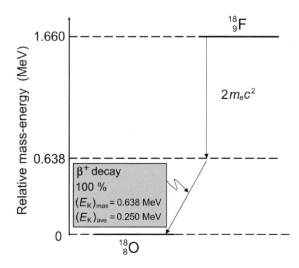

11.6   Electron Capture

11.6.1   Decay Energy in Electron Capture

Electron capture (EC) radioactive decay may occur when an atomic electron ventures inside the nuclear volume, is captured by a proton, and triggers a proton to neutron transformation. Of all atomic electrons, the K-shell electrons have the largest probability for venturing into the nuclear volume and thus contribute most often to the EC decay process. Typical ratios EC(K shell)/EC(L shell) are of the order of 10:1.

Electron capture can occur in proton-rich, unstable parent nuclei, when the mass $M(Z, A)$ of the parent nucleus combined with the mass of one electron m_e exceeds the mass of the daughter nucleus $M(Z - 1, A)$. The decay energy Q_{EC} for electron capture is given as

$$Q_{EC} = \{[M(Z, A) + m_e] - M(Z - 1, A)\}\, c^2$$
$$= \{M(Z, A) - [M(Z - 1, A) - m_e]\}\, c^2 \qquad (11.47)$$

in terms of nuclear masses M. Adding and subtracting $Z m_e$ to the right-hand side of (11.47) and neglecting the electron binding energies to the nucleus we obtain the decay energy Q_{EC} in terms of atomic masses $\mathcal{M}$

$$Q_{EC} = \{\mathcal{M}(Z, A) - \mathcal{M}(Z - 1, A)\}\, c^2. \qquad (11.48)$$

For electron capture to occur, the atomic mass of the parent $\mathcal{M}(Z, A)$ must exceed the atomic mass of the daughter $\mathcal{M}(Z - 1, A)$; i.e., $\mathcal{M}(Z, A) > \mathcal{M}(Z - 1, A)$. The atomic rest energy difference between the parent and the daughter gives the energy released to the neutrino and the daughter atom in an EC radioactive decay event.

Electron capture is a competing process to β^+ decay; however, the conditions on electron capture as far as relative atomic masses of parent and daughter are concerned are less restrictive than those imposed on β^+ decay that results in positron emission and subsequent positron annihilation with emission of annihilation quanta. The condition on EC decay is that the parent atomic mass $\mathcal{M}(P)$ simply exceeds the daughter atomic mass $\mathcal{M}(D)$, while the condition on β^+ decay is that the parent atomic mass exceeds that of the daughter by a minimum of two electron masses.

- When the condition $Q_{EC} > 0$ is satisfied but Q_{β^+} of (11.38) is negative, the β^+ decay will not happen because it is energetically forbidden and EC decay will happen alone.
- When $Q_{\beta^+} > 0$ then Q_{EC} is always positive and both decays (β^+ and EC) can happen. The branching ratios $\lambda_{EC}/\lambda_{\beta^+}$ vary considerably from one radionuclide to another, for example, from a low of 0.03 for fluorine-18 to several hundred for some other proton-rich radionuclides.
- In contrast to β^- and β^+ decay processes in which three decay products share the decay energy and produce a continuous spectral distribution, in the EC decay the two decay products do not have a continuous spectral distribution; rather they are given discrete (monoenergetic) energies. The monoenergetic neutrinos produce a line spectrum with energy E_ν, while the daughter has the recoil kinetic energy $(E_K)_D$, as discussed below.

11.6.2 Recoil Kinetic Energy of Daughter Nucleus in Electron Capture Decay

The recoil kinetic energy $(E_K)_D$ of the daughter nucleus in electron capture decay is determined in two steps:

First, we determine the momenta of the daughter $p_D = M(D)v_D$ and the neutrino $p_\nu = E_\nu/c$. The two momenta are identical in magnitude but opposite in direction, so we can write

$$p_D = M(D)v_D = p_\nu = \frac{E_\nu}{c}, \qquad (11.49)$$

where

E_ν is the neutrino energy.
$M(D)$ is the mass of the daughter nucleus.
v_D is the velocity of the daughter nucleus.

The recoil kinetic energy of the daughter (classically) is given as follows after inserting v_D from (11.49)

$$(E_K)_D = \frac{M(D)v_D^2}{2} = \frac{E_\nu^2}{2M(D)c^2}. \qquad (11.50)$$

The energy available for sharing between the daughter nucleus D and neutrino ν is equal to the electron capture decay energy Q_{EC} decreased by the binding energy E_B of the captured electron

$$Q_{EC} - E_B = E_\nu + (E_K)_D = E_\nu + \frac{E_\nu^2}{2M(D)c^2} \qquad (11.51)$$

or

$$\frac{E_\nu^2}{2M(D)c^2} + E_\nu - (Q_{EC} - E_B) = 0. \qquad (11.52)$$

Equation (11.52) results in the following expression for the energy of the monoenergetic neutrino emitted in electron capture

$$E_\nu = \left\{ -1 + \sqrt{1 + \frac{2(Q_{EC} - E_B)}{M(D)c^2}} \right\} M(D)c^2 \approx Q_{EC} - E_B. \qquad (11.53)$$

In the first approximation the recoil kinetic energy $(E_K)_D$ of the daughter is neglected and so is the binding energy E_B of the captured electron. The energy of the monoenergetic neutrino in electron capture is then approximated by the electron capture decay energy, i.e., $E_\nu \approx Q_{EC}$.

11.6.3 Electron Capture Decay of Beryllium-7 into Lithium-7

An example for EC decay is given in Fig. 11.7 that shows the decay scheme for beryllium-7 decaying through EC into lithium-7. Beryllium-7 has too many protons for nuclear stability, so it achieves better stability by transforming a proton into a neutron. However, it can do so only through EC and not through β^+ decay, because the atomic rest energy of beryllium-7 exceeds that of lithium-7 by only 0.86 MeV and not by a minimum of 1.02 MeV required for β^+ decay to be energetically feasible.

The decay energy for electron capture decay of beryllium-7 into lithium-7 is calculated using (11.48) with appropriate atomic masses $\mathcal{M}$ provided in Appendix A as follows

$$\begin{aligned} Q_{EC} &= \left\{ \mathcal{M}(_4^7\text{Be}) - \mathcal{M}\left(_3^7\text{Li}\right) \right\} c^2 \\ &= (7.016929\ \text{u} - 7.016004\ \text{u})\, c^2 \\ &= 0.000925\ \text{u} \times 931.494\ \text{MeV/u} = 0.862\ \text{MeV}. \end{aligned} \qquad (11.54)$$

Same result is obtained if we use (11.47) with appropriate nuclear masses given in Appendix A.

Fig. 11.7 Decay scheme for electron capture decay of berillium-7 into lithium-7. The relative mass-energy levels for the ground states of the two nuclides are calculated from the respective atomic masses listed in Appendix A

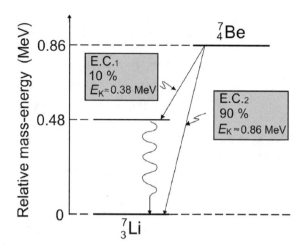

11.6.4 Decay of Iridium-192

Iridium-192 serves as an important radioactive source for use in brachytherapy with remote afterloading techniques. It decays with a half-life of 73.83 days into stable platinum-192 by β^- decay and into stable osmium-192 by electron capture decay. The source is produced through a neutron activation process on iridium-191 in a nuclear reactor (see Sect. 12.6.6). The natural abundance of stable iridium-191 is 37.3% in a mixture with 62.7% of stable iridium-193. The cross section σ for thermal neutron capture in Ir-192 is 954 b.

As shown in Fig. 11.8, iridium-192 has a very complicated γ-ray spectrum with 14 γ energies ranging from 0.2 MeV to $\sim$0.9 MeV and providing effective photon energy of 0.38 MeV. Because of the relatively short half-life, the iridium-192 source requires a source change in remote afterloading machines every 3–4 months.

Fig. 11.8 Decay scheme for decay of iridium-192 into platinum-192 through β^- decay and into osmium-192 through electron capture decay. The relative mass-energy levels for the ground states of the three nuclides are calculated from the respective atomic masses given in Appendix A

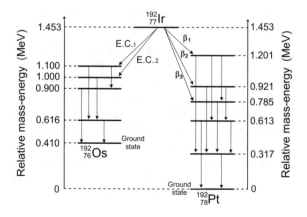

With $\mathcal{M}$ representing atomic masses, the β^- decay energy Q_{β^-} for iridium-192 decaying into platinum-192 is given as follows [see (11.26)]

$$
\begin{aligned}
Q_{\beta^-} &= \left\{ \mathcal{M}(^{192}_{77}\text{Ir}) - \mathcal{M}(^{192}_{78}\text{Pt}) \right\} c^2 \\
&= (191.962602 \text{ u} - 191.961035 \text{ u}) \, c^2 \\
&= 0.001567 \text{ u} \times 931.494 \text{ MeV/u} = 1.453 \text{ MeV} \qquad (11.55)
\end{aligned}
$$

The electron capture decay energy Q_{EC} for iridium-192 decaying into osmium-192 is [see (11.48)]

$$
\begin{aligned}
Q_{EC} &= \left\{ \mathcal{M}(^{192}_{77}\text{Ir}) - \mathcal{M}(^{192}_{76}\text{Os}) \right\} c^2 \\
&= (191.962602 \text{ u} - 191.961479 \text{ u}) \, c^2 \\
&= 0.001123 \text{ u} \times 931.5 \text{ MeV/u} = 1.043 \text{ MeV}. \qquad (11.56)
\end{aligned}
$$

Same results can be obtained for decay energies $Q_{\beta^-} = 1.453$ MeV and $Q_{EC} = 1.043$ MeV of the iridium-192 decay when appropriate nuclear masses are used in conjunction with (11.25) and (11.47), respectively.

11.7 Gamma Decay

11.7.1 General Aspects of Gamma Decay

The α decay as well as the three β decay modes may produce a daughter nucleus in an excited state without expending the full amount of the decay energy available. The daughter nucleus will reach its ground state (i.e., it will de-excite) through one of the following two processes:

1. Emit the excitation energy in the form of a γ photon in a decay process referred to as gamma (γ) decay.
2. Transfer the excitation energy to one of its associated atomic orbital electrons in a process called *internal conversion* (IC).

In most radioactive α or β decays the daughter nucleus de-excitation occurs instantaneously (i.e., within 10^{-12} s), so that we refer to the emitted γ-rays as if they were produced by the parent nucleus. For example, for the cobalt-60 β^- decay into nickel-60, the γ-rays following the β^- decay actually originate from nuclear de-excitations of nickel-60 (see Fig. 11.3), yet, for convenience, we refer to these γ-rays as the cobalt-60 γ-rays. Similarly, we refer to γ photons following the β^- decay of cesium-137 into barium-137m as cesium-137 γ-rays even though the γ photons actually originate from a transition in the barium-137 nucleus (see Fig. 11.4).

In certain α or β decays, the excited daughter nucleus does not immediately decay to its ground state; rather, it de-excites with a time delay of the order of several minutes or several hours:

- The excited state of the daughter is then referred to as a *metastable state* and the process of de-excitation is called an *isomeric transition*.The metastable states are characterized by their own half-lives $t_{1/2}$ and mean (average) lives τ.
- The nucleus in a metastable state is identified with a letter "m" next to the atomic mass number designation (e.g., barium-137m or $^{137m}_{56}$Ba with a half-life of 2.552 min; technetium-99m or $^{99m}_{43}$Tc with a half-life of 6.01 h).
- The term *isomer* is used for designation of nuclei that have the same atomic number Z and same atomic mass number A but differ in energy states.

In addition to α and β decay there are many other modes for producing nuclei in excited states that subsequently undergo γ decay. For example, excited states with energies up to 8 MeV may be produced with neutron capture (n, γ) reactions as well as with other nuclear reactions, such as (p, γ) and (α, γ), etc. Examples of γ-rays following α and β decays are given in Fig. 11.1 for α decay and Figs. 11.3 and 11.4 for β^- decay.

11.7.2 Emission of Gamma Rays in Gamma Decay

In general sense, γ decay stands for nuclear de-excitation either by emission of a γ-ray photon or by internal conversion. In a more narrow sense, γ decay only implies emission of γ photons. The energy of γ-rays emitted by a particular radionuclide is determined by the energy level structure of the radionuclides and can range from a relatively low value of 100 keV up to about 3 MeV.

The γ decay process may be represented as follows

$$^A_Z X^* \rightarrow {}^A_Z X + \gamma + Q_\gamma, \tag{11.57}$$

where $^A_Z X^*$ stands for an excited state of the nucleus $^A_Z X$ and Q_γ is the γ-decay energy.

11.7.3 Gamma Decay Energy

The decay energy Q_γ in γ emission is the sum of the γ-photon energy E_γ and the recoil kinetic energy of the daughter $(E_K)_D$ or

$$Q_\gamma = E_\gamma + (E_K)_D. \tag{11.58}$$

Since the magnitudes of the momenta of the daughter recoil nucleus $p_D = M(D)\upsilon_D$ and the γ photon $p_\gamma = E_\gamma/c$ are equal, i.e., $p_D = p_\gamma$, we can determine the partition of energy between $E_\gamma = p_\gamma c = M(D)\upsilon_D c$ and $(E_K)_D = \frac{1}{2}M(D)\upsilon_D^2$ as

$$(E_K)_D = \frac{M(D)v_D^2}{2} = \frac{E_\gamma^2}{2M(D)c^2},$$ (11.59)

where $M(D)$ and v_D are the rest mass and recoil velocity, respectively, of the daughter nucleus.

The γ-decay energy Q_γ may now be written as

$$Q_\gamma = E_\gamma + (E_K)_D = E_\gamma \left\{ 1 + \frac{E_\gamma}{2M(D)c^2} \right\}.$$ (11.60)

Equation (11.60) shows that the recoil kinetic energy of the daughter $(E_K)_D$ represents less than 0.1% of the γ-photon energy E_γ. The recoil energy of the daughter nucleus is thus negligible for most practical purposes. The label for *daughter* in γ decay is used in parallel with the same label used in other nuclear decays that are clearly defined with parent decaying into daughter. In γ decay the parent and daughter represent the same nucleus, except that the parent nucleus is in an excited state and the daughter nucleus is in a lower excited state or the ground state.

11.7.4   Resonance Absorption and Mössbauer Effect

The question of *resonance absorption* is of importance and deserves a brief discussion. The resonance absorption is a phenomenon in which a photon produced by a nuclear or atomic transition is re-absorbed by the same type of nucleus or atom, respectively. Since the photon shares the de-excitation energy with the atom or nucleus (recoil energy), it is quite possible that its energy will not suffice to allow triggering the reverse interaction and undergoing resonance absorption. However, if the recoil energy of the daughter atom or nucleus is not excessive, the resonance absorption is possible because of the natural width of the photon energy distribution and the finite lifetime of atomic and nuclear states, where the width and lifetime are governed by the uncertainty principle (see Sect. 1.24).

The photons' emission and absorption spectra differ because of the atomic or nuclear recoil energy that makes the emission energy slightly smaller than ΔE, the energy difference between the two states. However, if there is a region of overlap between the emission and absorption spectrum, resonance absorption is possible.

For atomic transitions that are of the order of eV to keV the resonance absorption is not hindered. On the other hand, for nuclear transitions that are of the order of 10 MeV, there is no overlap between the emission and the absorption photon spectrum and resonance absorption is not possible. However, there is a way around this problem. In 1957 *Rudolph Mössbauer* discovered that nuclear transitions occur with negligible nuclear recoil, if the decaying nucleus is embedded into a crystalline lattice. Here, the crystal as a whole rather than only the daughter nucleus absorbs the recoil momentum. This effect, called Mössbauer effect, minimizes the recoil energy and makes nuclear resonance absorption possible.

11.8 Internal Conversion

11.8.1 General Aspects of Internal Conversion

Nuclear de-excitation in which the de-ecitation energy is transferred from the parent nucleus almost in full to an orbital electron of the same atom is called *internal conversion* (IC). The process is represented as follows

$$
{}_Z^A X^* \rightarrow {}_Z^A X^+ + e^- + Q_{IC} \rightarrow {}_Z^A X, \tag{11.61}
$$

where

${}_Z^A X^*$ is the excited state of the nucleus most likely attained as a result of α or β decay.

${}_Z^A X^+$ is the singly ionized state of atom ${}_Z^A X$ following internal conversion decay.

Q_{IC} is the decay energy for internal conversion.

A small portion of the nuclear de-excitation energy Q_γ is required to overcome the binding energy E_B of the electron in its atomic shell, the remaining part of the decay energy Q_γ is shared between the conversion electron and the recoil daughter nucleus

$$
Q_{IC} = Q_\gamma - E_B = (E_K)_{IC} + (E_K)_D, \tag{11.62}
$$

where

Q_γ is the energy difference between two excited nuclear states, equal to the energy of a γ photon in γ decay.

$(E_K)_{IC}$ is the kinetic energy of the internal conversion electron ejected from the atom.

$(E_K)_D$ is the recoil kinetic energy of the daughter nucleus with nuclear mass $M(D)$.

The recoil kinetic energy $(E_K)_D$ of the daughter is much smaller than the kinetic energy $(E_K)_{IC}$ of the conversion electron and is usually neglected. It can be calculated with exactly the same approach that was taken for the β^+ decay to get

$$
(E_K)_D = \frac{m_e c^2}{M(D)c^2} (E_K)_{IC} + \frac{(E_K)^2_{IC}}{2M(D)c^2} \tag{11.63}
$$

and

$$
Q_{IC} = Q_\gamma - E_B = (E_K)_{IC} \left\{ 1 + \frac{m_e c^2}{M(D)c^2} + \frac{(E_K)_{IC}}{2M(D)c^2} \right\}, \tag{11.64}
$$

where $M(D)$ stands for the rest mass of the daughter nucleus.

11.8.2 Internal Conversion Factor

In any nuclear de-excitation both the γ-ray emission and the internal conversion electron emission are possible. The two nuclear processes are competing with one another and are governed essentially by the same selection rules. In contrast to the fluorescence yield ω (see Sect. 4.1.2) that is defined as the number of characteristic photons emitted per vacancy in a given atomic shell, the *total internal conversion factor* α_{IC} is defined as

$$\alpha_{IC} = \frac{\text{conversion probability}}{\gamma\text{-emission probability}} = \frac{N_{IC}}{N_\gamma}, \tag{11.65}$$

where

N_{IC} is the number of conversion electrons ejected from all shells per unit time.
N_γ is the number of γ photons emitted per unit time.

In addition to the total internal conversion factor α_{IC} one can define partial internal conversion factors according to the shell from which the electron was ejected

$$\frac{N_{IC}}{N_\gamma} = \frac{N_{IC}(K) + N_{IC}(L) + N_{IC}(M) + \cdots}{N_\gamma}$$
$$= \alpha_{IC}(K) + \alpha_{IC}(L) + \alpha_{IC}(M) + \cdots, \tag{11.66}$$

where α_{IC} (i) represents the partial internal conversion factors. Further distinction is possible when one accounts for subshell electrons.

The total internal conversion factors α_{IC} are defined with respect to N_γ so that α_{IC} can assume values greater or smaller than 1, in contrast to fluorescence yield ω that is always between 0 and 1. Since the K-shell electrons of all atomic electrons are the closest to the nucleus, most often the conversion electrons originate from the K atomic shell. The vacancy in the K shell, of course, is filled by a higher shell electron and the associated emission of characteristic photon or Auger electron, as discussed in Sect. 4.1.

An example for both the emission of γ photons and emission of conversion electrons is given in Fig. 11.4 with the β^- decay scheme for cesium-137 decaying into barium-137. Two channels are available for β^- decay of cesium-137:

1. 94.6% of disintegrations land in a barium-137 isomeric state (barium-137m) that has a half-life of 2.552 min and de-excitation energy of 662 keV.
2. 5.4% of disintegrations land directly in the barium-137 ground state.

The de-excitation energy of 0.662 MeV is emitted either in the form of a 662 keV γ photon or a conversion electron of kinetic energy $\sim$662 keV.

As shown in Fig. 11.4, for 100 disintegrations of cesium-137, 94.6 transitions land in barium-137m; of these 85 result in γ photons; 7.8 in K conversion electrons and 1.8 in higher shell conversion electrons. The internal conversion factor α_{IC} is $(7.8 + 1.8)/85 = 0.113$.

11.9 Spontaneous Fission

In addition to disintegrating through α and β decay processes, nuclei with very large atomic mass numbers A may also disintegrate by splitting into two nearly equal fission fragments and concurrently emit 2–4 neutrons. This decay process is called *spontaneous fission* (SF) and is accompanied by liberation of a significant amount of energy. It was discovered in 1940 by Russian physicists *Georgij N. Flerov* and *Konstantin A. Petržak* who noticed that uranium-238, in addition to α decay, may undergo the process of spontaneous fission.

Spontaneous fission follows the same process as neutron-induced nuclear fission, (See Sect. 12.8) except that it is not self-sustaining, since it does not generate the neutron fluence rate required to sustain a "chain reaction." In practice, SF is only energetically feasible for nuclides with atomic masses above 230 u or with $Z^2/A \geq 35$ where Z is the atomic number and A the atomic mass number of the radionuclide. SF can thus occur in thorium, protactinium, uranium and transuranic elements.

Transuranic (or transuranium) elements are elements with atomic numbers Z greater than that of uranium ($Z = 92$). All transuranic elements have more protons than uranium and are radioactive, decaying through β decay, α decay, or spontaneous fission. Generally, the transuranic elements are man-made and synthesized in nuclear reactions in a process referred to as nucleosynthesis. The nucleosynthesis reactions are generally produced in particle accelerators or nuclear reactors; however, neptunium ($Z = 93$) and plutonium ($Z = 94$) are also produced naturally in minute quantities, following the spontaneous fission decay of uranium-238.

The spontaneous fission neutrons emitted by U-238 can be captured by other U-238 nuclei thereby producing U-239 which is unstable and decays through β^- decay with a half-life of 23.5 min into neptunium-239 which in turn decays through β^- decay with a half-life 2.35 days into plutonium-239, or

$$^{238}_{92}\text{U} + \text{n} \xrightarrow{\quad\text{Neutron capture}\quad} {}^{239}_{92}\text{U} \xrightarrow{\beta^-} {}^{239}_{93}\text{Np} + \text{e}^- + \bar{\nu}_e \qquad (11.67)$$

$$^{239}_{93}\text{Np} \xrightarrow{\beta^-} {}^{239}_{94}\text{Pu} + \text{e}^- + \bar{\nu}_e.$$

Spontaneous fission is a competing process to α decay; the higher is A above uranium-238, the more prominent is the spontaneous fission in comparison with the α decay and the shorter is the half-life for spontaneous fission. For the heaviest nuclei, SF becomes the predominant mode of radioactive decay suggesting that SF is a limiting factor in how high in atomic number Z and atomic mass number A one can go in producing new elements.

- In uranium-238 the half-life for SF is $\sim 10^{16}$ a (years), while the half-life for α decay is 4.5×10^9 a. The probability for SF in uranium-238 is thus about 2×10^6 times lower than the probability for α decay.

- Fermium-256 has a half-life for SF of about 3 h making the SF in fermium-256 about 10 times more probable than α decay.
- Another interesting example is californium-256 (Cf-256) that decays essentially 100% of the time with SF and has a half-life of 12.3 min.
- For practical purposes, the most important radionuclide undergoing the SF decay is the transuranic californium-252 (Cf-252), used in industry and medicine as a very efficient source of fast neutrons (see Sect. 9.6.4). Californium-252 decays through α decay into curium-248 with a half-life of 2.65 a; however, about 3% of Cf-252 decays occur through SF producing on the average 3.8 neutrons per fission decay. The neutron production rate of Cf-252 is thus equal to $2.34 \times 10^6 \ (\mu g \cdot s)^{-1}$.

11.10 Proton Emission Decay

Proton-rich nuclides normally approach stability through β^+ decay or α decay. However, in the extreme case of a very large proton excess a nucleus may also move toward stability through emission of one or even two protons. Proton emission (PE) is thus a competing process to β^+ and α decay and is, similarly to α decay, an example of particle tunneling through the nuclear barrier potential.

Proton emission decay is much less common than are β^+ and α decay and is not observed in naturally occurring radionuclides. In PE decay the atomic number Z decreases by 1 and so does the atomic mass number A

$$\ce{^A_Z P} \rightarrow \ce{^{A-1}_{Z-1} D} + \text{p}. \tag{11.68}$$

- When a proton is ejected from a radionuclide P, the parent nucleus P sheds an orbital electron from its outermost shell to become a neutral daughter atom $\ce{^{A-1}_{Z-1} D}$.
- The energetic proton slows down in moving through the absorber medium and captures an electron from its surroundings to become a neutral hydrogen atom $\ce{^1_1 H}$.
- Since N, the number of neutrons does not change in proton emission decay, the parent P and daughter D are isotones.
- For lighter, very proton-rich nuclides with an odd number of protons Z, proton emission decay is likely.
- For lighter, very proton-rich nuclides ($A \approx 50$) with an even number of protons Z, a simultaneous two-proton emission may occur in situations where a sequential emission of two independent protons is energetically not possible (see example in Sect. 11.10.3).

11.10.1 Decay Energy in Proton Emission Decay

Decay energy Q_p released in proton emission decay appears as kinetic energy shared between the emitted proton and the daughter nucleus and is expressed as follows:

$$Q_p = \{\mathcal{M}(P) - [\mathcal{M}(D) + \mathcal{M}(H)]\}\, c^2 = \{M(P) - [M(D) + m_p]\}\, c^2, \quad (11.69)$$

where $\mathcal{M}(P)$, $\mathcal{M}(D)$, and $\mathcal{M}(_1^1H)$ are the atomic rest masses of the parent, daughter and hydrogen atom, respectively, and $M(P)$, $M(D)$ and m_p are nuclear rest masses of the parent, daughter and hydrogen nucleus (proton), respectively.

The total number of protons as well as the total number of neutrons does not change in the proton emission decay. Therefore, Q_p may also be expressed in terms of binding energies of the parent and daughter nucleus as follows:

$$Q_p = E_B(D) - E_B(P), \quad (11.70)$$

where

$E_B(D)$ is total binding energy of the daughter D nucleus.
$E_B(P)$ is total binding energy of the parent P nucleus.

The nuclear binding energy is defined in (1.25). For proton emission decay to be feasible, Q_p must be positive and this implies that the total binding energy of the daughter nucleus $E_B(D)$ must exceed the total binding energy of the parent nucleus $E_B(P)$; that is, $E_B(D) > E_B(P)$, or else that the rest mass of the parent nucleus must exceed the combined rest masses of the daughter nucleus and the proton, that is, $M(P) > M(D) + m_p$.

Two products are released in proton emission decay: a proton and the daughter product. For a decay of the parent nucleus at rest this implies that the proton and the daughter will acquire momenta p equal in magnitude but opposite in direction. The kinetic energy of the proton is $(E_K)_P = p^2/2m_p$ and of the daughter nucleus it is $(E_K)_D = p^2/2M(D)$.

The total decay energy Q_p must be positive for the proton emission decay and can be written as the sum of the kinetic energies of the two decay products:

$$Q_p = (E_K)_P + (E_K)_D = \frac{p^2}{2m_p} + \frac{p^2}{2M(D)}$$

$$= \frac{p^2}{2m_p}\left\{1 + \frac{m_p}{M(D)}\right\} = (E_K)_P\left\{1 + \frac{m_p}{M(D)}\right\}. \quad (11.71)$$

From (11.71) we determine the emitted proton kinetic energy $(E_K)_P$ as

$$(E_K)_P = Q_p\frac{1}{1 + \dfrac{m_p}{M(D)}}. \quad (11.72)$$

The kinetic energy of the recoil daughter $(E_K)_D$, on the other hand, is given as follows

$$(E_K)_D = Q_p - (E_K)_p = Q_p \frac{1}{1 + \dfrac{M(D)}{m_p}}. \tag{11.73}$$

The decay energy Q_{2p} released in two-proton emission decay appears as kinetic energy shared among the three emitted particles (two protons and the daughter nucleus) and may be calculated simply from the difference in binding energies E_B between the daughter D and the parent P nucleus

$$Q_{2p} = E_B(D) - E_B(P) \tag{11.74}$$

or from the following expression

$$Q_{2p} = \{\mathcal{M}(P) - [\mathcal{M}(D) + 2\mathcal{M}(^1_1H)]\}\, c^2 = \{M(P) - [M(D) + 2m_p]\}\, c^2, \tag{11.75}$$

where $\mathcal{M}$ stands for the atomic rest masses, M for nuclear rest masses and m_p for the proton rest mass.

11.10.2 Example of Proton Emission Decay

An example of proton emission decay is the decay of lithium-5 into helium-4 with a half-life of 10^{-21} s. The decay is schematically written as follows

$$^5_3Li \rightarrow {}^4_2He + p \tag{11.76}$$

and the decay energy may be calculated from (11.69) or (11.70). The required atomic and nuclear data are given in Appendix A as follows

$$\mathcal{M}(^5_3Li)c^2 = 5.012541 \text{ u} \times 931.5 \text{ MeV/u} = 4669.15 \text{ MeV} \tag{11.77}$$

$$\mathcal{M}(^4_2He)c^2 = 4.002603 \text{ u} \times 931.494 \text{ MeV/u} = 3728.4 \text{ MeV} \tag{11.78}$$

$$\mathcal{M}(^1_1H)c^2 = 1.007825 \text{ u} \times 931.494 \text{ MeV/u} = 938.78 \text{ MeV} \tag{11.79}$$

$$E_B(^5_3Li) = 26.32865 \text{ MeV} \tag{11.80}$$

$$E_B(^4_2He) = 28.29569 \text{ MeV} \tag{11.81}$$

We first notice that $\mathcal{M}(^5_3Li) > \mathcal{M}(^4_2He) + \mathcal{M}(^1_1H)$ and that $E_B(^4_2He) > E_B(^5_3Li)$. This leads to the conclusion that the proton emission decay is possible. Next we use (11.69) and (11.70) to calculate decay energy Q_p and get 1.97 MeV from both equations. Equations (11.72) and (11.73) give 1.57 MeV and 0.40 MeV for kinetic energies of the ejected proton $(E_K)_p$ and recoil helium-4 atom $(E_K)_{^4_2He}$, respectively, for a combined total energy of 1.97 MeV, as given by calculated Q_p value.

11.10.3 Example of Two-Proton Emission Decay

An example of two-proton emission decay is the decay of iron-45 (a highly proton rich radionuclide with $Z = 26$ and $N = 19$) which decays with a simultaneous emission of two protons at a half-life of 0.35 μs into chromium-43 (a proton-rich radionuclide with $Z = 24$ and $N = 19$). The decay is schematically written as follows

$$^{45}_{26}Fe \rightarrow {}^{43}_{24}Cr + 2p, \qquad (11.82)$$

and the decay energy Q_{2p} may be calculated from (11.74) or (11.75).

At first glance one could expect the iron-45 radionuclide to decay by a single proton emission into manganese-44; however, a closer inspection shows that the one-proton decay would produce negative decay energy Q_p from (11.69) and (11.70) and thus is not energetically feasible.

The atomic and nuclear data for radionuclides $^{45}_{26}Fe$, $^{44}_{25}Mn$, and $^{43}_{24}Cr$ are given as follows

$$\mathcal{M}(^{45}_{26}Fe)c^2 = 45.014564 \text{ u} \times 931.5 \text{ MeV/u} = 41930.79 \text{ MeV} \qquad (11.83)$$

$$\mathcal{M}(^{44}_{25}Mn)c^2 = 44.006870 \text{ u} \times 931.5 \text{ MeV/u} = 40992.14 \text{ MeV} \qquad (11.84)$$

$$\mathcal{M}(^{43}_{24}Cr)c^2 = 42.997711 \text{ u} \times 931.5 \text{ MeV/u} = 40052.11 \text{ MeV} \qquad (11.85)$$

$$\mathcal{M}(^1_1H)c^2 = 1.007825 \text{ u} \times 931.494 \text{ MeV/u} = 938.78 \text{ MeV} \qquad (11.86)$$

$$E_B(^{45}_{26}Fe) = 329.306 \text{ MeV} \qquad (11.87)$$

$$E_B(^{44}_{25}Mn) = 329.180 \text{ MeV} \qquad (11.88)$$

$$E_B(^{43}_{24}Cr) = 330.424 \text{ MeV} \qquad (11.89)$$

Inspection of (11.70) shows that one-proton emission decay of $^{45}_{26}Fe$ into $^{44}_{25}Mn$ is not possible, since it results in a negative Q_p. On the other hand, (11.74) results in positive decay energy Q_{2p} for a two-proton decay of $^{45}_{26}Fe$ into its isotone $^{43}_{24}Cr$. The decay energy Q_{2p} calculated from (11.74) and (11.75) then amounts to 1.12 MeV for the two-proton decay of $^{45}_{26}Fe$ into $^{43}_{24}Cr$.

11.11 Neutron Emission Decay

Neutron emission from a neutron-rich nucleus is a competing process to β^- decay but is much less common then the β^- decay and is not observed in naturally occurring radionuclides. In contrast to spontaneous fission that also produces neutrons, in neutron emission decay the atomic number Z remains the same but the atomic mass number A decreases by 1. Both the parent nucleus P and the daughter nucleus D are thus isotopes of the same nuclear species. The neutron emission (NE) decay relationship is written as follows:

$$_{Z}^{A}X \rightarrow \,_{Z}^{A-1}X + n. \tag{11.90}$$

11.11.1 Decay Energy in Neutron Emission Decay

The decay energy Q_n released in neutron emission decay appears as kinetic energy shared between the emitted neutron and the daughter nucleus and is expressed as follows

$$Q_n = \{\mathcal{M}(P) - [\mathcal{M}(D) + m_n]\}\, c^2 = \{M(P) - [M(D) + m_n]\}\, c^2, \tag{11.91}$$

where $\mathcal{M}(P)$ and $\mathcal{M}(D)$ are atomic masses of the parent and daughter atom, respectively; $M(P)$ and $M(D)$ are the nuclear masses of the parent and daughter, and m_n is the neutron rest mass.

The total number of protons Z as well as the total number of neutrons N does not change in the neutron emission decay. Therefore, Q_n may also be expressed in terms of binding energies of the parent and daughter nucleus as follows

$$Q_n = E_B\,(D) - E_B\,(P), \tag{11.92}$$

where

$E_B(D)$	is the total binding energy of the daughter D nucleus.
$E_B(P)$	is the total binding energy of the parent P nucleus.

For the neutron emission decay to be feasible, Q_n must be positive and this implies that the total binding energy of the daughter nucleus $E_B(D)$ must exceed the total binding energy of the parent nucleus $E_B(P)$; that is, $E_B(D) > E_B(P)$, or else that the rest mass of the parent nucleus $M(P)$ must exceed the combined rest masses of the daughter nucleus and the neutron; that is, $M(P) > M(D) + m_n$.

Two products are released in neutron emission decay: a neutron and the daughter product. For a decay of the parent nucleus at rest this implies that the neutron and the daughter will acquire momenta p equal in magnitude but opposite in direction. The kinetic energy of the neutron is $(E_K)_n = p^2/2m_n$ and of the daughter nucleus the kinetic energy is $(E_K)_D = p^2/[2M(D)]$.

For the NE decay to be possible the total decay energy Q_n must be positive and is expressed as follows

$$Q_n = (E_K)_n + (E_K)_D = \frac{p^2}{2m_n} + \frac{p^2}{2M(D)}$$

$$= \frac{p^2}{2m_n}\left\{1 + \frac{m_n}{M(D)}\right\} = (E_K)_n\left\{1 + \frac{m_n}{M(D)}\right\}. \tag{11.93}$$

From (11.93) we determine the emitted neutron kinetic energy $(E_K)_n$ as

$$(E_K)_n = Q_n\frac{1}{1 + \dfrac{m_n}{M(D)}}. \tag{11.94}$$

The kinetic energy of the recoil daughter $(E_K)_D$, on the other hand, is given as follows

$$(E_K)_D = Q_n - (E_K)_n = Q_n\frac{1}{1 + \dfrac{M(D)}{m_n}}. \tag{11.95}$$

11.11.2 Example of Neutron Emission Decay

An example of neutron emission decay is the decay of helium-5 into helium-4 with a half-life of 8×10^{-22} s. The decay is schematically written as follows:

$$\,^5_2\text{He} \rightarrow \,^4_2\text{He} + n, \tag{11.96}$$

and the decay energy may be calculated from (11.91) or (11.92). The required atomic and nuclear data obtained in Appendix A are as follows:

$$\mathcal{M}(^5_2\text{He})c^2 = 5.012221 \text{ u} \times 931.494 \text{ MeV/u} = 4668.85 \text{ MeV} \tag{11.97}$$

$$\mathcal{M}(^4_2\text{He})c^2 = 4.002603 \text{ u} \times 931.494 \text{ MeV/u} = 3728.4 \text{ MeV} \tag{11.98}$$

$$m_n c^2 = 1.008665 \text{ u} \times 931.5 \text{ MeV/u} = 939.56 \text{ MeV} \tag{11.99}$$

$$E_B(^5_2\text{He}) = 27.40906 \text{ MeV} \tag{11.100}$$

$$E_B(^4_2\text{He}) = 28.29567 \text{ MeV} \tag{11.101}$$

We first notice that $\mathcal{M}(P) > \mathcal{M}(D) + m_n$ and $E_B\left(^4_2\text{He}\right) > E_B\left(^5_2\text{He}\right)$ and conclude that neutron emission decay is possible. Next we use (11.91) and (11.92) to calculate decay energy Q_n and get 0.89 MeV from both equations. Equations (11.94) and (11.95) give 0.71 MeV and 0.18 MeV for the kinetic energies of the ejected neutron $(E_K)_n$ and recoil helium-4 atom $(E_K)_{^4_2\text{He}}$, respectively, for a combined total of 0.89 MeV, in agreement with the Q_n value calculated from (11.91) and (11.92).

11.12   Chart of the Nuclides (Segrè Chart)

All known nuclides are uniquely characterized by their number of protons Z (atomic number) and their number of neutrons $N = A - Z$ where A is the number of nucleons (atomic mass number). The most pertinent information on the 280 known stable nuclides and over 3000 known radioactive nuclides (radionuclides) is commonly summarized in a *Chart of the Nuclides* in such a way that it is relatively easy to follow the atomic transitions resulting from the various radioactive decay modes used by radionuclides to attain more stable configurations. Usually the ordinate (y) axis of the chart represents Z and the abscissa (x) axis represents N in a two-dimensional Cartesian plot with each nuclide represented by a unique square (pixel) that is placed onto the chart according to the N and Z value of the nuclide.

The chart of the nuclides is also referred to as the Segrè chart in honor of *Emilio Segrè* who was the first to suggest the particular arrangement in the 1940s. Similarly to the *Periodic Table of Elements* (see Sect. 3.2.3) that Mendeleev introduced in 1869 to represent conveniently the periodicity in chemical behavior of elements with increasing atomic number Z, Segrè chart of the nuclides presents an orderly formulation of all nuclear species (stable and radioactive) against both Z and N and, in addition, provides useful basic nuclear data for the nuclides and indicates the possible decay paths for radionuclides.

In addition to the number of protons Z and number of neutrons N for a given nuclide, the Segrè Chart usually provides the following nuclear data:

- *For stable nuclides* the atomic mass number A; the nuclear mass in unified atomic mass units u; the natural abundance; and, for example, cross sections for activation interactions.
- *For radionuclides* the atomic mass number A, nuclear mass in unified atomic mass units u, abundance, radioactive half-life, and mode of decay.

Nuclear charts are readily available in the literature and on the internet as a result of efforts by many authors, institutions, publishers, commercial vendors, and standards laboratories. A typical example of the chart of nuclides is shown in Fig. 11.9 displaying a condensed version of the "Karlsruher Nuklidkarte" (Karlsruhe Chart of Nuclides) issued by the Joint Research Center of the European Commission in Karlsruhe, Germany. The 7th edition of the chart, issued in 2007, contains data on 280 stable nuclides, 2962 radionuclides in the ground state, 692 isomeric radionuclides, and 8 radioactive decay modes. These numbers represent a significant increase over the nuclides that were listed in the first edition of the Karlsruhe chart which was issued in 1958 and presented data on 267 stable nuclides, 1030 radionuclides, and 4 radioactive modes. Of course, to get access to the data, smaller sections of the chart must be viewed and used; Fig. 11.9 simply presents a condensed version of the whole chart and allows identification of general features.

The chart in Fig. 11.9 covers all currently known stable and radioactive nuclides ranging in number of protons Z from 1 to 118 and in number of neutrons N from 0 to ~ 160. All pixels occupied by a nuclide form a "nuclide landscape" in the form

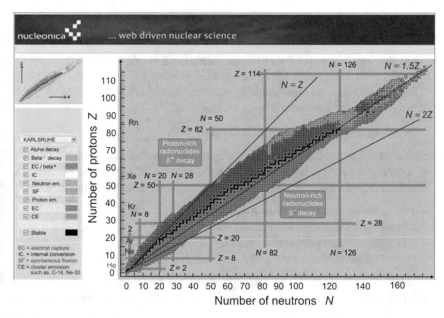

Fig. 11.9 *Chart of the Nuclides* also known as the *Segrè Chart*. Each known stable and radioactive nuclide is characterized by its unique combination of the number of protons Z and number of neutrons N, and assigned a pixel in a 2-dimensional chart displaying Z on the ordinate axis and N on the abscissa axis. The stable nuclides are shown by dark pixel squares, radioactive nuclides by colored pixel squares. The plot of stable nuclides forms a *"curve of stability,"* neutron-rich radionuclides are *below the curve* of stability and proton-rich radionuclides are *above the curve* of stability. The magic numbers for neutrons and protons are also indicated. Radionuclides are shown with colored pixels, each color representing the decay mode used by the particular radionuclide to attain a more stable configuration. The color code is displayed on the *left side-bar* to the chart. © *European Communities, 2009. Courtesy of Dr. Joseph Magill: Joint Research Centre of the European Commission, Karlsruhe, Germany. The "Karlsruher Nuklidkarte" is available on-line at*: www.nucleonica.net

of an elongated island oriented from South-West toward North-East: neutron-rich nuclides occupy the south shore; proton-rich nuclides occupy the north shore; and very heavy nuclides that are prone to spontaneous fission occupy the north-east tip of the island.

The magic numbers (see Sect. 1.16.2) for protons and neutrons are also identified on the chart; the stable nuclides are shown with black squares, the radionuclides with various colors depending on their decay mode. The color code used is shown in the left side-bar of the figure. The stable nuclides form the backbone of the island and follow a "curve of stability" which is defined by an optimum number of protons and neutrons. Nuclides with a magic number of protons Z or a magic number of neutrons N tend to exhibit maximum stability.

The following special features of the Segrè chart are noted for pixels representing a given element:

1. Horizontal rows give the list of known isotopes ($Z = $ const).
2. Vertical columns give the list of isotones ($N = $ const).
3. Diagonal lines (in direction roughly perpendicular to the "curve of stability") give the list of isobars for which $A = Z + N = $ const

The following features of nuclear stability are noted:

- Stable nuclides contain a balanced configuration of protons and neutrons because of a preference for pairing of nucleons which for the 280 known stable nuclides results in the following distributions of proton numbers Z and neutron numbers N:

  – 166 stable nuclides have even Z and even N.
  – 57 stable nuclides have even Z and odd N.
  – 53 stable nuclides have odd Z and even N.
  – Only 4 stable nuclides have odd Z and odd N.

- The stable nuclides follow a *curve of stability* on the Segrè chart. The curve of stability follows $Z \approx N$ for low Z nuclides and then slowly transforms into $N \approx 1.5Z$ with increasing Z.
- Elements with atomic number Z equal to or below $Z = 83$ (bismuth) have at least one stable isotope, with two notable exceptions: technetium ($Z = 41$) and promethium ($Z = 61$).
- Elements with $84 \leq Z \leq 92$ are unstable and present in nature either because they have a very long half-life or they are decay products of long-lived thorium or uranium series.
- A few low atomic number radionuclides with relatively short half-lives such as carbon-14 are produced by cosmic rays and their supply is continuously replenished.
- Below the curve of stability are neutron-rich radionuclides. Most of neutron-rich radionuclides undergo a nuclear transmutation by β^- decay but a few do so by direct neutron emission in situations where the neutron-proton inbalance is very large.
- Above the curve of stability are proton-rich radionuclides. Most of proton-rich radionuclides undergo a nuclear transmutation by β^+ decay or electron capture but a few do so by direct emission of one proton or even two protons in situations where the proton-neutron inbalance is very large.
- All nuclides with $Z > 82$ undergo α decay or spontaneous fission and some may also undergo β decay. α decay is also possible for $Z > 82$.
- Nuclides with atomic number Z exceeding 92 are called transuranic nuclides and are artificially produced either in a nuclear reactor or with a particle accelerator. Transuranic nuclides are also known as transuranium elements and all of them are unstable (radioactive) and undergo nuclear decay into other elements. To date all transuranium elements have been discovered in one of four national laboratories located one each in the USA, Russia, Germany, or Japan.

A small part of a simplified complete Segrè Chart is shown in Fig. 11.10 for proton numbers from $Z = 1$ to $Z = 12$, neutron numbers from $N = 0$ to $N = 18$, and atomic mass numbers from $A = 1$ to $A = 30$ (from hydrogen to magnesium). The stable nuclides are identified with black pixels (squares) indicating the curve of stability that is given by $Z \approx N$ for low atomic number elements; neutron-rich radionuclides and proton-rich radionuclides are shown in colored squares below and above the region of stability, respectively. A typical Segrè chart provides the relative abundance of stable nuclides as well as half-life and decay modes of radionuclides. Often cross sections for thermal neutron activation are also provided. In Fig. 11.10 these data are omitted and only the nuclide symbol and number of nucleons A are shown.

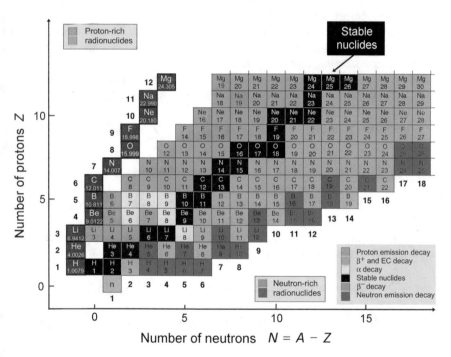

Fig. 11.10 A portion of the Chart of the Nuclides (Segrè Chart) for nuclides with proton numbers Z from 1 through 12, neutron numbers N from 0 through 18, and atomic mass numbers A from 1 through 30. Stable nuclides are shown with *black pixel squares*, radionuclides with *colored pixel squares*; the color code shown in the inset defines the particular decay. The *red squares* above the region of proton-rich radionuclides give the mean atomic mass in unified atomic mass units u for stable nuclides from hydrogen to magnesium. Each horizontal row ($Z = $ const) represents one element including all stable and all known radioactive isotopes of the particular chemical element; each vertical column ($N = $ const) represents nuclides with the same neutron number (isotones); each *diagonal line* perpendicular to the curve of stability represents nuclides with the same atomic mass number (isobars). Data were obtained from the Chart of Nuclides available from the Nuclear Data Center at: www.nndc.bnl.gov/chart/

Table 11.1 Isotopic composition, relative atomic mass and abundance of stable isotopes in natural oxygen

Stable oxygen isotope	Relative atomic mass (u)	Relative abundance (%)
O-16	15.994914	99.757
O-17	16.999132	0.038
O-18	17.999161	0.205

Data are from the NIST (http://www.physics.nist.gov/PhysRefData/Compositions/notes.html)

The red squares above the region of the proton-rich radionuclides present the first 12 elements of the *Periodic Table of Elements* along with their mean atomic mass $\overline{\mathcal{M}}$ in unified atomic mass units u. Natural elements generally contain a mixture of two or more isotopes of the element, as also evident from Fig. 11.10 where, of the 12 natural low atomic number elements, only three (beryllium, fluorine, and sodium) have only one stable isotope; six have two, and three have three stable isotopes.

The mean atomic mass $\overline{\mathcal{M}}$ of a given natural element is called the standard atomic weight of the natural element and is given as the weighted mean for all stable isotopes constituting the given element, as discussed in detail in Sect. 1.13.1.

$$\overline{\mathcal{M}} = \sum_i w_i \mathcal{M}_i \qquad \text{and} \qquad \sum_i w_i = 1, \qquad (11.102)$$

where w_i is the relative weight of stable isotope i and $\mathcal{M}_i$ is the atomic mass of stable isotope i constituting the given natural element.

For example, the NIST gives the standard atomic weight of natural oxygen as 15.99940 u. This value can be obtained from the basic NIST data (isotopic composition, atomic masses of the stable isotopes of oxygen, and relative abundance of the stable isotopes in natural oxygen) given in Table 11.1 as follows

$$\overline{\mathcal{M}} = \sum_i w_i \mathcal{M}_i = w_{O-16}\mathcal{M}_{O-16} + w_{O-17}\mathcal{M}_{O-17} + w_{O-18}\mathcal{M}_{O-18}$$

$$= 0.99757 \times 15.994915\,u + 0.00038 \times 16.999131\,u + 0.00205 \times 17999160\,u$$

$$= 15.99940\,u \qquad (11.103)$$

As shown in Fig. 11.10 for low atomic number Z nuclides, radionuclides below the curve of stability decay by β^- decay, possibly α decay, and neutron emission decay when the neutron-proton imbalance is very large. Above the curve of stability radionuclides decay by electron capture or β^+ decay, possibly α decay, and proton emission decay when the proton-neutron imbalance is very large. As suggested by the individual rows in the Segrè chart of Fig. 11.10 ($Z = $ const), a given atomic species or chemical element in general consists of one or more stable isotopes and several radioactive isotopes: neutron-rich isotopes to the right of the stable ones and proton-rich to the left.

In Fig. 11.10 proton is entered as stable isotope of hydrogen ${}_1^1\text{H}$ with $Z = 1$ and $N = 0$. Although it is not an element, neutron n appears in the chart of nuclides as neutral unstable "element" at position $Z = 0$ and $N = 1$. While neutrons bound in stable nuclei are stable, free neutrons are unstable undergoing β^- decay (see Sect. 11.4.3) with a half-life of 618 s (10.3 min) into a proton p, electron e^- and electron antineutrino $\bar{\nu}_e$.

Atomic masses $\mathcal{M}$ and nuclear masses M as well as the atomic rest energies $\mathcal{M}c^2$, nuclear rest energies Mc^2, and nuclear binding energies E_B for all nuclides discussed in this chapter were determined from the NIST data on atomic masses, as shown in Appendix A which lists the main attributes of all nuclides presented in this book. The data given in the table can be used to determine the various decay energies for the specific radioactive decay examples presented in this chapter.

The data were determined as follows:

1. Data for atomic masses $\mathcal{M}$ were obtained from the NIST and are given in unified atomic mass units u. The rest mass of the proton m_p, neutron m_n, electron m_e, and the unified atomic mass unit u are given by the NIST as follows:

$$m_p = 1.672\,621\,637 \times 10^{-27}\,\text{kg} = 1.007\,276\,467\,\text{u} = 938.272\,013\,\text{MeV}/c^2$$

$$m_n = 1.674\,927\,211 \times 10^{-27}\,\text{kg} = 1.008\,664\,916\,\text{u} = 939.565\,346\,\text{MeV}/c^2$$

$$m_e = 9.109\,382\,15 \times 10^{-31}\,\text{kg} = 5.485\,799\,094 \times 10^{-4}\,\text{u}$$

$$= 0.510\,998\,910\,\text{MeV}/c^2$$

$$u = 1.660\,538\,782 \times 10^{-27}\,\text{kg} = 931.494\,028\,\text{MeV}/c^2$$

2. For a given nuclide, its nuclear rest energy Mc^2 was determined by subtracting the rest energy of all atomic orbital electrons (Zm_ec^2) from the atomic rest energy $\mathcal{M}(u)c^2$ as follows

$$Mc^2 = \mathcal{M}(u)\,c^2 - Zm_ec^2 = \mathcal{M}(u) \times 931.494\,028\,\text{MeV/u}$$
$$- Z \times 0.510\,999\,\text{MeV}. \tag{11.104}$$

The binding energy of orbital electrons to the nucleus is ignored in (11.104).

3. The nuclear binding energy E_B for a given nuclide is determined using the mass deficit equation given in (1.25) to get

$$E_B = Zm_pc^2 + (A - Z)m_nc^2 - Mc^2, \tag{11.105}$$

with the nuclear rest energy Mc^2 given in (11.104) and the rest energy of proton m_pc^2, neutron m_nc^2, and electron m_ec^2 given in point (1) above.

11.13   Summary of Radioactive Decay Modes

Nuclear physics has come a long way since Ernest Rutherford's momentous discovery that most of the atomic mass is concentrated in the atomic nucleus which has the size of the order of 1 fm $= 10^{-15}$ m in comparison to the atomic size of the order of 1 Å $= 10^{-10}$ m. The atomic nucleus consists of nucleons – positively charged protons and neutral neutrons, and each nuclear species is characterized with a unique combination of the number of protons (atomic number) Z and number of neutrons N, the sum of which gives the number of nucleous $A = Z + N$ (atomic mass number).

Nucleons are bound together to form the nucleus by the strong nuclear force which, in comparison to the proton–proton Coulomb repulsive force, is at least two orders of magnitude larger but of extremely short range of only a few femtometers. To bind the nucleons into a stable nucleus a delicate equilibrium between the number of protons and the number of neutrons must exist. As evident from Figs. 11.9 and 11.10, for light (low A) nuclear species, a stable nucleus is formed from an equal number of protons and neutrons ($Z = N$). Above the nucleon number $A \approx 40$, more neutrons than protons must constitute the nucleus to form a stable configuration in order to overcome the Coulomb repulsion among the charged protons.

If the optimal equilibrium between protons and neutrons does not exist, the nucleus is unstable (radioactive) and decays with a specific decay constant into a more stable configuration that may also be unstable and decays further, forming a decay chain that eventually ends with a stable nuclide. As discussed in detail in this chapter, nine main processes are available to unstable nuclei (radionuclides) to advance toward a more stable nuclear configuration; for a given radionuclide generally only one type or at most two types of decay process will occur.

Nuclides with an excess number of neutrons are referred to as neutron-rich; those with an excess of protons are referred to as proton-rich.

- For a slight imbalance, radionuclides will decay by β decay characterized by transformation of a proton into a neutron in β^+ decay and a transformation of a neutron into a proton in β^- decay.
- For a large imbalance, the radionuclides will decay by emission of nucleons: α-particles in α decay, protons in proton emission decay, and neutrons in neutron emission decay.
- For very large atomic mass number nuclides ($A > 230$) spontaneous fission, which competes with α decay, is also possible.

Figure 11.11 shows schematically the decay paths (except for the spontaneous fission) possibly open to a parent radionuclide (N, Z) in its transition toward a more stable configuration. A very small segment of a typical Segrè chart is shown centered around the parent nucleus. Also shown are three special lines through the parent nucleus: isotope line for $Z = $ const, isotone line for $N = $ const, and isobar line for $A = $ const.

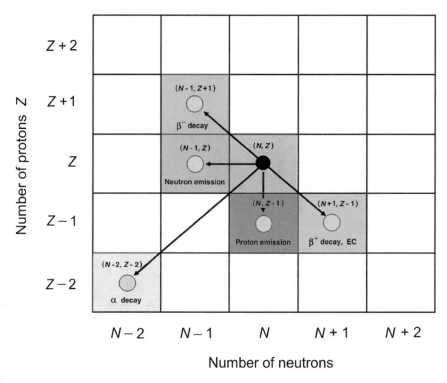

Number of neutrons

Fig. 11.11 Possible decay paths available in the Chart of the Nuclides (Segrè Chart) to a parent radionuclide (N, Z) in its quest to attain a more stable configuration. The parent radionuclide is shown by the *solid black circle*, daughter nuclides are shown by *open circles*. Three special lines through the parent nucleus are also indicated: isotope line for Z = const, isotone line for N = const, and isobar line for A = const. The parent and daughter are nearest neighbors on the isobar line (A = const) in β^- decay, electron capture, and β^+ decay; on the isotone line (N = const) in proton emission decay, and on the isotope line (Z = const) in neutron emission decay

The parent and daughter are nearest neighbors on:

1. Isobar line (A = const) in β^- decay, electron capture, and β^+ decay.
2. Isotone line (N = const) in proton emission decay.
3. Isotope line (Z = const) in neutron emission decay.

The following general features of radioactive decay processes are noted:

1. When radionuclide (Z, N) is below the curve of stability (i.e., is neutron-rich), the β^- decay and in extreme cases neutron emission are possible means to attain a more stable configuration. The resulting nucleus will be characterized by $(Z + 1, N - 1)$ for β^- decay and by $(Z, N - 1)$ for neutron emission decay. In β^- decay the atomic mass number A of the daughter is the same as that of the parent; in neutron emission decay A of the daughter decreases by 1.

Table 11.2 Main attributes of the nine decay modes available to an unstable nucleus for reaching a more stable configuration

Decay	Before decay	After decay	$\Delta Z = Z_b - Z_a$	$\Delta N = N_b - N_a$	$\Delta A = A_b - A_a$	Decay energy Q
α	P	D, α	+2	+2	4	$\{M(P) - [M(D) + m_\alpha]\}\, c^2$ $\{\mathcal{M}(P) - [\mathcal{M}(D) + \mathcal{M}(^4_2\text{He})]\}\, c^2$
β^-	P	D, e^-, $\bar{\nu}_e$	-1	+1	0	$\{M(P) - [M(D) + m_e]\}\, c^2$ $\{\mathcal{M}(P) - \mathcal{M}(D)\}\, c^2$
β^+	P	D, e^+, ν_e	+1	-1	0	$\{M(P) - [M(D) + m_e]\}\, c^2$ $\{\mathcal{M}(P) - [\mathcal{M}(D) + 2m_e]\}\, c^2$
Electron capture	P, e^-_{orb}	D, ν_e	-1	+1	0	$\{[M(P) + m_e] - M(D)\}\, c^2$ $\{\mathcal{M}(P) - \mathcal{M}(D)\}\, c^2$
γ	P*	P, γ	0	0	0	
Internal conversion	P*	P, e^-_{orb}	0	0	0	
Spontaneous fission	P	D_1, D_2	$\sim Z/2$	$\sim Z/2$	$\sim A/2$	
Proton emission	P	D, p	+1	0	-1	$\{M(P) - [M(D) + m_p]\}\, c^2$ $\{\mathcal{M}(P) - [\mathcal{M}(D) + \mathcal{M}(H)]\}\, c^2$
Neutron emission	P	D, n	0	+1	+1	$\{M(P) - [M(D) + m_n]\}\, c^2$ $\{\mathcal{M}(P) - [\mathcal{M}(D) + m_n]\}\, c^2$

P = parent nucleus; D = daughter nucleus; e_{orb} = orbital electron; M = nuclear rest mass; $\mathcal{M}$ = atomic rest mass; $\mathcal{M}(H)$ = rest mass of hydrogen atom (protium); $\mathcal{M}(^4_2\text{He})$ = rest mass of helium-4 atom; m_e = rest mass of electron; m_p = rest mass of proton; m_n = rest mass of neutron; m_α = rest mass of α-particle; b = before decay; a = after decay

2. When radionuclide (Z, N) is above the curve of stability (i.e., is proton-rich), the β^+ decay, electron capture or in extreme cases proton emission may be possible means to attain a more stable configuration. The resulting nucleus will be characterized by $(Z - 1, N + 1)$ for β^+ decay and electron capture, and by $(Z - 1, N)$ for proton emission decay. In β^+ decay and electron capture decay the atomic mass number A of the daughter nucleus is the same as that of the parent nucleus; in proton emission decay both Z and A decrease by 1.
3. Proton and neutron emission decay are much less common than α and β decays. The two nucleon emission decays are of no importance in medical physics and occur only in artificially produced radionuclides. The main characteristics of radionuclides which decay by proton or neutron emission are an extreme imbalance between the number of protons and the number of neutrons in their nuclei as well as very short half lives.
4. In addition to β decay the radionuclides (Z, N) with $Z > 83$ may decay by α decay or spontaneous fission. In α decay the resulting nucleus is characterized by $(Z - 2, N - 2)$, in contrast to spontaneous fission where the resulting nuclei are much lighter than the parent nucleus.
5. In γ decay and internal conversion decay the parent nucleus is excited and undergoes a de-excitation process by emitting a γ photon or a conversion electron, respectively. Both the parent nucleus and the daughter nucleus are characterized by (Z, N), since the number of protons as well as the number of neutrons does not change in the decay process.

the main characteristics of the eight most common radioactive decay modes is given in Appendix B. The appendix provides expressions for the basic decay relationship, the decay energy as well as the energy of the decay products (daughter nucleus and emitted particles) for eight radioactive decay modes (spontaneous fission is excluded). In radioactive decay the daughter recoil kinetic energy $(E_K)_D$ is generally ignored when determining the energy of the other, lighter decay products. However, we must keep in mind that in α decay as well as in proton and neutron emission decay $(E_K)_D$ is of the order of 100 keV, while in other radioactive decay modes, except for the spontaneous fission, it is of the order of 10 eV to 100 eV. Thus the daughter recoil kinetic energy in α decay and in proton and neutron emission decay is not negligible and should be accounted for, while for the other common radioactive decays it may be ignored.

Table 11.2 presents a summary of the main attributes of the nine modes of radioactive decay presented in this chapter, highlighting the changes in atomic number Z, neutron number N, and atomic mass number A as well as the expressions for the decay energy Q calculated using either the nuclear masses M or the atomic masses $\mathcal{M}$. The basic relationships governing the radioactive modes of decay are summarized in Appendix B.

Fission and Nuclear Chain Reaction

The discovery of the neutron-induced fission process in 1939, similarly to the discovery of x-rays, is one of the most significant scientific discoveries in human history. On the one hand, it represents the culmination of developments in modern physics that were unleashed with Röntgen's discovery of x-rays in 1895 and, on the other hand, it provided humanity with an effective tool for significant improvements in quality of life as well as for self-destruction, if not used wisely.

Many physicists and many scientific discoveries contributed to the long and painstaking process that ultimately lead to the discovery and practical use of fission. In addition to Röntgen, the most notable physicists and their discoveries that indirectly or directly lead to the discovery and understanding of fission are:

Henri Becquerel for discovery of natural radioactivity in 1896 at the Ecole Polytechnique in Paris.

Pierre Curie and Marie *Skłodowska-Curie* for discovery of radium-226 in 1898.

Ernest Rutherford for discovery of the atomic nucleus in 1911 at the University of Manchester.

James Chadwick for discovery of the neutron in 1932 at Cambridge University.

Jean-Frédéric Joliot and *Irène Joliot-Curie* for discovery of artificial radioactivity in 1934 at the "Institut du Radium" in Paris.

Enrico Fermi for experiments bombarding uranium with thermal neutrons in 1934 at the University of Rome.

Otto Hahn and *Friedrich Strassmann* for experimental discovery of the fission process in 1938 at the Kaiser Wilhelm Institute for Chemistry in Berlin.

Lise Meitner and *Otto Frisch* for the theoretical explanation of the fission process in 1938.

Enrico Fermi and *Leó Szilárd* for discovery of neutron multiplication in uranium in 1939 at Columbia University in New York.

Enrico Fermi and colleagues from the "Manhattan Project" for producing the first controlled nuclear chain reaction in 1942 and for developing the first atomic bomb that was used to force an end to Word War II in 1945.

The "Manhattan Project" ended the World War II and opened the door to the development of fission for peaceful purposes using controlled nuclear chain reaction for power generation and for production of artificial radionuclides that are used in science, industry, and medicine. In a fission chain reaction triggered by a thermal neutron, the fission of a fissile heavy nucleus, such as uranium-235, produces two radioactive fission fragments and at least one neutron which can propagate the chain reaction.

On next page, three generations of a nuclear fission chain reaction are shown schematically for a fissile radionuclide bombarded with a thermal neutron (Figure adapted from atomicarchive.com). Each generation in the example shown produces two fission fragments, two neutrons, and a significant amount of energy. Under these conditions the chain reaction can be made self-sustaining and used for neutron production and power generation.

Chapter 12
Production of Radionuclides

In 1896 Henri Becquerel discovered natural radioactivity and in 1934 Frédéric Joliot and Irène Joliot-Curie discovered artificial radioactivity. Most natural radionuclides are produced through one of four radioactive decay chains, each chain fed by a long-lived and heavy parent radionuclide. The vast majority of currently known radionuclides, however, are man-made and artificially produced through a process of nuclear activation which uses bombardment of a stable nuclide with a suitable energetic particle to induce a nuclear transformation. Various particles or electromagnetic radiation generated by a variety of machines are used for this purpose, most notably neutrons from nuclear reactors for neutron activation and protons from cyclotrons for proton activation.

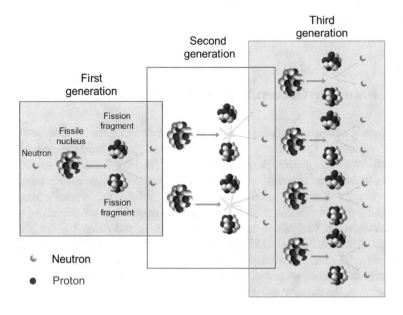

© Springer International Publishing Switzerland 2016
E.B. Podgoršak, *Radiation Physics for Medical Physicists*,
Graduate Texts in Physics, DOI 10.1007/978-3-319-25382-4_12

Three models for nuclear activation are discussed in this chapter and relevant medical physics examples are given for each model. Also discussed are two nuclear chain reactions: fission and fusion. While the fission chain reaction has been used for practical purposes for the past few decades, the fusion chain reaction has not left the laboratory environment yet. Also addressed is the production of the technetium-99m radionuclide with a molybdenum-technetium radionuclide generator for use in nuclear medicine imaging. The chapter concludes with a discussion of proton activation with protons obtained from a cyclotron.

12.1 Origin of Radioactive Elements (Radionuclides)

With respect to their origin radioactive nuclides (radionuclides) are divided into two categories:

1. *Naturally-occurring.*
2. *Man-made or artificially produced.*

There is no essential physical difference between the two categories of radioactivity; the division is mainly historical and related to the sequence of radioactivity-related discoveries. *Henri Becquerel* discovered natural radioactivity in 1896 when he noticed that uranium spontaneously produced an invisible, penetrating radiation that affected photographic plates. Almost 40 years later, in 1934 *Irène Joliot-Curie* and *Frédéric Joliot* discovered artificial radioactivity during a series of experiments in which they bombarded boron samples with naturally occurring α-particles and produced nitrogen that was unstable and emitted positrons through β^+ decay.

12.2 Naturally-Occurring Radionuclides

The naturally occurring radioactive elements are almost exclusively members of one of four radioactive series that all begin with very heavy and long-lived parents that have half-lives of the order of the age of the earth. A few long-lived light radionuclides are also found in nature and so is carbon-14, a carbon radioisotope produced by cosmic proton radiation. Cosmic rays are energetic particles that originate from outer space and strike the Earth's atmosphere. The vast majority of cosmic rays ($\sim$87%) are protons, some 12% are α-particles (helium ions), and about 1% are energetic electrons. Robert Millikan in 1925 coined the term "cosmic ray" and proved that the "rays" were of extraterrestrial origin. Subsequently, it was shown that the cosmic rays were mainly charged particles; however, the term "ray" continues to be used for designation of the extraterrestrial radiation.

The four naturally occurring series and their original parent radionuclide are named as follows:

- *Thorium series* originates with thorium-232.
- *Actinium series* originates with uranium-235.
- *Neptunium series* originates with neptunium-237.
- *Uranium series* originates with uranium-238.

The main characteristics of the four naturally occurring series are listed in Table 12.1. The series begin with a specific parent nucleus of very long half-life that decays through several daughter products to reach a stable lead isotope in the thorium, actinium, and uranium series and stable bismuth-209 nuclide in the neptunium series. For each of the four series most of the transitions toward the stable nuclides are α decays interspersed with few β decays. The number of α decays in each series varies between 6 and 8, as shown in Table 12.1 with the column labeled N_α.

The atomic mass numbers A for each member of the thorium series are multiples of 4 and, consequently, the thorium series is sometimes referred to as the $4n$ series. The atomic mass numbers of members of the neptunium series follow the rule $4n + 1$, the uranium series $4n + 2$, and the actinium series $4n + 3$. Therefore, these series are often referred to as the $4n + 1$, $4n + 2$, and $4n + 3$ series, respectively, as indicated in Table 12.1.

It is assumed that collapsing stars created all heavy radioactive elements in approximately equal proportions; however, these elements differ in their half-lives and this resulted in significant variations in today's abundance of radioactive heavy elements. Neptunium-237 has a significantly shorter half-life than the other three long-lived parent nuclei listed in Table 12.1. It does no longer occur naturally because it has completely decayed since the formation of the earth some 4.6×10^9 years ago. The other three parent nuclei (^{232}Th, ^{235}U, and ^{238}U) with much longer half-lives are still found in nature and serve as parents of their own series.

Cosmic ray protons continually produce small amounts of radioactive materials. The most notable example is carbon-14 that decays with a half-life of 5730 years

Table 12.1 The four naturally occurring radioactive series (N_α gives the number of steps in the decay chain required to reach the final stable nucleus)

Name of series		Parent	First decay	N_α	Found in nature	Half-life (10^9 years)	Stable end-product
Thorium	$4n$	$^{232}_{90}$Th	$^{228}_{88}$Ra $+ \alpha$	6	YES	14.05	$^{208}_{82}$Pb
Actinium	$4n + 3$	$^{235}_{92}$U	$^{231}_{90}$Th $+ \alpha$	7	YES	0.704	$^{207}_{82}$Pb
Neptunium	$4n + 1$	$^{237}_{93}$Np	$^{233}_{91}$Pa $+ \alpha$	7	NO	2.144×10^{-3}	$^{209}_{83}$Bi
Uranium	$4n + 2$	$^{238}_{92}$U	$^{234}_{90}$Th $+ \alpha$	8	YES	4.47	$^{206}_{82}$Pb

and is used for the so-called carbon dating of once-living objects not older than some 50 000 years.

A few naturally-occurring lighter-than-lead radioactive elements can be found in the earth. Most notable among them is potassium-40 (^{40}K) with a half-life of 1.277×10^9 years. Since it is present in all foods, ^{40}K accounts for the largest proportion of the naturally-occurring radiation load through ingestion among humans.

The existence of the four radioactive series with long-lived parent nuclei serves as the source of many short-lived daughters that are in transient or secular equilibrium with their parents. For example, radium-226 with its half-life of 1602 years would have disappeared long ago were it not for the uranium-238 decay series that provides constant replenishment of radium-226 in the environment.

12.3   Man-Made (Artificial) Radionuclides

Man-made (artificial) radionuclides are manufactured by bombarding stable nuclides or very long-lived radionuclides with energetic particles or energetic x-rays produced by special machines of various kinds. The process is referred to as radioactivation or nuclear activation. Since Irène Joliot-Curie and Frédéric Joliot discovered artificial radioactivity in 1934 over 3000 different artificial radionuclides have been synthesized and investigated. Thus, the current list of known nuclides contains some 280 stable nuclides and over 3000 radioactive nuclides (radionuclides). Some 200 radionuclides are used in industry and medicine, and most of them are produced through radioactivation.

A variety of particles may induce radioactivation; however, most commonly radioactivation is achieved by bombarding stable target nuclei with neutrons produced in nuclear reactors or by protons produced in cyclotrons. The following terminology is used in radioactivation:

1. Nuclear activation induced with thermal or fast neutrons from a nuclear reactor is called *neutron activation* or *neutron capture*.
2. Nuclear activation induced with protons (and possibly heavier ions such as deuterons, α-particles, and heavy ions) from a cyclotron and synchrotron is called *proton activation* or *proton capture*.
3. Nuclear activation induced by high-energy x-rays from a linear accelerator is referred to as *nuclear photoactivation*.

In addition to radioactivation, short-lived radionuclides used in nuclear medicine can be obtained from the so-called *radionuclide generators* that contain a relatively long-lived parent (produced through radioactivation) decaying into short-lived daughter that can be chemically extracted from the parent stored in a radionuclide generator.

Nuclear reactors are the main source of radionuclides used in medicine. These radionuclides are produced either through neutron activation of stable target nuclei placed into the reactor or by chemical separation from fission products resulting from the fission process induced in special targets or nuclear fuel whereby fissile nuclei, upon bombardment with thermal neutrons, split into two lighter fragments and two or three fission neutrons.

12.4 Radionuclides in the Environment

Over 60 radionuclides can be found in the environment and some of them pose a health hazard to humans. They are grouped into four categories as follows:

1. *Primordial* – originate from before the creation of the Earth.
2. *Secondary* – originate from decay of primordial radionuclides.
3. *Cosmogenic* – continually produced by cosmic radiation hitting the Earth.
4. *Man-made* or *artificial* – produced through the process of radioactivation mainly in nuclear reactors.

Pathways of radionuclides into environment:

1. *Atmospheric pathway* (through human activity, radioactive decay, cosmogenic reactions).
2. *Water pathway* (deposited in water from air or from ground through erosion, seepage, leaching, mining, etc.).
3. *Food chain pathway* (radionuclides in water and air may enter the food chain).

Pathways of radionuclides into human body:

1. *Ingestion*
2. *Inhalation*
3. *Through skin.*

12.5 General Aspects of Nuclear Activation

Several types of nuclear activation are known; for example: neutron activation, proton activation and nuclear photoactivation. In medical physics neutron activation is important in production of radionuclides used for external beam radiotherapy, brachytherapy, and molecular imaging; proton activation is important in production of positron emitters used in PET scanning; and nuclear photoactivation is important from a radiation protection point-of-view when components of high-energy radiotherapy machines become activated during patient treatment and pose a potential radiation hazard to staff using the equipment.

The three activation processes listed above are inherently different, yet there are several common features that govern the physics behind the processes, such as cross section, target thickness, Q value, and threshold.

12.5.1 Nuclear Reaction Cross Section

Particles of an incident beam striking a target can interact with the target nuclei through the following three processes:

1. *Scattering*
2. *Absorption*
3. *Nuclear reaction.*

In traversing a target the beam is attenuated in:

1. *Intensity or*
2. *Energy or*
3. *Both intensity and energy.*

In a simplistic approach we can estimate the probability for a reaction between the incident particle and a target nucleus by treating the incident particles as points and the target nuclei as projecting an area πR^2 defined by the nuclear radius R. The following assumptions are made:

- Any time an incident particle hits a nucleus, a reaction is assumed to happen; no reaction occurs when the particle misses the nucleus.
- This geometrical picture takes no account of the finite size of the incident particle nor does it consider the range of interaction forces that are in effect between the incident particle and the target nucleus.
- Rather than treating a geometrical cross sectional area πR^2 as a measure of interaction probability, we assign to the nucleus an effective area σ perpendicular to the incident beam such that a reaction occurs every time a bombarding particle hits any part of the effective disk area.
- This effective area is referred to as the reaction cross-section σ and is usually measured in barn, where 1 barn $= 1\,\mathrm{b} = 10^{-24}\,\mathrm{cm}^2$. The cross section σ is proportional to the reaction probability P.
- The range of reaction cross sections σ in nuclear physics varies from a low of 10^{-19} b to a high of 10^6 b with the lower limit in effect for weak neutrino interactions with nuclei and the upper limit in effect for thermal neutron capture in certain nuclides.

The target of thickness x_0 projects an area S to the incident particle beam. The target contains N nuclei, each characterized with a reaction cross section σ. The density of nuclei $n^\square$ represents the number of nuclei N per volume V of the target with $V = Sx_0$. To determine the reaction rate $\mathcal{R}$ (number of reactions per unit time) we consider two target options with regard to target thickness x_0: *thin targets* and *thick targets*.

12.5.2 Thin Targets

A *thin target* is thin enough so that no significant overlap between target nuclei occurs as the particle beam penetrates the target. This implies that negligible masking of target nuclei occurs in a thin target. The probability P for an incident particle to trigger a reaction in a thin target is the ratio of the effective area σN over the target area S

$$P = \frac{\sigma N}{S} = \frac{\sigma N x_0}{S x_0} = n^\square \sigma x_0. \tag{12.1}$$

If the number of incident particles per unit time is $\dot{N}_0$, then $\mathcal{R}$, the number of reactions per unit time, is given as follows

$$\mathcal{R} = P\dot{N}_0 = \dot{N}_0 n^\square \sigma x_0. \tag{12.2}$$

12.5.3   Thick Target

In comparison with a thin target, a thick target has a thickness x_0 that engenders considerable masking of target nuclei. In this case we assume that a thick target consists of a large number of thin targets. In each thin target layer of thickness dx the number of incident particles per unit time $\dot{N}$ diminishes by $d\dot{N}$, so that we can write $d\dot{N}(x)$ as

$$- d\dot{N}(x) = \dot{N}(x) n^\square \sigma dx \tag{12.3}$$

or

$$\int_{\dot{N}_0}^{\dot{N}(x_0)} \frac{d\dot{N}(x)}{\dot{N}(x)} = - \int_0^{x_0} n^\square \sigma dx, \tag{12.4}$$

where

$\dot{N}_0$ is the number of particles per unit time striking the target.
$\dot{N}(x_0)$ is the number of particles per unit time that traverse the thick target x_0.

The solution to (12.4) is

$$\dot{N}(x_0) = \dot{N}_0 e^{-n^\square \sigma x_0}, \tag{12.5}$$

and the number of reactions per unit time $\mathcal{R}$ in the thick target is now given by the following

$$\mathcal{R} = \dot{N}_0 - \dot{N}(x_0) = \dot{N}_0\{1 - e^{-n^\square \sigma x_0}\}. \tag{12.6}$$

Equation (12.6) reduces to thin target relationship of (12.2) for small thicknesses x_0 to give

$$e^{-n^\square \sigma x_0} \approx 1 - n^\square \sigma x_0 \qquad \text{or} \qquad 1 - e^{-n^\square \sigma x_0} \approx n^\square \sigma x_0 \tag{12.7}$$

to result in $\mathcal{R}$ given for a thin target in (12.2).

12.6 Nuclear Activation with Neutrons (Neutron Activation)

In practice the most commonly used nuclear activation process is triggered by *thermal neutrons* in a nuclear reactor, where a stable parent target P upon bombardment with neutrons is transformed into a radioactive daughter D that decays with a decay constant λ_D into a granddaughter G

$$P \rightarrow D \rightarrow G. \tag{12.8}$$

The situation in neutron activation is similar to the Parent $\rightarrow$ Daughter $\rightarrow$ Granddaughter decay series discussed in Sect. 10.3, except that λ_P in neutron activation does no longer apply, since the parent is stable or long-lived. Yet, we can use the decay formalism for the activation problem as long as we replace the parent decay constant λ_P by the product $\sigma_P \dot\varphi$ where:

σ_P is the probability for activation of the parent nucleus governed by the activation cross section usually expressed in barn/atom where 1 barn $= 1\,\mathrm{b} = 10^{-24}\,\mathrm{cm}^2$.

$\dot\varphi$ is the fluence rate of neutrons in the reactor usually expressed in neutrons per cm^2 per second, i.e., $\mathrm{cm}^{-2}{\cdot}\mathrm{s}^{-1}$. Typical modern reactor fluence rates are of the order of $10^{11}\,\mathrm{cm}^{-2}{\cdot}\mathrm{s}^{-1}$ to $10^{14}\,\mathrm{cm}^{-2}{\cdot}\mathrm{s}^{-1}$. An assumption is made that the neutron fluence rate $\dot\varphi$ remains constant for the duration of the activation process, and this is not always easy to achieve in practice, especially when activation times are long.

12.6.1 Infinite Number of Parent Nuclei: Saturation Model

The daughter nuclei are produced at a rate of $\sigma_P \dot\varphi N_P(t)$ and they decay with a rate of $\lambda_D N_D(t)$. The number of daughter nuclei is $N_D(t)$ and the overall rate of change of the number of daughter nuclei is dN_D/dt obtained by combining the production rate of daughter nuclei $\sigma_P \dot\varphi N_P(t)$ with the decay rate of daughter nuclei $\lambda_D N_D(t)$ to get

$$\frac{dN_D(t)}{dt} = \sigma_P \dot\varphi N_P(t) - \lambda_D N_D(t), \tag{12.9}$$

where $N_P(t)$ is the number of parent target nuclei. Two simplifying assumptions are usually made when dealing with neutron activation theory. The two assumptions are:

1. That in neutron activation a negligible fraction of the parent atoms is transformed, so that the number of residual target atoms $N_P(t_0)$ equals to $N_P(0)$, the initial number of target atoms placed into the reactor for activation purposes at time $t = 0$. The time t_0 is the total time the target is left in the reactor. The activation

model that neglects the depletion of the number of target nuclei is referred to as the *saturation model*.

2. That the neutron fluence rate $\dot{\varphi}$ at the position of the sample is contributed from all directions. The sample in the form of pellets is irradiated in a "sea" of thermal neutrons and we may assume that the sample is a thin target that does not appreciably affect the neutron fluence inside the pellet.

For the initial conditions $N_P(t = 0) = N_P(0)$ and $N_D(t = 0) = N_D(0) = 0$ as well as the general condition that $N_P(t) = N_P(0) = \text{const}$, the differential equation for dN_D/dt of (12.9) is written as

$$\frac{dN_D(t)}{dt} = \sigma_P \dot{\varphi} N_P(0) - \lambda_D N_D(t) \tag{12.10}$$

or in integral form as

$$\int_0^{N_D(t)} \frac{d\{\sigma_P \dot{\varphi} N_P(0) - \lambda_D N_D\}}{\sigma_P \dot{\varphi} N_P(0) - \lambda_D N_D} = -\lambda_D \int_0^t dt. \tag{12.11}$$

The solution of (12.11) is as follows

$$N_D(t) = \frac{\sigma_P \dot{\varphi} N_P(0)}{\lambda_D} \{1 - e^{-\lambda_D t}\}. \tag{12.12}$$

The daughter activity $\mathcal{A}_D(t)$ equals to $\lambda_D N_D(t)$, thus we can write $\mathcal{A}_D(t)$ as

$$\mathcal{A}_D(t) = \sigma_P \dot{\varphi} N_P(0)\{1 - e^{-\lambda_D t}\} = (\mathcal{A}_D)_{\text{sat}}\{1 - e^{-\lambda_D t}\}, \tag{12.13}$$

where we define $(\mathcal{A}_D)_{\text{sat}}$, the saturation daughter activity that can be produced by bombardment of the parent target with neutrons, as equal to $\sigma_P \dot{\varphi} N_P(0)$.

Equation (12.13) is a simple exponential relationship and its initial slope $d\mathcal{A}_D(t)/dt$ at $t = 0$ is defined as the *radioactivation yield* Y_D of the daughter produced in the radioactivation process. The radioactivation yield represents the initial rate of formation of new daughter activity that depends upon the irradiation conditions as well as the decay constant of the daughter λ_D, as seen from the following expression

$$Y_D = \left.\frac{d\mathcal{A}_D}{dt}\right|_{t=0} = \sigma_P \dot{\varphi} N_P(0)\lambda_D = \lambda_D(\mathcal{A}_D)_{\text{sat}} = \frac{(\mathcal{A}_D)_{\text{sat}}}{\tau_D}. \tag{12.14}$$

The build up of daughter activity $\mathcal{A}_D(t)$ in a target subjected to constant bombardment with neutrons in a reactor is illustrated in Fig. 12.1. The radioactivation yield Y_D is given by the initial slope of the growth curve at time $t = 0$. The extrapolation of the tangent to the growth curve at $t = 0$ intersects the asymptotic saturation activity line at a time $t = \tau_D = (t_{1/2})_D/\ln 2$, where $(t_{1/2})_D$ is the half-life of the daughter.

Fig. 12.1 Growth of
daughter activity $\mathcal{A}_D(t)$
normalized to saturation
activity $(\mathcal{A}_D)_{sat}$ and plotted
against time normalized to
the half-life of the daughter
$(t_{1/2})_D$. The slope of the
tangent on the
$\mathcal{A}_D(t)/(\mathcal{A}_D)_{sat}$ versus t
curve at $t = 0$, defined as the
activation yield Y_D, is also
shown

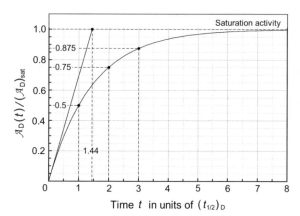

The following observations regarding the daughter activity growth curve, as given
in (12.13), can now be made:

1. Initially at small t, where $\exp(-\lambda_D t) \approx 1 - \lambda_D t$, the growth of $\mathcal{A}_D(t)$ is rapid and
   almost linear with time, since

$$\mathcal{A}_D(t) = (\mathcal{A}_D)_{sat}\{1 - e^{-\lambda_D t}\} \approx (\mathcal{A}_D)_{sat}\{1 - 1 + \lambda_D t - \cdots\}$$
$$\approx (\mathcal{A}_D)_{sat}\lambda_D t, \tag{12.15}$$

   but eventually at large times t the daughter activity $\mathcal{A}_D(t)$ becomes saturated (i.e.,
   reaches a steady-state) at $(\mathcal{A}_D)_{sat}$ and decays as fast as it is produced.
2. Equation (12.13) and Fig. 12.1 show that:

   (a) For an activation time $t = (t_{1/2})_D$, half the maximum activity $(\mathcal{A}_D)_{sat}$ is
       produced
   (b) For $t = 2(t_{1/2})_D$, 3/4 of $(\mathcal{A}_D)_{sat}$ is produced
   (c) For $t = 3(t_{1/2})_D$, 7/8 of $(\mathcal{A}_D)_{sat}$ is produced, etc.

3. Because of the relatively slow approach to saturation $(\mathcal{A}_D)_{sat}$, it is generally
   accepted that in practice activation times beyond $2(t_{1/2})_D$ are not worthwhile.

12.6.2 Finite Number of Parent Nuclei: Depletion Model

In situations where a measurable fraction of the target is consumed during the activation process, we can no longer assume that $N_P(t) = \text{const.}$ The fractional decrease
in the number of parent atoms depends on the activation cross section σ_P and on the
fluence rate $\dot{\varphi}$ of the reactor. The activation model that accounts for the depletion
of the number of the target nuclei during the radioactivation process is called the
depletion model.

In general, the rate of change in the number of parent atoms $N_P(t)$ with time t can be written as follows

$$\frac{dN_P(t)}{dt} = -\sigma_P \dot{\varphi} N_P(t), \tag{12.16}$$

similarly to the expression for radioactive decay given in (10.6) but replacing λ_P in (10.6) with the product $\sigma_P \dot{\varphi}$. The general solution for $N_P(t)$ of (12.16) is then

$$N_P(t) = N_P(0)e^{-\sigma_P \dot{\varphi} t}, \tag{12.17}$$

with $N_P(0)$ the initial number of parent nuclei placed into the reactor at time $t = 0$ and $N_P(t)$ the number of parent nuclei at time t.

The general expression for dN_D/dt, the rate of change in the number of daughter nuclei, is the number of parent nuclei transformed into daughter nuclei [governed by $N_P(t)$, σ_P, and $\dot{\varphi}$] minus the number of daughter nuclei that decay [governed by $N_D(t)$ and λ_D]

$$\frac{dN_D(t)}{dt} = \sigma_P \dot{\varphi} N_P(t) - \lambda_D N_D(t), \tag{12.18}$$

with $N_P(t)$ given in (12.17) in parallel to (10.25) for the P $\to$ D $\to$ G decay series.

The solution to (12.18), following the steps taken in the derivation of (10.34) for the P $\to$ D $\to$ G decay series and using the initial conditions for N_P and N_D

$$N_P(t = 0) = N_P(0) \tag{12.19}$$

and

$$N_D(t = 0) = N_D(0) = 0 \tag{12.20}$$

is now as follows

$$N_D(t) = N_P(0)\frac{\sigma_P \dot{\varphi}}{\lambda_D - \sigma_P \dot{\varphi}}\{e^{-\sigma_P \dot{\varphi} t} - e^{-\lambda_D t}\}. \tag{12.21}$$

Recognizing that $\mathcal{A}_D(t) = \lambda_D N_D(t)$ and assuming the validity of the depletion model, we get the following general expression for the growth of the daughter activity $\mathcal{A}_D(t)$

$$\mathcal{A}_D(t) = N_P(0)\frac{\sigma_P \dot{\varphi} \lambda_D}{\lambda_D - \sigma_P \dot{\varphi}}\{e^{-\sigma_P \dot{\varphi} t} - e^{-\lambda_D t}\}. \tag{12.22}$$

Since (12.22) for $\mathcal{A}_D(t)$ in neutron activation is identical in form to (10.35) for a radioactive decay series, we use the analysis presented with regard to the decay series to obtain solutions for the general daughter growth in neutron activation. Generally, in neutron activation $\sigma_P \dot{\varphi} < \lambda_D$ and this results in transient equilibrium dynamics, as discussed for the decay series in Sect. 10.3.4.

When $\sigma_P \dot{\varphi} \ll \lambda_D$, we are dealing with a special case of transient equilibrium called secular equilibrium for which (12.22) will simplify to an expression that was given in (12.13) for the saturation model and was derived under the assumption that the fraction of nuclei transformed from parent to daughter in neutron activation is negligible in comparison to the initial number of parent atoms $N_P(0)$. Equation (12.22) then reads

$$\mathcal{A}_D(t) = \sigma_P \dot{\varphi} N_P(0)\{1 - e^{-\lambda_D t}\} = (\mathcal{A}_D)_{sat}\{1 - e^{-\lambda_D t}\}, \qquad (12.23)$$

where $(\mathcal{A}_D)_{sat} = \sigma_P \dot{\varphi} N_P(0)$ is the saturation activity that is attainable by the target under the condition of secular equilibrium.

In the saturation model the activity $\mathcal{A}_D(t)$ approaches the saturation activity $(\mathcal{A}_D)_{sat}$ exponentially, as given in (12.23) and shown in Fig. 12.1. In saturation the production rate of the daughter equals the decay rate of the daughter resulting in a constant $N_D(t)$ and constant saturation activity $(\mathcal{A}_D)_{sat}$.

Usually, the growth of daughter in neutron activation is treated under the condition of secular equilibrium; however, with high enough reactor fluence rate $\dot{\varphi}$ and low enough daughter decay constant λ_D, this approximation may no longer be valid. The theoretical treatment then should recognize the radioactivation process as one of transient equilibrium for which account must be taken of the depletion of target nuclei, as given in (12.22). The following points should be noted:

- As discussed in detail in Sect. 10.3, the daughter activity $\mathcal{A}_D(t)$ in transient equilibrium cannot be assumed to reach saturation with increasing time t. Rather, the daughter activity $\mathcal{A}_D(t)$ is zero at time $t = 0$, and with increasing time first rises with t, reaches a maximum $(\mathcal{A}_D)_{max}$ at time $t = (t_{max})_D$, and then drops as t increases further until at $t = \infty$ it becomes zero again.
- The daughter activity will reach its maximum $(\mathcal{A}_D)_{max} = \mathcal{A}_D[(t_{max})_D]$ at the point of ideal equilibrium that occurs at a time $(t_{max})_D$ where $d\mathcal{A}_D(t_{max})/dt = 0$ and $\mathcal{A}_D[(t_{max})_D] = \sigma_P \dot{\varphi} N_P[(t_{max})_D]$. Note that in general $\mathcal{A}_D[(t_{max})_D] < (\mathcal{A}_D)_{sat}$.
- The time $(t_{max})_D$ is given as

$$(t_{max})_D = \frac{\ln \dfrac{\sigma_P \dot{\varphi}}{\lambda_D}}{\sigma_P \dot{\varphi} - \lambda_D}. \qquad (12.24)$$

Equation (12.24) with $\sigma_P \dot{\varphi}$ replaced by λ_P is identical in form to $(t_{max})_D$ that was calculated for a decay series in (10.37).

Defining new parameters m, $(x_D)_{max}$ and $(y_D)_{max}$ as well as variables x, y_P, and y_D, similarly to the approach we took in Sect. 10.4 for the radioactive decay series, we can understand better the dynamics resulting from the saturation and depletion models of the neutron activation process. The parameters and variables are for neutron activation defined as follows:

1. Factor m, now called *activation factor* in parallel with the decay factor m of the radioactive series decay:

$$m = \frac{\sigma_P \dot{\varphi}}{\lambda_D}, \qquad\qquad\qquad \text{compare with (10.44).} \qquad (12.25)$$

2. Variable x

$$x = \frac{\sigma_P \dot{\varphi}}{\ln 2} t = m \frac{\lambda_D}{\ln 2} t = m \frac{t}{(t_{1/2})_D}, \qquad \text{compare with (10.41).} \qquad (12.26)$$

3. Normalized number of parent nuclei y_P

$$y_P = \frac{N_P(t)}{N_P(0)} = e^{-\sigma_P \dot{\varphi} t} = e^{-x \ln 2} = \frac{1}{2^x}, \quad \text{see (10.42) and (10.47).} \qquad (12.27)$$

4. Normalized number of daughter nuclei y_D

$$y_D = \frac{A_D(t)}{\sigma_P \dot{\varphi} N_P(0)}, \qquad \text{see (10.43), (10.45) and (10.46).} \qquad (12.28)$$

5a. x coordinate of the $(y_D)_{max}$ point for $m > 0$ and $m \neq 1$

$$(x_D)_{max} = \frac{m}{m-1} \frac{\ln m}{\ln 2}, \qquad \text{see (10.48) and (10.50).} \qquad (12.29)$$

5b. x coordinate of the $(y_D)_{max}$ point for $m = 1$

$$(x_D)_{max} = \frac{1}{\ln 2} = 1.44, \qquad \text{see (10.51).} \qquad (12.30)$$

6a. Maximum point in y_D for $m > 0, m \neq 1$

$$(y_D)_{max} = y_P(x_{max}) = 2^{\left(\frac{m}{1-m}\right) \frac{\ln m}{\ln 2}} = \frac{1}{2^{x_{max}}},$$
$$\equiv e^{\frac{m}{1-m} \ln m} = e^{-(\ln 2)(x_{max})_D}, \qquad \text{see (10.53).} \qquad (12.31)$$

6b. Maximum point in y_D for $m = 1$

$$(y_D)_{max} = \frac{1}{e} = 0.368, \qquad \text{see (10.54).} \qquad (12.32)$$

Like in (10.45), the variable $y_D(x)$ for the depletion model is given by the following function, after inserting (12.25) and (12.28) into (12.22) to get

$$y_D = \frac{1}{1-m} \left\{ e^{-x \ln 2} - e^{-\frac{x}{m} \ln 2} \right\} = \frac{1}{1-m} \left\{ \frac{1}{2^x} - \frac{1}{2^{x/m}} \right\}. \qquad (12.33)$$

Equation (12.33) is valid for all positive m except for $m = 1$. For $m = 1$, $y_D(x)$ is given by the following function, as discussed in relation to (10.46), using L'Hôpital rule

$$y_D(m = 1) = \frac{x \ln 2}{2^x}. \qquad (12.34)$$

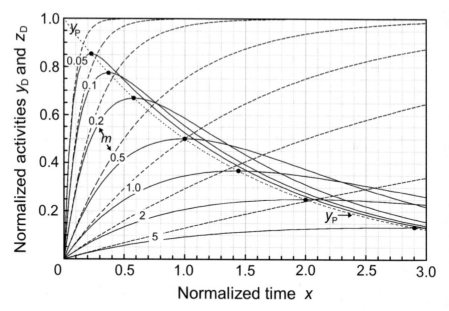

Fig. 12.2 Normalized daughter activities z_D of (12.35) for the saturation model (*dotted curves*) and y_D of (12.33) for the depletion model (*solid curves*) against the variable x for various decay factors m in the transient equilibrium region from 0.05 to 1.0 as well as for $m = 2$ and $m = 5$ in the non-equilibrium region where $m > 1$. The y_P curve is shown *dashed*. Points of ideal equilibrium specified for the depletion model by $(x_D)_{max}$ and $(y_D)_{max}$ are indicated with *heavy dots* on the y_P curve. Variables x and y_P are given by (12.26) and (12.27), respectively. Variables z_D for the saturation model and y_D for the depletion model ($m \neq 1$) are given by (12.35) and (12.33), respectively. Variable y_D for the depletion model with $m = 1$ is given by (12.34). The activation factor m is defined in (12.25)

For the saturation model of (12.13), on the other hand, the normalized daughter activity z_D is given for any $m > 0$ as follows, after inserting (12.25) and (12.28) into (12.23)

$$z_D = \frac{\mathcal{A}_D(t)}{(\mathcal{A}_D)_{sat}} = \frac{\mathcal{A}_D(t)}{\sigma_P \dot{\varphi} N_P(0)} = 1 - e^{-\lambda_D t} = 1 - e^{-\frac{\lambda_D}{\sigma_P \dot{\varphi}}(\ln 2)x}$$

$$= 1 - e^{-\frac{(\ln 2)x}{m}} = 1 - \frac{1}{2^{\frac{x}{m}}}. \tag{12.35}$$

To illustrate the general case of neutron activation for any m between zero (secular equilibrium) and one (start of non-equilibrium conditions) we show in Fig. 12.2 a plot of y_D and z_D, against x for various m in the range from 0.05 to 5 for both activation models: the saturation model of (12.35) with dashed curves and the depletion model of (12.33) with solid curves. For comparison we also show the y_P curve that indicates the depletion of the target nuclei during the neutron activation process.

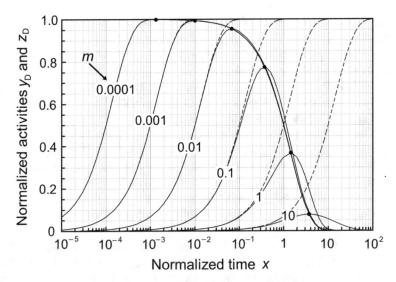

Fig. 12.3 Normalized daughter activities z_D of (12.35) for the saturation model (*dashed curves*) and y_D of (12.33) for the depletion model (*solid curves*) against the variable x for various activation factors m in the range from $m = 10^{-4}$ to $m = 10$. The *heavy dots* represent $(y_D)_{max}$, the maxima of y_D for given m and follow the normalized parent activity y_P of (12.27). Variables x and y_P are given by (12.26) and (12.27), respectively. Variables z_D for the saturation model and y_D for the depletion model ($m \neq 1$) are given by (12.35) and (12.33), respectively. Variable y_D for the depletion model with $m = 1$ is given by (12.34). The activation factor m is defined in (12.25)

The points of ideal equilibrium in the depletion model, where y_D reaches its maximum, are shown with dots on the y_P curve in Fig. 12.2. The expressions for $(x_D)_{max}$ and $(y_D)_{max}$ in terms of the activation factor $m \neq 1$ are given by (12.29) and (12.31), respectively, and for $m = 1$ by (12.30) and (12.32), respectively.

In Fig. 12.3 we plot the normalized daughter activities z_D for the saturation model and y_D for the depletion model from (12.35) and (12.33), respectively, against the variable x on a logarithmic scale to cover five orders of magnitude in the activation factor m ranging from 10^{-4} to 10. Some of the data presented in Fig. 12.3 have already been plotted in Fig. 12.2 that covers a much smaller range in m (from 0.05 to 5). The maxima $(y_D)_{max}$ in depletion model curves are indicated with heavy dots that also follow a trace of y_P, the normalized number of parent nuclei given in (12.27).

The following conclusions can now be reached with regard to Figs. 12.2 and 12.3:

1. In practical nuclear activation procedures the activation factor defined in (12.25) as $m = \sigma_P \dot{\varphi} / \lambda_D$ is generally small, justifying the use of the saturation model in studies of radioactivation dynamics. However, since m depends linearly on the fluence rate $\dot{\varphi}$, nuclear activation processes with very high fluence rates or relatively long activation times may invalidate the saturation model in favor of the depletion model.
2. The initial slope dy_D/dx at $t = 0$ is proportional to the activation yield Y defined for the saturation model in (12.14). A closer look at Figs. 12.2 and 12.3 reveals

that both the saturation model and the depletion model predict y_D with the same initial slopes equal to $(\ln 2/m)$ irrespective of the magnitude of m. This result can be obtained by taking the derivative dy_D/dx at $x = 0$ of (12.35) for the saturation model and (12.33) for the depletion model.

3. For all m in the saturation model z_D approaches its saturation value of 1.0 exponentially, while in the depletion model y_D reaches its peak value $(y_D)_{max}$ at $(x_D)_{max}$ and then decreases with increasing x.

4. In the saturation model, for a given m, the normalized daughter activity z_D approaches exponentially the saturation value $(z_D)_{sat} = 1$. The larger is m, the shallower is the initial slope, and the slower is the approach to saturation.

5. In the depletion model, for a given m, the normalized daughter activity y_D exhibits a maximum value $(y_D)_{max}$ that is smaller than the saturation value $(z_D)_{sat} = 1$. The larger is m, the larger is the discrepancy between the two models and the smaller is $(y_D)_{max}$ in comparison with $(z_D)_{sat} = 1$.

6. Parameter $(y_D)_{max}$ is the point of *ideal equilibrium* calculated from (12.31) for $m \neq 1$ and (12.32) for $m = 1$. It depends on $(x_D)_{max}$, as shown in (12.31). Parameter $(x_D)_{max}$ in turn depends on the activation factor m and is calculated from (10.50) and (12.29). As m decreases from $m = 1$ toward zero, $(x_D)_{max}$ decreases and $(y_D)_{max}$ increases, as shown by dots on the y_P curve in Figs. 12.2 and 12.3.

7. For $m > 10^{-3}$, parameter $(y_D)_{max}$ decreases with increasing m. Thus, in this region of m the depletion model should be used for determination of daughter activity.

8. For all $0 < m < 1$, variables y_P and y_D are said to be in transient equilibrium at $x \gg (x_D)_{max}$. For $m \geq 1$ no equilibrium between y_P and y_D exists at any x.

9. For $m < 10^{-2}$, y_P and y_D are in the special form of transient equilibrium called secular equilibrium.

10. For $m < 10^{-3}$, the saturation model and the depletion model give identical results, (i.e., $y_D = z_D$), for $x \leq (x_D)_{max}$ and attain a value of 1 at $x = (x_D)_{max}$. However, for $x > (x_D)_{max}$, z_D remains in saturation, while y_D decreases in harmony with y_P.

11. Using the expression for the normalized number of daughter nuclei (12.28), we can now express $(\mathcal{A}_D)_{max}$, the maximum daughter activity in the depletion model, as follows

$$(\mathcal{A}_D)_{max} = (y_{max})_D \, \sigma_P \dot{\varphi} N_P(0) = \sigma_P \dot{\varphi} N_P(0) \, 2^{-(x_{max})_D}$$
$$= \sigma_P \dot{\varphi} N_P(0) e^{-\frac{m}{1-m} \ln m}. \tag{12.36}$$

12. Equation (12.36) shows that the maximum daughter activity $(\mathcal{A}_D)_{max}$ depends on the saturation activity $(\mathcal{A}_D)_{sat} = \sigma_P \dot{\varphi} N_P(0)$ and on $(y_D)_{max}$ which approaches 1 for $m \to 0$, as shown in Fig. 10.4. However, as m increases toward 1, $(y_D)_{max}$ decreases, resulting in $(\mathcal{A}_D)_{max}$ that may be significantly smaller than $(\mathcal{A}_D)_{sat}$.

13. Since the normalized daughter activity $y_D(x)$ decreases with x for $x > (x_D)_{max}$, it is obvious that activation times beyond $(x_D)_{max}$ are counter-productive.

12.6.3 Maximum Attainable Specific Activities in Neutron Activation

As is evident from Fig. 12.3, $(y_D)_{max}$, the maximum normalized daughter activity for the depletion model decreases with the activation factor $m = \sigma_p \dot{\varphi}/\lambda_D$. In practice this means that, for a given daughter radionuclide, $(y_D)_{max}$ depends only on the particle fluence rate $\dot{\varphi}$, since the parameters σ_p and λ_D remain constant.

We now determine the maximum daughter specific activities that can be attained during the activation process, as predicted by the saturation model and the depletion model.

For the *saturation model* we use (12.35) to get the maximum daughter specific activity $(a'_D)_{max}$ as

$$
(a'_D)_{max} = (a_D)_{sat} = \frac{(A_D)_{sat}}{M_P} = (z_D)_{max}\,\sigma_P \dot{\varphi}\,\frac{N_P(0)}{M_P} = \left(\sigma_P \frac{N_A}{A_P}\right)\dot{\varphi}, \tag{12.37}
$$

where M_p and A_p are the atomic mass and the atomic mass number of the parent nucleus, respectively, and the parameter $(z_D)_{max}$ is equal to the saturation value of z_D equal to 1. Since $\sigma_p N_A/A_p$ is constant for a given parent nucleus, we note that $(a_D)_{max}$ is linearly proportional to $\dot{\varphi}$, the particle fluence rate. As $\dot{\varphi} \to \infty$ we get

$$
\lim_{\dot{\varphi} \to \infty} (a'_D)_{max} = \infty. \tag{12.38}
$$

This is obviously a problematic result, since we know that the maximum daughter specific activity produced through neutron activation cannot exceed the theoretical specific activity $(a_D)_{theor}$, given for the daughter in (10.2) as follows

$$
(a_D)_{theor} = \frac{A_D}{M_D} = \frac{\lambda_D N_D}{M_D} = \frac{\lambda_D N_A}{A_D}. \tag{12.39}
$$

Equation (12.37) shows that the saturation model is useful as an approximation only for relatively low particle fluence rates $\dot{\varphi}$; at high fluence rates the model breaks down and predicts a physically impossible result.

For the *depletion model* we use (12.33) to get the maximum daughter specific activity $(a_D)_{max}$ as

$$
(a_D)_{max} = \frac{(A_D)_{max}}{M_P} = (y_D)_{max}\,\sigma_P \dot{\varphi}\,\frac{N_P(0)}{M_P} = \left[\frac{\sigma_P N_A}{A_P}\right](y_D)_{max}\,\dot{\varphi}, \tag{12.40}
$$

where we use the identity $N_P/M_P = N_A/A_P$. The result of (12.40) is similar to $(a_D)_{max}$ given in (12.37) for the saturation model; however, it contains $(y_D)_{max}$, the normalized daughter activity that exhibits its own dependence on $\dot{\varphi}$, as shown in Fig. 12.3 and given in (12.31). Introducing the expression for $(y_D)_{max}$ of (12.31) into (12.40) and recognizing that the activation factor m of (12.25) is equal to $\sigma_p \dot{\varphi}/\lambda_D$, we get the following expression for the maximum daughter specific activity $(a_D)_{max}$

$$(a_D)_{max} = \left(\frac{\sigma_P N_A}{A_P}\right) \dot{\varphi} \, e^{\frac{m}{1-m} \ln m}. \tag{12.41}$$

At first glance, it seems that the depletion model of (12.41) also suffers the same catastrophe with $\dot{\varphi} \to \infty$, as shown in (12.37) for the saturation model. However, a closer look at $\lim_{\dot{\varphi} \to \infty} (a_D)_{max}$ for the depletion model produces a very logical result, namely that the maximum daughter specific activity $(a_D)_{max}$ will not exceed the theoretical specific activity $(a_D)_{theor}$, or

$$\lim_{\dot{\varphi} \to \infty} (a_D)_{max} = \left\{\frac{\sigma_P N_A}{A_P}\right\} \lim_{\dot{\varphi} \to \infty} \left\{\dot{\varphi} \exp\left[-\frac{\frac{\sigma_P \dot{\varphi}}{\lambda_D}}{\frac{\sigma_P \dot{\varphi}}{\lambda_D} - 1} \ln \frac{\sigma_P \dot{\varphi}}{\lambda_D}\right]\right\}$$
$$= \frac{\lambda_D N_A}{A_P} \approx (a_D)_{theor}. \tag{12.42}$$

The result of (12.42) is independent of the particle fluence rate $\dot{\varphi}$, irrespective of the magnitude of $\dot{\varphi}$ and depends only on the decay constant λ_D of the daughter and the atomic mass number A_P of the parent. Recognizing that $A_P \approx A_D$ at least for large atomic number activation targets, we can state that $\lambda_D N_A / A_P \approx (a_D)_{theor}$.

The depletion model, in contrast to the commonly used saturation model, thus adequately predicts (a_{theor}) as the limit for the maximum attainable daughter specific activity in neutron activation and should be taken as the correct model for describing the parent–daughter kinematics in radioactivation in general, irrespective of the magnitude of the particle fluence rate $\dot{\varphi}$ used in the radioactivation.

The saturation model is valid as a special case of the depletion model under one of the following two conditions:

1. For the activation factor $m = \sigma_p \dot{\varphi}/\lambda_D < 10^{-3}$.
2. For the activation time t_a short compared to $(t_{max})_D$, the time of ideal equilibrium between y_P and y_D.

The maximum attainable specific activities $(a_D)_{sat}$ and $(a_D)_{max}$ are plotted against the neutron fluence rate $\dot{\varphi}$ in Fig. 12.4 for the three most commonly used radionuclides in medicine: cobalt-60, iridium-192, and molybdenum-99. The theoretical specific activities $(a_{theor})_D$ for cobalt-60 and iridium-192 are also indicated in the figure. We note for iridium-192 and cobalt-60 that in the practical neutron fluence rate $\dot{\varphi}$ range from 10^{13} cm$^{-2}\cdot$s^{-1} to 10^{15} cm$^{-2}\cdot$s^{-1} the saturation model fails, while the depletion model approaches asymptotically the theoretical result. For molybdenum-99 in the neutron fluence range shown in Fig. 12.4 the maximum attainable specific activities are 5 to 6 orders of magnitude lower than $(a_{Mo})_{theor}$, so that the saturation and depletion model give identical results as a consequence of $\sigma_{MO} \dot{\varphi}/\lambda_{Mo} < 10^{-3}$.

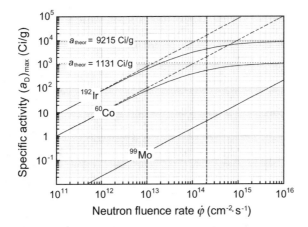

Fig. 12.4 Maximum attainable specific activities $(a_D)_{sat}$ of (12.40) and $(a_D)_{max}$ of (12.41) for the saturation model (*dashed curves*) and the depletion model (*solid curves*), respectively, plotted against neutron fluence rate $\dot{\varphi}$ for cobalt-60, iridium-192, and molybdenum-99 daughter products in neutron activation. The theoretical specific activities of cobalt-60 and iridium-192 are indicated with *horizontal dashed lines*. The *vertical dashed lines* at $\dot{\varphi} = 10^{13}$ cm$^{-2}\cdot$s^{-1} and 2×10^{14} cm$^{-2}\cdot$s^{-1} indicate data for the two neutron fluence rates of Table 12.3

Table 12.2 lists the important characteristics of cobalt-60, iridium-192, and molybdenum-99. The theoretical specific activity $(a_D)_{theor}$ is calculated from (10.2) assuming that the radioactive nuclide contains only the daughter nuclei, i.e., the source is carrier-free.

Table 12.3 lists the neutron activation characteristics for the saturation and depletion models applied to production of cobalt-60, iridium-192 and molybdenum-99 for two neutron fluence rates: $\dot{\varphi} = 10^{13}$ cm$^{-2}\cdot$s^{-1} and $\dot{\varphi} = 2\times10^{14}$ cm$^{-2}\cdot$s^{-1}. The two fluence rates are representative of rates used in activation processes with modern nuclear reactors. Of main interest in Table 12.3 are the maximum attainable specific activities $(a_D)_{sat}$ and $(a_D)_{max}$ predicted by the saturation model and the depletion model, respectively, and their comparison to the theoretical values $(a_D)_{theor}$ for the three daughter D products, also listed in the table.

Two interesting features of Fig. 12.4 and Table 12.3 are of note:

- For cobalt-60 $(a_{Co})_{theor} = 1131$ Ci/g, while at $\dot{\varphi} = 2\times10^{14}$ cm$^{-2}\cdot$s^{-1} the saturation model for cobalt-60 production predicts $(a_{Co})_{sat} = 2054$ Ci/g, a physically impossible result. On the other hand, the depletion model predicts that $(a_{Co})_{max} = 550$ Ci/g which is a realistic result that can be substantiated with experiment.
- A study of iridium-192 results in conclusions similar to those for cobalt-60 and this is understood, since the activation factors m for the two radionuclides are essentially identical. The activation factor m for molybdenum-99 for practical fluence rates, on the other hand, is so small that both models predict identical specific activities, both a miniscule fraction of $(a_{Mo})_{theor}$.

Table 12.2 Characteristics of three radionuclides of importance to medical physics and produced by thermal neutron activation in a nuclear reactor: *cobalt-60* as source in external beam radiotherapy, *iridium-192* as sealed source for brachytherapy, and *molybdenum-99* as source of technetium-99m for use in nuclear medicine imaging

Daughter nuclide	Cobalt-60	Iridium-192	Molybdenum-99
Half-life $(t_{1/2})_D$	5.27 y	73.8 d	66 h
Decay constant $\lambda_D(s^{-1})$	4.171×10^{-9}	1.087×10^{-7}	2.917×10^{-6}
Parent nuclide	Cobalt-59	Iridium-191	Molybdenum-98
Nuclear reaction	$^{59}_{27}$Co (n, γ) $^{60}_{27}$Co	$^{191}_{77}$Ir (n, γ) $^{192}_{77}$Ir	$^{98}_{42}$Mo (n, γ) $^{99}_{42}$Mo
Cross section (b)	37.2	954	0.13
a_{theor} (Ci/g)[a]	1.131×10^3	9.215×10^3	4.8×10^5
a_{pract} (Ci/g)[b]	~250	~450	~0.3
$\sigma_P N_A / A_P (cm^2/g)$	0.38	3.01	8×10^{-4}

[a]Theoretical specific activity: $a_{theor} = \lambda_D N_A / A_D$, assuming a carrier-free source
[b]Practical specific activity produced in a nuclear reactor

Table 12.3 Neutron activation characteristics for the saturation and depletion models applied to neutron activation of cobalt-59 into cobalt-60, iridium-191 into iridium-192, and molybdenum-98 into molybdenum-99 with neutron fluence rates of 10^{13} cm$^{-2}\cdot$s^{-1} and 2×10^{14} cm$^{-2}\cdot$s^{-1} (a = year; d = day; h = hour)

	Daughter nuclide	Cobalt-60		Iridium-192		Molybdenum-99	
1	$(a_D)_{theor}$ (Ci/g)	1131		9215		479 604	
	$\dot{\varphi}$ (cm$^{-2}\cdot$s^{-1})	10^{13}	2×10^{14}	10^{13}	2×10^{14}	10^{13}	2×10^{14}
2	$(a_D)_{sat}$ (Ci/g)	102.7	2054	813.5	16270	0.216	4.32
3	m	0.089	1.78	0.088	1.76	4.5×10^{-7}	8.9×10^{-6}
4	$(x_D)_{max}$	0.341	1.90	0.338	1.89	9.4×10^{-6}	1.5×10^{-4}
5	$(y_D)_{max}$	0.789	0.268	0.791	0.270	1.00	1.00
6	$(a_D)_{max}$ (Ci/g)	81.0	549.8	643.3	4398	0.22	4.32
7	t_{max}	12.18 a	5.61 a	284.0 d	79.3 d	1392.4 h	1107.2 h

1. $(a_D)_{theor} = \frac{\lambda_D N_A}{A_D}$ see (12.39)

2. $(a_D)_{sat} = \frac{\sigma_P N_A}{A_P} \dot{\varphi}$ see (12.37)

3. $m = \frac{\sigma_P \dot{\varphi}}{\lambda_D}$ see (12.25)

4. $(x_D)_{max} = \frac{m}{(m-1)} \frac{\ln m}{\ln 2}$ see (12.29)

5. $(y_D)_{max} = \frac{1}{2^{(x_D)_{max}}}$ see (12.31)

6. $(a_D)_{max} = \frac{\sigma_P N_A}{A_P} (y_D)_{max} \dot{\varphi}$ see (12.40)

7. $(t_{max})_D = \frac{(t_{1/2})_D}{m} (x_D)_{max} = \frac{\ln 2}{m \lambda_D} (x_D)_{max} = \frac{\ln m}{\lambda_D (m-1)}$ see (12.24)

Fig. 12.5 Time $(t_{max})_D/(t_{1/2})_D$ required for reaching the maximum specific activity $(a_D)_{max}$ plotted against neutron fluence rate $\dot{\varphi}$ for cobalt-60, iridium-192, and molybdenum-99. The data were calculated with the depletion model of radioactivation

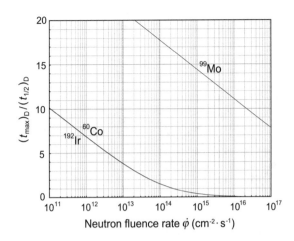

Of interest is also the activation time $(t_{max})_D$ required to obtain $(a_D)_{max}$ using the depletion model. From (12.26) and (12.29) we obtain

$$\frac{(t_{max})_D}{(t_{1/2})_D} = \frac{(x_D)_{max}}{m} = \frac{\ln m}{(\ln 2)\,(m-1)} = \frac{\lambda_D \ln \dfrac{\sigma_P \dot{\varphi}}{\lambda_D}}{(\ln 2)\,(\sigma_P \dot{\varphi} - \lambda_D)}, \qquad (12.43)$$

with roughly an inverse proportionality with fluence rate $\dot{\varphi}$. Thus, the higher is the particle fluence rate $\dot{\varphi}$, the shorter is the time required to reach the maximum specific activity $(a_D)_{max}$. For example, $(t_{max})_{Co}$ is 20.2 years at $\dot{\varphi} = 10^{13}$ cm^{-2}·s^{-1} and 5.61 years at $\dot{\varphi} = 2 \times 10^{14}$ cm^{-2}·s^{-1}, as also shown in Table 12.3.

The time $(t_{max})_D/(t_{1/2})_D$ of (12.43) is plotted against the neutron fluence rate $\dot{\varphi}$ for cobalt-60, iridium-192, and molybdenum-99 in Fig. 12.5. The curves for cobalt-60 and iridium-192 are essentially identical, because the activation factors m for the two radionuclides are fortuitously similar to one another as a result of similar ratios σ_P/λ_D for the two radionuclides.

12.6.4 Examples of Parent Depletion: Neutron Activation of Cobalt-59, Iridium-191, and Molybdenum-98

Using the general data of Fig. 12.3 we plot in Fig. 12.6 the specific activity a_D of cobalt-60 in part (a) and of iridium-192 in part (b) against activation time t for various neutron fluence rates $\dot{\varphi}$ in the range from 10^{13} cm^{-2}·s^{-1} to 2×10^{14} cm^{-2}·s^{-1}. The specific activity a_D is calculated for the saturation model (*dashed curves*) given by (12.35) and the depletion model (*solid curves*) given in (12.33). Both equations are

Fig. 12.6 Specific activity a_D of cobalt-60 in part (**a**) and of iridium-192 in part (**b**) plotted against activation time t for various neutron fluence rates $\dot{\varphi}$. The specific activity a_D is calculated for the saturation model (*dashed curves*) given by (12.35) and the depletion model (*solid curves*) given in (12.33). Both equations are used in conjunction with (12.26) to obtain a plot of a_D against activation time t rather than against the variable x. The *heavy dots* on the depletion model curves represent the time $(t_{max})_D$ at which the maximum specific activity $(a_D)_{max}$ occurs

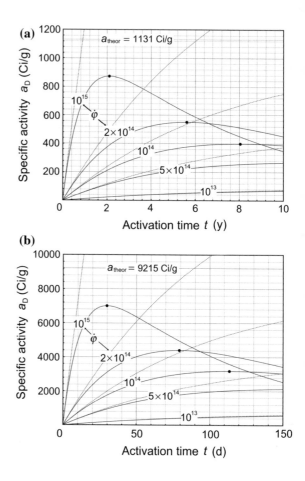

modified with incorporating (12.26) to obtain a plot of a_D against the activation time t rather than against the general variable x. The heavy dots on the depletion model curves represent the time $(t_{max})_D$ at which the maximum specific activity $(a_D)_{max}$ occurs. The theoretical specific activities a_{theor} of 1131 Ci/g and 9215 Ci/g for cobalt-60 and iridium-192, respectively, are indicated on the figure.

The discrepancy between the saturation model and the depletion model is evident, especially at high fluence rates and large activation times. An obvious breakdown of the saturation model occurs when it predicts a specific activity a_D of the daughter that exceeds the theoretical specific activity a_{theor} for a given radionuclide D.

Since both the saturation model and the depletion model show identical initial slopes, i.e., activation yields [see (12.14) and Fig. 12.1], one may use the saturation model as a simple yet adequate approximation to the depletion model at activation times short in comparison with $(t_{max})_D$. However, when the goal is to obtain optimal specific activities in the daughter of the order of the theoretical specific activity for a given radionuclide, such as the cobalt-60 source for external beam radiotherapy or iridium-192 source for industrial radiography, the saturation model fails and the depletion model should be used for estimation of the required radioactivation times and specific activities expected.

Equation (12.41) gives a relationship between the maximum attainable specific activity $(a_D)_{max}$ and neutron fluence rate $\dot{\varphi}$ for the depletion model. We now calculate the fraction f of the theoretical specific activity $(a_D)_{theor}$ that $(a_D)_{max}$ could reach at a given fluence rate $\dot{\varphi}$. The functional relationship between f and $(a_D)_{max}$ will allow us to estimate the maximum possible specific activity for a given parent–daughter combination in a radioactivation process with a given fluence rate $\dot{\varphi}$. We write $(a_D)_{max}$ as follows

$$(a_D)_{max} = f \times (a_D)_{theor} = f \frac{\lambda_D N_A}{A_P} = \frac{\sigma_P N_A}{A_P} \dot{\varphi} e^{\frac{m}{1-m} \ln m}, \qquad (12.44)$$

which gives

$$f = m e^{\frac{m}{1-m} \ln m} = m^{\frac{1}{1-m}}, \qquad (12.45)$$

where we used the following relationships: $m = \sigma_P \dot{\varphi}/\lambda_D$ of (12.25) and the approximation $A_D \approx A_P$.

We then introduce $m = \alpha \dot{\varphi}$, where α is defined as $\alpha = \sigma_P/\lambda_D$, to obtain the following expression for fraction f

$$f = (\alpha \dot{\varphi})^{\frac{1}{1-\alpha\dot{\varphi}}} \qquad (12.46)$$

and plot this expression in Fig. 12.7 for cobalt-60, iridium-192, and molybdenum-99 in the fluence rate $\dot{\varphi}$ range from 10^{11} cm^{-2}·s^{-1} to 10^{17} cm^{-2}·s^{-1}. Again, the data for cobalt-60 and iridium-192 are essentially the same for a given $\dot{\varphi}$, since the ratio σ_P/λ_D is almost identical for the two radionuclides. The molybdenum-99 fraction f data, on the other hand, are extremely small in comparison to those of the other two radionuclides indicating very low practical specific activities in the practical fluence rate range from 10^{12} cm^{-2}·s^{-1} to 10^{15} cm^{-2}·s^{-1}.

Data from Fig. 12.7 show that for cobalt-60 and iridium-192 the fraction f is 0.07 at $\dot{\varphi} = 10^{13}$ cm^{-2}·s^{-1} and 0.49 at $\dot{\varphi} = 2 \times 10^{14}$ cm^{-2}·s^{-1}. Same results are provided in Table 12.3 with the ratio $(a_D)_{max} / (a_D)_{theor}$. Thus, to obtain a higher specific activity in a cobalt-60 or iridium-192 target, we would have to surpass the currently available reactor fluence rates $\dot{\varphi}$. For example, to reach $f = 0.75$, i.e., $(a_{max})_D = 850$ Ci/g for a cobalt-60 source and 6900 Ci/g for an iridium-192 source, a $\dot{\varphi}$ of 10^{15} cm^{-2}·s^{-1} would be required. This would result in an activation factor m of 12.8 and, as shown in Fig. 12.5 and given by (12.43), the activation time t_{max} to reach this specific activity would be relatively short at 2.1 years for cobalt-60 and 30 days for iridium-192.

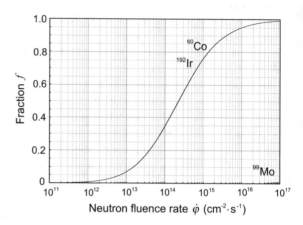

Fig. 12.7 Specific activity fraction f defined as $(a_D)_{max} / (a_D)_{theor}$ plotted against the neutron fluence rate $\dot{\varphi}$ for cobalt-60, iridium-192, and molybdenum-99. The data for molybdenum are visible only at very high fluence rates because the activation factor m at a given $\dot{\varphi}$ is several orders of magnitude smaller for molybdenum-99 in comparison with that of cobalt-60 and iridium-192

12.6.5 Neutron Activation of the Daughter: The Depletion–Activation Model

In the discussion of neutron activation above we have tacitly assumed that the daughter nuclide is not affected by exposure to activation particles. In situations where this assumption does not hold, account must be taken of the activation of the daughter radionuclide into a granddaughter that may or may not be radioactive. Ignoring the possibility of the granddaughter radioactivity, we account for the daughter activation by subtracting $\sigma_D \dot{\varphi} N_D(t)$ from the differential equation for dN_D/dt given in (12.18) to obtain

$$
\begin{aligned}
\frac{dN_D(t)}{dt} &= \sigma_P \dot{\varphi} N_P(t) - \lambda_D N_D(t) - \sigma_D \dot{\varphi} N_D(t) \\
&= \sigma_P \dot{\varphi} N_P(t) - [\lambda_D + \sigma_D \dot{\varphi}] N_D(t) \\
&= \sigma_P \dot{\varphi} N_P(t) - \lambda_D^* N_D(t),
\end{aligned}
\tag{12.47}
$$

where σ_P and σ_D are cross sections for activation of parent and daughter nuclei, respectively; $N_P(t)$ and $N_D(t)$ are numbers of parent and daughter nuclei, respectively; and $\dot{\varphi}$ is the neutron fluence rate. We now define a modified decay constant λ_D^* as follows

$$
\lambda_D^* = \lambda_D + \sigma_D \dot{\varphi}.
\tag{12.48}
$$

Using the same initial conditions as in (12.18), we get the following solution to (12.47)

$$
N_D(t) = N_P(0) \frac{\sigma_P \dot{\varphi}}{\lambda_D^* - \sigma_P \dot{\varphi}} \left\{ e^{-\sigma_P \dot{\varphi} t} - e^{-\lambda_D^* t} \right\}
\tag{12.49}
$$

and the following expression for the daughter activity $\mathcal{A}_D(t)$

$$\mathcal{A}_D(t) = \lambda_D N_D(t) = N_P(0) \frac{\sigma_P \dot{\varphi} \lambda_D}{\lambda_D^* - \sigma_P \dot{\varphi}} \left\{ e^{-\sigma_P \dot{\varphi} t} - e^{-\lambda_D^* t} \right\}$$

$$= \sigma_P \dot{\varphi} N_P(0) \frac{\dfrac{\lambda_D}{\lambda_D^*}}{1 - \dfrac{\sigma_P \dot{\varphi}}{\lambda_D^*}} \left\{ e^{-\sigma_P \dot{\varphi} t} - e^{-\lambda_D^* t} \right\}. \tag{12.50}$$

To obtain a general expression for the daughter activity in the *"parent depletion–daughter activation model"* we now introduce new parameters and variables, similarly to the approach we took in the discussion of the decay series with (10.41) through (10.44) and the radioactivation depletion model with (12.25) through (12.30), as follows

$$k^* = \frac{\sigma_P}{\sigma_D}, \tag{12.51}$$

$$\varepsilon^* = \frac{\lambda_D^*}{\lambda_D} = 1 + \frac{\sigma_D \dot{\varphi}}{\lambda_D}, \tag{12.52}$$

$$m = \frac{\sigma_P \dot{\varphi}}{\lambda_D}, \qquad \text{same as in (12.25)} \tag{12.53}$$

$$m^* = \frac{\sigma_P \dot{\varphi}}{\lambda_D^*} = \frac{m}{\varepsilon^*}, \tag{12.54}$$

$$x = \frac{\sigma_P \dot{\varphi}}{\ln 2} t = \frac{m^* \lambda_D^*}{\ln 2} t, \tag{12.55}$$

$$y_P = \frac{N_P(t)}{N_P(0)} = e^{-\sigma_P \dot{\varphi} t} = e^{-x \ln 2} = \frac{1}{2^x}, \tag{12.56}$$

$$y_D^* = \frac{\mathcal{A}_D(t)}{\sigma_P \dot{\varphi} N_P(0)} = \frac{1}{\varepsilon^* (1 - m^*)} \left[\frac{1}{2^x} - \frac{1}{2^{\frac{x}{m^*}}} \right]. \tag{12.57}$$

Equation (12.57) for the normalized daughter activity y_D^* of the depletion–activation model is similar to (12.28) for y_D of the depletion model, except for the factor ε^* which is larger than 1 and depends on $\dot{\varphi}$. In the depletion model $\sigma_D = 0$, $k^* = \infty$ and $\varepsilon^* = 1$, while for the depletion–activation model $\sigma_D \neq 0$ and $\varepsilon^* > 1$. Thus, we expect y_D^* of the depletion–activation model to behave in a similar manner to y_D of the depletion model: rise from 0 to reach a maximum $(y_D^*)_{max}$ at $x = (x_D^*)_{max}$ and then asymptotically decrease to zero at large x.

Like for (10.50) and (12.31), we find for $(x_D^*)_{max}$ the following expression

$$\left(x_D^* \right)_{max} = \frac{m^* \ln m^*}{(m^* - 1) \ln 2}, \tag{12.58}$$

and similarly to (10.53) and (12.31) we find the expression for $\left(y_D^* \right)_{max}$ as

$$\left(y_{\mathrm{D}}^{*}\right)_{\max} = \frac{1}{\varepsilon^{*}2^{(x_{\mathrm{D}}^{*})_{\max}}} = \frac{1}{\varepsilon^{*}}e^{-(\ln 2)(x_{\mathrm{D}}^{*})_{\max}} = \frac{1}{\varepsilon^{*}}e^{-\frac{m^{*}\ln m^{*}}{m^{*}-1}}. \tag{12.59}$$

The maximum specific activity $(a_{\mathrm{D}}^{*})_{\max}$ of the daughter, similar to (12.41), is expressed as

$$\left(a_{\mathrm{D}}^{*}\right)_{\max} = \left[\frac{\sigma_{\mathrm{P}}N_{\mathrm{A}}}{A_{\mathrm{P}}}\right](y_{\mathrm{D}}^{*})_{\max}\dot{\varphi} = \left[\frac{\sigma_{\mathrm{P}}N_{\mathrm{A}}}{A_{\mathrm{P}}}\right]\frac{\dot{\varphi}}{\varepsilon^{*}}e^{-\frac{m^{*}\ln m^{*}}{m^{*}-1}}. \tag{12.60}$$

Since both ε^{*} and m^{*} depend on $\dot{\varphi}$, the question arises about the behavior of $(a_{\mathrm{D}}^{*})_{\max}$ in the limit as the neutron fluence rate $\dot{\varphi}$ becomes very large, i.e., $\dot{\varphi} \to \infty$. We determine $\lim\limits_{\dot{\varphi}\to\infty}(a_{\mathrm{D}}^{*})_{\max}$ as follows

$$
\begin{aligned}
\lim_{\dot{\varphi}\to\infty}\left(a_{\mathrm{D}}^{*}\right)_{\max} &= \left[\frac{\sigma_{\mathrm{P}}N_{\mathrm{A}}}{A_{\mathrm{P}}}\right]\lim_{\dot{\varphi}\to\infty}\left\{\frac{\lambda_{\mathrm{D}}\dot{\varphi}}{\lambda_{\mathrm{D}}+\sigma_{\mathrm{D}}\dot{\varphi}}\exp\left[-\frac{\frac{\sigma_{\mathrm{P}}\dot{\varphi}}{\lambda_{\mathrm{D}}+\sigma_{\mathrm{D}}\dot{\varphi}}}{\frac{\sigma_{\mathrm{P}}\dot{\varphi}}{\lambda_{\mathrm{D}}+\sigma_{\mathrm{D}}\dot{\varphi}}-1}\ln\frac{\sigma_{\mathrm{P}}\dot{\varphi}}{\lambda_{\mathrm{D}}+\sigma_{\mathrm{D}}\dot{\varphi}}\right]\right\} \\
&= \left[\frac{\sigma_{\mathrm{P}}N_{\mathrm{A}}}{A_{\mathrm{P}}}\right]\left\{\frac{\lambda_{\mathrm{D}}}{\sigma_{\mathrm{D}}}\exp\left[-\frac{\frac{\sigma_{\mathrm{P}}}{\sigma_{\mathrm{D}}}}{\frac{\sigma_{\mathrm{P}}}{\sigma_{\mathrm{D}}}-1}\ln\frac{\sigma_{\mathrm{P}}}{\sigma_{\mathrm{D}}}\right]\right\} \\
&= \left[\frac{\lambda_{\mathrm{D}}N_{\mathrm{A}}}{A_{\mathrm{P}}}\right]\left\{k^{*}e^{-\frac{k^{*}\ln k^{*}}{k^{*}-1}}\right\} \approx g\times (a_{\mathrm{D}})_{\text{theor}}, \tag{12.61}
\end{aligned}
$$

where we used the definition of $k^{*} = \dfrac{\sigma_{\mathrm{P}}}{\sigma_{\mathrm{D}}}$ of (12.51) and defined the function $g\,(k^{*})$ as

$$g\,(k^{*}) = k^{*}e^{-\frac{k^{*}\ln k^{*}}{k^{*}-1}}. \tag{12.62}$$

A plot of the function $g\,(k^{*})$ against the parameter k^{*} in the range from $k^{*} = 10^{-3}$ to $k^{*} = 10^{3}$ is given in Fig. 12.8. We note several interesting features of $g\,(k^{*})$:

1. In general, $g\,(k^{*}) < 1$ for all finite k^{*}.
2. As $\sigma_{\mathrm{D}} \to 0$ that implies that the activation of the daughter is negligible, the parameter $k^{*} = \sigma_{\mathrm{P}}/\sigma_{\mathrm{D}}$ approaches infinity ∞ and function g approaches 1, as evident from the following

$$\lim_{k^{*}\to\infty}g\,(k^{*}) = \lim_{k^{*}\to\infty}k^{*}e^{-\frac{k^{*}\ln k^{*}}{k^{*}-1}} = 1. \tag{12.63}$$

3. For $k^{*} = \infty$ the maximum specific activity $(a_{\mathrm{D}}^{*})_{\max}$ for the depletion–activation model transforms into $(a_{\mathrm{D}})_{\text{theor}}$ given in (12.42) for the depletion model.
4. The function $g\,(k^{*}) = 1$ applies to the depletion model in which $\sigma_{\mathrm{D}} = 0$, $\varepsilon^{*} = 1$, and $k^{*} = \infty$.

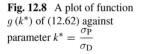

Fig. 12.8 A plot of function $g(k^*)$ of (12.62) against parameter $k^* = \dfrac{\sigma_P}{\sigma_D}$

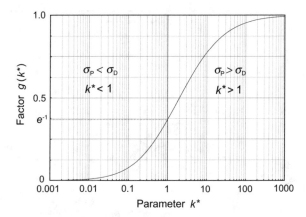

12.6.6 Example of Daughter Neutron Activation: Iridium-192

A closer investigation of the iridium radioactivation reveals a considerably more complicated picture than the one given in Sect. 12.6.5:

- Firstly, iridium has two stable isotopes: iridium-191 (Ir-191) with a natural abundance of 37.3% ($\sigma_P = \sigma_{191} = 954$ b) and iridium-193 with a natural abundance of 62.7% ($\sigma_P = \sigma_{193} = 100$ b). The Ir-191 isotope is of interest in industry and medicine, since iridium-192, the product of neutron activation has a reasonably long half-life of 73.8 days. In contrast, neutron activation of Ir-193 results in Ir-194 that decays with a short half-life of 19.3 h. Since the Ir-192 radionuclide is produced through the neutron activation of the Ir-191 stable nuclide, the natural mixture of Ir-191 (37.3%) and Ir-193 (62.7%) in the activation target will result in a lower final specific activity of the Ir-192 source in comparison with activation of a pure Ir-191 target.
- Secondly, iridium-192, the daughter product of iridium-191 neutron activation, itself has a significant cross section for neutron activation $\sigma_D = \sigma_{192} = 1420$ b in contrast to the parent cross section $\sigma_P = \sigma_{191} = 954$ b. As shown in Sect. 12.6.5, the activation of the daughter product will affect the specific activity of the iridium-192 source.

In Fig. 12.9 we plot the normalized activity functions for iridium-191: z_D of (12.35) for the saturation model; y_D of (12.33) for the depletion model; and y_D^* of (12.57) for the depletion–activation model. The functions are plotted against the variable x of (12.26) for two neutron fluence rates: $\dot{\varphi} = 10^{13}$ cm$^{-2} \cdot$s^{-1} and $\dot{\varphi} = 2 \times 10^{14}$ cm$^{-2} \cdot$s^{-1}.

The relevant parameters for these functions and three activation models are listed in Table 12.4. The following features are of note:

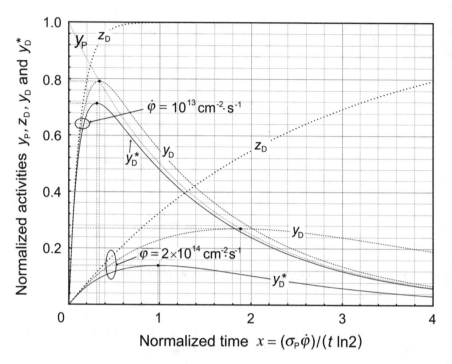

Fig. 12.9 Plot of normalized activity functions against x for iridium-191 neutron activation: z_D for the saturation model (*dashed curves*); y_D for the depletion model (*light solid curves*); and y_D^* for the depletion–activation model (*heavy solid curves*). The functions are plotted for two neutron fluence rates: $\dot\varphi = 10^{13}$ cm$^{-2}\cdot$s^{-1} and $\dot\varphi = 2\times10^{14}$ cm$^{-2}\cdot$s^{-1}

1. For all three functions (z_D, y_D, and y_D^*) the initial slopes at $x = 0$ are identical and equal to $(\ln 2)/m$.
2. For the saturation model z_D saturates at 1; for the depletion model y_D reaches its maximum of $(y_D)_{max}$ at $(x_D)_{max}$; for the depletion–activation model y_D^* reaches its maximum of $(y_D^*)_{max}$ at $(x_D^*)_{max}$.
3. $(x_D^*)_{max}$ and $(y_D^*)_{max}$ for the depletion–activation model decrease in comparison to $(x_D)_{max}$ and $(y_D)_{max}$ for the depletion model, respectively. The larger is $\dot\varphi$, the larger is the discrepancy between the two parameters.
4. $(y_D^*)_{max}$ no longer occurs at the point of ideal equilibrium where $y_P = y_D^*$, in contrast to $(y_D)_{max}$ of the daughter in the depletion model that occurs at the point of ideal equilibrium.

Figure 12.10 shows a plot of the maximum attainable specific activity for iridium-192 against neutron fluence rate $\dot\varphi$ for three activation models: (1) saturation model with straight line $(a_D)_{sat}$; (2) depletion model $(a_D)_{max}$; and (3) depletion–activation model $(a_D^*)_{max}$.

Table 12.4 Parameters of the depletion model and the depletion–activation model applied to neutron activation of iridium-191 nuclide into iridium-192 radionuclide

	Particle fluence rate		Definition	References
$\dot{\varphi}(\mathrm{cm}^{-2}\cdot\mathrm{s}^{-1})$	10^{13}	2×10^{14}		
$\lambda_D(\mathrm{s}^{-1})$	1.087×10^{-7}	1.087×10^{-7}	$\lambda_D = \dfrac{\ln 2}{(t_{1/2})_D}$	(10.13)
$\lambda_D^*(\mathrm{s}^{-1})$	1.229×10^{-7}	3.927×10^{-7}	$\lambda_D^* = \lambda_D + \sigma_D\varphi,\ \sigma_D = 1420\ \mathrm{b}$	(12.48)
ε	1.0	1.0	$\varepsilon = 1$	
ε^*	1.13	3.61	$\varepsilon^* = \dfrac{\lambda_D^*}{\lambda_D} = 1 + \dfrac{\sigma_D\dot{\varphi}}{\lambda_D}$	(12.52)
m	0.088	1.76	$m = \dfrac{\sigma_P\dot{\varphi}}{\lambda_D},\quad \sigma_P = 954\ \mathrm{b}$	(12.25)
m^*	0.078	0.49	$m^* = \dfrac{\sigma_P\dot{\varphi}}{\lambda_D^*} = \dfrac{\sigma_P\dot{\varphi}}{\lambda_D + \sigma_D\dot{\varphi}} = \dfrac{m}{\varepsilon^*}$	(12.54)
$(x_D)_{\max}$	0.338	1.89	$(x_D)_{\max} = \dfrac{m\ln m}{(m-1)\ln 2}$	(12.29)
$(x_D^*)_{\max}$	0.311	0.98	$(x_D^*)_{\max} = \dfrac{m^*\ln m^*}{(m^*-1)\ln 2}$	(12.58)
$(y_D)_{\max}$	0.793	0.270	$(y_D)_{\max} = \dfrac{1}{2^{(x_D)_{\max}}}$	(12.31)
$(y_D^*)_{\max}$	0.713	0.140	$(y_D^*)_{\max} = \dfrac{1}{\varepsilon^* 2^{(x_D^*)_{\max}}}$	(12.59)
$(a_D)_{\max}$	643.8	4,398	$(a_D)_{\max} = \dot{\varphi}\dfrac{\sigma_P N_A}{A_P}(y_D)_{\max}$	(12.40)
$(a_D^*)_{\max}$	580.0	2,275	$(a_D^*)_{\max} = \varphi\dfrac{\sigma_P N_A}{A_P}(y_D^*)_{\max}$	(12.60)
$\dfrac{(t_{\max})_D}{(t_{1/2})_D}$	3.84	1.07	$\dfrac{(t_{\max})_D}{(t_{1/2})_D} = \dfrac{(x_D)_{\max}}{m}$	(12.26)
$\dfrac{(t_{\max}^*)_D}{(t_{1/2})_D}$	3.53	0.56	$\dfrac{(t_{\max}^*)_D}{(t_{1/2})_D} = \dfrac{(x_D^*)_{\max}}{(m^*\varepsilon^*)}$	(12.55)

The cross section for neutron activation of Ir-191 is $\sigma\,(\mathrm{Ir}\text{-}191) = \sigma_P = 954$ b and for Ir-192 it is $\sigma\,(\mathrm{Ir}\text{-}192) = \sigma_D = 1420$ b

The saturation model saturates at $\sigma_P\dot{\varphi}N_A/A_P$, the depletion model saturates at the theoretical specific activity a_{theor} for iridium-192 at 9215 Ci/g, as also shown in Fig. 12.11, while the depletion–activation model saturates at $g\,(a_{\mathrm{Ir}-192})_{\mathrm{theor}} = 2742$ Ci/g, where $g = 0.3$, as given in (12.62) with k^* for iridium-192 equal to 0.672, as shown in Fig. 12.9 (*Note:* $k^* = \sigma_P/\sigma_D = 954/1420$).

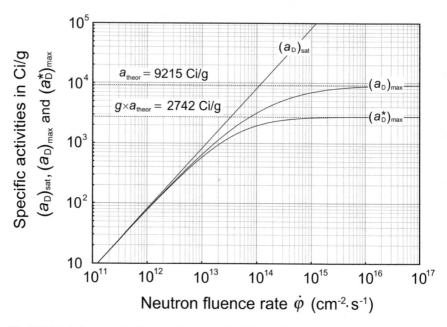

Fig. 12.10 Maximum attainable specific activity for iridium-192 against neutron fluence rate $\dot{\varphi}$ for three activation models: (*1*) saturation model shown with *straight line* $(a_D)_{sat}$; (*2*) depletion model $(a_D)_{max}$; and (*3*) depletion–activation model $(a_D^*)_{max}$. An assumption is made that the activation parent target contains pure iridium-191 rather than a natural mixture of iridium-191 and iridium-193 equal to 37.3% and 62.7%, respectively

Figure 12.10 also shows that when large specific activities of iridium-192 are produced with fluence rates $\dot{\varphi}$ of the order of 10^{13} cm$^{-2}\cdot$s^{-1} or higher, the best model for estimation of the specific activity of iridium-192 sample is the depletion–activation model.

In Fig. 12.11 we plot the specific activity of iridium-192 against activation time t normalized to $(t_{1/2})_D$, for three radioactivation models:

1. a_D' for the saturation model shown with *heavy dotted curves*.
2. a_D for the depletion model with *dotted curves*.
3. a_D^* for the depletion–activation model with *solid curves*.

The fluence rate $\dot{\varphi}$ in Fig. 12.11a is 10^{13} cm$^{-2}\cdot$s^{-1} and in Fig. 12.11b it is 2×10^{14} cm$^{-2}\cdot$s^{-1}. The maxima for the depletion curve $(a_D)_{max}$ and the depletion–activation curves $(a_D^*)_{max}$ are shown with heavy dots. The appropriate values for parameters $(a_D)_{max}$, $(t_{max})_D$, $(a_D^*)_{max}$, and $(t_{max}^*)_D$ are given in Table 12.4. Note that an assumption is made that the iridium activation sample contains only the iridium-191 stable nuclide rather than a natural mixture of iridium-191 and iridium-193. Thus, to get the specific activity for a natural sample of iridium, the natural abundance of iridium-191 in the sample would have to be taken into account.

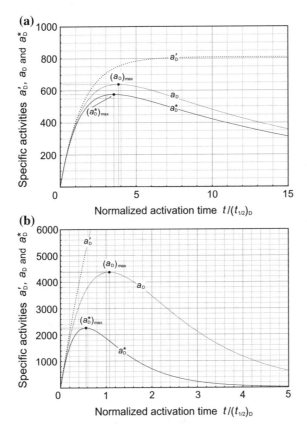

Fig. 12.11 Specific activity a_D of iridium-192 against activation time t normalized to $(t_{1/2})_D$ for iridium-192 for three radioactivation models: a_D' for the saturation model with *heavy dotted curves*; a_D for the depletion model with *dotted curves*; and a_D^* for the depletion–activation model with *solid curves*. The maxima for depletion curves $(a_D)_{max}$ and for the depletion–activation curves $(a_D^*)_{max}$ are indicated with *heavy dots*. Part **a** is for a fluence rate $\dot{\varphi}$ of 10^{13} cm^{-2}·s^{-1}; part **b** is for a fluence rate $\dot{\varphi}$ of 2×10^{14} cm^{-2}·s^{-1}. An assumption is made that the activation parent target contains pure iridium-191

Again we note that the activation of the daughter product iridium-192 has a significant effect on the daughter specific activity; this is especially pronounced at larger fluence rates, as shown in Fig. 12.12.

The following conclusions can now be made:

- The best model for description of the radioactivation kinematics is the depletion model when the daughter product is not activated by the exposure to radioactivation particles. An example for the use of this model is the activation of cobalt-59 into cobalt-60.

- The best model for describing the radioactivation kinetics in situations where the daughter product is activated by the radioactivation particles is the depletion–activation model. An example for the use of this model is the neutron activation of iridium-191 into iridium-192.
- The saturation model is only an approximation to the depletion and depletion–activation models. It is valid only at very short activation times or when $\sigma_P \dot{\varphi} / \lambda_D < 10^{-3}$. An example for the use of this model is the neutron activation of molybdenum-98 into molybdenum-99.

12.6.7 Practical Aspects of Neutron Activation

Thermal neutrons produced in nuclear reactors are the most common particles used for radioactivation. This type of the radioactivation process is referred to as *neutron activation* or *neutron capture*. Two types of neutron activation process occur commonly: (n, γ) and (n, p).

- The (n, γ) process results in neutron capture and emission of γ-rays. In the (n, γ) reaction the target nucleus ${}_{Z}^{A}X$ captures a neutron and is converted into an excited nucleus ${}_{Z}^{A+1}X^{*}$ that undergoes an immediate de-excitation to its ground state through emission of a γ-ray. Note that ${}_{Z}^{A}X$ and ${}_{Z}^{A+1}X$ are isotopes of the same chemical element. The schematic representation of the reaction is as follows:

$$ {}_{Z}^{A}X + n = {}_{Z}^{A+1}X^{*} + \gamma \qquad \text{or} \qquad {}_{Z}^{A}X \, (n, \, \gamma) \, {}_{Z}^{A+1}X. \qquad (12.64) $$

The ${}_{Z}^{A+1}X$ nucleus is neutron-rich as well as unstable and decays with a given half-life through β^{-} decay into a more stable configuration.

- The (n, p) process results in neutron capture and emission of a proton. It produces a new nucleus which is an isobar of the target nucleus. In the (n, p) reaction the target nucleus ${}_{Z}^{A}X$ captures a neutron and promptly ejects a proton to become converted into a new nucleus ${}_{Z-1}^{A}Y$. Note that ${}_{Z}^{A}X$ and ${}_{Z-1}^{A}Y$ do not represent the same chemical element; however, they possess the same atomic mass number A which means that they are isobars. Schematically the (n, p) reaction is represented as follows:

$$ {}_{Z}^{A}X + n = {}_{Z-1}^{A}Y + p \qquad \text{or} \qquad {}_{Z}^{A}X(n, p){}_{Z-1}^{A}Y \qquad (12.65) $$

Radionuclides produced by neutron activation in a nuclear reactor normally contain a mixture of stable parent nuclei in addition to radioactive daughter nuclei. The parent nuclei thus act as carrier of daughter nuclei and effectively decrease the specific activity of the source. The (n, γ) reaction is much more common than the (n, p) reaction and produces radioactive products that are not carrier-free. A chemical separation of the daughter nuclei from the parent nuclei is not feasible because the parent and daughter are isotopes of the same element; a physical separation, while possible, is difficult and expensive.

For example, the practical specific activity of cobalt-60 sources is limited to about 300 Ci/g or ∼25% of the carrier-free theoretical specific activity of 1133 Ci/g. This means that in a cobalt-60 teletherapy source ∼75% of the source mass is composed of stable cobalt-59 nuclei and only ∼25% of the source mass is composed of radioactive cobalt-60 nuclei. The reactor-produced molybdenum-99 is mixed with stable molybdenum-98 and the source has a practical specific activity that is significantly lower than the theoretical specific activity of molybdenum-99.

From a medical physics perspective the most important neutron activation processes are:

1. Production of *cobalt-60 sealed sources* for use in external beam radiotherapy with typical initial source activity of the order of 370 TBq (10^4 Ci).
2. Production of *iridium-192 sealed sources* for use in brachytherapy with typical activities of 0.37 TBq (10 Ci).
3. Production of *molybdenum-99 radionuclide* for generating the technetium-99m (^{99m}Tc) radionuclide in a radionuclide generator for use in nuclear medicine imaging.

12.7 Nuclear Fission Induced by Neutron Bombardment

When neutrons bombard certain heavy ($Z \geq 92$) nuclei, rather than undergoing neutron capture, i.e., inducing neutron activation, the neutrons may induce a process called nuclear fission in which the target nucleus fragments into two daughter nuclei of lighter mass. The fission process is accompanied with production of several fast neutrons, γ-rays, and neutrinos. The following nomenclature is used when dealing with nuclear fission:

- In general, a nucleus that can undergo induced fission when struck by a neutron is called a *fissionable nucleus*, while the narrower term *fissile nucleus* refers to fission induced by thermal neutrons.
- As fissile nuclei undergo the fission process, lighter, generally radioactive, nuclei called *fission fragments* are formed.
- Fission fragments combined with the nuclei that are subsequently formed through radioactive decay of fission fragments are called *fission products*.
- Nuclides that do not undergo fission themselves when bombarded with thermal neutrons but transform into fissile nuclides upon bombardment with thermal neutrons followed with two β^- decays are called *fertile nuclides*. Fertile nuclides thus serve as source of fissile nuclides.
- Fission can be considered a form of nuclear transmutation, since the fission fragments are not of the same nuclear species as was the original nucleus.

Fission is energetically possible only in heavy nuclei and the reason for this can be understood from an investigation of the diagram plotting, for all known elements, the binding energy per nucleon E_B/A against atomic mass number A, given in Fig. 1.3. For heavy nuclei $E_B/A \approx 7.5$ MeV while for intermediate nuclei $E_B/A \approx 8.4$ MeV. Thus, fission of a fissile heavy nucleus such as ^{235}U with a relatively low E_B/A of ~ 7.5 MeV into two tightly bound fragments with $E_B/A \approx 8.4$ MeV results in an energy release of 0.9 MeV per nucleon corresponding to energy release of the order of 235×0.9 MeV ≈ 200 MeV per ^{235}U nucleus or $(N_A/A) \times 2 \times 10^8 \times 1.6 \times 10^{-19}$ J $= \sim 8.2 \times 10^{10}$ J per 1 g of ^{235}U.

Nuclear fission is an exothermic (also referred to as exoergic) reaction. Of the 200 MeV per fission of a ^{235}U nucleus, about 10 MeV is carried away by neutrinos, about 10 MeV by γ-rays, 5 MeV by new neutrons produced in the fission process, and the remaining 175 MeV by the two fission fragments.

Considering that the chemical reaction of burning 1 g of coal produces at best an energy output six orders of magnitude smaller than the fission of 1 g of ^{235}U, the potential for power generation using fission is obvious and exciting, but also controversial. We can reach the same conclusion if we compare the energy release in a fission process with the energy release in a chemical reaction, for example, of combining hydrogen and oxygen into water which results in an energy release of about 3 eV per molecule.

Of course, the high energy yield in one fission event in itself is not of much practical importance; however, an important characteristics of fission is that each fission event produces on average 2–3 neutrons which under proper conditions can be used to sustain the nuclear reaction through a process referred to as *nuclear chain reaction* (see Sect. 12.8).

Many heavy nuclides are fissionable, however, only three fissile nuclei that can undergo thermal neutron induced fission are known. They are:

1. Uranium-235 (^{235}U) as the only naturally occurring fissile nuclide.
2. Plutonium-239 (^{239}Pu) artificially produced from uranium-238 bombarded with neutrons to get uranium-239 which undergoes a β^- decay with a half-life of 23.5 min into neptunium-239 which, in turn, undergoes a β^- decay with a half-life of 2.33 days into plutonium-239, or in short

$$^{238}_{92}\text{U} + \text{n} \rightarrow {}^{239}_{92}\text{U} \xrightarrow[23.5 \text{ min}]{\beta^-} {}^{239}_{93}\text{Np} \xrightarrow[2.33 \text{ d}]{\beta^-} {}^{239}_{94}\text{Pu}, \qquad (12.66)$$

3. Uranium-233 (^{233}U) artificially produced from thorium-232 bombarded with neutrons to get thorium-233 which undergoes a β^- decay with a half-life of 22 min into protactinium-233 which, in turn, undergoes a β^- decay with a half-life of 27 days into uranium-233, or in short

$$^{232}_{90}\text{Th} + \text{n} \rightarrow {}^{233}_{90}\text{Th} \xrightarrow[22 \text{ min}]{\beta^-} {}^{233}_{91}\text{Pa} \xrightarrow[27 \text{ d}]{\beta^-} {}^{233}_{92}\text{U}. \qquad (12.67)$$

Uranium-238 and thorium-232 do not undergo fission by thermal neutrons and therefore are not fissile. However, they are called fertile radionuclides, since they capture a neutron and transform into short-lived radionuclides (radioisotopes) that through two β^- decays transmute into artificial fissile nuclides plutonium-239 and uranium-233, respectively.

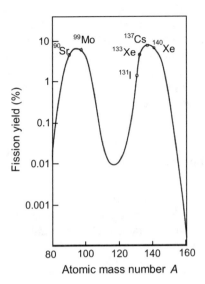

Fig. 12.12 Yield of fission fragments against atomic mass number A for uranium-235 bombarded with thermal neutrons. The yield for uranium-233 and plutonium-239 is similar in shape and distribution

The general equation for fission of uranium-235 is as follows

$$^{235}_{92}U + n \rightarrow {}^{236}_{92}U^* \rightarrow {}^{b}_{a}X + {}^{d}_{c}Y + fn, \qquad (12.68)$$

where the nucleus $^{235}_{92}U$ has been penetrated by a thermal neutron n to produce a compound nucleus $^{236}_{92}U^*$. The compound nucleus $^{236}_{92}U^*$ is unstable and divides by the fission process into two generally unstable nuclei (fission fragments) as well as several fast neutrons. The fission fragments ($^{b}_{a}X$ and $^{d}_{c}Y$) are of smaller Z and A than is the compound nucleus.

Any combination of lighter nuclei is possible for fission fragments as long as the fission process honors the conservation of atomic number Z and atomic mass number A. This means that $a + c = 92$ and $b + d + f = 236$, with f the number of fast neutrons produced in the fission process, typically equal to 2 or 3. Nuclei formed in fission as fission fragments range in atomic number from 30 to 64 and in atomic mass number from 60 to about 150. In general, the fission fragments have an asymmetrical mass distribution with daughter nuclei clustering around atomic mass numbers A of 95 and 140 with much higher probability than clustering around $A = 118$ for a symmetric distribution of fission yield.

A sketch of the yield of fission fragments from uranium-235 bombardment with thermal neutrons plotted against atomic mass number A is plotted in Fig. 12.12. The unstable fission fragments are generally neutron-rich and subsequently revert to stability by a succession of β^- decays forming chains of isobars, the most known being the chains with $A = 99$ (Mo-99) and $A = 140$ (Xe-140) and each occurring in about

6% of all uranium-235 fissions. The distributions of fission fragments for uranium-233 and plutonium-239 are similar to that shown for uranium-235 in Fig. 12.12.

Some of the fragments and fission products have half-lives of the order of thousands of years, making the handling and storage of radioactive waste produced by use of the fission process for power generation a serious concern for many future generations to come. This puts a damper on enthusiasm for commercial means of power generation based on nuclear fission.

12.8   Nuclear Chain Reaction

A nuclear chain reaction occurs when a nuclear reaction results in products that on the average cause one or more nuclear reactions of the same type as the original reaction and the process continues for several generations forming a chain of reactions. Two types of chain reaction are known:

1. Nuclear fission chain reaction based on fission of heavy nuclei such as uranium-235.
2. Nuclear fusion chain reaction based on fusion of light nuclei such as deuteron ^{2}H and triton ^{3}H.

The discovery and understanding of nuclear chain reactions is not only one of the most important scientific discoveries of all times, it is also one of the most controversial ones. On the one hand, nuclear chain reaction promises means for abundant and relatively inexpensive power generation when used appropriately and, on the other hand, it makes possible the destruction of humanity, if used for military purpose.

One of the most important characteristics of nuclear chain reactions is that the reactions are self-sustaining. For example, in fission of uranium-235 nucleus on average 2.5 neutrons are produced along with the two daughter fission fragments. These new neutrons can induce additional fission in the uranium-235 absorber and, in principle, a self-sustaining chain reaction becomes possible. Another important characteristic of chain reactions is that the individual reactions in the chain release an enormous amount of energy in comparison with standard chemical reactions. For example, the ratio between energy produced from the fission of one gram of uranium-235 to energy produced from burning of one gram of coal is $\sim 3 \times 10^6$, indicating several million times larger energy release in fission compared with a chemical reaction.

12.8.1   Nuclear Fission Chain Reaction

In a fission chain reaction the kinetic energy acquired by fission fragments is converted into heat that can be used in nuclear reactors in a controlled fashion for peaceful purposes: (1) in electric power generation and (2) in production of radionuclides through neutron activation. Unfortunately, uncontrolled chain reactions can be used for destructive purposes either directly in atomic bombs or indirectly as detonators of fusion-based hydrogen bombs.

Any nuclear chain reaction can be described by a parameter called effective neutron multiplication factor k, defined as the number of neutrons from a given fission in a given generation that can cause fission in the next generation. Generally, the factor k is smaller than the actual number of fission-generated neutrons because some of these neutrons either escape the system or undergo non-fission reactions. The value of k determines the fate of a chain reaction:

- For $k < 1$, the system cannot sustain a chain reaction, the power diminishes with time, and the mass of the fissile material is classified as *sub-critical*.
- For $k = 1$, every fission event causes a new fission on average. This leads to a constant power level, a steady-state chain reaction, and the classification of the fissile mass as *critical*. The critical mass is defined as the smallest amount of fissile material able to sustain a nuclear chain reaction. Nuclear power plants operate in this mode. The critical mass of a fissile material depends upon its nuclear fission cross section, density, enrichment, shape, and temperature. The shape with minimal critical mass is spherical since a sphere requires the minimum surface area per mass.
- For $k > 1$, every fission event causes an increase in number of subsequent fission events leading to a runaway chain reaction described as *super-critical*. Nuclear weapons operate in this mode.

12.8.2 Nuclear Reactor

In a nuclear reactor the nuclear chain reactions are controlled in such a way that only one new neutron on average is used for continuation of the chain reaction after each fission event. The first nuclear reactor was constructed in 1942 at the University of Chicago under the scientific leadership of *Enrico Fermi*. Since then, several hundred nuclear reactors have been constructed around the world, mainly for electric power generation but some also for research purposes and for production of radionuclides used in industry and medicine.

The principal component of any nuclear reactor is the *reactor core* shown schematically in Fig. 12.13 and containing: (1) fissile material also called fuel elements, (2) control rods, (3) moderator, and (4) coolant:

1. Three fissile radionuclides can be used in nuclear reactors: uranium-233, uranium-235, and plutonium-239. The fission-related physical characteristics of these materials are listed in Table 12.5. The fissile material that is most commonly used is uranium oxide with uranium-235 enriched using a suitable nuclide separation process to contain from 3% to 5% of $^{235}_{92}$U. Natural uranium contains 99.3% of $^{238}_{92}$U, 0.7% of $^{235}_{92}$U, and a minute amount (0.006%) of $^{234}_{92}$U. Some reactors use a much higher grade of enrichment operating at 20% or even higher. The

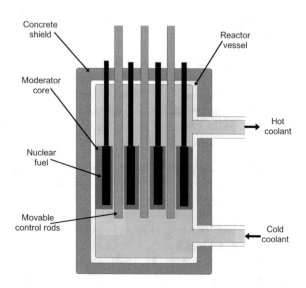

Fig. 12.13 Schematic diagram of nuclear reactor core highlighting the four main components: (*1*) fuel elements, (*2*) control rods, (*3*) moderator, and (*4*) coolant

Table 12.5 Main fission-related physical characteristics of the three fissile radionuclides

	Fissile nucleus	Means of production	Neutrons per fission	Critical mass (kg)	Half-life (years)	Decay process
1	$^{233}_{92}\text{U}$	$^{232}_{90}\text{Th} + \text{n}$	~2	16	160 000	α to $^{229}_{90}\text{Th}$
2	$^{235}_{92}\text{U}$	Naturally occurring	~2.5	52	0.7×10^9	α to $^{231}_{90}\text{Th}$
3	$^{239}_{94}\text{Pu}$	$^{238}_{92}\text{U} + \text{n}$	~3	10	24 110	α to $^{235}_{92}\text{U}$

$^{238}_{92}\text{U}$ uranium isotope that remains after the $^{235}_{92}\text{U}$ enrichment is called depleted uranium and is used for applications requiring very dense materials, for example, for collimators in teletherapy machines.

2. The neutron fluence rate, i.e., the number of neutrons available for inducing fission in the reactor core is controlled by movable *control rods* that are made of material with high cross section for absorption of neutrons, such as cadmium or boron compounds. This mechanism maintains a constant neutron multiplication factor k and ensures a constant criticality of the reactor and a constant power output.

3. The fission efficiency is the highest for thermal neutrons, but the new fission-generated neutrons have relatively large kinetic energies of the order of 1 MeV. *Moderators* are used to slow down the newly produced fast neutrons through elastic scattering events between neutrons and the nuclei of the moderator. Water serves as *moderator* material in most reactors; however, some reactors may use the

so-called heavy water (deuterium based), graphite, or beryllium for the purposes of moderation. Heavy water has a smaller probability for neutron absorption through the (n, γ) reaction than water; however, it is much costlier. Graphite also does not absorb many neutrons and scatters neutrons well. Beryllium is an excellent solid moderator with its low neutron absorption cross section and a high neutron scattering cross section.

4. In a nuclear power plant the reactor core is immersed in a suitable coolant. Fission occurs in the nuclear fuel and the fission energy in the form of kinetic energy of fission fragments and new neutrons is converted into heat. The *coolant* (usually water) is used to maintain a stable temperature in the reactor core and exits the core either as steam or as hot pressurized water, subsequently used to drive turbines connected to electric power generators.

12.8.3 Nuclear Power

Development of electric power generation with nuclear reactors has started in the early 1950 s and within two decades nuclear power became a viable alternative to power generation based on fossil fuels, such as oil, natural gas, and coal. The main advantage of nuclear power is that, in contrast to fossil fuels, it does not produce greenhouse gases and does not cause global warming; however, it has its own unique problems, real and imaginary, which detract from its widespread use. The main concerns causing opposition to nuclear power are:

1. Production of radioactive waste in the fission chain reaction and problems with the long-term disposal of the radioactive waste.
2. Potential for accidents resulting in contamination of environment making it inhabitable (area denial).
3. Increase in radiation exposure beyond the background level to general population living in the vicinity of nuclear power plants resulting in carcinogenesis (induction of cancer).

While it has been proven beyond doubt that nuclear reactors can be and are shielded so well that radiation exposure in the vicinity of nuclear power plants does not exceed the background radiation level, the real issues of radioactive waste and potential for accidents are still preventing a widespread use of nuclear power. The fission products generated in nuclear reactor chain reaction are generally long lived. Thus, the nuclear waste is bound to pose a health hazard to humans for thousands of years to come, and to date no solution to long-term safe storage of nuclear waste has been found. As far as accidents are concerned, accidents at the Three Mile Island nuclear power plant in the U.S. in 1979 and at the Chernobyl power plant in Ukraine in 1986 have proven that nuclear accidents with disastrous consequences can and do happen, no matter how unlikely they are deemed to be before they occur.

Waste from the nuclear fuel cycle has two components: front end and back end. Front-end waste is usually alpha particle emitting waste originating from the extraction of uranium and back-end waste consists mostly of spent reactor fuel rods containing fission products that emit beta and gamma radiation.

A partial solution to the current nuclear waste dilemma is a more efficient use of the nuclear fuel to minimize the amount of the nuclear waste generated. Two new developments in reactor technology are showing promise in this direction: the very high temperature (VHT) reactor and the sodium-cooled fast (SCF) reactor, both incorporating inherent safety features to prevent reactor core meltdown and release of radioactive fission products into the environment.

The core of the VHT reactor is made of graphite, a moderator material that remains strong and stable even at the very high operating temperatures. Helium gas is used for reactor cooling and for transport of heat to a heat exchanger where it is used to produce steam from water for running the turbines that generate electricity. The VHT reactor is operating at about 1000 °C in contrast to about 300 °C operating temperature of standard reactors and its nuclear fuel consists of carbon-coated uranium oxide to withstand the very high operating temperature.

The SCF reactors have been in use for several decades, however, modern designs introduce several important new features which increase efficiency. The operating temperature is about 500 °C and the reactors use sodium for the modulator material. In modern SCF reactors uranium is bombarded with fast neutrons which are neutrons with much higher energy than that of thermal neutrons used in standard reactors and this produces an almost two orders of magnitude increase in efficiency in comparison with the standard thermal reactors. Moreover, the SCF reactor can also burn spent nuclear fuel from standard reactors and can also run on depleted uranium left over from the uranium enrichment process, thereby significantly mitigating the nuclear waste problem.

12.8.4 Nuclear Fusion Chain Reaction

The binding energy per nucleon E_B/A plotted against atomic mass number A in Fig. 1.3 exhibits a maximum of ~ 8.8 MeV/nucleon at an atomic mass number $A \approx 60$ and falls to ~ 7.5 MeV/nucleon at very large A (heavy nuclei) and to even smaller values at small A (light nuclei). For example, E_B/A for deuteron is only ~ 1.1 MeV/nucleon. Just like this peculiar shape of the E_B/A curve lends itself to energy release in nuclear fission at large A, in principle, it also holds promise in nuclear fusion of two light nuclei (with relatively low E_B/A) into a heavier nucleus with larger E_B/A. The obvious problem with fusion is that the two light nuclei must be brought very close to each other for the short-range nuclear attractive potential to become effective and induce nuclear fusion, but, in doing so, the Coulomb repulsive potential must be overcome.

An estimate of the Coulomb repulsive potential for two light nuclei just touching results in a few MeV per nucleus, certainly an energy that can be achieved with modern accelerators. However, relying on collision of energetic machine-produced light particles is unlikely to enable a fusion chain reaction because most of the collision interactions will result in elastic scattering rather than fusion and the collision process will on average result in energy drain rather than gain.

Other means must thus be employed in the quest for using nuclear fusion as an inexpensive and "inexhaustible" energy source, and thermal energy seems the most practical option. However, the required temperature to provide the fusing nuclei with the thermal energy to overcome the Coulomb barrier of the order of a few MeV is extremely high and of the order of 10^7 K. (*Note:* the thermal energy kT at room temperature of about 300 K is 0.025 eV with k the Boltzmann constant).

Arthur S. Eddington in 1920s proposed proton–proton fusion to explain the energy source for the sun and other stars; however, the temperature of the sun was deemed too low for providing hydrogen nuclei with sufficient thermal energy to overcome the Coulomb barrier. Subsequent developments in quantum mechanics, specifically in understanding of the tunneling phenomena, have provided an answer for the possibility of fusion at temperatures lower than those estimated classically. Hans Bethe developed Eddington's work further, described nuclear mechanisms for fusion of hydrogen into helium, and in 1967 received the Nobel Prize in Physics *"for his theory of nuclear reactions, especially for his discoveries concerning the energy production in stars."*

Despite significant worldwide effort by many research groups to achieve controlled thermonuclear fusion, successes to date have been small. Several fusion reactions, albeit of very short duration, have already been observed under laboratory conditions; however, the energy expended to produce them always exceeded the energy output, so that chain reaction conditions so far have not been achieved. The main problem is containment of the nuclear fuel at the very high temperatures required to overcome the Coulomb repulsion between the two nuclei undergoing fusion. Of the two currently favored techniques, one uses magnetic confinement of hot plasma in tokamak machines and the other uses very intense laser beams or ion beams focused on very small volumes of nuclear fuel. Progress is slow, however, and practical power generation based on nuclear fusion is still far in the future.

In order to be useful as a potential source of energy, a fusion reaction should satisfy several criteria, such as:

- The reaction should involve low atomic number Z nuclei, such as deuteron d, triton t, helium-3 (helion), and helium-4 (Alpha particle).
- The reaction should be exoergic.
- The reaction should have two or more products for energy and momentum conservation.
- The reaction should conserve the number of protons and the number of neutrons.

A few examples of fusion reactions that meet the criteria listed above are given below. The Q values for the individual reactions are calculated using data provided in Appendix A. The energy given in the Q value is shared between the two reaction products in inverse proportion to their masses.

1. $d + d \rightarrow t + p + Q$

    $(1875.6129 + 1875.6129) \text{ MeV} = (2808.9209 + 938.2703) \text{ MeV} + Q$

    $Q = (3751.2250 - 3747.1912) \text{ MeV} = 4.03 \text{ MeV}$

    $Q = Q_t + Q_p = 0.75 \times 4.03 \text{ MeV} + 0.25 \times 4.03 \text{ MeV}$

    $\qquad = 3.02 \text{ MeV} + 1.01 \text{ MeV}$ (12.69)

2. $d + d \rightarrow {}_2^3\text{He} + n + Q$

    $(1875.6129 + 1875.6129) \text{ MeV} = (2808.3913 + 939.5654) \text{ MeV} + Q$

    $Q = (3751.2258 - 3747.9567) \text{ MeV} = 3.27 \text{ MeV}$

    $Q = Q_{{}_2^3\text{He}} + Q_n = 0.75 \times 3.27 \text{ MeV} + 0.25 \times 3.27 \text{ MeV}$

    $\qquad = 2.45 \text{ MeV} + 0.82 \text{ MeV}$ (12.70)

3. $d + t \rightarrow {}_2^4\text{He} + n + Q$

    $(1875.6129 + 2.808.9209) \text{ MeV} = (3727.3791 + 939.5654) \text{ MeV} + Q$

    $Q = (4684.5338 - 4666.9445) \text{ MeV} = 17.59 \text{ MeV}$

    $Q = Q_{{}_2^4\text{He}} + Q_n = 0.80 \times 17.59 \text{ MeV} + 0.20 \times 17.59 \text{ MeV}$

    $\qquad = 14.05 \text{ MeV} + 3.54 \text{ MeV}$ (12.71)

4. $d + {}_2^3\text{He} \rightarrow {}_2^4\text{He} + p + Q$

    $(1875.6129 + 2808.3913) \text{ MeV} = (3727.3791 + 938.2703) \text{ MeV} + Q$

    $Q = (4684.0042 - 4665.6494) \text{ MeV} = 18.36 \text{ MeV}$

    $Q = Q_{{}_2^4\text{He}} + Q_p = 0.80 \times 18.36 \text{ MeV} + 0.20 \times 18.36 \text{ MeV}$

    $\qquad = 14.69 \text{ MeV} + 3.67 \text{ MeV}$ (12.72)

12.9 Production of Radionuclides with Radionuclide Generator

Technetium-99m (Tc-99m) is the most widely used radionuclide for diagnostic imaging in nuclear medicine, being used in some 80% of all nuclear imaging tests. It emits 140.5 keV gamma rays with a physical half-life of 6.02 h and thus has properties which make it very suitable for nuclear imaging. It provides sufficiently high-energy gamma rays for imaging and has a half-life long enough for investigation of metabolic processes, yet short enough so as not to deliver an excessive total body dose to the patient.

In addition, since Tc-99m is characterized with a nuclear decay that produces only gamma rays and no α-particles or energetic β-particles, the total body patient dose is kept to a minimum. Technetium-99m does not target any specific organ or tissue in the human body; however, it can be incorporated as tracer into a wide range of biologically active substances for accumulation in a specific organ or tissue and subsequent imaging of the specific organ or tissue. Technetium-99m is well suited for its role as radioactive tracer because it emits readily detectable 140 keV gamma rays with a half-life of about 6 h.

12.9.1   Molybdenum–Technetium Decay Scheme

As shown in the simplified decay scheme of Fig. 12.14, the parent nucleus of Tc-99m is molybdenum-99 (Mo-99) which disintegrates with a half-life of 2.75 days (66 h) through β^- decay (Sect. 11.4) into several excited levels of Tc-99m. The Tc-99m radionuclide is unusual, since it undergoes a gamma decay into Tc-99 ground state with its own half-life of 6.02 h rather than decaying essentially instantaneously (within 10^{-16} s), as is the case with most excited radionuclides produced through β^- or β^+ decay. Tc-99m is thus the daughter product of the Mo-99 β^- decay exhibiting its own relatively long half-life of 6.02 h; hence the label 99 m with m designating an isomeric metastable state (Sect. 11.7.1).

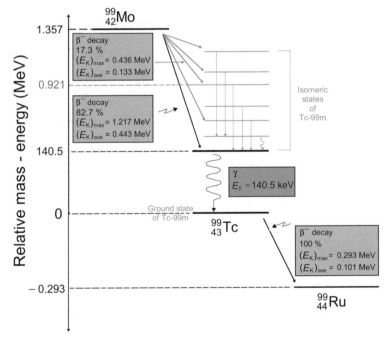

Fig. 12.14 Simplified scheme for nuclear series decay of molybdenum-99 into ruthenium-99, starting with β^- decay of molybdenum-99 into technetium-99m through gamma decay of technetium-99m into technetium-99 to β^- decay of technetium-99 into stable ruthenium-99. The relative mass-energy levels (not shown to scale) for the three nuclides are calculated using atomic masses listed in Appendix A

The diagram of Fig. 12.14 indicates a fairly complicated β^- decay scheme for the Mo-99 nucleus; however, all β^- transitions into the excited Tc-99m nucleus except the one to the 0.1405 MeV level are rare and are followed by internal conversion transitions to the 0.1405 MeV level which decays into ground state of Tc-99 through emission of a 140.5 keV gamma photon (88%) or internal conversion (12%) with associated internal conversion and Auger electrons and characteristic x-rays.

In the first approximation we assume that the Tc-99m gamma decay involves only one transition emitting a 140.5 keV gamma ray which is then used for diagnostic imaging. The Tc-99 nucleus, the daughter product of the Tc-99m gamma decay, is also radioactive and decays through β^- decay with a half-life of 2.13×10^5 years into stable ruthenium-99 (Ru-99).

The $^{99}_{42}\text{Mo} \xrightarrow[66\ \text{h}]{\beta^-} {}^{99\text{m}}_{43}\text{Tc} \xrightarrow[6\ \text{h}]{\gamma} {}^{99}_{43}\text{Tc} \xrightarrow[2.13 \times 10^5\ \text{y}]{\beta^-} {}^{99}_{44}\text{Ru}$ nuclear series decay thus proceeds through one β^- decay, one γ decay, and another β^- before ending with stable ruthenium-99 as follows

$$^{99}_{42}\text{Mo} \xrightarrow{\beta^-} {}^{99\text{m}}_{43}\text{Tc} + e^- + \bar{\nu}_e + Q_{\beta^-}(1.357\ \text{MeV}) \tag{12.73}$$

$$^{99\text{m}}_{43}\text{Tc} \xrightarrow{\gamma} {}^{99}_{43}\text{Tc} + \gamma(140.5\ \text{keV}) \tag{12.74}$$

$$^{99}_{43}\text{Tc} \xrightarrow{\beta^-} {}^{99}_{44}\text{Ru} + e^- + \bar{\nu}_e + Q_{\beta^-}(0.294\ \text{MeV}) \tag{12.75}$$

Decay energies Q_{β^-} for the β^- decay of $^{99}_{42}\text{Mo}$ into radioactive $^{99}_{43}\text{Tc}$ and the decay of $^{99}_{43}\text{Tc}$ into stable $^{99}_{44}\text{Ru}$ are determined, as shown in Sect. 11.4.2 for the β^- decay in general. We get

$$Q_{\beta^-}(^{99}_{42}\text{Mo}) = [\mathcal{M}(^{99}_{42}\text{Mo}) - \mathcal{M}(^{99}_{43}\text{Tc})]c^2 = (98.9077116 - 98.9062546)\text{u}c^2$$
$$= 0.001457 \times 931.494028\ \text{MeV} = 1.357\ \text{MeV} \tag{12.76}$$

and

$$Q_{\beta^-}(^{99}_{43}\text{Tc}) = [\mathcal{M}(^{99}_{43}\text{Tc}) - \mathcal{M}(^{99}_{44}\text{Ru})]c^2 = (98.9062546 - 98.9059393)\text{u}c^2$$
$$= 0.000315 \times 931.494028\ \text{MeV} = 0.294\ \text{MeV} \tag{12.77}$$

in agreement with the decay scheme shown in Fig. 12.14.

12.9.2 Molybdenum–Technetium Radionuclide Generator

The short half-life of only 6 h of technetium-99m is advantageous from the point-of-view of minimizing the patient dose resulting from the imaging test; however, it also makes the logistics of source production, delivery, and storage problematic. A method to circumvent the transportation and delivery problem was developed in 1950s at the Brookhaven National Laboratory in Upton, NY, whereby a supplier, rather than shipping the Tc-99m radionuclide, ships the longer-lived parent radionuclide Mo-99 in a device referred to as a radionuclide generator.

Originally, the device was named "radioisotope generator" and the old term is still often used even though the term "isotope" for designation of nuclear species has been replaced with the term "nuclide". Thus, the device should be called radionuclide generator not only to follow correct nomenclature but also to distinguish it from another type of "radioisotope generator" that is used for a broad class of power generators. These generators are based on heat produced by radioactive decay of certain long-lived radionuclides with preference for α emitters, such as plutonium-238, curium-244, and strontium-90. These generators can be considered as a type of battery and are used as power sources in satellites, space probes, and unmanned remote facilities, and have nothing in common with the radionuclide generators used in medicine.

Several radionuclide generators are used in nuclear medicine as a source of metastable radionuclides for imaging; however, the molybdenum-technetium (Mo–Tc) generator is by far the most common. In the molybdenum-technetium (Mo–Tc) radionuclide generator the Mo-99 radionuclide decays with a 65.94-h half-life to Tc-99m. Since the daughter's 6.02-h half-life is much shorter than the parent's half-life, the two radionuclides attain secular equilibrium and the daughter decays with the half-life of the parent, as discussed in Sect. 10.3.4. This effectively extends the life-time of the Tc-99m radionuclide and makes its transportation much less problematic than would be the shipping of carrier-free Tc-99m radionuclide.

The activities of Mo-99 and Tc-99m as well as the functional relationship between the two are illustrated in Fig. 10.2 with the curve labeled $m = 0.1$ where m is the decay factor given as $m = (t_{1/2})_D/(t_{1/2})_P = \lambda_P/\lambda_D$, and is defined in (10.44). The decay factor for the Mo–Tc decay is $m = (t_{1/2})_{Tc}/(t_{1/2})_{Mo} = \lambda_{Mo}/\lambda_{Tc} = 6.01/65.94 = 0.091$, so we can use the $m = 0.1$ curve in Fig. 10.2 as a reasonable approximation.

The parent Mo-99 decays exponentially, as one would expect, with its half-life of 66 h. The activity of the daughter Tc-99m, on the other hand, starts at zero rises rapidly to a maximum value which it attains when its activity is equal to the parent activity (point of ideal equilibrium as defined by Robley Evans), and then it decays with the activity of the parent. The time to reach maximum daughter activity (equilibrium activity) can be calculated using (10.48) and (10.50) to get

$$(t_{max})_D = \frac{m}{m-1}\frac{\ln m}{\ln 2}(t_{1/2})_P = \frac{0.091}{-0.90}\frac{\ln 0.091}{\ln 2}65.94 \text{ h} = 23.1 \text{ h}. \qquad (12.78)$$

As shown in (12.78), it takes 23.1 h (about four daughter half-lives) to reach equilibrium activity, but 50% of equilibrium activity will be reached within one daughter half-life and 75% within two daughter half-lives. Hence, removing the daughter nuclide from the generator in a simple procedure called the *elution process* (colloquially referred to as "milking the cow") can easily be done every 6 h to 12 h. Typically, about 80% of the available Tc-99m activity is extracted in a single elution process.

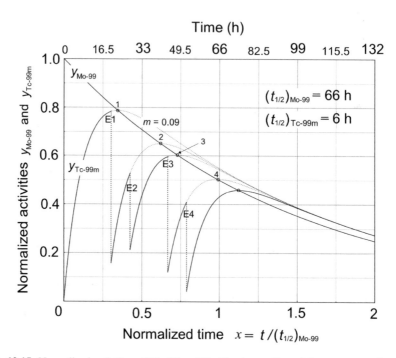

Fig. 12.15 Normalized activities of Mo-99 and Tc-99m in a radionuclide generator undergoing elution processes (E1, E2, E3, and E4) at times $t = 20, 28, 44$, and 52 h corresponding to normalized times x of 0.3; 0.42; 0.67; and 0.79, respectively. Points 1, 2, 3, 4, and 5 designate the appropriate points of ideal equilibrium on the Mo-99 exponential decay curve corresponding to the four elution processes. The branching decay ratio for Mo-99 decay is assumed equal to 100%, while in reality it is 86%, as discussed in Sect. 10.8

A large percentage of the Tc-99m generated by a Mo–Tc generator is produced in the first three half-lives of the parent or in about one week. Thus, regular delivery of one Mo–Tc radionuclide generator per week can ensure a steady supply of Tc-99m for a nuclear medicine department.

A typical record of Tc-99m activity in a Mo–Tc generator is shown in Fig. 12.15, starting at time $t = 0$ when the generator is loaded with Mo-99 and the activity of Tc-99m is zero. The generator is then shipped to the clinical site, the activity of Mo-99 decays exponentially, and the activity of Tc-99m increases toward the point of ideal equilibrium. If left alone, transient equilibrium between Mo-99 and Tc-99m would be reached in the generator, as discussed in Sect. 10.5.3, and the two activities would decrease exponentially, as given in (10.63) and shown in Fig. 10.2 for $m \approx 0.1$.

The activity trace given in Fig. 12.15 is significantly more complicated than that in Fig. 10.2 because of four elution processes (at times $t = 20$ h, 28 h, 44 h, and 52 h) that were carried out on the generator to obtain Tc-99m radionuclide for clinical nuclear medicine studies. After each elution, the activity of Tc-99m in the generator follows a similar pattern to the initial trace that is characterized with initial growth, peak at ideal equilibrium, and eventual transient equilibrium exponential decay, as indicated in Fig. 12.15.

Each elution process results in a different point of ideal equilibrium; however, all Tc-99m activity curves eventually converge to the initial Tc-99m activity curve which starts at zero at time $t = 0$. A simplifying assumption in the plot of Fig. 12.15 is that the branching decay fraction of Mo-99 into Tc-99m is assumed to be 100%, while in reality it is 86%, as discussed in Sect. 10.8.

12.9.3 Production of Molybdenum-99 Radionuclide

Two techniques, both based on nuclear reactor technology, are used for producing the parent radionuclide Mo-99 used in Mo–Tc generator for generating the Tc-99m radionuclide:

1. One method uses neutron activation of stable molybdenum-98 through the $^{98}\text{Mo}(n, \gamma)^{99}\text{Mo}$ reaction. The disadvantage of this technique is the relatively low specific activity of the resulting Mo-99 source, because the molybdenum radionuclide delivered to a clinic is not carrier free.
2. The second and more common technique uses fission of enriched uranium-235 foils to produce Mo-99 as one of the many fission fragments in the target foil. After irradiation with thermal neutrons, Mo-99 is radiochemically separated in a hot cell from other fission fragments and this results in a high specific activity Mo-99 source. The drawback of this technique is the large amount of radioactive waste it creates, much of it long-lived and problematic to dispose of.

12.10 Nuclear Activation with Protons and Heavier Charged Particles

Protons produced by cyclotrons are used in the production of proton-rich unstable radionuclides that decay through β^+ decay or electron capture into more stable configurations. When striking a target material, protons may cause nuclear reactions that produce radionuclides in a manner similar to neutron activation in a reactor. However, because of their positive charge, protons striking the target must have relatively high kinetic energies, typically 10 MeV to 20 MeV, to penetrate the repulsive Coulomb barrier surrounding the positively charged nucleus. Proton activation reactions are generally endoergic which means that energy must be supplied by the projectile for the reaction to occur. The minimum energy that will allow the reaction to occur is referred to as the threshold energy.

Proton capture by a target nucleus changes the atomic number from Z for the parent to $Z + 1$ for the daughter nucleus. This allows production of carrier-free radionuclides for use in medicine, because a chemical separation of the newly produced daughter radionuclide from the remaining parent nuclide is possible.

Positron emitters produced for use in medicine by proton activation in cyclotrons generally have much shorter half-lives (of the order of minutes) than radionuclides

Table 12.6 Main physical characteristics of four most common positron emitters produced in a cyclotron for use in medicine

Radionuclide	Specific activity	Target	Production reaction	Q value (MeV)	Half-life (min)
Carbon-11	8.4×10^8	Nitrogen-14	$^{14}_{7}N + p \rightarrow {}^{11}_{6}C + \alpha$	-2.92	20.4
Nitrogen-13	1.4×10^9	Oxygen-16	$^{16}_{8}O + p \rightarrow {}^{13}_{7}N + \alpha$	-5.22	10.0
Oxygen-15	6.0×10^9	Nitrogen-15	$^{15}_{7}N + p \rightarrow {}^{15}_{8}O + n$	-3.54	2.1
Fluorine-18	9.5×10^7	Oxygen-18	$^{18}_{8}O + p \rightarrow {}^{18}_{9}F + n$	-2.44	110.0

produced for use in medicine by neutron activation in nuclear reactors. This implies that cyclotrons used in production of radionuclides should be located close to the user of the radionuclides to minimize the transportation and delivery time.

Cyclotrons generally produce smaller quantities of radioactivity than do nuclear reactors because:

1. Cross sections for proton capture are lower by several orders of magnitude than those for neutron capture and they are strongly energy dependent.
2. Proton beam is monodirectional and is attenuated in the target.
3. Cyclotron particle fluence rates are generally lower than those produced by nuclear reactors.

For cyclotrons, rather than providing a fluence rate as is done for reactor produced neutrons, one provides a beam current, usually expressed in μA, where 1 μA of current is equal to 6.25×10^{12} electronic charges per second, i.e., 6.25×10^{12} electrons per second. Thus, a proton beam of 1 μA corresponds to 6.25×10^{12} protons per second; a helium He^{2+} beam corresponds to 3.125×10^{12} helium ions per second.

The cyclotron-produced radionuclides are positron emitters used in positron emission tomography (PET) scanners for diagnostic imaging; a non-invasive imaging technique that provides a functional image of organs and tissues, in contrast to CT scanning and MRI scanning that provide anatomic images of organs and tissues. For PET scanning the positron-emitting radionuclides are attached to clinically useful biological markers that are used in studies involving various metabolic processes in cancer diagnosis and treatment. The four most important positron emitting radionuclides used in medical PET imaging are: fluorine-18, carbon-11, nitrogen-13 and oxygen-15 and their main characteristics are given in Table 12.6.

All radionuclides currently used in PET scanning have short half-lives. This is advantageous for minimization of the total body dose that the patient receives during the diagnostic test but generally requires the presence of a cyclotron next to the PET machine making the operation of a PET scanner expensive. Only fluorine-18 with a half-life of 110 min can be manufactured at an offsite location.

12.10.1 *Nuclear Reaction Energy and Threshold Energy*

As discussed in Sect. 5.2, the *nuclear reaction energy* Q also known as the Q *value* for a nuclear reaction provides the energy release or energy absorption during the nuclear reaction. In general, the Q value is determined in one of the following two manners:

1. The sum of nuclear rest energies of the reaction products (i.e., the total rest energy after reaction) is subtracted from the sum of nuclear rest energies of the reactants (i.e., the total rest energy before reaction).
2. The sum of nuclear binding energies of the reactants (i.e., the total binding energy before reaction) is subtracted from the sum of nuclear binding energies of reaction products (i.e., the total binding energy after reaction).

For a given nuclear reaction the Q values obtained with the two methods should be identical and will be either positive or negative:

1. For $Q > 0$ the reaction is called *exoergic* and the excess energy is shared between the two reaction products.
2. For $Q < 0$ the reaction is called *endoergic* and for the reaction to occur, energy must be supplied in the form of the kinetic energy of the projectile.

By way of example we calculate the nuclear reaction energy Q for the activation of oxygen-18 into fluorine-18 in a proton cyclotron. The reaction is as follows

$$^{18}_{8}O + p \rightarrow {}^{18}_{9}F + n + Q\,(-2.44\ \text{MeV}) \tag{12.79}$$

and the Q value of -2.44 MeV is calculated using first the rest energy method and then the binding energy method with data from Appendix A.

1. Rest energy method:

$$Q = \left\{ \sum_i M_i c^2 \right\}_{\text{before}} - \left\{ \sum_i M_i c^2 \right\}_{\text{after}}$$
$$= \{M(^{18}_{8}O)c^2 + m_p c^2\} - \{M(^{18}_{9}F)c^2 + m_n c^2\}$$
$$= \{16762.0227 + 938.272\}\ \text{MeV} - \{16763.1673 + 939.5654\}\ \text{MeV}$$
$$= -2.44\ \text{MeV}. \tag{12.80}$$

2. Binding energy method

$$Q = \left\{ \sum_i (E_B)_i \right\}_{\text{after}} - \left\{ \sum_i (E_B)_i \right\}_{\text{before}}$$
$$= 137.3693\ \text{MeV} - 139.8071\ \text{MeV} = -2.44\ \text{MeV}. \tag{12.81}$$

Both methods produce the same result: -2.44 MeV. The production of fluorine-18 in a cyclotron is thus an endoergic reaction and energy must be supplied for the reaction to occur.

In general, for endoergic reactions to occur the projectiles must have a certain minimum *threshold kinetic energy* $(E_K)_{thr}$ that exceeds the absolute value of the Q value, so that the total momentum for before and after the interaction is conserved. The general relationship for the threshold of endoergic reactions was derived in (5.15), and based on that result we write the threshold kinetic energy $(E_K)_{thr}$ for the proton with rest mass m_p, activating target nuclide of rest mass M_t, as

$$(E_K)_{thr} = -Q \left\{ 1 + \frac{m_p}{M_t} \right\}. \tag{12.82}$$

For example, the threshold energy for fluorine-18 production from oxygen-18 in a proton cyclotron at 2.58 MeV is slightly higher than the absolute value of $|Q| = 2.44$ MeV. The threshold energy of 2.58 MeV for fluorine-18 is determined from (12.82) as follows

$$(E_K)_{thr} = -Q \left\{ 1 + \frac{m_p c^2}{M(^{18}_8 O)c^2} \right\}$$

$$= -(-2.44 \text{ MeV}) \left\{ 1 + \frac{938.272 \text{ MeV}}{16762.0227 \text{ MeV}} \right\} = 2.58 \text{ MeV}. \quad (12.83)$$

12.10.2   Targets in Charged Particle Activation

The targets used in charged particle activation are either thin or thick. The following conditions apply:

1. The thickness of a *thin target* is such that the target does not appreciably attenuate the charged particle beam.
2. Charged particles traversing a *thick target* lose energy through Coulomb interactions with electrons of the target and this affects the activation yield, since the cross section for activation depends on charged particle energy. The particle beam is completely stopped in a thick target or it is degraded in energy to a level below the threshold energy.
3. Similarly to the approach that one takes with thick x-ray targets assuming they consist of many thin x-ray targets, one may assume that a thick target in charged particle activation (CPA) consists of a large number of thin targets, each one characterized by a given charged particle kinetic energy and reaction cross section. The kinetic energy for each slice is determined from stopping power data for the given charged particle in the target material.
4. Target materials used in production of positron-emitting nuclides are either in a gaseous or liquid state.

5. Cyclotron targets are most commonly of the thick target variety resulting in complete beam absorption in the target material.
6. Essentially all energy carried into the target by the beam is transformed into heat because of charged particle Coulomb interactions with orbital electrons of the target atoms. Thus, targets are cooled with circulating helium gas.
7. Only a small fraction of one percent of the charged particle beam is used up for induction of activation, the rest is dissipated as heat.

The derivations presented in Sect. 12.4 for neutron activation could in principle be generalized to CPA; however, the issue of beam attenuation in thick targets that are routinely used for CPA of medical positron-emitting radionuclides complicates matters considerably. On the other hand, the specific activities produced by CPA are several orders of magnitude lower than specific activities produced in neutron activation, so that in general parent nuclide depletion is not of concern in CPA.

In Sect. 12.4 it was established that the depletion model should be used for activation factors m exceeding 10^{-3}. Since typical values of m in CPA are of the order of 10^{-7}, it is obvious that daughter activation in CPA can be calculated using the simple saturation model that accounts for the daughter decay during the activation procedure.

Assuming that there is no charged particle beam attenuation in the target (thin target approximation), the daughter activity $\mathcal{A}_D(t)$, similar to the neutron activation case of (12.23), can be written as follows

$$\mathcal{A}_D(t) = In^{\square}x\sigma_P \left(1 - e^{-\lambda_D t}\right) \qquad (12.84)$$

where

I	is intensity of charged particle beam in particles per unit time in (s^{-1}).
$n^{\square}$	is the number of target nuclei per volume in cm^{-3}.
x	is the target thickness in cm.
σ_P	is the reaction cross section of the parent nuclei at the energy of the charged particle beam in barn $(1\,b = 10^{-24}\,cm^2)$.

Acceleration Waveguide of a Linear Accelerator

Photograph on next page shows a cut-away view of a simple standing wave acceleration waveguide used in a medical linear accelerator (linac) to accelerate electrons to a kinetic energy of 6 MeV corresponding to an electron velocity of 99.7% of the speed of light in vacuum.

In contrast to x-ray tubes that use electrostatic potential between the anode (target) and the cathode (filament) for acceleration of electrons to a given kinetic energy of the order of 100 keV, a medical linac uses an acceleration waveguide in which electrons are accelerated with electromagnetic fields to much higher kinetic energies in the range from 4 MeV to 25 MeV.

Waveguides are evacuated or gas-filled structures of rectangular or circular cross sections used in transmission of microwaves. Two types of waveguides are used in linear accelerators: (1) radiofrequency power transmission waveguides (usually gas-filled) and (2) acceleration waveguides (always evacuated). The power transmission waveguides transmit the radiofrequency power from the power source to the acceleration waveguide in which electrons are accelerated. The electrons are accelerated in the acceleration waveguide by means of energy transfer from the high power radiofrequency field that is set up in the acceleration waveguide and produced by the radiofrequency power generator (either magnetron or klystron).

The simplest acceleration waveguide is obtained from a cylindrical uniform waveguide by adding a series of irises (disks) with circular holes at the center and placed at equal distances along the tube. These irises divide the waveguide into a series of cylindrical cavities that form the basic structure of the acceleration waveguide. The phase velocity of radiofrequency in a uniform waveguide exceeds that of the speed of light in vacuum and one of the roles of the irises is to slow down the radiofrequency below the speed of light in vacuum to allow electron acceleration. The irises also couple the cavities, distribute microwave power from one cavity to another, and provide a suitable electric field pattern for acceleration of electrons in the acceleration waveguide.

The waveguide cavities and the irises delineating them are clearly visible on the photograph of the waveguide: the accelerating cavities are on the central axis of the standing waveguide, the radiofrequency coupling cavities are offside. The source of electrons (electron gun) is on the left, the x-ray target on the right, both permanently embedded into the acceleration waveguide structure.

The electron gun is a simple electrostatic accelerator that produces electrons thermionically from a heated filament and accelerates them to a typical energy of 20 keV in preparation for further acceleration by the acceleration waveguide. The target is made of metal thicker than the 6 MeV electron range in the target material. The 6 MeV electrons are stopped in the target and a small portion (few %) of their kinetic energy is transformed into bremsstrahlung x-rays that form a photon spectrum, ranging in photon energies from 0 MeV to 6 MeV with an effective energy of about 2 MeV. The bremsstrahlung photon spectrum produced by 6 MeV electrons striking a target is referred to as a 6 MV x-ray beam.

Chapter 13
Waveguide Theory

This chapter is devoted to a discussion of theoretical aspects of disk-loaded waveguides used for acceleration of electrons with linear accelerators (linacs) in treatment of cancer with ionizing radiation. Linacs represent a significant technological advancement over x-ray machines and cobalt-60 units which were used for routine radiotherapy in the past before the advent of linacs.

Two research groups were involved in initial development of linear accelerator: William W. Hansen's group at Stanford University in the USA and D.W. Fry's group at the Telecommunications Research Establishment in the UK. Both groups were interested in linacs for research purposes and profited heavily from the microwave radar technology based on a design frequency of 2856 MHz and developed during World War II.

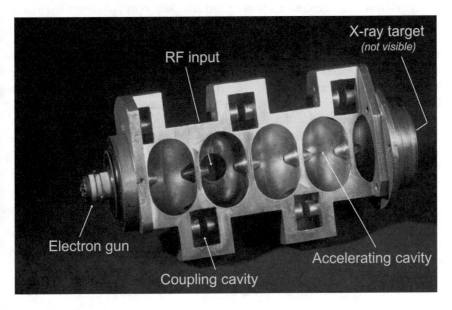

© Springer International Publishing Switzerland 2016

E.B. Podgoršak, *Radiation Physics for Medical Physicists*,

Graduate Texts in Physics, DOI 10.1007/978-3-319-25382-4_13

The potential for use of linacs in radiotherapy became apparent in the 1950s, and the first clinical linac was installed in the 1950s at the Hammersmith Hospital in London. During the subsequent years, the linac eclipsed the cobalt machine and became the most widely used radiation source in modern external beam radiotherapy. Technical and clinical aspects of linacs are discussed in Sect. 14.6; in this chapter we discuss the theoretical aspects of electron acceleration in an acceleration waveguide of a linac. First, the uniform waveguide theory is developed using Maxwell equations and then the results of the uniform waveguide theory are extrapolated to disk-loaded waveguides that are suitable for electron acceleration.

13.1 Microwave Propagation in Uniform Waveguide

In a disk-loaded acceleration waveguide of a linac, electrons are accelerated with electromagnetic (EM) radiofrequency (RF) fields in the microwave frequency range. Standard clinical linacs operate in the S-band (2856 MHz); miniature waveguides used in CyberKnife and Tomotherapy machines use the X band (10^4 MHz); and some research linacs run in the L band (10^3 MHz). In a clinical linac the electrons are accelerated to kinetic energy in the range from 4 MeV to 25 MeV and are used to produce two types of clinical beams:

1. Bremsstrahlung x-ray beams ranging in energy from 4 MV to 25 MV.
2. Electron beams in the kinetic energy range from 4 MeV to 25 MeV.

The disk-loaded acceleration waveguide evolved from a cylindrical uniform waveguide that is an efficient transmitter of radiofrequency (RF) power but is not suitable for charged particle acceleration. The theory of microwave propagation through disk-loaded acceleration waveguides is very complex, and often the final understanding and design of such a waveguide is achieved through empirical means and theoretical conclusions based on the theory of microwave propagation in uniform waveguides. The uniform waveguide theory is simpler, yet it provides an excellent basis for the final empirical steps in the study of disk-loaded waveguides.

In its most general form a waveguide consists of a metallic duct which has a uniform rectangular or circular cross section and is filled with a suitable dielectric gas or is evacuated to a very low pressure of the order of 10^{-6} torr. Waveguides are used either (1) for transmission of RF power or (2) for acceleration of charged particles such as electrons. The transmission waveguides have a rectangular or circular cross section; the acceleration waveguides have a circular cross section and are thus of cylindrical shape. The theory of acceleration waveguides uses cylindrical coordinates r, θ, and z with the axis of the cylinder coinciding with the z axis, as indicated in Fig. 13.1. The standard transformation equations between the Cartesian (x, y, and z) coordinates and the cylindrical (r, θ, and z) coordinates are given as follows

$$x = r\cos\theta, \qquad y = r\sin\theta, \qquad \text{and} \qquad z = z \tag{13.1}$$

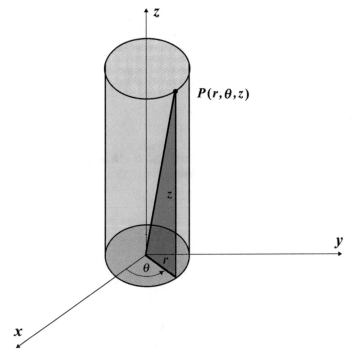

Fig. 13.1 Relationship between Cartesian coordinate system (x, y, z) and cylindrical coordinate system (r, θ, z)

or

$$r = \sqrt{x^2 + y^2}, \qquad \theta = \arctan \frac{y}{x}, \qquad \text{and} \qquad z = z, \qquad (13.2)$$

where $r \geq 0$; $0 \leq \theta \leq 2\pi$; and $0 < z \leq z_0$.

The propagation of microwaves through a uniform waveguide is governed by four Maxwell equations and appropriate boundary conditions. Maxwell equations for electric field vector $\mathcal{E}$ and magnetic field vector $\mathcal{B}$ were presented in general differential and integral forms in Sect. 1.29; for a uniform evacuated waveguide we write them in the differential form as follows

$$\nabla \cdot \mathcal{E} = 0, \qquad\qquad \text{Gauss law in electricity,} \qquad (13.3)$$

$$\nabla \cdot \mathcal{B} = 0, \qquad\qquad \text{Gauss law in magnetism,} \qquad (13.4)$$

$$\nabla \times \mathcal{E} = -\frac{\partial \mathcal{B}}{\partial t}, \qquad\qquad \text{Faraday law,} \qquad (13.5)$$

$$\nabla \times \mathcal{B} = \frac{1}{c^2} \frac{\partial \mathcal{E}}{\partial t}, \qquad\qquad \text{Ampère law,} \qquad (13.6)$$

recognizing that in vacuum the charge density ρ and current density $\mathbf{j}$, which appear in the general form of Maxwell equations, are equal to zero. Each of the four Maxwell equations also carries a specific name in honor of the physicist who discovered it independently from Maxwell.

13.2 Boundary Conditions

The boundary conditions on field vectors $\mathcal{E}$ and $\mathcal{B}$ at interfaces between two media (medium 1 and medium 2) where the dielectric and magnetic properties vary discontinuously are in general given by

$$(\mathcal{E}_2 - \mathcal{E}_1) \times \hat{\mathbf{n}} = 0 \tag{13.7}$$

and

$$(\mathcal{B}_2 - \mathcal{B}_1) \cdot \hat{\mathbf{n}} = 0, \tag{13.8}$$

where

$\hat{\mathbf{n}}$ is a unit vector normal at the surface.
$\mathcal{E}_1$ and $\mathcal{E}_2$ are the electric field vectors in media 1 and 2, respectively.
$\mathcal{B}_1$ and $\mathcal{B}_2$ are the magnetic field vectors in media 1 and 2, respectively.

A cylindrical evacuated uniform waveguide represents a special case in which medium 1 is vacuum and medium 2 is the waveguide wall made of a "perfect" conductor such as copper. The boundary conditions for the special case of a uniform cylindrical waveguide with radius a and the interior surface S separating vacuum (where field vectors $\mathcal{E}$ and $\mathcal{B}$ are non-zero and finite) from the copper wall (where $\mathcal{E}$ and $\mathcal{B}$ are equal to zero) may be written as follows

$$\hat{\mathbf{n}} \times \mathcal{E}\big|_S = 0 \quad \text{or} \quad \mathcal{E}_{\text{tang}}\big|_S = 0 \quad \text{or} \quad \mathcal{E}_z\big|_S = 0 \quad \text{or} \quad \mathcal{E}_z\big|_{r=a} = 0 \tag{13.9}$$

and

$$\hat{\mathbf{n}} \cdot \mathcal{B}\big|_S = 0 \quad \text{or} \quad \mathcal{B}_{\text{norm}}\big|_S = 0 \quad \text{or} \quad \frac{d\mathcal{B}_z}{dn}\bigg|_S = 0 \quad \text{or} \quad \frac{d\mathcal{B}_z}{dr}\bigg|_{r=a} = 0, \tag{13.10}$$

where

$\mathcal{E}_{\text{tang}}\big|_S$ is the tangential component of the electric field vector on the surface S at $r = a$.

$\mathcal{B}_{\text{norm}}\big|_S$ is the normal component of the magnetic filed vector on the surface S at $r = a$.

$\mathcal{E}_z$ is the z component of the electric field vector $\mathcal{E}$ with z along the cylinder axis.

$\mathcal{B}_z$ is the z component of the magnetic field vector $\mathcal{B}$ with z along the cylinder axis.

In mathematics the boundary condition $\mathcal{E}_z|_{r=a} = 0$ is referred to as the Dirichlet boundary condition, while the boundary condition $(d\mathcal{B}_z/dr)|_{r=a} = 0$ is referred to as the Neumann boundary condition.

The boundary conditions on $\mathcal{E}_z$ and $\mathcal{B}_z$ are different from one another, and in general cannot be satisfied simultaneously. Therefore, the fields inside a uniform waveguide are divided into two distinct modes: transverse magnetic (TM) and transverse electric (TE) with the following characteristics:

1. In the TM mode, $\mathcal{B}_z = 0$ everywhere and the boundary condition is of the Dirichlet type: $\mathcal{E}_z|_S = 0$.
2. In the TE mode, $\mathcal{E}_z = 0$ everywhere and the boundary condition is of the Neumann type: $\partial \mathcal{B}_z / \partial z|_S = 0$.

13.3 Differential Wave Equation in Cylindrical Coordinates

Applying the curl vector operator $(\nabla \times)$ on (13.5) and using the vector identity

$$\nabla \times \nabla \times \mathbf{A} = \nabla \nabla \cdot \mathbf{A} - \nabla^2 \mathbf{A}, \tag{13.11}$$

where

$\mathbf{A}$ is an arbitrary vector function.
∇ is the gradient vector operator often labeled as *grad*.
$\nabla\cdot$ is the divergence vector operator often labeled as *div*.
∇^2 is the vector Laplacian operator where $\nabla^2 \equiv \Delta = \nabla \cdot \nabla \equiv div\ grad$,

results in the following expression linking electric field vector $\mathcal{E}$ and magnetic field vector $\mathcal{B}$

$$\nabla \times \nabla \times \mathcal{E} = \nabla \nabla \cdot \mathcal{E} - \nabla^2 \mathcal{E} = -\frac{\partial}{\partial t} \nabla \times \mathcal{B}, \tag{13.12}$$

that, after inserting (13.3) and (13.6), evolves into a 3-dimensional partial differential wave equation for the electric field vector $\mathcal{E}$

$$\nabla^2 \mathcal{E} = \frac{1}{c^2} \frac{\partial^2 \mathcal{E}}{\partial t^2}. \tag{13.13}$$

In (13.13), $\mathcal{E}$ has three components $\mathcal{E}_r$, $\mathcal{E}_\theta$, and $\mathcal{E}_z$, and each of the three components is a function of spatial coordinates r, θ, and z as well as of the temporal coordinate t. Thus, we write

$$\mathcal{E} = [\mathcal{E}_r(r, \theta, z, t); \quad \mathcal{E}_\theta(r, \theta, z, t); \quad \mathcal{E}_z(r, \theta, z, t)]. \tag{13.14}$$

In a similar fashion, applying the curl vector operator ($\nabla \times$) on (13.6) and using the vector identity (13.11) results in the following expression for magnetic field vector $\mathcal{B}$ and electric field vector $\mathcal{E}$

$$\nabla \times \nabla \times \mathcal{B} = \nabla \nabla \cdot \mathcal{B} - \nabla^2 \mathcal{B} = -\frac{\partial}{\partial t} \nabla \times \mathcal{E}, \tag{13.15}$$

that, after incorporating (13.4) and (13.5), evolves into the following 3-dimensional partial differential wave equation for the magnetic field vector $\mathcal{B}$

$$\nabla^2 \mathcal{B} = \frac{1}{c^2} \frac{\partial^2 \mathcal{B}}{\partial t^2}. \tag{13.16}$$

In (13.16) $\mathcal{B}$ has three components $\mathcal{B}_r$, $\mathcal{B}_\theta$, and $\mathcal{B}_z$ and each of the three components is a function of spatial coordinates r, θ, and z as well as of the temporal coordinate t. Thus, we write

$$\mathcal{B} = [\mathcal{B}_r(r, \theta, z, t); \quad \mathcal{B}_\theta(r, \theta, z, t); \quad \mathcal{B}_z(r, \theta, z, t)]. \tag{13.17}$$

Wave equations (13.13) for electric field vector $\mathcal{E}$ and (13.16) for magnetic field vector $\mathcal{B}$ contain the vector Laplacian operator ∇^2 that, when operating on an arbitrary field vector $\mathbf{A}$ with components A_r, A_θ, and A_z, is expressed in cylindrical coordinate system as follows (see, for example: http://mathworld.wolfram.com/Laplacian.html)

$$\nabla^2 \mathbf{A} \equiv \begin{vmatrix} \dfrac{\partial^2 A_r}{\partial r^2} + \dfrac{1}{r}\dfrac{\partial A_r}{\partial r} + \dfrac{1}{r^2}\dfrac{\partial^2 A_r}{\partial \theta^2} + \dfrac{\partial^2 A_r}{\partial z^2} - \dfrac{2}{r^2}\dfrac{\partial A_\theta}{\partial \theta} - \dfrac{A_r}{r^2} \\[2ex] \dfrac{\partial^2 A_\theta}{\partial r^2} + \dfrac{1}{r}\dfrac{\partial A_\theta}{\partial r} + \dfrac{1}{r^2}\dfrac{\partial^2 A_\theta}{\partial \theta^2} + \dfrac{\partial^2 A_\theta}{\partial z^2} + \dfrac{2}{r^2}\dfrac{\partial A_r}{\partial \theta} - \dfrac{A_\theta}{r^2} \\[2ex] \dfrac{\partial^2 A_z}{\partial r^2} + \dfrac{1}{r}\dfrac{\partial A_z}{\partial r} + \dfrac{1}{r^2}\dfrac{\partial^2 A_z}{\partial \theta^2} + \dfrac{\partial^2 A_z}{\partial z^2} + 0 + 0 \end{vmatrix}$$

$$= \frac{1}{c^2}\frac{\partial^2 \mathbf{A}}{\partial t^2} \equiv \begin{vmatrix} \dfrac{1}{c^2}\dfrac{\partial^2 A_r}{\partial t^2} \\[2ex] \dfrac{1}{c^2}\dfrac{\partial^2 A_\theta}{\partial t^2} \\[2ex] \dfrac{1}{c^2}\dfrac{\partial^2 A_z}{\partial t^2} \end{vmatrix}. \tag{13.18}$$

As evident from (13.18), the individual relationships for the r and θ components of vector field $\mathbf{A}$ are quite complicated; however, the relationship for the z component of $\mathbf{A}$ retains the original form of the wave equation, expressed by the scalar Laplacian operator in the cylindrical coordinate system as

$$\frac{\partial^2 A_z}{\partial r^2} + \frac{1}{r}\frac{\partial A_z}{\partial r} + \frac{1}{r^2}\frac{\partial^2 A_z}{\partial \theta^2} + \frac{\partial^2 A_z}{\partial z^2} \equiv \frac{1}{r}\frac{\partial}{\partial r}\left(\frac{\partial A_z}{\partial r}\right) + \frac{1}{r^2}\frac{\partial^2 A_z}{\partial \theta^2} + \frac{\partial^2 A_z}{\partial z^2} = \frac{1}{c^2}\frac{\partial^2 A_z}{\partial t^2}.$$
(13.19)

Using the scalar Laplacian operator in cylindrical coordinates of (13.19), we now express the wave equations for $\mathcal{E}_z$ and $\mathcal{B}_z$ as follows

$$\nabla^2 \mathcal{E}_z \equiv \frac{\partial^2 \mathcal{E}_z}{\partial r^2} + \frac{1}{r}\frac{\partial \mathcal{E}_z}{\partial r} + \frac{1}{r^2}\frac{\partial^2 \mathcal{E}_z}{\partial \theta^2} + \frac{\partial^2 \mathcal{E}_z}{\partial z^2} = \frac{1}{c^2}\frac{\partial^2 \mathcal{E}_z}{\partial t^2}$$
(13.20)

and

$$\nabla^2 \mathcal{B}_z \equiv \frac{\partial^2 \mathcal{B}_z}{\partial r^2} + \frac{1}{r}\frac{\partial \mathcal{B}_z}{\partial r} + \frac{1}{r^2}\frac{\partial^2 \mathcal{B}_z}{\partial \theta^2} + \frac{\partial^2 \mathcal{B}_z}{\partial z^2} = \frac{1}{c^2}\frac{\partial^2 \mathcal{B}_z}{\partial t^2}.$$
(13.21)

Wave equations (13.13) and (13.16) describe the electric and magnetic fields, respectively, in a uniform waveguide. We first study the r, θ, and z components of the electric field vector $\mathcal{E}$ and the magnetic field vector $\mathcal{B}$, and then define the special RF modes that may prove useful for particle acceleration. Whereas radiofrequency can be transmitted through vacuum as well as various dielectric gases, vacuum is required for particle acceleration.

Equations (13.20) and (13.21) for $\mathcal{E}_z$ and $\mathcal{B}_z$, respectively, are known as 3-dimensional wave equations in cylindrical coordinates; they are linear partial differential equations of the second order in four variables (three spatial variables: r, θ, and z, and one temporal variable: t) with constant coefficients. The two equations have identical form and can in general be written as follows

$$\nabla^2 \eta \equiv \frac{1}{r}\frac{\partial}{\partial r}\left(r\frac{\partial \eta}{\partial r}\right) + \frac{1}{r^2}\frac{\partial^2 \eta}{\partial \theta^2} + \frac{\partial^2 \eta}{\partial z^2} = \frac{\partial^2 \eta}{\partial r^2} + \frac{1}{r}\frac{\partial \eta}{\partial r} + \frac{1}{r^2}\frac{\partial^2 \eta}{\partial \theta^2} + \frac{\partial^2 \eta}{\partial z^2} = \frac{1}{c^2}\frac{\partial^2 \eta}{\partial t^2},$$
(13.22)

with η a function of r, θ, z, and t representing both $\mathcal{E}_z(r, \theta, z, t)$ and $\mathcal{B}_z(r, \theta, z, t)$. The conditions imposed on $\eta(r, \theta, z, t)$ fall into two categories:

1. Those involving spatial coordinates r, θ, and z and governed by boundary conditions.
2. Those involving the temporal coordinate t and governed by initial conditions.

The most common approach to solving the 3-dimensional wave equation (13.22) is to apply the method of separation of variables. This method usually provides a solution to a partial differential equation in the form of an infinite series, such as a Fourier series, for example. We first separate out the time factor by defining $\eta\,(r, \theta, z, t)$ as a product of two functions: ϕ and T

$$\eta\,(r, \theta, z, t) = \phi\,(r, \theta, z)\,T\,(t)\,,\tag{13.23}$$

where

ϕ is a function of spatial coordinates r, θ, and z only.
T is a function of time t only.

Inserting (13.23) into (13.22) and dividing by $\phi(r, \theta, z)T(t)$ gives

$$\frac{\nabla^2\phi}{\phi} \equiv \frac{1}{c^2}\frac{1}{T}\frac{\partial^2 T}{\partial t^2}\,,\tag{13.24}$$

with the left hand side of (13.24) depending on spatial coordinates r, θ, and z only, and the right hand side depending on time t only. If this equality is to hold for all r, θ, z and t, it is evident that each side must be equal to a constant. This constant is identical for both sides and usually referred to as the separation constant Λ. From (13.24) we thus get two expressions

$$\nabla^2\phi = \Lambda\phi\tag{13.25}$$

and

$$\frac{\partial^2 T}{\partial t^2} = \Lambda c^2 T.\tag{13.26}$$

Equation (13.25) is referred to as the Helmholtz partial differential equation representing an eigenvalue problem in three dimensions with ϕ the eigenfunction, Λ the eigenvalue, and ∇^2 the Laplacian operator. The Helmholtz equation (13.25) results in three different types of solution depending on the value of separation constant Λ:

1. For $\Lambda > 0$ the solutions are exponential functions.
2. For $\Lambda = 0$ the solution is a linear function.
3. For $\Lambda < 0$ the solutions are trigonometric functions.

The Dirichlet boundary condition of (13.9) can be satisfied only for $\Lambda < 0$ and this will result in trigonometric solutions for function η. We now concentrate on finding solutions to the wave equation and set $\Lambda = -k^2$ to satisfy the usual periodicity requirement. Parameter k is called the *free space wave number* and is related to angular frequency ω through the standard relationship

$$k = \frac{\omega}{c}\,,\tag{13.27}$$

with c the speed of light in vacuum. Incorporating $\Lambda = -k^2$ into (13.25) and (13.26) yields the following equations for $\phi(r, \theta, z)$ and $T(t)$, respectively

$$\nabla^2 \phi + k^2 \phi = 0 \tag{13.28}$$

and

$$\frac{\partial^2 T}{\partial t^2} + k^2 c^2 T = \frac{\partial^2 T}{\partial t^2} + \omega^2 T = 0. \tag{13.29}$$

The solutions for $T(t)$ in (13.29) are either trigonometric or exponential but we reject the latter on physical grounds. Instead of using real trigonometric functions we express $T(t)$ as

$$T(t) \propto e^{-i\omega t}, \tag{13.30}$$

and assume that ω may be either positive or negative.

For the Helmholtz equation given in (13.28) we again use the method of separation of variables and express $\phi(r, \theta, z)$ as a product of three functions: $R(r)$, $\Theta(\theta)$, and $Z(z)$ to get

$$\phi(r, \theta, z) = R(r)\Theta(\theta)Z(z). \tag{13.31}$$

Insertion of (13.31) into (13.28) with the scalar Laplacian operator expressed in cylindrical coordinates as given in (13.22) and division by $\phi(r, \theta, z) = R(r)\Theta(\theta)Z(z)$ gives the following result

$$\frac{1}{R}\frac{\partial^2 R}{\partial r^2} + \frac{1}{rR}\frac{\partial R}{\partial r} + \frac{1}{r^2}\frac{1}{\Theta}\frac{\partial^2 \Theta}{\partial \theta^2} + \frac{1}{Z}\frac{\partial^2 Z}{\partial z^2} + k^2 = 0. \tag{13.32}$$

The first three terms of (13.32) are a function of r and θ alone while the fourth and fifth terms are independent of r and θ. This is possible only if the sum of the fourth and fifth term is equal to a constant that we designate as γ_n^2. Equation (13.32) now results in following two equations

$$\frac{1}{Z}\frac{\partial^2 Z}{\partial z^2} + k^2 = \gamma_n^2 \tag{13.33}$$

and

$$\frac{1}{R}\frac{\partial^2 R}{\partial r^2} + \frac{1}{rR}\frac{\partial R}{\partial r} + \frac{1}{r^2}\frac{1}{\Theta}\frac{\partial^2 \Theta}{\partial \theta^2} + \gamma_n^2 = 0. \tag{13.34}$$

We now express (13.33) as

$$\frac{\partial^2 Z}{\partial z^2} + (k^2 - \gamma_n^2)Z = \frac{\partial^2 Z}{\partial z^2} + k_g^2 Z = 0, \tag{13.35}$$

and get the following trigonometric solution

$$Z(z) \propto e^{+ik_g z},$$ (13.36)

where k_g is referred to as the *waveguide wave number* or the *propagation coefficient* defined as

$$k_g^2 = k^2 - \gamma_n^2.$$ (13.37)

Inserting (13.35) and (13.37) into (13.32) and multiplying the result by r^2 we get the following expression linking $R(r)$ and $\Theta(\theta)$

$$\frac{r^2}{R} \frac{\partial^2 R}{\partial r^2} + \frac{r}{R} \frac{\partial R}{\partial r} + \gamma_n^2 r^2 + \frac{1}{\Theta} \frac{\partial^2 \Theta}{\partial \theta^2} = 0,$$ (13.38)

with the first three terms of (13.38) depending on r alone and the fourth term depending on θ alone. Again, this is possible only if the fourth term is equal to a constant that must be negative to provide physically relevant solutions. We thus set the constant equal to $-m^2$ and get the following expression for the fourth term of (13.38)

$$\frac{\partial^2 \Theta}{\partial \theta^2} + m^2 \Theta = 0 \qquad (m = 0, 1, 2, \ldots).$$ (13.39)

Equation (13.39) has the following standard general trigonometric solution leading to trigonometric or complex exponential functions which serve as eigenfunctions

$$\Theta(\theta) = A \cos m\theta + B \sin m\theta.$$ (13.40)

Inserting (13.39) into (13.38) and multiplying the result by R/r^2 gives the following expression for $R(r)$

$$\frac{\partial^2 R}{\partial r^2} + \frac{1}{r} \frac{\partial R}{\partial r} + \left(\gamma_n^2 - \frac{m^2}{r^2} \right) R = 0,$$ (13.41)

representing the Bessel differential equation of order m or an eigenvalue equation with eigenvalue γ_n^2 when boundary conditions are imposed on $R(r)$. The physical conditions imposed on $\mathcal{E}_z$ and $\mathcal{B}_z$, and thus on $R(r)$ as well, stipulate that:

1. $R(r)$ must be finite at $r = 0$

and

2. $R(r = a)$ must satisfy either the Dirichlet boundary condition $R(r)|_{r=a} = 0$ of (13.9) or the Neumann boundary condition $dR/dr|_{r=a} = 0$ of (13.10).

The general solution to the Bessel equation (13.41) of order m consists of cylindrical functions; among these, given for non-negative integer values of m, the best known are the Bessel functions of the first kind $J_m(\gamma_n r)$ and Bessel functions of the second kind $N_m(\gamma_n r)$ (also known as Neumann functions).

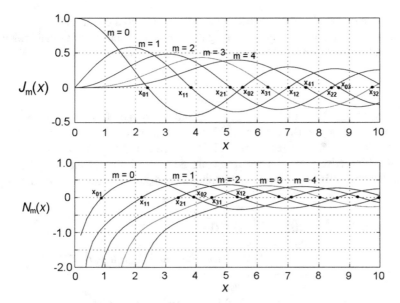

Fig. 13.2 Bessel functions of the first kind $J_m(x)$ and Neumann functions of the first kind $N_m(x)$ for integer values of m for $0 \leq m \leq 4$. Roots x_{mn} of $J_m(x)$ and $N_m(x)$ for $0 \leq m \leq 4$ and $0 \leq x \leq 10$ are also shown

In Fig. 13.2 we plot the Bessel functions $J_m(x)$ and $N_m(x)$ for non-negative integer values of m ranging from 0 to 4. A few important features of $J_m(x)$ and $N_m(x)$ are immediately apparent:

1. With increasing x, functions $J_m(x)$ and $N_m(x)$ oscillate about zero with a slowly diminishing amplitude and a decrease in separation between successive roots (zeros).
2. The two Bessel functions $J_m(x)$ and $N_m(x)$ possess an infinite number of roots, usually designated as x_{mn} and defined as those values of x at which the Bessel functions cross zero, i.e., where $J_m(x) = 0$ or $N_m(x) = 0$.
3. For $x = 0$, the Bessel functions of the first kind are finite; for integer $m > 0$ all Bessel functions of the first kind are equal to zero, i.e., $J_{m>0}(x)|_{x=0} = 0$ and for $m = 0$ the zero order Bessel function of the first kind equals to 1, i.e., $J_0(x)|_{x=0} = 1$.
4. For $x = 0$, the Bessel functions of the second kind (Neumann functions) exhibit a singularity, i.e., $\lim_{x \to 0} N_m(x) = -\infty$.
5. Roots x_{mn} of $J_m(x)$ and $N_m(x)$ for $0 \leq m \leq 4$ and $0 \leq x \leq 10$ are indicated in Fig. 13.2.

The general solution to the Bessel differential equation (13.41) is given as

$$R(r) = CJ_m(\gamma_n r) + DN_m(\gamma_n r), \tag{13.42}$$

where C and D are coefficients determined from the initial conditions.

Since the Neumann functions are singular at $r = 0$, to obtain a physically relevant solution to (13.41) we set $D = 0$ in (13.42) to get the following general solution for $R(r)$

$$R(r) = CJ_m(\gamma_n r), \tag{13.43}$$

where γ_n is a parameter defined in (13.37) and determined from the boundary conditions that are of the Dirichlet type, Neumann type or intermediate type. Combining solutions for $R(r)$, $\Theta(\theta)$, $Z(z)$, and $T(t)$ given in (13.43), (13.40), (13.36), and (13.30), respectively, we get the following general solution of the wave equation (13.22) representing the electric field component $\mathcal{E}_z$ and the magnetic field component $\mathcal{B}_z$

$$
\begin{aligned}
\eta(r, \theta, z, t) &= R(r)\Theta(\theta)Z(z)T(t) \\
&= \sum_{m=0}^{\infty}\sum_{n=1}^{\infty} J_m(\gamma_n r)\{A_{mn}\cos m\theta + B_{mn}\sin m\theta\}\, e^{+i\left(k_g z - \omega_{mn} t\right)} \\
&= \sum_{m=0}^{\infty}\sum_{n=1}^{\infty} J_m(\gamma_n r)\{A_{mn}\cos m\theta + B_{mn}\sin m\theta\}\, e^{+i\varphi}, \tag{13.44}
\end{aligned}
$$

with m the order of the Bessel function, n the rank order number of the given root of the Bessel function, and $(k_g z - \omega_{mn} t)$ usually referred to as the *phase of the wave* φ. Each pair of integers (m, n) corresponds to a particular characteristic mode of RF propagation through the uniform waveguide. The general solution (13.44) to the wave equation (13.22) is given as a superposition of all modes for $m = 0, 1, 2, \ldots$ and $n = 1, 2, 3, \ldots$.

13.4 Electric and Magnetic Fields in Uniform Waveguide

Based on (13.44) $\mathcal{E}_z$ and $\mathcal{B}_z$, the z components of the electric field vector $\mathcal{E}$ and magnetic field vector $\mathcal{B}$, respectively, can now be expressed in general terms as follows

$$\mathcal{E}_z(r, \theta, z, t) = \sum_{m=0}^{\infty}\sum_{n=1}^{\infty} J_m(\gamma_n r)\{A_{mn}\cos m\theta + B_{mn}\sin m\theta\}\, e^{i\left(k_g z - \omega_{mn} t\right)} \tag{13.45}$$

and

$$\mathcal{B}_z(r, \theta, z, t) = \sum_{m=0}^{\infty}\sum_{n=1}^{\infty} J_m(\gamma_n r)\{A'_{mn}\cos m\theta + B'_{mn}\sin m\theta\}\, e^{+i\left(k_g z - \omega_{mn} t\right)}, \tag{13.46}$$

where A_{mn}, B_{mn}, A'_{mn}, and B'_{mn} are constants to be determined from initial conditions, m stands for the order of the Bessel function of the first kind, and n represents the rank order number of the root of the Bessel function of the first kind.

$\mathcal{E}_z$ of (13.45) and $\mathcal{B}_z$ of (13.46) are determined by solving the wave equation given in (13.22). The r and θ components of $\mathcal{E}$ and $\mathcal{B}$, on the other hand, are calculated from Maxwell equations given for uniform waveguide in (13.6) and (13.5), respectively, in conjunction with the expressions for $\mathcal{E}_z$ of (13.45) and $\mathcal{B}_z$ of (13.46).

The curl operator operating on a vector field $\mathbf{A}$ is in cylindrical coordinates expressed as follows

$$\nabla \times \mathbf{A} = curl\,\mathbf{A} = \begin{pmatrix} \dfrac{\hat{\mathbf{r}}}{r} & \hat{\boldsymbol{\theta}} & \dfrac{\hat{\mathbf{z}}}{r} \\ \dfrac{\partial}{\partial r} & \dfrac{\partial}{\partial \theta} & \dfrac{\partial}{\partial z} \\ A_r & rA_\theta & A_z \end{pmatrix}$$

$$= \left\{ \frac{1}{r}\frac{\partial A_z}{\partial \theta} - \frac{\partial A_\theta}{\partial z} \right\} \hat{\mathbf{r}} + \left\{ \frac{\partial A_r}{\partial z} - \frac{\partial A_z}{\partial r} \right\} \hat{\boldsymbol{\theta}} + \frac{1}{r}\left\{ \frac{\partial (rA_\theta)}{\partial r} - \frac{\partial A_r}{\partial \theta} \right\} \hat{\mathbf{z}},$$

$$(13.47)$$

where

$\hat{\mathbf{r}}, \hat{\boldsymbol{\theta}}$, and $\hat{\mathbf{z}}$ are unit vectors.
A_r, A_θ, and A_z are components of vector field $\mathbf{A}$.

Noting from (13.45) and (13.46) that $\partial/\partial t = -i\omega$, we now use (13.6) in conjunction with (13.47) to express $\mathcal{E}_r$ and $\mathcal{E}_\theta$ as

$$\mathcal{E}_r = -\frac{c^2}{i\omega}\frac{1}{r}\frac{\partial \mathcal{B}_z}{\partial \theta} + \frac{c^2}{i\omega}\frac{\partial \mathcal{B}_\theta}{\partial z} \tag{13.48}$$

and

$$\mathcal{E}_\theta = -\frac{c^2}{i\omega}\frac{\partial \mathcal{B}_r}{\partial z} + \frac{c^2}{i\omega}\frac{\partial \mathcal{B}_z}{\partial r}. \tag{13.49}$$

Likewise, (13.5) allows us to express $\mathcal{B}_r$ and $\mathcal{B}_\theta$ as

$$\mathcal{B}_r = \frac{1}{i\omega}\frac{1}{r}\frac{\partial \mathcal{E}_z}{\partial \theta} - \frac{1}{i\omega}\frac{\partial \mathcal{E}_\theta}{\partial z} \tag{13.50}$$

and

$$\mathcal{B}_\theta = \frac{1}{i\omega}\frac{\partial \mathcal{E}_r}{\partial z} - \frac{1}{i\omega}\frac{\partial \mathcal{E}_z}{\partial r}. \tag{13.51}$$

The next step is to express r and θ components of vector fields $\mathcal{E}$ and $\mathcal{B}$ as a function of only the z components $\mathcal{E}_z$ and $\mathcal{B}_z$ given in (13.45) and (13.46), respectively. For example, we insert the expression for $\partial\mathcal{B}_\theta/\partial z$ that we obtain from (13.51) into (13.48) and, recognizing from (13.45) that $\partial/\partial z = ik_g$ and noting that $k^2 - k_g^2 = \gamma_n^2$ from (13.37), we get the following expression for $\mathcal{E}_r$

$$\mathcal{E}_r = \frac{1}{\gamma_n^2} \left\{ ik_g \frac{\partial E_z}{\partial r} + \frac{i\omega}{r} \frac{\partial B_z}{\partial \theta} \right\}. \tag{13.52}$$

In a similar manner, we determine the expressions for $\mathcal{E}_\theta$, $\mathcal{B}_r$, and $\mathcal{B}_\theta$ to get

$$\mathcal{E}_\theta = \frac{1}{\gamma_n^2} \left\{ \frac{ik_g}{r} \frac{\partial \mathcal{E}_z}{\partial \theta} - i\omega \frac{\partial B_z}{\partial r} \right\}, \tag{13.53}$$

$$\mathcal{B}_r = \frac{1}{\gamma_n^2} \left\{ -\frac{i\omega}{c^2} \frac{1}{r} \frac{\partial \mathcal{E}_z}{\partial \theta} + ik_g \frac{\partial B_z}{\partial r} \right\}, \tag{13.54}$$

$$\mathcal{B}_\theta = \frac{1}{\gamma_n^2} \left\{ \frac{i\omega}{c^2} \frac{\partial \mathcal{E}_z}{\partial r} + \frac{ik_g}{r} \frac{\partial B_z}{\partial \theta} \right\}. \tag{13.55}$$

The r, θ, and z components of the electric and magnetic field vectors $\mathcal{E}$ and $\mathcal{B}$ in a cylindrical uniform waveguide are thus determined by first solving the wave equations (13.20) and (13.21) for $\mathcal{E}_z$ and $\mathcal{B}_z$, respectively, and then applying $\mathcal{E}_z$ and $\mathcal{B}_z$ in conjunction with Maxwell equations (13.3) through (13.6) to determine the r and θ components.

13.5 General Conditions for Particle Acceleration

Six conditions that must be met for particle acceleration with radiofrequency fields in a waveguide are listed below; the first three are general conditions governing all methods of particle acceleration, the last three are specific to particle acceleration with radiofrequency fields:

1. Particle to be accelerated must be charged.
2. Electric field used for particle acceleration must be oriented in the direction of propagation of the charged particle.
3. Charged particle must be accelerated in vacuum rather than in a dielectric material to avoid deleterious interaction between the accelerated charged particle and atoms of the dielectric material through which the particle is moving.
4. The radiofrequency mode used for particle acceleration must provide a finite, non-zero value for $\mathcal{E}_z$ at $r = 0$, so as to enable the particle acceleration along the z axis of the cylindrical waveguide.
5. The radiofrequency mode should satisfy the Dirichlet boundary condition $\mathcal{E}_z = 0$ at $r = a$, i.e., $\mathcal{E}_z|_{r=a} = 0$ to obtain particle acceleration along the z axis of the acceleration waveguide.

6. The radiofrequency mode should produce $B_z = 0$ everywhere in the waveguide to thwart the interference of the magnetic field with the motion of the accelerated particle.

Modes satisfying conditions (4), (5), and (6) above are called transverse magnetic TM_{mn} modes. They are governed by the Dirichlet boundary condition on $\mathcal{E}_z$ and provide a finite electric field $\mathcal{E}_z$ in the direction of particle motion along the central axis of the cylindrical acceleration waveguide and may thus be useful for particle acceleration. In contrast, transverse electric TE_{mn} modes are characterized by the Neumann boundary condition on B_z and provide a non-zero magnetic field in the direction of propagation as well as $\mathcal{E}_z = 0$ everywhere. Obviously, the TE_{mn} modes with their $\mathcal{E}_z = 0$ everywhere characteristic cannot be used for particle acceleration along the z axis; however, they may be used for microwave transmission in uniform waveguides.

13.6   Dispersion Relationship

Parameter γ_n, which appears in expressions for the components of the electric and magnetic field vectors, is determined from the boundary condition on $R(r)$ at $r = a$, where a is the radius of the uniform cylindrical waveguide. The boundary condition on $R(r)$ in turn follows: (1) either from the Dirichlet boundary condition of (13.9) on electric field $\mathcal{E}_z|_{r=a}$ that stipulates that $\mathcal{E}_z|_{r=a} = 0$, thus $R(r = a) = 0$ or (2) from the Neumann boundary condition of (13.10) on the magnetic field $B_z|_{r=a}$ that stipulates that $dB_z/dr|_{r=a} = 0$, thus $dR/dr|_{r=a} = 0$.

For the Dirichlet boundary condition we have

$$R(r)|_{r=a} = 0 \Rightarrow J_m(\gamma_n a) = 0 \Rightarrow \gamma_n a = x_{mn} \Rightarrow \gamma_n = \frac{x_{mn}}{a} \qquad (13.56)$$

and for the Neumann boundary condition

$$\left. \frac{dR}{dr} \right|_{r=a} = 0 \Rightarrow \left. \frac{d}{dr} J_m(\gamma_n r) \right|_{r=a} = 0 \Rightarrow \gamma_n a = y_{mn} \Rightarrow \gamma_n = \frac{y_{mn}}{a}, \qquad (13.57)$$

where x_{mn} represents the nth root (zero) of the mth order Bessel function of the first kind and y_{mn} represents the nth root (zero) of the first derivative of the mth order Bessel function of the first kind.

Since we are interested in particle acceleration provided through TM_{mn} modes, we now consider only the Dirichlet boundary condition. From (13.37) we express γ_n as a function of $k = \omega/c$ and k_g, where k is the free space wave number and k_g is the waveguide wave number or propagation coefficient

$$\gamma_n^2 = k^2 - k_g^2 = \frac{\omega^2}{c^2} - k_g^2. \tag{13.58}$$

To satisfy the Dirichlet boundary condition parameter γ_n must be non-negative. Merging (13.58) and (13.56) we get the following expression

$$\gamma_n^2 = \left(\frac{x_{mn}}{a}\right)^2 = \frac{\omega^2}{c^2} - k_g^2, \tag{13.59}$$

that can be rearranged into the following hyperbolic $(\omega - k_g)$ dispersion relationship for a given TM$_{mn}$ mode for a cylindrical waveguide of radius a

$$\omega^2 = \omega_c^2 + c^2 k_g^2 \tag{13.60}$$

or

$$\omega = \sqrt{\omega_c^2 + c^2 k_g^2}, \tag{13.61}$$

where ω_c stands for the cutoff frequency, the lowest frequency that can propagate through the waveguide for a given m and given n since, as follows from (13.59),

$$\omega \geq \omega_c = \omega|_{k_g=0} = c\frac{x_{mn}}{a}. \tag{13.62}$$

From (13.62) we define a cutoff wavelength $(\lambda)_c$ as

$$(\lambda)_c = \frac{2\pi c}{\omega_c} = 2\pi\frac{a}{x_{mn}}, \tag{13.63}$$

such that only waves with frequency $\omega > \omega_c$ and free space wavelength $\lambda < (\lambda)_c$ can propagate without attenuation.

In Fig. 13.3 we show a typical $\omega - k_g$ diagram following the expression of (13.61). For an arbitrary point P on the hyperbola we define the phase velocity v_{ph} and group velocity v_{gr} of the radiofrequency wave as it propagates through a uniform waveguide. The line connecting point P and the origin of the coordinate system forms and angle α_{ph} with the abscissa (x) axis, while the tangent to the hyperbola at point P forms an angle α_{gr} with the abscissa (x) axis. Therefore, making use of (13.60), velocities v_{ph} and v_{gr} can be expressed as follows

$$v_{\text{ph}} = \frac{\omega}{k_g} = \tan \alpha_{\text{ph}} = \frac{c}{\sqrt{1 - \left(\frac{\omega_c}{\omega}\right)^2}} \tag{13.64}$$

and

$$v_{\text{gr}} = \frac{d\omega}{dk_g} = \tan \alpha_{\text{gr}} = c\sqrt{1 - \left(\frac{\omega_c}{\omega}\right)^2}. \tag{13.65}$$

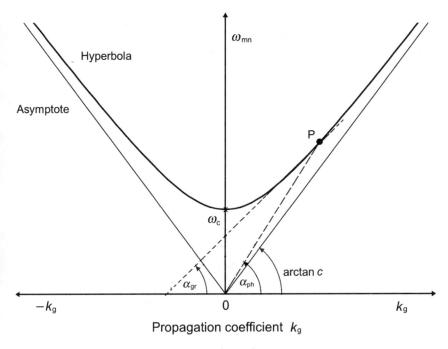

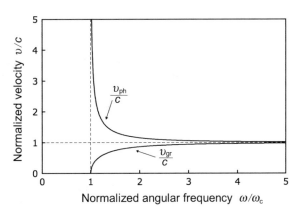

Fig. 13.3 Hyperbolic dispersion relationship $(\omega - k_g)$ for a uniform waveguide. Point P is an arbitrary point on the dispersion $\omega - k_g$ curve. Angles α_{ph} and α_{gr} are shown for point P. The asymptotes of the dispersion hyperbola form an angle arctan c with the abscissa (k_g) axis

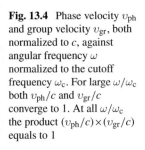

Fig. 13.4 Phase velocity v_{ph} and group velocity v_{gr}, both normalized to c, against angular frequency ω normalized to the cutoff frequency ω_c. For large ω/ω_c both v_{ph}/c and v_{gr}/c converge to 1. At all ω/ω_c the product $(v_{ph}/c) \times (v_{gr}/c)$ equals to 1

Phase velocity v_{ph} of (13.64) and group velocity v_{gr} of (13.65), both normalized to the speed of light in vacuum c, are plotted in Fig. 13.4 against normalized angular frequency ω/ω_c. The following characteristics of the hyperbolic dispersion relationship (13.61) as well as the phase and group velocities of (13.64) and (13.65), respectively, are notable from Figs. 13.3 and 13.4:

1. For $\omega \geq \omega_c$ the wave number k_g is real and waves can propagate through the uniform waveguide. For angular frequencies less than the cutoff frequency ω_c the propagation coefficient k_g is imaginary and the resulting modes, referred to as cutoff modes, cannot propagate through the uniform waveguide.

2. For a finite ω with $\omega > \omega_c$, as is evident from Fig. 13.4, $\tan \alpha_{ph} = v_{ph} > c$ and $\tan \alpha_{gr} = v_{gr} < c$.

3. As $\omega \to \infty$, both the phase velocity v_{ph} and group velocity v_{gr} approach the speed of light c in vacuum as evident from the following expressions for v_{ph} of (13.64) and v_{gr} of (13.65)

$$\lim_{\omega \to \infty} v_{ph} = \lim_{\omega \to \infty} \frac{c}{\sqrt{1 - \left(\dfrac{\omega_c}{\omega}\right)^2}} = c \tag{13.66}$$

and

$$\lim_{\omega \to \infty} v_{gr} = \lim_{\omega \to \infty} c\sqrt{1 - \left(\dfrac{\omega_c}{\omega}\right)^2} = c. \tag{13.67}$$

4. As $\omega \to \omega_c$, the phase velocity v_{ph} approaches ∞, while the group velocity v_{gr} approaches 0

$$\lim_{\omega \to \omega_c} v_{ph} = \lim_{\omega \to \omega_c} \frac{c}{\sqrt{1 - \left(\dfrac{\omega_c}{\omega}\right)^2}} = \infty \tag{13.68}$$

and

$$\lim_{\omega \to \omega_c} v_{gr} = \lim_{\omega \to \omega_c} c\sqrt{1 - \left(\dfrac{\omega_c}{\omega}\right)^2} = 0. \tag{13.69}$$

5. The range of phase velocity v_{ph} is from ∞ at $\omega = \omega_c$ to c at $\omega = \infty$, whereas the range of group velocity v_{gr} is from 0 at $\omega = \omega_c$ to c at $\omega = \infty$, i.e.,

$$c \text{ (at } \omega = \infty) \leq v_{ph} \leq \infty \text{ (at } \omega = \omega_c) \tag{13.70}$$

and

$$0 \text{ (at } \omega = \omega_c) \leq v_{gr} \leq \infty \text{ (at } \omega = \infty). \tag{13.71}$$

This implies that the phase velocity v_{ph} of the RF wave propagating in a uniform waveguide is always larger than c, whereas the group velocity v_{gr} is always smaller than c.

6. For all frequencies ω exceeding the cutoff frequency ω_c the following relationship holds between the phase velocity v_{ph} and the group velocity v_{gr}

$$v_{ph} v_{gr} = c^2. \tag{13.72}$$

Fig. 13.5 Behavior of the propagation coefficient k_g as a function of frequency ω for a uniform waveguide. Four modes are shown and each mode has its own values for m and n as well as cutoff frequency ω_c

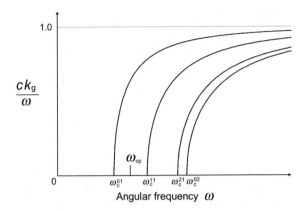

7. The asymptotes of the hyperbola (13.61) pass through the origin of the (k_g, ω) coordinate system and their angle with the abscissa (x) axis is equal to arctan c.

Equation (13.60) can be rearranged to give the waveguide propagation coefficient k_g as a function of frequency ω as follows

$$\frac{ck_g}{\omega} = \frac{k_g}{k} = \sqrt{1 - \left(\frac{\omega_c}{\omega}\right)^2}. \qquad (13.73)$$

A plot of (13.73) is given in Fig. 13.5 showing the behavior of the propagation coefficient k_g as a function of frequency ω. At any given frequency ω only a finite number of modes can propagate and each mode is characterized by its own cutoff frequency ω_c^{mn}. Since the propagation coefficient k_g is always less than the free space wave number k, the wavelength λ_g in the waveguide always exceeds the free space wavelength λ. The dimensions of the waveguide (for example, radius a) are usually chosen such that at the operating frequency ω_{op} only the lowest mode can occur, as shown schematically in Fig. 13.5.

13.7 Transverse Magnetic TM$_{01}$ Mode

The most important and the simplest transverse magnetic mode is the TM$_{01}$ mode for $m = 0$ and $n = 1$ with no azimuthal dependence and no roots (zeros) in $\mathcal{E}_z$ between $r = 0$ and $r = a$. In the TM$_{01}$ mode the electric field $\mathcal{E}_z$ of (13.45) simplifies to

$$\mathcal{E}_z = (\mathcal{E}_z)_0 \, J_0\left(2.405\frac{r}{a}\right) e^{i\varphi} \qquad (13.74)$$

where $(\mathcal{E}_z)_0$ is the amplitude of the electric field $\mathcal{E}_z$ and x_{01} is the first root (zero) of the zeroth order Bessel function of the first kind ($x_{01} = 2.405$, see Fig. 13.2).

Since, as shown in (13.62), the cut-off frequency ω_c is in general given as $\omega_c = c(x_{mn}a)$, we now get the following expressions for the cut-off frequency ω_c and cut-off wavelength $(\lambda)_c$, respectively, for the transverse magnetic TM_{01} mode

$$\omega_c = 2.405 \frac{c}{a} \tag{13.75}$$

and

$$(\lambda)_c = 2\pi \frac{a}{x_{01}} = 2.61a. \tag{13.76}$$

The other components of the electric field vector $\mathcal{E}$ and magnetic field vector $\mathcal{B}$ for the TM_{01} mode are obtained from (13.52) through (13.55) recognizing that in this case $\partial/\partial\theta = 0$ and $\mathcal{B}_z = 0$ everywhere. A summary of results for all components of the electromagnetic field is given below [*Note* φ is the phase of the radiofrequency wave and $\partial J_0(\alpha r)/\partial r = -\alpha J_1(\alpha r)$]

$$\mathcal{E}_r = i\frac{k_g}{\gamma_n^2}\frac{\partial \mathcal{E}_z}{\partial r} = -ik_g\frac{a}{2.405}(\mathcal{E}_z)_0 J_1\left(\frac{2.405}{a}r\right)e^{i\varphi} \tag{13.77}$$

$$\mathcal{E}_\theta = \frac{i}{\gamma_n^2}\left\{\frac{k_g}{r}\frac{\partial E_z}{\partial \theta} - \omega\frac{\partial \mathcal{B}_z}{\partial r}\right\} = 0 \quad \text{since} \quad \frac{\partial}{\partial\theta} = 0 \text{ and } \mathcal{B}_z = 0 \text{ everywhere,} \tag{13.78}$$

$$\mathcal{E}_z = (E_z)_0 J_0\left(\frac{2.405}{a}r\right)e^{i\varphi}, \qquad \text{see (13.45) for } m = 0 \text{ and } n = 1, \tag{13.79}$$

$$\mathcal{B}_r = \frac{i}{\gamma_n^2}\left\{-\frac{\omega}{c^2 r}\frac{\partial \mathcal{E}_z}{\partial \theta} + k_g\frac{\partial \mathcal{B}_z}{\partial r}\right\} = 0 \quad \text{since} \quad \frac{\partial}{\partial\theta} = 0 \text{ and } \mathcal{B}_z = 0 \text{ everywhere} \tag{13.80}$$

$$\mathcal{B}_\theta = \frac{i}{\gamma_n^2}\frac{\omega}{c^2}\frac{\partial \mathcal{E}_z}{\partial r} = -i\frac{\omega}{c^2}\frac{a}{2.405}(\mathcal{E}_z)_0 J_1\left(\frac{2.405}{a}r\right)e^{i\varphi} \tag{13.81}$$

$$\mathcal{B}_z = 0 \tag{13.82}$$

13.8 Relationship Between Radiofrequency Phase Velocity and Electron Velocity in Uniform Waveguide

In (13.44) the phase of the radiofrequency wave was defined as

$$\varphi = k_g z - \omega t, \tag{13.83}$$

where

ω is the angular frequency of the wave.
k_g is the waveguide wave number.
z is the coordinate along the central axis of the cylindrical uniform waveguide.

The angular frequencies of the wave as seen by a stationary observer and by a moving observer differ from one another and are given as follows:

1. The angular frequency ω seen by a stationary observer ($z = $ const) is given by

$$\omega = \frac{d\varphi}{dt}. \tag{13.84}$$

2. The angular frequency ω' seen by an observer (or accelerated electron) traveling with the radiofrequency wave is calculated as follows

$$\omega' = \frac{d\varphi}{dt'} = k_g \frac{dz}{dt'} - \omega \frac{dt}{dt'} = \frac{dt}{dt'} \left(k_g \frac{dz}{dt} - \omega \right) = \frac{dt}{dt'} (k_g v_{el} - \omega), \tag{13.85}$$

where

t' is the time measured in the reference frame of the moving observer.
v_{el} is the velocity of the accelerated electron ($v_{el} = dz/dt$) traveling in the waveguide.

Since $dt/dt' = \gamma = [1 - (v/c)^2]^{-1/2}$ in relativistic physics and $k_g = \omega/v_{ph}$ from (13.64), we can write (13.85) as

$$\omega' = \gamma(k_g v_{el} - k_g v_{ph}) = \gamma k_g v_{ph} \left[\frac{v_{el}}{v_{ph}} - 1 \right] = \gamma \omega \left[\frac{v_{el}}{v_{ph}} - 1 \right] \tag{13.86}$$

that is essentially the relativistic Doppler effect relationship.

For the electron to see continuously an accelerating electric field with a constant phase φ, the angular frequency ω' in the reference frame of the electron traveling with the radiofrequency wave must be equal to zero $(\omega' = 0)$. From (13.86) it is obvious that this is possible only if and when the electron velocity v_{el} is equal to wave phase velocity v_{ph}

$$v_{el} = v_{ph}. \tag{13.87}$$

In a uniform waveguide, for a finite angular frequency ω, as shown in (13.70), the phase velocity of the radiofrequency wave v_{ph} always exceeds the speed of light in vacuum c. On the other hand, the laws of relativistic mechanics stipulate that the electron velocity v_{el} cannot exceed c and this implies that an electron accelerated in a uniform waveguide cannot satisfy the condition (13.87) $v_{el} = v_{ph}$. This leads to the conclusion that uniform waveguides are *not* suitable for use in electron acceleration.

13.9 Relationship Between Velocity of Energy Flow and Group Velocity in Uniform Waveguide

The velocity of the energy flow v_{en} in a waveguide is of interest when the waveguide is to be used for charged particle acceleration. It is determined by finding the following ratio

$$v_{en} = \frac{\bar{P}}{W_{tot}}, \tag{13.88}$$

where

$\bar{P}$ is the mean power flowing through a transverse cross section of the waveguide.

W_{tot} is the total electromagnetic energy stored per unit length in the waveguide.

The mean power $\bar{P}$ is determined by integrating the Poynting vector S over the waveguide circular cross section with radius a as follows.

$$\bar{P} = \int_A S \, dA = \frac{1}{2} \text{Re} \int_A \frac{\mathcal{E}_T \times \mathcal{B}_T^*}{\mu_0} \, dA, \tag{13.89}$$

where $\mathcal{E}_T$ and $\mathcal{B}_T$ are the transverse components of the electric and magnetic field, respectively, and $\mathcal{B}_T^*$ is the complex conjugate of $\mathcal{B}_T$. Definition of the Poynting vector is provided in Sect. 1.30.

We now calculate $\bar{P}$ of (13.89) using the expressions provided for tangential electric and magnetic field components in the transverse magnetic TM$_{01}$ mode of (13.77) for $\mathcal{E}_r$ and (13.81) for $\mathcal{B}_\theta$ (*Note:* $\mathcal{E}_\theta = 0$ and $\mathcal{B}_r = 0$) and get

$$\bar{P} = \frac{1}{2} \int_0^a \int_0^{2\pi} \left\{ ik_g \left(\frac{a}{x_{01}} \right) (\mathcal{E}_z)_0 J_1 \left(\frac{x_{01}}{a} r \right) e^{i\varphi} \right\}$$

$$\times \left\{ \frac{i\omega}{\mu_0 c^2} \left(\frac{a}{x_{01}} \right) (\mathcal{E}_z)_0 J_1 \left(\frac{x_{01}}{a} r \right) e^{-i\varphi} \right\} r \, dr \, d\theta$$

$$= \pi \frac{k_g \omega}{\mu_0 c^2} \left(\frac{a}{x_{01}} \right)^2 (\mathcal{E}_z)_0^2 \int_0^a r J_1^2 \left(\frac{x_{01}}{a} r \right) dr$$

$$= \pi \sqrt{\frac{\varepsilon_0}{\mu_0}} k_g k \left(\frac{a}{x_{01}} \right)^2 (\mathcal{E}_z)_0^2 \int_0^a r J_1^2 \left(\frac{x_{01}}{a} r \right) dr. \tag{13.90}$$

The energy W_{mag} stored in the magnetic field per unit length is

$$W_{mag} = \frac{1}{4\mu_0} \int_A |\mathcal{B}_T|^2 \, dA, \tag{13.91}$$

while the energy W_{el} stored in the electric field per unit length is

$$W_{\mathrm{el}} = \frac{\varepsilon_0}{4} \int_A |\mathcal{E}_T|^2 \, dA. \tag{13.92}$$

The two energies W_{mag} and W_{el} stored per unit length are equal to one another ($W_{\mathrm{mag}} = W_{\mathrm{el}}$), so that the total energy W_{tot} stored in the waveguide per unit length is

$$W_{\mathrm{tot}} = W_{\mathrm{mag}} + W_{\mathrm{el}} = 2W_{\mathrm{mag}} = \frac{1}{2\mu_0} \int_A |\mathcal{B}_T|^2 \, dA. \tag{13.93}$$

The relationship between $\mathcal{E}_r$ and $\mathcal{B}_\theta$ is calculated using (13.77) and (13.81) to get

$$\frac{\mathcal{B}_\theta}{\mathcal{E}_r} = \frac{\frac{i\omega}{c^2}\left(\frac{a}{x_{01}}\right)(\mathcal{E}_z)_0 J_1\left(\frac{x_{01}}{a}r\right)e^{i\varphi}}{ik_{\mathrm{g}}\left(\frac{a}{x_{01}}\right)(\mathcal{E}_z)_0 J_1\left(\frac{x_{01}}{a}r\right)e^{i\varphi}} = \frac{\omega}{c^2 k_{\mathrm{g}}} = \frac{k}{ck_{\mathrm{g}}}. \tag{13.94}$$

The total EM energy W_{tot} stored per unit length calculated from (13.93) using (13.94) is

$$W_{\mathrm{tot}} = \frac{1}{2\mu_0} \int_A |\mathcal{B}_T|^2 \, dA = \frac{1}{2\mu_0} \frac{\omega^2}{c^4 k_{\mathrm{g}}^2} \int_A |\mathcal{E}_T|^2 \, dA$$

$$= \pi \frac{\omega^2}{\mu_0 c^4}\left(\frac{a}{x_{01}}\right)^2 (\mathcal{E}_z)_0^2 \int_0^a r J_1^2\left(\frac{x_{01}}{a}r\right) dr. \tag{13.95}$$

The velocity of energy flow v_{en} in the waveguide is now given as

$$v_{\mathrm{en}} = \frac{\bar{P}}{W_{\mathrm{tot}}} = \frac{\pi \sqrt{\frac{\varepsilon_0}{\mu_0}} k_0 k \left(\frac{a}{x_{01}}\right)^2 (\mathcal{E}_z)_0^2 \int_o^a r J_1^2\left(\frac{x_{01}}{a}r\right) dr}{\pi \frac{\omega^2}{\mu_0 c^4}\left(\frac{a}{x_{01}}\right)^2 (\mathcal{E}_z)_0^2 \int_o^a r J_1^2\left(\frac{x_{01}}{a}r\right) dr}$$

$$= \frac{\sqrt{\varepsilon_0 \mu_0} k_{\mathrm{g}} k \, c^4}{k^2 c^2} = c \frac{k_{\mathrm{g}}}{k} = \frac{c^2 k_{\mathrm{g}}}{\omega} = c\sqrt{1 - \left(\frac{\omega_c}{\omega}\right)^2}, \tag{13.96}$$

where we use $\bar{P}$ of (13.90) and W_{tot} of (13.95), and in the last step of (13.96) we use (13.60) to express k_{g} as a function of ω.

Since the last expression for v_{en} in (13.96) is identical to the expression for the group velocity v_{gr} of (13.65), one concludes that the velocity of energy flow v_{en} in a waveguide is equal to the group velocity v_{gr}

$$v_{\mathrm{en}} = v_{\mathrm{gr}}. \tag{13.97}$$

13.10   Disk-Loaded Waveguide

As discussed in Sect. 13.8, uniform waveguides cannot be used for charged particle
acceleration because they propagate radiofrequency waves with a phase velocity v_{ph}
that exceeds the speed of light in vacuum c. However, the theory of radiofrequency
wave propagation in uniform waveguides provides an excellent basis for finding
appropriate methods for circumventing the problem with ($v_{ph} = v_{el}$), the necessary
condition for charged particle acceleration. Since the electron velocity v_{el} cannot be
increased beyond the speed of light in vacuum c, the condition $v_{ph} = v_{el}$ can only
be satisfied by decreasing v_{ph} in a uniform waveguide below c. This is achieved in
practice by loading a uniform waveguide with periodic perturbations in the form
of disks or irises, as shown schematically in Fig. 13.6 with a comparison between
a uniform waveguide and a disk-loaded waveguide. The separation d between the
disks and the radius b of disk openings are chosen such that propagation of the TM_{01}
mode with $v_{ph} \leq c$ is achieved in the disk-loaded waveguide structure.

As an RF wave propagates through a disk-loaded waveguide, it is partially
reflected at each disk, the reflected fraction depending on the relative magnitudes
of the wavelength λ_g and the perturbation parameter $(a - b)$ with a the radius of
the uniform waveguide and b the radius of the disk opening, as shown in Fig. 13.6.
When radius b is comparable to a, i.e., $a - b \ll a$, the perturbation caused by the
disks is small, the reflection of the radiofrequency wave at the disk is negligible, and
the disk-loaded waveguide behaves much like uniform waveguide with radius a.

In general, when $\lambda_g \gg (a - b)$, corresponding to $k_g \ll (a - b)^{-1}$, the fraction
of wave reflection at disks is small. The dispersion relationship of the disk-loaded
waveguide then tends to that of a uniform waveguide (13.61) and the cutoff frequency
ω_c at $k_g = 0$ of the disk-loaded waveguide is identical to that of a uniform waveguide.

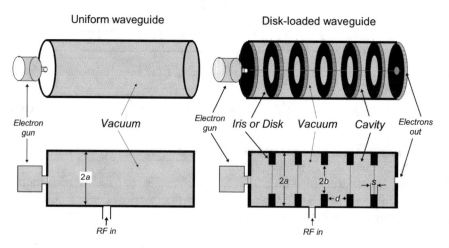

Fig. 13.6 Schematic comparison between a uniform waveguide and a disk-loaded waveguide

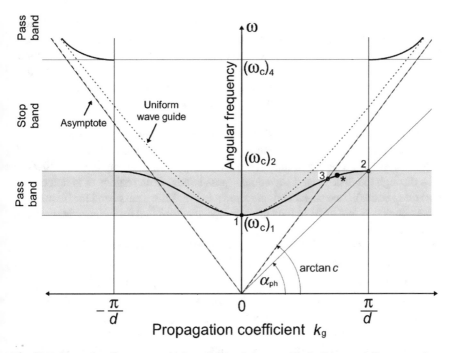

Fig. 13.7 Dispersion diagram $\omega - k_{\mathrm{g}}$ for a disk-loaded waveguide (*solid curves*). For comparison the dispersion hyperbola for a uniform waveguide of Fig. 13.3 is shown *dotted* in the background and its asymptotes are shown with *dashed lines*. The first pass band for frequencies between ω_{c_1} and ω_{c_2} is shown in *grey color*. The phase velocity v_{ph} for points between point 3 and point 2 on the disk-loaded dispersion curve is smaller than the speed of light c in vacuum

However, with increasing k_{g}, corresponding to a decreasing λ_{g} since $k_{\mathrm{g}} = 2\pi/\lambda_{\mathrm{g}}$, the fraction of the reflected wave at each disk steadily increases, and so does the interference between the incident and reflected wave, until at $\lambda_{\mathrm{g}} = 2d$ or $k_{\mathrm{g}} = \pi/d$ purely stationary waves are setup in each cavity defined by the disks. In this case, the cavities are in resonance, only stationary waves are present in the cavities, and there is no energy propagation possible from one cavity to another. This implies that $v_{\mathrm{gr}} = 0$ and the tangent to the $(\omega - k_{\mathrm{g}})$ dispersion relationship at $k_{\mathrm{g}} = \pi/d$ must be horizontal, in contrast to the uniform waveguide where the tangent to the dispersion relationship is horizontal only at $k_{\mathrm{g}} = 0$, and then with an increasing k_{g} its slope steadily rises to its limit of c as $k_{\mathrm{g}} \to \infty$.

A typical example of a dispersion relationship for a disk-loaded waveguide is shown in Fig. 13.7 with the solid curve, while the corresponding hyperbolic relationship for a uniform waveguide of Fig. 13.3 is shown with the dotted curve. The asymptotes of the uniform waveguide hyperbola are shown with dashed lines forming an angle arctan c with the abscissa (k_{g}) axis. For small k_{g} the two waveguides have identical $(\omega - k_{\mathrm{g}})$ diagrams, the same cutoff frequency ω_{c_1} (point 1 in Fig. 13.7) at $k_{\mathrm{g}} = 0$, and the same phase velocity v_{ph} which exceeds the speed of light in vacuum c.

For both the uniform as well as the disk-loaded waveguide the group velocity v_{gr} is zero at the cutoff frequency corresponding to the propagation coefficient $k_g = 0$ (point 1 in Fig. 13.7). As k_g increases from 0, the group velocity for uniform waveguide steadily increases until at $k_g = \infty$ it reaches a value of c. For a disk-loaded waveguide, on the other hand, with k_g increasing from zero, v_{gr} first increases, reaches a maximum smaller than c, and then decreases until at $k_g = \pi/d$ it reverts to $v_{gr} = 0$.

The dispersion curve for a disk-loaded waveguide thus deviates from that of a uniform waveguide and, as shown in Fig. 13.7, exhibits discontinuities at $k_g = n\pi/d$, with n an integer. The discontinuities in frequency ω separate regions of ω that can pass through the disk-loaded waveguide (pass bands) from regions of ω that cannot pass (stop bands). Two such bands are shown in Fig. 13.7: a pass band for frequencies ω between $(\omega_c)_1$ and $(\omega_c)_2$ in light grey color, and a stop band for frequencies between $(\omega_c)_2$ and $(\omega_c)_4$.

A closer look at the disk-loaded dispersion relationship curve of Fig. 13.7 shows the following features:

1. In the first pass band [$(\omega_c)_1 \leq \omega \leq (\omega_c)_2$]; frequencies in the region between points 3 and 2 on the dispersion plot have a phase velocity v_{ph} smaller than or equal to c as a result of $\alpha_{ph} \leq \arctan c$. Loading the uniform waveguide with disks thus decreases the phase velocity below c for certain angular frequencies ω, opening the possibility for electron acceleration with radiofrequency microwaves.

2. Frequency $(\omega_c)_2$ clearly has a phase velocity v_{ph} which is smaller than c; yet, the frequency $(\omega_c)_2$ would not be suitable for electron acceleration despite $v_{ph} \lesssim c$ because, simultaneously at frequency $(\omega_c)_2$, the group velocity of the wave is zero ($v_{gr} = 0$; tangent to dispersion curve at point 2 is horizontal). As shown in (13.97), zero group velocity makes energy transfer from the wave to the accelerated electrons impossible.

3. However, there are frequencies in the frequency pass band between $(\omega_c)_1$ and $(\omega_c)_2$, such as ω^* for point $*$ on the dispersion plot in Fig. 13.7, for which $v_{ph} \lesssim c$ and at the same time $v_{gr} > 0$, and these frequencies are suitable for electron acceleration.

4. In practice, frequencies which give v_{ph} smaller than yet close to c, i.e., $v_{ph} \lesssim c$, are used for electron acceleration in disk-loaded waveguides. The group velocities for these frequencies are non-zero but nonetheless very low, so that for a typical acceleration waveguide the phase velocity is about two orders of magnitude larger than the group velocity ($v_{ph}/v_{gr} \approx 100$).

The $(\omega - k_g)$ dispersion model is similar to models governing x-ray, neutron, and electron wave propagation in crystals. For example, the dispersion relationship for a uniform waveguide is analogous to the dispersion curve for a free electron model in which: (1) an electron moves through a periodic crystal lattice and (2) all effects of electron interactions with the lattice are ignored. More refined theories account for periodicity of the lattice as well as for variations in the strength of electron interactions and arrive at expressions for electron energy E that exhibits discontinuities when plotted against the wave number k of the electron.

The discontinuities occur at $k = n\pi/d$ where n is an integer and d the lattice constant, and are caused by Bragg reflections of the wave leading to energy gaps or band gaps. The region between $k = -\pi/d$ and $k = +\pi/d$ is called the first Brillouin zone; and the energy E versus wave number k diagram is called the Brillouin diagram.

Similarly to the Brillouin $(E - k)$ diagram, the $(\omega - k_g)$ disk-loaded waveguide dispersion relationship has an infinite number of branches alternating in frequency between pass bands where propagation is possible and stop bands with no propagation. However, only the part in the first Brillouin zone $(-\pi/d \leq k_g \leq \pi/d)$ is used for electron acceleration provided, of course, that $v_{ph} \leq c$ and $v_{gr} > 0$.

13.11 Capture Condition

In Sect. 13.7 it was shown that for electron acceleration in an acceleration waveguide the electron velocity v_{el} should be equal to the phase velocity v_{ph} of the wave. As shown in Fig. 13.7, in a disk-loaded acceleration waveguide there are certain frequencies for which $v_{ph} < c$ but close to c; however, the electron enters the acceleration waveguide from the electron gun with an initial velocity v_0 substantially smaller than c. Thus, the condition $v_{el} = v_{ph}$ cannot be fulfilled at the entrance side to the acceleration waveguide.

There are two possible solutions to this problem:

1. One option is to lower the phase velocity of the RF wave v_{ph} on the electron gun side of the acceleration waveguide to obtain $v_0 \approx v_{ph}$ and then gradually increase the phase velocity v_{ph} toward c as the accelerated charged particle gains kinetic energy. This approach is referred to as velocity modulation of the radiofrequency wave.

2. The other solution is to provide sufficiently large amplitude of the electric field $(\mathcal{E}_z)_0$ for the wave to capture the electron at the entrance to the acceleration waveguide despite its relatively low initial velocity v_0 which is smaller than the phase velocity of the radiofrequency wave v_{ph}.

Of the two, the first solution is more difficult as it involves modulation of the phase velocity v_{ph} by using non-uniform cavities in the entrance section of the acceleration waveguide and uniform cavities farther down the waveguide. Early linac designs contained many cavities with varying inner diameter, aperture radius, and axial spacing; more recently, only a few cavities were used for this purpose, and currently, a single half-cavity provides the phase modulation. The improved understanding of velocity modulation has resulted in a substantial lowering of the required gun injection voltage from historical levels of above 100 kV to current levels of around 25 kV.

The second solution is based on the calculation of the minimum amplitude of the electric field $[(\mathcal{E}_z)_0]_{min}$ which still allows the radiofrequency wave to capture the electron injected with a relatively low velocity v_{el} from the electron gun into the acceleration waveguide (capture condition).

We now derive the capture condition using two simplifying assumptions:

1. Wave propagates through the acceleration waveguide with a phase velocity equal to c (i.e., $v_{\mathrm{ph}} \approx c$)
2. Electric field is in the direction of propagation and has a sinusoidal behavior in time, such that

$$\mathcal{E}_z = (\mathcal{E}_z)_0 \sin \varphi, \tag{13.98}$$

with $(\mathcal{E}_z)_0$ the amplitude of the electric field and φ the phase angle between the wave and the electron, given in (13.44) as:

$$\varphi = k_{\mathrm{g}} z - \omega t, \tag{13.99}$$

where

ω is the angular frequency of the wave.
k_{g} is the waveguide wave number or propagation coefficient.
z is the coordinate along the waveguide axis.

The rate of change of phase φ with time t is from (13.99) given as

$$\frac{d\varphi}{dt} = k_{\mathrm{g}} \frac{dz}{dt} - \omega = k_{\mathrm{g}} v_{\mathrm{el}} - k_{\mathrm{g}} c = \frac{2\pi c}{\lambda_{\mathrm{g}}} (\beta - 1) \tag{13.100}$$

with $v_{\mathrm{el}} = dz/dt$, $v_{\mathrm{ph}} \approx c$ [assumption (1) above]; $k_{\mathrm{g}} = 2\pi/\lambda_{\mathrm{g}}$ where λ_{g} is the RF wavelength; and $\beta = v_{\mathrm{el}}/c$.

The relativistic equation of motion for the electron moving in the electric field $\mathcal{E}_z$ may be written as

$$F = \frac{dp}{dt} = \frac{d}{dt} m (v_{\mathrm{el}}) v_{\mathrm{el}} = \frac{d}{dt} \frac{m_{\mathrm{e}} \beta c}{(1 - \beta^2)^{1/2}} = e\mathcal{E}_z = e (\mathcal{E}_z)_0 \sin \varphi, \tag{13.101}$$

with

F force exerted on the electron by the electric field.
p electron momentum.
$m(v_{\mathrm{el}})$ mass of the electron at velocity v_{el}.
m_{e} electron rest mass ($0.511\ \mathrm{MeV}/c^2$).

Equations (13.100) and (13.101) are now simplified as follows

$$\frac{d\varphi}{dt} = a (\beta - 1) \tag{13.102}$$

and

$$\frac{d}{dt} \frac{\beta}{(1 - \beta^2)^{1/2}} = b \sin \varphi, \tag{13.103}$$

respectively, with the two acceleration waveguide parameters a and b given as: $a = 2\pi c/\lambda_g$ and $b = e(\mathcal{E}_z)_0/(m_e c)$.

Introducing $\beta = \cos\alpha$ into (13.102) and (13.103) we get, respectively,

$$\frac{d\varphi}{dt} = \frac{d\varphi}{d\alpha}\frac{d\alpha}{dt} = a(\cos\alpha - 1) \quad \text{or} \quad \frac{d\alpha}{dt} = a(\cos\alpha - 1)\frac{d\alpha}{d\varphi} \qquad (13.104)$$

and

$$\frac{d}{dt}\frac{\cos\alpha}{\sin\alpha} = \frac{d\cot\alpha}{d\alpha}\frac{d\alpha}{dt} = b\sin\varphi \quad \text{or} \quad \frac{d\alpha}{dt} = -b\sin^2\alpha\sin\varphi. \qquad (13.105)$$

After equating the two expressions above for $d\alpha/dt$, rearranging the terms, and integrating over φ from initial φ_0 to φ and over α from initial α_0 to α, we get

$$\frac{-b}{a}\int_{\varphi_0}^{\varphi}\sin\varphi d\varphi = \int_{\alpha_0}^{\alpha}\frac{\cos\alpha - 1}{\sin^2\alpha}d\alpha = \int_{\alpha_0}^{\alpha}\frac{\cos\alpha}{\sin^2\alpha}d\alpha - \int_{\alpha_0}^{\alpha}\frac{d\alpha}{\sin^2\alpha}, \qquad (13.106)$$

that results in

$$\frac{b}{a}[\cos\varphi - \cos\varphi_0] = \left[-\frac{1}{\sin\alpha} + \cot\alpha\right]_{\alpha_0}^{\alpha} = \left[\frac{\cos\alpha - 1}{\sqrt{1 - \cos^2\alpha}}\right]_{\alpha_0}^{\alpha} \qquad (13.107)$$

$$= \left[-\sqrt{\frac{1 - \cos\alpha}{1 + \cos\alpha}}\right]_{\alpha_0}^{\alpha} = \sqrt{\frac{1 - \cos\alpha_0}{1 + \cos\alpha_0}} - \sqrt{\frac{1 - \cos\alpha}{1 + \cos\alpha}}.$$

After inserting $\cos\alpha = \beta$ and $\cos\alpha_0 = \beta_0$, and recognizing that at the end of the acceleration $\beta \approx 1$, we obtain

$$\cos\varphi_0 - \cos\varphi = \frac{a}{b}\left(\frac{1 - \beta_0}{1 + \beta_0}\right)^{1/2} = \frac{2\pi}{\lambda_g e}\frac{m_e c^2}{(\mathcal{E}_z)_0}\left(\frac{1 - \beta_0}{1 + \beta_0}\right)^{1/2}, \qquad (13.108)$$

where $\beta_0 = v_0/c$ with v_0 the initial velocity of the electron injected into the acceleration waveguide from the electron gun. Since the left-hand side of (13.108) cannot exceed 2, we obtain the following relationship for the capture condition

$$(\mathcal{E}_z)_0 \geq \frac{\pi m_e c^2}{\lambda_g e}\left(\frac{1 - \beta_0}{1 + \beta_0}\right)^{1/2} = \frac{K}{\lambda_g}\sqrt{\frac{1 - \beta_0}{1 + \beta_0}}, \qquad (13.109)$$

where $K = \pi m_e c^2/e = 1.605$ MV is the capture constant for the electron. The minimum amplitude of the electric field $[(\mathcal{E}_z)_0]_{min}$ is thus expressed as follows

$$[(\mathcal{E}_z)_0]_{\min} = \frac{K}{\lambda_g}\sqrt{\frac{1-\beta_0}{1+\beta_0}}. \tag{13.110}$$

The capture condition must be satisfied if an electron with initial velocity v_0 is to be captured by the radiofrequency wave that has a phase velocity close to c. The well known relativistic relationship between the electron initial velocity β_0 and the electron initial kinetic energy $(E_K)_0$ is given as follows [see (2.7)]

$$\beta_0 = \frac{v_0}{c} = \sqrt{1 - \frac{1}{\left(1 + \dfrac{(E_K)_0}{m_e c^2}\right)^2}}, \tag{13.111}$$

allowing us to estimate $[(\mathcal{E}_z)_0]_{\min}$, the minimum amplitude of the radiofrequency field, for typical gun injection voltage potentials in the range from 20 keV to 100 keV.

Figures 13.8 and 13.9 display the capture condition given by the minimum electric field amplitude $[(\mathcal{E}_z)_0]_{\min}$ that the accelerating radiofrequency S-band field of 2856 MHz with phase velocity $v_{\rm ph} \approx c$ must possess to capture electrons injected

Fig. 13.8 Minimum electric field amplitude $[(\mathcal{E}_z)_0]_{\min}$ against initial electron velocity β_0 for the capture condition of (13.110)

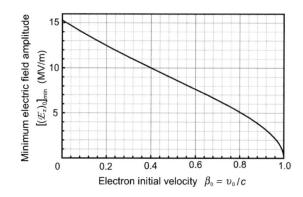

Fig. 13.9 Minimum electric field amplitude $[(\mathcal{E}_z)_0]_{\min}$ against initial electron kinetic energy $(E_K)_0$ for the capture condition of (13.110)

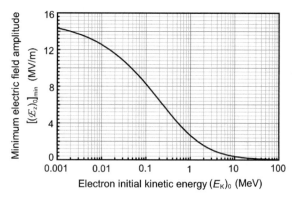

from the linac gun into the acceleration waveguide with a given initial velocity β_0 and initial kinetic energy $(E_K)_0$. Figure 13.8 plots $[(\mathcal{E}_z)_0]_{min}$ against electron initial velocity β_0, Fig. 13.9 plots $\left[(\mathcal{E}_z)_0\right]_{min}$ against electron initial kinetic energy E_K. The lower is the injected electron velocity υ_0 or kinetic energy $(E_K)_0$, the higher is the minimum required electric field amplitude $[(\mathcal{E}_z)_0]_{min}$ for capture of the injected electrons.

For example, assuming λ_g of 0.105 m, for the S band (2856 MHz), at electron initial kinetic energy of 20 keV $[(\beta_0 = 0.27$ from (13.111)] the minimum required amplitude of the RF electric field $[(\mathcal{E}_z)_0]_{min}$ is 11.6 MV/m and for 100 keV $[(\beta_0 = 0.55$ from 13.111)] electrons it is 8.2 MV/m. Since with current technology electric field amplitudes $(\mathcal{E}_z)_0$ of up to ~20 MV/m are possible, gun injection voltages of the order of 20 keV are adequate for injection of electrons into modern acceleration waveguides without requiring phase velocity modulation of the RF accelerating wave.

Treatment Planning in Radiotherapy

Figure on next page shows a conventional plan for radiation treatment of prostate cancer with seven stationary, conformal, and isocentric 18 MV x-ray beams superimposed on a transverse (axial) CT slice of a patient. The clinical target volume is delineated in magenta color, the planning target volume in orange, and the rectum in brown. Patient is placed supine into the treatment position such that the center of the target volume encompassing the prostate coincides with the machine isocenter. The dose distribution is shown with color-wash ranging from relatively low dose shown in blue (~15 Gy), through higher dose in green (~35 Gy), and high dose in brown of the order of 70 Gy.

Treatment planning is a very important component of the radiotherapy process, since it allows the physician to visualize the dose distribution that is to be delivered to the patient receiving radiotherapy. A good treatment plan provides 3-D dose distribution information not only for the target volume but also for the healthy tissues and organs surrounding the target. This provides the physician with appropriate tools to optimize the prescribed dose to the tumor without causing morbidity for the patient, fulfilling the basic objective of radiotherapy which is to maximize the tumor control probability (TCP) and to minimize the normal tissue complication probability (NTCP).

Radiation dosimetrists normally carry out the treatment planning on a computerized treatment planning system (TPS), and a radiation oncologist must approve a treatment plan before it is used in patient treatment. A medical physicist is responsible for the overall integrity of the computerized TPS and for the accuracy of all input data used in the treatment planning process.

The entire treatment planning process involves many steps, most of them in the realm of medical physics. The most important of these steps are: (1) Data acquisition for the treatment machine; (2) Data entry into the computerized TPS; (3) Patient data acquisition; (4) Treatment plan generation; and (5) Transfer of approved treatment plan to the radiotherapy machine. All hardware and software components of the treatment planning process in particular and radiotherapy in general are subject to stringent quality assurance protocols to ensure accurate and reliable execution of the planning process and optimal outcome of treatment.

Two approaches to treatment planning are used: (1) Forward or conventional planning and (2) Inverse planning. Forward planning represents the traditional approach whereby the plan is designed interactively and is based on experience and knowledge of dosimetrists, physicists, and physicians producing the plan. In forward planning several radiation beams of varying weights are aimed from different directions toward the target to give an acceptable dose distribution.

Inverse planning, on the other hand, is more objective and makes use of rigorous algorithm-based dose optimization techniques to satisfy the user specified criteria for the dose delivered to the target and critical structures. The inverse planning procedure starts with a prescribed dose distribution combined with various constraints imposed on dose to critical organs, and the inverse planning algorithm determines the optimal beam intensity modulation and beam orientations to achieve the prescribed dose distribution through a large number of mathematical iterations.

Figure on next page: Courtesy of William Parker, and Michael D.C. Evans, McGill University, Montreal. Reproduced with Permission.

Chapter 14
Particle Accelerators in Medicine

This chapter serves as introduction to particle accelerators used in medicine. Many types of particle accelerator were built for nuclear physics and particle physics research and most of them have also found some use in medicine, mainly for treatment of cancer. Two categories of particle accelerator are known: electrostatic and cyclic.

The best-known examples of electrostatic accelerator are the x-ray tube and the neutron generator. Three types of x-ray tube (Crookes tube, Coolidge tube, field emission tube) are discussed in this chapter; neutron generator is discussed briefly in Sect. 9.6.3. Cyclic accelerators fall into two categories: linear and circular. Many types of circular accelerator have been designed for research purpose and most are also used in medicine, such as the betatron, microtron, cyclotron, and synchrotron.

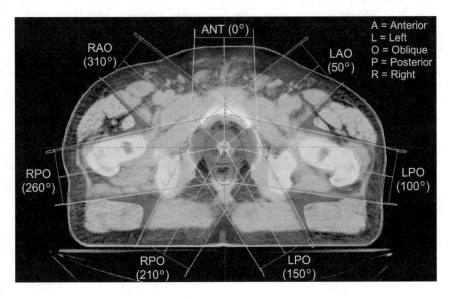

© Springer International Publishing Switzerland 2016
E.B. Podgoršak, *Radiation Physics for Medical Physicists*,
Graduate Texts in Physics, DOI 10.1007/978-3-319-25382-4_14

Of all cyclic accelerators, the linear accelerator is by far the most important and most widely used in medicine because of its versatility and compact design. Actually, one can say that modern radiotherapy achieved its successes as a result of the advances that were introduced during the past few years in the linear accelerator technology. In this chapter circular accelerators are discussed briefly, before the chapter undertakes a detailed discussion of the practical aspects of linear accelerators used clinically for cancer therapy. The theoretical aspects of electron acceleration in acceleration waveguides are discussed in Chap. 13.

14.1   Basic Characteristics of Particle Accelerators

Numerous types of accelerators have been built for basic research in nuclear physics and high-energy physics, and most of them have been modified for at least some limited use in radiotherapy. x-ray machine is the simplest accelerator and is widely used in medicine both for diagnosis of disease in diagnostic radiology and for treatment of disease in radiotherapy. In addition to megavoltage linear accelerators that are the most widely used machines in radiotherapy, other accelerators used in medicine are cyclotrons for proton and neutron radiotherapy as well as for production of positron emitting radionuclides for PET studies; betatrons and microtrons for x-ray and electron beam radiotherapy; and synchrotrons for hadron radiotherapy.

Irrespective of the accelerator type two basic conditions must be met for particle acceleration:

1. Particle to be accelerated must be charged.
2. Electric field must be provided in the direction of particle acceleration.

The various types of accelerators differ in the way they produce the accelerating electric field and in how the field acts on the particles to be accelerated. As far as the accelerating electric field is concerned there are two main classes of accelerators: electrostatic and cyclic.

In *electrostatic accelerators* the particles are accelerated by applying an electrostatic electric field through a voltage difference, constant in time, whose value fixes the value of the final kinetic energy of the accelerated particle. Since the electrostatic fields are conservative, the kinetic energy that the particle can gain depends only on the point of departure and point of arrival and, hence, cannot be larger than the potential energy corresponding to the maximum voltage drop existing in the machine. The kinetic energy that an electrostatic accelerator can reach is limited by the discharges that occur between the high voltage terminal and the walls of the accelerator chamber when the voltage drop exceeds a certain critical value (typically 1 MV).

The electric fields used in *cyclic accelerators* are variable and non-conservative, associated with a variable magnetic field and resulting in some closed paths along which the kinetic energy gained by the particle differs from zero. If the particle is made to follow such a closed path many times over, one obtains a process of gradual acceleration that is not limited to the maximum voltage drop existing in the accelerator. Thus, the final kinetic energy of the particle is obtained by submitting the charged particle to the same, relatively small, potential difference a large number of times, each cycle adding a small amount of energy to the total kinetic energy of the particle.

Cyclic accelerators fall into two main categories: *linear accelerators* and *circular accelerators*, depending on particle's trajectory during the acceleration. In a linear accelerator the particle undergoes rectilinear motion, while in a circular accelerator the particle's trajectory is circular. All cyclic accelerators except for the linear accelerator fall into the category of circular accelerator.

Examples of electrostatic accelerators used in medicine are: superficial and ortho-voltage x-ray machines and neutron generators. In the past, Van de Graaff accelerators have been used for megavoltage radiotherapy; however, their use was discontinued with the advent of first the betatron and then the linear accelerator. For medical use, the best-known example of a cyclic accelerator is the linear accelerator (linac); all other examples fall into the circular accelerator category and are the microtron, betatron, cyclotron, and synchrotron.

14.2 Practical Use of X-Rays

Roentgen's discovery of x-rays in 1895 is one of several important discoveries that occurred in physics at the end of the nineteenth century and had a tremendous impact on science, technology, and medicine in particular and modern society in general. Two other discoveries of similar significance are Becquerel's discovery of natural radioactivity in 1896 and discovery of radium by Marie Curie and Pierre Curie in 1898.

Studies in x-ray physics stimulated developments of modern quantum and relativistic mechanics and triggered the practical use of x-rays in medicine and industry. The use of x-rays in diagnosis of disease developed into modern diagnostic radiology, while the use of x-rays in treatment of cancer developed into modern radiotherapy. Concurrently with these two medical specialties, medical physics has evolved as a specialty of physics dealing with the physics aspects of diagnosis and treatment of disease, mainly but not exclusively with x-rays.

On a smaller scale x-ray research generated new research modalities, such as x-ray crystallography, x-ray spectroscopy, and x-ray astronomy. Scientific x-ray research to date resulted in 14 Nobel Prizes: eight of these in Physics, four in Chemistry and two in Medicine, as listed in Appendix E.

14.2.1 Medical Physics

Medical physics is a specialty of physics dealing with the application of physics to medicine, most generally in three areas: (1) Diagnostic imaging physics ($\sim$25% of total effort); (2) Nuclear medicine physics ($\sim$5% of total effort); and (3) Radiotherapy physics ($\sim$70% of total effort). While nuclear medicine concentrates mainly on application of unsealed radionuclides for diagnosis and treatment of disease, the use of x-rays forms an important component of diagnostic radiology as well as radiotherapy. The former uses x-rays in the photon energy range from 50 kVp to 150 kVp produced by x-ray tubes; the later uses x-rays in a much wider energy range extending from 50 kVp to 25 MV, produced by x-ray machines in the kilovolt range and linear accelerators in the megavolt range.

14.2.2 Industrial Use of X-Rays

X-rays for industrial use are produced by x-ray machines or linear accelerators and cover a wide variety of purposes dealing with safety and quality assurance issues, such as:

1. Inspection of luggage, shoes, mail, cargo containers, etc.
2. Nondestructive testing and inspection of welds, cast metals, parts of automobiles and airplanes, iron reinforcement bars, cracks and pipes inside concrete structures.
3. Food irradiation for sterilization and pest control.
4. Sterilization of surgical equipment and medical irradiation of blood.
5. Irradiation of small animals in radiobiological experiments.

14.2.3 X-Ray Crystallography

X-ray crystallography is a study of crystal structures through the use of x-ray diffraction techniques. X-rays are very suitable for this purpose because their wavelength in the 0.1 Å ($\sim$100 keV) to 1 Å ($\sim$10 keV) range is of the order of typical crystalline lattice separations. An x-ray beam striking a crystalline lattice is scattered by the spatial distribution of atomic electrons and the imaged diffraction pattern provides information on the atomic or molecular structure of the crystalline sample. In 1912 Max von Laue established the wave nature of x-rays and predicted that crystals exhibit diffraction phenomena. Soon thereafter, William H. Bragg and William L. Bragg analyzed the crystalline structure of sodium chloride, derived the Bragg relationship $2d \sin \phi = m\lambda$ (Fig. 1.8) linking the lattice spacing d with x-ray wavelength λ, and laid the foundation for x-ray crystallography. The crystal lattice of a sample acts as a diffraction grating and the interaction of x-rays with the atomic electrons creates a diffraction pattern which is related, through a Fourier transform, to the electron spectral distribution in the sample under investigation.

Instrumentation for x-ray diffraction studies consists of a monoenergetic x-ray source, a device to hold and rotate the crystal, and a detector suitable for measuring the positions and intensities of the diffraction pattern. Monoenergetic x-rays are obtained by special filtration of x-rays produced either by an x-ray tube or from an electron synchrotron storage ring. The basic principles of modern x-ray crystallography are essentially the same as those enunciated almost 100 years ago by von Laue and the Braggs; however, the technique received a tremendous boost by incorporation of computer technology after the 1970s, increasing significantly the accuracy and speed of the technique.

14.2.4   X-Ray Spectroscopy

X-ray spectroscopy is an analytical technique for determination of elemental composition of solid or liquid samples in many fields, such as material science, environmental science, geology, biology, forensic science, and archaeometry. The technique is divided into three related categories: the most common of them is the x-ray absorption spectrometry (also called x-ray fluorescence spectrometry), and the other two are x-ray photoelectron spectrometry and Auger spectrometry. All three techniques rely on creation of vacancies in atomic shells of the various elements in the sample under study as well as on an analysis of the effects that accompany the creation of vacancies (e.g., emission of photoelectron, emission of characteristic line spectrum, and emission of Auger electron). Like other practical emission spectroscopic methods, x-ray spectroscopy consists of three steps:

1. *Excitation of atoms in the sample* to produce fluorescence emission lines (or photoelectrons or Auger electrons) characteristic of the elements in the sample. The most common means for exciting characteristic x-ray photons for the spectroscopic analysis is by use of x-rays produced by x-ray machines; however, energetic electrons and heavy charged particles such as protons are also used for this purpose. Excitation by electrons is called the primary or impulse excitation; excitation by photons is called secondary or fluorescence excitation; excitation by heavy charged particles is called particle-induced x-ray emission (PIXE).

2. *Measurement of intensity and energy of the emitted characteristic lines* (or electrons). All methods for determining x-ray wavelengths λ use crystals as gratings and are based on the Bragg law $m\lambda = 2d \sin \phi$ where d is the lattice spacing and m is an integer. The dynamic range of these methods extends from 20 Å to 0.1 Å corresponding to photon energies of 6 keV to 130 keV, and the range of detectable elements in an unknown sample extends from beryllium ($Z = 4$) to uranium ($Z = 92$).

3. *Conversion of measured data to concentration or mass* with the nanogram range reached with standard spectrometers. The main disadvantage of the technique is that only a thin surface layer of the order of a few tenths of a millimeter can be analyzed because of absorption effects of the low energy fluorescence radiation. This requires a perfectly homogeneous sample for accurate results.

While x-ray spectroscopy was initially used to further the understanding of x-ray absorption and emission spectra from various elements, its role now is reversed and it is used as a non-destructive analytical tool for the purpose of chemical analysis of samples of unknown composition.

14.2.5   X-Ray Astronomy

X-ray astronomy is a relatively new branch of astronomy dealing with the study of x-ray emission from celestial objects, such as neutron stars, pulsars, and black holes. The specialty was born in 1962 when Italian–American astronomer Riccardo Giacconi discovered a cosmic x-ray source in the form of a compact star located in the constellation of Scorpius. For this discovery Giacconi received the 2002 Nobel Prize in Physics.

Since the x-rays emitted by celestial objects have relatively low energies of the order of a few kiloelectron volt, they cannot penetrate through the Earth's atmosphere to reach the surface of the Earth. Thus, to study these celestial rays, detectors must be taken above the Earth's atmosphere. Methods used to achieve this involve mounting x-ray detectors on rockets, balloons, or satellites. The x-ray detectors used for this purpose are either special charge-coupled devices (CCDs) or microcalorimeters.

14.3   Practical Considerations in Production of X-Rays

Chapter 4 dealt with general classical and relativistic relationships governing the emission of radiation by accelerated charged particles, including *bremsstrahlung* as the most important means and two more-specialized phenomena: the *synchrotron radiation* and the *Čerenkov radiation* In principle, all charged particles can emit radiation under certain conditions. In practice, however, the choice of charged particles that can produce measurable amounts of radiation of interest in medical physics, medicine, or industry is limited to light charged particles (electrons and positrons) that can undergo the following interactions:

1. Rapid deceleration of energetic electrons in targets through inelastic Coulomb collisions of electrons with nuclei of the target resulting in superficial, orthovoltage, or megavoltage x-rays (bremsstrahlung) for use in diagnosis (imaging) and treatment (radiotherapy) of disease.

2. Deceleration of electrons in retarding potentials resulting in microwave radiation. This process is used in magnetrons to produce radiofrequency photons and in klystrons to amplify radiofrequency photons. The radiofrequency used in standard clinical linear accelerators is 2856 MHz (S band); in miniature linear accelerator waveguide (tomotherapy and robotic arm mounting) it is at 10^4 MHz (X band).
3. Deceleration of electrons resulting in bremsstrahlung production in patients irradiated with photon or electron beams producing unwanted dose to the total body of the patient.
4. Acceleration of electrons in a linac waveguide (rectilinear motion of electrons) resulting in unwanted leakage radiation.
5. Circular motion of electrons in circular accelerators resulting in synchrotron radiation (sometimes referred to as magnetic bremsstrahlung) produced in high-energy circular accelerators and in storage rings. When charged particles pass through transverse magnetic fields, they experience an acceleration that, according to the Larmor relationship, results in emission of radiation that is typically of lower energy than bremsstrahlung. In comparison with synchrotron radiation, the accelerations in production of bremsstrahlung are random and also much stronger. Production of synchrotron radiation is still a very expensive undertaking, as it involves very expensive and sophisticated circular accelerators.
6. Deceleration of positrons (slowing down before annihilation) in positron emission tomography (PET) imaging studies of human organs resulting in unwanted stray radiation.
7. Atomic polarization effects when electrons move through transparent dielectric materials with a uniform velocity that exceeds the speed of light in the dielectric material result in visible light referred to as Čerenkov radiation. The efficiency for production of Čerenkov radiation is several orders of magnitude lower than the efficiency for bremsstrahlung production.
8. High energy electrons striking a nucleus may precipitate nuclear reactions (e, n) or (e, p) and transform the nucleus into a radioactive state thereby activating the treatment room and also the patient undergoing radiotherapy treatment.

14.4 Traditional Sources of X-Rays: X-Ray Tubes

Röntgen discovered x-rays in November of 1895 while investigating "cathode rays" produced in a Crookes tube The discovery of x-rays was just one of many important discoveries and advancements in physics that were engendered with experiments using the Crookes tube. Two other well-known experiments are Thomson's discovery of the electron in 1897 and Millikan's determination of electron charge in 1913. The Crookes tube thus occupies a very important place in the history of modern physics.

A typical x-ray machine used in medicine or industry has five basic components: (1) X-ray tube; (2) High voltage power supply; (3) Control console; (4) Imaged object; and (5) Image receptor. The x-ray tube is a vacuum tube that produces x-rays by accelerating electrons from a source (cathode) to a suitable target (anode) where a small amount of the electron kinetic energy is transformed into x-rays.

The most important component of an x-ray tube is the electron source. Three types of electron source are known:

1. Cold cathode stimulated by ionic bombardment in a Crookes x-ray tube (discussed in Sect. 14.4.1).
2. Hot cathode stimulated by high temperature in a Coolidge x-ray tube (discussed in Sect. 14.4.2)
3. Cold cathode stimulated by large electric field in a field emission x-ray tube (discussed in Sect. 14.4.3).

Generation of electrons with ionic bombardment of the cathode at room temperature (cold cathode) was used in Crookes cathode ray tubes as well as in the early x-ray tubes that are now referred to as Crookes x-ray tubes. Coolidge introduced the hot cathode design in 1913 (see Sect. 1.27) and this design, based on thermionic emission of electrons from a hot cathode, has since then been the most common practical electron source. Field emission from a cold cathode stimulated by a strong electric field (see Sect. 1.28) has shown promise for use in specialized x-ray tubes; however, since the efficiency of electron production with field emission currently cannot match the efficiency of thermionic emission generated by hot cathodes, field emission electron source is not used in standard x-ray tubes.

Electrons generated by the cathode bombard the target (anode) of an x-ray tube and a minute fraction of the electrons' kinetic energy (typically 1% or less) is transformed into x-rays (bremsstrahlung and characteristic radiation) and the rest into heat. The anode thus has three functions: (1) to define the positive potential in the x-ray tube; (2) to produce x-rays; (3) to dissipate the heat.

The anode material must have a high melting point to be able to withstand the high operating temperature and a relatively high atomic number for adequate x-ray production. Most common target materials for x-ray tubes are tungsten (also called wolfram) and molybdenum. Tungsten is used in x-ray tubes operating above 50 kVp and molybdenum for x-ray tubes operating at 50 kVp and below. Table 14.1 lists the main characteristics of molybdenum and tungsten of importance for use as target

Table 14.1 Important characteristics of target materials molybdenum and tungsten used as target (anode) in standard clinical and industrial x-ray tubes

Target material	Molybdenum Mo	Tungsten (Wolfram) W
Atomic number Z	42	74
Atomic mass A	95.94	183.84
Melting point ($^{\circ}$C)	2617	3422
K absorption edge (keV)	20	69.5
Energy of K_{α} line (keV)	$\sim$17.5	$\sim$58.6
Energy of K_{β} line (keV)	$\sim$19.6	$\sim$68.2

Table 14.2 Main characteristics of the Crookes x-ray tube, Coolidge x-ray tube, and field emission carbon nanotube (CNT) based x-ray tube

X-ray tube	Crookes X-ray tube	Coolidge tube	Field emission – carbon nanotube (CNT)
Source of electrons	COLD cathode: bombarded with positive ions of air to release electrons	HOT cathode: thermionic emission to produce electrons	COLD cathode: field emission in strong electric field
Air pressure	Intermediate vacuum 0.005 Pa to 0.1 Pa	High vacuum: $\sim 10^{-4}$ Pa	High vacuum: $\sim 10^{-4}$ Pa
X-ray output	Relatively low: depends on air pressure inside the tube	Relatively high: depends on cathode temperature	Relatively low: depends on cathode design and material
Period of use	1895 to $\sim$1920	1914 to present	Relatively new design

(anode) in an x-ray tube. Table 14.2 presents the main characteristics of Crookes, Coolidge, and field emission x-ray tubes.

14.4.1 Crookes Tube and Crookes X-Ray Tube

A Crookes tube is an electric discharge tube invented by British chemist and physicist *William Crookes* in the early 1870s. It consists of a sealed glass tube which is evacuated to an air pressure between 0.005 Pa and 0.1 Pa (4×10^{-5} torr and 7.5×10^{-4} torr) and incorporates two electrodes (cathode and anode) connected to an external DC power supply.

When high voltage is applied to the tube, electric discharge in the rarefied air inside the tube ionizes some air molecules. Positive ions move in the electric field toward the cathode and create more ions through collisions with air molecules. As positive ions strike the cathode, electrons are released from the cathode, move toward the anode in the electric field that is present between the cathode and the anode, and strike the anode.

During the first three decades after the invention of the Crookes tube, many important experiments were carried out with the tube and physicists soon established that unknown rays (referred to as cathode rays), originating in the cathode, were attracted by the anode. However, the exact nature of the "cathode rays" was not understood until Joseph J. Thomson in 1897 established that they were a new species of particle, negatively charged, and with mass of the order of 1800 times smaller than that of the hydrogen ion. He called the new particle electron and succeeded in measuring the ratio between its charge and mass.

In November 1895 Wilhelm Conrad Röntgen, a German physicist working at the University of Würzburg, discovered serendipitously that a Crookes tube, in addition to "cathode rays," generated a new kind of ray which penetrated the tube housing and behaved in a very peculiar fashion outside the tube. For example, the new rays were capable of exposing photographic film and also had the ability to penetrate opaque objects including hands, feet, and other parts of the human body. Röntgen named the unknown radiation x-rays and soon thereafter the new rays were introduced in medicine for diagnostic purposes. Röntgen's discovery ushered in the era of modern physics and revolutionized medicine by spawning three new specialties: diagnostic imaging and radiotherapy as specialties of medicine as well as medical physics as a specialty of physics. For his discovery Röntgen received many honors and awards that culminated in his receiving the inaugural Nobel Prize in Physics in 1901.

The exact nature of x-rays remained a mystery for a number of years until in 1912 Max von Laue, a German physicist, showed with a crystal diffraction experiment that x-rays were electromagnetic radiation similar to visible light but of much smaller wavelength. It then became apparent that when the "cathode ray" electrons strike the anode (target), they undergo interactions with orbital electrons and nuclei of the target and some of these interactions result in characteristic and bremsstrahlung photons, respectively, that form the x-ray spectrum.

For the first two decades after 1895, the x-ray tubes used for clinical work were of the Crookes tube type; very simple in design but suffering from severe practical problems related to the magnitude and reliability of the x-ray output. The situation improved significantly when in 1913 William Coolidge invented the "hot cathode" tube which turned out to be a vastly superior source of electrons in comparison with the "cold cathode" used in the Crookes type x-ray tubes. A schematic diagram of a Crookes x-ray tube is shown in Fig. 14.1a and the main characteristics of the Crookes tube are listed in Table 14.2.

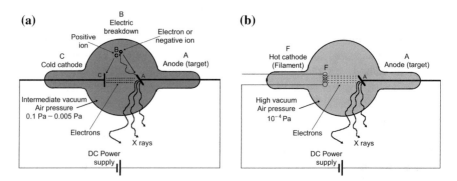

Fig. 14.1 Schematic diagram of **a** Crookes "cold cathode" x-ray tube and **b** Coolidge "hot cathode" x-ray tube

14.4.2   Coolidge X-Ray Tube

Low x-ray output combined with large output fluctuations and difficulties in controlling the output were the main drawbacks of Crookes x-ray tubes. In 1914 William Coolidge, an American physicist, introduced a new x-ray tube design based on a hot cathode which drastically improved the reliability and performance of clinical x-ray tubes. Almost 100 years later, Coolidge's hot cathode idea still provides the basis for design of modern x-ray tubes.

The hot cathode consists of a filament made of a high melting point metal, typically tungsten (melting point 3422 °C) or a tungsten based alloy, heated to a relatively high temperature to serve as source of electrons. The hot cathode emits electrons thermionically (Sect. 1.27) in contrast to the cold cathode of the Crookes x-ray tube in which positive air ions striking the cathode trigger the generation of electrons. Another important difference between the Coolidge tube and the Crookes tube is that the Coolidge tube operates under high vacuum of the order of 10^{-4} Pa to prevent collisions between electrons and molecules of air and also to prevent filament deterioration because of oxidation.

The main advantages of the Coolidge x-ray tube are its stability and its design feature which allows the external control of the x-ray output. The hotter is the filament, the larger is the number of emitted electrons. The filament is heated with electric current; increasing the filament current increases the filament temperature and this in turn results in an increase in number of thermionically emitted electrons. This number of emitted electrons is proportional to the number of electrons accelerated toward the anode (tube current) and this in turn is proportional to the number of x-rays produced in the anode (x-ray output). A schematic diagram of a typical Coolidge x-ray tube is shown in Fig. 14.1b and the main characteristics of the Coolidge tube are listed in Tables 14.2 and 14.3.

Table 14.3 Main characteristics of Coolidge-type x-ray tubes used in diagnostic radiology and radiotherapy

X-ray tube	Diagnostic radiology	Radiotherapy
Exposure time	Very short (of the order of 1 s)	Long (continuous operation)
Tube current	Large (up to 1 A)	Relatively small (~50 mA)
Anode (target)	Rotating (up to 10^4 rpm) To distribute heat	Stationary (water or oil cooled)
Anode material	Tungsten or Tungsten/rhenium	Block of tungsten embedded in copper
Focal spot	Point source (as small as possible)	Relatively large (diameter of few mm)
Instantaneous energy input	Large over very short time	Relatively small
Average energy input over long time	Relatively small	Large

Increasing the high voltage potential between the anode and the cathode increases the kinetic energy of the electrons striking the target (anode) and this increases the energy of the emitted x-rays.

In general, one can say that the x-ray output of a Coolidge x-ray tube is:

1. Proportional to the tube current for a given anode potential.
2. Proportional to the square of the anode potential for a given tube current.

As a metal is heated, electrons are "boiled off" its surface and form an electron cloud close to the surface. This space charge effectively prevents new electrons from leaving the metal. The current density j for the thermionic emission of electrons from the filament is governed by the filament temperature T and the work function $e\phi$ of the filament material, as given by the Richardson–Dushman relationship in (1.132) and discussed in Sect. 1.27. With a positive potential between the cathode (filament) and the anode (target) some of the electrons of the space charge are accelerated toward the positive anode and the space charge is replenished by emission from the hot filament.

As the anode voltage increases, the anode current I_a first increases linearly with anode voltage U_a, until it reaches saturation at very high voltages. The following observations can be made:

1. In the linear I_a versus U_a region the tube current depends on both the anode voltage U_a and filament temperature T, and the tube is said to operate in the space charge limited region.
2. In saturation, all electrons "boiled off" the filament are accelerated toward the anode, the tube is said to operate in the saturation mode, and the tube current depends only on the temperature of the filament. The tube current is said to be "temperature limited" or "filament-emission limited."

X-ray tubes come in many different forms and in many different designs depending on the purpose for which they are used; however, the main design criteria that Coolidge enunciated in 1913, namely high vacuum and hot cathode, are still valid today. Of interest in medical physics are two main categories of x-ray tube: diagnostic and therapy. The main characteristics of the two x-ray tube types are given in Table 14.3.

14.4.3   Carbon Nanotube Based X-Ray Tube

The standard source of electrons in x-ray machines, linear accelerators, and radiofrequency amplifiers is the hot cathode referred to as the filament from which electrons are emitted through the thermionic process (see Sect. 1.27). Coolidge invented the process in 1913 and Richardson explained it theoretically in 1928.

In comparison with the original cold cathode x-ray tube design based on the Crookes tube, the technology of hot cathode revolutionized x-ray tube design but has not changed much during the past century. Hot cathodes do have some drawbacks, so that for decades concurrently with improvements in hot cathode technology, search was on for alternative, more practical, and cheaper sources of electrons preferably based on cold cathode design.

The most serious drawback of the hot cathode is its required operating temperature around 1000 °C, limiting the choice of filament material to metals with very high melting point such as tungsten or tungsten alloys. The high operating temperature enables thermionic electron emission from the filament but, in comparison with cold cathode, makes the design of the electron source more complicated, consumes more power, and weakens the filament.

Field emission (see Sect. 1.28.2), which allows emission of electrons from the surface of a solid under the influence of a strong electric field, seems an excellent candidate for a practical and efficient cold cathode design. Attempts in this direction have been made for decades, however, the use of extremely small metal tips to achieve the large local electric fields always resulted in electrodes that were unreliable, relatively inefficient, and not durable enough for routine x-ray tube operation.

In original x-ray tubes using field-emission cold cathodes, the cathode was a metallic needle with a tip of about 1 μm in diameter. The electron field emission rate that was obtained with this arrangement, especially at relatively low voltages used in diagnostic radiology, could not produce sufficiently high tube currents for most standard radiological examinations. Thus, early attempts to manufacture field-emission x-ray tubes for clinical use were generally unsuccessful except for: (1) pediatric x-ray tubes where lower tube currents are acceptable and encouraged so as to limit the radiation exposure of pediatric patients and (2) chest-radiography x-ray tubes where high tube voltages (up to 300 kVp) are the norm and the high voltages significantly improve the field-emission efficiency.

During the past decade, a new generation of carbon based material called *carbon nanotube* (CNT) has been developed in nanotechnology laboratories and showed great promise for use as cold cathode-type electron source. Carbon nanotubes are ordered molecular structures formed by carbon, yet different from the two well-known carbon forms: graphite and diamond. They are molecular scale tubes with typical diameter of a few nanometers and a height of up to a few millimeters. The tubes have remarkable electronic properties and special physical characteristics that make them of great academic as well as potential commercial interest. They are extremely strong, yet flexible as well as light and thus hold promise for aerospace applications. Depending on their structure, they can behave like metal with conductivity higher than copper or like semiconductor potentially useful in design of nanoscale electronic devices. CNTs are mechanically, chemically, and thermally extremely robust and, since they also form atomically very sharp tips, they are also very efficient field emission materials for use as cold cathode electron source in x-ray tubes.

Miniature x-ray tubes using CNT cold cathode design are already commercially available. They generate electrons at room temperature and provide controllable as well as stable output currents and respectable life of the cathode. They can be used for "electronic brachytherapy" in medicine where they replace sealed radionuclide sources and in space exploration for performing remote mineralogical analyses on solid bodies of the solar system.

Use of cold cathode for high power x-ray tubes in medicine and industry, however, if it happens, is far in the future, since the technology of CNT production is still in a rudimentary stage and field emission cold cathodes are currently no match for the standard hot cathode x-ray tube design. The main characteristics of field emission x-ray tubes are summarized in Table 14.2 and compared with x-ray tubes of the Crookes type (see Sect. 14.4.1) and Coolidge type (see Sect. 14.4.2).

14.5 Circular Accelerators

With the exception of the linear accelerator, all cyclic particle accelerators used in science, industry, and medicine fall into the category of circular accelerators. Common to all circular accelerators is the circular motion of accelerated particles with either a constant radius or increasing radius. The particles are accelerated by an appropriate electric field oriented in the direction of motion; however, they are kept in circular orbit by a strong magnetic field. Because of their circular motion, the particles are constantly accelerated and emit part of their kinetic energy in the form of photons (see Larmor Law, discussed in Sect. 4.2.5). The emitted radiation is referred to as synchrotron radiation or magnetic bremsstrahlung (Sect. 4.3). Of interest and use in medicine are the following circular particle accelerators: betatron, cyclotron, microtron, and synchrotron. Each one of these machines is discussed briefly in this section, its schematic representation is shown in Fig. 14.2, and its basic characteristics are summarized in Table 14.4.

Table 14.4 Comparison of basic parameters of circular particle accelerators

Circular accelerator	Particles accelerated	Radiofrequency (RF) field	Magnetic field	Particle trajectory	Radius of orbit
Betatron	Electrons	Fixed 60 Hz–180 Hz	Variable	Circle	Const
Microtron	Electrons	Fixed 3 GHz or 10 GHz	Const	Spiral	Increases with energy
Cyclotron	Protons, ions	Fixed 10 MHz–30 MHz	Const	Spiral	Increases with energy
Synchro-cyclotron	Protons, ions	Variable	Const	Spiral	Increases with energy
Synchrotron	Electrons, protons	Variable, p: ~MHz e: few 100 MHz	Variable	Circle	Const

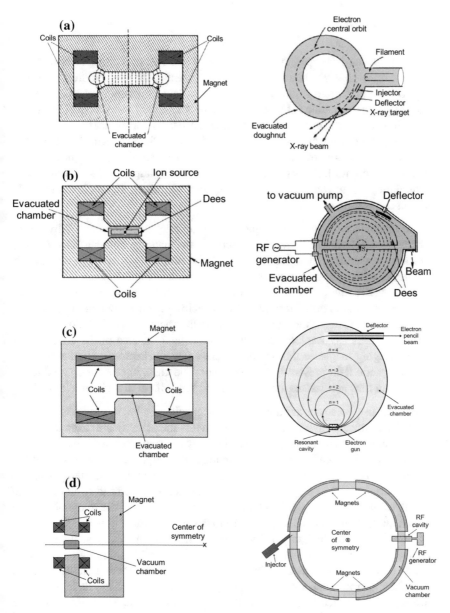

Fig. 14.2 Schematic diagrams of various circular particle accelerators used in medicine: **a** betatron (see Sect. 14.5.1); **b** cyclotron (see Sect. 14.5.2); **c** microtron (see Sect. 14.5.3); and **d** synchrotron (see Sect. 14.5.4). Vertical cross sections are on the *left*, top views on the *right*. A comparison of basic parameters of circular particle accelerators is provided in Table 14.4

14.5.1 Betatron

The betatron was developed in 1940 by *Donald W. Kerst* as a cyclic electron accelerator for basic physics research; however, its potential for use in radiotherapy was realized soon thereafter. In the 1950s betatrons played an important role in megavoltage radiotherapy, since at that time they provided the most practical means for production of megavoltage x-rays and electron beams for clinical use. However, the development of linacs pushed them into clinical oblivion because of the numerous advantages offered by linacs over betatrons, such as: much higher beam output (up to 10 Gy/min for linacs vs. 1 Gy/min for betatrons); larger field size; full isocentric mounting; more compact design; and quieter operation. The main features of a betatron are as follows:

1. The machine consists of a magnet fed by an alternating current of frequency between 50 Hz and 200 Hz. The electrons are made to circulate in a toroidal evacuated chamber (doughnut) that is placed into the gap between two magnet poles. A schematic diagram of a betatron is given in Fig. 14.2a.
2. Conceptually, the betatron may be considered an analog of a transformer: the primary current is the alternating current exciting the magnet and the secondary current is the electron current circulating in the vacuum chamber (doughnut).
3. The electrons are accelerated by the electric field induced in the doughnut by the changing magnetic flux in the magnet; they are kept in a circular orbit by the magnetic field present in the doughnut. The changing magnetic field in the doughnut thus accelerates the electrons and keeps them in their circular path.

For clinical use betatrons produce either megavoltage x-rays or clinical electron beams. To produce x-rays, electrons accelerated in the betatron are made to strike a thin target embedded in the doughnut and the x-ray beam so produced is flattened with a flattening filter, typically made of aluminum. A thin target produces an x-ray beam of the highest possible effective beam energy because electrons that interact with the target material are all of high energy. The electrons that traverse the thin target find themselves in a magnetic field and are swept into the wall of the doughnut before they can hit the flattening filter. In contrast to the betatron thin target one should note that linac targets are thick targets by necessity and thus produce less than optimal effective beam energies. It would be preferable also to use a thin target to produce x-ray beams in a linac; however, in a linac there is no magnetic field available to sweep the electrons transmitted through the target away from the flattening filter. Since these transmitted electrons would produce unwanted extrafocal x-rays in the flattening filter, one needs a target which is thick enough to stop all electrons striking it and this unfortunately degrades the x-ray beam so produced.

To produce clinical electron beams the accelerated electrons are brought through a window in the doughnut using a special device referred to as a peeler. Once outside the doughnut, the electron pencil beam traverses a thin scattering foil which scatters the electrons over a clinically useful field size.

14.5.2 Cyclotron

The cyclotron was developed in 1932 by *Ernest O. Lawrence* for acceleration of ions to a kinetic energy of a few MeV. The 1939 Nobel Prize in Physics was awarded to Lawrence for the invention and development of the cyclotron. Initially, the cyclotron was used for basic nuclear physics research but has later on found important medical uses in production of radionuclides for nuclear medicine as well as in production of proton and neutron beams for radiotherapy.

Lately, the introduction of PET/CT machines for use in radiotherapy and the increased interest in proton beam radiotherapy have dramatically increased the importance of cyclotrons in medicine. The PET/CT machines rely on glucose labeled with positron-emitting fluorine-18 as well as on other short-lived radionuclides that are produced by proton cyclotrons and are used for imaging and study of organ function (see Sect. 11.5.1). Proton beam radiotherapy is not yet widely available because of the relatively large cost involved in the infrastructure and maintenance of proton facilities in comparison with megavoltage x-ray installations.

While current conventional wisdom is that proton beam therapy is at least as good as modern x-ray therapy, the advantages of protons over x-rays have not been clearly demonstrated yet and the larger cost associated with protons is difficult to justify. It is agreed, however, that proton beam therapy offers a clear advantage over x-ray therapy in treatment of many pediatric tumors as a result of a significantly lower total body leakage dose produced by proton beam machines. This lower leakage dose translates into a lower rate of secondary cancer induction, an obviously important consideration in curative treatment of very young patients.

The salient features of a cyclotron are as follows:

1. In a cyclotron the particles are accelerated along a spiral trajectory guided inside two evacuated (of the order of 10^{-6} torr) half-cylindrical electrodes (referred to as dees because of their D-shape form) by a uniform magnetic field (of the order of 1 tesla) that is produced between the pole pieces of a large magnet. A diagram of the cyclotron is given in Fig. 14.2b.
2. A radiofrequency voltage with a constant frequency between 10 MHz and 30 MHz is applied between the two electrodes and the charged particle is accelerated while crossing the gap between the two electrodes. The frequency of operation is chosen such that protons are accelerated by a relatively small amount of energy each time they cross the gap between the dees.
3. Inside the metallic dee electrodes there is no electric field and the particle drifts under the influence of the magnetic field in a semicircular orbit with a constant speed, until it crosses the gap again. If, in the meantime, the electric field has reversed its direction, the particle will again be accelerated across the gap, gain a small amount of energy, and drift in the other electrode along a semicircle of a larger radius than the former one, resulting in a spiral orbit and a gradual increase in kinetic energy after a large number of gap crossings. Typical potential across the gap between the two dees is of the order of 150 kV.

The operation of a cyclotron is possible if the time required for the particles to describe each semicircle in a dee is constant and if the angular frequency ω of the RF generator is such that the transit time inside one of the dees is equal to half period of field oscillation. In standard cyclotron operation the Lorentz force F_L keeping the particle in circular orbit is equal to the centrifugal force

$$F_L = q\upsilon B = \frac{m\upsilon^2}{r}, \tag{14.1}$$

where

q is the charge of the accelerated charged particle.
υ is the velocity of the particle.
B is the magnetic field used for keeping the charged particle in circular orbit inside the dees.
m is the mass of the accelerated charged particle.
r is the radius of particle orbit during one revolution.

Equation (14.1) can be solved for the angular frequency $\omega = \upsilon/r$ and for r to get the following expressions for ω and r, respectively

$$\omega = \frac{qB}{m} = 2\pi\nu = \frac{2\pi}{T} = \omega_{cyc} \tag{14.2}$$

and

$$r = \frac{m\upsilon}{qB} = \frac{p}{qB} = \frac{\sqrt{2mE_K}}{qB}, \tag{14.3}$$

where p is the momentum of the particle and E_K is its kinetic energy. The angular frequency ω in (14.2) is referred to as the cyclotron frequency ω_{cyc} of a charged particle with a given ratio q/m in a given magnetic field B. It does not depend on the velocity υ of the accelerated particle moving in the constant magnetic field B; however, the radius r of the particle orbit given in (14.3) depends linearly on particle velocity. Slow particles move in relatively small circles, fast particles move in large circles, yet all orbits require the same period T to complete one revolution in the magnetic field B.

The final kinetic energy E_K of the accelerated particles in a cyclotron depends on the radius R of the dees. The velocity υ of the particle during the last acceleration across the gap between the dees is from (14.3) given as

$$\upsilon = \frac{qBR}{m}, \tag{14.4}$$

resulting in the following expression for E_K

$$E_K = \frac{mv^2}{2} = \frac{q^2 B^2 R^2}{2m}, \tag{14.5}$$

assuming the validity of classical mechanics for our calculation ($m = $ constant).

The cyclotron operates under the assumption that mass m of the accelerated charged particle in (14.1) is constant and this is true only under classical conditions where the kinetic energy E_K of the charged particle is much smaller than its rest energy E_0. For relativistic charged particles, mass m increases with particle velocity v and with kinetic energy E_K. This results in a decrease in cyclotron frequency ω_{cyc} of (14.2) and the particle loses its phase relationship with the constant frequency accelerating RF field. Consequently, electrons with their relatively small rest mass cannot be accelerated with a cyclotron, and heavier particles such as protons can only be accelerated to relatively low final kinetic energies. These kinetic energies of heavy charged particles are no longer useful for high-energy physics research; however, they are adequate for production of protons and heavier ions for use in medicine.

For relativistic particles expressions for ω_{cyc} and r of (14.2) and (14.3), respectively, are written as follows after inserting the relativistic mass γm_0 for the mass m to get

$$\omega_{cyc} = \frac{qB}{\gamma m_0} \tag{14.6}$$

and [see (1.64)]

$$r = \frac{p}{qB} = \frac{\gamma m_0 c^2 \beta}{qBc} = \frac{E_K}{qBc}\sqrt{1 + \frac{2m_0 c^2}{E_K}}, \tag{14.7}$$

respectively, where $\gamma = \sqrt{1 - (v/c)^2}$ is the Lorentz factor, $\beta = v/c$, and $m_0 c^2$ is the particle's rest mass. Equations (14.6) and (14.7) transform into (14.2) and (14.3), respectively, when we deal with classical mechanics where $\gamma = 1$ and $E_K \ll m_0 c^2$.

Higher heavy charged particle energies than those obtained with a standard cyclotron can be achieved with a special cyclotron in which the frequency of the RF accelerating field is modulated so as to remain in phase with the decreasing cyclotron frequency despite the increase in particle mass. These machines are called synchrocyclotron and are used in physics research but not in medicine.

14.5.3   Microtron

Microtron is an electron accelerator producing electrons in the energy range from 5 MeV to 50 MeV. It combines the features of a linac (resonant cavity for acceleration) and a cyclotron (constant magnetic field to keep accelerated particles in orbit) and is sometimes referred to as electron cyclotron. The concept of the microtron was proposed by *Vladimir I. Veksler* in 1944 and the first prototype unit was built in 1948 in Canada. The machine is used in modern radiotherapy, albeit to a much smaller extent than linear accelerator. Electrons are accelerated by a fixed frequency resonant cavity, make repeated passes through the same cavity, and describe circular orbits in a constant magnetic field.

Two types of microtron are in use: circular and racetrack. A schematic diagram of a circular microtron is shown in Fig. 14.2c.

In the circular microtron the electron gains energy from repeated transitions through a microwave resonant cavity. After each acceleration in a resonant cavity, electrons follow a circular orbit under the influence of a constant magnetic field B and return to the accelerating cavity to receive another boost in kinetic energy. The radius of the trajectory orbit increases with increasing electron energy. To keep the particle in phase with the microwave RF power, the cavity voltage, frequency, and magnetic field are adjusted in such a way that, after each passage through the cavity, the electrons gain an energy increment resulting in an increase in the transit time in the magnetic field equal to an integral number of microwave cycles.

The principle of operation of the racetrack microtron is similar to that of a circular microtron; however, in the racetrack model the magnet is split into two D-shaped pole pieces that are separated to provide greater flexibility in achieving efficient electron injection and higher energy gain per orbit through the use of multi-cavity accelerating structures similar to those used in linacs. The electron orbits consist of two semicircular and two straight sections.

14.5.4   Synchrotron

Synchrotron is the most recent and most powerful member of the circular accelerator family. In synchrotron, like in betatron, the particles follow a circular orbit of constant radius inside a vacuum chamber in the form of an evacuated circular tube but with a significantly larger radius than in a betatron. Because of its similarity with the betatron, the synchrotron is sometimes called the betatron for electrons. The evacuated chamber is placed into a magnetic field that changes in time to account for the increase in particle mass with energy. Unlike in betatron and cyclotron, in synchrotron there is no need for magnetic field within the whole circular orbit of the beam; instead, a narrow ring of magnets provides the guiding magnetic field. The particles are accelerated by an RF electric field which is produced in a resonant cavity (called resonator) placed at a certain point in the particle circular trajectory.

The particles pass through the resonant cavity a large number of times and gain a small amount of kinetic energy during each passage through the cavity.

Both light and heavy charged particles can be accelerated in a synchrotron; however, most synchrotrons are used for acceleration of protons. The frequency of operation is several MHz for acceleration of protons and several 100 MHz for acceleration of electrons. In order to avoid the need for a wide range of frequency modulation of the RF field, the particles are usually injected into the synchrotron after being accelerated to kinetic energy of a few MeV by means of an auxiliary electrostatic or linear accelerator called the injector. Synchrotrons are mainly used for high-energy physics research, but are also used clinically as source for proton beam radiotherapy. A schematic diagram of a synchrotron is given in Fig. 14.2d.

14.5.5 Synchrotron Light Source

As discussed in Sect. 4.3, a charged particle accelerated to a very high velocity in a synchrotron moves in a circular orbit under the influence of a magnetic field and is thus constantly accelerated. As a consequence of this acceleration, the particle emits part of its kinetic energy in the form of photons following Larmor law of (4.18). The radiation so emitted is called synchrotron radiation or magnetic bremsstrahlung and the radiation energy emitted, as given in (4.41), is proportional to the fourth power of the particle velocity and is inversely proportional to the square of the radius of the path.

A cyclic accelerator that keeps particles in a circular orbit such as a synchrotron may be used either for acceleration of charged particles to relativistic velocities or as source of high intensity photon beams (synchrotron radiation) produced by relativistic particles circulating in the accelerator. In the first instance we are dealing with a high energy cyclic accelerator for production of energetic particles, in the second instance we are dealing with a synchrotron light source also called a storage ring for production of intense photon beams.

In a storage ring a continuous or pulsed particle beam may be kept circulating for intervals up to few hours and the particles stored can be electrons, positrons, or protons. Most storage rings are used to store electrons for production of intense synchrotron radiation used in studies of various physical, chemical, and biological phenomena.

Broadband synchrotron radiation in the x-ray energy range that can be rendered monochromatic just above and just below the K absorption edge of iodine based contrast agent has been used successfully in cardiac imaging. The technique is called synchrotron radiation angiography (SRA) and is a novel tool for minimally invasive coronary artery imaging for detection of in-stent restenosis. It is based on acquisition and subtraction of two images, one taken just above the K edge of iodine and one taken just below the K edge of iodine after intravenous infusion of iodinated contrast agent. SRA is thus a novel form of the so-called digital subtraction angiography (DSA), an image subtraction technique based on standard x-ray machines.

14.6 Clinical Linear Accelerator

During the past few decades medical linear accelerators (linacs) have become the predominant machine in treatment of cancer with ionizing radiation. In contrast to linacs used for high-energy physics research, medical linacs are compact machines mounted isocentrically so as to allow practical radiation treatment aiming the beam toward the patient from various directions to concentrate the dose in the tumor and spare healthy tissues as much as possible. In this section we briefly discuss the practical aspects of clinical linear accelerators; the theoretical aspects of electron acceleration in acceleration waveguide are discussed in Chap. 13.

Medical linacs are cyclic accelerators which accelerate electrons to kinetic energies from 4 MeV to 25 MeV using non-conservative microwave radiofrequency (RF) fields in the frequency range from 10^3 MHz (L band) to 10^4 MHz (X band), with the vast majority running at 2856 MHz (S band).

In a linear accelerator the electrons are accelerated following straight trajectories in special evacuated structures called acceleration waveguides. Electrons follow a linear path through the same, relatively low, potential difference several times; hence, linacs also fall into the class of cyclic accelerators just like the other cyclic machines that provide curved paths for the accelerated particles (e.g., betatron and cyclotron).

Various types of linacs are available for clinical use. Some provide x-rays only in the low megavoltage range (4 MV or 6 MV) others provide both x-rays and electrons at various megavoltage energies. A typical modern high-energy linac will provide two photon energies (e.g., 6 MV and 18 MV) and several electron energies in the range from 4 MeV to 22 MeV.

14.6.1 Linac Generations

During the past 50 years, medical linacs have gone through five distinct generations, making the contemporary machines extremely sophisticated in comparison with the machines of the 1960s. Each generation introduced the following new features:

1. *Low energy megavoltage photons* (4 MV to 8 MV):
   straight-through beam; fixed flattening filter; external wedges; symmetric jaws; single transmission ionization chamber; isocentric mounting.
2. *Medium energy megavoltage photons* (10 MV to 15 MV) *and electrons*:
   bent beam; movable target and flattening filter; scattering foils; dual transmission ionization chamber; electron cones.
3. *High energy megavoltage photons* (18 MV to 25 MV) *and electrons*:
   dual photon energy and multiple electron energies; achromatic bending magnet; dual scattering foils or scanned electron pencil beam; motorized wedge; asymmetric or independent collimator jaws.
4. *High energy megavoltage photons* (18 MV to 25 MV) *and electrons*:
   computer-controlled operation; dynamic wedge; electronic portal imaging device; multileaf collimator (MLC).

5. *High energy megavoltage photons* (18 MV to 25 MV) *and electrons*:
   photon beam intensity modulation with multileaf collimator; full dynamic conformal dose delivery with intensity modulated beams produced with a multileaf
   collimator; on-board imaging for use in adaptive radiotherapy.

14.6.2 Components of Modern Linacs

The linacs are usually mounted isocentrically and the operational systems are distributed over five major and distinct sections of the machine:

1. *Gantry*
2. *Gantry stand or support*
3. *Modulator cabinet*
4. *Patient support assembly, i.e., treatment couch*
5. *Control console.*

A schematic diagram of a typical modern S-band medical linac is shown in
Fig. 14.3. Also shown are the connections and relationships among the various linac
components, listed above. The diagram provides a general layout of linac components; however, there are significant variations from one commercial machine to

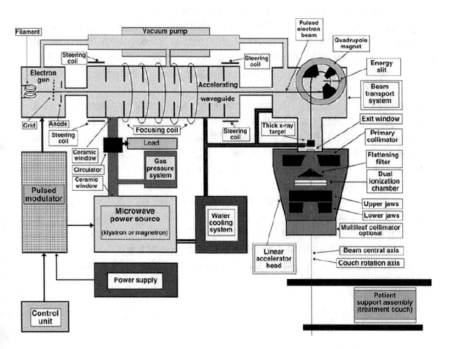

Fig. 14.3 Schematic diagram of a medical linear accelerator (linac)

another, depending on the final electron beam kinetic energy as well as on the particular design used by the manufacturer. The length of the acceleration waveguide depends on the final electron kinetic energy, and ranges from ~30 cm at 4 MeV to ~150 cm at 25 MeV.

The beam-forming components of medical linacs are usually grouped into six classes:

1. *Injection system*
2. *RF power generation system*
3. *Acceleration waveguide*
4. *Auxiliary system*
5. *Beam transport system*
6. *Beam collimation and beam monitoring system.*

The *injection system* is the source of electrons, essentially a simple electrostatic accelerator called an electron gun. Two types of electron gun are in use: diode type and triode type, both containing a heated cathode (at a negative potential of the order of -25 kV) and a perforated grounded anode. In addition, triode type gun also incorporates a grid placed between the cathode and anode. Electrons are thermionically emitted from the heated cathode, focused into a pencil beam and accelerated toward the perforated anode through which they drift into the acceleration waveguide.

The *radiofrequency* (RF) *power generating system* produces the high power microwave radiation used for electron acceleration in the acceleration waveguide and consists of two components: the RF power source and the pulsed modulator. The RF power source is either a magnetron or a klystron in conjunction with a low power RF oscillator. Both devices use electron acceleration and deceleration in vacuum for production of the high power RF fields. The pulsed modulator produces the high voltage, high current, short duration pulses required by the RF power source and the electron injection system.

Electrons are accelerated in the *acceleration waveguide* by means of an energy transfer from the high power RF field which is set up in the accelerating waveguide and produced by the RF power generator. The acceleration waveguide is in principle obtained from a cylindrical uniform waveguide by adding a series of disks (irises) with circular holes at the center, positioned at equal distances along the tube. These disks divide the acceleration waveguide into a series of cylindrical cavities that form the basic structure of the acceleration waveguide of a linac.

The *auxiliary system* of a linac consists of several basic systems that are not directly involved with electron acceleration, yet they make the acceleration possible and the linac viable for clinical operation. These systems are: the vacuum-pumping system, the water-cooling system, the air-pressure system, and the shielding against leakage radiation.

The *electron beam transport system* brings the pulsed high-energy electron beam from the acceleration waveguide onto the target in the x-ray therapy mode and onto the scattering foil in the electron therapy mode.

The *beam monitoring system* and the *beam collimation system* forms and essential system in a medical linac ensuring that radiation dose may be delivered to the patient as prescribed, with a high numerical and spatial accuracy.

14.6.3 Linac Treatment Head

The linac head contains several components, which influence the production, shaping, localizing, and monitoring of the clinical photon and electron beams. Electrons, originating in the electron gun, are accelerated in the acceleration waveguide to the desired kinetic energy and then brought, in the form of a pulsed pencil beam, through the beam transport system into the linac treatment head, where the clinical photon and electron beams are formed. The components found in a typical head of a modern linac include:

1. Several *retractable x-ray targets*
2. *Flattening filters* and *electron scattering foils* (also called scattering filters)
3. Primary and *adjustable secondary collimators*
4. *Dual transmission ionization chambers*
5. *Field defining light* and *range finder*
6. Optional *retractable wedges* or full *dynamic wedges*
7. Optional *multileaf collimator* (MLC).

Clinical Photon Beams

The clinical x-ray beams are produced in medical linacs with a target/flattening filter combination. The electron beam accelerated to a given kinetic energy in the acceleration waveguide is brought by the beam transport system onto an x-ray target in which a small fraction (of the order of 10%) of the electron pencil beam kinetic energy is transformed into bremsstrahlung x-rays. The intensity of the x-ray beam produced in the target is mainly forward peaked and a flattening filter is used to flatten the beam and make it useful for clinical applications. Each clinical photon beam produced by a given electron kinetic energy has its own specific target/flattening filter combination.

Photon beam collimation in a typical modern medical linac is achieved with three collimation devices: the primary collimator, the secondary movable beam defining collimator, and the multileaf collimator (MLC). The primary collimator defines a maximum circular field which is further truncated with the adjustable rectangular collimator consisting of two upper independent jaws and two lower independent jaws and producing rectangular or square fields with a maximum dimension of $40 \times 40 \, \text{cm}^2$ at the linac isocenter, 100 cm from the x-ray target.

The MLC is a relatively new addition to modern linac dose delivery technology. In principle, the idea behind an MLC is simple: MLC allows production of irregularly shaped radiation fields with accuracy as well as efficiency and is based on an array of narrow collimator leaf pairs, each leaf controlled with its own miniature motor. The building of a reliable MLC system presents a substantial technological challenge and current models incorporate up to 120 leaves (60 pairs) covering radiation fields up to $40 \times 40 \, cm^2$ and requiring 120 individually computer-controlled motors and control circuits.

Clinical Electron Beams

Megavoltage electron beams represent an important treatment modality in modern radiotherapy, often providing a unique option in the treatment of superficial tumors. They have been used in radiotherapy since the early 1950s, first produced by betatrons and then by microtrons and linacs. Modern high-energy linacs typically provide, in addition to two megavoltage x-ray energies, several electron beam energies in the range from 4 MeV to 25 MeV.

Electron beams used clinically are produced by retracting the target and flattening filter from the electron pencil beam path and (1) either scattering the pencil beam with a *scattering foil* or (2) deflecting and scanning the pencil beam magnetically to cover the field size required for electron treatment. Special cones (applicators) are used to collimate the electron beams. The electron pencil beam exits the evacuated beam transport system through a thin window usually made of beryllium, which, with its low atomic number, minimizes the pencil beam scattering and bremsstrahlung production.

Dose Monitoring System

In a medical linac an accurate measurement of dose delivered to the patient is of paramount importance. The dose monitoring system of a medical linac is based on transmission ionization chambers permanently imbedded in the linac clinical photon and electron beams. The chambers are used to monitor the beam output (patient dose) continuously during patient treatment. In addition to dose monitoring, the chambers are also used for monitoring the radial and transverse flatness of the radiation beam as well as its symmetry and energy.

For patient safety, the linac dosimetry system usually consists of two separately sealed ionization chambers with completely independent biasing power supplies and read out electrometers. If the primary chamber fails during patient treatment, the secondary chamber will terminate the irradiation, usually after an additional dose of only a few per cent above the prescribed dose has been delivered. Linacs are also equipped with backup timers. In the event of simultaneous failure of both the primary and secondary ionization chambers, the linac timer will shut the machine down with a minimal overdose to the patient.

Field Defining Light, Range Finder, and Laser Positioning Indicators

Accurate positioning of the patient into the radiation beam is very important for modern radiotherapy to ensure the spatial accuracy of dose delivery. The field defining light, range finder, and laser positioning indicators provide convenient visual methods for correctly positioning the patient for treatment using reference marks. The field light illuminates an area that coincides with the radiation treatment field on the patient's skin, while the range finder is used to place the patient at the correct treatment distance by projecting a centimeter scale whose image on the patient's skin indicates the vertical distance from the linac isocenter. Laser positioning devices are used for a practical and reliable indication of the position of the machine isocenter in the treatment room.

14.6.4 Configuration of Modern Linacs

At megavoltage electron energies the bremsstrahlung photons produced in the x-ray target are mainly forward-peaked and the clinical photon beam is produced in the direction of the electron beam striking the target.

In the simplest and most practical configuration, the electron gun and the x-ray target form part of the acceleration waveguide and are aligned directly with the linac isocentre, obviating the need for a beam transport system. A straight-through photon beam is produced and the RF power source is also mounted in the gantry. The simplest linacs are isocentrically mounted 4 MV or 6 MV machines with the electron gun and target permanently built into the acceleration waveguide, thereby requiring no beam transport nor offering an electron therapy option, as shown schematically in Fig. 14.4a.

Acceleration waveguides for intermediate (8 MeV to 15 MeV) and high (15 MeV to 30 MeV) electron energies are too long for direct isocentric mounting, so they are located either in the gantry, parallel to the gantry axis of rotation, or in the gantry stand. A beam transport system is then used to transport the electron beam from the acceleration waveguide to the x-ray target, as shown schematically in Fig. 14.4a, b. The RF power source in the two configurations is commonly mounted in the gantry stand. Various design configurations for modern isocentric linear accelerators are shown in Fig. 14.4 and a modern dual energy medical linac equipped with several imaging modalities for use in image guided radiotherapy (IGRT) and adaptive radiotherapy (ART) is shown in Fig. 14.5.

Image guided radiotherapy (IGRT) allows the imaging of patient's anatomy just before delivery of individual fractions of radiotherapy, thus providing precise knowledge of the location of the target volume on a daily basis and ensuring that the relative positions of the target volume and some reference point for each fraction are the same

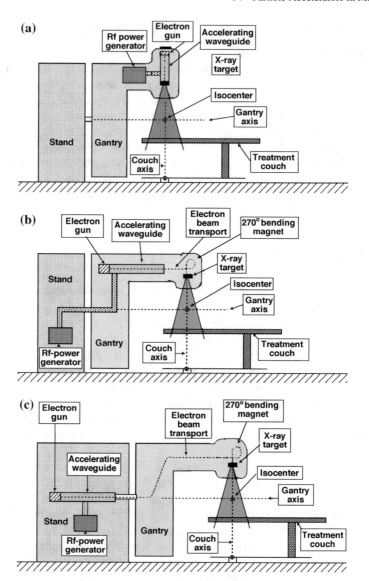

Fig. 14.4 Design configurations for isocentric medical linacs. **a** Straight-through beam design; the electron gun and target are permanently embedded into the acceleration waveguide; machine produces only x-rays with energies of 4 MV to 6 MV; the rf-power generator is mounted in the gantry. **b** Acceleration waveguide is in the gantry parallel to the isocenter axis; electrons are brought to the movable target through a beam transport system; the rf-power generator is located in the gantry stand; machine can produce megavoltage x-rays as well as electrons. **c** Acceleration waveguide and rf-power generator are located in the gantry stand; electrons are brought to the movable target through a beam transport system; machine can produce megavoltage x-rays as well as electrons

Fig. 14.5   Modern dual photon energy (6 MV and 20 MV) linear accelerator (Novalis TX, manufactured in collaboration between Varian Medical Systems, Palo Alto, CA and BrainLAB, Feldkirchen, Germany) equipped with several imaging modalities for use in image guided radiotherapy (IGRT), adaptive radiotherapy (ART), gated radiotherapy, and stereotactic cranial as well as extracranial radiosurgery. The machine incorporates a high definition multileaf collimator as well as a robotic treatment couch and allows static as well as dynamic intensity modulated treatments. *Courtesy of BrainLAB, Feldkirchen, Germany. Reproduced with Permission*

as in the treatment plan. Full implementation of IGRT leads to the concept of adaptive radiotherapy (ART) whereby the dose delivery for subsequent treatment fractions of a course of radiotherapy can be modified to compensate for inaccuracies in dose delivery that cannot be corrected for by simply adjusting the patient's positioning.

14.6.5   Pulsed Operation of Linacs

All particle accelerators operate in some sort of pulsed operation and the clinical electron linacs are no exception. The major linac components, including the source of radiofrequency (RF system), electron gun (injection system), acceleration of electrons in the acceleration waveguide, and production of bremsstrahlung x-rays in the linac target, operate in a pulsed mode with a relatively low duty cycle.

In telecommunications and electronics the duty cycle δ is defined as the fraction of time during which a particular system is in active state or

$$\delta = \frac{\tau}{T}, \tag{14.8}$$

where τ is the duration of the active pulse and T is the period of the periodic operation.

The operation of particle accelerators must be pulsed because of the large instantaneous power (of the order of megawatts) that is required to achieve the desired acceleration. The RF systems, operating at standard linac frequency of 2856 MHz or at $\sim 10^4$ MHz for miniature linacs, can supply such high power only in pulsed operation with a low duty cycle. The pulse repetition rates in typical clinical linacs range from a few pulses per second (pps) up to a few 100 pps.

Typical clinical linac pulsed operation proceeds as follows:

1. The RF power in the RF system is turned on and it takes about 1 μs to fill the acceleration waveguide and produce the field required for acceleration of electrons.
2. The electron gun is then tuned on, electrons are being injected into the acceleration waveguide for 1 μs to 2 μs, and accelerated toward the target.
3. The electron gun is then turned off and injection of electrons is stopped.
4. The RF system is switched off, the acceleration waveguide is completely de-excited in about 1 μs through dumping the residual energy into a dissipative load.
5. The system is ready for the next cycle.

A typical pulse sequence for electrons arriving at the x-ray target of a clinical linac which is run, for example, in the 10 MV x-ray mode is shown in Fig. 14.6. The pertinent parameters are as follows:

Pulse repetition rate (rep rate) in pulses per second (pps): $\rho = 100$ pps
Pulse period: $T = 10^4$ μs
Pulse width: $\tau = 2$ μs
Peak beam current: $I_{\text{peak}} = 50$ mA

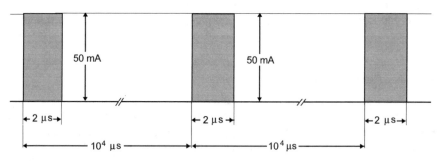

Fig. 14.6 Typical pulse sequence for electron arriving at the x-ray target of a megavoltage linac. The duty cycle δ for the example shown is $\delta = 2 \times 10^{-4}$

Nominal potential: $U = 10$ MV

Charge q carried per pulse: $q = \tau I_{\text{peak}} = (2 \times 10^{-6}\,\text{s}) \times (50 \times 10^{-3}\,\text{A}) = 10^{-7}\,\text{C}$

Charge deposited in the target per second: $q\rho = 10^{-5}\,\text{C}$

Duty cycle: $\delta = \tau/T = 2 \times 10^{-4}$

Number of electrons arriving per second in the x-ray target:

$$n_e = \delta I_{\text{peak}} e^{-1} = 2 \times 10^{-4} \times 50 \times 10^{-3}\,\text{A}/(1.6 \times 10^{-19}\,\text{As}) = 6.25 \times 10^{13}\,\text{s}^{-1}$$

Average beam current: $\bar{I} = \delta I_{\text{peak}} = 10\,\mu\text{A}$ or $\bar{I} = q\rho = 10\,\mu\text{A}$.

Kinetic energy carried by each electron: $E_K = 10^7\,\text{eV} = 1.6 \times 10^{-12}\,\text{J}$

Several interesting conclusions can be drawn from this example:

1. The fraction of the linac active on-time is fairly low considering that during the pulse duration (period) of $10^4\,\mu\text{s}$ the linac is active only for $2\,\mu\text{s}$ and is off for the remaining $9998\,\mu\text{s}$. Despite this very low duty cycle of $\delta = 2 \times 10^{-4}$, a clinical linac produces an average (mean) dose rate of $1\,\text{cGy/min}$ to $10\,\text{cGy/min}$ at the linac isocenter without difficulty.

2. Because of the low duty cycle, the average beam current in the waveguide is also very low, so that the actual power consumption by a typical clinical linac is not excessive.

3. The power P delivered to the linac x-ray target under normal operating conditions is also relatively low and can be estimated by multiplying the average beam current $\bar{I} = 10^{-5}\,\text{A}$ with the nominal potential $U = 10^7$ V to obtain $P = 100\,\text{W}$.

4. We can obtain the same result for the power P delivered to the x-ray target as in point (3) above, if we multiply n_e, the number of electrons arriving per second in the x-ray target with the kinetic energy E_K carried by each electron

$$P = n_e E_K = (6.25 \times 10^{13}\,\text{s}^{-1}) \times (1.6 \times 10^{-12}\,\text{J}) = 100\ \text{W} \qquad (14.9)$$

The power of 100 W delivered to the linac target is obviously much lower than the power delivered to an x-ray target of an x-ray tube operated in the diagnostic radiology range of 50 kVp to 150 kVp. For the same photon beam output, the substantially lower power input into a linac target compared to an x-ray tube target is a result of the forward-peaked x-ray production combined with a much higher efficiency for x-ray production in the megavoltage energy range in comparison with superficial and orthovoltage x-ray tube range.

14.6.6 Practical Aspects of Megavoltage X-Ray Targets and Flattening Filters

Traditionally, the requirements for target properties, established during the early days of x-ray technology development, were quite straightforward with two main stipulations:

1. The targets should be made of high atomic number Z material to maximize efficiency for x-ray production.
2. Targets should have a high melting point to minimize damage to the target from the energetic electron beam used for x-ray production.

Tungsten satisfies well both conditions and is thus the material of choice in most x-ray tubes. With the advent of megavoltage linear accelerators (linacs), it also seemed prudent to adopt tungsten as the target material in clinical linacs and the approach worked well for linac energies below 15 MV; however, at energies above 15 MV high Z targets did not prove optimal, as shown by research work carried out on a 25-MV linac at the Princess Margaret Hospital in Toronto in the early 1970s. The hospital had been using a 25-MV betatron for many years and purchased one of the first commercially available 25-MV linear accelerators when they became available in 1970.

The new linac incorporated a tungsten target and tungsten flattening filter for the production of the clinical megavoltage x-ray beam. It was installed under the assumption that it would provide several improvements over the betatron (such as, better output, isocentric mounting, quieter operation) and produce a clinical x-ray beam comparable to that of the betatron when operated at 25 MeV. However, during the linac commissioning process it became apparent that the linac provided a significantly less penetrating beam than did the betatron when both machines were operated at 25 MeV; actually, the 25-MeV linac produced an x-ray beam with tissue-penetrating properties that matched the betatron beam when the betatron was operated at 16 MV. This was a significant energy difference considering the extra cost in building a linac running at 25 MeV rather than at 16 MV.

The cause of the discrepancy between the linac beam and the betatron beam was traced to the target/flattening filter design and atomic number Z in the two machines. By virtue of its design, the betatron uses a thin target that inherently produces a more penetrating photon beam in comparison to linac's thick transmission tungsten target. Early betatrons also used an aluminum flattening filter in comparison to linac's tungsten filter, and aluminum with its low atomic number will soften the megavoltage x-ray beam less than does a high atomic number filter.

A thin target may be used in betatrons because the target is immersed in a strong magnetic field that engulfs the doughnut, keeps the electrons in circular motion, and sweeps the low-energy electrons transmitted through the target into the doughnut wall before they can strike the flattening filter and produce unwanted off-focus x-rays. In linacs the targets are not immersed in a strong magnetic field so they must be of the thick variety to prevent electrons from traversing the target and striking the flattening filter where they would produce extrafocal bremsstrahlung radiation. The 25-MV linac beam was thus formed with a thick high Z target and a high Z flattening filter, while the betatron beam was formed with a thin target and a low Z flattening filter.

A study of unfiltered linac x-ray beams at machine potentials above 15 MeV has shown that a low Z thick target produces the same quality x-ray beam as a thin betatron target. Thus, a conclusion can be made that x-ray targets in this energy range should be made of low atomic number material to produce the most penetrating photon beam. However, there is a practical problem with this stipulation: it is difficult to find a low Z target that also has a high mass density (i.e., engenders a relatively short range of megavoltage electrons) to make it compact for use in linacs.

For example, the required target thickness for 25-MeV electrons is 1 cm of lead ($Z = 82$, $\rho = 11.3 \, \text{g} \cdot \text{cm}^{-3}$), 0.5 cm of tungsten ($Z = 74$, $\rho = 19.25 \, \text{g} \cdot \text{cm}^{-3}$), or 4 cm of aluminum ($Z = 13$, $\rho = 2.7 \, \text{g} \cdot \text{cm}^{-3}$). From the atomic number Z point of view aluminum is an excellent choice of target material; however, its low mass density precludes its use as a practical target material in high-energy linear accelerators. In modern high-energy linacs a compromise between low Z requirement and concurrent high-density requirement is reached by the use of copper ($Z = 29$ and $\rho = 8.9 \, \text{g} \cdot \text{cm}^{-3}$) rather than tungsten ($Z = 74$ and $\rho = 19.25 \, \text{g} \cdot \text{cm}^{-3}$) as target material.

The low Z target recommendation goes against the target high Z requirement for maximizing the x-ray production; however, it turns out that in the megavoltage energy range the x-ray production in the forward direction is essentially independent of target atomic number and for practical radiotherapy one uses only photons projected in the forward direction defined by the electron pencil beam striking the target. The x-ray yield, of course, depends on the atomic number Z of the target (the higher is Z, the higher is the yield); however, this yield is stated for the 4π geometry and in megavoltage radiotherapy one uses only photons projected in the forward direction for which the yield is independent of Z. It is actually advantageous to have a lower x-ray yield in directions outside the useful radiotherapy beam, because this lowers the required shielding against leakage radiation produced in the linac target.

The traditional requirement on a high melting point of the target material is not as stringent for high-energy linacs in comparison with diagnostic x-ray tubes. At high photon energies used in radiotherapy, the efficiency for x-ray production is of the order of 10% to 20% rather than below 1% as is the case with diagnostic x-ray tubes. Therefore, the electron beam energy deposition and target cooling is of much less concern in megavoltage linacs as compared to diagnostic range x-ray tubes.

The Toronto target/flattening filter study revealed that depth doses for thebreak 25-MV x-ray beam from the linac could be made identical to those measured for the 25-MV betatron beam with the use of an aluminum thick target and an aluminum flattening filter for the shaping of the linac beam. Other beam energies were also studied and Fig. 14.7 shows the results by plotting percentage depth doses against electron kinetic energy in the range from 10 MeV to 32 MeV at various depths in water for four target/flattening filter combinations using high Z (lead) and low Z (aluminum) for target and flattening filter material.

As evident from Fig. 14.7, the most penetrating x-ray beam was obtained from a low Z target/low Z flattening filter combination for electron kinetic energies above 15 MeV and from a high Z target/low Z flattening filter combination for electron kinetic energies below 15 MeV. With both machines operating at 25 MV, the linac x-ray beam produced with an aluminum target and aluminum flattening filter exhibited the same depth dose characteristics as the betatron with its thin target and aluminum flattening filter.

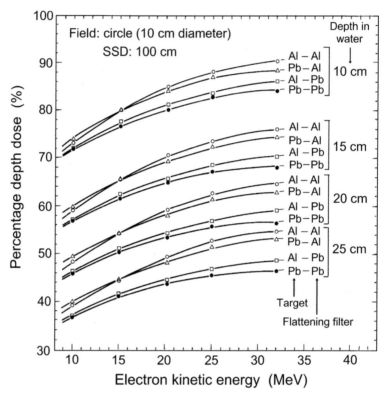

Fig. 14.7 Percentage depth dose at various depths in water against electron beam kinetic energy for various target and flattening filter combinations for megavoltage linacs. Circular field of 10 cm diameter at an SSD of 100 cm

As far as flattening filters are concerned, low atomic number materials are preferable in the range above 15 MV because, with their lower pair production cross section, they cause less beam softening than high atomic number materials; however, similarly to the situation with target materials, space constraints in linac heads limit the practical choices available.

Table 14.5 Best and worsts target/flattening filter combinations for megavoltage linear accelerators

Electron beam kinetic energy E_K (MeV)	BEST target/flattening filter combination		WORST target/flattening filter combination	
	Target	Flattening filter	Target	Flattening filter
$E_K < 15$ MeV	High Z	Low Z	High Z	High Z
$E_K = 15$ MeV	High or Low Z	Low Z	High Z	High Z
$E_K > 15$ MeV	Low Z	Low Z	High Z	High Z

Aluminum, with its relatively low mass density ($\rho = 2.7\,\mathrm{g} \cdot \mathrm{cm}^{-3}$), was a good choice for flattening the betatron beam, since the field size produced by the machine was limited to a $20 \times 20\,\mathrm{cm}^2$ field. Modern linear accelerators, however, deliver fields of up to $40 \times 40\,\mathrm{cm}^2$ at $100\,\mathrm{cm}$ from the target and these field sizes cannot be supported by aluminum flattening filters because of the associated required large size (height) of the filter.

The results of the Toronto target/flattening filter study are summarized in Table 14.5 and show that for linac potential below 15 MV the best target/flattening filter combination is high Z target and low Z flattening filter, while at potentials above 15 MV the best target/flattening filter combination is low Z target and low Z flattening filter. Because of spatial constraints in linac heads, intermediate atomic number materials provide a reasonable compromise for choice of flattening filter and target material for megavoltage linacs. Since the price of a linac increases with its maximum electron beam energy, it is important to optimize the design of the clinical beam shaping components to produce the highest effective x-ray energy for a given machine potential defined by the kinetic energy of electrons striking the target.

Momentous Discoveries in Radiation Physics during the last 5 Years of 19th Century

Wilhelm Conrad Röntgen, a German physicist, **discovered x-rays in 1895** at the University of Würzburg in Germany, where he held a position of physics chair from 1888 to 1900. On November 8, 1895, while investigating electric current flow in a Crookes tube, he noticed that a barium platino-cyanide plate on a bench a few meters from the tube was fluorescing whenever the tube was in operation. He did not plan the experiment with the plate but serendipitously concluded that his Crookes tube must be emitting a new unknown kind of ray. It turned out that his supposition was correct and he almost overnight became one of the most famous physicists ever. For his discovery of x-rays he was in 1901 awarded the inaugural Nobel Prize in Physics: *"in recognition of the extraordinary services he has rendered by the discovery of the remarkable rays subsequently named after him."*

Antoine-Henri Becquerel, a French physicist, **discovered natural radioactivity in 1896** at École Polytechnique in Paris, France where he held a position of physics chair from 1895 to 1908. In early March of 1896, he was investigating phosphorescence stimulated in uranium salts by solar light and hypothesized that the just-discovered Röntgen's new kind of rays may also be emitted by phosphorescing uranium salts. By accident he found that there was no need to trigger phosphorescence with solar light in uranium salts: the salts could expose a photographic plate in complete darkness and this indicated that the salts, rather than through phosphorescence, spontaneously produce a new kind of radiation. This unplanned experiment was the first example of an autoradiograph, ushered in the nuclear age, and brought Becquerel half of the shared 1903 Nobel Prize in Physics *"in recognition of the extraordinary services he has rendered by his discovery of spontaneous radioactivity."*

Marie Skłodowska-Curie, a Polish-born French physicist, and **Pierre Curie**, a French physicist, worked as a team on investigations of radioactivity that Becquerel discovered in 1896. At that time Pierre Curie was already a well-known physicist that distinguished himself with the discovery of piezoelectricity, investigation of magnetic properties of materials, and research on crystal symmetry. For her doctoral thesis at École Polytechnique in Paris, Marie Skłodowska-Curie as Becquerel's student decided to investigate the peculiar characteristics of "Becquerel's rays" and her husband Pierre Curie joined her on the difficult, yet novel, research project.

Marie Skłodowska-Curie started her thesis work by chemically extracting uranium from uranium ore with the goal of isolating pure uranium. However, once uranium was extracted from the ore, she noticed that the residual ore was more radioactive than uranium itself. She postulated that the residual ore must contain other radioactive elements and jointly with Pierre Curie began to process the residual ore. **By 1898 they discovered two new elements: radium and polonium.**

In 1903 Marie Skłodowska-Curie obtained her doctorate for a thesis on radioactive processes and together with Pierre Curie received the second half of the 1903 Nobel Prize *"in recognition of the extraordinary services they have rendered by their joint researches on the radiation phenomena discovered by Professor Henri Becquerel."*

The following quote by Marie Skłodowska-Curie seems even more important today than 100 years ago:

"We must not forget that when radium was discovered no one knew that it would prove useful in hospitals. The work was one of pure science. And this is proof that scientific work must not be considered from the point of view of the direct usefulness of it. It must be done for itself, for the beauty of science, and then there always is a chance that a scientific discovery may become, like radium, a benefit for mankind."

Chapter 15
Fundamentals of Radiation Dosimetry

Ionizing radiation is used in many aspects of modern life: in science, industry, power generation, and medicine. In medicine it is used in diagnosis of disease with x-rays and gamma rays (imaging) as well as in treatment of malignant disease with high energy x-rays, electrons, and heavy charged particles.

Measurement of the quantity and quality of ionizing radiation, generally referred to as radiation dosimetry, is the most important aspect of medical physics, essentially its "raison d'être". It has played a pivotal role in physics since the birth of radiation physics during the last five years of the 19th century. Initially, development of techniques in ionizing radiation measurement was stimulated by the rapid expansion of radiation physics research; later on, however, with the realization of the tremendous potential of x-rays in medicine, and radioactivity in medicine as well as industry and defense, radiation dosimetry became and still is the most important component of radiation physics.

Pioneers in radiation physics research did not pay attention to detrimental effects of ionizing radiations on human tissues and many of them suffered severe consequences as a result of their lack of understanding. The situation today is significantly different and the use of ionizing radiation is strictly controlled, monitored, and regulated.

Wilhelm Conrad RÖNTGEN	**Henri-Antoine BECQUEREL**	**Marie SKŁODOWSKA-CURIE & Pierre CURIE**
1895	1896	1898
Discovery of	Discovery of	Discovery of
x-rays	**natural radioactivity**	**radium** and **polonium**
Nobel Prize: 1901	Nobel Prize: 1903	Nobel Prize: 1903

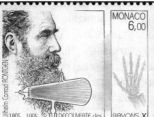  

© Springer International Publishing Switzerland 2016
E.B. Podgoršak, *Radiation Physics for Medical Physicists*,
Graduate Texts in Physics, DOI 10.1007/978-3-319-25382-4_15

Use of ionizing radiation in science, industry or medicine requires an ability to measure accurately the quantity and quality of ionizing radiation produced by a radioactive source or radiation-producing machine. Accuracy is especially important in medical use of ionizing radiation for diagnosis (x-ray imaging and nuclear medicine) or treatment (radiotherapy) of disease. In imaging the goal is to obtain optimal image quality with minimal radiation dose; in radiotherapy it is imperative to deliver the prescribed dose to the target volume with optimal numerical and geometrical accuracy.

The principal physical quantity determined in radiation dosimetry is the absorbed dose, i.e., energy per unit mass, imparted to an absorber; however, several other physically relevant quantities, characteristic of the radiation source or the radiation beam, may also be of interest. A few of these are: radiation source activity, beam fluence, air-kerma in air, and effective dose in an animate medium such as the human body.

Radiation dosimetry is divided into two categories: absolute and relative. Absolute radiation dosimetry, discussed in Chap. 16, refers to measurement of radiation dose directly without requiring calibration in a known radiation field. Only three absolute radiation dosimetry systems have been developed to date: (i) calorimetric absolute radiation dosimetry, (ii) chemical (Fricke) absolute radiation dosimetry, and (iii) ionometric radiation dosimetry. Relative radiation dosimetry is based on dosimeters that require calibration of their signal in a known radiation field. Many relative radiation dosimetry systems are known and the most important of them are discussed in Chap. 17.

This chapter presents an introduction to radiation dosimetry, starting with the basic principles of radiation dosimetry in Sect. 15.1 and quantities used for describing a radiation beam in Sect. 15.2. The chapter then introduces the concept of charged particle equilibrium (CPE) in Sect. 15.3 and a comparison between kerma and absorbed dose in Sect. 15.4. The theoretical basis of radiation dosimetry is introduced in Sect. 15.7 through a discussion of three basic cavity theories used in radiation dosimetry. The chapter concludes with a description of basic aspects of radiotherapy machine output calibration in Sect. 15.10 and radiation dosimetry protocols in Sect. 15.10.

15.1   Ionizing Radiation Beams

The term "radiation dosimetry" was originally used to describe the measurement of radiation dose delivered to an absorber by directly or indirectly ionizing radiation, but the meaning of the term has subsequently been expanded to include the determination or calculation of dose distribution in the human body, irradiated either by one or multiple external beams or by an internal radiation beam. An external beam is defined as an ionizing radiation beam originating in a radiation source located outside the patient, while the term internal beam implies that the radiation source is placed inside the patient.

External radiation beams are used in diagnostic radiology imaging and are produced by x-ray equipment. They are also used in standard external beam radiotherapy and are commonly produced by cobalt machines, x-ray tubes, and linear accelerators. Internal radiation beams, on the other hand, are produced either in nuclear medicine imaging with the use of unsealed radiation sources deposited into human tissues or in brachytherapy with the use of sealed radiation sources placed into body cavities or embedded in tissue.

In all applications of ionizing radiation in medicine the absorbed dose delivered to the patient must be known. In radiotherapy this is of utmost importance in order to ensure that the absorbed dose delivered to the patient matches as closely as possible the prescribed dose (see Sect. 1.12.5) to maximize the tumor control probability (TCP) and minimize the normal tissue complication probability (NTCP). The ICRU has recommended an overall accuracy in tumor dose delivery of $\pm 5\%$, based on an analysis of dose response data and on an evaluation of errors in dose delivery in a clinical setting. Considering all uncertainties involved in radiation dose delivery to the patient, the $\pm 5\%$ recommendation on accuracy is by no means easy to attain and implies that the accuracy of the basic physical calibration of radiotherapy machine output be of the order of $\pm 2\%$ or better.

In diagnostic and nuclear medicine imaging the knowledge of the accurate dose to the patient is not as critical as it is in radiotherapy. However, even here, in order to optimize image quality and simultaneously minimize the possible deleterious effects of ionizing radiation on human tissues, the radiation dose should be known to better than $\pm 10\%$. The same requirement must typically be fulfilled with respect to doses determined for radiation protection and health physics purposes.

15.2 Quantities Used for Describing a Radiation Beam

The International Commission on Radiation Units and Measurements (ICRU) in its Report 33 recommends an extensive list of quantities and units for general use in radiation sciences. Special emphasis is given to photon beams that are described with three distinctly different categories of quantities: radiometric quantities, interaction coefficients, and dosimetric quantities.

1. Radiometric quantities describe the radiation beam in terms of the number and energy of particles constituting the radiation beam.
2. Interaction coefficients deal with quantities related to photon interactions with matter (photoelectric effect, Compton scattering, pair production, etc.).
3. Dosimetric quantities describe the amount of energy the radiation beam deposits in a given medium, such as air, water, tissue, etc. For photon beams they are given as a product of radiometric quantities and interaction coefficients.

15.2.1 Important Radiometric Quantities

The definitions in this section are valid for all ionizing radiations:

1. Directly ionizing (charged) particles such as electrons, positrons, protons, etc.
2. Indirectly ionizing (neutral) particles such as photons and neutrons.

Particle fluence φ is defined as the ratio of dN/dA, where dN is the number of particles that enter a sphere of cross-sectional area dA. The SI unit of particle fluence φ is m^{-2}; the traditional unit, still in common use, is cm^{-2}, where $1\ m^{-2} = 10^{-4}\ cm^{-2}$ or $1\ cm^{-2} = 10^4\ m^{-2}$.

Particle fluence rate $\dot{\varphi}$ is defined as the ratio $d\varphi/dt$, where $d\varphi$ is the increment of particle fluence in the time interval dt. The SI unit of particle fluence rate φ is $m^{-2}\cdot s^{-1}$; the traditional unit, still in common use, is $cm^{-2}\cdot s^{-1}$, where

$$1\ m^{-2}\cdot s^{-1} = 10^{-4}\ cm^{-2}\cdot s^{-1} \text{ or}$$

$$1\ cm^{-2}\cdot s^{-1} = 10^4\ m^{-2}\cdot s^{-1}$$

$$\varphi = \frac{dN}{dA} \quad \text{and} \quad \dot{\varphi} = \frac{d\varphi}{dt}. \tag{15.1}$$

Energy fluence ψ is defined as the ratio dE_v/dA, where dE_v is the radiant energy incident on a sphere of cross-sectional area dA. The SI unit of energy fluence ψ is $J\cdot m^{-2}$; the traditional unit, still in common use, is $MeV\cdot cm^{-2}$, where

$$1\ J\cdot m^{-2} = 6.242\times10^8\ MeV\cdot cm^{-2} \text{ or}$$

$$1\ MeV\cdot cm^{-2} = 1.602\times10^{-9}\ J\cdot m^{-2} = 1.602\times10^{-13}\ J\cdot cm^{-2}.$$

For a monoenergetic photon beam dE_v equals the number of photons dN multiplied with their energy E_v to give

$$\psi = \frac{dE_v}{dA} = E_v\frac{dN}{dA} = E_v\varphi. \tag{15.2}$$

Energy fluence rate $\dot{\psi}$ is defined as the ratio $d\psi/dt$, where $d\psi$ is the increment of energy fluence in the time interval dt. The SI unit of photon energy fluence rate ψ is $W\cdot m^{-2}$; the traditional unit, still in common use, is $MeV\cdot cm^{-2}\cdot s^{-1}$, where

$$1\ W\cdot m^{-2} = 6.242\times10^8\ MeV\cdot cm^{-2}\cdot s^{-1} \text{ or}$$

$$1\ MeV\cdot cm^{-2}\cdot s^{-1} = 1.602\times10^{-9}\ J\cdot m^{-2}\cdot s^{-1} = 1.602\times10^{-13}\ J\cdot cm^{-2}\cdot s^{-1}.$$

$$\psi = \frac{dE_v}{dA} \quad \text{and} \quad \dot{\psi} = \frac{d\psi}{dt}. \tag{15.3}$$

15.2.2 Important Photon Interaction Coefficients

Mass attenuation coefficient μ/ρ of an absorber material for photons is defined as the ratio of dN/N over $\rho\,dl$, where dN/N is the fraction of photons that undergo interactions in traveling a distance dl in an absorber of mass density ρ (see Sect. 8.1)

$$\frac{\mu}{\rho} = \frac{1}{\rho N}\frac{dN}{dl}. \tag{15.4}$$

The SI unit of mass attenuation coefficient μ/ρ is $\mathrm{m^2 \cdot kg^{-1}}$; the traditional unit, still in use, is $\mathrm{cm^2 \cdot g^{-1}}$, where $1\ \mathrm{m^2 \cdot kg^{-1}} = 10\ \mathrm{cm^2 \cdot g^{-1}}$ or $1\ \mathrm{cm^2 \cdot g^{-1}} = 0.1\ \mathrm{m^2 \cdot kg^{-1}}$ (see Sect. 8.1).

Mass energy transfer coefficient μ_{tr}/ρ of an absorber material for photons is given as a product of the mass attenuation coefficient μ/ρ and $\bar{E}_{\mathrm{tr}}/E_\nu$, where E_ν is incident photon energy $h\nu$; $\bar{E}_{\mathrm{tr}}$ is the mean energy transferred from incident photon to charged particles (electrons and positrons) averaged over all possible photon interactions; and $\bar{f}_{\mathrm{tr}} = \bar{E}_{\mathrm{tr}}/E_\nu$ is the mean energy transfer fraction (Sect. 8.2)

$$\frac{\mu_{\mathrm{tr}}}{\rho} = \frac{\mu}{\rho}\frac{\bar{E}_{\mathrm{tr}}}{E_\nu} = \frac{\mu}{\rho}\bar{f}_{\mathrm{tr}}. \tag{15.5}$$

The SI unit of mass energy transfer coefficient μ_{tr}/ρ is $\mathrm{m^2 \cdot kg^{-1}}$; the traditional unit, still in use, is $\mathrm{cm^2 \cdot g^{-1}}$, where $1\ \mathrm{m^2 \cdot kg^{-1}} = 10\ \mathrm{cm^2 \cdot g^{-1}}$ or $1\ \mathrm{cm^2 \cdot g^{-1}} = 0.1\ \mathrm{m^2 \cdot kg^{-1}}$.

Mass energy absorption coefficient $\bar{\mu}_{\mathrm{ab}}/\rho$ of an absorber material for photons is the product of the mass energy transfer coefficient μ_{tr}/ρ and $(1 - \bar{g})$, where $\bar{g}$ is the mean radiation fraction, i.e., fraction of energy of secondary charged particles that is lost to bremsstrahlung and/or in-flight positron annihilation in the absorber (Sect. 8.3). We can also write $\bar{\mu}_{\mathrm{ab}}/\rho$ as a product of the mass attenuation coefficient μ/ρ and total mean energy absorption fraction $\bar{f}_{\mathrm{ab}}$. Thus, we have the following expressions for $\bar{\mu}_{\mathrm{ab}}/\rho$

$$\frac{\mu_{\mathrm{ab}}}{\rho} = \frac{\mu_{\mathrm{tr}}}{\rho}(1 - \bar{g}) \qquad \text{or} \qquad \frac{\mu_{\mathrm{ab}}}{\rho} = \frac{\mu}{\rho}\bar{f}_{\mathrm{ab}} = \frac{\mu}{\rho}\frac{\bar{E}_{\mathrm{ab}}}{E_\nu}. \tag{15.6}$$

The SI unit of the mass energy absorption coefficient $\bar{\mu}_{\mathrm{ab}}/\rho$ is $\mathrm{m^2 \cdot kg^{-1}}$; the traditional unit, still in use, is $\mathrm{cm^2 \cdot g^{-1}}$, where $1\ \mathrm{m^2 \cdot kg^{-1}} = 10\ \mathrm{cm^2 \cdot g^{-1}}$ or $1\ \mathrm{cm^2 \cdot g^{-1}} = 0.1\ \mathrm{m^2 \cdot kg^{-1}}$.

Note: the ICRU uses notations μ_{tr}/ρ and μ_{en}/ρ for the mass energy transfer coefficient and mass energy absorption coefficient, respectively. We use the notation of Johns and Cunningham: $\bar{\mu}_{\mathrm{tr}}/\rho$ and $\bar{\mu}_{\mathrm{ab}}/\rho$ for the two coefficients because they are easier to distinguish from each other; "tr" stands for energy transferred, while "ab" stands for energy absorbed.

15.2.3   Important Dosimetric Quantities

Exposure X is defined as the ratio dQ/dm, where dQ stands for the absolute value of the total charge of the ions of one sign produced in air when all charged particles (electrons and positrons) released by photons in air of mass dm are completely stopped in air. The SI unit of exposure X is $C \cdot kg^{-1}$; however, the old unit of exposure, the röntgen R, is still often used and corresponds to $1\ R = 2.58 \times 10^{-4}\ C \cdot kg^{-1}$ or $1\ C/kg = 3876\ R$.

The seemingly arbitrary choice of $1\ R = 2.58 \times 10^{-4}\ C/kg$ is traced back to the original definition of one röntgen (1 R) as being equal to one "electrostatic unit of electricity" (1 esu) of charge collected in $1\ cm^3$ of air at STP (standard temperature of $T = 0\ °C = 273.16\ K$ and standard pressure of $p = 760\ torr = 101.3\ kPa$). Since $1\ C = 3 \times 10^9$ esu or 1 esu $= 3.333 \times 10^{-10}\ C$ and mass density of air at STP is $1.293 \times 10^{-3}\ g/cm^3$ or $1\ cm^3$ of air at STP contains $1.293 \times 10^{-6}\ kg$ of air, we get:

$$1\ R = 1\ esu/cm^3 = [(3 \times 10^9\ esu/C) \times (1.293 \times 10^{-6} kg)]^{-1} = 2.58 \times 10^{-4}\ C/kg.$$

Exposure X is a measure of the ability of photons to ionize air and is of limited value in modern radiation physics because it is only defined for air as absorbing medium and only for photons in a relatively narrow energy $h\nu$ range of 1 keV $<$ $h\nu < 3$ MeV. It cannot be used for neutrons and charged particle beams, but is of some historical significance because: (i) it was the first radiation quantity defined by the ICRU in 1928 to quantify radiation beams and (ii) it was subsequently for several decades used as an intermediate step in the determination of absorbed dose in absorbing medium such as water and tissue.

Kerma K_{med} (acronym for "kinetic energy released in matter") is given by the ratio $d\bar{E}_{\text{tr}}$ over dm, where $d\bar{E}_{\text{tr}}$ is the mean energy transferred from indirectly ionizing radiation (photon and neutron) to secondary charged particles released by photons and neutrons in absorber medium of mass dm. The SI unit of kerma is $J \cdot kg^{-1}$ and the special unit is gray (Gy), with $1\ Gy = 1\ J \cdot kg^{-1}$.

For monoenergetic photons kerma may also be expressed as a product of photon energy fluence ψ and the mass energy transfer coefficient $(\mu_{\text{tr}}/\rho)_{\text{med}}$ of the absorber medium

$$K_{\text{med}} = \frac{d\bar{E}_{\text{tr}}}{dm} = \Psi \left(\frac{\mu_{\text{tr}}}{\rho} \right)_{\text{med}} = \varphi E_\nu \left(\frac{\mu_{\text{tr}}}{\rho} \right)_{\text{med}} = \varphi F_{\text{K}}, \qquad (15.7)$$

with $\psi = E_\nu \varphi$ and $F_K = E_\nu (\mu_{\text{tr}}/\rho)_{\text{med}}$ the so-called kerma factor. For a beam of photons with energy E_ν and photon fluence φ, inserting (15.5) into (15.7) and using (7.19) allows us to express kerma as follows

$$K_{\text{med}} = \frac{d\bar{E}_{\text{tr}}}{dm} = \Psi \left(\frac{\mu_{\text{tr}}}{\rho} \right)_{\text{med}} = \varphi E_\nu \left(\frac{\mu}{\rho} \right)_{\text{med}} \bar{f}_{\text{tr}} = \varphi E_\nu \left(\frac{\mu}{\rho} \right)_{\text{med}} \frac{\bar{E}_{\text{tr}}}{E_\nu} = \varphi \left(\frac{\mu}{\rho} \right)_{\text{med}} \bar{E}_{\text{tr}},$$
$$(15.8)$$

where $(\mu/\rho)_{\mathrm{med}}$ is the mass attenuation coefficient of the absorber medium and $\bar{E}_{\mathrm{tr}}$ is the mean energy transferred to charged particles in the medium at each interaction. The product $\varphi \times (\mu/\rho)$ gives the number of photon interactions per unit mass of absorber medium irradiated by photon fluence φ.

Kerma rate $\dot{K}_{\mathrm{med}}$ is defined as the ratio dK_{med}/dt, where dK_{med} is the increment of kerma in the time interval dt. The SI unit of kerma rate $\dot{K}_{\mathrm{med}}$ is $\mathrm{J \cdot kg^{-1} \cdot s^{-1}}$ or $\mathrm{Gy \cdot s^{-1}}$.

Absorbed dose D_{med} is the most important quantity in radiation dosimetry. It is defined as the ratio $d\bar{E}_{\mathrm{ab}}/dm$, where $d\bar{E}_{\mathrm{ab}}$ is the mean energy imparted by ionizing radiation to matter of mass dm. The SI unit of absorbed dose is $\mathrm{J \cdot kg^{-1}}$ and the special unit is gray (Gy), with $1 \mathrm{~Gy} = 1 \mathrm{~J \cdot kg^{-1}}$.

For indirectly ionizing radiations the mean energy $\bar{E}_{\mathrm{ab}}$ is imparted to absorbing medium through a two-step process: in the first step, energy is transferred from a neutral particle (photon or neutron) to energetic charged particles (resulting in kerma K_{med}); in the second step, the liberated charged particles, as they travel through the absorber, gradually impart a portion of their kinetic energy to absorbing medium (resulting in absorbed dose D_{med}).

Absorbed dose rate $\dot{D}_{\mathrm{med}}$ is defined as the ratio dD_{med}/dt, where dD_{med} is the increment of absorbed dose in the time interval dt. The SI unit of absorbed dose rate $\dot{D}_{\mathrm{med}}$ is $\mathrm{J \cdot kg^{-1} \cdot s^{-1}}$ or $\mathrm{Gy \cdot s^{-1}}$.

15.3 Concept of Charged Particle Equilibrium

As discussed in detail in Chap. 7, when photons traverse an absorbing medium, they may interact with absorber atoms through various interactions (effects) in which secondary charged particles (electrons and positrons) are released in the medium and the interacting photon either disappears or is scattered.

The concept of charged particle equilibrium (CPE) is an important concept in radiation physics. It states that under CPE each secondary charged particle that leaves volume-of-interest $\mathcal{V}$ with kinetic energy E_K is replaced by another secondary charged particle that enters $\mathcal{V}$ with same kinetic energy E_K and expends it inside $\mathcal{V}$. This means that under CPE a perfect energy balance in volume $\mathcal{V}$ is maintained and the kinetic energy that charged particles carry out of volume $\mathcal{V}$ is replenished by same amount of kinetic energy carried by other charged particles into volume $\mathcal{V}$.

The concept of CPE is illustrated schematically in Fig. 15.1 that depicts a block of material irradiated by 10 MeV photons. Secondary charged particles (only electrons are shown) released in photon interactions with absorber atoms are shown with straight arrows representing electron tracks. The volume-of-interest $\mathcal{V}$ is shown in yellow color, the block of absorbing material in grey color. The numbers at track origins (i.e., at points of photon interaction) indicate initial kinetic energy $(E_K)_0$ of secondary electrons in MeV. Numbers at track crossing the boundary of $\mathcal{V}$ indicate electron kinetic energy in MeV at specific track crossings. Electron kinetic energy at track-ends is zero.

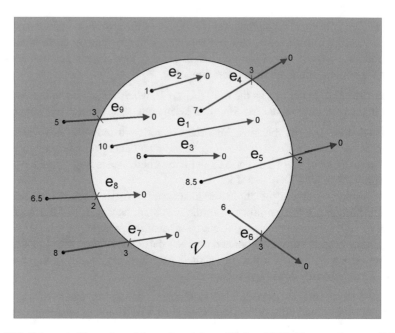

Fig. 15.1 Schematic illustration of charged particle equilibrium (CPE). Photons of energy 10 MeV (not shown) are irradiating a block of material (phantom) shown in *grey color* and containing a volume-of-interest $\mathcal{V}$ in *yellow color*. Secondary charged particles (only electrons are shown) released in photon interactions with absorber atoms are shown with *straight arrows*, representing electron tracks. There are 9 electron tracks, each starting with a *black dot* (•) and the initial electron kinetic energy in MeV is indicated. The tracks end when electron kinetic energy is zero at points indicated with 0. Electron kinetic energy in MeV entering or exiting $\mathcal{V}$ is indicated at the boundary of $\mathcal{V}$

Nine electron tracks are shown in Fig. 15.1:

1. Three electrons (e_1, e_2, and e_3) originate inside $\mathcal{V}$ and expend all their kinetic energy through collision losses inside $\mathcal{V}$.
2. Three electrons (e_4, e_5, and e_6) originate in $\mathcal{V}$ but carry a portion of their kinetic energy out of $\mathcal{V}$.
3. Three electrons (e_7, e_8, and e_9) originate outside of $\mathcal{V}$ and carry a portion of their kinetic energy into $\mathcal{V}$.
4. Electron e_4 carries kinetic energy of 3 MeV out of $\mathcal{V}$. Energy loss of 3 MeV is compensated by electron e_9 that brings kinetic energy of 3 MeV into $\mathcal{V}$.
5. Electron e_5 carries kinetic energy of 2 MeV out of $\mathcal{V}$. Energy loss of 2 MeV is compensated by electron e_8 that brings kinetic energy of 2 MeV into $\mathcal{V}$.
6. Electron e_6 carries kinetic energy of 3 MeV out of $\mathcal{V}$. Energy loss of 3 MeV is compensated by electron e_7 that carries kinetic energy of 3 MeV into $\mathcal{V}$.

Since all kinetic energy, lost to volume-of-interest $\mathcal{V}$ by electrons released in $\mathcal{V}$ and leaving $\mathcal{V}$ with kinetic energy E_K, is compensated by an equal amount of kinetic energy E_K entering $\mathcal{V}$, we can assume that: (i) all energy released in $\mathcal{V}$ is actually absorbed in $\mathcal{V}$ and (ii) conditions for charged particle equilibrium (CPE) are satisfied in volume-of-interest $\mathcal{V}$ in Fig. 15.1.

For simplicity, in Fig. 15.1 only electron tracks are shown, implying that no pair production photon interactions occur in the absorber. Moreover, an assumption is made that electrons experience only collision losses and no radiation losses occur inside volume-of-interest $\mathcal{V}$. These would be valid assumptions for photons below 1 MeV; at higher photon energies, however, pair production interactions as well as radiation losses (bremsstrahlung and in-flight positron annihilation) would increase in importance with photon energy and would have to be taken into consideration. However, the basic concepts of CPE would still apply.

15.4 Kerma Versus Absorbed Dose

Both kerma K_{med} in absorber medium and absorbed dose D_{med} in absorber medium are defined by the ratio dE/dm, where, for kerma, dE stands for mean energy transferred from incident particle to liberated charged particle in absorber medium of mass dm and, for absorbed dose, dE stands for mean energy transferred from the liberated charged particle to absorber medium of mass dm.

Kerma K_{med} occurs at the point of interaction between the incident particle (photon or neutron) and the absorber atom, while absorbed dose D_{med} is spread over a larger mass element dm. For kerma K_{med} the choice of mass element dm size is not important; for absorbed dose D_{med}, on the other hand, the choice of dm size is very important for two reasons:

(i) To avoid statistical fluctuations that occur when dm is too small
(ii) To achieve adequate spatial resolution that is less than optimal when dm is too large. For large dm the absorbed dose represents the mean dose $\bar{D}_{med}$ over the absorber mass dm.

Kerma K_{med} is conceptually easy to understand, relatively easy to calculate, but difficult to measure. It is defined only for beams of neutral particles (photons and neutrons) that are, because of their two-step dose delivery process, referred to as indirectly ionizing radiations, in contrast to charged particles that fall into the category of directly ionizing radiations (Sect. 1.10).

15.4.1 Absorbed Dose for Photons

For photons the two-step dose delivery process proceeds as follows:

1. In the first step the photon interacts with an absorber atom through one of the following effects: photoelectric effect, Compton scattering, nuclear pair production, and triplet production (see Chap. 7). An energetic secondary electron and/or an electron-positron pair is liberated in the absorber and the mean energy $\bar{E}_{tr}$ transferred from the photon to energetic charged particles per unit mass results in kerma K_{med}.
2. In the second step the energetic charged particles (electrons and positrons) travel through the absorber medium and undergo multiple Coulomb interactions with the atoms of the absorber thereby slowly losing their kinetic energy in two possible manners:

   - Coulomb interaction between energetic secondary charged particles (electrons and positrons) and orbital electrons of absorber atoms is described as collision loss and results in energy transfer from secondary charged particle to absorber. This energy per unit mass is called collision kerma K_{med}^{col} and is, under the condition of electronic equilibrium, called absorbed dose D_{med} in the absorber.
   - Coulomb interaction between the energetic secondary charged particles and nuclei of absorber atoms results in radiation loss in the form of either bremsstrahlung photons (for electrons and positrons) or annihilation photons following in-flight annihilation of a positron of non-zero kinetic energy. This radiation loss per unit mass results in radiation kerma K_{med}^{rad} that is assumed to escape the volume-of-interest and thus does not contribute to radiation absorbed dose D_{med} in the absorber.

For photons, kerma K_{med} is a sum of two components: the smaller and often negligible component called radiation kerma K_{med}^{rad} that escapes the volume of interest V and the larger component called collision kerma K_{med}^{col} that, under the condition of charged particle equilibrium (CPE), is equal to absorbed dose D_{med} in the volume-of-interest V

$$K_{med} = K_{med}^{col} + K_{med}^{rad} \tag{15.9}$$

and

$$D_{med} = K_{med}^{col} \text{ provided the CPE condition is satisfied.} \tag{15.10}$$

Equations (15.7) for kerma K_{med} and (15.10) for absorbed dose D_{med} suggest that measuring absorbed dose for indirectly ionizing radiations should be simple. However, both K_{med} and D_{med} depend on particle (photon or neutron) fluence that is difficult to measure with high accuracy and precision, and this makes radiation dosimetry based on particle fluence difficult in practice; especially so in a clinical setting. Therefore, other more suitable and practical techniques have been developed based on cavity theories, as discussed in Sect. 15.7.

15.4.2 Absorbed Dose for Neutrons

For neutrons the two-step dose delivery process is simpler than that for photons and is characterized as follows:

1. In the first step the neutron interacts with an absorber atom through one of the following interactions (Sect. 9.2): inelastic scattering, neutron capture, spallation, and nuclear fission. In many of these nuclear reactions a proton or heavier charged particle is released and the energy E_{tr} transferred from neutron to the secondary charged particles contributes to kerma K_{med} at a point-of-interest in the absorber.
2. In the second step the energetic secondary charged particles (protons and heavier ions) travel through the absorber and undergo multiple Coulomb interactions with the orbital electrons and nuclei of the absorber atoms. Interactions with orbital electrons result in collision losses (Sect. 6.4) that contribute to collision kerma K_{med}^{col}. Two simplifying assumptions can be made:

   - Since radiation loss of heavy charged particles in Coulomb interactions with absorber nuclei is negligible, neutron kerma has only one component

$$K_{med} = K_{med}^{col} \tag{15.11}$$

   - Since the range of secondary charged particles (protons and heavier ions) in absorber media is generally short, standard CPE conditions apply in most situations of interest in neutron dosimetry and absorbed dose D_{med} is generally equal to kerma K_{med}

$$D_{med} \equiv K_{med} = K_{med}^{col}. \tag{15.12}$$

15.4.3 Example of Kerma and Absorbed Dose Calculation

By way of example we now take a closer look at Sect. 8.7.1 in which we examined the interaction of photons of energy $E_\nu = 2$ MeV with lead absorber. We established that in a 2 MeV photon interaction with lead absorber on average the following conditions apply:

1. Mean energy transferred to secondary charged particles (electrons and positrons) is $\bar{E}_{tr} = 1.13$ MeV
2. Energy of scattered photon is $E_\nu' = 0.87$ MeV

Of the $\bar{E}_{tr} = 1.13$ MeV transferred to secondary charged particles:

1. Mean energy absorbed in lead is $\bar{E}_{ab} = 1.04$ MeV.
2. Energy emitted in the form of bremsstrahlung photons and in-flight annihilation photons is $E_\nu'' = 0.09$ MeV.

We also determined the mass attenuation coefficient $(\mu/\rho)_{Pb}$, mass energy transfer coefficient $(\mu_{tr}/\rho)_{Pb}$, and mass energy absorption coefficient $(\mu_{ab}/\rho)_{Pb}$ and got the following results:

$$\left(\frac{\mu}{\rho}\right)_{Pb} = 0.0453 \ \frac{cm^2}{g}, \ \left(\frac{\mu_{tr}}{\rho}\right)_{Pb} = 0.0256 \ \frac{cm^2}{g}, \ \text{and} \ \left(\frac{\mu_{ab}}{\rho}\right)_{Pb} = 0.0235 \ \frac{cm^2}{g}.$$

$$(15.13)$$

The mean radiation fraction $\bar{g}$ (Sect. 8.3.1) was determined as $\bar{g} = 0.08$. It has two components: the predominant bremsstrahlung component $\bar{g}_B$ and the smaller, often neglected, in-flight annihilation g_A component.

To illustrate the difference between kerma K_{Pb} and absorbed dose D_{Pb} we show in Fig. 15.2a schematic diagram of the 2-MeV photon interaction presented in Sect. 8.7.1 with a lead atom at point A. The volume-of-interest $\mathcal{V}$ containing mass of lead dm is shown with a circle and dm is indicated in yellow color. The following facts related to Fig. 15.2 are of interest:

1. Interaction at point A can be photoelectric effect, Rayleigh scattering, Compton scattering, nuclear pair production or electronic (triplet) production. As shown in Sect. 8.7.1, on the average over a large number of such interactions, mean energy $\bar{E}_{tr}$ transferred from 2-MeV photon to charged particles (electrons and positrons) is $\bar{E}_{tr} = 1.13$ MeV, while $E'_v = 0.87$ MeV goes to scattered photons (Rayleigh and Compton).

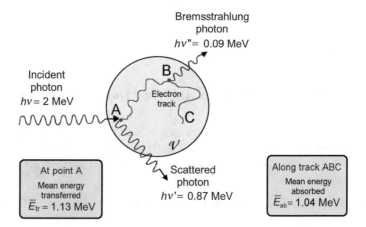

Fig. 15.2 Photon of energy 2 MeV interacts with lead atom at point A. In the interaction the photon is scattered and a secondary charged particle (electron and possibly positron) are released. Volume-of-interest $\mathcal{V}$ containing mass of lead dm is shown with *yellow circle*. Electron track ABC is shown in *blue color*. Radiation loss in the form of bremsstrahlung photons and annihilation photons produced in in-flight positron annihilation is shown with one bremsstrahlung photon originating at point B

2. The secondary charged particles are represented by a single electron track ABC. Starting with initial kinetic energy $(\bar{E}_K)_0 = 1.13$ MeV at point A the secondary charged particle undergoes numerous collision (ionization) losses as well as radiation losses (bremsstrahlung and in-flight positron annihilation) progressing from point A through point B to point C where its kinetic energy is fully expended $(E_K = 0)$.

3. Bremsstrahlung and in-flight annihilation photons that are produced through radiation losses in track ABC are represented in Fig. 15.2 by a single bremsstrahlung photon of energy $E_v'' = 0.09$ MeV originating at point B and escaping volume-of-interest $\mathcal{V}$.

4. Mean energy $\bar{E}_{tr} = 1.13$ MeV, transferred from photon to secondary electron on a single atom at point A, forms part of kerma K_{Pb} that has two components: collision kerma K_{Pb}^{col} and radiation kerma K_{Pb}^{col}. Under the condition of charged particle equilibrium, collision kerma K_{Pb}^{col} is equal to absorbed dose D_{Pb}.

5. The secondary charged particle loses all of its initial kinetic energy $(E_K)_0 = 1.13$ MeV in the volume-of-interest $\mathcal{V}$ through numerous interactions with lead atoms along the track ABC:

   • 1.04 MeV of 1.13 MeV is lost through collision (ionization) losses and represents mean energy $\bar{E}_{ab}$ that is absorbed in $\mathcal{V}$ and contributes to absorbed dose D_{Pb};
   • 0.09 MeV of 1.13 MeV is lost through radiation losses E_v'' and escapes $\mathcal{V}$.

6. To get an idea about the size of the volume-of-interest $\mathcal{V}$ containing mass dm we now estimate the CSDA range (Sect. 6.8) of a secondary electron with initial kinetic energy of $\bar{E}_{tr} = (E_K)_0 = 1.13$ MeV. The NIST gives a range of ~ 1 g/cm² that corresponds to ~ 0.9 mm of lead. This is a relatively short distance, yet sufficiently long to allow the electron to expend all of its initial kinetic energy through a large number of ionizations and excitations of lead atoms. Since the ionization energy IE(Pb) of lead is 7.4 eV, a 1.13 MeV electron can create of the order of 10^5 ionizations and excitations during its ~ 0.9 mm travel through lead.

Let us now assume that the photon fluence φ of the 2 MeV photon beam is $\varphi = 5 \times 10^{10}$ cm⁻² and the charged particle equilibrium (CPE) is satisfied in the lead absorber.

Determine the following dosimetric quantities in volume of interest $\mathcal{V}$: 1. Kerma; 2. Collision kerma; 3. Radiation kerma; and 4. Absorbed dose.

The calculation of the dosimetric quantities proceeds as follows:

1. Kerma will be determined using (15.8) to get

$$K_{Pb} = \varphi \left(\frac{\mu}{\rho}\right)_{Pb} \bar{E}_{tr} = (5 \times 10^{10} \text{ cm}^{-2}) \times \left(0.0453 \frac{\text{cm}^2}{\text{g}}\right) \times (1.13 \text{ MeV})$$

$$\times \left(1.602 \times 10^{-13} \frac{\text{J}}{\text{eV}}\right) \times \left(\frac{10^3 \text{ g}}{\text{kg}}\right) = 0.4100 \frac{\text{J}}{\text{kg}} = 0.4100 \text{ Gy} = 41 \text{ cGy}.$$

$$(15.14)$$

2. Collision kerma K_{Pb}^{col} is calculated from kerma K_{Pb} and mean radiation fraction $\bar{g}$

$$K_{Pb}^{col} = K_{Pb}(1 - \bar{g}) = (0.4100 \text{ Gy}) \times (1 - 0.08) = 0.3772 \text{ Gy}. \qquad (15.15)$$

3. Radiation kerma K_{Pb}^{rad} is calculated from kerma K_{Pb} and mean radiation fraction $\bar{g}$ as

$$K_{Pb}^{rad} = K_{Pb}(\bar{g}) = (0.4100 \text{ Gy}) \times 0.08 = 0.0328 \text{ Gy}. \qquad (15.16)$$

4. Since an assumption is made that the charged particle equilibrium (CPE) is satisfied, we can determine absorbed dose D_{Pb} in lead simply by stating that absorbed dose equals collision kerma K_{Pb}^{col} of (15.15) or else we can use the following expression

$$D_{Pb} = \varphi \left(\frac{\mu}{\rho}\right) \bar{E}_{ab} = (5 \times 10^{10} \text{ cm}^{-2}) \times \left(0.0453 \frac{\text{cm}^2}{\text{g}}\right) \times (1.04 \text{ MeV})$$

$$\times \left(1.602 \times 10^{-13} \frac{\text{J}}{\text{eV}}\right) \times \left(\frac{10^3 \text{g}}{\text{kg}}\right) = 0.3774 \frac{\text{J}}{\text{kg}} = 0.3774 \text{ Gy} = 37.74 \text{ cGy}.$$

or $\qquad\qquad\qquad\qquad\qquad\qquad\qquad\qquad\qquad\qquad\qquad\qquad\qquad\qquad (15.17)$

$$D_{Pb} = \varphi \left(\frac{\mu}{\rho}\right) \bar{E}_{tr}(1 - \bar{g}) = (5 \times 10^{10} \text{ cm}^{-2}) \times \left(0.0453 \frac{\text{cm}^2}{\text{g}}\right) \times (1.13 \text{ MeV})$$

$$\times (1 - 0.08) \times \left(1.602 \times 10^{-13} \frac{\text{J}}{\text{eV}}\right) \times \left(\frac{10^3 \text{ g}}{\text{kg}}\right)$$

$$= 0.3772 \frac{\text{J}}{\text{kg}} = 0.3772 \text{ Gy} = 37.72 \text{ cGy}.$$

where $\bar{E}_{ab}$ is the portion of the mean energy $\bar{E}_{tr}$ transferred to charged particles that contributes to energy absorbed in the absorber. The slight discrepancy between (15.17) and (15.15) results from rounding errors.

15.5 Radiation Dosimetry Systems

Radiation-related quantities, such as dose absorbed in medium, are measured with devices referred to as radiation dosimetry system. These devices consist of a radiation dosimeter (detector) and associated electronics (reader). The radiation dosimeter responds to a certain physical characteristic of the ionizing radiation under investigation and produces a suitable physical or chemical signal that is measured by the reader.

Radiation dosimetry generally relies on measurement of a radiation-induced physical signal that is emitted by a radiation dosimeter and measured by a reader of a dosimetry system. Typical physical signals measured in modern radiation dosimetry cover a wide variety of physical quantities, such as, for example:

1. Temperature rise in calorimetric radiation dosimetry (discussed in Sect. 16.1).
2. Ion current in ionization chamber dosimetry (discussed in Sect. 16.3).
3. Emission of ultraviolet or visible light (thermally activated phosphorescence) in thermoluminescence (TL) dosimetry (discussed in Sect. 17.2.4).
4. Emission of ultraviolet or visible light in optically stimulated luminescence (OSL) dosimetry (discussed in Sect. 17.2.5).
5. Transmission of visible light in radiographic and radiochromic film dosimetry (discussed in Sect. 17.4).

15.5.1 Active Versus Passive Radiation Dosimetry System

With respect to the connection between the dosimeter and reader of a dosimetry system two categories of dosimetry system are in use: 1. Active or 2. Passive.

1. A dosimetry system that allows direct measurement of absorbed dose rate as well as the accumulated absorbed dose is referred to as an active or electronic dosimetry system. It is characterized with a direct connection between the dosimeter that responds to ionizing radiation and the reader that measures the dosimeter response during the dosimeter's exposure to radiation.
2. A dosimetry system in which the dosimeter and reader are not connected during dosimeter exposure to ionizing radiation is called a passive dosimetry system. In this category of dosimetry system the dosimeter and reader are not connected during dosimeter's exposure to radiation; the dosimeter accumulates a suitable signal proportional to absorbed dose during exposure to radiation and the reader measures the accumulated dose in the dosimeter upon completion of radiation exposure after connection between the dosimeter and reader is established.

15.5.2 Absolute Versus Relative Radiation Dosimetry System

A radiation dosimetry system consists of a detector–reader combination, with the detector (dosimeter) producing a suitable signal M in response to ionizing radiation and the reader measuring the signal M that is related to mean dose $\bar{D}_{cav}$ deposited in the dosimeter's sensitive volume V by ionizing radiation. With regard to dose determination two major categories of dosimetry system are known: absolute dosimetry systems and relative dosimetry systems.

1. *Absolute radiation dosimetry system* is based on a dosimeter that produces a signal from which the absorbed dose in its sensitive volume can be determined directly without requiring calibration in a known radiation field. As discussed in Chap. 16, three types of absolute radiation dosimetry system are in use:

   (i) Calorimetric radiation dosimetry, based on carbon or water calorimeters.
   (ii) Chemical radiation dosimetry, based on ferrous sulfate chemical dosimeters.
   (iii) Ionometric radiation dosimetry, based on ionization chambers.

   Note that ionization chamber-based systems are used not only in absolute dosimetry, they are also commonly used as relative dosimetry systems.
2. *Relative dosimetry systems* are based on a dosimeter that requires calibration of its signal in a known radiation field. Many relative dosimetry systems have been developed to date, ranging from systems of significant practical value and usefulness (e.g., ionization chambers, radiographic film, and thermoluminescence dosimetry) down to systems of purely scientific interest and of little practical value in radiation dosimetry (e.g., radioelectret and thermally activated exoelectron emission).

15.5.3 Main Characteristics of Radiation Dosimetry Systems

Many dosimetry systems have been developed since Röntgen and Becquerel discovered ionizing radiation in 1895 and 1896, respectively. The basic requirements that a radiation dosimetry system (detector–reader combination) must satisfy are as follows:

(1) *High sensitivity to dose*. The signal produced by the radiation detector in response to the radiation beam should be of adequate magnitude to match the reader specifications even at relatively low doses.
(2) *Large range* of absorbed dose coverage.
(3) *Linear response to dose* over the large range of the dosimetry system to avoid difficulties with supralinear response in the intermediate dose range and signal saturation of response at high doses. In supralinear response the detector signal exceeds linear response with increasing dose (example: thermoluminescence dosimetry); in saturation at high doses the signal attains a constant level or even diminishes with increasing dose (example: Geiger–Müller counter).
(4) *High accuracy and high precision*. Accuracy is stipulated by the degree of agreement between the measured value and the true (i.e., expectation) value of a physical quantity, while precision is inferred from the reproducibility of results measured at same conditions. Thus, accuracy is associated with systematic errors, precision with random errors.
(5) *Dose rate independence of the detector signal*. Ideally, for the same dose the detector signal should be independent of dose rate; however, in practice this is

generally not the case and the dose rate effect must be known and accounted for when one is interested in accurate determination of absorbed dose.

(6) *Detector signal dependence on beam energy* must be known and accounted for when the measured user's beam quality differs from that of the calibration beam.

(7) *Special issues*, such as *beam direction* striking the detector, *spatial resolution* of the detector, and various *spurious effects* affecting the detector response, must also be considered when accurate dosimetry is the objective of the dose measurement process.

15.6   Radiation Dosimeters

A radiation dosimeter typically consists of two major components: (1) Dosimeter cavity and (2) Dosimeter wall.

1. "Cavity" is a shorthand term for the radiation sensitive medium producing the radiation-induced dosimeter signal and its volume is called the radiation sensitive volume. The signal that the cavity produces in response to ionizing radiation contains information on the dose imparted to the cavity and is based on a wide variety of physical or chemical quantities that are proportional to the dose imparted to the dosimeter.
2. "Wall" is a shorthand term for the dosimeter component that defines, supports, and contains gaseous and liquid cavities. Solid cavities do not need a wall for containment and are in this context more practical than gaseous and liquid cavities. The wall can be of various sizes and shapes, such as cylinder, vial, pillbox, sphere, etc. and its thickness is made as thin as possible to minimize its effect on the cavity signal, yet thick enough to provide mechanical support for the cavity.

Since direct non-invasive measurements with a radiation dosimeter are difficult to accomplish on a patient, they are normally carried out in a suitable tissue-equivalent medium referred to as a phantom. Water is the most commonly used soft tissue substitute material for absorbed dose measurements in photon and electron beams. However, use of water in dosimetry is not always practical and dosimetric measurements are often carried out in more practical solid phantom materials, such as polystyrene, Lucite, and "Solid Water" that closely approximate water and soft tissue in terms of absorption and scattering properties. Three physical parameters of a given material are important in determining water equivalency: mass density, number of electrons per gram, and effective atomic number.

Determination of absorbed dose at a given point A in a patient is accomplished by measuring the dose at the corresponding point A in a tissue-equivalent phantom. As shown schematically in Fig. 15.3, a radiation dosimeter, consisting of a radiation sensitive cavity and wall, is placed into the tissue-equivalent phantom in such a way that its reference point (typically the center of the cavity volume) corresponds with the point-of-interest A at depth z on the central axis of the radiation beam. The

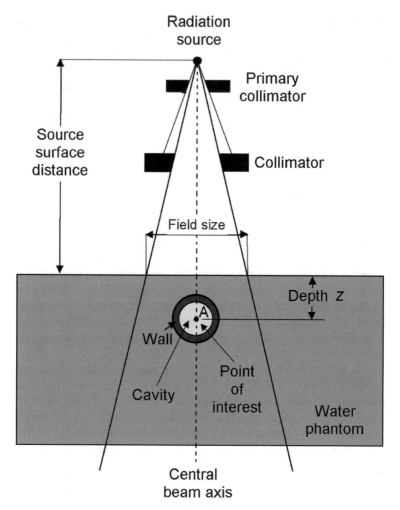

Fig. 15.3 Measurement of dose to point A in a tissue-equivalent phantom with a radiation dosimeter consisting of: (i) cavity filled with radiation sensitive medium and (ii) wall to contain the radiation sensitive medium. The reference point of the dosimeter, typically but not necessarily at the center of the cavity volume, is placed into point A in the phantom

radiation beam is produced by an external radiation source that is characterized by beam type and energy as well as by the source-surface distance (SSD) and field size A at the surface of the water phantom.

The absorbed dose measurement involves many experimental and calculation steps as well as some approximations and corrections. Assuming that an appropriate phantom was chosen and experimental setup was completed, as depicted in Fig. 15.3, the absorbed dose (or absorbed dose rate) determination proceeds as follows:

1. Dosimeter cavity produces a radiation-induced signal that is converted to dose-to-cavity D_{cav} (or dose rate-to-cavity $\dot{D}_{cav}$) using corrections for influence quantities and parameters appropriate for the dosimetry technique used in D_{cav} measurement. Influence quantities are quantities that affect the dosimeter response and may arise from environmental causes, dosimeter itself, or radiation field. Some influence quantities can be controlled during the measurement, the effect of other influence quantities on measurement result must be accounted for with appropriate correction factors.
2. Dose-to-cavity D_{cav} is converted into dose-to-medium D_{med} where the subscript "med" refers to the tissue-equivalent phantom into which the dosimeter is introduced and in which the cavity radiation-induced signal is measured. D_{med} is determined from D_{cav} using a cavity theory appropriate for the conditions in which D_{cav} was measured (see Sect. 15.7).

15.7 Cavity Theories for Radiation Dosimetry

To determine absorbed dose D_{med} in a medium such as a tissue-equivalent (water) phantom, a radiation dosimeter (cavity) is introduced into the phantom and absorbed dose D_{cav} to the cavity is measured. Generally, the radiation sensitive material of the cavity differs from the medium in which dose D_{med} is to be determined. Therefore, the dosimeter placed into the phantom may perturb the photon fluence as well as the charged particle fluence and, consequently, the measured D_{cav} may not reflect the fluence conditions that prevail in the phantom in the absence of the dosimeter. This makes the relationship between D_{cav} and D_{med} very complex and involves several computational steps, parameters, corrections, and approximations.

Various theories deal with the D_{cav} to D_{med} transition. They are referred to as cavity theories of radiation dosimetry and are governed by the cavity size they address. Somewhat vaguely, the cavities used in radiation dosimetry are categorized into small (Bragg–Gray) cavities, intermediate (Burlin) cavities, and large size cavities depending on the relative magnitude of their size d compared to the range R of secondary charged particles released through photon interactions in the phantom and traversing the cavity (dosimeter).

Figure 15.4 shows a schematic representation of the three cavity sizes: small cavity in (b), intermediate cavity in (c), and large cavity in (d). Part (a) shows the phantom with an incident photon undergoing an interaction at point P releasing an energetic secondary electron whose range R in the phantom is indicated by an arrow. Point A represents the point-of-interest at which the absorbed dose is to be determined with a radiation dosimeter. The reference point of the dosimeter is positioned in such a way that it coincides with point A in the phantom.

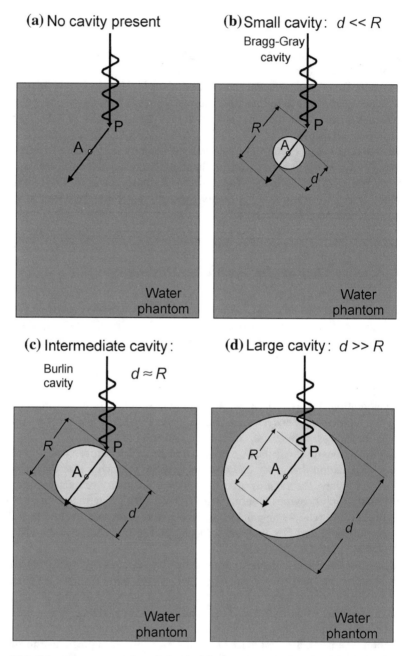

Fig. 15.4 Schematic representation of absorbed dose measurement at point A in water phantom. Part **a** shows the phantom with an incident photon undergoing an interaction at point P releasing an energetic secondary electron whose range R in the phantom is indicated by a straight arrow through point A. Parts **b**, **c**, and **d** show the measurement of dose at point A with dosimeters of various cavity sizes: Bragg–Gray (small cavity) in (**b**); Burlin (intermediate) cavity in (**c**); and large cavity in (**d**)

15.7.1 Small Cavity and Bragg–Gray Cavity Theory for Photon Beams

For $d \ll R$, the cavity size is classified as small (Fig. 15.4b) and called a Bragg–Gray cavity in honor of William H. Bragg and Louis H. Gray who developed the first cavity theory, now referred to as the Bragg–Gray cavity theory. The theory was subsequently refined and expanded into the so-called Spencer-Attix cavity theory, but its basic tenets enunciated a century ago in the format of two specific conditions are still valid today. The two Bragg–Gray conditions for photon beams of energy $hv > 200$ keV are as follows:

1. The radiation cavity is small to give a reasonable spatial resolution of the measurement as well as to avoid perturbing the secondary charged particle fluence produced by photon interactions in the phantom. This means that $\varphi_{\text{cav}} \approx \varphi_{\text{med}}$ where φ_{cav} and φ_{med} are secondary charged particle fluences in the cavity and in phantom medium surrounding the cavity, respectively.
2. Dose D_{cav} absorbed in the Bragg–Gray cavity is deposited solely by secondary charged particles traversing it; photon interactions in the cavity are rare and therefore ignored.

The following two relationships hold for D_{cav} in the cavity and D_{med} at the same point A in the phantom in the absence of the cavity, respectively

$$D_{\text{cav}} = \varphi_{\text{cav}}(\bar{S}_{\text{col}}/\rho)_{\text{cav}} \quad \text{and} \quad D_{\text{med}} = \varphi_{\text{med}}(\bar{S}_{\text{col}}/\rho)_{\text{med}}, \tag{15.18}$$

where $(\bar{S}_{\text{col}}/\rho)_{\text{cav}}$ and $(\bar{S}_{\text{col}}/\rho)_{\text{med}}$ are the spectrum averaged mean mass collision stopping powers for cavity (dosimeter) and surrounding medium (phantom), respectively (see Sect. 6.5).

From the first Bragg–Gray condition ($\varphi_{\text{cav}} \approx \varphi_{\text{med}}$) and (15.18) we now get the following Bragg–Gray relationship between the cavity dose D_{cav} and phantom dose D_{med}

$$\frac{D_{\text{cav}}}{D_{\text{med}}} = \frac{\varphi_{\text{cav}}(\bar{S}_{\text{col}}/\rho)_{\text{cav}}}{\varphi_{\text{med}}(\bar{S}_{\text{col}}/\rho)_{\text{med}}} \approx \frac{(\bar{S}_{\text{col}}/\rho)_{\text{cav}}}{(\bar{S}_{\text{col}}/\rho)_{\text{med}}} \tag{15.19}$$

or

$$D_{\text{med}} \approx D_{\text{cav}} \frac{(\bar{S}_{\text{col}}/\rho)_{\text{med}}}{(\bar{S}_{\text{col}}/\rho)_{\text{cav}}} \tag{15.20}$$

for a small Bragg–Gray cavity embedded in a tissue-equivalent phantom irradiated by an ionizing photon beam. *Note*: The Spencer-Attix relationship that currently constitutes the recommended approach to radiation dosimetry is based on the Bragg–Gray theory except that, instead of unrestricted mean mass collision stopping powers $\bar{S}_{\text{col}/\rho}$ of Bragg–Gray equations, it uses mean restricted mass collision stopping powers $\bar{L}/\rho$ and kinetic energy threshold Δ of 10 keV (see Sect. 6.10).

15.7.2 Large Cavity in Photon Beam

For $d \gg R$, the cavity is classified as a large cavity (Fig. 15.4d) and the dose D_{cav} absorbed in the cavity is attributed to secondary electrons released mainly by photon interactions inside the cavity. The contribution to D_{cav} from electrons that originated in the phantom medium outside the cavity is negligible and therefore ignored.

Absorbed dose-to-cavity D_{cav} and absorbed dose-to-phantom D_{med}, respectively, can be expressed as

$$D_{\mathrm{cav}} = \psi_{\mathrm{cav}}(\bar{\mu}_{\mathrm{ab}}/\rho)_{\mathrm{cav}} \quad \text{and} \quad D_{\mathrm{med}} = \psi_{\mathrm{med}}(\bar{\mu}_{\mathrm{ab}}/\rho)_{\mathrm{med}}, \qquad (15.21)$$

where ψ_{cav} and ψ_{med} are photon energy fluences (in MeV·cm^{-2}) in the cavity and phantom in the absence of the cavity, respectively, while $(\bar{\mu}_{\mathrm{ab}}/\rho)_{\mathrm{cav}}$ and $(\bar{\mu}_{\mathrm{ab}}/\rho)_{\mathrm{med}}$ are mass energy absorption coefficients (in cm^2·g^{-1}) averaged over the photon fluence spectra for the cavity and phantom medium, respectively (see Sect. 8.3).

Assuming that the cavity density ρ_{cav} is similar to the density ρ_{med} of the tissue-equivalent phantom ($\rho_{\mathrm{cav}} \approx \rho_{\mathrm{med}}$), we stipulate that the photon energy fluences ψ_{cav} for the cavity and ψ_{med} for the phantom are similar (i.e., $\psi_{\mathrm{cav}} \approx \psi_{\mathrm{med}}$) to get the following relationship between D_{cav} and D_{med} from (15.21)

$$\frac{D_{\mathrm{med}}}{D_{\mathrm{cav}}} = \frac{\psi_{\mathrm{med}} \cdot (\bar{\mu}_{\mathrm{ab}}/\rho)_{\mathrm{med}}}{\psi_{\mathrm{cav}} \cdot (\bar{\mu}_{\mathrm{ab}}/\rho)_{\mathrm{cav}}} = \frac{(\bar{\mu}_{\mathrm{ab}}/\rho)_{\mathrm{med}}}{(\bar{\mu}_{\mathrm{ab}}/\rho)_{\mathrm{cav}}} \qquad (15.22)$$

or

$$D_{\mathrm{med}} \approx D_{\mathrm{cav}} \frac{(\bar{\mu}_{\mathrm{ab}}/\rho)_{\mathrm{med}}}{(\bar{\mu}_{\mathrm{ab}}/\rho)_{\mathrm{cav}}} \qquad (15.23)$$

for a large cavity embedded in a tissue-equivalent phantom irradiated by ionizing photon beam.

15.7.3 Intermediate Cavity and Burlin Cavity Theory

Radiation dose measurements based on cavity theory are most often carried out with the use of a Bragg–Gray cavity embedded in a tissue-equivalent (water) phantom. The main stipulation of the Bragg–Gray cavity theory is that the cavity dimensions d are small compared to the range R of secondary charged particles traversing the cavity. This means that: (i) cavity is small to prevent perturbation of the secondary charged particle fluence in the phantom and (ii) dose absorbed in the cavity is produced solely by charged particles released by photon interactions in the phantom; photon interactions in the small cavity are negligible.

Meeting the Bragg–Gray conditions in practice is not always possible, such as, for example, for $d \gg R$ in the case of large cavities and for $d \approx R$ where the cavity dimensions d are comparable to the range R of secondary charged particles in the cavity. The $d \approx R$ case deals with the so-called intermediate cavity (Fig. 15.4c) that is also known as Burlin cavity in honor of T. E. Burlin who investigated this problem both experimentally and theoretically in the 1960s.

Burlin proposed a semi-empirical equation for D_{cav}/D_{med}, the ratio of cavity dose D_{cav} versus phantom dose D_{med} in the absence of the cavity in the phantom. The Burlin equation is

$$\frac{D_{cav}}{D_{med}} = \Delta \frac{(\bar{S}_{col}/\rho)_{cav}}{(\bar{S}_{col}/\rho)_{med}} + (1 - \Delta)\frac{(\bar{\mu}_{ab}/\rho)_{cav}}{(\bar{\mu}_{ab}/\rho)_{med}}, \tag{15.24}$$

where Δ is a parameter related to the size d of the cavity and

$(\bar{S}_{col}/\rho)_{cav}$ and $(\bar{S}_{col}/\rho)_{med}$ are the mean mass collision stopping powers of the cavity medium and phantom medium, respectively.

$(\bar{\mu}_{ab}/\rho)_{cav}$ and $(\bar{\mu}_{ab}/\rho)_{med}$ are the mean mass energy absorption coefficients of the cavity medium and phantom medium, respectively.

Burlin equation in principle covers the whole range of dosimetric cavities from small (Bragg–Gray) cavities where $d \ll R$ through intermediate (Burlin) cavities where $d \approx R$ to large size cavities where $d \gg R$. Parameter Δ in Burlin equation is related to the cavity size, approaching $\Delta \to 0$ for large cavities in agreement with (15.23) and $\Delta \to 1$ for small (Bragg–Gray) cavities in agreement with (15.20).

15.8 Media Used for Cavity, Wall, and Phantom

The dosimeter that is placed into a phantom to determine D_{med} in phantom consists of two major components: radiation sensitive cavity and wall that contains the cavity. As shown in Fig. 15.3, there are three components of major importance in dose-to-medium (phantom) D_{med} measurement with a dosimeter at point A in a phantom:

(i) Radiation sensitive cavity C filled with solid, liquid, or gas radiation sensitive medium;
(ii) Wall W of the cavity for liquid and gaseous cavities (*Note*: solid cavities do not require a wall for containment); and
(iii) Phantom P into which the radiation dosimeter is embedded.

In general, each one of these components is made of its own suitable material that, in its own way, affects the fluence of photons as well as the fluence of secondary charged particles released by photon interactions in the phantom. The dosimeter placed into a phantom to determine D_{med} in phantom, consists of a cavity filled with a radiation sensitive material and a wall.

The goal of D_{med} measurement is to determine D_{med} at point A in phantom in the absence of the dosimeter; however, the dosimeter measures the mean dose-to-cavity $\bar{D}_{cav}$ at point A from which D_{med} is inferred by applying an appropriate cavity theory, as discussed in Sect. 15.7.

To simplify the transition from $\bar{D}_{cav}$ to D_{med} one must coordinate the choice of material for each of the three components: phantom, wall, and cavity, accounting for relevant constraints. For example, in most clinical dosimetry situations the phantom must be tissue-equivalent and the wall for containment of liquid or gaseous cavities must be rigid. For simplicity, it is desirable for the wall and cavity to be tissue-equivalent and of same material, i.e., cavity and wall are homogeneous and matched. Obviously, the constraints are stringent, difficult to satisfy, and somewhat contradictory. Therefore, in practice compromises must be made and accounted for in the determination of D_{med} from $\bar{D}_{cav}$.

Some of the parameters of importance for matching cavity and wall to phantom medium are: atomic composition, effective atomic number, density (solid, liquid, or gas), mass collision stopping power, and mass energy absorption coefficient. The four most obvious media matches are: **P = W = C**. In an ideal situation, the phantom P, wall W, and cavity C are made of same material. The phantom-dosimeter combination is said to be homogeneous and its three components are perfectly matched. In practice, finding perfect match and tissue equivalency is difficult, essentially impossible, considering that one must match with the water phantom the dosimeter with its radiation sensitive cavity and the wall.

P ≈ W ≈ C. In this more realistic approach, the components are not matched perfectly; however, an attempt is made to match most of the pertinent parameters as closely as possible, so that field perturbation that the dosimeter introduces into the phantom are small and can be ignored.

P ≠ W ≈ C. The wall and cavity of the dosimeter are almost perfectly matched (i.e., they are homogeneous), but the dosimeter is not matched to the phantom. Thus, $D_{wall} \approx \bar{D}_{cav}$, and a correction must be applied to D_{wall} in the transition from dose-to-wall D_{wall} to dose-to-medium (phantom) D_{med}.

W ≠ C. In a general case where the wall and cavity media are not matched (i.e., the dosimeter is not homogeneous), one can invoke the Burlin relationship for intermediate cavities see (15.24) and express the relationship between the mean dose-to-cavity $\bar{D}_{cav}$ and dose-to-wall D_{wall} as

$$\frac{\bar{D}_{cav}}{D_{wall}} = \Delta \frac{(\bar{S}_{col}/\rho)_{cav}}{(\bar{S}_{col}/\rho)_{wall}} + (1 - \Delta) \frac{(\bar{\mu}_{ab}/\rho)_{cav}}{(\bar{\mu}_{ab}/\rho)_{wall}}, \qquad (15.25)$$

where Δ is a parameter related to cavity size and

$(\bar{S}_{col}/\rho)_{cav}$ and $(\bar{S}_{col}/\rho)_{wall}$ are the mean mass collision stopping powers of the cavity medium and wall medium, respectively.

$(\bar{\mu}_{ab}/\rho)_{cav}$ and $(\bar{\mu}_{ab}/\rho)_{wall}$ are the mean mass energy absorption coefficients of the cavity medium and wall medium, respectively.

Note that irrespective of the value of parameter Δ we get a perfect match between the cavity medium and wall medium when both the mass collision stopping powers $(\bar{S}_{col}/\rho)$ and mass energy absorption coefficients $\bar{\mu}_{ab}/\rho$ for cavity and wall are matched, i.e., $(\bar{S}_{col}/\rho)_{cav} = (\bar{S}_{col}/\rho)_{wall}$ and $(\bar{\mu}_{ab}/\rho)_{cav} = (\bar{\mu}_{ab}/\rho)_{wall}$. If the match between the cavity and wall is bad, then a reasonable approximation for parameter Δ must be estimated and (15.25) used to obtain D_{wall} from the measured $\bar{D}_{cav}$ and subsequently D_{med} from D_{wall}. This can become very difficult to manage in practice; finding reasonable matches from an assortment of suitable materials seems to be a more promising approach.

15.9 Basic Calibration of Radiotherapy Machine Output

Output of ionizing radiation beams produced by external beam radiotherapy treatment machines must be determined accurately before the machine is used clinically and, moreover, it must also be verified on a regular basis during clinical use to ensure accurate delivery of the prescribed dose to the patient. Radiotherapy beams that need accurate calibration in modern radiotherapy include photon beams of standard field sizes in the range from 4×4 cm^2 to 40×40 cm^2, electron beams of standard field sizes of 4×4 cm^2 to 30×30 cm^2, photon and electron beams of small field sizes (less than 4×4 cm^2), as well as heavy charged particle beams such as proton beams and heavier particle beams. Brachytherapy sealed radionuclide sources such as iridium-192 and cesium-137 require their own calibration before clinical use. Each beam type has its own special demands and presents its own special constraints, making accurate absorbed dose determination a very complex undertaking.

The basic output calibration of a radiotherapy beam (to $\pm2\%$ or better) is but one, albeit the essential one, of the links constituting the chain representing an accurate prescribed dose delivery to the patient during the radiotherapy treatment. The other links deal with:

(i) Procedures for measurement and use of relative dose data, such as percentage depth dose distribution, surface dose, exit dose, depth of dose maximum, etc. (see Sect. 1.12);
(ii) Equipment commissioning and quality assurance;
(iii) Treatment planning consisting of calculation of dose distribution in tissues of the target volume and tissues surrounding the target volume;
(iv) Actual set up of the patient on the treatment machine.

The basic machine output calibration can in principle be carried out with one of the three known absolute radiation dosimetry systems: (i) calorimetric, (ii) Fricke chemical, and (iii) ionometric. These systems are discussed in detail in Chap. 16; however, they are quite complicated and cumbersome to use, so that, in practice, they are relegated to use in radiation standards laboratories that define primary radiation standards.

Rather than with one of the three absolute radiation dosimetry systems, clinical machines are calibrated with a suitable relative dosimetry system that traces its calibration coefficient to an accredited National or International Standards Laboratory. This hybrid approach to clinical machine output calibration can be regarded as the second tier absolute radiation dosimetry, since it provides absolute machine output combining an absolute radiation dosimetry technique with a suitable relative dosimetry system. The relative dosimetry system so used is most commonly based on an ionization chamber with a calibration coefficient traceable to a national standards laboratory. One should note that when an absolute dosimetry system is used for radiation beam calibration in a standards laboratory, it relies on its own accuracy, in contrast to a hybrid field-calibration with a relative dosimetry system that traces its calibration coefficient to a standard that is common to many other users and originated in a standards laboratory.

Not all National and International Standards Laboratories deal with radiation standards; however, the ones that do so, typically provide absorbed dose calibration coefficients on two distinct levels: clinical level and high dose level.

1. Clinical level. For cobalt-60 gamma beams in the absorbed dose range from $\sim$1 Gy to $\sim$10 Gy that are used in radiotherapy where the cobalt-60 calibration coefficient of a dosimeter, traceable to a standards laboratory, is used in dosimetry protocols (see Sect. 15.10) for calibration of megavoltage x-ray beams (4 MeV to 25 MV) and megavoltage electron beams (4 MeV to 30 MeV).
2. High dose level. For cobalt-60 gamma beams as well as megavoltage photon and electron beams in the dose range from $\sim$0.5 kGy to $\sim$1 MGy used in radiation processing and food irradiation.

15.10 Dosimetry Protocols

The procedures to be followed when carrying out a basic output calibration of a clinical photon or electron beam produced by a radiotherapy machine are prescribed in international, national, or regional radiation dosimetry protocols or dosimetry codes of practice. The choice of the dosimetry protocol to be used is largely left to individual radiotherapy departments or is prescribed by a national governmental agency. Output of radiotherapy equipment is usually quoted as absorbed dose or dose rate under specific reference conditions. Water was chosen as the standard reference medium, because of its absorption and scattering characteristics that are similar to those of tissue.

The best known international radiation dosimetry protocol is the IAEA TRS–398 protocol issued by the Vienna-based International Atomic Energy Agency (IAEA) in 2000 entitled: *"Absorbed Dose Determination in External Beam Radiotherapy"*;

however, many other dosimetry protocols are also available from national or regional organizations, most notably from the American Association of Physicists in Medicine (AAPM) for North America; Institution of Physics and Engineering in Medicine and Biology (IPEMB) for the UK; Deutsches Institut für Normung (DIN) for Germany; Nederlandse Commissie voor Stralingsdosimetrie (NCS) for Holland and Belgium; and Nordic Association of Clinical Physics (NACP) for Scandinavia.

Radiation Dosimetry in Radiation Medicine

Radiation dosimetry, defined as measurement of ionizing radiation dose or some other dose-related physical quantity, is a branch of medical physics that plays a very important role in the use of ionizing radiation in diagnosis and treatment of disease. In addition, radiation dosimetry forms the foundation of radiation protection services that deal with:

1. Recommendations on, and enforcement of, safe practices in the use of ionizing radiation in medicine, science, industry, and power generation.
2. Estimation of absorbed dose to general public resulting from peaceful use of ionizing radiation in medicine, science, industry and power generation.
3. Estimation of absorbed dose to individuals involved in radiation accidents.

Diagnosis of disease (imaging) with ionizing radiation is carried out in two specialties of radiation medicine: diagnostic radiology using x-rays and nuclear medicine (also called molecular imaging) with unsealed radioactive sources. Both of these specialties strive to obtain the optimal image quality with a minimum dose to the patient and staff.

Treatment of disease with ionizing radiation usually implies treatment of cancer with ionizing radiation; a specialty of radiation medicine, commonly called radiotherapy or radiation oncology. It is carried out with a variety of external or internal radiation sources producing x-rays, gamma rays, electrons, protons, or heavier particles. The goal of radiotherapy is to deliver the prescribed tumoricidal (oncolytic) dose to the tumor with a high degree of spatial and numerical accuracy and minimal total body patient dose as well as minimal dose to staff providing the radiotherapy treatment to the patient.

Radiotherapy involves directly a team of four professionals sharing responsibilities as follows:

1. Radiation oncologist diagnoses the patient's malignant disease, decides on course of treatment, and prescribes the type of radiation, dose fractionation, and treatment dose.
2. Medical physicist: (i) organizes equipment maintenance and servicing, (ii) calibrates the radiation beams produced by radiotherapy equipment, such as x-ray machines, cobalt units, and linear accelerators, and (iii) deals with governmental regulatory agencies and ensures that hospitals and clinics meet regulatory requirements to make the use of ionizing radiation in diagnosis and treatment of disease safe for patients and staff.
3. Medical dosimetrist, in cooperation with medical physicist and radiation oncologist, produces a detailed computerized treatment plan and determines the optimal dose distribution for patient's treatment.
4. Radiation therapist (radiotherapy technologist) delivers the prescribed radiation dose to the patient on the prescribed radiotherapy machine according to the treatment plan and a prescribed fractionation of typically 20 to 30 daily fractions.

As evident from Chaps. 16 and 17 that deal with absolute and relative radiation dosimetry, respectively, radiation dosimetry is an important, well-developed and sophisticated field of physics, mainly applied to medicine. Professionals who work in this field must carry out their work with utmost accuracy and care to ensure that the patient treatment is safe for the patient as well as staff and that the desired diagnostic or treatment outcome is attained.

Chapter 16
Absolute Radiation Dosimetry

This chapter deals with one of two major categories of radiation dosimetry referred to as absolute radiation dosimetry. The other major category is the relative radiation dosimetry, discussed in Chap. 17.

Absolute radiation dosimetry is defined as direct measurement of radiation dose in the dosimeter material (cavity) without any need for calibration of the dosimeter response in a known radiation field. Three types of absolute dosimetry are known and described in this chapter: calorimetric absolute radiation dosimetry, discussed in Sect. 16.1; chemical (Fricke) absolute radiation dosimetry, discussed in Sect. 16.2; and ionometric absolute radiation dosimetry, discussed in Sect. 16.3.

16.1 Calorimetric Absolute Radiation Dosimetry

16.1.1 Introduction to Calorimetry

In principle, the most fundamental measurement of ionizing radiation dose absorbed in a given medium is based on calorimetry, a technique that measures the thermal (heat) energy transferred from an ionizing radiation beam to absorbing medium as a result of the beam interaction with the absorbing medium. In practice, however, calorimetric radiation dosimetry does not enjoy the same status and widespread use as do ionometric, i.e., ionization chamber-based, dosimetry techniques because it has a relatively low sensitivity and is not practical for routine clinical use.

The apparatus used in calorimetry is referred to as a calorimeter and its operation is based on the following four physical concepts:

1. Heat is a form of energy akin to mechanical and electrical energy. The so-called "mechanical equivalent of heat", introduced by physicists Julius von Mayer and James P. Joule during 1840s, was a precursor to the principle of conservation of energy as well as to the establishment of thermodynamics as a branch of physics.

© Springer International Publishing Switzerland 2016
E.B. Podgoršak, *Radiation Physics for Medical Physicists*,
Graduate Texts in Physics, DOI 10.1007/978-3-319-25382-4_16

2. In energy transfer from one body to another as a result of temperature difference between the two bodies, the amount of energy transfer is referred to as thermal energy or heat.

3. In thermal energy transfer from one body to another, the hotter body loses thermal energy and the cooler body gains the same amount of thermal energy. The thermal energy transfer continues until both bodies reach the same temperature that is called the equilibrium temperature.

4. Thermal energy can be transferred from one location to another by three means:

   (i) *Conduction*, with thermal energy travelling through the heated object (applicable to solids).

   (ii) *Convection*, with heat transported by the movement of a heated substance (applicable to liquids and gases).

   (iii) *Radiation*, with heat transported through space in the form of electromagnetic (EM) radiation and governed by Stefan–Boltzmann law.

16.1.2   Basic Aspects of Absorbed Dose Calorimetry

Thermal energy transfer ΔQ in the calorimeter is given by the following expression

$$\Delta Q = mC\Delta T, \tag{16.1}$$

where

m is the mass of the reference medium of interest.

ΔT is the change in temperature of the reference medium of interest, i.e., $\Delta T = T_{final} - T_{initial}$.

C is the specific heat capacity (also called specific heat) of the reference medium of interest with units $J \cdot kg^{-1} \cdot K^{-1}$ and in general dependent on temperature, pressure and volume.

A portion of energy carried by an ionizing radiation beam traversing an absorbing medium is deposited in the absorbing medium and manifests itself as a temperature rise of the absorbing medium. Equation (16.1) suggests that calorimetric techniques that measure energy directly can be used for measuring ionizing radiation related quantities, such as dose absorbed in reference medium. Indeed, absorbed dose calorimeters have been developed; however, they are impractical for routine clinical use because of the extremely small temperature rises that occur when clinical radiation doses of the order of 1 Gy are used to irradiate the calorimeter.

For example, from (16.1) we get the following temperature rise ΔT for a dose of 1 Gy absorbed in calorimeter core made of water

$$\frac{\Delta Q}{m} = \frac{E}{m} = \bar{D} = C_{\text{water}} \Delta T \tag{16.2}$$

or

$$\Delta T = \frac{\bar{D}}{C_{\text{water}}} = \frac{1 \text{ J·kg}^{-1}}{4186 \text{ J·kg}^{-1}\text{·K}^{-1}} = 2.39 \times 10^{-4} \text{ K} = 0.239 \text{ mK} \tag{16.3}$$

where we assumed that all of the energy absorbed in the sensitive water core was converted into heat and we used the definition of dose as energy absorbed per unit mass ($\bar{D} = E/m$) with $\bar{D}$ the mean absorbed dose in the core (cavity) medium. Specific heat of water is given by the NIST as $C_{\text{water}} = 4186 \text{ J·kg}^{-1}\text{·K}^{-1}$, when measured at 15 °C and a standard air pressure of 101.3 kPa.

It is obvious from (16.3) that the temperature rise ΔT of 0.239 mK for a dose of 1 Gy in water is extremely small and therefore difficult to measure. This clearly reveals the biggest disadvantage of calorimetric radiation dosimetry, namely, the limitation on the sensitivity of the technique because of the small temperature differences ΔT that must be measured when using radiation doses of the order of doses prescribed in radiotherapy (typically of the order of 2 Gy).

Compared to most media that are or could be used as reference material in calorimeter core, water has a relatively high specific heat capacity C. Therefore, substituting water in the calorimeter core with a medium of lower specific heat would increase the temperature rise per unit dose, since, as shown in (16.3), ΔT is inversely proportional to the specific heat C. However, in terms of energy deposition in tissue, water is an excellent and practical tissue equivalent material and, as such, it is often used in absorbed dose calorimeters.

For example, specific heat of graphite, that serves as reference medium in some absorbed dose calorimeters, is $C_{\text{graphite}} = 710 \text{ J·kg}^{-1}\text{·K}^{-1}$, almost 6 times smaller than that of water, producing an almost 6 times larger temperature rise ΔT per unit dose (0.239 mK/Gy for water versus 1.41 mK/Gy for graphite). Unfortunately, graphite despite having an atomic number close to water, does not match the tissue equivalence as well as water does, and mean cavity dose $\bar{D}_{\text{cav}}$ determined in a graphite calorimeter must be converted into dose-to-water D_{med} with the help of various scaling factors that lead to possible errors and uncertainty in the final result.

In contrast to the minute temperature rise for clinical dose in water of the order of gray, for typical industrial and food radiation processing doses of 20 kGy delivered to water, the temperature rise amounts to about 5 K, as shown in (16.4)

$$\Delta T = (0.239 \times 10^{-3} \text{ K·Gy}^{-1}) \times (20 \times 10^3 \text{ Gy}) \approx 5 \text{ K}, \tag{16.4}$$

representing a much easier to measure temperature difference ΔT.

16.1.3   Properties of Thermistors

To be useful as absolute dose calorimeter the device must measure absorbed dose
to a statistical uncertainty of 0.5% or better. Thus, in absolute dose calorimetry, a
temperature increase in the calorimeter core of ~ 7 μK for graphite and ~ 1.2 μK
for water should be detectable and this is by no means a simple proposition.
The associated thermometry technique must incorporate a thermal detector that is:
(i) very sensitive in order to detect the small temperature rise that results in energy
transfer from the radiation beam to the calorimeter core and (ii) of small dimensions
so as to interfere minimally with energy transfer from the radiation beam to the
calorimeter core (dosimeter cavity).

Of the two most suitable thermometry techniques available for calorimetric
dosimetry, one is based on thermocouples and the other on thermistors. Thermistors are about an order of magnitude more sensitive than thermocouples and are
therefore universally used in absorbed dose calorimetry.

Thermistors (thermal resistors) are temperature sensitive semiconductors fabri-
cated from oxides of metals such as iron, cobalt, nickel, manganese, and copper.
Depending on their composition, thermistors exhibit either increasing or decreasing
electrical resistance with increasing temperature. The latter type is more
common and referred to as negative temperature coefficient (NTC) thermistor.
Resistance for typical thermistors changes by a factor of ~ 300 in the temperature
range from -50 °C to $+100$ °C, amounting to a change of $\sim 3\%$ to $\sim 6\%$
per °C. Typical thermistor probe sizes range from a fraction of a millimeter to about
2 cm in diameter and come in the form of beads, rods, or chips encapsulated in epoxy
or glass.

16.1.4   Measurement of Thermistor Resistance

Resistance R of a thermistor is measured with a standard Wheatstone bridge circuit,
shown schematically in Fig. 16.1 with the following main components: a DC power
supply U, two resistors of known resistance (R_1 and R_2), one precision variable
resistor (potentiometer) R_p, thermistor with resistance R_T, and a null detector. All
resistors are chosen to be similar in resistance and sufficiently large so that the power
dissipated in the thermistor R_T is negligible.

The null detector is a device that is connected to the bridge output at points P_1 and
P_2 in the bridge circuit of Fig. 16.1. It indicates zero reading when the two points P_1
and P_2 are at identical potentials irrespective of the input voltage U that the power
supply provides to the circuit.

The procedure for determination of thermistor resistance R_T at temperature T
requires that the bridge be balanced and this is achieved by varying the potentiometer
R_p until the null detector reads zero, indicating that points P_1 and P_2 are at the same
potential, i.e., there is no current flow through the null detector between the two

Fig. 16.1 Schematic diagram of a Wheatstone bridge with the following main components: DC power supply, two resistors (R_1 and R_2), precision variable resistor (potentiometer) R_P, thermistor with resistance R_T, and null detector

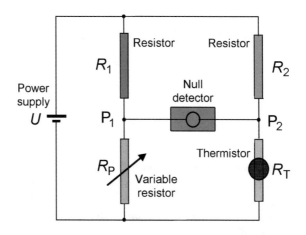

points. Thus, a null detector can be a low impedance galvanometer reading zero current or a high impedance voltmeter reading zero voltage.

The Wheatstone equation for a balanced bridge is expressed as follows

$$\frac{R_1}{R_P} = \frac{R_2}{R_T} \quad \text{or} \quad R_T = R_P \frac{R_2}{R_1}, \tag{16.5}$$

providing simple means to determine the unknown resistance R_T of the thermistor at temperature T from known resistances of resistors R_1 and R_2 as well as potentiometer R_p.

16.1.5 Resistance versus Temperature Relationship for Thermistor

The relationship between resistance R and temperature T of a thermistor is highly non-linear and this stimulated several attempts at developing empirical expressions for describing the $R(T)$ curve. The best-known approximation is the polynomial logarithmic equation that was proposed by John S. Steinhart and Stanley R. Hart in 1968 and is expressed as follows

$$\frac{1}{T} = a_0 + a_1 \ln R + a_2 \ln R^2 + a_3 \ln R^3, \tag{16.6}$$

where T is the thermistor temperature in degrees kelvin, R is the thermistor resistance in ohms at temperature T, and $a_0, a_1, a_2 = 0$ and a_3 are the Steinhart–Hart curve fitting parameters.

Inserting two new parameters (β and T_0) into (16.6), the Steinhart–Hart equation can be simplified to get the so-called beta formula for $R(T)$

$$\frac{1}{T} = \frac{1}{T_0} + \frac{1}{\beta}\ln\frac{R}{R_0} \quad \text{or} \quad R = R_0 e^{-\beta\left(\frac{1}{T_0}-\frac{1}{T}\right)}, \tag{16.7}$$

where

$$a_0 = \frac{1}{T_0} - \frac{1}{\beta}\ln R_0, \qquad a_1 = \frac{1}{\beta}, \qquad \text{and} \qquad a_2 = a_3 = 0, \tag{16.8}$$

with T_0 a reference temperature (usually 25 °C) and R_0 the resistance R at T_0.

A special parameter called α is used to describe the slope dR/dT of the $R(T)$ non-linear curve at any given point on the $R(T)$ curve. The slope gives the rate of change in resistance R of the thermistor at a given temperature T and is from (16.7) expressed as

$$\left.\frac{dR}{dT}\right|_T = -\beta\frac{R_0}{T^2}e^{-\beta\left(\frac{1}{T_0}-\frac{1}{T}\right)} = -\frac{\beta}{T^2}R, \tag{16.9}$$

where R_0 and R are thermistor resistances at reference temperature T_0 and arbitrary temperature T, respectively, and β is determined from (16.9) as

$$\beta = \frac{\ln\frac{R}{R_0}}{\frac{1}{T} - \frac{1}{T_0}}. \tag{16.10}$$

Equation (16.10) can be linearized to read

$$y = \beta x + \kappa, \tag{16.11}$$

with $y = \ln R$; $x = 1/T$; and $\kappa = \ln R_0 - \beta/T_0$.

The Steinhart–Hart approximation (16.6) provides satisfactory agreement with the non-linear R versus T thermistor curve provided that parameters $a_0, a_1, a_2 = 0$, and a_3 of the approximation are determined through a careful measurement of the thermistor $R(T)$ curve in the temperature range of interest. Thus the thermistor resistance must be calibrated either in-house or in a standards laboratory, such as the NIST.

The calibration of thermistors is carried out with resistance temperature detectors (RTDs) that consist of fine wire made of pure metal such as platinum, nickel, or copper wrapped around a ceramic or glass core. In comparison with thermistors, RTDs are less sensitive to small temperature changes, have a slower response time, and are larger in size. However, they are also more stable, usable over a wider temperature range, and have an essentially linear response with change in temperature. The higher sensitivity and smaller size ensure that thermistors are the temperature sensor of choice in absorbed dose calorimetry; the wider temperature range and linear temperature response make RTDs the sensor of choice in calibration laboratories offering temperature standards.

16.1.6 Practical Aspects of Calorimetric Radiation Dosimetry

In its simplest form an absorbed dose calorimeter is a vessel containing a small spherical core (cavity) of the reference medium in which dose is to be determined, surrounded by, but thermally insulated from, a shell (also called mantle) that is typically of the same material as the core and thermally insulated from the surrounding air by a special shield. The core also contains a temperature sensor, most commonly a thermistor. The cavity medium is usually a tissue equivalent solid such as graphite or tissue equivalent liquid such as water.

The role of the core is to define the small sensitive volume in which the dose is determined and the mantle surrounding the core ensures that conditions of charged particle equilibrium as well as conditions of thermal equilibrium are met. The theoretical equation (16.2) is in practice modified slightly to account for thermal leakage from the core as well as for endothermic and exothermic chemical reactions that accompany the energy deposition in the core

$$\bar{D} = \frac{C \Delta T}{1 - \kappa},$$ (16.12)

where

$\bar{D}$ is the mean dose deposited in the calorimeter core (cavity).
C is the specific heat of the core material.
ΔT is the measured rise in temperature in the core.
κ is the so-called heat defect accounting for the fraction of energy E that is absorbed in the calorimeter core but does not contribute to ΔT, the measured rise in temperature. It can be expressed as $\kappa = 1 - E_H/E_A$, where E_A is energy absorbed by calorimeter core and E_H is energy that appears as heat.

Standards laboratories use either graphite core calorimeters or water core calorimeters for primary dosimetry standards in cobalt-60 radiotherapy beams. Graphite calorimeters have been in use much longer than water calorimeters. Each of the two core materials has some advantages and some disadvantages, and it is impossible to state which of the two is clearly better suited for absorbed dose calorimetry. It is clear, however, that water is a better tissue equivalent material than graphite.

Graphite calorimeters measure dose-to-graphite while water calorimeters measure dose-to-water. Since the end result for both calorimeters must be dose-to-water, one could conclude that the necessity of a complicated dose conversion from graphite to water in graphite calorimetry would give water calorimetry a significant advantage. However, water, in comparison with graphite, presents its own drawbacks with regard to various radiation-induced effects not observed in graphite, such as issues with temperature dependent convection, impurities, dissolved gases, and radiolysis (see Sect. 16.2.2), that all adversely affect the heat defect κ and are difficult to quantify.

The basic approach to absorbed dose calorimetry can be stated as follows: mean absorbed dose $\bar{D}_{cav}$ in the calorimeter core (cavity) can be determined adiabatically without any need for energy calibration or knowledge of the sensitive mass of the core provided that in an absorbed dose calorimeter the following conditions are satisfied (*Note*: an adiabatic process is a process that occurs without transfer of heat between a system and its surroundings):

   (i) Specific heat C and heat defect κ of the core are known.
  (ii) Thermal leakage of the core is negligible.
 (iii) Thermistor is calibrated to read the core temperature.

During the past 100 years many elegant and practical variations have been developed on the theme above, but the basics of absorbed dose calorimetry have remained logical and simple; the problems only arise in dealing with the details related to the quest for extreme accuracy that is required from a primary dose standard.

Some examples of modifications to basic ideas of absorbed dose calorimetry are:

1. The amount of energy that radiation deposits in the sensitive core volume is determined by subsequent dissipation of measured amount of electrical energy in a heating coil to produce the same temperature rise ΔT in the core. *Advantage*: Measurement of dose does not depend on the knowledge of specific heat C of the calorimeter core.

2. In some calorimeters thermistors are used not only for measurement of the temperature rise ΔT in the core and mantle, they are also used for controlled heating of calorimeter components. *Advantage*: Use of heaters that introduce foreign materials into the calorimeter core is avoided thereby reducing the heat defect problems.

3. Some calorimeters incorporate a modification to the Wheatstone bridge and employ two identical thermistors, one in each arm of the Wheatstone bridge. *Advantage*: Doubling of sensitivity compared to a one-thermistor bridge.

4. Some water calorimeters operate at +4 °C to minimize heat transfer from the core by convection in water. *Advantage*: At +4 °C water density reaches maximum value, justifying the assumption that heat transfer by convection in water at +4 °C is negligible.

5. When isothermic heating is used, energy absorbed in the core produces a change of state rather than a change in core temperature. The absorbed energy is determined from the measured amount of material that changes its state. For example, ice melted into water at 0 °C results in volume change that is proportional to energy absorbed in the core. *Advantage*: Isothermal operation.

6. Based on low thermal diffusivity of water that allows water to keep a relatively stable temperature map, resulting from energy transfer from ionizing radiation to water absorber, Steve Domen in early 1980s designed a simple water calorimeter that dispensed with a defined sensitive volume (core) and associated wall and vacuum gap problems. A small bead thermistor was sandwiched between two thin films submersed in the calorimeter water bath. The two films had a threefold purpose:

(i)   To hold the thermistor at the desired depth in distilled water.
(ii) To electrically insulate the thermistor and connecting wires.
(iii) To prevent convection currents in the vicinity of the thermistor.

Ignoring the possible heat defect, the dose at the position of the thermistor was determined using the product $C_{water} \Delta T$, as given in (16.2); however, problems were encountered with the heat defect caused by impurities in water, dissolved gases such as nitrogen and oxygen in water, and radiolysis of water. *Advantage*: Simple design of calorimeter.

16.1.7   Calorimetric Absolute Radiation Dosimetry: Summary

Of the three known absolute radiation dosimetry techniques, calorimetric radiation dosimetry or radiation calorimetry is unequivocally the most absolute, either intrinsically or through electrical heating calibration. The basic premises of radiation calorimetry are simple: as ionizing radiation interacts with medium, part of its energy is transferred to the medium and manifests itself as a temperature rise that is proportional to the dose absorbed in the medium. In contrast to the other two known absolute dosimetry techniques, radiation calorimetry has the following attributes:

• It comes closest to determining directly the energy transferred from ionizing radiation and absorbed in the medium.
• It is best suited for the measurement of dose-to-water D_{med} in phantom without the need for troublesome correction factors.
• Its dose rate independence extends to very high dose rates.
• It exhibits no LET dependence.
• It adds up absorbed doses from various types of ionizing radiations.

The main disadvantages of calorimetric radiation dosimetry are that it has a relatively low sensitivity and that it is impractical for routine clinical or fieldwork.

Absorbed dose calorimetry, graphite-based or water-based, provides a primary standard for calibration of cobalt-60 radiotherapy beams. The principles of absorbed dose calorimetry, as shown above, are well understood, as are the practical problems that arise when absorbed dose calorimeters are used in absorbed dose measurements of ionizing radiation beams. Each primary standards laboratory developed its own unique solution to calorimetric radiation dosimetry, many years ago with graphite calorimeters and more recently with water calorimeters after Steve Domen from the NIST introduced a practical water calorimeter in the early 1980s.

Current research in calorimetric radiation dosimetry is carried out with the goal of extending the primary absorbed dose standard from cobalt-60 beams to other megavoltage x-ray and electron beams as well as from standard fields to various other unconventional irradiation conditions, such as small-size photon beams (smaller than 2×2 cm^2) and charged particle radiotherapy. Progress in these new areas of interest would allow determination of dose to water with much lower uncertainties than is currently possible with other absolute dosimetry techniques.

16.2   Fricke Chemical Absolute Radiation Dosimetry

16.2.1   Introduction to Fricke Chemical Absolute Radiation Dosimetry

Radiation dosimetry relies on measurement of a suitable radiation induced physical or chemical signal that is emitted by a radiation dosimeter and measured by a reader of a radiation dosimetry system. Most radiation dosimetry systems are based on physical signals; however, a category of dosimetry systems based on various well-defined chemical reactions that ionizing radiation triggers in absorbing medium (dosimeter cavity) are also available. Quantification of chemical reaction products with the goal of determining the dose absorbed in medium is referred to as chemical dosimetry and is carried out with various techniques, such as (i) titration, (ii) ultraviolet spectrophotometry, and (iii) electron paramagnetic resonance (EPR). All chemical dosimetry systems are of the passive type and all but one fall into the category of relative radiation dosimetry.

The Fricke ferrous sulfate chemical dosimetry technique is by far the best-known and oldest chemical dosimetry technique, developed and understood so well that it is considered one of the three known absolute radiation dosimetry techniques, the other two being calorimetric dosimetry and ionometric dosimetry. Other chemical systems used in radiation processing as well as for reference and transfer standard dosimetry are based on alanine or various aqueous chemical solutions, such as iodide-iodate, ceric-cerous sulfate, dichromate, and ethanol-chlorobenzene solutions. However, these chemical techniques must be calibrated in known radiation fields and therefore belong into the category of relative dosimetry techniques.

Most chemical dosimetry systems are based on aqueous solutions with water serving as solvent for radiation sensitive chemicals that serve as the solute. Since the concentration of radiation sensitive chemicals dissolved in water is relatively low, an assumption is made that the chemical dosimetry system measures absorbed dose in water. Thus, to understand chemical radiation dosimetry one must understand not only the physics but also the underlying chemistry of the interaction of ionizing radiation with water.

The dissociation of water molecules induced by ionizing radiation is called radiolysis, in contrast to dissociation caused by visible light called photolysis and dissociation caused by direct electric current passing through water referred to as electrolysis. The branch of chemistry that studies chemical effects induced by ionizing radiation is called radiation chemistry.

16.2.2   Radiolysis of Water

Radiation chemistry traces its beginnings to the end of 19th century soon after Röntgen's discovery of x-rays and Becquerel's discovery of natural radioactivity. Since
then, much effort has been expended on unraveling the mechanisms of radiation
induced formation of free radicals and molecular products in water and other media
irradiated with ionizing radiation. Modern radiolytic techniques use very short pulses
(of the order of 1 ns or shorter) of megavoltage electron beams produced by linear
accelerators. They are referred to as pulse radiolysis and were introduced in 1960s
by English scientists John Keene and Jack Boag.

Current model divides the radiolysis of water into three somewhat overlapping
stages:

(i)   Physical stage
(ii)   Intermediate stage
(iii)   Chemical stage

Physical Stage of Water Radiolysis

The physical stage of water radiolysis is of extremely short duration (~ 1 fs) and is
characterized by standard radiation interactions resulting in ionization and excitation
of water molecules. For photon beams these interactions comprise the photoelectric
effect, Compton scattering, and triplet production that ionizing photons undergo with
water molecules; for charged particle beams they comprise Coulomb interactions
between charged particles and water molecules.

Excitation of a water molecule results in an unstable excited water molecule
H_2O^*, while ionization results in an unstable water cation (positive ion) H_2O^+ and
a free electron e^- called a primary electron. The free electron propagates through
water, and may be energetic enough to excite or ionize water on its own, releasing
low energy secondary electrons in the process. These electrons propagate through
water, eventually becoming thermalized and incapable to ionize or excite other water
molecules. According to the NIST, ionization energy (IE) of a water molecule is
~ 12.6 eV, while the minimum excitation energy of a water molecule is ~ 7.5 eV.
In summary, the physical stage of water radiolysis produces unstable excited water
molecules H_2O^*, unstable positively charged ionized water molecules H_2O^+, as well
as a large number of thermalized electrons e^-.

Intermediate Stage of Water Radiolysis

During the intermediate stage of water radiolysis, lasting ~ 1 ps, the unstable molecules H_2O^+ and H_2O^* undergo various processes resulting in highly reactive hydrogen and hydroxyl radicals $H^\bullet$ and $HO^\bullet$, respectively, while the thermalized free
primary and secondary electrons e^- may recombine with ionized water molecules
or undergo a process referred to as solvation. The chemical processes occurring in
the intermediate stage are as follows:

(i)   The ionized water molecule H_2O^+ interacts with a neighboring neutral water
        molecule H_2O and transforms into an $HO^\bullet$ radical by transferring a proton

Table 16.1 Comparison of characteristics of reducing agents with those of oxidizing agents

Reducing agent	Oxidizing agent
Loses electrons	Gains electrons
Is electron donor	Is electron acceptor
Is oxidized	Is reduced
Is electropositive	Is electronegative
Reduces substances with which it interacts	Oxidizes substances with which it interacts

onto the neutral H_2O molecule. This chemical process is called protonation and transforms the neutral water molecule into a hydronium cation H_2O^+ (also known as oxonium or aqueous cation)

$$H_2O^+ + H_2O \rightarrow HO^\bullet + H_3O^+. \tag{16.13}$$

(ii) The excited water molecule H_2O^* can revert to ground state H_2O through emission of heat or it can undergo one of several dissociative processes of bond breakage, the most common being dissociation into a hydrogen radical $H^\bullet$ and a hydroxyl radical $HO^\bullet$

$$H_2O^* \rightarrow H^\bullet + HO^\bullet. \tag{16.14}$$

(iii) The thermalized primary and secondary electrons e^- interact with dipoles of water molecules and may become trapped into a reactive cluster of typically six water molecules forming an entity called a solvated electron and designated as e_{aq}^-. Like hydrogen radical $H^\bullet$, a solvated electron e_{aq}^- (also known as hydrated electron or aqueous electron) is a strong reducing agent, in contrast to the hydroxyl radical $HO^\bullet$ that is a powerful oxidizing agent. A general comparison of chemical properties of reducing and oxidizing agents is provided in Table 16.1.

In summary, the intermediate stage of water radiolysis produces chemically reactive radicals $H^\bullet$ and $HO^\bullet$ as well as solvated electrons e_{aq}^- from the three initial species H_2O^+, H_2O^*, and energetic e^- produced during the physical stage. These radicals are highly chemically reactive as a result of having an unpaired valence electron and play an important role in further chemical reactions that occur during the chemical stage of water radiolysis.

Chemical Stage of Water Radiolysis

During the chemical stage of water radiolysis, lasting ~ 1 μs, the free radicals $H^\bullet$ (hydrogen radical), $HO^\bullet$ (hydroxyl radical), and e_{aq}^- (solvated electron radical) that were produced in the intermediate stage, undergo one of several dozen possible chemical reactions. These reactions result in various chemical products in the form of oxidative and reductive radicals, ions, and molecules that drive the reactions of importance in chemical radiation dosimetry. In addition to the three radicals ($H^\bullet$,

Table 16.2 List of most important radicals, ions, and molecules produced in radiolysis of water

Radicals	Ions	Molecules
H$^\bullet$ hydrogen	H$^+$ hydrogen - proton	H$_2$ dihydrogen
OH$^\bullet$ hydroxyl	OH$^-$ hydroxyl	O$_2$ oxygen
O$_2^\bullet$ superoxide	HO$_2^-$ hydroperoxy	H$_2$O$_2$ hydrogen peroxide
HO$_2^\bullet$ hydroperoxy	H$_2$O$^+$ hydronium	
e$_{aq}^-$ solvated electron		

HO$^\bullet$, and e$_{aq}^-$) of the intermediate stage, Table 16.2 also lists the most important chemical products produced during the chemical stage of water radiolysis.

Summary of Water Radiolysis

A summary of the three stages of water radiolysis is given in Table 16.3:

1. During the extremely short physical stage ($\sim$1 fs) of water radiolysis, ionizing radiation interacts with a water molecule H$_2$O* resulting in either an excited water molecule through excitation of the water molecule or in an ionized water molecule H$_2$O$^+$ and an energetic electron e$^-$ through ionization of the water molecule. The energetic electron propagates through water, loses its kinetic energy through Coulomb interactions with other water molecules, and eventually becomes thermalized.
2. During the intermediate stage ($\sim$1 ps) the unstable excited water molecule H$_2$O* dissociates into a hydrogen H$^\bullet$ radical and hydroxyl radical HO$^\bullet$. On the other hand, the positive water ion H$_2$O$^+$ interacts with a neutral water molecule, resulting in a hydroxyl radical HO$^\bullet$ and a hydronium ion H$_3$O$^+$. The thermalized free electrons can attach themselves to a cluster of six water molecules and form a reactive species called solvated electron.
3. The radical species are extremely reactive and contribute to several dozen chemical reactions ($\sim$1 μs) during the chemical stage of water radiolysis. Ten most common of these reactions that eventually remove the chemically reactive species from water are also listed in Table 16.3, many more are available in the literature.

16.2.3 Radiolytic Yield in Chemical Dosimetry

Radiolytic yield $G(X)$ of a radiation-induced entity X that can be a molecule, ion, radical, etc. is an important parameter in chemical radiation dosimetry. In the original definition $G(X)$ is the so-called G value representing the number of entities X produced per 100 eV of energy imparted by ionizing radiation to a chemical dosimeter, i.e., $G(X) = G/(100\ eV)$, where G is the number of entities produced per 100 eV of absorbed energy. Modern nomenclature measures the radiolytic yield $G(X)$ in units of mol/J and the relationship between $G(X)$ expressed in G/(100 eV) and $G(X)$ expressed in units of mol/J is determined as follows

Table 16.3 Radiolysis is defined as dissociation of molecules induced by ionizing radiation

The table provides a schematic representation of the 3 stages of radiolysis of water: physical stage, intermediate stage, and chemical stage. The physical stage lasts about 1 fs and is characterized by excitation and ionization of water molecules. The intermediate stage that lasts about 1 ps consists of dissociation of excited water molecules, protonation water molecules, and solvation of free electrons. During the chemical stage that lasts about 1 μs various reactions occur among free radicals, ions, and molecules that were produced during the intermediate stage. Examples of most important reactions are given in the table

$$G(X) = \frac{G}{100\,\text{eV}} = \frac{G}{(100\,\text{eV})} \times \frac{\text{mol}}{\text{mol}}$$

$$= \frac{G \times \text{mol}}{6.022 \times 10^{23} \times (100\,\text{eV}) \times (1.602 \times 10^{-19}\,\text{J/eV})}$$

$$= G \times (1.037 \times 10^{-7}\,\text{mol/J}), \tag{16.15}$$

where G is the number of entities produced per 100 eV of energy absorbed in dosimeter cavity. Typically, $G(X)$ in aqueous chemical dosimeters is of the order of 10^{-7} mol/J to 10^{-6} mol/J.

16.2.4 Absorbed Dose in Chemical Dosimeter

Determination of dose absorbed in the dosimeter cavity is straightforward provided that:

(i) Radiolytic yield $G(X)$ in mol/J for product X is known.
(ii) Radiation-induced molar concentration ΔM in mol/ℓ of product X in the cavity solution has been measured.

Mean energy $\Delta \bar{E}_{ab}$ absorbed in the cavity per unit volume ΔV of the chemical solution can be expressed by the ratio $\Delta M / G(X)$, i.e.,

$$\frac{\Delta \bar{E}_{ab}}{\Delta V} = \frac{\Delta M}{G(X)}. \tag{16.16}$$

Dividing (16.16) by the density ρ_{sol} of the chemical solution of the cavity, we get the following relationship for the cavity dose $\bar{D}_{cav}$ averaged over the sensitive volume of the cavity

$$\bar{D}_{cav} = \frac{\Delta \bar{E}_{ab}}{\Delta m} = \frac{\Delta \bar{E}_{ab}}{\rho_{sol} \Delta V} = \frac{\Delta M}{\rho_{sol} G(X)}, \tag{16.17}$$

where m is the mass of the chemical dosimeter cavity.

16.2.5 Background to Fricke Ferrous Sulfate Chemical Radiation Dosimetry

Ferrous sulfate radiation dosimetry has been studied since late 1920s and is today the best-known chemical radiation dosimetry technique. In honor of Hugo Fricke, a Danish-born American physicist and pioneer in this field, ferrous sulfate dosimetry is also known as Fricke dosimetry. The technique is based on radiolysis of water; a process that, as discussed in Sect. 16.2.2, produces a variety of radicals, ions, and molecular products in water under the influence of ionizing radiation. In an irradiated aqueous solution of ferrous (Fe^{2+}) ions, some of these radicals and molecular products oxidize the Fe^{2+} (ferrous) ions into Fe^{3+} (ferric) ions and the number of Fe^{3+} ions produced in the Fricke solution is proportional to dose absorbed in the solution (dosimeter cavity).

16.2.6   Composition of Fricke Solution

The standard Fricke solution is prepared from high purity solutes and contains 0.001 M ferrous sulfate (FeSO$_4$) and 0.4 M sulfuric acid (H$_2$SO$_4$) dissolved in air-saturated (i.e., fully aerated) triply distilled water. The molarity M of a solute is defined as the number of moles of the solute contained in 1000 cm^3 (1 liter) of solution. Thus, a 0.001 M solution of FeSO$_4$ contains 0.1519 g of FeSO$_4$ in 1000 cm^3 of solution and 0.4 M solution of H$_2$SO$_4$ contains 39.22 g of H$_2$SO$_4$ in 1000 cm^3 (1 liter) of solution. Note that the standard molecular weights of ferrous sulfate FeSO$_4$ and sulfuric acid H$_2$SO$_4$ are 98.08 g/mol and 151.9 g/mol, respectively. To reduce the deleterious effect of organic impurities often 0.001 M NaCl (sodium chloride) is added to the Fricke solution.

16.2.7   Components of Fricke Dosimeter

A Fricke dosimeter has two components: (i) Fricke solution that forms the dosimeter cavity and (ii) container in the form of a cylindrical pancake-shaped vial with inner diameter of $\sim$3 cm and inner height of $\sim$0.5 cm, defining a sensitive volume of $\sim$4 cm^3. To minimize the perturbation of the dosimeter reading and to obviate the need for correcting the cavity dosimeter reading the vial wall is made as thin as possible and of a plastic, tissue-equivalent material, such as polystyrene or Lucite. When glass or quartz vials are used, a correction factor is applied to account for the cavity signal perturbation caused by the wall material. The relatively large sensitive volume of $\sim$4 cm^3 is used because of:

(i) Low inherent sensitivity of Fricke dosimetry systems.
(ii) Desire to obtain more than one dose reading from a single irradiation.

16.2.8   Oxidation of Ferrous Ions to Ferric Ions

The Fricke dosimetry system offers a reliable means for measurement of absorbed dose-to-water based on radiolysis of water combined with various chemical processes that result in oxidation of ferrous Fe^{2+} ions to ferric Fe^{3+} ions. Three products of water radiolysis: hydroxyl radical HO$^\bullet$, hydrogen peroxide molecule H$_2$O$_2$, and hydrogen radical H$^\bullet$ drive the Fe$^{2+} \rightarrow$ Fe^{3+} oxidation reactions in an aerated Fricke aqueous solution. The three channels available for Fe^{3+} production are as follows:

1. The hydroxyl HO$^\bullet$ channel produces one Fe^{3+} ion per HO$^\bullet$ radical

$$HO^\bullet + Fe^{2+} \rightarrow Fe^{3+} + HO^-. \tag{16.18}$$

2. The hydrogen peroxide H_2O_2 channel produces two Fe^{3+} ions per H_2O_2 molecule

$$H_2O_2 + Fe^{2+} \rightarrow Fe^{3+} + HO^\bullet + HO^+$$
$$HO^\bullet + Fe^{2+} \rightarrow Fe^{3+} + HO^-$$
(16.19)

or, in shorter format after combining the two steps of the H_2O_2 channel (16.19)

$$H_2O_2 + 2Fe^{2+} \rightarrow 2Fe^{3+} + HO^-$$
(16.20)

3. The hydrogen radical $H^\bullet$ channel produces three Fe^{3+} ions per $H^\bullet$ radical as follows: For this channel to be fully effective oxygen molecules O_2 must be dissolved in the water solution (i.e., solution must be fully aerated), so that the hydroperoxyl radical $HO_2^\bullet$ can be produced from the ($H^\bullet$, O_2) reaction. The $HO_2^\bullet$ radical triggers one $Fe^{2+} \rightarrow Fe^{3+}$ oxidation reaction and the reaction product HO_2^- produces a hydrogen peroxide molecule H_2O_2 that in turn, as shown in channel 2, oxidizes two Fe^{2+} ions into two Fe^{3+} ions. Thus, in fully aerated Fricke water solutions each $H^\bullet$ radical produces three Fe^{3+} ions, one Fe^{3+} ion through the $HO_2^\bullet$ radical plus two Fe^{3+} ions through the H_2O_2 molecule

$$H^\bullet + O_2 \rightarrow HO_2^\bullet \quad \text{(hydroperoxyl radical)}$$
$$HO_2^\bullet + Fe^{2+} = Fe^{3+} + HO_2^- \quad \text{(hydroperoxyl anion)}$$
$$HO_2^- + H^+ \rightarrow H_2O_2 \quad \text{(hydrogen peroxide)}$$
$$H_2O_2 + 2Fe^{2+} \rightarrow 2Fe^{3+} + 2HO^-$$
(16.21)

Radiolytic yield $G(Fe^{3+})$ for an aerated standard Fricke solution is a sum of three components, one for each of the three $Fe^{2+} \rightarrow Fe^{3+}$ oxidation channels based on hydroxyl, peroxide, and hydrogen radicals

$$G(Fe^{3+}) = G(HO^\bullet) + 2G(H_2O_2) + 3G(H^\bullet),$$
(16.22)

where $G(HO^\bullet)$, $G(H_2O_2)$, and $G(H^\bullet)$ and are the primary radiolytic yields that depend on the molarity and aeration of the Fricke solution as well as on the type and energy of the ionizing radiation beam. The factors preceding the radiolytic yield $G(Fe^{3+})$ in (16.22) correspond to the number of Fe^{3+} ions produced in each of the three channels that contribute to the Fe^{3+} production in aerated Fricke solutions.

By way of example, the following primary radiolytic yields $G(X)$ are used for standard aerated Fricke solutions in a cobalt-60 gamma ray beam: $G(^\bullet OH) = 2.90$; $G(H_2O_2) = 0.83$ and $G(H^\bullet) = 3.63$ to get, after using (16.22), a radiolytic yield of $G(Fe^{3+}) = 15.5$ for production of ferric Fe^{3+} ions per 100 eV of absorbed energy in the Fricke solution. As shown in (16.15), the G value of 15.5 ions per 100 eV of absorbed energy corresponds to $G(Fe^{3+}) = 15.5 \times (1.037 \times 10^{-7} \, \text{mol/J}) = 1.607 \times 10^{-6} \, \text{mol/J}$.

Since the Fricke solution is kept in sealed vials during irradiation, the dissolved oxygen gets gradually depleted with dose at high doses with a corresponding decrease in $G(Fe^{3+})$. This results in a non-linear response of the dosimeter at high doses and sets an upper limit of 400 Gy for dose linearity in standard Fricke solutions. The lower limit is around 40 Gy and is set by the relatively low sensitivity of the Fricke dosimetry systems. Codes of practice for Fricke dosimetry recommend that, for best results, the Fricke solution should be prepared a short time before irradiation and, if storage is required, the Fricke solution should be kept in dark and cold storage.

16.2.9 Measurement of Radiation-Induced Ferric Concentration

Based on (16.17) we note that the mean dose $\bar{D}_{cav}$ absorbed in Fricke dosimeter cavity is directly proportional to ΔM, the change in concentration of Fe^{3+} ions in the Fricke solution that results from exposure to ionizing radiation. The radiation-induced change ΔM in concentration of in a Fricke solution is determined by evaluating the Fe^{3+} concentrations before and after the irradiation of the Fricke dosimeter.

The reading before irradiation is important because it establishes a baseline for the radiation-induced change in concentration measurement. It turns out that the $Fe^{2+} \rightarrow Fe^{3+}$ oxidation, in addition to being triggered by ionizing radiation, also occurs spontaneously in an un-irradiated Fricke solution, albeit with a significantly lower, yet not negligible, yield. This means that the baseline reading will depend on the time the solution was kept in storage before being irradiated. For best results, Fricke dosimetry protocols recommend preparation of Fricke solutions a short time before irradiation rather than using solutions stored for long periods of time.

Two methods are available for the determination of ΔM. While a simple chemical titration can do the job, a more practical and sensitive method based on spectrophotometry is usually used for this purpose. Spectrophotometry is a term used to describe a quantitative measurement of transmission of ultraviolet (uv) or visible light through an absorber such as a chemical solution. The transmission of light in a spectrophotometer through a Fricke solution sample depends upon:

(i) Wavelength λ of the beam passing through the sample.

(ii) Absorbed dose to which the sample was exposed. This dose is proportional to the number of Fe^{3+} ions produced in the radiation induced $Fe^{2+} \rightarrow Fe^{3+}$ oxidation process; the larger is the radiation induced concentration of Fe^{3+} ions in the sample, the lower is the transmitted uv beam intensity and the higher is the mean radiation dose $\bar{D}_{cav}$ absorbed in the sample.

In Fricke spectrophotometry special containers, called cuvettes, are used for measurements of light transmission. Fricke solution is irradiated in a sealed vial in a water phantom; however, prior to measurement of ferric concentration $C_{Fe^{3+}}$ in a spectrophotometer, the irradiated solution is transferred to a cuvette. The role of the cuvettes is to contain samples filled with Fricke solution and to define with great

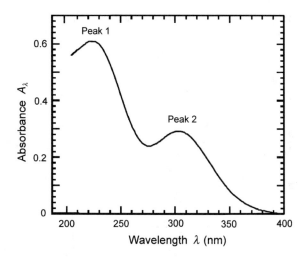

Fig. 16.2 Absorbance spectrum for ferric (Fe^{3+}) ion in standard Fricke solution in the ultraviolet spectral region plotted for wavelengths λ in the region from 200 nm to 400 nm. Absorbance peak 1 is at $\lambda = 224$ nm, peak 2 at $\lambda = 303$ nm

precision the length of the light path L through the Fricke solution (typical length of L is 1 cm). The spectrophotometer is used to determine the ratio of light intensity I transmitted through a cuvette filled with irradiated sample of Fricke solution to light intensity I_0 transmitted through the same cuvette but filled with un-irradiated sample from the same Fricke solution batch.

The logarithm with base 10 of the ratio I_0/I is defined as absorbance A_λ of a solution. We thus have the following expressions for absorbance A_λ

$$A_\lambda = \log_{10}\frac{I_0}{I} = -\log_{10}\frac{I}{I_0} = -\log_{10}T \quad \text{or} \quad \frac{I}{I_0} = T = 10^{-A_\lambda}, \quad (16.23)$$

where T is defined as the transmittance of the light beam through the Fricke solution.

Figure 16.2 plots the absorbance spectrum for Fe^{3+} ion in standard Fricke solution in the form of absorbance A_λ against wavelength λ in the uv spectral region from 200 nm to 400 nm. The Fe^{3+} absorption spectrum exhibits two maxima: peak 1 at $\lambda = 224$ nm and peak 2 at $\lambda = 303$ nm. Absorbance at peak 1 is about double that at peak 2 and this implies that the sensitivity of Fricke dosimeter would be doubled if measurements were carried out at 224 nm. However, peak 2 is usually used in Fricke dosimetry because, in comparison with peak 1, it is less affected by impurities in the Fricke solution. In addition, the ratio of absorbances $A_\lambda(Fe^{3+})/A_\lambda(Fe^{2+})$ is significantly larger for peak 2 in comparison with that of peak 1, highlighting the advantage of using peak 2 for dosimetry.

The absorbance A_λ of a Fricke solution is directly proportional to the following two parameters:

(i) Ferric ion concentration $C_{Fe^{3+}}$ in the solution.
(ii) Length of the light path L through the Fricke solution (optical path length) that is equal to the thickness of the cuvette in which $C_{Fe^{3+}}$ is measured.

Table 16.4 Relationship between transmittance T and absorbance A_λ in the transmittance range from $T = 0$ to $T = 1$

	Transmittance		Absorbance A_λ
	$T = \frac{I}{I_0} = 10^{-A_\lambda}$	$\frac{I_0}{I} = \frac{1}{T} = 10^{A_\lambda}$	$A_\lambda = \log\frac{I_0}{I} = -\log T$
1	0	∞	–
2	0.0001	10 000	4.0
3	0.001	1 000	3.0
4	0.01	100	2.0
5	0.05	20	1.301
6	0.1	10	1.0
7	0.2	5	0.699
8	0.3	3.33	0.523
9	0.5	2	0.301
10	0.75	1.33	0.125
11	0.9	1.11	0.046
12	1.0	1.00	0

Absorbance can thus, in conjunction with (16.23), be expressed as follows

$$A_\lambda = \varepsilon_{Fe^{3+}} C_{Fe^{3+}} L = \log_{10}\frac{I_0}{I}, \tag{16.24}$$

where $\varepsilon_{Fe^{3+}}$ is the proportionality constant called molar absorption coefficient that measures the intrinsic ability of a chemical species to absorb ultraviolet or visible light at a given wavelength. In the past $\varepsilon_{Fe^{3+}}$ was known as molar absorptivity or molar extinction coefficient; however, the IUPAC is discouraging the use of older nomenclature to avoid confusion.

For Fricke solutions the molar absorption coefficient $\varepsilon_{Fe^{3+}}$ is usually expressed in units of $\ell \cdot mol^{-1} \cdot cm^{-1}$ or cm^2/mol (*Note:* 1 $\ell \cdot mol^{-1} \cdot cm^{-1} = 10^3$ cm^2/mol). It is normalized to a readout temperature of $T_\varepsilon = 25\,°C$ and decreases by as $\sim 0.7\%/K$ as T_ε decreases from $T_\varepsilon = 25\,°C$. Equation (15.49) is in physics and chemistry known as the Beer–Lambert law in honor of German physicists August Beer and Johann H. Lambert who were among the first to study in the 19th century the attenuation of light in relation to properties of the material through which light is propagating.

Table 16.4 displays the relationship between transmittance T and absorbance A_λ in the transmittance range of $0 \le T \le 1$, from no transmittance through the sample ($T = 0$) to full transmittance through the sample ($T = 1$). The transmittance through a sample is measured as a ratio I/I_0, therefore it varies exponentially with path length L and concentration C of the solute. Absorbance A_λ of a sample, on the other hand, is measured as a logarithm of I_0/I, therefore, it is directly proportional to the optical path length L and concentration C of the solute.

16.2.10 Molar Absorption Coefficient Versus Molecular Cross Section

Molar absorption coefficient ε (expressed in units of $\ell\cdot\text{mol}^{-1}\cdot\text{cm}^{-1}$ or cm^2/mol where $\ell\cdot\text{mol}^{-1}\cdot\text{cm}^{-1} = 10^3\ \text{cm}^2/\text{mol}$) is directly related to the molecular cross section $_{\text{mol}}\sigma$ in units of $\text{cm}^2/\text{molecule}$. The derivation of the relationship proceeds as follows: merging (16.23) and (16.24) we get

$$\frac{I}{I_0} = 10^{-A_\lambda} = 10^{-\varepsilon CL}, \tag{16.25}$$

with A_λ the absorbance; ε molar absorption coefficient; C concentration in mol/ℓ; and L optical path length through the solution in cm. We can also express the ratio I/I_0 in the format used for describing exponential attenuation of a photon beam (see Sect. 7.1.1) to get

$$\frac{I}{I_0} = e^{-\sigma L} = e^{-\frac{\sigma}{\rho}\rho L} = e^{-_{\text{mol}}\sigma n^\square L}, \tag{16.26}$$

where

σ	is linear attenuation coefficient with unit of cm^{-1}.
σ/ρ	is mass attenuation coefficient expressed in units of cm^2/g.
$_{\text{mol}}\sigma$	is molecular cross section expressed in units of $\text{cm}^2/\text{molecule}$.
$n^\square$	is concentration of molecules (number of molecules per unit volume): $n^\square = \rho N_A/M$.
ρ	is mass density in g/cm^3.
N_A	is the Avogadro number: $N_A = 6.022\times10^{23}$ molecules/mole.
M	is molecular mass expressed in g/mol.

Merging (16.25) and (16.26) we now write $\ln(I/I_0)$ as

$$\ln\frac{I}{I_0} = -\varepsilon CL \ln 10 = -_{\text{mol}}\sigma n^\square L, \tag{16.27}$$

yielding the following simple relationship between the molecular cross section $_{\text{mol}}\sigma$ and molar absorption coefficient ε

$$_{\text{mol}}\sigma = \frac{\varepsilon C \ln 10}{n^\square} = \frac{\ln 10}{N_A}\times\frac{CM}{\rho}\varepsilon = \left(3.824\times10^{-24}\frac{\text{mol}}{\text{molecule}}\right)\times\frac{CM}{\rho}\varepsilon. \tag{16.28}$$

Since $CM/\rho \approx 1$ in (16.28), the relationship between $_{\text{mol}}\sigma$ and ε becomes very simple suggesting a direct proportionality between the two attenuation coefficients with the proportionality constant $(\ln 10)/N_A$, i.e.,

$$_{mol}\sigma = \frac{\ln 10}{N_A}\varepsilon, \tag{16.29}$$

where $_{mol}\sigma$ is given in cm^2/molecule and ε in $\ell \cdot mol^{-1} \cdot cm^{-1} = 10^3$ cm^2/mol.

By way of example we now calculate the molecular cross section $_{mol}\sigma_{Fe^{3+}}$ in cm^2/molecule for the ferric ion at wavelength $\lambda = 303$ nm in the spectropho-tometer where the molar absorption coefficient $\varepsilon_{Fe^{3+}}$ is 2187 $\ell \cdot mol^{-1} \cdot cm^{-1} = 2.187 \times 10^6$ cm^2/mol. Inserting $\varepsilon_{Fe^{3+}}$ into (16.29) we get

$$_{mol}\sigma_{Fe^{3+}} = \frac{\ln 10}{N_A}\varepsilon = \frac{(\ln 10) \times (2.187 \times 10^6 \ cm^2/mol)}{6.022 \times 10^{23} \ molecules/mol} = 8.362 \times 10^{-18} \frac{cm^2}{molecule} \tag{16.30}$$

Note that this molecular cross section is several orders of magnitude larger than is a typical atomic cross section for x-rays and gamma rays.

16.2.11 Dependence of Radiolytic Yield on Irradiation Conditions and Readout Temperature

The measured radiation-induced concentration of ferric ions $C_{Fe^{3+}}$ in a Fricke solu-tion depends not only on ionizing radiation beam type and energy but also on the temperature T_G at which the irradiation is carried out. As for the temperature depen-dence, $C_{Fe^{3+}}$ is usually corrected for irradiation of the Fricke solution at $T_G = 25\,°C$. It has been shown experimentally that the radiolytic yield $G(Fe^{3+})$ decreases by $\sim 0.15\%/K$ as the temperature T_G of the Fricke solution during irradiation decreases from 25 °C. Since this correction is quite small, the ICRU recommends that irradia-tion be carried out in the temperature range between 20 °C to 25 °C.

Data from ICRU Reports 14 and 21 for photons and electrons, respectively, reveal that, for standard Fricke solution irradiated at 20 °C < T_G < 25 °C, the radiolytic yield expressed in number of Fe^{3+} ions per 100 eV of absorbed energy:

(i) For photon beams ranges from for 15.3 ± 0.3 for Cs-137 γ-rays through 15.4 ± 0.3 for 2 MV x-rays and 15.5 ± 0.2 for Co-60 γ-rays to 15.7 ± 0.6 for 30 MV x-rays.

(ii) For electron beams between 1 MeV and 30 MeV is assumed constant at 15.7 ± 0.6.

16.2.12 Determination of Cavity Dose in Fricke Chemical Dosimetry

Mean dose $\bar{D}_{cav}$ to Fricke cavity was in (16.17) shown to be directly proportional to the change in molar concentration ΔM that in turn was shown to be equal to the radiation-induced change in concentration $C_{Fe^{3+}}$ in the Fricke dosimeter expressed from (16.24) as follows

$$C_{Fe^{3+}} = \frac{A_\lambda}{\varepsilon_{Fe^{3+}} L}. \tag{16.31}$$

Combining (16.17) with (16.24) we now get the following expression for the mean dose $\bar{D}_{cav}$ to Fricke cavity, i.e., Fricke solution

$$\bar{D}_{cav} = \frac{\Delta M}{\rho_F G(Fe^{3+})} = \frac{C_{Fe^{3+}}}{\rho_F G(Fe^{3+})} = \frac{A_\lambda}{\rho_F \varepsilon_{Fe^{3+}} G(Fe^{3+}) L} = \frac{\log_{10}(I_0/I)}{\rho_F \varepsilon_{Fe^{3+}} G(Fe^{3+}) L}, \tag{16.32}$$

where

ρ_F is the mass density of the standard Fricke solution: $\rho_F = 1.024$ kg/ℓ.

$G(Fe^{3+})$ is the radiolytic yield of Fe^{3+} for standard Fricke solution: $G(Fe^{3+}) = 1.607 \times 10^{-6}$ mol/J for low LET radiations such as cobalt-60 gamma rays irradiated at 20 °C $< T_G <$ 25 °C.

$\varepsilon_{Fe^{3+}}$ is the molar absorption coefficient for Fe^{3+} ions: at $\lambda = 303$ nm and $T = 25$ °C the molar absorption coefficient of Fricke solution is $\varepsilon_{Fe^{3+}} = 2187$ $\ell \cdot mol^{-1} \cdot cm^{-1}$

L is the thickness of the spectrophotometer cuvette, i.e., optical path length usually of the order of 1 cm.

A_λ is the absorbance measured for the Fricke sample with the spectrophotometer and defined as $\log_{10}(I_0/I)$ with I_0 and I the ultraviolet light intensities measured before and after exposure of the Fricke dosimeter to ionizing radiation, respectively.

Inserting the parameters given above, we can now simplify (16.32) to get the following simple equation for the mean dose to Fricke cavity $\bar{D}_{cav}$ for a standard Fricke solution exposed to cobalt-60 gamma radiation

$$\bar{D}_{cav} = \frac{A_\lambda}{(1.024 \text{ kg}/\ell) \times (2187 \text{ } \ell \cdot mol^{-1} \cdot cm^{-1}) \times (1.607 \times 10^{-6} \text{ mol/J}) \times (1 \text{ cm})}$$

$$= (277.9 \text{ Gy}) \times A_\lambda = (277.9 \text{ Gy}) \times \log_{10}\frac{I_0}{I}. \tag{16.33}$$

16.2.13   Determination of Dose to Water from Mean Dose to Cavity

The Fricke dosimeter consists of the Fricke solution contained in the dosimeter cavity by a vial with a rigid wall that defines the cavity volume and insulates the Fricke solution from the surrounding water phantom. As shown schematically in Fig. 16.3a, the dosimeter is placed into the water phantom in such a way that its reference point corresponds with the point-of-interest A at depth z on the central axis of the radiation beam. The radiation beam is produced by an external radiation source that is characterized by beam type and energy as well as by the source-surface distance SSD and field size A at the surface of the water phantom.

The measurement provides the mean dose $\bar{D}_{cav}$ to the dosimeter cavity placed at point A in the water phantom. This dose is subsequently used to calculate D_{med}, dose to point A in the water phantom in the absence of the radiation dosimeter. The transition from $\bar{D}_{cav}$ to D_{med} is in general a complicated process, since one needs to account for several factors that may adversely affect the accuracy of the final result. Three issues that deserve our attention with regard to Fricke cavity are: (i) Size of the cavity; (ii) Density and effective atomic number of the cavity material; and (iii) Density, thickness, and effective atomic number of the vial that serves as the wall of the dosimeter.

Fricke Cavity Size

Of all known cavity theories used in radiation dosimetry, the most practical and reliable is the Bragg–Gray theory (see Sect. 15.7) governed by the two standard Bragg–Gray conditions. However, not all dosimetry techniques can meet the two Bragg–Gray conditions and Fricke dosimetry is one of these. The relatively low sensitivity of the Fricke solution and cavity material in liquid form are two notable characteristics of Fricke dosimetry that preclude the use of small Bragg–Gray cavities ($d \ll R$) and demand a relatively large cavity volume of the intermediate ($d \approx R$) or large ($d \gg R$) variety (*note*: d stands for chamber dimensions, R for the range of secondary charged particles in the cavity medium).

Figure 16.3b, c depict schematically an intermediate (Burlin) cavity ($d \approx R$) embedded into a water phantom: part (b) is for the cavity alone and part (c) for the cavity and its wall. The arrows with labels 1, 2, and 3 represent the primary energetic electrons, released by photon interactions in the phantom, wall, or cavity. The length of the arrows represents the range of secondary electrons. Photons release three varieties of energetic charged particles (electrons):

Type 1 electrons are released in the cavity and expend all their energy in the cavity.
Type 2 electrons are released in the wall, enter the cavity, and end their tracks in the cavity.
Type 3 electrons are released in the cavity, enter the wall, and end their tracks in the wall.

In an ideal situation, the radiation sensitive medium of the cavity and the wall of the cavity are made of the same tissue-equivalent material as the phantom. Three conclusions can then be made:

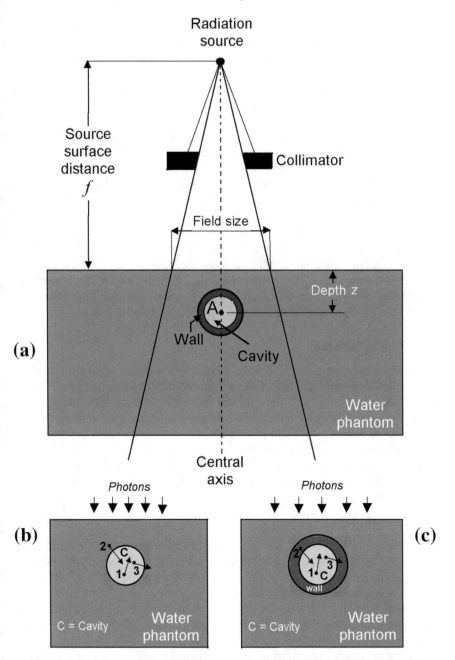

Fig. 16.3 Determination of dose-to-medium (water) D_{med} from mean dose-to-cavity (Fricke solution) $\bar{D}_{cav}$. **a** Reference point of Fricke dosimeter is placed into point A on beam central axis at depth z in a water phantom. Radiation source is at source-surface distance f from the phantom surface. Fricke dosimeter consists of dosimeter (Burlin intermediate) cavity filled with standard Fricke solution and surrounded with dosimeter wall. **b** Dosimeter cavity alone in water phantom. **c** Cavity and wall in water phantom

1. Dosimeter with its wall and cavity does not perturb the photon spectrum and the charge particle fluence in the phantom. This means that the photon spectrum and the charged particle fluence existing in the phantom in the absence of the dosimeter do not change when the dosimeter is introduced into the phantom.

2. Mean dose in the cavity $\bar{D}_{cav}$ is produced by electrons in the cavity volume under the condition of charged particle equilibrium (Sect. 15.3). To satisfy the charged particle equilibrium (CPE) condition each electron released in the cavity that crosses the cavity boundary and carries kinetic energy E_K into the wall (type 3 electron in Fig. 16.3c) there must be an electron released in the wall that carries the same energy into the cavity (type 2 electron in Fig. 16.3c). Then, an assumption can be made that the dosimetric signal from the ferrous sulfate solution in the cavity is produced in its entirety by electrons released and stopped in the cavity (type 1 electrons in Fig. 16.3c).

3. For the cavity, wall, and phantom made of tissue-equivalent materials no corrections are necessary to the measured mean cavity dose $\bar{D}_{cav}$ because the cavity and wall are made of same material as the phantom. Furthermore, the translation from dose-to-cavity $\bar{D}_{cav}$ to dose-to-phantom D_{med} does not require any corrections, since the cavity medium corresponds to the phantom medium. Thus, in this case the $\bar{D}_{cav} \rightarrow \bar{D}_{med}$ translation is trivial, since $D_{med} = \bar{D}_{cav}$.

Fricke Cavity Material

The Fricke cavity contains the Fricke solution that is based on oxygen-saturated water and sulfuric acid (0.4 M) as solvents in which the ferrous sulfate (0.001 M) $FeSO_4$ solute is dissolved. It turns out that the Fricke solution is essentially water ($\sim$96%), so that its density and effective atomic number are only slightly different from those of water ($\rho_F = 1.024$ g/cm^3), thereby causing a negligible perturbation in measurement of $\bar{D}_{cav}$. An assumption is therefore made that the slight differences in density as well as concentration of ferrous sulfate between the Fricke solution and pure water can be ignored in the determination of dose-to-water D_{med} from the mean dose to cavity $\bar{D}_{cav}$.

Fricke Cavity Wall

Two types of material are used in fabrication of vials that contain the Fricke solution during irradiation: (i) water-equivalent materials, such as polystyrene ($\rho \approx$ 1.04 g/cm^3), Lucite ($\rho = 1.18$ g/cm^3), and polyethylene ($\rho \approx 0.96$ g/cm^3) or (ii) regular glass and quartz glass ($\rho \approx 2.2$ g/cm^3).

In contrast to glass vials, plastic vials have the advantage of being almost water equivalent but are much more difficult to clean to ensure that bothersome organic impurities do not contaminate the Fricke solution. Glass vials offer better purity; however, they have a larger effect on the cavity signal because of their higher density and atomic number.

The wall effect is mainly caused by increased or decreased scattering of secondary electrons in the wall, perturbing the electron fluence inside the cavity thereby causing an increase or decrease in the measured $\bar{D}_{cav}$. Experiments and Monte Carlo calculations have shown that glass vials increase $\bar{D}_{cav}$ by up to ~2% at high photon energies from no correction at cobalt-60, while plastic vials decrease $\bar{D}_{cav}$ by less than 0.5% even at high photon energies.

When used in photon beams, the wall is made thicker than the range of secondary charged particles released in the wall material. In electron beams, on the other hand, to minimize wall effects and obviate the need for a wall correction of $\bar{D}_{cav}$, walls of the vial are usually made as thin as possible and of plastic tissue-equivalent materials rather than glass. When glass vials are used, an appropriate wall correction factor $p_{wall} < 1$ must be used in the determination of $\bar{D}_{cav}$.

By way of example, we now take a closer look at wall effects in four Fricke dosimeters, each with a different wall material (polystyrene, Lucite, polyethylene, and Pyrex glass) and cavity filled with standard Fricke solution. The dosimeters are irradiated to the same dose in a water phantom by external cobalt-60 γ-ray beam, as indicated schematically in Fig. 16.3a. The objective of this example is to determine the relationship between the measured mean cavity dose $\bar{D}_{cav}$ (dose-to-Fricke solution) and the dose-to-medium (water) D_{med} in the phantom in the absence of the dosimeter.

In our experiment D_{med} will be the same for the four irradiations; however, the ratio $D_{med}/\bar{D}_{cav}$ will not necessarily be the same for all four irradiations because of the likely variations in the wall perturbation of the photon spectrum as well as the charged particle spectrum in the dosimeter cavity. To describe the two dose ratios $\bar{D}_{cav}/D_{wall}$ and D_{wall}/D_{med} of interest in our study we will use the general Burlin equation given in (15.24) as follows

$$\frac{\bar{D}_{cav}}{D_{wall}} = \Delta \frac{(\overline{S}_{col}/\rho)_{cav}}{(\overline{S}_{col}/\rho)_{wall}} + (1-\Delta)\frac{(\overline{\mu}_{ab}/\rho)_{cav}}{(\overline{\mu}_{ab}/\rho)_{wall}} \tag{16.34}$$

and

$$\frac{D_{wall}}{D_{med}} = \Delta \frac{(\overline{S}_{col}/\rho)_{wall}}{(\overline{S}_{col}/\rho)_{med}} + (1-\Delta)\frac{(\overline{\mu}_{ab}/\rho)_{wall}}{(\overline{\mu}_{ab}/\rho)_{med}}, \tag{16.35}$$

where $(\overline{S}_{col}/\rho)$ and $(\overline{\mu}_{ab}/\rho)$ are defined in conjunction with (15.24) and

$\bar{D}_{cav}$ is the mean dose-to-cavity with "cav" standing for the standard Fricke solution.

D_{wall} is dose-to-wall where the wall is of a thickness larger than the range of secondary electron in the wall material, yet thin enough not to disrupt the photon energy fluence in the cavity.

D_{med} is dose-to-water phantom at point-of-interest A in the absence of the Fricke dosimeter.

Assuming that D_{wall}/D_{med} of (16.35) meets conditions for large cavity, we insert $\Delta = 0$ into (16.35) and get the following expression for the ratio D_{wall}/D_{med}

$$\frac{D_{\text{wall}}}{D_{\text{med}}} = \frac{(\overline{\mu}_{\text{ab}}/\rho)_{\text{wall}}}{(\overline{\mu}_{\text{ab}}/\rho)_{\text{med}}}. \tag{16.36}$$

Next, we express the ratio $\overline{D}_{\text{cav}}/D_{\text{med}}$ as

$$\frac{\overline{D}_{\text{cav}}}{D_{\text{med}}} = \frac{\overline{D}_{\text{cav}}}{D_{\text{wall}}} \frac{D_{\text{wall}}}{D_{\text{med}}} = \Delta \frac{(\overline{S}_{\text{col}}/\rho)_{\text{cav}}}{(\overline{S}_{\text{col}}/\rho)_{\text{wal}}} \frac{(\overline{\mu}_{\text{ab}}/\rho)_{\text{wall}}}{(\overline{\mu}_{\text{ab}}/\rho)_{\text{med}}} + (1 - \Delta) \frac{(\overline{\mu}_{\text{ab}}/\rho)_{\text{cav}}}{(\overline{\mu}_{\text{ab}}/\rho)_{\text{med}}}, \tag{16.37}$$

where we merged (16.34)–(16.36) to get the product $(\overline{D}_{\text{cav}}/D_{\text{wall}}) \times (\overline{D}_{\text{wall}}/D_{\text{med}})$.

Equation (16.37) will let us evaluate the wall effect on the measured mean dose $\overline{D}_{\text{cav}}$ for the four wall materials; however, before embarking on the calculation of $\overline{D}_{\text{cav}}/D_{\text{med}}$, we must make reasonable approximations of the photon and secondary electron spectra in the cavity to be able to use the NIST data on mean mass energy absorption coefficient $\overline{\mu}_{\text{ab}}/\rho$ and mean mass stopping power $\overline{S}_{\text{col}}/\rho$. This means that we estimate the mean photon energy in the photon spectrum and the mean energy in the secondary electron spectrum in the cavity for use in the NIST tables for mass energy absorption coefficients μ_{ab}/ρ and mass stopping powers S_{col}/ρ.

Mean photon energy. In its beta minus decay into nickel-60 a cobalt-60 γ-source emits two energetic photons of energies 1.17 MeV and 1.33 MeV (Sect. 11.4.4). Since the two gamma photon energies are very close in magnitude, it is safe to take the mean energy of $E_\gamma = 1.25$ MeV for energy of cobalt-60 γ-rays. Therefore, in (16.37) we use the NIST tabulated values for mass energy absorption coefficient μ_{ab}/ρ at $E_\gamma = 1.25$ MeV.

Mean electron kinetic energy. When cobalt-60 photons of $E_\gamma = 1.25$ MeV interact with absorbing media, the predominant mode of photon interaction with absorber atoms is the Compton effect (see Fig. 8.5) in which the photon is scattered and transfers some of its energy to a "free" atomic electron. The mean energy transfer fraction $\overline{f}_C$ for energy transfer from 1.25 MeV photon to Compton electron in Compton effect is $\overline{f}_C = \overline{E}_K^C/E_\gamma = 0.47$, resulting in mean initial kinetic energy $\overline{E}_K^C$ of Compton electrons of $\overline{E}_K^C = 0.47 E_\gamma = 0.588$ MeV ≈ 0.6 MeV. In the first approximation we assume that at any point in the irradiated phantom-dosimeter combination, the mean kinetic energy $\overline{E}_K$ in the electron spectrum ranging from 0 to $\overline{E}_K^C \approx 0.6$ MeV will be equal to $\overline{E}_K = \frac{1}{2} \overline{E}_K^C \approx 0.3$ MeV. Therefore, in (16.37) we use the NIST tabulated values for mass collision stopping power at S_{col}/ρ at $E_K = 0.3$ MeV.

We are now ready to use (16.37) in output calibration of a cobalt-60 radiotherapy machine with a Fricke ferrous sulfate dosimeter in a water phantom. To evaluate wall effects on the dosimeter signal four different vials are studied and results for $\overline{D}_{\text{cav}}/D_{\text{med}}$ of (16.37) are summarized in Tables 15.5 and 15.6. The tables also show the NIST data for μ_{ab}/ρ at photon energy $E_\gamma = 1.25$ MeV and S_{col}/ρ at electron kinetic energy $\overline{E}_K = \frac{1}{2} \overline{E}_K^C \approx 0.3$ MeV.

Table 16.5 Wall effects in four Fricke chemical dosimeter, each with a different wall material (polyethylene, polystyrene, Lucite, and Pyrex glass) and dosimeter cavity filled with standard Fricke solution

(1)	(2)	(3)	(4)	(5)	(6)	(7)	(8)	(9)
(2)	Component	Material	μ_{ab}/ρ (cm^2/g)	S_{col}/ρ $\left(\frac{\text{MeV·cm}^2}{\text{g}}\right)$	$\frac{\bar{D}_{Fricke}}{D_{wat}}$ using (15.25)	$\frac{\bar{D}_{Fricke}}{D_{wat}}$ $\Delta = 1$	$\frac{\bar{D}_{Fricke}}{D_{wat}}$ $\Delta = 0$	Range (6) to (7) in %
(3)	Cavity	Fricke solution	0.02955	2.342	–	–	–	
(4)	Wall	Polyethylene	0.03049	2.497	0.997 – 0.033Δ	0.964	0.997	96.7–100
(5)		Polystyrene	0.02874	2.305	0.997 – 0.012Δ	0.985	0.997	98.8–100
(6)		Lucite	0.02882	2.292	0.997	0.997	0.997	100–100
(7)		Pyrex glass	0.02650	1.962	0.997 – 0.068Δ	0.929	0.997	93.2–100
(8)	Medium	Water	0.02965	2.355	–	–	–	

The Fricke dosimeters are irradiated to the same dose at same place in a water phantom by external cobalt-60 gamma ray beam. Equation (16.37) is used to determine the ratio between the measured mean cavity dose $\bar{D}_{cav} = \bar{D}_{Fricke}$ and the dose to medium $D_{med} = D_{water}$ in the phantom in the absence of the dosimeter. Data for μ_{ab}/ρ at photon energy $h\nu = 1.25$ MeV and S_{col}/ρ at electron kinetic energy $\bar{E}_K = \frac{1}{2}E_K^C \approx 300$ keV are from the NIST

A sample calculation of $\bar{D}_{cav}/D_{med}$ for the vial made of Pyrex glass is shown below

$$\frac{\bar{D}_{cav}}{D_{med}} = \frac{\bar{D}_{Fricke}}{D_{wat}} = \Delta \frac{(\bar{S}_{col}/\rho)_{Fricke}}{(\bar{S}_{col}/\rho)_{Pyrex}} \times \frac{(\bar{\mu}_{ab}/\rho)_{Pyrex}}{(\bar{\mu}_{ab}/\rho)_{wat}} + (1 - \Delta) \frac{(\bar{\mu}_{ab}/\rho)_{Fricke}}{(\bar{\mu}_{ab}/\rho)_{wat}}$$

$$= \Delta \frac{2.342}{1.962} \times \frac{0.02650}{0.02965} + (1 - \Delta) \frac{0.02955}{0.02965} = 1.065\Delta + (1 - \Delta) \times 0.997$$

$$= 0.997 - 0.068\Delta \tag{16.38}$$

Several interesting features can be gleaned from Tables 15.5 and 15.6.

1. Ratio $\bar{D}_{Fricke}/D_{wat}$ given in (16.37) depends on parameter Δ whose values range from $\Delta = 0$ for large cavities to $\Delta = 1$ for small cavities. For a given wall material, as shown in columns (5), (6), and (7) of Table 15.5, the ratio $\bar{D}_{Fricke}/D_{wat}$ ranges from $(\bar{D}_{Fricke}/D_{wat})_{min}$ at $\Delta = 1$ to $(\bar{D}_{Fricke}/D_{wat})_{max}$ at $\Delta = 0$.
2. For all wall materials $(\bar{D}_{Fricke}/D_{wat})_{max} = 0.997$, reflecting a minute difference in density and effective atomic number between the Fricke solution and water. This small difference is generally ignored and an assumption is made that $(\bar{D}_{Fricke}/D_{wat})_{max} \approx 1$.
3. Ratio $(\bar{D}_{Fricke}/D_{wat})_{min}$ varies significantly among the four wall materials ranging from 0.929 for Pyrex glass to 0.997 for Lucite. A conclusion can be reached that Lucite is an excellent match for water as far as dosimeter wall is concerned,

Table 16.6 Wall effects in four Fricke chemical dosimeter, each with a different wall material (polyethylene, polystyrene, Lucite, and Pyrex glass) and dosimeter cavity filled with Fricke solution

(1)	(2)	(3)	(4)	(5)	(6)	(7)
(2)	Component	Material	μ_{ab}/ρ (cm^2/g)	S_{col}/ρ $\left(\dfrac{\text{MeV}\cdot\text{cm}^2}{\text{g}}\right)$	$\dfrac{(\mu_{ab}/\rho)_x}{(S_{col}/\rho)_x}$	$\dfrac{\frac{(\mu_{ab}/\rho)_{\text{wat}}}{(S_{col}/\rho)_{\text{wat}}}}{\frac{(\mu_{ab}/\rho)_x}{(S_{col}/\rho)_x}}$
(3)	Cavity	Fricke solution	0.02955	2.342	0.01262	0.998
(4)	Wall	Polyethylene	0.03049	2.497	0.01221	1.031
(5)		Polystyrene	0.02874	2.305	0.01247	1.010
(6)		Lucite	0.02882	2.292	0.01258	1.001
(7)		Pyrex glass	0.02650	1.962	0.01351	0.912
(8)	Medium	Water	0.02965	2.355	0.01259	1.000

The Fricke dosimeters are irradiated to the same dose at same place in a water phantom by external cobalt-60 gamma ray beam. Equivalency between phantom medium (water) and wall medium (x) is established when the ratios $(\mu_{ab}/\rho)_x/(S_{col}/\rho)_x$ for water and wall are matched. Data for μ_{ab}/ρ at photon energy $h\nu = 1.25$ MeV and S_{col}/ρ at electron kinetic energy $\bar{E}_K = \frac{1}{2}E_K^C \approx 300$ keV are from the NIST

while a glass wall requires a 7% correction. Polystyrene and polyethylene are less problematic than glass, yet for accurate dosimetry they require a 1.2% and 3.3% correction, respectively.

4. Of the four materials in our study, Lucite is almost perfectly matched with water, while the other three common dosimetric materials show various degrees of deviation from water equivalency. Since in mass density Lucite deviates from water significantly more than do polystyrene and polyethylene, the question arises on what parameter to use for establishing dosimetric water equivalency for wall materials. From Burlin equation (16.37) we note that $\bar{D}_{\text{Fricke}}/\bar{D}_{\text{wat}} \approx 1$ under the assumption that: (i) Fricke solution is water equivalent and (ii) $(S_{col}/\rho)_{\text{wat}}/(\mu_{ab}/\rho)_{\text{wat}} = (S_{col}/\rho)_{\text{wall}}/(\mu_{ab}/\rho)_{\text{wall}}$. The wall material and cavity material (water) are then perfectly matched and the size of the cavity does not affect the dose in the cavity, as shown by Lucite in rows (6) of Tables 16.5 and 16.6.

16.2.14 Fricke Chemical Absolute Radiation Dosimetry: Summary

Fricke dosimetry is the best known, most studied, and oldest chemical radiation dosimetry system. Of the relatively large number of chemical dosimetry techniques, Fricke dosimetry is the only one that qualifies for the select group of absolute dosimetry techniques. Its absoluteness is governed by the accuracy with which the radiolytic yield $G(Fe^{3+})$ is known when dose is measured for a given type and energy of an ionizing radiation field.

Dilute aqueous ferrous sulfate solutions have an effective atomic number Z_{eff}, mass density ρ_{Fe}, and mass energy absorption coefficient μ_{ab}/ρ that are close to those of water. This means that Fricke solutions used in Fricke dosimeter cavities can be assumed water equivalent, obviating any need for corrections to measured mean dose-to-cavity $\bar{D}_{cav}$ as a result of mismatch between the cavity medium and water.

The relatively low sensitivity of Fricke dosimetry combined with the mass density of Fricke solutions that is of the order of that of water precludes the use of Bragg–Gray cavity ($d << R$) theory in Fricke dosimetry; instead, the Burlin cavity theory that is based on intermediate cavity size ($d \approx R$) is applied (d represents cavity dimensions; R represents the range of secondary charged particles in cavity medium). The dose-to-cavity $\bar{D}_{cav}$ determined with the Burlin theory represents the mean dose over the cavity sensitive volume.

The standard Fricke solution (0.001 M ferrous sulfate and 0.4 M sulfuric acid in water solution) covers the dose range from ~40 Gy to ~400 Gy. The lower limit of this range can be decreased by an order of magnitude by increasing the spectrophotometer's cuvette thickness from 1 cm to 10 cm; the upper limit can be extended to ~4000 Gy by raising the ferrous sulfate molarity from 0.001 M to 0.05 M and by aerating the Fricke solution during irradiation.

The useful dose range of Fricke dosimetry is too high for use in personnel dosimetry and barely practical for use in clinical dosimetry; however, its high dose range is very suitable for calibration of photon and electron machines used for radiation processing, food irradiation, and equipment sterilization.

16.3   Ionometric Absolute Radiation Dosimetry

16.3.1   Introduction to Cavity Ionization Chamber

The oldest and still the simplest means for detection and measurement of ionizing radiation is based on passing ionizing particles (including photons, neutrons, and charged particles) through a suitable radiation sensitive medium. This medium is most commonly some type of gas; however, liquids and solids are also used for this purpose under special circumstances. During the passage through the medium the particles ionize molecules along their tracks and the ions thereby created are collected and counted, providing information on the presence of ionizing radiation and on measurable quantities associated with its passage through the medium. The science that deals with production and measurement of ionization is referred to as ionometry; measurement of ionizing radiation with ionometric techniques is called ionometric radiation dosimetry.

The simplest of all gas-filled radiation detectors is the cavity ionization chamber filled with ambient air. It relies on collection of ions that the passage of ionizing radiation produces in the chamber sensitive volume (cavity) and it achieves this

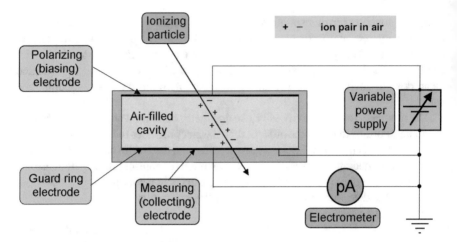

Fig. 16.4 Schematic diagram of a parallel-plate (also called end-window, pancake, and parallel-plate) ionization chamber system with three main components: power supply, electrometer, and cavity ionization chamber. The cavity chamber volume is defined with three electrodes: polarizing (biasing) electrode, measuring (collecting) electrode, and guard electrode

through an application of an electrostatic field in the chamber cavity. While more sophisticated radiation detectors (such as proportional counters and Geiger–Müller counters) usually operate in the pulse mode, ionization chambers are commonly used in the current mode as DC devices.

16.3.2 Ionization Chamber Dosimetry Systems

As shown schematically in Fig. 16.4, the basic design of an ionization chamber dosimetry system is quite simple and is composed of only three main components: cavity ionization chamber, power supply, and electrometer:

1. Cavity ionization chamber contains a cavity filled with the radiation sensitive material, most commonly in gaseous form, such as ambient air.
2. Variable power supply establishes an electric field in the sensitive volume of the cavity and enables the collection of charges that ionizing radiation produces in the dosimeter cavity. The power supply should have a polarity reversal option to allow measurement of the ionization chamber polarity effect.
3. Electrometer measures the small ionization charge produced in the radiation sensitive volume or the associated small ionization current (of the order of ~ 1 fA to ~ 1 pA). It is often calibrated separately in a standards laboratory in terms of charge or current per scale division and should have a digital display with four-digit (0.1%) resolution.

16.3.3   Electrodes of an Ionization Chamber

Typical ionization chambers incorporate three electrodes: polarizing, measuring, and guard:

1. *Polarizing electrode* (also called biasing electrode) is connected directly to the power supply and its function is to establish the electric field in the ionization chamber cavity.
2. *Measuring electrode* (also called collecting electrode) is connected to ground potential through the low impedance electrometer to measure the charge or current produced in the chamber sensitive volume.
3. *Guard electrode* is grounded directly and has two functions: it helps to define the chamber sensitive volume and it prevents the bulk of the chamber leakage current from being measured as part of the radiation-induced signal. (*Note*: measuring and guard electrodes are kept at approximately the same potential to minimize the leakage current between them).

Like capacitor, ionization chamber comes in three types with regard to the geometry of electrodes:

1. *Parallel-plate ionization chamber* (IC), also called flat IC, plane-parallel IC, or end-window IC, consists of two parallel conductive electrodes, usually disk-shaped and held apart by suitable insulators, as shown in Fig. 16.4.
2. *Cylindrical ionization chamber* (also called thimble ionization chamber) consists of two coaxial cylindrical electrodes.
3. *Spherical ionization chamber* consists of two concentric spherical electrodes.

16.3.4   Configuration of Ionization Chamber-Based Dosimetry System

From a practical standpoint, the most common configuration of an ionization chamber-based dosimetry system is depicted schematically in Fig. 16.5. The power supply and electrometer are merged into one unit and a triaxial cable is used to connect the ionization chamber with the electrometer/power supply unit. Two distinct ionization chamber configurations are in use: 1. Grounded electrometer and 2. Floating electrometer.

1. Grounded electrometer configuration is shown in Fig. 16.5a:
   Measuring electrode is connected through the core of the triaxial cable and through the electrometer to ground potential.
   Guard electrode is connected through the inner shield of the triaxial cable directly to ground potential.
   Polarizing electrode is connected through the outer sheath of the triaxial cable to the HV (high voltage) side of the power supply.

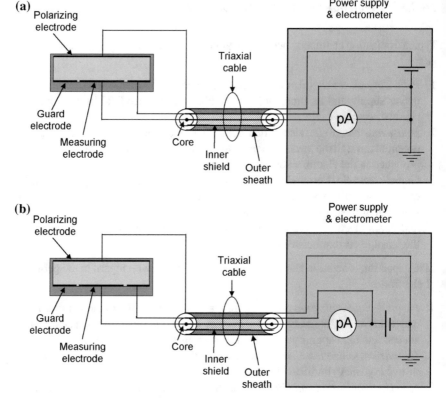

Fig. 16.5 Schematic diagram of a common configuration for ionization chamber system with power supply and electrometer merged into one unit and a triaxial cable used to connect the ionization chamber with the electrometer/power supply unit. **a** Grounded electrometer configuration, high voltage on outer shield of triaxial cable. **b** Grounded polarizing electrode, high voltage on core of triaxial cable

2. Floating electrometer configuration is shown in Fig. 16.5b:
   Measuring electrode is connected through the core of the triaxial cable and through the electrometer to the HV side of the power supply.
   Guard electrode is connected through the inner shield of the triaxial cable directly to the HV side of the power supply.
   Polarizing electrode is connected through the outer sheath of the triaxial cable to ground potential.

The grounded electrometer configuration is the older approach and seems more practical and logical. However, it places the high potential of the ionization chamber onto the outer sheath of the triaxial cable creating a shock hazard. The floating electrometer configuration is safer, since it keeps the outer sheath of the triaxial cable at ground potential, but the electrometer must be able to withstand operation at the high potential used for biasing of ionization chambers.

16.3.5 Ion Pairs Produced in an Ionization Chamber

The neutral atoms and molecules forming a gas are in constant thermal motion that manifests itself by random atomic and molecular collisions and a mean free path (MFP) travelled between two consecutive collisions. The MFP in gases at atmospheric pressure and temperature is of the order of 10^{-5} cm and, in general, is proportional to temperature and inversely proportional to pressure.

While passage of ionizing radiation through solids creates electron-hole pairs in the solid, the corresponding process in gases and liquids ionizes atoms and molecules and creates positive ions and secondary free electrons. The positive ion and free electron are referred to as an ion pair and serve as the basic component of the radiation induced signal obtained from the cavity ionization chamber. Under ideal conditions this signal is related to the total number of ion pairs created in the cavity that, in turn, is related to the energy transferred from ionizing radiation to cavity gas. The positive ions and free electrons created by ionizing radiation in the chamber cavity as ion pairs are also subjected to thermal motion as well as collisions in the cavity gas and are affected by the electric field present in the chamber cavity.

Two distinct charge collection mechanisms occur in ionization chambers:

1. If the chamber cavity is filled with an electronegative gas, i.e., gas that has a high affinity for free electrons, the free electrons forming an ion pair will tend to attach themselves to neutral gas atoms and create negative ions. Thus, in the cavity electric field, positive ions will drift toward the negative electrode and negative ions (rather than free electrons) will drift toward the positive electrode. Examples of highly electronegative gas are oxygen O_2, fluorine F, Freon, sulfur-hexafluoride SF_6, and air (because of oxygen content).

   Since ambient air is the most common medium used in an ionization chamber and air contains oxygen, a known electronegative gas, the charges collected on electrodes of an air-filled ionization chamber are positive and negative ions and **not** positive ions and free electrons.

2. When a non-electronegative gas (i.e., gas that does not attract electrons) is used in a gas-filled radiation detector, free electrons of the ion pairs remain free as they drift in the electric field toward the positive electrode. Examples of gases that have low electron affinity (non-electronegative behavior) are nitrogen N_2, hydrogen H_2, carbon dioxide CO_2, noble (inert) gases, methane, ethylene, and butane.

Sometimes, elements with very low electronegativity are referred to as electropositive elements, since they are characterized as electron donors in certain reactions and, as a result of losing an electron, become positive ions. As far as radiation sensitive gas in ionization chambers is concerned, we are dealing either with electronegative gases (strong affinity for free electrons) or non-electronegative gases that have no affinity for electrons but are not electron donors either.

Since in a gas at standard temperature and pressure electron drift velocity is some three orders of magnitude larger than ionic drift velocity, the free electrons are cleared out of the chamber cavity much faster than positive ions, and this affects the ion recombination in the cavity gas (see Sect. 16.3.6).

16.3.6   Fate of Ions in an Ionization Chamber

An ionization chamber can be regarded as a capacitor designed to give very low intrinsic leakage current and as large as possible radiation-induced current. The radiation-induced current is produced in the radiation sensitive cavity volume as a result of interactions between ionizing radiation and neutral gas molecules or atoms present in the chamber cavity. In these interactions, ion pairs consisting of positive and negative ions in electronegative gases or positive ions and free electrons in non-electronegative gases are created in chamber cavity. The charge carriers drift toward appropriate electrode in the electric field that is present in the chamber sensitive volume.

As ions strike an electrode they revert to their original neutral state. On the one hand, positive ions land on the negative electrode, collect an electron from the electrode, and revert to neutral state. On the other hand, free electrons or negative ions land on the positive electrode; the free electron enters the electrode while the negative ion transfers its extra electron to the electrode.

The electron exchange on electrodes creates movement of electrons in the ionization chamber circuitry enabled by the power supply to keep the capacitor (ionization chamber) charged to the level allowed by the intrinsic capacitance of the chamber. This current (or charge) is measured by the electrometer and is proportional to the dose rate (or dose, respectively) absorbed in the cavity medium of the ionization chamber provided that all charges that ionizing radiation produces in the cavity sensitive volume are actually collected. An ionization chamber is thus a simple air-filled capacitor in which ionizing radiation produces a leakage current by ionizing the air molecules. The power supply keeps the capacitor charged to the level in effect when no radiation is present.

16.3.7   Ion Recombination

As positive and negative ions in an electronegative gas or positive ions and free electrons in a non-electronegative gas drift in the electric field present in the ionization chamber cavity, they experience numerous random collisions with neutral atoms and molecules of the cavity gas. In addition, they also experience collisions with other ions and free electrons drifting in the electric field. While in collisions of ions with neutral atoms or molecules charge transfer can occur, the final result of the collision does not change the charge picture in the chamber cavity. However, when both collision partners are charged and of opposite polarity, the situation can become more complicated:

1. In a positive ion collision with a negative ion in an electronegative gas, the electron may transfer from negative ion to positive ion and the interaction, called an ion–ion recombination, results in two neutralized ions (two neutral atoms).
2. In a positive ion collision with a free electron in a non-electronegative gas, the positive ion may capture the free electron and become a neutral atom in a reaction called an ion–electron recombination.

The two collision mechanisms listed above between charged particles in an ionization chamber may result in neutral atoms or molecules and the process is then called ion recombination. In either of the two cases, ion recombination represents loss of charge that was produced through ionizing radiation interactions in the cavity ionization chamber. This charge loss decreases the measured radiation-induced output signal of the ionization chamber and, if the goal of measurement is accurate ionization chamber-based radiation dosimetry, one must account for this charge recombination loss. The larger is the loss of charges to recombination, the larger is the recombination correction factor. Determination of this recombination correction factor is based on recombination theories and is by no means a trivial process.

16.3.8 Collection Efficiency and Saturation Curve of Ionization Chamber

When voltage V applied to the polarizing electrode of an ionization chamber exposed to constant radiation intensity is increased from zero to a high value, the ionization signal collected on the measuring electrode increases from zero (assuming that the intrinsic leakage current in the chamber is negligible), at first linearly with voltage V, then more slowly, until it finally approaches asymptotically a saturation value for the given radiation intensity. The ionization signal is either the ionization current j or the ionization charge Q.

The saturation ionization signal is in principle attained at a relatively high voltage when all ion pairs that ionizing radiation produces in the chamber cavity are collected on the measuring electrode and none are lost to recombination or any other charge loss mechanism. The saturation signal is represented either by the saturation current j_{sat} (proportional to radiation dose rate in the chamber cavity) or saturation charge Q_{sat} (proportional to dose to chamber cavity).

The curve relating the measured ionization current j or the measured ionization charge Q to applied voltage potential V (or electric field $\mathcal{E}$) is called the saturation curve of the ionization chamber for a given radiation beam intensity. A typical example of saturation curves for a typical ionization chamber is given in Fig. 16.6 depicting three saturation curves for various doses (100 cGy, 200 cGy, and 300 cGy) delivered to an ionization chamber air cavity. The following observations can be made:

1. With applied potential increasing from $V = 0$ the collected charge Q for the same dose increases first linearly with V, then more gradually with V until saturation signal Q_{sat} appears at relatively high V.

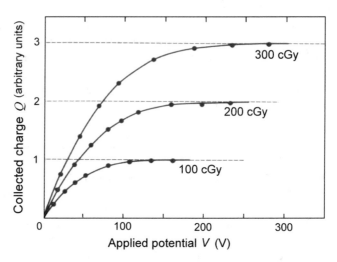

Fig. 16.6 Saturation curves for a typical ionization chamber exposed to ionizing radiation displayed as charge Q (typical units are nC) collected on the measuring electrode against applied potential V (typical units V). Saturation curves for three absorbed doses (100 cGy, 200 cGy, and 300 cGy) are shown. At low voltages V, collected charge Q first increases linearly with V, then more gradually with increasing V, until it asymptotically approaches saturation Q_{sat} at relatively high voltages. Saturation charge Q_{sat} is proportional to delivered dose

2. Saturation signal Q_{sat} depends linearly on dose – doubling of dose results in doubling of the saturation signal.
3. The onset of the saturation region moves to higher potential as the dose increases.
4. Another way to depict a saturation curve is to plot measured current j (typical units nA or pA) collected on measuring electrode against applied potential V for various dose rates (typical units cGy/min).

The ratio of the current j or charge Q measured at an arbitrary voltage V to saturation current j_{sat} or saturation charge Q_{sat}, respectively, is called collection efficiency $f(V)$. The collection efficiency $f(V)$ is thus defined as the ratio of radiation signal (j or Q) collected on the measuring electrode at voltage V to the saturation signal (j_{sat} or Q_{sat}, respectively) produced in the chamber cavity by ionizing radiation

$$f(V) = \frac{j(V)}{j_{sat}} \qquad \text{or} \qquad f(V) = \frac{Q(V)}{Q_{sat}}. \qquad (16.39)$$

Defined as in (16.39), $f(V)$ lies between 0 and 1, where 1 indicates full saturation. Often collection efficiency $f(V)$ is also given as percentage of full saturation that occurs at $f(V) = 100\%$. Several distinct regions of the saturation region can be specified, such as: low collection efficiency region where $f < 50\%$; near-saturation region where $70\% \leq f \leq 99\%$; extreme near-saturation region where $99\% < f(V) < 100\%$; and full saturation where $f(V) = 100\%$.

Full saturation (above 99%) is difficult to attain in an ionization chamber experimentally, because, when collection efficiency reaches 99% and applied voltage continues to increase in search of full saturation, the recombination becomes negligible but two other phenomena begin to occur that may cause a spurious increase in nondosimetric ionization signal: 1. Breakdown and 2. Charge multiplication.

1. *Mechanical or electrical breakdown* of insulating material that separates the chamber electrodes and can make a significant contribution to the measured ionization signal. This contribution is not radiation-induced and as such should not be counted as part of the radiation-induced signal.
2. *Charge multiplication*, in which free electrons, as they drift in the chamber electric field, can gain enough kinetic energy between consecutive collisions with atoms of the cavity gas to be able to ionize other atoms with which they subsequently collide. Thus, a single ion, as it travels toward collecting electrode, can give rise to a cascade of ions. This too increases the measured ionization signal but the charge multiplication is not radiation-induced and thus should not be counted as part of the radiation-induced dosimetric signal.
   It should be noted that, while charge multiplication is clearly unwanted in ionization chamber dosimetry, the operation of more sophisticated radiation sensing devices, such as proportional counters and Geiger-Müller counters, is based to a large extent on charge multiplication. However, we must note that these devices are not suitable for absolute radiation dosimetry.

The investigation of ion recombination and saturation curves of parallel plate ionization chambers has a long history that started with Ernest Rutherford and John J. Thomson in 1896, soon after Röntgen discovered x-rays and measurement of x-ray intensity became of importance. During the subsequent several decades many physicists, most notably Gustav Mie, John S. Townsend, and Jack Boag, proposed various theories to deal with the recombination mechanisms and saturation behavior of gases; however, none of the proposed theories succeeded in covering the complete saturation curve and, in addition, most of them are very complicated.

Based on Mie's quite elaborate but complicated theory, Boag has developed a simple recombination theory for air-filled ionization chambers governed by collection of negative and positive ions. The Boag theory fails to describe the saturation curve at low collection efficiencies but has three redeeming features: (i) it is simple, (ii) it works well in the near-saturation region where collection efficiency exceeds 70%, and (iii) it ignores the spurious non-radiation induced effects in the saturation region where measurement of saturation signal (current j_{sat} or charge Q_{sat}) is unreliable.

Since the region above collection efficiency of 70% is of most relevance to medical physics and radiation dosimetry, national and international dosimetry protocols rely on Boag theory when dealing with charge recombination issues in accurate radiation dosimetry, such as the determination of saturation current j_{sat} or saturation charge Q_{sat} and the required correction for loss of charge as a result of ion recombination.

16.3.9   Charge Loss in Ionization Chamber
for Continuous Beams

A closer look at the charge loss mechanism in an ionization chamber reveals that there are two types of recombination loss: initial recombination and general recombination. In addition to the two recombination losses, there is one more charge loss mechanism: diffusion of charge carriers against electric field in the dosimeter cavity. Thus, at voltages V below the saturation voltage, the lack of saturation, i.e., incomplete collection of charge in an ionization chamber, may be attributed to the following three distinct mechanisms, listed in decreasing order of importance:

1. General (volume) recombination where recombination occurs between charge carriers originating from different ionizing tracks.
2. Initial (columnar) recombination where positive and negative charge carriers formed in the track of a single ionizing particle collide and recombine.
3. Diffusion loss caused by the diffusion of ions onto the measuring electrode against the electric field.

The three distinct charge loss mechanisms, listed above, are affected by the electric field in the chamber cavity, properties of the cavity gas, as well as the type of the radiation beam. Both the general recombination and diffusion loss are affected by the density of radiation tracks, i.e., dose rate; initial recombination does not depend on the dose rate.

Three types of radiation beams are considered in recombination theories:

1. Continuous beam, such as a radiation beam emitted by a gamma source or x-ray machine.
2. Pulsed beam, such as a beam emitted by a standard clinical linear accelerator that uses a flattening filter for production of flat radiation fields.
3. Pulsed-scanned beam, such as a beam emitted by a linear accelerator that uses magnetically scanned electron pencil beam for producing large radiation fields.

Pulse-scanned beams are not common and are listed here only for completeness; in further discussions we will refer only to the first two common options: either continuous radiation beam or pulsed radiation beam. Examples of machines for producing continuous beams are cobalt-60 machines and kilovoltage x-ray machines; best example for a machine producing a pulsed beam is a linear accelerator (linac).

General (Volume) Recombination in a Continuous Radiation Beam

General recombination (also referred to as volume recombination) is defined as recombination between charge carriers that originate in different ionizing tracks. In continuous radiation beams the general recombination is by far the most important charge loss mechanism, so that, in comparison with general recombination, initial recombination and diffusion against electric field in the cavity chamber are usually ignored. Under these conditions the Boag simple theory provides the following expression for collection efficiency f_{gen}^c for a constant dose rate and an electronegative cavity gas

$$f_{gen}^c = \frac{Q(V)}{Q_{sat}} = \frac{1}{1 + \Lambda_{gen}^c/V^2}, \tag{16.40}$$

where V is the potential applied on the chamber cavity in volt (V) and Λ_{gen}^c is a parameter that is proportional to dose rate but also depends on chamber geometry and properties of ions of the cavity gas, such as ion mobility and recombination coefficient. *Note* that the labels "gen" and "c" stand for "general recombination" and "continuous beam", respectively. For the purposes of radiation dosimetry protocols, the detailed information on parameter Λ_{gen}^c is not important. The relationship (16.40) was found to be valid in the near saturation region ($f > 0.7$) and may also be written in the following linearized form related to $f_{gen}^c = Q(V)/Q_{sat}$

$$\frac{1}{Q(V)} = \frac{1}{Q_{sat}} + \frac{\lambda_{gen}^c}{V^2}, \tag{16.41}$$

where Λ_{gen}^c and λ_{gen}^c are related through the following relationship $\lambda_{gen}^c = \Lambda_{gen}^c/Q_{sat}$, indicating that parameter λ_{gen}^c is independent of dose rate since both Λ_{gen}^c and saturation charge Q_{sat} are directly proportional to dose rate.

Equation (16.41) implies that when Q, measured as a function of V, is plotted in the form $1/Q$ against $1/V^2$ a straight line should result that, when extrapolated to $1/V^2 = 0$, yields $1/Q_{sat}$, the reciprocal of the saturation charge Q_{sat}. Thus, the extrapolation of the straight line to $1/V^2 = 0$ should provide a reliable method to correct for general recombination in a continuous radiation beam, provided that the data points do not fall below $f_{gen}^c = Q(V)/Q_{sat} = 0.7$ and do not extend to very large potentials above the near saturation region where electrical breakdown or charge multiplication would make a spurious non-dosimetric contribution to the collected charge.

A simplified version of the straight-line approach in the near-saturation region is the so-called *two-voltage technique* in which collected charge is measured only at two voltages, V_H and V_L assuming, of course, that the linear expression (16.41) is valid in the region spanned by the two voltage points (high H and low L). If Q_H and Q_L are the charges measured at V_H and V_L, respectively, then f_{gen}^c, the collection efficiency at applied potential V_H, can be expressed in terms of Q_H, Q_L, V_H and V_L as

$$f_{gen}^c(V_H) = \frac{Q_H}{Q_{sat}} = \frac{\left(\frac{V_H}{V_L}\right)^2 - \frac{Q_H}{Q_L}}{\left(\frac{V_H}{V_L}\right)^2 - 1}, \tag{16.42}$$

where, as recommended in some dosimetry protocols, the ratio V_H/V_L must be greater than 2 for the determination of f_{gen}^c and Q_{sat}. Equation (16.42) is derived by writing (16.41) for Q_H and Q_L, and then using both equations to derive (16.42).

The ion recombination factor $P_{ion}^c(V_H)$ to be applied to the measured Q_H for a continuous radiation beam is defined as $[f_{gen}^c(V_H)]^{-1}$ and given as

$$P_{\text{ion}}^{c}(V_{\text{H}}) = \frac{1}{f_{\text{gen}}^{c}(V_{\text{H}})} = \frac{Q_{\text{sat}}}{Q_{\text{H}}} = \frac{\left(\dfrac{V_{\text{H}}}{V_{\text{L}}}\right)^{2} - 1}{\left(\dfrac{V_{\text{H}}}{V_{\text{L}}}\right)^{2} - \dfrac{Q_{\text{H}}}{Q_{\text{L}}}}, \tag{16.43}$$

while the saturation charge Q_{sat} is given as

$$Q_{\text{sat}} = Q_{\text{H}} P_{\text{ion}}^{c}(V_{\text{H}}) = \frac{Q_{\text{H}}}{f_{\text{gen}}^{c}(V_{\text{H}})} = Q_{\text{H}} \frac{\left(\dfrac{V_{\text{H}}}{V_{\text{L}}}\right)^{2} - 1}{\left(\dfrac{V_{\text{H}}}{V_{\text{L}}}\right)^{2} - \dfrac{Q_{\text{H}}}{Q_{\text{L}}}}. \tag{16.44}$$

By way of example, we now calculate the saturation current j_{sat} with the two-voltage technique for measurement of saturation current using a dosimetry system based on an ionization cavity chamber exposed to a cobalt-60 gamma ray beam. *Note*: The gamma ray beam from a cobalt-60 source is classified as a continuous photon beam. The pertinent experimental data are as follows:

Standard (high) biasing potential for the ionization chamber: $V_{\text{H}} = 300$ V
Lower biasing potential for the two-voltage technique: $V_{\text{L}} = 120$ V
Current measured at standard potential: $j_{\text{H}} = 1.500$ pA at $V_{\text{H}} = 300$ V
Current measured at lower potential: $j_{\text{L}} = 1.390$ pA at $V_{\text{L}} = 120$ V

Based on information above, we now calculate the saturation current j_{sat} and the ion recombination factor $P_{\text{ion}}^{c}(V_{\text{H}})$ that must be applied to the current j_{H} measured at standard potential of 300 V. We get j_{sat} using (16.44) and $P_{\text{ion}}^{c}(V_{\text{H}})$ using (16.43) with the following parameters: $V_{\text{H}}/V_{\text{L}} = 300/120 = 2.5$ and $j_{\text{H}}/j_{\text{L}} = 1.500/1.390 = 1.079$. Hence, the saturation current j_{sat} is calculated as follows

$$j_{\text{sat}} = j_{\text{H}} P_{\text{ion}}^{c}(V_{\text{H}}) = \frac{j_{\text{H}}}{f(V_{\text{H}})} = j_{\text{H}} \frac{\left(\dfrac{V_{\text{H}}}{V_{\text{L}}}\right)^{2} - 1}{\left(\dfrac{V_{\text{H}}}{V_{\text{L}}}\right)^{2} - \left(\dfrac{j_{\text{H}}}{j_{\text{L}}}\right)}$$

$$= (1.500 \text{ pA}) \times \frac{6.250 - 1}{6.250 - 1.079} = 1.523 \text{pA} \tag{16.45}$$

and the ion recombination factor $P_{\text{ion}}^{c}(V_{\text{H}})$ is

$$P_{\text{ion}}^{c}(V_{\text{H}}) = \frac{j_{\text{sat}}}{j_{\text{H}}} = \frac{1.523}{1.500} = 1.015. \tag{16.46}$$

We also note that the collection efficiency f_{gen}^{c} for the current j_{H} measured at the standard potential of $V_{\text{H}} = 300$ V is $f_{\text{gen}}^{c} = j_{\text{H}}/j_{\text{sat}} = 1/P_{\text{ion}}^{c}(V_{\text{H}}) = 98.5\%$.

Results of the j_{sat} calculation are illustrated in Fig. 16.7 with $1/j$ versus $1/V^{2}$ graph where $1/j$ is the reciprocal of ionization current j and $1/V^{2}$ is the square of the reciprocal of applied potential V. The straight line through the two data points (ionization currents j_{H} and j_{L} at biasing voltages of 300 V and 120 V, respectively)

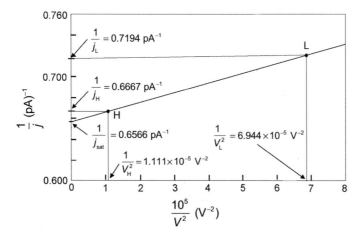

Fig. 16.7 Illustration of the two-voltage method for determination of saturation current j_{sat} with ionization chamber in continuous radiation beam. For standard voltage point H at $V_H = 300$ V, the measured current is $j_H = 1.500$ pA; for low voltage point L at $V_L = 120$ V, the measured current is $j_H = 1.390$ pA. For the two measured points (H and L), a graph of $1/j$ against $1/V^2$ is plotted and a straight line through the two points H and L extrapolated to $1/V^2 \rightarrow 0$ (corresponding to $V \rightarrow \infty$) gives $1/j_{sat} = 0.6566$ pA^{-1}

is extrapolated to $1/V^2 = 0$ and the intersection point with the ordinate axis yields $1/j_{sat}$, the reciprocal of the saturation current j_{sat}.

In our calculation we used the theory that Jack Boag developed for the general ion recombination in a continuous radiation beam, such as a cobalt-60 beam or kilovoltage x-ray beam. In continuous radiation beams, initial recombination as well as diffusion loss amount to less than 0.2% of the charge loss caused by general recommendation and are largely ignored, unless extreme accuracy is desired.

We will now assume that we have data for the saturation curve of Fig. 16.7 for biasing potentials ranging from 115 V to 600 V and we plot the data in Fig. 16.8 on the $1/j$ versus $1/V^2$ graph. The two-voltage technique depicted in Fig. 16.7 seems clear and straightforward; however, there is an obvious question that must be answered with respect to the two-voltage technique: Do we know for sure that the two measured $1/j$ points (reciprocals of ionization currents j_H and j_L for biasing voltages of 300 V and 120 V, respectively) fall into the region $0.70 < f < 0.99$ in which Boag theory is valid and where the $1/j$ versus $1/V^2$ graph yields a straight line? This question can only be answered with a measurement of the saturation curve in two saturation curve regions: near-saturation where data fall onto a straight line $(0.70 < f < 0.99)$ and extreme near-saturation $(f > 0.99)$ where breakdown and charge multiplication effects may add an unwanted signal to that resulting from the general recombination.

In Fig. 16.8 the reciprocals of ionization currents j_H and j_L of Fig. 16.7 for square of reciprocals of voltage points at 300 V and 120 V, respectively, are shown with black dots; reciprocals of currents for all other voltage points, ranging from 115 V to

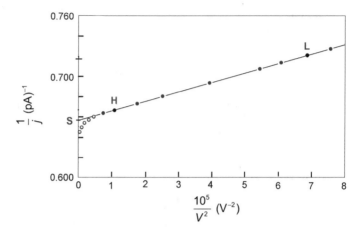

Fig. 16.8 Plot of complete saturation data for the two-voltage experiment of Fig. 15.11 plotted in the form of $1/j$ against $1/V^2$. The two data points H and L of Fig. 15.11 are shown with black dots, the other data points that fit onto the straight line defined by H and L points are shown with red dots. In the extreme near-saturation region at very large voltages, the data points no longer fit onto the straight line and they are shown with hollow red dots. The deviation of these data points from the straight line are attributed to non-dosimetric effects resulting from the onset of charge multiplication. S indicates $1/j_{\text{sat}}$, the inverse of the saturation current j_{sat} calculated from Boag theory

600 V, are shown with solid or hollow red dots. All points between 115 V to 350 V are shown with solid red dots and fall on the straight line defined by the two black dots H and L of the two-voltage technique, shown in Fig. 16.7.

In the region between 115 V and 350 V in Fig. 16.8, Boag theory and the two-voltage technique are valid. However, it is obvious that reciprocals of currents for voltage points above 350 V (shown with hollow red dots) deviate from the straight line and the reason for this deviation is that the measured currents j are too large (i.e., current reciprocals are too low) – the higher is the voltage, the larger is the deviation from the straight line. We conclude that the discrepancy between Boag theory and measurement at $V > 350$ V arises from charge multiplication effects in the cavity gas. These effects are not of a dosimetric origin and must therefore be ignored when the measurement is used for radiation dosimetry.

The high-voltage non-linearity of the saturation curve on a $1/j$ versus $1/V^2$ plot clearly shows that direct measurement of j_{sat} is difficult. On the other hand, the two-voltage technique is simple and reliable provided, of course, that reciprocals of currents j_H and j_L measured for voltages V_H and V_L, respectively, fall on the straight portion of the measured saturation curve, plotted in the format $1/j$ versus $1/V^2$. If not, the magnitude of V_H must be adjusted accordingly (lowered) to ensure that constraints imposed on the two-voltage technique are respected.

The discussion above is based on the theory of ionization chambers that Jack Boag developed in 1960s but is still in general use in national and international dosimetry protocols. The theory introduced several simplifying, yet reasonable, assumptions that lead to the following steps for determining the collection efficiency $f(V_H)$, saturation charge Q_{sat} (or saturation current j_{sat}), and ion recombination factor $P_{ion}^{cont}(V_H)$ from measured ionization charge Q (or ionization current j) at standard applied potential V_H in continuous radiation beams:

1. Initial recombination and ion diffusion against electric field in the chamber cavity are ignored and the predominance of general recombination is assumed in continuous radiation beams.
2. A sufficiently high polarizing voltage V_H (yet, in order to avoid charge multiplication, not too high) is applied to the chamber cavity to ensure that all ionization charges Q (or ionization currents j) are measured in the near saturation region where the collection efficiency is between $\sim 70\%$ and $\sim 99\%$.
3. Collection efficiency $f(V_H)$ is determined using the two-voltage technique that was derived from the assumed linear relationship in the near saturation region between $1/Q$ and $1/V^2$, where the higher voltage point is the standard voltage V_H and the lower voltage point V_L is chosen such that $V_H/V_L > 2$.
4. Saturation charge Q_{sat} or saturation current j_{sat} is calculated from the ratio $Q(V_H)/f(V_H)$ or $j(V_H)/f(V_H)$, respectively.

Initial Recombination for Continuous and Pulsed Radiation Beams

Initial recombination (also called columnar recombination) is defined as recombination of charge carriers that were formed in the same charged particle track. Thus, in contrast to general recombination, initial recombination is independent of dose rate. Moreover, it makes only a small contribution to total recombination in ionization chambers operating at atmospheric pressure for low LET radiations and continuous radiation beams where general recombination predominates. However, initial recombination can make a significant contribution to recombination for ionizing radiations that form densely ionizing tracks. Examples of these are high LET radiations, such as α-particles in standard pressure air or electron tracks in high-pressure gas.

Russian-American physicist George C. Yaffé was the first to present a theoretical treatment of initial recombination reducing recombination to a problem of simple Brownian motion under the influence of two forces: (i) Coulomb attraction between oppositely charged ions produced in the same charged particle ionization track and (ii) applied electric field in an ionization chamber.

Jaffé found that the collection efficiency for initial recombination f_{init} in an ionization chamber for large polarizing potential V applied to chamber cavity may be expressed in the following form

$$f_{init} = \frac{Q(V)}{Q_{sat}} = \frac{1}{1 + \Lambda_{init}/V}, \tag{16.47}$$

where Λ_{init} is a dose rate-independent parameter incorporating various chamber as well as cavity gas parameters and $Q(V)$ is the ionization charge collected at applied potential V. The equation was initially developed for electric field normal to the ionization track; however, it retains its general form when it is extended to arbitrary angles between the applied electric field and ionization track.

Equation (16.47) is valid in the near saturation region where it can be linearized to get

$$\frac{1}{Q} = \frac{1}{Q_{\text{sat}}} + \frac{\lambda_{\text{init}}}{V},\tag{16.48}$$

with $\lambda_{\text{init}} = \Lambda_{\text{init}}/Q_{\text{sat}}$ a parameter dependent on properties of the cavity gas, such as mobility, diffusion coefficient, and recombination coefficient. According to (16.48), a plot of $1/Q$ versus $1/V$ will provide a straight line that, when extrapolated to $1/V = 0$ corresponding to $V \to \infty$, results in $1/Q_{\text{sat}}$.

For continuous low LET radiation beams and chamber cavity gas at ambient temperature and pressure, the initial recombination is negligible in comparison with general recombination. The linear relationship in $1/Q$ versus $1/V$ plot can be noticed only at high potentials where the general recombination is no longer present but initial recombination still exists. However, this is also the area of the onset of charge multiplication that makes it very difficult to distinguish the minute contribution of initial recombination from the contribution of unwanted charge multiplication.

In pulsed beams, in contrast to continuous radiation beams, the initial recombination contributes to the charge loss process, since both the general recombination and initial recombination are governed by the $1/Q$ versus $1/V$ relationship. Therefore, there is no point in separating the two contributions from each other because a single plot of $1/Q$ versus $1/V$ relationship covers combined effects of both charge loss mechanisms.

Thermal Ion Diffusion Against Applied Electric Field

In addition to the general and initial recombination, some charge loss in a polarized ionization chamber may result from thermal diffusion of ions against the applied electric field in the chamber cavity. The diffusion of ions in the chamber is a natural result of Brownian motion in which identical gas particles become evenly distributed throughout the cavity volume. French physicist Paul Langevin worked on this problem and found that the collection efficiency f_{dif} for thermal diffusion may be expressed as

$$f_{\text{dif}} = \frac{1}{1 + \Lambda_{\text{dif}}/V},\tag{16.49}$$

Equation (16.49), in same fashion as (16.47) may be linearized to get

$$\frac{1}{Q} = \frac{1}{Q_{\text{sat}}} + \frac{\lambda_{\text{dif}}}{V},\tag{16.50}$$

where the diffusion parameter Λ_{dif} is given as $\Lambda_{\text{dif}} = 2kT/e = \lambda_{\text{diff}} Q_{\text{sat}}$, with k the Boltzmann constant, T the temperature, e and the elementary charge (1.602×10^{-19} C). At room temperature $kT \approx 0.05$ eV and $\Lambda_{\text{dif}} \approx 0.05$ V. Equation (16.50), similarly to (16.48), implies that a plot of $1/Q$ versus $1/V$ results in a straight line that, when extrapolated to $1/V = 0$ corresponding to $V \to \infty$, gives $1/Q_{\text{sat}}$.

Similarly to charge loss in initial recombination, charge loss in diffusion for an ionization chamber exposed to low LET continuous radiation is negligible compared to charge loss in general recombination. In pulsed radiation, on the other hand, ion diffusion contributes to the total charge loss; however, since all three dosimetric charge loss mechanisms in the near saturation region exhibit linearity of $1/Q$ versus $1/V$, an extrapolation of ionization data to $1/V=0$ accounts for all three charge loss mechanisms and yields $1/Q_{sat}$ that is assumed valid for determination of dose or dose rate delivered to the chamber cavity gas.

16.3.10 Charge Loss in Ionization Chambers for Pulsed Radiation Beams

For purposes of radiation dosimetry a pulsed beam is a beam with a pulse width τ significantly shorter than the drift time t_{dr} of ions between the ionization chamber electrodes. The pulse repetition rate (rep rate) of the linac must be low enough, so that ions can clear out of the chamber cavity between consecutive pulses.

The most common source of pulsed beams is a linear accelerator (linac). All linacs operate in a pulsed mode with a relatively low duty cycle (see Sect. 14.6.5). The pulse width τ is defined as the time period in a pulse during which radiation is produced in an accelerator, while pulse duration T is the time interval between corresponding points on successive pulses and is, thus, the reciprocal of the pulse repetition rate (also known as rep rate or pulse repetition frequency). Duty cycle δ is defined as the ratio τ/T between the pulse width τ and pulse duration T.

As discussed in Sect. 14.6.5 and shown in Fig. 14.6, the typical linac pulse width is $\tau = 2 \, \mu s$ and the typical rep rate is 100 pps (pulses per second). This means that the typical pulse duration is $T = 10^{-2} \, s = 10^4 \, \mu s$ and the duty cycle $\delta = 2 \times 10^{-4}$.

To clarify the definition of pulsed beams in radiation dosimetry we now estimate with an example the drift time t_{drift} of nitrogen and oxygen ions (N_2^+ and O_2^-, respectively) in an ionization chamber open to ambient air. Recall that N_2^+ ion originates with ionization of N_2 molecule by ionizing radiation particle (photon or charged particle), while O_2^- ion is formed when a O_2 molecule (because of its electronegativity) attracts a nearby free electron that was produced in an ionization event in air.

Consider a typical parallel-plate ionization chamber with a plate separation d of 2 mm and an applied potential U of 300 V, resulting in an electric field $\mathcal{E}$ of (300 V)/$(2 \times 10^{-3} \, m) = 1.5 \times 10^5 \, V/m$ in the chamber cavity.

By definition the drift velocity v_{drift} of ions in an electric field $\mathcal{E}$ is given as $v_{drift} = k\mathcal{E}$ where k stands for mobility defined as the drift velocity in m/s per 1 V/m. The SI unit of mobility is thus $m^2/(V \cdot s)$. The mobility k_+ of the N_2^+ ion in air is $k_+ = 0.18 \times 10^{-3} \, m^2/(V \cdot s)$; the mobility of the O_2^- ion in air is $k_- = 0.33 \times 10^{-3} \, m^2/(V \cdot s)$. For our estimate of the drift time t_{drift} between the two electrodes of the ionization chamber ($d = 2$ mm) we use the mean mobility of N_2^+ and O_2^- ions given as $\bar{k} = \frac{1}{2}(k_+ + k_-) = 0.25 \times 10^{-3} \, m^2/(V \cdot s)$.

The ion drift velocity v_{drift} for our ionization chamber is given as

$$v_{\text{drift}} = k\mathcal{E} = \left(0.25 \times 10^{-3} \, \frac{\text{m}^2}{\text{V} \cdot \text{s}}\right) \times \left(1.5 \times 10^5 \, \frac{\text{V}}{\text{m}}\right) = 37.5 \text{ m/s}, \qquad (16.51)$$

resulting in the following drift time t_{drift} of N_2^+ and O_2^- ions for a typical chamber electrode separation of 2 mm

$$t_{\text{drift}} \approx \frac{d}{v_{\text{drift}}} = \frac{2 \times 10^{-3} \text{ m}}{37.5 \text{ m/s}} = 5.3 \times 10^{-5} \text{ s} = 53 \text{ } \mu\text{s}. \qquad (16.52)$$

Since for a typical clinical linac the pulse width $\tau = 2$ µs and pulse duration $T = 10^4$ µs (for a typical rep rate of 100 pps), we conclude that the ion drift time t_{drift} is both: (i) much larger than pulse width τ and (ii) much smaller than pulse duration T. Thus, conditions for pulsed operation $\tau \ll t_{\text{drift}} \ll T$ are met and three conclusions important for radiation dosimetry can be made:

1. If $\tau \ll t_{\text{drift}}$, the production of charge ($\tau = 2$ µs) in the ionization chamber can be assumed to be almost instantaneous in comparison with the ion drift time ($t_{\text{drift}} \approx 50$ µs).
2. If $t_{\text{drift}} \ll T$, the ions produced by a given pulse have ample time ($t_{\text{drift}} \approx 50$ µs) to get cleared out from the cavity volume during the duration $T = 10^4$ µs of the pulse.
3. If, on the other hand, $\tau > t_{\text{drift}}$, then the theory of general recombination in continuous radiation beams applies.

General Recombination for Pulsed Radiation Beams in Electronegative Gas

In addition to having made in the 1960s an important contribution to the theory of general recombination in continuous radiation beams that is still relevant today, Jack Boag also worked on recombination theory for pulsed ionizing radiation beams. He proposed his theory on collection efficiency of ionization chambers exposed to pulsed radiation beams in the early 1950s and the theory still forms the basis of modern dosimetry protocols for calibration of radiation beams produced by linear accelerators.

Boag has shown that the collection efficiency $f_{\text{gen}}^{\text{P}}$ for general recombination in a pulsed radiation beam for electronegative gas may be expressed as

$$f_{\text{gen}}^{\text{P}} = \frac{Q(V)}{Q_{\text{sat}}} = \frac{1}{u} \ln(1 + u), \qquad (16.53)$$

where u is a quantity defined as $u = \Lambda_{\text{gen}}^{\text{P}}/V$ with V the applied potential in the chamber cavity and $\Lambda_{\text{gen}}^{\text{P}}$ a parameter that depends upon:

(i) initial charge density of ions created by the radiation pulse per volume of cavity gas,
(ii) chamber geometry (equivalent electrode spacing),

(iii) relevant properties of the cavity gas, such as mobility of positive and negative ions (k_+ and k_-) as well as recombination coefficient α.

For $f_{gen}^P > 0.9$ and small initial charge densities per pulse the quantity u becomes sufficiently small $u < 0.1$ to allow the function $\ln(1 + u)$ expanded as

$$\ln(1 + u) = u - \frac{1}{2}u^2 + \frac{1}{3}u^3 + \frac{1}{4}u^4 + \cdots ,\tag{16.54}$$

resulting in the following expression for (16.53) when only the first two terms of the expansion (16.54) are used

$$\frac{1}{f_{gen}^P} = \frac{Q_{sat}}{Q(V)} = \frac{u}{\ln(1 + u)} \approx \frac{u}{u - \frac{1}{2}u^2} = 1 + \frac{1}{2}u.\tag{16.55}$$

Equation (16.55) can be linearized to read

$$\frac{1}{Q(V)} = \frac{1}{Q_{sat}} + \frac{\lambda_{gen}^P}{V},\tag{16.56}$$

where $\lambda_{gen}^P = \Lambda_{gen}^P/(2Q_{sat})$. For charge losses below 5% the linear approximation (16.56) deviates from the exact formula (16.53) by less than 0.1%. Thus, a plot of the reciprocal of charge $Q(V)$ against the reciprocal of applied potential V results in a straight line that, when extrapolated to $1/V = 0$ corresponding to $V \to \infty$, yields the reciprocal of saturation charge Q_{sat}.

A simplified version of this extrapolation is the so-called two-voltage technique that is used to determine the collection efficiency f_{gen}^P and saturation current Q_{sat} with ionization chamber currents measured in pulsed radiation beams at only two applied potential points (standard voltage V_H and a lower voltage V_L) in the near-saturation region for $\sim 0.90 < f_{gen}^P < \sim 0.99$. If Q_H and Q_L are charges measured at applied potentials V_H and V_L, respectively, the collection efficiency $f_{gen}^P(V_H)$ at applied potential V_H is given by

$$f_{gen}^P(V_H) = \frac{\frac{V_H}{V_L} - \frac{Q_H}{Q_L}}{\frac{V_H}{V_L} - 1},\tag{16.57}$$

where, as recommended in some dosimetry protocols, the ratio V_H/V_L must be at greater than 2 for the determination of f_{gen}^P and Q_{sat}. Equation (16.57) is derived by writing (16.56) for Q_H and Q_L, and then using both equations to derive (16.57).

The ion recombination factor $P_{ion}^P(V_H)$ to be applied to the measured Q_H for a pulsed radiation beam is defined as the reciprocal of collection efficiency $f_{gen}^P(V_H)$ and given as

$$P_{\text{ion}}^{\text{p}}(V_{\text{H}}) = \frac{1}{f_{\text{gen}}^{\text{p}}(V_{\text{H}})} = \frac{Q_{\text{sat}}}{Q_{\text{H}}} = \frac{\dfrac{V_{\text{H}}}{V_{\text{L}}} - 1}{\dfrac{V_{\text{H}}}{V_{\text{L}}} - \dfrac{Q_{\text{H}}}{Q_{\text{L}}}}, \tag{16.58}$$

while the saturation charge Q_{sat} is given as

$$Q_{\text{sat}} = Q_{\text{H}} P_{\text{ion}}^{\text{p}}(V_{\text{H}}) = \frac{Q_{\text{H}}}{f_{\text{gen}}^{\text{p}}(V_{\text{H}})} = Q_{\text{H}} \frac{\dfrac{V_{\text{H}}}{V_{\text{L}}} - 1}{\dfrac{V_{\text{H}}}{V_{\text{L}}} - \dfrac{Q_{\text{H}}}{Q_{\text{L}}}} \tag{16.59}$$

By way of example we now determine, using the two-voltage technique, the following dosimetric quantities for a pulsed radiation beam:

1. Collection efficiencies $f_{\text{gen}}^{\text{p}}(V_{\text{H}})$ and $f_{\text{gen}}^{\text{p}}(V_{\text{L}})$ calculated with (16.57).
2. Ion recombination factor $P_{\text{gen}}^{\text{p}}(V_{\text{H}})$ and $P_{\text{gen}}^{\text{p}}(V_{\text{L}})$ calculated with (16.58).
3. Saturation charge Q_{sat} calculated with (16.59).
4. Parameters $\lambda_{\text{gen}}^{\text{p}}$ and $\Lambda_{\text{gen}}^{\text{p}}$ in Boag recombination equation (16.56) for pulsed radiation beams.

Radiation beam: 18 MV x-ray beam from a clinical linear accelerator. Field size: $10 \times 10 \text{ cm}^2$.

Ionization chamber is of cylindrical shape with cavity volume of 0.6 cm^3 open to ambient air through a rubber sheath. The chamber is positioned in a water phantom at a depth of 5 cm on the beam central axis. Surface of the water phantom is at 100 cm from the linac target.

Standard applied potential in chamber cavity for the two-voltage technique: $V_{\text{H}} = 250 \text{ V}$.

Lower applied potential in chamber cavity for two-voltage technique: $V_{\text{L}} = 100 \text{ V}$.

In this example, the two-voltage technique was used to determine the saturation charge Q_{sat} for a pulse-irradiated parallel-plate ionization chamber. Two separate measurements of charge (Q_{H} and Q_{L}) collected on the measuring electrode of the ionization chamber were made at the same dose per pulse but with different applied potentials: first at standard potential V_{H} and next at lower potential V_{L}.

Basic measured charge data obtained for the two-voltage technique on the ionization chamber:

$$Q_{\text{H}} = Q(V_{\text{H}}) = 41.82 \text{ nC} \quad \text{and} \quad Q_{\text{L}} = Q(V_{\text{L}}) = 40.85 \text{ nC}. \tag{16.60}$$

The applicable ratios $V_{\text{H}}/V_{\text{L}}$ and $Q_{\text{H}}/Q_{\text{L}}$ for calculation of dosimetric quantities $f_{\text{gen}}^{\text{p}}(V_{\text{H}})$, $P_{\text{ion}}^{\text{p}}(V_{\text{H}})$, and Q_{sat}, given in (16.57)–(16.59), respectively, are as follows

$$\frac{V_{\text{H}}}{V_{\text{L}}} = \frac{250 \text{ V}}{100 \text{ V}} = 2.50 \quad \text{and} \quad \frac{Q_{\text{H}}}{Q_{\text{L}}} = \frac{Q(V_{\text{H}})}{Q(V_{\text{L}})} = \frac{41.82 \text{ nC}}{40.85 \text{ nC}} = 1.02375 \tag{16.61}$$

The calculation results are as follows:

1. Collection efficiency $f_{gen}^P(V_H)$ is calculated using (16.57)

$$f_{gen}^P(V_H) = \frac{Q_H}{Q_{sat}} = \frac{\dfrac{V_H}{V_L} - \dfrac{Q_H}{Q_L}}{\dfrac{V_H}{V_L} - 1} = \frac{2.50 - 1.02375}{2.50 - 1.0} = 0.9842 \qquad (16.62)$$

It is easy to show that $f_{gen}^P(V_L)$ is given by

$$f_{gen}^P(V_L) = \frac{Q_L}{Q_{sat}} = \frac{\dfrac{V_L}{V_H} - \dfrac{Q_L}{Q_H}}{\dfrac{V_L}{V_H} - 1} = \frac{0.400 - 0.9768}{0.400 - 1.00} = \frac{-0.5768}{-0.600} = 0.9613$$

$$(16.63)$$

2. Ion recombination factor $P_{ion}^P(V_H)$ is calculated using (16.58)

$$P_{ion}^P(V_H) = \frac{Q_{sat}}{Q_H} = \frac{1}{f_{gen}^P(V_H)} = \frac{\dfrac{V_H}{V_L} - 1}{\dfrac{V_H}{V_L} - \dfrac{Q_H}{Q_L}} = \frac{2.50 - 1.0}{2.50 - 1.02375} = 1.0161$$

$$(16.64)$$

Since $P_{ion}^P(V_L) = [f_{gen}^P(V_L)]^{-1}$, we calculate $P_{ion}^P(V_L)$ as follows

$$P_{ion}^P(V_L) = \frac{Q_{sat}}{Q_L} = \frac{\dfrac{V_L}{V_H} - 1}{\dfrac{V_L}{V_H} - \dfrac{Q_L}{Q_H}} = \frac{0.400 - 1.00}{0.400 - 0.9768} = \frac{-0.600}{-0.5768} = 1.040$$

$$(16.65)$$

3. Saturation charge Q_{sat} is calculated using (16.59)

$$Q_{sat} = \frac{Q_H}{f_{gen}^P(V_H)} = Q_H \frac{\dfrac{V_H}{V_L} - 1}{\dfrac{V_H}{V_L} - \dfrac{Q_H}{Q_L}} = (41.82 \text{ nC}) \times \frac{2.50 - 1.0}{2.50 - 1.02375}$$

$$= 42.493 \text{ nC} \qquad (16.66)$$

4. Parameter λ_{gen}^P is determined using (16.56) either in conjunction with appropriate values of $Q_H = Q(V_H)$ from (16.60) and Q_{sat} from (16.66)

$$\lambda_{gen}^P = \left[\frac{1}{Q(V_H)} - \frac{1}{Q_{sat}} \right] V_H = \left[\frac{1}{41.82 \text{ nC}} - \frac{1}{42.493 \text{ nC}} \right] \times (250 \text{ V})$$

$$= 0.0947 \text{ V/nC} \qquad (16.67)$$

or in conjunction with appropriate values of $Q_L = Q(V_L)$ from (16.60) and Q_{sat} from (16.66)

$$\lambda^p_{gen} = \left[\frac{1}{Q(V_L)} - \frac{1}{Q_{sat}} \right] V_H = \left[\frac{1}{40.85 \text{ nC}} - \frac{1}{42.493 \text{ nC}} \right] \times (100 \text{ V}) = 0.0947 \text{ V/nC}$$
$$(16.68)$$

Equations (16.55) and (16.56) show that parameters Λ^p_{gen} and λ^p_{gen} are related through $\Lambda^p_{gen} = 2Q_{sat}\lambda^p_{gen}$ and we use this relationship to determine Λ^p_{gen} as follows

$$\Lambda^p_{gen} = 2Q_{sat}\lambda^p_{gen} = 2 \times (42.493 \text{ nC}) \times \left(0.0947 \frac{\text{V}}{\text{nC}} \right) = 8.05 \text{ V}. \quad (16.69)$$

Results for this example on the two-voltage technique for determination of saturation charge in an ionization chamber exposed to a pulsed radiation beam are presented in Fig. 16.9. The figure shows experimental data with $1/Q$ against $1/V$ plot for the charge Q collected at two applied voltages: V_H and V_L, with V_H the standard chamber operating potential holding the chamber close to saturation and V_L a lower potential, such that $V_H/V_L = 2.5$ in our example.

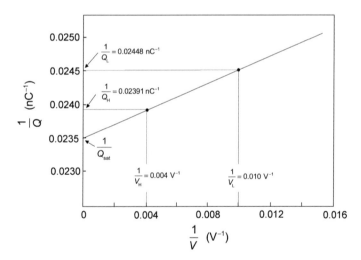

Fig. 16.9 Illustration of the two-voltage method for determination of saturation charge Q_{sat} with ionization chamber in a 18 MV pulsed x-ray beam. For standard voltage point H at $V_H = 250$ V, the measured collected charge is $Q_H = 41.82$ nC; for low voltage point L at $V_L = 100$ V, the measured collected charge is $Q_L = 40.85$ nC. For the two measured points (H and L), a graph of $1/Q$ against $1/V$ is plotted and a straight line through the two points H and L extrapolated to $1/V \to 0$ (corresponding to $V \to \infty$) gives $1/Q_{sat} = 0.02353$ nC^{-1}

Table 16.7 Summary of data of Fig. 16.9 for the two-voltage method of saturation charge determination for a typical ionization chamber irradiated in a water phantom with a pulsed ionizing radiation beam

Point	$V(V)$	$\frac{1}{V}(V^{-1})$	$Q(V)$ (nC)	$\frac{1}{Q(V)}$ (nC^{-1})	$f_{gen}^{p}(V)$	$P_{ion}^{p}(V)$
V_L	**100**	0.010	**40.85**	0.02448	0.9613[1]	1.040[3]
V_H	**250**	0.004	**41.82**	0.02391	0.9842[2]	1.016[4]
V_∞	∞	0	42.493	0.02353	1.000	1.000

Measured data for the two-voltage technique are shown in bold
$V_H/V_L = 2.5$ $Q_H/Q_L = Q(V_H)/Q(V_L) = 1.0237$ $Q_{sat} = Q(V_\infty) = 42.493$ nC
[1]see (16.63) [2]see (16.62) [3]see (16.65) [4]see (16.64)

Three assumptions have been made when drawing the diagram:

1. That the collection efficiencies $f_{gen}^{p}(V_H)$ and $f_{gen}^{p}(V_L)$ for voltages V_H and V_L, respectively, exceed 90%.
2. That the charges $Q_H = Q(V_H)$ and $Q_L = Q(V_L)$ for the two voltage points (shown with black solid dots in Fig. 16.9) fall onto the straight portion of the saturation curve plotted in $1/Q$ against $1/V$ format.
3. That an extrapolation of the straight line through the two voltage points in Fig. 16.9 to $1/V = 0$ corresponding to $1/V \rightarrow \infty$ yields the reciprocal of the saturation charge Q_{sat} from which Q_{sat} can easily be calculated.

Saturation charge Q_{sat}, determined graphically in Fig. 16.9, agrees well with Q_{sat} calculated in (16.66), showing that both methods can be used reliably in output calibration of pulsed radiation beams. The data shown in Fig. 16.9 are summarized in Table 16.7 with measured data for the two voltage points shown in bold print.

The two-voltage technique for determination of saturation signal can also be used with the basic Boag equation (16.53) and for our example discussed above the calculation would proceed as follows:

1. Set up (15.78) for the two voltage points: $Q(V_H) = 41.82$ nC and $Q(V_L) = 40.85$ nC

$$Q_H = Q(V_H) = Q_{sat}\frac{1}{u_H}\ln(1 + u_H) = 41.82 \text{ nC} \qquad (16.70)$$

and

$$Q_L = Q(V_L) = Q_{sat}\frac{1}{u_L}\ln(1 + u_L) = 40.85 \text{ nC} \qquad (16.71)$$

where $u_H = \text{const}/V_H$ and $u_L = \text{const}/V_L$ allowing us to write the following ratio $u_H/u_L = V_L/V_H$.
2. Write out the ratio Q_H/Q_L based on the two equations (16.70) and (16.71) to get

$$\frac{Q_H}{Q_L} = \frac{Q(V_H)}{Q(V_L)} = \frac{u_L \ln(1 + u_H)}{u_H \ln(1 + u_L)} = \frac{V_H}{V_L}\frac{\ln(1 + u_H)}{\ln(1 + u_H V_H/V_L)}. \qquad (16.72)$$

Equation (16.72) can in general be written as

$$\frac{Q_H}{Q_L} \ln\left(1 + u_H \frac{V_H}{V_L}\right) = \frac{V_H}{V_L}\ln(1 + u_H) \tag{16.73}$$

or for our example with known ratios $Q_H/Q_L = 1.02375$ and $V_H/V_L = 2.50$

$$\frac{1.02375}{2.50} \ln\left(1 + u_H \frac{V_H}{V_L}\right) \equiv 0.4095 \ln\left(1 + u_H \frac{V_H}{V_L}\right) = \ln(1 + u_H). \tag{16.74}$$

Equation (16.74) can be solved for u_H either numerically or graphically. Once u_H is determined, it is inserted into (16.70) to find the saturation signal Q_{sat}. The method of plotting $1/Q$ against $1/V$ and extrapolation of the straight line to $1/V = 0$ that we used above seems simpler and yields the same result as the more general Boag equation used in (16.72), provided that both charges for the two voltage points are measured within 5% of saturation.

16.3.11 Mean Energy to Produce an Ion Pair in Gas

Absolute dosimetry based on ionization chamber relies on ion pairs that ionization chamber produces in the chamber cavity filled with a gas. To determine the dose absorbed in cavity gas we must know:

1. Mass of gas in the dosimeter cavity exposed to radiation.
2. Number of ion pairs or total charge that ionizing radiation produced in cavity gas.
3. Mean energy $\bar{W}_{\text{gas}}/e$ that ionizing radiation dissipates in the cavity gas per ion pair liberated.

Mass of gas in the chamber cavity is in principle known from the general design of the ionization chamber and geometrical data. However, its determination with great accuracy is difficult and often is achieved at a standards laboratory that issues a calibration coefficient for the chamber for a specific beam type and energy. The calibration coefficient is proportional to the mass of gas in the cavity chamber, so it may be used to extract information about the cavity effective volume and mass of gas contained in the cavity.

Number of ion pairs or total ionization charge can be determined from the measured ionization current j or ionization charge Q in the air cavity. The elementary charge $e = 1.602 \times 10^{-19}$ C is used for this purpose.

Mean energy $\bar{W}/e$ dissipated by ionizing radiation to produce an ion pair in cavity gas is very important in ionization chamber dosimetry, because it directly affects the accuracy of the dosimetric measurement. It exceeds the ionization energy of the atom making up the cavity gas, because some of the energy of the interacting particle is expended for excitation of the atom and for ionization of higher-level atomic shell electrons. Thus, $\bar{W}/e$ represents the mean value of a large number of possible interactions that a particle can have with an atom of a given gas, making its calculation from first principles difficult and the result uncertain.

$\bar{W}/e$ of gases and liquids for use in radiation dosimetry is usually determined experimentally; however, the experiment is difficult and the $\bar{W}/e$ that the ICRU recommends for use in radiation dosimetry comes from the weighted mean of available experimental data obtained mainly from absolute dose measurements using two parallel methods of absorbed dose measurement: (i) with graphite calorimeter and (ii) with graphite wall ionization chamber embedded in a graphite phantom. Since both dosimetric methods must result in the same dose-to-graphite D_{graph} for same dose delivery conditions, we get (see Sect. 15.7)

$$\bar{D}_{\text{graph}}^{\text{cal}} = D_{\text{graph}}^{\text{ion ch}} = \frac{Q_{\text{cor}}}{m_{\text{air}}} \left(\frac{\bar{W}}{e}\right)_{\text{air}} \frac{(\bar{S}/\rho)_{\text{graph}}}{(\bar{S}/\rho)_{\text{air}}}, \qquad (16.75)$$

resulting in the following expression for $(\bar{W}/e)_{\text{air}}$

$$\left(\frac{\bar{W}}{e}\right)_{\text{air}} = \frac{\bar{D}_{\text{graph}}^{\text{cal}}}{\dfrac{Q_{\text{cor}}}{m_{\text{air}}} \dfrac{(\bar{S}/\rho)_{\text{graph}}}{(\bar{S}/\rho)_{\text{air}}}}, \qquad (16.76)$$

where Q_{corr} is the charge collected in air mass m_{air} of the ionization chamber cavity and corrected for influence quantities such as air temperature, air pressure, and ion recombination; $(\bar{S}_{\text{col}}/\rho)_{\text{graph}}$ and $(\bar{S}_{\text{col}}/\rho)_{\text{air}}$ are the mean mass collision stopping powers of graphite or air, respectively, calculated for the photon or electron beam energy used.

$\bar{W}/e$ depends on the atomic number of the gas as well as on the type and energy of directly or indirectly ionizing radiation; however, the dependence on these quantities is relatively feeble and $\bar{W}/e$ of most gases amounts to roughly 30 eV per ion pair to 40 eV/i.p. for electrons and α-particles, irrespective of energy.

Since the cavity of ionization chambers used in radiation dosimetry is usually filled with ambient air, $(\bar{W}/e)_{\text{air}}$ is an important dosimetric "constant" and a few of its salient attributes are as follows:

1. Currently recommended value of $(\bar{W}/e)_{\text{air}}$ for dry air is 33.97 eV/i.p. or 33.97 J/C. The ICRU has modified the recommended value of $(\bar{W}/e)_{\text{air}}$ several times during the past 50 years, as indicated in Table 16.8.
2. It is generally assumed that a constant value of $(\bar{W}/e)_{\text{air}}$ can be used for the complete photon and electron energy range used in radiotherapy.
3. In general, $(\bar{W}/e)_{\text{air}}$ for humid air is about 0.997 of the recommended value for dry air over the range of relative humidity from 15% to 75% and, in most circumstances, this minute correction applicable to humid air is neglected.

Table 16.8 $(\bar{W}/e)_{\text{air}}$, mean energy required to produce an ion pair in air in J/C or eV/ion pair as recommended by the ICRU, the International Commission on Radiation quantities and Units

ICRU report	10 b	31	37
Year	1962	1979	**1984**
$(\bar{W}/e)_{\text{air}}$ J/C or eV/ion pair	33.73 ± 0.15	33.85 ± 0.15	$\mathbf{33.97 \pm 0.06}$
R to cGy conversion in air	$D_{\text{air}} = 0.869$ cGy/R	$D_{\text{air}} = 0.873$ cGy/R	$D_{\mathbf{air}} = \mathbf{0.876\,cGy/R}$

Year of initial recommendation and the ICRU Report in which the recommendation was first given are also shown. The currently recommended values are shown in bold

4. It is known that the $(\bar{W}/e)_{\text{air}}$ value at a temperature of 20 °C, pressure of 101.3 kPa and 50% relative humidity is 0.6% lower than that for dry air at the same temperature and pressure, resulting in a value of 33.77 J/C instead of 33.97 J/C. Thus, for the same amount of energy available for creating charge, 0.6% more charge will be created in air at 50% relative humidity than in dry air at 20 °C and 101.3 kPa.

16.3.12 Dose to Ionization Chamber Cavity Gas

We are now ready to determine dose D_{cav} in cavity air from the saturation charge Q_{sat} in an ionization chamber cavity of volume $\mathcal{V}_{\text{cav}}$ filled with mass m_{air} of ambient air. As discussed in Sect. 15.4, under charged particle equilibrium (CPE) D_{cav} is simply equal to collision air-kerma in air $(K_{\text{air}}^{\text{col}})_{\text{air}}$ and expressed as

$$D_{\text{cav}} \equiv (K_{\text{air}}^{\text{col}})_{\text{air}} = \frac{Q_{\text{sat}}}{m_{\text{air}}} \left(\frac{\bar{W}}{e}\right)_{\text{air}}, \tag{16.77}$$

where Q_{sat} is the total ionization charge that ionizing radiation produces in the cavity of the ionization chamber. Note that (16.77) is dimensionally correct, since on the left hand side the unit of D_{cav} is gray (Gy), where 1 Gy $= 1$ J/kg and the unit on the right hand side is also Gy, since Q_{sat} is given in coulomb (C), mass in kg, and $(\bar{W}/e)_{\text{air}}$ in J/C, resulting in J/kg.

The raw measured data that we have for use in (16.77) are the measured ionization charge Q_{H} and cavity volume $\mathcal{V}_{\text{cav}}$, while for (16.77) we need saturation charge Q_{sat} and cavity air mass m_{air}. We make the transition from the raw data to required data as follows:

1. Saturation charge Q_{sat} exceeds measured charge Q_{H} because of recombination loss in the cavity of the ionization chamber. As shown in (16.43) for continuous radiation and (16.58) for pulsed radiation, Q_{sat} and Q_{H} are related through the following expression

$$Q_{\text{sat}} = P_{\text{ion}}(V_{\text{H}})Q_{\text{H}}, \tag{16.78}$$

where

V_{H} is the standard applied potential in the chamber cavity

$P_{\text{ion}}(V_{\text{H}})$ is the ion recombination factor at standard applied potential V_{H}, measured separately for a given ionization chamber.

2. Air mass m_{air} of (16.77) is determined from the known cavity volume $\mathcal{V}_{\text{cav}}$ and ambient air density $\rho_{\text{air}}(T, p)$ at the time of measurement. Considering air as an ideal gas, the density $\rho_{\text{air}}(T, p)$ at an arbitrary $T(°C)$ temperature and pressure $p(\text{kPa})$ is given by the following relationship, normalized to air density $\rho_{\text{STP}} = 1.293 \text{ kg/m}^3$ at STP (standard temperature $T_s = 0\,°C = 273.16$ K and standard pressure $p_s = 101.325$ kPa)

$$\rho_{\text{air}}(T, p) = \rho_{\text{STP}} \times \frac{273.16 \text{ K}}{T(\text{K})} \times \frac{p \text{ (kPa)}}{101.325 \text{ kPa}}, \tag{16.79}$$

resulting in the following expression for m_{air} for the cavity air mass

$$m_{\text{air}} = \rho_{\text{air}}(T, p)\mathcal{V}_{\text{cav}}. \tag{16.80}$$

In (16.80) we assume that $\mathcal{V}_{\text{cav}}$ is constant, i.e., independent of temperature T and pressure p. The effective volume of the cavity $\mathcal{V}_{\text{cav}}$ is determined either (i) directly by measurement making the chamber an absolute dosimeter under special circumstances or (ii) indirectly through calibration of the chamber response in a known calibration field making the chamber a relative dosimeter.

The question now arises on the special case of dose absorbed in air for an exposure of one roentgen (1 R). We can write this special situation using (16.77) with air as cavity medium and recognizing that the ratio Q/m represents exposure (Sect. 15.2). Hence, we now have

$$D_{\text{air}} = (1 \text{ R}) \cdot (\bar{W}/e)_{\text{air}} = (2.58 \times 10^{-4} \text{ C/kg}_{\text{air}}) \times (33.97 \text{ J/C})$$
$$= 0.00876 \text{ J/kg}_{\text{air}} = 0.876 \text{ cGy} \tag{16.81}$$

Thus, absorbed dose in air for an exposure of 1 R is 0.876 cGy. In general terms, dose-to-air in cGy for an exposure of X roentgens is 0.876 X cGy. According to the ICRU, C/kg is the modern unit of exposure and roentgen should no longer be used for this purpose. Correspondingly, exposure of X C/kg corresponds to absorbed dose-to-air of 33.97X Gy.

Ionization chambers are usually embedded in a medium (water or water-equivalent phantom) in which absorbed dose D_{med} is to be determined. The subsequent conversion of the cavity air dose D_{cav} into dose-to-medium (usually water) D_{med}, as shown in Sect 15.7, is based on cavity theories, most commonly on the Bragg–Gray cavity theory or on its refinement, the Spencer–Attix cavity theory.

16.3.13 *Absolute Dosimetry with Ionization Chamber*

Ionization chamber, with its numerous types and models as well as relatively simple design and long history, is the most practical and widely used dosimeter for measurement of either radiation exposure (air-kerma in air) or absorbed dose-to-medium. While exposure, characterized as the ability of photons to ionize air, is limited to only ionizing photons interacting with air, dose is defined more broadly as energy of any ionizing radiation absorbed per unit mass of any absorbing medium.

National and regional standards laboratories involved with radiation standards employ ionization chambers as one of three known absolute dosimetry techniques: the other two techniques are calorimetric dosimetry discussed in Sect. 16.1 and chemical Fricke dosimetry discussed in Sect. 16.2. Three types of ionization chamber are used in standards laboratories for the purpose of absolute dosimetry:

1. Standard free air ionization chamber for exposure (air-kerma in air) standard in the kilovoltage photon energy range from $\sim$50 kVp to $\sim$400 kVp x-rays.
2. Graphite cavity ionization chamber with accurately known sensitive volume for absorbed dose-to-water standard, mainly for cobalt-60 γ-rays and, optionally, also for other beam types and qualities, such as megavoltage photon and electron beams, kilovoltage x-rays, protons, and heavier ions.
3. Extrapolation chamber for absolute dose-to-water standard for low energy x-rays in the kilovoltage energy range as well as water phantom-embedded extrapolation chamber for megavoltage x-ray and electron beams.

The three ionization chamber types listed above are designed for absolute dosimetry but are too cumbersome and impractical for use in a clinical environment. Therefore, measurement of absolute dose in radiotherapy clinics around the world is carried out with the second tier of absolute dosimetry that is essentially relative dosimetry and is almost exclusively based on specially designed cavity ionization chambers suitable for use in radiotherapy clinics. These chambers are usually calibrated in a cobalt-60 γ-ray beam at a primary standards laboratory and the calibration is carried out either in terms of air-kerma in air in older dosimetry protocols or in terms of absorbed dose-to-water in newer protocols. Thus, the calibration coefficients of these chambers are traceable to a National or Regional Standards Laboratory and are used in clinics in conjunction with a suitable dosimetry protocol or code of practice for measurement of photon and electron beam output of clinical megavoltage radiotherapy machines.

In addition to use in output measurement of radiotherapy machines, cavity ionization chambers are also used in general commissioning as well as in quality assurance of ionizing radiation-emitting equipment installed in hospitals for diagnosis and treatment of disease. They are also used for calibration of radioactive sources, for environmental monitoring, and as survey meters for radiation safety around all equipment producing ionizing radiation in hospitals, clinics, research institutes, university laboratories, and nuclear installations.

16.3.14 Standard Free-Air Ionization Chamber

Standard free-air ionization chambers are designed for measurement of exposure X_p in C/kg of air at a point of interest P in air or air-kerma in air $(K_{air}^p)_{air}$ in Gy at a point of interest P in air. The measurements are based on the definition of exposure $X_p = \Delta Q / \Delta m_{air}$ that stipulates the collection of all ions formed in air (total charge Q) along the tracks of secondary electrons that were liberated by photon interactions in a small, well-defined volume (cavity) $\mathcal{V}_S$ filled with air of mass Δm_{air} and encircling the point of interest P.

The possible photon interactions with air molecules, as discussed in detail in Chap. 7, are: (i) Photoelectric effect releasing photoelectron; (ii) Compton effect releasing Compton (recoil) electron, and (iii) Pair production (nuclear and electronic) producing free electron and positron.

Note 1: Exposure is defined for photon energies in the keV range; therefore, pair production (with threshold energy of 1.022 MeV) is ignored in discussions of exposure.

Note 2: In photon interactions the x-ray photon spectrum releases a spectrum of secondary electrons (photoelectrons and Compton electrons) of kinetic energy E_K between 0 and a maximum kinetic energy $(E_K)_{max}$ approximately equal to the maximum photon energy $(E_v)_{max}$ in the photon spectrum; i.e., $(E_K)_{max} \approx (E_v)_{max}$.

Simplified Model of Free-Air Ionization Chamber

The exposure measurement is illustrated in Fig. 16.10 depicting a simplified model of a free-air ionization chamber. In its basic form the free-air chamber consists of two concentric air volumes: small air cavity volume $\mathcal{V}_S$ and large air volume $\mathcal{V}_L$, each volume serving its own special purpose: (i) $\mathcal{V}_S$ filled with a known mass of air Δm_{air} serves as the source of energetic secondary electrons and (ii) $\mathcal{V}_L$ enables secondary electrons to expend their total kinetic energy from initial kinetic energy $(E_k)_i$ down to 0 through formation of ion pairs in air.

We now carry out the following "gedanken" (thought) experiment on exposure measurement:

1. Imagine a narrow diverging x-ray beam of low or intermediate energy traversing air and collimated by a circular collimator, as depicted in Fig. 16.10.
2. Place the point of interest P at which exposure X_p is to be measured onto the central axis of the x-ray beam and define a small spherical air volume $\mathcal{V}_s$ of radius r_s surrounding point P.

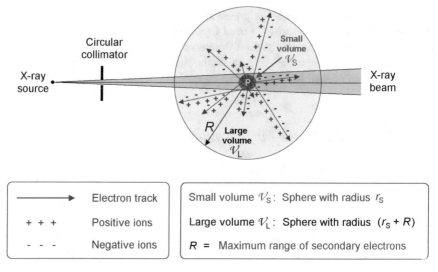

Fig. 16.10 Simplified model of a free air ionization chamber with two defined volumes of air. Small volume (cavity volume) $\mathcal{V}_S$ is a sphere with radius r_S; large volume $\mathcal{V}_L$ a sphere with radius $(r_S + R)$ where R is the maximum range of secondary electrons in air

3. Assume that photon interactions with air molecules can only occur in the small cavity volume $\mathcal{V}_s$ that is filled with air. There are no photon interactions outside of cavity volume $\mathcal{V}_s$.
4. Imagine a large sphere with volume $\mathcal{V}_L$ and radius $(r_s + R)$ filled with air and, like $\mathcal{V}_s$, also centered at point P. *Note: R* stands for the range in air of the most energetic secondary electron released by photon interactions in the small air volume $\mathcal{V}_s$.
5. From 3. we recall that all secondary electrons released in photon interactions with air originate in volume $\mathcal{V}_s$, while from 4. we note that all secondary electrons released in photon interactions expend their total initial kinetic energy in large volume $\mathcal{V}_L$ through Coulomb interactions with air molecules, resulting first in a large number of ion pairs consisting of positive ion and free electron. Since oxygen of air is a strongly electronegative gas, the free electrons attach themselves to oxygen atoms thereby forming negative ions that are, rather than free electrons, collected in the free-air ionization chamber (see Sect. 16.3.5).
6. Assume that there is no recombination loss in large air volume $\mathcal{V}_L$. The ratio of total charge Q produced in volume $\mathcal{V}_L$ to the mass of air m_{air} contained in small air cavity volume $\mathcal{V}_S$ fulfills the definition of exposure X_p at point P. Thus, an accurate determination of small cavity volume $\mathcal{V}_S$ and total charge Q produced in large volume $\mathcal{V}_L$ provide excellent means for measurement of exposure X_p and this conclusion forms the basis for design and operation of practical standard free-air ionization chambers used in ionizing radiation dosimetry.

Practical Standard Free-Air Ionization Chamber

As discussed above, the role of the standard free-air ionization chamber is to collect all positive and negative ions produced in large air volume $\mathcal{V}_L$ by charged particles that photons liberate in small air cavity volume $\mathcal{V}_S$. This is achieved by defining the two volumes $\mathcal{V}_S$ and $\mathcal{V}_L$ as well as aiming the x-ray beam centrally between two polarized electrodes that are oriented in parallel with the central axis of the x-ray beam. The electrodes define the electric field in both the chamber cavity $\mathcal{V}_S$ and the large air volume $\mathcal{V}_L$, thereby enabling collection and counting of liberated ionization charge carriers. Obviously, the situation is much more complicated in practice than the simple picture discussed in the "gedanken" experiment above and depicted in Fig. 16.10.

Many practical and theoretical constraints impose significant difficulties on the design, construction, and use of free-air ionization chambers, precluding a routine clinical use of these chambers and relegating them to national standards laboratories and a few large specialized laboratories. However, it is clear that a measurement with a free-air ionization chamber provides the most convenient means for an accurate determination of exposure X_p in C/kg and air-kerma in air $(K_{air})_{air}$ in Gy in the low to intermediate kilovoltage energy x-ray region from $\sim$50 kVp to $\sim$400 kVp.

A schematic diagram of a typical standard free-air ionization chamber is shown in Fig. 16.11. Part (a) represents a vertical cut containing the x-ray beam central axis, x-ray source S, point of interest P, as well as two parallel-plate electrodes. Part (b) is a vertical cut, perpendicular to the x-ray beam central axis through point P. The whole ionization chamber is usually encased in a lead-lined box (not shown in Fig. 16.11) to shield the chamber cavity volume $\mathcal{V}_S$ from stray and scattered radiation. As shown in Fig. 16.11a, the x-ray beam enters the chamber from left to right and is collimated by a circular collimator before entering the ionization chamber.

The operation of the free-air ionization chamber is based on the thought experiment discussed above (see Fig. 16.10); however, the practical free-air chamber is significantly more elaborate so as to ensure that its cavity volume $\mathcal{V}_S$ is well delineated and accurately known, and that the large chamber volume $\mathcal{V}_L$ that encompasses cavity $\mathcal{V}_S$ allows unhindered ionization of air molecules to all secondary electrons released in small cavity volume $\mathcal{V}_S$.

Several important features of the standard free-air ionization chamber are notable:

1. Polarizing electrode is made of a single sheet of metal and connected to a power supply; collecting electrode consists of two separate electrodes: guard electrode connected directly to ground and measuring electrode grounded through an electrometer.

2. All electrodes are located at a distance from cavity volume $\mathcal{V}_S$ that exceeds R, the maximum range of secondary electrons that x-ray photons release in air. Typical ranges in air of electrons in the keV energy range, obtained from the NIST, are listed in Table 16.9. It is obvious that to satisfy this requirement the dimensions of the ionization chamber become quite large as the x-ray energy increases, placing a practical energy limit for exposure measurement with standard free-air ionization chambers at below $\sim$400 kVp.

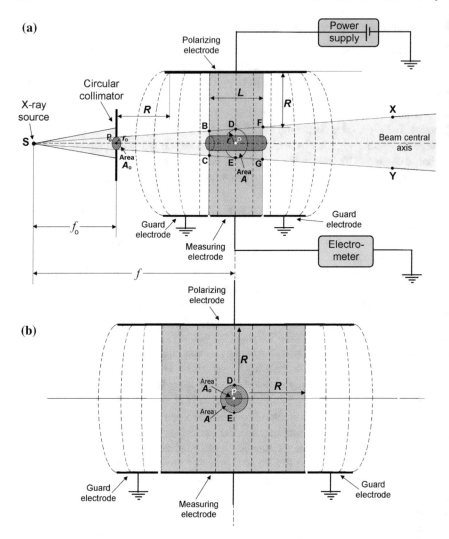

Fig. 16.11 Schematic diagram of a typical standard free-air ionization chamber. **a** Vertical cut containing the x-ray beam central axis, x-ray source S, point-of-interest P on the x-ray beam central axis, and the two parallel-plate electrodes. **b** Vertical cut, perpendicular to the x-ray beam central axis through point P. The divergent x-ray beam (SXY) enters the chamber from left to right and is collimated by a circular collimator before entering the ionization chamber. R is the range of maximum energy secondary electrons in air, L is the width of the measuring electrode. Conical segment BFGC defines the cavity volume (small cavity volume $\mathcal{V}_S$); measuring electrode defines the large volume $\mathcal{V}_L$ (ion collection volume)

Table 16.9 Range in air for electrons with kinetic energy E_K in the range from 100 keV to 3 MeV

Electron kinetic energy E_K (MeV)	Range in air	
	(g/cm^2)	(cm)
0.1	0.016	12.6
0.2	0.051	39.3
0.3	0.095	73.7
0.4	0.146	112.6
0.5	0.200	154.7
0.8	0.372	287.7
1.0	0.491	380.0
1.5	0.790	611.0
2.0	1.085	839.1
3.0	1.658	1282.3

Air density: $\rho_{air}(STP) = 1.293 \times 10^{-3}$ g/cm^3. Data are from the NIST

3. Cavity volume $\mathcal{V}_S$ is at a distance larger than R from the circular collimator to ensure that no secondary electron released in photon interactions with the collimator can reach and ionize air in the cavity $\mathcal{V}_S$ thereby negatively disrupting measurement of X_p.

4. Chamber cavity volume $\mathcal{V}_S$ is indicated with a conical section BFGC in Fig. 16.11a. The volume is defined by the conical section that is delineated by the x-ray beam and by two vertical planes perpendicular to the x-ray beam central axis, one plane containing points B and C, the other plane points F and G. The two planes are defined by the narrow gaps separating the measuring electrode and the guard electrode. The height of the conical section is equal to the width L of the measuring electrode, as shown in Fig. 16.11a. The large volume $\mathcal{V}_L$ (ion collection volume) is defined by the width and length of the measuring electrode as well as the separation between the measuring and polarizing electrodes.

5. The circular collimator of the free-air chamber is at a distance of at least 1 m from the x-ray source. This means that the beam divergence is small and the conical segment defining $\mathcal{V}_S$ through points BFGC in Fig. 16.11a can be approximated by a cylinder of circular cross sectional area A and radius r, containing point P at its center and located at a distance f from the x-ray source. The cylinder approximating cavity volume $\mathcal{V}_S$ is depicted in Fig. 16.12.

6. While the chamber cavity volume $\mathcal{V}_S$ is a cylinder of radius r and height L, as shown in Fig. 16.12, the large volume $\mathcal{V}_L$, as indicated in Figs. 16.11 and 16.12, is a rectangular parallelepiped (sometimes referred to as cuboid) of length L and height as well as width equal to $\sim 2(r + R)$. The requirement on accuracy of $\mathcal{V}_S$ is quite stringent because $\mathcal{V}_S$ defines the mass of air m_{air} in the defining equation of exposure $X_P = \Delta Q / \Delta m_{air}$ For the cuboid $\mathcal{V}_L$, on the other hand, only its length L must be well defined, while its height and width must only exceed $\sim 2(r + R)$ to allow an accurate determination of the total charge Q to satisfy the definition of exposure.

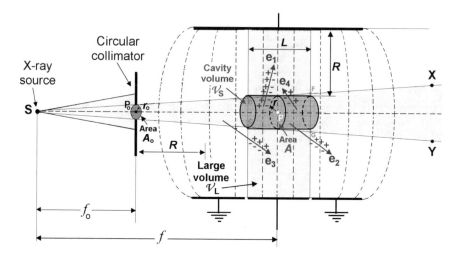

Fig. 16.12 Schematic diagram of a standard free air ionization chamber of Fig. 16.11. Actual cavity volume shown as conical segment BFGC is for simplicity approximated as a cylinder of radius r, height L, and volume $\mathcal{V}_S = \pi r^2 L$. Large volume $\mathcal{V}_L$ (ion collection volume) is defined by the width L of the measuring electrode. Source-collimator distance is f_0; source-point P distance is f

7. In the determination of exposure X_P at point P the important parameter is the mass of air m_{air} in the cavity volume $\mathcal{V}_S$ expressed as

$$m_{air} = \rho_{air}\mathcal{V}_S = \rho_{air}A\cdot L = \rho_{air}\pi r^2 L, \qquad (16.82)$$

where ρ_{air} is the density of cavity air given in (15.104) as a function of air temperature T and pressure p.

In the simple thought experiment depicted in Fig. 16.10 all secondary electrons contributing to ionizations in the large volume $\mathcal{V}_L$ of air originated in the small cavity air volume $\mathcal{V}_S$ and contributed fully to measured total charge Q. In a real-life free-air ionization chamber, the situation is more complicated, as shown in Fig. 16.12. Here, some secondary electrons, such as e_1 and e_4, originate in $\mathcal{V}_S$ and expend all their kinetic energy on ionizations in $\mathcal{V}_L$, thereby fully contributing to the measured total charge Q.

However, we also have secondary electrons, such as e_3, that originate outside (in front) of $\mathcal{V}_S$, produce ionizations outside both $\mathcal{V}_S$ and $\mathcal{V}_L$, eventually penetrate $\mathcal{V}_L$, and make a partial contribution to ionizations in despite not originating in the cavity volume $\mathcal{V}_L$. This partial contribution to ionizations in $\mathcal{V}_L$, if included in total charge Q, will cause an erroneous increase in the measured total charge.

In addition, there also are secondary electrons, such as e_2, that originate in $\mathcal{V}_S$ but do not fully contribute to ionizations in $\mathcal{V}_L$ because, after partially contributing to ionizations in $\mathcal{V}_L$, they carry some of their kinetic energy out of $\mathcal{V}_L$, thereby not making their full contribution to measured total charge Q.

The various modes of secondary electron contributions to total charge Q make for a seemingly confusing situation; however, in general, the number of ions that are gained in $\mathcal{V}_L$ from type e_3 secondary electrons will be balanced by the number of ions that are lost in $\mathcal{V}_L$ from type e_2 secondary electrons. As a result of this charged particle equilibrium, discussed in more detail in Sect. 15.3, we can assume that all secondary electrons producing ions in $\mathcal{V}_L$ originated in cavity volume $\mathcal{V}_S$ and that the charge Q collected on the measuring electrode of the free-air ionization chamber is the charge to be used for determination of exposure X_P. Of course, the charge Q measured in free-air ionization chamber, like in any other ionization chamber type, must be corrected for recombination loss, as discussed in Sect. 16.3.9.

Exposure X_P at point P located in the free-air ionization chamber, as shown in Figs. 16.11 and 16.12, can now be determined from: (i) measured charge Q corrected for various influence quantities to get the saturation charge Q_{sat} and (ii) known mass of air m_{air} in the cavity volume $\mathcal{V}_S$ given in (15.106) to get

$$X_P = \frac{Q_{sat}}{m_{air}} = \frac{Q_{sat}}{\rho_{air} A \cdot L} = \frac{Q_{sat}}{\rho_{air} \pi r^2 \cdot L}. \qquad (16.83)$$

It is easy to show that (16.83) is dimensionally correct, since it provides units of C/kg (coulomb per kilogram of air) for exposure X_P at point P in air.

Inverse Square Law

In practice it is more convenient to determine exposure X at a point in space outside the free-air ionization chamber and standards laboratories usually choose the center of the circular collimator of the free-air chamber for such special reference point on the central axis of the x-ray beam (see point P_o in Figs. 16.11 and 16.12). This is then the point on the beam central axis at which cavity chambers are centered for determination of their calibration coefficients in kilovoltage x-ray beams.

The relationship between the exposure X_{P_0} at point P_0 at a distance f_0 from the source S and exposure X_P of (16.83) given for point P at a distance f from the source S is governed by the so-called inverse square law that is derived as follows:

Photon sources are commonly assumed to be point sources and the beams they produce are divergent beams, as shown schematically in Fig. 16.13. Let us assume that we have an x-ray point source S and two circular radiation fields, one with radius r_0 and area A_0 at a distance f_0 from the source S and the other with radius r and area A at a distance f from the source S where $f_0 < f$. The two x-ray fields: A_0 and A resemble fields A_0 and A in the free-air ionization chamber depicted in Figs. 16.11 and 16.12. As shown in Fig. 16.13, they are related geometrically as follows

$$\tan\theta = \frac{r_0}{f_0} = \frac{r}{f} \quad \text{or} \quad \frac{r_0}{r} = \sqrt{\frac{A_0}{A}} = \frac{f_0}{f}, \qquad (16.84)$$

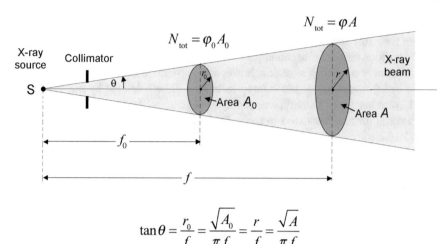

$$\tan\theta = \frac{r_0}{f_0} = \frac{\sqrt{A_0}}{\pi f_0} = \frac{r}{f} = \frac{\sqrt{A}}{\pi f}$$

Fig. 16.13 Schematic illustration of inverse square law for a divergent circular x-ray or gamma ray beam originating in a point source S. At a distance f_0 from the source the field size is $A_0 = \pi r_0^2$; at a distance f from the source the field size is $A = \pi r^2$. An assumption is made that the total number N_{tot} of photon passing through area A_0 is equal to the number of photons passing through area A; i.e., photon interactions in air between $A_0(f_0)$ and $A(f)$ are negligible and $N_{tot} = \varphi_0(A_0) = \varphi A$, where φ_0 and φ are photon fluences at f_0 and f, respectively

where θ is the angle between the radiation beam central axis and the geometric edge of the divergent radiation beam.

The x-ray source S emits x-ray photons and produces fluence φ_0 at distance f_0 from the source and fluence φ at distance f from the source. Since the total number N_{tot} of x-ray photons crossing area A_0 is equal to the total number N_{tot} of x-ray photons crossing area A (under the reasonable assumption that photon interactions occurring in air between areas A_0 and A are negligible), we can write, in conjunction with (16.84)

$$N_{tot} = \varphi_0 A_0 = \varphi A \qquad \text{and} \qquad \frac{\varphi_0}{\varphi} = \frac{A}{A_0} = \frac{\pi r^2}{\pi r_0^2} = \frac{f^2}{f_0^2}. \qquad (16.85)$$

Since, at a given point P in air, exposure X_P and air-kerma in air $(K_{air})_{air}$ are directly proportional to photon fluence φ at point P and photon fluence φ, in turn, is proportional to the square of the distance f from the source, it is reasonable to conclude that both exposure X_P and air-kerma in air $(K_{air})_{air}$ follow the inverse square law behavior, i.e.,

$$\frac{X_P}{X_{P_0}} = \frac{X(f)}{X(f_0)} = \frac{(K_{air}^P)_{air}}{(K_{air}^{P_0})_{air}} = \frac{f_0^2}{f^2}. \qquad (16.86)$$

The inverse square law plays an important role in physics and applies to physical quantities that are inversely proportional to the square of the distance from the source of the given physical quantity.

We now return to the relationship between exposure at point P_0 to exposure at point P in the free-air ionization chamber and, based on (16.86) and neglecting the x-ray beam attenuation in the air path from P_0 to P, conclude that exposure at P_0 will exceed that at P by the inverse square factor $(f/f_0)^2$. Hence, we get the following result for exposure X_{P_0} at P_0 related to exposure X_P at P given by (15.107)

$$X_{P_0} = X_P \frac{f^2}{f_0^2} = \frac{f^2}{f_0^2} \frac{Q_{\text{sat}}}{\rho_{\text{air}} AL} = \frac{Q_{\text{sat}}}{\rho_{\text{air}} A_0 L} = \frac{Q_{\text{sat}}}{\rho_{\text{air}} \pi r_0^2 L}, \qquad (16.87)$$

where we used (16.85) to write $f^2/(f_0^2 A) = 1/A_0$. The small cavity volume $A_0 L$ is shown in Fig. 16.11a with the dark grey cylinder. Point P_0 is at the center of the circular collimator at the beam entrance side of the free-air ionization chamber; point P is at the center of the free-air chamber.

As suggested in (16.77), (16.87) can be written in terms of collision air-kerma in air $(K_{\text{air}}^{\text{col}})_{\text{air}}$ as follows

$$(K_{\text{air}}^{\text{col}})_{\text{air}} = X_{p_0} \left(\frac{\bar{W}}{e} \right)_{\text{air}} = \frac{Q_{\text{sat}}}{\rho_{\text{air}} A_0 L} \left(\frac{\bar{W}}{e} \right)_{\text{air}}, \qquad (16.88)$$

with $(\bar{W}e)_{\text{air}}$ the mean energy dissipated by ionizing radiation to produce an ion pair in air. For dry air the current value of $(\bar{W}e)_{\text{air}}$ is 33.97 J/C. Recognizing that the radiation fraction $\bar{g}$ is negligible at photon energies below 400 keV, we now express air-kerma in air $(K_{\text{air}})_{\text{air}}$ at point P in a standard free-air ionization chamber as follows

$$(K_{\text{air}})_{\text{air}} \approx (K_{\text{air}}^{\text{col}})_{\text{air}} = X_{p_0} \left(\frac{\bar{W}}{e} \right)_{\text{air}} = \frac{Q_{\text{sat}}}{\rho_{\text{air}} A_0 L} \left(\frac{\bar{W}}{e} \right)_{\text{air}} = \frac{Q_{\text{sat}}}{\rho_{\text{air}} \pi r_0^2 L} \left(\frac{\bar{W}}{e} \right)_{\text{air}}. \qquad (16.89)$$

16.3.15 Standard Bragg–Gray Cavity Ionization Chamber

The free-air ionization chamber fulfills well the definition of exposure and air-kerma in air; however, the large volume of air required to allow secondary electrons to fully dissipate in air their initial kinetic energy makes the free-air chamber excessively large and limits its usefulness to x-ray energies below ~400 kVp. Since exposure and air-kerma in air are defined up to ~3 MeV, there is a need for measurement techniques that cover exposure and air-kerma in air at least up to 1.5 MeV to include cesium-137 and cobalt-60 gamma rays with photon energies of 0.662 MeV and ~1.5 MeV, respectively.

At primary standards laboratories exposure and air-kerma in air measurement in the photon energy $h\upsilon$ range $\sim$0.4 MeV $< h\upsilon < \sim$1.5 MeV is carried out with Bragg–Gray cavity ionization chambers that make use of the Bragg–Gray cavity theory (Sect. 15.7.1). In contrast to large and bulky free-air ionization chamber, discussed in Sect. 16.3.14, the standard Bragg–Gray cavity ionization chamber is small and practical. It consists of a small Bragg–Gray cavity filled with ambient air and of volume $\mathcal{V}_{cav}$ known to high degree of accuracy.

A graphite wall with a thickness of the order of the range R of secondary electrons in graphite surrounds and defines the chamber cavity. This means that the graphite wall is thick enough to provide full buildup of secondary electrons and the chamber cavity fulfills both Bragg–Gray conditions (Sect. 15.7.1): (i) Chamber cavity does not perturb the secondary charged particles fluence that photons produce in the chamber graphite wall and (ii) photon interactions in the small air-filled cavity are rare and therefore neglected.

Many parameters and correction factors are used in the process of exposure and air-kerma in air measurement with a standard cavity ionization chamber at primary standards laboratories. The final expression for air-kerma in air $(K_{air})_{air}$, as will be seen below, is inversely proportional to the volume $\mathcal{V}$ of the chamber cavity that is related to cavity air mass m_{air} through $m_{air} = \rho_{air}\mathcal{V}$. This makes the accurate knowledge of the chamber sensitive volume of great importance to primary standards laboratories, but the problem is not as simple as it seems at first glance.

Often an assumption is made that the geometrical or mechanical volume of the chamber provides adequate approximation to the sensitive volume of an ionization chamber and much effort is expended with blueprints, radiographic images, and micro-CT scans of a given chamber to determine its mechanical volume. However, as far as charge collection in an ionization chamber is concerned, it is the electric field lines produced by the specific electric field in the cavity volume that define the cavity effective volume and it turns out that this volume is not necessarily equal to the mechanical volume. The two volumes may differ by a certain percentage and, moreover, the difference may vary with chamber voltage and exposure rate. Thus, the cavity mechanical volume gives a good approximation to the chamber effective volume; however, the true effective cavity volume is determined through a combination of experimental and theoretical work including measurement of chamber response in a known radiation field, modeling of electric field in the cavity, and Monte Carlo calculations.

The primary air-kerma in air $(K_{air})_{air}$ standard in the range from $\sim$0.4 MeV to $\sim$1.5 MeV at primary standards laboratories is provided through the use of standard Bragg–Gray cavity chambers and the following relationship

$$(K_{air})_{air} = \frac{Q_{sat}}{m_{air}}\left(\frac{\bar{W}}{e}\right)_{air}\frac{1}{1-\bar{g}}\frac{(\bar{S}_{col}/\rho)_{graph}}{(\bar{S}_{col}/\rho)_{air}}\frac{(\bar{\mu}_{ab}/\rho)_{air}}{(\bar{\mu}_{ab}/\rho)_{graph}}\prod_i k_i, \qquad (16.90)$$

where

Q_{sat} is chamber saturation charge calculated from the measured charge Q corrected for recombination loss in the chamber cavity.

m_{air} is the mass of cavity air determined from the known cavity volume V through $m_{air} = \rho_{air} V$ with ρ_{air} the density of ambient air.

$\bar{g}$ is the mean radiation fraction, i.e., the mean fraction of secondary electron energy lost to radiation processes such as bremsstrahlung and in-flight annihilation.

$(\bar{S}_{col}/\rho)$ is the mean mass collision stopping power for either graphite or air.

$(\bar{\mu}_{ab}/\rho)$ is the mean mass energy absorption coefficient for either air or graphite.

$\prod_i k_i$ stands for the product of correction factors for influence quantities.

Equation for air-kerma in air $(K_{air})_{air}$ given in (16.90) is derived as follows:

1. According to the Bragg–Gray equation (15.20) the relationship between dose-to-air D_{air} in the cavity of known volume V and dose-to-graphite wall (medium) D_{graph} in which secondary electron spectrum is being built up is expressed as

$$D_{graph} = D_{air} \frac{(\bar{S}_{col}/\rho)_{graph}}{(\bar{S}_{col}/\rho)_{air}} = \frac{Q_{sat}}{\rho_{air} V} \frac{(\bar{S}_{col}/\rho)_{graph}}{(\bar{S}_{col}/\rho)_{air}}, \qquad (16.91)$$

where we used (15.102) for dose-to-air D_{air} in the chamber cavity.

2. As discussed in Sect. 15.1.4, under the condition of charged particle equilibrium (CPE) dose to cavity air D_{air} is equal to collision air-kerma in air $(K_{air}^{col})_{air}$ that in turn can be expressed as a product of photon energy fluence ψ_{air} and mean mass energy absorption coefficient $(\bar{\mu}_{ab}/\rho)_{air}$

$$D_{air} = (K_{air}^{col})_{air} = \psi_{air} \cdot (\bar{\mu}_{ab}/\rho)_{air} \qquad (16.92)$$

We get a similar relationship for the dose-to-graphite wall D_{graph}

$$D_{graph} = \psi_{graph} \cdot (\bar{\mu}_{ab}/\rho)_{graph}. \qquad (16.93)$$

Recognizing that photon energy fluence ψ_{graph} in graphite is equal to photon energy fluence ψ_{air} in the air cavity, we get after combining (15.116) and (15.117)

$$\psi_{air} = \psi_{graph} = \frac{(K_{air}^{col})_{air}}{(\bar{\mu}_{ab}/\rho)_{air}} = \frac{D_{graph}}{(\bar{\mu}_{ab}/\rho)_{graph}}. \qquad (16.94)$$

From (16.94) we now get the following relationship between collision air-kerma in air $(K_{air}^{col})_{air}$ and dose-to-graphite wall D_{graph}

$$(K_{air}^{col})_{air} = D_{graph} \frac{(\bar{\mu}_{ab}/\rho)_{air}}{(\bar{\mu}_{ab}/\rho)_{graph}}. \qquad (16.95)$$

3. In general, as discussed in Sect. 15.4, kerma K consists of two components: collision kerma K^{col} and radiation kerma K^{rad} i.e., $K = K^{col} + K^{rad}$. Thus, air-kerma in air $(K_{air})_{air}$ can be expressed as a sum of two components: collision kerma and radiation kerma, as follows

$$(K_{air})_{air} = (K_{air}^{col})_{air} + (K_{air}^{rad})_{air} \tag{16.96}$$

or, after solving for $(K_{air}^{col})_{air}$

$$(K_{air}^{col})_{air} = (K_{air})_{air} - (K_{air}^{rad})_{air} = (K_{air})_{air}\left[1 - \frac{(K_{air}^{rad})_{air}}{(K_{air})_{air}}\right]$$
$$= (K_{air})_{air}[1 - \bar{g}], \tag{16.97}$$

where $\bar{g} = (K_{air}^{rad})_{air}/(K_{air})_{air}$ stands for the mean radiation fraction, i.e., mean fraction of secondary electron energy that the secondary electron loses to radiation processes such as bremsstrahlung and in-flight annihilation.

After inserting (16.95) and (16.91) into (16.97), we get the following result for air-kerma in air $(K_{air})_{air}$ at the point of interest P in the standard Bragg–Gray cavity ionization chamber with all parameters defined in conjunction with (16.90).

$$(K_{air})_{air} = \frac{Q_{sat}}{m_{air}}\left(\frac{\bar{W}}{e}\right)_{air}\frac{1}{1 - \bar{g}}\frac{(\bar{S}_{col}/\rho)_{graph}}{(\bar{S}_{col}/\rho)_{air}}\frac{(\bar{\mu}_{ab}/\rho)_{air}}{(\bar{\mu}_{col}/\rho)_{graph}}, \tag{16.98}$$

The result (16.98) is similar to that stated in (16.90) except that (16.90) contains the product $\prod_i k_i$ of various correction factors. Primary standards laboratories use these correction factors to account for the effects of:

(i) photon attenuation and scattering in the chamber graphite wall,
(ii) radiation source non-uniformity,
(iii) minor deviations from Bragg–Gray or Spencer–Attix cavity theories.

16.3.16 Practical Ionization Chambers in Radiotherapy Department

By far the most prevalent application of ionization chambers is found in external beam radiotherapy, where they are used in acceptance testing, commissioning, and routine calibration of major radiotherapy machines. These machines produce kilo-voltage x-rays with x-ray machines and megavoltage x-rays and electrons with linear accelerators. It is imperative that these machines be properly calibrated and maintained to ensure that radiation doses delivered to patients actually comply with doses prescribed by the physician.

(a)

(b)

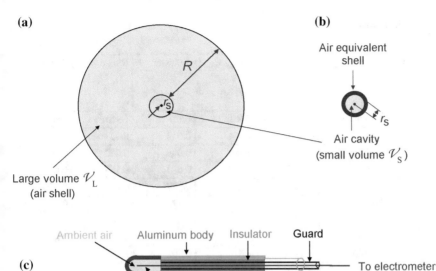

(c)

Fig. 16.14 Practical ionometric radiation dosimetry: **a** Two concentric spherical air volumes: small volume $\mathcal{V}_S$ for cavity with radius r_s and large volume $\mathcal{V}_L$ with radius $r_s + R$ where R is the maximum range of secondary electrons in air; **b** Spherical air shell compressed into a shell of density ~ 1 g/cm^3, and **c** Schematic diagram of a thimble ionization chamber with an air wall compressed to a density of ~ 1 g/cm^3

Absolute ionometric dosimetry was discussed above based on two types of absolute ionization chamber: standard free-air chamber (Sect. 16.3.14) applicable in energy range from 50 kVp to 400 kVp and standard Bragg–Gray cavity chamber (Sect. 16.3.15) applicable in energy range from 400 kVp to 1.5 MeV. Unfortunately, the two standard ionization chambers are impractical for use in clinical environment and, moreover, their energy range does not extend above 1.5 MeV where most of modern radiotherapy is carried out. Therefore, practical solutions had to be found for accurate dose measurement in clinical radiotherapy departments covering the full clinical photon range from 50 kVp to 30 MeV and electron range from 4 MeV to 20 MeV.

For photon energies below 1.5 MeV (including cobalt-60 gamma rays at 1.25 MeV) the solution is relatively simple and depicted in Fig. 16.14a, as follows:

1. Imagine two concentric spherical air volumes: small volume $\mathcal{V}_S$ for cavity with radius r_S and large volume $\mathcal{V}_L$ with radius $r_S + R$ where R is the maximum range of secondary electrons generated by photon interactions in air volume $\mathcal{V}_L$. This ensures that for the cavity $\mathcal{V}_S$ conditions of electronic equilibrium apply (see Sect. 15.3).

Table 16.10 Range in air for electrons with kinetic energy E_K in the range from 100 keV to 3 MeV

Electron kinetic energy E_K (MeV)	Range R_{CSDA} in air		Range R_{CSDA} in water		Range R_{CSDA} in lead	
	(g/cm^2)	(cm)	(g/cm^2)	(cm)	(g/cm^2)	(cm)
0.1	0.016	12.6	0.014	0.014	0.031	0.003
0.2	0.051	39.3	0.049	0.049	0.092	0.008
0.3	0.095	73.7	0.084	0.084	0.167	0.015
0.4	0.146	112.6	0.129	0.129	0.294	0.026
0.5	0.200	154.7	0.177	0.177	0.336	0.030
0.8	0.372	287.7	0.330	0.330	0.605	0.053
1.0	0.491	380.0	0.437	0.437	0.784	0.069
1.5	0.790	611.0	0.708	0.708	1.219	0.107
2.0	1.085	839.1	0.979	0.979	1.629	0.144
3.0	1.658	1282.3	1.514	1.514	2.381	0.210

Densities: air $\rho_{air}(STP) = 1.293 \times 10^{-3}$ g/cm^3; water $\rho_{water} = 1$ g/cm^3; lead $\rho_{lead} = 11.34$ g/cm^3. Data are from the NIST

2. Immerse the large volume into a photon beam and assume that we can collect and measure the ionization charge produced in cavity $\mathcal{V}_S$ by electrons released in the large air volume $\mathcal{V}_L$.
3. Assume that we know the air mass contained in the cavity $\mathcal{V}_S$ and that we know the total charge of one sign collected in $\mathcal{V}_S$. With this information we can thus determine the charge Q_{sat} per unit air mass m_{air} (exposure) at the center C of the cavity.

Thus far, the discussion seemed logical and plausible. However, when we try to estimate the required radius $r_S + R$ of the large air sphere $\mathcal{V}_L$, we note, as shown in Table 16.10 providing the range of electrons in air, water and lead, that we would need excessively large air masses to provide electronic equilibrium in the cavity $\mathcal{V}_S$. For example, for cobalt-60 gamma rays, $r_S + R$ would exceed 400 cm, an obviously untenable proposition. However, if the air shell of density $\rho_{air}(STP) = 1.293 \times 10^{-3}$ g/cm^3 and thickness R surrounding the cavity $\mathcal{V}_S$ were compressed into a solid air-equivalent shell with density of ~1 g/cm^3, as shown schematically in Fig. 16.14b, the result would be the so-called thimble chamber that, for cobalt-60 gamma rays (energy ~1.25 MeV), would require an air-equivalent wall thickness of about 0.45 cm. We estimate this by comparing densities of air and low atomic number plastics such as Lucite ($\rho_{PMMA} = 1.18$ g/cm^3) getting a density ratio $\rho_{air}/\rho_{PMMA} \approx 10^{-3}$ and an air-equivalent wall thickness that is about 3 orders of magnitude smaller than the actual thickness of air. The required wall thicknesses at photon energies below that of cobalt-60 gamma rays are even smaller than 0.45 cm and at very low photon energies amount to only a fraction of a millimeter.

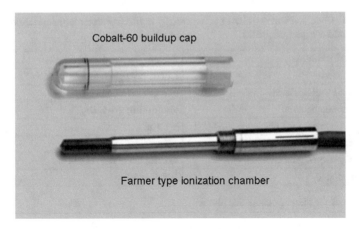

Fig. 16.15 Farmer type ionization chamber and its cobalt-60 buildup cap. Typical volume of Farmer type thimble chamber cavity is 0.6 cm^3; typical central electrode: material – pure aluminum, diameter – 1 mm, length – 2 cm; typical thimble (cavity wall) material is pure graphite

To cover a large range of energies general purpose thimble chambers are made with a thin air-equivalent wall (typically $\sim$0.4 mm) and additional sleeves (called build-up caps), usually made of Lucite (PMMA), are available as attachment to increase the wall thickness to that appropriate for given photon energy. A schematic diagram of a thimble chamber is depicted in Fig. 16.14c, while Fig. 16.15 shows a Farmer type thimble chamber and its cobalt-60 build-up cap.

Thimble Ionization Chambers. The vast majority of ionization chambers used in external beam radiotherapy are cylindrical with a thimble-shaped wall and a cavity with nominal ambient air volume of 0.6 cm^3. Thimble chambers became commercially available during 1950s with the advent of high-energy radiotherapy carried out on cobalt-60 machines. They were designed by UK medical physicist Frank T. Farmer, manufactured by the Baldwin Instrument Company in London, UK, and referred to as Baldwin-Farmer ionization chambers. During 1960s other companies began to offer thimble chambers following Farmer's design and these chambers are now called Farmer type ionization chamber.

The basic design of the chamber, such as its nominal volume of 0.6 cm^3, has not changed during the past 50 years; however, some useful modifications have been introduced with the passage of time. Most notable new features currently commercially available are: (i) Waterproof design allowing chamber use in water phantoms; (ii) Chamber electrical connection to electrometer and power supply with a triaxial cable; (iii) Chamber venting through a flexible tube surrounding the triaxial cable; and (iv) Shell wall and central collecting electrode built with air-equivalent or tissue-equivalent Shonka conductive plastics (SCP).

Table 16.11 General characteristics of commercially available Farmer type ionization chambers (entries in **bold** are most common)

Nominal cavity volume	0.6 cm^3
Shell wall material	**Lucite (PMMA); graphite**; SCP(*); delrin
Nominal shell wall thickness	**0.4 mm** to 0.5 mm
Central electrode material	**aluminum; graphite**; SCP
Central electrode nominal diameter	$\sim$**1 mm**
Central electrode nominal length	$\sim$**2 cm**
Build-up cap (Co-60) material	**Lucite**; bras; SCP; delrin
Polarizing voltage	**300 V** to 1000 V

(*) SCP stands for Shonka conductive plastics

Table 16.11 provides a summary of important characteristics of commercially available Farmer type thimble ionization chambers. For certain characteristics that vary from one manufacturer to another the most common ones are shown in bold. We note that the nominal cavity volume is 0.6 cm^3 for all Farmer type thimble chamber models and the shell wall thickness is between 0.3 mm and 0.5 mm. In most cases the collecting electrode is made either of aluminum or graphite with a diameter of 1 mm and length of 2 cm. The shell wall is mainly made of Lucite or graphite. Since Lucite is an insulator, the interior of Lucite thimble is painted with graphite dag to provide a conductive surface for charge collection. Most often the cobalt-60 build-up caps are made of Lucite, but some manufacturers also offer caps made of other "air-equivalent" materials.

Thimble Ionization Chamber and Photon Energy Below 1.5 MeV

We are now ready to determine air-kerma in air $(K_{air}^P)_{air}$ at point P in air with a thimble ionization chamber in a photon beam of energy less than 1.5 MeV. As discussed above, we replaced the standard free-air chamber (Sect. 16.3.14) as well as the standard Bragg–Gray cavity chamber (Sect. 16.3.15) with a thimble ionization chamber composed of an ambient air cavity and an air-equivalent shell wall (thimble) with appropriate thickness. In Sect. 16.3.12 we saw that for standard free air ionization chamber $(K_{air}^P)_{air}$ is expressed by (16.77) as follows

$$(K_{air}^P)_{air} = \frac{Q_{sat}}{m_{air}} \left(\frac{\bar{W}}{e} \right)_{air}, \tag{16.99}$$

where Q_{sat} is the charge produced in the chamber cavity, m_{air} is the mass of ambient air in the chamber cavity, and $(\bar{W}/e)_{air}$ is the mean energy required to produce an ion pair in air.

Based on similarity of basic principles behind the standard free-air chamber and thimble chamber, we conclude that (16.99) cannot only be used in thimble chamber dosimetry, it could also qualify the thimble chamber for absolute dosimetry. However, a closer look at (16.99) reveals that the ratio Q_{sat}/m_{air} is problematic and may cause difficulties with respect to absolute dosimetry.

On the one hand, Q_{sat}, the saturation charge collected on the measuring electrode is relatively easy to measure to a high degree of accuracy with a calibrated electrometer. On the other hand, m_{air}, the mass of air in the thimble cavity, is defined by the cavity effective volume V_S that is not necessarily equal to the cavity geometrical volume, and the discrepancy between the two can amount to several percent.

In contrast to the relatively easy to determine geometrical volume of the thimble cavity, the effective volume of the cavity is difficult to ascertain because it is governed by possibly non-uniform electric field lines that run between the polarizing and measuring electrodes and may also depend on the chamber polarizing voltage as well as radiation beam energy and dose rate. Consequently, the uncertainty in m_{air} resulting from uncertainty on thimble cavity effective volume has an adverse effect on the accuracy of the ratio Q_{sat}/m_{air} and precludes the direct use of (16.99) for absolute dosimetry with thimble chambers.

The standard method for obviating the problem with the direct determination of the Q_{sat}/m_{air} ratio of (16.99) is to calibrate the given cavity chamber at a primary standards laboratory that measures and issues a calibration coefficient for the chamber at a given photon energy up to cobalt-60 gamma rays. An assumption can be made that this calibration coefficient indirectly provides the effective volume of the chamber cavity at a given photon energy.

A thimble chamber used for calibration of kilovoltage x-ray beams usually possesses calibration coefficients covering a range of x-ray beams between 80 kVp and 400 kVp with various beam filtrations that affect the effective energy of the x-ray beam. Since the calibration coefficients vary slowly with effective beam energy, we conclude that effective cavity volumes related to m_{air} of (16.99) also vary with energy, while the geometrical volumes obviously do not. Thus, based on an accurate determination of m_{air} through the calibration coefficient, a claim can be made that calibrated thimble chambers can be used for absolute dosimetry in clinical radiotherapy departments.

16.3.17 Absolute Dosimetry of Megavoltage X-Ray Beams and Electron Beams

In principle, direct absolute dose measurement with ionization chamber in water for high-energy x-ray and electron beams is quite feasible; however, there are many practical difficulties that prevent a reliable use of the direct approach to clinical ionization chamber dosimetry in practice.

The Spencer–Attix air cavity relationship for dose D_{med} in phantom medium provides the theoretical basis for measurement of absolute dose in medium, such as water, in high-energy x-ray and electron beams. It is given as (see (15.20) in Sect. 15.7.1 and (16.77) in Sect. 16.3.12)

$$D_{med} = D_{cav} \frac{(\bar{L}/\rho)_{med}}{(\bar{L}/\rho)_{air}} = \frac{Q_{sat}}{m_{air}} \left(\frac{\bar{W}}{e}\right)_{air} \frac{(\bar{L}/\rho)_{med}}{(\bar{L}/\rho)_{air}}, \qquad (16.100)$$

where $(\bar{L}/\rho)_{med}$ and $(\bar{L}/\rho)_{air}$ are the mean restricted mass collision stopping powers of medium and air, respectively, for the electron spectrum at the position of the cavity in the medium and

Q_{sat} is the saturation charge, determined from charge Q collected on the measuring electrode of the ionization chamber and corrected for recombination loss.

m_{air} is the mass of ambient air in the chamber cavity given as $m_{air} = \rho_{air} \mathcal{V}_{eff}$ with ρ_{air} the air density in the cavity and $\mathcal{V}_{eff}$ the effective volume of the chamber cavity.

$(\bar{W}/e)_{air}$ is the mean energy dissipated by ionizing radiation in producing an ion pair in air.

In (16.100) an assumption is made that the sensitive air mass in the chamber cavity satisfies the Bragg–Gray cavity conditions and does not perturb the secondary charged particle spectrum in the medium. The Spencer–Attix cavity relationship currently represents the recommended approach to radiation dosimetry and uses restricted mass collision stopping powers (Sect. 6.10) averaged over the slowing-down spectrum of all generations of electrons with kinetic energy in the range between a low energy limit Δ (typically 10 keV) and maximum electron energy. The ratio $(\bar{L}/\rho)_{med}/(\bar{L}/\rho)_{air}$ in (16.100) is not very sensitive to choice of Δ that is generally taken as the minimum kinetic energy required by an electron just to traverse a typical Bragg–Gray cavity ($\sim$2 mm).

The Spencer–Attix equation of (16.100) provides a simple linear relationship between the dose D_{med} to a point P in the medium and the ratio Q_{sat}/m_{air}. Unlike the charge Q that is related to Q_{sat} and easily measured to a high degree of accuracy with a calibrated electrometer, the effective mass m_{air} related to effective volume $\mathcal{V}_{eff}$ of the cavity air is difficult to determine experimentally with a high degree of accuracy.

Two methods are known for dealing with the problem of determination of the ratio Q_{sat}/m_{air} in the Spencer–Attix cavity equation (16.100):

1. Standard method by which a chamber used in output measurement of megavoltage photon and electron beams possesses a cobalt-60 calibration coefficient traceable to a primary standards laboratory. This calibration coefficient is used in conjunction with a suitable dosimetry protocol to determine dose to water at the reference point in phantom.
2. Possible method for direct determination of the ratio Q_{sat}/m_{air} with an uncalibrated extrapolation chamber in a water equivalent solid phantom (see Sect. 16.3.18).

The standard method for obviating the problem with the direct determination of the Q_{sat}/m_{air} ratio of (16.100) is to obtain a calibration coefficient for the given cavity chamber in a cobalt-60 gamma ray beam at, or trace its calibration coefficient to, a primary standards laboratory. An assumption can be made that this calibration coefficient indirectly provides the effective volume of the chamber cavity at cobalt-60 gamma ray energy. However, since this chamber effective volume was determined in a known radiation field at the primary standards laboratory, a claim that the chamber provides an option for absolute calibration of a radiation beam is somewhat ambiguous. On the other hand, the cobalt-60 calibration coefficient indirectly provides the effective cavity volume with a high degree of accuracy and if, in addition, we know the chamber composition, we can use (16.100) and claim that conditions for absolute dosimetry were met. Therefore, calibration of megavoltage machine output in clinical radiotherapy departments with thimble chambers possessing cobalt-60 calibration coefficient traceable to a primary standards laboratory falls into the category of absolute dosimetry.

The primary standards laboratory usually provides a chamber calibration coefficient for cobalt-60 gamma rays either in terms of air-kerma in air (traditional option) or in terms of absorbed dose to water (modern option). Either one of these two cobalt-60 options is then used in conjunction with a suitable radiation dosimetry protocol for determination of absorbed dose to water phantom at a reference point in the user's megavoltage radiation beam based on the measured charge Q collected on the measuring electrode of the thimble chamber.

The exact procedures to be followed in determining dose to water D_{med} at reference point P in water phantom from the measured charge Q and the chamber cobalt-60 calibration coefficient are prescribed in international, national, or regional dosimetry protocols or dosimetry codes of practice (Sect. 15.10). It is customary that the clinical user chooses the protocol to follow; however, there are some countries or regions where government agencies prescribe a dosimetry protocol to be used in their jurisdiction.

The recent trend at primary standards laboratories aims at extending absorbed dose chamber calibration procedures from cobalt-60 gamma rays to high-energy x-rays and electron beams, thereby favoring the advantages of using the same radiation quantity (absorbed dose to water), beam quality, and experimental conditions as does the user. This approach by-passes the numerous correction factors used in dosimetry protocols thereby reducing uncertainties in the final dose-to-water result; unfortunately, these services are not yet widely available.

The dosimetry protocol (code of practice) lays out the road map for the procedure to be used to determine the desired dose-to-water from the measured charge Q. The protocol incorporates the chamber calibration coefficient as well as numerous correction factors that introduce various levels of uncertainty into the final result. These factors account for:

(i) Chamber calibration coefficient (air-kerma in air or absorbed dose to water).
(ii) Calibration mode (cobalt-60 or megavoltage vs and electrons).
(iii) Effects of chamber dimensions and wall composition.
(iv) Ion collection efficiency.
(v) Disruption of the photon fluence and secondary electron fluence caused by the presence of the chamber in the phantom.

16.3.18   Standard Extrapolation Chamber

At primary standards laboratories absolute ionometric dosimetry for low energy x-rays (∼50 kVp to ∼400 kVp) is carried out with standard free air ionization chambers (Sect. 16.3.14) and for intermediate x-ray energies (∼400 kVp to ∼1.5 MeV) with standard Bragg–Gray cavity chambers (Sect. 16.3.15). In the megavoltage x-ray and electron range the situation is less clear-cut. No absolute megavoltage ionometric dosimetry technique has been put into practice to date; however, the possibility of using extrapolation ionization chambers for this purpose has been known and discussed for several decades. Extrapolation chambers are circular parallel-plate ionization chambers that feature a variable air cavity volume and operate in the Bragg–Gray cavity region.

Since 1937 when Italian-born American Physicist Gioacchino Failla designed the first extrapolation ionization chamber, these chambers have been used mainly for determination of relative surface dose in kilovoltage and megavoltage photon beams and also in dosimetry of beta rays and superficial x-rays. In 1955 American medical physicist John S. Laughlin who used an extrapolation chamber embedded in a polystyrene phantom was the first physicist to investigate the measurement of a cobalt-60 machine output with an extrapolation chamber. More recently, in 1991, Stanley Klevenhagen used a Lucite extrapolation chamber immersed in a water phantom for absolute dosimetry of megavoltage x-ray and electron beams. However, both polystyrene and Lucite chamber components required troublesome corrections of the chamber readout Q to get the final result of absorbed dose to water and this was perceived a major hindrance to developing the technique further.

In 1995 staff and students at McGill University in Montreal built an extrapolation chamber directly into a "Solid Water" phantom and proposed their extrapolation chamber–phantom configuration as an option for absolute megavoltage dosimetry. Solid Water is dosimetric water-equivalent phantom material made of special epoxy resins and powders, designed and manufactured for use as a solid substitute of water in radiation dosimetry.

The Solid Water phantom–chamber configuration is simple and straightforward. Its main advantage is that both the chamber and phantom are made of identical material that is designed to be a close match to water in density and in terms of electron mass collision stopping powers as well as photon mass energy attenuation coefficients. Therefore, at least in principle, there is no need for any cumbersome correction factors to the signal Q measured in Solid Water when determining dose to water. Nevertheless, since the absorbed dose measurement is not carried out in water directly and the match of Solid Water to regular water is not absolutely perfect, there is concern that the technique is not suitable for use in primary standards laboratories. It is possible that continued refinements in solid water technology will change this perception in the future.

Figure 16.16 shows a schematic diagram of a typical phantom-embedded extrapolation chamber: a variable air-cavity cylindrical parallel-plate ionization chamber that forms an integral part of a water-equivalent solid phantom. A 7 cm diameter, 10 cm

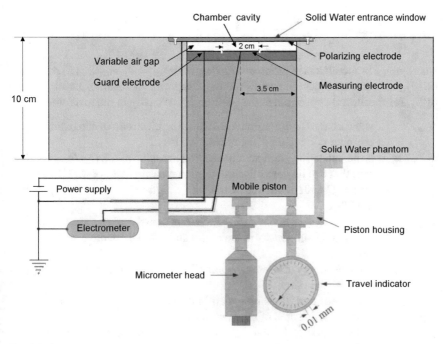

Fig. 16.16 Schematic diagram of the cylindrical, parallel-plate, variable air-volume extrapolation chamber embedded in Solid Water phantom. Nominal collecting electrode diameter is 2 cm

height Solid Water piston was fashioned to move inside a cylindrical aperture bored along the center of the Solid Water phantom ($30 \times 30 \times 10$ cm^3). The 0.5 mm thick polarizing electrode is fixed to the phantom at the top of the aperture; the measuring electrode with nominal radius of 1 cm and the 2.5 cm wide guard ring are attached to the movable piston. Graphite dag of thickness ~ 0.1 mm, spray-painted directly onto the Solid Water material, forms the electrodes. A micrometer that is mounted on the phantom body controls the electrode separation, determined by the displacement of the piston. A mechanical distance travel indicator monitors the movement of the piston.

For sufficiently small cavity air mass m_{air} the ratio Q_{sat}/m_{air} of the Spencer–Attix equation (16.100) is constant as a function of m_{air} and may be replaced by the easier-to-measure derivative dQ_{sat}/dm_{air} in (16.100) to get the following modified Spencer–Attix cavity equation

$$D_{med} = \frac{dQ_{sat}}{dm_{air}} \left(\frac{\bar{W}}{e} \right)_{air} \frac{(\bar{L}/\rho)_{med}}{(\bar{L}/\rho)_{air}}, \tag{16.101}$$

or, for a cylindrical parallel-plate extrapolation chamber

$$D_{\mathrm{med}} = \frac{1}{\rho_{\mathrm{air}} A_{\mathrm{eff}}} \frac{\mathrm{d}Q_{\mathrm{sat}}}{\mathrm{d}z} \left(\frac{\bar{W}}{e}\right)_{\mathrm{air}} \frac{(\bar{L}/\rho)_{\mathrm{med}}}{(\bar{L}/\rho)_{\mathrm{air}}} = k_{\mathrm{ec}} \frac{\mathrm{d}Q_{\mathrm{sat}}}{\mathrm{d}z},\qquad(16.102)$$

where $\mathrm{d}m_{\mathrm{air}}$ is the air mass differential given as $\mathrm{d}m_{\mathrm{air}} = \rho_{\mathrm{air}} A_{\mathrm{eff}} \mathrm{d}z$, $(\bar{L}/\rho)_{\mathrm{med}}$ and $(\bar{L}/\rho)_{\mathrm{air}}$, are the mean restricted collision stopping powers of medium and air, respectively, for the electron spectrum at the position of the cavity in the medium, and

Q_{sat} is the charge collected on the measuring electrode of the extrapolation chamber corrected for recombination loss.

$\mathrm{d}Q_{\mathrm{sat}}/\mathrm{d}z$ is the ionization gradient representing the slope of the $Q_{\mathrm{sat}}/m_{\mathrm{air}}$ relationship that can be determined through a measurement of $\Delta Q/\Delta m$ where both ΔQ and Δm are measured with a high degree of accuracy (within $\pm 0.2\%$).

m_{air} is the mass of ambient air in the chamber cavity given as $m_{\mathrm{air}} = \rho_{\mathrm{air}} \mathcal{V}_{\mathrm{eff}}$ with ρ_{air} the air density in the cavity and $\mathcal{V}_{\mathrm{eff}}$ the effective volume of the chamber cavity.

$(\bar{W}/e)_{\mathrm{air}}$ is mean energy dissipated by ionizing radiation in producing an ion pair in air.

ρ_{air} is the density of the ambient air in the cavity.

A_{eff} is the effective area of the measuring electrode.

z is the separation between the polarizing and measuring electrodes.

k_{ec} is the proportionality constant for an extrapolation chamber at a given photon or electron energy $k_{\mathrm{ec}} = (\rho_{\mathrm{air}} A_{\mathrm{eff}})^{-1} (\bar{W}/e)_{\mathrm{air}} [(\bar{L}/\rho)_{\mathrm{med}}/(\bar{L}/\rho)_{\mathrm{air}}]$.

Equation (16.102) forms the basis for absolute megavoltage radiation dosimetry with an extrapolation chamber showing that absorbed dose D_{med} in phantom is linearly proportional to the measured ionization gradient $\mathrm{d}Q_{\mathrm{sat}}/\mathrm{d}z$. The basic assumptions are that: (i) chamber cavity satisfies the Bragg–Gray cavity conditions and (ii) the four parameters forming the extrapolation chamber constant k_{ec} are known to a high degree of accuracy.

As far as the Bragg–Gray conditions are concerned (see Sect. 15.7.1), the extrapolation chamber must operate at relatively small electrode separation z typically in the range between 1 mm and 2.5 mm in order to remain in the Bragg–Gray cavity domain where $\mathrm{d}Q_{\mathrm{sat}}/\mathrm{d}z$ is constant.

As for the extrapolation chamber constant k_{ec} of (16.102)

$$k_{\mathrm{ec}} = \frac{1}{\rho_{\mathrm{air}} A_{\mathrm{eff}}} \left(\frac{\bar{W}}{e}\right)_{\mathrm{air}} \frac{(\bar{L}/\rho)_{\mathrm{med}}}{(\bar{L}/\rho)_{\mathrm{air}}},\qquad(16.103)$$

mean ionization energy $(\bar{W}/e)_{\mathrm{air}}$ of air as well as the ratio $(\bar{L}/\rho)_{\mathrm{med}}/(\bar{L}/\rho)_{\mathrm{air}}$ for a given photon or electron beam energy are readily available from the literature, and the ambient air density ρ_{air} can be easily determined for a given temperature and pressure [see (16.79)].

Since all ionization chambers possess an inherent capacitance C, one can determine the effective area A_{eff} of the measuring electrode of the extrapolation chamber to a high degree of accuracy using a simple chamber capacitance measurement as follows:

1. For a sufficiently large guard electrode the capacitance C of the parallel-plate ionization chamber is given by

$$C = \frac{\Delta Q}{\Delta V} = \varepsilon_0 \frac{A_{eff}}{z},\tag{16.104}$$

where

ε_0 is the electric constant (8.85×10^{-12} As/Vm).

z is the given separation between the polarizing and measuring electrodes of the extrapolation chamber.

ΔQ is the change in the charge measured by the electrometer when the polarizing voltage is changed by ΔV.

2. First write (16.104) as $\Delta Q = C \Delta V$ and, for a given z, plot ΔQ against ΔV to get a straight line with slope $C(z)$. Repeat for several different z.
3. Next, write (16.104) as $z = (\varepsilon_0 A_{eff})/C$ and plot z against $1/C$ to get a straight line with a slope of $\varepsilon_0 A_{eff}$. Divide the slope $\varepsilon_0 A_{eff}$ by ε_0 to arrive at the effective area A_{eff} of the measuring electrode.

In the modified Spencer–Attix equation (16.102) of an extrapolation chamber the derivative dQ_{sat}/dz (ionization gradient) represents the slope of the Q_{sat}/m_{air} relationship and can be determined through a measurement of $\Delta Q_{sat}/\Delta z$ where both ΔQ_{sat} and Δz are measured with a high degree of accuracy. As discussed above, the effective area A_{eff} of the measuring electrode can be determined accurately through a measurement of chamber capacitance C as a function of relative electrode separation z measured with a commercial high-accuracy mechanical travel distance indicator. With the particular extrapolation chamber design, it is not necessary to measure z in an absolute manner; only the easy-to-measure relative electrode separations that can be determined with high accuracy are required for absorbed dose calculations.

Un-calibrated, variable air-volume extrapolation chambers, built as an integral part of the phantom in which absorbed dose is measured, can serve as radiation dosimeters in output calibrations of megavoltage x-ray and electron beams in radiotherapy. In contrast to dosimetry with calibrated thimble chambers, the dosimetry with extrapolation chambers is simple and requires no correction factors to account for chamber wall properties, for perturbation of the secondary charged particle fluence, and for the lack of access to high-energy photon and electron calibrations at primary standards laboratories. The main drawback of extrapolation chambers is that currently available water-equivalent solid phantoms still are not perfectly water-equivalent.

16.4   Absolute Radiation Dosimetry: Summary

Quantities related to ionizing radiation are measured with radiation dosimetry systems that consist of a radiation dosimeter (detector) and a reader. In general, the radiation dosimeter has two components: a cavity composed of a radiation sensitive medium and a wall enclosing the cavity. The cavity produces a signal in response to ionizing radiation, the wall contains the cavity medium, and the reader measures the signal.

The radiation quantities most often measured in radiation dosimetry are: (i) dose absorbed in the dosimeter cavity and (ii) particle fluence in the dosimeter cavity. With regard to absorbed dose in the cavity two major categories of radiation dosimetry are known: absolute and relative.

An absolute dosimetry system incorporates a dosimeter cavity that produces a signal from which the absorbed dose in the cavity can be determined directly without any need for calibration of dosimeter response in a known radiation field. A relative dosimetry system, on the other hand, is a system that requires calibration of its signal response in a known radiation field.

Only three absolute dosimetry systems have been developed to date, but there are many known relative radiation dosimetry systems, ranging from very useful and practical systems, through practical yet of limited use, all the way down to systems that are not practical but are of some academic interest. The three known absolute radiation dosimetry systems are: (i) Calorimetric absolute dosimetry, (ii) Chemical (Fricke) absolute dosimetry, and (iii) Ionometric (ionization chamber-based) absolute dosimetry.

The three absolute dosimetry systems are discussed in this chapter; a few of the most practical and best-known relative dosimetry systems used in medicine for diagnosis and treatment of disease are discussed in Chap. 17.

16.4.1   Calorimetric Absolute Radiation Dosimetry System

Calorimetry is in general defined as the process of measuring heat generated or absorbed in a chemical reaction or physical process. Calorimetric absolute radiation dosimetry systems rely on direct measurement of heat (energy) absorbed in the core of the calorimeter (dosimeter cavity) as a result of standard physical interactions of ionizing radiation with the dosimeter cavity.

As shown in (16.12), the mean dose $\bar{D}_{cav}$ to the cavity (calorimeter core) is directly proportional to the rise in temperature ΔT of the cavity as well as to the specific heat C of the core material and is inversely proportional to $(1 - \kappa)$ where κ is the so-called heat defect that accounts for the fraction of energy that is absorbed in the calorimeter core but does not contribute to ΔT

$$\bar{D}_{cav} = \frac{C \Delta T}{1 - \kappa}.$$

(16.105)

Calorimetry has been applied successfully to a variety of physical and chemical processes; however, its application to radiation dosimetry, despite being logical and simple in principle, has had limited success. The main problem is that dosimetric cavities, applied in calorimetric radiation dosimetry, absorb a relatively small amount of heat from ionizing radiation. This small amount of heat produces a very weak radiation-induced signal that can be measured only with extremely sensitive signal detection, making the experimental procedures very cumbersome and difficult.

Use of calorimetric methods in radiation dosimetry has a long history. Calorimeters have been investigated for detection of ionizing radiation almost immediately after Röntgen's and Becquerel's discovery of ionizing radiations; however, calorimetric radiation dosimetry has always been relegated to primary standards laboratories where it provides a primary absorbed dose standard or is used in conjunction with ionometric absolute dosimetry systems.

16.4.2 Fricke Chemical Absolute Radiation Dosimetry System

Radiation dosimetry generally implies measurement of a radiation-induced signal that results from a physical interaction between ionizing radiation and dosimeter cavity. In addition, radiation dosimetry also encompasses a special dosimetry category that relies on a variety of well-defined chemical reactions that ionizing radiation triggers in certain absorbing media in dosimeter cavity. All but one of these radiation-induced chemical reactions fall into the category of relative dosimetry; however, the Fricke ferrous sulfate chemical dosimetry system, as a result of its high reliability and accuracy in measurement of absorbed dose to water, is considered one of the three known absolute dosimetry systems.

Hugo Fricke, a Danish-American physicist and pioneer in chemical dosimetry, introduced the ferrous sulfate based system in 1920s. It relies on radiolysis of water, a process that under the influence of ionizing radiation results in a variety of highly reactive radicals, ions, and molecular products in water. In an irradiated aqueous solution of ferrous (Fe^{2+}) ions in a Fricke dosimeter cavity, some of these radicals and molecular products oxidize Fe^{2+} ions into ferric Fe^{3+} ions and the number of Fe^{3+} ions produced in the Fricke dosimetric solution is proportional to dose absorbed in the cavity of the Fricke dosimeter.

In Fricke dosimetry the mean dose $\bar{D}_{cav}$ to the cavity of the Fricke dosimeter is inversely proportional to: (i) density of the Fricke ferrous sulfate solution ρ_F (usually assumed to be 1.024 g/cm^3) and (ii) radiolytic yield $G(Fe^{3+})$ of ferric Fe^{3+} ions where, in general, the radiolytic yield $G(X)$ of entity X is defined as the number of entities X produced per 100 eV of absorbed energy. $\bar{D}_{cav}$ is also directly proportional to the change in molar concentration ΔM of the Fe^{3+} ion in aqueous ferrous sulfate solution as a result of exposure to ionizing radiation

$$\bar{D}_{cav} = \frac{\Delta M}{\rho_F G(Fe^{3+})}. \tag{16.106}$$

Since $G(X)$ depends on ionizing beam energy, mode, and ionization density the beam produces in cavity medium, it is difficult to measure. This means that the absoluteness of Fricke dosimetry depends strongly on the accepted value for radiolytic yield $G(X)$. *For example*, the agreed upon value of $G(X) = G(Fe^{3+})$ for cobalt-60 radiation is 15.5 ferric (Fe^{3+}) ions produced per 100 eV of energy absorbed in water of the Fricke dosimeter cavity.

The change in molar concentration of Fe^{3+} ions in an irradiated Fricke dosimeter is determined with a spectrophotometer that measures transmission of ultraviolet (uv) light ($\lambda = 303$ nm) through a Fricke solution. The larger is the concentration of Fe^{3+} ions in the sample, the lower is the transmitted uv beam intensity and the higher was the mean radiation dose absorbed in the dosimeter cavity.

16.4.3   Ionometric Absolute Radiation Dosimetry System

The oldest and still the simplest means of detecting ionizing radiation is based on passing ionizing particles (photons, neutrons, or charged particles) through a gas and collecting the ions that the particles release in gas by ionizing atoms and molecules of the gas. Radiation detectors used for this purpose are referred to as ionization chambers.

The simplest of all gas-filled radiation detectors is the cavity ionization chamber filled with ambient air. A DC power supply establishes an electric field in the air-filled chamber cavity and enables the collection of charges that ionizing radiation produces in the dosimeter cavity. A sensitive electrometer is used for measuring the collection of ions on the measuring electrode of the ionization chamber. The basic design of an ionization chamber has not changed much since the pioneers in ionizing radiation research started in the late 1890s developing devices for measuring the presence and quantity of ionizing radiation emitted either by x-ray tubes or radioactive substances.

Ionization chambers are used for both absolute and relative radiation dosimetry. An ionization chamber made of known material and having a cavity of known volume can be considered an absolute ionization chamber. As far as absolute radiation dosimetry is concerned, four types of absolute ionization chambers are in use, all using ambient air as the radiation sensitive gas and each covering a different energy range and mode of ionizing radiation:

1. Standard free air ionization chamber is the primary radiation standard for photons in the range from $\sim$50 kVp to $\sim$400 kVp.
2. Standard Bragg–Gray cavity ionization chamber serves as primary radiation standard for photons in the range from $\sim$400 kVp to $\sim$1.5 MeV including cobalt-60 gamma rays at 1.17 MeV and 1.33 MeV.
3a. Thimble ionization chamber having an air-equivalent shell and possessing a calibration coefficient issued by, or traceable to, a primary standards laboratory for photons from 80 kVp to 1.5 MeV.

3b. Thimble ionization chamber of known composition, possessing a cobalt-60 calibration coefficient issued by, or traceable to, a primary standards laboratory for photons from cobalt-60 to 35 MV, electrons from 10 MeV to ~30 MeV as well as clinical proton and heavy ion beams. The chamber cavity volume should be between 0.1 cm^3 and 1 cm^3; most common choice is a Farmer type chamber with a 0.6 cm^3 cavity volume. The cobalt-60 calibration coefficient is used in conjunction with an appropriate dosimetry protocol.

4. Parallel-plate ionization chamber has a more restricted usefulness in radiotherapy departments. Its use is optional for photon beams and electron beams of kinetic energy above 10 MeV; however, its use is mandatory for electron beams below 10 MeV. For use in absolute dosimetry the chamber must possess at least a cobalt-60 calibration coefficient in terms of absorbed dose to water from a primary standards laboratory and, if possible, a calibration coefficient in terms of absorbed dose to water in an electron beam of user's quality, also from a primary standards laboratory. The reference point in a parallel-plate chamber is taken to be on the inner surface of the entrance window.

5. Extrapolation chamber is added to this list of four ionization chambers for academic interest. It could potentially be used (but is not used) to provide a standard for photons from cobalt-60 gamma rays to 35 MeV and for electrons from ~4 MeV to ~35 MeV.

The dosimetric information provided by the five types of ionization chambers listed above is based on the Spencer–Attix cavity theory (see Sect. 15.1.7) and the basic equation given in (16.77)

$$\bar{D}_{\text{cav}} = \frac{Q_{\text{sat}}}{m_{\text{air}}} \left(\frac{\bar{W}}{e} \right)_{\text{air}}, \tag{16.107}$$

where

Q_{sat} is the saturation charge calculated from the measured charge Q corrected for recombination loss.

m_{air} is the air mass in the chamber cavity.

$(\bar{W}/e)_{\text{air}}$ is the mean energy required to produce an ion pair in air.

16.4.4 "Absoluteness" of Absolute Radiation Dosimetry Systems

Table 16.12 lists the main characteristics of the three absolute radiation dosimetry systems: calorimetric, chemical and ionometric. As far as absoluteness of absorbed dose measurements is concerned, it is reasonable to state that calorimetric radiation dosimetry is the most absolute of the three known absolute dosimetry techniques. This conclusion is justified, since calorimetric dosimetry measures heat (energy) absorbed

Table 16.12 Main characteristics of the three known absolute radiation dosimetry systems: (i) calorimetric, (ii) chemical (Fricke), and (iii) ionometric

Absolute radiation dosimetry systems	Calorimetric radiation dosimetry	Chemical radiation dosimetry	Ionometric radiation dosimetry
Cavity material	Water or Graphite	Ferrous sulfate solution in water	Air
Reaction in cavity	Ionization and excitation	Oxidation $Fe^{2+} \rightarrow Fe^{3+}$	Ionization and production of positive and negative ions
Dosimetric signal in cavity	ΔT Change in cavity temperature	$\Delta M = C_{Fe^{3+}}$ Change in molar concentration	Ionization charge or ionization current
Absorbed dose $\bar{D}_{cav}$ in cavity	$\bar{D}_{cav} = \dfrac{C\Delta T}{1 - \kappa}$ see (16.12)	$\bar{D}_{cav} = \dfrac{C_{Fe}^{3+}}{\rho_F\, G(Fe^{3+})}$ see (16.32)	$\bar{D}_{cav} = \dfrac{Q_{sat}}{m_{air}} \left(\dfrac{\bar{W}}{e}\right)_{air}$ see (16.77)
Important parameter for absolute radiation dosimetry	–	$G(X) = G/(100 \text{ eV})$ $G = G(X) \times 1.037 \times 10^{-7}$ mol/J	$\left(\dfrac{\bar{W}}{e}\right)_{air} = 33.97$ J/C
Measurement instrument (reader) of the dosimetry system	**Thermistor** with **Wheatstone bridge** to measure rise in temperature	**Spectrophotometer** to measure transmission of u.v. light (303 nm) through Fricke solution	Electrometer to measure ionization charge or ionization current
Usable dose range (Gy)	$\sim 10^2$ to $\sim 10^5$	~ 20 to ~ 400	–

in dosimeter cavity directly, in contrast to ionometric dosimetry and Fricke chemical dosimetry that determine absorbed dose in the cavity relying on specific conversion coefficients. The conversion coefficient for ionometric dosimetry is $(\bar{W}/e)_{air}$, the mean energy required to produce an ion pair in air; for Fricke chemical dosimetry it is $G(Fe^{3+})$, the radiolytic yield or G value defined as the number of ferric (Fe^{3+}) ions created from ferrous (Fe^{2+}) ions per 100 eV of absorbed energy. In ionometric dosimetry and Fricke chemical dosimetry the accuracy of the end result is linked to the accuracy of the measured conversion coefficient. When the recommended value of the conversion coefficient changes, the end result for the whole calibration chain changes. Table 16.8 traces the evolution of $(\bar{W}/e)_{air}$ values since 1962 when the ICRU issued its first recommendation on the $(\bar{W}/e)_{air}$ value to be used in radiation dosimetry. During the past 50 years the value of $(\bar{W}/e)_{air}$ increased by about 1%.

Antozonite, Thermoluminescence, and Free Elemental Fluorine

If an ionic crystal is exposed to ionizing radiation and subsequently heated, a tiny fraction of the energy that the crystal absorbed from radiation is emitted in the form of visible or ultraviolet light. This phenomenon of thermally activated phosphorescence is called thermoluminescence (TL) and is the most spectacular and widely known of a number of radiation induced thermally activated phenomena observed in ionic crystals. Its practical applications are in two seemingly unrelated areas: in archaeological pottery dating and in dosimetry of ionizing radiation.

One of the best-known TL materials is calcium fluoride (CaF_2), an important industrial mineral known as fluorite or fluorspar that crystallizes in cubic lattice structure and is used in numerous chemical, ceramic, and metallurgical processes. Fluorite also has historical significance as a result of the 1852 discovery by George G. Stokes (of the "Stokes theorem" fame) that fluorite produces a blue-violet glow when illuminated with ultraviolet light. Stokes coined the term fluorescence to describe this instantaneous emission of light by a crystalline solid under the influence of some type of stimulation. In contrast, the post-stimulation emission of light with a time delay exceeding 10^{-8} s is called phosphorescence.

The photograph on the next page shows spectacular TL emission of visible light from a pulverized fluorite mineral heated to 300 °C on a hot plate. The mineral shown in the photograph is a unique variety of fluorite that is radioactive, originates in Wölsendorf, a village in Bavaria, Germany, and was discovered some 200 years ago. It is officially known as antozonite but is colloquially called stinkspat or fetid fluorite because of the pungent smell that it emits when crushed. Radioactive uranium-238 and its daughter products that are mixed naturally with the Wölsendorf antozonite mineral generated the ionizing radiation that started the TL process shown in the photograph on the next page.

It is interesting to note that, in addition to emission of TL light, the natural radioactivity found in antozonite is also responsible for triggering another unique natural phenomenon, namely, the only known natural source of free elemental fluorine (F_2). Until lately, the conventional wisdom was that F_2 is the most reactive chemical molecule and as such cannot exist freely in nature. Florian Kraus, a German chemist, proved this contention wrong in 2012 with an NMR spectroscopy experiment that confirmed the presence of free F_2 trapped in inclusions inside the antozonite mineral. It is assumed that the ionizing particles produced in the uranium-238 decay series can induce oxidation of F^- lattice ions into free F_2 molecules that become trapped in antozonite inclusions. When antozonite is crushed, the trapped fluorine reacts with atmospheric oxygen and water vapor resulting in ozone (O_3) and hydrogen fluoride (HF).

Photo on next page: Courtesy of Prof. Robert Schwankner, Radiolab Munich and Michael Dönhöfer (B. Eng.), Technischer Überwachungsverein AG, Munich, Germany. Reproduced with permission.

Chapter 17
Relative Radiation Dosimetry

Ionizing radiation induces various effects in matter with which it interacts and in many of these effects the radiation-induced signal is proportional to absorbed dose and can be used to measure absorbed dose in the given material with a device referred to as a radiation dosimetry system.

Since humans cannot sense ionizing radiation, the ability to detect ionizing radiation and to measure its quantities is an important aspect of radiation physics. Measurement of ionizing radiation is referred to as radiation dosimetry and is divided into two major categories: absolute and relative. Absolute radiation dosimetry systems do not require calibration in a known radiation field and are discussed in detail in Chap. 16; relative dosimetry systems must be calibrated in a known radiation field and are dealt with in this chapter.

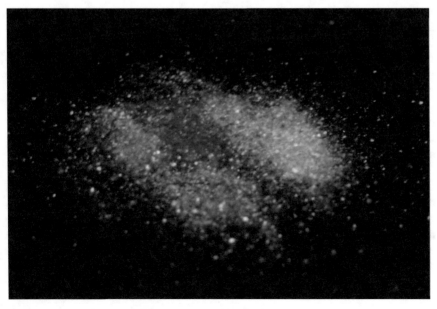

© Springer International Publishing Switzerland 2016
E.B. Podgoršak, *Radiation Physics for Medical Physicists*,
Graduate Texts in Physics, DOI 10.1007/978-3-319-25382-4_17

As discussed in Chap. 16, there are only three known categories of absolute radiation dosimetry: (i) calorimetric, (ii) chemical (Fricke), and (iii) ionometric. On the other hand, a large variety of relative dosimetry techniques are known, some very practical and generally useful, others of limited or very specialized use, and some not at all practical and studied only for academic interest. A relative radiation dosimetry system is defined as a system whose response to ionizing radiation must be calibrated in a known radiation field before its radiation induced (dosimetric) signal can be used to provide absorbed dose or dose rate in the dosimeter chamber cavity.

The most practical of the relative dosimetry systems used in medicine fall into four major categories and are discussed in this chapter. Each of the four major categories is subdivided into many subcategories, as indicated in Table 17.1, and each subcategory is based on a specific radiation induced signal representing a measurable physical quantity that is proportional to the radiation dose absorbed in the dosimeter cavity. Some of the relative dosimetry systems can be used in both the active and passive mode, while other systems only support one of the two modes. The four major categories of relative radiation dosimetry as applied to medicine are:

(i) Relative ionometric dosimetry discussed in Sect. 17.1.
(ii) Luminescence dosimetry discussed in Sect. 17.2.
(iii) Semiconductor dosimetry discussed in Sect. 17.3.
(iv) Film dosimetry discussed in Sect. 17.4.

Each of the four broad relative dosimetry categories listed above and discussed in this chapter focuses on the most practical and best known radiation dosimetry systems in the given category; however, in addition to these systems, many systems of lesser practical value or of only academic interest are listed in Table 17.1 *in italic font* but are not discussed in this chapter.

The chapter begins with a discussion of four relative ionometric radiation dosimetry techniques in Sect 17.1: (a) area survey meters for use in radiation protection in Sect. 17.1.1; (b) re-entrant well type ionization chambers in Sect. 17.1.2 used in calibration of brachytherapy sealed sources and in activity measurement of radiopharmaceuticals; (c) radiation detectors used in quality assurance of megavoltage linear accelerators in Sect. 17.1.3, and (d) liquid ionization chambers in Sect. 17.1.4.

Luminescence dosimetry is covered in Sect. 17.2 with three subsections: basics of energy band structure in solids in Sect. 17.2.1; basics of luminescence dosimetry in Sect. 17.2.2; thermoluminescence (TL) dosimetry in Sect. 17.2.4; and optically stimulated luminescence (OSL) in Sect. 17.2.7.

Semiconductor dosimetry provides in Sect. 17.3 a short introduction to semiconductor physics in Sect. 17.3.1; and covers the p-n junction in Sect. 17.3.2; diode radiation dosimeter in Sect. 17.3.3; and diamond radiation dosimeter in Sect. 17.3.4.

Film dosimetry addresses in Sect. 17.4 two types of film use in radiation dosimetry: radiographic film dosimetry in Sect. 17.4.2 and radiochromic film dosimetry in Sect. 17.4.4. The chapter on relative dosimetry systems concludes in Sect. 17.5 with a summary of the four major categories of relative dosimetry.

Table 17.1 Relative radiation dosimetry techniques. (Note: entries shown in *italic font* are of lesser importance and are not discussed in this book)

1	Relative ionometric dosimetry	Parallel-plate ionization chamber
		Thimble ionization chamber
		Area survey meter
		Re-entrant well ionization chamber
2	Luminescence dosimetry	Thermoluminescence (TL)
		Optically stimulated luminescence (OSL)
		Thermally activated current
		Thermally activated depolarization
		Thermally stimulated exoelectron dosimetry
		Thermal electret dosimetry
3	Semiconductor dosimetry	Diode
		Diamond detector
		MOSFET
4	Film dosimetry	Radiographic film
		Radiochromic film
5	*Relative Chemical Dosimetry*	*Ceric-cerous dosimetry*
		Alanin-EPR dosimetry
		Ethanol-chlorobenzene dosimetry
		Gel dosimetry

17.1 Relative Ionometric Radiation Dosimetry

Ionometry is defined as measurement and collection of ions, while ionometric radiation dosimetry refers to measurement of absorbed dose or other dose related quantities with ionization chambers. These devices are based on electrostatic collection of ions and provide one of three techniques for absolute radiation dosimetry, discussed in detail in Sect. 16.3. In addition, ionometric techniques also play an important role in the ionometric category of relative radiation dosimetry. The basic characteristics of ionization chambers are discussed in detail in Chap. 16 in the context of absolute ionometric radiation dosimetry; however, the general discussion applies equally well to all ionization chambers, irrespective of their classification either as absolute or relative dosimeter.

Four categories of relative dosimetry make use of ionization chambers:

(i) Area survey meters for use in radiation protection (Sect. 17.1.1).
(ii) Re-entrant well ionization chambers used in calibration of brachytherapy sealed sources and measurement of activities of radiopharmaceuticals in nuclear medicine (Sect. 17.1.2).
(iii) Radiation detectors used in quality assurance programs for clinical megavoltage x-ray and electron beams in external beam radiotherapy (Sect. 17.1.3).
(iv) Liquid ionization chambers (Sect. 17.1.4).

Ionization chambers play a primary and invaluable role in acceptance testing, commissioning, and calibration of clinical megavoltage beams as well as in calibration of brachytherapy radioactive sealed sources; moreover, in radiation protection, ionization chambers complement other useful devices, such as proportional counters, Geiger counters and scintillation detectors.

To fulfill their assigned role in radiation dosimetry, ionization chambers used in the four categories listed above vary significantly in geometrical design (parallel-plate, cylindrical, spherical), size (cavity volume from 1 mm^3 to 1 l), and cavity medium used (ambient air, pressurized gas, or suitable liquid). For effective use, these chambers must be periodically calibrated in a primary standards laboratory or they must at least trace their calibration coefficients to a standards laboratory. However, no such calibration is required for ionization chambers that are used solely as relative dosimeters, since the data they provide are not critical from the point of view absolute dosimetry.

17.1.1 Area Survey Meter

Area survey meters are used for detection and measurement of radiation:

   (i) To check personnel, equipment, facilities, and environment for radioactive contamination or presence of ionizing radiation.
  (ii) To survey radiation producing equipment for radiation leakage.
 (iii) To evaluate adequacy of shielding in radiation producing facilities.
  (iv) To verify compliance with radiation safety regulations.

Area survey meters are designed to be hand-held, portable, and battery-operated. They have easy to read analog or digital display of count rate and dose rate in differential mode or of total counts and total dose in integral mode. Most of them feature audible indication of count rate that can be switched ON and OFF, as desired.

Various types of radiation detectors are used in area survey meters, such as ionization chambers, proportional counters, Geiger tubes, and scintillation counters. All except ionization chambers use charge multiplication to amplify the radiation signal, thereby optimizing sensitivity of radiation detection. However, the flat energy response of ionization chambers is one of the advantages of ionization chambers over the charge multiplication type detectors and many survey meters use an ionization chamber as their radiation detector for measurement of exposure (air-kerma in air) despite its relatively weak signal.

Area survey meters based on ionization chamber usually have relatively large volume cylindrical cavity either open to ambient air or sealed and filled with pressurized gas. Large cavity volume and gas under pressure are used to increase the detector sensitivity. As for radiation quantities, ionization chamber based survey meters are suited for measurement of exposure rate (air-kerma rate in air) as well as for estimation of dose rate. They can be used for measurements of α- and β-particles as well

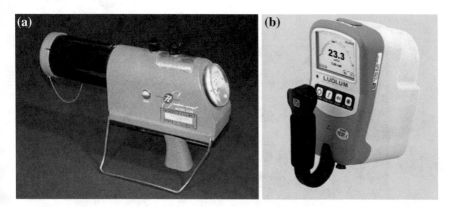

Fig. 17.1 Examples of ionization chamber based area survey meters. **a** Original "Cutie Pie" area survey meter manufactured by Nuclear Chicago in 1957. **b** Modern digital area survey meter manufactured by Ludlum Instruments, Inc., Sweetwater, Texas, USA

as γ-rays and x-rays. Two area survey meters based on ionization chamber detector, one shown in (a) designed some 50 years ago (type: Cutie Pie) and the other shown in (b) commercially available now (manufactured by Ludlum), are shown in Fig. 17.1.

17.1.2 Re-Entrant Well Ionization Chamber

Well type ionization chambers, also called re-entrant chambers (where re-entrant stands for "pointing inward"), are ideally suited for calibration of brachytherapy sealed sources, such as iridium-192 and cobalt-60. These sources, used in remote afterloading brachytherapy machines, can be calibrated with thimble chambers that are routinely used in calibration of kilovoltage and megavoltage external beams generated by x-ray machines and linear accelerators, respectively. However, calibration of brachytherapy sources with thimble chambers in open-air geometry is very cumbersome as well as time consuming and well type ionization chambers provide a much more reliable, simpler, and significantly faster method for this purpose. Well type chambers can be used with standard DC power supplies and conventional electrometers that are readily available for other dosimetric purposes in radiotherapy departments.

Computer driven remote afterloading systems have been developed to optimize the efficiency of brachytherapy procedures and to minimize radiation exposure to medical and support staff delivering the procedures. With respect to dose rate, brachytherapy machines are classified as either low dose rate (LDR) or high dose rate (HDR) machines. Because of the large difference in activity between an LDR and HDR source, the currents that these sources generate in a given well chamber vary significantly. It is therefore customary to use a larger volume well type chamber to calibrate LDR sources, and a smaller cavity well chamber to calibrate HDR sources.

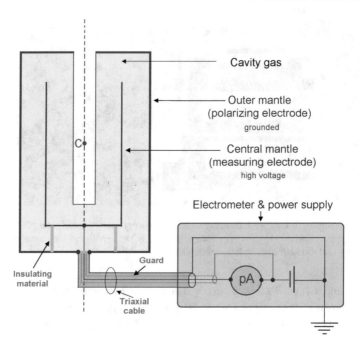

Fig. 17.2 Schematic diagram of a re-entrant well ionization chamber used for calibration of brachytherapy sources. The chamber is cylindrically symmetrical and incorporates three concentric mantles: inner, central, and outer. The outer and inner mantles are connected and play the role of grounded polarizing electrode; the cental electrode is connected to a power supply through an electrometer

Currently the most common source used in afterloading machines is iridium-192 (Ir-192) because of its effective gamma ray energy of ~380 keV and its high specific activity. Its relatively short half-life of ~74 days is a distinct disadvantage, since the use of Ir-192 source in remote afterloading machines requires frequent replacement of the source (3 to 4 times per year). Typical activity of an Ir-192 source is 10 Ci to 20 Ci (37 GBq to 74 GBq) delivering treatment dose rates exceeding 2 Gy/min (see Sect. 11.6.4).

With respect to their use, there are three types of well re-entrant ionization chambers:

(i) Conventional well type chambers designed for measuring activities of radio-pharmaceuticals used in nuclear medicine.
(ii) Brachytherapy well type chambers designed for measuring activities of LDR sources.
(iii) Brachytherapy well type chambers designed to measure HDR brachytherapy sources.

As shown schematically in Fig. 17.2, the typical well re-entrant ionization chamber is cylindrically symmetrical and consists of three concentric cylinder mantles: inner, central, and outer. The following features of the cylinder mantles are notable:

1. The inner and outer cylinder mantles are metallic, mechanically and electrically connected, and serve as the polarizing electrode of the chamber. They are connected to ground potential, define the chamber cavity, and contain the cavity gas.
2. The outer cylinder mantle has a relatively thick wall and, in addition to serving as polarizing electrode, by virtue of being at ground potential also serves as the chamber housing.
3. The central cylinder mantle is the measuring electrode and is connected through the floating electrometer to the high voltage end of the power supply.
4. The inner and central metallic cylinder mantles are relatively thin to transmit charged particles as well as x-rays and gamma rays emitted by the sealed source undergoing calibration with the well chamber, yet thick enough to provide electronic equilibrium in the cavity gas.
5. The special design of the polarizing electrode ensures that the well chamber provides almost 4π collection geometry when the source is placed at the center C of the chamber.
6. The chamber cavity gas in well chambers is either ambient air that requires air density correction or sealed pressurized argon gas that does not require temperature and pressure correction. Ambient air contains oxygen that is electronegative and attracts free electrons to create negative ions. This delays collection of negative charges in the cavity thereby increasing recombination loss in the chamber. Argon, a non-electronegative gas, does not suffer this problem.

17.1.3 Practical Ionization Chambers in Radiotherapy Department

From the perspective of relative ionometric radiation dosimetry, ionization chambers are used in all branches of radiation medicine. As discussed above, they are used in radiation safety, brachytherapy, and nuclear medicine; however, by far the most common and important application of ionization chambers is found in many technical aspects of external beam radiotherapy, such as acceptance testing, commissioning, and routine quality assurance of major radiotherapy machines. These machines produce kilovoltage x-rays with x-ray machines and megavoltage x-rays and electrons with linear accelerators. It is imperative that these machines be properly calibrated and maintained to ensure that radiation doses delivered to patients actually comply with doses prescribed by the physician.

In Chap. 16 absolute ionometric dosimetry was discussed based on four types of ionization chamber: standard free-air chamber (Sect. 16.3.14), standard Bragg-Gray cavity chamber (Sect. 16.3.15), thimble ionization chamber (Sect. 16.3.16), and parallel-plate ionization chamber. The first two chambers are cumbersome to use and limited to absolute dosimetry in primary standards laboratories; the latter two are used in clinical departments not only for absolute dosimetry but also for relative dosimetry.

When thimble and parallel-plate chambers are used for absolute dosimetry, they must possess a calibration coefficient from a primary standards laboratory; when used for relative dosimetry, no calibration coefficient is required.

Quality assurance (QA) in radiotherapy is a very important component of the radiotherapy process. It encompasses all procedures that ensure consistency of the medical prescription of the radiation dose as well as safe and reliable fulfillment of that prescription with respect to dose absorbed in the target volume, combined with minimal dose to normal tissue, minimal exposure to personnel, and adequate patient follow up. QA programs on radiotherapy machines, in turn, play a major role in the radiotherapy QA, and relative ionization chamber dosimetry represents one of the most important aspects of the equipment QA programs.

A radiotherapy machine QA program typically consists of a list of mandatory procedures, their frequency (daily, weekly, etc.) and required action level. Many of the procedures involve radiation tests and most of these require measurements with a suitable ionization chamber. Machine output calibration is typically carried out 2–3 times per week on each linear accelerator beam and twice per year on a cobalt machine. These tests fall into the category of absolute dosimetry tests. As discussed in Sect. 16.3.17, a Farmer type ($0.6\,cm^3$) ionization chamber with a cobalt-60 calibration coefficient is the most suitable chamber for this purpose.

Commissioning of a radiotherapy machine is carried out upon installation of the machine and subsequently on a yearly basis. This is an important and time-consuming undertaking, since each radiation beam available from the machine must be thoroughly tested and a clinical linear accelerator typically produces two x-ray beams (10 MV and 18 MV) as well as about six electron beams in the range from 4 MeV to 20 MeV. Many of the tests involve measuring a set of data for each radiation beam and these measurements are carried out with a suitable thimble chamber that for this purpose is used as relative dosimeter. To allow measurements in water phantoms the ionization chambers should be waterproof and have relatively small cavity volumes to provide a reasonable spatial resolution for tests involving small radiation beams. A typical data set for megavoltage x-ray beams consists of:

1. Percent depth dose distributions measured at source-surface distance (SSD) of 100 cm in water phantom from phantom surface to a depth of 35 cm for field sizes of $2\times2\,cm^2$ to $40\times40\,cm^2$ in increments of 2 cm.
2. Beam profiles measured in water phantom at source-surface distance (SSD) of 100 cm at depths of dose maximum, 5 cm, 10 cm, 20 cm, and 30 cm for field sizes from $2\times2\,cm^2$ to $40\times40\,cm^2$ in increments of 2 cm.

In addition to a set of thimble ionization chambers, a radiotherapy machine QA program also requires access to parallel-plate ionization chambers with a thin front window to deal with some specialized tasks that thimble chambers cannot address. Some of these specialized tasks are:

1. Absolute output calibration of megavoltage electron beams with kinetic energy of 10 MeV and below.
2. Absolute output calibration of superficial x-ray beams at 50 kVp and below.

3. Measurement of surface dose for megavoltage photon and electron beams.
4. Measurement of percent depth dose of megavoltage photon beams in the buildup region (see Sect. 1.12.1).
5. Measurement of percent depth dose of electron beams with kinetic energy of 10 MeV and below.
6. Measurement of exit dose for kilovoltage and megavoltage photon beams.

17.1.4   Liquid Ionization Chamber

The great success of air-filled ionization chambers in radiation dosimetry has stimulated a search for non-gaseous forms of dosimetric cavity media. The obvious advantage of a suitable liquid or solid chamber cavity over an air-filled cavity would be the potential for tissue equivalence of the cavity medium and a significant increase in the magnitude of the dosimetric signal for the same cavity volume. Since liquids and solids used as cavity medium in ionization chambers must be strong insulators and, in addition, must generate charge carriers with high mobility in the chamber electrical field, the choice of suitable materials is limited.

A liquid-filled ionization chamber (also known as liquid ionization chamber or LIC) offers some exciting advantages over air-filled ionization chamber; however, despite a significant research effort, it is currently used as relative dosimeter and is not yet ready for use in absolute dosimetry. Its principle of operation is similar to that of an air-filled ionization chamber, but its three orders of magnitude larger cavity mass density provides a significant increase in signal magnitude over that obtained from an air filled chamber for same cavity volume. This allows a corresponding decrease in cavity volume that results in a high spatial resolution in dynamic measurements involving very small radiation fields in radiotherapy, such as those used in stereotactic radiosurgery and intensity modulated radiotherapy (IMRT). Use of cavity volumes as small as 1 mm^3 has been reported. Liquids used in liquid ionization chambers must be strong electrical insulators, such as isooctane, and of high purity grade to minimize leakage currents.

In principle, some notable advantages of liquid ionization chambers over gas filled ionization chambers are as follows:

1. Compared to air-filled chambers of same cavity volume, liquid ionization chambers provide a much stronger signal. Conversely, for same signal the liquid chamber cavity is of much smaller volume so that liquid ionization chamber offers a much higher spatial resolution.
2. Liquid ionization chamber exhibits excellent energy response in the whole photon energy range of interest in medical physics.
3. Liquid ionization chambers, unlike air-filled chambers, require no temperature and pressure correction of signal response.

17.1.5   Relative Ionometric Dosimetry: Conclusions

Ionization chambers are used in many areas of science, medicine, and technology. In medicine they play an invaluable role in application of ionizing radiation to diagnosis and treatment of disease where they are used both as absolute as well as relative dosimeters. They are also used in radiation protection as radiation detectors and in general public safety in fire and smoke detectors.

The basic design of ionization chambers is quite simple and the essential principles of their operation are well understood. However, the in-depth theory of saturation characteristics of ionization chambers is still not complete, yet it is sufficiently developed to allow reliable use of ionization chambers in absolute radiation dosimetry with standard free-air chamber, standard Bragg-Gray cavity chamber, Farmer type thimble chamber, and dosimetry level parallel-plate chamber.

In comparison with absolute ionization chambers, the relative ionization chambers come in a larger variety of sizes, shapes, and cavity sensitive media. As for geometrical shape, parallel-plate, cylindrical (thimble), and spherical ionization chambers are in use. As for cavity media, in contrast to absolute ionization chambers that all use ambient air, relative ionization chambers use a variety of options ranging from ambient air in clinical physics work, through sealed cavities filled with pressurized non-electronegative gas, such as argon, to increase collection efficiency and signal magnitude, to a sealed cavity filled with a suitable liquid, such as isooctane.

17.2   Luminescence Dosimetry

Luminescence dosimetry, as the name implies, is the measurement of ionizing radiation based on the process of luminescence. When ionizing radiation interacts with matter, most of the absorbed energy ultimately goes into heat while a small fraction is used to break chemical bonds. In some solids referred to as phosphors a very minute fraction of the absorbed energy is stored in metastable energy states often referred to as traps. The traps are characterized with a trap activation energy E_a (also called trap depth) and trap lifetime τ. Trap depth or its activation energy E_a is defined as the energy required to empty a filled trap; trap lifetime τ is defined as the mean time delay between energy deposition in the phosphor by ionizing radiation (filling of empty traps) and emission of light (emptying of filled traps).

The energy stored in traps is subsequently released in the form of ultraviolet, visible or infrared light and the phenomenon of stored energy release in the form of light is called luminescence. Depending on the trap lifetime τ, two types of luminescence are known: fluorescence for $\tau < 10^{-8}$ s and phosphorescence for $\tau > 10^{-8}$ s.

Phosphors usually contain a series of traps of different activation energies and emit radiation of different wavelengths; the larger is the activation energy E_a of a given trap, the larger is its lifetime τ. Some traps are so shallow that the phosphor exhibits fluorescence during interaction with ionizing radiation, other traps are deeper resulting in phosphorescence that may be delayed by minutes, days, or even years.

The process of phosphorescence may be accelerated with a suitable excitation of the pre-irradiated phosphor with energy either in the form of heat or in the form of ultraviolet or visible light:

1. If the activation agent is heat, the phenomenon of thermally accelerated phosphorescence is called thermoluminescence (TL), and the phosphor emitting light upon heating is called TL phosphor. *Note*: TL is a phenomenon clearly different from black body radiation emitted by material heated to high temperatures. The TL light is superimposed onto the black body spectrum of the phosphor, but is emitted only upon heating of previously irradiated phosphor.
2. If the activation agent is ultraviolet or visible light, the phenomenon of optically accelerated phosphorescence is called optically stimulated luminescence (OSL), and the phosphor emitting light upon optical stimulation is called OSL phosphor.

Except for the difference in agents activating phosphorescence, the physical processes underlying TL and OSL are similar and based on the energy band structure of crystalline solids, discussed below.

17.2.1 Energy Band Structure in Solids

Binding energies of electrons in a single and isolated atom are arranged into allowed discrete energy levels, as discussed in Chap. 3. Atoms in solids, on the other hand, are arranged into periodic structures (lattice) and held together by bonds, such as the ionic bond and covalent bond. Energies of electrons in solids are governed by electronic band theory that predicts a band structure for electron energies with alternating allowed and forbidden energy bands. Thus, electrons in a solid can only have certain values that are within allowed energy bands.

Electrons in free atoms have an infinite number of electronic energy levels available to them; however, they tend to occupy the lowest possible energy level available. Similarly, electrons in solids have an infinite number of allowed bands available to them and they tend to occupy the lowest possible band available. In all solids the lowest lying bands are completely filled and an electron occupies each available energy level. No electric current can propagate through a filled energy band.

As far as electrical conductivity of solids is concerned, there are three possible band categories in solids, resulting in three classes of solid, as shown schematically in Fig. 17.3:

1. *Insulator*: If the allowed bands in a solid are either completely filled or completely empty, no electrons can move in an electric field and the solid behaves as an insulator. In an insulator the highest allowed band that is completely filled is called the valence band. There is a relatively large energy gap E_g of the order of

10 eV between the top of the valence band and the bottom of the next allowed band that is normally empty and called the conduction band.

2. *Intrinsic semiconductor*: If all lowest allowed bands are completely filled except for one or two top bands that are minimally filled or minimally empty, the solid is called a semiconductor. In a semiconductor the top almost filled band is called the valence band and the bottom of the next allowed band that is minimally filled by electrons is called the conduction band. In a semiconductor the energy gap E_g between the top of the valence band and the bottom of the conduction band is of the order of 1 eV.

3. *Metal*: If one or more bands are partially, yet substantially filled, the solid will act as a metal and it will be a good conductor of electricity. In a metal the outer band that is only partially filled is called the conduction band.

The Fermi energy E_F in Fig. 17.3 has a special meaning and at $T = 0$ K represents the energy demarcation between filled electron states and empty electron states. In metals the band structure features a partially filled outer band referred to as conduction band allowing electrons to move freely within the metal. The band structure of insulator and semiconductor are similar to each other with the exception of energy band gap E_g that in an insulator is of the order of 10 eV and in semiconductor of the order of 1 eV.

Figure 17.4 depicts two energy band diagrams applicable to insulators:

(a) For a perfect insulator showing three allowed bands; the lower two bands are completely filled (the higher of the two filled bands is called valence band) and the top band (called conduction band) is completely empty.

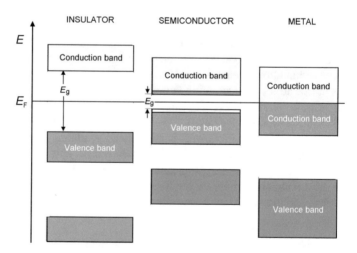

Fig. 17.3 Schematic diagram showing electron occupancy of allowed energy bands for an insulator, semiconductor, and metal. Energy is plotted on the y (ordinate) axis, position on the x (abscissa) axis. E_F stands for Fermi energy, E_g for the width of the forbidden band (energy gap) that separates the conduction band from the valence band. Shaded areas indicate regions filled with electrons

(b) For a typical insulator, showing bands like in part (a) but, in addition, also showing energy levels in the forbidden band gap between the valence band and the
conduction band. These energy levels represent traps, i.e., bound states for charge
carriers that play a role in luminescence, and are attributed to imperfections in
the crystal lattice (atomic or ionic vacancies and impurities), either intrinsic or
introduced on purpose through doping.

As shown in Fig. 17.4b, energy levels just below the conduction band represent
electron traps; levels just above the valence band represent hole traps. Electron trap
activation energy (electron trap depth) E_e is defined as energy difference between the
lower edge of the conduction band and the trap level. Similarly, hole trap activation
energy (hole trap depth) E_h is defined as the energy difference between the trap level
and the upper edge of the valence band that is usually taken as zero energy for energy
band diagrams of solids.

The three electron traps and the three hole traps shown in Fig. 17.4b are examples
of a variety of traps found in insulators, ranging from shallow traps (relatively small
activation energy) through normal traps to deep-seated traps with large activation
energy. Since E_g of insulators is of the order of 10 eV, the trap depths of insulators
are of the order of a few tenths of eV for shallow traps to a few eV for deep-seated
traps.

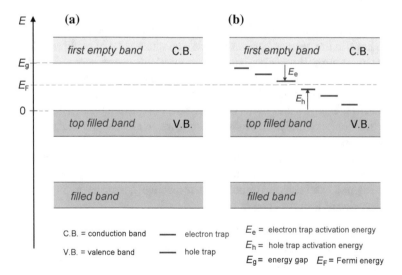

Fig. 17.4 Schematic representation of energy band diagrams of insulators. **a** For perfect insulator
with no imperfections in crystal lattice. **b** For a typical insulator with imperfections in crystal lattice.
The imperfections introduce electron and hole traps into the forbidden energy gap that separates
the conduction band from the valence band

17.2.2 Basic Aspects of Luminescence Dosimetry

An electron cannot move in an insulator unless it acquires enough energy to cross the band gap E_g separating the valence band from the conduction band. Various stimulating modalities can supply this energy leading to emission of light either as fluorescence or phosphorescence. The process usually carries a name that combines the stimulating modality with the term "luminescence", such as, for example, bioluminescence, chemi–, electro–, radio–, tribo–, etc. Ionizing radiation (photons, neutrons, or charged particles) can also be used to supply energy to electrons bound in an insulator and this leads to TL and OSL processes that are used in luminescence dosimetry.

As discussed in detail in Chaps. 7 and 8, the primary photon interactions with atoms of an insulator can be with the entire atom (photoelectric effect – PE), or with an orbital electron of an atom (Compton effect – CE, triplet production – TP) or with an atomic nucleus (nuclear pair production – NPP). The cross sections for each of these effects depend strongly on the photon energy $h\nu$ and atomic number Z of the absorber.

The highly energetic secondary charged particles, generally electrons (from PE and CE) and to a lesser degree positrons (originating from less common TP and NPP), that are released in the primary photon interactions with the insulator, move through the insulator and release numerous low energy free electrons by ionizing atoms and ions of the insulator. If the free electron receives energy exceeding the gap energy E_g of the insulator, it is lifted into the conduction band and leaves an electron vacancy, called a hole, in the valence band. A hole is a quantum state of positive polarity that migrates to the top of the valence band and its migration causes successive transitions of electrons to lower energy levels in the valence band. The free electron lifted to the the conduction band, on the other, on the other hand, migrates through progressively lower electron energy states to the bottom of the conduction band. Thus, in contrast to air where ionizing radiation produces ion pairs, in an insulator ionizing radiation creates electron-hole pairs.

The free electrons and holes released by energetic charged particles migrate through the insulator in their respective energy bands and they either recombine or become trapped in an electron or hole trap, respectively, somewhere in the insulator. Two categories of trap are known:

1. First category is called a *storage trap* and its purpose is to trap free charge carriers during irradiation and release them during subsequent stimulation either with heat (TL process) or with light (OSL process).
2. Second category acts as a *recombination center* or *luminescence center* where a charge carrier released from a storage trap recombines with a trapped charge carrier of opposite sign. The recombination energy is at least partially emitted in the form of visible or ultraviolet light. The nature of the recombination process depends strongly on the relative magnitudes of the capture cross section of the unfilled storage traps and filled recombination centers.

Any real crystal contains a wide variety of imperfections (lattice defects) and impurities that may act as storage traps and luminescence centers. The luminescence centers are mainly due to impurities and differ among themselves by the wavelength of light emitted in the recombination process.

The traps are nominally empty before irradiation. This means that the hole traps contain electrons and the electron traps do not. During irradiation of the insulator the secondary charged particles lift electrons into the conduction band either (i) from the valence band leaving a free hole in the valence band or (ii) from an empty hole trap thereby filling the hole trap. The system may approach thermal equilibrium in various ways:

1. Two free charge carriers meet and recombine (electron-hole recombination); recombination energy is converted into heat.
2. Free charge carrier recombines with a charge carrier of opposite sign trapped at a luminescence (recombination) center; recombination energy is emitted as optical fluorescence.
3. Free charge carrier becomes trapped at a storage trap, eventually resulting in natural phosphorescence or accelerated phosphorescence called TL when the accelerating agent is heat and OSL when the accelerating agent is visible or ultraviolet light.

Both thermoluminescence (TL) and optically stimulated luminescence (OSL) are thus two-step processes:

1. In the first step, a fraction of energy that ionizing radiation deposits in an insulator (phosphor) is stored in metastable states (traps) present in the insulator.
2. Subsequently to irradiation, in the second step the stored energy is read out either by heating the insulator (TL dosimetry) or exposing it to light (OSL dosimetry).

Figure 17.5 illustrates the two major stages of the TL and OSL dosimetry processes. Parts (a) and (c) show "Step 1: Irradiation" that consists of the following components:

1. Interaction (PE, CE, TP, and NPP) of photon with an atom of insulator (phosphor) and release of energetic charged particle (photoelectron, Compton electron, PP electron and positron).
2. Propagation of the energetic charged particle through insulator and creation of electron-hole (e-h) pairs through ionization of atoms or ions of the insulator. (*Note*: only one e-h creation is shown; however, each energetic charged particle creates thousands of electron-hole pairs).
3. If the electron of the e-h pair was supplied with kinetic energy E_K exceeding the energy gap E_g, the electron is lifted from the valence band into the conduction band (*Note*: if $E_K < E_g$, then both the electron and hole remain in the valence band, but they are bound together into a neutral entity called exciton, and migrate through the valence band).
4. Migration of free electron through conduction band of the insulator.
5. Trapping of electron into an empty electron trap.

6. Migration of hole of the e-h pair through valence band of the insulator.
7. Trapping of hole into an empty hole trap.

Parts (b) and (d) of Fig. 17.5 show "Step 2: Readout" of the TL and OSL processes. Part (b) is for $E_e < E_h$, indicating that the electron trap is a storage trap and the hole trap is a luminescence center; part (d) is for $E_e > E_h$, indicating that in this situation the electron trap acts as a luminescence center and the hole trap acts as a storage trap. Part (b) has the following components:

1. Ejection of trapped electron from the electron storage trap into conduction band as a result of heating (TL) or exposure to light (OSL) of the previously irradiated insulator (phosphor).
2. Migration of free electron through conduction band of the insulator.
3. Recombination of free electron with a trapped hole at a luminescence center.
4. Emission of visible or ultraviolet light (TL or OSL).

Part (d) of Fig. 17.5 has the following components:

1. Ejection of trapped hole from hole storage trap into valence band as a result of heating (TL) or exposure to light (OSL) of the previously irradiated insulator.
2. Migration of free hole through valence band of the insulator.
3. Recombination of free hole with a trapped electron at a luminescence center.
4. Emission of visible or ultraviolet light (TL or OSL).

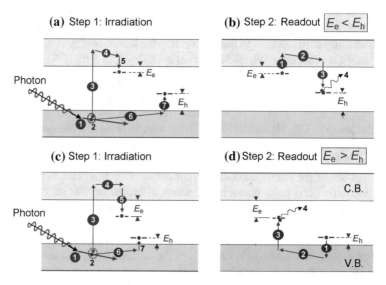

Fig. 17.5 Two-step processes of thermoluminescence (TL) and optically stimulated luminescence (OSL). Parts (**a**) and (**c**) illustrate step 1, parts (**b**) and (**d**) illustrate step 2

17.2.3 Kinetics of Thermoluminescence

The method of thermal glow, introduced by Urbach and applied to any one of the numerous thermally activated processes in insulators, is a powerful tool for determining charge carrier trapping parameters. Almost all early work was done on thermoluminescence but in recent years the method has been extended to include other thermally activated phenomena, such as thermally activated currents, photoconductivity, and ionic space charge electrets.

A very broad range of different specific models of reaction kinetics yield expressions for the time and temperature dependence of the measured quantity (usually current) that are either of the Randall-Wilkins form or of the Garlick-Gibson form depending, respectively, upon whether the decay rate processes are dominated by first order (linear) kinetic terms or second order (bilinear) kinetic terms.

In studies of thermally activated phenomena some type of current is recorded as a function of temperature. For a constant heating rate $\beta = dT/dt$ the plot of current against temperature is called a thermogram. In TL one plots the photomultiplier tube (PMT) current (Sect. 17.2.5) against the temperature of the phosphor and obtains a TL thermogram that is more commonly referred to as a glow curve. A glow curve usually consists of a number of minima and maxima with the maxima referred to as peaks. Each peak can be characterized by the temperature at which the maximum is observed (peaking temperature T_M) and by its full width at half maximum (FWHM) measured in K. From a glow curve one can estimate the thermal activation energies of various peaks and gain information about the type of kinetics involved in the decay process.

In this section we discuss the phenomenological theory of glow curve (thermogram) shapes for single, well resolved glow curve peaks with well defined full widths at half maximum (FWHM). The first phenomenological assumption is that the glow curve peak is produced from certain filled storage traps whose concentration we denote by $n(t)$ at time t. The PMT current $I(t)$ or TL intensity we measure is proportional to the rate of decay of the filled traps

$$I(t) = -\eta \frac{dn}{dt}, \tag{17.1}$$

where η is temperature and time independent numerical factor related to detection efficiency. The integrated intensity (i.e., charge release) observed upon tracing out a thermogram peak, is obtained by integrating (17.1) over time

$$\phi = \int_0^\infty I(t)dt = \eta n_0, \tag{17.2}$$

where n_0 is the initial concentration of filled storage traps (in m^{-3}).

TL processes have been described with three different phenomenological models:

(i)   First order (linear) Randall-Wilkins model.
(ii) Second order (bilinear) Garlick-Gibson model.
(iii) General order May-Partridge model.

Randall-Wilkins First Order Glow Curve Model

The Randall-Wilkins (RW) first order linear model of TL kinetics assumes:

- Presence of storage traps and recombination centers in a phosphor.
- No possibility of re-trapping a charge carrier released from a filled storage trap.

p is the probability that a charge carrier will undergo recombination at a luminescence center. In RW glow curve model there is no possibility of re-trapping, hence $p = 1$.

N is the total concentration in cm^{-3} of storage traps in the phosphor.

$n(t)$ is the concentration in cm^{-3} of filled storage traps in the phosphor at time t.

n_0 is the initial concentration (in cm^{-3}) of filled storage traps at time $t = 0$.

The TL readout process is depicted in Fig. 17.5b for an electron trapped in a storage trap that upon release from storage trap recombines with a hole trapped in a recombination center. In Fig. 17.5d the roles of electron and hole are reversed. Since re-trapping is ignored ($p = 1$), a first order kinetic process dominates the decay rate of the concentration of filled traps $n(t)$ at time t and dn/dt is given by

$$\frac{dn}{dt} = -pn(t)W(T) = -n(t)W_0 e^{-\frac{E_a}{kT}}, \quad \text{or} \quad \int_{n_0}^{n} \frac{dn}{n} = -W_0 e^{-\frac{E_a}{kT}} \int_0^t dt \quad (17.3)$$

where $W(T)$ is a temperature dependent decay rate proportional to a simple thermal activation form

$$W(T) = W_0 e^{-\frac{E_a}{kT}}, \quad (17.4)$$

with E_a the thermal activation energy of the filled traps and W_0 a pre-exponential frequency factor proportional to the frequency of fundamental lattice vibrations in solids ($\sim 10^{11 \pm 2}\ s^{-1}$).

The *intensity of phosphorescence* can now, after combining the solution to (17.3) with (17.1) for a constant temperature of the TL phosphor, be expressed as

$$I(t) = -\eta \frac{dn}{dt} = \eta n(t) W_0 e^{-\frac{E_a}{kT}} = \eta n_0 W_0 e^{-\frac{E_a}{kT}} \exp\left\{-W_0 e^{-\frac{E_a}{kT}} t\right\}, \quad (17.5)$$

where n is the concentration of filled traps in cm^{-3}.

TL intensity for a TL phosphor following first order kinetics and heated at a constant heating rate $\beta = dT/dt$ is calculated from (17.3) as follows (*note*: $dt = dT/\beta$)

$$\frac{dn}{n} = -W_0 e^{-\frac{E_a}{kT}} dt = -\frac{W_0}{\beta} e^{-\frac{E_a}{kT}} dT \quad \text{and} \quad \int_{n_0}^{n} \frac{dn}{n} = \ln \frac{n}{n_0} = -\frac{W_0}{\beta} \int_{T_0}^{T} e^{-\frac{E_a}{kT}} dT,$$

(17.6)

where T_0 is the temperature at which the phosphor was irradiated and is also the temperature at which the TL readout procedure starts. Equation (17.6) finally gives the following expression for concentration of filled traps n

$$n = n_0 \exp \left[-\frac{W_0}{\beta} \int_{T_0}^{T} e^{-\frac{E_a}{kT}} dT \right],$$

(17.7)

where n_0 is the concentration of filled traps at initial temperature T_0. The temperature dependence of the TL emission is, after inserting (17.7) into (17.1), expressed as

$$I(T) = -\eta \frac{dn}{dt} = \eta n W(T) = \eta n W_0 e^{-\frac{E_a}{kT}} = \eta n_0 W_0 e^{-\frac{E_a}{kT}} \exp \left[-\frac{W_0}{\beta} \int_{T_0}^{T} e^{-\frac{E_a}{kT}} dT \right].$$

(17.8)

The peak temperature T_M related to activation energy E_a of the filled trap can be obtained by setting $dI(T)/dT \big|_{T=T_M} = 0$ to get

$$\frac{d}{dT} \ln W(T) \bigg|_{T=T_M} = \frac{W(T)}{\beta} \bigg|_{T=T_M},$$

(17.9)

or, after inserting (17.4) into (17.9)

$$\frac{\beta E_a}{k T_M^2} = W(T_M) = W_0 e^{-\frac{E_a}{k T_M}}.$$

(17.10)

Equation (17.10) shows that the peak temperature T_M depends on the heating rate β, trap activation energy E_a, and frequency factor W_0, but does not depend on the initial number of filled traps n_0. It is a transcendental equation that does not have a closed form solution; however, it can be solved with numerical methods or with suitable simplifying approximations.

Typical TL glow curve temperatures are of the order of a few hundred degrees kelvin. This gives thermal energy kT anywhere from 10^{-2} eV (115 K) to 7×10^{-2} eV (800 K). Typical trap activation energies E_a in the same temperature range are from ~ 0.3 eV to ~ 2 eV, suggesting that kT is always much smaller than E_a and $E_a/(kT) \gg 1$. Rewriting (17.10) in the following form

$$\frac{E_a}{k T_M} = \ln \frac{W_0 T_M}{\beta} - \ln \frac{E_a}{k T_M}$$

(17.11)

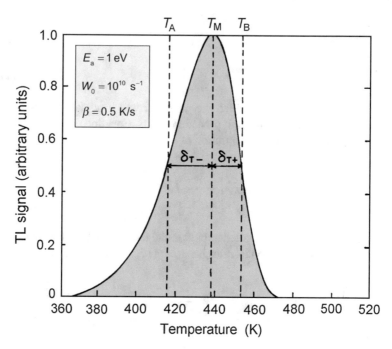

Fig. 17.6 Typical Randall-Wilkins (RW) glow curve (thermogram) calculated from the Randall-Wilkins (17.8) for trap activation energy $E_a = 1$ eV, frequency factor $W_0 = 10^{10}$ s^{-1}, and constant heating rate $\beta = dT/dt = 0.5$ K/s. The glow curve has a peak temperature at 438 K and is asymmetric with asymmetry ratio of $\delta T_-/\delta T_+ \approx 1.5$. The full width at half maximum *FWHM* is $\Delta T = \delta T_- + \delta T_+ = T_B - T_A = 39$ K

leads to a significant simplification of the transcendental (17.10) that, after assuming that $E_a/kT_M \gg \ln E_a/(kT_M)$, reads

$$\frac{E_a}{kT_M} \approx \ln \frac{W_0 T_M}{\beta}. \tag{17.12}$$

For a narrow glow curve we thus have a simple approximation for estimation of trap activation energy E_a from known W_0, β, and T_M

$$E_a \approx kT_M = \ln \frac{W_0 T_M}{\beta}. \tag{17.13}$$

Figure 17.6 shows a typical glow curve calculated from the Randall-Wilkins equation (17.8) with the following parameters: trap activation energy $E_a = 1$ eV, frequency factor $W_0 = 10^{10}$ s^{-1}, and constant heating rate $\beta = 0.5$ K/s, normalized to $I(T_M) = I_M = 1.0$. The important parameters of the glow curve $I(T)$ are the temperature T_M of the glow curve peak and temperatures T_A and T_B at which the current $I(T)$ attains half-intensity on each side of T_M, i.e., on the ascending side with half-width $\delta T_- = T_M - T_A$ and the on the descending side with half-width $\delta T_+ = T_B - T_M$, respectively. Thus, the full-width-at-half-maximum ΔT is given as the sum $\delta T_- + \delta T_+ = T_B - T_A$. In Fig. 17.6 one immediately notes the well-known characteristic of the RW glow curve, namely its 50% low-temperature asymmetry with δT_- about 50% larger than δT_+ and resulting in an asymmetry ratio $\delta T_-/\delta T_+ \approx 1.5$.

The peak temperature T_M of the RW glow curve in Fig. 17.6 is 438 K. Inserting $T_M = 438$ K into (17.13) gives $E_a \approx 1.12$ eV, showing that (17.13) provides a simple, yet reasonable, first approximation for a quick estimation of storage trap activation energy E_a. In the second approximation we insert into (17.11) the first approximation value $E_a \approx 1.12$ eV and get $E_a = 0.997$ eV within a fraction of percent from the actual $E_a = 1$ eV.

Garlick-Gibson Second Order Glow Curve Model

The Garlick-Gibson glow curve model relates to traps whose decay is dominated by second order bilinear kinetics and assumes:

- Presence of storage traps and recombination centers in a phosphor.
- Strong possibility of re-trapping a charge carrier released from a filled storage trap.

p is the probability that a charge carrier released from a storage trap will undergo recombination at a luminescence center with a trapped charge carrier of opposite sign. The probability p is given as $p = n/N$.

N is the total concentration in cm^{-3} of storage traps in the phosphor.

$n(t)$ is the concentration in cm^{-3} of filled storage traps in the phosphor at time t.

n_0 is the initial concentration (in cm^{-3}) of filled storage traps at time $t = 0$.

The TL readout process is depicted in Fig. 17.5b for an electron trapped in a storage trap that upon release recombines with a hole trapped in a recombination center. In Fig. 17.5d the roles of electron and hole are reversed. Since there is a strong possibility of re-trapping the charge carrier released from a storage trap, a second order (bilinear) kinetic process dominates the decay rate of the concentration $n(t)$ of filled traps at time t.

For second order kinetics the rate of change dn/dt is given as

$$\frac{dn}{dt} = -pn(t)W(T) = -\frac{[n(t)]^2}{N}W_0 e^{-\frac{E_a}{kT}} = -[n(t)]^2 W_0' e^{-\frac{E_a}{kT}} , \qquad (17.14)$$

where

p is the probability for recombination accounting for re-trapping $(p = n/N)$.

$W(T)$ is a temperature dependent decay rate proportional to a simple thermal activation form

$$W(T) = W_0 e^{-\frac{E_a}{kT}} \quad \text{and} \quad \frac{W(T)}{N} = \frac{W_0}{N} e^{-\frac{E_a}{kT}} = W_0' e^{-\frac{E_a}{kT}} \qquad (17.15)$$

with W_0' a pre-exponential factor $(W_0' = W_0/N)$ in units of $cm^3 \cdot s^{-1}$. Integration of (17.14) at a constant temperature gives

$$\int_{n_0}^{n} \frac{dn}{n^2} = -W_0' e^{-\frac{E_a}{kT}} \int_0^t dt \quad \text{and} \quad n = \frac{n_0}{1 + W_0' n_0 t e^{-\frac{E_a}{kT}}} \tag{17.16}$$

Intensity of phosphorescence $I(t)$ at constant temperature T_0 is, after combining (17.16) with (17.1), written as

$$I(t) = -\eta \frac{dn}{dt} = n^2 W_0' e^{-\frac{E_a}{kT}} = \frac{\eta n_0^2 W_0' e^{-\frac{E_a}{kT}}}{\left[1 + W_0' n_0 e^{-\frac{E_a}{kT}} t\right]^2}. \tag{17.17}$$

TL intensity for a phosphor following second order kinetics and heated at a constant heating rate $\beta = dT/dt$ is calculated from (17.14) as follows (*note*: $dt = dT/\beta$)

$$\int_{n_0}^{n} \frac{dn}{n^2} = -\frac{W_0'}{\beta} \int_{T_0}^{T} e^{-\frac{E_a}{kT}} dT \quad \text{and} \quad n = \frac{n_0}{\left[1 + \frac{W_0' n_0}{\beta} \int_{T_0}^{T} e^{-\frac{E_a}{kT}} dT\right]}, \tag{17.17}$$

where n_0 is the initial concentration of filled storage traps at initial temperature T_0. The temperature dependence of TL emission is, after inserting (17.17) into (17.1), given as

$$I(T) = -\eta \frac{dn}{dt} = \eta [n(t)]^2 W_0' e^{-\frac{E_a}{kT}} = \eta \frac{n_0^2 W_0' e^{-\frac{E_a}{kT}}}{\left[1 + \frac{W_0' n_0}{\beta} \int_{T_0}^{T} e^{-\frac{E_a}{kT}} dT\right]^2}. \tag{17.18}$$

Figure 17.7 shows a typical glow curve calculated from the Garlick-Gibson equation (17.18) with the following parameters: storage trap activation energy $E_a = 1\,eV$, the product of frequency factor × initial number of filled storage traps $W_0' \times n_0 = 10^{10}\,s^{-1}$, and constant heating rate $\beta = 0.5\,K/s$, normalized to $I(T_M) = I_M = 1.0$. The important parameters of the glow curve are the temperature T_M of the glow curve peak and temperatures T_A and T_B at which the current attains half-intensity on each side of T_M, i.e., on the ascending side $\delta T_- = T_M - T_A$ and the descending side $\delta T_+ = T_B - T_M$ of T_M, respectively. In Fig. 17.7 one immediately notes the well-known characteristic of the GG glow curve, namely its symmetry, with half-width δT_- equal to half-width δT_+.

Figure 17.8 shows two TL glow curves for comparison: a Randall-Wilkins (RW) first order kinetics curve, solid and calculated from (17.8) and a Garlick-Gibson (GG) second order kinetics curve, dashed and calculated from (17.18). Both curves are normalized to $I(T_M) = I_M = 1$ and both are calculated for equal values of pertinent parameters: activation energy $E_a = 1\,eV$ of storage traps, frequency factor $W_0 = n_0 W_0' = n_0 W_0/N = 10^{10}\,s^{-1}$, and heating rate $\beta = 0.5\,K/s$. A comparison

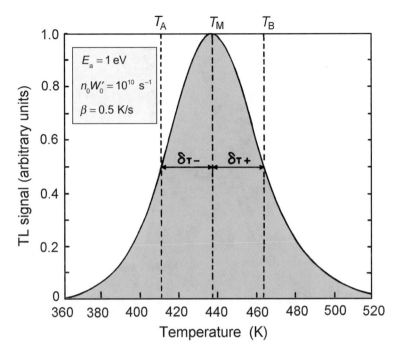

Fig. 17.7 Typical Garlick-Gibson glow curve (thermogram) calculated from the Garlick-Gibson (GG) (17.18) for trap activation energy $E_a = 1$ eV, product of initial concentration of filled storage traps $n_0 \times$ frequency factor W_0, i.e., $n_0 W_0 = 10^{10}$ cm$^{-3} \cdot$s^{-1}, and constant heating rate $\beta = dT/dt = 0.5$ K/s. The glow curve has a peak temperature at 438 K and is symmetric ($\delta T_- = \delta T_+$). The full width at half maximum $FWHM$ is $\Delta T = \delta T_- + \delta T_+ = T_B - T_A = 47$ K

of the two glow curves shows that the RW curve is asymmetric with ascending side half-width δT_- about 50% larger than the descending side half-width δT_+. The FWHM of the RW curve is $\Delta T_{RW} = 39$ K. The GG curve is symmetric with a FWHM ΔT_{GG} of 47 K.

The difference between the shapes of the two curves is explained by re-trapping of charge carriers in a storage trap upon release: GG model allows re-trapping while RW model does not. Re-trapping will delay recombination and shift the curve to higher temperatures. This is not immediately obvious from Fig. 17.8 because both glow curves are normalized to 1. If they were plotted on an absolute scale, then the RW curve peak would be higher than the GG curve peak, the leading edges of both curves would be similar, and the delayed TL emission of the GG curve compared to the RW curve would be clearly discernable.

May-Partridge General-Order TL Model

For situations where a measured glow curve matches neither the first order Randall-Wilkins model nor the second order Garlick-Gibson model, May and Partridge have

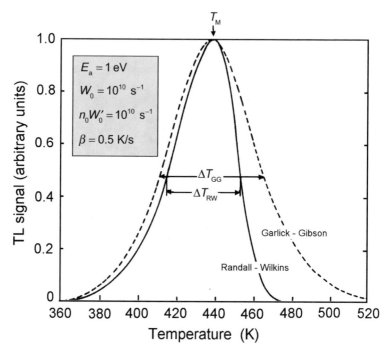

Fig. 17.8 Comparison between a Randall-Wilkins (RW) glow curve (*dashed*) and a Garlick-Gibson (GG) glow curve calculated from (17.8) and (17.18), respectively, with $W_0 = 10^{10}$ s^{-1}, $n_0 W_0 = 10^{10}$ cm^{-3}·s^{-1}, $\beta_{RW} = \beta_{GG} = 0.5$ K/s, and $E_a = 1$ eV. The RW glow curve is asymmetric with $\delta T_-/\delta T_+ \approx 1.5$ and $\Delta T = 39$ K; the GG glow curve is symmetric and $\Delta T = 47$ K

proposed a general-order empirical relationship to describe the TL glow curve. The model is called the May-Partridge general-order kinetic model and is expressed as follows

$$I(T) = -\eta \frac{dn}{dt} = \eta \frac{n_0 W_0'' e^{-\frac{E_a}{kT}}}{\left[1 + \frac{(b-1)W_0''}{\beta} \int_{T_0}^{T} e^{-\frac{E_a}{kT}} dT\right]^{\frac{b}{b-1}}} \tag{17.19}$$

where b is the kinetic order, a parameter with a typical value between 1 and 2, $W_0'' = W_0(n_0/N)^{b-1}$ is an empirical parameter playing the role of an effective frequency factor and, similarly to Randall-Wilkins and Garlick-Gibson kinetic models,

N is the total concentration in cm^{-3} of storage traps in the phosphor.
$n(t)$ is the concentration in cm^{-3} of filled storage traps in the phosphor at time t.
n_0 is the initial concentration (in cm^{-3}) of filled storage traps at time $t = 0$.

17.2.4 Thermoluminescence Dosimetry System

From a practical perspective the TL process consists of two steps:

(i) Exposure of TL phosphor to ionizing radiation – results in trapping of charge carriers (electrons and holes) in storage traps and recombination centers present in the phosphor (signal storage).

(ii) Thermal activation of the luminescence process (heating of TL phosphor) – results in release of charge carriers trapped in storage traps and their recombination with charge carriers of opposite sign trapped in recombination centers. Recombination energy is released in the form of ultraviolet or visible light (signal readout).

The two-step TL process was illustrated on a microscopic scale in Fig. 17.5 and the basic TL models were discussed in Sect. 17.2.3. The discussion of TL continues here with practical aspects of TL dosimetry. In principle TL measurements are simple and, as shown in Fig. 17.9, a TL dosimetry reader comprises only three major components:

(i) Reliable heating system to heat the TL phosphor during the readout cycle.

(ii) Sensitive light sensor to measure the intensity of the emitted TL light as a function of temperature of the TL phosphor or time.

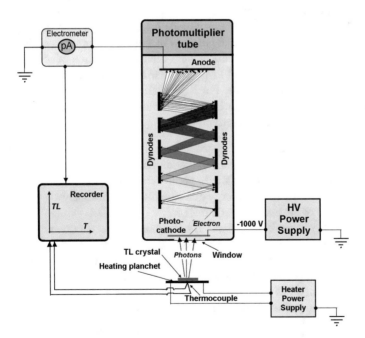

Fig. 17.9 Schematic diagram of a typical thermoluminescence (TL) reader, showing the main components: photomultiplier tube (PMT) with high voltage power supply, heating system with thermocouple, and electrometer with recorder

(iii) Recording system to plot, record, and analyze the TL signal against temperature of the TL phosphor or against time.

Heating System

The TL phosphor is usually irradiated at room temperature and the heating system must be capable of heating the phosphor from room temperature to at least 400 °C to get a reliable reading of the TL signal. Many heating methods have been used for this purpose and the three most common methods are: (i) Ohmically heated metallic planchet, (ii) Pre-heated nitrogen gas, and (iii) Focused laser light.

A thermocouple is used to measure the sample temperature. Thermal contact between the thermocouple and the sample can be quite troublesome and ensuring that the thermocouple measures the actual temperature of the phosphor is not a trivial task. The heater is also expected to follow in a reproducible fashion a constant heating rate or a pre-set heating schedule and heating rate.

Light Sensor

The signal emitted from a TL phosphor during the readout heating cycle is most often measured with a photomultiplier tube (PMT); however, simpler photodiodes are also sometimes used for this purpose. PMT is an important component of a TL reader especially when low doses in the radiation protection range are measured because of extremely faint signal produced even by most sensitive TL phosphors. A PMT is a very sophisticated device that for many decades has been providing highly precise photometric capability to many areas of medicine, nuclear, particle, and high-energy physics, as well as the oil industry. It is a glass vacuum tube that survived the transition from electronic valves to solid-state devices after the invention of solid-state diode and transistor during the 1950s. A PMT has three types of electrode: a photocathode, 10 to 14 dynodes, and an anode. For its operation a PMT relies on two physical phenomena: (i) surface photoelectric effect occurring on the photocathode and (ii) secondary emission occurring on dynodes.

(i) Surface photoelectric effect is a quantum phenomenon (see Sect. 1.26)) whereby a bound electron from the photocathode surface absorbs a photon; the photon disappears and the electron called a photoelectron is ejected from the photocathode surface with a kinetic energy that equals the photon energy $h\nu$ less the work function $e\phi$ of the photocathode material.

(ii) In secondary emission an electron of certain kinetic energy, when striking a dynode, ejects a few electrons of lesser kinetic energy and these electrons are accelerated to the next dynode in the dynode chain, where the multiplication process gets repeated. The current gain after electrons progress through all 10 to 14 dynodes of a PMT results in a typical current gain of up to 10^8. The electron multiplication process is illustrated in Fig. 17.9.

PMTs typically operate at a potential of 1 kV to 2 kV: the photocathode is held at a negative potential (-1000 V to -2000 V) and the anode is grounded through an electrometer. The dynodes are held at intermediate voltages between the photocathode and the anode set by a resistive voltage divider. The current flowing from the anode to ground is directly proportional to, and many orders of magnitude larger than, the miniscule photoelectric current generated by the photocathode and attracted to the first dynode of the PMT dynode chain. The current gain comes from the secondary emission of electrons at successive dynodes (each incident electron releases from 3 to 5 electrons at each dynode) that produces a total PMT gain of up to a factor of 10^8.

17.2.5 TL Thermogram or TL Glow Curve

A plot of the TL light intensity transformed into anode current by the PMT of a TL reader versus the phosphor temperature measured by the reader's thermocouple is called a TL thermogram or glow curve. The temperature scan is begun at the temperature at which the crystal was exposed to ionizing radiation, most commonly room temperature. Of course, for research purposes irradiation of TL phosphors can be carried out also at lower temperatures.

The maximum temperature of the scan is limited either by the melting point of the phosphor or the heater capability. A TL thermogram usually exhibits a series of intensity maxima and minima that result from the distribution of storage traps and luminescence centers in the TL phosphor. Maxima in the TL signal are referred to as the TL glow peak. From the TL glow curve shape and the peak temperature one can get an estimate of the thermal activation energy of the trap involved and a glimpse of the physical process behind a particular glow peak emission of light, as discussed in Sect. 17.2.3.

A typical example of a TL glow curve is given in Fig. 17.10 with a plot of TL intensity against temperature for LiF:Mg,Ti (lithium fluoride doped with magnesium and titanium), a well known TL phosphor developed by Harshaw Chemical Co. during 1960s and sold commercially under the name TLD-100. LiF has been studied extensively at the University of Wisconsin in Madison where during 1950s and 1960s Farrington Daniels and John R. Cameron with their students carried out most of the pioneering work on TL applied to radiation dosimetry. LiF in its pure form exhibits very little thermoluminescence; however, addition of Mg (300 ppm) and Ti (15 ppm) impurities make LiF:Mg,Ti one of the most practical and useful TL dosimeters (*note*: ppm stands for parts per million).

The glow curve of Fig. 17.10 results from exposure of TLD-100 phosphor to 100 kVp x-rays at room temperature and subsequently heated to 250 °C. Over the given temperature range the glow curve exhibits 6 peaks with peaks 4 and 5 usually serving as "dosimetry peaks", that is, peaks that yield reliable information on the dose delivered to the phosphor sample. It is obvious that a typical glow curve is much more complex than the ideal glow curves that are discussed in Sect. 17.2.3 and, for ease of theoretical manipulation, feature only one TL peak.

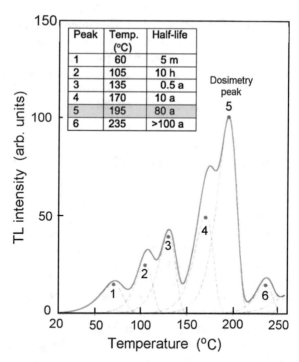

Peak	Temp. (°C)	Half-life
1	60	5 m
2	105	10 h
3	135	0.5 a
4	170	10 a
5	195	80 a
6	235	>100 a

Fig. 17.10 Glow curve measured for LiF:Mg,Ti (lithium fluoride doped with magnesium and titanium, TLD-100) in the temperature range from room temperature to 250 °C. Six glow peaks can be identified and isolated from one another. Peaks 4 and 5 are considered the dosimetry peaks and the area under the two peaks combined is used for indication of the TL signal. The *insert* shows the peak temperature and half-life for each of the six glow peaks

If the glow peaks are not too close together in a measured glow curve, it is possible to separate them into individual peaks (labeled with numbers from 1 through 6), as shown with the grey curves in Fig. 17.10. The dosimetric peaks must occur at relatively high temperatures (of the order of 150 °C to 250 °C) to ensure reasonable stability of the TL signal after irradiation.

The insert to Fig. 17.10 contains a table listing the half-life of the six LiF thermoluminescence peaks. It is clear that peaks close to room temperature undergo significant phosphorescence when stored at room temperature. Crystal lattice vibrations can cause random release of trapped charge carriers resulting in charge recombination at luminescence centers – the closer is the peak temperature to the temperature at which the irradiated phosphor is stored, the higher is the probability of undesired phosphorescence. This effect is referred to as TL signal fading. From Fig. 17.10 we note that LiF peak 1 has a half-life of 5 minutes, peak 2 ten hours, and the dosimetry peaks 4 and 5 ten years and 80 years, respectively.

17.2.6 Practical Considerations in Thermoluminescence Dosimetry

TL dosimetry falls into the category of relative dosimetry since the response of a thermoluminescence dosimeter must be first calibrated in a known radiation field before it is used in radiation dosimetry. Moreover, TL dosimetry is considered passive dosimetry since the dosimeter is not attached to the reader during irradiation. In addition, there are several peculiarities of the TL process that must be considered before TL is used for radiation dosimetry.

The response of a TL phosphor is affected by its previous radiation history and heating history. Residual effects must be erased with special procedures that involve heating the phosphor at a prescribed temperature before irradiation (pre-irradiation annealing) or after irradiation (post-irradiation annealing) but before readout.

For example, the standard pre-irradiation annealing procedure for LiF (TLD-100) is 1 hour of heating in an oven at 400 °C and then 24 hours at 80 °C. This removes peaks 1 and 2 from the glow curve (see Fig. 17.10). The same result, namely removal of peaks 1 and 2, can be achieved with a 10-minute post-irradiation annealing in an oven at 100 °C. Similar empirical annealing schedules were determined for other TL phosphors through tedious trial and error experiments.

Another concern is the onset of supralinear response of the TL material, usually at relatively high doses. In supralinear response a detector signal exceeds linear response with increasing dose. Supralinearity is not of importance in standard personnel dosimetry; however, it is of concern in radiation accidents and in food irradiation. The supralinear response depends on chemical composition of the phosphor, on concentration of impurities, as well as on the heating, cooling, and irradiation history of the TL material.

In general, the response of TL dosimeters is not affected by radiation dose rate but can be affected by radiation quality. The effective atomic number of LiF is 8.2 compared to 7.5 for water. This makes LiF almost water equivalent, even for x-rays below 100 keV where phosphors of higher effective atomic numbers show a significant increase in TL signal response because of increase in their photoelectric cross section.

TL dosimeters come in many solid forms such as single crystals, chips, rods, and ribbons. They are also available in powder form for use with special dispensers to ensure reproducibility in dosimeter mass for the readout procedure.

Many insulators in crystalline form emit luminescence upon exposure to ionizing radiation but only a few satisfy the full list of properties that are deemed essential for use in radiation dosimetry. Four of the best-known TL phosphors used in radiation dosimetry and radiation medicine are: lithium fluoride (LiF), lithium borate ($Li_2B_4O_7$), calcium fluoride (CaF_2), and calcium sulfate ($CaSO_4$) and their main characteristics are listed in Table 17.2. Of note are three additional materials exhibiting luminescence: aluminum oxide (also known as alumina, sapphire, and corundum Al_2O_3), beryllium oxide (BeO), and strontium sulfide (SrS).

Table 17.2 Characteristics of most common TL materials used in TL dosimetry

TL phosphor	LiF:Mg,Ti	Li$_2$B$_4$O$_7$:Mn	CaF$_2$:Mn	CaSO$_4$:Dy
	Lithium fluoride	Lithium borate	Calcium fluoride	Calcium sulfate
Mass density (g/cm^3)	2.64	2.40	3.18	2.96
Effective atomic number	8.2	7.4	16.3	15.3
Temperature of dosimetry peak (°C)	195	200	260	220
Sensitivity 30 keV/cobalt-60	1.25	0.9	13	12
TL emission spectrum max (nm)	400	605	500	480

Doping of TL materials with a controlled amount of impurities is a special skill, as is growing and cutting of solid crystals for use in dosimetry.

The TL dosimetric signal is superimposed onto the Planck black body emission of the heating planchet and TL dosimeter. Generally, the black body non-dosimetric signal is significantly lower than the dosimetric TL signal, and thus of no concern. However, at low radiation doses and high temperatures of the TL dosimeter the black body signal can become appreciable and the need for separating the dosimetric signal from the sum of the two signals should not be neglected.

TL dosimetry is a very useful relative dosimetry technique but not an easy one to use reliably and reproducibly. Many variables that affect readout results must be understood and kept under control for effective use of the technique. With considerable care and effort one can obtain a precision of about 2%.

The main advantage of TL dosimetry is that TL dosimeters can be very small. Since the dosimeters are not attached to the reader during irradiation, they can be inserted directly into tissues and body cavities for in-vivo patient dosimetry. Since TL detectors can integrate low doses for extended periods of time, they are often used in dose monitoring for radiation protection.

17.2.7 Optically Stimulated Luminescence Dosimetry

Luminescence dosimetry comprises two relative dosimetry techniques: thermoluminescence (TL) and optically stimulated luminescence (OSL). Of the two, TL dosimetry is older and better known, having been in commercial use since 1960s; OSL dosimetry, in commercial use since 1990s, is less known than TL dosimetry but has several features that give it an advantage over TL dosimetry. Both the TL and

OSL phenomena became known soon after the discovery of ionizing radiation during 1890s, but remained limited to academic interest until 1950s when their potential for practical use in radiation dosimetry for measurement of absorbed dose and in archaeology for dating of archaeological artifacts was established.

In many respects TL and OSL are similar: they both have the same theoretical background, they both can be described as a process of stimulated phosphorescence occurring in a previously irradiated crystalline mineral referred to as a phosphor, and in both the stimulation results in emission of visible light proportional to the dose absorbed in the phosphor. The major difference between the two techniques is in the agent triggering acceleration of phosphorescence: in the TL process the stimulating agent is heat while the OSL process is stimulated with visible light.

17.2.8 Optically Stimulated Luminescence (OSL) Reader

The two most important components of an OSL reader depicted schematically in Fig. 17.11a are: (i) light sensor and (ii) light source:

Light sensor is a photomultiplier tube (PMT) with the photocathode matched as close as possible to the light that is emitted by the OSL dosimeter and filtered into a narrower spectral band by filtration at the PMT entrance window.

Light source is a laser or a light emitting diode (LED) matched with the wavelength that stimulates the phosphorescence in the OSL dosimeter, previously exposed to ionizing radiation.

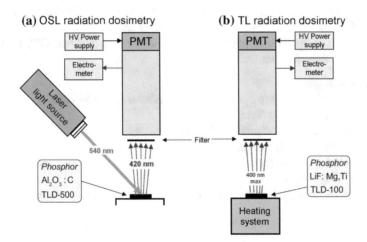

Fig. 17.11 Schematic diagram of readout equipment for two luminescence dosimetry techniques: **a** Optically stimulated luminescence (OSL) dosimetry and **b** Thermoluminescence (TL) dosimetry

17.2.9 Optically Stimulated Luminescence (OSL) Dosimeter

Currently, the best known OSL dosimeter material is aluminum oxide (also known as alumina, sapphire, and carborund) doped with carbon (Al_2O_3:C). The material has already been investigated during 1960s as a promising TL material but its sensitivity to visible light was found a serious disadvantage that precluded its future as a TL phosphor. However, the sensitivity of alumina to visible light eventually became recognized as an advantage for OSL and engendered the development of alumina into the best-understood OSL phosphor. Compared to TL dosimetry, OSL dosimetry does not feature such a broad array of available phosphors. In addition to alumina, beryllium oxide BeO and potassium bromide doped with europium (KBr:Eu) are known to exhibit the OSL behavior that may be useful for OSL dosimetry.

For the Al_2O_3:C OSL dosimeter the stimulation light can be a broad spectrum of visible light with a peak at 540 nm; however, a green laser light is more convenient and is used in commercial equipment. The Al_2O_3:C dosimeter emits the stimulated luminescence as blue light with spectral maximum at 420 nm. The intensity of luminescence emitted by the OSL dosimeter is proportional to the dose absorbed in the dosimeter as well as to the intensity of, and exposure time to, the stimulation light. Clearly, OSL dosimetry, like TL dosimetry, requires dosimeter/reader response calibration in a known radiation field prior to clinical use.

A notable feature of the OSL dosimetry is its fast and essentially non-destructive signal readout that depletes only a small fraction of 1% of the stored dosimetric signal, in contrast to TL dosimetry where the readout erases the major part of the TL signal. This means that in OSL dosimetry multiple readouts of same dosimetric signal and after-readout storage of dosimeter for permanent dose record are possible.

Re-use of OSL dosimeters is possible through optical signal annealing in strong white light for a few minutes to erase the record of any previous irradiation.

A short list of the most important characteristics of the Al_2O_3:C OSL dosimeter is as follows: *Signal response is*:

  (i) Reproducible.
 (ii) Linear with dose up to 500 cGy and supralinear above 500 cG.
(iii) Independent of dose rate.
 (iv) Isotropic, showing no angular dependence.
  (v) Fast and essentially non-destructive.
 (vi) Independent of temperature of irradiation in the room temperature range from $\sim$10 °C to $\sim$40 °C.
(vii) Unstable during first 10 minutes after irradiation and stable afterwards.

17.2.10 TL Dosimetry and OSL Dosimetry: Summary

Luminescence dosimetry has two major categories: thermoluminescence (TL) and optically stimulated luminescence (OSL). During the past few years it became clear that there is a rivalry between the two commercially available luminescence dosimetry techniques. On the one hand, TL dosimetry is a well-established modality of relative radiation dosimetry with several decades of clinical service in radiotherapy and radiation protection. On the other hand, OSL is relatively new to radiation dosimetry but it is making steady in-roads into areas where TL dosimetry played a leading role during the past four decades. Schematic diagrams for the two techniques are shown in Fig. 17.11; part (a) is for OSL dosimetry, part (b) for TL dosimetry. A short comparison of the two techniques reveals the following:

1. Compared to TL dosimetry that has become well established and developed commercially during the past four decades, OSL dosimetry is a relatively new luminescence dosimetry technique that needs more commercial development as well as a variety of suitable dosimetry materials.
2. Compared to TL readout that requires heating of the dosimeter, OSL readout is simpler and significantly faster, requiring only illumination of dosimeter with visible laser or LED light rather than heating to high temperatures.
3. In TL dosimetry, the readout procedure erases the stored signal. In OSL dosimetry, the readout uses only a small fraction of 1 per cent of the stored signal, making multiple readouts and storage of permanent dose records possible.
4. OSL dosimeter signal fading is smaller than signal fading observed with TL dosimeters.
5. Procedures for pre- and post-irradiation optical annealing of OSL dosimeters are much simpler than procedures for thermal pre- and post-irradiation annealing of TL dosimeters.
6. Non-dosimetric Planck black body emission of the dosimeter support assembly and of the dosimeter itself does not interfere with the OSL dosimetric signal, while at low doses and high temperatures it may interfere with the TL dosimetric signal.

17.3 Semiconductor Radiation Dosimetry

Detectors based on silicon (Si) and germanium (Ge), the two best known semiconductor crystals, have been used in physics research for energy spectrometry and as particle detectors since the early days of semiconductor physics research. More recently, the use of semiconductor detectors has entered relative radiation dosimetry with semiconductor diodes, metal-oxide-semiconductor field effect transistors (MOSFETs), and diamond detectors.

Since their radiation sensitive medium is about three orders of magnitude denser than air, the semiconductor detectors significantly exceed the sensitivity of gas-filled ionization chambers on a per volume basis. In addition, semiconductor dosimetry is based on creation of electron-hole pairs in the cavity volume and energy required to create an electron-hole pair in a semiconductor (few eV) is almost an order of magnitude smaller than mean energy required to produce an ion pair in air (33.97 eV). This also has a favorable effect on the sensitivity of semiconductor dosimeters.

17.3.1 Introduction to Semiconductor Physics

A semiconductor is an insulator that in thermal equilibrium with its environment has some mobile charge carriers: electrons in the conduction energy band and holes in the valence energy band (see Sect. 17.2.1). In comparison to insulators that have an energy band gap E_g of the order of 10 eV, the energy band gap of semiconductors is an order of magnitude smaller with Si at 1.14 eV and Ge at 0.67 eV. Both Si and Ge crystallize in the diamond structure configuration that is supported with covalent bonds. The peculiar properties of semiconductors result from thermal lattice vibrations, impurities in the crystal lattice, and lattice defects.

Crystal imperfections and impurities affect drastically the electronic properties of semiconductors. For example, one boron (B) atom to 10^5 Si atoms increases the intrinsic conductivity of pure Si by three orders of magnitude at room temperature.

The deliberate addition of impurities into a semiconductor crystal is referred to as doping. Two types of semiconductor impurity are known: 1. Donors and 2. Acceptors. Both are substitutional atomic impurities that replace directly the four-valent atom in the semiconductor covalent lattice.

1. *Donors* are pentavalent atoms such as phosphorus (P), arsenic (As), and antimony (Sb) that introduce occupied energy levels into the forbidden energy band gap right below the conduction band, as shown schematically in Fig. 17.12a. Electrons can easily be excited from these levels into the conduction band. Semiconductors containing a predominance of donor impurities have many more free electrons in the conduction band than free holes in the valence band and conduction occurs mainly by electrons in the conduction band. These crystals are then referred to

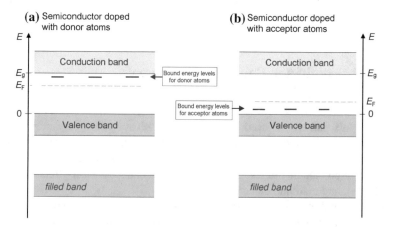

Fig. 17.12 Typical energy band diagram for a semiconductor: **a** Doped with pentavalent donor atoms introducing occupied electron energy levels in the forbidden energy band *right* below the conduction band and **b** Doped with trivalent acceptor atoms atoms introducing vacant electron energy levels in the forbidden energy band *right* above the valence band

as n-type semiconductors since they carry a preponderance of negative charge carriers (electrons). In n-type semiconductors electrons are described as majority carriers and holes as minority carriers.

2. *Acceptors*, on the other hand, are trivalent atoms such as boron (B), aluminum (Al), and gallium (Ga) that introduce vacant energy levels in the forbidden energy band gap right above the valence energy band, as shown schematically in Fig. 17.12b. Electrons can be excited to these energy levels from the valence band thereby producing free holes in the valence band. Semiconductors containing a predominance of acceptor impurities have many more free holes in the valence band than free electrons in the conduction band and conduction occurs mainly by holes in the valence band. These crystals are referred to as p-type semiconductors, since positive charge carriers (holes) predominate. In p-type semiconductors holes play the role of majority carriers and electrons are minority carriers.

Commonly both types of impurities are present in a doped semiconductor crystal; however, one type or the other type predominates and the type that predominates determines the type of conductivity in the crystal.

17.3.2 Semiconductor p-n Junction

By themselves, the n-type and p-type semiconductors are of little practical interest; however, when a mixture of two or more of them was put on the same semiconductor crystal during 1950s, a whole new world of scientific and technological advancement opened up. This started a major revolution in science and technology and the process is still ongoing seven decades later. The first step in this revolution was the p-n junction that dealt with a simple unification of an n-type and p-type semiconductor on a single crystal. Such junctions are called a diode and have important electric properties, including diode rectification, solar cells, photodiode, etc., and have also proved useful as relative dosimeters in medical radiation dosimetry.

The p-n junction is produced with special doping procedures. If the diode is to be of the p-type variety, a Si crystal is doped in two stages, first with acceptor atoms and then with diffusion of higher concentration of donor atoms to superimpose an n-type region over a portion of the p-type crystal. For an n-type diode the doping procedure is reversed and starts with donor doping first and continues with diffusion of a higher concentration of acceptor atoms. Both p-type and n-type diodes end up with a narrow junction region that contains no charge carriers and is called a p-n transition region or depletion region separating the n-type region of the crystal from the p-type region of the crystal.

After the physical formation of the p-n junction some of the electrons from the n-side diffuse into the p-side across the physical boundary and recombine with holes, thereby forming negative acceptor ions. These ions are locked into lattice positions close to the physical p-n boundary. Furthermore, this diffusion of electrons leaves behind positive donor ions locked into their respective positions on the n-side close to

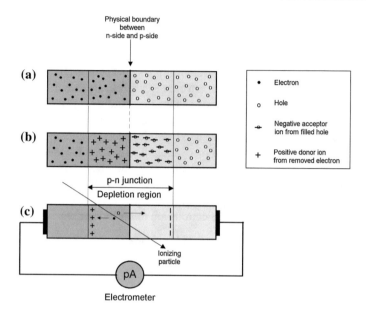

Fig. 17.13 Semiconductor p-n junction: **a** Without charge carrier diffusion over the physical junction barrier; **b** With charge carrier diffusion over the physical junction barrier, and **c** Connected in short-circuit mode as diode radiation detector

the physical boundary. The negative acceptor ions from the crystal p-side combined with the positive donor ions from the crystal n-side create a dipole charge layer that is depleted of free charge carriers and defines the so-called p-n junction also known as depletion region or diode space charge region.

The formation of p-n junction is illustrated in Fig. 17.13, in part (a) without charge carrier diffusion and in part (b) with charge carrier diffusion. The typical width of the depletion region is 10^{-4} cm to 10^{-3} cm. The dipole layer of the depletion region creates an electric field in the depletion region from positive donor-side polarity to negative acceptor-side polarity and, for a Si diode, a potential difference of ~ 0.7 V across the depletion region. When this equilibrium is reached, no more electrons from the n-side can move across the physical p-n boundary to the p-side because they are repelled by the negative acceptor ions on the p-side and attracted by the positive donor ions on the n-side.

17.3.3   Diode Radiation Dosimeter

The p-type and the n-type sides of a p-n junction diode are connected to metallic plates to which external wires are attached, as shown schematically in Fig. 17.13c. In an electronic circuit there are three possible configurations for a diode connection:

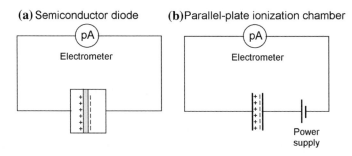

Fig. 17.14 Radiation detectors: **a** Semiconductor diode connected in short-circuit mode and **b** Parallel-plate ionization chamber connected to external power supply (see Fig. 16.4)

1. Short circuit or no bias mode that leaves the p-n junction in equilibrium, characterized in a Si diode with a polarity of 0.7 V across the depletion region.
2. Forward bias mode in which the p-side is connected to positive terminal and n-side to negative terminal thereby reducing the width of the depetion region from the equilibrium width.
3. Reversed bias mode in which the p-side is connected to negative terminal and the n-side to positive terminal thereby increasing the width of the depetion region from the equilibrium width.

An ideal diode has zero resistance in the forward bias mode and infinite resistance in the reverse bias mode.

The diode detectors used in radiation dosimetry are connected in the short circuit diode mode (no bias) through an electrometer that measures the current flowing in the external circuit. This current is proportional to the dose rate of ionizing radiation to which the diode is exposed and its time integral is proportional to the dose absorbed in the diode.

As shown in Fig. 17.14, the diode based dosimetry system can be regarded as an analog to a parallel-plate ionization chamber system except that a diode dosimeter provides its own biasing in the depletion region, while an ionization chamber relies on an external power supply to provide electric field inside the chamber cavity volume. An ionization chamber acts like a perfect capacitor in which ionizing radiation induces a "leakage current" that can be measured with an electrometer in an external circuit and this "leakage signal" is proportional to absorbed dose or dose rate. The diode depletion region plays the role of the dosimeter cavity. Like in all radiation detectors, ionizing radiation releases energetic charged particles that ionize atoms and molecules of the cavity sensitive medium through collision losses. In gaseous and liquid cavity media, these ionizations produce ions pairs; in condensed cavity media such as semiconductor diode (and TL dosimeters, etc.), they produce electron-hole pairs.

When ionizing radiation strikes a diode, electron-hole pairs may be created in the depletion region; electrons are attracted to the n-side of the diode, holes to the p-side. The electric field inside the depletion region sweeps the electrons and holes out of the depletion region and their recombination is measured by the electrometer connected to the diode in the external "short-circuit" diode mode, as shown in Fig. 17.13c.

17.3.4 Silicon Diode Dosimeter: Practical Issues

Silicon diodes are used in radiation dosimetry in two areas: (i) quality assurance of radiotherapy equipment (measurement of parameters of radiation beams produced by high technology radiotherapy equipment, such as x-ray machines, cobalt units, and linear accelerators) and (ii) in-vivo patient dosimetry (measurement of dose to patient during radiotherapy treatment).

The advantages of silicon diode dosimeters in comparison with air-filled ionization chamber are:

1. *Higher sensitivity per unit volume*: Cavity mass density of silicon at 2.33 g/cm^3 compared to that of air at $\sim 1.3 \times 10^{-3}$ g/cm^3; mean energy to produce an electron-hole pair in silicon is 3.5 eV compared to mean energy to produce an ion pair in air at ~ 34 eV.
2. *Higher spatial resolution* makes diodes excellent for measurements in high dose gradient radiation fields such as those in brachytherapy, radiosurgery, and in penumbral regions of radiation beams.
3. *Mechanical stability and strength*: Diodes are much more rugged and strong than ionization chambers with their brittle cavity walls.
4. *No biasing is required for diodes* compared to high voltage biasing required for ionization chambers.

Disadvantages of diodes in comparison with ionization chambers:

1. *Effect of accumulated dose on detector sensitivity*: Diodes lose sensitivity with accumulated dose, while ionization chambers are immune from this phenomenon. The diode loss of sensitivity is caused by radiation damage to semiconductor cavity of the diode and is most pronounced at the beginning of the diode "service" as radiation detector. After prolonged use the sensitivity loss levels off when the accumulated dose history reaches a level of a few kGy. To minimize the fractional sensitivity loss during clinical use some manufacturers actually pre-irradiate the diodes to high radiation doses before they are sold.
2. *Temperature effect*: The diode response is affected by diode temperature and the effect amounts to about 0.2% to 0.4% per °C. While in routine relative dose measurements this is not of concern, the effect should be considered when diodes are used for in-vivo patient dosimetry.
3. *Energy response*: Silicon diodes (atomic number Z of Si is 14) are not water or tissue equivalent, therefore their response varies with energy of both photon as well as electron beams. Since atomic number Z of diodes is about double that of water ($Z_{wat} \approx 7.5$), one can expect a diode to increase its signal per dose to water by almost an order of magnitude, when response to photons at 1 MeV is compared

Fig. 17.15 Mass collision stopping power ratios of water/silicon, water/carbon, and water/air for electrons with kinetic energy from 10 MeV to 100 MeV

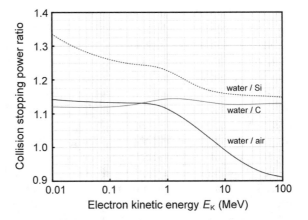

to response at around 50 keV photons. This increase in response is a result of photoelectric effect, its Z^3 dependence, and its predominance at relatively low photon energies below 100 keV. In electron beams, silicon diodes show a slight energy response as a result of a variation in collision stopping power ratio water/Si with electron energy (see Fig. 17.15). However, the variation is relatively small and is often ignored under the assumption that the diode measures relative dose in water.

4. *Directional effect*: Diode response varies with the angle between the radiation beam central axis and the axis of symmetry of the diode. The sensitivity drops with increasing angle and the drop can amount to as much as 5% at large angles, depending on the diode. Small size ionization chambers do not exhibit this effect (Fig. 17.15).

17.3.5 Diamond Radiation Dosimeter

Within a few years after discovery of x-rays and natural radioactivity at the end of 19-th century air-filled ionization chamber became the most widely used and best understood radiation dosimeter. It is safe to say that, despite the wide armamentarium of radiation dosimeters currently available in radiation dosimetry, even today, air-filled ionization chambers still occupy one of the top spots in the ranking of radiation dosimeters both in the category of absolute dosimetry as well as in the category of relative dosimetry.

Since the density of air in the ionization chamber cavity is some three orders of magnitude smaller than that of tissue, replacing cavity air in an ionization chamber with a denser tissue-equivalent liquid or solid insulator should increase the ionization chamber sensitivity and this, in turn, could be used to improve the spatial resolution of the dosimeter. However, finding suitable insulators that would allow sufficient charge carrier mobility in a liquid or electron and hole mobility in a solid proved elusive. As for liquid ionization chambers (see Sect. 17.1.4) their feasibility has been studied since 1970s but they are still not widely used in medical physics.

Natural diamond was among the first materials used as solid cavity medium in ionization chambers since 1940; however, for many decades the large cost as well as variations in quality of natural diamonds precluded a widespread use of diamonds for routine radiation detection in particle and medical physics. Moreover, the development of silicon and germanium semiconductor radiation detectors during 1950s offered a better option for solid-state dosimetry and sidetracked dosimetry research based on natural diamond serving as dosimeter cavity.

The situation changed in the early 1980s when well-controlled synthetic manufacturing of diamonds became available. The technology, referred to as chemical vapor deposition (CVD), is relatively simple as well as inexpensive. The synthetic diamonds so produced are called CVD diamonds and are, in comparison with natural diamonds, more suitable for use in radiation dosimetry because they are of uniform quality and of higher purity. These two attributes combined with a few other advantages of diamond over silicone diode have during the past two decades made diamond dosimeter a serious contender for primacy in solid-state radiation dosimetry.

Using the CVD technology a synthetic diamond is grown on a substrate layer that can be silicone or metal from a hydrocarbon gas mixture composed of about 1% of methane (CH_4) and 99% of hydrogen (H_2). Methane serves as source of carbon, while hydrogen etches off the non-diamond carbon that is also produced during the CVD process.

Diamond is one of the forms (allotropes) of carbon with atoms arranged in a face centered cubic crystal structure called diamond lattice. The other common allotropes of carbon are: graphite (often used for ionization chamber walls and central electrode, Sect. 16.3.16), graphene, carbon nanotube (used in cold cathode x-ray tubes, Sect. 14.4.3), and fullerene. Each of these allotropes of carbon exhibits its own set of physical properties that make them particularly useful for a specific scientific or practical purpose.

A solid-state ionization chamber based on synthetic diamond cavity has a number of advantages over a silicon diode dosimeter that is currently the most widely used solid-state dosimeter in medical radiation dosimetry. A comparison of basic characteristics of importance in medical radiation dosimetry for diamond, silicon, and air is given in Table 17.3. Based on these data and a few other known characteristics of these three cavity media we conclude that the most notable advantages of synthetic diamond dosimeter over silicon diode dosimeter are as follows:

1. Diamond with $Z = 6$ is much closer to tissue and water equivalence ($Z \approx 7.5$) than silicon with $Z = 14$.
2. Diamond dosimeter cavity, because of its high resistivity, produces a much lower leakage (dark) current than silicon diode.
3. In contrast to silicon diode, diamond cavity exhibits much higher radiation hardness (i.e., is much less affected by radiation exposure history).
4. In comparison with silicon, diamond has higher charge carrier mobility.

Table 17.3 Characteristics of importance for radiation dosimetry with dosimeter cavity made of diamond, silicon, or air

Cavity medium	Atomic number Z	Mass density (g/cm³)	Energy gap (eV)	$\bar{W}$ (eV)	Resistivity (Ω·cm)
Diamond	6	3.51	5.45	13	$10^{13} - 10^{16}$
Silicon	14	2.33	1.14	3.6	2.3×10^5
Air @ STp	~7.5	1.293×10^{-3}	–	33.97	–

Note: $\bar{W}$ stands for mean energy to produce an ion pair in air or an electron-hole pair in diamond and silicon; STp stands for standard temperature $0\,°C$ and pressure 101.3 kPa

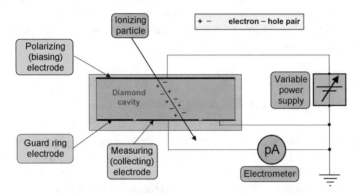

Fig. 17.16 Schematic diagram of a diamond dosimeter configuration. The diagram is similar to the one shown in Fig. 16.4 for an air-filled ionization chamber, except that the air-filled cavity in the ionization chamber is replaced by a diamond cavity. Note that in an air-filled cavity ionizing radiation creates ion pairs while in a solid diamond cavity ionizing radiation creates electron-hole pairs. Electrodes in both types of chamber are ohmic electrodes

5. Diamond, because of its physical hardness, is less prone to mechanical damage in comparison to silicon diode.
6. Because of its mechanical sturdiness and high melting point, diamond dosimeter can be safely sterilized for in-vivo use.

As shown in Fig. 17.16, the basic configuration of a diamond dosimeter is the same as that for an air-filled ionization chamber, shown in Fig. 16.4, except that ambient air in the dosimeter cavity of the air-filled ionization chamber is replaced with a diamond to get a diamond solid-state dosimeter. One can thus say that in principle a diamond dosimeter is a solid-state ionization chamber. The electrodes used in diamond dosimeter are ohmic electrodes that provide electric field in the diamond cavity. The typical potential applied to diamond detector is of the order of several hundred volts.

Of course, the cavity mass density in the air-filled dosimeter is more than three orders of magnitude smaller than is the cavity mass density in the diamond dosimeter ($\rho_{\text{air, STp}} = 1.293 \times 10^{-3}$ g/cm³ compared to $\rho_{\text{diamond}} = 3.51$ g/cm³). This means that the diamond dosimeter will, on a per volume basis, supply a much larger signal than the air-filled dosimeter. Or else, we may state that to get the same signal from the diamond detector as from the air-filled detector, we could use a much smaller volume diamond cavity thereby significantly increasing the spatial resolution of the dosimeter.

It was well accepted since 1940s that diamond has very attractive properties for use as cavity medium in a standard ionization chamber. The introduction of the CVD process into diamond manufacturing has solved the only major problem with diamond dosimetry, namely, the diversity in natural diamonds originating from different sources. Since synthetic diamonds do not suffer this drawback and, in addition, as listed above, offer some significant advantages over the diode, the use of synthetic diamond in dosimetry is growing rapidly. It is reasonable to expect that in the future diamond dosimetry will eclipse diode dosimetry and become the most widely used category of solid-state dosimetry.

17.3.6 Semiconductor Dosimetry: Summary

Semiconductor radiation detectors have been in use in particle physics and nuclear physics since the introduction of semiconductor devices in electronic technology during 1950s. These detectors are either in the form of forward or reverse biased p-n junction diodes made of silicon or germanium semiconductors or in the form of lithium drifted silicon or germanium semiconductors cooled to low temperatures to diminish leakage currents. They are used for particle counting and photon spectroscopy but are much too sophisticated for use in medical dosimetry. However, the experience with semiconductor detectors of ionizing radiation in particle and nuclear physics research has during the past two decades engendered a new category of relative radiation dosimetry that was found very practical, simple to use, and useful in quality assurance procedures on intermediate energy x-ray machines, cobalt units, and megavoltage linear accelerators used in radiotherapy.

The two most popular semiconductor relative dosimetry systems are: (i) silicon diode connected in the short circuit (no bias) mode and (ii) diamond dosimeter that effectively is an analog to a gas-filled ionization chamber in which diamond replaces air as cavity radiation sensitive medium.

Since the diode dosimeter in the "no bias mode" provides its own internal bias that establishes electric field in the diode depletion region, one can say that the diode dosimeter is, like the diamond dosimeter, an analog to an air filled ionization chamber. Of course, the diamond dosimeter, like an air filled ionization chamber, uses an external power supply to establish electric field inside the cavity volume.

We must note that, on the one hand, charge carriers in an ionization chamber are ion pairs created in chamber gas (positive and negative ions in electronegative gas; positive ions and free electrons in a non-electronegative gas, see Sect. 16.3.6). On the other hand, in semiconductor dosimetry, the charge carriers are electron-hole pairs produced in the semiconductor cavity medium. Energy to create an electron-hole pair in a semiconductor is only a few electronvolts, while energy to create an ion pair in air is an order of magnitude larger at ~ 34 eV

Advantages of semiconductor dosimeters over air-filled ionization chambers are as follows: (i) higher sensitivity per unit cavity volume, (ii) higher spatial resolution, (iii) higher mechanical stability and strength. Since diamond dosimeter has several advantages over silicon diode dosimeter, one can conclude that in many respects diamond dosimeter offers several advantages of ionization chamber; however, it is not ready yet for absolute dosimetry.

17.4 Film Radiation Dosimetry

Medical film radiation dosimetry falls into the group of relative radiation dosimetry techniques and comprises two major film categories: (i) radiographic film and (ii) radiochromic film.

Both types of dosimetric film serve the same purpose and give similar results; however, the use of radiographic film in dosimetry is waning while radiochromic film is gaining in importance. The main reason for this shift is the significant difference in film processing required by the two film types; processing of radiographic film requires access to sophisticated developers and fixers, while radiochromic film is self-processing and requires no additional film processing equipment.

In the past film developers were readily available in clinics and hospitals that were using radiographic films for diagnostic procedures and they could also be used for radiation film dosimetry. During the past decade, however, the use of analog radiographic film for diagnostic imaging has been discontinued in the developed world in favor of digital computerized radiographic techniques. This shift caused the disappearance of film developers from hospitals and clinics and provided a strong economic incentive for the use of radiochromic film for medical film radiation dosimetry.

17.4.1 Absorbance and Optical Density

Film-based radiation dosimetry depends strongly on transmission of light through processed film, be it radiographic or radiochromic. The transmission is affected by film opacity that can be measured in terms of optical density OD with devices, such as film densitometer, laser densitometer, automatic film scanner, and spectrophotometer. Similarly to absorbance A_λ of a Fricke solution, discussed in Sect. 16.2.9 and measured with a spectrophotometer, optical density OD of film depends on absorbed dose and is defined as

$$OD = \log \frac{I_0}{I} \quad \text{or} \quad I = I_0 10^{-OD} = I_0 e^{-(2.302)OD}, \qquad (17.20)$$

where

I_0 is the initial light intensity emitted from a light source, such as a light emitting diode (LED), laser, or white light fluorescent source.

I is the light intensity transmitted through film and measured by a sensor such as a photodiode, photomultiplier tube (PMT), or linear charge coupled device (CCD).

For example, if a processed film transmits 10% of the incident light beam, then $I/I_0 = 0.1$ or $I_0/I = 10$, and the optical density of the film is $OD = \log 10 = 1$. Similarly, an optical film density $OD = 2$ corresponds to film transmission fraction $I/I_0 = \exp(-2.302 \times 2) = 0.01$.

In clinical practice, optical densities range from less than 1 to over 3. The advantage of the logarithmic definition of optical density given in (17.20) is that it makes optical densities additive. Optical density of two layers of film or two exposures of the same film are additive and the total optical density is the sum of two densities ($OD_{tot} = OD_1 + OD_2$). A densitometer that has been calibrated by a standard strip of film containing regions of known optical density can provide a direct reading of optical density.

A graph of optical density OD against absorbed dose of a typical dosimetric film is referred to as the sensitometric curve and is also called characteristic curve of film or the H&D sensitometric curve in honor of Ferdinand Hurter and Vero C. Driffield who were the first to investigate the relationship. A typical H&D characteristic curve for a dosimetric film is shown in Fig. 17.17. The curve has four regions:

   (i) Base + fog at very low doses.
  (ii) Toe in transition region between base and the linear region.
 (iii) Linear region at intermediate doses.
  (iv) Shoulder in the transition from the linear region to saturation.
   (v) Saturation at high doses.

Note: The dosimetric quantity of interest is usually the net optical density (netOD) or OD_{net} that is defined as the measured optical density OD_{meas} less the base optical density OD_{base}, or

$$OD_{net} = OD_{meas} - OD_{base}. \tag{17.21}$$

General characteristics of radiographic and radiochromic film dosimetry:

1. While film is not useful for absolute dosimetry, film dosimetry possesses several attributes that make it very useful and practical in special areas of radiation dosimetry as well as in acceptance testing, commissioning, and quality assurance of high technology equipment used in radiation medicine.
2. Film has an excellent spatial resolution and is often used to measure 2D dose distribution maps, especially in regions of high dose gradient and regions of dynamic dose.
3. Film is especially useful for gaining qualitative data that provide fast general information on radiation dose distribution.

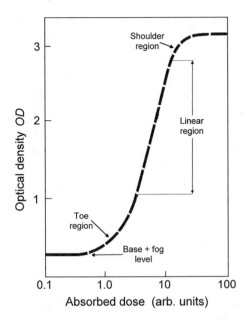

Fig. 17.17 A typical H&D characteristic curve for a dosimetric film showing five regions: (i) base + fog, (ii) toe, (iii) linear region, (iv) shoulder, and (v) saturation

17.4.2 Radiographic Film Dosimetry

Radiographic film has been in use for x-ray detection since Wilhelm K. Röntgen discovered x-rays in 1895. The discovery of x-rays was translated into hospitals and clinics around the world for medical purposes within a few weeks of their discovery and x-ray images produced with steady improvements in film technology have formed the backbone of radiology for more than a century.

Composition of Radiographic Film

Modern radiographic film is typically about 0.25 mm thick and comprises several layers, as shown in Fig. 17.18(a). The radiographic film layers and their functions are as follows:

  (i) Film base – made of polyester resin that provides support for active layers of film.
 (ii) Adhesive layer – binds emulsion to film base.
(iii) Emulsion – active layer of film; comprises gelatin and uniformly distributed small silver halide crystals.
(iv) Protective coating – protects the active film layer.

Note: Layers listed above form single sided films; to increase sensitivity films are double sided containing the same layers on each side of the film base (in same order starting from the film base).

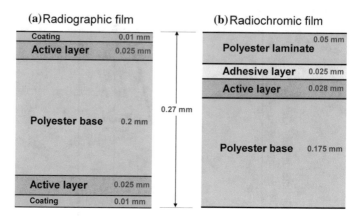

Fig. 17.18 Transverse section through two typical film types used in film radiation dosimetry: **a** Double-sided radiographic film with silver halide based active layer and **b** Single sided EBT2 radiochromic film with monomeric di-acetylene based active layer

Emulsion forms the active film layer of which silver halide crystals (~95% silver bromide AgBr and ~5% silver iodide AgI) form the most important component. The silver halide crystals (also called grains) are ionic crystals composed of silver ions Ag^+ and halogen ions Br^- and I^-. The microscopic crystals, arranged into a lattice structure, are not perfect nor are they chemically pure. Rather, they contain imperfections called sensitivity specks that are either naturally occurring (edge dislocations) or are introduced into the crystals through doping (impurities) to increase sensitivity. It is estimated that each crystal grain contains about 10^{10} atoms and is about 1 to 2 μm in diameter.

Another important imperfection in the AgBr crystal is the so-called Frenkel defect that is characterized by silver ions Ag^+ located interstitially in the lattice with their corresponding negatively charged silver ion vacancies. The interstitial Ag^+ ion is extremely mobile through the Ag^+Br^- ionic lattice and serves as the basis for forming the latent image during the exposure of film to visible light or ionizing radiation.

The photographic effect as well as the utility of radiographic film in radiation dosimetry arises from the lattice defects that occur in imperfect AgBr crystals. These lattice defects serve as shallow traps for free electrons that are located in the energy band below the conduction band in the energy gap separating the conduction band from the valence band (see Sect. 17.2.1).

Formation of Latent Image in Radiographic Film

During film exposure, ionizing radiation releases energetic free electrons (photoelectrons, Compton electrons, etc.) in the film and these electrons through ionization of atoms and ions of the film components release free electrons into the conduction band of the AgBr crystals. Some of the free electrons, while migrating through silver halide crystals get close to a sensitivity speck and may get trapped into a shallow electron trap.

An electron trapped in an electron trap (sensitivity speck) attracts a nearby interstitial silver ion Ag^+ from a Frenkel defect and reduces it to neutral Ag^0 atom $(Ag^+ + e^- \rightarrow Ag^0)$. If the reduction process is repeated at least another 3 times on a given crystal, a latent image that can be made visible through film processing is formed. The extent to which a given area of film is affected by ionizing radiation depends on the number of particles (photons or charged particles) incident upon that area.

After exposure to ionizing radiation, radiographic film contains a latent image that is not visible before the film undergoes appropriate chemical processing that consists of two steps: (i) film development and (ii) film fixation.

Radiographic Film Processing

During development the silver bromide crystals that have been sufficiently affected during exposure to radiation (typically 4 hits or more) are converted into opaque black silver specks with all Ag^+ ions reduced to Ag^0 atoms and all Br^- ions removed. The crystals that have not been affected during exposure remain in their original state in the emulsion. A chemical suitable as film developer must, on the one hand, be able to donate rapidly electrons to crystals containing a latent image and, on the other hand, be much slower in donating electrons to crystals that do not contain a latent image. The duration of the development process must be carefully timed to prevent eventual complete reduction of all silver ions to atomic silver irrespective of whether or not they contain latent image.

During fixation the unaffected AgBr crystals in the emulsion are dissolved leaving behind the record of radiation exposure in the form of a corresponding pattern of opaque black silver specks distributed over the gelatin. The metallic silver is unaffected by the fixer but causes darkening of the developed film and the degree of darkening depends on radiation exposure of the film that can be measured with a film densitometer.

17.4.3 General Characteristics of Radiographic Film

Radiographic film has a long and illustrious history dating back to the discovery of x-rays in 1895. Since then it was used both in the clinic for imaging as well as in radiation dosimetry as detector of ionizing radiation.

Dosimetry based on radiographic film consists of the following four steps:

1. Film manufacture – crystals of silver halide (mainly AgBr) of appropriate size and containing a desirable concentration of lattice defects (sensitivity specks) are suspended uniformly in gelatin to produce film emulsion. The emulsion is bound onto polyester fim base and coated with a protective layer.
2. Exposure to ionizing radiation (photons, electrons, protons) to create a latent image in the active film layer.
3. Radiographic film processing consists of three stages: (a) Development of exposed film to convert the latent image to atomic silver Ag^0; (b) Fixation of developed film to dissolve and remove the AgBr remaining on the developed film and to

harden the gelatin; (c) Washing the fixated film in pure water and air-drying the fixated film.

4. Measurement of optical film density to determine absorbed dose based on measurement of the film H&D sensitometric curve.

When used in radiation dosimetry, radiographic film is characterized by several advantages and disadvantages in comparison with other dosimetric techniques. The significant advantages of radiographic film are:

- Very high spatial resolution limited by the physical size of the light source of the densitometer rather than by the film itself.
- Permanent storage of processed image and dosimetric information allowing multiple non-destructive readouts.
- Convenient physical shape of the dosimeter cavity (small thickness and arbitrarily large area) and flexibility allowing easy use in radiation field mapping and general quality assurance measurements of photon and electron beams.
- Until recently, widespread commercial availability of a variety of radiographic film types and shapes. The advent of digital radiography has significantly curtailed the manufacturing of radiographic films and severely limited access to radiographic film developers, so that this feature can no longer be considered an advantage of radiographic films.

Some of the notable disadvantages of radiographic film are:

- Radiographic film can only be used as a relative dosimetry technique, since the response of a radiographic film dosimetry system must be calibrated in a known radiation field before use for measurement of absorbed dose.
- Radiographic film is not tissue equivalent because it contains silver bromide that increases its effective atomic number above that of water and tissue.
- Radiographic film exhibits strong energy dependence of film response for photons with energy below $\sim$100 keV as a result of the high atomic number of silver ($Z = 47$) and predominance of photoelectric effect at low photon energies (see Sect. 8.2.3, Fig. 8.5). For example, the relative response per unit of x-ray exposure normalized to cobalt-60 γ-rays for a typical radiographic film is about 20 times higher for 50 keV x-rays and the ratio of mass energy absorption coefficients for typical radiographic film emulsion to water at 50 keV x-rays is $\sim$100 to 1.
- Radiographic film processing involves wet chemical development and fixation that both require strong quality control and meticulous equipment maintenance to obtain results that are in line with standards of radiation dosimetry.
- Radiographic film is sensitive to visible light, so any work involving open film must be carried out in a dark room.

17.4.4   Radiochromic Film Dosimetry

Radiochromic dosimetry is a chemical dosimetry technique based on radiochromic reactions defined as reactions that are: (i) triggered by ionizing radiation and (ii) result in a color change in the radiation sensitive medium. The radiochromic dosimetry has been developed during 1960s for high dose measurements in the kGy range for industrial applications of ionizing radiation.

As the sensitivity of radiochromic techniques was improved to the level of medical dosimetry, it became apparent that radiochromic dosimeters could be fashioned into the format used for radiographic film dosimetry. The goal was to develop radiochromic film systems with characteristics that match, or improve upon, those attributed to radiographic film. However, unlike radiographic films that are based on silver halides, radiochromic films would be based on radiochromic reactions producing a color change in the irradiated detector that is proportional to the dose absorbed in the detector. Hence, measured color change in radiochromic film exposed to ionizing radiation would be used for relative radiation dosimetry.

Radiochromic Film Based on Polymerization of Di-Acetylene

In the past, many organic materials and several chemical processes were used in radiochromic radiation detectors; however, none of these techniques were sensitive enough for medical applications. This changed during 1980s with the development of a new radiochromic detector in the format of radiographic film. The active radiation sensitive medium of the new radiochromic film is based on di-acetylene (C_4H_2), a highly unsaturated hydrocarbon monomer that contains three single bonds and two triple bonds ($H - C \equiv C - C \equiv C - H$). The film consists of thin microcrystalline monomeric di-acetylene emulsion coated on a flexible polyester film base. A schematic diagram of a typical radiochromic film is given in Fig. 17.18b for comparison with a typical radiographic film shown in Fig. 17.18a. Pioneers in the development of medical radiochromic film dosimetry are William McLaughlin, David F. Lewis, and Christopher G. Soares.

The radiochromic film is translucent before irradiation. Under exposure to ionizing radiation the di-acetylene monomers undergo progressive chain-growth photopolymerization. This results in formation of colored polymer chains that grow in length and the film turns to a shade of blue that becomes darker as the absorbed dose increases.

According to the IUPAC (International Union of Pure and Applied Chemistry), polymerization is the process of converting a monomer or a mixture of monomers into a polymer and photo-polymerization is chain-growth polymerization initiated by visible light, ultraviolet light, or ionizing radiation.

It is notable that no physical, chemical or thermal processing is required to develop or fix the radiochromic film image – radiochromic film is self-developing; however, after exposure to ionizing radiation, polymerization slowly continues for about 24 hours post-irradiation and then color stability sets in.

Several companies are involved in production of radiochromic films for various applications. The majority of these films require doses that are much higher than those used in clinical work and the films are then used for commercial purpose such as food irradiation, medical equipment sterilization, and waste management. The best-known medical radiochromic film is the GAFchromic EBT2 film produced by International Specialty Products, Wayne, NJ, USA.

Absorption Spectrum of Radiochromic Film

The dosimetric signal in medical radiochromic film dosimetry is obtained from the color change or film darkening that is proportional to the dose absorbed in the film active layer. The amount of light that is transmitted through (or absorbed in) the film during the readout procedure depends on the dose of ionizing radiation and the wavelength λ of the light used in the readout device.

Each type of radiochromic film has its own absorbance spectrum and an example of such spectrum is given in Fig. 17.19 for the EBT2 radiochromic film in the wavelength region from 350 nm to 800 nm. Two absorption curves are shown: one for unexposed film and one for film exposed to 200 cGy of cobalt-60 gamma rays. The curves show three absorbance peaks: two dosimetric peaks (Peak 2 at 583 nm and Peak 3 at 635 nm) and one broad non-dosimetric peak (labeled Peak 1) at about 420 nm. The non-dosimetric Peak 1 is attributed to the addition of yellow dye to the active film layer for inhibition of film sensitivity to visible and ultraviolet light.

Measurement of the absorption spectrum of a given radiochromic film type is important, since it allows optimization of the measured film sensitivity during the film analysis. Generally, the light source of the film readout device is selected such that it matches as close as possible the absorption maximum of a dosimetry peak of the given radiochromic film. For example, a helium-neon laser with emission of red light at 633 nm has been found a good match for the dosimetry peak at 635 nm of the EBT2 film.

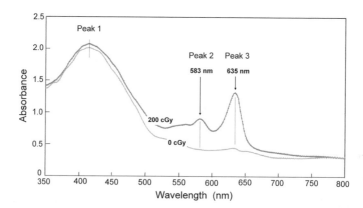

Fig. 17.19 Absorbance spectrum (adapted from David F. Lewis, ISP Corporation) for GAFchromic™ EBT2 film in the wavelength range from 350 nm to 800 nm for: (i) unexposed film and (ii) film exposed to 200 cGy of cobalt-60 gamma rays. Exposed film clearly shows two radiation induced absorbance peaks (dosimetric peaks), peak 2 at 583 nm and peak 3 at 635 nm. The non-dosimetric absorbance peak at ∼ 420 nm (peak 1) is attributed to the addition of yellow dye into the active layer of the radiochromic film. The yellow dye significantly decreases the film sensitivity to ambient fluorescent light; however, the absorbance peak is not of much value for dosimetry

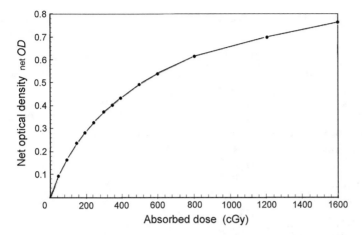

Fig. 17.20 Typical example of an H&D sensitometric curve of EBT2 radiochromic film. The plot "net*OD*" against dose would serve as a calibration curve in radiochromic film dosimetry using the same radiochromic film batch, same measuring equipment, and same radiation source

Calibration of Radiochromic Film

Since medical film dosimetry is a relative dosimetry technique, it is imperative that the radiochromic film technique be calibrated before use for measurement of unknown doses. The calibration procedure consists of the determination of the H&D sensitometric relationship for the radiochromic film and must encompass all components of the measuring system. Moreover, the calibration must be carried out on the specific radiation source that will be used for measuring unknown doses. The measured H&D calibration curve should encompass the dose region of future interest; extrapolation out of the calibration region is not acceptable.

A typical H&D sensitometric calibration curve for EBT2 radiochromic film is shown in Fig. 17.20 with a plot of net*OD* against absorbed dose. The plot would serve as the calibration curve for the specific film batch, measuring equipment, and radiation source. The measured net*OD* of an unknown dose can be used in conjunction with the calibration curve to determine the actual dose delivered to the radiochromic film.

17.4.5 *Comparison Between Radiographic and Radiochromic Radiation Dosimetry*

Radiographic film has for over a century been used in medicine for clinical imaging as well as radiation dosimetry. Its use in both areas has been waning during the past decade as a result of the advent of digital radiography techniques in clinical x-ray imaging and advances made in radiochromic film technology in terms of greater film sensitivity and decreased cost for use in film dosimetry. Clinical use of radiographic films has essentially disappeared in the developed world in favor of digital imaging

Table 17.4 Comparison of characteristics of radiographic film dosimetry and radiochromic film dosimetry

Characteristics	Radiographic film	Radiochromic film
Radiation sensitive medium	Silver halide AgBr (90%) + AgI (10%)	Di-acetylene C_4H_2 monomer
Radiation induced process	Reduction $Ag^+ + e^- \rightarrow Ag^0$	Polymerization of di-acetylene monomer into polymer
Post-irradiation film processing	Dark room Film processing	**NO**
Darkroom required	YES	**NO**
Film spatial resolution	High	High
Water equivalence	NO	$\sim$ **YES**
Signal low-energy dependence	Severe	**Slight**
Dose rate dependence	NO	NO
Visible light sensitivity of unexposed film	YES	**NO**
Ultraviolet light sensitivity of unexposed film	YES	YES
Allows direct water immersion during irradiation	NO	**YES**

techniques. This has caused the disappearance of darkrooms and well-maintained film processors from hospitals and clinics and severely curtailed the availability of radiographic film dosimetry in radiology and radiotherapy departments.

It is simply economically prohibitive to maintain a film processor in a standard medical physics department just for occasional use in film dosimetry, especially if another technique, such as radiochromic film dosimetry that, in addition, features a few other advantages, requires no film processing. Evidently, the significant physical advantages of radiochromic films over radiographic films, combined with economic considerations, have given medical radiochromic film dosimetry a clear advantage over radiographic film dosimetry. Table 17.4 provides a comparison of various pertinent characteristics of traditional radiographic film dosimetry and the novel polymeric radiochromic film dosimetry.

A brief comparison of several important dosimetric parameters for the two known film radiation dosimetry techniques yields the following results:

1. *Effective atomic number* Z_{eff}: In contrast to radiographic film that relies on dosimetric signal produced in silver halides embedded in active film layer, all medical radiochromic films have effective atomic number Z_{eff} close to and slightly lower than that of water, since their radiation sensitive material is organic and contains mainly hydrogen and carbon. For example, Z_{eff} of the EBT2 radiochromic film is $\sim$6.8 compared to $\sim$7.5 for water. This means that the active sensor medium has a mass collision stopping power similar to that of water and mass energy absorption coefficient for photons of energy $h\nu$ exceeding 100 keV also similar to that of water.

2. *Water equivalency*. Based on Z_{eff} one may conclude that radiochromic film is almost water equivalent and its response is essentially independent of energy for photons above 100 keV. For photon energies below 100 keV the situation is not as clear: for some radiochromic films the response is actually up to 40% below that at 1 MeV and for other films it is slightly above. However, the low energy response of radiochromic films is far less pronounced than that of radiographic films where the over-response at low energies exceeds the response at 1 MeV by a factor of one order of magnitude or more.

3. *Dose rate dependence*. Similarly to radiographic film, radiochromic films do not exhibit a dependence on dose rate.

4. *Environmental factors*, such as ambient temperature and humidity during irradiation and readout, affect the radiochromic film response and have to be strictly controlled for best results.

5. *Sensitivity to visible and ultraviolet photons*. Radiochromic film is **not** sensitive to photons above 300 nm but is affected by ultraviolet light. This means that exposure to incandescent lighting is fine; however, ambient fluorescent light and sunlight must be avoided so as not to trigger photo-polymerization with the ultraviolet light that would be of non-dosimetric origin.

6. *Post-irradiation processing*. In contrast to radiographic film, the radiochromic film requires no post-irradiation processing, no chemicals, and no darkroom. Radiochromic film is self-developing through the process of radiation-induced polymerization of di-acetylene monomers that starts within a fraction of a second after the start of irradiation and stabilizes within about 24 hours post-irradiation.

7. *Cutting and immersion in water*. Radiochromic film can be cut to desired shape at room temperature in ambient light in preparation for use in anthropomorphic phantoms. It can also be safely immersed in water.

8. *Polarization effects*. New radiochromic films exhibit film orientation dependent polarization effects. Measured optical density may vary as the polarization plane of the analyzing light changes with film orientation. The effect is most severe when laser beam is used for readout. Film mounting convention on the readout device should be consistent with calibration procedure.

17.4.6 Main Characteristics of Radiochromic Film

Radiochromic radiation dosimetry is a relative chemical dosimetry technique that relies on radiation induced radiochromic reactions for measurement of absorbed dose. These reactions produce in the radiochromic film (radiation dosimeter cavity) a color change that is proportional to the dose absorbed in the dosimeter and can be measured by determining the optical density of the radiochromic film with a film densitometer.

Radiochromic film dosimetry has been known for several decades but has only during the past two decades become a serious competitor to radiographic film dosimetry, largely because of significant improvements in radiochromic film technology in terms of increased film sensitivity, improved understanding of the chemistry behind the radiochromic process, and decreased film production cost. Radiochromic film, similarly to radiographic film (see Sect. 17.4.3), has several features of importance to radiation dosimetry, such as high spatial resolution, permanent storage of processed image allowing multiple non-destructive readouts, and convenient physical shape and flexibility of the dosimeter cavity. In addition as discussed in Sect. 17.4.5, radiochromic film, in comparison with radiographic film, offers several important advantages, such as better water equivalence, insensitivity to visible light, and self-development after exposure.

17.5 Relative Radiation Dosimetry: Summary

Radiation dosimetry systems comprising a radiation sensitive dosimeter (cavity) and a reader are divided into two major categories: absolute and relative. An absolute dosimetry system is defined as a system whose response can be used to infer directly the dose absorbed in the dosimeter cavity without any need for calibration of the dosimeter signal in a known radiation field. A relative dosimetry system, on the other hand, is a system that requires calibration of its response in a known radiation field.

Three absolute dosimetry systems are known: calorimetric, chemical (Fricke), and ionometric and they were discussed in detail in Chap. 16. In contrast, a large variety of relative dosimetry systems are known ranging from very useful and practical, through mundane, all the way to impractical and of only academic interest. This chapter deals with relative dosimetry systems that are of practical interest to medical physics and are used in radiation medicine for calibration of high technology equipment used for diagnostic imaging or treatment of disease with ionizing radiation.

Four categories of relative dosimetry are used in radiation medicine: ionographic, luminescence, semiconductor, and film. Each of these broad categories is split into several subcategories and each of the subcategories can be classified as either and

Table 17.5 Four categories of relative radiation dosimetry systems discussed in this chapter

	Category of dosimetry system	Dosimeter	Mode
1.	Relative Ionometric Dosimetry	Parallel-plate ionization chamber	A
		Thimble (cylindrical) ionization chamber	A
		Area survey meter (ionization chamber)	A
		Re-entrant well ionization chamber	A
2.	Luminescence Dosimetry	Thermoluminescence (TL)	P
		Optically stimulated luminescence (OSL)	P
3.	Semiconductor Dosimetry	Diode (p-n junction)	A
		Diamond dosimeter	A
4.	Film Dosimetry	Radiographic film	P
		Radiochromic film	P

For each category the dosimeters discussed in this chapter as well as their mode of operation (active A or passive P) are also presented

Table 17.6 Basic characteristics of the three absolute dosimetry techniques

	Absolute dosimetry	Cavity medium	Radiation induced process
1.	Calorimetric	Water or carbon	Heat transfer
2.	Fricke Chemical ferrous sulfate	Ferrous sulfate solution	Oxidation of ferrous sulfate (Fe^{2+}) ion into ferric (Fe^{3+}) ion
3.	Ionometric	Ambient air	Production of ion pairs

Table 17.7 Basic characteristics of most important relative dosimetry techniques used in medicine

Relative dosimetry		Cavity medium	Radiation induced process
Ionometric	Ionization chamber	Air	Production of ion pair
	Ionization chamber	Liquid	Production of ion pair
Luminescence	Thermoluminescence (TL)	Crystalline solid	Phosphorescence emission accelerated by heat
	Optically stimulated luminescence (OSL)	Crystalline solid	Phosphorescence emission accelerated by visible light
Semiconductor	Silicon diode	Si p-n junction	Production of electron-hole pair
	Diamond detector	Diamond (carbon)	Production of electron-hole pair
Film	Radiographic	Silver bromide	Reduction of silver ion Ag^+ into silver atom Ag^0
	Radiochromic	Di-acetylene	Polymerization od di-acetylene monomer into polymer

active or passive dosimetry system. An active dosimeter is electrically connected with the reader and its signal is read out directly during exposure to radiation. A passive dosimeter stores the radiation-induced signal during irradiation and is readout with the reader subsequently to irradiation. An active system can measure absorbed dose or absorbed dose rate during irradiation, a passive system can only measure absorbed dose and does this subsequently to irradiation when the dosimeter is connected to the reader for the readout process. Table 17.5 lists the four categories of relative dosimetry systems presented in this chapter and for each category lists several examples of dosimetry systems used in radiation medicine. For each dosimetry system the table also gives their mode of operation (active or passive).

Essential characteristics of the three absolute dosimetry techniques and the most important relative dosimetry techniques discussed in this chapter are summarized in Tables 17.6 and 17.7, respectively.

Appendix A
Main Attributes of Nuclides Presented in This Book

Data given in Table A.1 can be used to determine the various decay energies for the specific radioactive decay examples as well as for the nuclear activation examples presented in this book. M stands for the nuclear rest mass; $\mathcal{M}$ stands for the atomic rest mass. The data were obtained from the NIST and are based on CODATA 2010 as follows:

1. Data for atomic masses $\mathcal{M}$ are given in atomic mass constants u and were obtained from the NIST at: ⟨http://www.physics.nist.gov/pml/data/comp.cfm⟩.
2. Rest mass of proton m_p, neutron m_n, electron m_e, and of the atomic mass constant u are from the NIST (http://www.physics.nist.gov/cuu/constants/index.html) as follows:

$$m_p = 1.672\,621\,777 \times 10^{-27}\ \text{kg} = 1.007\,276\,467\ \text{u} = 938.272\,046\ \text{MeV}/c^2 \tag{A.1}$$
$$m_n = 1.674\,927\,351 \times 10^{-27}\ \text{kg} = 1.008\,664\,916\ \text{u} = 939.565\,379\ \text{MeV}/c^2 \tag{A.2}$$
$$m_e = 9.109\,382\,91 \times 10^{-31}\ \text{kg} = 5.485\,799\,095 \times 10^{-4}\ \text{u} = 0.510\,998\,928\ \text{MeV}/c^2 \tag{A.3}$$
$$1\ \text{u} = 1.660\,538\,922 \times 10^{-27}\ \text{kg} = 931.494\,060\ \text{MeV}/c^2 \tag{A.4}$$

3. For a given nuclide, its nuclear rest energy was determined by subtracting the rest energy of all atomic orbital electrons (Zm_ec^2) from the atomic rest energy $\mathcal{M}(\text{u})c^2$ as follows

$$Mc^2 = \mathcal{M}(\text{u})c^2 - Zm_ec^2 = \mathcal{M}(\text{u}) \times 931.494\,060\ \text{MeV/u} - Z \times 0.510\,999\ \text{MeV}. \tag{A.5}$$

Binding energy of orbital electrons to the nucleus is ignored in (A.5).
4. Nuclear binding energy E_B for a given nuclide was determined using the mass deficit equation given in (1.25) to get

$$E_B = Zm_pc^2 + (A - Z)m_nc^2 - Mc^2, \tag{A.6}$$

with Mc^2 given in (A.5) and the rest energy of the proton, neutron, and electron given in (A.1), (A.2), and (A.3), respectively.
5. For a given nuclide the binding energy per nucleon E_B/A is calculated by dividing the binding energy E_B of (A.6) with the number of nucleons equal to the atomic mass number A of a given nuclide.

© Springer International Publishing Switzerland 2016
E.B. Podgoršak, *Radiation Physics for Medical Physicists*,
Graduate Texts in Physics, DOI 10.1007/978-3-319-25382-4

Table A.1 Main attributes of nuclides presented in this book (in the entry for half-life: a = year, d = day, h = hour, min = minute, s = second)

Element and its nuclides		Z	A	Atomic mass M(u)	Nuclear rest energy Mc^2 (MeV)	Binding energy E_B (MeV)	$\dfrac{E_B \text{ (MeV)}}{\text{nucleon}}$	Half-life $t_{1/2}$
Hydrogen	H	1	1	1.007825	938.2720	–	–	Stable
Deuterium	D	1	2	2.014102	1875.6130	2.2244	1.1122	Stable
Tritium	T	1	3	3.016049	2808.9206	8.4821	2.8274	12.3 a
Helium	He	2	3	3.016029	2808.3910	7.7184	2.5728	Stable
	He	2	4	4.002603	3727.3788	28.2959	7.0740	Stable
	He	2	5	5.012220	4667.8310	27.4091	5.4818	8×10^{-22} s
	He	2	6	6.018889	5605.5372	29.2682	4.8780	0.801 s
Lithium	Li	3	5	5.012540	4667.6181	26.3287	5.2657	10^{-21} s
	Li	3	6	6.015123	5601.5182	31.9939	5.3323	Stable
	Li	3	7	7.016004	6533.8328	39.2446	5.6064	Stable
Beryllium	Be	4	7	7.016929	6534.1835	37.6006	5.3715	53 d
	Be	4	8	8.005305	7454.8498	56.4996	7.0625	8.19×10^{-17} s
	Be	4	9	9.012182	8392.7497	58.1651	6.4628	Stable
	Be	4	12	12.026921	11200.9611	68.6497	5.7208	21.49 ms
	Be	4	13	13.035691	12140.6243	68.5518	5.2732	2.7×10^{-21} s
Boron	B	5	9	9.013329	8393.3071	56.3143	6.2571	8.5×10^{-19} s
	B	5	10	10.012937	9324.4360	64.7508	6.4751	Stable
	B	5	11	11.009305	10252.5469	76.2053	6.9278	Stable
Carbon	C	6	11	11.011434	10254.0190	73.4398	6.6763	20.33 min
	C	6	12	12.000000	11174.8623	92.1618	7.6802	Stable
	C	6	13	13.003355	12109.4815	97.1080	7.4698	Stable
	C	6	14	14.003242	13040.8703	105.2845	7.5203	5730 a

(continued)

Table A.1 (continued)

Element and its nuclides		Z	A	Atomic mass M(u)	Nuclear rest energy Mc^2 (MeV)	Binding energy E_B (MeV)	$\dfrac{E_B \text{ (MeV)}}{\text{nucleon}}$	Half-life $t_{1/2}$
Nitrogen	N	7	13	13.005739	12111.1912	94.1050	7.2388	10 min
	N	7	14	14.003074	13040.2028	104.6587	7.4756	Stable
	N	7	15	15.000109	13968.9350	115.4919	7.6995	Stable
	N	7	17	17.008450	15839.6935	123.8646	7.2862	4.173 s
Oxygen	O	8	15	15.003066	13971.1784	111.9551	7.4637	122.24 s
	O	8	16	15.994915	14895.0798	127.6191	7.9762	Stable
	O	8	17	16.999132	15830.5023	131.7624	7.7507	Stable
	O	8	18	17.999160	16762.0221	139.8075	7.7671	Stable
Fluorine	F	9	18	18.000938	16763.1673	137.3690	7.6316	1.83 h
	F	9	19	18.998403	17692.3009	147.8013	7.7790	Stable
Neon	Ne	10	20	19.992440	18617.7285	160.6451	8.0323	Stable
Aluminum	Al	13	27	26.981539	25126.4995	224.9516	8.3315	Stable
Phosphorus	P	15	30	29.978314	27916.9555	250.6049	8.3535	2.498 min
Chromium	Cr	24	43	42.997710	40039.8461	330.4238	7.6843	21 ms
Manganese	Mn	25	44	44.006870	40979.3616	329.1803	7.4814	0.1 μs
	Mn	25	45	44.994513	41899.3452	348.7621	7.7503	70 ns
Iron	Fe	26	45	45.014560	41917.5078	329.3061	7.3179	0.35 μs
Cobalt	Co	27	57	56.936291	53022.0207	498.2859	8.7419	271.736 d
	Co	27	59	58.933200	54882.1269	517.3085	8.7679	Stable
	Co	27	60	59.933822	55814.2003	524.8005	8.7467	5.26 a
Nickel	Ni	28	60	59.930791	55810.8659	526.8415	8.7807	Stable
	Ni	28	63	62.929669	58604.3058	552.0998	8.7635	100.17 a

(continued)

Table A.1 (continued)

Element and its nuclides	Z	A	Atomic mass M (u)	Nuclear rest energy Mc^2 (MeV)	Binding energy E_B (MeV)	$\dfrac{E_B (\text{MeV})}{\text{nucleon}}$	Half-life $t_{1/2}$
Krypton	36	82	81.913484	76283.5252	714.2732	8.7106	Stable
	36	89	88.917632	82807.8494	766.9094	8.6170	3.15 min
	36	91	90.923451	84676.2538	777.6357	8.5454	8.57 min
	36	92	91.926156	85610.2731	783.1817	8.5129	1.84 s
	36	94	93.934363	87480.9176	791.6680	8.4220	210 ms
Rubidium	37	82	81.918209	76287.4155	709.0895	8.6474	1.2575 min
	37	90	89.914802	83736.1973	776.8335	8.6315	2.633 min
	37	96	95.934270	89343.2988	807.1242	8.4075	203 ms
Strontium	38	90	89.907738	83729.1065	782.6310	8.6959	28.808 a
	38	94	93.915361	87462.1839	807.8150	8.5938	1.255 min
Zirconium	40	96	95.908273	89317.5477	828.9953	8.6354	Stable
Molybdenum	42	95	94.905842	88382.7670	821.6240	8.6487	Stable
	42	98	97.905408	91176.8409	846.2430	8.6351	Stable
	42	99	98.907711	92110.4849	852.1677	8.6078	65.92 h
	42	100	99.907478	93041.7604	860.4575	8.6046	Stable
Technetium	43	99	98.906255	92108.6129	852.7430	8.6136	2.11×10^5 a
Cadmium	48	113	112.904402	105145.2523	963.5555	8.5270	Stable
	48	114	113.903359	106075.7748	972.5984	8.5316	Stable
Iodine	53	125	124.904210	116320.4427	1056.6789	8.4534	59.407 d
	53	127	126.904473	118183.6758	1072.5765	8.4455	Stable
	53	131	130.906125	121911.1907	1103.3230	8.4223	8.02 h

(continued)

Table A.1 (continued)

Element and its nuclides		Z	A	Atomic mass M (u)	Nuclear rest energy Mc^2 (MeV)	Binding energy E_B (MeV)	$\dfrac{E_B \, (\text{MeV})}{\text{nucleon}}$	Half-life $t_{1/2}$
Xenon	Xe	54	131	130.905082	121909.7082	1103.5122	8.4238	Stable
	Xe	54	140	139.921641	130308.7287	1160.7287	8.2909	13.6 s
Cesium	Cs	55	137	136.907084	127500.0262	1149.2929	8.3890	30.2 a
	Cs	55	138	137.911017	128435.1888	1153.7002	8.3602	33.42 min
	Cs	55	142	141.924299	132173.5374	1173.6131	8.2649	1.689 s
Barium	Ba	56	137	136.905821	127498.3387	1149.6870	8.3919	Stable
	Ba	56	139	138.908841	129364.1457	1163.0154	8.3670	1.3861 h
	Ba	56	141	140.914412	131232.3218	1173.9700	8.3260	18.27 min
	Ba	56	142	141.916454	132165.7185	1180.1387	8.3108	10.6 min
	Ba	56	144	143.922951	134034.7607	1190.2273	8.2655	11.5 s
Lanthanum	La	57	139	138.906353	129361.3169	1164.5508	8.3781	Stable
Samarium	Sm	62	152	151.919732	141480.6412	1253.1048	8.2441	Stable
Europium	Eu	63	151	150.919850	140548.7460	1244.1412	8.2393	$\geq 1.7 \times 10^{18}$ a
	Eu	63	152	151.921745	141482.0053	1250.4474	8.2266	13.54 a
	Eu	63	153	152.921230	142413.0196	1258.9984	8.2287	Stable
Gadolinium	Gd	64	152	151.919791	141479.6741	1251.4852	8.2335	1.08×10^{14} a
Osmium	Os	76	192	191.961479	178772.1354	1526.1178	7.9485	Stable
Iridium	Ir	77	191	190.960591	177839.3032	1518.0913	7.9481	Stable
	Ir	77	192	191.962602	178772.6704	1524.2894	7.9390	73.8 d
	Ir	77	193	192.962924	179704.4644	1532.0607	7.9381	Stable
Platinum	Pt	78	192	191.961035	178770.6998	1524.9667	7.9425	Stable
Gold	Au	79	197	196.966552	183432.7980	1559.4019	7.9157	Stable
	Au	79	198	197.968242	184365.8743	1565.8975	7.9090	2.695 d

(continued)

Table A.1 (continued)

Element and its nuclides		Z	A	Atomic mass M(u)	Nuclear rest energy Mc^2(MeV)	Binding energy E_B (MeV)	$\dfrac{E_B(\text{MeV})}{\text{nucleon}}$	Half-life $t_{1/2}$
Thallium	Tl	81	208	207.982019	193692.6254	1632.2134	7.8472	3.05 min
Lead	Pb	82	205	204.974482	190890.6040	1614.2387	7.8743	1.73×10^7 a
	Pb	82	206	205.974465	191822.0821	1622.3258	7.8754	Stable
	Pb	82	207	206.975881	192754.8952	1629.0781	7.8699	Stable
	Pb	82	208	207.976636	193687.0925	1636.4462	7.8675	Stable
	Pb	82	212	207.976652	197413.9072	1667.8998	7.8675	Stable
Bismuth	Bi	83	209	208.980383	194621.5658	1640.2449	7.8481	Stable
	Bl	83	212	211.991286	197426.2120	1654.3017	7.8033	1.01 h
Polonium	Po	84	210	209.982873	195554.8683	1645.2144	7.8344	15.6 min
	Po	84	211	210.986653	196489.8918	1649.7632	7.8188	0.516 s
	Po	84	216	216.001915	201161.5783	1675.9036	7.7588	0.145 s
	Po	84	218	218.008973	203031.1325	1685.4730	7.7315	3.167 min
Radon	Rn	86	220	220.011394	204895.3622	1697.7946	7.7172	0.9267 min
	Rn	86	222	222.017578	206764.1021	1708.1781	7.6945	3.8 d
Radium	Ra	88	224	224.020212	208628.5302	1720.3014	7.6799	3.657 d
	Ra	88	226	226.025403	210496.3452	1731.6097	7.6620	1602 a
	Ra	88	228	228.031070	212364.6211	1742.4720	7.6424	5.739 a
Actinium	Ac	89	228	228.031021	212364.0642	1741.7356	7.6392	6.139 h
Thorium	Th	90	228	228.028741	212361.4305	1743.0760	7.6451	1.91286 a
	Th	90	232	232.038050	216096.0679	1766.6924	7.6151	1.4×10^{16} a

(continued)

Table A.1 (continued)

Element and its nuclides	Z	A	Atomic mass $M(u)$	Nuclear rest energy Mc^2 (MeV)	Binding energy E_B (MeV)	$\dfrac{E_B \text{ (MeV)}}{\text{nucleon}}$	Half-life $t_{1/2}$
Uranium	U	92	233	233.039628	217028.0099	1771.7291	7.6040	1.6×10^6 a
	U	92	235	235.043923	218894.9987	1783.8710	7.5909	0.7×10^9 a
	U	92	236	236.045568	219828.0342	1790.4086	7.5865	2.3×10^7 a
	U	92	238	238.050783	221695.8708	1801.6949	7.5701	4.5×10^9 a
	U	92	239	239.054288	222630.6297	1806.5013	7.5586	23.5 min
	U	92	240	240.056592	223564.2793	1812.4250	7.5518	14.11 h
Neptunium	Np	93	239	239.052913	222628.8379	1806.9998	7.5607	2.35 d
Plutonium	Pu	94	239	239.052157	222627.6227	1806.9217	7.5603	24×10^3 a
	Pu	94	240	240.053814	223560.6690	1813.4486	7.5560	6567.1 a
	Pu	94	244	244.064204	227296.3238	1836.0554	7.5248	7.93×10^7 a
Curium	Cm	96	248	248.072349	231028.8557	1859.1901	7.4967	0.35×10^6 a
Californium	Cf	98	252	252.081626	234762.4513	1881.2693	7.4654	2.65 a
	Cf	98	256	256.093440	238499.4321	1902.5499	7.4318	12.3 min
Fermium	Fm	100	256	256.091767	238496.8517	1902.5436	7.4318	158 min

Appendix B
Basic Characteristics of the Main Radioactive Decay Modes

The following decay modes are presented: α decay, β^- decay, β^+ decay, electron capture, γ decay, internal conversion, proton emission decay, and neutron emission decay. For each decay mode the table gives the basic relationship, the decay energy Q, and the kinetic energy E_K of the decay products. P stands for the parent nucleus or atom; D for the daughter nucleus or atom. M represents the nuclear rest mass, $\mathcal{M}$ the atomic rest mass, m_e the electron rest mass, m_p the proton rest mass, and m_n the neutron rest mass.

© Springer International Publishing Switzerland 2016
E.B. Podgoršak, *Radiation Physics for Medical Physicists*,
Graduate Texts in Physics, DOI 10.1007/978-3-319-25382-4

Table B.1 Alpha (α) Decay

Basic relationship: (see (11.2))	
$$^A_Z\text{P} \to {}^{A-4}_{Z-2}\text{D} + \alpha + Q_\alpha$$	(B.1)
Decay energy: (see (11.3) and (11.4))	
$$\begin{aligned} Q_\alpha &= \{M(\text{P}) - [M(\text{D}) + m_\alpha]\}\,c^2 \\ &= \left\{\mathcal{M}(\text{P}) - [\mathcal{M}(\text{D}) + \mathcal{M}({}^4_2\text{He})]\right\}c^2 \\ &= E_\text{B}(\text{D}) + E_\text{B}(\alpha) - E_\text{B}(\text{P}) = (E_\text{K})_\alpha + (E_\text{K})_\text{D} \end{aligned}$$	(B.2)
Kinetic energy of α-particle: (see (11.7))	
$$(E_\text{K})_\alpha = \frac{Q_\alpha}{1 + \dfrac{m_\alpha}{M(\text{D})}} \approx \frac{A_\text{P} - 4}{A_\text{P}}Q_\alpha$$	(B.3)
Daughter recoil kinetic energy: (see (11.8))	
$$(E_\text{K})_\text{D} = \frac{Q_\alpha}{1 + \dfrac{M(\text{D})}{m_\alpha}} \approx \frac{4}{A_\text{P}}Q_\alpha$$	(B.4)

Table B.2 Beta minus (β^-) Decay

Basic relationship: (see (11.15))	
$$^A_Z\text{P} \to {}^A_{Z+1}\text{D} + e^- + \bar{\nu}_e + Q_{\beta^-}$$	(B.5)
Decay energy: (see (11.19), (11.25), and (11.26))	
$$\begin{aligned} Q_{\beta^-} &= \{M(\text{P}) - [M(\text{D}) + m_e]\}\,c^2 \\ &= \{\mathcal{M}(\text{P}) - \mathcal{M}(\text{D})\}\,c^2 \\ &= (E_{\beta^-})_\text{max} + (E_\text{K})_\text{Dmax} = (E_\beta)_\text{max}\left\{1 + \frac{m_e c^2 + \dfrac{1}{2}(E_\beta)_\text{max}}{M(\text{D})c^2}\right\} \end{aligned}$$	(B.6)
Daughter maximum recoil kinetic energy: (see (11.23))	
$$(E_\text{K})_\text{Dmax} = \frac{m_e}{M(\text{D})}(E_{\beta^-})_\text{max}\left\{1 + \frac{(E_{\beta^-})_\text{max}}{2M(\text{D})c^2}\right\}$$	(B.7)
Combined energy given to electron/antineutrino	
$$(E_{\beta^-})_\text{max} = Q_{\beta^-} - (E_\text{K})_\text{Dmax} \approx Q_{\beta^-}$$	(B.8)

Table B.3 Beta plus (β^+) Decay

Basic relationship: (see (11.16)	
$$^A_Z P \rightarrow {}^A_{Z-1} D + e^+ + \nu_e + Q_{\beta^+}$$	(B.9)
Decay energy: (see (11.19), (11.33), and (11.34))	
$$\begin{aligned} Q_\beta^+ &= \{M(P) - [M(D) + m_e]\}\, c^2 \\ &= \{\mathcal{M}(P) - \mathcal{M}(D) + 2m_e\}\, c^2 \\ &= (E_{\beta^+})_{max} + (E_K)_{Dmax} = (E_\beta)_{max}\left\{1 + \dfrac{m_e c^2 + \dfrac{1}{2}(E_\beta)_{max}}{M(D)c^2}\right\} \end{aligned}$$	(B.10)
Daughter maximum recoil kinetic energy: (see (11.24))	
$$(E_K)_{Dmax} = \frac{m_e}{M(D)}(E_{\beta^+})_{max}\left\{1 + \frac{(E_{\beta^+})_{max}}{2mM(D)c^2}\right\}$$	(B.11)
Combined energy given to positron/neutrino	
$$(E_{\beta^+})_{max} = Q_{\beta^+} - (E_K)_{Dmax} \approx Q_{\beta^+}$$	(B.12)

Table B.4 Electron Capture (EC)

Basic relationship: (see (11.17))	
$$^A_Z P + e^- = {}^A_{Z-1} D + \nu_e + Q_{EC}$$	(B.13)
Decay energy: (see (11.42) and (11.43))	
$$\begin{aligned} Q_{EC} &= \{[M(P) + m_e] - M(D)\}\, c^2 \\ &= \{M(P) - [M(D) - m_e]\}\, c^2 \\ &= \{\mathcal{M}(P) - \mathcal{M}(D)\}\, c^2 = (E_K)_D + E_{\nu_e} \end{aligned}$$	(B.14)
Daughter recoil kinetic energy: (see (11.45))	
$$(E_K)_D = \frac{E_\nu^2}{2M(D)c^2} \approx \frac{Q_{EC}}{2M(D)c^2}$$	(B.15)
Energy given to neutrino: (see (11.48))	
$$E_\nu = \left\{-1 + \sqrt{1 + \frac{2(Q_{EC} - E_B)}{M(D)c^2}}\right\} M(D)c^2 \approx Q_{EC} - E_B$$	(B.16)

Table B.5 Gamma (γ) Decay

Basic relationship: (see (11.52))	
$$^A_Z P^* \rightarrow\, ^A_Z P + \gamma + Q_\gamma$$	(B.17)
Decay energy: (see (11.53))	
$$Q_\gamma = E^* - E = E_\gamma + (E_K)_D \\ = E_\gamma \left\{ 1 + \frac{E_\gamma}{2M(D)c^2} \right\}$$	(B.18)
Daughter recoil kinetic energy: (see (11.54))	
$$(E_K)_D = \frac{E_\gamma^2}{2M(D)c^2}$$	(B.19)
Energy of gamma photon: (see (11.55))	
$$E_\gamma = Q_\gamma - (E_K)_D = Q_\gamma \left\{ 1 - \frac{E_\gamma}{2M(D)c^2} \right\} \approx Q_\gamma$$	(B.20)

Table B.6 Internal Conversion (IC)

Basic relationship: (see (11.56))	
$$^A_Z P^* \rightarrow\, ^A_Z P^+ + e^- + Q_{IC}$$	(B.21)
Decay energy: (see (11.57))	
$$Q_{IC} = (E^* - E) - E_B = (E_K)_{IC} + (E_K)_D$$	(B.22)
Daughter recoil kinetic energy: (see (11.58))	
$$(E_K)_D = \frac{m_e}{M(D)}(E_K)_{IC} + \frac{(E_K)_{IC}^2}{2M(D)c^2} \\ = \frac{(E_K)_{IC}}{M(D)c^2} \left\{ m_e c^2 + \frac{1}{2}(E_K)_{IC} \right\}$$	(B.23)
Kinetic energy of internal conversion electron	
$$(E_K)_{IC} = Q_{IC} - (E_K)_D \approx Q_{IC}$$	(B.24)

Table B.7 Proton emission decay

Basic relationship: (see (11.63))	
$$^A_Z P \rightarrow {}^{A-1}_{Z-1} D + p + Q_p$$	(B.25)
Decay energy: (see (11.64) and (11.65))	
$$\begin{aligned} Q_P &= \{M(P) - [M(D) + m_p]\}\, c^2 \\ &= \{\mathcal{M}(P) - [\mathcal{M}(D) + \mathcal{M}(^1_1 H)]\}\, c^2 \\ &= E_B(D) - E_B(P) = (E_K)_p + (E_K)_D \end{aligned}$$	(B.26)
Kinetic energy of the emitted proton: (see (11.67))	
$$(E_K)_p = \frac{Q_p}{1 + \dfrac{m_p}{M(D)}}$$	(B.27)
Daughter recoil kinetic energy: (see (11.68))	
$$(E_K)_D = \frac{Q_p}{1 + \dfrac{M(D)}{m_p}}$$	(B.28)

Table B.8 Neutron emission decay

Basic relationship: (see (11.85))	
$$^A_Z P \rightarrow {}^{A-1}_{Z} D + n + Q_n$$	(B.29)
Decay energy: (see (11.86) and (11.87))	
$$\begin{aligned} Q_n &= \{M(P) - [M(D) + m_n]\}\, c^2 \\ &= \{\mathcal{M}(P) - [\mathcal{M}(D) + m_n]\}\, c^2 \\ &= E_B(D) - E_B(P) = (E_K)_n + (E_K)_D \end{aligned}$$	(B.30)
Kinetic energy of the emitted neutron: (see (11.89))	
$$(E_K)_n = \frac{Q_n}{1 + \dfrac{m_n}{M(D)}}$$	(B.31)
Daughter recoil kinetic energy: (see (11.90))	
$$(E_K)_D = \frac{Q_n}{1 + \dfrac{M(D)}{m_n}}$$	(B.32)

Appendix C
Short Biographies of Scientists Whose Work is Discussed in This Book

The biographical data were obtained mainly from two sources:

1. Book by William H. Cropper: *"Great Physicists: The Life and Times of Leading Physicists from Galileo to Hawking"* published by Oxford University Press in 2001.
2. The website: www.Nobelprize.org that contains biographies and Nobel lectures of all Nobel Prize winners in Physics, Chemistry, Physiology or Medicine, Literature, Peace, and Economic Sciences from 1901 to date.

ANDERSON, Carl David (1905–1991)

American physicist, educated at the California Institute of Technology (Caltech) in Pasadena (B.Sc. in engineering physics in 1927; Ph.D. in engineering physics in 1930). He spent his entire professional career at Caltech, becoming Professor of Physics in 1939, Chairman of the Physics, Mathematics and Astronomy division (1962–1970), and Professor Emeritus in 1976.

Early in his career Anderson concentrated on studies of x-rays, later on on studies of cosmic rays with cloud chambers that lead to the discovery of the positron in 1932. Positron was the first known particle in the category of antimatter. Paul A.M. Dirac enunciated its existence in 1928 with his relativistic quantum theory for the motion of electrons in electric and magnetic fields. Dirac's theory incorporated Albert Einstein's special theory of relativity and predicted the existence of an antiparticle to the electron (same mass, opposite charge). In 1933 Anderson succeeded in producing positrons by gamma radiation through the effect of pair production. In 1936 Anderson, in collaboration with his graduate student Seth Neddermeyer, discovered, again while studying cosmic radiation, the muon (μ meson), the first known elementary particle that is not a basic building block of matter.

In 1936 Anderson shared the Nobel Prize in Physics with Victor Franz Hess, an Austrian physicist. Anderson received the Prize *"for his discovery of the positron"* and Hess *"for his discovery of cosmic radiation"*.

© Springer International Publishing Switzerland 2016
E.B. Podgoršak, *Radiation Physics for Medical Physicists*,
Graduate Texts in Physics, DOI 10.1007/978-3-319-25382-4

AUGER, Pierre Victor (1899–1993)

French physicist who was active as a basic scientist in atomic, nuclear and cosmic ray physics but also made important contributions to French and international scientific organizations. The world's largest cosmic ray detector, the Pierre Auger observatory, is named after him. Auger is also credited with the discovery in 1925 of radiation*less* electronic transitions in atoms that are followed by emission of orbital electrons. The process is named after him as the Auger effect and the emitted electrons are called Auger electrons. Lise Meitner actually discovered the radiation *less* atomic transition process in 1923, two years before Auger; nonetheless, the process is referred to as the Auger effect.

AVOGADRO, Amedeo (1776–1856)

Italian lawyer, chemist, physicist, best known for the "Avogadro principle" and "Avogadro number". The Avogadro's principle states that "equal volumes of all gases at the same temperature and pressure contain the same number of molecules". The concepts of gram–atom and gram-mole were introduced long after Avogadro's time; however, Avogadro is credited with introducing the distinction between the molecule and the atom. The number of atoms per gram–atom and number of molecules per gram–mole is constant for all atomic and molecular entities and referred to as Avogadro constant (number) $\left(N_A = 6.022 \times 10^{23} \text{ atom/mol}\right)$ in honor of Avogadro's contributions to chemistry and physics.

BALMER, Johann Jakob (1825–1898)

Swiss mathematician who studied in Germany at the University of Karlsruhe and the University of Berlin before receiving a doctorate at the University of Basel. He then spent his professional life teaching mathematics at the University of Basel.

Balmer is best known for his work on spectral lines emitted by the hydrogen gas. In 1885 he published a formula that predicted the wavelengths of the lines in the visible part of the hydrogen spectrum. The formula predicted the lines very accurately but was empirical rather than based on any physical principles. Several other scientists subsequently proposed similar empirical formulas for hydrogen lines emitted in other portions of the photon spectrum (Lymann in the ultraviolet and Paschen, Brackett and Pfund in the infrared). In 1913 Niels Bohr derived from first principles the general relationship for spectral lines of hydrogen. The relationship is governed by n, the principal quantum number, and contains a constant that is now referred to as the Rydberg constant $\left(R_\infty = 109\,737 \text{ cm}^{-1}\right)$. The spectral line series for $n = 1$ is called the Lymann series; for $n = 2$ the Balmer series; for $n = 3$ the Paschen series; for $n = 4$ the Brackett series; for $n = 5$ the Pfund series, and for $n = 6$ the Humphreys series.

BARKLA, Charles Glover (1877–1944)

British physicist, educated in mathematics and physics at the University College in Liverpool from where he graduated in 1898. He worked as research assistant with Joseph J. Thomson in the Cavendish Laboratory in Cambridge and as academic physicist at the University of London. In 1913 he was appointed Chair of Natural Philosophy at the University of Edinburgh and held the position until his death in 1944.

Barklas's most important research involved studies of the production of x-rays and of their interactions with matter. He is credited with the discovery of characteristic (fluorescence) radiation and the polarization of x-rays.

In 1917 Barkla was awarded the Nobel Prize in Physics *"for his discovery of the characteristic Röntgen radiation of the elements"*.

BECQUEREL, Henri Antoine (1852–1908)

French physicist, educated at the École Polytechnique in basic science and at the École des Ponts et Chaussées becoming an *ingénieur* in 1877. In 1888 he acquired the degree of docteur-ès-sciences. In 1895 he became Professor of Physics at the École Polytechnique in Paris, the foremost French "grande école" of engineering, founded in 1794.

Becquerel was active in many areas of physics investigating polarization of visible light, naturally occurring phosphorescence in uranium salts, and terrestrial magnetism. In 1896, shortly after Wilhelm Röntgen's discovery of x-rays, Becquerel accidentally discovered natural radioactivity while investigating phosphorescence in uranium salts upon exposure to light. He observed that when the salts were placed near a photographic plate covered with opaque paper, the developed plate was nonetheless fogged. Becquerel concluded that the uranium salts were emitting penetrating rays that were emanating from uranium atoms. He subsequently showed that the rays were causing ionization of gases and that, in contrast to Röntgen's x-rays, they were deflected by electric and magnetic fields.

In 1903 Bq shared the Nobel Prize in Physics with Pierre Curie and Marie Skłodowska-Curie. He was awarded the prize *"in recognition of the extraordinary services he has rendered by his discovery of spontaneous radioactivity"* and the Curies received their prize *"in recognition of the extraordinary services they have rendered by their joint researches on the radiation phenomena discovered by Professor Henri Becquerel"*.

Becquerel and his work are honored by the SI unit of radioactivity named Becquerel (Bq). In addition, there are Becquerel craters on the moon and Mars.

BERGER, Martin Jacob (1922–2004)

Austrian-born American physicist, educated at the University of Chicago where he received his degrees in Physics: B.Sc. in 1943, M.Sc. in 1948, and doctorate in 1951. In 1952 Berger joined the Radiation Theory Section at the National Bureau of Standards (NBS), now National Institute of Science and Technology (NIST) in Washington D.C. In 1964 he became the Section Chief and later, as well, Director of the Photon and Charged-Particle Data Center at the NBS/NIST, a position he held until his retirement in 1988.

Berger is best known for his early work on the transport of gamma rays and applications of Monte Carlo calculations in complex media involving boundaries and inhomogeneities. He also worked on charged-particle transport with emphasis on electrons and protons, and developed algorithms for use in charged particle Monte Carlo codes. His ETRAN code, first published in the 1960s, became the industry standard for coupled electron-photon transport. Berger, in collaboration with Stephen Seltzer, also developed cross-section data for electron and heavy charged particle interactions as well as for electron bremsstrahlung production. He was also involved in applications of Monte Carlo calculations to important problems in radiological physics and radiation dosimetry.

BETHE, Hans Albrecht (1906–2005)

German-born American physicist, educated at the Universities of Frankfurt and Munich. He received his doctorate in theoretical physics under Arnold Sommerfeld in 1928. For four years he worked as Assistant Professor at the University of Munich, then spent a year in Cambridge and a year in Rome with Enrico Fermi. He returned to Germany as Assistant Professor at the University of Tübingen but lost the position during the rise of Nazism. He first emigrated to England and then in 1935 moved to Cornell University in Ithaca, New York as Professor of Physics. He stayed at Cornell essentially all his professional life, but also served as Director of Theoretical Physics on the Manhattan project at Los Alamos (1943–1946).

Bethe made important theoretical contributions to radiation physics, nuclear physics, quantum mechanics, and quantum electrodynamics. He was also a strong advocate for peaceful use of atomic energy, despite having been involved with the Manhattan project as well as with the development of the hydrogen bomb. In collision theory Bethe derived the stopping power relationships that govern inelastic collisions of fast particles with atoms. With Heitler, he developed the collision theory for relativistic electrons interacting with atomic nuclei and producing bremsstrahlung radiation in the process. Bethe's work in nuclear physics lead to the discovery of the reactions that govern the energy production in stars.

In 1967 Bethe was awarded the Nobel Prize in Physics *"for his theory of nuclear reactions, especially his discoveries concerning the energy production in stars"*.

BHABHA, Homi Jehandir (1909–1966)

Indian nuclear physicist, educated in Mumbai (Bombay) and Cambridge (U.K.) where he first studied engineering and later-on physics and received his Ph.D. in physics in 1935 studying cosmic rays. He was already well respected in the international physics community when he returned to India in 1939. He took a post in theoretical physics at the Indian Institute of Science in Bangalore under C.V. Raman and carried out experimental work in cosmic radiation and theoretical work in mathematics. In 1945 he became director of the newly established Tata Institute of Fundamental Research (TIFR) in Mumbai and remained in the position until 1966 when he died in an airplane crash.

Under Bhabha's leadership TIFR became a leading nuclear science institute committed to peaceful use of nuclear energy. He was very influential in Indian nuclear policy and developed a close personal relationship with India's first Prime Minister Jawaharlal P. Nehru. He was instrumental in getting the Indian Constituent

Assembly to pass the Indian Atomic Energy Act and creating the Indian Atomic Energy Commission.

Bhabha's important contributions to nuclear physics are recognized by the term Bhabha scattering which defines positron scattering on electrons. He was also elected Fellow of the Royal Society and his contribution to Indian nuclear science was recognized in 1967 by renaming the TIFR into Bhabha Atomic Research Centre (BARC).

BLOCH, Felix (1905–1983)

Swiss-born American physicist, educated at the Eidgenössische Technische Hoch - schule in Zürich (ETHZ) and at the University of Leipzig where he received his doctorate in physics in 1928. During the next few years he held various assistantships and fellowships that gave him the opportunity to work with the giants of modern physics (Pauli, Heisenberg, Bohr, and Fermi) and to further his understanding of solid state physics in general and stopping powers of charged particles in particular. In 1933 Bloch left Germany and in 1934 accepted a position at Stanford University where he got involved with experimental physics of neutron momenta and polarized neutron beams. During the war years he worked on the Manhattan project at Los Alamos and on radar technology at Harvard where he became familiar with modern techniques of electronics. This helped him upon return to Stanford in 1945 with development of new techniques for measuring nuclear moments that culminated in 1946 with the invention of the nuclear magnetic resonance (NMR) technique, a purely electromagnetic procedure for the study of nuclear moments in solids, liquids, and gases. At Harvard Edward M. Purcell with students Robert Pound and Henry C. Torrey invented the NMR technique independently and at about the same time as Bloch.

In 1952 Bloch and Purcell received the Nobel Prize in Physics *"for their development of new methods for nuclear magnetic precision measurements and discoveries in connection therewith"*. Since the late 1970s NMR provided the basis for magnetic resonance imaging (MRI), which is widely used as a non-invasive diagnostic imaging technique.

BOAG, John (Jack) Wilson (1911–2007)

Scottish born medical physicist who spent most of his professional life in London, England. He grew up in Glasgow and received his undergraduate degree in electrical engineering from the University of Glasgow. He then received a scholarship to St. John's College in Cambridge that enabled him to work in the Cavendish laboratory. In 1934 he began his Ph.D. studies in Braunschweig in Germany but the start of World War II forced him to return to U.K. before completion of his studies.

In 1942 Boag joined the new radiotherapy department at Hammersmith Hospital in London where he designed and built a 2 MeV van de Graaff. In 1955 he moved to the radiotherapy department in St. Bartolomew's Hospital in London and in 1958 to the radiobiology unit at Mount Vernon Hospital in Middlesex where he got involved in radiation dosimetry of clinical kilovoltage and megavoltage radiation beams. During the Mount Vernon period of his career he designed and built many types of ionization chambers, derived recombination theory for both continuous as well as pulsed ionizing radiation beams, developed the pulsed radiolysis technique for radiobiological studies, and discovered the solvated electron, an essential product of water radiolysis. In 1964 Boag became head of Medical Physics department at the Royal Marsden Hospital in London and retired from this position in 1976.

Boag received the Barclay Medal from the British Institute of Radiology (BIR) in 1974, the Gray Medal from the International Commission on Radiation Units and Measurements (ICRU) in 1975, and an honorary Doctorate from the University of Glasgow in 1954.

After retirement Boag dedicated his time to promote peaceful uses of scientific research and to reduce the risk of nuclear conflict during the cold war era. During 1980s he served as the British secretary of the Pugwash movement that grew out of an international conference of scholars organized by Sir Bertrand Russell. The conference was held in 1957 in the small village of Pugwash, Nova Scotia in Canada and hosted by Pugwash's native son, steel magnate Cyrus Eaton.

The first Pugwash conference brought together scientists from both sides of the cold war divide to state their opposition to nuclear weapons of mass destruction and to express their support for the Russell–Einstein Manifesto that the two eminent scientists issued in 1955. The Pugwash group remained active after the first Pugwash conference and continued to organize conferences on the same theme on an annual basis at various locations around the world, attended by several hundred scientists who espouse the mission of the Pugwash movement. The conferences are now known as the "Pugwash Conferences on Science and World Affairs" and Boag remained involved with them until his death in 2007.

BOHR, Niels Henrik David (1885–1962)

Danish physicist, educated at the University of Copenhagen where he obtained his M.Sc. degree in physics in 1909 and doctorate in physics in 1911. Between 1911 and 1916 Bohr held various academic appointments in the U.K. and Copenhagen. In 1911 he worked in Cambridge with Joseph J. Thomson and in 1912 he worked in Manchester with Ernest Rutherford. He was a lecturer in physics at the University of Copenhagen in 1913 and at the University of Manchester between 1914 and 1916. In 1916 he was appointed Professor of Theoretical Physics and in 1920 he became the first Director of the Institute of Theoretical Physics (Niels Bohr Institute) at the University of Copenhagen. He remained in both positions until his death in 1962.

Bohr was an exceptionally gifted theoretical physicist who made important contributions to atomic, nuclear, and quantum physics. He is best known for his expansion in 1913 of the Rutherford's atomic model into the realm of Planck's quantum physics to arrive at a model that is now called the Rutherford-Bohr atomic model. With four postulates that merged simple classical physics concepts with the idea of quantization of angular momenta for electrons revolving in allowed orbits about the nucleus, he succeeded in explaining the dynamics of one-electron structures and in predicting the wavelengths of the emitted radiation.

Bohr is also known as the author of the principle of complementarity which states that a complete description of an atomic scale phenomenon requires an evaluation from both the wave and particle perspective. In 1938 he proposed the so-called liquid drop nuclear model and in 1939 he succeeded in explaining the neutron fission of natural uranium in terms of fissionable uranium-235 (a uranium isotope with an abundance of only 0.7 % in natural uranium) and the much more abundant non-fissionable uranium-238.

During World War II Bohr worked on the Manhattan project in Los Alamos but his contribution to the development of atomic weapons was only minor. After the war he used his considerable credibility and influence to promote peaceful use of the atomic energy and in 1954 helped found the CERN (Centre Européen de Recherche Nucléaire) in Geneva, touted as the world's largest particle physics laboratory and recognized as the birthplace of the *worldwide web*. In addition to producing his theoretical masterworks, Bohr was also keenly interested in politics and advised Presidents Roosevelt and Truman as well as Prime Minister Churchill on nuclear matters. Only Albert Einstein and Marie Curie among scientists of the 20th century have attained such esteem from physics colleagues, world leaders, and the general public.

In tribute to Bohr's contributions to modern physics the element with atomic number 107 is named bohrium (Bh). Bohr received the 1922 Nobel Prize in Physics *"for his services in the investigation of the structure of atoms and of the radiation emanating from them"*.

BORN Max (1882–1970)

German mathematician and physicist, educated at universities of Breslau (1901), Heidelberg (1902), Zürich (1903), and Göttingen where he received his doctorate in 1907. In 1909 he was appointed lecturer at the University of Göttingen and in 1912 he moved to the University of Chicago. In 1919 he became Professor of Physics at the University of Frankfurt and then in 1921 Professor of Physics at the University of Göttingen. From 1933 until 1936 he lectured at the University of Cambridge and from 1936 until 1953 at the University of Edinburgh.

Born is best known for his work on relativity in general and the relativistic electron in particular. He was also working on crystal lattices and on quantum theory, in particular on the statistical interpretation of quantum mechanics. He is best known for his formulation of the now-standard interpretation of the probability density for $\psi^*\psi$ in the Schrödinger equation of wave mechanics.

In 1954 Born shared the Nobel Prize in Physics with Walther Bothe. Born received his half of the prize *"for his fundamental research in quantum mechanics, especially for his statistical interpretation of the wavefunction"* and Bothe *"for the coincidence method and his discoveries made herewith"*.

BRAGG, William Henry (1862–1942)

British physicist, educated at King William College on Isle of Man and at the Trinity College at Cambridge where he graduated in 1884. His first academic appointment was at the University of Adelaide in Australia from 1885 until 1909. In 1909 he returned to England and worked as Professor of Physics at the University of Leeds from 1909 until 1915 and at the University College in London from 1915 until 1923. From 1923 until 1942 he was Director of the Royal Institution in London.

Henry Bragg is best known for the work he carried out in collaboration with his son Lawrence on the diffraction of x-rays on crystalline structures. Von Laue discovered the diffraction of x-rays on crystals; however, it was the father-son Bragg team that developed the discipline of x-ray crystallography based on the Bragg crystal spectrometer, a very important practical tool in solid state physics and analytical chemistry.

The 1915 Nobel Prize in Physics was awarded to William Henry Bragg and his son William Lawrence Bragg *"for their services in the analysis of crystal structure by means of x-rays"*.

BRAGG, William Lawrence (1890–1971)

Australian-born British physicist, educated at Adelaide University where he graduated at age 18 with an honors B.A. degree in mathematics. He then entered Trinity College in Cambridge, continued his studies in mathematics but switched to physics the second year and graduated in physics in 1912. He first worked as lecturer at the Cavendish Laboratory in Cambridge but from 1915 spent three years in the army. He became Langworthy Professor of Physics at the University of Manchester in 1919. During 1938 he was Director of National Physical Laboratory in Teddington and then worked in Cambridge as the Cavendish Professor of Experimental Physics from 1939 until 1954 and as Director of the Royal Institution from 1954 until 1966.

In 1912 William L. Bragg became interested in the great debate on the nature of x-rays: were they waves or particles? Following the experiments of von Laue and colleagues he developed an ingenious way of treating the phenomenon of x-ray diffraction on crystalline structures. He pointed out that the regular arrangement of atoms in a crystal defines a large variety of planes on which the atoms effectively lie. This means that the atoms in a regular lattice simply behave as if they form reflecting planes. The well-known Bragg equation is then expressed as $2d \sin \phi = n\lambda$, with d the separation between two atomic planes, ϕ the angle of incidence of the x-ray beam, λ the x-ray wavelength, and n an integer. The basis of a Bragg spectrometer is then as follows: For a known d, an x-ray spectrum can be analyzed by varying ϕ and observing the intensity of the reflected x-rays that are scattered through and angle $\theta = 2\phi$ from the direction of the incident collimated beam. On the other hand, if mono-energetic x-rays with a known λ are used, it is possible to determine various effective values of d in a given crystal and hence the basic atomic spacing a. With the knowledge of a one may determine the Avogadro constant N_A with great accuracy.

The 1915 Nobel Prize in Physics was awarded to William Lawrence Bragg and his father William Henry Bragg *"for their services in the analysis of crystal structure by means of x-rays"*.

CHADWICK, James (1891–1974)

British physicist, educated at Manchester University (B.Sc. in 1911 and M.Sc. in 1913) before continuing his studies in the Physikalisch Technische Reichanstalt at Charlottenburg. In 1919 he moved to Cambridge to work with Ernest Rutherford on nuclear physics research. He remained in Cambridge until 1935 when he became the Chairman of Physics at the University of Liverpool. From 1943 to 1946 he was the Head of the British Mission attached to the Manhattan project.

Chadwick is best known for his 1932 discovery of the neutron, a constituent of the atomic nucleus that in contrast to the proton is devoid of any electrical charge. In recognition of this fundamental discovery that paved the way toward the discovery of nuclear fission, Chadwick was awarded the 1935 Nobel Prize in Physics *"for the discovery of the neutron"*.

COMPTON, Arthur Holly (1892–1962)

American physicist, educated at College of Wooster (B.Sc. in 1913) and Princeton University (M.A. in 1914 and Ph.D. in 1916). He worked as physics instructor at the University of Minnesota, research engineer at Westinghouse in Pittsburgh, and research fellow at Cambridge University. Upon return to the U.S. in 1920 he worked as Chairman of the Physics department at the Washington University in St. Louis and in 1923 he moved to the University of Chicago as Professor of Physics.

Compton is best known for his experimental and theoretical studies of x-ray scattering on atoms that lead to his discovery, in 1922, of the increase in wavelength of x-rays scattered on essentially free atomic electrons. This effect illustrates the corpuscular nature of photons and is now known as the Compton effect. As Chairman of the National Academy of Sciences Committee to Evaluate Use of Atomic Energy in War, Compton was instrumental in developing the first controlled uranium fission reactors and plutonium-producing reactors.

In 1927 Compton was awarded the Nobel Prize in Physics *"for the discovery of the effect that bears his name"*. The co-recipient of the 1927 Nobel Prize was C.T.R. Wilson for his discovery of the cloud chamber.

COOLIDGE, William David (1873–1975)

American physicist and inventor, educated at the Massachusetts Institute of Technology (MIT) in Boston (B.Sc. in electrical engineering in 1896) and the University of Leipzig (doctorate in physics in 1899). In 1899 he returned for five years to Boston as a research assistant in the Chemistry department of the MIT. In 1905 Coolidge joined the General Electric (GE) Company in Schenectady, and remained with the company until his retirement in 1945. He served as director of the GE Research Laboratory (1932–1940) and as vice president and director of research (1940–1944).

During his 40-year career at General Electric, Coolidge became known as a prolific inventor and was awarded 83 patents. He is best known for his invention of ductile tungsten in the early years of his career. He introduced ductile tungsten for use as filament in incandescent lamps in 1911 producing a significant improvement over Edison's design for incandescent lamps. In 1913 he introduced ductile tungsten into x-ray tubes and revolutionized x-ray tube design that at the time was based on three major components: cold cathode, low pressure gas, and anode (target). The role of the low pressure gas was to produce ions which produced electrons upon bombardment of the cold aluminum cathode. This x-ray tube design was based on the Crookes device for studying cathode rays, and is now referred to as the Crookes tube. The performance of the Crookes x-ray tube was quite erratic and Coolidge introduced a significant improvement when he replaced the cold aluminum cathode with a hot tungsten filament and replaced the low pressure gas with high vacuum. Coolidge's x-ray tube design is now referred to as the Coolidge tube and is still used today for production of superficial and orthovoltage x-rays. In the Coolidge x-ray tube the electrons are produced by thermionic emission from the heated filament cathode and accelerated in the applied electric field toward the anode (target).

In honor of Coolidge's contribution to radiology and medical physics through his hot filament innovation, the highest award bestowed annually by the American Association of Physicists in Medicine is named the William D. Coolidge Award.

CORMACK, Allen MacLeod (1924–1990)

South African-born American physicist, educated in x-ray crystallography at the University of Cape Town where he obtained his B.Sc. in 1944 and M.Sc. in 1945. For a year he continued his studies in nuclear physics at the Cavendish Laboratory in Cambridge, and then returned to a lectureship in the Physics department at the University of Cape Town. On a part time basis he assumed responsibilities for supervising the use of radioactive nuclides in the Groote Shuur hospital, thus learning about medical physics in a radiotherapy department. In 1956 Cormack took a sabbatical at Harvard and developed there a crude theory for the x-ray absorption problem to be used in future CT algorithms. From Harvard he returned to Cape Town for a few months and carried out actual experiments on a crude cylindrical CT phantom. In 1957 Cormack moved to Tufts University in Boston and continued intermittent work on his tomography idea.

During 1963 and 1964 Cormack published two seminal CT papers in the "Journal of Applied Physics". The two papers were largely ignored, but earned him the 1979 Nobel Prize in Medicine and Physiology which he shared with Godfrey N. Hounsfiled *"for the development of computer assisted tomography"*.

COULOMB, Charles–Augustin (1736–1806)

French physicist, educated at the Collège des Quatre–Nations in Paris and in Ecole du Génie at Mézières from where he graduated in 1761 as military engineer. For 20 years after graduation he held various military posts in France and Martinique related to engineering and structural design. During 1770s he wrote several theoretical works in mathematics and produced prize-winning work in applied physics, most notably on torsion balance for measuring very small forces and on friction. In the early 1780s he became recognized as eminent scientist, was elected to the Académie des Sciences, and produced seminal work on electricity and magnetism. After the French Revolution in 1789 the Académie des Sciences was abolished and replaced by the "Institut de France" to which Coulomb was elected in 1795. During the last years of his life he was involved with education as inspector general of public education and as such was responsible for setting up the system of lycées across France. The system is still in use today with the lycée representing the second and last stage of secondary education and completed with the exit exam referred to as the "baccalauréat".

Coulomb is considered the father of the renaissance in French physics and is best known for the Coulomb law of electrostatics which states that the force between two electrical charges is proportional to the product of the charges and inversely proportional to the square of the distance between them. He is also honored by the SI unit of charge called the coulomb C.

CROOKES, William (1832–1919)

British chemist and physicist, educated in the Royal College of Chemistry in London where he also served as assistant from 1850–1854. Upon leaving the Royal College, he first worked as a superintendent at the Radcliffe Observatory in Oxford and then became a lecturer in chemistry at the Chester College in Chester. In 1880 he moved to London where he built and equipped his own laboratory and from then on devoted his life to his versatile research interests carrying out his research projects in his own private laboratory.

During his professional career Crookes was active as researcher in many areas of chemistry and physics, as member and officer of various scientific organizations, and as founder and long-time editor of the Chemical News. He discovered the rare earth element thallium; carried out pioneering work in the field of radioactivity, especially on radium; and invented the radiometer for measurement of radiant energy and the spinthariscope for counting single alpha particles. He is best known for his most important invention, the Crookes tube which he invented in the 1870s and which toward the end of 19th century became a very important device in physics laboratories around the world for studies of "cathode rays" in particular and atomic physics in general. Most notable experiments based on the Crookes tube research are: Röntgen's serendipitous discovery of x-rays in 1895; Thomson's discovery of the electron in 1897; and Millikan's determination of the charge of the electron in 1913.

The Crookes tube not only resulted in one of most important discoveries of all times, namely the discovery of x-rays, it also served as a precursor to modern cathode ray TV tubes. To recognize the importance of Crookes's experimental work the European Physical Society (EPS) established the William Crookes Prize in plasma physics which is awarded to a mid-career (10–20 years post Ph.D.) researcher judged to have made a major contribution to plasma physics.

CURIE, Pierre (1859–1906)

French physicist and chemist, educated in Paris where, after obtaining his "licence ès sciences" (equivalent to M.Sc.) at the age of 18, he was appointed a laboratory assistant at the Sorbonne. In 1882 he was appointed supervisor at the École de Physique et Chimie Industrielle in Paris and in 1895 obtained his doctorate. In 1900 he was appointed lecturer and in 1904 Professor of Physics at the Sorbonne.

Pierre Curie's contributions to physics have two distinct components clearly separated by the date of his wedding to Maria Skłodowska-Curie in 1895. Before that date, he was involved in crystallography and magnetism discovering the piezoelectric effect as well as showing that magnetic properties of a given substance change at a certain temperature that is now referred to as the Curie point. To carry out his experiments he constructed delicate devices that proved very useful in his collaborative studies of radioactivity with his wife Marie Curie. After their discovery of polonium and radium, Pierre Curie concentrated on investigating the physical properties of radium while Marie concentrated on preparing pure compounds.

Pierre Curie and one of his students are credited with making the first observation of nuclear power through measuring the continuous production of heat in a sample of radium. He was also the first to report the decay of radioactive materials and the deleterious biological effects of radium after producing a radium burn and wound on his own skin.

In his honor the 1910 Radiology Congress accepted the definition of the curie (Ci), a unit of activity, as the activity of 1 g of radium–226 or 3.7×10^{10} s^{-1}. The curie is still defined as 3.7×10^{10} s^{-1}, however, subsequent measurements have shown that the specific activity of radium–226 is 0.988 Ci/g. In tribute to the work of Pierre and Marie Curie the element with atomic number 96 was given the name curium (Cm).

Pierre and Marie Curie shared the 1903 Nobel Prize in Physics with Henri Becquerel "*in recognition of the extraordinary services they have rendered by their joint researches on the radiation phenomena discovered by Professor Henri Becquerel*". Becquerel was awarded his share of the Nobel Prize "*in recognition of the extraordinary services he has rendered by his discovery of spontaneous radioactivity*".

SKŁODOWSKA–CURIE, Marie (1867–1934)

Polish-born French physicist and chemist, educated at the Sorbonne in Paris where she obtained a "licence ès sciences" (equivalent to M.Sc.) in physical sciences (1893) and mathematics (1894) and her doctorate in physics in 1903. Curie spent her professional life at various institutions in Paris. In 1906 she was appointed lecturer in physics at the Sorbonne and was promoted to Professor of Physics in 1908.

In 1914 Marie Curie helped found the "Radium Institute" in Paris dedicated to scientific disciplines of physics, chemistry and biology applied to prevention, diagnosis and treatment of cancer. The institute had two divisions: the Curie Laboratory dedicated to research in physics and chemistry of radioactivity and the Pasteur Laboratory devoted to studies of biological and medical effects of radioactivity. The Curie Laboratory was headed by Marie Curie; the Pasteur Laboratory by Claudius Regaud who is regarded as the founding father of both radiotherapy and radiobiology. In 1920, the Curie Foundation was inaugurated to raise funds to support the activities of the Radium Institute. In 1970 the Radium Institute and the Curie Foundation were merged into the Curie Institute mandated to carry out cancer research, teaching and treatment.

After obtaining her "licence" at the Sorbonne, Curie, looking for a doctoral degree subject, decided to investigate the phenomenon of radiation emission from uranium discovered by Henri Becquerel in 1896. She coined the name "radioactivity" for the spontaneous emission of radiation by uranium and established that radioactivity was an atomic rather than chemical phenomenon process. She then investigated if the peculiar property of uranium could be found in any other then-known element and discovered that thorium is also an element which exhibits radioactivity. Noticing that some minerals (for example, pitchblende uranium ore) exhibited a much larger rate of radioactivity than warranted by their uranium or thorium content, she surmised that the minerals must contain other highly radioactive unknown elements. In collaboration with her husband Pierre Curie, Marie Curie discovered miniscule amounts of new elements radium and polonium after sifting through several tons of pitchblende uranium ore. In tribute to the work of Pierre and Marie Curie the element with atomic number 96 was given the name curium (Cm).

The discovery of the new radioactive elements in 1898 earned Marie Curie a doctorate in physics and, in addition, both Marie Skłodowska-Curie and Pierre Curie shared, with Henry Becquerel, the 1903 Nobel Prize in Physics *"in recognition of the extraordinary services they have rendered by their joint researches on the radiation phenomena discovered by Professor Henri Becquerel"*.

In 1911 Marie Curie was awarded another Nobel Prize, this time in Chemistry, *"in recognition of her services to the advancement of chemistry by the discovery of the elements of radium and polonium, by the isolation of radium and the study of the nature and compounds of this remarkable element"*.

Marie Skłodowska-Curie contribution to science has been enormous not only in her own work but also in the work of subsequent generations of physicists whose lives she touched and influenced. She was the first woman to teach at the Sorbonne, the first woman to receive a Nobel Prize, and the first scientist to have received two Nobel Prizes.

ČERENKOV, Pavel Alekseevič (1904–1990)

Russian physicist, educated at the Voronež State University in Voronež in Central Russia, where he graduated with a degree in mathematics and physics in 1928. In 1930 he accepted a post as senior scientific officer in the Peter N. Lebedev Institute of Physics in the Soviet Academy of Sciences (now the Russian Academy of Sciences in Moscow) under the directorship of Sergei I. Vavilov. In 1940 Čerenkov was awarded a doctorate in physics and in 1953 he became Professor of Experimental Physics. In 1970 he became an Academician of the USSR Academy of Sciences.

Čerenkov is best known for his studies of the visible light emitted by energetic charged particles which move through a transparent medium with a velocity that exceeds c/n, the speed of light in the medium, where c is the speed of light in vacuum and n is the index of refraction. In 1934 Čerenkov and Sergei I. Vavilov observed that gamma rays from a radium source, besides causing luminescence in solutions, also produce a faint light from solvents. Their subsequent research lead to two important conclusions: firstly, the emitted light was not a luminescence phenomenon and secondly, the light they observed was not emitted by photons, rather, it was emitted by high energy electrons released in the medium by photon interactions with orbital electrons of the medium. The effect is now referred to as the Čerenkov effect (or sometimes as the Čerenkov-Vavilov effect) and the blue light emitted by energetic charged particles is called Čerenkov radiation. Ilja Frank and Igor Tamm, also from the Lebedov Institute, explained the Čerenkov effect theoretically in 1937 showing that Čerenkov radiation originates from charged particles that move through the medium faster then the speed of light in the medium. The Čerenkov effect is used in Čerenkov counters in nuclear and particle physics for determination of particle energy and velocity.

The 1958 Nobel Prize in Physics was awarded to Čerenkov, Frank, and Tamm *"for the discovery and the interpretation of the Čerenkov effect"*.

DAVISSON, Clinton Joseph (1881–1958)

American physicist, educated at the University of Chicago (B.Sc. in 1908) and Princeton University where he received his doctorate in physics in 1911. He spent most of his professional career at the Bell Telephone Laboratories. Upon retirement from Bell Labs he became Visiting Professor of Physics at the University of Virginia in Charlottesville.

Davisson is best known for his work on electron diffraction from metallic crystals. In 1927 he was studying elastic electron scattering on a nickel single crystal in collaboration with Lester H. Germer. When they analyzed the angular distribution of scattered electrons they discovered that electrons produced diffraction patterns similar to those produced by x-rays. The diffraction patterns were governed by the Bragg formula with a wavelength λ given by the de Broglie equation: $\lambda = h/p$ with h the Planck's constant and p the momentum of the electron. The experiment, now known as the Davisson-Germer experiment, confirmed the hypothesis formulated in 1924 by Louis de Broglie that electrons exhibit dual nature, behaving both as waves and as particles. George P. Thomson, a physicist at the University of Aberdeen in Scotland, confirmed the de Broglie's hypothesis with a different experiment. He studied the behavior of electrons as they traversed very thin films of metals and also observed that electrons under certain conditions behave as waves despite being particles. Thomson's apparatus is referred to as an electron diffraction camera and produces a series of rings when a narrow electron beam is made to traverse a thin metallic foil.

In 1937 Davisson and Thomson shared the Nobel Prize in Physics *"for their experimental discovery of the diffraction of electrons by crystals"*.

De BROGLIE, Louis (1892–1987)

French theoretical physicist, educated at the Sorbonne in Paris, first graduating with an arts degree in 1909 and then with Licence ès Sciences (equivalent to M.Sc.) in 1913. De Broglie spent the war years 1914–1918 in the army and in 1920 resumed his studies in theoretical physics at the Sorbonne. He obtained his doctorate in theoretical physics in 1924, taught physics at the Sorbonne for two years and became Professor of Theoretical Physics at the Henri Poincaré Institute. From 1932 to his retirement in 1962 he was Professor of Theoretical Physics at the Sorbonne.

De Broglie is best known for his theory of electron waves based on the work of Max Planck and Albert Einstein. The theory, presented in his doctorate work, proposed the wave-particle duality of matter. De Broglie reasoned that if x-rays behave as both waves and particles, then particles in general and electrons in particular should also exhibit this duality. De Broglie's theory was confirmed experimentally by Clinton J. Davisson and Lester H. Germer in the United States and by *George P. Thomson* in the U.K. The theory was subsequently used by Erwin Schrödinger to develop wave mechanics.

The 1929 Nobel Prize in Physics was awarded to de Broglie *"for his discovery of the wave nature of electrons"*.

DIRAC, Paul Adrien Maurice (1902–1984)

British physicist, educated at the University of Bristol where he obtained his Bachelor's degree in electrical engineering in 1921 and at the St. John's College in Cambridge where he received his doctorate in mathematics in 1926. In 1927 he became a Fellow of the St. John's College and from 1932 until 1969 he was Lucasian Professor of Mathematics in Cambridge. In 1969 Dirac moved to Florida to become Professor of Physics at the Florida State University.

Dirac was an extremely productive and intelligent theoretical physicist, mainly involved with mathematical and theoretical aspects of quantum mechanics. Quantum mechanics, dealing with dimensions of the order of the atomic size, introduced the second revolution in physics, the first one being Albert Einstein's special theory of relativity that deals with velocities of the order of the speed of light in vacuum.

In 1926 Dirac developed his version of quantum mechanics that merged the "matrix mechanics" of Werner Heisenberg with the "wave mechanics" of Erwin Schrödinger into a single mathematical formalism. In 1928 he derived a relativistic equation for the electron that merged quantum mechanics with relativity and is now referred to as the Dirac equation. The equation predicts the existence of an antiparticle (same mass, opposite charge) to the electron and infers the electron quantum spin. Dirac also predicted that in an electron/anti-electron encounter the charges cancel, and the two particles annihilate with the combined mass transforming into radiation according to Albert Einstein's celebrated equation $E = mc^2$. Four year later, in 1932 Carl D. Anderson discovered the anti-electron, a new particle which is now called the positron. In 1931 Dirac showed theoretically that the existence of a magnetic monopole would explain the observed quantization of the electrical charge (all charges found in nature are multiples of the electron charge). No monopoles have been found in nature so far.

The 1933 Nobel Prize in Physics was awarded to Paul M. Dirac and Erwin Schrödinger *"for their discovery of new productive forms of atomic theory"*.

DUANE, William (1872–1935)

American physicist, educated at the University of Pennsylvania and Harvard, where he received a B.A. degree in 1893 and a M.A. degree in 1895. From 1895 to 1897 he held the Tyndall Fellowship of Harvard University and studied physics in Göttingen and Berlin receiving the Ph.D. degree from Berlin in 1897. From 1898 to 1907 he held a position of Professor of Physics at the University of Colorado. He then moved to Paris and worked for 6 years with Marie Skłodowska-Curie at the Sorbonne on various projects involving radioactivity.

During his Paris period Duane also got interested in the application of radium and x-rays in medicine and in 1913, when the newly formed Harvard Cancer Commission was formed, he accepted a job offer of Assistant Professor of Physics at Harvard and Research Fellow in Physics at the Harvard Cancer Commission. By 1917 he was promoted to Professor of Biophysics, probably filling the first such position in North America, and remained with Harvard and the Cancer Commission till his retirement in 1934. In view of his hospital appointment and significant contributions to imaging and cancer therapy one can conclude that Duane was among the first medical physicists in North America.

Duane is best known for the Duane-Hunt law that he discovered with his Ph.D. student Franklin Hunt in 1915. This law states that there is a sharp upper limit to the x-ray frequencies emitted from a target stimulated by the impact of energetic electrons. He also established that the Duane-Hunt law could be used as a very accurate method of determining Planck's constant h and the ratio h/e, where e is the charge of the electron. He was also the first to discover that the total intensity produced by an x-ray target depends linearly on the atomic number of the target.

Duane is one of the most important early contributors to radiation dosimetry of gamma rays and x-rays used in treatment of cancer. He developed the technical details for measurement of radiation dose with ionization chambers and was instrumental in gaining national and international acceptance of 1 unit of x-ray intensity as "that intensity of radiation which produces under saturation conditions one electrostatic unit of charge per cm^3 of air under standard temperature and pressure". Duane's unit of x-ray intensity was subsequently named rontgen (R). During his professional career, Duane received numerous awards for his scientific work and was awarded honorary Sc.D. degrees by the University of Pennsylvania in 1922 and University of Colorado in 1923.

EINSTEIN, Albert (1879–1955)

German-born theoretical physicist, educated at the Eidgenössische Technische Hochschule in Zürich (ETHZ) from which he graduated in 1900 as a teacher of mathematics and physics. He did not succeed in obtaining an academic post after graduating and spend two years teaching mathematics and physics in secondary schools. From 1902 until 1909 he worked as a technical expert in the Swiss Patent Office in Bern. In 1905 he earned a doctorate in physics from the University of Zürich.

Following publication of three seminal theoretical papers in 1905 and submission of his "Habilitation" thesis in 1908, Einstein's credibility in physics circles rose dramatically; he started to receive academic job offers and entered a period of frequent moves and changes in academic positions. In 1908 he became lecturer at the University of Bern and in 1909 Professor of Physics at the University of Zürich. During 1911 he was Professor of Physics at the Karl-Ferdinand University in Prague and in 1912 he moved back to Zürich to take a chair in theoretical physics at the ETHZ. Finally, in 1914 he moved to Berlin to a research position without teaching responsibilities at the then world-class center of physics at the University of Berlin.

During the Berlin period (1914–1933) Einstein produced some of his most important work, became an international "star" physicist and scientist, got involved in political issues, and traveled a great deal to visit physics colleagues and present invited lectures on his work. In 1932 he moved to the United States to become Professor of Theoretical Physics at the Institute for Advanced Study in Princeton, one of the world's leading centers for theoretical research and intellectual inquiry.

Einstein was an extremely gifted physicist and his contribution to modern physics is truly remarkable. His three papers published in Volume 17 of the "Annalen der Physik" each dealt with a different subject and each is now considered a masterpiece. The first of the three papers dealt with the photoelectric effect and contributed to quantum theory by assuming that light under certain conditions behaves like a stream of particles (quanta) with discrete energies. The second paper dealt with statistical mechanics and lead to an explanation of Brownian motion of molecules. The third paper addressed the connection between the electromagnetic theory and ordinary motion and presented Einstein's solution as the "special theory of relativity". In 1916, after a decade of futile attempts, Einstein completed his "general theory of relativity" based on the "equivalence principle" stating that uniform acceleration of an object is equivalent to a gravitational field. The gravitational field causes curvature of space-time as observed experimentally by measuring the precession of the mercury perihelion and the bending by the sun of light from the stars.

At the end of the Berlin period and during his American period from 1933 until his death in 1955 Einstein concentrated on developing a unified field theory, unsuccessfully attempting to unify gravitation, electromagnetism and quantum physics. Throughout his life Einstein was a pacifist detesting both militarism as well as nationalism. In tribute to Einstein's contributions to modern physics the element with atomic number 109 is named einsteinium (Es).

In 1921 the Nobel Prize in Physics was awarded to Einstein *"for his services to Theoretical Physics and especially for his discovery of the law of the photoelectric effect"*.

In recognition of Einstein's tremendous contribution to modern physics the year 2005, the centenary of Einstein's "annus mirabilis," was proclaimed the world year of physics, a worldwide celebration of physics and its impact on humanity.

EVANS, Robley (1907–1995)

American nuclear and medical physicist, educated at the California Institute of Technology (Caltech) where he studied physics and received his B.Sc. in 1928, M.Sc. in 1929, and Ph.D. under Robert A. Millikan in 1932. After receiving his doctorate he studied biological effects of radiation as post-doctoral fellow at the University of California at Berkeley before accepting a faculty position at the Massachusetts Institute of Technology (MIT) in Boston. He remained an active member of the MIT faculty for 38 years and retired in 1972 to become a special project associate at the Mayo Clinic in Rochester, Minnesota.

At the MIT Evans was instrumental in building the first cyclotron in the world for biological and medical use. He established the Radioactivity Center in the Physics department at the MIT for research in nuclear physics related to biology, introduced the first iodine radionuclide for diagnosis and treatment of thyroid disease, and built the first total body counter to measure the uptake and body burden of radium in the human body. In 1941 he established one ten-millionth of a gram of radium ($0.1\ \mu Ci$) as the maximum permissible body burden. The standard is still internationally used and has been adapted for other radioactive substances including plutonium-239 and strontium–90. Evans's book "The Atomic Nucleus" was first published in 1955 and remained the definitive nuclear physics textbook for several decades and is still considered an important nuclear physics book.

In 1985 Evans received the William D. Coolidge Award from the American Association of Physicists in Medicine in recognition to his contribution to medical physics and in 1990 he received the Enrico Fermi Award in recognition of his contributions to nuclear and medical physics.

FANO, Ugo (1912–2001)

American physicist of Italian descent, educated in mathematics and physics at the University of Torino. He received postdoctoral training from Enrico Fermi at the University of Rome (1934–1936) and from Werner Heisenberg at the University of Leipzig (1936–1937). In 1940 Fano started his American career in radiation biology and physics at the Carnegie Institution at Cold Spring Harbor. In 1946 he joined the staff of the National Bureau of Standards (the predecessor of the National Institute of Standards and Technology) and during his two decades there made outstanding contributions to radiation physics and basic solid state physics.

From 1966 to 1982 Fano was on staff at the University of Chicago where he worked in atomic and molecular physics and continued his lifetime interest in radiation physics. He published over 250 scientific papers and made major contributions in radiation physics and radiation dosimetry. He developed the first general theory of the ionization yield in a gas; characterized statistical fluctuations of ionization by the now known Fano factor; developed methods for dealing with the transport of photons and charged particles in matter; and demonstrated the cavity principle of radiation equilibrium under general conditions. He was also a big proponent of the use of synchrotron radiation for spectroscopic studies.

Fano was recognized with many honors, most notably with the Enrico Fermi Award by the U.S. government; membership in the National Academy of Sciences of the United States; and several honorary doctorates.

FERMI, Enrico (1901–1954)

Italian-born physicist who graduated from the University of Pisa in 1921. He was a lecturer at the University of Florence for two years and then Professor of Theoretical Physics at the University of Rome from 1923 to 1938. In 1938 he moved to the United States and worked first for four years at Columbia University in New York and from 1942 till his death in 1954 at the University of Chicago.

Fermi is recognized as one of the great scientists of the 20th Century. He is best known for his contributions to nuclear physics and quantum theory. In 1934 he developed the theory of the beta nuclear decay that introduced the neutrino and the weak force as one of the basic forces in nature. The existence of neutrino was actually enunciated by Wolfgang Pauli in 1930 and experimentally confirmed only in 1956.

In 1934, while at the University of Rome, Fermi began experiments bombarding various heavy elements with thermal neutrons. He discovered that the thermal neutrons bombarding uranium were very effective in producing radioactive atoms, but did not realize at the time that he succeeded in splitting the uranium atom.

Otto Hahn and Fritz Strassmann in 1938 repeated Fermi's experiments and discovered that uranium bombarded with thermal neutrons splits into two lighter atoms. Lise Meitner and Otto Frisch explained the process theoretically and named it nuclear fission.

Upon his move to the United States Fermi continued his fission experiments at Columbia University and showed experimentally that uranium fission results in two lighter by-products, releasing several neutrons and large amounts of energy. In 1942 he was appointed Director of the Manhattan project at the University of Chicago with a mandate to develop an "atomic bomb". With his team of scientists Fermi produced the first nuclear chain reaction and developed the atomic bombs that were dropped on Hiroshima and Nagasaki by the United States at the end of the World War II.

In 1938 Fermi was awarded the Nobel Prize in Physics *"for his demonstrations of the existence of new radioactive elements produced by neutron irradiation, and for his related discovery of nuclear reactions brought about by slow neutrons"*.

Fermi's name is honored by the unit of length that is of the order of the size of the atomic nucleus (1 fermi = 1 femtometer = 1 fm = 10^{-15} m). One of the American national laboratories is named Fermi National Laboratory (Fermilab), and the oldest and most prestigious science and technology prize awarded in the United States is the Enrico Fermi Award. A common name for particles with half-integer spin, such as electron, neutron, proton and quark, is fermion; the artificially produced element with atomic number Z of 100 is fermium (Fm); and the quantum statistics followed by fermions is known as the Fermi-Dirac statistics, after its inventors.

FLEROV, Georgij Nikolaevič (1913–1990)

Russian nuclear physicist, educated in physics at the Polytechnical Institute of Leningrad (now Sankt Petersburg) from where he graduated in 1938. He started his scientific career at the Leningrad Institute of Physics and Technology and was involved in basic research in a number of fundamental and applied areas of nuclear physics. From 1941 to 1952 Flerov, together with Igor V. Kurčatov, participated in investigations linked with the use of atomic energy for military purposes and nuclear power industry. From 1960 to 1988 he was the director of the Nuclear Reactions Laboratory of the Joint Institute for Nuclear Research in Dubna.

Flerov is best known for his discovery in 1940 (in collaboration with Konstantin A. Petržak) of the spontaneous fission of uranium–238. With colleagues in Dubna Flerov carried out research that resulted in the synthesis of new heavy elements (nobelium No–102, rutherfordium Rf–104, dubnium Db–105), production of a large number of new nuclei on the border of stability, and discovery of new types of radioactivity (proton radioactivity) and new mechanisms of nuclear interactions.

FRANCK, James (1882–1964)

German-born American physicist, educated at the University of Heidelberg and the University of Berlin where he received his doctorate in physics in 1906. He worked at the University of Berlin from 1911 to 1918 and at the University of Göttingen until 1933 when he moved to the United States to become Professor at Johns Hopkins

University in Baltimore. From 1938 to 1947 he was Professor of Physical Chemistry at the University of Chicago.

Franck is best known for the experiment he carried out in 1914 at the University of Berlin in collaboration with Gustav Hertz. The experiment is now known as the Franck–Hertz experiment and it demonstrated the existence of quantized excited states in mercury atoms. This provided the first experimental substantiation of the Bohr atomic theory which predicted that atomic electrons occupied discrete and quantized energy states.

In 1925 James Franck and Gustav Hertz were awarded the Nobel Prize in Physics *"for their discovery of the laws governing the impact of electron upon an atom"*. In addition to the Nobel Prize, Franck was also honored with the 1951 Max Planck medal of the German Physical Society and was named honorary citizen of the university town of Göttingen.

GAMOW, George (1904–1968)

Ukranian-born American physicist and cosmologist, educated at the Novorossia University in Odessa (1922–1923) and at the Leningrad University (1923–1928) where he received his doctorate in physics in 1928. After a fellowship with Niels Bohr at the Institute for Theoretical Physics in Copenhagen and a short visit to Ernest Rutherford at the Cavendish Laboratory in Cambridge, he returned to SSSR in 1931 to become a Professor of Physics at the University of Leningrad. From 1934 until 1956 he was Chair of Physics at the George Washington University in Washington D.C. and from 1956 until his death in 1968 he was a Professor of Physics at the University of Colorado in Boulder. During World War II he was involved with the Manhattan nuclear weapons project in Los Alamos.

Gamow is best known for his (1928) theory of the alpha decay based on tunneling of the alpha particle through the nuclear potential barrier. He was also a proponent of the Big-Bang theory of the Universe and worked on the theory of thermonuclear reactions inside the stars that is still today relevant to research in controlled nuclear fusion. His name is also associated with the beta decay in the so-called Gamow-Teller selection rule for beta emission. Gamow was also well known as an author of popular science books and received the UNESCO Kalinga Prize for popularization of science.

GEIGER, Hans (1882–1945)

German physicist, educated in physics and mathematics at the university of Erlangen where he obtained his doctorate in 1906. From 1907 to 1912 he worked with Ernest Rutherford at the University of Manchester where, with Ernest Marsden, he carried out the α-particle scattering experiments that lead to the Rutherford–Bohr atomic model. He also discovered, in collaboration with John M. Nuttall, an empirical linear relationship between $\log \lambda$ and $\log R_\alpha$ for naturally occurring α emitters with the decay constant λ and range in air R_α (Geiger–Nuttall law). In collaboration with Walther Müller he developed a radiation detector now referred to as the Geiger-Müller counter.

GERLACH, Walther (1889–1979)

German physicist, educated at the University of Tübingen where he received his doctorate in physics in 1912 for a study of blackbody radiation and the photoelectric effect. He worked at the University of Göttingen and University of Frankfurt before returning in 1925 to Tübingen as Professor of Physics. From 1929 to 1952 he was Professor of Physics at the University of Munich.

Gerlach made contributions to radiation physics, spectroscopy and quantum physics. He is best known for his collaboration with Otto Stern in 1922 at the University of Frankfurt on an experiment that demonstrated space quantization using a beam of neutral silver atoms that, as a result of passage through an inhomogeneous magnetic field, split into two district components, each component characterized by a specific spin (angular momentum) of the silver atoms.

GERMER, Lester H (1896–1971)

American physicist, educated at Columbia University in New York. In 1927 he worked as graduate student at Bell Laboratories under the supervision of Clinton T. Davisson on experiments that demonstrated the wave properties of electrons and substantiated the Louis de Broglie's hypothesis that moving particles exhibit particle-wave duality. The electron diffraction experiments on metallic crystals are now referred to as the Davisson-Germer experiment.

HAHN, Otto (1879–1968)

German chemist, educated at University of Munich and University of Marburg. In 1901 he obtained his doctorate in organic chemistry at the University of Marburg. He spent two years as chemistry assistant at the University of Marburg, and then studied radioactivity for one year under William Ramsay at the University College in London and for one year under Ernest Rutherford at McGill University in Montreal. In 1905 he moved to the Kaiser Wilhelm Institute (now Max Planck Institute) for chemistry in Berlin and remained there for most of his professional life. From 1928–1944 he served as the Director of the Institute.

Early in his career in Berlin he started a life-long professional association with Austrian-born physicist Lise Meitner; a collaboration that produced many important discoveries in radiochemistry and nuclear physics. Hahn's most important contribution to science is his involvement with the discovery of nuclear fission. In 1934 the Italian physicist Enrico Fermi discovered that uranium bombarded with neutrons yields several radioactive products. Hahn and Meitner, in collaboration with Friedrich Strassmann, repeated Fermi's experiments and found inconclusive results. In 1938, being Jewish, Meitner left Germany for Stockholm to escape persecution by the Nazis; Hahn and Strassmann continued with the neutron experiments and eventually concluded that several products resulting from the uranium bombardment with neutrons were much lighter than uranium suggesting that the neutron bombardment caused uranium to split into two lighter components of more or less equal size. Hahn communicated the findings to Meitner in Stockholm, who, in cooperation with Otto Frisch, explained the theoretical aspects of the uranium splitting process and called it nuclear fission. The discovery of nuclear fission led to the atomic bomb and to modern nuclear power industry.

In 1944 Hahn alone was awarded the Nobel Prize in Chemistry *"for his discovery of the fission of heavy nuclei"*. In 1966 Hahn, Strassmann and Meitner shared the Enrico Fermi Prize for their work in nuclear fission. It is now universally accepted that four scientists are to be credited with the discovery of the nuclear fission process: Hahn, Strassmann, Meitner and Frisch.

HARTREE, Douglas (1897–1958)

British mathematician and physicist, educated in Cambridge where he obtained a degree in Natural Sciences in 1921 and a doctorate in 1926. In 1929 he was appointed Professor of Applied Mathematics at the University of Manchester and in 1937 he moved to a Chair of Theoretical Physics at the University of Manchester. In 1946 he was appointed Professor of Mathematical Physics at Cambridge University and held the post until his death in 1958.

Hartree was both a mathematician and physicist and he is best known for applying numerical analysis to complex physics problems such as calculations of wave functions for multi-electron atoms. Hartree approached the problem by using the method of successive approximations, treating the most important interactions in the first approximation and then improving the result with each succeeding approximation. Hartree's work extended the concepts of the Bohr theory for one-electron atoms or ions to multi-electron atoms providing reasonable, albeit not perfect, approximations to inter-electronic interactions in multi-electron atoms.

HEISENBERG, Werner (1901–1976)

German theoretical physicist, educated in physics at the University of Munich and the University of Göttingen. He received his doctorate in physics at the University of Munich in 1923 and successfully presented his habilitation lecture in 1924. During 1924–1926 he worked with Niels Bohr at the University of Copenhagen. From 1927 until 1941 Heisenberg held an appointment as Professor of Theoretical Physics at the University of Leipzig and in 1941 he was appointed Professor of Physics at the University of Berlin and Director of the Kaiser Wilhelm Institute for Physics in Berlin. From 1946 until his retirement in 1970 he was Director of the Max Planck Institute for Physics and Astrophysics in Göttingen. The institute moved from Göttingen to Munich in 1958.

In 1925 Heisenberg invented matrix mechanics which is considered the first version of quantum mechanics. The theory is based on radiation emitted by the atom and mechanical quantities, such as position and velocity of electrons, are represented by matrices. Heisenberg is best known for his Uncertainty Principle stating that a determination of particle position and momentum necessarily contains errors the product of which is of the order of the Planck's quantum constant h. The principle is of no consequence in the macroscopic world, however, it is critical for studies on the atomic scale.

In 1932 Heisenberg was awarded the Nobel Prize in Physics *"for creation of quantum mechanics, the application of which has, inter alia, led to the discovery of allotropic forms of hydrogen"*.

HERTZ, Gustav (1887–1975)

German physicist, educated at universities of Göttingen, Munich and Berlin, and graduating with a doctorate in physics in 1911. During 1913–1914 he worked as research assistant at the University of Berlin. Hertz alternated work in industry (Philips in Eindhoven; Siemens in Erlangen) with academic positions at universities of Berlin, Halle and Leipzig.

Hertz made many contributions to atomic physics but is best known for the experiment in which he studied, in collaboration with James Franck, the impact of electrons on mercury vapor atoms. The experiment is now referred to as the Franck–Hertz experiment and demonstrated the existence of quantized excited states in mercury atoms, thereby substantiating the basic tenets of the Bohr atomic theory.

In 1925 James Franck and Gustav Hertz were awarded the Nobel Prize in Physics *"for their discovery of the laws governing the impact of an electron upon an atom"*. Hertz was also the recipient of the Max Planck Medal of the German Physical Society.

HOFSTADTER, Robert (1915–1990)

American physicist, educated at the College of the City of New York (B.Sc., 1935) and Princeton University in Princeton, New Jersey (M.A. and Ph.D. in physics, 1938). During 1938 he was postdoctoral fellow at Princeton working on photoconductivity of willemite crystals and then a year at the University of Pennsylvania where he helped to construct a large Van de Graaff accelerator for nuclear research.

During the war years Hofstadter first worked at the National Bureau of Standards (NBS) and later at Norden Laboratory. In 1945 he returned to Princeton as Assistant Professor of Physics and got involved in radiation detection instrumentation, discovering in 1948 that sodium iodide activated with thallium made an excellent scintillation counter that could also be used as spectrometer for measurement of energy of gamma rays and energetic charged particles. In 1950 he moved to Stanford University in Stanford, California to become Associate Professor and later on Professor of Physics carrying out important research work on development of radiation detectors and electron scattering by nuclei using W.W. Hansen's invention of electron linear accelerator. Hofstadter's work on using high-energy electrons to probe the nucleus resulted in much of the current knowledge on scattering form factors, nuclear charge distribution and size as well as charge and magnetic moment distributions of the proton and neutron.

Hofstadter was elected to the National Academies (USA) and was named California Scientist of the Year in 1959. In 1961 he shared the Nobel Prize in Physics with Rudolf L. Mössbauer. Hofstadter received the prize *"for his pioneering studies of electron scattering in atomic nuclei and for his thereby achieved discoveries concerning the structure of the nucleons"*, Mössbauer for the discovery of gamma ray resonance in connection with the effect which bears his name.

HOUNSFIELD, Godfrey Neubold (1919–2004)

British electrical engineer and scientist, educated at the Electrical Engineering College in London from which he graduated in 1951. The same year he joined the research staff of the EMI in Middlesex. He remained associated with the EMI throughout his professional career.

Hounsfield made a significant contribution to early developments in the computer field and was responsible for the development of the first transistor-based solid-state computer in the U.K. He is best known, however, for the invention of computed tomography (CT), an x-ray-based diagnostic technique that non-invasively forms two-dimensional cross sections through the human body. Originally, the technique was referred to as computer assisted tomography (CAT), now the term computed tomography (CT) is more commonly used.

Following his original theoretical calculations, he first built a laboratory CT model to establish the feasibility of the approach, and then in 1972 built a clinical prototype CT-scanner for brain imaging. From the original single slice brain CT-scanner the technology evolved through four generations to the current 64 slice body and brain CT-scanners. Röntgen's discovery of x-rays in 1895 triggered the birth of diagnostic radiology as an important medical specialty; Hounsfield's invention of the CT-scanner placed diagnostic radiology onto a much higher level and transformed it into an invaluable tool in diagnosis of brain disease in particular and human malignant disease in general. In 1979 Hounsfield shared the Nobel Prize in Medicine and Physiology with *Allan M. Cormack "for the development of computer assisted tomography"*. Cormack derived and published the mathematical basis of the CT scanning method in 1964.

Hounsfield's name is honored with the Hounsfield scale which provides a quantitative measure of x-ray attenuation of various tissues relative to that of water. The scale is defined in hounsfield units (HU) running from air at -1000 HU, fat at -100 HU, through water at 0 HU, white matter at ~ 25 HU, grey matter at ~ 40 HU, to bone at $+400$ HU or larger, and metallic implants at $+1000$ HU.

HUBBELL, John Howard (1925–2007)

American radiation physicist, educated at the University of Michigan in Ann Arbor in engineering physics (B.Sc. in 1949, MSc. in 1950). In 1950 he joined the staff of the National Bureau of Standards (NBS) now known as the National Institute of Science and Technology (NIST) in Washington D.C. and spent his professional career there, directing the NBS/NIST x-ray and Ionizing Radiation Data Center from 1963 to 1981. He retired in 1988.

Hubbell's collection and critical evaluation of experimental and theoretical photon cross section data resulted in the development of tables of attenuation coefficients and energy absorption coefficients, as well as related quantities such as atomic form factors, incoherent scattering functions, and atomic cross sections for photoelectric effect, pair production and triplet production. Hubbell's most widely known and important work is the "National Standard Reference Data Series Report 29: Photon Cross Sections, Attenuation Coefficients and Energy Absorption Coefficients from 10 keV to 100 GeV".

JOHNS, Harold Elford (1915–1998)

Born in Chengtu, China to Canadian parents who were doing missionary work in China, Johns obtained his Ph.D. in Physics from the University of Toronto and then worked as physicist in Edmonton, Saskatoon, and Toronto. His main interest was diagnosis and therapy of cancer with radiation and his contributions to the field of medical physics are truly remarkable. While working at the University of Saskatchewan in Saskatoon in the early 1950s, he invented and developed the cobalt-60 machine which revolutionized cancer radiation therapy and had an immediate impact on the survival rate of cancer patients undergoing radiotherapy.

In 1956 Johns became the first director of the Department of Medical Biophysics at the University of Toronto and Head of the Physics division of the Ontario Cancer Institute in Toronto. He remained in these positions until his retirement in 1980 and built the academic and clinical departments into world-renowned centers for medical physics. With his former student John R. Cunningham, Johns wrote the classic book "The Physics of Radiology" that has undergone several re-printings and is still considered the most important textbook on medical physics.

In 1976 Johns received the William D. Coolidge Award from the American Association of Physicists in Medicine.

JOLIOT-CURIE, Irène (1897–1956)

French physicist, educated at the Sorbonne in Paris where she received her doctorate on the alpha rays of polonium in 1925 while already working as her mother's (Marie Skłodowska-Curie) assistant at the Radium Institute. In 1927 Irène Curie married Frédéric Joliot who was her laboratory partner and Marie Curie's assistant since 1924. In 1932 Joliot-Curie was appointed lecturer and in 1937 Professor at the Sorbonne. In 1946 she became the Director of the Radium Institute.

Joliot-Curie is best known for her work at the "Institut du Radium" in Paris, in collaboration with her husband Frédéric Joliot, on the production of artificial radioactivity through nuclear reactions in 1934. They bombarded stable nuclides such as boron-10, aluminum-27, and magnesium-24 with naturally occurring α-particles and obtained radionuclides nitrogen-13, phosphorus-30, and silicon-27, respectively, accompanied by release of a neutron. The discovery of artificially produced radionuclides completely changed the periodic table of elements and added several hundred artificial radionuclides to the list. In 1938 Joliot-Curie's research of neutron bombardment of uranium represented an important step in eventual discovery of uranium fission by Otto Hahn, Friedrich Strassmann, Lise Meitner, and Otto Frisch.

The 1935 Nobel Prize in Chemistry was awarded to Frédéric Joliot and Irène Joliot-Curie "*in recognition of their synthesis of new radioactive elements*".

JOLIOT, Jean Frédéric (1900–1958)

French physicist, educated at the École de Physique et Chimie Industriele in Paris where he received an engineering physics degree in 1924. Upon graduation he became Marie Skłodowska-Curie assistant at the "Institut du Radium" in Paris. He married Irène Curie, Marie Curie's daughter, in 1927 and worked on many nuclear physics projects in collaboration with his wife. In 1930 he obtained his doctorate in physics and in 1937 he became Professor of Physics at the Collège de France in Paris.

In 1934 Joliot discovered artificial radioactivity with Irène Curiie and in 1939 he confirmed the fission experiment announced by Otto Hahn and Friedrich Strassmann. He recognized the importance of the experiment in view of a possible chain reaction and its use for the development of nuclear weapons. In 1935 Joliot and Irène Joliot-Curie shared the Nobel Prize in Chemistry *"in recognition of their synthesis of new radioactive elements"*.

KERST, Donald William (1911–1993)

American physicist, educated at the University of Wisconsin in Madison where he received his doctorate in physics in 1937. From 1938 to 1957 he worked through academic ranks to become Professor of Physics at the University of Illinois. He then worked in industry from 1957 to 1962 and from 1962 to 1980 he was Professor of Physics at the University of Wisconsin. Kerst made important contributions to the general design of particle accelerators, nuclear physics, medical physics, and plasma physics. He will be remembered best for this development of the betatron in 1940, a cyclic electron accelerator that accelerates electrons by magnetic induction. The machine found important use in industry, nuclear physics and medicine during the 1950 and 1960s before it was eclipsed by more practical linear accelerators.

KLEIN, Oskar (1894–1977)

Swedish-born theoretical physicist. Klein completed his doctoral dissertation at the University of Stockholm (Högskola) in 1921 and worked as physicist in Stockholm, Copenhagen, Lund and Ann Arbor. He is best known for introducing the relativistic wave equation (Klein-Gordon equation); for his collaboration with Niels Bohr on the principles of correspondence and complementarity; and for his derivation, with Yoshio Nishina, in 1929 of the equation for Compton scattering (Klein-Nishina equation). Klein's attempts to unify general relativity and electromagnetism by introducing a five-dimensional space-time resulted in a theory now known as the Kaluza-Klein theory.

LARMOR, Joseph (1857–1942)

Irish physicist, educated at Queen's University in Belfast where he received his B.A. and M.A. In 1877 he continued his studies in mathematics at the St. Johns College in Cambridge. In 1880 he returned to Ireland as Professor of Natural Philosophy at Queens College Galway. In 1885 he moved back to Cambridge as lecturer and in 1903 he became the Lucasian Chair of Mathematics succeeding George Stokes. He remained in Cambridge until retirement in 1932 upon which he returned to Ireland.

Larmor worked in several areas of physics such as electricity, dynamics, thermodynamics, and, most notably, in ether, the material postulated at the end of the 19th century as a medium pervading space and transmitting the electromagnetic radiation. He is best known for calculating the rate at which energy is radiated from a charged particle (Larmor law); for explaining the splitting of spectral lines by a magnetic field; and for the Larmor equation $\omega = \gamma B$, where ω is the angular frequency of a precessing proton, γ the gyromagnetic constant, and B the magnetic field.

LAUE, Max von (1879–1960)

German physicist, educated at the University of Strassbourg where he studied mathematics, physics and chemistry, University of Göttingen and University of Berlin where he received his doctorate in physics in 1903. He then worked for two years at the University of Göttingen, four years at the Institute for Theoretical Physics in Berlin, and three years at the University of Munich, before starting his series of Professorships in Physics in 1912 at the University of Zürich, 1914 at the University of Frankfurt, 1916 at the University of Würzburg and 1919 at the University of Berlin from which he retired in 1943.

Von Laue is best known for his discovery in 1912 of the diffraction of x-rays on crystals. Since the wavelength of x-rays was assumed to be of the order of inter-atomic separation in crystals, he surmised that crystalline structures behave like diffraction gratings for x-rays. Von Laue's hypothesis was proven correct experimentally and established the wave nature of x-rays and the regular internal structure of crystals. The crystalline structure essentially forms a three-dimensional grating, presenting a formidable problem to analyze. William L. Bragg proposed a simple solution to this problem now referred to as the Bragg equation. Von Laue also made notable contributions to the field of superconductivity where he collaborated with Hans Meissner who with Robert Ochsenfeld established that, when a superconductor in the presence of a magnetic field is cooled below a critical temperature, all of the magnetic flux is expelled from the interior of the sample.

The 1914 Nobel Prize in Physics was awarded to von Laue *"for his discovery of the diffraction of x-rays by crystals"*.

LAUTERBUR, Paul Christian (1929–2007)

American chemist, educated at the Case Institute of Technology in Cleveland (B.Sc. in chemistry in 1951) and University of Pittsburgh (Ph.D. in chemistry in 1962). His first academic position was at Stony Brook University as Associate Professor and from 1969 until 1985 as Professor of Chemistry. From 1985 until 1990 he was Professor of Chemistry at the University of Illinois at Chicago and from 1985 to 2007 he was Professor and Director of the Biomedical MR Laboratory at the University of Illinois at Urbana-Champaign.

Being trained in nuclear magnetic resonance (NMR), Lauterbur started his academic career in this area. However, in the early 1970s when investigating proton NMR relaxation times of various tissues obtained from tumor-bearing rats, he observed large and consistent differences in relaxation times from various parts of the sacrificed animals. Some researchers were speculating that relaxation time measurements might supplement or replace the observations of cell structure in tissues by pathologists but Lauterbur objected to the invasive nature of the procedure. He surmised that there may be a way to locate the precise origin of the NMR signals in complex objects, and thus non-invasively form an image of their distribution in two or even three dimensions. He developed the method of creating a two dimensional image by introducing gradients into the NMR magnetic field, analyzing the characteristics of the emitted radio waves, and determining the location of their source. To allay fears by the general public of everything nuclear, the NMR imaging became known as magnetic resonance imaging or MRI.

Lauterbur shared the 2003 Nobel Prize in Medicine with Peter Mansfield *"for their discoveries concerning magnetic resonance imaging"*.

LAWRENCE, Ernest Orlando (1900–1958)

American physicist, educated at the University of South Dakota (B.A. in chemistry in 1922), University of Minnesota (M.A. in chemistry in 1923) and Yale University (Ph.D. in physics in 1925). He first worked at Yale as research fellow and Assistant Professor of Physics and was appointed Associate Professor at the University of California at Berkeley in 1928 and Professor of Physics in 1930. In 1936 he was appointed Director of the University's Radiation Laboratory and remained in these posts until his death in 1958.

The reputation of the Berkeley Physics department as an eminent world-class center of physics is largely based on Lawrence's efforts. He was not only an excellent physicist, he was also an excellent research leader, director of large-scale physics projects, and government advisor. Lawrence is best known for his invention of the cyclotron (in 1930), a cyclic accelerator that accelerates heavy charged particles to high kinetic energies for use in producing nuclear reactions in targets or for use in cancer therapy. During World War II Lawrence worked on the Manhattan project developing the atomic fission bomb. His research interests were also in the use of radiation in biology and medicine.

In 1939 Lawrence was awarded the Nobel Prize in Physics *"for the invention and development of the cyclotron and for results obtained with it, especially with regard to artificial radioactive elements"*. Lawrence's name is honored by Lawrence Berkeley Laboratory in Berkeley, Lawrence Livermore National Laboratory in Livermore, California, and lawrencium, an artificial element with an atomic number 103.

LICHTENBERG, Georg Christoph (1742–1799)

German physicist and philosopher, educated at the University of Göttingen, where he also spent his whole professional life, from 1769 until 1785 as Assistant Professor of Physics and from 1785 until his death in 1799 as Professor of Physics.

In addition to physics, Lichtenberg taught many related subjects and was also an active researcher in many areas, most notably astronomy, chemistry, and mathematics. His most prominent research was in electricity and in 1777 he found that discharge of static electricity may form intriguing patterns in a layer of dust, thereby discovering the basic principles of modern photocopying machines and xeroradiography. High voltage electrical discharges on the surface or inside of insulating materials often result in distinctive patterns that are referred to as Lichtenberg figures or "trees" in honor of their discoverer.

Lichtenberg is credited with suggesting that Euclid's axioms may not be the only basis for a valid geometry and his speculation was proven correct in the 1970s when Benoit B. Mandelbrot, a Polish-American mathematician, introduced the techniques of fractal geometry. Coincidentally, these techniques also produce patterns that are now referred to as Lichtenberg patterns.

Lichtenberg was also known as a philosopher who critically examined a range of philosophical questions and arrived at intriguing, interesting and often humorous conclusions. Many consider him the greatest German aphorist and his "Waste Books" contain many aphorisms and witticisms that are still relevant to modern societies.

LORENTZ, Hendrik Antoon (1853–1928)

Dutch physicist, educated at the University of Leiden where he obtained a B.Sc. degree in mathematics and physics in 1871 and a doctorate in physics in 1875. In 1878 he was appointed to the Chair of Theoretical Physics at the University of Leiden and he stayed in Leiden his whole professional life.

Lorentz made numerous contributions to various areas of physics but is best known for his efforts to develop a single theory to unify electricity, magnetism and light. He postulated that atoms were composed of charged particles and that atoms emitted light following oscillations of these charged particles inside the atom. Lorentz further postulated that a strong magnetic field would affect these oscillations and thus the wavelength of the emitted light. In 1896 Pieter Zeeman, a student of Lorentz, demonstrated the effect now known as the Zeeman effect. In 1904 Lorentz proposed a set of equations that relate the spatial and temporal coordinates for two systems moving at a large constant velocity with respect to each other. The equations are now called the Lorentz transformations and their prediction of increase in mass, shortening of length, and time dilation formed the basis of Albert Einstein's special theory of relativity.

In 1902 Lorentz and Zeeman shared the Nobel Prize in Physics *"in recognition of the extraordinary service they rendered by their researches into the influence of magnetism upon radiation phenomena"*.

MANDELBROT, Benoit (1924–2010)

Polish-born American mathematician, educated in France at the École Polytechnique in Paris and the California Institute of Technology (Caltech) in Pasadena. Mandelbrot received his doctorate in mathematics from the University of Paris in 1952. From 1949 until 1957 he was on staff at the Centre National de la Recherche Scientifique. In 1958 he joined the research staff at the IBM T. J. Watson Research Center in Yorktown Heights, New York and he remained with the IBM until his retirement in 1987 when he became Professor of Mathematical Sciences at Yale University.

Mandelbrot is best known as the founder of fractal geometry, a modern invention in contrast to the 2000 years old Euclidean geometry. He is also credited with coining the term "fractal". Man-made objects usually follow Euclidean geometry shapes, while objects in nature generally follow more complex rules designs defined by iterative or recursive algorithms. The most striking feature of fractal geometry is the self-similarity of objects or phenomena, implying that the fractal contains smaller components that replicate the whole fractal when magnified. In theory the fractal is composed of an infinite number of ever diminishing components, all of the same shape.

Mandelbrot discovered that self-similarity is a universal property that underlies the complex fractal shapes, illustrated its behavior mathematically and founded a completely new methodology for analyzing these complex systems. His name is now identified with a particular set of complex numbers which generate a type of fractal with very attractive properties (Mandelbrot Set).

MANSFIELD, Peter (born in 1933)

British physicist, educated at the Queen Mary College in London where he obtained his B.Sc. in physics in 1959 and doctorate in physics in 1962. He spent 1962–1964 as research associate at the University of Illinois in Urbana and 1964–1978 as lecturer and reader at the University of Nottingham. In 1979 he was appointed Professor of Physics at the University of Nottingham and since 1994 he is Emeritus Professor of Physics at the University of Nottingham.

Mansfield's doctoral thesis was on the physics of nuclear magnetic resonance (NMR), at the time used for studies of chemical structure, and he spent the 1960s perfecting his understanding of NMR techniques. In the early 1970s Mansfield began studies in the use of NMR for imaging and developed magnetic field gradient techniques for producing two-dimensional images in arbitrary planes through a human body. The term "nuclear" was dropped from NMR imaging and the technique is now referred to as magnetic resonance imaging or MRI. Mansfield is also credited with developing the MRI protocol called the "echo planar imaging" which in comparison to standard techniques allows a much faster acquisition of images and makes functional MRI (fMRI) possible.

Mansfield shared the 2003 Nobel Prize in Medicine and Physiology with Paul C. Lauterbur *"for their discoveries concerning magnetic resonance imaging"*.

MARSDEN, Ernest (1889–1970)

New Zealand-born physicist who made a remarkable contribution to science in New Zealand and England. He studied physics at the University of Manchester and as a student of Ernest Rutherford, in collaboration with Hans Geiger, carried out the α-particle scattering experiments that inspired Rutherford to propose the atomic model, currently known as the Rutherford-Bohr model of the atom. In 1914 he returned to New Zealand to become Professor of Physics at Victoria University in Wellington. In addition to scientific work, he became involved with public service and helped in setting up the New Zealand Department of Scientific and Industrial Research. During World War II, he became involved with radar technology in defense work and in 1947 he was elected president of the Royal Society of New Zealand. He then returned to London as New Zealand's scientific liaison officer and "ambassador" for New Zealand science. In 1954 he retired to New Zealand and remained active on various advisory committees as well as in radiation research until his death in 1970.

MAXWELL, James Clerk (1831–1879)

Scottish mathematician and theoretical physicist, educated at the University of Edinburgh (1847–1850) and Trinity College of Cambridge University. From 1856–1860 he held the position of Chair of Natural Philosophy of Marischal College of Aberdeen

University and from 1860–1865 he was Chair of Natural Philosophy at King's College in London where he established a regular contact with Michael Faraday.

In 1865 Maxwell resigned his position at King's College and for the following five years held no academic appointments. In 1871 he became the first Cavendish Professor of Physics at Cambridge and was put in charge of developing the Cavendish Laboratory. He remained at Cavendish until his death in 1879.

While at King;s College, Maxwell made his most important advances in the theory of electromagnetism and developed the famous set of four Maxwell equations, demonstrating that electricity, magnetism and light are all derived from electromagnetic field. His work on electromagnetic theory is considered of the same importance as Newton's work in classical mechanics and Einstein's work in relativistic mechanics.

Maxwell also worked on the kinetic theory of gases and in 1866 derived the Maxwellian distribution describing the fraction of gas molecules moving at a specified velocity at a given temperature. Together with Willard Gibbs and Ludwig Boltzmaun Maxwell is credited with developing statistical mechanics and paving the way for quantum mechanics and relativistic mechanics.

MEITNER, Lise (1878–1968)

Austrian-born physicist who studied physics at the University of Vienna and was strongly influenced in her vision of physics by Ludwig Boltzmann, a leading theoretical physicist of the time. In 1907 Meitner moved to Berlin to work with Max Planck and at the University of Berlin she started a life-long friendship and professional association with radiochemist Otto Hahn. At the Berlin University both Meitner and Hahn were appointed as scientific associates and progressed through academic ranks to attain positions of professor.

During her early days in Berlin, Meitner discovered the element protactinium with atomic number $Z = 91$ and also discovered, two years before Auger, the non-radiative atomic transitions that are now referred to as the Auger effect. Meitner became the first female physics professor in Germany but, despite her reputation as an excellent physicist, she, like many other Jewish scientists, had to leave Germany during the 1930s. She moved to Stockholm and left behind in Berlin her long-term collaborator and friend Otto Hahn, who at that time was working with Friedrich Strassmann, an analytical chemist, on studies of uranium bombardment with neutrons. Their experiments, similarly to those reported by Irene Joliot-Curie and Pavle Savić were yielding surprising results suggesting that in neutron bombardment uranium was splitting into smaller atoms with atomic masses approximately half of that of uranium. In a letter Hahn described the uranium disintegration by neutron bombardment to Meitner in Stockholm and she, in collaboration with Otto Frisch, succeeded in providing a theoretical explanation for the uranium splitting and coined the term nuclear fission to name the process.

The 1944 Nobel Prize in Chemistry was awarded to Hahn *"for the discovery of the nuclear fission"*. The Nobel Committee unfortunately ignored the contributions by Strassmann, Meitner and Frisch to the theoretical understanding of the nuclear fission process. Most texts dealing with the history of nuclear fission now recognize the four scientists: Hahn, Strassmann, Meitner, and Frisch as the discoverers of the fission process.

Despite several problems that occurred with recognizing Meitner's contributions to modern physics, her scientific work certainly was appreciated and is given the same ranking in importance as that of Marie Skłodowska-Curie. In 1966 Meitner together with Hahn and Strassmann shared the prestigious Enrico Fermi Award. In honor of Meitner's contributions to modern physics the element with atomic number 109 was named meitnerium (Mt).

MENDELEEV, Dmitri Ivanovič (1834–1907)

Russian physical chemist, educated at the University of St. Petersburg where he obtained his M.A. in chemistry in 1856 and doctorate in chemistry in 1865. The years between 1859 and 1861 Mendeleev spent studying in Paris and Heidelberg. He worked as Professor of Chemistry at the Technical Institute of St. Petersburg and the University of St. Petersburg from 1862 until 1890 when he retired from his academic posts for political reasons. From 1893 until his death in 1907 he was Director of the Bureau of Weights and Measures in St. Petersburg.

While Mendeleev made contributions in many areas of general chemistry as well as physical chemistry and was an excellent teacher, he is best known for his 1869 discovery of the Periodic Law and the development of the Periodic Table of Elements. Until his time elements were distinguished from one another by only one basic characteristic, the atomic mass, as proposed by John Dalton in 1805. By arranging the 63 then-known elements by atomic mass as well as similarities in their chemical properties, Mendeleev obtained a table consisting of horizontal rows or periods and vertical columns or groups. He noticed several gaps in his Table of Elements and predicted that they represented elements not yet discovered. Shortly afterwards elements gallium, germanium and scandium were discovered filling three gaps in the table, thereby confirming the validity of Mendeleev's Periodic Table of Elements. Mendeleev's table of more than a century ago is very similar to the modern 21st century Periodic Table, except that the 111 elements of the modern periodic table are arranged according to their atomic number Z in contrast to Mendeleev's table in which the 63 known elements were organized according to atomic mass. To honor Mendeleev's work the element with atomic number Z of 101 is called mendelevium.

MILLIKAN, Robert Andrews (1868–1952)

American physicist, educated at Oberlin College (Ohio) and Columbia University in New York where he received a doctorate in physics in 1895. He then spent a year at the universities of Berlin and Götingen, before accepting a position at the University of Chicago in 1896. By 1910 he was Professor of Physics and remained in Chicago until 1921 when he was appointed Director of the Norman Bridge Laboratory of Physics at the California Institute of Technology (Caltech) in Pasadena. He retired in 1946.

Millikan was a gifted teacher and experimental physicist. During his early years at Chicago he authored and coauthored many physics textbooks to help and simplify the teaching of physics. As a scientist he made many important discoveries in electricity, optics and molecular physics. His earliest and best known success was the accurate determination, in 1910, of the electron charge with the "falling-drop method" now commonly referred to as the Millikan experiment. He also verified experimentally the Einstein's photoelectric effect equation and made the first direct photoelectric determination of Planck's quantum constant h.

The 1923 Nobel Prize in Physics was awarded to Millikan *"for his work on the elementary charge of electricity and on the photoelectric effect"*.

MØLLER, Christian (1904–1980)

Danish theoretical physicist, educated at the University of Copenhagen where he first studied mathematics and then theoretical physics. In 1929 he obtained his M.Sc. degree and in 1933 his doctorate under Niels Bohr on passage of fast electrons through matter. He first worked as lecturer at the Bohr institute in Copenhagen and from 1943 until retirement in 1975 he was Professor of Mathematical Physics at the University of Copenhagen.

Møller began his theoretical physics work in nuclear and high-energy physics and was influenced by the many eminent physicists who were visiting the Bohr institute during the 1930s, such as George Gamow, Nevill Mott, and Yoshio Nishima. He made important contributions in nuclear and high-energy physics problems combined with the theory of relativity, most notably in alpha decay, α-particle scattering, electron-positron theory and meson theory of nuclear forces. He is best known for his work on electron scattering on atomic orbital electrons which in his honor is termed Møller scattering.

MÖSSBAUER, Rudolf Ludwig (born in 1929)

German physicist, educated at the Technische Hochschule (Technical University) in Munich, where he received his doctorate in physics in 1958, after carrying out the experimental portion of his thesis work in Heidelberg at the Institute for Physics of the Max Planck Institute for Medical Research. During 1959 Mössbauer worked as scientific assistant at the Technical University in Munich and from 1960 until 1962 as Professor of Physics at the California Institute of Technology (Caltech) in Pasadena. In 1962 he returned to the Technical Institute in Munich as Professor of Experimental Physics and stayed there his whole professional career except for the period 1972–1977 which he spent in Grenoble as the Director of the Max von Laue Institute.

Mössbauer is best known for his 1957 discovery of recoil-free gamma ray resonance absorption; a nuclear effect that is named after him and was used to verify Albert Einstein's theory of relativity and to measure the magnetic field of atomic nuclei. The Mössbauer effect involves the emission and absorption of gamma rays by atomic nuclei. When a free excited nucleus emits a gamma photon, the nucleus recoils in order to conserve momentum. The nuclear recoil uses up a minute portion of the decay energy, so that the shift in the emitted photon energy prevents the absorption of the photon by another target nucleus. While working on his doctorate thesis in Heidelberg, Mössbauer discovered that by fixing emitting and absorbing nuclei into a crystal lattice, the whole lattice gets involved in the recoil process, minimizing the recoil energy loss and creating an overlap between emission and absorption lines thereby enabling the resonant photon absorption process and creating an extremely sensitive detector of photon energy shifts.

Mössbauer received many awards and honorable degrees for his discovery; most notably, he shared with Robert Hofstadter the 1961 Nobel Prize in Physics *"for his researches concerning the resonance absorption of gamma radiation and his discovery in this connection of the effect which bears his name"*. Hofstadter received his share of the 1961 Nobel Prize for his pioneering studies of electron scattering in atomic nuclei.

MOSELEY, Henry Gwen Jeffreys (1887–1915)

British physicist, educated at the University of Oxford where he graduated in 1910. He began his professional career at the University of Manchester as lecturer in physics and research assistant under Ernest Rutherford.

Based on work by *Charles Barkla* who discovered characteristic x-rays and work of the team of William Bragg and Lawrence Bragg who studied x-ray diffraction, Moseley undertook in 1913 a study of the K and L characteristic x-rays emitted by then-known elements from aluminum to gold. He found that the square root of the frequencies of the emitted characteristic x-ray lines plotted against a suitably chosen integer Z yielded straight lines. Z was subsequently identified as the number of positive charges (protons) and the number of electrons in an atom and is now referred to as the atomic number Z. Moseley noticed gaps in his plots that corresponded to atomic numbers Z of 43, 61, and 75. The elements with $Z = 43$ (technetium) and $Z = 61$ (promethium) do not occur naturally but were produced artificially years later. The $Z = 75$ element (rhenium) is rare and was discovered only in 1925. Moseley thus found that the atomic number of an element can be deduced from the element's characteristic spectrum (non-destructive testing). He also established that the periodic table of elements should be arranged according to the atomic number Z rather than according to the atomic mass number A as was common at his time.

There is no question that Moseley during a short time of two years produced results that were very important for the development of atomic and quantum physics and were clearly on the level worthy of Nobel Prize. Unfortunately, he perished during World War I shortly after starting his professional career in physics.

MOTT, Nevill Francis (1905–1996)

British physicist, educated at Clifton College in Bristol and St. John's College in Cambridge where he studied mathematics and physics, received a baccalaureate degree in 1927, and carried out his first work in theoretical physics studying scattering of electrons on nuclei. During 1928 he continued his physics studies under Niels Bohr in Copenhagen and Max Born in Göttingen. He spent the 1929–30 academic year as lecturer in Manchester where William L. Bragg introduced him to solid atate physics. In 1930 Mott returned to Cambridge, obtained his M.Sc. degree in Physics, and continued his work on particle scattering on atoms and nuclei in Rutherford's laboratory. His contributions to collision theory are recognized by the description of electron–nucleus scattering as Mott scattering.

The period from 1933 to 1954 Mott spent in Bristol, first as Professor of Theoretical Physics and from 1948 as Chairman of the Physics Department. His work concentrated on solid state physics and resulted in many important publications and several books. In 1954 Mott became Cavendish Professor of Physics at Cambridge. He continued his work in solid state physics, concentrating on amorphous semiconductors and producing research for which he shared the 1977 Nobel Prize in Physics with Philip W. Anderson and John H. Van Vleck *"for their fundamental theoretical investigations of the electronic structure of magnetic and disordered systems"*.

In addition to his contributions to experimental and theoretical physics, Mott has also taken a leading role in science education reform in the U.K. and served on many committees that dealt with science education.

NISHINA, Yoshio (1890–1951)

Japanese physicist, educated at the University of Tokyo where he graduated in 1918. He worked three years as an assistant at the University of Tokyo and then spent several years in Europe: 1921–1923 at the University of Cambridge with Ernest Rutherford and 1923–1928 at the University of Copenhagen with Niels Bohr. From 1928 to 1948 he worked at the University of Tokyo.

Nishina is best known internationally for his collaboration with Oskar Klein on the cross section for Compton scattering in 1928 (Klein-Nishina formula). Upon return to Japan from Europe, Nishina introduced the study of nuclear and high-energy physics in Japan and trained many young Japanese physicists in the nuclear field. During World War II Nishina was the central figure in the Japanese atomic weapons program that was competing with the American Manhattan project and using the same thermal uranium enrichment technique as the Americans. The race was tight; however, the compartmentalization of the Japanese nuclear weapons program over competing ambitions of the army, air force and the navy gave the Americans a definite advantage and eventual win in the nuclear weapons competition that resulted in the atomic bombs over Hiroshima and Nagasaki and Japanese immediate surrender.

PAULI, Wolfgang (1900–1958)

Austrian-born physicist, educated at the University of Munich where he obtained his doctorate in physics in 1921. He spent one year at the University of Göttingen and one year at the University of Copenhagen before accepting a lecturer position at the University of Hamburg (1923–1928). From 1928 to 1958 he held an appointment of Professor of Theoretical Physics at the Eidgenössische Technische Hochschule in Zürich in (ETHZ). From 1940 to 1946 Pauli was a visiting professor at the Institute for Advanced Study in Princeton.

Pauli is known as an extremely gifted physicist of his time. He is best remembered for enunciating the existence of the neutrino in 1930 and for introducing the exclusion principle to govern the states of atomic electrons in general. The exclusion principle is now known as the Pauli Principle and contains three components. The first component states that no two electrons can be at the same place at the same time. The second component states that atomic electrons are characterized by four quantum numbers: principal, orbital, magnetic and spin. The third component states that no two electrons in an atom can occupy a state that is described by exactly the same set of the four quantum numbers. The exclusion principle was subsequently expanded to other electronic and fermionic systems, such as molecules and solids.

The 1945 Nobel Prize in Physics was awarded to Pauli *"for his discovery of the Exclusion Principle, also called the Pauli Principle"*.

PLANCK, Max Karl Ernst (1858–1947)

German physicist, educated at the University of Berlin and University of Munich where he received his doctorate in physics in 1879. He worked as Assistant Professor at the University of Munich from 1880 until 1885, then Associate Professor at the University of Kiel until 1889 and Professor of Physics at the University of Berlin until his retirement in 1926.

Most of Planck's work was on the subject of thermodynamics in general and studies of entropy and second law of thermodynamics in particular. He was keenly interested in the blackbody problem and the inability of classical mechanics to predict the blackbody spectral distribution. Planck studied the blackbody spectrum in depth and concluded that it must be electromagnetic in nature. In contrast to classical equations that were formulated for blackbody radiation by Wien and Rayleigh, with Wien's equation working only at high frequencies and Rayleigh's working only at low frequencies, Planck formulated an equation that predicted accurately the whole range of applicable frequencies and is now known as Planck equation for blackbody radiation. The derivation was based on the revolutionary idea that the energy emitted by a resonator can only take on discrete values or quanta, with the quantum energy ε equal to $h\nu$, where ν is the frequency and h a universal constant now referred to as the Planck's constant. Planck's idea of quantization has been successfully applied to the photoelectric effect by Albert Einstein and to the atomic model by Niels Bohr.

In 1918 Planck was awarded the Nobel Prize in Physics *"in recognition of the services he rendered to the advancement of Physics by his discovery of energy quanta"*. In addition to Planck constant and Planck's formula, Planck's name and work are honored with the Max Planck Medal that is awarded annually as the highest distinction by the German Physical Society (Deutsche Physikalische Gesellschaft) and the Max Planck Society for the Advancement of Science that supports basic research at 80 research institutes focusing on research in biology, medicine, chemistry, physics, technology and humanities.

POYNTING, John Henry (1852–1914)

British physicist, born in Monton near Manchester and educated at Owens College in Manchester (B.Sc. in 1876) and Trinity College in Cambridge (Sc.D in 1887). He started his academic career in Manchester (1876–1878) where he met Joseph J. Thomson with whom he completed "A Textbook of Physics". In 1878 he became a Fellow of Trinity College in Cambridge and for two years worked in Cavendish Laboratory under J.C. Maxwell. In 1880 he moved to the University of Birmingham as professor of physics and stayed there for the rest of his professional life.

Poynting was an excellent theoretical as well as experimental physicist. His greatest discovery was the Poynting theorem in electromagnetism from which comes the definition of the Poynting vector. He is also remembered for many other contributions to physics, such as an accurate measurement of Newton's gravitational constant; determination of the mean density of the Earth; discovery of the Poynting-Robertson effect (small particles in the orbit about Sun spiral into the Sun); and method for determining absolute temperature of celestial objects.

The Poynting theorem deals with the conservation of energy for the electromagnetic field and can be derived from the Lorentz force in conjuction with Maxwell's equations. The Poynting vector represents the energy flow through a given area for electromagnetic field and is usually written as a vector product $\mathbf{S} = \mathcal{E} \times \mathcal{B}/\mu_0$ where $\mathcal{E}$ is the electric field, $\mathcal{B}$ the magnetic field, and μ_0 the permeability of vacuum.

Poynting was held in high esteem by his peers and received numerous awards for his work. He was President of Physical Society and was elected Fellow of the Royal Society. In 1905 he received the Royal Medal from the Royal Society.

PURCELL, Edward Mills (1912–1997)

American physicist, educated at Purdue University in Indiana where he received his Bachelor's degree in electrical engineering in 1933 and Harvard where he received his doctorate in physics in 1938. After serving for two years as lecturer of physics at Harvard, he worked at the Massachusetts Institute of Technology on development of new microwave techniques. In 1945 Purcell returned to Harvard as Associate Professor of Physics and became Professor of Physics in 1949.

Purcell is best known for his 1946 discovery of nuclear magnetic resonance (NMR) with his students Robert Pound and Henry C. Torrey. NMR offers an elegant and precise way of determining chemical structure and properties of materials and is widely used not only in physics and chemistry but also in medicine where, through the method of magnetic resonance imaging (MRI), it provides non-invasive means to image internal organs and tissues of patients.

In 1952 Purcell shared the Nobel Prize in Physics with Felix Bloch *"for their development of new methods for nuclear magnetic precision measurements and discoveries in connection therewith"*.

RAYLEIGH, John William Strutt (1842–1919)

English mathematician and physicist who studied mathematics at the Trinity College in Cambridge. Being from an affluent family he set up his physics laboratory at home and made many contributions to applied mathematics and physics from his home laboratory. From 1879 to 1884 Rayleigh was Professor of Experimental Physics and Head of the Cavendish Laboratory at Cambridge, succeeding James Clark Maxwell. From 1887 to 1905 he was Professor of Natural Philosophy at the Royal Institution in London.

Rayleigh was a gifted researcher and made important contributions to all branches of physics known at his time, having worked in optics, acoustics, mechanics, thermodynamics, and electromagnetism. He is best known for explaining that the blue color of the sky arises from scattering of light by dust particles in air and for relating the degree of light scattering to the wavelength of light (Rayleigh scattering). He also accurately defined the resolving power of a diffraction grating; established standards of electrical resistance, current, and electromotive force; discovered argon; and derived an equation describing the distribution of wavelengths in blackbody radiation (the equation applied only in the limit of large wavelengths).

In 1904 Rayleigh was awarded the Nobel Prize in Physics *"for his investigations of the densities of the most important gases and for his discovery of the noble gas argon in connection with these studies"*. He discovered argon together with William Ramsey who obtained the 1904 Nobel Prize in Chemistry for his contribution to the discovery.

RICHARDSON, Owen Willans (1879–1959)

British physicist, educated at Trinity College in Cambridge from where he graduated in 1890 as a student of Joseph J. Thomson at the Cavendish Laboratory. He was appointed Professor of Physics at Princeton University in the United States in 1906 but in 1914 returned to England to become Professor of Physics at King's College of the University of London.

Richardson is best known for his work on thermionic emission of electrons from hot metallic objects that enabled the development of radio and television tubes as well as modern x-ray (Coolidge) tubes. He discovered the equation that relates the rate of electron emission to the absolute temperature of the metal. The equation is now referred to as the Richardson law or the Richardson-Dushman equation.

In 1928 Richardson was awarded the Nobel Prize in Physics *"for his work on the thermionic phenomenon and especially for the law that is named after him"*.

RÖNTGEN, Wilhelm Konrad (1845–1923)

German physicist, educated at the University of Utrecht in Holland and University of Zürich where he obtained his doctorate in physics in 1869. He worked as academic physicist at several German universities before accepting a position of Chair of Physics at the University of Giessen in 1979. From 1888 until 1900 he was Chair of Physics at the University of Würzburg and from 1900 until 1920 he was Chair of Physics at the University of Munich.

Röntgen was active in many areas of thermodynamics, mechanics and electricity but his notable research in these areas was eclipsed by his accidental discovery in 1895 of "a new kind of ray". The discovery occurred when Röntgen was studying cathode rays (now known as electrons, following the work of Joseph J. Thomson) in a Crookes tube, a fairly mundane and common experiment in physics departments at the end of the 19th century. He noticed that, when his energized Crookes tube was enclosed in a sealed black and light-tight envelope, a paper plate covered with barium platinocyanide, a known fluorescent material, became fluorescent despite being far removed from the discharge tube. Röntgen concluded that he discovered an unknown type of radiation, much more penetrating than visible light and produced when cathode rays strike a material object inside the Crookes tube. He named the new radiation x-rays and the term is generally used around the World. However, in certain countries x-rays are often called Röntgen rays. In 1912 Max von Laue showed with his crystal diffraction experiments that x-rays are electromagnetic radiation similar to visible light but of much smaller wavelength. In tribute to Röntgen's contributions to modern physics the element with the atomic number 111 was named röntgenium (Rg).

In 1901 inaugural Nobel Prize in Physics was awarded to Röntgen *"in recognition of the extraordinary services he has rendered by the discovery of the remarkable rays subsequently named after him"*.

RUTHERFORD, Ernest (1871–1937)

New Zealand-born nuclear physicist, educated at the Canterbury College in Christchurch, New Zealand (B.Sc. in mathematics and physical science in 1894) and at the Cavendish Laboratory of the Trinity College in Cambridge. He received his science doctorate from the University of New Zealand in 1901. Rutherford was one of the most illustrious physicists of all time and his professional career consists of three distinct periods: as MacDonald Professor of Physics at McGill University in Montreal (1898–1907); as Langworthy Professor of Physics at the University of Manchester (1908–1919); and as Cavendish Professor of Physics at the Cavendish Laboratory of Trinity College in Cambridge (1919–1937).

With the exception of his early work on magnetic properties of iron exposed to high frequency oscillations, Rutherford's career was intimately involved with the advent and growth of nuclear physics. Nature provided Rutherford with α-particles, an important tool for probing the atom, and he used the tool in most of his exciting discoveries that revolutionized physics in particular and science in general.

Before moving to McGill in 1898, Rutherford worked with Joseph J. Thomson at the Cavendish Laboratory on detection of the just-discovered x-rays (Wilhelm Röntgen in 1895) through studies of electrical conduction of gases caused by x-ray ionization of air. He then turned his attention to the just-discovered radiation emanating from uranium (Henri Becquerel in 1896) and radium (Pierre Curie and Marie Skłodowska-Curie in 1898) and established that uranium radiation consists of at least two components, each of particulate nature but with different penetrating powers. He coined the names α and β-particles for the two components.

During his 10 years at McGill, Rutherford published 80 research papers, many of them in collaboration with Frederick Soddy, a chemist who came to McGill from Oxford in 1900. Rutherford discovered the radon gas as well as gamma rays and speculated that the gamma rays were similar in nature to x-rays. In collaboration with Soddy he described the transmutation of radioactive elements as a spontaneous disintegration of atoms and defined the half-life of a radioactive substance as the time it takes for its activity to drop to half of its original value. He noted that all atomic disintegrations were characterized by emissions of one or more of three kinds of rays: α, β, and γ.

During the Manchester period Rutherford determined that α-particles were helium ions. He guided Hans Geiger and Ernest Marsden through the now-famous α-particle scattering experiment. Based on the experimental results Rutherford in 1911 proposed a revolutionary model of the atom which was known to have a size of the order of 10^{-10} m. He proposed that most of the atomic mass is concentrated in a miniscule nucleus with a size of the order of 10^{-15} m) and that the atomic electrons are distributed in a cloud around the nucleus. In 1913 Niels Bohr expanded Rutherford's nuclear atomic model by introducing the idea of the quantization of electrons' angular momenta and the resulting model is now called the Rutherford-Bohr atomic model. During his last year at Manchester, Rutherford discovered that nuclei of nitrogen, when bombarded with α-particles, artificially disintegrate and produce protons in the process. Rutherford was thus first in achieving artificial transmutation of an element through a nuclear reaction.

During the Cambridge period Rutherford collaborated with many world-renowned physicists such as John Cocroft and Ernest Walton in designing a proton linear accelerator now called the Cocroft-Walton generator, and with James Chadwick in discovering the neutron in 1932. Rutherford's contributions to modern physics are honored with the element of atomic number 104 which was named rutherfordium (Rf).

In 1908 Rutherford was awarded the Nobel Prize in Chemistry *"for his investigations into the disintegration of the elements and the chemistry of radioactive substances"*.

RYDBERG, Johannes (1854–1919)

Swedish physicist, educated at Lund University. He obtained his Ph.D. in mathematics in 1879 but worked all his professional life as a physicist at Lund University where he became Professor of Physics and Chairman of the Physics department.

Rydberg is best known for his discovery of a mathematical expression that gives the wavenumbers of spectral lines for various elements and includes a constant that is now referred to as the Rydberg constant $(R_\infty = 109\,737\,\text{cm}^{-1})$. In honor of Rydberg's work in physics the absolute value of the ground state energy of the hydrogen atom is referred to as the Rydberg energy $(E_R = 13.61\,\text{eV})$.

SCHRÖDINGER, Erwin (1887–1961)

Austrian physicist, educated at the University of Vienna where he received his doctorate in Physics in 1910. He served in the military during World War I and after the war moved through several short-term academic positions until in 1921 he accepted a Chair in Theoretical Physics at the University of Zürich. In 1927 he moved to the University of Berlin as Planck's successor. The rise of Hitler in 1933 convinced Schrödinger to leave Germany. After spending a year at Princeton University, he accepted a post at the University of Graz in his native Austria. The German annexation of Austria in 1938 forced him to move again, this time to the Institute for Advanced Studies in Dublin where he stayed until his retirement in 1955.

Schrödinger made many contributions to several areas of theoretical physics, however, he is best known for introducing wave mechanics into quantum mechanics. Quantum mechanics deals with motion and interactions of particles on an atomic scale and its main attribute is that it accounts for the discreteness (quantization) of physical quantities in contrast to classical mechanics in which physical quantities are assumed continuous. Examples of quantization were introduced by Max Planck who in 1900 postulated that oscillators in his blackbody emission theory can possess only certain quantized energies; Albert Einstein who in 1905 postulated that electromagnetic radiation exists only in discrete packets called photons; and Niels Bohr who in 1913 introduced the quantization of angular momenta of atomic orbital electrons. In addition, Louis de Broglie in 1924 introduced the concept of wave-particle duality.

Schrödinger's wave mechanics is based on the so-called Schrödinger's wave equation, a partial differential equation that describes the evolution over time of the wave function of a physical system. Schrödinger and other physicists have shown that many quantum mechanical problems can be solved by means of the Schrödinger equation. The best known examples are: finite square well potential; infinite square well potential; potential step; simple harmonic oscillator; and hydrogen atom.

In 1933 Schrödinger shared the Nobel Prize in Physics with Paul A.M. Dirac *"for the discovery of new productive forms of atomic theory"*.

SEGRÈ, Emilio Gino (1905–1989)

Italian-born American nuclear physicist, educated at the University of Rome, where he received his doctorate in physics as Enrico Fermi's first graduate student in 1928. In 1929 he worked as assistant at the University of Rome and spent the years 1930–1931 with Otto Stern in Hamburg and Pieter Heman in Amsterdam. In 1932 he became Assistant Professor of Physics at the University of Rome and in 1936 he was appointed Director of the Physics Laboratory at the University of Palermo. In 1938 Segrè came to Berkeley University, first as research associate then as physics lecturer. From 1943 until 1946 he was a group leader in the Los Alamos Laboratory of the Manhattan Project and from 1946 until 1972 he held an appointment of Professor of Physics at Berkeley. In 1974 he was appointed Professor of Physics at the University of Rome.

Segrè is best known for his participation with Enrico Fermi in neutron experiments bombarding uranium-238 with neutrons thereby creating several elements heavier than uranium. They also discovered thermal neutrons and must have unwittingly triggered uranium-235 fission during their experimentation. It was Otto Hahn and colleagues, however, who at about the same time discovered and explained nuclear fission. In 1937 Segrè discovered technetium, the first man-made element not found in nature and, as it subsequently turned out, of great importance to medical physics in general and nuclear medicine in particular. At Berkeley Segrè discovered plutonium-239 and established that it was fissionable just like uranium-235. Segrè made many other important contributions to nuclear physics and high-energy physics and, most notably, in collaboration with Owen Chamberlain discovered the antiproton. Segrè and Chamberlain shared the 1959 Nobel Prize in Physics *"for their discovery of the antiproton"*.

SELTZER, Stephen Michael (born in 1940)

American physicist, educated at the Virginia Polytechnic Institute where he received his B.S. in physics in 1962 and at the University of Maryland, College Park where he received his M.Sc. in physics in 1973. In 1962 he joined the Radiation Theory Section at the National Bureau of Standards (NBS), now the National Institute of Standards and Technology (NIST), and has spent his professional career there, becoming the Director of the Photon and Charged-Particle Data Center at NIST in 1988 and the Leader of the Radiation Interactions and Dosimetry Group in 1994. He joined the International Commission on Radiation Units and Measurements (ICRU) in 1997.

Seltzer worked with Martin Berger on the development of Monte Carlo codes for coupled electron-photon transport in bulk media, including the transport-theoretical methods and algorithms used, and the interaction cross-section information for these radiations. Their ETRAN codes, underlying algorithms and cross-section data have been incorporated in most of the current radiation-transport Monte Carlo codes. Seltzer was instrumental in the development of extensive data for the production of bremsstrahlung by electrons (and positrons), electron and positron stopping powers, and a recent database of photon energy-transfer and energy-absorption coefficients. His earlier work included applications of Monte Carlo calculations to problems in space science, detector response, and space shielding, which led to the development of the SHIELDOSE code used for routine assessments of absorbed dose within spacecraft.

SIEGBAHN, Karl Manne Georg (1886–1978)

Swedish physicist, educated at the University of Lund where he obtained his Doctorate in Physics in 1911. From 1907 to 1923 he lectured in physics at the University of Lund, first as Assistant to Professor J. R. Rydberg, from 1911 to 1915 as Lecturer in Physics, and from 1915 to 1923 as Professor of Physics. In 1923 he moved to the University of Uppsala where he stayed as Professor of Physics until 1937 when he became a Research Professor of Experimental Physics and the first Director of the Physics Department of the Nobel Institute of the Royal Swedish Academy of Sciences in Stockholm. He remained with the Academy till 1975 when he retired.

Siegbahn's main contribution to physics was in the area of x-ray spectroscopy and covered both the experimental and theoretical aspects. He made many discoveries related to x-ray emission spectra from various target materials and also developed equipment and techniques for accurate measurement of x-ray wavelengths. He built his own x-ray spectrometers, produced numerous diamond-ruled glass diffraction gratings for his spectrometers, and measured x-ray wavelengths of many target elements to high precision using energetic electrons to excite the characteristic spectral emission lines.

To honor Siegbahn's significant contributions to x-ray spectroscopy the notation for x-ray spectral lines that are characteristic to elements is referred to as Siegbahn's notation. The notation has been in use for many decades and only recently the International Union of Pure and Applied Chemistry (IUPAC) proposed a new notation referred to as the IUPAC notation that is deemed more practical and is slated to replace the existing Siegbahn notation.

In 1924 Siegbahn received the Nobel Prize in Physics *"for discoveries and research in the field of x-ray spectroscopy"*.

SODDY, Frederick (1877–1956)

British chemist, educated at Morton College in Oxford where he received his degree in chemistry in 1898. After graduation he spent two years as research assistant in Oxford, then went to McGill University in Montreal where he worked with Ernest Rutherford on radioactivity. In 1902 Soddy returned to England to work with William Ramsay at the University College in London. He then served as lecturer in physical chemistry at the University of Glasgow (1910–1914) and Professor of Chemistry at the University of Aberdeen (1914–1919). His last appointment was from 1919 until 1936 as Lees Professor of Chemistry at Oxford University.

Soddy is best known for his work in the physical and chemical aspects of radioactivity. He learned the basics of radioactivity with Ernest Rutherford at McGill University in Montreal and then collaborated with William Ramsay at the University College. With Rutherford he confirmed the hypothesis by Marie Skłodowska-Curie that radioactive decay was an atomic rather than chemical process, postulated that helium is a decay product of uranium, and formulated the radioactive disintegration law. With Ramsay he confirmed that the alpha particle was doubly ionized helium atom. Soddy's Glasgow period was his most productive period during which he enunciated the so-called displacement law and introduced the concept of isotopes. The displacement law states that emission of an alpha particle from a radioactive element causes the element to transmutate into a new element that moves back two places in the Periodic Table of Element. The concept of isotopes states that certain elements exist in two or more forms that differ in atomic mass but are chemically indistinguishable.

Soddy was awarded the 1921 Nobel Prize in Chemistry *"for his contributions to our knowledge of the chemistry of radioactive substances, and his investigations into the origin and nature of isotopes"*.

STERN, Otto (1888–1969)

German-born physicist educated in physical chemistry at the University of Breslau where he received his doctorate in 1912. He worked with Albert Einstein at the University of Prague and at the University of Zürich before becoming an Assistant Professor at the University of Frankfurt in 1914. During 1921–1922 he was an Associate Professor of Theoretical Physics at the University of Rostock and in 1923 he was appointed Professor of Physical Chemistry at the University of Hamburg. He remained in Hamburg until 1933 when he moved to the United States to become a Professor of Physics at the Carnegie Institute of Technology in Pittsburgh.

Stern is best known for the development of the molecular beam epitaxy, a technique that deposits one or more pure materials onto a single crystal wafer forming a perfect crystal; discovery of spin quantization in 1922 with Walther Gerlach; measurement of atomic magnetic moments; demonstration of the wave nature of atoms and molecules; and discovery of proton's magnetic moment.

Stern was awarded the 1943 Nobel Prize in Physics *"for his contribution to the development of the molecular ray method and his discovery of the magnetic moment of the proton"*.

STRASSMANN, Friedrich Wilhelm (1902–1980)

German physical chemist, educated at the Technical University in Hannover where he received his doctorate in 1929. He worked as an analytical chemist at the Kaiser Wilhelm Institute for Chemistry in Berlin from 1934 until 1945. In 1946 Strassmann became Professor of Inorganic Chemistry at the University of Mainz. From 1945 to 1953 he was Director of the Chemistry department at the Max Planck Institute.

Strassmann is best known for his collaboration with Otto Hahn and Lise Meitner on experiments that in 1938 lead to the discovery of neutron induced fission of uranium atom. Strassmann's expertise in analytical chemistry helped with discovery of the light elements produced in the fission of uranium atoms. In 1966 the nuclear fission work by Hahn, Strassmann and Meitner was recognized with the Enrico Fermi Award.

SZILÁRD Leó (1898–1964)

Hungarian born American physicist, educated in engineering first at the Budapest Technical University and at the "Technische Hochschule" in Berlin. After the basic training in engineering he switched to physics and received his Ph.D. in Physics in 1923 from the Humboldt University in Berlin. He worked as physics instructor and inventor at the University of Berlin. In 1933 he moved to Britain where he worked till 1938 on various nuclear physics and engineering projects in London and Oxford. His main interests during that time were the practical use of atomic energy and the nuclear chain reaction process. He received a British patent for proposing that if any neutron-driven process released more neutrons than the number required to start it, an expanding nuclear chain reaction would result in a similar fashion to chain reactions known in chemistry.

In 1938 Szilárd moved to Columbia University in New York City where he was soon joined by Enrico Fermi who moved to the U.S. from Italy. In 1939 Szilárd and Fermi learned about nuclear fission experiment carried out by Hahn and Strassmann and concluded that uranium would be a good material for sustaining a chain reaction through the fission process and neutron multiplication.

The use of nuclear chain reaction for military purpose became obvious and Szilárd was instrumental in the creation of the Manhattan project whose purpose was to develop nuclear weapons for use in the World War II against Germany and Japan. In 1942 both Szilárd and Fermi moved to the University of Chicago and in December of 1942 they set off the first controlled nuclear chain reaction and subsequently many well-known theoretical and experimental physicists became involved with the project. Since uranium-235 was one of the two fissionable nuclides of choice (the other was plutonium) for the bomb, several laboratories around the U.S.A. were working on techniques for a physical separation of U-235 from the much more abundant U-238 in natural uranium. Atomic bombs became available for actual military use in 1945; they were used on Hiroshima and Nagasaki in Japan and are credited with accelerating the rapid surrender of Japan.

In 1955 Szilárd, with Enrico Fermi, received a patent for a nuclear fission reactor in which nuclear chain reactions are initiated, controlled, and sustained at a steady observable rate. These reactions are used today as source of power in nuclear power plants.

Szilárd was very conscious socially and had great concern for the social consequences of science. He believed that scientists must accept social responsibility for unexpected detrimental consequences of their discoveries. After the military use of the atomic bombs that caused enormous civilian casualties, Szilárd became a strong promoter of peaceful uses of atomic energy and control of nuclear weapons.

THOMSON, George Paget (1892–1975)

British physicist, educated in mathematics and physics at the Trinity College of the University of Cambridge. He spent the first world war years in the British army and after the war spent three years as lecturer at the Corpus Christi College in Cambridge. In 1922 he was appointed Professor of Natural Philosophy at the University of Aberdeen in Scotland and from 1930 until 1952 he held an appointment of Professor of Physics at the Imperial College of the University of London. From 1952 until 1962 he was Master of the Corpus Christi College in Cambridge.

In Aberdeen Thomson carried out his most notable work studying the passage of electrons through thin metallic foils and observing diffraction phenomena which suggested that electrons could behave as waves despite being particles. This observation confirmed Louis de Broglie's hypothesis of particle-wave duality surmising that particles should display properties of waves and that the product of the wavelength of the wave and momentum of the particle should equal to the Planck's quantum constant h. Clinton J. Davisson of Bell Labs in the United States noticed electron diffraction phenomena with a different kind of experiment.

In 1937 Thomson shared the Nobel Prize in Physics with Clinton J. Davisson "*for their experimental discovery of the diffraction of electrons by crystals*".

THOMSON, Joseph John (1856–1940)

British physicist, educated in mathematical physics at the Owens College in Manchester and the Trinity College in Cambridge. In 1884 he was named Cavendish Professor of Experimental Physics at Cambridge and he remained associated with the Trinity College for the rest of his life.

In 1987 Thomson discovered the electron while studying the electric discharge in a high vacuum cathode ray tube. In 1904 he proposed a model of the atom as a sphere of positively charged matter in which negatively charged electrons are dispersed randomly ("plum-pudding model of the atom").

In 1906 Thomson received the Nobel Prize in Physics "*in recognition of the great merits of his theoretical and experimental investigations on the conduction of electricity by gases*". Thomson was also an excellent teacher and seven of his former students also won Nobel Prizes in Physics during their professional careers.

WIEN Wilhelm (1864–1928)

German physicist educated in mathematics and natural sciences at Universities of Göttingen and Berlin. In 1886 he completed his doctorate under Hermann von Helmholtz on the diffraction of light at the University of Berlin. Most of Wien's professional life was related to study of thermal radiation but he was also active in many other areas of contemporary physics at the end of the 19th century and the beginning of 20th century. He excelled both as experimental as well as theoretical physicist and played an important role in Planck's work on the quantum theory of black body radiation.

During his professional career Wien advanced through increasingly more advanced academic positions. He was professor of Physics at the University of Aachen from 1896 to 1899 and at the University of Giessen from 1899 to 1900. From 1900 to 1920 he was chairman of Physics at the University of Würzburg, succeeding Wilhelm Röntgen who moved to chairmanship at the University of Munich, and from 1920 to 1928 he was chairman of Physics at the University of Munich, again succeeding Röntgen.

Wien is best known for the so-called displacement law that carries his name and establishes a relationship between the temperature T of a Planck blackbody object and the frequency ν_{max} (in the frequency domain) or wavelength λ_{max} (in the wavelength domain) at which the spectral energy density of emitted blackbody radiation exhibits a maximum. The displacement relationships are given as: $\nu_{max} = C_\nu T$ and $\lambda_{max} = C_\lambda T^{-1}$, where C_ν and C_λ are constants.

Wien also carried out in-depth research using various refinements on Crookes tubes generating cathode rays and canal rays. Thomson in 1897 identified cathode rays as electrons of negative charge and of mass about 1840 times smaller than the mass of hydrogen atom. German physicist Eugen Goldstein in 1886 determined that canal rays (also known as anode rays) transmitted through a specially modified Crookes tube equipped with a perforated cathode consist of positive particles generated in the rarefied gas of theCrookes tube. Wien measured the charge – mass ratio of canal rays and found a positive particle equal in mass to that of hydrogen atom. Based on this finding Wien is credited with discovering the hydrogen nucleus, i.e., proton, in 1898 and laying the foundation for mass spectroscopy.

In 1911 the Nobel Prize in Physics was awarded to Wilhelm Wien "*for his discoveries regarding the laws governing the radiation of heat*".

Appendix D
Electronic Databases of Interest in Nuclear and Medical Physics

Atomic Weights and Isotopic Compositions

J.S. Coursey, D.J. Schwab, and R.A. Dragoset

The atomic weights are available for elements 1 through 112, 114, and 116, and isotopic compositions or abundances are given when appropriate.
www.physics.nist.gov/PhysRefData/Compositions/index.html

Bibliography of Photon Attenuation Measurements

J.H. Hubbell

This bibliography contains papers (1907–1995) reporting absolute measurements of photon (XUV, x-ray, gamma ray, bremsstrahlung) total interaction cross sections or attenuation coefficients for the elements and some compounds used in a variety of medical, industrial, defense, and scientific applications. The energy range covered is from 10 eV to 13.5 GeV.

www.physics.nist.gov/PhysRefData/photoncs/html/attencoef.html

Elemental Data Index and Periodic Table of Elements

M.A. Zucker, A.R. Kishore, R. Sukumar, and R.A. Dragoset

The Elemental Data Index provides access to the holdings of NIST Physics Laboratory online data organized by element. It is intended to simplify the process of retrieving online scientific data for a specific element.

www.physics.nist.gov/PhysRefData/Elements/cover.html

Fundamental Physical Constants

CODATA

CODATA, the Committee on Data for Science and Technology, is an interdisciplinary scientific committee of the International Council for Science (ICSU), which works to improve the quality, reliability, management and accessibility of data of importance

© Springer International Publishing Switzerland 2016
E.B. Podgoršak, *Radiation Physics for Medical Physicists*,
Graduate Texts in Physics, DOI 10.1007/978-3-319-25382-4

to all fields of science and technology. The CODATA committee was established in 1966 with its secretariat housed at 51, Boulevard de Montmorency, 75016 Paris, France. It provides scientists and engineers with access to international data activities for increased awareness, direct cooperation and new knowledge. The committee was established to promote and encourage, on a world-wide basis, the compilation, evaluation, and dissemination of reliable numerical data of importance to science and technology. Today 23 countries are members, and 14 International Scientific Unions have assigned liaison delegates.

www.codata.org

Fundamental Physical Constants

The NIST Reference on Constants, Units, and Uncertainty.

www.physics.nist.gov/cuu/constants/

Ground Levels and Ionization Energies for the Neutral Atoms

W.C. Martin, A. Musgrove, S. Kotochigova, and J.E. Sansonetti

This table gives the principal ionization energies (in eV) for the neutral atoms from hydrogen ($Z = 1$) through rutherfordium ($Z = 104$). The spectroscopy notations for the electron configurations and term names for the ground levels are also included.

www.physics.nist.gov/PhysRefData/IonEnergy/ionEnergy.html

International System of Units (SI)

The NIST Reference on Constants, Units, and Uncertainty

The SI system of units is founded on seven SI base units for seven base quantities that are assumed to be mutually independent. The SI base units as well as many examples of derived units are given.

www.physics.nist.gov/cuu/Units/units.html

Mathematica

Wolfram MathWorld

Wolfram MathWorld™ is web's most extensive mathematical resource, provided as a free service to the world's mathematics and internet communities as part of a commitment to education and educational outreach by Wolfram Research, makers of Mathematica, an extensive technical and scientific software. Assembled during the past decade by Eric W. Weisstein, MathWorld emerged as a nexus of mathematical information in mathematics and educational communities. The technology behind MathWorld is heavily based on Mathematica created by Stephen Wolfram. In addition to being indispensable in the derivation, validation, and visualization of MathWorld's content, Mathematica is used to build the website itself, taking advantage of its advanced mathematical typesetting and data-processing capabilities.

www.mathworld.wolfram.com

Nuclear Data

National Nuclear Data Center

The National Nuclear Data Center (NNDC) of the *Brookhaven National Laboratory* (BNL) in the USA developed a software product (NuDat 2) that allows users to search and plot nuclear structure and nuclear decay data interactively. The program provides an interface between web users and several databases containing nuclear structure, nuclear decay and some neutron-induced nuclear reaction information. Using NuDat 2, it is possible to search for nuclear level properties (energy, half-life, spin-parity), gamma ray information (energy, intensity, multipolarity, coincidences), radiation information following nuclear decay (energy, intensity, dose), and neutron-induced reaction data from the BNL-325 book (thermal cross section and resonance integral). The information provided by NuDat 2 can be seen in tables, level schemes and an interactive chart of nuclei. The software provides three different search forms: one for levels and gammas, a second one for decay-related information, and a third one for searching the Nuclear Wallet Cards file.

www.nndc.bnl.gov

Nuclear Data Services

International Atomic Energy Agency (IAEA)

The Nuclear Data Section (NDS) of the *International Atomic Energy Agency* (IAEA) of Vienna, Austria maintains several major databases as well as nuclear databases and files, such as: ENDF – evaluated nuclear reaction cross section libraries; ENSDF – evaluated nuclear structure and decay data; EXFOR – experimental nuclear reaction data; CINDA – neutron reaction data bibliography; NSR – nuclear science references; NuDat 2.0 – selected evaluated nuclear data; Wallet cards – ground and metastable state properties; Masses 2003 – atomic mass evaluation data file; Thermal neutron capture gamma rays; Q-values and Thresholds.

www-nds.iaea.org

Nuclear Energy Agency Data Bank

Organisation for Economic Cooperation and Development (OECD)

The nuclear energy agency data bank of the Organization for Economic Cooperation and Development (OECD) maintains a nuclear database containing general information, evaluated nuclear reaction data, format manuals, preprocessed reaction data, atomic masses, and computer codes.

www.nea.fr/html/databank/

Nucleonica

European Commission: Joint Research Centre

Nucleonica is a new nuclear science web portal from the European Commission's Joint Research Centre. The portal provides a customizable, integrated environment and collaboration platform for the nuclear sciences using the latest internet "Web 2.0" dynamic technology. It is aimed at professionals, academics and students working with radionuclides in fields as diverse as the life sciences, the earth sciences, and the more traditional disciplines such as nuclear power, health physics and radiation protection, nuclear and radiochemistry, and astrophysics. It is also used as a knowledge management tool to preserve nuclear knowledge built up over many decades by creating modern web-based versions of so-called legacy computer codes. Nucleonica also publishes and distributes the Karlsruhe Nuklidkarte (Karlsruhe Chart of the Nuclides).

www.nucleonica.net/unc.aspx

Photon Cross Sections Database: XCOM

M.J. Berger, J.H. Hubbell, S.M. Seltzer, J.S. Coursey, and D.S. Zucker

A web database is provided which can be used to calculate photon cross sections for scattering, photoelectric absorption and pair production, as well as total attenuation coefficients, for any element, compound or mixture ($Z \leq 100$) at energies from 1 keV to 100 GeV.

www.physics.nist.gov/PhysRefData/Xcom/Text/XCOM.html

Stopping-Power and Range Tables for Electrons, Protons, and Helium Ions

M.J. Berger, J.S. Coursey, and M.A. Zucker

The databases ESTAR, PSTAR, and ASTAR calculate stopping-power and range tables for electrons, protons, or helium ions, according to methods described in ICRU Reports 37 and 49. Stopping-power and range tables can be calculated for electrons in any user-specified material and for protons and helium ions in 74 materials.

www.physics.nist.gov/PhysRefData/Star/Text/contents.html

X-Ray Form Factor, Attenuation, and Scattering Tables

C.T. Chantler, K. Olsen, R.A. Dragoset, A.R. Kishore, S.A. Kotochigova, and D.S. Zucker

Detailed Tabulation of Atomic Form Factors, Photoelectric Absorption and Scattering Cross Section, and Mass Attenuation Coefficients for Z from 1 to 92. The primary interactions of x-rays with isolated atoms from $Z = 1$ (hydrogen) to $Z = 92$ (uranium) are described and computed within a self-consistent Dirac-Hartree-Fock framework. The results are provided over the energy range from either 1 or 10 eV to 433 keV, depending on the atom. Self-consistent values of the $f1$ and $f2$ components of the atomic scattering factors are tabulated, together with the photoelectric attenuation coefficient τ/ρ and the K-shell component τ_K/ρ, the scattering attenuation coefficient σ/ρ (coh + inc), the mass attenuation coefficient μ/ρ, and the linear attenuation coefficient μ, as functions of energy and wavelength.

www.physics.nist.gov/PhysRefData/FFast/Text/cover.html

X-Ray Mass Attenuation Coefficients and Mass Energy-Absorption Coefficients

J. H. Hubbell and S. M. Seltzer

Tables and graphs of the photon mass attenuation coefficient μ/ρ and the mass energy-absorption coefficient μ_{en}/ρ are presented for all elements from $Z = 1$ to $Z = 92$, and for 48 compounds and mixtures of radiological interest. The tables cover energies of the photon (x-ray, gamma ray, bremsstrahlung) from 1 keV to 20 MeV.

www.physics.nist.gov/PhysRefData/XrayMassCoef/cover.html

X-ray Transition Energies

R.D. Deslattes, E.G. Kessler Jr., P. Indelicato, L. de Billy, E. Lindroth, J. Anton, J.S. Coursey, D.J. Schwab, K. Olsen, and R.A. Dragoset

This x-ray transition table provides the energies and wavelengths for the K and L transitions connecting energy levels having principal quantum numbers $n = 1$, 2, 3, and 4. The elements covered include $Z = 10$, neon to $Z = 100$, fermium. There are two unique features of this database: (1) all experimental values are on a scale consistent with the International System of measurement (the SI) and the numerical values are determined using constants from the Recommended Values of the Fundamental Physical Constants: 1998 and (2) accurate theoretical estimates are included for all transitions.

www.physics.nist.gov/PhysRefData/XrayTrans/index.html

Appendix E
Nobel Prizes for Research in X-Rays

	Year	Field	Scientists and Justification for Nobel Prize
(1)	1901	Physics	**Wilhelm Conrad ROENTGEN** *"for his discovery of the remarkable rays subsequently named after him"*
(2)	1911	Physics	**Wilhelm WIEN** *"for his discoveries regarding the laws governing the radiation of heat"*
(3)	1914	Physics	**Max von LAUE** *"for his discovery of the diffraction of x-rays by crystals"*
(4)	1915	Physics	**William Henry BRAGG and William Lawrence BRAGG** *"for their services in the analysis of crystal structure by means of x-rays"*
(5)	1917	Physics	**Charles Glover BARKLA** *"for his discovery of the characteristic Rontgen radiation of the elements"*
(6)	1924	Physics	**Karl Manne Georg SIEGBAHN** *"for discoveries and research in the field of x-ray spectroscopy"*
(7)	1927	Physics	**Arthur Holly COMPTON** *"for the discovery of the effect that bears his name"*
(8)	1936	Chemistry	**Peter J.W. DEBYE** *"for his contributions to our knowledge of molecular structure through his investigations on dipole moments and on the diffraction of x-rays and electrons in gases"*
(9)	1962	Chemistry	**Max Ferdinand PERUTZ and John Cowdery KENDREW** *"for their studies of the structures of globular proteins"*

(continued)

© Springer International Publishing Switzerland 2016
E.B. Podgoršak, *Radiation Physics for Medical Physicists*,
Graduate Texts in Physics, DOI 10.1007/978-3-319-25382-4

	Year	Field	Scientists and Justification for Nobel Prize
(10)	1962	Medicine	**Francis CRICK, James WATSON, and Maurice WILKINS** *"for their discoveries concerning the molecular structure of nucleic acids and its significance for information transfer in living material"*
(11)	1979	Medicine	**A. McLeod CORMACK and G. Newbold HOUNSFIELD** *"for the development of computer assisted tomography"*
(12)	1981	Physics	**Kai M. SIEGBAHN** *"for his contribution to the development of high-resolution electron spectroscopy"*
(13)	1985	Chemistry	**Herbert A. HAUPTMAN and Jerome KARLE** *"for their outstanding achievements in the development of direct methods for the determination of crystal structures"*
(14)	1988	Chemistry	**Johann DEISENHOFER, Robert HUBER and Hartmut MICHEL** *"for the determination of the three dimensional structure of a photosynthetic reaction centre"*
(15)	2002	Physics	**Raymond DAVIS, Jr., Masatoshi KOSHIBA, Riccardo GIACCONI** *"for pioneering contributions to astrophysics, in particular for the detection of cosmic neutrinos and discovery of cosmic x-ray sources"*

Bibliography

F.H. Attix, *Introduction to Radiological Physics and Radiation Dosimetry* (Wiley, New York, 1986)

V. Balashov, *Interaction of Particles and Radiation with Matter* (Springer, Heidelberg, 1997)

British Journal of Radiology, Suppl. 25: *Central Axis Depth Dose Data for Use in Radiotherapy* (British Institute of Radiology, London, 1996)

J.R. Cameron, J.G. Skofronick, R.M. Grant, *The Physics of the Body*, 2nd edn. (Medical Physics Publishing, Madison, 1999)

S.R. Cherry, J.A. Sorenson, M.E. Phelps, *Physics in Nuclear Medicine*, 3rd edn. (Saunders, Philadelphia, 2003)

W.H. Cropper, *Great Physicists: The Life and Times of Leading Physicists from Galileo to Hawking* (Oxford University Press, Oxford, 2001)

R. Eisberg, R. Resnick, *Quantum Physics of Atoms, Molecules, Solids, Nuclei and Particles* (Wiley, New York, 1985)

R.D. Evans, *The Atomic Nucleus* (Krieger, Malabar, 1955)

H. Goldstein, C.P. Poole, J.L. Safco, *Classical Mechanics*, 3rd edn. (Addison Wesley, Boston, 2001)

D. Greene, P.C. Williams, *Linear Accelerators for Radiation Therapy*, 2nd edn. (Institute of Physics Publishing, Bristol, 1997)

J. Hale, *The Fundamentals of Radiological Science* (Thomas Springfield, Illinois, 1974)

W. Heitler, *The Quantum Theory of Radiation*, 3rd edn. (Dover Publications, New York, 1984)

W. Hendee, G.S. Ibbott, *Radiation Therapy Physics* (Mosby, St. Louis, 1996)

W.R. Hendee, E.R. Ritenour, *Medical Imaging Physics*, 4th edn. (Wiley, New York, 2002)

International Commission on Radiation Quantities and Units (ICRU), Average Energy Required to produce an Ion Pair, ICRU Report 31 (ICRU, Bethesda, 1979)

International Commission on Radiation Quantities and Units (ICRU), Radiation Quantities and Units, ICRU Report 33 (ICRU, Bethesda, 1980)

International Commission on Radiation Units and Measurements (ICRU), *Electron Beams with Energies Between 1 and 50 MeV*, ICRU Report 35 (ICRU, Bethesda, 1984)

International Commission on Radiation Units and Measurements (ICRU), *Stopping Powers for Electrons and Positrons*, ICRU Report 37 (ICRU, Bethesda, 1984)

J.D. Jackson, *Classical Electrodynamics*, 3rd edn. (Wiley, New York, 1999)

H.E. Johns, J.R. Cunningham, *The Physics of Radiology*, 4th edn. (Thomas Springfield, Illinois, 1984)

C.J. Karzmark, C.S. Nunan, E. Tanabe, *Medical Linear Accelerators* (McGraw-Hill Inc, New York, 1993)

F. Khan, *The Physics of Radiation Therapy*, 3rd edn. (Williams & Wilkins, Baltimore, 2003)

© Springer International Publishing Switzerland 2016

E.B. Podgoršak, *Radiation Physics for Medical Physicists*,

Graduate Texts in Physics, DOI 10.1007/978-3-319-25382-4

S.C. Klevenhagen, *Physics and Dosimetry of Therapy Electron Beams* (Medical Physics Publishing, Madison, 1993)

K.S. Krane, *Modern Physics* (Wiley, New York, 1996)

D.R. Lide, *CRC Handbook of Chemistry and Physics*, 87th edn. (CRC Press, Boca Raton, 2005)

P. Marmier, E. Sheldon, *Physics of Nuclei and Particles* (Academic Press, New York, 1969)

P. Mayles, A. Nahum, J.C. Rosenwald (eds.), *Handbook of Radiotherapy Physics: Theory and Practice* (Taylor and Francis, New York, 2007)

P. Metcalfe, T. Kron, P. Hoban, *The Physics of Radiotherapy X-Rays from Linear Accelerators* (Medical Physics Publishing, Madison, 1997)

E. Persico, E. Ferrari. S.E. Segre, Principles of Particle Accelerators (W.A. Benjamin, Inc. New york, 1968)

E.B. Podgoršak Compendium to Radiation Physics for Medical Physicists: 300 Problems and Solutions (Springer, Heidelberg, 2014)

B. Povh, K. Rith, C. Scholz, F. Zetsche, *Particles and Nuclei: An Introduction to the Physical Concepts*, 5th edn. (Springer, Heidelberg, 2006)

J.W. Rohlf, *Modern Physics from α to Z0* (Wiley, New York, 1994)

R.A. Seaway, C.J. Moses, C.A. Moyer, *Modern Physics* (Saunders College Publishing, Philadelphia, 1989)

P. Sprawls, *Physical Principles of Medical Imaging* (Medical Physics Publishing, Madison, 1995)

R.L. Sproull, *Modern Physics: The Quantum Physics of Atoms, Solids, and Nuclei* (Krieger Pub. Co., Malabar, 1990)

J. Van Dyk (ed.), *The Modern Technology of Radiation Oncology* (Medical Physics Publishing, Madison, 1999)

Index

© Springer International Publishing Switzerland 2016   885
E.B. Podgoršak, *Radiation Physics for Medical Physicists*,
Graduate Texts in Physics, DOI 10.1007/978-3-319-25382-4

Printed in the United States
By Bookmasters